FUNDAMENTALS OF GENERAL, ORGANIC, AND BIOLOGICAL CHEMISTRY

FUNDAMENTALS OF GENERAL, ORGANIC, AND BIOLOGICAL CHEMISTRY

SIXTH EDITION

JOHN R. HOLUM

Augsburg College

JOHN WILEY & SONS, INC.

New York / Chichester / Weinheim
Brisbane / Singapore / Toronto

Cover Photos

inset: © Bob Sacha
background: Courtesy Affymetrix, Santa Clara, California

Acquisitions Editor   Clifford Mills
Marketing Manager   Kimberly Manzi
Senior Production Editor   Elizabeth Swain
Designer   Madelyn Lesure
Photo Editor   Hilary Newman
Illustration Editor   Edward Starr

This book was set in Caslon 224 Book by Progressive Information Technologies, Inc. and printed and bound by Von Hoffmann Press. The cover was printed by The Lehigh Press, Inc.

Recognizing the importance of preserving what has been written, it is a policy of John Wiley & Sons, Inc. to have books of enduring value published in the United States printed on acid-free paper, and we exert our best effort to that end. The paper in this book was manufactured by a mill whose forest management programs include sustained yield harvesting of its timberlands. Sustained yield harvesting principles ensure that the number of trees cut each year does not exceed the amount of new growth.

Holum, John R.
    Fundamentals of general, organic, and biological chemistry / John
R. Holum. — 6th ed.
        p.   cm.
    Includes index.
    ISBN 0-471-17574-9 (cloth :alk. paper)
    1. Chemistry.   I. Title.
QD31.2.H62   1997
540—dc21   97-15451
                                                            CIP

Printed in the United States of America
10   9   8   7   6   5   4   3   2   1

PREFACE

This book is about nature and about human life at their molecular levels. It's meant for two kinds of students. Many plan on post-university careers in any one of the health sciences. Others are liberal arts or general education students who sense that *chemical* knowledge underlies the current and upcoming advances in the sciences that will most affect their health and inform their business, political, and ecological judgments.

A course built around this text is more truly a *general* chemistry course than any other because students experience the excitement of learning during the first year about how life works at the molecular level. And they'll also be introduced to how genetic engineering is done and how the genetic code will help biochips become major information carriers.

SIXTH EDITION FEATURES

"This Chapter in Context" Units Weave an *Awareness* of Relevance into Each Chapter   To many students, the topics in the first several chapters seem remote from the molecular basis of life, and it's hard for them to stay motivated. Teachers know that *relevance motivates,* and students want relevance. The first section of Chapter 1 gives an overview of the entire book. Beginning with Chapter 2, each chapter then opens with a short unit called "This Chapter in Context." A new feature in this edition, it broadly surveys "where we have been" and "where we are going."

Students readily grant that basic concepts and vocabulary usually form an essential background in any field, so frequent reminders of relevance are helpful. Moreover, many interactions between health and the environment require little if any knowledge of biochemistry. When sports enthusiasts, for example, first learn how buffers are essential to breathing, their respect improves for all of the concepts that are background for the study of buffers. (One senior physical education major even told me that he wished he had gone into chemistry! Imagine that.)

Interaction Units Apply Chemistry to Life   This would be a good time to look at a special index, the index to Interactions. We used to call these units "Special Topics," but the new term better suggests that chemistry does interact with many topics of current interest. We've had many such topics in the past. Among the *new* Interactions of this edition are the following.

1.1	Scurvy, Lime Juice, Ascorbic Acid, and the Scientific Method
6.1	Air Bags
6.2	Breathing at High Altitude
6.4	Surfactant Replacement Therapy for Preterm Babies
7.1	The Environment Friendly Solvent, Liquid CO_2
10.6	Chornobyl—An Explosion That Shook the World
11.2	Enlisting Microbes Against Oil Spills
13.1	Ethyl Alcohol and Alcoholism
19.3	Fake Fat—How Olestra Escapes Digestion

Environmentally significant Interactions that have carried over with updating include discussions of the greenhouse effect, acid rain, ozone in smog, and the ozone hole.

Among the deleted Special Topics are those on solubility products and on the interrelating calculations involving pH, $[H^+]$, and K_a.

Traditional General Chemistry Topics Make Up Chapters 1 to 10   We cannot go into the molecular basis of life without knowing about molecules as well as several fundamental concepts concerning the structure and properties of matter in general. As long as students retain confidence that a topic, no matter how seemingly remote, relates somehow to their interest in life and health, their motivation is enhanced. Acids, bases, and buffers are studied, for example, because the acid–base status of the body is a matter of life and death. Any teacher realizes, of course, that acids, bases, and buffers cannot be studied without a good background in formulas, structures, equations, solutions, and equilibria.

For future health care professionals there is no more important topic in the entire book than the acid–base status of the blood and how this relates to the respiratory gases. It's a topic that not only is used as background in later courses, it is also the only chemistry topic that truly carries over beyond formal schooling to careers. Virtually all medical emergencies seen by health care professionals in emergency rooms, operating rooms, and critical care units involve the blood gases, respiration, and blood buffers. Ask topflight nurses if this is right (or read almost any issue of *The American Journal of Nursing*); that's how I first became aware of the importance of this topic.

We've changed the first ten chapters in the following ways. All changes reflect a desire to accomplish two goals: to control the book's length, and to make the level more appropriate for current students. Stoichiometry has been moved later, from being Chapter 3 to being Chapter 5. This enables a more unified treatment of atomic and molecular structure. The treatment of redox reactions has been shortened and made simpler by removing old Chapter 10 and consolidating the basic vocabulary of such reactions in Chapter 4.

Old Section 1.6 ("Accuracy, Error, Uncertainty, and Precision") has been shortened and folded into a new section (1.4, "Physical Quantities, Measurement, and Significant Figures.") This places the topic on significant figures earlier where it better supports the discussion of scientific notation.

Old Section 2.4 ("Heat and the Thermal Properties of Matter") has also been shortened (and renamed "Heat"). The discussion of heat capacity has been eliminated. The concept of specific heat is retained. Heats of vaporization and fusion are now in chapter 6 ("States of Matter and the Kinetic Theory").

A new section (2.5, "Heat and Molecular Kinetic Energy") delivers on requests turned up by reviewers, namely, getting the concept of molecular kinetic energy earlier than in the chapter on gases.

I have deleted the topics of hybrid atomic orbitals both here and in the chapters on organic chemistry. VSEPR theory is all that we need to account for molecular geometry at this level. And I have always believed that the chemistry of the double bond can be adequately taught in a course such as this without the mention of pi bonds. (It's easy to point out that a double bond must be a region of higher electron density than a single bond and so would be more attractive to an electron-seeking reactant, like the hydrogen ion.)

The mathematical treatment of the gas laws has been shortened as I have shifted to the use of the combined gas law as the only one that students need for the more relevant gas law calculations. The discussion of Dalton's law is also reduced by omitting its application to the collection of gases over water.

In Chapter 7 ("Solutions and Colloids"), following an excellent suggestion by a reviewer, I have added a summary of all of the forces of attraction that we've studied so students can review these in one place.

Chapters 11 to 17 Survey the Functional Groups of Organic Chemistry Deemed Most Essential to the Study of Biochemistry   The wedding between the theme of the course and the limitation of time results in a very abbreviated survey of organic chemistry. Some of the major topics developed in even a one-term course of "regular" organic chemistry have had to be excluded, topics like the theory of resonance, nucleophilic substitution reactions, the Grignard synthesis, and many others. I have stressed only those functional groups that occur widely among the molecules of life and their reactions with four kinds of compounds: acids, bases, oxidizing agents, and reducing agents. I have provided some mechanisms because organic reactions otherwise seem too much like magic, and their learning becomes merely rote. Some mechanisms (those of acid catalyzed alcohol dehydration, aldol condensation, and Claisen ester condensation) have been moved to Appendix D. Teachers who want to omit them anyway will find it easier now. Teachers who want to include them may do so whenever they deem them most appropriate. (For example, I've preferred to discuss the aldol condensation not in the aldehyde chapter but in the chapter on the metabolism of carbohydrates where the reaction actually applies.)

Chapters 18 to 29 Constitute One Illustration after Another of the Molecular Basis of Life   Carbohydrates, lipids, and proteins begin this closing section of the book. Because of their importance to all that follows, I next take up enzymes (Chapter 21). Hormones and neurotransmitters come next, this time in a separate chapter (22). Then come the extracellular fluids of the body (Chapter 23). The chapter on nucleic acids (24) has been considerably updated in its discussion of viruses, recombinant DNA technology, and the applications of genetic engineering to medicine (including biochip technology).

Biochemical energetics (Chapter 25), after an overview, first takes up the citric acid cycle. This time I've given descriptive *chemical* labels, like ""dehydration" "oxidation," and so forth, to the various steps. The discussion of the respiratory chain has been rewritten in an effort to make it easier. Then come treatments of the metabolism of carbohydrates (Chapter 26), lipids (Chapter 27), and proteins (Chapter 28). With all of this background, the study of nutrition (Chapter 29) is made much easier.

Many Design Features Aid Students   There are frequent **margin comments** to restate a point, offer data, or simply remind.

Key terms are highlighted in boldface at those places where they are defined and then discussed. A complete **glossary** of these terms plus a few others appears at the end of the book. The ***Study Guide*** that accompanies this book also has individual chapter glossaries.

Each section of a chapter begins with a **headline**. This is *not* a one-sentence summary of the section but rather a lead-in to the beginning of the section that tries to state the section's major point.

Each chapter has a **Summary** that uses key terms in a narrative manner. The summaries are not necessarily organized in the same order in which the material occurs in the various chapter sections. The summaries assume that the sections have been studied so that the needed vocabulary is in place. The summaries thus illustrate that the pedagogy for first-time learning is not necessarily the same as that for reviewing.

The chapters in the first two-thirds of the book have several **worked examples.** In those involving calculations, the **factor-label** method is exploited. These examples generally have labeled parts, such as "Problem," "Analysis," "Solution," and "Check." Thus, immediately after the statement of the *problem* comes the *analysis.* What is the problem really asking? In a multistep solution, what must be done first? Then comes the *solution.* We want to encourage students to see that *solving* a problem (figuring out what to do) occurs *before* the calculations. Following the "Solution" section of an example there is often a "Check" section. "Does the *size* of the answer make sense?" This takes the student back over the problem and encourages the use of the mind (as opposed to a mechanical use of factor-labels) to see the sense of the analysis and the solution. Among problems in the organic chapters, "Check" sections help students to learn how to double-check their answers.

Nearly all worked examples are followed by **Practice Exercises,** which encourage immediate reinforcements of skills learned in the examples. Answers to all Practice Exercises are in Appendix E. A copious number of **Review Exercises** closes each chapter, including some that are "additional." These are not identified by topic, and some require the use of material from earlier chapters. Thus you will find stoichiometry problems scattered throughout the book.

SUPPLEMENTARY MATERIALS FOR STUDENTS AND TEACHERS

The complete package of supplements that are available to help students study and teachers teach includes the following.

Laboratory Manual for Fundamentals of General, Organic, and Biological Chemistry, sixth edition.   This is a revised edition prepared by Dr. Sandra Olmsted, Augsburg College. An *Instructor's Manual* to this laboratory manual is a section of the general Teachers' Manual described below.

Study Guide for Fundamentals of General, Organic, and Biological Chemistry, sixth edition.   Each chapter of this softcover book contains a discussion of what are the "must study and master" topics. There are also chapter glossaries, additional worked examples and exercises, sample examinations, and the answers to all of the Review Exercises.

Teachers' Manual for Fundamentals of General, Organic, and Biological Chemistry, sixth edition.   This softcover supplement is available to teachers. It contains all the usual services for *both the text and the laboratory manual.*

Test Bank.   Available in both hard copy and software (Macintosh© and IBM© compatible) versions, this test resource contains roughly 1000 questions.

Transparencies.   Instructors who adopt this book may obtain from Wiley, without charge, a set of color transparencies that duplicate key illustrations from the text.

ACKNOWLEDGMENTS

My wife Mary has been my strongest supporter, and I am deeply grateful to this wonderful woman now stricken by a brain aneurysm. Our daughters, Liz, Ann, and Kathryn, have also been strong champions, and I thank them for what they have meant to Mary and me. These daughters have sometimes asked me "How did you do it all these years?" My answer has always been, "It wouldn't have been remotely possible without the strong support of your mother."

At Augsburg College, I always enjoyed unstinting encouragement from the Chair of the Chemistry Department, who was then Dr. Earl Alton (and is now Assistant Academic Dean), and from Dr. Charles Anderson, then President. Dr. Arlin Gyberg, Dr. Joan Kunz (current Chair), and Dr. Sandra Olmsted of the Chemistry Department all have been important sources of suggestions and corrections.

Extraordinarily nice people abound at John Wiley & Sons. I think particularly of my Editor, Clifford Mills, Editorial Assistant, Alicia Solis, and Supplements' Editor, Jennifer Yee.

The overall design was the responsibility of Maddy Lesure, with whom I have worked with pleasure on this and other books. Edward Starr has been skillful, artistic, and faithful in handling the line drawing art work. Hilary Newman, Photo Editor, produced such a rich supply of outstanding choices for photographs that my choosing became difficult, yet exciting and pleasurable. Elizabeth Swain has long served splendidly in managing the copy editing, and now has capably supervised the production, surely one of the most difficult jobs in textbook publishing.

Two outstanding proofreaders saved me from innumerable embarrassments—Connie Parks and Dr. Sandra Olmsted. It's hard to imagine that any errors remain but, based on experience, no doubt some do. They are now entirely my responsibility. Please use a letter to my Chemistry Editor to let me know about them.

The professional critiques of many teachers are part of the process of preparing a manuscript. I am most pleased to acknowledge and to thank the following people for their work.

Robert Ake
Old Dominion University

Margaret Asirvatham
University of Colorado/Boulder

David Ball
Cleveland State University

Muriel Bishop
Clemson University

Ronald Bost
North Central Texas College

Lorraine Brewer
University of Arkansas

Sybil Burgess
University of North Carolina, Wilmington

Jack Dalton
Boise State University

Henry Fisher
University of Rhode Island

Arlin Gyberg
Augsburg College

Donald Harriss
University of Minnesota/Duluth

Larry Jackson
Montana State University

Herman Knoche
University of Nebraska—Lincoln

John Meisenheimer
Eastern Kentucky University

Sandra Olmsted
Augsburg College

Nancy Paisley
Montclair State University

Richard Petersen
University of Memphis

Fred Schell
University of Tennessee/Knoxville

Ram Singhal
Wichita State University

Joan Stover
South Seattle Community College

Kent Thomas
Kansas-Newman College

Atilla Tuncay
Indiana University Northwest

Ruiess Van Fossen Bravo
Indiana University of Pennsylvania

Justine Walhout
Rockford University

Leslie Wynston
*California State University/
Long Beach*

JOHN R. HOLUM
Saint Paul, MN

CONTENTS

Index to Interactions

FUNDAMENTALS OF GENERAL, ORGANIC, AND BIOLOGICAL CHEMISTRY

GOALS, METHODS, AND MEASUREMENTS

1

When the living body is in health, all of the forces of nature, even at the molecular level of life, are more or less in balance, as these two hikers no doubt unconsciously assume as they view the sunset over the Lysefjord, Norway.

1.1 CHEMISTRY AND THE MOLECULAR BASIS OF LIFE

The theme of this book is the molecular basis of life.

One of the wonders of nature, if you reflect for a moment, is that all animals have so much in common. We infer this particularly in what they take in. Virtually all animals use air and water, for example, and those of a variety of species eat the same kinds of food. Many species around the world prosper simply by eating grass and hay. Vultures, lions, and humans can all obtain essential nourishment from cattle. Both a kitten and a human baby prosper on milk. Somewhere, it seems, at some deep level of existence, there must be a common pool of parts that all species can tap and then put back together in their own unique ways. What are these common parts?

The shared parts are evidently not whole organs or tissues, but much, much smaller things that are the building blocks of everything else. They are extremely tiny particles, called molecules, which are made of even smaller particles, called atoms. All of life, whether plant or animal, has a *molecular* basis, and chemistry has been the route to this discovery. **Chemistry** is the study of that part of nature that bears on substances, their compositions and structures, and their abilities to be changed into other substances. There are so many different substances, however, that we must develop a plan of study or lose our way.

■ Well over 9 million chemical substances are known.

Molecules, Like Maps, Can Be Read When the Keys or Map Signs Are Known   Life at the molecular level involves molecules and chemical reactions that often are complicated. The symbols we use for molecules, however, are actually less complex than many symbol systems you have already mastered, such as map symbols. The symbols for molecules are like those of maps because the same pieces of molecules, like molecular "map signs," occur over and over again in different situations. When you learn these "signs" among simple substances, you'll be amazed (and relieved!) to see how easy it is to study some of nature's most complicated molecules.

■ The atoms of all matter are made of varying combinations of three extremely tiny particles: electrons, protons, and neutrons.

Molecules, as we said, are made of atoms, so to understand molecules we must first learn about atoms and how their own (even tinier) parts get reorganized into molecules. Hence, the study of atomic and molecular structure occurs mainly in the first few chapters of the book. Here also is essential background about several common substances that are highly important to all living systems, such as acids, bases, salts, and solutions.

As you study chemistry, keep before you a major goal, namely, to learn how nature works at the molecular level. You'll be surprised at how enjoyable knowing this can be (of all things!). People who know how nature works have a window on the inner beauty of nature of which others are not even aware. So expect to be surprised by beauty in unexpected places. Expect also to learn how chemistry is in service to society in a surprising number of ways—and thus is in service to many careers. They include medicine, nursing, dentistry, veterinary science, dietetics, nutrition, inhalation therapy, physical and occupational therapy, public health, science education, pharmacy, clinical lab work, crime lab work, consumer products safety, agriculture, forestry, home economics, engineering, and many others. Little wonder that chemistry is often referred to as the *central science*.

1.2 FACTS, HYPOTHESES, AND THEORIES IN SCIENCE

Scientific theories speak to one general question: "How does nature work?"

One of the most common activities of scientists is the gathering of information and facts by observing nature. Some facts are reproducible and some aren't. Both kinds are important in science, but chemistry normally deals only with *reproducible facts*, those that can be observed over and over by independent observers. Chemical experiments, for example, can usually be repeated to check their results.

■ In science, the most reliable facts are those that can be obtained in repeated observations or measurements.

Hypotheses and Theories Are Used To Explain Facts Scientists are forever interested in underlying causes. "How does nature work?" "What must be true about what we cannot see in nature to account for what we do see?" Facts and observations, therefore, are seldom interesting all by themselves. They are valued, instead, as raw material for *hypotheses*.

A **hypothesis** is a conjecture that appears to explain a set of facts in terms of a common cause. A hypothesis also serves as the basis for designing additional tests that can disclose the truth about the hypothesis. In other words, is the hypothesis right or wrong? In medicine, a preliminary diagnosis is an example of a hypothesis, and it's based on observed or reported facts. The diagnosis usually suggests what new information should be sought, perhaps by further questions of the patient or by additional lab tests.

■ The aim in testing a hypothesis is not to prove the hypothesis but to discover the truth about it. Is the hypothesis right or wrong?

Sometimes hypotheses are considered on a large scale. In the history of nutrition science, for example, a number of quite different ailments simply did not fit the notion that *all* diseases are caused by germs. In searching for other causes, trace substances in certain foods were discovered that are vital to health, and the *vitamin theory* developed. After much research, it became so solidly based on experiments and tests that nobody calls it a theory anymore. It's simply another fact about our world, namely, we need various vitamins to be healthy. What clinched the promotion from vitamin *theory* to *fact* were the discoveries of how vitamins work chemically inside cells. This and similar success stories are behind the flowering throughout the 20th century of an important truth: *We cannot claim to know what a disease really is until we know its chemistry*.

A *theory* differs in scope from a hypothesis. A **theory** is an explanation for a large number of facts, observations, and hypotheses in terms of one or a few basic convictions about what the world is like. For example, the *kinetic-molecular theory*, one of the broadest, grandest theories in science, is that the molecules of gases and liquids move around chaotically, bumping into each other and their containers. (Even molecules in solids jiggle somewhat.) We will use this theory to explain many observations about the behavior of things at the molecular level of life.

The *Scientific Method* Bases Conclusions on Evidence The use of isolated facts and human reason to construct testable hypotheses and theories has occurred often in the history of science, and it is popular now to call this approach to questions the **scientific method**. It's the way by which scientists operate, but you no doubt have also used the scientific method often without even realizing it. Anytime you ask, "What's going on here?" you have posed the kind of question that launches the scientific method for solving your "puzzle." The method generally involves the following steps.

1. You're faced with a puzzle; "What's going on here?" Or, "How did that happen?" [The engine of my car is pinging.]

2. You shrewdly come up with a hypothesis that tentatively answers the question *in terms of what you already know*. [Most likely, a spark plug is faulty.]

3. You think about the hypothesis and soon realize that, if it's true, you ought to observe something else if *you try an experiment*. [Not knowing which plug might be at fault, you replace each, one after the other, and listen to the engine.]

4. The experiment either tells you that a plug is no good, or it tells you that your hypothesis is wrong. If wrong, you use your knowledge about engines to devise another hypothesis. (Or, if you're like this author, who has almost no practical knowledge about engines, you let a mechanic have the fun of devising and testing hypotheses about the pinging.)

The engine-pinging problem used above to illustrate the scientific method is trivial, of course. A more serious example involving sailors, scurvy, and the vitamin theory is described in Interaction 1.1.

"How?" versus "Why?" In any scientific study, asking the right question of nature is critical all in itself. Asking "*How* does nature work?" leads, for example, to more progress in understanding nature than asking "*Why* does nature work?" If we ask, for example, "*Why* do we get sick?" instead of "*How* do we get sick?" we can get bogged down in speculations that people have never resolved to everyone's satisfaction. Not that asking "Why?" isn't important, or that this question has no answer, but the value of the "*How?*" question is that it gets at *mechanism*. Knowledge of the physical or chemical *mechanisms* of various illnesses doesn't answer all of the serious questions in life, but it has certainly helped to reduce pain and suffering. Scientists, of course, still use the language of "Why?" After all, both "Why?" and "How?" are ways of asking "What causes . . . ?" But almost always "Why?" means "How?" in science.

1.3 PROPERTIES AND THE STATES OF MATTER

A physical property differs from a chemical property by being observable without changing a substance into something else.

A **property** is any characteristic of something that we can use to identify and recognize the thing when we see it again. Properties such as color, height, or mass that can be observed without changing the object into something different are called **physical properties**. For example, some physical properties of liquid water are that it is colorless and odorless; that it dissolves sugar and table salt but not butter; that it makes a thermometer read 100 °C (212 °F) when it boils (at sea level); and that if it is mixed with gasoline it will sink, not float. If you were handed a glass containing a liquid having these properties, your initial hypothesis undoubtedly would be that it is water. Think of how often each day you recognize things (and people) by simply observing physical properties.

The observations of some properties, however, change an object or a sample of a substance into something else. We can measure, for example, how much gasoline it takes to drive a car 100 miles, but this measurement uses up the gasoline. As it burns it changes into water and carbon dioxide (the fizz in soda pop). A property that, when observed, causes a substance to change into new substances is called a **chemical property**, and what is being observed is called a **chemical reaction**. A chemical property of iron, for example, is that it rusts in moist air; it changes slowly into iron oxide, a reddish solid quite unlike metallic iron. *Chemistry* is the study of

INTERACTION 1.1
SCURVY, LIME JUICE, ASCORBIC ACID, AND THE SCIENTIFIC METHOD

Before the 1500s, sailors who went on long voyages without being resupplied with fresh vegetables and fruit developed bleeding gums, weakened and spongy gum tissue, and the loss of teeth. The victims also bruised easily and became susceptible to killing diseases. A high percentage died. "What was going on here?" The question could not be satisfactorily answered because science had not yet advanced enough, but the disease could at least be named. It was called *scurvy*.

Does Citrus Juice Kill Germs? As early as the mid-1500s, the Dutch knew that the diets of sailors should include citrus fruits. Two centuries later, James Lind (1716–1794), a Scottish physician, realized that citrus fruit juice specifically prevents or cures scurvy, as if by magic, and published a treatise on it in 1753. Scurvy had long been a scourge of the British navy; in a voyage around the world (1740–1744), 1051 of 1955 sailors died, chiefly of scurvy. Formal orders were finally issued by the British Admiralty in 1795 to include daily issues of lemon juice to all sailors. Lemons were called limes then, and British sailors were soon popularly called "limeys."

Still, a question persisted. "What's going on here with lemon juice?" What does lemon juice have that prevents scurvy? As chemistry advanced and improved its methods for isolating, purifying, and identifying chemicals, the antiscurvy substance turned out to be ascorbic acid or vitamin C. But this advance still begged the question. "How does ascorbic acid do what it does?" Perhaps germs cause scurvy and ascorbic acid kills them. At one time this would have been a reasonable hypothesis, *one that could be tested*. Or does vitamin C work in another way?

Nature Bats Last You can see that the use of the scientific method begins with a puzzle, "How does something work?" Or "What is behind this or that observation?" Then a possible explanation, a hypothesis, is put forward, one that is reasonable in the light of the observations and the knowledge existing at the time. The hypothesis, if well made, suggests an approach, perhaps an experiment, that would be a test for it. That scurvy might be caused by germs, like so many other diseases, was reasonable, but no germs could be found. So the hypothesis got nowhere. Despite how reasonable it once might have been, the hypothesis was wrong, which neatly illustrates that the test of reason isn't alone sufficient in science. A hypothesis must be aligned with nature as it really is. Scientists always know that "nature bats last."

By the early 20th century, partly as the result of the research of a Polish chemist, Casimir Funk, we came to know that not all diseases are caused by germs, but that many arise from the absence of specific chemicals in the diet, chemicals that are not carbohydrates, proteins, or edible fats and oils. Believing that these chemicals were all ammonia-like, Casimir Funk called them "vitamines" (a contraction of "vital amines"), but today we drop the "e," and name them *vitamins*. Ascorbic acid is a vitamin, and scurvy is a *vitamin-deficiency disease*, not a germ-caused condition.

Ascorbic Acid and Collagen We still have the persistent, altogether human question, "How does ascorbic acid do what it does?" It wasn't until several decades into the 20th century that ascorbic acid was shown to be essential to the proper formation of a protein called collagen. Collagen functions something like strong flexible reinforcing rods in bones, teeth, cartilage, tendons, skin, blood vessels, and certain ligaments. The deterioration of gum tissue, which relies on collagen, is simply an early sign that collagen is not being made properly. But we're not finished yet. "Just exactly how does ascorbic acid work in the manufacture of collagen?" In short, ascorbic acid is needed to make an enzyme without which proper collagen cannot be made. This answer lands us finally at the molecular basis of scurvy, but to appreciate it we need to know more basic chemistry.

The juice from lemons was known to prevent scurvy as early as the 16th century.

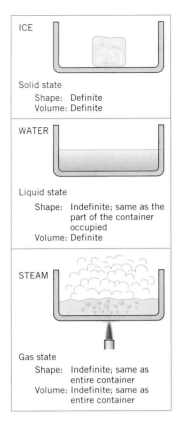

FIGURE 1.1   The three states of matter, as illustrated by water.

     2 = a number
2 yards = a physical quantity

these kinds of changes in substances, how they occur, and how atoms and molecules become reorganized as they happen.

States of Matter   Substances are able to exist as solids, liquids, or gases, which we call the three **physical states** of matter. You're familiar with water as solid ice, as a liquid, and as water vapor (see Figure 1.1). A **solid**, like an ice cube, has both a *definite* volume and a *definite* shape. A **liquid** has a definite volume but no fixed shape; it takes whatever shape its container has. A **gas** has no definite volume and no fixed shape. It will spread out and occupy any container into which it is released. Physical changes brought about by heating or cooling convert a given sample of matter into its different states.

Don't confuse "*states* of matter" with "*kinds* of matter." The kinds of matter, yet to be studied, are elements, compounds, and mixtures.

1.4   PHYSICAL QUANTITIES, MEASUREMENT, AND SIGNIFICANT FIGURES

The most fundamental quantities of measurement are called *base quantities* and each has an official *standard of reference* for one unit.

You may have noticed when we described water's physical properties how much our observations depend on human senses, our abilities to see, taste, and feel, and to sense hotness or coldness. Our senses, however, are limited, so inventors have developed instruments that extend the senses and make possible finer and sharper observations. These devices are equipped with scales or readout panels, and the data we obtain by using measuring instruments are called *physical quantities*.

A **physical quantity** is a physical property to which we can assign both a numerical value *and a unit*. Your own height is a simple example. If it is, say, 5.5 feet, its numerical value, 5.5, and its unit, feet, together tell us at a glance how much greater your height is than an agreed-upon reference of height, a 1-foot measuring stick.

The unit in a physical quantity is just as important as the number. If you said that your height is "two," people would ask, "Two what?" Thus, we can't describe a physical property by a physical quantity without giving both a number and a unit.

$$\text{Physical quantity} = \text{number} \times \text{unit}$$

Physical Quantities Are Obtained by Measurements   A **measurement** is an operation by which we compare an unknown physical quantity with one we know. Maybe, as you were growing up, someone measured your height by comparing it with how many 1-foot rulers it took to equal your height. Usually the number did not match your height exactly, so fractions called inches (each with their own fractions) were also used. Somebody has decided what the inch, the foot, and the yard are, and the rest of us have agreed to the definitions. That's all they are, *definitions*.

Mass, Length, Time, and Temperature Are Base Quantities   The most fundamental measurements in chemistry are those of *length*, *mass*, *temperature*, *time*, and *quantity of chemical substance*. These are called **base quantities** because, as the term implies, all other measurements are based on them or are derived from them. They are truly basic and fundamental.

Length, for example, is a physical quantity that describes how far an object extends into space, or it is the distance between two points. The **volume** of an object is the space it occupies, but volume itself, although important, is not a base quantity. Instead, volume is an example of a **derived quantity** because it can be defined in terms of the base quantity length. The volume of a cube, for example, is the product of (length) × (length) × (length), or (length)³.

Mass, a particularly important base quantity in chemistry, is the measure of the **inertia** of an object, its inherent resistance to any kind of change in motion. When something has a high inertia, such as a train engine, a massive boulder, or an ocean liner, it is very hard to get it into motion or, if it is in motion, it is difficult to slow it down or make it change course. Mass is our way of describing inertia quantitatively, that is, with a number and a unit. A large inertia means a large *mass.*

We must draw an important distinction between *mass* and *weight.* A large mass doesn't always mean a large weight, because your mass does not depend on where you are in the universe, but your weight does. The **weight** of an object is a measure of the gravitational force of attraction experienced by the object. The gravitational force is less on the moon, which is a smaller, less massive object than the earth—about six times less. Thus, an astronaut's *weight* on the moon is one-sixth of its value on earth, but the astronaut's *mass,* the fundamental resistance to any change in motion, is the same in both locations.

When we use a laboratory balance to *weigh* something, we are actually measuring mass because we are comparing two weights *at the same place on the earth* and, therefore, under the same gravitational influence (see Figure 1.2). One weight is the quantity being measured, and the other is a "weight" (or set of weights) built into the weighing balance. Although we commonly call the result of the measurement a "weight," we'd more properly call it the *mass* of the object or the sample.

The base quantity **time** is our measure of how long events last. We need it to describe how rapidly the heart beats, for example, or how fast some chemical reaction occurs.

Temperature is the base quantity we use to describe the hotness or coldness of an object.

All these base quantities are necessary to all sciences, but chemistry has a special base quantity called the *mole* that describes a certain amount of a chemical substance. (We'll study it in Chapter 5.)

Every Base Quantity Has a Base Unit and a *Reference Standard* To measure and report an object's mass, its temperature, or any of its other base or derived physical quantities we obviously need some reference standards and units. By international treaties among the countries of the world, the base units and their reference standards are decided by a diplomatic organization called the General Conference of Weights and Measures, headquartered in Sèvres, a suburb of Paris, France.

Volume of a cube = (*l*)³

The large mass of logs has a large inertia, requiring a powerful tug to move it.

(a)

(b)

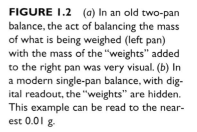

FIGURE 1.2 (*a*) In an old two-pan balance, the act of balancing the mass of what is being weighed (left pan) with the mass of the "weights" added to the right pan was very visual. (*b*) In a modern single-pan balance, with digital readout, the "weights" are hidden. This example can be read to the nearest 0.01 g.

■ The other two SI base quantities are *electric current* and *luminous intensity*. Their base units are called the *ampere* and the *candela*, respectively.

■ An alloy is a mixture of two or more metals made by mixing them in their molten states.

■ We use the period in the abbreviation of inch (in.) to avoid any confusion with the preposition "in," which has the same spelling.

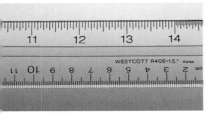

Common ruler marked in inches and centimeters. Notice that there are 10 1-millimeter spaces marked between the centimeter divisions.

■ One cubic meter holds a little over 250 gallons.

The General Conference has defined a unit called a **base unit** for each of seven base quantities, but we need units only for five: length, mass, time, temperature, and mole. We also need units for important derived quantities, such as volume, density, pressure, and energy. The reference standards and the definitions of base and derived quantities and units together make up the **International System of Units** or the **SI** (after the French name, *Système Internationale d'Unites*).

A **reference standard** is the physical description or embodiment of a base unit. The SI *base unit* of length, for example, is called the **meter**, abbreviated **m**, and its *reference standard* is called the *standard meter*. Until 1960, the standard meter was very simple to visualize and understand, being the distance separating two thin scratches on a bar of platinum-iridium alloy stored in an underground vault in Sèvres. The bar, of course, could have been lost or stolen, so the new reference for the meter is based on a property of light, something available everywhere, in all countries, and that obviously can't be lost or damaged, nor can it suffer corrosion. This change in reference standard did not change the actual length of the meter. It only changed its official reference.[1]

In the United States, older but still common units are legally defined in terms of the meter. For example, the yard (yd), roughly nine-tenths of a meter, is defined as 0.9144 m (exactly). The foot (ft), roughly three-tenths of a meter, is defined as 0.3048 m (exactly). The inch (in.) is about two and a half centimeters; more exactly,

$$1 \text{ in.} = 2.54 \text{ cm (exactly)}$$

Table 1.1 gives several relationships among various units of length.

In chemistry, the meter is usually too long for convenience, and submultiples are often used, particularly the **centimeter**, or **cm**, and the **millimeter**, or **mm**. Expressed mathematically, these are defined as follows:

$$1 \text{ m} = 100 \text{ cm} \quad \text{or} \quad 1 \text{ cm} = 0.01 \text{ m}$$
$$1 \text{ m} = 1000 \text{ mm} \quad \text{or} \quad 1 \text{ mm} = 0.001 \text{ m}$$
$$1 \text{ cm} = 10 \text{ mm} \quad \text{or} \quad 1 \text{ mm} = 0.1 \text{ cm}$$

Notice that the subunits are in fractions based on 10. The millimeter, for example, is one-tenth of a centimeter. As we'll often see, letting subunits be divisible by 10 makes many calculations much easier.

The SI unit of volume, a derived unit, is the cubic meter, m^3, called the *stere*, but this is much too large for convenience in chemistry. An older unit, the **liter**, abbreviated **L**, is accepted as a *unit of convenience*. The liter occupies a volume of 0.001 m^3 (exactly), and 1 liter is almost the same as one liquid quart; 1 quart (qt) = 0.946 L.

TABLE 1.1   Some Common Measures of Length[a]

SI	U.S. Customary
1 kilometer (km) = **1000** meters (m)	1 mile (mi) = **5280** feet (ft)
1 meter = **100** centimeters (cm)	= **1760** yards (yd)
1 centimeter = **10** millimeters (mm)	1 yard = **3** feet (ft)
	1 foot = **12** inches (in.)

Other Relationships	
1 meter = 39.37 inches	1 inch = **2.54** centimeters

[a] Numbers in boldface are exact.

[1] The SI now defines the standard meter as how far light will travel in 1/299,792,458 of a second. It is thus based on the speed of light as measured by an "atomic clock."

TABLE 1.2   Some Common Measures of Liquid Volume[a]

SI

1 cubic meter (m^3) = **1000** liters (L)
1 liter = **1000** milliliters (mL)
1 milliliter = **1000** microliters (μL)

U.S. Customary

1 gallon (gal) = **4** liquid quarts (liq qt)
1 liquid quart = **2** liquid pints (liq pt)
1 liquid pint = **16** liquid ounces (liq oz)

Other Relationships

1 cubic meter = 264.2 gallons	1 liter = 1.057 liquid quarts
1 liquid quart = 946.4 milliliters	1 liquid ounce = 29.57 milliliters

[a] Numbers in boldface are exact.

FIGURE 1.3 Some apparatus used to measure liquid volumes. From left to right we see a thin pipet, a volumetric flask, a graduated cylinder, a buret with its stopcock, two more volumetric flasks, a bulb pipet, another graduated cylinder, and another pipet.

Even the liter is often too large for convenience in chemistry, and two submultiples are used, the **milliliter (mL)** and the **microliter (μL)**, which are related as follows.

$$1 \text{ L} = 1000 \text{ mL} \quad \text{or} \quad 1 \text{ mL} = 0.001 \text{ L}$$
$$1 \text{ mL} = 1000 \ \mu L \quad \text{or} \quad 1 \ \mu L = 0.001 \text{ mL}$$

The milliliter is the unit you will most often encounter. Table 1.2 gives several other relationships among units of volume. Figure 1.3 shows apparatus used to measure volumes in the lab.

The SI base unit of mass is the **kilogram**, abbreviated **kg**. Its reference standard is the *standard kilogram mass*, a cylindrical block of platinum-iridium alloy prepared in 1889 and housed at Sèvres under the most noncorrosive conditions possible (Figure 1.4). It is the only SI reference that could still be lost or stolen or suffer corrosion. No alternative has yet been officially adopted, but teams of scientists around the world are hard at work to find one. Duplicates made as much as possible like the original kilogram mass (nicknamed "Le Grand K") are stored in other countries. One kilogram has a mass roughly equal to 2.2 pounds, and Table 1.3 gives some useful relationships among various quantities and units of mass.

FIGURE 1.4 The SI standard kilogram mass. Shown here is the U.S. copy of the reference standard, "Le Grand K," kept at the International Bureau of Weights and Measures in France. The U.S. copy is at the National Institute of Standards and Technology near Washington, D.C. It is made out of a very corrosion-resistant alloy of platinum and iridium.

TABLE 1.3   Some Common Measures of Mass[a]

SI

1 kilogram (kg) = **1000** grams (g)
1 gram = **1000** milligrams (mg)
1 milligram = **1000** micrograms (μg, γ, or mcg)[b]

U.S. Customary (avoirdupois)[c]

1 short ton = **2000** pounds (lb avdp)
1 pound = **16** ounces (oz avdp)

Other Relationships

1 kilogram = 2.205 lb	1 lb avdp = 453.6 grams

[a] Numbers in boldface are exact.
[b] The microgram was once called a *gamma* in medicine and biology.
[c] These are the common units in the United States.

1 kg of butter

Paper clip, 0.4 g
Penny, 3.4 g

One drop of water is about 60 mg.

■ The *kelvin* is named after William Thomson, Baron Kelvin of Largs (1824–1907), a British scientist.

The most often used units of mass in chemistry are the kilogram (kg), the **gram (g)**, the **milligram (mg)**, and the **microgram (μg)**, which are defined as follows.

$$1 \text{ kg} = 1000 \text{ g} \qquad \text{or} \qquad 1 \text{ g} = 0.001 \text{ kg}$$
$$1 \text{ g} = 1000 \text{ mg} \qquad \text{or} \qquad 1 \text{ mg} = 0.001 \text{ g}$$
$$1 \text{ mg} = 1000 \text{ μg} \qquad \text{or} \qquad 1 \text{ μg} = 0.001 \text{ mg}$$

Lab experiments in chemistry usually involve grams or milligrams of substances.

The SI unit of time is the **second**, abbreviated **s**. The *duration* of the second is 1/18,400 of a mean solar day. (Although the duration is basically the same, the actual SI *definition* of the second is different from this, but it involves complexities of atomic physics unneeded in our study.) Decimal-based multiples and submultiples of the second are used in science, but so also are such deeply entrenched old units as minute, hour, day, week, month, and year.

The SI unit for degree of temperature is called the **kelvin, K.** (Be sure to notice that the abbreviation is K, not °K.) The kelvin is essentially the same as the **degree Celsius (°C)** and the *degree centigrade* (also °C). All three are identical in size. Only the *numbers* assigned to points on the scales differ (see Figure 1.5). Think of each kelvin or each degree centigrade as 1/100 of the interval between the freezing point and boiling point of water. The most extreme coldness possible, 0 K on the **Kelvin scale**, is called **absolute zero**. It is −273.15 °C on the Celsius scale.

Because 0 K corresponds to −273.15 °C, we have the following simple relationships between kelvins and degrees Celsius (where we follow common practice of rounding 273.15 to 273).

$$°C = K - 273$$
$$K = °C + 273$$

■ PRACTICE EXERCISE 1   Normal body temperature is 37 °C. What is this in kelvins?

The Kelvin scale is used in chemistry mostly to describe temperatures of gases. The Celsius scale is more popular for most other uses, including medicine. The de-

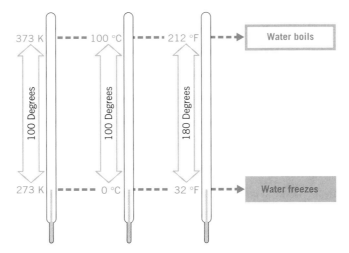

FIGURE 1.5   Relationships among the Kelvin, Celsius, and Fahrenheit scales of temperature.

gree Fahrenheit (°F), with which most United States citizens are still most familiar, is five-ninths the size of the degree Celsius. To convert a Celsius temperature, t_C, to a Fahrenheit temperature, t_F, we can use either of the following equations.

$$t_C = \frac{5\,°\text{C}}{9\,°\text{F}}(t_F - 32\,°\text{F})$$

$$t_F = \frac{9\,°\text{F}}{5\,°\text{C}}t_C + 32\,°\text{F}$$

Table 1.4 gives some common temperatures in both °C and °F.

In Taking Measurements, We Strive for Both Accuracy and Precision Most people use the terms *accuracy* and *precision* as if they mean the same thing, but they don't. The same is true about the terms *error* and *uncertainty*.

Accuracy refers to the closeness of a measurement to the true value. The **error** is the difference between an experimental value and the correct value. In an accurate measurement, the error is small. Perfect accuracy is an ideal, however. In the real world of measurements, we nearly always have to estimate the last digit of a physical quantity. You have probably noticed this in reading a speedometer or a mercury thermometer. Thus, there is always some element of *uncertainty* in a measurement because we are unsure of the last digit and try to estimate it.

We describe **uncertainty** by the *range* in values that we judge best encloses the true value. Suppose, for example, you want to test the *accuracy* of an automobile or bicycle odometer (mileage gauge) against mile posts set up by the highway department. If you've ever done this, you no doubt sensed the *uncertainty* of estimating the reading beyond the first decimal place. You might, therefore, record the reading as 5.1 ± 0.1 mi as you pass the 5-mile post, because you judge that you cannot read the odometer more closely than to a tenth of a mile. The symbol ± stands for "plus or minus," and what follows this symbol indicates how much uncertainty is carried in the last digit. By recording 5.1 ± 0.1 mi, the mileage is said to be between 5.0 and 5.2 (but closest to 5.1), so the range of uncertainty is 2 in the tenths position. Thus, here the tenths position holds the first uncertain digit. You can see that *uncertainty* does not directly supply information about the *accuracy* of the measurement. *Uncertainty* is only an estimate of how finely the number could be read at the time it was taken. *Accuracy*, we repeat, is the closeness of a measurement to the true value.

When a measurement can be repeated, we might take several as carefully as we can. Then we use the resulting data to calculate the **precision** of the measurement, a statement of how *reproducible* the measurements are. The results of the measurements are averaged, and the *precision* is calculated by some statistical index, such

■ We assume here that you are able to begin the test with your body exactly opposite the 0-mile post and that your body is exactly opposite the 5-mile post when you take your odometer reading. We also assume that the highway department has accurately positioned the posts. You can see that there are many sources of uncertainty in even an ordinary measurement such as this.

TABLE 1.4   Some Common Temperature Readings in °C and °F

	°F	°C
Room temperature	68	20
Very cold day	−20	−29
Very hot day	100	38
Normal body temperature	98.6[a]	37
Hottest temperature the hands can stand	120	49

[a] A revision in this value is currently underway. In some healthy people, the normal temperature is as low as 98.2 °F and in others as high as 99 °F.

as the average absolute deviation from the mean or by a standard deviation. These indices will not concern us, because seldom in routine laboratory work is more than one measurement taken of the same quantity. Our intent in defining precision has simply been to point out that *precision* and *accuracy* are not the same concepts. An average of several measurements, all agreeing very closely with each other (and so of high *precision*), might still have an average value greatly different from the accurate value (and so of great *error*); the instrument might have been inaccurately manufactured, for example. Figure 1.6 illustrates the difference between accuracy and precision.[2]

The Number of Significant Figures in a Physical Quantity Is the Number of Digits That Are Known To Be Accurate Plus One More   As we just noted, there is always some uncertainty in a measured quantity because we have to estimate the value of the last decimal place. How, then, should we handle derived quantities, which are

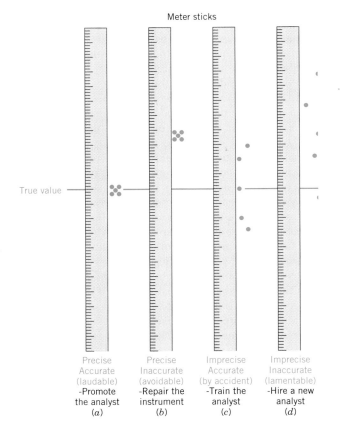

FIGURE 1.6   Accuracy and precision. Each dot represents one attempt at measuring a person's height. (*a*) High precision and great accuracy; the dots are tightly clustered by the true value. (*b*) High precision (tightly clustered dots) but poor accuracy (large error); perhaps the meter stick was not made correctly. (*c*) Poor precision (dots are not tightly clustered) but, by accident, high accuracy; the average of the measurements would be close to the true value. (*d*) No precision and no accuracy.

calculated from the (somewhat uncertain) values of base quantities? How do we deal with the uncertainty in the calculated result? Let's work a very simple case to illustrate the problem.

Suppose that you read a report that refers to "4.57 mg of antibiotic" but are told to weigh out a third of this quantity for the purposes of some experiment. We are certain that the "4" and the "5" in "4.57 mg" are accurate, but we're equally sure that the last digit was estimated. However, we can't tell from the number itself how closely the estimate was made. In other words, we have no information about the

[2] Suggested additional reading: Charles J. Guare, "Error, Precision, and Uncertainty," *Journal of Chemical Education*, August 1991, page 649.

range of uncertainty in this quantity, that is, about what should be in place of the question mark in "4.57 ± ? mg." The rule that we'll use in this text is that, unless told otherwise, the uncertainty is assumed to be plus or minus *one unit of the last decimal place.* Thus, in our example, 4.57 mg means 4.57 ± 0.01 mg.

To carry out the experiment, you next get out your pocket calculator and divide 4.57 by 3 and see 1.5233333 appear on the panel. Now what? The calculator knows nothing about uncertainty, but you instinctively know that even if you had a weighing balance accurate to the seventh decimal place (and there aren't any), trying to measure a dose of precisely 1.5233333 mg would be foolish. This kind of problem arises all the time in lab work, however, so scientists have devised rules for handling it, the *rounding-off rules.*

The goal of rounding off the numerical part of a physical quantity, like 1.5233333 mg, is to leave it with a definite and realistic number of *significant figures.* The number of **significant figures** in a physical quantity is the number of digits known with complete certainty to be accurate plus one. The quantity "4.57 mg," therefore, has three significant figures, two certain digits plus one. The first two digits, 4 and 5, are certain, but in the last digit there is a small uncertainty, assumed, by our rule, to be ±0.01 mg. By recording a value of 4.57 mg, the analyst is saying that the actual measurement of the mass of this one sample of antibiotic was closer to 4.57 mg than to 4.56 mg or 4.58 mg.

If the mass had been reported as 4.570 mg, then the quantity carries four significant figures. Three digits, 4, 5, and 7, are presumed to be certainly accurate, but some uncertainty resides in the last digit, 0. The measurement, as reported, means a value closer to 4.570 mg than to 4.569 mg or 4.571 mg. Thus "4.570 mg" implies a greater certainty or fineness of measurement than 4.57 mg.

Before we apply the concept of significant figures to rounding off, we need to know how to determine the number of significant figures in a physical quantity. This is easy, provided we have an agreement on how to treat zeros. Are all the zeros counted as *significant* in such quantities as 4,500,000 people, or 0.0004500 L, or 400,005 mi? We will use the following rules to decide.

Rules Governing Significant Figures

1. **Zeros sandwiched between nonzero digits are always counted as significant.** Thus, both 400,005 mi and 400.005 mi have six significant figures; the zeros occur between nonzero digits and so are counted. Similarly, both 4056 g and 4506 g have four significant figures.

2. **Zeros that do no more than set off the decimal point on their *left* are never counted as significant figures.** Thus, such quantities as 0.045 mL, 0.0045 mL, and 0.00045 mL all have only two significant figures. Although their zeros are necessary to convey the general *size* of a quantity, they don't say anything about the *certainty* of the measurement.

3. **Trailing zeros to the *right* of the decimal point are always significant.** Trailing zeros are any that come to the right of a decimal point at the very end of the quantity, as in the three trailing zeros in 4.56000 g. Because the three are to the *right* of the decimal point, all are significant. The quantity 4.56000 g thus has six significant figures and represents considerable certainty or fineness of measurement.

4. **Trailing zeros that are to the *left* of the decimal point are counted as significant only if the author of the book or article has somewhere said or implied so.** *In this text, if we leave trailing zeros before the decimal point, count them as significant figures.* Thus, 200 mL has three significant figures.

Rule 4 is the only tricky rule, but scientific notation helps deal with ambiguities, as we'll see next.

1.5 SCIENTIFIC NOTATION

Scientific notation expresses very large or very small numbers in exponential form to make comparisons and calculations easier.

■ Appendix A has a review of exponential numbers.

The typical human red blood cell has a diameter of 0.000008 m. Whether we want to write it, say it, or remember it, 0.000008 m is an awkward quantity, and to make life easier scientists have developed a method for recording very small or very large numbers called *scientific notation* or, by some, *exponential notation*. In **scientific notation**, a number is written as the product of two numbers, the first being a decimal number, usually between 1 and 10, followed by a times (×) sign and then the second number, which is 10 with an exponent. For example, we can rewrite the number 4000 as follows:

$$4000 = 4 \times 1000 = 4 \times 10 \times 10 \times 10$$
$$= 4 \times 10^3$$

■ When the decimal point is omitted, we assume that it is after the last digit in the number.

Notice that the exponent 3 is the number of places to the left that we have to move the decimal point in 4000 to get to 4, a number in the desirable range between 1 and 10.

$$4\,000$$
$$3\ 2\ 1$$

If our large number is 42,195, the number of meters in a marathon distance, we can rewrite it as follows after figuring out that we have to move the decimal point four places to the left to get a decimal number between 1 and 10.

$$42{,}195 \text{ m} = 4.2195 \times 10^4 \text{ m}$$

■
4 3 2 1

In rewriting numbers smaller than 1 in scientific notation, we have to move the decimal point to the *right* to get a number in the acceptable range of 1 to 10. The number of rightward moves is the value of the *negative* exponent of 10. For example, we can rewrite 0.000008 as

$$0.000008 = 8 \times 10^{-6}$$

■ 0.000008
1 2 3 4 5 6

You should not continue until you are satisfied that you can change large or small numbers into scientific notation. For practice, do the following exercises.

■ PRACTICE EXERCISE 2   Express each number in scientific notation.[3] Let the decimal part be a number between 1 and 10.

(a) 545,000,000   (b) 5,670,000,000,000   (c) 6454

(d) 25   (e) 0.0000398   (f) 0.00426

(g) 0.168   (h) 0.00000000000987 (See footnote 3.)

---

[3] Some of the numbers in this exercise illustrate a small problem that the SI is trying to get all scientists to handle in a uniform way. In part (h), for example, you might become dizzy trying to count closely spaced zeros. The SI recommends—and most European scientists have accepted the suggestion—that the digits in numbers having four or more digits be grouped in threes separated by thin

Prefixes to the Names of SI Base Units Are Used To Specify Fractions or Multiples of These Units   If we rewrite 3000 m as 3×10^3 m and try to pronounce the result, we have to say "three times ten to the third meters." There's nothing wrong with this, but it's clumsy. This is why the SI has names for several exponential expressions, not independent names but prefixes that can be attached to the name of any unit. For example, 10^3 has been assigned the prefix *kilo-*, abbreviated *k*. Thus 1000 or 10^3 meters can be called 1 kilometer. Abbreviated, this becomes 10^3 m = 1 km.

With just a few exceptions, the prefixes defined by the SI go with exponentials that involve powers of 3, 6, 9, 12, 15, and 18 or powers of -3, -6, -9, -12, -15, and -18. These are all divisible by 3 (see Table 1.5). You'll meet those given in boldface so often that you should memorize them now.

Notice in Table 1.5 that there are four prefixes that do not go with powers divisible by 3. The SI hopes their usage will gradually fade away, but this hasn't happened yet. The two in boldface have to be learned. However, *centi* is used almost entirely in just one physical quantity, the centimeter. *Deci* is limited almost completely to another physical quantity, the deciliter (100 mL or 1/10 L), and you won't see it often in strictly chemical situations. (Clinical chemists often use the deciliter because it saves space on clinical report sheets to abbreviate 100 mL to 1 dL.)

To take advantage of the SI prefixes, we sometimes have to modify a rule used in converting a large or small number into scientific notation. The goal in this conversion will now be to get the exponential part of the number to match one with an SI prefix even if the decimal part of the number isn't between 1 and 10. For example,

■ 1 dL = 1 $\times$ 10^{-1} L = 1/10 liter.
But 1/10 liter = 100 mL.
Therefore, **1 dL = 100 mL.**

TABLE 1.5   SI Prefixes for Multiples and Submultiples of Base Units[a]

Relationship	Prefix	Symbol
1 000 000 000 000 000 000 = 10^{18}	exa	E
1 000 000 000 000 000 = 10^{15}	peta	P
1 000 000 000 000 = 10^{12}	tera	T
1 000 000 000 = 10^{9}	giga	G
1 000 000 = 10^{6}	**mega**	**M**
1 000 = 10^{3}	**kilo**	**k**
100 = 10^{2}	hecto	h
10 = 10^{1}	deka	da
0.1 = 10^{-1}	**deci**	**d**
0.01 = 10^{-2}	**centi**	**c**
0.001 = 10^{-3}	**milli**	**m**
0.000 001 = 10^{-6}	**micro**	**μ**
0.000 000 001 = 10^{-9}	nano	n
0.000 000 000 001 = 10^{-12}	pico	p
0.000 000 000 000 001 = 10^{-15}	femto	f
0.000 000 000 000 000 001 = 10^{-18}	atto	a

[a] The most commonly used prefixes and their symbols are in boldface. Thin spaces instead of commas are used to separate groups of three zeros to illustrate the format being urged by the SI (but not yet widely adopted in the United States).

spaces. For large numbers, just omit the commas. Thus 545,000,000 would be written as 545 000 000. The number 0.00000000000987 becomes 0.000 000 000 009 87. You will not soon see this as common usage in the United States, but when you do you'll know what it means. Incidentally, European scientists use a comma instead of a period to locate the decimal point. You might see this yourself soon when you first weigh something in the lab. If the weighing balance was made in Europe, a reading of 1,045 g means 1.045 g.

we know that the number 545,000 can be rewritten as 5.45×10^5, but 5 isn't divisible by 3, and there isn't an SI prefix to go with 10^5. If we counted 6 spaces to the left, however, we could use 10^6 as the exponential part.

$$5\underset{6}{.}4\underset{5}{5}\underset{4}{0}\underset{3}{0}\underset{2}{0}\underset{1}{0} = 0.545 \times 10^6$$

■ We usually put a zero in front of a decimal point in numbers that are less than 1, such as in 0.545. The zero helps us to remember that the decimal point is there.

Now we could rewrite 545,000 m as 0.545×10^6 m or 0.545 Mm (megameter), because the prefix *mega*, abbreviated M, goes with 10^6.

We also could have rewritten 545,000 as 545×10^3, and then 545,000 m could have been written as 545 km (kilometers) because *kilo* goes with 10^3.

EXAMPLE 1.1   Rewriting Physical Quantities Using SI Prefixes

Bacteria that cause pneumonia have diameters roughly equal to 0.0000009 m. Rewrite this using the SI prefix that goes with 10^{-6}.

ANALYSIS   In straight exponential notation, 0.0000009 m is 9×10^{-7} m, but -7 is not divisible by 3 and no SI prefix goes with 10^{-7}. If we move the decimal six places instead of seven to the right, however, we get 0.9×10^{-6} m, and -6 is divisible by 3.

SOLUTION   The prefix for 10^{-6} is *micro* with the symbol μ, so

$$0.0000009 \text{ m} = 0.9 \times 10^{-6} \text{ m} = 0.9 \ \mu\text{m}$$

The diameter of one of these bacteria is 0.9 micrometers (0.9 μm).

■ PRACTICE EXERCISE 3   Complete the following conversions to exponential notation by supplying the exponential parts of the numbers.

(a) $0.0000398 = 39.8 \times$ _____
(b) $0.000000798 = 798 \times$ _____
(c) $0.000000798 = 0.798 \times$ _____
(d) $16500 = 16.5 \times$ _____

■ PRACTICE EXERCISE 4   Write the abbreviation of each of the following.

(a)   milliliter
(b)   microliter
(c)   deciliter
(d)   millimeter
(e)   centimeter
(f)   kilogram
(g)   microgram
(h)   milligram

■ PRACTICE EXERCISE 5   Write the full name that goes with each of the following abbreviations.

(a)   kg
(b)   cm
(c)   dL
(d) μg
(e)   mL
(f)   mg
(g)   mm
(h) μL

■ PRACTICE EXERCISE 6   Rewrite the following physical quantities using the standard SI abbreviated forms to incorporate the exponential parts of the numbers.

(a) 1.5×10^6 g
(b) 3.45×10^{-6} L
(c) 3.6×10^{-3} g
(d) 6.2×10^{-3} L
(e) 1.68×10^3 g
(f) 5.4×10^{-1} m

■ PRACTICE EXERCISE 7 Express each of the following physical quantities in a way that uses an SI prefix.

(a) 275,000 g   (b) 0.0000625 L   (c) 0.000000082 m

Scientific Notation Removes Ambiguities As we said, rule 4 for determining the number of significant figures in a physical quantity is really the only tricky rule. The zeros in 4,500,000 people, for example, are trailing zeros to the left of the decimal point, but are they really *significant*, as the term is used in science? Suppose this quantity stands for the population of a city, which changes constantly as people are born and die, and as they move in and out. No one could claim to know a population is *exactly* 4,500,000 people—not 4,499,999 and not 4,500,001, but exactly 4,500,000 people. More likely, the census bureau might be sure that the population is known to be closer to 4,500,000 people than to 4,400,000 or to 4,600,000 people. In other words, the "4" is certain; the "5" is uncertain (and so are all of the zeros). The number of certain figures in 4,500,000 people (one, namely, the "4") plus one more is 2, so only two significant figures should be in the result. Thus 4,500,000 people would be rewritten in scientific notation as 4.5×10^6 people to express two significant figures. If we know the population to three significant figures, namely, as 4.50×10^6 people, we are now sure that both the "4" and the "5" are correct but the trailing "0" *after* the decimal point is uncertain. Similarly, there are four significant figures in 4.500×10^6 people.

A Few Rules Govern the Rounding Off of Calculated Physical Quantities Finally, we're ready to deal with the matter of dividing 4.57 mg by 3, the calculator result being 1.5233333 mg. The uncertain digit in 4.57 mg is the "7," and notice that it stands in the second decimal place, which we can call the *position of uncertainty* in the quantity. When we mathematically rework a physical quantity by multiplying, dividing, adding, or subtracting, *we have to round the result so that it expresses the same amount of uncertainty allowed by the data.* Normally, such rounding is done at the *end* of a calculation (unless specified otherwise) to minimize the errors that can accumulate and grow when we round at intermediate steps in a multistep calculation. We will use four simple rules for rounding calculated quantities.[4]

■ Resist the impulse that some owners of new calculators have of keeping all the digits they paid for.

Rules for Rounding Calculated Results

1. **Multiplication or Division** When we multiply or divide quantities, the result is allowed no more significant figures than carried by the least certain quantity (the one with the fewest significant figures).

2. **Addition or Subtraction** When we add or subtract numbers, the result is allowed no more decimal places than are in the number having the fewest decimal places.

3. **When to Round Up** When the first of the digits to be removed by rounding is 5 or higher, round the digit to its left *upward* by one unit. Otherwise, drop it and all others after it.

4. **Exact Numbers** Treat *exact numbers* as having an infinite number of significant figures.

---

[4] When the actual range in uncertainty is known from experiment for each piece of data, the uncertainty in a calculated result can be expressed with greater care. See the reference in footnote 2.

An **exact number** referred to in rule 4 is any that we define to be so, and we usually encounter exact numbers in statements relating units. For example, all of the numbers in the following expressions are exact and, for purposes of rounding calculated results, have an infinite number of significant figures.

$$1 \text{ in.} = 2.54 \text{ cm (exactly, as } \textit{defined by law}\text{)}$$

$$1 \text{ L} = 1000 \text{ mL (exactly, by the } \textit{definition} \text{ of mL)}$$

The significance of having an infinite number of significant figures is that we don't let such numbers affect how we round results. It would be silly to say that the "1" in "1 L" has just one significant figure when we intend, by definition, that it be an exact number. To return to our example, that of taking a third of 4.57 mg, the divisor, "3," is an exact number; the "4.57 mg" has three significant figures. When we divide 4.57 mg by 3 and get 1.5233333, we must round the result to the second decimal place so that it also has only three significant figures. Thus you would try to weigh out 1.52 mg of the antibiotic. (The "3" that follows the "2" in 1.5233333 is less than five, so by rule 3 we drop it and all the rest.) Now we'll work some examples to give you experience in correctly rounding off.

EXAMPLE 1.2   Rounding the Result of a Multiplication or a Division

A floor is measured as 11.75 m long and 9.25 m wide. What is its area, correctly rounded by our rules?

SOLUTION

$$\text{Area} = (\text{length}) \times (\text{width})$$
$$= 11.75 \text{ m} \times 9.25 \text{ m}$$
$$= 108.6875 \text{ m}^2 \text{ (not rounded)}$$

But the measured width, 9.25 m, has only three significant figures whereas the length, 11.75 m, has four. By our rules, we have to round the calculated area to three significant figures.

$$\text{Area} = 109 \text{ m}^2 \text{ (correctly rounded)}$$

EXAMPLE 1.3   Rounding the Result of an Addition or a Subtraction

Samples of a medication having masses of 1.12 g, 5.1 g, and 0.1657 g are mixed. How should the total mass of the resulting sample be reported?

SOLUTION   The sum of the three values, obtained with a calculator, is 6.3857 g, which shows four places following the decimal point. However, one mass is precise only to the first decimal place, so by our rules we have to round to this place. The final mass should be reported as 6.4 g. Notice that the value of the second sample mixed, 5.1 g, says nothing about the second or third decimal places. We don't know whether the mass is 5.09 g or 5.11 g, or what; the sample just wasn't measured precisely. This is why we can't know anything beyond the first decimal place in the sum.

■ **PRACTICE EXERCISE 8** The following numbers are the numerical parts of physical quantities. After the indicated mathematical operations are carried out, how must the results be expressed?

(a) 16.4×5.8

(b) $5.346 + 6.01$

(c) 0.00467×5.6324

(d) $2.3000 - 1.00003$

(e) $16.1 + 0.004$

(f) $(1.2 \times 10^2) \times 3.14$

(g) $9.31 - 0.00009$

(h) $\dfrac{1.0010}{0.0011}$

1.6 FACTOR-LABEL METHOD IN CALCULATIONS

In calculations involving physical quantities, the units are multiplied or canceled as if they were numbers.

Many people have developed a mental block about any subject that requires the use of mathematics. They know perfectly well how to multiply, divide, add, and subtract, but the problem is in knowing *when*, a decision that no pocket calculator can make. We said earlier that 1 in. equals 2.54 cm, but, if asked to convert, say, 75.0 cm into inches, some people are stymied over whether to divide or multiply. Science teachers have worked out a method called the *factor-label method* for correctly setting up such a calculation and *knowing* that it is correct.

The **factor-label method** begins with a relationship between units, stated as an equation (such as 1 in. = 2.54 cm) and then restates the relationship in the form of a fraction, called a **conversion factor**. The quantity that we want to convert, like 75.0 cm, is then multiplied by the conversion factor. Identical units (the "labels") are multiplied or canceled as if they were numbers. If the units that remain for the answer are right, we know that the calculation was correctly set up. We can best learn how this works by completing our example of changing 75.0 cm into inches, but first let's see how to construct conversion factors.

The relationship, 1 in. = 2.54 cm, can be restated in either of the following two ways, and both are examples of conversion factors.

$$\frac{2.54 \text{ cm}}{1 \text{ in.}} \quad \text{or} \quad \frac{1 \text{ in.}}{2.54 \text{ cm}}$$

If we read the divisor line as "per," then the first conversion factor says "2.54 cm per 1 in." and the second says "1 in. per 2.54 cm." These expressions are thus alternative ways of saying that "1 in. equals 2.54 cm." *Any relationship between two units can be restated as two conversion factors.* For example,

$$1 \text{ L} = 1000 \text{ mL} \qquad \frac{1000 \text{ mL}}{1 \text{ L}} \quad \text{or} \quad \frac{1 \text{ L}}{1000 \text{ mL}}$$

$$1 \text{ lb} = 453.6 \text{ g} \qquad \frac{453.6 \text{ g}}{1 \text{ lb}} \quad \text{or} \quad \frac{1 \text{ lb}}{453.6 \text{ g}}$$

■ Some call the factor-label method the cancel-unit or the factor-unit method.

■ When we divide both sides of the equation 2.54 cm = 1 in. by 2.54 cm, we get

$$\frac{2.54 \text{ cm}}{2.54 \text{ cm}} = \frac{1 \text{ in.}}{2.54 \text{ cm}}$$

This only restates the relationship of the centimeter and the inch; it doesn't change it. The use of a conversion factor changes just *units*, not actual quantities.

■ **PRACTICE EXERCISE 9** Restate each of the following relationships in the forms of their two possible conversion factors.

(a) $1 \text{ g} = 1000 \text{ mg}$ (b) $1 \text{ kg} = 2.205 \text{ lb}$

To convert 75.0 cm into inches, let's call "75.0 cm" the *given*. We multiply the given by whichever one of the two conversion factors relating inches to centimeters that lets us cancel the unit no longer wanted and leaves the unit we want. Thus,

$$75.0 \; \cancel{\text{cm}} \times \frac{1 \text{ in.}}{2.54 \; \cancel{\text{cm}}} = 29.5 \text{ in. (rounded correctly from 29.527559)}$$

Notice how the units of "cm" cancel. Only "in." remains, and it is on top in the numerator where it has to be. Suppose we had used the wrong conversion factor.

$$75.0 \text{ cm} \times \frac{2.54 \text{ cm}}{1 \text{ in.}} = 190.5 \; \frac{\text{cm}^2}{\text{in.}} \text{ (not yet correctly rounded)}$$

■ The arithmetic is correct, but the result is still all wrong.

That's right. We *must* do to the units exactly what the times sign and the divisor line tell us, and (cm) times (cm) equals (cm)^2 just as $2 \times 2 = 2^2$. Of course, the resulting units, (cm)^2/in., make no sense, so we know with certainty that we can't set up the solution this way. The reliability of the factor-label method lies in this use of the units (the "labels") as a guide to setting up the solution. Now let's work an example.

EXAMPLE 1.4   Using the Factor-Label Method

How many grams are in 0.230 lb?

ANALYSIS In Table 1.3 we find that 1 lb = 453.6 g, so we have our pick of the following conversion factors.

$$\frac{453.6 \text{ g}}{1 \text{ lb}} \qquad \text{or} \qquad \frac{1 \text{ lb}}{453.6 \text{ g}}$$

To change the given, 0.230 lb, into grams, we want "lb" to cancel and we want "g" in its place in the numerator. Therefore, we pick the first conversion factor; it's the only one that can give this result.

SOLUTION

$$0.230 \text{ lb} \times \frac{453.6 \text{ g}}{1 \text{ lb}} = 104 \text{ g (correctly rounded)}$$

There are 104 g in 0.230 lb. (We rounded from 104.328 g to 104 g because the given value, 0.230 lb, has only three significant figures. Remember that the "1" in "1 lb" has to be treated as an exact number because it's in a definition.)

■ PRACTICE EXERCISE 10   The *grain* is an old unit of mass still used by some pharmacists and physicians, and 1 grain = 0.0648 g. How many grams of aspirin are in a tablet containing 5.00 grain of aspirin?

Often there is no single conversion factor that does the job, and two or more have to be used. For example, we might want to find out how many kilometers are in, say, 26.22 miles, but our tables don't have a direct relationship between kilometers and miles. However, if we can find in a table that 1 mile = 1609.3 m and that 1 km = 1000 m, we can still work the problem. We'll see in the next example how we can string two (or more) conversion factors together before the calculation of the final answer.

EXAMPLE 1.5 Using the Factor-Label Method. Stringing Conversion Factors

---

How many kilometers are there in 26.22 miles, the distance of a marathon race? Use the fact that 1 mile equals 1609.3 meters and any other relationships that are available.

ANALYSIS The fact that 1 mile equals 1609.3 meters gives us the following conversion factors.

$$\frac{1 \text{ mile}}{1609.3 \text{ m}} \quad \text{or} \quad \frac{1609.3 \text{ m}}{1 \text{ mile}}$$

We're given 26.22 miles, so we know that if we use the second conversion factor, the following calculation would convert miles into meters.

$$26.22 \text{ mile} \times \frac{1609.3 \text{ m}}{1 \text{ mile}}$$

If we paused to carry out this calculation, the answer would not be in kilometers (km). Therefore, *before doing this calculation*, we look for another conversion factor that, if possible, directly relates meters to kilometers and so would let us cancel "m" and replace it by "km." Knowing that "kilo-" stands for 1000, we know that the relationship, 1 km = 1000 m, is what we need.

SOLUTION

$$26.22 \text{ mile} \times \frac{1609.3 \text{ m}}{1 \text{ mile}} \times \frac{1 \text{ km}}{1000 \text{ m}} = 42.20 \text{ km (correctly rounded)}$$

The marathon distance is 42.20 km.

■ PRACTICE EXERCISE 11 Using the relationships between units given in the exercises or in tables in this chapter, carry out the following conversions. Be sure that you give the answers in the correct number of significant figures.

(a) How many milligrams are in 0.324 g (the aspirin in one normal tablet)?

(b) A long-distance run of 10.0×10^3 m is how far in feet? (This is the 10-km distance.)

(c) A prescription calls for 5.00 fluidrams of a liquid. What is this in milliliters? (8 fluidram = 1 liquid ounce.)

(d) One drug formulation calls for a mass of 10.00 drams. If only an SI balance is available, how many grams have to be weighed out? (16 drams = 1 ounce.)

(e) How many microliters are in 0.00478 L?

On page 11, equations were given relating degrees Celsius and degrees Fahrenheit. The use of these equations illustrates further examples of how to cancel units no longer wanted, as you can demonstrate by using the equations to work the following practice exercises.

■   PRACTICE EXERCISE 12   A child has a temperature of 104 °F. What is this in degrees Celsius?

■   PRACTICE EXERCISE 13   If the water at a beach is reported as 15 °C, what is this in degrees Fahrenheit? (Would you care to swim in it?)

1.7   DENSITY AND SPECIFIC GRAVITY

One of the important physical properties of a liquid is its density, its amount of mass per unit volume.

TABLE 1.6   Densities of Some Common Substances at 25 °C

Substance	Density (g/cm³)
Aluminum	2.70
Bone	1.7–2.0
Butter	0.86–0.87
Cement, set	2.7–3.0
Cork	0.22–0.26
Diamond	3.513
Glass	2.4–2.8
Gold	19.3
Iron	7.86
Marble	2.6–2.8
Mercury	13.534
Milk	1.028–1.035
Wood, balsa	0.11–0.14
ebony	1.11–1.33
maple	0.62–0.75
teak	0.98

■   The density of mercury changes only from 13.60 g/mL to 13.35 g/mL when its temperature changes from 0 °C to 100 °C, a density change of only about 2%.

Properties Are Called *Extensive* or *Intensive* According to Their Dependence on Sample Size   Both the mass of a chemical sample and its volume are examples of **extensive properties**, those that are directly proportional to the *size* of the sample. Length is also an extensive property.

An **intensive property** is independent of the sample's size. Temperature and color are intensive properties, for example. Generally, intensive properties disclose some essential quality of a substance that is true for any sample size, and this is why scientists find intensive properties particularly useful.

An Object's Density Is the Ratio of Its Mass to Its Volume   One useful intensive property of a substance, particularly if it is a fluid, is its density. **Density** is the mass per unit volume of a substance.

$$\text{Density} = \frac{\text{mass}}{\text{volume}}$$

The density of mercury, the silvery liquid used in some thermometers, is 13.5 g/mL (at 25 °C), making mercury one of the most dense substances known. In contrast, the density of liquid water at 25 °C is 1.0 g/mL. Table 1.6 gives the densities of several common substances.

Don't make the mistake of confusing *heaviness* with *denseness*. A pound of mercury is just as heavy as a pound of water or a pound of feathers, because a pound is a pound. But a pound of mercury occupies only 1/13.5 the volume of a pound of water.

The density of a substance varies with temperature, because for most substances the volume of a sample but not the sample's mass changes with temperature. Most substances expand in volume when warmed and contract when cooled, but the effect of such changes in volume on densities isn't great for liquids or solids. Table 1.7 gives the density of water at several temperatures. Notice that, when rounded to two significant figures, the density of water is 1.0 g/mL in the (liquid) range of 0 to 30 °C (32 to 86 °F).

One of the uses of density is to calculate what volume of a liquid to take when the problem or experiment specifies a certain mass. Often it is easier (and sometimes safer) to measure a volume than a mass, as we will note in the next example.

EXAMPLE 1.6   Using Density To Calculate Volume from Mass

Concentrated sulfuric acid is a thick, oily, and very corrosive liquid that no one would want to spill on the pan of an expensive balance, to say nothing of the skin. It is an example of a liquid that is usually measured by volume instead of by mass, but suppose an experiment called for 25.0 g of sulfuric acid. What volume (in mL) should be taken to obtain this mass? The density of sulfuric acid is 1.84 g/mL.

ANALYSIS The given value of density means that 1.84 g acid = 1.00 mL acid. This gives two possible conversion factors:

$$\frac{1.84 \text{ g acid}}{1 \text{ mL acid}} \quad \text{or} \quad \frac{1 \text{ mL acid}}{1.84 \text{ g acid}}$$

The "given" in our problem, 25.0 g acid, should be multiplied by the second of these conversion factors to get the unit we want, mL.

SOLUTION

$$25.0 \text{ g acid} \times \frac{1 \text{ mL acid}}{1.84 \text{ g acid}} = 13.6 \text{ mL acid}$$

Thus if we measure 13.6 mL of acid, we will obtain 25.0 g of acid. (The pocket calculator result is 13.58695652, but our rules require us to round this result to three significant figures.)

TABLE 1.7 Density of Water at Various Temperatures

Temperature (°C)	Density (g/mL)
0	0.9987
3.98	1.0000
10	0.9973
20	0.99823
25	0.99707
30	0.99567
35	0.99406
45	0.99025
60	0.98324
80	0.97183
100	0.95838

■ PRACTICE EXERCISE 14 An experiment calls for 16.8 g of methyl alcohol, the fuel for fondue burners, but it is easier to measure this by volume than by mass. The density of methyl alcohol is 0.810 g/mL, so how many milliliters have to be taken to obtain 16.8 g of methyl alcohol?

■ PRACTICE EXERCISE 15 After pouring out 35.0 mL of corn oil for an experiment, a student realized that the mass of the sample also had to be recorded. The density of the corn oil is 0.918 g/mL. How many grams are in the 35.0 mL?

The Concept of Specific Gravity Occurs in Clinical Work The **specific gravity** of a liquid is the ratio of the mass contained in a given volume to the mass in the same unit of the identical volume of water at the same temperature. Because the density of water is simple, namely, 1.0 g/mL (or extremely close to this over a wide temperature range), when we divide the density of anything else in units of g/mL by the density of water it's like dividing by 1, *but all the units cancel*. Specific gravity thus has no units, and a value of specific gravity is numerically so close to its density that we usually say they are numerically the same.

In clinical work, the concept of specific gravity surfaces most commonly in connection with urine specimens. Normal urine has a specific gravity in the range of 1.010 to 1.030. It's slightly higher than water because the addition of wastes to water usually increases the mass of a sample more rapidly than its volume. Thus, the more wastes contained in 1 mL of urine, the higher is its specific gravity, which makes measuring a urine specimen's specific gravity clinically important.

One of the important functions of the kidneys is to remove chemical wastes from the bloodstream and put them into the urine being made. The kidneys' mechanism for doing this does not remove from the blood those substances that ought to remain in the blood. The clinical significance, therefore, of a change in the concentration of substances dissolved in the urine is that it indicates a change in the activity of the kidneys. This might be the result of a kidney disease having caused substances that should stay in the blood to leak into the urine being made. Or it might mean that wastes are being generated somewhere else in the body more rapidly than the kidneys can remove them.

Figure 1.7 shows the traditional method of using a urinometer to measure the specific gravity of a urine specimen. This method, however, has largely been supplanted by another instrument, a refractometer, which needs only one or two drops of urine for a measurement. (*How* the refractometer does this is beyond the scope of our study.)

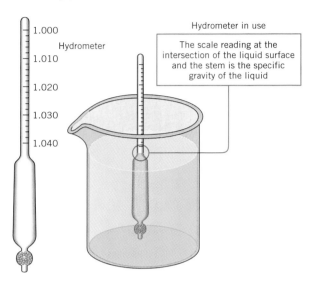

FIGURE 1.7 A hydrometer designed to serve as a urinometer.

Hydrometer in use

The scale reading at the intersection of the liquid surface and the stem is the specific gravity of the liquid

Hydrometer

1.000
1.010
1.020
1.030
1.040

SUMMARY

Chemistry and the molecular basis of life Down at the level of nature's tiniest particles, we find the "parts"—molecules—that nature shuffles from organism to organism in the living world. One of the many ways of looking at life is to examine its molecular basis, the way in which health depends on chemicals and their properties.

Scientific facts and physical quantities Scientific facts are usually capable of being checked by independent observers, and these facts often are physical quantities that pertain to physical properties, whether these are extensive or intensive properties. Physical properties—those that can be studied without changing the substance into something else—include mass, volume, time, temperature, color, and density. Physical properties distinguish the three states of matter, solid, liquid, and gas.

For our purposes, the important base quantities are (with the names of the SI base units given in parentheses) mass (kilogram), length (meter), time (second), temperature degree (kelvin), and quantity of chemical substance (mole). Except for temperature, all are extensive properties, those that depend on the size of a sample.

The liter is an important derived unit for volume, and volume is another extensive property. An important intensive physical property is the density of something, the ratio of its mass to its volume. It is often reported in units of grams per milliliter (g/mL).

Special prefixes can be attached to the names of the base units to express multiples or submultiples of these units. To select a prefix we have to be able to put very large or very small numbers into scientific notation.

Accuracy, error, uncertainty, and precision An accurate measurement is one that is true. The error in the measurement is the difference between the true value and what is observed. The range of uncertainty in a direct measurement is determined by the quality of the measuring device, how accurately it is set, how reliable (or steady) is its gauge, scale, or readout panel, and how acute is the observer. In this text, the range of uncertainty is assumed to be plus or minus one unit in the last decimal place allowed by the number of significant figures. The number of significant figures in a physical quantity equals the number of digits known to be true plus one more. When we add or subtract physical quantities, the decimal places in the result can be no more than the least number of decimal places among the original quantities. When we multiply or divide physical quantities, we have to round the result to show the same number of significant figures as are in the least precise original quantity.

Factor-label method The units of the physical quantities involved in a calculation are multiplied or canceled as if they were numbers. To convert a physical quantity into its equivalent in other units, we multiply the quantity by a conversion factor that permits the final units to be correct. The conversion factor is obtained from a defined relationship between the units.

REVIEW EXERCISES

The answers to Review Exercises whose numbers are in color are found in Appendix E. The answers to the other Review Exercises are found in the Study Guide that accompanies this book. The more challenging questions are marked with asterisks.

Molecular Basis of Life

1.1 Scientists who are *chemists* are most curious about what aspects of nature?

1.2 At what level of complexity—for example, whole organism, tissue, cell, or molecular—are different species most often able to exchange intact "parts?" What are these parts, in most general terms?

Hypotheses, Theories, and the Scientific Method

1.3 You're driving down a dry highway and your car starts to shake a little and to respond poorly to the steering wheel. You'll probably devise what? (What's the better term here, hypothesis or theory?) Explain.

1.4 What was initially called the vitamin *theory* eventually became established fact. Briefly, what made this shift possible?

1.5 What makes a fact a *reproducible* fact?

1.6 To a physician, nurse, or chemist, an observed fact has value for what purpose, if the scientific method is being applied?

1.7 What goal is sought when someone constructs a *hypothesis* involving several facts or observations?

*1.8 Which of the following two questions is more appropriate after you have devised a hypothesis?
(a) How can I prove the hypothesis?
(b) Is the hypothesis true or false?

1.9 What is the chief role of reason or logic in the application of the scientific method?

1.10 The question, "How do we get sick?" is more often successfully put to nature than the question, "Why do we get sick?" Explain.

Properties and the States of Matter

1.11 What are the names of the three states of matter?

1.12 A substance with an indefinite shape and volume is in what state?

1.13 When we speak of the *properties* of some substance, what is meant?

1.14 What is the basis for distinguishing between *chemical* and *physical* properties?

Physical Quantities and Measurements

1.15 What marks the difference between a *physical property* and a *physical quantity*?

1.16 In a word, what operation must be done to obtain a value for a physical quantity?

1.17 What is the general name we give to those fundamental quantities in terms of which all other physical quantities are defined? Name five of these physical quantities that are defined or mentioned in this chapter.

1.18 What is meant by the *inertia* of an object, and what is the name of the physical quantity used to describe this property?

1.19 Your *weight* on the moon would be less than your weight on earth, yet your *mass* is the same at both locations. Explain.

1.20 What makes it possible for us to say that we determine the *mass* of an object when the actual operation we use is *weighing* (and the instrument is a two-pan balance)?

1.21 In order for a physical quantity to have any meaning or any usefulness, what must be defined for it?

1.22 What is the *name* of the base unit for the following physical quantities?
(a) length   (b) time   (c) mass   (d) temperature
(e) quantity of chemical substance

1.23 In general terms, the *definition* of a base unit involves what kind of a standard? What organization has been responsible for these definitions?

1.24 Why is *volume* not called a *base* quantity?

1.25 What are the *names* of the SI base unit of length and the corresponding reference standard?

1.26 What are the *names* of the SI base unit of mass and its corresponding reference standard?

1.27 Examine each pair of quantities and state which is larger.
(a) meter and yard   (b) inch and centimeter
(c) gram and ounce   (d) millimeter and centimeter
(e) pound and kilogram   (f) kilogram and ton
(g) liter and quart   (h) microliter and milliliter
(i) ounce and pound   (j) gram and kilogram

1.28 How many milliliters are in 1 liter?

1.29 How many micrograms are in 1 milligram?

1.30 How many grams are in 1 kilogram?

1.31 Rewrite the following physical quantities using abbreviated units.
(a) 6.0 kilograms of sugar
(b) 500 milligrams of vitamin C
(c) 65 micrograms of melatonin
(d) 75.0 milliliters of Zantac solution
(e) 2.54 centimeters per inch
(f) 30 millimeter gap

1.32   Rewrite the following physical quantities with their units written out in full.
(a) 250 mL of coffee   (b) 25.0 kg of flour
(c) 45 μL venom   (d) 45.5 mg salt
(e) 65 dL Ringer's solution   (f) 36 cm = 360 mm

Degrees and Scales of Temperature

1.33   What is the value given to the point on a mercury-filled thermometer where the mercury level eventually comes to rest after the thermometer is immersed in an ice–water slush on each of the following scales of temperature?
(a) Celsius   (b) Kelvin   (c) Fahrenheit

1.34   When a mercury-filled thermometer is immersed into a container of boiling water (at sea level), what is the value given to the level the mercury eventually reaches on each of the following scales?
(a) Celsius   (b) Kelvin   (c) Fahrenheit

1.35   How many degree divisions (arbitrarily) separate the mercury levels for the freezing and the boiling points of water on each of the following scales?
(a) Celsius   (b) Kelvin   (c) Fahrenheit

1.36   Which is the larger degree, the Celsius degree or the Fahrenheit degree? By how much is it larger?

1.37   Which is the larger degree, the kelvin or the Fahrenheit degree? By how much is it larger?

1.38   When expressed in degrees Celsius, the zero point on the Kelvin scale has what value?

1.39   What is true about a temperature of 0 K?

1.40   A Canadian weather report said that the temperature at one remote reporting station was −40 °C. What is this in °F?

*1.41   If you read a Kelvin thermometer in your room as 278 K, would you be comfortable without a coat or sweater? (Do calculations as support for your answer)

1.42   An American visitor to Germany wanted to set a room thermostat for the degree Celsius equivalent of 68 °F. What should the setting be in °C?

*1.43   A clinical thermometer was used to take the temperature of a patient, and it registered 40 °C. Did the patient have a fever? (Do the calculation. Normal body temperature has traditionally been taken to be 98.6 °F, but many healthy individuals have normal temperatures slightly lower or slightly higher than this.)

Accuracy, Precision, and Error

*1.44   When a scale was used to take six successive measurements of a person's mass, the following data were recorded

   59.85 kg, 59.70 kg, 59.91 kg, 59.73 kg, 59.94 kg, 59.91 kg

The balance had earlier been tested against a set of official reference standard masses and found to be working exceptionally well. The true value of the mass was verified as 59.86 kg.
(a) Can the measurements be described as *accurate*? Explain.
(b) What, if anything, do the data disclose about their *uncertainty*?

1.45   What is the specific problem when a measurement is known to be in *error*?

1.46   How many significant figures are in each quantity?
(a) 65.0010   (b) 600.50   (c) 6000   (d) 0.00600

1.47   Round off the following physical quantity to the specified number of significant figures.

   160543 mi

(a) 1   (b) 2   (c) 3   (d) 4   (e) 5

Scientific Notation

*1.48   Restate the following physical quantities in scientific notation in which the decimal part of the number is between 1 and 10.
(a) 0.130 L   (b) 3568.5 m
(c) 0.00000420 g   (d) 0.0045 g

*1.49   How would the following physical quantities be re-expressed in scientific notation in which the decimal part of the number is between 1 and 10?
(a) 6324 g   (b) 0.000 000 006 78 L
(c) 8746 000 m   (d) 0.001 004 93 g

*1.50   Use a suitable SI prefix to express each of the quantities in Exercise 1.48.

*1.51   Use a suitable SI prefix to restate each of the quantities in Exercise 1.49.

1.52   The number of fans attending a professional football game was reported as 65,356. Restate this number in scientific notation, but retain only two significant figures.

1.53   The number of autos sold by a dealership in 1 year was reported as 1080. Give this number in scientific notation, retaining two significant figures.

1.54   The population of the United States in 1790, according to one almanac, was 3,939,214. Rewrite this figure in scientific notation, using three significant figures.

*1.55   Rewrite the following number according to the number of significant figures specified in each part. Express your answers in scientific notation.

   16,560,010.01

(a) one   (b) two   (c) three
(d) four   (e) five   (f) six

*1.56   Rewrite the following number according to the number of significant figures specified in each part. Give your answers in scientific notation.

   199,898.9091

(a) three   (b) four   (c) five
(d) six   (e) eight   (f) nine

1.57 The relationship between the milliliter and the microliter is

$$1 \text{ mL} = 1000 \ \mu\text{L}$$

How many significant figures are considered to be in each number?

*1.58 Consider that the following mathematical operations are calculations that involve physical quantities. (The units have been omitted.) Determine the significant figures that can be retained in the answer in each part according to our rules, and express the results of the calculations in the proper way. Use scientific notation; let the decimal part of the number be between 1 and 10.
(a) $4.665 \times 3.2 \times 10^{-5}$ (b) $6.3 \times 5.6000 \times 10^{3}$
(c) $4.005 \times 6.23 \times 10^{23}$ (d) $4.5 + 62.003$
(e) $6.004 - 3.2$ (f) $45.0023 + 0.023$
(g) $90.00 \div 3.0$ (h) $0.00050 \div 0.005$
(i) $6.40 \div 3.200$

Converting Between Units

1.59 Conversion factors relate identical physical amounts but that are expressed in different units. Write each of the following relationships between units in the forms of two conversion factors.
(a) 39.37 in. = 1 m   (b) 1 L = 1.057 qt
(c) 1 g = 1000 mg   (d) 1 kg = 2.205 lb
(e) 1 oz = 23.35 g   (f) $1 \ \mu\text{L} = 1000$ mL

*1.60 Given the relationships expressed in Review Exercise 1.59, which of the two quantities that follow the symbol, ≈ ("equals approximately"), most closely matches the quantity in the first column. You should develop the skill to make these kinds of judgments without doing an actual calculation using conversion factors.
(a) 0.50 m ≈ 20 in. or 80 in.
(b) 0.5 lb ≈ 4.4 kg or 0.23 kg
(c) 6 g ≈ 140 oz or 0.25 oz
(d) 4500 mg ≈ 4.5 g or 4.5×10^{6} g

1.61 Make the following conversions using relationships found in tables in this chapter. Do all calculations to three significant figures.
(a) Convert 163 cm into inches (the height of an adult female).
(b) Convert 154 lb into kilograms (the mass of an adult male).

1.62 Make the following conversions using relationships found in tables in this chapter. Do all calculations to three significant figures.
(a) Convert 111.5 lb into kilograms (the mass of an adult female).
(b) Convert 192 cm into inches (the height of an adult male).

1.63 The normal content of cans of popular soft drinks is 12.0 liquid ounces. How much is this in milliliters (to three significant figures)?

1.64 The popular-size bottles of mineral water hold 296 mL. How much is this in liquid ounces (to the proper number of significant figures)?

1.65 The gasoline tank of a small car holds 12.0 U.S. gallons. How much is this in liters?

*1.66 A physician prescribed 0.50 g of valinomycin. The pharmacy dispenses valinomycin in 250-mg tablets. How many tablets are needed for one prescribed dose?

*1.67 Valium is available in tablets containing 5 mg of this medication. How many tablets must be administered to give a dose of 0.015 g of Valium?

1.68 The highest mountain in the world, Mount Everest in Nepal, is 8847.7 m. What is this in feet?

1.69 The highest mountain in the United States is Alaska's Mount Denali, 20,322 ft. How high is it in meters?

1.70 A diamond rated as 2.50 carats has a mass of how many grams? (1 carat = 200 mg)

Density and Specific Gravity

1.71 Mass is called an extensive property, but density is an intensive property. Explain the difference.

*1.72 This book has dimensions of about $8.25 \times 10.0 \times 1.60$ in. Its mass is about 3.75 lb.
(a) Calculate its density in units of g/cm^{3}.
(b) What would be the mass of this book in kilograms and in pounds if it were made of solid lead? (Assume that the density of lead is $11.50 \ \text{g/cm}^{3}$.)

1.73 Corrosive chemical solutions are usually more safely measured by volume than by mass. To obtain 35.2 g of sulfuric acid (density, 1.84 g/mL), how many milliliters should be measured?

*1.74 Corn oil has a density of 7.60 lb/gal.
(a) Calculate its density in g/mL.
(b) How many milliliters of corn oil must be taken to get 250 g?

1.75 Specific gravity is defined such that the numerical value of something's density is virtually the same as its specific gravity. Explain.

1.76 If a urine specimen has an abnormally low value of specific gravity, what is known about the specimen?

1.77 A urinometer float rides higher in what kind of fluid, one with a high density or one with a low density?

1.78 Scarcely any change in volume occurs when 3.00 g of sugar is dissolved in 100 mL of water. Assuming no volume change, what is the specific gravity of this solution?

1.79 If the density of mercury is 13.5 g/mL at 25 °C, what is its specific gravity at the same temperature? (The density of water at 25 °C is 0.99707 g/mL.)

Vitamin C (Interaction 1.1)

1.80   In broad terms, how is ascorbic acid used in the body to prevent scurvy?

1.81   We could say that the ascorbic acid–scurvy story is just one question after the other. In your own words, state this series of questions. Where do they end?

Additional Exercises

1.82   What observations of nature led chemists to conclude that living things *must* have many common features?

1.83   You are talking by long-distance phone to a good friend when suddenly the line "goes dead." What is the most likely explanation that you would make, your first hypothesis to explain this observation of a dead line? Your hypothesis might be logical *based on previous experiences*, but is it therefore *necessarily* correct or incorrect? What would you do to test your hypothesis? Why would it be pretentious to dub your hypothesis a *theory*? (The last question has to do with how best to use the term)?

1.84   Scientists regard the SI reference standard for length as far more satisfactorily specified than the SI reference standard for mass? Give the reasons.

1.85   In testing an American recipe, a French baker had to decide how to set the French oven for a recipe specification of 320 °F. What oven setting in °C was needed?

1.86   Study the following numbers.
(A) 4.55×10^8 (B) 0.0455   (C) 45,500
(D) 0.00455   (E) 4550   (F) 4.550×10^{-3}
(G) 4.550   (H) 0.45500   (I) 4.5500×10^7
(a) Which of these numbers has three significant figures? (Identify them by their letters.)
(b) Which has four significant figures?
(c) Which has five significant figures?

1.87   The following mathematical operations are calculations that involve physical quantities. (The units have been omitted.) Determine how many significant figures can be retained, and express the results of the calculations in the proper way. Use scientific notation in which the decimal part of the number is between 1 and 10.
(a) $9600.00 \div 320.0000$ (b) $45.0 \div 1.50$
(c) $45.0 + 1.50$ (d) 45.0×1.50
(e) $45.0 - 1.50$ (f) 0.000009×1.1

*1.88   Given the relationships that can be found among the tables in this chapter, what quantity most nearly matches what is given in a different unit before the symbol $\approx$?
(a) 1 cup $\approx$ 50 mL or 200 mL (1 qt = 4 cups)
(b) 1 mile $\approx$ 1.6 km or 0.66 km
(c) 10 mm $\approx$ 2.5 in. or 0.25 in.
(d) 1000 mL $\approx$ 1 μL or 1 L

*1.89   While driving on a country road in a European country you come to a bridge limited to 1.4×10^3 kg. Your vehicle has a mass of 4.5×10^3 lb. Should you cross? (Do the calculation.)

*1.90   An IV (intravenous) solution contains 25 mg of a drug per 5.0 mL of solution. You are to administer 0.75 g of the drug. What volume of the solution should be used?

*1.91   A vial of a medication carries a label instructing the user to add 7.50 mL of water to the contents of the vial to obtain a solution containing 25.0 mg of the active drug per mL of the solution. How many milliliters of this solution must be taken to obtain 0.175 g of the drug?

1.92   One pound of butter can be made into 128 equal-sized pats of butter. What is the mass of each pat in grams?

1.93   The density of aluminum is 2.70 g/cm^3. A block of aluminum with a volume of 250 mL (about 1 cup) has a mass of how many grams? Pounds?

1.94   Liquids and solutions generally expand in volume as they are warmed.
(a) Assuming no loss by evaporation, does the mass of a sample change as it is warmed?
(b) When a liquid sample is warmed, does its density increase, decrease, or stay the same?

1.95   What is the specific gravity of water?

1.96   Which of the two members of each of the following pairs is the larger physical quantity?
(a) 3 ft or 1 yd   (b) 2 qt or 2 L
(c) 10^3 μL or 10^2 mL   (d) 15 mm or 5 cm
(e) 10 in. or 250 mm   (f)   10 kelvin or 12 degrees of Fahrenheit

1.97   Reexpress the following physical quantity with the specified number of significant figures.

$$6.0604 \times 10^3 \text{ g}$$

(a) 1   (b) 2   (c) 3   (d) 4

*1.98   Rehydration therapy is a life-saving procedure for victims of cholera, who typically lose large amounts of fluid. An English physician, Thomas Latta, was the first to use this procedure during the cholera epidemic in London in 1832. The solution he had the victims drink contained 3.0 drachmas of sodium chloride and 2.0 scruples of sodium bicarbonate per 6.0 pints of water. [1 drachma = 60 grains; 1 ounce (avdp) = 480 grains; 1 ounce (avdp) = 28.35 g; 1 scruple = 20 grains]
(a) How many milligrams of sodium chloride were in each 6-pint unit of the solution? How many would be in 1 L of solution?
(b) How many milligrams of sodium bicarbonate were in each 6-pint unit of the solution? How many would be in 1 L of solution?

MATTER AND ENERGY

2

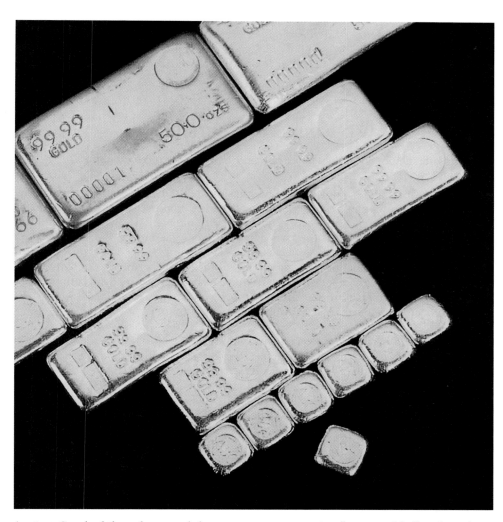

Ancient Greek philosophers used the term *atomas,* meaning "not cuttable," to describe the smallest gold particle that could exist. Today we call it the gold *atom.* The pieces made by breaking a gold atom are electrons, protons, and neutrons, not smaller pieces of gold. We'll learn about atoms in this chapter.

THIS CHAPTER IN CONTEXT

Chapter 1 gave us bearings for our travels into the molecular basis of life. Before we can say much about molecules, however, we need more systematic knowledge about matter and energy, the chief topics of this chapter. Be sure to pick up the idea of molecules in motion, a concept that carries forward to most chapters. We also need tools that simplify chemical descriptions of life at the molecular level, so we'll learn here about atomic symbols, symbols for compounds, and equations for describing chemical reactions. Thus, this chapter is basically about matter and energy and certain tools.

2.1 STATES AND KINDS OF MATTER

Elements and compounds have definite compositions, but mixtures do not.

Matter is anything that occupies space and has mass. It includes literally everything, and because there is such a huge variety of matter, making sense out of it might seem impossible. Throughout history, however, we humans have had a powerful impulse to sort and classify whenever we face what is very complex. Biologists, for example, created kingdoms, phyla, species, subspecies, and groups for plants and animals. One reason for sorting and classifying is to focus our minds on patterns in what we see around us that suggest underlying causes. We've already mentioned one classification of matter, its three *states,* namely, solids, liquids, and gases. Now let's turn to an even more basic classification, the *kinds* of matter.

The Three *Kinds* of Matter Are Elements, Compounds, and Mixtures   Any sample of matter consists of one or more *pure substances* of which there are basically two kinds, *elements* and *compounds.*[1] *Mixtures,* the third kind of matter, are simply combinations of elements or compounds jumbled together in no particular proportion and without having chemically reacted. In this chapter we'll define elements, compounds, and mixtures and return to them for a deeper study in later chapters.

Elements Cannot Be Changed into Simpler Pure Substances   As their name implies, elements are elementary. An **element** is a pure substance that cannot be broken down into simpler pure substances. Very familiar elements include aluminum, iron, copper, silver, and gold, as well as the chief components of air, oxygen and nitrogen. Water is not an element because we can break water down into two elements, oxygen and hydrogen, and we can cause them to recombine to give water.

A table of the 109 known elements is inside the front cover of this book, but throughout our study we'll be concerned with only about a dozen. Of the bulk mass of the human body, 99% consists of substances made from only four elements, carbon, nitrogen, hydrogen, and oxygen. The remainder of the body is made from at least 21 other elements (possibly 24), but all are vitally important. Many are the "trace elements" of nutrition and must be regularly replenished by the diet.

Ninety elements occur naturally; the rest have been made by scientists using special equipment. The synthetic elements plus a few that occur naturally exhibit **radioactivity,** the emission of one or more kinds of radiation. We will return to their special properties in a later chapter.

■ Uranium, plutonium, and radium are naturally occurring radioactive elements.

---

[1] To chemists (but not always to nonchemists), the word "pure" in *pure substance* is unnecessary. When chemists say "substance," they mean *pure* substance. However, we will often use the two-word term for the sake of emphasis.

At room temperature, 2 elements are liquids, 11 are gases, and the rest are solids. All but about 20 elements are *metals*. **Metals** have shiny surfaces (when polished), can be hammered into sheets and drawn into wires, and are generally good conductors of electricity and heat.

Sometimes two or more metals are melted, mixed together, and allowed to cool to give a solid mixture of metals called an **alloy.** Steel, for example, is actually the name for a family of alloys in which the principal element is iron. The many iron alloys include chromium steel and nickel steel, which have exceptional strength and which resist corrosion far better than does iron alone. A few alloys are used to replace bones or to strengthen them, so they must be unusually resistant to corrosion.

Several of the solid elements, like carbon and sulfur, plus all the gaseous elements are **nonmetals.** Nonmetals cannot be worked into sheets or wires, and they do not conduct electricity or heat as well as the metals. Metals and nonmetals have somewhat opposite *chemical* properties, and large numbers of compounds, like table salt, are made by combining metals with nonmetals.

Compounds Are Made From Elements   Chemical **compounds** are pure substances made from two or more elements *always* combined in a proportion by mass that is both definite and unique for the compound. When water, for example, is broken down into hydrogen and oxygen, the elements are invariably obtained in a mass ratio of 2.0 g of hydrogen to 16.0 g of oxygen. Yet water is not just a *mixture* of hydrogen and oxygen taken in this mass ratio. Such a mixture can be prepared (but only by very experienced chemists), and at room temperature it is in the gaseous state, not the liquid state. But this mixture is extremely unstable; given the slightest spark or exposure to ultraviolet light, it explodes and water forms. The vitamin C in lemon and orange juice as well as their citric acid and fruit sugar are all chemical compounds. Their constituent elements are carbon, hydrogen, and oxygen, but the proportions are different for each.

Table salt is also a compound. It's made from the elements sodium and chlorine, and it is somewhat amazing that salt, so essential to life, is made from two very dangerous elements. Sodium (see Figure 2.1) is a shiny metal (but only when freshly cut); it combines vigorously with both the oxygen and the moisture in humid air. Chlorine (see Figure 2.2) is a greenish-yellow, poisonous gas. When mixed, sodium and chlorine combine violently (see in Figure 2.3) to give salt, sodium chloride, a compound needed by all animals.

■ Mercury is a metallic element but a liquid at room temperature.

■ The current U.S. nickel coin is actually an alloy of copper (75%) and nickel (25%).

■ Diamonds consist of one form of pure carbon. Graphite is another form.

FIGURE 2.1   The shiny, metallic luster of sodium will soon fade, because this soft, easily cut metal reacts quickly with both oxygen and moisture in air.

FIGURE 2.2   Chlorine, a pale yellowish-green gas, is a poison. Warring sides used it in World War I as a weapon.

FIGURE 2.3   When a small piece of sodium metal, speared on the tip of a pointed glass rod, is thrust into chlorine, the reaction produces an instant shower of light and heat as sodium chloride forms.

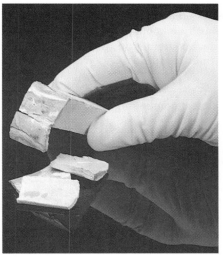

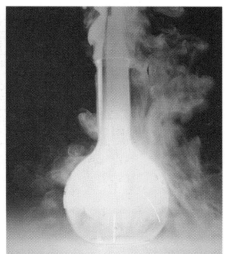

FIGURE 2.1   **FIGURE 2.2**   **FIGURE 2.3**

■ In chemistry, the opposite of *pure* is not *impure,* because *impure* carries an extra biological meaning, such as "dangerous to human health." The opposite of *pure substance* in chemistry is *mixture.*

Mixtures Have Variable Compositions   A **mixture** consists of two or more pure substances that are present in a proportion that can vary considerably. Each pure substance contributes something to the mixture's overall properties. Freshly squeezed orange juice, for example, consists of water, vitamin C, a little fruit sugar, citric acid, and other substances. If you remove any of them, you would notice the change (although the removal of some minor components, like vitamin C, would not be noticed right away). You probably can sense that if we want to understand any mixture, like orange juice, we have to study its individual pure substances. To recapitulate, there are three *states* of matter: solid, liquid, and gas. There are three *kinds* of matter: elements, compounds, and mixtures. There are two kinds of *pure substances:* elements and compounds.

Chemical Reactions Change Substances into Other Substances   Compounds are made by **chemical reactions,** events in which substances called **reactants** change into different substances called **products.** In lab work we tell reactants and products apart because they almost always have at least some quite different physical properties. Sodium and chlorine, two reactants, and sodium chloride, a product of their reaction, certainly illustrate such differences. Chemical reactions always feature at least some changes in physical properties as reactants change over into products.

■ Huge quantities of both sodium and chlorine are made annually by passing electricity through molten salt (sodium chloride).

Laws of Chemical Combination Govern Chemical Composition   Sodium chloride, as we said, is a *compound,* not just a mixture of the elements. A stable *mixture* of sodium metal and chlorine gas cannot actually be prepared. Yet it is possible to break down sodium chloride into its elements. (It's done by passing a current of electricity through *molten* salt in such a way that the sodium and chlorine emerge into separate containers.) The breakdown of sodium chloride is also a chemical reaction, because the initial substance changes over into different substances.

Every time sodium chloride is changed into its elements, or every time sodium and chlorine are allowed to react to form sodium chloride, the ratio of chlorine to sodium is invariably 1.5421 g of chlorine to 1.0000 g of sodium. A sample of sodium chloride obtained from any place in the world would have this ratio of chlorine to sodium. No matter in what proportion you might initially mix sodium and chlorine, the reacting proportion would be exactly 1.5421 g of chlorine to 1.0000 g of sodium. You might have either chlorine or sodium left over, depending on how carefully you mixed these elements, but for every 1.0000 g of sodium that reacts, 1.5421 g of chlorine would react as well, to form 2.5421 g of sodium chloride.

All compounds display definite proportions of their elements. This truth is so universal that we put it into our *definition* of a compound. A **compound** is a substance made from two or more elements chemically combined in a definite proportion by mass. This is one of the laws of chemical combination, the **law of definite proportions,** and the first of the scientific laws that we study.

> **Law of Definite Proportions**   In a given chemical compound, the elements are always combined in the same proportion by mass.

Lemonade is little more than lemon-flavored sugar-water, and its sweetness can be varied from nearly sour to syrupy sweet.

Elements also obey this law, because an element consists of 100% of itself. Chemists call elements and compounds **pure substances** because their compositions obey the law of definite proportions. The expression "pure substance" is thus actually redundant. When we call something (in this text) a *substance,* we mean a *pure* substance (element or compound), one that obeys the law of definite proportions.

Mixtures, in contrast, can be prepared in widely varying proportions of their components. For example, we can prepare mixtures (solutions) of sugar and water, two compounds, in almost any proportion we please. Moreover, we can separate the two simply by letting the water evaporate—a physical change because only a change in physical state occurs. Mixtures, in general, require only physical changes or operations to be separated into their components.

Another law of chemical combination was suggested when we pointed out that 2.5421 g of sodium chloride forms when 1.0000 g of sodium combines with 1.5421 of chlorine. The mass of the product is the sum of the masses of the reactants. This mass relationship between reactants and products has always been observed for chemical reactions, and these observations are behind the **law of conservation of mass.**

Law of Conservation of Mass   In any chemical reaction, the sum of the masses of the reactants always equals the sum of the masses of the products.

Any widespread regularity in nature generally suggests some very basic truth about the world in which we live, and the laws of chemical combination are examples of regularity that draw us to the question: "What must be true about substances to explain these laws?"

2.2   ATOMS, CHEMICAL SYMBOLS, AND EQUATIONS

Dalton saw that the laws of chemical combination virtually compel the belief that atoms exist.

John Dalton (1766–1844), an English scientist, was the first to find a reasonable explanation for both definite proportions in compounds and the conservation of mass in reactions.

Dalton's Atomic Theory Proposed Indestructible Atoms   Dalton reasoned that matter must be made of very tiny, individual particles that undergo a variety of chemical reactions *without breaking apart or losing any mass.* In order to explain the *definite* compositions of compounds, he said that these tiny particles simply cannot exist as major fragments of themselves. Each particle is an unbreakable unit.

The idea of a tiny, invisible, unbreakable particle had been around for centuries, because ancient Greek philosophers had proposed it. The Greek word for "not cut" is *atomos,* and from this term came our word *atom.* Dalton revived this ancient belief in "not cuttable" particles with the enormously important difference that he had solid evidence, namely, the laws of chemical combination.

The chief postulates of **Dalton's atomic theory** are the following.

1. Matter consists of definite particles called **atoms.**

2. Atoms are indestructible.

3. All atoms of one particular element are identical in mass.

4. Atoms of different elements have different masses.

5. By becoming bound together in different ways, atoms form compounds in definite ratios by *atoms.*

John Dalton

Not all of Dalton's postulates turned out to be correct, as we'll see in later chapters. His theory, however, was never meant to explain everything about atoms and compounds, only something. The only way, said Dalton, that we can observe definite ratios *by mass* in compounds is that they possess definite ratios of atoms, each kind of atom having its own unique mass. Let's see how Dalton's theory works for compounds of iron and sulfur.

The elements iron and sulfur can be made to combine in the ratio of 1.000 g of iron to 0.574 g of sulfur to form one kind of iron sulfide with the formal name iron(II) sulfide. Dalton reasoned that if the *atoms* of these two elements combine in a ratio of 1 *atom* of iron to 1 *atom* of sulfur, each atom with its own mass, *then the mass ratio cannot help but be a constant* (see Figure 2.4).

■   Sulfur forms more than one compound with iron, so we need a special name to designate one of them. The "(II)" in the formal name, iron(II) sulfide, uniquely defines the sulfide described here.

55.8g Fe
(6.02 × 10²³ atoms)
32.1g S
(6.02 × 10²³ atoms)
87.9g FeS
(6.02 × 10²³ formula units)

FIGURE 2.4   Mass ratio versus atom ratio in iron(II) sulfide. If the mass of 1 sulfur atom is 0.574 times the mass of 1 iron atom, then combining sulfur and iron atoms in a simple 1 to 1 ratio (by atoms), regardless of how much this is scaled up, must result in a constant ratio by mass.

The Discovery of the Law of Multiple Proportions Compelled Belief in Atoms
Powerful evidence for Dalton's atomic theory came from the study of different compounds that can be made from the same elements but in different mass ratios. The mineral pyrite is another compound that can be made from iron and sulfur, but the ratio by mass in pyrite is 1.000 g of iron to 1.148 g of sulfur. In iron(II) sulfide, described earlier, the mass ratio is 1.000 g of iron to 0.574 g of sulfur. Notice that 1.148 is exactly twice the size of 0.574. There is a *whole-number* ratio between the grams of sulfur combined with 1.000 g of iron in the two compounds. If Dalton is right, there *must* be a whole-number ratio because atoms combine as whole units, as whole particles. Figure 2.5 shows why this is so. It shows several possible combinations of intact iron and sulfur atoms. Three of the four compounds in Figure 2.5 are known, namely the first, second, and the fourth.

■   Pyrite is often found as golden crystals embedded in rock samples. Many a novice gold miner felt an increased heartbeat on finding what more experienced miners called "fool's gold."

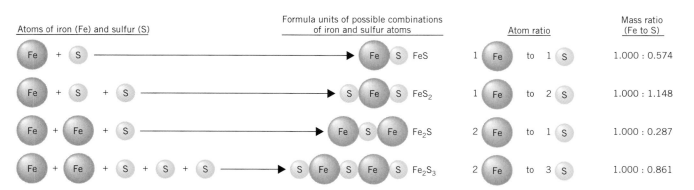

FIGURE 2.5   Possible multiple proportions for combinations of iron atoms and sulfur atoms. If atoms remain essentially intact (and suffer no detectable loss in mass) when they form compounds, then they must assemble in *whole-number ratios by atoms*, regardless of the masses of the individual atoms.

Two compounds of tin and oxygen further illustrate the special relationship among compounds made from the same elements. In one compound of tin and oxygen, 1.000 g of oxygen is combined with 3.710 g of tin. In another, 1.000 g of oxygen is combined with 7.420 g of tin. Now compare 7.420 g of tin with 3.710 g of tin.

$$\frac{7.420}{3.710} = 2$$

The ratio of the quantities of tin in the two compounds that combine with the same mass of oxygen is a ratio of small, whole numbers, 2 to 1. These and several other examples led to the third law of chemical combination, the **law of multiple proportions.**

> **Law of Multiple Proportions**   Whenever two elements form more than one compound, the different masses of one that combine with the same mass of the other are in the ratio of small whole numbers.

Our examples illustrating this law, plus many others, combined with Dalton's shrewd interpretations, convinced scientists that atoms exist. So much additional evidence for atoms has accumulated that their existence is taken as fact, not theory.

Chemical Formulas Describe a Substance's Composition   Chemists use **chemical formulas** as symbols for compounds because such symbols, unlike, say, nicknames, carry important information. They tell us, for example, what elements are combined in the compound and in what ratio by atoms.

To construct formulas bearing the information we want them to have, each element has been assigned an **atomic symbol** consisting of one or two letters. Those with which we will most often work are given in Table 2.1, and the complete list of atomic symbols appears in a table inside the front cover of this book.

Many elements, like those in the first column of Table 2.1, have single-letter symbols, usually (but not invariably) the capitalized first letter. Because there are more elements than letters, several elements have names beginning with the same letter—for example, carbon, calcium, chlorine, chromium, cobalt, and copper. Many atomic symbols, therefore, consist of the first two letters with the first letter *always* capitalized and the second letter always lower cased. Examples are in the second column of Table 2.1. The third column shows how the first letter and a letter occurring beyond the second place in the name are combined to make a symbol. Thus chlorine has the symbol Cl and chromium is Cr. The last column of Table 2.1 lists some elements named long ago when Latin was the almost universal language of educated people, and so the symbols of some elements were derived from Latin names,

■ Tin is the metal used to coat the inner surfaces of "tin" cans. Tin, unlike iron and less expensive steels, won't corrode in an environment of food juices.

■ Among the trickier, but still common, pairs of chemical symbols are

P = phosphorus
K = potassium

S = sulfur
Na = sodium

I = iodine
Fe = iron

TABLE 2.1   Names and Symbols of Some Common Elements[a]

C Carbon	Al Aluminum	Cl Chlorine	Ag Silver (*argentum*)
H Hydrogen	Ba Barium	Mg Magnesium	Cu Copper (*cuprum*)
O Oxygen	Br Bromine	Mn Manganese	Fe Iron (*ferrum*)
N Nitrogen	Ca Calcium	Pt Platinum	Pb Lead (*plumbum*)
S Sulfur	Li Lithium	Zn Zinc	Hg Mercury (*hydrargyrum*)
P Phosphorus	Si Silicon	As Arsenic	K Potassium (*kalium*)
I Iodine	Co Cobalt	Cs Cesium	Na Sodium (*natrium*)
F Fluorine	Ra Radium	Cr Chromium	Au Gold (*aurum*)

[a] The names in parentheses in the last column are the Latin names from which the atomic symbols were derived.

as shown. Students often find the symbols for sodium (Na) and potassium (K) the trickiest, so be sure to take some extra time to fix these firmly in mind.

Empirical Formulas Give the Ratios of Atoms Combined in Compounds   The simplest chemical formula is the **empirical formula.** In it the atomic symbols of the elements making up the compound are displayed in the *smallest* possible whole number ratio that corresponds to the compound's actual atom ratio. The formula FeS for iron(II) sulfide, for example, is an empirical formula. The ratio that distinguishes this compound from all other iron–sulfur compounds is 1 Fe atom to 1 S atom. Notice that the two atomic symbols, one each of Fe and S, are written together without spaces between. We must now note that because atoms are so amazingly tiny even the tiniest speck of FeS has billions of Fe and S atoms. The formula FeS says only that their *ratio* is 1:1.

In pyrite, another iron–sulfur compound, the combining ratio is 1 atom of iron to 2 atoms of sulfur. Thus the empirical formula of pyrite is written as FeS_2. The 2 is called a *subscript,* and subscripts in formulas always *follow* the symbols to which they refer. In Fe_2S_3, the empirical formula of still another compound of iron and sulfur (given in Figure 2.5), the ratio is 2 atoms of Fe to 3 atoms of S. You can see that empirical formulas give definitive chemical information about a compound's composition (but say nothing about the pattern of their arrangement).

The subscript 1 is always "understood" in chemical formulas. We write the formula of iron(II) sulfide as FeS, not as Fe_1S_1. We do not write the formula as Fe_2S_2 or as Fe_3S_3 either, although both 2:2 and 3:3 are equivalent to a 1:1 ratio. By convention, however (not by law of nature), empirical formulas have the *smallest* whole numbers that specify the ratio by atoms.

We might have conveyed the identical information about FeS by writing the formula as SFe, but the standard convention is that the symbols of metal elements are generally placed before those of nonmetals.

Generally speaking, empirical formulas are used for compounds between metals and nonmetals, like FeS or NaCl. Many, in fact, most compounds are made of two or more nonmetals, like carbon dioxide, CO_2, and methane (natural gas), CH_4. Notice what is a general rule; when one of the nonmetals is carbon its symbol comes first.

It is too early in our study to go more deeply into the rules for writing formulas and names of compounds, so don't worry about this phase of our study yet. For the present, just be sure you can spot the difference between say, FeS and fes, or that you can tell that CO can't possibly be the symbol for an element (the second letter is not lower cased) but that Co might be. Let's now work an example to illustrate how to use the rules and to become more familiar with chemical formulas and the use of subscripts.

■ The formal name of Fe_2S_3 is iron(III) sulfide. Why the "(III)" is used will be explained in a later chapter.

■ There are also *molecular formulas* and *structural formulas,* each differing in the amount of chemical information it carries.

■ CO is the formula of carbon monoxide—a compound made of carbon, C, and oxygen, O—and Co is the atomic symbol of cobalt.

EXAMPLE 2.1   Writing Chemical Formulas from Atomic Compositions

Aluminum, a metal with the symbol Al, and sulfur, a nonmetal with the symbol S, form a compound in which the atom ratio is 2 atoms of Al to 3 atoms of S. Write the formula.

ANALYSIS   The symbol for aluminum has to come first because aluminum is a metal.

SOLUTION   The formula is Al_2S_3 in which "2" goes with Al and "3" with S.

EXAMPLE 2.2 Writing Chemical Formulas from Atomic Compositions

Carbon, a nonmetal with the symbol C, and chlorine, another nonmetal but with the symbol Cl, form a compound in which the ratio of atoms is 1 of C to 4 of Cl. Write the formula.

ANALYSIS By convention, the symbol of carbon comes first.

SOLUTION The formula is CCl_4. (The subscript, 1, of carbon is understood.)

■ PRACTICE EXERCISE 1   Write the formula of the compound between sodium, a metal, and sulfur in which the atom ratio is 2 atoms of sodium to 1 atom of sulfur.

■ PRACTICE EXERCISE 2   Give the names of the elements that are present and the ratio of their atoms in K_2CO_3.

A Formula Unit Is a Real or Hypothetical Particle Having the Composition Given by the Formula of the Compound   If Fe stands for an atom, what kind of particle does FeS stand for? The most *general* name we can give to the particle having the composition of the formula of a compound is **formula unit.** There are special names for certain kinds of formula units, such as *atom, molecule,* or *set of ions* or *ion group,* but *formula unit* embraces all of them. One formula unit of FeS is made from *one* iron atom and *one* sulfur atom. One formula unit of FeS_2 is made from *one* iron atom and *two* sulfur atoms.

As we indicated, atoms are *formula units,* too. The chemical formula for sodium, for example, is its atomic symbol, Na; hence one formula unit of sodium is one atom of sodium.

The formula units of most substances, particularly those made only of nonmetals, are discrete, individual units called *molecules.* A **molecule** is an actual particle made from two or more atoms that are joined together and *that then exist together as a single, independent "package."* Water, for example, consists of water molecules (H_2O), each one being made from two atoms of hydrogen and one of oxygen, each one capable of moving around. No such *independent* packages of atoms that can move around exist in sodium chloride, NaCl, or iron(II) sulfide, FeS. In these substances, the atoms making them up occur in vast networks, like enormous jungle gyms made of endlessly repeating parts in which the ratios of the parts are in simple, whole numbers. The parts, being locked together, are not free to migrate independently, although they do jiggle about their fixed positions.

Chemical Equations Use Formulas to Describe Reactions   Now that we know something about chemical formulas we can learn how to use them in describing chemical reactions by means of equations. A **chemical equation** is a shorthand description of a reaction that groups the symbols of the reactants, separated by plus signs, on one side of an arrow and places the symbols of the products, also separated by plus signs, on the arrowhead side of the arrow. A very simple example is the equation for the formation of FeS from iron and sulfur. (The "translation" of the equation appears beneath it.)

$$Fe + S \longrightarrow FeS$$

Iron reacts with   to give   iron(II) sulfide
sulfur in a ratio
of 1 atom of Fe to
1 atom of S

Read the + sign separating reactants as "reacts with"; the arrow means "to give."

A more complicated example of an equation describes the formation of aluminum sulfide, Al_2S_3, from aluminum and sulfur.

$$2Al + 3S \longrightarrow Al_2S_3$$

Aluminum reacts   to give   aluminum
with sulfur in a   sulfide
ratio of 2 Al atoms
to 3 S atoms

The numbers in front of the formulas, called **coefficients,** specify the proportions of the formula units involved in the reaction. As with subscripts, whenever a coefficient is 1, the 1 is not written; it's understood.

Our objective here is very limited. It's simply to recognize equations and translate them, not to write them. (That will come later.) An essential feature of any chemical equation, however, is that it be a **balanced equation,** one in which all atoms present in the reactants occur somewhere among the products. For example, in the equation for the formation of Al_2S_3, there are 2 Al atoms on the left and 2 on the right in Al_2S_3. Similarly, there are 3 S atoms on the left and 3 on the right. The unbreakability of atoms by chemical means and the conservation of mass in chemical reactions ensure this kind of balance. As Dalton said so long ago, when reactions occur the atoms of the reactants rearrange; they do not break up or disappear.

Chemical Reactions Always Convert Substances into Different Substances The absolutely necessary and sufficient condition that determines whether an event is a chemical reaction is that substances change into other substances. To tell whether this has happened, we rely on changes in physical appearances or physical properties. As a chemical reaction occurs, some or all of the physical properties associated with the reactants disappear and those of the products emerge. Such changes might be changes in color, odor, or physical state—a gas might bubble out, or a solid might separate. If, once started, the event occurs without any further intervention and heat is released, then the change is usually *chemical*. Most chemical reactions that "go by themselves" release heat. If none of these tests decides the matter, then more sophisticated measures must be used, steps that positively identify different compounds. The involvement of heat energy in so many reactions, as we have intimated, means that we must now pay some attention to energy as a factor at the molecular level of life.

■

■ Always be *very* cautious when checking odors of materials. A deep enough whiff of the fumes of ammonia or hydrogen sulfide ("rotten eggs" odor), for example, could hurt you.

2.3 KINETIC AND POTENTIAL ENERGY

Chemical energy is a form of potential energy that can be changed to other forms when reactions occur.

Energy is the ability to cause change. Sound energy, for example, changes the sound level; light energy changes the illumination level. Thus, we say that things "have energy" when they have the ability to cause change.

Moving Objects Have Kinetic Energy In the most general terms, energy can be possessed in two ways, as *kinetic energy* and as *potential energy*. **Kinetic energy** is the energy associated with motion. A moving car has kinetic energy; so does an

■ *Kinetic* is from the Greek *kinetikos,* meaning "of motion."

avalanche, or a falling star, or a running child. The kinetic energy (KE) of a moving object is calculated by the equation

$$\text{KE} = \tfrac{1}{2} mv^2 \qquad (2.1)$$

where m is the mass of the object and v is its velocity. Thus the kinetic energy of a moving object is directly proportional to its mass and to the *square* of its velocity. If you double the velocity of a moving car, its kinetic energy increase by a factor of four (because $4 = 2^2$). The energy quadruples, not doubles, and this is why a small increase in vehicle velocity can be much more dangerous than the numbers might indicate.

■ Velocity, a derived unit, is distance per period of time, and in the SI its units are meters per second (m/s).

The SI unit of energy, the *joule,* is based on Equation 2.1. An object with a mass of exactly 2 kg moving at a velocity of exactly 1 m/s (meter per second) has one **joule (J)** of kinetic energy.

■ The *joule* is named in honor of J. P. Joule (1818–1889), an English physicist.

$$1 \text{ J} = \tfrac{1}{2} \times (2 \text{ kg}) \times (1 \text{ m/s})^2$$

The joule is a small amount of energy; if you dropped a 4.4 pound object only 4 inches, it would acquire 1 joule of energy. The **kilojoule (kJ)** is often used for larger amounts of energy.

$$1 \text{ kJ} = 1000 \text{ J}$$

Potential Energy Is Stored Energy   There are several ways to put energy into storage. When you wind the spring of a windup toy, you transfer some of your own energy to the coiled spring. There this energy remains until you release the spring and the stored energy appears as the kinetic energy of the moving toy. The stored energy is called **potential energy,** symbolized as **PE,** because it is potentially available to be changed to kinetic energy when you switch on the toy.

■ Unlike kinetic energy, no simple equation exists for calculating the amount of potential energy in something.

When water is pumped into a town's water tank, the stored water has potential energy because of gravitational attraction. Some of this stored energy enables a flow of water that moves wastes through the pipes of the town's sanitary system.

When battery makers arrange the right chemicals in a battery, the package contains potential energy. At the turn of a switch, chemical reactions occur in the battery that give kinetic energy to electrons to push them through a wire. Thus *chemicals themselves, simply by virtue of their compositions, have potential energy,* usually called **chemical energy.** When you put gasoline into a car, you know that you will be able to use the chemical energy in the gasoline, released when it burns in the engine, to make the car move (and so have kinetic energy).

■ Electrons are present in all atoms and so in all matter. They are extremely tiny particles, and when electricity "flows," electrons are moving in one direction.

A piece of paper has stored or chemical energy because of the chemical nature of paper. You can hold a crumpled newspaper in your hands comfortably, until someone puts a match to it. Now the paper burns, giving off heat and light (and maybe a little sound). (Have you ever thought of how remarkable this is? How can the paper have all this energy?)

When you eat, you replenish your reserves of chemicals having chemical energy. Some of the chemical energy might be changed into the mechanical energy of moving arms and legs; some might be transformed into sound energy for speaking; some will appear as heat needed to maintain a steady body temperature.

The chemical energy in paper becomes heat energy.

Total Energy Is Conserved as Changes Occur   When you toss a ball into the air you give its mass a certain velocity, so the ball gains kinetic energy. As the ball rises, however, its velocity decreases and it loses kinetic energy. Eventually, at the top of its ascent, the velocity of the ball becomes zero, and it momentarily has no kinetic energy. Where did the energy go? It became potential energy, PE. At the moment of no more upward motion, all of the ball's kinetic energy has changed to potential energy. As the ball falls, the reverse process occurs. The ball's potential energy changes into an increasing amount of kinetic energy as the velocity increases. As you might infer, the *total energy* of the ball never changes, only the form of the energy does. The total energy is the sum of KE and PE.

The experience with the ball illustrates one of the most important of all natural laws, the **law of conservation of energy.**

> **Law of Conservation of Energy**   The total energy of the universe is constant and can neither be created nor destroyed; it can only be transformed.

A great deal of play energy comes from the chemical energy in a sandwich.

The total energy of the universe is the sum of all potential and kinetic energies everywhere. This law is also called the *first law of thermodynamics*.

The law of conservation of energy is an example of a *law of nature*. Like all such laws, this law is not *known* to be true *without exception*. It would be impossible to carry out measurements everywhere to see whether there are exceptions or violations. The law, instead, expresses the universal experience of those who have made measurements.

The substances in a sandwich or in a piece of paper have chemical energy not because of their locations in the universe, like the tossed ball, but because of their chemical nature. It is in the nature of the way in which the atoms of food, paper, and oxygen are organized that they can undergo energy-releasing chemical reactions and liberate their chemical or stored energy in other forms such as heat.

2.4   HEAT

Heat can change an objects' temperature or its physical state.

Heat is the energy that transfers from one object to another when the two are at different temperatures and in some kind of contact. We say that heat *flows* from the object with the higher temperature to the one with the lower temperature. If left to itself, the flow continues until both objects reach the same intermediate temperature. *Heat is thus a temperature-changing capacity possessed by an object.* To get heat to flow out of an object, all we have to do is put the object next to one with a lower temperature.

Heat is also a physical state–changing capacity. A block of ice at 0 °C in contact with a warm radiator will not itself undergo a change in temperature. The ice will simply melt, that is, change its physical state from solid to liquid. As long as the freshly melted water is in contact with some ice, its temperature is the same as that of the ice. The temperature at which a solid changes into a liquid is called the **melting point** of the solid.

If you put a pan of water at 100 °C on a hot burner, the flame's higher temperature won't raise the temperature of the water. Instead, the water will boil at a constant temperature and change its state from liquid to gas (vapor). The temperature at which the change from liquid to vapor occurs is called the **boiling point**. Thus, when an object at a higher temperature is in contact with one at a lower temperature, either a change in state or a change in temperature occurs. In either case, heat flows.

■   Turning up a burner on a pan of boiling water won't cook the potatoes faster, because it doesn't raise the temperature. It just boils the water away faster.

The Calorie Is a Unit of Energy Commonly Used for Heat   The **thermal properties** of matter are physical properties related to the ability of a substance to handle heat without undergoing a chemical change. Water, for example, readily absorbs heat when it's in contact with something hotter (such as the bottom of a pan holding the water on a hot burner). Substances, however, differ widely in their ability to absorb energy, so scientists have singled out and defined one thermal property, *specific heat*, to describe the differences. **Specific heat** is the quantity of energy required to change the temperature of a 1-gram sample of something by 1 °C. It is useful to think of specific heat as the heat-holding ability, the heat-absorbing ability, or the heat-delivering ability of 1 g of a substance.

Water is so common that its heat-absorbing ability was used originally to *define* an older unit of heat, the **calorie (cal)**: 1 cal is the amount of heat that, when added to 1 g of water at 14.5 °C, makes the temperature increase to 15.5 °C.[2] If the 1-g sample of water is cooled so that its temperature decreases from 15.5 to 14.5 °C, then the water has given up 1 cal.

One degree is a small change, and 1 gram of water isn't much—about 16 drops—so the calorie is a small quantity of heat. Scientists, therefore, often use a multiple of the calorie called the **kilocalorie, kcal**.

$$1 \text{ kcal} = 1000 \text{ cal}$$

The relationship between the calorie and the joule is known because kinetic energy can be changed quantitatively into heat. The kinetic energy possessed by a paddle rotating below the surface of a sample of water, for example, is changed entirely to heat, which increases the temperature of the water. Roughly 4.2 J are the equivalent of 1 cal. More exactly, the SI now *defines* the calorie in terms of the joule.

$$1 \text{ cal} = 4.184 \text{ J (exactly)}$$

Increasingly the joule and the kilojoule (kJ) are replacing the calorie and the kilocalorie as the preferred units for any form of energy, including heat. The change of usage, however, is occurring slowly, particularly in the life sciences, so we will normally use the calorie and the kilocalorie as our units of energy.

■ PRACTICE EXERCISE 3   The relationship between the calorie and the joule can be restated in the form of two conversion factors.

(a)   What are these factors?

(b)   Use the proper conversion factor to calculate how many joules and how many kilojoules of energy are equivalent to the 79.7 cal needed to melt 1.00 g of ice.

Now that we have energy units, we can put specific heat on a more general basis. Specific heat is defined by

$$\text{Specific heat} = \frac{\text{cal}}{\text{g } \Delta t_C} \tag{2.2}$$

where cal = calories, g = mass in grams, and Δt_C = the *change* in temperature in Celsius degrees. The units in Equation 2.2 give us the common units for specific heat, cal/g °C (calories per gram per degree Celsius).

■ In nearly all popular books on nutrition and diet, the word *calorie* actually means *kilocalorie*. The nutritionist's "Calorie" (capital C) is 1 kcal.

■ The symbol Δ is the Greek capital *delta*. Pronounce Δt as "delta tee." When Δ is in front of any other symbol, it means a *change* in the value of whatever the other symbol represents. Thus ΔE refers to a *change* in energy; Δm is a change in mass.

[2] Although the specific Celsius degree, the one between 14.5 and 15.5 °C, is specified in this formal definition, a 1-degree transition anywhere between the 0 °C and 100 °C marks requires virtually an identical quantity of heat.

■ Specific heat could also be in the units of

$$\frac{J}{g\ ^\circ C}$$

■ To two significant figures, water's specific heat is 1.0 cal/g °C over the entire range of 0 to 100 °C.

The specific heats of several substances are given in Table 2.2. Notice that metals have very low values. The specific heat of iron (about 0.1 cal/g °C), for example, is only one-tenth that of water. Thus, only 0.1 cal is needed to make the temperature of a 1 g sample of iron increase by 1 °C. Put another way, the 1 cal that raises the temperature of only 1 g of water by just 1 °C can raise the temperature of the same mass of iron by 10 °C, 10 times as much. Iron thus undergoes a much larger change in temperature than the same mass of water by the gain or loss of a relatively small amount of heat.

■ A large iron nail has a mass of about 25.4 g.

TABLE 2.2   Specific Heats of Some Substances

Substance	Specific Heat (cal/g °C)[a]
Ethyl alcohol	0.58
Gold	0.031
Granite	0.192
Iron	0.12
Olive oil	0.47
Water (liquid)	1.00

[a] These values are valid in the temperature range of several degrees Celsius on either side of room temperature.

■ The specific heat is known from Table 2.2 to only two significant figures, so we round the answer accordingly.

EXAMPLE 2.3   Using Specific Heat Data

When a 25.4 g piece of iron at 20.0 °C receives 115 cal of heat, what does its temperature change to?

ANALYSIS
What the question really calls for is a value of Δt, and we can calculate Δt with Equation 2.2 using the specific heat of iron, 0.12 cal/g °C (Table 2.2). *It is essential that we carry along all of the units as we place data into Equation 2.2, because we have to be sure that the units will cancel properly to leave the answer in the correct unit.* NEVER OMIT THE UNITS OF PHYSICAL QUANTITIES DURING CALCULATIONS UNTIL THEY CANCEL OR MULTIPLY PROPERLY TO GIVE THE CORRECT FINAL UNITS. THE UNITS ARE OUR PRIMARY CHECK ON SETTING UP A SOLUTION CORRECTLY.

SOLUTION
Using the data and Equation 2.2, we have

$$\text{Specific heat of iron} = \frac{0.12\ \text{cal}}{g\ ^\circ C} = \frac{115\ \text{cal}}{25.4\ g \times \Delta t}$$

To solve this for Δt we have to cross multiply. If you think this mathematical procedure is something you can't do, turn to Appendix A where it is described using this problem. Cross multiplication gives us

$$\Delta t = \frac{(115\ \text{cal})(g\ ^\circ C)}{(25.4\ g) \times (0.12\ \text{cal})}$$

Notice that we cross multiply units just like numbers. Notice also that all of the units cancel, except °C. After doing the arithmetic, we get

$$\Delta t = 38\ ^\circ C\ (\text{rounded by our rules from } 37.72965879)$$

In other words, 115 cal of heat will *increase* the temperature of a 25.4 g piece of iron by 38 °C. So its new temperature is 20.0 °C + 38 °C = 58 °C.

■ PRACTICE EXERCISE 4   Suppose that the same amount of heat used in Example 2.3, 115 cal, was absorbed by 25.4 g of water instead of iron, with the initial temperature also 20.0 °C. What will be the final temperature of the water in degrees Celsius? This exercise demonstrates the superior ability of water to absorb heat, compared to iron, without experiencing a large change in temperature.

Chemical Reactions That Give Off Heat Are Said To Be *Exothermic* Three common sources of heat are solar energy coming from the sun; nuclear energy, coming from the unique properties of radioactive elements; and chemical reactions. All of them supply heat and all can be used to generate electricity. Combustion or burning releases the chemical energy in the fossil fuels—gas, oil, and coal—or the stored energy in wood, charcoal, and wastes. A reaction such as combustion that continuously releases heat is called an **exothermic** reaction, and most (but not all) spontaneous reactions are exothermic. *Spontaneous events* are those that, once arranged or started, continue with no further human intervention. Even a small bolt of lightning or the flame from a tiny match can initiate a gigantic forest fire.

■ *exo,* out
endo, in
therm, heat

Some chemical reactions can be made to take place only if we continuously supply the reactants with heat or some other form of energy. Reactions that require a continuous input of heat, such as changing sodium chloride back into its elements, are called **endothermic** reactions. Another endothermic process is *photosynthesis* in plants, with the necessary energy coming from the sun. **Photosynthesis** is the use of solar energy to convert simple compounds with little chemical energy—carbon dioxide and water—into large molecules with considerable chemical energy, like starch. Even the chemical energy in the fossil fuels originated in the sun, because these fuels are the altered remains of plants that flourished in sunlight eons ago.

Our Water Content Gives Us a Large Thermal Cushion Because the adult body is about 60% water and because of the relatively high specific heat of water, our bodies can absorb or release considerable heat without resulting in more than a 1 degree change in body temperature. The body thus has a thermal "cushion" to help it maintain a steady temperature despite major swings in outside temperature.

■ If your body's core temperature changes even a few degrees from normal, which is 37.0 °C, you could die.

Events inside the body, the chemical reactions of *metabolism*, also threaten the body's steady temperature. **Metabolism** consists of the entire collection of the many spontaneous, generally exothermic reactions inside cells. The reactions of metabolism convert the chemical energy in the foods we eat into other forms of energy, about half as heat. The body uses the rest of the energy to run itself and perform the tasks you ask of it. Interaction 2.1 discusses some of the serious, life-threatening consequences of overheating or overcooling the body.

■ A person using 2000 Calories (the nutritionist's Calorie being 1000 cal) generates 1×10^6 cal per day of heat.

2.5 HEAT AND MOLECULAR KINETIC ENERGY

An object holds heat energy because its formula units individually possess kinetic energy.

Heat or thermal energy is actually a form of kinetic energy because each individual tiny particle of a substance, such as a molecule, moves or jiggles about and so has kinetic energy. The sum total of these tiny energies is called the **molecular kinetic energy** of the substance.

It's not that the entire sample of the substance spontaneously moves or jiggles about but rather that its tiniest particles do, and they move randomly in every direction. In a gas, for example, the molecules move more or less freely, bumping into each other and changing directions constantly. Because of the nature of the collisions, some being direct hits, others glancing blows, and some mere taps, individual molecules do not all move at the same speed. Some are moving slowly, some rapidly.

■ So tiny are molecules that in only 1 mL of a gas under ordinary conditions there are over 10^{19} molecules.

Remember that the amount of kinetic energy possessed by a moving object depends on both its mass and the square of its velocity; $KE = \frac{1}{2}mv^2$. So within a gas sample the value of v varies from molecule to molecule, and the molecules of a gas have a wide range of kinetic energies, some being low values and others relatively

INTERACTION 2.1
METABOLISM AND BODY TEMPERATURE

Basal Metabolism. The body's *basal activities* are the minimum activities inside the body that must take place just to maintain muscle tone, control body temperature, circulate the blood, breathe, make compounds or break them down, and otherwise operate tissues and glands during periods of rest. The sum total of the chemical reactions that supply the energy for basal activities is called the body's *basal metabolism.* The rate at which chemical energy is used for basal activities is called the *basal metabolic rate,* which is customarily in units of kcal/min or kcal/kg h (kilocalories per kilogram of body weight per hour).

Measurements of basal metabolic rates are taken when the person is lying down, has done no vigorous exercise for several hours, has eaten no food for at least 14 hours, and is otherwise awake but at complete rest. A 70-kg (154-lb) adult male has a basal metabolic rate of 1.0 to 1.2 kcal/min. The rate for a 58-kg (128-lb) woman is 0.9 to 1.1 kcal/min. When a person is more active, the metabolic rate is higher than the basal rate. It can go as high as 12 kcal per minute when carrying a heavy load uphill or sprinting in a race.

Body Temperature, Metabolic Rate, Hyperthermia, and Hypothermia. If you eat, say, 2400 food "calories," meaning 2400 kcal, then about 1200 kcal of heat must be released from the body. If all of this heat stayed in the body, the body temperature would soar by 24 °C! Happily, we have several mechanisms for getting rid of heat, including perspiration.

One reason that the body tries to maintain a steady temperature is that even small changes in temperature can greatly affect the rates of chemical reactions, including those of metabolism. If the body's core temperature increases—a condition called *hyperthermia*—the reactions of metabolism speed up. To sustain this, the body would need more oxygen—about 7% more for every 1 degree Fahrenheit increase. To deliver this oxygen, the heart would have to work harder, so a sustained condition of hyperthermia creates problems for the heart.

The opposite of hyperthermia is *hypothermia*—a condition of a lower than normal body temperature. Under this condition, the rates of metabolism eventually slow down, including those reactions that keep vital functions working normally. The initial response of the body to a decrease in its temperature is uncontrolled shivering, however, which *accelerates* the metabolic rate and releases thermal energy from exothermic reactions. This response, however, cannot overcome more than a 2–3 °F decrease in the temperature of the body's core. At ever larger decreases, amnesia sets in, the muscles become more rigid, the heart rate becomes erratic, and the victim loses consciousness and eventually dies. To prevent this, a victim *must* be made dry, given shelter, and warmed—by drinking warm fluids if conscious and by blankets or sleeping bags in any case. Give no alcoholic beverages; alcohol dilates (enlarges) capillaries, which enables the heart to force a flood of chilled blood rapidly from the capillaries at the skin's surface into the body's core, chilling it further.

■   Because of its extremely small mass, an individual gas molecule's kinetic energy is very small; but all the energies do add up.

high. There will, of course, be an *average molecular kinetic energy,* and it is this average that relates to the temperature of a substance. *The temperature of an object is a measure of its average molecular kinetic energy.* At a high temperature, the average molecular kinetic energy is high. At a low temperature, this average is low.

All Kinetic Molecular Motions Cease at 0 K When something becomes colder, its average molecular kinetic energy becomes smaller because all of its molecules are slowing down. We can therefore imagine a state in which all such motions have stopped, when the value of v in $\frac{1}{2}mv^2$ becomes zero. At this temperature, the kinetic molecular energy must also be zero. (Substitute v equal to 0 in the kinetic energy equation to see this.) The absence of molecular kinetic energy occurs at absolute zero or 0 K. We're more interested, however, in the concept of molecules in collision rather than at rest, because hard enough collisions result in the reorganizations of the parts of molecules that we call chemical reactions.

SUMMARY

Matter Matter, anything with mass that occupies space, can exist in three physical states: solid, liquid, and gas. Broadly, the three kinds of matter are elements, compounds, and mixtures. Elements and compounds are called pure substances, and they obey the law of definite proportions. Elements can be classified as metals or nonmetals. Mixtures, which do not obey this law, can be separated by operations that cause no chemical changes, but to separate the elements that make up a compound requires chemical reactions.

A chemical reaction is an event in which substances change into different substances with different formulas.

Dalton's atomic theory The law of definite proportions and the law of conservation of mass in chemical reactions led John Dalton to the idea—now regarded as well-established fact—that all matter consists of discrete, noncuttable particles called atoms. The atoms of the same element, said Dalton, all have the same mass, and those of different elements have different masses. When atoms of different elements combine to form compounds, they combine as *whole* atoms; they do not break apart. Dalton realized that when different elements combine in different proportions by atoms, the resulting compounds must display a pattern now summarized by the law of multiple proportions.

Symbols, formulas, and equations Every element is given a one- or two-letter symbol, and it can stand either for the element or for one atom of the element. The symbol for a compound is called a formula. The empirical formula (one kind of formula) consists of the symbols of the atoms in one formula unit, with the smallest whole-number subscripts used to show the proportions of the different atoms present.

The empirical formula is most often the one used for compounds made by combining a metal with a nonmetal, like FeS or NaCl. When a compound is made entirely of nonmetals, like water or carbon dioxide, the formula unit is nearly always a discrete, individual package of atoms called a molecule.

To describe a chemical reaction, the symbols of the substances involved as reactants, separated by plus signs, are on one side of an arrow that points to the symbols of the products, also separated by plus signs. The equation is balanced when all atoms showing in the formulas of the reactants are present in like numbers in the formulas of the products. Coefficients, numbers standing in front of formulas, are employed as needed to achieve the correct balance. In both formulas and equations, the numbers used for subscripts and coefficients are generally the smallest whole numbers that show the correct proportions.

Forms of energy When something is able to cause a change in motion, position, illumination, sound, or chemical composition, it has energy of one form or another—kinetic energy (energy of motion), light, sound, potential energy, and chemical energy (a special form of potential energy). Energy is neither created nor does it disappear into nothing; it can only be transformed from one type into another or from one place to another. Spontaneous chemical reactions are usually exothermic—they release heat—but many reactions can be made to occur by continuously heating the reactants. These are endothermic reactions.

Heat energy Heat is the form of energy that transfers because of temperature differences. This transfer either causes changes in the temperature or causes changes in physical state, such as melting or freezing, boiling or condensing, or it causes chemical change.

The heat that changes the temperature of 1 gram of a substance by 1 degree Celsius is called the substance's specific heat. When the substance is water, this quantity of heat is defined as the calorie. Because the specific heat of water is about the highest in nature, and because we're at least 60% water, our bodies can gain or release considerable heat without large swings in body temperature.

Heat is actually molecular kinetic energy. The motions or jigglings of the molecules of a substance add up to a total amount of heat, and the average molecular kinetic energy is proportional to the substance's temperature. At 0 K, molecular motions cease.

REVIEW EXERCISES

The answers to Review Exercises whose numbers are in color are found in Appendix E. The answers to the other review exercises are found in the Study Guide that accompanies this book. The more challenging questions are marked with asterisks.

States and Kinds of Matter

2.1   It is important to distinguish between the states of matter and the *kinds* of matter.
(a) What are the names of the states of matter?
(b) What are the names of the kinds of matter?

2.2   Among which, the states of matter or the kinds of matter, are the distinctions basically in terms of *physical* properties?

2.3   Ice, water, and steam do not look like each other at all, yet we say that they are *chemically* the same. What does this mean?

2.4   What have we learned in this chapter about the distinctions between elements and compounds?

2.5   Sulfur is an element that can be obtained as bright yellow, shiny pieces. The pieces are brittle, however, and they do not conduct electricity. Is it a metal or a nonmetal?

2.6   Tungsten, a solid element, can be polished to a shiny finish, and it is a good conductor of electricity. Is it a metal or a nonmetal?

2.7   One element is a liquid at room temperature, and its surface is bright and shiny, like silver. It conducts electricity well. Is it a metal or a nonmetal? What is unusual about the properties described here?

2.8   Changes can be either physical or chemical. What must happen if the change is to be classified as *chemical*? Will physical changes also occur? Explain.

2.9   Which of the following events are chemical changes (as well as being physical changes)?
(a) When heated in a pan, sugar turns brown (caramelizes).
(b) When stirred in water, table salt seems to disappear.
(c) When struck with a hammer, an ice cube shatters.
(d) A bleaching agent causes a colored fabric to lose its color.

2.10   A glass of water and a solution of sugar in water look identical, but the word *substance* properly belongs only to one. Which? Why?

2.11   What is the most basic distinction between a compound and a mixture?

2.12   What are alloys?

2.13   Some elements are said to be *radioactive*. What is true about them to earn this title?

2.14   At room temperature, in which *state* are most of the elements found?

2.15   Which term best fits the concept of an element, *substance, state of matter*, or *particle of matter*?

2.16   How many elements are liquids at room temperature, 0, 1, 2, 20, or 40?

2.17   Roughly how many elements are there, 50, 100, 150, 200, or 250?

2.18   How are compounds different from elements?

2.19   How are compounds different from mixtures?

2.20   Sodium reacts with chlorine to give sodium chloride (table salt).
(a) Physical changes obviously occur (as seen in Figures 2.1–2.3), so why is this event also called a *chemical* reaction?
(b) Which substances are the *reactants* and which are the *products*?
(c) What is always true about the *ratio of the masses* of sodium and chlorine that combine to form sodium chloride?

2.21   The fact that 2.0 g of hydrogen combines with 16.0 g of oxygen, no more, no less, to give 18.0 g of water illustrates what important law of chemical combination?

2.22   What law of chemical combination most surely distinguishes compounds from mixtures?

Dalton's Atomic Theory

2.23   What are the postulates of Dalton's atomic theory?

2.24   What law of chemical combination contributed most to Dalton's postulate that compounds include their elements in definite ratios *by atoms*?

2.25   The fact that the ratio of the elements *by mass* in compounds made of just two elements is almost never 1 : 1 suggested what postulate of Dalton's?

2.26   Dalton's postulate that atoms react as *whole* units and do not break up into pieces of atoms when they react is strongly supported by which law of chemical combination?

*2.27   Copper can combine with oxygen to give two compounds. In one, called cuprite, 1.00000 g of oxygen is combined with 7.94454 g of copper. In the other, called tenorite, the ratio is 1.00000 g of oxygen to 3.97265 g of copper. Do these ratios illustrate the law of multiple proportions? (Do the appropriate calculation.)

Chemical Symbols, Formulas, and Equations

2.28   Of the symbols NO and No, which stands for an element and which for a compound?

2.29   Write the symbols of the following elements.
(a) phosphorus   (b) calcium   (c) bromine
(d) platinum   (e) carbon   (f) barium

2.30   What are the symbols of the following elements?

(a) lead   (b) mercury   (c) fluorine
(d) potassium   (e) hydrogen   (f) iron

2.31 Write the symbols of the following elements.
(a) sulfur   (b) iodine   (c) manganese
(d) sodium   (e) copper   (f) magnesium

2.32 What are the symbols of the following elements?
(a) nitrogen   (b) oxygen   (c) silver
(d) zinc   (e) lithium   (f) chlorine

2.33 What are the names of the elements with the following symbols?
(a) I (b) Pb (c) Li
(d) N (e) Zn (f) Ba
(g) C (h) Ca (i) Cl
(j) F (k) Cu (l) Fe

2.34 Write the names of the elements with these symbols.
(a) H (b) Mg (c) Al
(d) Hg (e) Mn (f) Na
(g) O (h) Ag (i) Br
(j) P (k) K (l) Pt

2.35 The symbol Fe represents the *element* iron. It also represents what else?

2.36 One formula unit of a constituent of smog consists of one atom of nitrogen and two atoms of oxygen. Which of the following formulas best represents it?
(a) N_2O (b) 2NO (c) NO_2 (d) $2NO_2$ (e) N_2O_4

2.37 The formula of sodium chloride is NaCl.
(a) Why would it not be a violation of anything fundamental about sodium chloride to write its formula as Na_2Cl_2?
(b) What is improper about writing Na_2Cl_2 as the formula of sodium chloride?
(c) What kind of formula is represented by NaCl?
(d) What is the name we use for the particle made out of just 1 sodium atom and 1 chlorine atom?

*2.38 We say that H_2O represents a *molecule* of water.
(a) Explain why the term "molecule" fits this situation.
(b) Does the symbol H_2O also represent a formula unit of water? Explain.
(c) Is the formula H_2O also an empirical formula? Explain.
(d) Why don't we say that NaCl represents a *molecule* of sodium chloride?

2.39 The following sentences describe chemical reactions. Convert them into balanced equations.
(a) Calcium reacts with sulfur to give calcium sulfide, CaS.
(b) Sodium reacts with sulfur to give sodium sulfide, Na_2S.

Energy

2.40 Compare and contrast the definitions of *matter* and *energy*.

2.41 What is the name of the form of energy associated with the motion itself of a moving object? What is the equation for this form of energy?

2.42 What is the name and symbol of the unit of energy

based on the equation that defines kinetic energy?

*2.43 If the *mass* of an object moving at a velocity of 5 m/s is doubled (with no change in velocity), by what factor is the energy of motion of this object changed?

*2.44 If the *velocity* of a moving object changes from 8 m/s to 16 m/s, by what factor is its energy of motion changed?

2.45 When we say that "energy is conserved," do we mean kinetic energy, potential energy, or something else?

*2.46 Consider the units used for energy.
(a) Using the equation that defines kinetic energy, what are the *units* of the joule?
(b) How much kinetic energy in joules is possessed by a vehicle with a mass of 2.00×10^3 kg (about the mass of a normal-sized station wagon) when it travels with a velocity of 25.0 m/s (about 56 mi/h)?
(c) When expressed in calories, how much energy does this vehicle have? How much in kilocalories? (1 cal = 4.184 J)
(d) What would the velocity of the vehicle have to become (in m/s) to have twice the kinetic energy it has in part (b)?

2.47 When an earthquake occurs, a huge amount of kinetic energy is generated. If energy is supposed to be conserved, what happens to this energy?

2.48 Where do you suppose that the light energy released when you turn on a flashlight existed before you closed the switch? (In what form was this energy?)

2.49 When you pick up a book you have to make muscles move. In what form did the kinetic energy of these muscles exist before you caused this to happen?

Thermal Properties of Matter and Heat

2.50 If no heat transfers between two objects in contact, what must be true about them?

2.51 Suppose that two objects, A and B, are in contact and that A has a higher temperature than B.
(a) In which direction, A to B or B to A, will heat transfer?
(b) If by this heat transfer the temperature of B does *not* change, what might be the explanation?

2.52 What name do we give the temperature reading:
(a) At which a solid changes to its liquid state?
(b) At which a liquid changes to its vapor state?

2.53 What is the name we give the amount of heat needed to raise the temperature of 1 g of water by 1°C? If your finger received this much heat all at once, do you think the skin would be burned?

2.54 When you read that a certain food portion has, say, "100 calories," what does the term "calorie" mean in this context?

2.55 When we say that the human body has a "substantial thermal cushion," what is meant? What property of water contributes most to this thermal cushion?

2.56 What is meant when heat is described as a form of kinetic energy?

2.57 With what physical property is the average molecular kinetic energy of a substance associated?

2.58 What happens to the average molecular kinetic energy of a substance when it cools? What happens to the aggregate of the molecular kinetic energies?

Basal Metabolism (Interaction 2.1)

2.59 What is the difference between *basal metabolism* and *basal metabolic rate*?

2.60 How do *basal* activities differ from other kinds?

2.61 Describe the basal activities of the body.

2.62 What is the average basal metabolic rate in kcal/min for each (give ranges)?
(a) adult females
(b) adult males

2.63 Calculate how much energy (in kilocalories) an adult with a basal metabolic rate of 1.00 kcal/min expends just on basal activities alone over a 24.00-hour period.

2.64 When the body loses heat more rapidly than it can replace the losses, what condition concerning the body's core temperature results?

2.65 In a hyperthermic condition, the body usually breathes more rapidly. Why?

2.66 Why does the heart beat faster (initially) when hyperthermia occurs?

Additional Questions

*2.67 If the ratio *by mass* of two elements, *X* and *Y*, in the hypothetical compound *XY* were exactly 1:1, what would be true in Dalton's theory about the relative masses of the *atoms* of *X* and *Y*?

*2.68 Hydrogen and oxygen form two compounds, water and hydrogen peroxide. In water, 1.00000 g of hydrogen is combined with 7.93655 g of oxygen. In hydrogen peroxide, 1.00000 g of hydrogen is most likely combined with how much oxygen, 9.45386 g of oxygen, 13.0521 g of oxygen, or 15.8731 g of oxygen? How can you tell?

2.69 The balanced equation for the formation of NaCl from sodium and chlorine, as we will see in the next chapter, is properly written as follows.

$$2Na + Cl_2 \longrightarrow 2NaCl$$

(a) The *formula unit* of sodium is evidently how many atoms of sodium?

(b) The *formula unit* of chlorine is apparently made out of how many atoms of chlorine?

(c) What is the general name for the numbers that stand in front of formulas in equations?

(d) How many Cl atoms are represented by 2NaCl?

*2.70 Use conversion factors made from relationships among units given in Chapter 1 for this question.
(a) What is the equivalent in miles per hour of a velocity of 30.0 m/s?
(b) How does a velocity of 30.0 m/s compare with the normal walking pace of most people (generally taken to be 3.00 mi/h)?
(c) How much energy in kilojoules is possessed by 70.0 kg adult moving at 30.0 m/s in a vehicle? How much is possessed by the same person moving at a walking pace of 3 mi/h?

*2.71 Referring to Table 2.2, which sample has more heat energy at 25°C, 100 g of iron or 100 g of water? How can you tell?

*2.72 The National Academy of Sciences uses the following conversion factors for the energy content of foods:

Proteins	4.0 kcal/g
Carbohydrates	4.0 kcal/g
Food fat	9.0 kcal/g

One serving of a commercial preparation of baked beans contains 16 g of protein, 49 g of carbohydrate, and 8.0 g of food fat. Calculate the kilocalories of energy in this serving.

*2.73 One serving of elbow macaroni (prepared without cheese sauce) contains 7.0 g of protein, 41 g of carbohydrate, and 1.0 g of food fat. Using the information given in Review Exercise 2.72, calculate the kilocalories of energy in this serving.

*2.74 A large box of buttered popcorn bought at a theater contains 8.0 g of protein, 32 g of carbohydrate, and 25 g of food fat. Using the information in Review Exercise 2.72:
(a) What is the energy content of this popcorn in kilocalories?
(b) If you walk at a speed of 3.5 mi/h and need 5.0 kcal/min to sustain this activity, how many hours must you walk to "work off" the kilocalories in the popcorn? How many miles long must the walk be?

2.75 It is easy to poke one's finger through the surface of liquid water but not through frozen water. We haven't studied a reason for this yet, but offer a possible explanation (a hypothesis) that explains this at the "molecular level."

ATOMIC THEORY AND THE PERIODIC TABLE

3

The arrival of each row of wave crests at the beach is more or less a periodic function of time. Some of the properties of the elements are a periodic function of atomic number.

THIS CHAPTER IN CONTEXT

In Chapters 1 and 2 we sharpened our knowledge about *what matter is.* We now move toward an understanding of *how matter behaves.* As we've said, we cannot say much about the molecular basis of life without knowing about molecules, and to understand molecules we must know about atoms. This chapter is about atoms, what they are made of, and how their subatomic pieces are organized. You'll be amazed at how much of nature "makes sense" when you have this background, because *changes in atomic organizations are at the heart of the reactions of molecules.*

When it comes to studying the elements, nature has handed us a nice simplification. The 109 elements can be grouped into only a few families, displayed by the periodic table, whose members have similar chemical properties. Electron configurations explain how the periodic table is possible, and we'll use it the way chemists do, to make life easier. So this chapter is also about the elements, focusing on their atomic structures and introducing their families.

3.1 THE NUCLEAR ATOM

The protons and neutrons of an atom occur together in a core, or *atomic nucleus,* and the electrons are outside this core.

Dalton reasoned that atoms are indestructible, but on this postulate he was wrong. His error was unimportant to the idea of an atom because enough energy to break atoms apart is never available in purely *chemical* reactions. Powerful atom smashing machines are needed, and they produce debris consisting of **subatomic particles.** Remarkably, the same small set of particles is obtained *regardless of the element.* All elements, indeed, all of matter, including you(!), are made of the same subatomic pieces with variations only in the numbers used and their arrangements.

Of all of the subatomic particles, only three, the **electron,** the **proton,** and the **neutron,** are important enough to the chemical properties of matter to warrant our studying them further. We'll first consider two of their physical properties, mass and electrical charge. The existence of charge will help us understand how particles are able to attract or to repel each other and so either stick together or fly apart. Only the electron and the proton have charge, however; the neutron does not.

When particles have electrical charge they are able to exert pushes and pulls on each other called forces of repulsion and attraction, even when not touching. When two objects carry *opposite* electrical charges, they *attract* each other, mimicking the behavior of the opposite poles of magnets. We say, in short, that *unlike charges attract.* An electron and a proton thus attract each other because they bear opposite charges. A proton, however, exerts a force of repulsion on another proton. An electron also repels another electron. In short, *like charges repel.* These two elegantly simple laws about charged particles explain much about how atoms of different elements are able to join to make compounds. A lot of chemistry depends on these two laws.

The charge on a proton has the same *intensity* as that on an electron, but it is *opposite* in character. The amount of charge on either a proton or an electron is defined as *one unit of charge,* and we assign plus and minus signs to signify opposite character. Thus, we arbitrarily say that a proton has a charge of 1+, and an electron, a charge of 1−.

All three subatomic particles have mass. The masses in grams are extremely small (Table 3.1), however, so there's an advantage to reexpressing them in a new

■ The proton and the neutron appear to be made of still smaller particles that physicists have named *quarks.*

Like charges repel.

■ The numerical values of charge, 1+ and 1−, are *relative* values, not the values in SI units. The charges are *equal* but *opposite.*

TABLE 3.1 Properties of Subatomic Particles

Particle	Mass (g)	Mass (u)	Electrical Charge	Symbol
Electron	$9.1093897 \times 10^{-28}$	0.0005485712	1−	$_{-1}^{0}e$
Proton	$1.6726430 \times 10^{-24}$	1.0072725	1+	$_{1}^{1}H, _{1}^{1}p$
Neutron	1.674954×10^{-24}	1.008664	0	$_{0}^{1}n$

unit of mass, the **atomic mass unit** or **u.**

$$1 \text{ u} = 1.6605665 \times 10^{-24} \text{ g}$$

Defining 1 u this way lets the mass of the proton or the neutron be restated, when rounded to two significant figures, as 1.0 u, a much simpler number. The mass of the proton, for example, when converted from grams to atomic mass units, becomes

$$1.6726430 \times 10^{-24} \frac{\text{g}}{\text{proton}} \times \frac{1 \text{ u}}{1.6605665 \times 10^{-24} \text{ g}} = 1.0072725 \text{ u/proton}$$

This rounds to 1.0 u, for the mass of one proton.

The mass of the electron, a much lighter particle, is about 1/1836th u. The practical result of this much smaller mass is that we can ignore electron masses when working with the masses of whole atoms. Instead, what is important about an atom's electrons is their arrangement or *electron configuration*.

An Atom's Protons and Neutrons Make Up Its Atomic *Nucleus* Early in the 20th century, British scientists led by Ernest Rutherford (1871–1937) discovered that atoms are mostly empty space. Essentially all of the atom's mass is in an extremely dense central core, a particle that they named the **nucleus.** It contains all of the atom's heavy, subatomic particles, its protons and neutrons, so the nucleus has essentially all of the atom's mass and all of its positive charge. Thus, the total nuclear charge simply is the sum of the protons.

■ No known atom has more than 109 electrons, and they contribute only about 0.02% to the mass of such an atom.

■ Rutherford's discovery of the nucleus earned him the 1908 Nobel Prize in Chemistry and, in 1930, the title Baron Rutherford of Nelson.

> Charge on an atomic nucleus = number of protons

Each Element Has a Unique *Atomic Number* We can now identify something special about a given element. *All atoms of the same element have identical nuclear charges,* meaning identical numbers of protons. *Atoms of different elements have different nuclear charges* or different numbers of protons. Thus, each of the elements owns a unique number, called its **atomic number,** the number of protons in one of its atoms.

> Atomic number of element = positive charge on its atomic nuclei
> = number of protons per atom

The Number of an Atom's Electrons Also Equals the Atomic Number *All atoms are electrically neutral.* For this to be true, the positive charge on an atom's nucleus must be balanced by the total negative charge contributed by the atom's electrons, all located outside the nucleus. The number of protons in an atom, therefore, equals the number of its electrons, so that each charge of 1+ is neutralized in an electrical sense by each charge of 1−. The atomic number, therefore, also tells us how many electrons an atom has.

Ernest Rutherford

> Atomic number of an element = number of protons
> = number of electrons

We'll return to this in a later section because *the arrangements of electrons within atoms determine the chemical properties of the elements.*

Each Kind of Atom Has a Unique Mass Number   The sum of an atom's neutrons and protons is its **mass number.**

> Mass number = protons + neutrons

Because each neutron and proton has a mass of 1.0 u, the mass number is the same as the mass of the atom in atomic mass units, u. Only when a precision requiring three or more significant figures is needed would we have to modify this statement.

Summary about Atoms   Atoms are tiny *neutral* particles consisting of nuclei and electrons, and atomic nuclei consist of protons and neutrons. Each element has its own *atomic number,* which equals its number of protons or number of electrons. Each kind of atom has a *mass number,* which equals the sum of its numbers of protons and neutrons.

■   PRACTICE EXERCISE 1   What are the mass numbers of atoms that have the following nuclear compositions?

(a) 7 protons and 8 neutrons   (b) 12 protons and 12 neutrons

(c) 11 protons and 13 neutrons

■   PRACTICE EXERCISE 2   How many neutrons are in each of these atoms?

(a) Atomic number 4, mass number 9

(b) Atomic number 17, mass number 35

(c) Atomic number 17, mass number 37

Nearly Every Element Consists of a Mixture of a Small Number of Its *Isotopes*
According to Dalton's third postulate, "all atoms of a particular element are identical in mass." This was another error that, fortunately, didn't matter to the basic idea of an atom. However, nearly any element we care to name is actually a mixture of atoms that differ in mass numbers. When such atoms are of the same element, their atomic numbers are identical but their numbers of neutrons are different. Substances whose atoms are identical in atomic number *but different in mass number* are called **isotopes.** About 250 isotopes occur in nature. More than a thousand have been made by nuclear transformations. Many isotopes are useful in medicine.

■   Chlorine is used to disinfect water in the United States.

Chlorine (atomic number 17) illustrates an element with isotopes. It consists chiefly of two, called chlorine-35 and chlorine-37, where 35 and 37 are *mass numbers*. The atoms of both possess 17 protons (the shared atomic number), but an atom of chlorine-35 has 18 neutrons (35 − 17) and the chlorine-37 atom has 20 neutrons (37 − 17). Wherever chlorine is found in nature, regardless of its state of chemical combination, the *ratio* of chlorine-35 atoms to chlorine-37 atoms is very close to 3 : 1. The ratios of the isotopes for each of the other elements are different from 3 : 1, and many elements have more than two isotopes. *For any given element, however, the proportions of its isotopes are essentially constant every-*

where in nature.

As we'll see, the atoms of the isotopes of the same element have identical electron configurations, and so an element's isotopes have identical *chemical* properties. This is why the existence of isotopes could not have affected Dalton's theory. Dalton's postulates were based on *chemical* properties and on the mass relationships observed in *chemical* reactions. Isotopes have no bearing on these. We need to know about isotopes for one reason, to understand an important physical property, an element's *atomic mass.*

The *Atomic Mass* or *Atomic Weight* of an Element Reflects the Relative Abundances of Its Isotopes   As we said, the two chlorine isotopes occur *everywhere,* all over the earth, in the same ratio of close to 3:1. For every four chlorine atoms in nature, three are chlorine-35 and one is chlorine-37. To find the *average* mass of these four atoms, the only kind of mass that could be obtained by purely *chemical* experiments, we'll assume that the ratio is exactly 3:1.

$$\text{Mass of 3 atoms of chlorine-35: } 3 \times 35 \text{ u} = 105 \text{ u}$$

$$\text{Mass of 1 atom of chlorine-37: } 1 \times 37 \text{ u} = 37 \text{ u}$$

$$\text{Mass of these four atoms} = 142 \text{ u}$$

Dividing 142 u by 4 gives us 35.5 u as the average mass of the chlorine atoms. The ratio of atoms is not exactly 3 to 1 so the actual average, 35.4527 u, isn't exactly 35.5 u. The average mass of the atoms of the various isotopes of any given element, as they occur in their natural proportions, is called the **atomic mass** or the **atomic weight** of the element. Thus, the existence of isotopes causes atomic masses not to be simple, whole numbers in the way that mass numbers are. The Table of Atomic Weights and Numbers inside the front book cover clearly illustrates this fact.

■ Because of historical precedent (and human inertia) atomic masses are still widely called atomic weights.

Summary About Isotopes   It is important to realize that isotopes are *substances.* Isotopes are not particles; rather, they consist of particles (atoms) having identical numbers of protons but different numbers of neutrons. Thus, an element, which is one of the three kinds of *matter* (and not one of a few kinds of particles), usually consists of two or more *substances,* its *isotopes,* physically mixed. The terms *atomic mass* and *atomic weight* apply to an element as it occurs with its natural proportions of isotopes. The term *mass number* has meaning only in connection with a single isotope.

■ PRACTICE EXERCISE 3   Suppose that element number 25 consists of two isotopes in a 50:50 ratio. One has 25 neutrons and the other has 30. What is the atomic mass of this element?

We now know enough about elements and their atoms to go into one of the great simplifications concerning elements uncovered by chemists, the periodic table. Having some knowledge of the periodic table will ease our study of electron configurations.

3.2 PERIODIC LAW AND PERIODIC TABLE

When the elements are arranged in their order of increasing atomic number, several properties recur periodically.

Dimitri Mendeleev, a Russian scientist, was one of the first to notice that the elements can be grouped into a small number of families whose members have much in common. His discovery came as he tried to make chemistry easier for his students and during the preparation of a chemistry textbook, published in 1869.

Many Properties of the Elements Vary Periodically with Their Atomic Numbers Mendeleev observed that the physical and chemical properties of the elements seem to go through cycles as we move through the elements from lowest to highest atomic mass. (Atomic numbers were not known in his day.) The *boiling points* of the elements, for example, roughly illustrate what Mendeleev noted, as seen in Figure 3.1*a*, a plot of boiling points through element 20. Notice that the boiling point temperatures do not continuously increase but, instead, fluctuate with higher and higher elements. The fluctuations aren't perfect, but there are definite ups and downs. One general improvement came with the discovery of atomic numbers a few decades after Mendeleev's work. The properties of the elements fit into a periodic pattern better if atomic numbers instead of atomic masses are the basis for ordering the elements.

■ A substance's boiling point is the temperature at which it boils.

The *ionization energies* of the elements display a similar rising and falling in values (see Figure 3.1*b*). An element's ionization energy is the energy needed to remove an electron from each atom in a large standard number of atoms. Notice that helium, neon, and argon are at the bottoms of cycles in the plot of boiling points versus atomic numbers, but they are at the tops in the plot of ionization energies. In other words, helium, neon, and argon seem to form a set or a family.

■ The product of *ionization*, besides the electron, is an electrically charged particle called an *ion*.

The *combining abilities* of the atoms of one element for atoms of another also go through a cyclical rise, fall, rise, fall pattern. For example, most of the first 20 elements form binary compounds with hydrogen. A **binary compound** is one made of only two elements.

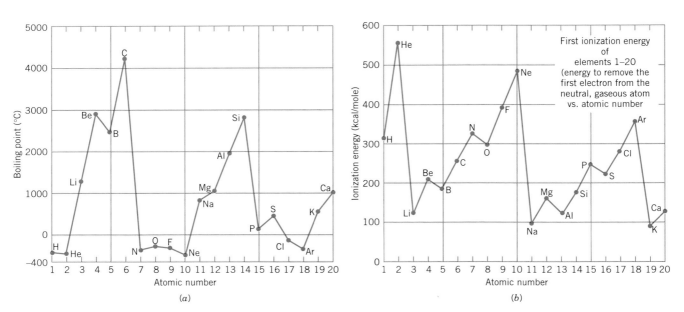

FIGURE 3.1   Two properties of elements 1 through 20 that show periodic fluctuations. (*a*) Boiling points versus atomic numbers. (*b*) Ionization energies versus atomic numbers.

An atom of atomic number:	3	4	5	6	7	8	9	10
can bind these many H atoms:	1	2	3	4	3	2	1	0
The formulas are:	LiH	BeH$_2$	BH$_3$	CH$_4$	NH$_3$	H$_2$O	HF	—

This pattern repeats itself as we go to still higher atomic numbers.

An atom of atomic number:	11	12	13	14	15	16	17	18
can bind these many H atoms:	1	2	3	4	3	2	1	0
The formulas are:	NaH	MgH$_2$	AlH$_3$	SiH$_4$	PH$_3$	H$_2$S	HCl	—

■ CH$_4$ is methane, the natural gas used for heating and cooking. NH$_3$ is ammonia.

In both series, we see an increase from 1 to 4 in the number of hydrogens and then we see the number fall back again. The number of hydrogens does not keep increasing to ever higher values as we go to elements of higher atomic numbers.

Just as helium (2), neon (10), and argon (18) seem to be similar with respect to boiling points and ionization energies, they also similarly form no compound with hydrogen. These three seem to be in a family. Notice that atoms of elements 6 and 14 are alike in their ability to bind four hydrogen atoms, and these two elements similarly share high boiling points (Figure 3.1*a*). In these examples we thus see that periodically, with increasing atomic number, physical and chemical properties recur—more or less. This is the essence of the **periodic law,** one of the important laws of nature.

■ *Periodic* means the recurrence of something in a regularly repeating way, such as lampposts or sunsets.

> **Periodic Law** The properties of the elements are a periodic function of atomic numbers.

The Periodic Table Organizes the Elements to Show Off Their Periodic Properties

The heart of Mendeleev's discovery was that elements of similar properties line up in vertical columns when horizontal rows, made of atomic symbols arranged in order of increasing atomic *mass* (Mendeleev's criterion), were broken at the right places. The result was a table of the elements called the **periodic table.** Its modern form, which is based on atomic numbers rather than atomic masses, is shown inside the front book cover. The table, of course, incorporates elements discovered since the time of Mendeleev. Each horizontal row is called a **period,** and each vertical column is called a **group.**

In constructing his periodic table, Mendeleev had the boldness to leave blanks in the columns whenever this seemed necessary to get elements to line up vertically in families. He even went so far as to declare that these blanks represented elements that had not yet been discovered, and he was right. Such a blank space prompted the search for and the discovery of the element germanium (atomic number 32), for example.

In order to achieve the best vertical sorting into families, Mendeleev even switched some pairs of elements from their order of increasing atomic masses. He listed, for example, tellurium (atomic mass 127.6) *before* iodine (atomic mass 126.9), because iodine seemed to fit far better with fluorine, chlorine, and bromine in group VIIA than with oxygen, sulfur, and selenium in group VIA. When atomic numbers were discovered, it was gratifying to learn that placing tellurium *before* iodine had put these elements in the correct order according to increasing atomic numbers.

The horizontal rows or periods are not all the same length; several are broken. This is necessary if the highest priority is to be the chemical similarities of the elements in vertical columns or groups. Thus period 1 is very short, containing only hydrogen and helium, and like the next two periods it is separated into two parts.

Dimitri Mendeleev
(1834–1907)

■ Notice in the periodic table that Co (at. no. 27) has a higher atomic mass than Ni (at. no. 28).

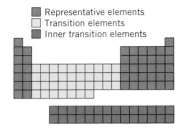

Representative elements
Transition elements
Inner transition elements

The (vertical) groups have both numbers and letters. Some have roman numerals followed by the letter A—IA, IIA, IIIA, and so forth up to VIIA. These A groups plus group 0 are called the **representative elements.** The other groups, clustered near the middle of the periodic table, use roman numerals followed by the letter B (except for a cluster in the middle designated as VIII). The B series plus group VIII are called the **transition elements.** There are 10 such elements in each of periods 4, 5, and 6. The two rows of elements placed outside the table are the **inner transition elements.** (The table would not fit well on the page if the inner transition elements were not handled in this way.) Elements 58 through 71 constitute the *lanthanide series,* named after element 57, lanthanum, which just precedes it. Elements 90 to 103 make up the *actinide series,* named after actinium, element 89. These two series each have 14 elements.

We are using here the column labels still widely employed in the United States. The International Union of Pure and Applied Chemistry, or IUPAC, has urged different designations for the vertical columns, but chemists in the United States have resisted the new numbers because they do not correlate electron configurations as easily with the column numbers. In the periodic table, the IUPAC numbers for the groups are given in parentheses beneath the designations we will normally use. Thus group VIIA is the same as the IUPAC group number 17. (The IUPAC group number will sometimes be given in parentheses in this chapter.)

Several of the groups among representative elements also have names. Except for hydrogen, the elements in group IA (IUPAC 1) are called the **alkali metals,** because they all react with water to give an alkaline or caustic (skin-burning) solution.

The elements in group IIA (IUPAC 2) are called the **alkaline earth metals,** because they are commonly found in "earthy" substances. Calcium carbonate or $CaCO_3$, for example, is a compound of calcium, an alkaline earth metal of group IIA, and $CaCO_3$ is the chief substance in limestone.

The elements in group VIIA (IUPAC group 17) are called the **halogens** after a Greek word signifying salt-forming ability. Chlorine of group VIIA, for example, is present in a chemically combined form in table salt, NaCl (sodium chloride).

The elements in group 0 (IUPAC 18), are all gases discovered after Mendeleev's first table. Except for a few compounds that xenon and krypton form with fluorine and oxygen, these elements chemically react with nothing. For this reason they are called the **noble gases** (*noble* signifying, perhaps, aloofness from change).

Other groups of representative elements are named simply after the first member—for example, the **boron family** (group IIIA, IUPAC 3), the **carbon family** (group IVA, IUPAC 14), the **nitrogen family** (group VA, IUPAC 15), and the **oxygen family** (VIA, IUPAC 16).

■ The IUPAC is an organization made of representatives from the world's several chemical societies, such as the American Chemical Society.

■ Hydrogen is a nonmetal, physically and chemically unlike the alkali metals, and it could be left to stand alone in the periodic table. It is put in group IA only because of its electron configuration (as we will soon see).

■ PRACTICE EXERCISE 4   Referring to the periodic table, pick out the symbols of the elements as specified.

(a) A member of the carbon family: Sr, Sn, Sm, S

(b) A member of the halogen family: C, Ca, Cl, Co

(c) A member of the alkali metals: Rn, Ra, Ru, Rb

(d) A member of the alkaline earth metals: Mg, Mn, Mo, Md

(e) A member of the noble gas family: Ac, Al, Am, Ar

Metals and Nonmetals Are Separated in the Periodic Table   An interesting feature of the periodic table is the location of the metals and nonmetals. As you can see in Figure 3.2, the great majority of all elements are metals and that, except for hydro-

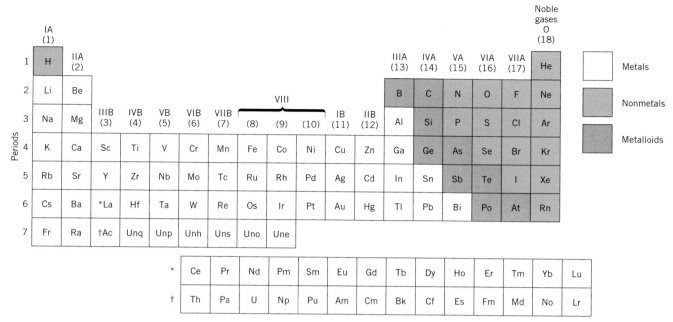

FIGURE 3.2 Locations of metals, nonmetals, and metalloids in the periodic table.

gen, the nonmetals are all clustered in the upper right hand corner. A few elements lie along the borderline between metals and nonmetals. Sometimes called **metalloids,** they have properties that are partly metallic and partly nonmetallic. Thus, carbon (as graphite) conducts electricity, like a metal, but shatters when struck, like a solid nonmetal.

The Members of a Group in the Periodic Table Form Compounds with Similar Formulas and Chemical Properties   As we saw, perhaps the most noteworthy common property of the elements in the same group, particularly among the representative elements, is that their compounds have like formulas and properties. For example, all of the alkali metals of group IA react with water in the same way. If we let M represent any alkali metal, the general equation for the reaction is

$$2M + 2H_2O \longrightarrow 2MOH + H_2$$

alkali   alkali metal
metal   hydroxide

When M is sodium, for example, sodium hydroxide (lye) forms.

$$2Na(s) + 2H_2O \longrightarrow 2NaOH(aq) + H_2(g)$$

sodium   sodium hydroxide

All of the group IA metal hydroxides, MOH, have the common property of being caustic substances capable of causing a chemical burn to the skin. Thus, if we knew that lye (NaOH) is caustic but had never handled potassium hydroxide, KOH, we would be very careful with it. KOH, like lye, is the hydroxide of an alkali metal and is very likely to be just as caustic as lye (which is true).

All of the binary compounds between the halogens (group VIIA) and hydrogen have the common formula HX (where X can be F, Cl, Br, or I). If we know that an aqueous solution of HCl gives a chemical burn to the skin, we would be very care-

■ To designate a physical state in an equation we use (s) for solid, (l) for liquid, (g) for gas, and (aq) for aqueous solution (a solution in water).

■ NaOH flakes are an ingredient in one kind of drain cleaner. Use it very carefully, wearing protective gloves, and keep it out of reach of children.

ful when handling similar solutions of HF, HBr, and HI. We would expect them to be like HCl.

■ We have much more to learn about acids, so consider the definition here to be a partial definition.

All of the aqueous HX solutions are *acids*. **Acids** are substances capable of destroying or *neutralizing* the caustic properties of the alkali metal hydroxides according to the following general equation:

$$HX(aq) + MOH(aq) \longrightarrow MX(aq) + H_2O$$

For example, hydrochloric acid, HCl(*aq*), and sodium hydroxide, NaOH(*aq*), react as follows:

$$HCl(aq) + NaOH(aq) \longrightarrow NaCl(aq) + H_2O$$

■ Pure HCl is a gas called *hydrogen chloride*. The name of its solution in water is *hydrochloric acid*, HCl(*aq*).

Water and sodium chloride (table salt) form in this reaction, and we know from experience that neither causes a chemical burn to the skin. Because the reaction of an acid with an alkali destroys a characteristic property of both, it is an important kind of reaction and so is given a special name, **neutralization.** Substances that can neutralize acids are generally called **bases,** so the reaction of hydrochloric acid with sodium hydroxide is an example of an *acid–base neutralization*. Similarly, hydrobromic acid, HBr(*aq*), reacts with and neutralizes potassium hydroxide, KOH(*aq*).

$$HBr(aq) + KOH(aq) \longrightarrow KBr(aq) + H_2O$$

Notice in our examples of chemical properties how we have been able to write *general* equations that apply to several reactions. General equations release us from having to learn each and every reaction of a family, because members of the same family have similar properties.

The existence of the periodic law and the periodic table begs several questions. Why? Why do the elements in the same group have similar properties? Why are the properties periodic? Why are certain periods broken up? Why are there transition elements and why 10 of them per period? Why are there 14 elements in each of the sets of inner transition elements? Answers to these questions lie in the finer details of atomic structure, namely, electron configurations of atoms.

3.3   ELECTRON CONFIGURATIONS, AN OVERVIEW

The electrons in an atom are confined to particular energy shells outside the nucleus.

When elements combine to form compounds, some electrons of the elements' atoms become relocated relative to atomic nuclei. To understand this, we have to learn where an atom's electrons are *initially*. Then we can think about how they might relocate to give more stable arrangements in compounds.

An atom's electrons do not occur simply at random in the space near its nucleus. They are constrained to particular patterns, and the specific arrangement of electrons about a nucleus is called the atom's **electron configuration.**

The Bohr Atom Was an Early Atomic Model   Two postulates about electron configurations that have survived to this day were proposed in 1913 by Danish physicist Niels Bohr (1885–1962), only 2 years after Rutherford's discovery of the nucleus. *Bohr's first postulate* was that electrons are confined to what came to be called *allowed energy states*. Electrons, said Bohr, cannot be just anywhere, like buzzing mosquitoes. They can only be in particular places, much as tennis balls on a stairway can only be on the steps, not suspended in air between them. Bohr's allowed en-

Niels Bohr

ergy states, in fact, are commonly called *energy levels* or *energy shells*. Like the steps of a stairway, lower energy states or levels are more stable places to be than are higher states. In other words, each of the allowed energy states of an atom corresponds to a different value of energy.

Bohr's second postulate was that as long as the atom's electrons remain in allowed energy states, the atom neither radiates nor absorbs any energy associated with the electrons' movements. We're not saying that electrons cannot move from one allowed state to another. This occurs, for example, when an iron fireplace poker is heated until it glows in the dark. Before being heated, the iron atoms are nearly all in their *ground state* arrangement of electrons; all their electrons are in the lowest energy states available. The heat causes electrons in iron atoms to shift to higher energy states called *excited states*. This is partly how the iron actually soaks up heat energy. As soon as atoms have become excited in this way, the electrons begin to shift back to lower states, *and the difference in energy between the excited state and the lower state is emitted as light*. Scientists had known long before Bohr that such emitted light did not possess every conceivable value of light energy, but had only certain values. This is what led Bohr to believe that atoms had only certain allowed energy states.

The precise quantity of energy that is emitted when one electron changes from a higher to a lower energy state is called a **quantum** of energy or a **photon** of energy. Sometimes electricity rather than heat is used to generate excited atoms that then emit light. Sodium vapor lamps or mercury vapor lamps along highways work this way, and you know that sodium vapor lamps have a characteristic yellow color. Excited sodium atoms don't emit all colors of light, just yellow light, and the yellow light from sodium lamps is characterized by photons of just two, very nearly equal values of energy.

To help people understand his postulates, Bohr suggested an analogy. He pictured an atom's electrons as being in very rapid motion around the nucleus, and that they follow paths, called *orbits*, much as planets move in orbits around the sun. This picture of an atom is an example of a scientific **model**, a mental or visual construction used to explain a number of facts. The **Bohr model** of the atom was quickly dubbed the "solar system" model, and it is still commonly used in the communications media in almost any discussion of things atomic.

The solar system view of the atom is now obsolete, because it has failed to explain too many observations. Yet Bohr's two fundamental postulates behind the model are still true.

Heisenberg's Uncertainty Principle Gives Us Nature's Limit on What We Can Know about an Atom's Electrons   The Bohr model worked well only for one element, hydrogen. The problem, as the German physicist Werner Heisenberg (1901–1976) soon realized, is that calculations based on the model assumed that the location of an electron and its energy can *both* be precisely known at the *same* instant, which is untrue.

The electron is small enough that any act of measuring its location gives it a nudge and changes its energy, and any attempt to find its energy changes its location. So the location and energy of an electron cannot *both* be precisely known at the same time. Some uncertainty about either an electron's location or its energy always exists, which is one way of stating what came to be called the **Heisenberg uncertainty principle.**

Because of Heisenberg's insight, scientists gave up the idea that an electron moves in a fixed orbit. Instead, the locations of electrons are described in terms of *probabilities* of their being at certain places. The question became, "In what particu-

■ Fireflies use *chemical* energy, not heat, to promote electrons to excited states in molecules.

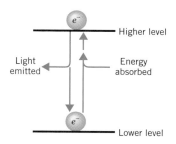

Electrons shift to higher levels when the correct amount of energy is absorbed, and they emit this energy when they drop back again.

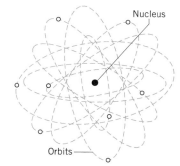

The Bohr "solar system" model

■ Heisenberg won the 1932 Nobel Prize in Physics.

■ The idea that the very act of measuring something actually alters what is being measured has profound implications for the fields of psychology, sociology, poll taking, and even television news.

TABLE 3.2   The Principal Energy Levels and the Number of Subshells

	1	2	3	4	5	6	7
Principal level number[a]	1	2	3	4	5	6	7
Maximum number of electrons actually observed in nature	2	8	18	32	32	18	8
Number of subshells in theory	1	2	3	4	5	6	7

[a] Principal quantum number, n.

lar parts of the space surrounding a nucleus is it *likely* that an electron will be?" It's like asking: "If I go out a certain distance from the nucleus in one particular direction, what are the chances or the probability of an electron being there?" The answer depends on what energy state the electron is in, as we'll see next.

Electrons in Atoms Are Confined to Principal Energy Levels   The specific energy states are called the atom's **principal energy levels.** They roughly correspond to certain successive distances from the atom's center. These levels are given numbers called the *principal quantum numbers,* symbolized by n. The value of n can't be just any number, like 0.467 or 12.689, but n is limited to *whole* numbers beginning with n equal to 1 for the lowest energy level. Level 1 happens to be the level nearest the nucleus, but it's better to associate a lower value of n with a lower value of energy rather than a location. An electron in level 1 has the least quantity of energy it can have in the atom. Just as nature has a powerful tendency to take up positions having the lowest possible energy, so an atom's electrons nearly always are in the lowest allowed levels—in the ground state.

There Are Limits to the Numbers of Electrons in Each Principal Energy Level   When an atom has three or more electrons, not all can crowd into level 1. Electrons, remember, are like-charged and so they repel each other. Each principal energy level has a limit to its number of electrons, and the limit for level 1 is only two. At $n = 2$, the limit is 8 electrons. Table 3.2 summarizes the maximum number of electrons that can be in the various principal energy levels.

The Principal Energy Levels Are *Electron Shells*   If you've ever studied a whole onion, you know that it exists in sections, each one a hollow sphere with thick walls, each successively larger than the one just inside it. The sections all have a common center, so they are called *concentric* spheres. Thus, we can roughly think of an atom's principal energy levels as concentric spheres, each with a definite thickness. Hence, the principal energy levels are often called **electron shells.** The first principal energy level is thus the first shell.

Summary about Principal Energy Levels   The electrons of an atom reside in principal energy levels or shells that occur concentrically around the atom's nucleus. Each shell has a unique, whole-number value of the principal quantum number, n, and each shell has a unique value for the number of electrons it can hold (Table 3.2).

3.4   ELECTRON CONFIGURATIONS AND ATOMIC ORBITALS

The principal energy levels have *subshells* made up of regions called *atomic orbitals.*

Electron Shells Have *Subshells*   The thickness of an electron shell allows for some fine structure. All electron shells except the first have a small number of **subshells,** the allowed number equaling the value of n for the main level. When $n = 1$, there is 1 subshell so we could call the entire shell a whole subshell. When $n = 2$, there are 2 subshells; at $n = 3$, there are 3 subshells, and so forth. It's as though an atom is an apartment house for electrons (with the nucleus in the basement) and each floor is a principal energy level or shell. Each floor can have one, two, three, four, or five apartments, called subshells.

Subshells Have Regions Called *Orbitals*   An atom's electrons are in subshells but these, like apartments, have further structure. Each subshell has a particular number of spaces, called *atomic orbitals,* where individual electrons can reside. **Atomic orbitals** are particularly shaped spaces within subshells that can hold up to two electrons apiece, no more. It's as though each apartment (subshell) on a given floor (main shell) has a certain number of rooms (orbitals) for electrons, but no room can hold more than two electrons. To specify the location of an electron, we must name its main energy shell, its subshell, and its orbital. When we do this, we specify all that we can know about an electron's location relative to the atomic nucleus.

Heisenberg said that if we give up wanting to know precisely *where* an electron is within an orbital, we can know what is more important, the *energy* of the electron. Energy is important because of a major fact about our world and the way it works: *Nature, given the opportunity, tends to change in whichever direction results in a more stable, lower energy arrangement of things.* When we know that one arrangement of electrons and nuclei is more stable (has lower energy) than another, we then know which arrangement nature prefers. When something gets into the less stable arrangement, it sooner or later will change, perhaps undergoing a chemical reaction leading to what is more stable. These basic principles of the way nature works are at the heart of chemical reactions, because they exemplify nature's preference for the most stable arrangement of its parts.

An Orbital Can Hold Two Electrons Only If They Are Spinning in Opposite Directions   The final complexity in electron configurations is that electrons are *spinning* particles. Like the earth, an electron spins about an axis, but, unlike the earth, an electron has the option of spinning in either of two opposite directions. When an atomic orbital holds two electrons, they can be present only if one electron spins in a direction opposite that of the other. Part of the reason is that the actual spinning makes an electron behave as a tiny magnet. By spinning *oppositely*, the two electrons have a magnetic *attraction* for each other. This helps to overcome the electrical repulsion between two like-charged electrons when they are in the same space.

Wolfgang Pauli (1900–1958), an Austrian-born physicist, was the first to realize the limitations on the number and spins of electrons in the same orbital, and we call the rule the **Pauli exclusion principle.**

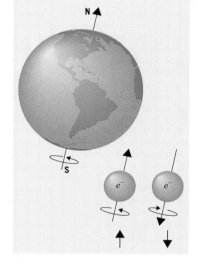

Electron spin

■ Wolfgang Pauli won the 1945 Nobel Prize in Physics.

> **Pauli Exclusion Principle**   An orbital can hold as many as two electrons, but only if they have opposite spins.

Orbitals Have Unique Shapes   Now let's look more closely at the kinds of "rooms," the atomic orbitals, available in the atom's electron apartments (subshells). As we said, when $n = 1$, we're at the first main level where there is only one subshell. The subshell at $n = 1$ has only one orbital named the 1s orbital, where "1" is the value of n, and "s" comes from a German word of no interest here. Think of "s" as meaning "spherical."

■ The orbital, subshell, and level are identical when $n = 1$.

Each atomic orbital has a particular shape deduced by calculating the probabilities of finding electrons at definite points relative to the atom's nucleus. The shape of an orbital is simply that of an imaginary envelope wrapped around enough of a particular space to enclose a region of high probability, say, 90%, of having a particular electron somewhere in it. Each point on the envelope's surface has the same probability as any other point.

Figure 3.3 shows the shape of the 1s orbital. When viewed from the outside, it looks like a sphere with the nucleus its center. The surface of the sphere encloses a space within which the probability of finding an electron belonging to level 1 is greater than 90%. An electron in a 1s orbital moves about very rapidly, but we can't know exactly how or where because we would rather know about the 1s electron's energy. It is more important to know whether an electron is in a 1s state than to know exactly where within the state it is.

The rapid movement of an electron within an orbital gives us another useful image, that of an **electron cloud.** An electron moves so rapidly that the influence of its negative charge is distributed throughout its orbital much as water molecules are distributed throughout a cloud.

Principal level 2 has 2 subshells. One consists of only one orbital named the 2s orbital. It corresponds to slightly less energy than that of the other subshell at $n = 2$. The shape of the 2s orbital looks like that of the 1s orbital when "viewed" from the outside, which explains why "s" is used in the name 2s. The second subshell at $n = 2$ holds three orbitals called the 2p orbitals. They have identical energies, and each 2p orbital has two lobes. The long axes of these orbitals, each axis symmetrically piercing both of the orbital's lobes, project at right angles along the x, y, and z coordinate axes (see Figure 3.4). Each 2p orbital is characterized by a subscript to indicate the direction of projection, so we write $2p_x$, $2p_y$, and $2p_z$. In atoms with two or more electrons, a p orbital electron has more energy than an s orbital electron in the same main level. Thus in atoms that normally have an electron at level two, the electron is in a slightly more stable location in the 2s orbital than in any of the 2p orbitals.

When $n = 3$, we're at main level 3, so it has 3 subshells. (Remember, the value of n is the same as the number of subshells.) One is an orbital all by itself, the 3s or-

■ The atomic orbitals that we are describing are based on calculations done on the hydrogen atom. Ample evidence exists that the orbitals of other atoms are like those of hydrogen.

■
$\overline{2p_x}$ $\overline{2p_y}$ $\overline{2p_z}$

$\overline{2s}$

Subshells at main level 2

■ 3d __ __ __ __ __
3p __ __ __
3s __

Subshells at main level 3

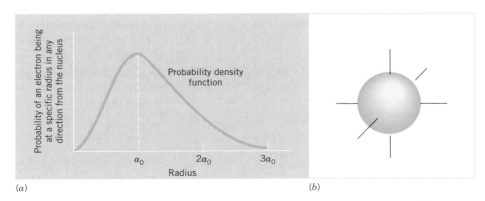

(a)   (b)

FIGURE 3.3   The 1s orbital. (a) Imagine that the space around a nucleus is made up of layer upon layer of extremely thin, concentric shells. At each distance away from the nucleus there is a point on the curve that indicates the probability of finding a 1s electron at this distance. Notice that the probability is zero for a zero distance—the electron is not on the nucleus. The probability reaches a maximum at the radius marked a_0, which is 52.9 pm (1 pm = 10^{-12} m). (b) One of the thin spheres described in part (a) encloses a space within which the total probability of finding an electron is large, say, 90%. This sphere is the "envelope" discussed in the text, and its shape is the shape of the 1s orbital.

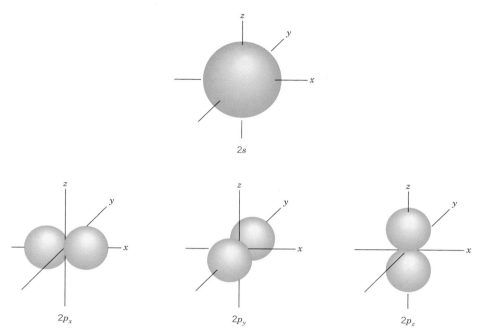

FIGURE 3.4   The orbitals at principal energy level 2. From the outside, the 2s orbital looks like a 1s orbital, seen in Figure 3.3. However, its radius is larger. Each of the 2p orbitals has the same shape and energy as the others, but they are oriented along the three different axes.

bital. From the outside, it looks like the other s orbitals, only it has a larger radius. Another subshell at $n = 3$ has three orbitals of identical energy but with right-angle orientations, the $3p_x$, $3p_y$, and the $3p_z$. They resemble in shape the two-lobed p orbitals at main level 2. An electron in a $3p$ orbital has more energy than one in the $3s$ state. The third subshell at $n = 3$ consists of five orbitals, called the $3d$ orbitals, which correspond to more energy than the $3p$ orbitals.

At $n = 4$, the fourth main level, there are 4 subshells, the $4s$ (1 orbital), $4p$ (3 orbitals), $4d$ (5 orbitals), and $4f$ (7 orbitals). We will not study the shapes and names of d or f orbitals.

Now that we know the main accommodations available to electrons, we are ready to describe the electron configurations of the elements. We can figure these out from atomic numbers because the configurations of atoms of successively higher atomic numbers follow a simple set of rules. In going to elements from lower to higher atomic number, we follow the **aufbau principle** (from the German for "building up"), namely, as we add a proton to the nucleus we place an electron in whichever of the *available* orbitals that corresponds to the lowest energy. Knowing what constitutes an "available orbital" is critical, of course, but only a few rules are needed. We've already given one, the Pauli exclusion principle.

■ 4f __ __ __ __ __ __ __
  4d __ __ __ __ __
  4p __ __ __
  4s __
Subshells at main level 4

Electrons Spread Out among Orbitals of the Same Subshell   Another rule concerning "available orbitals" governs where electrons go when orbitals *of the same energy* are present at some value of n, like the three p orbitals at one of the main levels. **Hund's rule** handles this question.

Hund's Rule Electrons *at the same subshell* spread out among the subshell's orbitals as much as possible.

This rule makes sense because electrons are like charged and they tend to be as far from each other as possible if it makes no difference in terms of the orbital energies. It is also true that when electrons do spread out like this, they have the same spins. Thus, the distribution on the left is more stable than the one on the right.

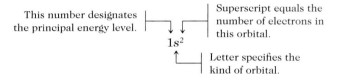

This configuration with unpaired spins is more stable.   This configuration with paired spins is less stable.

As a symbol for the direction of spin, we use an arrow, which can point either down (↓) or up (↑). However, because two electrons in the same orbital *must* have opposite spins, we seldom need such arrows. Usually, the symbol used for a pair of electrons in the same orbital is a right superscript as in $1s^2$.

This number designates the principal energy level.   Superscript equals the number of electrons in this orbital.

$1s^2$

Letter specifies the kind of orbital.

Beginning with the filling of orbitals at the $3d$ subshell and higher, what constitutes a lower energy subshell does not correlate well with the value of n. For example, the $4s$ subshell is used *before* any electrons are put into any orbitals of the $3d$ subshell, because the energy of the $4s$ subshell is less than that of the $3d$ despite its having a larger value of n. Figure 3.5 displays the relative energies of all of the subshells, and you can see several overlaps of subshells among different main levels.

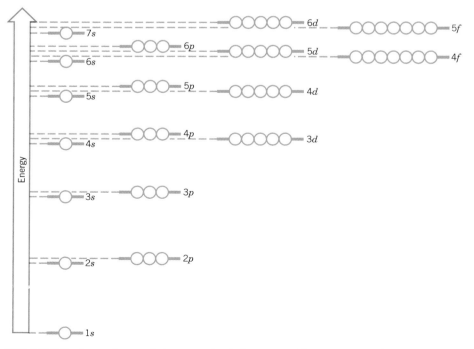

FIGURE 3.5 Approximate relative energy levels for atoms with two or more electrons.

One easily reconstructed device for knowing the order in which subshells are filled is given in Figure 3.6. Just follow the arrows, starting at the top. Thus, Figure 3.6 tells us that the order of filling orbitals is first $1s$, then $2s$, $2p$, $3s$, $3p$, $4s$, $3d$, $4p$, $5s$, $4d$, etc.

The *Aufbau Principle* Lets Us Construct the Electron Configurations of Elements from Atomic Numbers A *hydrogen* atom has one electron. We have learned that the 1s orbital has the lowest associated energy, so this is where the electron resides.

H $1s^1$

meaning $1s$ ↑

Helium has atomic number 2 and therefore its atoms have two electrons. Both can (and must) go into the 1s orbital.

He $1s^2$

meaning $1s$ ⇅

Lithium, atomic number 3, has three electrons in each atom. The first two fill the 1s orbital. According to Figure 3.5, the third electron must go into the next lowest orbital, the 2s.

Li $1s^2 2s^1$

meaning $2p$ __ __ __
$2s$ ↑
$1s$ ⇅

Beryllium, atomic number 4, has four electrons per atom. The first two fill the 1s orbital, and the last two fill the 2s orbital. None enters a 2p orbital, because these orbitals are at a higher energy level (Figure 3.5), and we must fill the lower energy orbitals first.

Be $1s^2 2s^2$

meaning $2p$ __ __ __
$2s$ ⇅
$1s$ ⇅

Boron, atomic number 5, has five electrons per atom. The first four fill the 1s and the 2s orbitals, exactly as in beryllium (number 4), and boron's fifth electron enters a 2p orbital. We don't know (or need to know) which of the three takes the electron, so we just arbitrarily assign it to the $2p_x$ orbital.

B $1s^2 2s^2 2p_x^1$

meaning $2p$ ↑ __ __
$2s$ ⇅
$1s$ ⇅

The next element, carbon (atomic number 6), is important at the molecular level of life, because its atoms make up the "backbones" of most if not quite all of the molecules, other than water, that are in living cells.

C $1s^2 2s^2 2p_x^1 2p_y^1$

meaning $2p$ ↑ ↑ __
$2s$ ⇅
$1s$ ⇅

Notice that carbon illustrates the application of Hund's rule. The last two electrons go into *separate* orbitals at the 2p subshell.

■ Hydrogen has three isotopes and this is the electron configuration of all three.

■ Helium is a gas used to fill dirigibles such as the Goodyear blimp. It is much less dense than air, and it won't burn.

■ Think of the three lines at the 2p subshell as representing, in order, the $2p_x$, $2p_y$, and $2p_z$ orbitals, but they are empty in Li and Be.

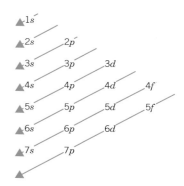

FIGURE 3.6 The order in which sublevels are filled. Each horizontal row names the subshells at a given main energy level. In a many-electron atom, the filling of sublevels begins with the 1s subshell (first slanting arrow at the top); then the 2s subshell receives electrons (second slanting arrow from the top). Hereafter, to find the further order of subshell filling, follow the successive slanting arrows in the directions they point. The order is thus $1s \rightarrow 2s \rightarrow 2p \rightarrow 3s \rightarrow 3p \rightarrow 4s \rightarrow 3d \rightarrow 4p \rightarrow 5s \rightarrow 4d \rightarrow$ etc.

■ Air is about 79% nitrogen in the form of N_2.

Nitrogen, number 7, another very important element among biological chemicals, also illustrates Hund's rule.

$$\textbf{N} \quad 1s^2 2s^2 2p_x^1 2p_y^1 2p_z^1$$

meaning   2p ↑ ↑ ↑
           2s ⇅
           1s ⇅

■ Air is about 21% oxygen in the form of O_2.

Oxygen (number 8) illustrates that we don't start with level 3 until level 2 is filled. Oxygen's eighth electron goes into a $2p$ orbital, not the $3s$.

$$\textbf{O} \quad 1s^2 2s^2 2p_x^2 2p_y^1 2p_z^1$$

meaning   2p ⇅ ↑ ↑
           2s ⇅
           1s ⇅

■ Fluorine is so reactive that it burns with water.

Fluorine (atomic number 9) has nine electrons, and we continue to fill the $2p$ orbitals.

$$\textbf{F} \quad 1s^2 2s^2 2p_x^2 2p_y^2 2p_z^1$$

meaning   2p ⇅ ⇅ ↑
           2s ⇅
           1s ⇅

■ Neon is the gas in "neon" lights.

With the next element, neon (atomic number 10), we complete the filling of all the atomic orbitals at level 2.

$$\textbf{Ne} \quad 1s^2 2s^2 2p_x^2 2p_y^2 2p_z^2$$

meaning   2p ⇅ ⇅ ⇅
           2s ⇅
           1s ⇅

Level 2 now has its maximum of eight electrons, two at the $2s$ subshell and six at the $2p$ subshell.

With element 11, sodium, we start to fill the third main level.

$$\textbf{Na} \quad 1s^2 2s^2 2p_x^2 2p_y^2 2p_z^2 3s^1$$

meaning   3s ↑
           2p ⇅ ⇅ ⇅
           2s ⇅
           1s ⇅

We could also write the configuration of the Na atom as $1s^2 2s^2 2p^6 3s^1$, because when *all* orbitals at a p subshell are filled we can save space by writing p^6.

The aufbau process described for elements 1–11 continues through the remainder of period 3, just as we have shown. Period 3 ends with argon (atomic number 18), with the configuration

Argon $1s^2 2s^2 2p^6 3s^2 3p^6$

As we move into period 4, remember (from Figures 3.5 and 3.6) that the $4s$ orbital accepts electrons *before* the $3d$. Therefore the electron configurations of potassium and calcium, the first two elements in period 4, are

Potassium $1s^2 2s^2 2p^6 3s^2 3p^6 4s^1$

Calcium $1s^2 2s^2 2p^6 3s^2 3p^6 4s^2$

EXAMPLE 3.1   Writing an Electron Configuration

Phosphorus, atomic number 15, is another element important at the molecular level of life. Write its electron configuration.

ANALYSIS   With 15 electrons, we know that both levels 1 and 2 are filled; they take $2 + 8 = 10$ electrons. The remaining five must be in level 3.

SOLUTION $1s^2 2s^2 2p^6 3s^2 3p_x^1 3p_y^1 3p_z^1$

meaning

$3p$	↑	↑	↑
$3s$	⇅		
$2p$	⇅	⇅	⇅
$2s$	⇅		
$1s$	⇅		

■ **PRACTICE EXERCISE 5**   Using the one-line representation rather than the system with arrows, write the electron configuration of each of the following elements. Use the table inside the front cover to find out their atomic numbers.

(a) aluminum   (b) chlorine   (c) silicon   (d) calcium

Abbreviated Electron Configurations Focus Attention on the Electrons of the Highest Occupied Level   Still another way to represent the electron configurations of elements like sodium and the others in the third period of the periodic table is in the abbreviated form illustrated for Na and the next element, magnesium, as follows.

$$\text{Sodium} \qquad [Ne]3s^1$$
$$\text{Magnesium} \qquad [Ne]3s^2$$

"[Ne]" stands for $1s^2 2s^2 2p^6$, the electron configuration of neon. In other words, we let "[Ne]" represent all of the inner electrons, the **core electrons,** those in main shells with n less than that of the highest *occupied* shell. Following the symbol for the core electrons, we finish the configuration by specifying in more detail the **outside shell electrons,** those in orbitals associated with the highest value of n at which electrons occur. The outside shell electrons are generally those involved in chemical reactions; the core electrons usually are undisturbed.

Elements in period 4, elements 19 through 35 following argon, have the argon configuration for the inner core electrons, $1s^2 2s^2 2p^6 3s^2 3p^6$. This allows us to write the following abbreviated configurations for potassium and calcium.

$$\text{Potassium} \qquad [Ar]4s^1$$
$$\text{Calcium} \qquad [Ar]4s^2$$

The $3d$ orbitals of potassium and calcium are empty like those of argon.

After calcium, we move into the first series of 10 transition elements. We begin to place electrons in the five $3d$ orbitals, which correspond to lower energy than the $4p$ state (see Figure 3.5). From element 21 (scandium) through element 30 (zinc), the $3d$ orbitals fill until at zinc the $3d$ orbitals hold a total of 10 electrons. Only after the $3d$ orbitals have filled do electrons go into the $4p$ orbitals, beginning with element 31 (gallium). The filling of the $3d$ orbitals is a general pattern for transition elements in any period. As we move through any series of transition elements from left

■ The *d* orbitals come in sets of five, each capable of holding 2 electrons, so 5 × 2 or 10 is the maximum number of electrons that can be held by a *d* orbital subshell.

to right in the periodic table, *d* orbitals are filling to their maximum of 10 electrons. Because it takes a series of 10 elements to fill the 5 *d* orbitals, each set of transition elements numbers 10.

The period 5 elements have a krypton core; krypton is the noble gas element immediately *preceding* period 5. Therefore, the first element in period 6, rubidium (an alkali metal), has the electron configuration [Kr]5s^1. Period 6 elements all have xenon cores, and the elements in period 7 have radon cores. We will not carry this study to greater detail, but you can find all of the electron configurations in Appendix B, where you can see that the atoms of the inner transition elements correspond to the filling of the *f* orbitals. Because there are seven *f* orbitals at any value of *n*, it takes 14 electrons to fill them, which is why there are 14 elements in each series of inner transition elements. The elements of most common significance at the molecular level of life are found among the first 20, although a number of transition elements are vitally important as the "trace elements" of nutrition.

When We Write the Electron Configuration of an Element, We Do So for Any of Its Isotopes   Atoms of the different isotopes of the same element differ only in their numbers of neutrons, but these are in the nucleus and have nothing to do with electron configurations. *It is because all isotopes of any given element have the same electron configurations, that they all have the same chemical properties.*

Elements in the Same Family Have the Same Outside Shell Configurations   Among the representative elements of groups IA–VIIA, the roman numeral of the group equals the number of outside shell electrons in the atoms. Thus, all atoms of group IA have one electron in the outside shell. The atoms of the group IIA elements have two outside shell electrons, and so on.

■

Group number and name		Outside shell electrons
IA	Alkali metals	1
IIA	Alkaline earth metals	2
IIIA	Boron family	3
IVA	Carbon family	4
VA	Nitrogen family	5
VIA	Oxygen family	6
VIIA	Halogen family	7
0	Noble gases	8

Group IA		Group IIA	
Lithium	[He] 2s^1	Beryllium	[He] 2s^2
Sodium	[Ne] 3s^1	Magnesium	[Ne] 3s^2
Potassium	[Ar] 4s^1	Calcium	[Ar] 4s^2
Rubidium	[Kr] 5s^1	Strontium	[Kr] 5s^2
Cesium	[Xe] 6s^1	Barium	[Xe] 6s^2
Francium	[Rn] 7s^1	Radium	[Rn] 7s^2

■ We note in passing that somehow an outside level of 8 electrons confers unusual chemical stability (to be continued in Chapter 4).

Similarly, the group IIIA atoms all have 3 outside shell electrons. The atoms in group IVA have 4. Helium, the first element in group 0, has just two electrons in its outside shell, but this is shell 1 and so could not hold more than 2 anyway. Otherwise, after helium, all noble gas elements have 8 outside shell electrons. For understanding chemical properties, the *total* number of electrons in the outside shell is the important number obtained from electron configurations. This number can found at a glance at the periodic table when the element is a representative element, because it's the same as the roman numeral of the group.

EXAMPLE 3.2   Finding Information in the Periodic Table

How many electrons are in the outside shell of an atom of iodine?

ANALYSIS   We need the atomic number of iodine to find it most easily in the periodic table, which will then tell us whether iodine is a representative element. A search of the Table of Atomic Masses and Numbers (inside the front cover) tells us that the atomic number of iodine is 53. Now we use the periodic table, and find that iodine is in group VIIA.

SOLUTION   Being one of the A-type elements, we know that iodine is a *representative* element, which means that its group number is the same as the number of outside shell electrons, 7.

■   PRACTICE EXERCISE 6   How many electrons are in the outside shell of an atom of each of the following elements?

(a) potassium   (b) oxygen   (c) phosphorus   (d) chlorine

Most Elements with 4–8 Outside Shell Electrons Are Nonmetals   If you compare the locations of the elements in Figure 3.2 with their electron configurations, you will see that all of the nonmetals, except hydrogen and helium, have four to eight electrons in their outside shells. All the metallic elements have atoms with one, two, or three outside shell electrons, occasionally four. Only among some elements with high atomic numbers do we find metals with more than three outside-shell electrons. Tin and lead of group IVA are common examples.

SUMMARY

Atomic structure   Atoms, which are electrically neutral particles, are the smallest representatives of an element that can display the element's chemical properties. Each atom has one nucleus—a hard inner core—surrounded by enough electrons to balance the positive charge on the nucleus. All the atom's protons and neutrons (and thus all of its mass and positive charge) are in the nucleus. The proton has a charge of 1+, the electron's charge is 1−, and the neutron is electrically neutral. In atoms, the atomic number equals both the number of nuclear protons and the number of extranuclear electrons. The mass number of an isotope is the sum of the protons and neutrons in its atoms' nuclei. An element's atomic mass or atomic weight is the average mass in atomic mass units (u) of the element's individual masses taking into account the proportions of the isotopes as they occur naturally.

Periodic properties   Because many properties of the elements are periodic functions of atomic numbers, the elements fall naturally into vertical groups or families in the periodic table. The atoms of a group of representative elements, those in groups IA–VIIA, have the same number of outside shell electrons, a number corresponding to the group number itself. Of the group 0 elements, all except helium have outside shells with 8 electrons; helium atoms have 2.

The horizontal rows of the periodic table are called periods. In the long periods, there are transition and inner transition elements which involve the systematic filling of inner *d* or *f* orbitals. The nonmetallic elements are in the upper right-hand corner of the table, and the metals, the great majority of the elements, make up the rest of the table. At the border between metals and nonmetals occur the metalloids, which have both metallic and nonmetallic properties.

The metal hydroxides of the group IA alkali metals have the general formula, *M*OH and are all bases. They neutralize acids, such as hydrochloric acid or the other H*X* acids of group VIIA, in the same way.

Atomic orbitals   The places where electrons can be in an atom are organized as principal energy shells, which consist of subshells made up of orbitals. An atomic orbital is a volume of space near the nucleus where there is a high probability of finding a particular electron. The shape of an orbital comes from wrapping an imaginary envelope about that much of the space within which the overall probability of finding an electron is at least 90%.

Principal level or shell 1 has only one subshell. (It is its own subshell.) And this subshell has just one orbital. It is its own orbital, the 1s orbital, the 1 standing for principal level 1, and the s specifying the shape of the orbital (spherical). At level (shell) 2, there are 2 subshells—the s and the p types. Any s subshell, no matter at which principal level, has just 1 orbital. A p-type subshell always has 3 orbitals, designated as p_x, p_y, and p_z, to correspond to the 3 coordinate axes along which the 3 are aligned. Each p orbital consists of two lobes of equal size and shape. Level 3 has s, p, and d subshells. (Subshell d consists of 5 orbitals.) Levels 4 and higher have all these plus an f subshell (with 7 orbitals).

Each occupied orbital can be viewed as an electron "cloud." It isn't possible to obtain precise information simultaneously about an electron's location and energy, but the knowledge of where electrons most likely are—information obtained from electron configurations—is enough for understanding chemical properties.

As long as electrons remain in their orbitals, an atom neither absorbs nor radiates energy. However, an atom can absorb the energy that corresponds exactly to the difference in energy between two orbitals, provided the orbital of higher energy has a vacancy. The absorbed energy makes an electron move to the higher orbital. When it drops back down, this energy is radiated as a quantum or photon of light.

Electron configurations   To write an electron configuration of an atom, we use the aufbau principle and follow certain rules. We place the atom's electrons one by one into the available orbitals, starting with the one of lowest energy, the 1s orbital. According to the Pauli exclusion principle, each orbital can hold two electrons (if their spins are opposite), but where two or more orbitals are available at the same subshell, electrons spread out (Hund's rule).

REVIEW EXERCISES

The answers to Review Exercises whose numbers are in color are found in Appendix E. The answers to the other Review Exercises are found in the Study Guide that accompanies this book. The more challenging questions are marked with asterisks.

The Nuclear Atom

3.1   Dalton said that atoms are indestructible, but we now know that they can be broken up into three (actually more) particles.
(a) What is the general name for such particles?
(b) What are the names and electrical conditions of the three particles of greatest interest in understanding the chemical properties of substances?
(c) Two of these particles attract each other. What are their names?
(d) Two simply stated rules govern the behaviors of electrically charged particles toward each other. What are they?

3.2   The atomic mass unit has approximately what size, 10^{-230} g, 10^{-23} g, 10^{-3} g, or 10^{23} g?

3.3   What fact lets us ignore the masses of electrons in an atom when we compute an atomic mass from its subatomic particles?

3.4   To five significant figures, calculate the mass in grams of a sample of 6.0220×10^{23} hydrogen atoms, each one having just one proton. Ignore the mass of the electron. How does the result compare with the atomic mass of hydrogen? (To two significant figures, are the results the same or different?)

3.5   Who discovered the existence of atomic nuclei and what did he conclude about the nature of an atom?

3.6   What causes the positive charge on an atomic nucleus, and why is it always a *whole* number?

3.7   When we know the specific arrangement of the electrons around an atomic nucleus, what do we know about the atom? (What short term is used?)

3.8   When we say that nearly all elements consist of mixtures of substances, what general name do we give to these substances?

3.9   Which of the following are pairs of isotopes? (Use the hypothetical symbols for your answer.)

M has 12 protons and 13 neutrons
Q has 13 protons and 13 neutrons
X has 12 protons and 12 neutrons
Z has 13 protons and 12 neutrons

3.10   The members of which of the following pairs of elements would be expected to have identical chemical properties and why?

Pair 1   A (10 protons and 9 neutrons)
         B (10 protons and 10 neutrons)
Pair 2   C (7 protons and 6 neutrons)
         D (6 protons and 6 neutrons)

3.11   Carbon, atomic number 6, has three isotopes that are found in nature. The most abundant (98.89%) has a mass number of 12. The isotope with a mass number of 13 makes up 1.11% of naturally occurring carbon. The third isotope is carbon-14, which is obviously present in the

merest trace, because 98.89% + 1.11% = 100.00%. But carbon-14 makes possible the dating of ancient artifacts.
(a) What is the same about these isotopes?
(b) In what specific feature of atomic structure do they differ?

3.12 Later, when we want to describe some *chemical* reaction of sulfur, which consists principally of two isotopes (mass numbers 32 and 34), we will use the symbol S. Why won't we have to specify which isotope?

The Periodic Table

3.13 Describe in a general way how the boiling points of the first 20 elements vary as their atomic numbers increase.

3.14 What is meant by the *first ionization energy* of an element? In general terms, how does the value of the first ionization energy change as you move from atomic number 1 to 20?

3.15 The binary hydrides of the first 20 elements display a periodic change in what specific way?

*3.16 What are the likely formulas of the binary hydrides of the following elements?
(a) sodium   (b) selenium   (c) gallium
(d) germanium   (e) arsenic   (f) calcium
(g) bromine

3.17 What general name is given to the elements in the same vertical column in the periodic table? What name is given to a horizontal row of elements in the periodic table?

3.18 In the form of the periodic table used in this textbook, vertical columns are assigned roman numerals followed by either A or B.
(a) Which designates a family of *representative* elements, IIIA or IIIB?
(b) Are the elements in group 0 regarded as representative or transition elements?
(c) Are the elements in the B groups likely to be metals or nonmetals?
(d) Are the elements in period 2 representative or transition elements or a mixture of both?

*3.19 Suppose element *X* forms a compound HX.
(a) To what *group* does *X* most likely belong and what is the family name?
(b) Is HX an acid or a base?
(c) What reaction will HX give with KOH? Write the equation.
(d) What *name* is given to the reaction of part (c)?

3.20 Give the group number and the chemical family name of the set of elements to which each of the following belongs.
(a) bromine   (b) lithium
(c) selenium   (d) barium

3.21 Give the group number and the chemical family name of the set of elements to which each of the following belongs.
(a) chlorine   (b) phosphorus
(c) calcium   (d) aluminum

Electron Configurations

3.22 Niels Bohr suggested two postulates about the arrangements of electrons in atoms. What are they? Are they still true?

3.23 When light is emitted from, say, a red-hot bar of iron, what happens at the atomic level?

3.24 What analogy did Bohr suggest to illustrate his view of atomic structure? Do scientists still use this analogy?

3.25 What two facts about an electron in an atom cannot be known precisely *at the same time*? Who first realized this? How was the Bohr model affected?

3.26 Instead of trying to describe the orbits taken by electrons in atoms, we instead say that a given electron is in a particular *energy state*. Each such allowed energy state has what two general names (synonyms of each other)?

3.27 When all of the electrons of an atom are in their lowest energy states, what name do we give to this condition or state?

3.28 What does *principal quantum number* refer to? What relationship exists between the number of subshells and the corresponding principal quantum number?

3.29 Complete the following table to describe the *numbers* of subshells and orbitals associated with principal energy levels.

Principal Energy Level Number	Number of Subshells	Number of Orbitals
1	_____	_____
2	_____	_____
3	_____	_____
4	_____	_____

3.30 What are the symbols used for the individual orbitals at each location?
(a) Principal level 1, subshell 1
(b) Principal level 2, subshell 1
(c) Principal level 2, subshell 2
(d) Principal level 3, subshell 2

3.31 When all lower orbitals are filled and one electron is in principal level 3, what subshell and what orbital is it normally in? Give the combination symbol that summarizes this.

*3.32 When we describe the geometry of the 1s atomic orbital as that of a sphere:

(a) What is said about the region inside the sphere?

(b) What is true about the surface of the sphere?

(c) Is it possible for a 1s electron to be outside its 1s sphere?

3.33 What does the cross section of a 2p orbital look like, roughly?

3.34 Of the three p orbitals at principal level 2, what is different about them?

3.35 What is the maximum number of electrons that are allowed in each?

(a) Principal energy levels 1, 2, and 3

(b) The p orbitals of principal levels 2 and 3

(c) The s orbitals of principal levels 1, 2, and 3

(d) The d orbitals of principal levels 4 and higher

(e) The f orbitals where they occur

3.36 What is Hund's rule, and what is a likely reason for it?

3.37 What does the Pauli exclusion principle tell us about writing electron configurations?

3.38 How many electrons are represented in the following electron configurations, and what is the name of the associated element?

(a) $1s^2 2s^2 2p_x^1 2p_y^1 2p_z^1$

(b) $1s^2 2s^2 2p_x^2 2p_y^2 2p_z^2 3s^2 3p_x^1 3p_y^1$

3.39 How many electrons are represented in the following electron configurations? Name the associated element.

(a) $[\text{Ne}]\, 3s^2 3p_x^1 3p_y^1 3p_z^1$

(b) $[\text{Ar}]\, 4s^2$

3.40 Write the electron configuration of sulfur in each of the following ways.

(a) The unabbreviated one-line mode

(b) The condensed one-line mode

3.41 Write the electron configuration of potassium in each of the following ways.

(a) The unabbreviated one-line mode

(b) The condensed one-line mode

*3.42 The electron configuration of an atom of the metal zinc is $1s^2 2s^2 2p^6 3s^2 3p^6 3d^{10} 4s^2$. Using only this information, answer the following questions.

(a) What is the atomic number of zinc? How can you tell?

(b) Is principal energy level number 3 completely filled?

(c) Does the zinc atom have electrons with unpaired spins? How can you tell?

(d) How would the electron configuration of zinc be written in the condensed mode? (Let your answer reflect the subshell filling order of Figure 3.6, i.e., 4s fills before 3d.)

(e) Using the condensed mode, what would be the electron configuration of an element with four more electrons than zinc? What is its atomic number? What is its atomic symbol?

(f) In answering part (e), how was Hund's rule used?

(g) In answering part (e), how was the Pauli exclusion principle used?

3.43 Using the condensed mode, write the electron configuration of neon, atomic number 10.

*3.44 Using only the following atomic numbers and the aufbau principle, write the electron configurations of the following atoms in both the noncondensed and abbreviated forms.

(a) 3 (b) 8 (c) 12 (d) 16

*3.45 Suppose that an atom has the following electron configuration.

$$1s^2 2s^2 2p^6 3s^2 3p^6 3d^{10} 4s^2 4p^3$$

Without consulting the periodic table, answer the following questions.

(a) To what family does it belong (using the roman numeral–letter designation)? How can you tell?

(b) Is it likely a metal or a nonmetal? How can you tell?

Additional Exercises

*3.46 In Chapter 4 we will learn that particles having the following compositions exist.

Particle X: 11 protons, 12 neutrons, and 10 electrons
Particle Y: 17 protons, 18 neutrons, and 18 electrons

(a) What is the *net* electrical charge carried by particle X?

(b) What is the *net* electrical charge carried by particle Y?

(c) Would the particles X and Y attract each other, repel each other, or be indifferent to each other? How can you tell?

(d) What are the mass numbers of X and Y?

3.47 It costs energy to make an electron leave an atom, but which of the following changes would cost the *least* energy, and why?

(a) Removal of an electron from a particle having 12 protons and 12 electrons

(b) Removal of an electron from a particle having 12 protons and 11 electrons

3.48 Rutherford found that essentially all of the mass of an atom is in its nucleus. The following calculations will give you an idea of what this means.

(a) Calculate the density in g/cm^3 of the nucleus of a hydrogen atom from the following data. Assume that the nucleus is a perfect sphere so that you can calculate its volume by the standard equation, volume of sphere = $(4/3)\pi r^3$, where r = radius and π = 3.14. The radius of the hydrogen nucleus is 5×10^{-14} cm and the mass of the nucleus is 1.67×10^{-24} g.

(b) Convert the answer to part (a) into units of metric tons per cubic centimeter (1 metric ton = 1000 kg).

3.49 An atom has a nucleus with 13 protons and 14 neutrons. What is its atomic number? What is its mass number?

3.50 In what way would period 5 of the periodic table be different if the elements were arranged in their order of increasing atomic mass instead of increasing atomic number?

Why did Mendeleev insist on the arrangement we presently see despite the fact that it violated his version of the periodic law?

3.51 In this chapter we looked ahead to the use we will make of an electron configuration. In general terms, what is this?

3.52 Given the opportunity and freedom to change, nature generally tends to assume a posture of lowest energy. What rule used in connection with the aufbau principle reflects this?

*3.53 Examine the following possible electron configurations and answer the questions about them.

$$1s^2 2s^2 2p_x^2 2p_y^2 2p_z^2 3s^2 3p_x^1 \qquad 1s^2 2s^2 2p_x^2 2p_y^2 2p_z^2 3s^2 4s^1$$

$$1 \qquad\qquad\qquad 2$$

(a) Are these electron configurations for atoms of the same element or of different elements? How can you tell?
(b) Which is the more stable configuration, **1** or **2**? How can you tell?
(c) Which configuration corresponds to the *higher* energy state for the atoms? Why?
(d) What has to happen to change an atom from one configuration to the other?

3.54 Using only the following atomic numbers as well as the aufbau principle, write the electron configurations of the following atoms in both the noncondensed and abbreviated forms.
(a) 5 (b) 7 (c) 14 (d) 17

*3.55 Suppose that an atom has the following electron configuration.

$$1s^2 2s^2 2p^6 3s^2 3p^6 3d^{10} 4s^2 4p^6 4d^{10} 5s^2 5p^2$$

Without consulting the periodic table, answer the following questions. When you write condensed electron configurations, let your answers reflect the subshell filling order of Figure 3.6, namely, 5s fills before 4d.
(a) What is the group number of this element? How can you tell without referring to the periodic table?
(b) When written in the condensed form, what is its electron configuration?
(c) Using the condensed form, write the electron configuration of the element standing immediately to its right in the periodic table.
(d) Using the condensed form, write the electron configuration of the element standing immediately to its left in the periodic table.
(e) Using the condensed form, write the electron configuration of the element standing immediately above it in the periodic table. (The group 0 element nearest this one is argon, Ar.)

*3.56 The following table is a section from the periodic table where the numbers are atomic numbers. The numbers of one row have been given hypothetical atomic symbols. You should be able to answer the following questions without referring to the actual periodic table.

5	6	7	8	9
13 a	14 b	15 g	16 d	17 e
31	32	33	34	35

(a) Give the atomic numbers of the elements in the same period as g.
(b) What are the atomic numbers of the elements in the same group as b?
(c) Give the atomic numbers of the elements in the same family as e.
(d) Above each box of the top row of elements, write the group numbers of the elements, including the A or B designation.
(e) How many electrons are in the highest occupied principal energy level of the element that would stand immediately to the left of a?
(f) How many electrons would be in the outside shell of the element standing immediately below d in the periodic table?
(g) Element g forms a binary hydride with hydrogen with the formula gH_3. What are the likely formulas of the binary hydrides of elements 33 (give it the symbol X) and 7 (give it the symbol Z)?
(h) Which element is more likely to be a nonmetal, element 9 or 31?

*3.57 Using Table 3.1 data, convert the neutron's mass from g to u.

4

CHEMICAL COMPOUNDS AND CHEMICAL BONDS

Crystals of table salt (NaCl) all have the same, regular, cubic shapes. We'll see why in this chapter.

THIS CHAPTER IN CONTEXT

With our background in atomic structure we can now raise a fundamental question: What holds things together? When we have the answer we'll be able to understand why rock salt is hard, why water is mobile, and why air moves so readily out of our way as we walk. The question arises because atoms and molecules are electrically *neutral*, so how do they become stuck together strongly enough to account for the existence of solids and liquids? The major part of the answer concerns net electrical forces of attraction. This chapter is mostly about how such forces arise from neutral particles.

The principles laid down here are crucial to life at the molecular level. They speak to related questions for later study, such as how do cells keep their insides in? Why are bones strong? What stabilizes the molecules of genes? How can molecules become organized into strong muscle fibers?

4.1 ELECTRON TRANSFERS AND IONIC COMPOUNDS

Strong forces of attraction exist between oppositely charged ions in a large number of chemical compounds.

The Two Principal Kinds of Compounds Are Ionic and Molecular   To get net electrical forces of attraction out of neutral atoms, electrons and nuclei have to reorganize. This happens in basically two ways. One leads to a *molecule*, a package of two or more atomic nuclei located among enough electrons to make the whole particle neutral. In *molecular elements*, like Cl_2 or O_2, the nuclei are from the same element. In *molecular compounds* the nuclei are from different elements, as in the molecules of water (H_2O), ammonia (NH_3), sugar ($C_{12}H_{22}O_{11}$), or cholesterol ($C_{27}H_{46}O$).

The second way to reorganize the electrons and nuclei of atoms into compounds produces tiny particles of opposite electrical charge called *ions*, and these strongly attract each other. Compounds consisting of oppositely charged ions are called *ionic compounds*. Let's consider them next.

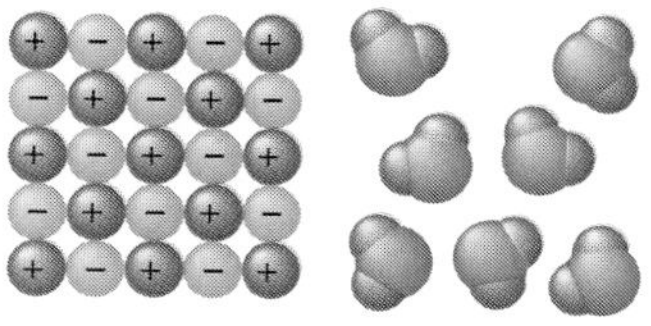

Electron Transfers between Atoms Produce Ions   Sodium chloride (table salt) is a typical ionic compound. Its parent elements, sodium and chlorine, cannot be stored in each other's presence because they react violently. Their atoms undergo the following changes in electron configurations. To simplify this discussion, we're going to ignore the fact that chlorine consists of molecules, Cl_2, instead of individual atoms. The circles shown next stand for atomic nuclei, which contain protons (p^+) and neutrons (n). Notice that a 3s electron from a sodium atom transfers to the $3p_z$ orbital of a chlorine atom. By losing an electron, the sodium atom gives up its electrical neutrality as well and becomes positively charged. When a chlorine atom accepts the electron, its net charge changes to $1-$.

■ *Molecule* is from a Greek term meaning *little mass*.

An ionic substance   A molecular substance

■ For purposes of illustration, we have picked the sodium-23 and chlorine-35 isotopes.

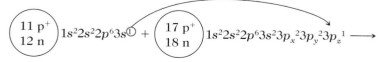

One sodium *atom*
Na

One chlorine *atom*
Cl

$$\left[\overset{\text{Outer}}{\text{octet}} \quad \binom{11\ p^+}{12\ n}\ 1s^2 2s^2 2p^6 \overbrace{} \right]^+ + \left[\binom{17\ p^+}{18\ n}\ 1s^2 2s^2 2p^6 \overbrace{3s^2 3p_x^2 3p_y^2 3p_z^2}^{\text{Outer octet}} \right]^-$$

One sodium *ion*
Na$^+$

One chloride *ion*
Cl$^-$

■ *Ion* is from the Greek *ienai*, to go or to move. Ions, unlike atoms, can move in response to electrical forces.

■ A 0.1-mg sample of NaCl has about 10^{18} pairs of ions.

Crystal of sodium chloride

■ Sodium chloride melts at 804 °C.

Electrically charged particles at the atomic level of size are called **ions,** with positively charged ions designated **cations** and negatively charged ions called **anions.** Ions of either charge having *one* atomic nucleus are **monatomic ions.** In the reaction of sodium with chlorine, electrons relocate relative to atomic nuclei to give sodium ions and chloride ions. These ions, not intact atoms, constitute sodium chloride.

Ionic Compounds Have Ionic Bonds between Symmetrically Organized Ions   It isn't physically possible to arrange a chemical meeting between just one atom of sodium and one of chlorine. Any visible sample, even the tiniest speck, has upward of at least 10^{18} atoms. Therefore, when actual samples of these elements are mixed, a storm of electron transfers occurs, and countless numbers of oppositely charged ions form.

Because like charges repel, the new sodium ions repel each other and the new chloride ions also repel each other. Being unlike charged, however, sodium ions and chloride ions attract each other. To maximize the forces of attraction and to minimize the forces of repulsion, Na$^+$ and Cl$^-$ ions *cannot help but assemble in a symmetrical array.* Spontaneously, Na$^+$ and Cl$^-$ nestle together as nearest neighbors, and like-charged ions stay just a little farther apart (see Figure 4.1). This is why crystals of sodium chloride have regular shapes.

What forms as oppositely charged ions come together is an **ionic compound,** an orderly aggregation of oppositely charged ions. The force of attraction between the ions is called the **ionic bond.** Ionic bonds are very strong, so crystals of ionic compounds remain rigid up to high temperatures, after which the crystals melt. Ionic crystals, such as salt crystals, can be pulverized without too much difficulty, however. A sharp blow makes one layer of ions shift slightly, causing like-charged ions suddenly to become closest neighbors. Now the net force—at least along this layer in the crystal—is one of repulsion, not attraction, and the crystal splits apart.

The Formula Unit of NaCl Is a Hypothetical Particle   The ionic bond does not extend in any single, unique direction. The force of attraction that Na$^+$ has for a negative charge radiates equally in all directions, like light from a light bulb. Therefore, one particular Na$^+$ ion doesn't belong to any particular Cl$^-$ ion. Except in our imagination, therefore, no separate, discrete *molecule* exists that consists of just one Na$^+$ ion and one Cl$^-$ ion. Instead, we speak of the *formula unit* of NaCl a hypothetical particle made of one of each of its ions. Thus, NaCl is an empirical formula, not a molecular formula.

Monatomic Ions Are Named after the Parent Elements   We'll pause now to learn how monatomic ions and the simpler ionic compounds are named before going further.

All ions derived from metals have the same name as the element (plus the word *ion*). The sodium ion, Na$^+$; the potassium ion, K$^+$; the magnesium ion, Mg^{2+}; and the calcium ion, Ca^{2+}, are particularly important. Notice that the formula of an ion, when set separately, always includes the electrical charge as a right superscript.

All metals form monatomic *cations*. The transition metals generally can form two cations that differ in the amount of charge. Iron forms two ions, for example, Fe^{2+} and Fe^{3+}. To name them, we indicate the amount of charge by a roman numeral. Thus, Fe^{2+} is named the iron(II) ion, and Fe^{3+} is the iron(III) ion. Long ago, the iron ions were called the ferrous ion (Fe^{2+}) and the ferric ion (Fe^{3+}). Copper can likewise form two cations, Cu^+ or the copper(I) ion (cuprous ion), and Cu^{2+}, the copper(II) ion (cupric ion). We'll now ignore the older names because they are almost never used except occasionally on bottles of chemicals.

Monatomic ions derived from nonmetals are generally *anions,* and their names end in *-ide,* as in the anion of chlorine, the *chloride ion,* Cl^-. Other important monatomic anions are the fluoride ion, F^-; bromide ion, Br^-; iodide ion, I^-; oxide ion, O^{2-}; and sulfide ion, S^{2-}.

The names, symbols, and electrical charges of several common ions are given in Table 4.1, *and they must be learned now.* Ions and ionic compounds are everywhere about and in us, as Interaction 4.1 discusses. Generally speaking, expect a compound to be *ionic* if its formula includes the atomic symbol of a metal.[1]

Binary Ionic Compounds Are Named After Their Ions   A compound made of the ions of just two elements is called a *binary ionic compound,* like sodium chloride, NaCl; calcium oxide, CaO; or calcium chloride, $CaCl_2$. As you can see, they are easy to name. Just write the name of the cation, then the name of the anion, omitting the "ion," of course. A binary compound of the iron(III) ion and the bromide ion is thus called iron(III) bromide.

[1] One common exception involves a cation made of N and H with the formula NH_4^+, the ammonium ion. Other exceptions are found among organic compounds.

■ The *metal* elements in groups IVA and VA can form positively charged ions.

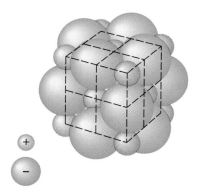

FIGURE 4.1   The structure of a sodium chloride crystal. The sodium ions are surrounded by chloride ions as nearest neighbors, and like-charged ions are just a little farther apart.

TABLE 4.1   Some Important Ions

Group	Element	Symbol for Neutral Atom	Symbol for its Common Ion	Name of Ion
IA	Lithium	Li	Li^+	Lithium ion
	Sodium	Na	Na^+	Sodium ion
	Potassium	K	K^+	Potassium ion
IIA	Magnesium	Mg	Mg^{2+}	Magnesium ion
	Calcium	Ca	Ca^{2+}	Calcium ion
	Barium	Ba	Ba^{2+}	Barium ion
IIIA	Aluminum	Al	Al^{3+}	Aluminum ion
VIA	Oxygen	O	O^{2-}	Oxide ion
	Sulfur	S	S^{2-}	Sulfide ion
VIIA	Fluorine	F	F^-	Fluoride ion
	Chlorine	Cl	Cl^-	Chloride ion
	Bromine	Br	Br^-	Bromide ion
	Iodine	I	I^-	Iodide ion
Transition Elements	Silver	Ag	Ag^+	Silver ion
	Zinc	Zn	Zn^{2+}	Zinc ion
	Copper	Cu	Cu^+	Copper(I) ion (cuprous ion)[a]
			Cu^{2+}	Copper(II) ion (cupric ion)
	Iron	Fe	Fe^{2+}	Iron(II) ion (ferrous ion)
			Fe^{3+}	Iron(III) ion (ferric ion)

[a] The names in parentheses are older names but are still in use.

INTERACTION 4.1
IONS AND IONIC COMPOUNDS IN FOODS, MEDICINES, AND THE HOME

Many familiar substances besides sodium chloride are ionic compounds. Sodium bicarbonate in baking soda, sodium carbonate in washing soda, calcium sulfate in plaster and plaster of paris, and sodium hydroxide in lye and drain cleaners are other examples (Figure 1). Some drain cleaners also carry visible bits of aluminum metal. When the NaOH of the drain cleaner dissolves, it is able to react with the aluminum in the following way.

$$2Al + 2NaOH + 2H_2O \longrightarrow 2NaAlO_2 + 3H_2$$

FIGURE 1 Ionic compounds in the home.

As the hydrogen gas forms, its bubbles generate a turbulence that aids the rest of the sodium hydroxide to attack greasy matter, which normally is responsible for a clogged drain.

Ionic Compounds in the Body. Every fluid in every living thing, whether plant or animal, contains dissolved ions. At the molecular level of life, ions are everywhere. People in professional health care fields speak of the "electrolyte status" of this or that body fluid. An **electrolyte** is any substance that can furnish ions in a solution in water, and all body fluids include dissolved electrolytes. These electrolytes in blood, for example, furnish the Na^+, K^+, Cl^-, and HCO_3^- ions. All of them are vital to health at low levels, but their concentrations must be maintained between narrow values or you're in a serious medical emergency. For example, despite the blood's need for Na^+ and Cl^- ions, if you were to drink too much seawater, their levels would rise and you would die. Everything is toxic in the wrong amount.

When prepared foods are advertised as "low in sodium," or if your physician tells you that your blood's sodium level is normal, "sodium" refers to the sodium *ion,* certainly not the extremely reactive sodium atom.

This, incidentally, hints at how profoundly chemical properties differ between atoms and their ions. Something "rich in calcium," like milk, is rich in calcium *ions*, not calcium atoms.

Correct Formulas Must Reflect the Electrical Neutrality of Ionic Compounds The ratio of the ions in an empirical formula must be one that permits all opposite charges to cancel, because compounds are electrically neutral. When calcium ions, Ca^{2+}, and oxide ions, O^{2-}, form an ionic compound, they can combine *only* in a 1:1 ratio. Only in this ratio can the 2+ on the calcium ion cancel the 2− on the oxide ion. Thus the formula is CaO. By convention, the cation is written first, and the charges are omitted. They are "understood" in empirical formulas.

Calcium chloride, a compound of calcium ions and chloride ions, must have *two* Cl^- ions for every Ca^{2+} ion because it takes two Cl^- ions to give enough negative charge to cancel the charge of 2+ of a calcium ion. To show this ratio, the formula of calcium chloride is written as $CaCl_2$.

■ *Empirical* is from the Latin *empiricus,* something experienced (here signifying "from experimental data").

■ CaO is also called "quick lime" and is an ingredient in cement.

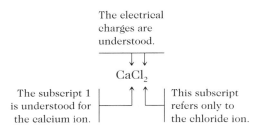

The electrical charges are understood.

$CaCl_2$

The subscript 1 is understood for the calcium ion.

This subscript refers only to the chloride ion.

The subscripts used are generally the smallest whole numbers that express the correct ratio. We don't write Ca_2Cl_4 for calcium chloride, even though it gives the correct ratio because $2:4$ is equivalent to $1:2$.

EXAMPLE 4.1   Writing Formulas from Names of Ionic Compounds

Write the formula of aluminum oxide.

ANALYSIS We must use the name, aluminum oxide, to infer the ions *and their electrical charges*. Unless this shows that a 1:1 ratio works, we use the charges to determine their lowest common multiple and thus to see what is the smallest number of each ion that will have a total charge numerically equal to this multiple. Finally, we assemble the ions in the correct order for a formula using the numbers just determined as subscripts.

SOLUTION From the name "aluminum oxide," we infer that the ions are Al^{3+} and O^{2-}. We know that $3+$ isn't canceled by $2-$, so we can't simply write AlO. When a 1:1 ratio of ions cannot work, we use the lowest common multiple of the ionic charges to obtain subscripts. With charges of $3+$ and $2-$, the least common multiple (ignoring the signs) is 6 ($2 \times 3 = 6$). An electrically neutral balance is thus obtainable if we have $6+$ total charge balanced by $6-$ total charge. So we can pick the smallest number of aluminum ions that give a total of $6+$ and the smallest number of oxide ions that give a total of $6-$. This requires $2\ Al^{3+}$, because $[2 \times (3+) = 6+]$, and it requires $3\ O^{2-}$, because $[3 \times (2-) = 6-]$. The ratio in aluminum oxide, therefore, is $2\ Al^{3+}$ to $3\ O^{2-}$, and writing the aluminum first in the formula gives us the answer

$$Al_2O_3$$

■ Aluminum oxide is a buffing powder for polishing metals.

The strategy in Example 4.1 to find and use the lowest common multiple of (numerically unequal) charges on the ions always works. Try Practice Exercise 1 to develop experience in using this approach.

■ When a ratio of 1:1 does work, the calculation of a least common multiple is obviously unnecessary.

■ PRACTICE EXERCISE 1   Write the formulas of the following compounds.

(a)   Silver bromide (a light-sensitive chemical used in photographic film)

(b)   Sodium oxide (a very caustic substance that changes to lye in water)

(c)   Iron(III) oxide (the chief component in iron rust)

(d)   Copper(II) chloride (an ingredient in some laundry-marking inks)

EXAMPLE 4.2   Writing Names from Formulas of Ionic Compounds

Write the name of the ionic compound $FeCl_2$, using both the modern and the older forms.

ANALYSIS The symbol Fe stands for one of the two ions of iron, but which? What charge does the iron ion bear in $FeCl_2$? Here we must use our knowledge of the charge on one ion, Cl^-, to figure out the charge on the other. Because there are two Cl^- ions in $FeCl_2$, there must be a total negative charge of $2-$ in each formula unit of $FeCl_2$. The lone Fe part of $FeCl_2$ must therefore provide $2+$ in charge, so we must be dealing with the Fe^{2+} ion, the iron(II) ion.

SOLUTION   The compound is named iron(II) chloride.

■   **PRACTICE EXERCISE 2**   Write the names of each of the following compounds.

(a)   CuS   (b)   NaF   (c) FeI_2 (d) $ZnBr_2$ (e) Cu_2O

The technique used in Example 4.2 of finding the electrical charge on one ion from the known charge on the other in an ionic compound greatly reduces the number of ions that have to be memorized. The next exercise gives some practice in this.

■   **PRACTICE EXERCISE 3**   What are the charges on the *metal* ions in each of these substances?

(a) Cr_2O_3 (a green pigment in stained glass)

(b)   HgS ("Chinese red," a bright, scarlet-red pigment)

(c) $CoCl_2$ (an ingredient in invisible ink)

The Relative Sizes of Ions Are Important at the Molecular Level of Life   When involved in the processes of life, such as the workings of enzymes, hormones, neurotransmitters, genes, many medications, and many poisons, *the shapes and sizes of ions and molecules are as important as anything else.* Ions, for example, have to move through tiny channels in cell membranes, and the diameters of these channels might let a small ion, like Na^+, pass through but not its larger relative, K^+. We'll consider first the simplest principles governing size and geometry, those involving monatomic ions.

■   The principal ions in blood are Na^+ and Cl^-, but many others are also present.

Atoms and the ions made from them have quite different sizes. As you can see in Figure 4.2, *cations are always smaller than their parent atoms; anions are always larger.* When a cation forms from a metal atom, the remaining electron cloud shrinks. The excess positive charge in the cation pulls its electron cloud more tightly inward, so metal ions are always smaller than their parent atoms. The sodium *ion*, for example, has a radius (95 pm) only about half as large as the sodium *atom* (186 pm).

■   Monatomic ions are spherical, and the *radius* of a sphere is the distance from its center to its surface.

When an anion forms from an atom, one or more electrons are added to the atom's electron cloud. The electron cloud is thus larger in the anion than in the parent atom. The radius of the Cl^- *ion*, 181 pm, for example, is almost twice as large as the radius of the Cl *atom*, 99 pm.

■   The unit "pm" is the *picometer;* 1 pm $= 10^{-12}$ meter.

Notice also in Figure 4.2 that as we descend a group, the sizes of both atoms and associated ions increase. We'd expect this because electrons must populate orbitals at increasing distances from the nuclei.

The Sizes and Relative Numbers of Ions Determine Crystal Structure   The sizes of the Na^+ and Cl^- ions are such that the best "fit" using a 1:1 ratio causes the final shape of the sodium chloride crystal to be a cube. We say that these ions form a *cubic crystal.* So do K^+ and Cl^- ions. Other combinations of ions form crystals with different but still regular shapes. The shapes *must* be regular because ions can maximize interionic attractions, cations to anions, only this way.

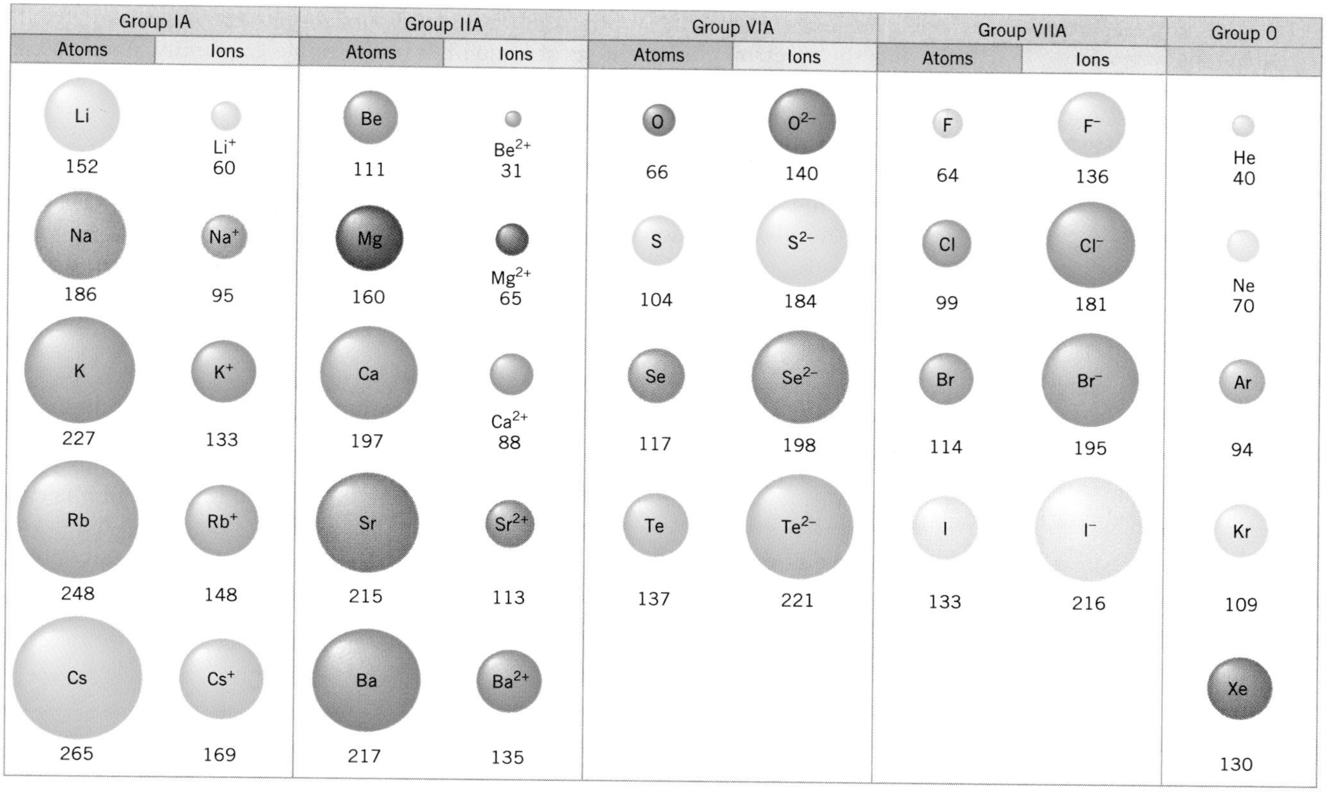

| Group IA | | Group IIA | | Group VIA | | Group VIIA | | Group 0 |
Atoms	Ions	Atoms	Ions	Atoms	Ions	Atoms	Ions	
Li 152	Li$^+$ 60	Be 111	Be^{2+} 31	O 66	O^{2-} 140	F 64	F$^-$ 136	He 40
Na 186	Na$^+$ 95	Mg 160	Mg^{2+} 65	S 104	S^{2-} 184	Cl 99	Cl$^-$ 181	Ne 70
K 227	K$^+$ 133	Ca 197	Ca^{2+} 88	Se 117	Se^{2-} 198	Br 114	Br$^-$ 195	Ar 94
Rb 248	Rb$^+$ 148	Sr 215	Sr^{2+} 113	Te 137	Te^{2-} 221	I 133	I$^-$ 216	Kr 109
Cs 265	Cs$^+$ 169	Ba 217	Ba^{2+} 135					Xe 130

FIGURE 4.2 Atomic and ionic radii of some representative elements. The radii are given in picometers, pm (10^{-12} meter).

4.2 MONATOMIC IONS AND THE OCTET RULE

Atoms and ions whose outside energy levels hold eight electrons are substantially more stable than those whose outside levels do not.

The formation of sodium and chloride ions raises several questions. Does sodium always form singly charged ions, Na$^+$? Why not Na^{2+} ions? We could ask similar questions about chlorine. Indeed, we could ask, why do atoms form ions at all? Not all elements do.

Atoms of the Noble Gases Have the Most Stable Electron Configurations   The group 0 elements, the noble gases, are the least reactive, most stable of all elements, forming neither positive nor negative ions. There is evidently something quite stable about the two kinds of electron configurations found among noble gas atoms. One is the **outer octet**, eight electrons in whichever principal level happens to be the *outside* level. The other is a *filled* level 1 when it is the *outside* level (as in helium). These two noble gas configurations, conditions of unusual chemical stability, will help us understand why ions have the charges they do and why the elements in the same chemical family, if they form ions at all, form ions of the *same* charge.

■ Just why noble gas configurations are so stable is still not fully understood.

Monatomic Ions of the Representative Elements Have Noble Gas Configurations   A sodium atom, in group IA, has one electron in its outside level—level 3—but after losing this electron, it has a new outside level—the former inner level, number 2,

■ The *atoms* in sodium metal re-act violently with water. The sodium *ion* is exceptionally stable, gives no reaction with water (or with much of anything else for that matter), and is present in all body fluids.

■ The *element* chlorine is a poiso-nous gas. The chloride *ion* is unusu-ally stable, has very few reactions, and is present in all body fluids.

■ G. N. Lewis (1875–1946) was chiefly responsible for the develop-ment of the octet rule.

■ "Valence" is from the Latin *valere,* meaning "to be strong." Here it refers to "combining strength" or "combining ability."

■ One of the many roles of Mg^{2+} in the body is to activate enzymes, substances that control how rapidly reactions occur.

which has eight electrons. As the sodium *ion* the particle has an outer octet, and it also has the electron configuration of a member of the noble gas family, neon. What-ever causes an outer octet to lend stability to the neon *atom* evidently is at work in the sodium *ion*, because the sodium ion is an unusually stable particle, too. (It's sta-ble provided that there is something oppositely charged nearby, so that the overall system is electrically neutral.)

A chlorine atom, in group VIIA, has seven electrons in its highest occupied en-ergy level, number 3. When this level accepts one electron, the Cl atom acquires an outer octet and becomes Cl^-. Thus the chloride *ion* has the same electron configura-tion as an atom of one of the noble gases, argon. The chloride ion, provided that an oppositely charged system is nearby, is also a particle of unusual chemical stability, like argon.

Electron Transfers Lead to Noble Gas Configurations   When we put sodium and chlorine together, the electron transfers between sodium and chlorine generate vastly more stable particles, those having noble gas configurations. The atoms sacri-fice neutrality for the increased stabilities of ions as they nestle together in a crystal of NaCl.

We haven't explained *why* a noble gas configuration is stable. We are only point-ing out that for some reason it is. The pattern we noted for the formation of sodium chloride is so general among representative metals and nonmetals that it amounts to a law of nature traditionally called the **octet rule** or the **noble gas rule.**

OCTET RULE (NOBLE GAS RULE)   The atoms of the reactive representa-tive elements tend to undergo those chemical reactions that most directly give them electron configurations of the nearest noble gas.

The Charges on Monatomic Ions of Representative Elements Correlate with the Pe-riodic Table   The phrase "most directly" in the octet rule refers to an electron transfer path involving the fewer number of electron losses or gains. Thus a sodium atom does not accept seven more electrons to give it an outer octet at level 3. In-stead it gives up just one electron to get an outer octet at level 2. Giving up one elec-tron is thus the most direct path to a noble gas configuration.

All metal atoms have just a few outside level electrons, so changing to ions with noble gas configurations "most directly" for all metal atoms means *losing* a few elec-trons rather than gaining many. All metals, therefore, have *positively* charged ions. The metals of the *representative* elements always give up only as many electrons as needed to strip the electron configuration to that of the noble gas nearest the metal (and preceding it) in the periodic table.

Notice that the "action" leading to a monatomic cation involves the outside *level* of the original atom. However, because "level" in some situations means a spe-cific orbital, chemists often use the term **valence shell** for an atom's highest occu-pied principal energy level. For the representative elements, the number of valence shell electrons equals the group number. Therefore, *all members of the same repre-sentative family of elements gain or lose electrons in the same way and form ions of identical charges.*

For representative metal elements, *the positive charge on the ion is the same as the group number.* For example, the ions of the group IA metals all have charges of 1+: Li^+, Na^+, K^+, Rb^+, and Cs^+. Two ions, Na^+ and K^+, are particularly important in the fluids of living systems. The ions of the group IIA metals have charges of 2+, for example, Mg^{2+}, Ca^{2+}, and Ba^{2+}. We will be interested in only one element in group

IIIA, aluminum, and its ion has a charge of 3+, Al^{3+}. The electrical charge on an ion—both the number and the sign of the charge—is sometimes called the ion's *electrovalence* because this charge ("electro-") is responsible for the ionic bonding ("valence") in ionic crystals.

For the ions of the transition metals there is no simple correlation between their group numbers and the size of their positive charges. Generally, the charges vary from 1+ to 3+, and several transition metals can exist as ions of more than one charge, as we have noted.

Nonmetal atoms of the *reactive* elements (not the noble gas elements) have 4, 5, 6, or 7 valence shell electrons. Being already close to valence shell octets, atoms with valence shells of 6 or 7 electrons can achieve noble gas configurations most directly by *gaining* 2 or 1 electrons rather than by losing 6 or 7. Group VIA or VIIA nonmetal atoms thus *accept* electrons to acquire octets. Atoms of group VIA (six valence shell electrons) accept two electrons and form monatomic ions with charges of 2−, like O^{2-} and S^{2-} (see Table 4.1). Atoms of group VIIA elements (seven valence shell electrons) need just one more electron for outer octets. So their monatomic ions all have charges of 1−, for example, F^-, Cl^-, Br^-, and I^-.

The group IVA and VA elements evidently have valence shells too far from octets. Their atoms would have to gain more electrons (four or three) than their nuclear charges can attract *and hold*. The result is that the nonmetal elements in groups IVA and VA exist as anions in so few compounds that we will ignore them. When we encounter an exception, we will note it.

■ The nitride ion (N^{3-}), a group VA anion, occurs with ions of group IIA metals in saltlike compounds; e.g., Mg_3N_2, magnesium nitride.

EXAMPLE 4.3 **Using the Noble Gas (Octet) Rule**

When nutritionists speak of the *calcium requirement* of the body, they always mean the calcium *ion* requirement. Calcium has atomic number 20. What charge does the calcium ion have, and what is its symbol?

ANALYSIS There are two methods for solving this kind of problem, and you should learn both. The first is to write the electron configuration of the atom, determine the number of electrons in its valence shell, and then decide how many electrons must be gained or lost to acquire an outer octet in the most direct manner.

The second approach is to use the periodic table to find out in which *group* the element is. Then the group number tells how many electrons must be gained or lost to acquire an outer octet in the most direct manner.

SOLUTION Using the rules, we write the electron configuration of element 20, remembering that $4s$ fills before electrons go into $3d$.

$$1s^2 2s^2 2p^6 3s^2 3p^6 4s^2$$

The atom has two electrons in the valence shell (level 4) and eight electrons in the next lower level (level 3 with two $3s$ electrons plus six $3p$ electrons). Only by losing *both* of the $4s$ electrons can the calcium atom get a new outside level that holds an octet. Losing *one* electron won't do. Neither will losing three or more. Therefore the only stable ion that calcium forms has the following electron configuration.

$$1s^2 2s^2 2p^6 3s^2 3p^6$$

By losing two electrons the net charge becomes 2+, so the symbol of the calcium ion is Ca^{2+}.

Using the second approach, we locate calcium in group IIA in the periodic table. All group IIA atoms are representative and have two valence shell electrons, so all most directly acquire configurations of the nearest noble gases by losing *two* electrons. Thus all of the group IIA ions bear charges of 2+: Be^{2+}, Mg^{2+}, Ca^{2+}, Sr^{2+}, Ba^{2+}, and Ra^{2+}.

EXAMPLE 4.4   Using the Noble Gas (Octet) Rule

Oxygen exists as the oxide ion in such substances as calcium oxide, an ingredient in stucco and mortar. What is the symbol for the oxide ion, including its electrical charge?

ANALYSIS   As in Example 4.3, we will solve this from both the electron configuration of oxygen and its location in the periodic table.

SOLUTION   The electron configuration of oxygen (atomic number 8) is

$$1s^2 2s^2 2p_x^2 2p_y^1 2p_z^1$$

The valence shell, level 2, has a total of six electrons, just two short of an octet and a neon configuration. We could also say that the oxygen atom has six too many electrons to have the helium configuration ($1s^2$). However, gaining two electrons to become like neon ($1s^2 2s^2 2p^6$) is much simpler than losing six, so oxygen achieves a noble gas configuration most directly by accepting two electrons from some metal atom donor. The configuration of the anion of oxygen, then, is

$$1s^2 2s^2 2p_x^2 2p_y^2 2p_z^2 \qquad \text{or} \qquad 1s^2 2s^2 2p^6$$

The two extra electrons give the anion a charge of 2−, so the symbol of the oxide ion is O^{2-}.

Using the periodic table, we locate oxygen in group VIA, so its valence shell has six electrons. It must pick up two electrons, not just one and not more than two, to have an outer octet. These two extra electrons give the particle a charge of 2−, so we can write O^{2-} directly.

EXAMPLE 4.5   Using the Noble Gas (Octet) Rule

Hardly any element is involved in more compounds than carbon. (Roughly six million carbon compounds are known.) Can carbon atoms change to ions? If so, what is the symbol of the ion?

ANALYSIS   When we find carbon's place in the periodic table, we see that it is in group IVA, so carbon atoms have four valence shell electrons. A carbon atom either must lose these four electrons (and become helium like) or gain four electrons (and become neon like) to achieve a noble gas configuration. In one or two very rare situations, carbon can do the latter—become the C^{4-} ion. Because this is so rare we ignore it.

■ The methanide ion, C^{4-}, apparently occurs in Be_2C, beryllium methanide, a brick-red solid.

SOLUTION   By ignoring anions of groups IVA and VA, we need not write their configurations or symbols.

■ PRACTICE EXERCISE 4 Write the electron configuration of an atom of each of the following elements, and deduce from it the charge on the corresponding ion. If the atom isn't expected to have a corresponding ion, state so. The numbers in parentheses are atomic numbers. Do not use the periodic table for this practice exercise.

(a) potassium (19)   (b) sulfur (16)   (c) silicon (14)

■ PRACTICE EXERCISE 5 Write the electron configurations of the *ions* of the elements in Practice Exercise 4 that can form ions.

■ PRACTICE EXERCISE 6 Relying on their locations in the periodic table, write the symbols of the ions of each of the following elements. Always remember that no symbol of an ion is complete without its electrical charge. (The numbers in parentheses are atomic numbers.)

(a) cesium (55)   (b) fluorine (9)   (c) phosphorus (15)   (d) strontium (38)

4.3 REDOX REACTIONS

Energy-producing reactions at the molecular level of life are redox reactions.

Reactions in Which Electrons Transfer Are Called Oxidation–Reduction Reactions

The reaction between sodium and chlorine studied earlier illustrates an important family of reactions, *redox reactions,* or **oxidation–reduction reactions.** These occur, for example, when we use oxygen from air to obtain chemical energy from foods.

A **redox reaction** is one in which oxidation numbers change. **Oxidation numbers** are numbers assigned to each atom in a formula according to a set of rules. The rules are simple for elements and monatomic ions.

Oxidation Number Rules for Elements and Monatomic Ions

1. The oxidation number of any *element* is zero.

2. The oxidation number for any *monatomic ion* is the same as the charge on the ion.

Thus the oxidation number of Na in sodium metal is 0, but the oxidation number of sodium in NaCl is $+1$ because there is a charge of $1+$ on the sodium ion. The oxidation number of chlorine in NaCl is -1 because the chloride ion has a charge of $1-$, but Cl has an oxidation number of 0 in elemental Cl_2. In $FeCl_3$, the oxidation number of iron is $+3$. The oxidation numbers in MgS are $+2$ for magnesium and -2 for sulfur.

When a sodium atom transfers an electron to a chlorine atom, the oxidation number of Na changes from 0 to $+1$. Any change that makes an oxidation number more positive is defined as an **oxidation.** We say that the sodium atom *is oxidized* to the sodium ion. Of course, this occurs here only because an electron actually leaves the Na atom, so in electron-transfer reactions involving elements or monatomic systems, *oxidation means the loss of electrons.*

When a chlorine atom accepts an electron and goes from Cl to Cl^-, the oxidation number of chlorine changes from 0 to -1. We define any change that makes an oxidation number more negative as a **reduction.** This can't take place without the particle accepting an electron, so for elements and monatomic systems *reduction means the gain of electrons.*

■ The chemical reactions involved in respiration, photosynthesis, and rusting all include oxidation–reduction reactions.

■ The earliest examples of oxidation involved oxygen itself as the oxidizing agent; hence the name.

■ People used to call the conversion of ores, like iron ore, to the metal a *reduction* of the ore to the metal. This is where we got the general name, reduction.

A reduction cannot occur without an oxidation of something else. Electrons don't just leave from or go to outer space; they *transfer*. A reaction that involves a reduction also *must* involve an oxidation, and any such reaction is a *redox reaction*.

The reactant that causes a reduction is called the **reducing agent.** Sodium is the reducing agent in the reaction with chlorine because it causes the reduction of Cl to Cl^-.

The reactant that oxidizes something else is called an **oxidizing agent.** Chlorine is the oxidizing agent in the reaction with sodium because it causes the oxidation of Na to Na^+. Using the correct formula for chlorine, Cl_2, we have

$$2Na + Cl_2 \longrightarrow 2Na^+ + 2Cl^-$$

$$\underset{\text{agent}}{\text{reducing}} \quad \underset{\text{agent}}{\text{oxidizing}}$$

Notice that an oxidizing agent is always itself reduced in a redox reaction, and that a reducing agent is always itself oxidized.

EXAMPLE 4.6   Analyzing a Redox Reaction

The reaction of calcium with sulfur is a redox reaction in which calcium ions and sulfide ions form to make calcium sulfide, CaS.

$$Ca + S \longrightarrow CaS$$

Determine the oxidation numbers of calcium and sulfur in CaS, and decide what is the oxidizing agent and what is the reducing agent in this redox reaction.

ANALYSIS   The reactants are both elements, so they both have oxidation numbers of 0. From Table 4.1 (and soon, it must be said, from memory) we know that the charge on Ca in CaS is 2+ and that the charge on S in CaS is 2−. Calcium, therefore, has an oxidation number of +2 in CaS, and sulfur's oxidation number in this compound is −2.

SOLUTION   We can see that calcium's oxidation number becomes more positive (0 to +2), so calcium is oxidized. Sulfur's oxidation number becomes more negative (0 to −2), so sulfur is reduced. Calcium is the reducing agent and sulfur is the oxidizing agent.

■   PRACTICE EXERCISE 7   Identify by their chemical symbols what is oxidized and what is reduced in the following reactions. Also identify what is the oxidizing agent and what is the reducing agent.

(a) $Mg + S \rightarrow MgS$

(b) $CuCl_2 + Zn \rightarrow ZnCl_2 + Cu$

(*Hint:* Both $CuCl_2$ and $ZnCl_2$ are compounds of the chloride ion. From this information, you should be able to deduce the charges on the metal ions in these compounds.)

4.4 ELECTRON SHARING AND MOLECULAR COMPOUNDS

When electron density becomes sufficiently concentrated between two atomic nuclei there is a chemical bond—a covalent bond—between them.

Nature has ways besides electron transfers to develop net forces of attraction or chemical bonds between atoms. Our purpose in this section is to learn about a second method. It most often occurs when atoms of nonmetals form compounds.

Compounds of Nonmetals Are Not Ionic   Several million compounds, probably over 90% of all compounds, involve only nonmetal atoms. These cannot be ionic compounds, however. To have *ionic* compounds, ions of both positive and negative charge are needed. But atoms of nonmetals do not become positive ions because too many electrons have to be lost from their valence shells to achieve noble gas configurations. The nonmetals in groups IVA and VA, moreover, almost never become negative ions, because they would have to gain too many electrons to get outer octets.

The result is that compounds made entirely of nonmetals, instead of being ionic, consist of discrete *molecules*. A **molecule,** as we have said, is a small, electrically neutral particle consisting of at least two nuclei and enough electrons to make the whole system neutral. A compound that consists of molecules is called a **molecular compound.** Generally speaking, if the formula of a compound does *not* contain the atomic symbol of a metal, it is a molecular compound. (An exception is NH_4^+; see footnote 1.). Our question now is, "What holds molecules together?"

Electron Clouds Can become Dense Between Suitable Atoms   We'll consider the simplest molecule first, that of H_2. Imagine two isolated hydrogen *atoms* moving directly toward each other (see Figure 4.3*a*). As they draw closer, the 1s electron cloud of each atom begins to sense the presence of the positively charged nucleus of the other atom *and experiences an attraction toward it.* The attraction distorts both of the two electron clouds, and they bulge toward the sides of the atoms that are nearing each other (see Figure 4.3*b*). As the atoms continue toward collision, their electrons spend more and more of their time on facing sides of the atoms. Each nucleus, however, begins more and more to sense the like-charged nature of the other, so a collision never reaches the actual touching of the nuclei. Instead, the atoms brake so that their two nuclei are a short distance apart (see Figure 4.3*c*).

The atoms do not rebound, like two billiard balls, because the electron density between their two nuclei is now too great. The nuclei are attracted toward this concentrated electron cloud, and the cloud itself is attracted to the two nuclei. Because the pair of electrons has become concentrated more or less between the nuclei, the two nuclei now have a bond between them, and a molecule is born, H_2.

■ A region of *high electron density* is one with a thick or dense electron cloud.

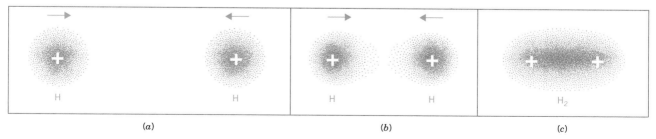

(a)   (b)   (c)

FIGURE 4.3   Formation of the covalent bond in hydrogen, H_2. See text for explanation.

■   Co- comes from cooperative;
-valent from the Latin valere, to be
strong, signifying strong binding.

The bonding pair of electrons is called the *shared pair*. Neither of the shared electrons any longer belongs exclusively to just one of the nuclei, as each did before the atoms came together. The two electrons now belong to both nuclei, and we say that the nuclei *share* the two electrons. The bond created by the sharing of electron pairs between atomic nuclei is called a **covalent bond**, sometimes an *electron pair bond*.

The whole new package, the two protons and two electrons of H_2, is electrically neutral, as molecules *must* be. Unlike a single pair of ions, like Na^+ and Cl^- in NaCl, a molecule is a discrete and separate particle that can enjoy independent existence. It can move around as a unit.

The covalent bonds in larger molecules are like those in H_2. They involve the electron density of a pair of electrons that becomes concentrated between two nuclei. The nuclei are attracted into this region and held there at a very small distance apart.

Atomic Orbitals Overlap to Form Molecular Orbitals   An atomic orbital is associated with only one atomic nucleus. In a widely held view of the covalent bond, the bond's shared pair of electrons resides in a space associated with more than one nucleus. In the H_2 molecule, for example, the 1s orbitals of the previously separated atoms partially merge—it's called the **overlapping of orbitals**—to create a new space for the electrons to be shared. The new space created by the overlapping of atomic orbitals is called a **molecular orbital,** and it surrounds two nuclei (sometimes more). Like an atomic orbital, a molecular orbital can hold a maximum of two electrons, provided their spins are opposite. The shared electron pair of a covalent bond thus resides in a molecular orbital whose space encloses both nuclei held by the covalent bond.

The atomic orbitals that overlap are generally those of valence shell electrons. Figure 4.4 shows how orbital overlapping happens between two fluorine atoms. On the left in the figure, we see two separated fluorine atoms, but only their half-filled p_z orbitals are pictured. Their nuclei are centered at the *nodes,* where pairs of lobes touch. Imagine that these two atoms move toward each other. Eventually, the spaces

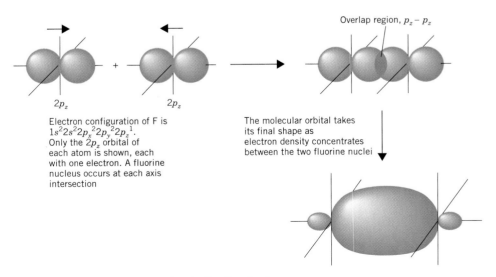

Overlap region, $p_z - p_z$

$2p_z$ $2p_z$

Electron configuration of F is $1s^2 2s^2 2p_x^2 2p_y^2 2p_z^1$. Only the $2p_z$ orbital of each atom is shown, each with one electron. A fluorine nucleus occurs at each axis intersection

The molecular orbital takes its final shape as electron density concentrates between the two fluorine nuclei

FIGURE 4.4   The covalent bond in the F—F molecule.

occupied by facing lobes of the p_z orbitals start to overlap. A molecular orbital forms, and the former two p_z electrons take up residence in it, become the shared pair, and thus electron density becomes concentrated between the two nuclei. The covalent bond in F—F has formed

4.5 LEWIS STRUCTURES AND THE OCTET RULE

The electron pairs of covalent bonds are counted with the other valence electrons in computing octets.

The stability of a noble gas configuration influences the *number* of covalent bonds that an atom can form. To study how, we'll need a new kind of symbol, the *electron-dot structure.*

Electron-Dot Structures Show Only Valence Shell Electrons in Atoms We can draw attention to valence shell electrons by representing them as dots placed around the atomic symbol, and the result is called an **electron-dot structure.** The first four dots are evenly distributed around the sides of an imaginary square. The fifth, sixth, seventh, and eighth electrons are placed so as to create pairs of electrons. Because of the great contributions of G. N. Lewis, an American chemist, to the theory of the covalent bond, electron-dot structures are often called **Lewis structures.**

The group numbers of the representative elements tell us how many electrons are in their valence shells and thus how many electron dots must be used. All group IA elements require just one electron dot. Those in group VIIA need seven. (Remember that those of group 0, the noble gases, need eight, with the exception of helium, which needs two.) The electron-dot structure of hydrogen is simply H· and that of helium is ·He·. The electron-dot or Lewis structures of the atoms in the second period of the periodic table are as follows.

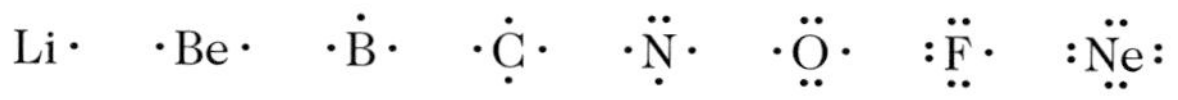

$$\text{Li·} \quad \text{·Be·} \quad \text{·}\overset{\displaystyle\cdot}{\text{B}}\text{·} \quad \text{·}\overset{\displaystyle\cdot}{\text{C}}\text{·} \quad \text{·}\overset{\displaystyle\cdot}{\text{N}}\text{·} \quad \text{·}\overset{\displaystyle\cdot\cdot}{\text{O}}\text{·} \quad \text{:}\overset{\displaystyle\cdot\cdot}{\text{F}}\text{·} \quad \text{:}\overset{\displaystyle\cdot\cdot}{\underset{\displaystyle\cdot\cdot}{\text{Ne}}}\text{:}$$

Notice how, after the first four dots have gone into place, additional dots are paired with others. More importantly, notice how the number of dots by atoms in groups IA–VIIA always equals the group number of the atom.

■ It doesn't matter which sides of the symbols get the pairs.

EXAMPLE 4.7 Writing the Electron-Dot Symbol of a Representative Atom

What is the electron-dot symbol of sulfur, S?

ANALYSIS Sulfur is in group VIA of the periodic table, so its atoms must have six electrons in their valence shells. We write the first four electron dots at the sides of an imaginary square, and we add the last two to create two pairs. (It doesn't matter which of the possible pairs is created.)

SOLUTION

$$\text{·}\overset{\displaystyle\cdot\cdot}{\underset{\displaystyle\cdot\cdot}{\text{S}}}\text{·}$$

■ **PRACTICE EXERCISE 8** Write the electron-dot or Lewis symbols for the atoms of the period 3 elements of the periodic table.

■ **PRACTICE EXERCISE 9** Write the electron-dot or Lewis symbol for antimony, Sb, atomic number 51.

The Formation of Ions Can Be Shown by Electron Dot Symbolism The electron transfer that occurs when sodium and chlorine react can be represented by electron-dot symbolism as follows.

$$\text{Na} \overset{\frown}{} + \cdot \overset{..}{\underset{..}{\text{Cl}}} : \longrightarrow \text{Na}^+ + \left[: \overset{..}{\underset{..}{\text{Cl}}} : \right]^-$$

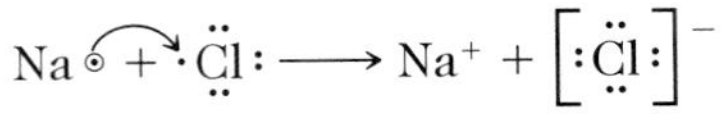

The valence shell of sodium loses its electron, so no dot remains. (Only valence shells, the *original* outside levels, are given electron dots.) The valence shell of chlorine accepts this electron and an octet forms. To show that this electron is fully the property of chlorine, we put brackets about the symbol of the ion.

The reaction between magnesium and chlorine can similarly be represented.

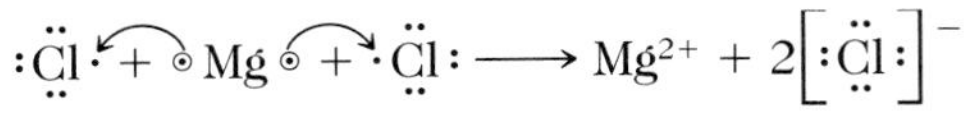

$$: \overset{..}{\underset{..}{\text{Cl}}} \cdot \overset{\frown}{} + \text{Mg} \overset{\frown}{} + \cdot \overset{..}{\underset{..}{\text{Cl}}} : \longrightarrow \text{Mg}^{2+} + 2\left[: \overset{..}{\underset{..}{\text{Cl}}} : \right]^-$$

Each chlorine atom has room in its valence shell for only one more electron, so two chlorine atoms are needed to take care of the two electrons of one magnesium atom.

EXAMPLE 4.8 **Representing a Reaction in Electron-Dot Symbolism**

Sodium sulfide is an ionic compound of the sodium ion, Na^+, and the sulfide ion, S^{2-}. How can its formation be represented in electron-dot symbolism?

ANALYSIS We first write the Lewis symbols for sodium and sulfur.

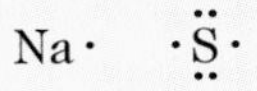

$$\text{Na} \cdot \qquad \cdot \overset{..}{\text{S}} \cdot$$

To achieve an octet, S needs two electrons, so we need two Na atoms for one S atom.

SOLUTION We can therefore write (remembering the brackets),

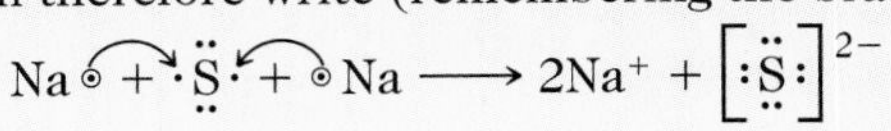

$$\text{Na} \overset{\frown}{} + \cdot \overset{..}{\underset{..}{\text{S}}} \cdot \overset{\frown}{} + \text{Na} \longrightarrow 2\text{Na}^+ + \left[: \overset{..}{\underset{..}{\text{S}}} : \right]^{2-}$$

■ **PRACTICE EXERCISE 10** Represent the formation of calcium oxide by electron-dot symbolism. It consists of Ca^{2+} and O^{2-} ions.

Lewis Structures of Molecules Display Noble Gas Configurations for the Atoms Present The formation of H_2 can be represented in electron-dot symbolism as follows:

$$\text{H} \cdot \overset{\frown}{} + \overset{\frown}{} \cdot \text{H} \longrightarrow \text{H} : \text{H}$$

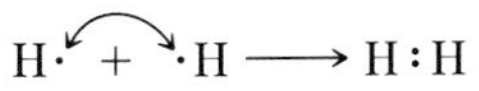

The shared pair of electrons is shown between the two atomic symbols. Since they are shared, they both count for each hydrogen when assessing whether a noble gas configuration is achieved. Each hydrogen atom acquires the helium configuration by this sharing of the electron pair.

What we see here is a general rule. Noble gas configurations dominate the formation of covalent bonds as they do of ions. Atoms that form covalent bonds tend, by the sharing of electron pairs, to acquire as many electrons as needed to achieve noble gas configurations.

We can represent the formation of the diatomic halogen molecules in electron-dot symbolism as follows. We will also carry the symbolism one step further in these examples. It is customary to represent any shared pair of electrons—any covalent bond—by a short line, a "dash" bond. Any pairs of valence shell electrons that are not used for covalent bonds are called **unshared pairs.** Unshared electron pairs are often omitted from structures when they are not involved in chemical reactions. The structures following the arrows, next, thus illustrate these options.

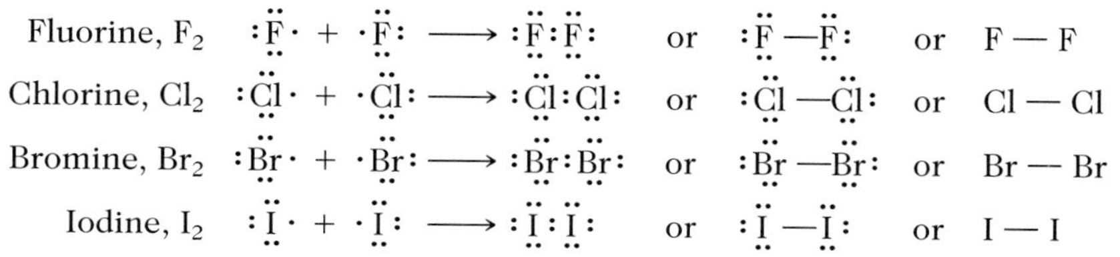

Fluorine, F_2 $:\ddot{F}\cdot + \cdot\ddot{F}: \longrightarrow :\ddot{F}:\ddot{F}:$ or $:\ddot{F}-\ddot{F}:$ or $F-F$

Chlorine, Cl_2 $:\ddot{Cl}\cdot + \cdot\ddot{Cl}: \longrightarrow :\ddot{Cl}:\ddot{Cl}:$ or $:\ddot{Cl}-\ddot{Cl}:$ or $Cl-Cl$

Bromine, Br_2 $:\ddot{Br}\cdot + \cdot\ddot{Br}: \longrightarrow :\ddot{Br}:\ddot{Br}:$ or $:\ddot{Br}-\ddot{Br}:$ or $Br-Br$

Iodine, I_2 $:\ddot{I}\cdot + \cdot\ddot{I}: \longrightarrow :\ddot{I}:\ddot{I}:$ or $:\ddot{I}-\ddot{I}:$ or $I-I$

■ Remember, only the valence shell electrons are shown. The core electrons are "understood."

By counting the shared electron pairs of the covalent bonds for either atom, we can see that each halogen atom in these diatomic molecules has achieved a noble gas configuration, an outer octet. Each molecule has one covalent bond.

Hydrogen chloride, HCl, is a diatomic molecule with one covalent bond, and its formation can be represented as follows.

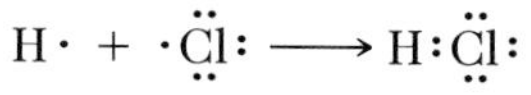

$$H\cdot + \cdot\ddot{Cl}: \longrightarrow H:\ddot{Cl}:$$

Again we see that noble gas configurations are achieved for H and Cl by the sharing of an electron pair.

The atoms of the nonmetals of groups IVA, VA, and VIA must form more than one covalent bond to acquire octets. Of particular importance at the molecular level of life are atoms of carbon, nitrogen, oxygen, and sulfur. Their Lewis symbols are

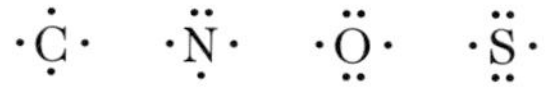

$$\cdot\dot{\underset{\cdot}{C}}\cdot \qquad \cdot\ddot{N}\cdot \qquad \cdot\ddot{O}\cdot \qquad \cdot\ddot{S}\cdot$$

You can see at a glance from these symbols how many electrons each must get by sharing to acquire octets. Carbon has four electrons so it needs four more, and if it combines with hydrogen (which has just one electron per atom to share), it *must* have four hydrogen atoms to get the necessary four additional electrons.

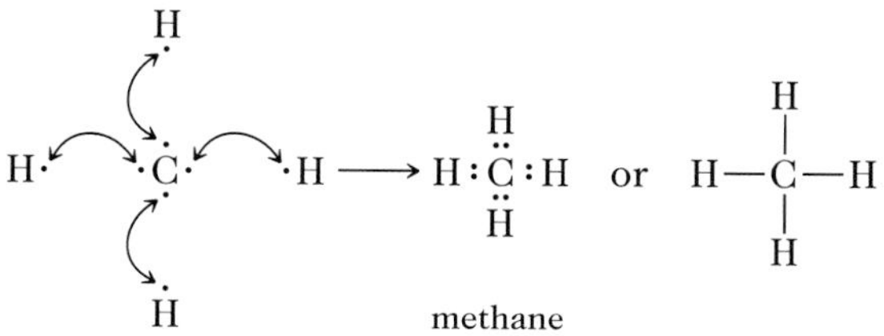

$$H \cdot \overset{H}{\underset{H}{\cdot\underset{}{C}\cdot}} \cdot H \longrightarrow H:\overset{H}{\underset{H}{\ddot{C}}}:H \quad \text{or} \quad H-\overset{H}{\underset{H}{\overset{|}{\underset{|}{C}}}}-H$$

methane

■ Methane is the chief constituent of natural gas.

Similarly, a nitrogen atom, which has five electrons and needs a share of three more to have an octet, combines with three hydrogen atoms. And oxygen and sulfur combine with two.

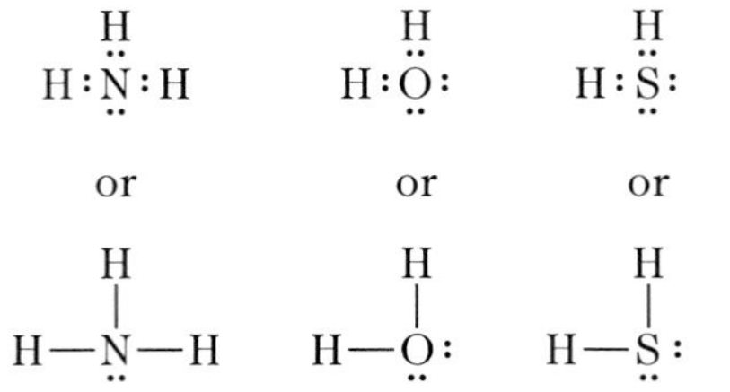

$$
\begin{array}{ccc}
H & H & H \\
H:\overset{\cdot\cdot}{N}:H & H:\overset{\cdot\cdot}{O}: & H:\overset{\cdot\cdot}{S}: \\
\overset{\cdot\cdot}{} & \overset{\cdot\cdot}{} & \overset{\cdot\cdot}{}
\end{array}
$$

or   or   or

$$
\begin{array}{ccc}
H & H & H \\
| & | & | \\
H-\overset{\cdot\cdot}{N}-H & H-\overset{\cdot\cdot}{O}: & H-\overset{\cdot\cdot}{S}: \\
\end{array}
$$

ammonia   water   hydrogen sulfide

■ Hydrogen sulfide is what gives rotten eggs their odor.

The symbols for methane, ammonia, water, and hydrogen sulfide that show the sequence in which the atoms are joined together are called **structural formulas** or simply **structures**. A formula of a molecular substance, such as H_2O, which gives the composition of one molecule is called a **molecular formula**. You can see that structures are more informative than other kinds of formulas.

EXAMPLE 4.9   Using the Octet Rule and Electron-Dot Structures to Figure Out Structural Formulas

---

Phosphine is a very poisonous compound of phosphorus and hydrogen. What is its most likely structure?

ANALYSIS We need the electron-dot structures for the atoms first in order to see how outer octets are to be satisfied. Phosphorus, like nitrogen, is in group VA.

$$H\cdot \quad \cdot\overset{\cdot\cdot}{P}\cdot$$

Phosphorus needs three more electrons for an outer octet. To get them using H atoms, P *must* have *three* H atoms.

■ Phosphine is present in the very unpleasant odor of decaying fish.

SOLUTION The structural formula of phosphine, therefore, must be

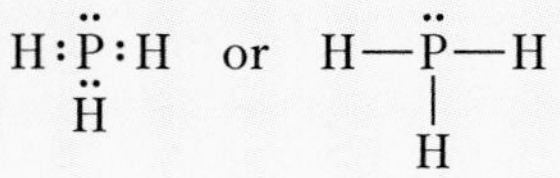

$$
H:\overset{\cdot\cdot}{P}:H \quad \text{or} \quad H-\overset{\cdot\cdot}{P}-H
$$
$$
\overset{\cdot\cdot}{H} \qquad\qquad \overset{|}{H}
$$

■ PRACTICE EXERCISE 11   Silicon and hydrogen form a simple compound called silane that has one Si atom per molecule. What is the structural formula of silane?

In Many Molecules, Two or Three Pairs of Electrons Are Shared in Double or Triple Bonds Ethylene has the molecular formula C_2H_4. Its structure is

■ Ethylene is used to make polyethylene plastics.

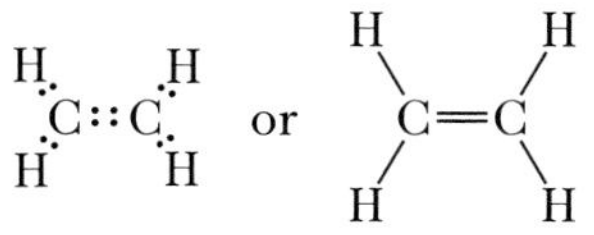

$$
\begin{array}{cc}
H \quad H & H \qquad H \\
\overset{\cdot\cdot}{C}::\overset{\cdot\cdot}{C} \quad \text{or} & \diagdown \quad \diagup \\
H \quad H & C=C \\
& \diagup \quad \diagdown \\
& H \qquad H
\end{array}
$$

Each bond with one shared pair, like the bond from H to C, is called a **single bond.**

Each carbon supplies four electrons, as its location in group IVA requires, and each carbon atom has an octet. This would not have been possible without placing two pairs of electrons between the carbon symbols. When two pairs of electrons are shared, the result is called a **double bond.** Such bonds are extremely prevalent in nature.

The nitrogen molecule has a **triple bond,** because three pairs of electrons are shared between the nitrogen nuclei. We can think of N_2 forming as follows.

$$:\overset{..}{N}\cdot\longleftrightarrow\cdot\overset{..}{N}:\ \longrightarrow\ :N:::N:\quad\text{or}\quad:N\equiv N:$$

8 electrons

All six valence shell electrons between the two nitrogen nuclei count toward the octet of each N. The triple bond is not as common as single and double bonds, but acetylene, C_2H_2, the fuel for oxyacetylene torches, has one.

$$H:C:::C:H\quad\text{or}\quad H\!\!-\!\!C\equiv C\!\!-\!\!H$$

acetylene

Many molecules have two double bonds. Carbon dioxide, CO_2, is a common example. Its Lewis structure is

$$:\overset{..}{O}::C::\overset{..}{O}:\quad\text{or}\quad:\overset{..}{O}\!=\!C\!=\!\overset{..}{O}:$$

carbon dioxide

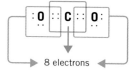

8 electrons

Note how each atom in CO_2 has an octet.

The *Covalence Number* of a Nonmetal Is the Number of Covalent Bonds Its Atoms Have in Molecules   We can summarize part of the foregoing discussion in terms of the bond-forming abilities of the nonmetal elements. Each can be assigned a number, called its **covalence number,** that equals the number of bonds its atoms have in molecules. The covalence number of an element thus also equals the number of additional electrons that its atoms must acquire by sharing to have a noble gas configuration. The common covalence numbers of several elements are given in Table 4.2; they apply to the atoms given as they occur in electrically neutral particles. Some elements have more than one covalence number.

We have studied bond-*forming* events, but covalent bonds can be broken if enough energy is supplied. Such reactions occur, for example, in the stratosphere where bond-breaking reactions provide us with protection from the sun's ultraviolet radiation (see Interaction 4.2).

TABLE 4.2   Common Covalences of Nonmetals

Periodic Table Group Number									
IA		IVA		VA		VIA		VIIA	
H	1	C	4	N	3	O	2	F	1
		Si	4	P	3	S	2	Cl	1
								Br	1
								I	1

OZONE IN THE STRATOSPHERE AND THE OZONE "HOLE"

The Ozone Hole Covalent bonds break when a molecule absorbs light of the proper energy. The occurrence of such ruptures in the rarefied regions of our stratosphere is the molecular basis of the ozone "hole."

The *stratosphere* is an envelope of space surrounding the earth between the altitudes of 10 and 50 km (6 to 31 miles). It holds ultrasmall concentrations of a number of substances, like oxygen, nitrogen, methane, nitrogen dioxide, and ozone.

Ozone is a form of oxygen in which the oxygen atoms exist as triatomic molecules, O_3. It is an extremely reactive substance, attacking almost anything in living systems. Smog contains some ozone, which is one reason that breathing smoggy air is unhealthy. Ozone in the stratosphere, however, is one of the silent, unseen protectors of all life!

Light is a stream of tiny massless particles called *photons,* each being a packet of energy. Some photons in sunlight make up what we call *ultraviolet* or UV light, which is invisible, and they carry more energy than those of visible light. UV photons have enough energy to break chemical bonds.

Roughly 7% of the solar energy streaming toward the earth is carried by UV photons. This may seem small, but neither plant nor animal life, including human life, would be possible without the removal of its most dangerous portions, called UV-B radiation. UV-B causes sunburn, for example. Sufficiently prolonged or severe exposure to UV-B damages genes in the cells of the skin, which can lead to one of two kinds of skin cancer. Fortunately, nearly all incoming UV-B radiation is absorbed in the stratosphere, but a small amount does survive to reach us at the earth's surface.

Most of the UV-B photons are removed by chemical reactions in the stratosphere. Such photons have sufficient energy to break oxygen molecules into oxygen atoms by the following reaction.

$$O_2 \xrightarrow{\text{UV radiation}} 2O \qquad (1)$$

Stratospheric Ozone Cycle The breaking of an O_2 molecule launches a series of reactions called the *stratospheric ozone cycle.* The region of the stratosphere where the ozone cycle occurs is called the *ozone layer.* Overall, the ozone cycle converts UV-B radiation into heat. By "cycle" we mean a *chemical chain reaction* whereby the reactants needed for an early step are continuously generated by a later step thus making possible further occurrences of the early step.

The oxygen atoms generated by equation 1 are generally *electronically excited,* meaning that they have a valence-shell electron in a higher than normal energy state. An excited O atom—we'll symbolize it by O*—is a very reactive species. It reacts with O_2 to make O_3 if it manages to collide with the O_2 molecule at the surface of a neutral particle, *M*. *M* can be another molecule of O_2 or an N_2 molecule, for example. (The function of *M* is to absorb some of the energy of the collision between O and O_2 so that the new O_3 molecule has insufficient energy to break up as soon as it forms.) Thus the first step of the ozone cycle itself is the following reaction.

$$O* + O_2 + M \longrightarrow O_3 + M + \text{heat} \qquad (2)$$

The newly formed ozone is next broken apart by the absorption of UV radiation, and the second step of the cycle occurs.

$$O_3 + \text{UV energy} \longrightarrow O_2 + O* \qquad (3)$$

Notice that the O* *product* of the second reaction, 3, is the necessary *reactant* for the preceding step, 2. Reaction 3, therefore, now takes over from reaction 1 as the supplier of O* for reaction 2. When 2 occurs again, reaction 3 can take place once more. Thus 2 and 3 together constitute a *chemical chain reaction.* The cycle repeats itself sometimes hundreds of times before other events terminate a chain.

If we add equations 2 and 3, canceling identical species on opposite sides of the arrows, the net result is simply

$$\text{UV energy} \longrightarrow \text{heat} \qquad (4)$$

This is the net effect of the ozone cycle in the stratosphere, the conversion of dangerous UV radiation into heat.

One of the many reactions that can terminate a chain of the ozone cycle is the recombination of two O* atoms to form molecular oxygen. When O* is thus lost, equation 2 cannot generate the ozone needed for further occurrences of 3, so the cycle is broken.

The overall result of the chain initiation (reaction 1), the cycle itself (reactions 2 and 3), and any natural chain terminations is normally a somewhat steady concentration of ozone in the stratosphere. Because this concentration varies widely with latitude, altitude, and the month of the year, no single figure can be cited to describe the stratospheric ozone level. At latitudes reaching 60° north or south from the equator, the ozone concentration in 1974 varied roughly between 260 and 360 Dobson units, reaching as high as 500 Dobson units. The *Dobson unit* equals 2.7×10^{16} molecules of O_3 in a column of air 1 cm^2 in area at its base (the earth's surface) and extending through the stratosphere. In 1993, the ozone level over the South Pole dropped to its all time low, 85 Dobson units. In 1996, the low reading in October was 111 Dobson units.

Chlorofluorocarbons and the Ozone Cycle The chlorofluorocarbons (CFCs) are a family of volatile, nonflammable, chemically stable, and essentially odorless and tasteless compounds. CFC-11, for example, is $CFCl_3$, boiling at 24 °C, and CFC-12 is CCl_2F_2, boiling at −30 °C. They have been widely used as the fluids in air conditioners, refrigerators, and freezers; as cleaning solvents for computer parts; and as aerosol propellants. At one time, $CFCl_3$ was used in 50–60% of all aerosol cans sold.

Eventually, all of the CFCs migrate throughout the entire atmosphere, become globally distributed, and work their way into the stratosphere. In 1974, American chemists M. J. Molina and F. S. Rowland warned that the CFCs could reduce our protection from UV radiation by interfering with the stratospheric ozone cycle. Molina and Rowland shared the 1995 Nobel Prize in chemistry with P. Crutzen (Germany) for this work. According to Molina and Rowland, the CFCs in the stratosphere absorb UV radiation, which breaks their C—Cl bonds and generates chlorine *atoms*.

$$CCl_3F \xrightarrow{\text{UV radiation}} CCl_2F + Cl$$

$$CCl_2F_2 \xrightarrow{\text{UV radiation}} CClF_2 + Cl$$

Atomic chlorine, Cl, is able to destroy ozone and so disrupt the ozone cycle by the following chain reaction.

$$Cl + O_3 \longrightarrow ClO + O_2 \quad (5)$$
$$\underline{ClO + O \longrightarrow Cl + O_2} \quad (6)$$
$$\text{Net:} \quad O_3 + O \longrightarrow 2O_2$$

Notice that this is also a *chain reaction;* the Cl atoms made by reaction 6 feed back to reaction 5. Each occurrence of reactions 5 and 6 destroys ozone. So the breakup of only one CFC molecule can initiate a cycle that causes the destruction of thousands of ozone molecules.

Because of their inertness, CFCs endure for several decades. They break up naturally only in the stratosphere. Their influence on the stratospheric ozone level will continue throughout the 21st century even if no more are released. Other processes can also interfere with the ozone layer, but the CFCs constitute a major threat that humans can control. The threat is serious, so in 1987, 36 nations signed a treaty called the Montreal Protocol on Substances that Deplete the Ozone Layer, which called for cutting the worldwide CFC production in half by 1998. Two years later, more deeply alarmed by additional research, 80 nations agreed to ban CFC production by 2000. Developed countries stopped production at the end of 1995.

The Antarctic Ozone "Hole" The British Antarctic Survey, which had been measuring the ozone density over the Antarctic since the 1960s, reported in 1985 a puzzling and disturbing trend from 1970 on: a decline in the ozone concentration occurred during each Antarctic spring. (Spring in the southern hemisphere occurs during the fall of the northern hemisphere.) The ozone level reaches its lowest levels in October each year. In 1984, the loss by mid-October was 30%; in 1989, the reduction was 70%. The result is that a huge, continent-broad column of the atmosphere over the Antarctic becomes significantly less able to destroy UV radiation during this precipitous decline in ozone level. The ozone-poor column of the atmosphere over the South Pole is called the *Antarctic ozone hole*. A similar "hole" occurs seasonally over the North Pole, but it has not been as severe.

Slowly, as spring turns to summer in the southern hemisphere, between November and March, a significant recovery in the ozone level in the Antarctic ozone hole occurs, but another decline takes place the following spring.

Why over the Antarctic? Why in the Antarctic Spring? A circulating wind pattern called the *Antarctic vortex* develops over the South Pole and the Antarctic continent during the Antarctic winter (see Figure 1). It is generated by the rotation of the earth and the sharply uneven heating of the atmosphere. Wind velocities as high as 80 m/s (180 mph) are seen in the jet stream encircling the South Pole, but very little wind is in the center of the vortex. These winds help to confine the chemical reactions that destroy ozone largely to this vortex. The circulation pattern thus answers the question, "Why over the Antarctic (or over the Arctic)?"

Following summer and fall, the Antarctic winter sets in again, and it is a period of *total* darkness during June, July, and August. The stratospheric temperature drops below 195 K (−78 °C), low enough for vast, yet thin, clouds to form from traces of water vapor and other substances. These polar stratospheric clouds (PSCs) include microcrystals of water containing sulfur oxides (SO_2 and SO_3) injected into the

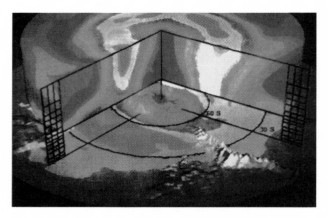

FIGURE 1 The Antarctic polar vortex and the ozone hole, which is highlighted by the purple column extending upward from the middle. The wind speeds are indicated by color code in the upper part of the figure (the cylinder missing a wedge). The darkest blue indicates essentially zero wind velocity and the red areas correspond the velocities up to 80 m s^{-1}. The surface plot is color coded for total ozone where the darkest purple colors are around the South Pole and signify values below 200 Dobson units. Latitudes are drawn in at 30° and 60° south. Extending to the lower right is the Andes chain of mountains of South America. Extending to the lower left is the tip of Africa. (Used by permission from M. R. Schoeberl and D. L. Hartmann, *Science*, January 4, 1991, page 47.)

stratosphere from volcanoes (or migrating as *anthropogenic*—human-produced—air pollutants). Another type of cloud involves crystals of a form of nitric acid ($HNO_3 \cdot 3H_2O$).

During the Antarctic winter, the polar stratospheric clouds collect and store chlorine, Cl_2. This forms by the following reaction of hydrogen chloride with chlorine nitrate (a compound made in the stratosphere).

$$HCl + ClNO_3 \rightarrow Cl_2 + HNO_3$$

The hydrogen chloride forms from the reaction of *atomic* chlorine (taken from the ozone-destroying cycle, reactions 5 and 6) with CH_4 to give CH_3 (methyl) and HCl.

$$CH_4 + Cl \rightarrow HCl + CH_3$$

(Methane, CH_4, is an air pollutant produced both naturally and by pipeline leaks. Methyl, CH_3, is another very unstable species that can exist only in a low-pressure situation, and then not for long.)

The chlorine and nitric acid molecules stay on the cloud crystals until the Antarctic winter is over and the sun reappears. With the subsequent warmup, the Cl_2 is released and starts to be broken apart by UV radiation into chlorine atoms.

$$Cl_2 + UV \rightarrow 2Cl$$

Just as sunlight returns, therefore, and the Antarctic spring commences, a large supply of chlorine atoms is released. They quickly destroy ozone by the cycle of equations 5 and 6, and the Antarctic ozone hole once again appears. Thus we see why the severity of the ozone depletion—ozone hole—peaks in the Antarctic *spring*.

Recovery of the Ozone Layer The implementation of the Montreal Protocol in the mid 1990s appears to be working. The average concentration of ozone-destroying substances in the lower atmosphere reached a peak in 1994. By mid 1995, the level was dropping by 1% per year. According to one prediction, the ozone layer will recover to its mid 1970s value by the mid 2040s. Had the production of the CFCs continued unabated, the Antarctic ozone hole would have been a serious factor in an accelerating rate of skin cancer throughout all of the 21st century. The entire experience illustrates several important truths, particularly the "law of unintended consequences." The development of the CFCs was meant to provide an utterly safe fluid for refrigeration and air conditioning equipment as well as for aerosols. They worked splendidly well, but an unknown and unexpected consequence of their extraordinary chemical stability surfaced. Rowland and Molina's early warning about what the CFCs might be doing far away from us, in the stratosphere, eventually led to international action to control the problem and eventually see it disappear.

4.6 STRUCTURES OF POLYATOMIC IONS

Many important ions are electrically charged clusters of atoms held together by covalent bonds.

A **polyatomic ion** is a cluster of atoms with a net electrical charge and held together by covalent bonds. The ammonium ion, NH_4^+, is a particularly important example, because it and substances like it have functions at the molecular level of life. To understand this ion, we must expand our understanding of the covalent bond.

A Shared Electron Pair Can Originate from One Atom   The electron-dot structure of ammonia, NH_3, is

$$H:\overset{\cdot\cdot}{\underset{\overset{|}{\ddot{H}}}{N}}:H \qquad or \qquad H-\overset{\cdot\cdot}{\underset{\overset{|}{H}}{N}}-H$$

Now notice that the nitrogen atom in this molecule has one unshared pair in the valence shell. Both electrons of this pair hold a fourth nucleus of hydrogen in the ammonium ion. The nucleus of a hydrogen atom, of course, is just a bare proton, and we can symbolize it here as H^+, a *hydrogen ion,* because it's a hydrogen atom minus its $1s$ electron. Then we can visualize the formation of NH_4^+ from NH_3 and H^+ as follows.

$$H-\overset{\cdot\cdot}{\underset{\overset{|}{H}}{N}}-H + H^+ \longrightarrow \left[H-\overset{\overset{\textstyle H}{|}}{\underset{\overset{|}{H}}{N}}-H\right]^+ \quad or \quad \left[H-\overset{\overset{\textstyle H}{|}}{\underset{\overset{|}{H}}{N}}-H\right]^+ \quad or \quad NH_4^+$$

ammonia   hydrogen   ammonium
       ion   ion

The new bond to the fourth hydrogen is a covalent bond like the other bonds, because it's an electron-pair bond. Sometimes it's useful, however, to have a special name for a covalent bond for which *both* shared electrons come from one atom; we call the bond a **coordinate covalent bond.**

The ammonium ion bears a net charge of 1+ because we have added the 1+ charge of H^+ to a neutral particle (zero charge), NH_3. The resulting cluster of atoms, all held together by covalent bonds, is therefore an example of a *polyatomic ion.* It can exist in solution and move around as one unit, or it can be present as a unit in a solid ionic compound, like ammonium chloride, NH_4Cl.

In the structure of the ammonium ion, the nitrogen still has an octet, only now all four electron pairs of the octet are involved in covalent bonds. All four of these bonds are equivalent. The molecule cannot remember which bond formed in which way, so you can see that a coordinate covalent bond and a covalent bond are not different *once they have formed.*

Before we consider other polyatomic ions we want to say more about two families of compounds, *acids* and *bases,* that are not just sources of polyatomic ions, but are also involved in supplying or in combining with the hydrogen ion. We do this to facilitate some of your laboratory work, but we have much more to learn about acids and bases later. (In fact, the acid–base status of body fluids is one of the most important topics that we study.)

Compounds That Furnish Hydrogen Ions Are Called Acids   If you're wondering where we can obtain a hydrogen ion, H^+, to make NH_4^+, it is supplied by any of a large family of **acids.** For example, hydrochloric acid is actually a 1:1 mixture of hydrogen ions and chloride ions in water. Sulfuric acid, H_2SO_4, furnishes hydrogen ions in water plus both the hydrogen sulfate ion, HSO_4^-, and the sulfate ion, SO_4^{2-}. Nitric acid provides, besides the hydrogen ion, the nitrate ion, NO_3^-.

Dilute solutions of these acids all have very tart tastes (but don't experiment with them unless your instructor shows you what to do; some acids are poisons and can harm teeth!). The tartness of lemon juice is caused by citric acid, and the tart-

■ Even dilute solutions of nitric, sulfuric, and hydrochloric acids will quickly eat holes in blue jeans.

ness of vinegar is caused by acetic acid. Another property common to acids is that they react with most metals such as iron. *The reason why acids have properties in common is that all acids are sources of hydrogen ions.*

■ Ammonia, a gas under ordinary conditions, is available in the lab as solutions in water called *aqueous ammonia.*

Bases are compounds or ions that combine with hydrogen ions. Ammonia, which can react with and bind hydrogen ions, is an example of a base, but it's just one of several. If we represent hydrochloric acid by its separated ions, H^+ and Cl^-, then its reaction with ammonia, a base, can be written as follows:

$$NH_3 \quad + [H^+ + Cl^-] \quad \longrightarrow [NH_4^+ + Cl^-]$$

ammonia   hydrochloric acid   ammonium chloride

Generally, when an acid and a base react, one product is a *salt,* which is a general term. A **salt** is an ionic compound formed from any cation except H^+ and any anion except OH^- or O^{2-}. Sodium chloride is thus just one of thousands of known salts. Ammonium chloride (above) is another salt.

Parentheses Are Sometimes Needed in Chemical Formulas Involving Polyatomic Ions Table 4.3 lists several of the important polyatomic ions. *Their names and formulas should be learned now.* Many common substances involve polyatomic ions, for example:

NH_4Cl	ammonium chloride, an ingredient in smelling salts
$NaOH$	sodium hydroxide, a raw material for making soap
NH_4NO_3	ammonium nitrate, a fertilizer
$NaNO_2$	sodium nitrite, a preservative in bacon and bologna
Na_3PO_4	sodium phosphate, a powerful cleaning agent
Na_2CO_3	sodium carbonate, washing soda
$NaHCO_3$	sodium bicarbonate, baking soda (not baking *powder*)

All are examples of salts, except NaOH, because its anion is OH^-.

Whenever a formula has more than one polyatomic ion, we place parentheses about it and put a subscript *outside* the closing parenthesis. One example is ammonium sulfate.

TABLE 4.3   Some Important Polyatomic Ions

Name	Formula	Name	Formula
Ammonium ion	NH_4^+	Dihydrogen phosphate ion	$H_2PO_4^-$
Hydronium ion[a]	H_3O^+	Nitrate ion	NO_3^-
Hydroxide ion	OH^-	Nitrite ion	NO_2^-
Acetate ion	$C_2H_3O_2^-$	Hydrogen sulfite ion[d]	HSO_3^-
Carbonate ion	CO_3^{2-}	Sulfite ion	SO_3^{2-}
Bicarbonate ion[b]	HCO_3^-	Cyanide ion	CN^-
Sulfate ion	SO_4^{2-}	Permanganate ion	MnO_4^-
Hydrogen sulfate ion[c]	HSO_4^-	Chromate ion	CrO_4^{2-}
Phosphate ion	PO_4^{3-}	Dichromate ion	$Cr_2O_7^{2-}$
Monohydrogen phosphate ion	HPO_4^{2-}		

[a] This ion is known only in a water solution.
[b] Formal name: hydrogen carbonate ion.
[c] Common name: bisulfate ion.
[d] Common name: bisulfite ion.

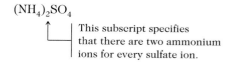

This subscript specifies that there are two ammonium ions for every sulfate ion.

Other compounds whose formulas require parentheses are the following (and all are salts).

$Ca(NO_3)_2$ calcium nitrate

$Mg_3(PO_4)_2$ magnesium phosphate

$Al_2(SO_4)_3$ aluminum sulfate

■ PRACTICE EXERCISE 12   Spend some time learning the names and formulas of the polyatomic ions that your instructor has assigned, and then drill yourself by writing the formulas of the following compounds.

(a)   potassium bicarbonate

(b)   sodium monohydrogen phosphate

(c)   ammonium phosphate

■ PRACTICE EXERCISE 13   Write the name of each of the following compounds.

(a)   NaCN   (b) KNO_3 (c) $NaHSO_3$ (d) $(NH_4)_2CO_3$ (e) $NaC_2H_3O_2$

4.7   WRITING LEWIS STRUCTURES OF MOLECULES AND POLYATOMIC IONS

As much as possible, Lewis structures are written so that all atoms have valence shell octets.

Learning how to write Lewis structures is one way to gain a better sense of where electrons are in molecules and polyatomic ions. This is useful to our study, because chemical reactions rearrange valence electrons as bonds break and reform.

Chemists have reduced the writing of a Lewis structure to a few simple steps. The first step is to write a *skeletal structure,* one that roughly groups the atoms of the molecular formula as they are arranged in the structure, but without any valence electrons. This is often the trickiest step, and sometimes you have to make an educated guess. For example, sometimes the molecular formula suggests what might be the *central atom,* the atom around which the others are grouped. Hydrogen is never the central atom because once a covalent bond extends from H, no more bonds can extend from it. In H_2O, therefore, oxygen must be the central atom. Otherwise, oxygen is seldom the central atom. Here are the correct skeletal structures for H_2O, CO_2, and NO_2. (For these, the central atom is the one that has the highest covalence number.)

H O H   O C O   O N O

■ NO_2 is the air pollutant that gives the red-brown color to smog.

Now consider carbonic acid, H_2CO_3 which has more than three atoms. Which of the following groupings is correct is not obvious.

O
H O C O H   or   H O C O O H   or   O C O O H
correct   incorrect   H
                                                       incorrect

■ Carbonic acid forms by a reaction of CO_2 with water. It isn't a stable acid, and it cannot be isolated and made pure.

What we need is a guideline, a rule of thumb that can help us out in most situations like this and lead us to the identity of the central atom. The key word in carbonic acid is *acid*, plus the fact that its formula, H_2CO_3, contains both oxygen and hydrogen besides another nonmetal. Such acids are classified as *oxoacids*, and in all those of our study, H is always joined covalently to O and O is never the central atom. The central atom in carbonic acid is thus C. Around it, as shown above, can be grouped all of the O atoms of H_2CO_3.

EXAMPLE 4.10   Writing Skeletal Structures

Write the skeletal structure of nitric acid, HNO_3.

ANALYSIS This is an acid with both H and O, so we know that one oxygen atom holds the hydrogen atom and that the central atom is N. We can now write the skeletal structure by grouping the O atoms around the N and placing H near one of the O atoms. (It does not matter which one.)

SOLUTION

$$O$$
$$H \quad O \quad N \quad O$$

■ PRACTICE EXERCISE 14   Write the skeletal structures of sulfuric acid, H_2SO_4, probably the most important acid used in industry, and phosphoric acid, H_3PO_4.

The second step in writing a Lewis structure is to add up all of the valence electrons. The total equals the number of dots that must be used. Use the locations of the elements in the periodic table to find the number of their valence electrons.

EXAMPLE 4.11   Adding Up Valence Electrons

How many dots must appear in the Lewis structures of H_2CO_3, HNO_3, and H_2SO_4?

ANALYSIS We use the group number of the element in the periodic table to determine quickly how many electrons one of its atoms has in its valence shell. Then we add up the valence shell electrons over all atoms in the formula.

SOLUTION

H_2CO_3 H has 1 valence electron, so for 2 H $2 \times 1 = 2$

C (group IVA) has 4, so for 1 C $1 \times 4 = 4$

O (group VIA) has 6, so for 3 O $3 \times 6 = \underline{18}$

Total valence electrons 24

HNO$_3$	H has 1 valence electron, so for 1 H	$1 \times 1 =$	1
	N (group VA) has 5, so for 1 N	$1 \times 5 =$	5
	O (group VIA) has 6 each, so for 3 O	$3 \times 6 =$	$\underline{18}$
	Total valence electrons		24
H$_2$SO$_4$	H has 1 electron, so for 2 H	$2 \times 1 =$	2
	S (group VIA) has 6, so for 1 S	$1 \times 6 =$	6
	O (group VIA) has 6, so for 4 O	$4 \times 6 =$	$\underline{24}$
	Total valence electrons		32

■ PRACTICE EXERCISE 15 How many valence electrons are in H$_3$PO$_4$, phosphoric acid?

In the third step, we start to place the electrons *by pairs* into the skeletal structure. Do this in the following order. Put *one* pair in each bond. Then use pairs to complete the octets of all atoms attached to the central atom. Finally, if necessary, complete the octet of the central atom using as many pairs as needed. (Remember to limit the electrons by any H atom to two.)

EXAMPLE 4.12   Placing Electron Pairs into Skeletal Lewis Structures

Fill in the electron pairs for the skeletal structure of sulfuric acid.

ANALYSIS   In Practice Exercise 14, the skeletal structure was found to be the following.

$$
\begin{array}{c}
O \\
H \ O \ S \ O \ H \\
O
\end{array}
$$

(You might have placed the H atoms by different oxygens. This would not matter.) In Example 4.11 we calculated that there are 32 valence electrons—32 dots—to distribute.

SOLUTION   First we put electron pairs into each bond.

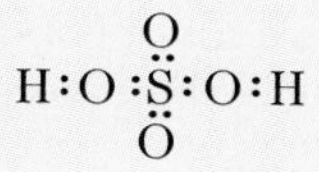

$$
\begin{array}{c}
O \\
H{:}O{:}\overset{\cdot\cdot}{S}{:}O{:}H \\
\overset{\cdot\cdot}{O}
\end{array}
$$

This used up 12 electrons. Now we finish the octets for the oxygen atoms.

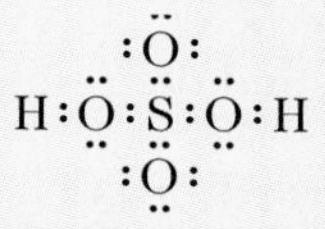

$$
\begin{array}{c}
:\overset{\cdot\cdot}{O}: \\
H{:}\overset{\cdot\cdot}{O}{:}\overset{\cdot\cdot}{S}{:}\overset{\cdot\cdot}{O}{:}H \\
:\underset{\cdot\cdot}{O}:
\end{array}
$$

This uses up 20 more electrons. We have used the 32 electrons we were required to place, 12 + 20. We see that the central atom has an octet. So we have answered the problem, but we can still change this structure to one with dash bonds and write the Lewis structure of sulfuric acid as follows.

■ The Lewis structure of sulfuric acid (and phosphoric acid), as well as those for the anions, will be modified in a later chapter.

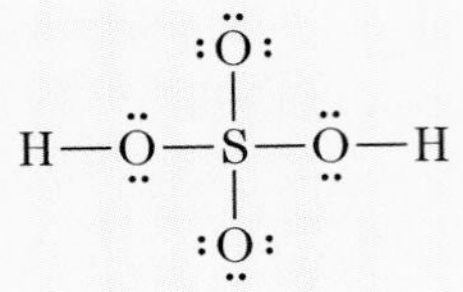

$$
\begin{array}{c}
:\!\ddot{O}\!: \\
| \\
H\!-\!\ddot{O}\!-\!S\!-\!\ddot{O}\!-\!H \\
| \\
:\!\ddot{O}\!:
\end{array}
$$

■ **PRACTICE EXERCISE 16** One of the oxoacids of chlorine is chloric acid, $HClO_3$. Chlorine is its central atom. Write its skeletal structure, calculate the total number of valence electrons, fill them in (remembering to complete the octet of the central atom), and display the Lewis structure using dash bonds.

In many applications of the steps just given, you run out of electrons before you complete all of the octets. This is a sign that *double or triple bonds have to be created*. (Beryllium and boron are exceptions, as will be discussed at the end of this section.)

EXAMPLE 4.13   Creating Double Bonds in Writing a Lewis Structure

Write the Lewis structure of carbon dioxide, CO_2.

ANALYSIS   Its correct skeleton was given earlier,

$$O \quad C \quad O$$

Its total number of valence electrons is 16. (Two O atoms contribute $2 \times 6 = 12$ electrons, and the C contributes 4 electrons for a total of 16.) We proceed to place these 16 electrons around the atomic symbols of the skeleton structure and check to see that each atom has an octet. If it does, we have solved the problem; otherwise, we must consider an additional procedure.

SOLUTION   Filling in the electrons according to our guidelines gives us the following:

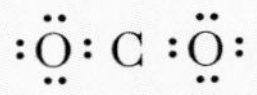

$$:\!\ddot{O}\!:\ C\ :\!\ddot{O}\!:$$

We have now used up all 16 electrons, but the central atom does not have an octet. There are only 4 electrons around C, and we've run out of valence electrons.

The procedure now is to move a *nonbonding* pair of electrons on an adjacent oxygen atom into an existing carbon–oxygen bond.

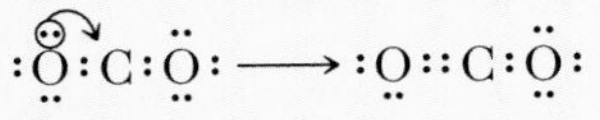

$$:\!\overset{\curvearrowright}{\ddot{O}}\!:\!C\!:\!\ddot{O}\!: \longrightarrow :\!O\!::\!C\!:\!\ddot{O}\!:$$

But carbon still doesn't have an octet. So we move another nonbonding pair, taking it from the other oxygen, and the result is two double bonds in CO_2.

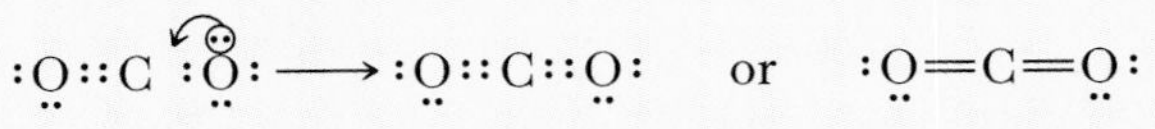

$$:\!O\!::\!C\ :\!\overset{\curvearrowleft}{\ddot{O}}\!: \longrightarrow :\!O\!::\!C\!::\!O\!: \quad \text{or} \quad :\!O\!\!=\!\!C\!\!=\!\!O\!:$$

carbon dioxide

Here O is the "adjacent atom." These shifts of electron pairs cannot be done when a halogen atom is the adjacent atom. Halogen atoms (F, Cl, Br, or I) generally do not engage in double or triple bonds. S and N, however, can.

■ PRACTICE EXERCISE 17   Write the Lewis structure for sulfur trioxide, SO_3, an air pollutant.

Lewis Structures of Polyatomic Ions Also Show Noble Gas Configurations. We need only a small modification of our preceding steps to construct Lewis structures of polyatomic ions. These modifications concern the number of valence shell electrons that are to be inserted into a skeletal structure.

1. For each negative charge on the ion, we have to add an electron to the count. A charge of 1− means adding one more electron. A charge of 2− means adding two more electrons, and so forth.
2. For each positive charge on an ion, we subtract an electron.

Let's see how this works with the sulfate ion, SO_4^{2-}.

EXAMPLE 4.14   Writing the Lewis Structure of a Polyatomic Negative Ion

What is the Lewis structure of SO_4^{2-}?

ANALYSIS   We first have to total the valence shell electrons and then draw a skeletal structure, using S as the central atom. Next we insert electron pairs into the covalent bonds, place remaining electrons so as to provide for complete octets, and then see whether double bonds are needed.

SOLUTION   First, we total the valence shell electrons.

For S (group VIA), 6 electrons	$1 \times 6 = 6$
For each O (group VIA), 6 electrons	$4 \times 6 = 24$
For each negative charge, add 1 electron	$2 \times 1 = \underline{2}$
Total valence shell electrons	32

Next we write a reasonable skeletal structure. We use S as the central atom and arrange the O atoms around it symmetrically.

$$
\begin{array}{c}
O \\
O \; S \; O \\
O
\end{array}
$$

Next, we insert electron pairs into the covalent bonds. This will use up 8 electrons out of the 32.

$$
\begin{array}{c}
O \\
O : \overset{..}{\underset{..}{S}} : O \\
O
\end{array}
$$

Now we place electron pairs around the oxygens to give each an octet. This will use up all the remaining 24, so this gives the final structure of the sulfate ion. We don't have to create double bonds. To show that the ion carries a charge of 2−, we'll place brackets about it and place the 2− charge as a right superscript.

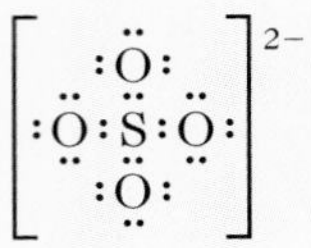

$$\left[\begin{array}{c} :\ddot{O}: \\ :\ddot{O}:S:\ddot{O}: \\ :\ddot{O}: \end{array}\right]^{2-} \quad \text{the sulfate ion, } SO_4{}^{2-}$$

■ PRACTICE EXERCISE 18   Write the Lewis structures of the following ions.

(a) H_3O^+ (b) OH^-

■ PRACTICE EXERCISE 19   Write the Lewis structures of the following ions. Each will require that you form a double bond. Remember, it won't matter which electron pair is moved into position as the second bond of the double bond so long as proper octets are realized.

(a) $CO_3{}^{2-}$ (b) $NO_3{}^-$

Sometimes the Octet Rule Fails   The octet rule is not really a law of nature because it fails sometimes. We will just cite, by structures, some examples of failures without making anything big out of them. We will seldom encounter failures of the octet rule when only the row 1 and the row 2 nonmetal elements of the periodic table are involved. This means that H, C, N, and O will never give us trouble on this score. Their valence shells can never hold more than eight electrons (just two for hydrogen, of course).

Some failures of the octet rule involve more than eight valence shell electrons in the structure. These can occur when a "central" nonmetal atom is from period 3 (or higher) of the periodic table. These atoms have their valence electrons at principal energy level 3 (or higher), *which can hold more than eight electrons*. Phosphorus pentachloride, PCl_5, and sulfur hexafluoride, SF_6, are examples of octet rule failures of this type.

■ SF_6 is a colorless, odorless, tasteless, nonflammable, nontoxic, and unusually stable gas used to insulate high-voltage generators and switches.

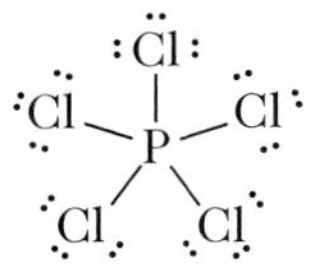

10 electrons
around P and
8 electrons
around each Cl

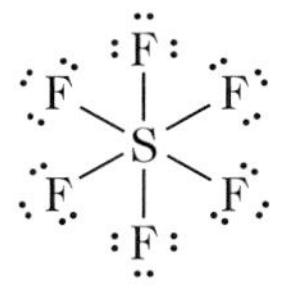

12 electrons
around S and
8 electrons
around each F

A few failures of the octet rule involve fewer than eight electrons around a central atom. The classic examples are beryllium chloride, $BeCl_2$, and boron trichloride, BCl_3. The Lewis structure of Be is simply $\cdot Be\cdot$, because Be is in group IIA. We can represent the formation of $BeCl_2$ as follows:

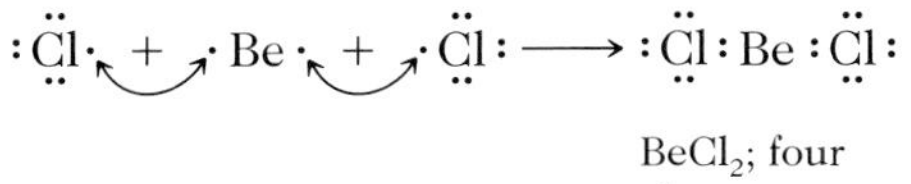

$$:\ddot{Cl}\cdot + \cdot Be\cdot + \cdot\ddot{Cl}: \longrightarrow :\ddot{Cl}:Be:\ddot{Cl}:$$

$BeCl_2$; four
electrons
around Be

Only four valence electrons are around the Be atom in this structure. Recall that we cannot shift electron pairs from Cl to make multiple bonds and so give Be an octet in $BeCl_2$. Halogen atoms do not participate in multiple bonds.

The formation of BCl_3 is represented as follows.

$$\cdot \overset{\cdot}{B} \cdot \; + \; 3 \cdot \overset{..}{\underset{..}{Cl}} : \; \longrightarrow \; \overset{\overset{..}{:Cl:}}{\underset{..}{:Cl:B:Cl:}}$$

BCl_3; B has
six outside-
level electrons.

Boron has only six electrons in its valence shell in BCl_3.

■ Boron is the first member of group IIIA.

4.8 SHAPES OF MOLECULES

The overall shape of a molecule is forced by the repulsions of electron clouds of valence-shell electron pairs.

We said earlier that molecular shape is as important as anything else at the molecular level of life. We'll see how such shapes arise in this section.

The water molecule is not a linear molecule, with all its atoms lined up in a straight line. It's a bent molecule, and the bond angle is known, 104.5°. The angle formed by two bonds from the same atom is called the **bond angle**.

■ *Linear* means in a straight line.

$$---H—O—H---$$
$$\underset{180°}{\smile}$$

linear molecule
(incorrect)

$$\overset{O}{\underset{H \quad \quad H}{\diagdown \quad \diagup}}$$
$$\underset{104.5°}{\smile}$$

bent molecule
(correct)

The VSEPR Theory Is a Simple Explanation for Bond Angles Why is the water molecule bent and not linear? The easiest way to answer this question is by a theory with a long but descriptive name, the **valence-shell electron-pair repulsion theory**, or the **VSEPR theory** for short. The *valence shell* is where covalent bonds originate. *Electron pair* refers to the electrons of the valence shell. *Repulsion* refers to the effect that an electron cloud of one pair has on the electron cloud of another in the same valence shell.

VSEPR theory says that the shape of a molecule is largely determined by the efforts of the valence shell electron clouds to stay out of each other's way as much as they can. Imagine that each electron cloud is a balloon with an imaginary line or axis running from where it's tied off to the opposite surface. (See the figure in the margin.) Now imagine how four identical balloons *must* arrange themselves most comfortably (and so in the most stable way) if we tie them all close together at one point (see Figure 4.5). The balloons are least crowded when their axes make angles of 109.5° with each other. This array is called **tetrahedral** because the four axes point to the corners of a regular tetrahedron. Figure 4.6 indicates how VSEPR theory deals with the bond angles in molecules of water, ammonia, and methane.

The valence shell of O in H_2O has four electron pairs. Two pairs are unshared, and the other two carry and hold one hydrogen nucleus apiece. These hydrogen nuclei shrink the associated clouds somewhat, so the clouds repel each other less. Consequently, the actual bond angle in H_2O is not quite as large as the true tetrahedral angle of 109.5°, but it is very close.

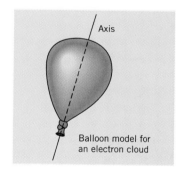

Axis

Balloon model for
an electron cloud

FIGURE 4.5 When four identical balloons are tied off at a common point, their axes will point to the corners of a tetrahedron.

Number of Pairs in Bonds	Number of Lone Pairs	Structure	
4	0		Tetrahedral (Example, CH_4) All bond angles are 109.5°.
3	1		Trigonal pyramidal (Pyramid shaped) (Example, NH_3)
2	2		Nonlinear, bent (Example, H_2O)

FIGURE 4.6   Shapes of molecules that have four pairs of electrons around a central atom.

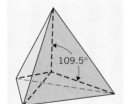

A regular tetrahedron is a four-sided space bounded by identical equilateral triangles. Any two lines from the corners to the midpoint of the space make an angle of 109.5°.

The central atom is N in NH_3. Once again the VSEPR theory predicts a tetrahedral bond angle of 109.5°, because there are four valence-shell electron pairs. The actual bond angle in ammonia is 107.3°, also quite close.

In methane, CH_4, C is the central atom. VSEPR theory again predicts bond angles of 109.5° for each of the H—C—H bonds, and the bond angles in methane are all 109.5°.

VSEPR Theory Is Unusually Successful for Systems Like $BeCl_2$, BCl_3, PCl_5, and SF_6, Too   What if there are fewer than four electron pairs in the valence shell, as in $BeCl_2$ and BCl_3? What if there are more than four pairs as around P in PCl_5 and around S in SF_6?

Figure 4.7 illustrates the shapes of molecules with anywhere from two to six valence-shell electron pairs. Each shape is exactly what VSEPR theory predicts. When only two electron pairs are in the valence shell—think of just two balloons tied together—the axes of their clouds must point oppositely to give the most comfortable room, so the bond angle in $BeCl_2$ is 180°. When there are just three electron clouds, their axes will be in a plane and point to the corners of a regular triangle, so the bond angle in BCl_3 is 120°. The bond angles and shapes of the rest of the systems in Figure 4.7 follow from the same arguments.

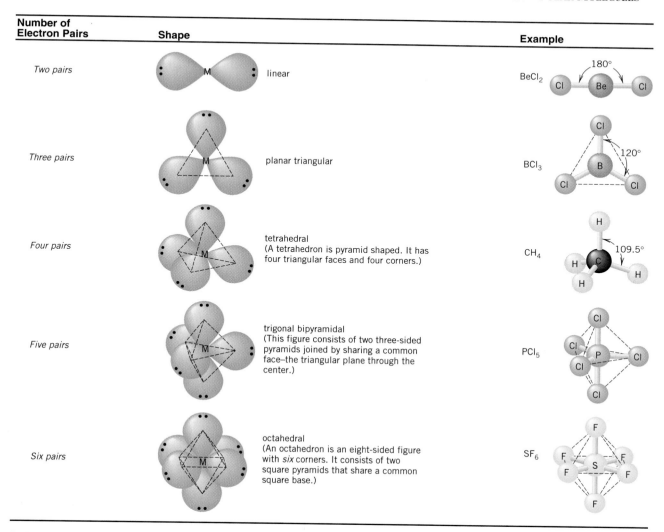

Number of Electron Pairs	Shape		Example
Two pairs		linear	$BeCl_2$
Three pairs		planar triangular	BCl_3
Four pairs		tetrahedral (A tetrahedron is pyramid shaped. It has four triangular faces and four corners.)	CH_4
Five pairs		trigonal bipyramidal (This figure consists of two three-sided pyramids joined by sharing a common face–the triangular plane through the center.)	PCl_5
Six pairs		octahedral (An octahedron is an eight-sided figure with *six* corners. It consists of two square pyramids that share a common square base.)	SF_6

FIGURE 4.7   Molecular shapes to be expected for different numbers of pairs of valence shell electrons.

4.9 POLAR MOLECULES

Even electrically neutral molecules can attract each other if they are polar.

Now that we know quite a lot about molecules we can consider an important question about molecular compounds. If molecules are neutral, how can they stick together? Sugar molecules, for example, gather together naturally to make beautiful crystals that are not easy to melt. What holds sugar molecules together in such crystals? We'll introduce the answer here. The same principles will apply to the cohesion of the molecules of protein or DNA at the molecular level of life.

Shared Pairs between Unlike Atoms Are Usually Not Equally Shared   When atoms have *unequal* nuclear charges, the electron cloud of the shared pair usually is drawn toward the nucleus with the larger charge. This is particularly true when the bond holds atoms of the *same period* in the periodic table. Such atoms have the same number and arrangement of core electrons, so atomic nuclei from the same period are shielded in identical ways by their core electrons. Hence, the stronger the nuclear charge becomes as we move from left to right across a period, the greater will be the ability of the nucleus to attract a valence-shell electron cloud. Thus the nucleus with the larger positive charge pulls some electron density of the shared electron cloud away from the other nucleus, *somewhat exposing its positive charge influence*. The positive charge of the nucleus of *lower* charge is therefore not entirely canceled *where this nucleus is in the molecule;* its associated electron cloud is too thin in negative charge. In other words, where the electron cloud is too thin, a fraction of a nuclear positive charge is still exerting whatever attracting or repelling influences any charge can exert. A fractional charge is called a *partial charge,* and we use the Greek lowercase letter delta, δ, to stand for *partial*. Thus a partial positive charge has the symbol $\delta+$.

The hydrogen flouride molecule, for example, has a $\delta+$ charge at its hydrogen end. The hydrogen nucleus has only a charge of $1+$, but the fluorine nucleus has a charge of $9+$, nine times as great. Moreover, because fluorine's other electrons are not concentrated between the two nuclei, they cannot entirely shield the influence of fluorine's nuclear charge from the shared pair. This is why the electron cloud of the shared pair is pulled toward the fluorine end of the H—F molecule. Consequently, the hydrogen end retains insufficient electron density to neutralize fully the positive charge of the hydrogen nucleus *where it is*. Thus the hydrogen end of the H—F molecule has a $\delta+$ charge. Now let's look at the other end.

The fluorine nucleus of the H—F molecule pulls some electron density of the shared pair toward itself. This causes the total electron density at the fluorine end of H—F to be more than enough to neutralize the positive charge of the F nucleus. In other words, the electron cloud is thicker than necessary at F in H—F, so this end has a partial negative charge, $\delta-$.

We can't say precisely what the sizes of the fractions represented by $\delta+$ and $\delta-$ are. However, the molecule as a whole is electrically neutral, so the algebraic sum of $\delta+$ and $\delta-$ must be zero.

Polar Bonds Have Opposite Partial Charges at Either End   When a covalent bond has a $\delta+$ at one end and a $\delta-$ at the other, it is called a **polar bond.** We can symbolize the electrical polarity of the bond in hydrogen fluoride in either of two ways, as seen in structures **1** and **2**.

$$\overset{\delta+\quad\delta-}{\text{H—F}} \qquad \overset{\longrightarrow}{\text{H—F}}$$

$$\textbf{1} \qquad\qquad \textbf{2}$$

In **2**, the arrow points toward the end of the bond that is richer in electron density. At the other end, there is a hint of the positive character by the merger of the arrow with a plus sign. Because there are two partial, opposite charges in H—F, this molecule is sometimes said to have an **electrical dipole.**

A magnet with two *magnetic* poles—labeled north and south—is a good analogy for describing the consequences of *electrical* polarity. Perhaps you have played with toy magnets and know that two magnets can stick to each other *if they are lined up properly*. In fact, if you have a great many magnets, and line all of them up correctly, you can make them all cling together. You just have to make sure that poles of opposite kind are nearest neighbors and that poles that are alike are as far apart as possible.

Molecules that are electrically polar can stick to each other just like magnets. Given the freedom to move, polar molecules will line up automatically the way mag-

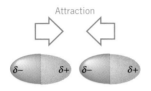

Two polar molecules can attract each other.

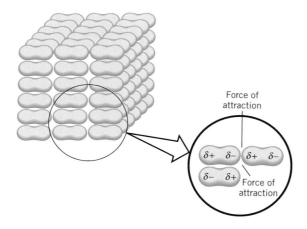

FIGURE 4.8 Polar molecules attract each other in a crystal of a molecular substance.

nets can (see Figure 4.8). This is how neutral molecules are able to adhere to each other. How tightly they stick depends on the sizes of the partial charges and on the shapes of the molecules. In some large molecules with complex shapes, partial charges can be shielded by neighboring sections of the molecules. This interferes with the ability of $\delta+$ and $\delta-$ sites on different molecules to get close to each other and so weakens forces of attraction between the molecules.

When Bonded Atoms Have Different Electronegativities, the Bond Is Polar The relative ability of an element's atoms to draw away electron density of a covalent bond is called the **electronegativity** of the element.

Fluorine has the highest electronegativity of all of the elements, because its atoms have the highest nuclear positive charge while being shielded by only level 1 and level 2 electrons. Oxygen, which stands just to the left of fluorine in the periodic table, has the next highest electronegativity. Its atoms have one less positive charge on their nuclei than fluorine atoms and also are shielded by only level 1 and level 2 electrons. The element with the third highest electronegativity, you might now guess, lies just to the left of oxygen. It is nitrogen with atoms that have one less charge on their nuclei than oxygen atoms.

Figure 4.9 shows the relative electronegativities of several elements, metals and nonmetals, and their locations in the periodic table. Notice that carbon isn't the element with the fourth highest electronegativity; chlorine ranks fourth. The chlorine atom has a large positive nuclear charge, 17+. This doesn't make chlorine even more electronegative than fluorine, however, evidently because chlorine has a sufficient number of additional core electrons to shield the chlorine nucleus. Moreover, the covalent bond in a molecule such as H—Cl is a longer bond than it is in H—F. The shared pair has its electron density farther from the chlorine nucleus to start with, and this also makes it harder for this nucleus to pull electron density toward itself. On balance, chlorine is less electronegative than fluorine, but more so than carbon.

Notice in Figure 4.9 that metals have the lowest electronegativities. In fact, the general trend is that as you move to the right in the same period or as you move upward in the same group, the electronegativities become larger. The most electronegative element, as we said, is fluorine; and the least electronegative is cesium, the last element in group IA (not pictured in Figure 4.9, but below rubidium, Rb). The *trends* should be learned, but not specific values of electronegativities. We'll be working so often with oxygen, nitrogen, carbon, and hydrogen, however, that you should memorize the order of their electronegativities, $O > N > C > H$. We'll see shortly how knowing this can be useful.

IA						
H 2.20	IIA	IIIA	IVA	VA	VIA	VIIA
Li 0.97	Be 1.47	B 2.01	C 2.50	N 3.07	O 3.50	F 4.10
Na 1.01	Mg 1.23	Al 1.47	Si 1.74	P 2.06	S 2.44	Cl 2.83
K 0.91	Ca 1.04				Se 2.48	Br 2.74
Rb 0.89	Sr 0.99				Te 2.01	I 2.21

FIGURE 4.9 Relative electronegativities.

■ Group IA and IIA elements are so weakly electronegative that they give up electrons entirely to group VIA or VIIA elements and form ions, not polar bonds.

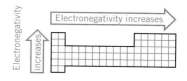

In electronegativity, metals are lowest and nonmetals are highest.

Polar Bonds Make Molecules Polar if the Bond Polarities Do Not Cancel   It's easy to tell whether a *bond* is polar; it *always* is if the atoms that the bond joins have different electronegativities. For diatomic molecules such as H—F and H—Cl that have only one bond, when the bond is polar so is the molecule. Because such diatomics as H—H and F—F involve identical atoms and one bond, we can tell right away that these molecules are nonpolar.

Whether larger molecules are polar in an overall sense depends not just on the presence of polar bonds but also on the *geometry* of the molecule. It's possible for the polarities of individual bonds to cancel each other. Consider, for example, the carbon dioxide molecule, **3**.

■ Nonbonding valence-shell electrons are often omitted from Lewis structures in general discussions.

$$O=C=O \qquad \underset{H}{\overset{O}{\diagup}}\overset{}{\diagdown}_{H}$$

3 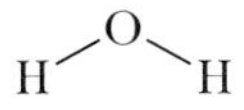   4

Because oxygen is more electronegative than carbon, each carbon–oxygen double bond must be polar. But the two dipoles in CO_2 point in exactly opposite directions, so in an overall sense they cancel each other. This leaves the CO_2 molecule overall nonpolar. The water molecule, **4**, on the other hand, is bent. Its two individual O—H bond polarities thus cannot cancel, and the water molecule is polar, quite polar, in fact.

Another useful way of thinking about the polarity of a molecule uses the idea of a *center of density of charge,* something like a "balance point" for electrical charge. A **polar molecule** is one in which the center of density of positive charge is not at the same place as the center of density of negative charge. We don't have to be able to pinpoint these centers exactly to know whether they are in the same place or not. The *symmetry* will usually tell us. For example, the symmetry of the carbon dioxide molecule, **3**, tells us that all the positive charges on the three nuclei balance around the center of the carbon nucleus. Similarly, all of the negative charges contributed by all the electrons must also balance *around the identical point*. Because these two centers are in the same place, the molecule is nonpolar. Wherever these two centers are in the water molecule, **4**, we know that the molecule's angularity prevents their being at the same place. Hence, we know that the H_2O molecule must be polar.

EXAMPLE 4.15   Predicting Molecular Polarity

Place δ+ and δ− signs at the correct ends of each of the bonds in the following structures (whose correct geometries are shown). Then decide whether each molecule as a whole is polar or nonpolar. (Use information in Figure 4.9 as needed.) A three-dimensional view of the carbon tetrachloride molecule is given. This molecule is entirely symmetrical.

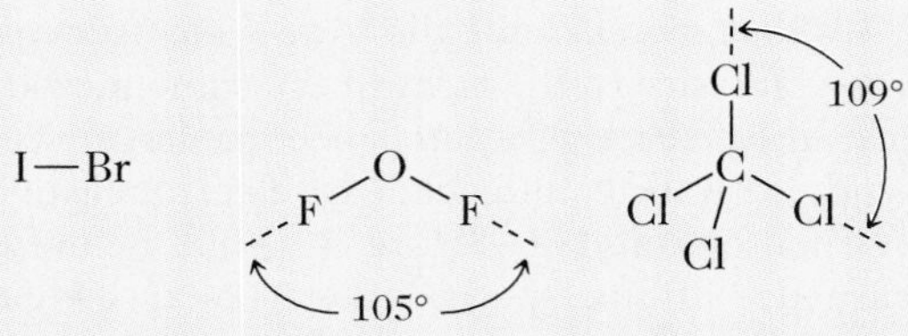

I—Br

$$\underset{H}{\overset{O}{\diagup}}$$

F ⋯⋯ F   105°

Cl   109°
Cl—C—Cl
Cl

iodine
bromide

oxygen
difluoride

carbon
tetrachloride

ANALYSIS There are six nonmetal atoms involved in these structures, and we must use their relative electronegativities to decide where partial charges are to be placed. Because Br stands *above* I in group VIIA, Br is more electronegative than I. This makes the Br end of the bond in I—Br the site with the $\delta-$ charge. Because F stands to the right of O in period 2, F is more electronegative than O and must have the $\delta-$ charge in the bonds of oxygen difluoride. From Figure 4.9 we learn that Cl is more electronegative than C, so we must place $\delta-$ at the Cl ends of each C—Cl bond in CCl_4. Once we have the partial charges correctly placed, we consider the geometry of each molecule to decide if it has an overall polarity.

SOLUTION

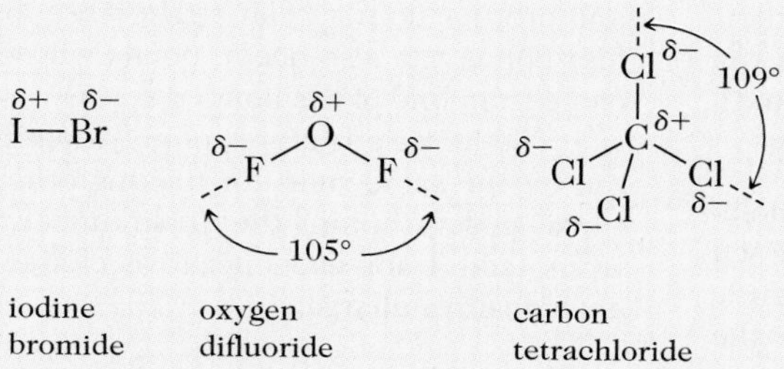

iodine bromide   oxygen difluoride   carbon tetrachloride

The linear iodine bromide molecule has a polar bond and so must be a polar molecule. Because the oxygen fluoride molecule is angular, the centers of density of positive and negative charge cannot be at the same location, so the molecule must be (and is) polar. Although each bond in CCl_4 is polar, this molecule is *symmetrical*. Therefore, the balance point—the center of charge density—for all positive charge has to be at the center of the carbon nucleus. Similarly, the center of all negative charge density has to be in the identical place—the symmetrical disposition of the chlorine atoms about this center guarantees this result. Hence, the CCl_4 molecule as a whole is not polar.

■ PRACTICE EXERCISE 20   The structure of chloroform is just like that of carbon tetrachloride (in Example 4.15, above) except that one Cl has been replaced by H. Using the structure of carbon tetrachloride as a model, make the needed changes to draw a structure of chloroform and then place $\delta+$ and $\delta-$ signs by each atom. Finally, decide whether the molecule as a whole is polar.

SUMMARY

Ionic bonds and ionic compounds   A reaction between a metal and a nonmetal usually goes by the transfer of an electron from the metal atom to the nonmetal atom. The metal atom changes into a positively charged ion (cation), and the nonmetal atom becomes a negatively charged ion (anion). The oppositely charged ions aggregate in whatever whole-number ratio ensures that the product is electrically neutral. The electrical force of attraction between the ions is called an ionic bond, and compounds made of ions are called ionic compounds.

Octet rule (noble gas rule)   The ions of the representative elements have electron configurations of the nearest noble gases, usually outer octets. Metal atoms lose exactly enough valence shell electrons to achieve (new) noble gas configurations, and nonmetal atoms gain exactly as many valence shell electrons needed to acquire noble gas configurations.

Redox reactions   Electron transfers mean changes in oxidation numbers and are called redox reactions. Elements have zero oxidation numbers. Monatomic ions have oxidation numbers that equal their electrical charges. An atom that loses electrons and whose oxidation number becomes more positive is oxidized, and one that gains electrons and a more negative oxidation number is reduced. Anything that causes an oxidation is called an oxidizing agent; a reducing agent is anything that causes a reduction. Metals tend to be reducing agents and nonmetals are oxidizing agents.

Formulas of ionic compounds   The empirical formulas of ionic compounds begin with the symbol of the cation (without the sign indicating the charge). Subscripts are used to give the ratios of the ions. When a formula must include two or more polyatomic ions, parentheses enclose their symbols. The names of ionic compounds are based on the names of the ions (except that the word *ion* is omitted). The cation is named first, then the anion.

Molecular compounds   Atoms of nonmetals form molecules by sharing valence shell electrons in pairs. Each shared pair constitutes one covalent bond, and each pair is counted as the joint property of both atoms when the structure is checked to see if it adheres to the octet rule. The shared pair creates a region of relatively high electron density between the two atoms toward which the nuclei are electrically attracted, and this attraction is called a covalent bond. The covalent bond–forming abilities of atoms are their covalence numbers.

Molecular orbitals form by the partial overlapping of atomic orbitals. This creates a space between the two nuclei in which the electron pair of the single covalent bond resides.

Lewis structures   Molecules have unique structures, which display the sequence in which the atoms are joined. Lewis structures (electron-dot structures) display the valence-shell electrons, both the shared and unshared pairs, in molecules. (The shared pairs are usually replaced by dash bonds.) In nearly all molecules the atoms have noble gas configurations, but a few have fewer than four electron pairs and a few have more than four.

To write a Lewis structure, the skeletal structure must be known as well as the total number of valence shell electrons present. First, electron pairs are placed into the bonds. Next, additional electron pairs are used to give noble gas configurations to all atoms, doing this for the central atom last. If there aren't enough for the central atom, double (or triple) bonds have to be made by moving electron pairs from other atoms, like C, N, or O (but not halogens), toward the central atom.

Polyatomic ions   The atoms of polyatomic ions are joined by covalent bonds, but the overall numbers of electrons and protons do not balance. In these ions, coordinate covalent bonds can also occur, bonds whose electron pairs both came from one of the bonded atoms.

The Lewis structures of polyatomic ions can be figured out by the same rules for making other electron-dot structures. In determining the total number of valence shell electrons, add 1 for each negative charge and subtract 1 for each positive charge.

Acids and bases   Acids are substances that can furnish H^+ ions, and bases are compounds that can combine with H^+ ions. Acids and bases react to neutralize each other.

Molecular shapes and the VSEPR theory   Valence-shell electron pairs in molecules stay out of each other's way as much as possible. When there are four pairs, the result is a tetrahedral geometry, or close to it. If only three pairs are present, the geometry is that of a triangle with the three atoms at the corners.

Polar molecules   Even though molecules are electrically neutral, they can still be polar. If individual bond polarities, caused by electronegativity differences between the joined atoms, do not cancel, the molecule is polar and can adhere to adjacent polar molecules much as magnets can stick together.

REVIEW EXERCISES

The answers to Review Exercises whose numbers are in color are found in Appendix E. The answers to the other Review Exercises are found in the Study Guide that accompanies this book. The more challenging questions are marked with asterisks.

Ions, Ionic Compounds, and Their Names and Formulas

4.1 No matter what kind of bond we chose to describe, what kind of force is responsible for it? What law of nature is particularly important for understanding the chemical bond?

4.2 When a chemical bond forms between two atoms, what parts of the atoms become reorganized relative to each other? Are the two *atoms* still present once the bond forms? Explain.

4.3 Based on the way that electrons and nuclei can become reorganized, depending on the atoms involved, what two fundamental kinds of compounds are recognized?

4.4 What is the name of the fundamental *particle* that carries all of the chemical properties of a molecular compound?

4.5 Which kinds of elements, metals or nonmetals, largely make up molecular compounds?

4.6 Use electron configurations, including the compositions of nuclei, as we did in Section 4.1 for the reaction of sodium with chlorine, to show how a potassium atom and a fluorine atom can change to particles that attract each other. What are the names of the particles that form?

4.7 In terms of their fundamental structures, what is the difference between a potassium atom and a potassium ion?

4.8 Write the symbols with the correct charges for the following ions.
(a) fluoride ion   (b) magnesium ion
(c) cupric ion   (d) calcium ion
(e) barium ion   (f) aluminum ion
(g) sulfide ion   (h) lithium ion
(i) bromide ion   (j) chloride ion
(k) potassium ion   (l) zinc ion

4.9 What are the names of the following ions?
(a) I^- (b) Cu^{2+} (c) Cl^- (d) Cu^+
(e) Zn^{2+} (f) S^{2-} (g) Ba^{2+} (h) Na^+
(i) Al^{3+} (j) Mg^{2+} (k) Li^+ (l) Fe^{2+}

4.10 What are the formulas of the following compounds?
(a) sodium iodide   (b) iron(II) bromide
(c) aluminum oxide   (d) calcium chloride
(e) barium oxide   (f) copper(I) sulfide

4.11 Write the names of the following compounds.
(a) KI   (b) $BaCl_2$ (c) CaS
(d) $FeCl_2$ (e) Al_2O_3 (f) AgI

4.12 Which has the smaller radius, the atom or the ion of a given metal? Explain.

4.13 Which has the smaller radius, the oxygen atom or the oxide ion? Explain.

Ions and the Octet Rule

4.14 Show how calcium atoms and chlorine atoms can interact to form ions that will aggregate in the correct ratio. (Follow the directions given in Review Exercise 4.6.)

4.15 If M is the symbol of a representative element and M^{3+} is the symbol of its ion, to what group in the periodic table does this element most likely belong?

4.16 What electric charge is carried by the ions of the elements in the following groups?
(a) group IA   (b) group IIA   (c) group IIIA
(d) group VIIA   (e) group VIA

4.17 The atoms of which group in the periodic table have exceptionally stable electron configurations? Describe these configurations in your own words.

4.18 Can the atom with the following electron configuration be changed to a stable ion? If it can, what would be the electrical charge on the ion?
$$1s^2 2s^2 2p_x^2 2p_y^2 2p_z^1$$

4.19 Study the following electron configuration. Can the atom with this configuration be changed to a reasonably stable ion? If so, what is the charge on the ion?
$$1s^2 2s^2 2p^6 3s^2 3p^6 4s^1$$

4.20 Can an atom with the following electron configuration be changed to a stable ion? If it can, what would be the charge on this ion?
$$1s^2 2s^2 2p_x^2 2p_y^1 2p_z^1$$

4.21 Are nonrepresentative elements likelier to have positively or negatively charged ions? Explain.

*4.22 Using the atomic numbers only (and the corresponding electron configurations), without consulting the periodic table, write the symbols and the electron configurations of the *ions* of the following elements. (The atomic symbols are hypothetical.)
(a) M, atomic number 3
(b) Q, atomic number 19
(c) Z, atomic number 16

Oxidation and Reduction

4.23 What are the oxidation numbers of the following ions?
(a) mercury(II) ion   (b) copper(II) ion
(c) oxide ion   (d) aluminum ion
(e) chloride ion   (f) potassium ion

4.24 Determine the oxidation number of the metal component in each of the following compounds.
(a) PbF_4 (b) V_2O_5 (c) Cr_2O_3
(d) $AuCl_3$ (e) $ZrOCl_2$ (f) $GaCl_3$

4.25 What would be the formula of an oxide of manganese, Mn, if the oxidation number of Mn were $+7$? (The oxidation number of oxygen is always -2.)

4.26 The following reaction between aluminum and oxygen seals an invisible coating of aluminum oxide, an ionic compound, over any aluminum surface exposed to air.
$$4Al + 3O_2 \longrightarrow 2Al_2O_3$$
Write the chemical symbol for the following species.
(a) the substance reduced
(b) the substance oxidized
(c) the reducing agent
(d) the oxidizing agent

4.27 Calcium reacts with chlorine to form calcium chloride, an ionic compound, as follows.
$$Ca + Cl_2 \longrightarrow CaCl_2$$
Write the chemical symbol for the following substances in this reaction.
(a) the oxidizing agent
(b) the substance oxidized
(c) the reducing agent
(d) the substance reduced

Molecules, Molecular Compounds, and Lewis Structures

4.28   In what ways are a molecule and an atom alike and in what ways are they different?

4.29   How are a molecule and an ion alike and how do they differ?

4.30   How do molecular elements and molecular compounds differ?

4.31   How do molecular compounds and ionic compounds differ?

4.32   What kind of force of attraction holds the two nuclei in one hydrogen molecule, H_2, quite near each other? What special name do we give to this force of attraction when it operates within molecules?

4.33   In your own words, describe how the force of attraction arises as two hydrogen atoms combine to form a hydrogen molecule.

4.34   In what ways is a molecular orbital different from an atomic orbital?

4.35   In what ways is a molecular orbital similar to an atomic orbital?

4.36   How do we envision the formation of a molecular orbital from atomic orbitals?

4.37   How do we envision the formation of the covalent bond in the hydrogen molecule, H_2?

4.38   How does the octet rule work for molecular compounds?

4.39   Using the periodic table, write the electron-dot structures of the following.
(a) rubidium (atomic number 37)
(b) strontium (atomic number 38)
(c) gallium (atomic number 31)
(d) tellurium (atomic number 52)

4.40   Use Lewis structures to diagram the formation of $CaCl_2$ from neutral atoms.

Polyatomic Ions and Formulas of Compounds Involving Them

4.41   Write the names of the following ions.
(a) NO_3^- (b) CO_3^{2-} (c) SO_4^{2-} (d) H_3O^+
(e) HSO_4^- (f) HSO_3^- (g) $H_2PO_4^-$ (h) MnO_4^-
(i) $C_2H_3O_2^-$ (j) HCO_3^- (k) CN^- (l) OH^-

4.42   Write the formulas of the following ions.
(a) phosphate ion   (b) ammonium ion
(c) sulfate ion   (d) hydronium ion
(e) hydroxide ion   (f) acetate ion
(g) carbonate ion   (h) nitrate ion
(i) sulfite ion   (j) cyanide ion
(k) bicarbonate ion   (l) monohydrogen phosphate ion

4.43   Write the formulas of the following compounds.
(a) potassium phosphate   (b) sodium carbonate
(c) calcium sulfate   (d) ammonium cyanide
(e) lithium nitrite   (f) sodium hydrogen sulfite
(g) calcium dichromate   (h) magnesium acetate

4.44   Write the names of the following compounds.
(a) Na_2CO_3 (b) NH_4NO_3 (c) $Mg(OH)_2$
(d) $Ca(C_2H_3O_2)_2$ (e) $KHCO_3$ (f) $BaSO_4$
(g) $NaNO_2$ (h) $(NH_4)_3PO_4$

4.45   What is the total number of atoms of all kinds in one formula unit of each of the following compounds?
(a) $(NH_4)_2CO_3$ (b) $Al(C_2H_3O_2)_3$ (c) $Ba(H_2PO_4)_2$

Acids and Bases

4.46   Solutions in water of any of the common acids all contain what ion? (Give its name and formula.)

4.47   What is the formula of nitric acid, and what are the principal ions present in a solution of this in water? (Give their names and formulas.)

4.48   Hydrochloric acid is an aqueous solution in which the principal ions are what? (Give their names and formulas.)

4.49   What are two properties of solutions that contain acids? What chemical species is responsible for both?

4.50   What is the general family name for the substances that can react with acids to neutralize them? When these substances neutralize acids, do they react with the positive ion, the negative ion of the acid, or both?

4.51   Using the structural formula of the ammonia molecule in which each bond is represented by a line and the unshared pair of outer level electrons on nitrogen is shown by a pair of dots, diagram how the ammonia molecule neutralizes a hydrogen ion.

4.52   Using the periodic table, determine the group numbers of arsenic (As), selenium (Se), and germanium (Ge). On the basis of these locations, write the most likely molecular formulas and the Lewis structures of the compounds of these elements with hydrogen.

4.53   Write the skeletal structures of the following species.
(a) $PbCl_4$ (b) OF_2 (c) NCl_3
(d) NH_3 (e) HSO_4^- (f) HSO_3^-

4.54   How many dots must appear in the structures of Review Exercise 4.53?

4.55   Draw the Lewis structures of the compounds of Review Exercise 4.53.

VSEPR Theory and the Shapes of Molecules

4.56   What, in general terms, is true at the molecular level of life that makes the study of bond angles useful?

4.57 What does the acronym *VSEPR* represent? In general terms, what factor is used in VSEPR theory to explain bond angles?

4.58 Describe the geometric arrangements assumed by the *axes* of the electron clouds of the valence shell of a central atom in a molecule under each circumstance given.
(a) when the atom has four valence-shell electron pairs
(b) when the atom has three valence-shell electron pairs
(c) when the atom has two valence-shell electron pairs

4.59 The central atom, Be, in $BeCl_2$ holds two atoms just like the central atom, O, in H_2O. The bond angles at these two central atoms are different, however.
(a) What are the two bond angles?
(b) Why are these angles different?

Polar Molecules

4.60 What is the underlying cause of the polarity of a covalent bond?

4.61 Atoms X and Y are of elements in the same period of the periodic chart. Atom X has the larger atomic number. Which has the higher electronegativity? Briefly explain.

4.62 Suppose that X and Y form a diatomic molecule $X—Y$, and that X is less electronegative than Y.
(a) Is the $X—Y$ molecule polar?
(b) If so, where are the $\delta+$ and the $\delta-$ charges located?

4.63 Describe the ways in which relative electronegativities change in each circumstance.
(a) within the same family in the periodic table
(b) within the same period in the periodic table
(c) as one moves across the table from left to right

Common Ionic Substances (Interaction 4.1)

4.64 If you see a proper chemical name or a chemical formula on the ingredients list of a supermarket product, what is a good clue as to whether the compound is ionic or not?

4.65 Give the formulas of (a) lye and (b) baking soda.

4.66 Clinical chemists can measure the *potassium level* of the blood. What exactly does this refer to?

4.67 What kinds of compounds are more likely to be electrolytes, ionic or molecular compounds?

Ozone in the Stratosphere (Interaction 4.2)

4.68 Write the equation by means of which the ozone cycle is *initiated*.

4.69 What two equations represent the ozone cycle? Why is the cycle called a chemical *chain reaction?*

4.70 What happens to the UV radiation as a result of the ozone cycle?

4.71 By means of equations, explain how CFC-11 can reduce the stratospheric ozone level.

4.72 What is the polar vortex? The Antarctic ozone hole?

4.73 Polar stratospheric clouds provide reservoirs for the subsequent release of chlorine atoms. How is this done? (Include equations.)

4.74 How does the reappearance of the sun at the onset of the Antarctic spring initiate the decline in ozone levels in the ozone hole?

Additional Exercises

4.75 Why don't nonmetal atoms readily form positively charged ions?

4.76 What are the formulas of the following ions?
(a) oxide ion   (b) silver ion
(c) iodide ion   (d) copper(I) ion
(e) iron(III) ion   (f) sodium ion
(g) iron(II) ion   (h) lead(IV) ion

4.77 What are the names of the following ions?
(a) Br^- (b) Ca^{2+} (c) Ag^+ (d) O^{2-}
(e) K^+ (f) F^- (g) Fe^{3+} (h) Pb^{2+}

4.78 Write the formulas of each of the following compounds.
(a) calcium oxide   (b) iron(III) chloride
(c) copper(II) sulfide   (d) lithium oxide
(e) sodium bromide   (f) magnesium fluoride

4.79 What are the names of the following compounds?
(a) ZnO   (b) Li_2O (c) $FeBr_3$
(d) $MgCl_2$ (e) NaF   (f) $CuBr_2$

4.80 Sulfur and oxygen are in the same family in the periodic table. Which atom has the smaller radius, S or O? Explain.

4.81 Using the atomic numbers only (and the corresponding electron configurations), without consulting the periodic table, write the symbols and the electron configurations of the *ions* of the following elements. (The atomic symbols are hypothetical.)
(a) X, atomic number 20
(b) Y, atomic number 17
(c) Z, atomic number 10

4.82 Write diagrams that show how lithium atoms and oxygen atoms can cooperate to form particles that aggregate in a particular ratio to form an ionic compound. (Follow the directions given in Review Exercise 4.6.)

4.83 An atom of a representative nonmetal X can, in some circumstances, accept two electrons and become an ion. To what group in the periodic table does X most likely belong?

4.84 Study the following electron configuration. Can the atom with this configuration be changed to a reasonably stable ion? If so, what is the charge on the ion?

$$1s^2 2s^2 2p_x^{\,1} 2p_y^{\,1}$$

4.85 Can an atom with the following electron configuration be changed to a stable ion? If it can, what would be the charge on this ion?

$$1s^2 2s^2 2p^6 3s^2 3p^6 3d^{10} 4s^2 4p_x{}^2 4p_y{}^2 4p_z{}^2$$

4.86 In the compound RbH, rubidium hydride, rubidium exists as its normal ion, Rb^+. What is the charge on H in this ionic compound? Write the electron configuration of this ion. In what way does this ion agree with the noble gas rule?

4.87 Tungsten, W, has an oxidation state of 5+ in one of its oxides. Oxygen always has an oxidation number of 2− in oxides, so what is the formula of this tungsten oxide?

*4.88 Iron(III) oxide reacts with hydrogen as follows.

$$3H_2 + Fe_2O_3 \longrightarrow 2Fe + 3H_2O$$

Write the chemical symbols of the following species in this reaction.
(a) the species whose oxidation number becomes less positive
(b) the species whose oxidation number becomes more positive
(c) the reducing agent
(d) the substance reduced
(e) the substance oxidized
(f) the oxidizing agent

4.89 What atomic orbitals partially overlap to form the covalent bond in the chlorine molecule, Cl_2?

4.90 What atomic orbitals partially overlap to form the covalent bond in the hydrogen fluoride molecule?

4.91 The diatomic molecule, He_2, does not exist. Offer an explanation for its failure to form.

4.92 Use Lewis structures to diagram the formation of the following molecules from neutral atoms.
(a) HF   (b) H_2S (c) SiH_4

4.93 The terms *bond length* and *bond distance* are used in connection with the (indirectly) measured distance between two atomic nuclei in a covalent bond. Consider now the distance between two carbon nuclei in ethane and ethylene.

ethane   ethylene

In which molecule is the carbon–carbon bond distance shorter? Explain.

4.94 Write the names of the following ions.
(a) $HPO_4{}^{2-}$ (b) $CrO_4{}^{2-}$ (c) $NH_4{}^+$
(d) $NO_2{}^-$ (e) $PO_4{}^{3-}$ (f) $Cr_2O_7{}^{2-}$

4.95 Write the formulas of the following ions.
(a) hydrogen sulfate ion   (b) dihydrogen phosphate ion

(c) dichromate ion   (d) hydrogen sulfite ion
(e) chromate ion   (f) nitrite ion

4.96 What are the formulas of the following compounds?
(a) ammonium carbonate
(b) sodium monohydrogen phosphate
(c) aluminum hydroxide
(d) sodium hydrogen sulfate
(e) lithium bicarbonate
(f) calcium nitrate
(g) ammonium dihydrogen phosphate
(h) sodium permanganate

4.97 What are the names of the following compounds?
(a) $KHSO_4$ (b) Li_2HPO_4 (c) $Ca(CN)_2$
(d) $Na_2Cr_2O_7$ (e) Na_2SO_3 (f) $BaCrO_4$
(g) $Al_2(SO_4)_3$ (h) $KMnO_4$

4.98 One formula unit of each of the following compounds has how many atoms of all kinds?
(a) $Al_2(CO_3)_3$ (b) $Ca(NO_3)_2$ (c) $(NH_4)_3PO_4$

4.99 The *Merck Index,* an encyclopedia of chemicals, drugs, and biologicals, gives the formula of iron(II) gluconate, a hematinic agent (promotes the formation of red blood cells), as $Fe[HOCH_2(CHOH)_4CO_2]_2$. How many atoms of all kinds are in one formula unit of this compound?

4.100 What is the formula of sulfuric acid? In a solution of this acid in water, what are the names and formulas of the principal ions present?

4.101 An unshared pair of electrons on the oxygen atom of a water molecule can form a coordinate covalent bond to a hydrogen ion. The product is the hydronium ion. Using the structural formula of the water molecule in which each bond is shown by a line and each unshared pair of electrons on the oxygen is given by a pair of dots, diagram how the water molecule forms this new bond to H^+ to give H_3O^+.

*4.102 Draw the Lewis structure of the acetylide ion, $C_2{}^{2-}$.

*4.103 Dinitrogen tetroxide, N_2O_4, is a constituent in smog.
(a) What is its empirical formula?
(b) Propose a Lewis structure for it. (It is a symmetrical molecule with a single bond between the two nitrogen atoms.)

4.104 Write the electron-dot structure of indium. What would be a reasonable electrovalence for it?

4.105 What bond angles would be predicted by VSEPR theory for silane, SiH_4?

4.106 Arrange the atomic symbols, C, F, H, N, and O in the order of the relative electronegativities of the corresponding elements by placing the symbols in the correct order, left to right, with the most electronegative being on the left.

QUANTITATIVE RELATIONSHIPS IN CHEMICAL REACTIONS

5

If you were told that the net mass of the jelly beans in this jar is 500 g, and that the average mass of a jelly bean is, say 1 g, you'd soon realize that the jar holds 500 jelly beans. In other words, you'd have power to *count* jelly beans by weighing them. That's much of what we'll be trying to do as we study atomic and molecular masses in this chapter, namely, to be able to count molecules in samples of substances by weighing the samples.

THIS CHAPTER IN CONTEXT

The atoms of elements are extremely small, and they cannot exist as major fragments of themselves. When atoms combine to make the formula units of compounds they do so as whole atoms. These basic facts about our world dominate experimental work in chemistry and any of its applications, including all aspects of clinical analysis.

The problem is this. If, for example, we want two atoms of hydrogen and one of oxygen to make a water molecule and we don't want any hydrogen or oxygen atoms left over, we have to figure out how to count them. But atoms are too small to count directly. The chief purpose of this chapter is to see how we can "count" atoms and molecules by *weighing* samples of substances. In principle this is as simple as "counting" the pennies in a sack by weighing the pile and dividing by the average mass of a single penny. The same sort of operation is done by chemists to get atoms in desired ratios, and the field is called **stoichiometry**. It deals with the ratios by *atoms* of the elements in compounds and the ratios by *formula units* of the reactants and products in chemical reactions.

■ "Stoy-kee-ah-meh-tree" is obtained from the Greek *stoicheion*, meaning "element," and *-metron*, meaning "measure."

5.1   MOLE CONCEPT

The SI unit for quantity of chemical substance is the *mole*.

The smallest scale on which we could imagine the reaction of hydrogen with oxygen to form water, as we've already indicated, is

$$2 \text{ atoms H} + 1 \text{ atom O} \longrightarrow 1 \text{ molecule H}_2\text{O}$$

If we mistakenly begin with a 3 to 1 ratio of H to O atoms, one H atom will be left over. Thus,

$$3 \text{ atoms H} + 1 \text{ atom O} \longrightarrow 1 \text{ molecule H}_2\text{O} + 1 \text{ H atom (left over)}$$

At a larger scale, say 1 dozen molecules of H_2O, then we need the following reaction, but the *ratio* of H to O is the same, 2 to 1.

$$2 \text{ dozen atoms H} + 1 \text{ dozen atoms O} \longrightarrow 1 \text{ dozen molecules H}_2\text{O}$$

If we want 1 gross (144) molecules of H_2O, then we need

$$2 \text{ gross atoms H} + 1 \text{ gross atoms O} \longrightarrow 1 \text{ gross molecules H}_2\text{O}$$

You can see the point. Regardless of the *scale* of the reaction, whether the target is one, a dozen, or a gross of H_2O molecules, the combining ratio of atoms to make H_2O, with nothing left over, is always 2 H to 1 O. We could say that this is the *stoichiometric law* for water. Even when we correct for the diatomic natures of hydrogen and oxygen, the key ratio is still 2 to 1, only this time 2 *molecules* of H_2 and 1 *molecule* of O_2.

$$2 \text{ molecules H}_2 + 1 \text{ molecule O}_2 \longrightarrow 2 \text{ molecules H}_2\text{O}$$

Alternatively,

$$2H_2 + 1O_2 \longrightarrow 2H_2O$$

In this balanced equation we can count 4 atoms of H plus 2 atoms of O on the left, a ratio of 4:2 or 2:1.

Avogadro's Number Is a Chemical "Counting Unit" The smallness of atoms and formula units necessitates a reference number larger than dozen or gross. Chemists have adopted 6.02×10^{23} for this purpose and call it **Avogadro's number** to honor one of the scientists who pioneered in stoichiometry.

■ Amadeo Avogadro (1776–1856) was an Italian scientist.

$$\text{Avogadro's number} = 6.02 \times 10^{23}$$

Avogadro's number seems unnecessarily awkward, but there is sense to it. It's scaled large enough so that a sample of Avogadro's number of hydrogen atoms, the element with the lightest atoms, will weigh at least 1 g.

No one really comprehends the magnitude of 10^{23}. If you had 10^{23} dollars and distributed dollars equally among the roughly 6 billion or 6×10^9 people of the earth's population, each person could receive over $5000 *per second* for an entire century before the money ran out!

■ Avogadro's number is known to eight significant figures: 6.0221367×10^{23}, but we'll round to 6.02×10^{23}.

Avogadro's Number Provides One Mole When we have a sample of any pure chemical substance that contains Avogadro's number of its formula units, we have the *SI unit for amount of substance* called one **mole** (abbreviated **mol**). The first of two important meanings for *mole* is that it stands for a pure number, namely, Avogadro's number. The following equations for a much larger scale synthesis of H_2O are identical, therefore.

■ *Mol* stands for both the plural and the singular.

$$2 \text{ mol H atoms} \quad + \quad 1 \text{ mol O atoms} \quad \longrightarrow \quad 1 \text{ mol } H_2O \text{ molecules}$$
$$2 \times [6.02 \times 10^{23} \text{ H atoms}] \; + \; 1 \times [6.02 \times 10^{23} \text{ O atoms}] \longrightarrow 1 \times [6.02 \times 10^{23} \; H_2O \text{ molecules}]$$

As we've said, to count (or, at least, to estimate) a large number of small things, like pennies, we weigh the pile then divide its mass by the mass of one penny.

$$\text{grams of pennies} \times \frac{1 \text{ penny}}{\text{grams for one penny}} = \text{number of pennies}$$

The conversion factor in this equation is a connection, like all conversion factors. Here the connection is between grams of pennies and number of pennies. What we now need is the connection—the conversion factor—between the *mass* of a chemical sample and the *number* of formula units present. For this, we must learn more about the concept of atomic mass.

5.2 ATOMIC, FORMULA, AND MOLECULAR MASSES

Atoms of the carbon-12 isotope, are used to define the mole in mass units.

We learned in Section 3.1 that the average mass of the atoms of an element as they occur in their natural abundance is the element's **atomic mass** or **atomic weight** in units of u, the atomic mass unit. The atomic mass of carbon, for example, is 12.011 u, a little over 12 u. Nearly all carbon atoms are of the carbon-12 isotope (six protons and six neutrons per nucleus). A tiny percentage are carbon-13 atoms (six protons and seven neutrons). Any sample of carbon obtained directly from nature has the proportions of carbon-12 and carbon-13 atoms having an *average* mass of 12.011 u.

■ In terms of percentages, naturally occurring carbon is 98.89% carbon-12 and 1.11% carbon-13.

■ $1 \text{ u} = 1.6605665 \times 10^{-24} \text{ g}$

The SI Definition of the *Mole* Is Based on Carbon-12 The SI uses carbon-12 as the basis for *defining* Avogadro's number or one mole.

FIGURE 5.1 Avogadro's number or I mole of atoms is present in each quantity of these elements: I mol of iron (55.8 g; the paper clips), I mol of liquid mercury (200.6 g), I mol of copper (63.5 g; the wire), and I mol of sulfur (32.1 g).

> One **mole** of any element or compound has the identical number of formula units (Avogadro's number) as there are atoms in 0.012 kg or 12 g (exactly) of carbon-12.

Thus if we weigh out exactly 12 g of just the carbon-12 isotope, we have Avogadro's number of atoms (of carbon-12). This explains why Avogadro's number is awkward. If, *purely for our convenience*, we want a sample of carbon-12 atoms to have a mass *numerically equal to something familiar about this isotope*, namely its mass number, we are stuck with whatever number of atoms are in such a sample.

Naturally occurring chlorine (element 17) consists of two isotopes, chlorine-35 (75.53%) and chlorine-37 (24.47%). The average mass of the atoms of these isotopes as they occur in nature is 35.45 u, the atomic mass or atomic weight of chlorine. The average chlorine atom is thus heavier than the carbon-12 atom, 35.45/12 times as heavy, so if we take Avogadro's number or 1 mole of average chlorine atoms the sample has a greater mass, 35.45 g, than that of 1 mole of carbon-12 atoms.

$$1 \text{ mol Cl} = 35.45 \text{ g Cl}$$

Thus we have the second, far more important, meaning of *mole* than just "Avogadro's number." *When we have 1 mol of an element we not only have Avogadro's number of its atoms, we have an amount of the element equaling its atomic mass in grams.* Samples of four elements, each sample having Avogadro's number of atoms or 1 mol, are shown in Figure 5.1. The *masses* of the samples are different, however, because the atomic masses are different.

The table of the elements inside the front cover gives the atomic masses of the elements. The footnotes to this table explain why the numbers of significant figures in atomic masses vary so much among the elements. So before we work an example, we need a policy for rounding atomic masses when they are used in calculations.

> **Policy on Rounding Atomic Masses** Round atomic masses to the first decimal place *before* using them in any calculation. (Round the atomic mass of H, however, to 1.01.)

EXAMPLE 5.1   Relating Masses to Avogadro's Number

How many carbon atoms are in 6.00 g of naturally occurring carbon?

ANALYSIS The given units in this problem, "atoms" and "grams," suggest what we need, a conversion factor that connects them. The connection here is Avogadro's number. *Virtually all problems in chemistry can be analyzed by returning to basic definitions and connections. Whenever you are stuck and don't know what to do, go back to definitions and look for connections among units. These connections are sources of conversion factors.*

We solve this problem by working with the basic meaning of Avogadro's number. Applied to elements, Avogadro's number gives us the number of atoms present in however many grams of the element are equal to its atomic mass. So we first look up the atomic mass of carbon (12.011) and round it by our rule to the first decimal place, 12.0. Now we can write the connecting relationship between mass and atoms of carbon.

$$12.0 \text{ g of C} = 6.02 \times 10^{23} \text{ atoms of C}$$

As we said, whenever we have an equation connecting two quantities, we can devise two conversion factors, in this example, the following.

$$\frac{6.02 \times 10^{23} \text{ atoms C}}{12.0 \text{ g C}} \quad \text{or} \quad \frac{12.0 \text{ g C}}{6.02 \times 10^{23} \text{ atoms C}}$$

SOLUTION   If we multiply what is given, 6.00 g C, by the first conversion factor, the units "g C" will cancel, and our answer will be in the units we want, namely, atoms of carbon.

$$6.00 \text{ g C} \times \frac{6.02 \times 10^{23} \text{ atoms C}}{12.0 \text{ g C}} = 3.01 \times 10^{23} \text{ atoms C}$$

Thus 6.00 g of carbon contains 3.01×10^{23} atoms of carbon.

CHECK   Does the answer make sense? Yes; 6.00 g of C is *less* than a whole mole of C atoms (exactly half as much, in fact). So the answer should be *less* than Avogadro's number (exactly half, in fact), and 3.01×10^{23} is half of 6.02×10^{23}. We will sometimes make "sense checks" like this; it's a habit that you will want to develop not just to avoid embarrassing mistakes but also to further your understanding.

■ PRACTICE EXERCISE 1   How many atoms of gold are in 1.00 oz of gold? Note, 1.00 oz = 28.4 g.

The *Formula Mass* or *Molecular Mass* of a Compound Is the Sum of the Atomic Masses in the Compound's Formula   Because atoms lose no weighable quantity of mass when they combine to form compounds, we can expand the idea of an atomic mass to that of a *formula mass* for every compound. The **formula mass** of a compound is simply the sum of the atomic masses of all of the atoms present in one formula unit, no matter what kind of formula unit it represents. For example, the formula mass of ordinary salt, NaCl, is calculated as follows from its formula. We round atomic masses by our rule. We will also follow a common practice of letting the unit of u be "understood."

1 atom of Na in NaCl gives	23.0
1 atom of Cl in NaCl gives	35.5
Formula mass of NaCl	58.5

Thus the formula mass of NaCl is 58.5, and one formula unit of NaCl has a mass of 58.5 u. This means that

$$1 \text{ mol NaCl} = 58.5 \text{ g NaCl}$$

The idea of a *formula mass* is general; it applies to anything with a definite formula, including elements as well as compounds. The formula of sodium, for example, is Na, so we might just as well say that its formula mass is 23.0 as to say that this is its atomic mass. The formula of chlorine is Cl_2 so its formula mass is twice the atomic mass of Cl, or $2 \times 35.5 = 71.0$. Thus *subscripts in formulas are multipliers of the atomic masses of the atoms to which the subscripts belong*.

Many compounds have parentheses within their formulas because polyatomic ions occur as units. The formula for calcium nitrate, $Ca(NO_3)_2$, for example has a subscript "2" outside the parenthesis. It's a multiplier for everything within the parentheses, so in one formula unit of $Ca(NO_3)_2$ there are 2 N, (3×2) or 6 O, and 1 Ca.

The terms **molecular mass** and **molecular weight**, are preferred by many scientists, particularly for molecular compounds. However, *formula mass* is a more general term because it applies to any substance with a chemical formula, whether the substance consists of atoms, ion groups, or molecules.

Chemists who study substances with unusually large formula masses, like those of proteins and genes, frequently attach the word *daltons* to the formula mass. The **dalton**, abbreviated **D**, is a synonym of *atomic mass unit* but is easier to say. A gene might be described as having a mass of 2.5×10^9 daltons or 2.5×10^9 D, for example.

EXAMPLE 5.2   Calculating a Formula Mass

Some baking powders contain ammonium carbonate, $(NH_4)_2CO_3$. Calculate its formula mass.

ANALYSIS   A formula mass is the sum of the atomic masses of all the atoms expressed in the formula. So we must look up and write down the atomic masses of all of the elements present, rounding each to the first decimal place (except H is 1.01).

$$N, 14.0 \qquad H, 1.01 \qquad C, 12.0 \qquad O, 16.0$$

Each atomic mass must be multiplied by the number of times the atom appears in the formula, and then the resulting numbers are added.

SOLUTION   In each formula unit of $(NH_4)_2CO_3$, N occurs 2 times, H occurs 8 times, C occurs 1 time, and O occurs 3 times. Therefore,

2 N	+8 H	+1 C	+3 O	$= (NH_4)_2CO_3$
2×14.0	$+8 \times 1.01$	$+1 \times 12.0$	$+3 \times 16.0$	$= 96.1$ (correctly rounded)

The formula mass of ammonium carbonate is 96.1. This means that one formula unit has a mass of 96.1 u, and it means that Avogadro's number of these units has a total mass of 96.1 g. It also means that

$$1 \text{ mol } (NH_4)_2CO_3 = 96.1 \text{ g } (NH_4)_2CO_3$$

■   PRACTICE EXERCISE 2   Calculate the formula masses of the following compounds.

(a) $C_9H_8O_4$ (aspirin)   (b) $Mg(OH)_2$ (milk of magnesia)

(c) $Fe_4[Fe(CN)_6]_3$ (ferric ferrocyanide or Prussian blue, an ink pigment)

Think of the Mole as the Lab-Sized Unit for Amount of Chemical Substance   The mole is a quantity of a substance large enough to be manipulated experimentally, and it can be taken in fractions or in multiples. For example, the formula mass of H_2O is 18.0, so 1 mol of H_2O has a mass of 18.0 g. If we wished, we could weigh out a smaller sample, say 1.80 g, and then we would have 0.100 mol of water, because 1.80 is one-tenth of 18.0. Or we could take 36.0 g of H_2O, and then have 2.00 mol, because 36.0 is 2 times 18.0. Figure 5.2 pictures 1-mole samples of four compounds. Each sample has Avogadro's number of formula units, but the masses of the samples vary according to their formula masses.

One of the SI prefixes is often used for the mole, the prefix *milli-* signifying one-thousandth. The abbreviation of *millimole* is *mmol*.

$$1 \text{ mmol} = 0.001 \text{ mol}$$

$$1000 \text{ mmol} = 1 \text{ mol}$$

Two kinds of calculations are particularly helpful in fixing the mole concept in mind. They are also the two most often used calculations in experimental work—calculating grams from moles and moles from grams.

■ The moles-to-millimoles conversion factors are $\dfrac{1000 \text{ mmol}}{1 \text{ mol}}$ and $\dfrac{1 \text{ mol}}{1000 \text{ mmol}}$.

EXAMPLE 5.3   Converting Moles to Grams

About 21% of the air we breathe is oxygen, O_2. How many grams of O_2 are in 0.250 mol of O_2?

ANALYSIS   The problem comes down to the following question.

$$0.250 \text{ mol } O_2 = \ ? \text{ g } O_2$$

What *always* connects grams to moles is a formula mass. So we first must calculate the formula mass of O_2. It is twice the atomic mass of O, 16.0, or $2 \times 16.0 = 32.0$. This tells us that

$$1 \text{ mol } O_2 = 32.0 \text{ g } O_2$$

This equation connects moles and grams of O_2 and so gives us the following conversion factors.

$$\frac{1 \text{ mol } O_2}{32.0 \text{ g } O_2} \quad \text{or} \quad \frac{32.0 \text{ g } O_2}{1 \text{ mol } O_2}$$

If we now multiply what is given, 0.250 mol of O_2, by whichever conversion factor lets us cancel "mol O_2," we'll have the answer in the desired final unit, "g O_2." The second conversion factor is what we need.

SOLUTION

$$0.250 \text{ mol } O_2 \times \frac{32.0 \text{ g } O_2}{1 \text{ mol } O_2} = 8.00 \text{ g } O_2$$

Thus 0.250 mol of O_2 has a mass of 8.00 g of O_2.

CHECK   Does the size of the answer, 8.00 g O_2, make sense? Of course. A whole mole of O_2 has a mass of 32.0 g, so a quarter of a mole of O_2 is one-fourth of 32.0 g, or 8.00 g.

FIGURE 5.2   Avogadro's number or 1 mole of formula units is present in each of these samples of compounds: white sodium chloride (1 mol NaCl = 58.5 g), blue copper sulfate pentahydrate (1 mol $CuSO_4 \cdot 5H_2O$ = 249.7 g), yellow sodium chromate (1 mol Na_2CrO_4 = 162.0 g), and liquid water (1 mol H_2O = 18.0 g).

■ PRACTICE EXERCISE 3   An experiment calls for 24.0 mol of NH_3. How many grams is this?

The next worked example shows how to use a formula mass to convert grams to moles.

EXAMPLE 5.4   Converting Grams to Moles

A student was asked to use 12.5 g of NaCl in an experiment. How many moles of NaCl are in 12.5 g NaCl?

ANALYSIS   Here the question is

$$12.5 \text{ g NaCl} = ? \text{ mol NaCl}$$

The connection between grams and moles of NaCl is given by the formula mass of NaCl, which we already found to be 58.5. This tells us that

$$58.5 \text{ g NaCl} = 1 \text{ mol NaCl}$$

Therefore we have these two possible conversion factors.

$$\frac{58.5 \text{ g NaCl}}{1 \text{ mol NaCl}} \quad \text{or} \quad \frac{1 \text{ mol NaCl}}{58.5 \text{ g NaCl}}$$

If we multiply the given, 12.5 g NaCl, by the second ratio, the units will cancel properly and we'll have the number of moles of NaCl in 12.5 g NaCl.

SOLUTION

$$12.5 \text{ g NaCl} \times \frac{1 \text{ mol NaCl}}{58.5 \text{ g NaCl}} = 0.214 \text{ mol of NaCl (from 0.2136752137)}$$

Thus 12.5 g of NaCl consists of 0.214 mol of NaCl.

CHECK   Is the answer reasonable? Yes. The sample is about two-tenths of one mole (58.5 g NaCl), so the amount must certainly be *less*, not more, than 58.5 g NaCl.

■   PRACTICE EXERCISE 4   A student was asked to prepare 6.84 g of aspirin, $C_9H_8O_4$. How many moles is this? How many millimoles?

5.3   BALANCED CHEMICAL EQUATIONS AND STOICHIOMETRY

The coefficients of a balanced equation give the ratios by moles in which the substances react.

One of the most important properties of substances involves their mole relationships when they react. We have just learned that atoms combine only in definite *ratios* by atoms or by moles when they form compounds. Likewise, compounds react only in definite *ratios* by formula units or moles. The coefficients of the reaction's balanced equation give us such ratios. This is the major reason that balanced equations are so important in experimental work in chemistry.

Balanced Equations Use Formulas and Coefficients to Tell What Reacts, What Forms, *and in What Proportions by Moles*   We've already learned to read a balanced equation. We'll now carry this another step forward and learn how to write and balance equations when we are given the formulas of the reactants and products.

Recall that a chemical equation is a **balanced equation** when all the atoms given among the reactants appear in identical numbers among the products. In most equations one or more of the formulas is multiplied by some whole number in order to show the correct balance. These multipliers of formulas in chemical equations are called **coefficients**. For example, the formation of water from hydrogen and oxygen has the following equation (in which we now use the formulas H_2 and O_2).

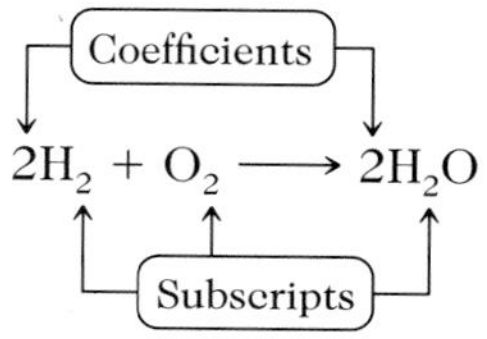

$$2H_2 + O_2 \longrightarrow 2H_2O$$

Because of the coefficient of 2 in $2H_2O$, there are $2 \times 2 = 4$ atoms of H on the right side of the equation, and $2 \times 1 = 2$ atoms of O. These figures match the 4 H atoms and 2 O atoms on the left. The equation is thus balanced.

We are never allowed to change subscripts, once we have the right formulas, just to get an equation to balance. For example, changing H_2O to H_2O_2 makes a change from the formula for water to the formula for hydrogen peroxide, an entirely different substance. We can adjust coefficients, however, to balance an equation, as we will see in the next worked example.

■ H_2O_2 cannot be made by a direct combination of H_2 and O_2.

EXAMPLE 5.5 Balancing a Chemical Equation

Sodium, Na, reacts with chlorine, Cl_2, to give sodium chloride, NaCl. Write the balanced equation for this reaction.

ANALYSIS The first step is to set down all of the correct formulas in the format of an equation. *Never worry about the coefficients until the correct formulas are down.* Then, *never change the formulas.*

SOLUTION After the first step, we have

$$Na + Cl_2 \longrightarrow NaCl \quad \text{(unbalanced)}$$

So far, we see two chlorine atoms on the left (in Cl_2) but only one on the right. We can't fix this by writing $NaCl_2$, because this isn't the correct formula for sodium chloride. The only way we are allowed to get two Cl atoms on the right is to put a coefficient of 2 in front of NaCl.

$$Na + Cl_2 \longrightarrow 2NaCl \quad \text{(unbalanced)}$$

Of course, writing 2 in front of NaCl makes it a multiplier for both Na and Cl, so now we have two Na atoms on the right and just one on the left. To fix this, we write a 2 before the Na.

$$2Na + Cl_2 \longrightarrow 2NaCl \quad \text{(balanced)}$$

Notice particularly how we used a subscript, the 2 in Cl_2, to suggest a coefficient for another formula on the other side of the arrow. This is a standard strategy in balancing equations.

■ $NaCl_2$ doesn't even exist.

EXAMPLE 5.6   Balancing a Chemical Equation

Iron, Fe, can be made to react with oxygen, O_2, to form an oxide with the formula Fe_2O_3. Write the balanced equation for this reaction.

ANALYSIS   We first write down the correct formulas in the format of an equation. Then we use the subscripts to suggest coefficients.

SOLUTION   After the first step, the unbalanced equation is

$$Fe + O_2 \longrightarrow Fe_2O_3 \quad \text{(unbalanced)}$$

Next we exploit subscripts to suggest coefficients. Oxygen has a subscript of 2 in O_2 and a subscript of 3 in Fe_2O_3. To get a balance, we use the 3 as a coefficient for O_2 and the 2 as a coefficient for Fe_2O_3. This is cross-switching the numbers.

$$Fe + 3O_2 \longrightarrow 2Fe_2O_3 \quad \text{(unbalanced)}$$

Now there are six oxygen atoms on the left (in $3O_2$) and six on the right (in $2Fe_2O_3$). Of course, the coefficient of 2 in the formula on the right also means that there are 4 Fe atoms on the right. To fix this, we simply use a coefficient of 4 on the left, for Fe.

$$4Fe + 3O_2 \longrightarrow 2Fe_2O_3 \quad \text{(balanced)}$$

■   PRACTICE EXERCISE 5   In the presence of an electrical discharge like lightning, oxygen, O_2, can be changed into ozone, O_3. Write the balanced equation for this reaction.

■   PRACTICE EXERCISE 6   Aluminum, Al, reacts with oxygen to give aluminum oxide, Al_2O_3. Write the balanced equation for this change.

■   (g) gas

(l) liquid

(s) solid

(aq) aqueous solution

(solution in water)

Physical States Are Often Included in Equations   Following the formula of a substance in an equation, a chemist often adds a symbol in parentheses to specify whether the substance is in the gaseous, liquid, or solid state or is in solution in water. The symbols are shown in the margin. Thus, the equation for the reaction of iron with oxygen (Example 5.6) could be written as follows.

$$4Fe(s) + 3O_2(g) \longrightarrow 2Fe_2O_3(s) \quad \text{(balanced)}$$

The practice is not always used, but we get more chemical information when it is.

Sometimes, when we adjust coefficients to balance an equation, we get an equation with coefficients all divisible by the same whole number. Suppose, for example, that we had obtained the following as we tried to balance the equation for the reaction of sodium with chlorine in Example 5.5.

$$4Na + 2Cl_2 \longrightarrow 4NaCl$$

The equation is surely balanced and all of its formulas are correct, so there is nothing basically wrong with it. Chemists, however, generally (but not always) write balanced equations using the set of *smallest* whole numbers as coefficients. We will normally follow this rule. In the equation above, the coefficients are all divisible by 2.

When the formulas in an equation include groups of atoms inside parentheses, and when it is obvious that the groups do not themselves change, treat the groups as whole units in balancing equations.

EXAMPLE 5.7 **Balancing Equations Involving Polyatomic Groups**

When water solutions of $(NH_4)_2SO_4(aq)$ and $Pb(NO_3)_2(aq)$ are mixed, a white solid separates that has the formula $PbSO_4(s)$. The other product is $NH_4NO_3(aq)$, but it remains dissolved as indicated by the (aq). Represent this reaction by a balanced equation, showing the physical states.

ANALYSIS As usual, we start by simply writing the correct formulas in the format of an equation.

$$(NH_4)_2SO_4(aq) + Pb(NO_3)_2(aq) \longrightarrow PbSO_4(s) + NH_4NO_3(aq)$$

Because polyatomic groups are involved, we next examine the formulas to see whether any such groups change or if they all appear to react as whole units. We can see here that they do remain as intact units. The subscript of 2 in $(NH_4)_2SO_4$ suggests that we use 2 as the coefficient in the formula on the right where NH_4 occurs.

SOLUTION Placement of 2 as a coefficient for $NH_4NO_3(aq)$ gives

$$(NH_4)_2SO_4(aq) + Pb(NO_3)_2(aq) \longrightarrow PbSO_4(s) + 2NH_4NO_3(aq)$$

This automatically brought into balance the units of NO_3 on each side of the arrow. The equation is now balanced.

■ PRACTICE EXERCISE 7 Balance each of the following equations.

(a) $Ca + O_2 \rightarrow CaO$

(b) $KOH + H_2SO_4 \rightarrow H_2O + K_2SO_4$

(c) $Cu(NO_3)_2 + Na_2S \rightarrow CuS + NaNO_3$

(d) $AgNO_3 + CaCl_2 \rightarrow AgCl + Ca(NO_3)_2$

(e) $Al + H_2SO_4 \rightarrow Al_2(SO_4)_3 + H_2$

(f) $CH_4 + O_2 \rightarrow H_2O + CO_2$

An Equation's Coefficients Provide Connections among the Mole Quantities of Reactants and Products With the concept of a mole, we can now think about the coefficients in a balanced equation at two levels at the same time. For example, in the following equation for the reaction of Fe with S, notice that each formula has a coefficient of 1. Beneath each formula in the equation we can see various ways of interpreting these coefficients.

Fe	+ S	$\longrightarrow$ FeS
1 atom of Fe	+ 1 atom of S	$\longrightarrow$ 1 formula unit of FeS
1 dozen atoms of Fe	+ 1 dozen atoms of S	$\longrightarrow$ 1 dozen formula units of FeS
6.02×10^{23} atoms Fe	+ 6.02×10^{23} atoms S	$\longrightarrow 6.02 \times 10^{23}$ formula units of FeS
1 mol of Fe	+ 1 mol of S	$\longrightarrow$ 1 mol of FeS

Notice that the proportions all remain the same, provided we work with *formula units* of one kind or another. The most important relationship in all of stoichiometry is the following.

> Equal numbers of moles contain equal numbers of formula units.

All that changes as we move through the equations for the formation of FeS is the *scale* of the reaction, namely, the actual numbers of formula units used, not their proportions in relationship to each other. *The coefficients in a balanced equation give us the proportions of substances in moles*, not the scale of a particular experiment. Thus, using an earlier equation,

$$2Na + Cl_2 \longrightarrow 2NaCl$$

we can now interpret it to mean that for every *2 mol* of Na that reacts, 1 *mol* of Cl_2 also reacts and *2 mol* of NaCl forms.

One kind of stoichiometric calculation is to use an equation's coefficients to find how many moles of one substance must be present if a certain number of moles of another are involved. *An equation's coefficients are the connection between moles of one substance and moles of another*. Each connection provides two conversion factors, as we'll see. But, first, we must introduce a substitute for the equals sign when we want to specify a moles-to-moles connection between two *different* substances. We let $\Leftrightarrow$ stand for "is chemically equivalent to." You'll see how this symbol is used in the next example.

EXAMPLE 5.8   Using the Mole Concept with Equations

How many moles of oxygen are needed to combine with 0.500 mol of hydrogen in the reaction that produces water by the following equation?

$$2H_2 + O_2 \longrightarrow 2H_2O$$

ANALYSIS   We need the connection between moles of H_2 (the given) and moles of O_2. The coefficients tell us that 2 mol of H_2 combines with 1 mol of O_2 in this particular reaction. They tell us that in this reaction 2 mol of H_2 *is chemically equivalent to* 1 mol of O_2. So we can symbolize this connection as follows.

$$2 \text{ mol } H_2 \Leftrightarrow 1 \text{ mol } O_2$$

It would not be right to say that "2 mol of H_2 *is equal to* 1 mol of O_2" and so write 2 mol H_2 = 1 mol O_2. Hydrogen and oxygen are not "equal." Their ability to react in a 2 to 1 ratio, however, tells us that 2 mol of H_2 *requires* exactly 1 mol of O_2, so *for this reaction only* we know that 1 mol of O_2 is the chemical equivalent of 2 mol of H_2. A connection, of course, is a connection, whether it is symbolized by = or $\Leftrightarrow$, and we can use it to construct conversion factors. The connection we just made, 2 mol $H_2 \Leftrightarrow 1$ mol O_2, lets us construct and select between the following conversion factors.

$$\frac{2 \text{ mol } H_2}{1 \text{ mol } O_2} \quad \text{or} \quad \frac{1 \text{ mol } O_2}{2 \text{ mol } H_2}$$

We have to choose one of these ratios to multiply by the given quantity, 0.500 mol of H_2, to find out how much O_2 is needed.

SOLUTION   The correct conversion factor is the second.

$$0.500 \text{ mol } H_2 \times \frac{1 \text{ mol } O_2}{2 \text{ mol } H_2} = 0.250 \text{ mol } O_2$$

In other words, 0.500 mol of H_2 requires 0.250 mol of O_2 for this reaction.

CHECK Is the size of the answer sensible? Yes. We can see by the balanced equation that half as many moles of O_2 are needed for the given moles of H_2, and half of 0.500 mol is 0.250 mol.

■ PRACTICE EXERCISE 8   How many moles of H_2O are made from the 0.250 mol of O_2 in Example 5.8? How many millimoles of H_2O are thus made?

■ PRACTICE EXERCISE 9   Nitrogen and oxygen combine at high temperature in an automobile engine to produce nitrogen monoxide, NO, an air pollutant. The equation is $N_2 + O_2 \rightarrow 2NO$. To make 8.40 mol of NO, how many moles of N_2 are needed? How many moles of O_2 are also needed?

■ PRACTICE EXERCISE 10   Ammonia, an important nitrogen fertilizer, is made by the following reaction: $3H_2 + N_2 \rightarrow 2NH_3$. In order to make 300 mol of NH_3, how many moles of H_2 and how many moles of N_2 are needed?

With the ability to make the mole calculations involving a balanced equation, we move to study a very common problem that arises in the laboratory—how many grams of one substance are needed to make a given mass of another according to some equation? The next worked example illustrates how this is handled.

EXAMPLE 5.9   Mole Calculations Using Balanced Equations

How many grams of aluminum are needed to make 24.4 g of Al_2O_3 by the following equation?

$$4Al + 3O_2 \longrightarrow 2Al_2O_3$$

ANALYSIS *All problems of this nature in stoichiometry must first be worked at the mole level, because the coefficients refer to moles, not masses.* Thus we must first find out how many *moles* are in 24.4 g of AL_2O_3. Then we use the connection, given by the equation's coefficients,

$$2 \text{ mol } Al_2O_3 \Longleftrightarrow 4 \text{ mol } Al$$

to calculate how many *moles* of Al are chemically equivalent to this much Al_2O_3 *for this reaction.* When we know the number of moles of Al, we use the connection,

$$1 \text{ mol } Al = 27.0 \text{ g } Al$$

to calculate the number of grams of Al needed. Figure 5.3 outlines the calculation "flow." In short, our calculation "trip" is

$$24.4 \text{ g of } Al_2O_3 \longrightarrow \text{? mol of } Al_2O_3 \longrightarrow \text{? mol of } Al \longrightarrow \text{? g of } Al$$

Knowing that we'll be needing formula masses, it's usually a good idea at the start of a problem such as this to compute any needed formula masses. The atomic mass of Al is 27.0; the formula mass of Al_2O_3 is 102.0.

■ Aluminum oxide, Al_2O_3, is sometimes used as a white filler for paints.

■ The *equals* sign in 1 mol Al = 27.0 g Al is suitable because both sides refer to a quantity of the identical substance but in different units. It's like writing, 1 in. = 2.54 cm; both sides refer to the identical length but in different units.

■ 2Al: 2 × 27.0 = 54.0
   3O: 3 × 16.0 = 48.0
Formula mass Al_2O_3 = 102.0

SOLUTION   To determine the number of moles of Al_2O_3 in 24.4 g Al_2O_3, we devise a conversion factor based on the connection between moles and mass: 1 mol Al_2O_3 = 102.0 g Al_2O_3. Then we carry out the following calculation.

$$24.4 \text{ g } Al_2O_3 \times \frac{1 \text{ mol } Al_2O_3}{102.0 \text{ g } Al_2O_3} = 0.239 \text{ mol of } Al_2O_3$$

Now we can use the connection, given by the equation's coefficients, between the number of moles of Al_2O_3 and the number of moles of Al, 2 mol Al_2O_3 ⟺ 4 mol Al, to calculate how many moles of Al are needed. The connection gives us the following conversion factors.

$$\frac{4 \text{ mol Al}}{2 \text{ mol } Al_2O_3} \qquad \text{or} \qquad \frac{2 \text{ mol } Al_2O_3}{4 \text{ mol Al}}$$

So we multiply 0.239 mol of Al_2O_3 by the first factor.

$$0.239 \text{ mol } Al_2O_3 \times \frac{4 \text{ mol Al}}{2 \text{ mol } Al_2O_3} = 0.478 \text{ mol Al}$$

The problem calls for the answer in grams of Al, not moles of Al, so we next have to convert 0.478 mol of Al into grams of Al. We use the formula mass of Al, 1 mol Al = 27.0 g Al, to devise the correct conversion factor.

$$0.478 \text{ mol Al} \times \frac{27.0 \text{ g Al}}{1 \text{ mol Al}} = 12.9 \text{ g Al}$$

This is the answer; it takes 12.9 g of Al to prepare 24.4 g of Al_2O_3 according to the given equation.

CHECK   It's harder to do a "head check" of the sense of this answer, but we can ask if the *size* of the answer makes sense. The answer we found is close to *half* a mole of Al. The equation's coefficients tell us that half again of this amount of Al is chemically equivalent to Al_2O_3, which has a formula mass of 102.0. In other words, half a mole of Al is chemically equivalent to a quarter mole of Al_2O_3. Is this what we began with? Yes, a quarter of a mole of Al_2O_3 is about 25 g, close to the 24.4 g of Al_2O_3 actually given.

FIGURE 5.3   All calculations involving masses of reactants and products that participate in a chemical reaction must be worked out at the mole level. There is no one-step route from grams of one substance to grams of another, although one can always use a string of conversion factors.

■   PRACTICE EXERCISE 11   How many grams of oxygen are needed for the experiment described in Example 5.9? Use a diagram of the solution in the style of Figure 5.3 as you work out the answer.

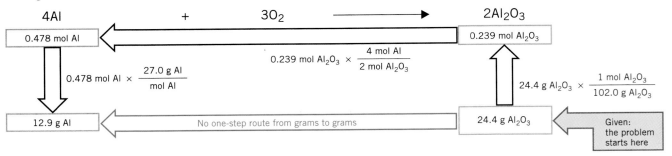

■ **PRACTICE EXERCISE 12**   If 28.4 g of Cl_2 are used up in the following reaction, how many grams of Na are also used up, and how many grams of NaCl form?

$$2Na + Cl_2 \longrightarrow 2NaCl$$

When the Ideal Stoichiometric Ratio of Reactants Is Not Used, One Reactant *Limits* How Much of the Products Can Be Made   Exact stoichiometric ratios by moles of the reactants, the ratios given by the coefficients, are seldom supplied among the reactions occurring in nature and in our cells. A living cell, for example, may have to operate in a short supply of oxygen, or its protein-making machinery shuts down because one "building block" molecule (an amino acid) is unavailable when needed.

To illustrate what "short supply" means with a simple example, consider again the reaction in Example 5.9 and in Practice Exercise 11.

$$4Al + 3O_2 \longrightarrow 2Al_2O_3$$

The ideal stoichiometric ratio of the reactants is 4 mol Al to 3 mol O_2, because 4 and 3 are the respective coefficients. But suppose that in the lab we actually measured out the reactants in a ratio of only 2 mol Al to 3 mol O_2. What then? Now we do not have enough Al for all of the O_2 supplied, so the O_2 taken cannot all be used up. Some O_2 will remain left over from the reaction. (How many moles of O_2 will this leftover amount be?) An excess of O_2 is present, so all of the Al will be used up.

When the ideal mole ratio of reactants is not taken, the reactant that can be entirely used up is called the **limiting reactant**. It's said to be "limiting" because it is this reactant that limits how many moles of product can form. When Al is the limiting reactant, and we start with 2.0 mol of it, by the equation's coefficients we can only obtain 1.0 mol of Al_2O_3, because

$$2.0 \text{ mol Al} \times \frac{2 \text{ mol Al}_2O_3}{4 \text{ mol Al}} = 1.0 \text{ mol Al}_2O_3$$

The concept of a limiting reactant is important in all areas at the molecular level of life. A short supply of vitamins or amino acids, for example, can have injurious or fatal consequences.

5.4   REACTIONS IN SOLUTION

Virtually all of the chemical reactions studied in the lab and that occur in living systems take place in aqueous solutions.

If the particles of one substance are to react with those of another, they must have enough freedom to move about to find each other. Such freedom exists in the gaseous and liquid states but not in the solid state. To get one solid to react with another, chemists usually dissolve them in something. This puts them into a liquid state, and their particles can move about. In order to learn in the next section about mass relationships when reactants are in solution, we must first learn the common terms used to describe solutions.

A Solution Is Made of a Solvent and One or More Solutes   A **solution** is a uniform mixture of particles that are exceedingly small, at the atomic size or somewhat larger depending on the formula units involved. A minimum of two substances is needed to have a solution. One is called the *solvent* and all of the others are called the *solutes*.

■ Both solids and gases can function as solvents. Air is a solution chiefly of oxygen and nitrogen. An *alloy* is a hardened solution made by stirring together two or more molten metals.

The **solvent** is the medium into which the other substances are mixed or dissolved. The solvent is usually a liquid, like water. Unless we state otherwise, we will always be dealing with aqueous solutions, meaning that water is the solvent.

A **solute** is anything that is dissolved by the solvent. In an aqueous solution of sugar, the solute is sugar and the solvent is water. The solute can be a gas. Club soda is a solution of carbon dioxide in water. The solute can be a liquid. Antifreeze, for example, is mostly an aqueous solution of the liquid ethylene glycol.

Solutions Can Be Dilute or Concentrated   Several terms are used to describe a solution. A **dilute solution** is one in which the ratio of solute to solvent is very small, such as a few crystals of sugar dissolved in a glass of water. Most of the aqueous solutions in living systems have more than two solutes and are dilute in all of them. In a **concentrated solution**, the ratio of solute to solvent is large. Syrup, for example, is a concentrated solution of sugar in water.

Solutions Can Be Unsaturated, Saturated, or Supersaturated   Some solutions are **saturated solutions**, which means that it isn't possible to dissolve more of the solute in them (assuming that the temperature of the solution is kept constant). If more solute is added to a solution already saturated in this solute, the extra solute will just remain separate. If the solute is a solid, it will generally sink to the bottom and lie there.

An **unsaturated solution** is one in which the ratio of solute to solvent is lower than that of the corresponding saturated solution. If more solute is added to an unsaturated solution, at least some of it will dissolve.

It isn't easy, but sometimes a **supersaturated solution** can be made. This is an unstable system in which the ratio of dissolved solute to solvent is actually higher than that of a saturated solution. We can sometimes make a supersaturated solution by carefully cooling a saturated solution. The ability of most solutes to dissolve in water decreases with temperature, so when a saturated solution is cooled some of the now excess solute should separate. But this doesn't always happen. If the excess solute does not separate, then we have a supersaturated solution. If we now scratch the inner wall of the container with a glass rod, or if we add a crystal—a "seed" crystal—of the pure solute to the system, the excess solute will usually separate immediately. This event can be dramatic and pretty to watch (see Figure 5.4). The separation of a solid from a solution is called **precipitation**, and the solid is referred to as the **precipitate**.

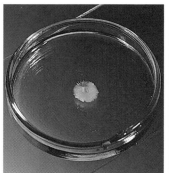   

FIGURE 5.4 Supersaturation. A seed crystal has been added to a supersaturated solution, left photo, and whatever solute was present in solution in excess quickly separates. Any solution that still remains in contact with the crystals (right photo) is now a saturated solution.

Solubility Is Sometimes Reported as Grams of Solute per 100 g of Solvent The amount of solute needed to give a saturated solution in a given quantity of solvent at a specific temperature is called the **solubility** of the solute in the given solvent. It is the maximum quantity of the solute that can dissolve and form a stable solution at the given temperature in the given quantity of solvent. If you take more than this maximum by even the smallest quantity, like a single crystal of a solid, the extra will simply remain undissolved in the system. The presence of an extra amount of solute, in fact, is a sign that the solution is saturated (provided enough time has been allowed for the solution to form).

Solubilities vary widely from substance to substance, as the data in Table 5.1 show. Notice particularly that a saturated solution can still be quite dilute. For example, only a very small mass of barium sulfate can dissolve in 100 g of water.

The solubilities of most solids increase with temperature, as you can see in Table 5.1. All gases become less and less soluble in water as the temperature increases, assuming that the measurements are made under the same pressure.

5.5 MOLAR CONCENTRATION

The unit of *moles per liter* is the most useful unit of concentration when working with the stoichiometry of reactions in solution.

The **concentration** of a solution is the ratio of the quantity of solute to some given unit of the solution. The units can be anything we wish, but for the stoichiometry of reactions in solution, the best units are those of moles of solute per liter of solution. This ratio is called the solution's **molar concentration**, or **molarity**, abbreviated M. The molar concentration of a solution, its molarity M, is the number of moles of solute per liter of solution.

$$ M = \frac{\text{mol solute}}{\text{L solution}} = \frac{\text{mol solute}}{1000 \text{ mL solution}} $$

■ M = moles/liter

mol = moles

TABLE 5.1 Solubilities of Some Substances in Water

Solute	Solubilities (g/100 g water)			
	0 °C	20 °C	50 °C	100 °C
Solids				
Sodium chloride, $NaCl$	35.7	36.0	37.0	39.8
Sodium hydroxide, $NaOH$	42	109	145	347
Barium sulfate, $BaSO_4$	0.000115	0.00024	0.00034	0.00041
Calcium hydroxide, $Ca(OH)_2$	0.185	0.165	0.128	0.077
Gases				
Oxygen, O_2	0.0069	0.0043	0.0027	0
Carbon dioxide, CO_2	0.335	0.169	0.076	0
Nitrogen, N_2	0.0029	0.0019	0.0012	0
Sulfur dioxide, SO_2	89.9	51.8	4.3	1.8[a]
Ammonia, NH_3			28.4	7.4[b]

[a] At 90 °C.
[b] At 96 °C.

A bottle might, for example, have the label "0.10 M NaCl." If so, we know that the solution in this container has a concentration of 0.10 mol of NaCl per liter of solution (or per 1000 mL of solution). There might be a small or large amount of solution in the bottle, but in any case the *ratio* is the same, 0.10 mol NaCl per liter of solution. A value of molarity always gives two conversion factors. In our example, they are

$$\frac{0.10 \text{ mol NaCl}}{1000 \text{ mL NaCl solution}} \quad \text{and} \quad \frac{1000 \text{ mL NaCl solution}}{0.10 \text{ mol NaCl}}$$

A Volumetric Flask Is Used to Make a Solution of Known Molarity Figure 5.5 shows how to make a solution having a known molarity. The mass of solute corresponding to the moles of solute we want is weighed out. The sample is then placed in a *volumetric flask*, a special piece of glassware pictured in Figure 5.5. These flasks are available in several fixed capacities, so the one selected must allow a final volume of solution that gives the solute the molar concentration we want. The solvent is added until the solute all dissolves and the liquid level exactly reaches the etched mark on the flask, as described in the figure legend.

The concept of molarity will become clearer as we study how to do some of the calculations associated with it. In the next worked example we'll see what kinds of calculations have to be done in order to go into the lab and prepare a certain volume of a solution that has a given molar concentration.

■ Volumetric flasks as large as 5 L can be purchased as well as many of smaller sizes down to 1 mL.

FIGURE 5.5 The preparation of a solution of known molarity. The volumetric flask has an etched line on its neck that marks the liquid level at which the flask will hold the specified volume. (*a*) The solute, accurately weighed, has been placed in the flask. (*b*) Some water (distilled or deionized) is added. (*c*) The flask is agitated so that the solute dissolves. (*d*) Enough water is added to bring the level to the etched line. (*e*) After the flask is stoppered, it is shaken so that the solution will be uniform.

EXAMPLE 5.10 Preparing a Solution of Known Molar Concentration

How many grams of sodium bicarbonate, $NaHCO_3$, are needed to prepare 500 mL of 0.125 M $NaHCO_3$?

ANALYSIS The desired concentration, 0.125 M, indirectly refers to *moles*, but the question asks for the answer in grams. Before we can calculate the grams needed, we have to find out how many moles of $NaHCO_3$ are required to prepare this solution. The given concentration, 0.125 M, provides the conversion factor.

(*a*)

(*b*)

(*c*)

(*d*)

(*e*)

$$\frac{0.125 \text{ mol NaHCO}_3}{1000 \text{ mL NaHCO}_3 \text{ solution}} \quad \text{or} \quad \frac{1000 \text{ mL NaHCO}_3 \text{ solution}}{0.125 \text{ mol NaHCO}_3}$$

If we next multiply the given volume, 500 mL of $NaHCO_3$ solution, by the first conversion factor, the units will correctly cancel.

SOLUTION

$$500 \text{ mL NaHCO}_3 \text{ solution} \times \frac{0.125 \text{ mol NaHCO}_3}{1000 \text{ mL NaHCO}_3 \text{ solution}} = 0.0625 \text{ mol NaHCO}_3$$

In other words, the 500 mL of solution has to contain 0.0625 mol of $NaHCO_3$. Now we make the connection between moles of $NaHCO_3$ to what we can measure, namely, grams of $NaHCO_3$. The moles-to-grams connection is always a formula mass. Here, we need one of the conversion factors that the formula mass of $NaHCO_3$, 84.0, makes available.

$$\frac{84.0 \text{ g NaHCO}_3}{1 \text{ mol NaHCO}_3} \quad \text{or} \quad \frac{1 \text{ mol NaHCO}_3}{84.0 \text{ g NaHCO}_3}$$

If we multiply 0.0625 mol of $NaHCO_3$ by the first of these, the units will work out right. (Draw in the cancel lines yourself.)

$$0.0625 \text{ mol NaHCO}_3 \times \frac{84.0 \text{ g NaHCO}_3}{1 \text{ mol NaHCO}_3} = 5.25 \text{ g NaHCO}_3$$

Thus to prepare 500 mL of 0.125 *M* $NaHCO_3$, we have to weigh out 5.25 g of $NaHCO_3$, dissolve it in some water in a 500 mL volumetric flask, and then carefully add water until its level reaches the mark, making sure that the contents become well mixed.

CHECK Is the *size* of the answer, 5.25 g $NaHCO_3$, reasonable? Yes. The molarity is 0.125 *M* $NaHCO_3$, and the formula mass of $NaHCO_3$, is 84 g. If the molarity were 0.100 *M* instead of 0.125 *M*, then we'd need 8.4 g of solute for a whole liter, and 4.2 g for the specified half a liter. But the molarity is larger than 0.100 *M*, so we need a somewhat larger mass of $NaHCO_3$ than 4.2 g, which is what the answer is. (By checking the *size* of the answer, you get an idea if you multiplied when you should have divided something, or vice versa.)

■ Sodium bicarbonate is the "bicarb" of home medicine cabinets. It is also an ingredient in baking powders.

■ PRACTICE EXERCISE 13 How many grams of each solute are needed to prepare the following solutions?

(a) 250 mL of 0.100 *M* H_2SO_4 (b) 100 mL of 0.500 *M* glucose ($C_6H_{12}O_6$)

Another calculation is to find the volume of a solution of known molar concentration that will deliver a certain quantity of its solute. The next worked example shows how this is done.

EXAMPLE 5.11 Using Solutions of Known Molar Concentration

In an experiment to see whether mouth bacteria can live on mannitol ($C_6H_{14}O_6$), a student needed 0.100 mol of mannitol. It was available as a

■ Mannitol is the sweetening agent in some sugarless chewing gums.

0.750 M solution. How many milliliters of this solution must be used in order to obtain 0.100 mol of mannitol?

ANALYSIS The molarity is the connection between moles of solute and volume of solution. Thus we have two conversion factors from 0.750 M.

$$\frac{0.750 \text{ mol mannitol}}{1000 \text{ mL mannitol solution}} \quad \text{and} \quad \frac{1000 \text{ mL mannitol solution}}{0.750 \text{ mol mannitol}}$$

We must multiply the given, 0.100 mol of mannitol, by the second factor to calculate the answer.

SOLUTION

$$0.100 \text{ mol mannitol} \times \frac{1000 \text{ mL mannitol solution}}{0.750 \text{ mol mannitol}} = 133 \text{ mL mannitol solution}$$

Thus 133 mL of 0.750 M mannitol solution holds 0.100 mol of mannitol.

CHECK If 1000 mL of solution holds 0.750 mol of solute, we need roughly one-seventh as much $\left(\dfrac{0.100}{0.750} \text{ is about one-seventh}\right)$ to hold 0.100 mol, and one-seventh of 1000 is about 140 mL.

■ PRACTICE EXERCISE 14   To test sodium carbonate, Na_2CO_3, as an antacid, a scientist needed 0.125 mol of Na_2CO_3. It was available as 0.800 M Na_2CO_3. How many milliliters of this solution are needed for 0.125 mol of Na_2CO_3?

Once solutions of known molar concentration have been prepared, then the most common kind of calculation involves the stoichiometry of some reaction when at least one reactant is in solution. In the next worked example, we'll see how we can do stoichiometric calculations for such a reaction.

EXAMPLE 5.12   Stoichiometric Calculations That Involve Molar Concentrations

Potassium hydroxide, KOH, reacts with hydrochloric acid as follows:

$$HCl(aq) \quad + \quad KOH(aq) \longrightarrow KCl(aq) \quad + H_2O$$

| hydrochloric acid | potassium hydroxide | potassium chloride | |

■ We usually do not specify the liquid state for water, as in $H_2O(l)$, because this is the state water normally is in.

How many milliliters of 0.100 M KOH are needed to react with the acid in 25.0 mL of 0.0800 M HCl?

ANALYSIS Always remember, *no matter in what form the quantities are given, they must be converted into moles before we can use the balanced equation's coefficients.*

The volume and the molarity of the HCl solution will enable us to calculate how many moles of HCl were used. The equation's coefficients will let us relate the number of moles of HCl to the number of moles of KOH. Finally, we will use the number of moles of KOH and the molarity of the KOH solution to find the volume of its solution.

SOLUTION We can find the number of moles of HCl in 25.0 mL of 0.0800 M HCl by using the first of the following conversion factors obtained from the fundamental meaning of M, moles per liter or moles per 1000 mL.

$$\frac{0.0800 \text{ mol HCl}}{1000 \text{ mL HCl soln}} \quad \text{or} \quad \frac{1000 \text{ mL HCl soln}}{0.0800 \text{ mol HCl}}$$

We multiply the given, 25.0 mL of HCl solution, by the first factor:

$$25.0 \text{ mL HCl soln} \times \frac{0.0800 \text{ mol HCl}}{1000 \text{ mL HCl soln}} = 0.00200 \text{ mol HCl}$$

The next step is to find out how many moles of KOH are needed to react with 0.00200 mol of HCl. The coefficients of the balanced equation tell us that the mole ratio is 1:1, which means 1 mol of HCl $\Longleftrightarrow$ 1 mol of KOH in this reaction. Thus, 0.00200 mol of HCl requires 0.00200 mol of KOH.

Finally, we have to calculate the volume (in mL) of the KOH solution that contains 0.00200 mol of KOH. The molarity of the KOH solution gives us the option of the following conversion factors:

$$\frac{0.100 \text{ mol KOH}}{1000 \text{ mL KOH soln}} \quad \text{or} \quad \frac{1000 \text{ mL KOH soln}}{0.100 \text{ mol KOH}}$$

If we multiply 0.00200 mol of KOH by the second conversion factor, we'll have the right units:

$$0.00200 \text{ mol KOH} \times \frac{1000 \text{ mL KOH soln}}{0.100 \text{ mol KOH}} = 20.0 \text{ mL KOH soln}$$

Thus 20.0 mL of 0.100 M KOH solution provides exactly the right amount of KOH to react with the acid in 25.0 mL of 0.0800 M HCl. Check this out yourself.

The next worked example shows how to solve a problem in which the relevant mole ratio in the chemical equation is not 1:1.

EXAMPLE 5.13   Stoichiometric Calculations That Involve Molar Concentrations

---

Sodium hydroxide, NaOH, reacts with sulfuric acid by the following equation:

$$H_2SO_4(aq) + 2NaOH(aq) \longrightarrow Na_2SO_4(aq) + 2H_2O$$

sulfuric	sodium	sodium
acid	hydroxide	sulfate

■ Sodium hydroxide, NaOH, is commonly known as "lye" and is an ingredient in some oven cleaners.

How many milliliters of 0.125 M NaOH provide enough NaOH to react completely with the sulfuric acid in 16.8 mL of 0.118 M H_2SO_4 by the given equation?

ANALYSIS Again, *always remember to shift given data to the moles level first.* We start by asking how many moles of sulfuric acid are in 16.8 mL of 0.118 M H_2SO_4. Then, we can relate this number of moles to the number of moles of NaOH that match it according to the coefficients. For this equation, we know from its coefficients that 1 mol $H_2SO_4 \Leftrightarrow$ 2 mol NaOH. Finally, we will find out how many milliliters of the NaOH solution hold this calculated number of moles of NaOH. Figure 5.6 has a flowchart of these steps.

SOLUTION   First, the number of moles of H_2SO_4 that react:

$$16.8 \text{ mL } H_2SO_4 \text{ soln} \times \frac{0.118 \text{ mol } H_2SO_4}{1000 \text{ mL } H_2SO_4 \text{ soln}} = 0.00198 \text{ mol } H_2SO_4$$

Next, the moles of NaOH that chemically match 0.00198 mol of H_2SO_4 based on the fact that 1 mol $H_2SO_4 \Leftrightarrow$ 2 mol NaOH:

$$0.00198 \text{ mol } H_2SO_4 \times \frac{2 \text{ mol NaOH}}{1 \text{ mol } H_2SO_4} = 0.00396 \text{ mol NaOH}$$

Finally, the volume of 0.125 M NaOH solution that holds 0.00396 mol NaOH:

$$0.00396 \text{ mol NaOH} \times \frac{1000 \text{ mL NaOH soln}}{0.125 \text{ mol NaOH}} = 31.7 \text{ mL NaOH soln}$$

Thus 31.7 mL of 0.125 M NaOH solution is needed to react with all of the sulfuric acid in 16.8 mL of 0.118 M H_2SO_4. As a check, look again at the conversion factors used and if units have properly canceled.

■ PRACTICE EXERCISE 15   Sodium bicarbonate reacts with sulfuric acid as follows:

$$2NaHCO_3(aq) + H_2SO_4(aq) \longrightarrow Na_2SO_4(aq) + 2CO_2(g) + 2H_2O$$

How many milliliters of 0.112 M H_2SO_4 will react with 21.6 mL of 0.102 M NaHCO_3 *according to this equation?*

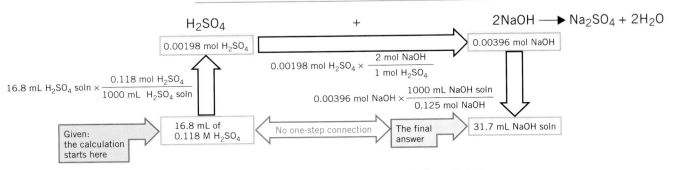

FIGURE 5.6   The calculation flow diagram for Example 5.13.

5.6   PREPARING DILUTE SOLUTIONS FROM CONCENTRATED SOLUTIONS

The dilution of a fixed volume of a concentrated solution changes only the concentration, not the moles of solute.

Often in the lab we have a relatively concentrated solution of known molarity, but we'd rather use a more dilute solution in some experiment. We'll study here how to do the calculations needed for the preparation of dilute solutions of known molarity from concentrated solutions. This or a similar calculation sometimes occurs in a clinical situation when a medication must be diluted. One basic idea guides these calculations. The mass or moles of solute in the final volume of the dilute solution will be the same as were present in the concentrated solution. We add only *solvent*, not solute.

The calculation begins with three facts. We know what final concentration we want, we know what final volume we wish to have, and we know the concentration of the initial concentrated solution. The calculation is to tell us what volume of the concentrated solution must be taken. Because the moles of actual solute are the same in both solutions, we can calculate this amount in the usual way from the volume and molarity data *for both solutions*. In the dilute solution,

■ To save shipping costs, acids are often shipped in very concentrated solutions, which are then diluted to the desired concentration at the laboratory.

$$\text{mol solute} = \text{liters}_{\text{dil soln}} \times \frac{\text{mole solute}}{\text{liter}_{\text{dil soln}}} = \text{liters}_{\text{dil soln}} \times M_{\text{dil soln}}$$

In the concentrated solution,

$$\text{mol solute} = \text{liters}_{\text{concd soln}} \times \frac{\text{mole solute}}{\text{liter}_{\text{concd soln}}} = \text{liters}_{\text{concd soln}} \times M_{\text{concd soln}}$$

These two expressions for the moles of solute equal each other, as we have said. Therefore,

$$\text{liters}_{\text{dil soln}} \times M_{\text{dil soln}} = \text{liters}_{\text{concd soln}} \times M_{\text{concd soln}}$$

Actually, the unit of liters isn't required. We can use any volume unit that we please provided that it is the same unit on both sides of the equation. Normally the mL unit is used in the lab, so our equation can be restated as follows:

■ dil = dilute
concd = concentrated
soln = solution

$$\text{mL}_{\text{dil soln}} \times M_{\text{dil soln}} = \text{mL}_{\text{concd soln}} \times M_{\text{concd soln}} \qquad (5.1)$$

The following example shows how this equation is used.

EXAMPLE 5.14   Doing the Calculations for Making Dilutions

Hydrochloric acid can be purchased at a concentration of 1.00 M HCl. How can we prepare 500 mL of 0.100 M HCl?

ANALYSIS   What the question really asks is how many milliliters of 1.00 M HCl would have to be diluted to a final volume of 500 mL to make a solution with a concentration of 0.100 M HCl? Equation 5.1 is meant for this problem.

■ Hydrochloric acid can be purchased in hardware stores as "muriatic acid."

SOLUTION   We first assemble the known data:

$$\text{mL}_{\text{dil soln}} = 500 \text{ mL} \qquad \text{mL}_{\text{concd soln}} = ?$$
$$M_{\text{dil soln}} = 0.100 \, M \qquad M_{\text{concd soln}} = 1.00 \, M$$

Now we use the equation:

$$\text{mL}_{\text{dil soln}} \times M_{\text{dil soln}} = \text{mL}_{\text{concd soln}} \times M_{\text{concd soln}}$$
$$500 \text{ mL} \times 0.100 \, M = \text{mL}_{\text{concd soln}} \times 1.00 \, M$$

■ From here on, as you judge their usefulness, draw in the cancel lines in calculations involving conversion factors.

Rearranging terms to solve for mL$_{conced\ soln}$ gives us

$$mL_{conced\ soln} = \frac{500\ mL \times 0.100\ M}{1.00\ M} = 50.0\ mL$$

In other words, if we take 50.0 mL of 1.00 M HCl, place this in a 500-mL volumetric flask, and add water to the mark, we will have 500 mL of 0.10 M HCl. Figure 5.7 illustrates how a dilution is carried out.

■ **PRACTICE EXERCISE 16**   Calculate the volume of 0.200 M $K_2Cr_2O_7$ needed to prepare 100 mL of 0.0400 M $K_2Cr_2O_7$. Figure 5.7 shows the steps for doing this dilution.

■ **PRACTICE EXERCISE 17**   The concentrated sulfuric acid that can be purchased from chemical supply houses is 18 M H_2SO_4. How could we use this to prepare 250 mL of 1.0 M H_2SO_4?

(a)

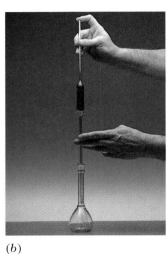
(b)

(c)

(d)

FIGURE 5.7   Preparing a dilute solution by diluting a concentrated solution. The long glass tube with a bulge in its middle is a volumetric pipet. Like a volumetric flask, it has an etched line on its upper narrow section so that when the liquid level is at this line, the pipet contains the volume printed on it. Notice that suction is being supplied by a suction bulb, not by mouth. (a) The calculated volume of the more concentrated solution is withdrawn and (b) placed in a volumetric flask. This flask would already contain some of the additional water to be added if the solution in the pipet were a concentrated acid or base. (c) Now additional water is added slowly as the new solution is swirled to promote mixing until the final volume is reached. (d) Then the new solution is transferred to a dry bottle and labeled.

SUMMARY

Mole concept   The number of atoms in 12 g (exactly) of the carbon-12 isotope is called Avogadro's number, and to three significant figures it equals 6.02×10^{23} atoms. This many formula units of any pure chemical substance constitutes one mole of the substance. Equal numbers of moles contain identical numbers of formula units.

Formula masses   The sum of the atomic masses of the atoms shown in a formula is the formula mass (formula weight) of the substance. For a substance of exceptionally large formula mass, the unit *dalton* is sometimes added to the number. An amount of a substance equal to its formula mass taken in grams is one mole of the substance.

Stoichiometry   For calculation purposes, the relationship

1 mol of an element = atomic mass of element in grams

connects amount of an element by particles with quantity by mass. The coefficients in a balanced equation give the proportions of the chemicals involved in moles. When working problems involving balanced equations and quantities of substances, solve them at the mole level where the coefficients can be used. Then, as needed, convert moles to grams.

When reactants are mixed in mole ratios that do not conform to the reaction's coefficients, the reactant that can be entirely used up is called the limiting reactant because its mole quantity limits the moles of products.

Solutions   A solution has a solvent and one or more solutes, and its concentration is the ratio of quantity of solute to some unit quantity of solvent or of solution.

A solution can be described as dilute or concentrated according to its ratio of solute to solvent being small or large. Whether a solution is unsaturated, saturated, or supersaturated depends on its ability to dissolve any more solute (at the same temperature).

Each substance has a particular solubility in a given solvent at a specified temperature, and this is often expressed as the grams of solute that can be dissolved in 100 g of the solvent. The solubility is the maximum quantity of solute that can dissolve and form a stable solution at the given temperature in the given solvent.

Molar concentration   The ratio of the moles of solute per liter (or 1000 mL) of solution is the molar concentration or the molarity of the solution. When we have to prepare one solution by diluting a more concentrated solution, the equation that we use is

$$mL_{\text{dil soln}} \times M_{\text{dil soln}} = mL_{\text{concd soln}} \times M_{\text{concd soln}}$$

REVIEW EXERCISES

The answers to Review Exercises whose numbers are in color are found in Appendix E. The answers to the other Review Exercises are found in the Study Guide that accompanies this book. The more challenging questions are marked with asterisks.

The Mole Concept and Avogadro's Number

5.1   What two important aspects of chemistry involve *stoichiometry*?

5.2   What is it about chemical substances and chemical reactions that makes the study of the *numbers* of formula units in a given mass of a substance important?

5.3   When rounded to one significant figure, what is Avogadro's number?

5.4   When applied to a chemical substance, what is another name for Avogadro's number?

5.5   Describe the stoichiometric relationships given by the formula $CaCl_2$.

5.6   The atomic mass unit has approximately what size, 10^{-240} g, 10^{-24} g, 10^{-4} g, or 10^{24} g?

5.7   Avogadro's number is actually known to eight significant figures. Why did chemists pick such an ungainly number to stand for one unit quantity of chemical substance?

5.8   How many atoms are in 16.00 g of oxygen-16? (Express the answer in two ways.)

5.9   How many atoms are in 5.89 g of cobalt?

Formula Masses[1]

5.10   What law of chemical combination permits us simply to add atomic masses to calculate formula masses?

5.11   Calculate the formula masses of the following substances.

(a) HCl   (b) KOH   (c) Na_2CO_3
(d) H_2SO_4 (e) $NaHCO_3$ (f) $Ba(NO_3)_2$
(g) $(NH_4)_2HPO_4$ (h) $Ca(C_2H_3O_2)_2$ (i) $C_6H_{12}O_6$

---

[1] Remember that the policy is to round values of atomic masses to their first decimal point before starting any calculations (except that we round the atomic mass of H to 1.01).

Moles of Chemical Substances

*5.12 What is the fundamental reason that we have to calculate the amount of *mass* in a mole in connection with running chemical reactions in the laboratory?

5.13 How is the quantity of mass in one mole of some substance calculated?

5.14 After calculating the formula mass of sodium hydroxide, NaOH, what two conversion factors can we prepare for chemical calculations?

5.15 Calculate the number of grams in 0.125 mol of each of the substances in Review Exercise 5.11.

5.16 Calculate the number of moles in 50.0 g of each of the compounds in Review Exercise 5.11.

5.17 How many *molecules* of N_2 are there in 1.00 g of N_2, roughly the amount of nitrogen in one liter of air?

5.18 How many *molecules* of water are in one drop, which we can assume has a volume of 0.0625 mL and a density of 1.00 g/mL?

5.19 A "5-grain" aspirin tablet holds about 180 mg of aspirin ($C_9H_8O_4$). How many moles and how many molecules are in 180 mg of aspirin?

Balanced Equations

5.20 State the information given in the following equation in words.

$$S + O_2 \longrightarrow SO_2 \text{ (sulfur dioxide)}$$

5.21 Write in your own words what is said by the following equation for the reaction of nitrogen monoxide with oxygen to give nitrogen dioxide.

$$2NO + O_2 \longrightarrow 2NO_2$$

5.22 The following is a balanced equation, but what would be a more acceptable way to write it?

$$4H_2SO_4 + 8NaOH \longrightarrow 4Na_2SO_4 + 8H_2O$$

5.23 Balance the following equations.
(a) $Mg + O_2 \rightarrow MgO$
(b) $CaO + HCl \rightarrow CaCl_2 + H_2O$
(c) $MgCl_2 + AgNO_3 \rightarrow Ca(NO_3)_2 + AgCl$
(d) $HBr + Ca(OH)_2 \rightarrow CaBr_2 + H_2O$
(e) $Na_2CO_3 + HCl \rightarrow NaCl + CO_2 + H_2O$

Stoichiometry and Its Use with Balanced Equations

*5.24 Propane, C_3H_8, is a common heating and cooking fuel used where natural gas (mostly CH_4) is unavailable. When propane burns in a plentiful supply of oxygen, the products are CO_2 and H_2O.
(a) Write the balanced equation for this reaction.
(b) What pairs of conversion factors express the mole rela-

tionships between each of the following pairs of substances?
  C_3H_8 and O_2
  C_3H_8 and CO_2
  C_3H_8 and H_2O

5.25 In one brand of stomach antacid the active ingredient is calcium hydroxide, $Ca(OH)_2$. The stomach acid is hydrochloric acid, HCl, which is neutralized (destroyed) by the following reaction:

$$Ca(OH)_2 + 2HCl \longrightarrow CaCl_2 + 2H_2O$$

What conversion factors describe the *mole* relationship between $Ca(OH)_2$ and HCl?

5.26 Aluminum metal is vigorously attacked by sulfuric acid according to the equation

$$2Al + 3H_2SO_4 \longrightarrow Al_2(SO_4)_3 + 3H_2$$

What conversion factors describe the mole relationships between the two substances in each of the following pairs?

  Al and H_2SO_4
  Al and H_2
  H_2SO_4 and $Al_2(SO_4)_3$

5.27 When glucose ($C_6H_{12}O_6$) is used in the body as a source of energy, the overall reaction is with oxygen, and the products are CO_2 and H_2O.
(a) Write the balanced equation for the reaction of glucose with oxygen.
(b) What conversion factors express the mole relationship between glucose and oxygen?

5.28 Gasohol is a fuel consisting of various hydrocarbons and ethyl alcohol, C_2H_6O. The ethyl alcohol burns in oxygen to give only carbon dioxide and water.
(a) Write the balanced equation for this reaction.
(b) The burning of 7.00 mol of ethyl alcohol uses up how many moles of oxygen?
(c) How many moles of carbon dioxide are produced by the burning of 0.500 mol of ethyl alcohol?

*5.29 The rusting of iron involves the reaction of oxygen with iron. Although the process is complicated, the following equation can be used to represent the overall results.

$$4Fe + 3O_2 \longrightarrow 2Fe_2O_3$$

(a) If 0.148 mol of iron is changed in this way, how many moles of oxygen are consumed?
(b) How many moles of Fe_2O_3 are produced from 0.148 mol of iron?

5.30 Butane, C_4H_{10}, the fuel in lighters, burns according to the following equation.

$$2C_4H_{10} + 13O_2 \longrightarrow 8CO_2 + 10H_2O$$

(a) If 5.56 mol of O_2 are to be consumed by this reaction, how many moles of butane will be used up?

(b) To produce 2.66 mol of CO_2 by this reaction requires how many moles of butane and how many moles of oxygen?

*5.31 Aluminum metal is made industrially by passing a current of electricity through a solution of aluminum oxide, Al_2O_3, in a special solvent. The other product is molecular oxygen.
(a) Complete and balance the following equation for this reaction.

$$Al_2O_3 \xrightarrow{\text{electric current}}$$

(b) How many grams of aluminum can be made from 100 g of aluminum oxide?
(c) How many grams of oxygen are produced from 100 g of aluminum oxide?
(d) What is the total mass of aluminum plus oxygen produced from 100 g of aluminum oxide? Compare the answer with the amount of aluminum oxide used. What law of chemical combination is illustrated by this?

5.32 One chemical reaction that is used industrially to make iron from iron oxide is the reduction of iron(III) oxide, Fe_2O_3, by carbon monoxide according to the following equation.

$$Fe_2O_3 + 3CO \longrightarrow 2Fe + 3CO_2$$

(a) How many grams of iron can be made from 646 g of Fe_2O_3?
(b) How many grams of carbon monoxide are needed to reduce 646 g of Fe_2O_3 by this reaction?
(c) How many grams of carbon dioxide are produced by this reaction from 646 g of iron(III) oxide?

5.33 When a small amount of an acid is accidentally spilled onto a laboratory bench, it should be promptly destroyed (neutralized) before further cleanup is tried. One common way to do this that poses little danger is to sprinkle the acid spill with sodium carbonate until the fizzing caused by escaping carbon dioxide stops. Sulfuric acid, for example, reacts as follows with sodium carbonate.

$$H_2SO_4 + Na_2CO_3 \longrightarrow Na_2SO_4 + CO_2 + H_2O$$

Suppose that a spill of 45.0 g of sulfuric acid occurs. What is the minimum number of grams of sodium carbonate needed to destroy the acid?

Solutions

5.34 A well-stirred mixture of finely divided sand in water is not properly called a *solution*. Explain.

5.35 When water is the dissolving medium for something like sugar, the water itself is designated in what way? How is the sugar designated? What general term can be used to describe any solution for which water is the dissolving medium?

5.36 Sugar (sucrose) is very soluble in water; 100 g of water will dissolve 200 g of sugar.
(a) A solution made up at this concentration would be described as *supersaturated*, *saturated*, or *unsaturated*?
(b) A solution made up at this concentration would be described as *dilute* or *concentrated*?

5.37 Using the information in a table in this chapter, name a compound (not a gas) that can form the most dilute solution that could still be called saturated.

5.38 What laboratory operation *not* involving the use of any added solute or solvent could be used to convert a saturated solution of sodium hydroxide into a supersaturated solution? Into an unsaturated solution?

Molar Concentrations

5.39 What is another term for *molar concentration*?

5.40 Distinguish between the terms *molar concentration*, *molarity*, *mole*, and *molecule*.

5.41 Do the units of molar concentration refer to moles per liter of solvent or to moles per liter of solution?

*5.42 Calculate the number of moles and the number of grams of solute needed to prepare the given volumes of the following solutions.
(a) 500 mL of 0.125 M NaCl
(b) 250 mL of 0.100 M $C_6H_{12}O_6$
(c) 100 mL of 0.250 M H_2SO_4
(d) 125 mL of 0.500 M Na_2CO_3

5.43 How many milliliters of 0.150 M HNO_3 contain 0.0100 mol of HNO_3?

5.44 To obtain 0.150 mol of H_2SO_4, how many milliliters of 0.250 M H_2SO_4 would have to be taken?

5.45 The stock solution of hydrochloric acid is 5.00 M HCl. If 100 mL of this solution is taken, how many moles of HCl are taken?

5.46 A student obtained 25.0 mL of 4.00 M Na_2CO_3 solution. How many moles of Na_2CO_3 are in this quantity?

5.47 How many moles of glucose, $C_6H_{12}O_6$, are in 250 mL of 0.0100 M glucose solution?

*5.48 If a stock solution of nitric acid, HNO_3, has a concentration of 4.00 M, how many milliliters of this solution are needed to obtain 5.00 g of HNO_3?

5.49 To obtain 7.50 g of HCl, how many milliliters of 12.0 M HCl have to be taken?

Stoichiometry of Reactions in Solution

5.50 The label on a reagent bottle reads "0.125 M HNO_3." What two conversion factors are available from this information? (Base these factors on the milliliter unit for the volume.)

*5.51   Barium sulfate, the ingredient in a "barium cocktail" given to patients about to undergo an X ray of the intestinal tract, can be made by the following reaction.

$$Ba(NO_3)_2(aq) + Na_2SO_4(aq) \longrightarrow BaSO_4(s) + 2NaNO_3(aq)$$

The desired product, as you can see, is a water-insoluble compound that can be separated from the other substances by filtration (that is, by letting the mixture flow through filter paper). To prepare 3.00 g of $BaSO_4$, how many milliliters of 0.400 M $Ba(NO_3)_2$ and how many milliliters of 0.350 M Na_2SO_4 must be mixed together?

*5.52   Gold is attacked by very few chemicals. A mixture of concentrated nitric acid and hydrochloric acid, called *aqua regia* ("royal water"), however, dissolves gold by the following equation.

$$Au(s) + 3HNO_3(aq) + 4HCl(aq) \longrightarrow$$
$$HAuCl_4(aq) + 3NO_2(g) + 3H_2O$$

To dissolve 28.4 g of Au (1.00 oz) by this reaction, what is the minimum number of milliliters of 12.0 M HCl and of 16.0 M HNO_3 needed?

*5.53   Vinegar is a 5.00% solution of acetic acid ($HC_2H_3O_2$) in water, which corresponds to 0.837 M $HC_2H_3O_2$. If mixed with aqueous sodium carbonate (Na_2CO_3), the following reaction occurs.

$$Na_2CO_3(aq) + 2HC_2H_3O_2(aq) \longrightarrow$$
$$2NaC_2H_3O_2(aq) + CO_2(g) + H_2O$$

(a) If the acetic acid in 35.0 mL of vinegar is to react entirely with 0.450 M Na_2CO_3 by this equation, how many milliliters of the sodium carbonate solution are needed?
(b) If solid sodium carbonate is used instead of the solution, how many grams of this solid are needed to react with the acetic acid in a spill of 48.0 mL of 0.837 M acetic acid?

*5.54   The active ingredient in milk of magnesia, an over-the-counter antacid, is finely divided magnesium hydroxide slurried in water. The acid it destroys in the stomach by the following equation is roughly 0.1 M HCl. (A minimum recommended dose of milk of magnesia is 2 tablespoons or 30 mL.)

$$Mg(OH)_2(s) + 2HCl(aq) \longrightarrow MgCl_2(aq) + 2H_2O$$

How many milliliters of 0.100 M HCl can be destroyed by 30.0 mL of milk of magnesia when this medication contains 1.20 g of solid $Mg(OH)_2$ per 15.0 mL of milk of magnesia slurry? (Normally, roughly 2 L of 0.1 M HCl is secreted per day into the stomach.)

Preparing Dilute Solutions from Concentrated Solutions

5.55   The commercially available concentrated phosphoric acid is 15.0 M H_3PO_4. How many milliliters of this acid are needed to prepare 250 mL of 0.500 M H_3PO_4? Describe how one would prepare this dilute solution.

5.56   The concentrated aqueous ammonia that is commercially available is 15.0 M NH_3. If you dissolved 50.0 mL of this solution in water and made up the final volume to equal 250 mL, what would be the molarity of the resulting solution?

Additional Exercises

5.57   Which has more mass, 0.50 mol of NH_3 or 2.0 mol of He?

*5.58   Iodized salt contains a trace amount of calcium iodate, $Ca(IO_3)_2$, to help prevent a thyroid condition called *goiter*. How many moles of iodine, I, are in 0.100 mol of $Ca(IO_3)_2$?

*5.59   Baking soda is sodium bicarbonate, $NaHCO_3$. If a sample of baking soda is large enough to contain 2.50 g of C, then how many grams of Na, H, and O are also present?

*5.60   An acid spill involved 25.60 mL of 12.0 M HNO_3, concentrated nitric acid. In an attempt to neutralize the acid, 20.0 g of Na_2CO_3 was stirred into the spill. The equation is

$$2HNO_3 + Na_2CO_3 \longrightarrow 2NaNO_3 + H_2O + CO_2$$

(a) Was enough Na_2CO_3 added to neutralize all of the acid?
(b) What is the formula of the limiting reactant in this situation?
(c) Which reactant was left over? How many moles of this reactant remained?
(d) How many moles and how many grams of CO_2 formed?

5.61   How many atoms are in 1.00 oz of silver? (1 oz = 28.4 g)

5.62   Calculate the formula masses of the following compounds.
(a) $MgBr_2$ (b) HNO_3 (c) $(NH_4)_3PO_4$
(d) $Mg_3(PO_4)_2$ (e) $Al(C_2H_3O_2)_3$ (f) $Ca(ClO_4)_2$
(g) $(NH_4)_2SO_3$ (h) $K_2Cr_2O_7$ (i) $Fe_4(OH)_2(SO_4)_5$

*5.63   The *mole* is the SI base unit for amount of chemical substance. All other SI base units, like the meter and the kilogram mass, have constant values, but the mass of one mole varies from substance to substance. Explain. Is there any feature about a mole that is *constant* from substance to substance?

5.64   Chemical laboratory balances generally read in *grams*, not in *moles*. Explain why.

5.65   How many grams are in 0.750 mol of each of the compounds in Review Exercise 5.62?

5.66   Calculate the number of moles in 1.50 g of each of the compounds in Review Exercise 5.62.

*5.67   At a level of only 0.5 μg of ozone, O_3, in one cubic meter of air, the air is considered dangerous for active children to breathe. How many molecules of ozone are in 0.5 μg?

5.68   Balance each of the following equations.
(a) $PCl_3 + O_2 \rightarrow POCl_3$

(b) $Fe_2O_3 + HCl \rightarrow FeCl_3 + H_2O$
(c) $KHSO_3 + H_2SO_4 \rightarrow K_2SO_4 + SO_2 + H_2O$
(d) $HNO_3 \rightarrow N_2O_5 + H_2O$
(e) $C_3H_6 + O_2 \rightarrow CO_2 + H_2O$

*5.69 The high explosive TNT (trinitrotoluene) decomposes (breaks down) according to the following equation when it explodes.

$$2C_7H_5N_3O_6 \longrightarrow 3N_2 + 7CO + 5H_2O + 7C$$

(a) How many moles of N_2 are produced for each mole of TNT that decomposes?
(b) How many moles of CO per mole of TNT are produced?
(c) The water forms as high-temperature water vapor. How many moles of water per mole of TNT are produced?
(d) What is the total number of moles of gases of all three kinds, N_2, CO, and H_2O vapor, produced from 1 mol of TNT? (The carbon forms as finely divided soot.)
(e) TNT is a solid at room temperature, and 1.00 mol of TNT occupies a volume of about 0.14 L. If the gases produced by the explosion occupy a volume of roughly 31 L/mol, what is the ratio of the volume of gases produced to the volume of TNT taken? (This huge expansion, occurring suddenly, is what makes TNT such an effective explosive. Rapidly produced and expanding gases push things out of the way!)

*5.70 The chemical reaction that causes silver to tarnish is between silver metal, oxygen in the air, and traces of hydrogen sulfide, also in the air. The black tarnish consists of silver sulfide.

$$4Ag + 2H_2S + O_2 \longrightarrow 2Ag_2S + 2H_2O$$

(a) If 4.68 mg of Ag tarnish by this reaction, how many milligrams of hydrogen sulfide are needed?
(b) How many milligrams of silver sulfide form from 4.68 mg of silver?

*5.71 The deeply purple colored permanganate ion, MnO_4^-, can be made by the oxidation of the Mn(II) ion using sodium bismuthate, $NaBiO_3$, and nitric acid, HNO_3,

$$2Mn(NO_3)_2 + 5NaBiO_3 + 14HNO_3 \longrightarrow$$
$$2NaMnO_4 + 5Bi(NO_3)_3 + 3NaNO_3 + 7H_2O$$

Consider an experiment in which 12.6 g of sodium permanganate, $NaMnO_4$, is to be made by this reaction.
(a) How many grams of manganese(II) nitrate, $Mn(NO_3)_2$, are needed?
(b) How many grams of sodium bismuthate are required?
(c) This preparation will also require how many grams of nitric acid?
(d) How much bismuth(III) nitrate (in grams) will also be produced?

5.72 Suppose that for a series of experiments you needed to have on hand a saturated solution of sodium chloride in water. How could such a solution be prepared in the certain knowledge that it is saturated without actually weighing out the solute?

*5.73 How many moles and how many grams of solute are needed to prepare the stated volume of each of the following solutions?
(a) 500 mL of 0.200 M $NaC_2H_3O_2$
(b) 250 mL of 0.125 M HNO_3
(c) 100 mL of 0.100 M NaOH
(d) 50.0 mL of 0.250 M $NaHCO_3$

5.74 An experiment called for 0.0250 mol of Na_2CO_3, which was available in a solution with a concentration of 0.112 M. How many milliliters of this solution are required?

*5.75 The stock supply of sulfuric acid, H_2SO_4, has a concentration of 0.125 M. To obtain 0.500 g of sulfuric acid in the form of this solution, how many milliliters have to be taken?

*5.76 The concentration of sodium bicarbonate in pancreatic juice, one of the digestive juices, can go as high as 0.120 M $NaHCO_3$. It reacts by the following equation with the hydrochloric acid delivered in the stomach contents as they move into the upper intestinal tract.

$$NaHCO_3(aq) + HCl(aq) \longrightarrow NaCl(aq) + CO_2(g) + H_2O$$

How many liters of 0.120 M $NaHCO_3$ solution provide enough solute to react with the solute in 2.00 L of 0.100 M HCl?

6

STATES OF MATTER AND THE KINETIC THEORY

Near wintry Lake Superior's edge, beneath Minnesota's Split Rock Lighthouse, water is in its three states, solid, liquid, and vapor. We'll be using the kinetic theory of matter in this chapter to understand them better.

THIS CHAPTER IN CONTEXT

So far we have developed a background in atoms, ions, and molecules, the submicroscopic particles of substances in any physical state, and have learned that above 0 K they are in incessant motion, possessing molecular kinetic energy. These facts affect the physical properties of all things, including the air in which we live and breathe. Moreover, our bodies are about 60% water, a liquid. And what keeps us from dribbling away are solids. Clearly, no understanding of the molecular basis of life is possible without a good background in the states of matter.

We begin this chapter with the gaseous state because it is the easiest. We'll then be in a position to start our survey of the molecular basis of respiration.

6.1 GASEOUS STATE

The four properties that can be used to define completely the physical state of any gas are pressure, temperature, volume, and amount.

Unlike liquids and solids, all gases obey a small number of laws, *regardless of their chemical identities*. These *gas laws* are expressed in terms of the following four physical quantities.

1. Pressure (P). You became aware of gas pressure when you first pumped up a bicycle tire or blew up a balloon. We'll study the units of pressure shortly.

2. Volume (V). A liquid or a solid might occupy only part of its container, but the volume of a confined gas is always the container's *total* volume. The reason is that gases easily spread out or diffuse. Gas volumes are usually given in liters (L) or milliliters (mL).

3. Temperature (T). We have to use the Kelvin scale to simplify the gas laws and make the calculations easy.

4. Amount (n). The amount of gas is in moles when it appears in a gas law. The mole is preferred to a mass unit, like gram, because when we compare two gases consisting of the same number of moles we can be sure that they have the same number of molecules.

Pressure Is Defined as Force per Unit Area   Air is matter, so it has mass, and like anything with mass it is subject to the earth's gravitational attraction. This causes air to exert a force on each unit of the area of the earth's surface, which we call the *atmospheric pressure*. The force per unit area exerted by any mass is called **pressure.**

A column of air 1 in.² in cross section that extends from sea level to the edge of outer space exerts a force that averages 14.7 pounds, so the air pressure at sea level is 14.7 lb/in.² At higher elevations, less air is in the column, so the air pressure is less. From the top of Mount Everest, for example, earth's highest mountain, the column of air to outer space is so relatively short that the air pressure is about one-third that at sea level.

One way to observe atmospheric pressure is with a Torricelli barometer (Figure 6.1). At 0 °C and sea level, the pressure of the atmosphere supports a column of mercury in the barometer averaging 760 mm high, a value now defined as one **standard atmosphere**, abbreviated **atm.**

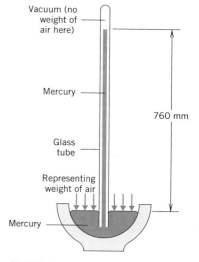

FIGURE 6.1 The Torricelli barometer (named after Evangelista Torricelli, 1608–1647, an Italian physicist). The long tube, sealed at one end, is filled with mercury and inverted into a dish of mercury. Some mercury runs out of the tube, but the remainder is held up by air pressure. No air is above the mercury inside the tube, so no air pressure acts inside to oppose the air pressure outside. The column of mercury in the tube stands 760 mm high at sea level and 0 °C.

■ By *outer space* we mean space beyond the earth's atmosphere where the atmosphere is so thin as to be almost nonexistent.

■ The height fluctuates with the temperature and weather.

$$1 \text{ atm} = 760 \text{ mm Hg (at 0 °C)}$$

■ TV and radio weather reports in the United States use *inches* of mercury, not mm Hg, as the pressure unit.

I atm = 29.9 in. Hg

In scientific work and medicine, gas pressures are often expressed in a smaller unit, the **millimeter of mercury**, abbreviated **mm Hg**, defined by the following equation:

$$1 \text{ mm Hg} = \frac{1}{760} \text{ atm}$$

The air pressure in the lab might, for example, be 740 mm Hg (0.974 atm) one day and 747 mm Hg (0.983 atm) on another, depending on the weather. At the top of Mount Everest, the pressure is close to 250 mm Hg.

The SI unit of pressure is the **pascal (Pa).**

$$1 \text{ mm Hg} = 133.3224 \text{ Pa}$$
$$1 \text{ atm} = 101,325.024 \text{ Pa}$$

■ I kPa = 1000 Pa

This makes the standard atmosphere about 101 kilopascals (kPa).[1]

The mm Hg unit is a strange unit when you think about it. Why should a unit of length, the mm, be used for something that is not length but is force per unit area? So most scientists have adopted an exact synonym for "mm Hg," namely the **torr.**

$$1 \text{ mm Hg} = 1 \text{ torr}$$
$$760 \text{ mm Hg} = 760 \text{ torr} = 1 \text{ atm}$$

Most people in the health care fields, however, still favor the mm Hg, which is why we'll do so too.

6.2   PRESSURE–VOLUME–TEMPERATURE RELATIONSHIPS FOR A FIXED AMOUNT OF GAS

The product of the pressure and volume of a gas divided by its temperature is a constant.

■ Any physical quantity that a lab worker is free to change is called a variable.

Most of the gas laws concern a fixed amount of gas, a specific number of moles. Let's now see how such a sample responds to changes in the other three variables, namely, pressure, volume, and temperature.

The Volume of a Gas Sample Varies Inversely with Its Pressure at Constant Temperature   When he trapped a gas sample in a device similar to that shown in Figure 6.2, English scientist Robert Boyle (1627–1691) discovered that an increase in gas pressure reduced the gas volume proportionately. If the pressure is doubled, the volume is cut in half, for example.

Whenever one quantity decreases in proportion to an increase in the other, we say that the quantities are *inversely proportional*. In time, other scientists found that all gases had this property, so we now have a law of nature called the **pressure–volume law,** or **Boyle's law.**

> **Boyle's Law (Pressure–Volume Law)**   The volume of a given amount of gas at constant temperature varies inversely with the pressure.

---

[1] Although most professional organizations urge health care specialists to switch from mm Hg to kPa to report pressures of respiratory gases, relatively few have made the change.

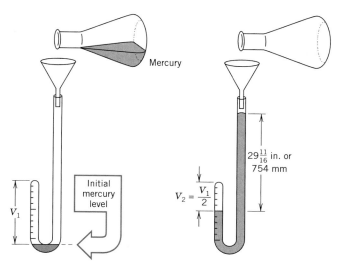

FIGURE 6.2 J-tube apparatus for pressure–volume data. On the left, the pressures in the two arms of the tubes are the same and equal, say, 754 mm Hg. A volume of gas, V_1, has been trapped in the shorter arm with the sealed end. On the right, enough mercury has been added to make the mercury column in the longer arm extend 754 mm above the top of the shorter mercury column. In other words, the pressure on the entrapped gas is now 754 + 754 mm Hg, or twice the initial value. This has squeezed the volume of the gas to half its original size. Thus, doubling the pressure halves the volume, as Robert Boyle discovered.

Put into mathematical form, the law says the following.

$$V \propto \frac{1}{P} \qquad (T \text{ and mass are constant}) \tag{6.1}$$

■ The symbol $\propto$ stands for *is proportional to.*

The proportionality sign, $\propto$, in Equation 6.1 can be replaced by introducing a proportionality constant, C.

$$V = \frac{1}{P} \times C$$

Rearranging gives

$$PV = C \tag{6.2}$$

Let us now imagine that we have a fixed amount of gas at constant temperature and that some change is made in the sample's volume or pressure. Using subscripts to distinguish the data, let the initial values of pressure and volume be P_1 and V_1 and the final values P_2 and V_2. Initially, before the change, Equation 6.2 tells us that

$$P_1V_1 = C \tag{6.3}$$

After the change,

$$P_2V_2 = C \tag{6.4}$$

■ The same constant, C, applies. This is at the heart of Boyle's law.

The constant C, of course, *must be the same for both Equations 6.3 and 6.4.* Otherwise we could not call C a *constant.* Two terms that equal the same thing, C in this instance, are equal to each other, so

$$P_1V_1 = P_2V_2 \tag{6.5}$$

Equation 6.5 is probably the most useful way to express Boyle's law, because when we know any three of its four values, we can calculate the fourth without the need to figure out the Boyle's law constant, C.

EXAMPLE 6.1 Calculating with Boyle's Law

A sample of nitrogen occupies a volume at 25 °C of 5.65 L when the gas pressure is 740 mm Hg. If, at the same temperature, the pressure is changed to 760 mm Hg, what is the final volume?

ANALYSIS The problem is about the interaction of pressure with volume for a fixed amount of gas at constant temperature, all of the specifications of Boyle's law.

SOLUTION You will generally find gas law calculations easier if you assemble all pertinent data first.

$$V_1 = 5.65 \text{ L} \qquad P_1 = 740 \text{ mm Hg}$$
$$V_2 = ? \text{ L} \qquad P_2 = 760 \text{ mm Hg}$$

Now we apply Boyle's law using Equation 6.5. After inserting the given data and rearranging, we have

$$V_2 = 5.65 \text{ L} \times \frac{740 \text{ mm Hg}}{760 \text{ mm Hg}} = 5.50 \text{ L}$$

CHECK Is the answer reasonable? Yes; 5.50 L is less than 5.65 L, and the calculated *decrease* in volume is mandated by the *increase* in pressure; it's Boyle's law.

■ PRACTICE EXERCISE 1   If 2.5 L of a gas is at a pressure of 760 mm Hg, and its pressure changes to 730 mm Hg, what is the new volume (assuming a constant temperature)?

A Gas That Obeys the Gas Laws Exactly Is an *Ideal Gas*   Most gases closely follow Boyle's law rather well at ordinary temperatures and pressures, like those in our calculations, but no gas "obeys" the law *exactly* over a wide range of pressures. A gas most poorly obeys the Boyle's law equation when it is under high pressure (10 atm or more) or at a low temperature (below 200 K), or some combination of these. Under such extreme conditions, most gases are close to changing over to their liquid state where the gas laws are not applicable.

Although real gases do not *exactly* obey Boyle's law or any of the other gas laws, we can yet imagine a hypothetical gas that does. It's called an **ideal gas** and is *defined* as one that obeys the gas laws exactly. A real gas increasingly shows ideal gas behavior as its pressure is decreased and its temperature is increased, changes that lower the tendency of the gas to change to a liquid.

Gas Volume Varies Directly with the Kelvin Temperature at Constant Pressure
Jacques Charles discovered how a gas changes its volume when its temperature is changed while both the pressure and number of moles are kept constant. His discovery is now called the **temperature–volume law,** or **Charles' law.**

Temperature–Volume Law (Charles' Law)   The volume of a fixed amount of any gas is directly proportional to its Kelvin temperature, if the pressure is kept constant.

Charles' law can be expressed in mathematical form as follows.

$$V \propto T$$

When two things are *directly* proportional to each other, as V and T, we say that there is a *linear* (straight line) relationship, and the plots of Figure 6.3 bear this out for a gas under Charles' law conditions. Notice that the plots all converge at 0 K, absolute zero.

By rearranging terms and introducing a constant of proportionality we can reexpress Charles' law as follows.

$$\frac{V}{T} = \text{a constant} \qquad (6.6)$$

A useful way to restate Equation 6.6, where we use the subscripts 1 and 2 as we did in Equation 6.5, is

$$\frac{V_1}{T_1} = \frac{V_2}{T_2} \qquad \text{(constant } P \text{ and } n) \qquad (6.7)$$

Charles' law calculations are handled much as those of Boyle's law. Remember that *temperatures in degrees Celsius must first be changed to kelvins.* The volumes, however, may be in any unit provided it's the same for both V_1 and V_2.

■ Using laser technology, scientists at Colorado's Joint Institute for Laboratory Astrophysics were able in 1991 to cool a sample of cesium vapor to within 10^{-6} degree of 0 K.

■ $\dfrac{V_1}{T_1} = C'$ and

$\dfrac{V_2}{T_2} = C'$. Hence

$\dfrac{V_1}{T_1} = \dfrac{V_2}{T_2}$

which is Equation 6.7.

■   **PRACTICE EXERCISE 2**   A sample of cyclopropane, an anesthetic, with a volume of 575 mL at a temperature of 30 °C, was cooled to 15 °C at the same pressure. What was the new volume?

Gas Pressure Varies Directly with Kelvin Temperature   Joseph Gay-Lussac discovered the way in which an increase in the temperature of a confined gas changes its pressure, a relationship now called the **temperature–pressure law**, or **Gay-Lussac's law**.

Temperature–Pressure Law (Gay-Lussac's Law)   The pressure of a fixed amount of gas is directly proportional to its Kelvin temperature if the volume is kept constant.

Gay-Lussac's law can be written in mathematical form as follows.

$$P \propto T \qquad (V \text{ and } n \text{ are constant})$$

The direct proportionality of P and T is why an aerosol can must not be put into an incinerator. The can is sealed, and it still holds residual gas. As its temperature increases so does its internal pressure. Eventually, the can explodes, which may damage the incinerator or possibly hurt bystanders.

The General Gas Law Incorporates the Laws of Boyle, Charles, and Gay-Lussac
When gases are under conditions in which they obey the gas laws, a remarkable relationship exists. The ratio of PV to T for a given amount of gas is itself a constant. In equation form,

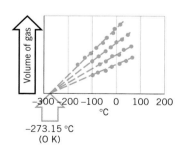

FIGURE 6.3 Plots of temperature–volume data obtained at constant pressure. For different masses of the same gas, the volume is directly proportional to the temperature. The plots all converge at the baseline at a temperature of 0 K, absolute zero.

$$\frac{PV}{T} = \text{constant} \qquad \text{(for a fixed amount of gas)}$$

■ Because $\frac{P_1V_1}{T_1}$ and $\frac{P_2V_2}{T_2}$ equal the same constant, they must equal each other.

If we assign before and after subscripts in the usual way, we get the following equation, called the **general gas law** or sometimes the **combined gas law**.

$$\frac{P_1V_1}{T_1} = \frac{P_2V_2}{T_2} \qquad \text{(constant amount)} \tag{6.8}$$

As we have said, Equation 6.8 includes the laws of Boyle, Charles, and Gay-Lussac as special cases. To prove this, notice that when T_1 equals T_2, that is, when T is constant (Boyle's law conditions), the temperatures cancel, leaving $P_1V_1 = P_2V_2$, an expression of Boyle's law. When the pressure is constant (Charles' law condition), P_1 and P_2 cancel from Equation 6.8, leaving Equation 6.7, the Charles' law equation. When the volume is held constant, V_1 and V_2 cancel leaving the equation for Gay-Lussac's law (which we did not develop above).

EXAMPLE 6.2　Using the General Gas Law

A sample of an anesthetic gas with a volume of 925 mL at 20.0 °C and 750 mm Hg pressure is warmed to 37.0 °C at a pressure of 745 mm Hg. What is the final volume of the gas?

ANALYSIS　It's best to collect the data first to isolate what is to be calculated.

$$V_1 = 925 \text{ mL} \qquad\qquad V_2 = \text{?}$$
$$P_1 = 750 \text{ mm Hg} \qquad\quad P_2 = 745 \text{ mm Hg}$$
$$T_1 = 293 \text{ K (20.0 °C)} \qquad T_2 = 310 \text{ K (37.0 °C)}$$

Because n is a constant but P, V, and T all change, the appropriate equation is the general gas law, Equation 6.8.

SOLUTION　Using Equation 6.8,

$$\frac{750 \text{ mm Hg} \times 925 \text{ mL}}{293 \text{ K}} = \frac{745 \text{ mm Hg} \times V_2}{310 \text{ K}}$$

Solving for V_2, we find

$$V_2 = \frac{750 \text{ mm Hg} \times 925 \text{ mL} \times 310 \text{ K}}{745 \text{ mm Hg} \times 293 \text{ K}} = 985 \text{ mL}$$

CHECK　Notice that there is an *increase* in volume from 925 mL to 985 mL. Is this result sensible? Yes, the pressure itself *decreases*, and Boyle's law tells us that the volume should therefore increase. The temperature *increases*, and Charles' law says that this also should work to increase the volume. (A "head check" like this is trickier when the individual effects tend to work oppositely.)

■　PRACTICE EXERCISE 3　A fire occurred in a storage room where a steel cylinder of oxygen-enriched air was kept. The pressure of the gas in the cylinder was 300 atm at 20.0 °C. To what value does the pressure change if the temperature increases to 200 °C but the volume of the cylinder does not change?

6.3 IDEAL GAS LAW

The ratio of *PV* to *nT* is a constant, the universal gas constant, *R*.

Under Identical Pressure and Temperature, Equal Volumes of Gases Contain the Same Number of Moles   As we said, the gas laws so far studied assume a constant amount of gas. In fact, it isn't possible to increase the number of moles of a gas in a container unless something else changes, namely, *P*, *V*, or *T*. The volume must increase, for example, if we want to add more gas but still keep the pressure and temperature constant. In fact, at constant pressure and temperature, the volume of a gas is directly proportional to the number of moles.

$$V \propto n \qquad \text{(at constant } P \text{ and } T\text{)}$$

This relationship becomes an equation when we insert a proportionality constant, *C*:

$$V = Cn \qquad \text{(at constant } P \text{ and } T\text{)}$$

Amedeo Avogadro discovered an important fact about this equation, namely, that *C* *is the same for all gases*. A 45-L sample of oxygen, for example, has the same number of moles as a 45-L sample of nitrogen or any other gas (under the same temperature and pressure). Thus, we have another important gas law, now called **Avogadro's principle.**

■ Conditions of constant temperature and pressure are assumed throughout this discussion.

Avogadro's Principle   When measured at the same temperature and pressure, equal volumes of gas contain equal numbers of moles.

$$V \propto n \qquad \text{(at constant } T \text{ and } P\text{)}$$

At 273 K and 1 Atm, One Mole of a Gas Occupies 22.4 L   The actual volume occupied by one mole of any gas varies, of course, with pressure and temperature. To make comparisons, scientists have therefore agreed to reference conditions called the **standard conditions of temperature and pressure** or **STP**. Standard pressure is 1 atm and standard temperature is 273 K (0 °C). At STP, one mole of any gas (to three significant figures) occupies 22.4 L, a value called the **standard molar volume** of a gas (see Table 6.1).

Before we move on to the most general equation about gases, one that combines all four gas laws studied, we ought to reflect on the enormous volume-increasing effect of changing a liquid or solid to its vapor. One mole of water, for example, has a mass of only 18 g. That corresponds to the volume of a finger. If it behaved as an ideal gas at STP, its volume would be over 22 L, or over 1200 times as much. In the operation of an air bag during a vehicle accident, a solid chemical, sodium azide, changes suddenly to a large volume of gas, nitrogen, as discussed further in Interaction 6.1.

The Ratio *PV/nT* Is the Same for All Gases   As scientists worked more and more with the laws of Boyle, Charles, Gay-Lussac, and Avogadro, they eventually discovered that the result of multiplying *P* and *V* and dividing by *T* is proportional to the moles of the gas, *n*.

$$\frac{PV}{T} \propto n$$

Using a proportionality constant, *R*, we have

$$\frac{PV}{T} = nR$$

TABLE 6.1 Molar Volumes of Some Gases at STP

Gas	Formula	Molar Volume (L)
Helium	He	22.398
Argon	Ar	22.401
Hydrogen	H_2	22.410
Nitrogen	N_2	22.413
Oxygen	O_2	22.414
Carbon dioxide	CO_2	22.414

INTERACTION 6.1
AIR BAGS

Federal legislation has mandated that automobiles used in the United States, from the 1998 model year and thereafter, must all be equipped with dual air bags (Figure 1). They work on simple chemical principles, particularly the huge volume-expanding result of converting a solid chemical into a gas, nitrogen. The National Transportation Safety Administration estimates that air bags have saved about 1100 lives, to say nothing about how many serious injuries have been prevented. However, the bags fly out at about 200 miles/hour, and people have been killed when airbags have deployed. Although properly strapped in car seats facing the rear, over two dozen infants and children have died in crashes when the enlarging air bag forced the car seat violently into the back of the vehicle seat. (These data were reported near the end of 1996.) Short adults have also died when an air bag hit the face and not the chest. Research continues to try to develop less dangerous air bags, and parents are urged to have all children sit in rear seats. (Many states may require this by law.)

There are four operating units to an air bag system; namely, a nylon bag into which the nitrogen expands, a gas generator, a minicomputer or microprocessor, and electronic sensors. The sensors, located at the front of the vehicle, receive the first signal that a rapid loss in speed is occurring. The microprocessor evaluates the sensors' signals, comparing them with its stored data on crash patterns. (Gen-

erally, it takes a precrash velocity of 20 mph to activate the air bags.) The microprocessor then launches the mechanism for generating nitrogen. The nitrogen expands into the nylon bag, but the bag is just porous enough to allow some to escape as it enters so that when the bag meets your head it feels not like a brick wall but more like a good pillow. All of this activity must occur within several thousandths of a second after the crash starts.

The gas generator contains a powdered mixture of sodium azide (NaN_3), potassium nitrate (KNO_3), and silicon dioxide (SiO_2). Sodium azide is the chief source of nitrogen. NaN_3 is an explosive, but it's safe to have it in the system because it takes either a high temperature (300 °C) or an electrical impulse to make it decompose, an impulse that the microprocessor initiates. Sodium azide decomposes according to the following equation.

$$2NaN_3(s) \xrightarrow{300\ °C} 2Na(g) + 3N_2(g)$$

The sodium vapor would be very hazardous if it leaked out, but the potassium nitrate now serves its purpose. The sodium is changed to sodium oxide by the reaction of the following equation (and a little more nitrogen is generated).

$$10Na(g) + 2KNO_3(s) \longrightarrow K_2O(s) + 5Na_2O(s) + N_2(g)$$

The oxides of sodium and potassium, Na_2O and K_2O, are still hazardous; with moisture they change to the very caustic, skin-harming hydroxides, sodium and potassium hydroxide. But these are "neutralized" by the third chemical in the system, silicon dioxide, and are converted into a silicate "glass," a safe and stable compound. The equation for the reaction is

$$K_2O(s) + Na_2O(s) + SiO_2 \longrightarrow$$
$$\text{sodium-potassium silicate ("glass")}$$

(A definite formula for the glass cannot be written; pure sodium silicate, however, is Na_2SiO_3. In a school project you may have used "water glass," an aqueous solution of sodium silicate.)

Most air bags are never used, which means that junked cars carry an explosive substance. It's a problem not yet fully resolved, but it no doubt will yield to some kind of recycling operation.

A Reference A. Madlung, "The Chemistry Behind the Air Bag," *Journal of Chemical Education,* April, 1996, page 347.

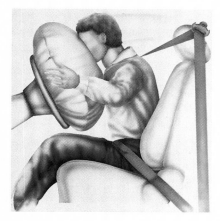

FIGURE 1 When both a seat belt and an air bag operate during a vehicle crash, injuries to the head and the thorax are sharply reduced and often eliminated. Children, however, should be harnassed into a rear seat.

or

$$PV = nRT \qquad (6.9)$$

Equation 6.9 is called the **universal gas law**, and the constant, **R**, is the **universal gas constant.**

The value of R is found by rearranging Equation 6.9 and inserting the values for the standard pressure, temperature, and molar volume. Thus, when n is 1 mol, V is 22.4 L, P is 1.00 atm, and T is 273 K, we have

$$R = \frac{PV}{nT} = \frac{(1.00 \text{ atm})(22.4 \text{ L})}{(1.00 \text{ mol})(273 \text{ K})}$$

$$= 0.0821 \frac{\text{atm L}}{\text{mol K}}$$

Or, arranging the units in their most commonly given order,

$$R = 0.0821 \text{ L atm/mol K}$$

As you've probably noticed, the value of R does depend on the *units* used for pressure and volume. We've picked atm and L and will stay with them.[2]

EXAMPLE 6.3 **Using the Ideal Gas Law**

A sample of oxygen at 24.0 °C and 745 torr was found to have a volume of 455 mL. How many grams of O_2 were in the sample?

ANALYSIS In this question we are given pressure, volume, and temperature of a gas and then asked to determine the amount of gas. These quantities are related by the ideal gas law, so we need the equation $PV = nRT$. We can use the equation to calculate the number of moles of O_2, and then we can use the formula mass of O_2 as a tool to calculate the mass of O_2 in grams.

SOLUTION First, let's solve the ideal gas law for n.

$$n = \frac{PV}{RT}$$

To use the value of 0.0821 L atm/mol K for R, we must have all the data in matching units: V in liters, P in atmospheres, and T in kelvins. Gathering the data and making the necessary unit conversions, we have

$P = 0.980$ atm   From: $745 \text{ torr} \times \dfrac{1 \text{ atm}}{760 \text{ torr}}$

$V = 0.455$ L   From: 455 mL

$T = 297$ K   From: $24.0 + 273$

---

[2] If we choose 760 mm Hg (the same as 1 atm) and 22.4×10^3 mL (the same as 22.4 L) to calculate R, we get one common alternative value of R.

$$R = \frac{(760 \text{ mm Hg})(22.4 \times 10^3 \text{ mL})}{(1.00 \text{ mol})(273 \text{ K})}$$

$$= 6.24 \times 10^4 \text{ mL mm Hg/mol K}$$

■ Show that the identical answer, 0.0183 mol O_2, is obtained using the alternative value of R.

$$R = 6.24 \times 10^4 \frac{mL \ mm \ Hg}{mol \ K}$$

Substituting these values gives

$$n = \frac{(0.980 \ \text{atm})(0.455 \ \text{L})}{(0.0821 \ \text{L atm/mol K})(297 \ \text{K})}$$

$$= 0.0183 \ \text{mol of} \ O_2$$

The formula mass of O_2 is 32.0. Therefore,

$$0.0183 \ \text{mol} \ O_2 \times \frac{32.0 \ \text{g} \ O_2}{1 \ \text{mol} \ O_2} = 0.586 \ \text{g} \ O_2$$

The sample of oxygen must have had a mass of 0.586 g.

■   PRACTICE EXERCISE 4   What volume in milliliters does a sample of nitrogen with a mass of 0.245 g occupy at 21 °C and 750 mm Hg?

6.4   DALTON'S LAW OF PARTIAL PRESSURES

The total pressure of a mixture of gases is the sum of the partial pressures of the individual gases.

The molecular basis of respiration begins with air, a mixture principally of nitrogen and oxygen but with small traces of a few other gases. In 100 L of dry air there are 21 L of oxygen and 79 L of nitrogen, so on a volume basis air is 21% oxygen and 79% nitrogen. *These percentages are true at all altitudes of the earth's atmosphere.* What changes with altitude is not the *volume* ratio of oxygen to nitrogen but their individual pressures called their *partial pressures.*

■ For every 100.000 L of pure dry air, there is 0.934 L of argon, so air is nearly 1% in argon on a volume basis.

The **partial pressure** of one particular gas in a gas mixture is the contribution that the gas makes to the total pressure. It's the pressure that the gas would exert if all of the other gases were removed from the container to leave the one gas all alone in the same container (at the same temperature). When dry air at 0 °C is at a total pressure of 760 mm Hg, the partial pressure of oxygen is 160 mm Hg and that of nitrogen is 600 mm Hg. Notice that the sum (160 mm Hg + 600 mm Hg) is 760 mm Hg.

John Dalton (of atomic theory fame) was the first to discover that all of the partial pressures of individual gases add up to the total pressure of a gas mixture.

Dalton's Law of Partial Pressures   The total pressure exerted by a mixture of gases is the sum of their individual partial pressures.

$$P_{\text{total}} = P_a + P_b + P_c + \text{etc.}$$

P_a, P_b, and so on refer to the partial pressures of the individual gases, a, b, etc. When the gases are known, then their formulas can be substituted for the small letters. Thus, PO_2 refers to the partial pressure of oxygen. The value of PO_2 in the atmosphere decreases with altitude until oxygen-enriched air eventually becomes essential for breathing (see Interaction 6.2).

■ You'll sometimes see a partial pressure like PO_2 written as P_{O_2}.

INTERACTION 6.2
BREATHING AT HIGH ALTITUDE

When the total pressure of dry air is 760 mm Hg, the value of PO_2 is 160 mm Hg. At an altitude of 5.5 km (3.4 miles; 18,000 ft), the total pressure is one half as much, 380 mm Hg, and so the partial pressure of oxygen is also one half as much, one-half of 160 mm Hg or 80 mm Hg.

For the lung systems of most people, a partial pressure of 80 mm Hg for oxygen in air is not high enough to force atmospheric oxygen out of the lungs and into the bloodstream at a rate fast enough to support life. Those who challenge this fact develop altitude sickness, which is life threatening. The only cure is supplemental, oxygen-enriched air or, without a supply of this, *prompt* return to low altitude. At 18,000 ft, air must be enriched to 42% oxygen to give its PO_2 the same value as it has at sea level. Nearly all people have to use pressure tanks that deliver oxygen-enriched air to the lungs when they operate even above 14,000 ft (Figure 1), although with conditioning at altitude, experienced mountain climbers are able, unaided by such air, to handle peaks this high and higher. Even Mount Everest has been climbed by several people without supplemental oxygen.

FIGURE 1 To survive and operate at high altitude, most people need to breathe oxygen-enriched air to beef up the effective partial pressure of the oxygen.

EXAMPLE 6.4   Using Dalton's Law of Partial Pressures

Suppose you want to fill a pressurized tank having a volume of 4.00 L with oxygen-enriched air for use in diving, and you want the tank to contain 50.0 g of O_2 and 150 g of N_2. What must be the total gas pressure at 25 °C?

ANALYSIS   If we calculate the pressure that each gas will have if it is *alone* in the tank and then add the two partial pressures together, we'll have the answer. To find the individual partial pressures, we use the ideal gas law, $PV = nRT$, for a temperature of 25 °C, or 298 K, and a volume of 4.00 L. To use this law, we must first express the amounts of the gases in *moles*.

SOLUTION   For moles of oxygen,

$$n = 50.0 \text{ g O}_2 \times \frac{1 \text{ mol O}_2}{32.0 \text{ g O}_2} = 1.56 \text{ mol O}_2$$

For moles of nitrogen,

$$n = 150 \text{ g N}_2 \times \frac{1 \text{ mol N}_2}{28.0 \text{ g N}_2} = 5.36 \text{ mol N}_2$$

Next, we calculate the partial pressures. For each, $P_a = \dfrac{nRT}{V}$, where V is the volume of the cylinder.

$$PO_2 = \frac{1.56 \text{ mol} \times 0.0821 \text{ L atm/K mol} \times 298 \text{ K}}{4.00 \text{ L}}$$

$$= 9.54 \text{ atm}$$

$$PN_2 = \frac{5.36 \text{ mol} \times 0.0821 \text{ L atm/K mol} \times 298 \text{ K}}{4.00 \text{ L}}$$

$$= 32.8 \text{ atm}$$

For the total pressure, using Dalton's law, we have

$$P_{total} = 9.54 \text{ atm} + 32.8 \text{ atm}$$
$$= 42.3 \text{ atm}$$

A relatively high pressure, 42.3 atm, is needed for the tank to hold only 50.0 g of O_2 and 150 g N_2.

■   PRACTICE EXERCISE 5   How many grams of oxygen are present at 25 °C in a 5.00-L tank of oxygen-enriched air under a total pressure of 30.0 atm when the only other gas is nitrogen at a partial pressure of 15.0 atm?

■   PRACTICE EXERCISE 6   The atmospheric pressure atop Mount Everest (8.8 km) is 250 mm Hg and the partial pressure of nitrogen is 198 mm Hg. What is the partial pressure of oxygen, assuming no other gas is present? (The value is too low to force O_2 from the lungs into the bloodstream at a rate fast enough to sustain human activities for nearly all people.)

■   Alveolar air is air inside *alveoli*, thin-walled air sacs enmeshed in beds of fine blood capillaries. We have roughly 300 million such air sacs in our lungs, providing an enormous surface area for the exchange of O_2 and CO_2 between the alveolar air and the bloodstream.

The Air We Exhale Includes Water and Carbon Dioxide   Table 6.2 gives partial pressure data for inhaled and exhaled air as well as for air in the lung's air sacs (alveoli). The air we inhale is generally not dry air, and its water vapor makes a small contribution to the total pressure. The warm, moist environment in the alveoli loads the exhaled air with water vapor, as the data in Table 6.2 also show. But notice that the total pressure, wherever the air sample, is still the sum of the partial pressures.

Notice, also in Table 6.2, the decrease in PO_2 between inhaled and exhaled air and the relatively large increase in PCO_2. The PO_2 decreases, of course, because oxygen is removed from the air in the lungs and is absorbed by the blood. The value of PCO_2 increases because carbon dioxide, a gaseous waste product of metabolism, is removed from the blood at the lungs and transferred to the air about to be exhaled.

TABLE 6.2   The Composition of Air during Breathing

Gas	Partial Pressure (in mm Hg)		
	Inhaled Air	**Exhaled Air**	**Alveolar Air**
Nitrogen	594.70	569	570
Oxygen	160.00	116	103
Carbon dioxide	0.30	28	40
Water vapor	5.00[a]	47	47
Totals	760.00	760	760

[a] A partial pressure of water vapor of 5.00 mm Hg corresponds to air with a relative humidity of about 20%, a condition in which air is holding 20% of the maximum amount of water vapor it is able to hold at the given temperature.

6.5 KINETIC THEORY OF GASES REVISITED

All the gas laws can be explained in terms of the model of an ideal gas described by the kinetic theory.

As the gas laws unfolded over the decades, more and more scientists asked the question, "What must gases really be like given the *experimental* facts about gases (the gas laws)?"

A Model of an Ideal Gas Led to Theoretical Conclusions That Matched the Gas Laws   Scientists used the behavior of real gases to postulate the following features of an ideal gas.

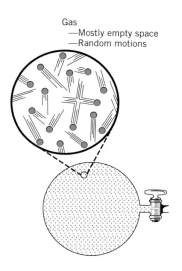

Gas
—Mostly empty space
—Random motions

Model of an Ideal Gas

1. The ideal gas consists of a large number of extremely tiny particles in chaotic, utterly random motion.

2. The particles are perfectly hard, and when they collide they lose no energy because of friction.

3. The particles neither attract nor repel each other.

4. The particles move in accordance with the known laws of motion.

These postulates and the calculations based on them make up the **kinetic theory of gases.** The truth about gases, according to this theory, is that they consist of tiny particles in random motion, each particle possessing kinetic energy. Let us now see how the kinetic model explains the gas laws.

We've already looked at one relationship. In Section 2.5 we learned that *the temperature of a gas is directly proportional to the average molecular kinetic energy of its molecules.* In other words, when we heat a gas and raise its temperature, we increase the average molecular kinetic energy of its particles. When we cool a gas, we reduce the average molecular kinetic energy. Because there's no such thing as *negative* kinetic energy, there ought to be a lower limit to which something can be cooled. As we know, this limit is −273.15 °C, absolute zero.

■ The term "molecular" in "molecular kinetic energy" can be stretched to include the monatomic, noble gases.

The Kinetic Theory Explains Gas Pressure and Boyle's Law   Theorists could demonstrate that the pressure exerted by a gas arises from the innumerable collisions per second that the particles make with each unit of area of the walls of the container. If we imagine, then, that the volume of the container is made smaller, the wall area battered by the gas particles is also less. More hits *per unit wall area* therefore occur after the volume has been reduced. Moreover, the reduction in volume does not affect the average speed at which the particles travel—only the temperature affects speed—and the particles don't have to travel as far to hit a wall. Thus, reducing the volume of the gas increases the frequency with which its particles hit the walls without reducing the strengths of each tiny collision, so the gas pressure must increase (see Figure 6.4).

■ Pressure is related to the ratio $\dfrac{\text{number of hits per second}}{\text{area hit}}$; so, as the denominator (area hit) grows smaller, the ratio (pressure) *increases.*

When the theoretical calculations on which the preceding description is based were carried out, the result was identical with Boyle's pressure–volume law for gases. This kind of agreement and others like it give us confidence that the model of an ideal gas closely describes real gases, too.

The Kinetic Theory Explains Charles' Law   If we heat a gas to make the gas molecules move with a higher average speed and energy, they will hit the container's

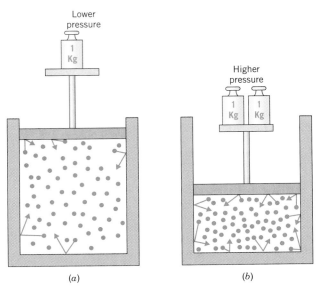

FIGURE 6.4 The kinetic theory and Boyle's law. The pressure of the gas is proportional to the frequency of collisions per unit area. When the gas volume is made smaller in (*b*), the frequency of the collisions per unit area of the container's walls increases. This is how the pressure increase occurs.

walls more frequently and with greater force. If we want to prevent the pressure from increasing—and remember that constant pressure is a condition of Charles' law—we have to let the gas volume expand. The theoretical calculations along these lines showed that gas volume at constant pressure is, in theory, proportional to the Kelvin temperature, just as Charles and others had earlier shown by experiment.

■ This is the law behind the warning not to incinerate aerosol cans.

The Kinetic Theory Also Explains Gay-Lussac's Law If we don't let the volume expand when we heat a confined gas, then the pressure of the gas must increase. The calculations based on the model of an ideal gas predicted the same relationship between P and T that had long been known from Gay-Lussac's law (pressure–temperature law).

The Kinetic Theory Explains Gas Diffusion Another property of gases that the kinetic theory explains is the ability of a gas to diffuse. **Diffusion** is the spontaneous spreading out of molecules because of their kinetic motions. Gas molecules in random motion eventually find their way into all the space available. Almost everyone has experienced the diffusion of perfume, cologne, or aftershave fragrances throughout a room.

6.6 LIQUID STATE

Molecules in liquids experience attractions for each other but retain the freedom to move about.

Liquid
Densely packed
Random motions

The lack of large empty spaces between molecules in liquids and solids prevents the existence of general laws, such as the gas laws, for these physical states. Yet, the kinetic theory helps us understand some of their features. Let's see how, starting with the liquid state.

Substances with Polar Molecules Are More Likely to Be Liquids than Those with Nonpolar Molecules of About the Same Size   Recall (Section 4.9) that when the molecules of a substance are polar they have permanent $\delta+$ and $\delta-$ sites. Remember also that opposite charges attract, even when they are partial charges. The attraction between permanent $\delta+$ and $\delta-$ sites on separate polar molecules is called a **dipole–dipole attraction,** and it is one origin of the electrical force that draws molecules together and enables a substance to be in the liquid state. When the partial charges are very weak, however, this attractive force cannot outbalance the repelling force at work as the electron clouds of gas molecules draw near each other on a collision course. Electron clouds repel each other. This is why gas molecules, being too weakly polar in relationship to the energies of their collisions, remain in the gaseous state.

To change a gas into a liquid we must cool it, exert pressure on it, or do some combination of both. Chilling a gas slows its molecules down and so weakens the force of repulsion caused by colliding electron clouds. Exerting a pressure squeezes the gas molecules together, and when they are packed tightly enough, forces of attraction between gas molecules are strong enough to enable the gas to change over to the liquid state.

London Forces of Attraction Exist between Nonpolar Molecules   Even the most completely nonpolar substances, like the noble gases, are changed into liquids and solids by lowering the temperature enough. This behavior raises the question, "How do electrical forces of attraction develop between *nonpolar* atoms or molecules?" The answer lies in the electron cloud encompassing any atom or molecule.

As one nonpolar particle closely approaches another, their two somewhat "soft" electron clouds distort each other (see Figure 6.5). ("Like charges repel.") The distortion gives to the approaching particles a *temporary* polarity; we say that polarization is *induced.* The induced polarization of molecules (or atoms) gives *temporary dipoles* even to particles lacking permanent dipoles.

In a sample with billions and billions of particles without *permanent* dipoles, we can still easily imagine that there are innumerable, temporarily polarized molecules. Existing between these molecules, therefore, is a net electrical force of attraction, which is called the **London force** (after physicist Fritz London). At sufficiently low temperatures, London forces are strong enough to cause even such totally nonpolar substances as the noble gases to change from a gas to the liquid state or from the liquid to the solid state.

London forces are related to the polarization of electron clouds, so the larger the overall electron cloud per atom or molecule, the more induced polarization a particle can receive. Thus, *substances with larger molecules or atoms generally have larger London forces than substances with smaller molecules or atoms.*

Boiling Points Reflect Attractive Forces   An excellent indication of the attractive forces between particles is simply the temperature required to make the substance boil, a temperature called the *boiling point.* When a liquid boils, its molecules are made to separate from each other. A high boiling point indicates that relatively large forces of attraction must be overcome before such separation can occur. Because greater London forces are possible with the larger electron clouds of bigger atoms or molecules, the rule of thumb for the boiling points of nonpolar substances is that *the higher the formula mass, the higher the boiling point.* The boiling points of the noble gases, given in the margin, illustrate this rule.

With sufficient molecular size, even nonpolar substances can be liquids and solids at room temperature and pressure. Examples are gasoline and paraffin wax,

■   When the charges are the full charges of ions, attractions between the ions are called ion–ion attractions, and they are much stronger than dipole–dipole attractions.

■   Each gas has a particular temperature called its *critical temperature* above which no amount of pressure can liquefy the gas. The critical temperature of water is 374.1 °C; for carbon dioxide it's 31 °C.

■   Helium becomes a liquid only at about 4 degrees above absolute zero (4.18 K), and changing helium to a solid requires an extra 26 atm of pressure.

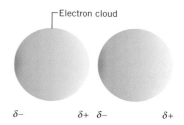

FIGURE 6.5 London forces. Molecules can be temporarily polarized by coming close to each other, and the force of attraction between such temporary dipoles is called a London force.

■

Noble Gas	Formula Mass	Boiling Point (°C)
He	4.00	−269
Ne	20.2	−246
Ar	39.9	−186
Kr	83.8	−152
Xe	131.3	−107

compounds made entirely of C and H atoms. The electronegativities of C and H are nearly the same, so neither C—C nor C—H bonds are very polar. Hence, molecules with only these bonds are nearly nonpolar. Yet London forces are great enough so that substances with such molecules can be liquids and solids under ordinary conditions.

6.7 VAPOR PRESSURE, DYNAMIC EQUILIBRIA, AND CHANGES OF STATE

When two opposing changes occur at the same rate, the system is in a state of *dynamic equilibrium.*

The Kinetic Theory Helps to Explain Vapor Pressure   Suppose we add some fixed quantity of water to a glass bulb "sealed" by a movable column of mercury (see Figure 6.6). There is "free space" in the bulb for the water to change between its liquid and vapor states. Initially, for a very brief moment, we can imagine that no water molecules are in the air space above the sample, so the pressure inside and outside the bulb is initially the same, namely, atmospheric pressure. This situation quickly changes, however, because some water molecules escape and move around in the space above the liquid. The change from liquid to vapor is called **evaporation.** As more and more of the liquid water evaporates, the pressure inside the bulb increases and pushes on the mercury column. In other words, water is exerting its vapor pressure. **Vapor pressure** is the partial pressure exerted by a vapor in contact with its liquid. A liquid's vapor pressure can be regarded as the escaping tendency of its molecules.

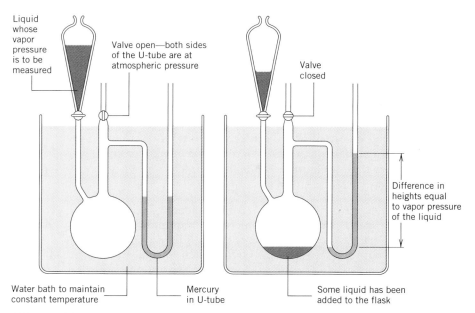

FIGURE 6.6   Measuring the vapor pressure of a liquid. When the liquid is added to the closed glass bulb, its vapor forces the mercury column to rise. The difference in the heights of the columns in the U-tube on the right is the equilibrium vapor pressure of the liquid. (From J. E. Brady and J. R. Holum, *Chemistry. The Study of Matter and Its Changes,* 2nd edition, 1996, John Wiley & Sons, Inc., New York. Used by permission.)

Eventually, molecules of water vapor, suffering random collisions with each other and with the liquid surface, begin to return to the liquid state. The change of a vapor to its liquid form is called **condensation.** Now, water molecules are coming and going, some leaving the liquid for the vapor state and others returning. The two processes, evaporation and condensation, are said to *oppose* each other because one reverses the effect of the other. Eventually, for every molecule leaving the liquid state there is one entering it. The two opposing processes then occur at identical rates and so exactly cancel each other's effect. After this, no further changes occur in the vapor pressure or in the actual masses of material in each of the two physical states. The system in the closed bulb has reached a steady state.

Equilibrium Exists When the Rates of Evaporation and Condensation Are Equal
When a dynamic process reaches a steady state in which opposing changes occur at identical rates and no further net change takes place, **dynamic equilibrium** exists for the system. We say "dynamic" because there is considerable coming and going; and we say "equilibrium" because there is no *net* change. An equilibrium is like a town with a stable population; there are births and deaths as well as departures and arrivals, but no net change occurs in number of people.

Dynamic Equilibria Involve Forward and Reverse Processes   To represent a dynamic equilibrium, we use an equation with double arrows. For convenience, we use the format of a chemical change even for one that is physical, and we may even refer to materials in two different physical states with the language of "reactant" and "product." Thus the equilibrium involving water is illustrated as follows.

$$\text{water}(l) \rightleftharpoons \text{water}(g) \tag{6.10}$$

This is an example of an *equilibrium expression* or an **equilibrium equation.** The change from left to right is the **forward reaction** or change, and the opposing change, from right to left, is the **reverse reaction.** The double arrows $\rightleftharpoons$ signify that equilibrium exists between the materials on both sides of the arrows.

At equilibrium between its liquid and vapor states, each liquid has an *equilibrium vapor pressure.* For a given liquid, it always increases with temperature, as seen in Figure 6.7, where the equilibrium vapor pressures of several liquids are plotted against temperature. Notice that ether develops high vapor pressures even at low temperatures. Such liquids are said to be **volatile;** they readily evaporate from open containers at room temperature. In contrast, propylene glycol (in Figure 6.7) and salad oil have very low vapor pressures at room temperature. They are called **nonvolatile** liquids because they do not evaporate at room temperature. Water is more volatile than salad oil, but less volatile than ether.

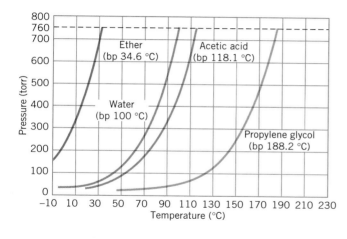

FIGURE 6.7 Equilibrium vapor pressure versus temperature. (Ether was once widely used as an anesthetic. Acetic acid is the sour component in vinegar. Propylene glycol is in several brands of antifreeze mixtures.)

Water Has an Unusually High Heat of Vaporization   Sometimes we include heat in an equilibrium expression as if it were either a reactant or a product. Just from boiling water to make coffee, we know that heat is needed to change a liquid into its vapor. When the vapor condenses the same amount of heat is released. We can therefore place "heat" on the *left* in the following equation, as if it were a reactant, because it is "consumed" by the forward change (evaporation) and released in the identical amount by the reverse change (condensation).

$$\text{water}(l) + \text{heat} \rightleftharpoons \text{water}(g) \tag{6.11}$$

Evaporation is thus *endothermic*, and condensation is *exothermic*.

The heat needed to change 1 g of a substance from its liquid to its gaseous form is called its **heat of vaporization.** The heat of vaporization of water at its boiling point, for example, is 539.6 cal/g, but this value is different at different temperatures. Evaporation, as you well know, can occur at any temperature. At body temperature (37 °C), for example, the heat of vaporization of water is about 580 cal/g.

Water's heat of vaporization is unusually high among substances, and this explains why letting a little water evaporate from the skin or from the lungs rids the body of considerable heat. Evaporation carries away about 40% of the heat that must be released each day from the body.

Dynamic Equilibria Can Be Shifted   Once an equilibrium is established, no net change occurs *spontaneously*, but this doesn't mean that we cannot cause a change. By including heat on the appropriate side of the equilibrium equation, it's easy to predict how an equilibrium will *shift* in response to an increase or a decrease in temperature. If, for example, we add heat to liquid water in equilibrium with its vapor in the glass bulb of Figure 6.6, the rate of evaporation will increase. For a time, the system will no longer be in equilibrium because the forward rate is now greater than the reverse. We say that the addition of heat *upsets the equilibrium*.

Once we stop adding heat but yet maintain the temperature of the (closed) system at a constant but higher value, the rate at which vapor molecules return to the liquid will catch up. The two opposing rates of evaporation and condensation will become the same again. Both will be faster than at the lower temperature, but when both are *equally* fast there is no further net change. Once again, the system is in dynamic equilibrium. At the higher temperature, of course, the vapor pressure of the liquid is higher because we have more of the gas in the same enclosed space.

Generally, any equilibrium shifts in response to a disturbance, like the addition or removal of heat. A disturbance to an equilibrium is called a **stress,** which is any factor that makes one of the two opposing changes faster than the other, at least for a while. Whichever change, forward or reverse, becomes faster is said to be *favored*. If the forward reaction becomes faster, we say that the equilibrium shifts to the right and at least some of the materials before the arrow are changed into materials after the arrow. Thus, adding heat to liquid water that is in equilibrium with its vapor places a stress on the equilibrium. In response, some liquid water evaporates, and more water vapor forms.

On the other hand, an equilibrium shifts to the left when the reverse reaction is favored for any reason. Thus, the removal of heat by cooling the water–water vapor system slows the rate of evaporation and increases the rate of condensation. However, as more and more vapor condenses, the *rate* of this change slows down so, eventually, the rates of evaporation and condensation become equal, and once again there will be equilibrium. Less water mass will remain as a vapor and more will be in the liquid state, and the vapor pressure is less, but equilibrium again reigns.

Shifts in Equilibria Can Be Predicted by Le Châtelier's Principle   Heat—its addition or removal—is just one stress that can affect an equilibrium, but it neatly illus-

■ The heat of vaporization of gasoline is only about 78 cal/g; that of ethyl alcohol is 204 cal/g.

■ "Spontaneous" here means "happening without the addition or removal of anything (or heat) to the system at equilibrium."

■ By **system** we simply mean something we have chosen to study, like the water–vapor equilibrium. Everything else in the universe would be called the **surroundings** to the system. Between the system and the surroundings stands a **boundary,** the walls of the container.

trates a major principle about how nature works. The principle is called **Le Châtelier's principle,** after Henri Louis Le Châtelier (1850–1936), a French chemist.

> **Le Châtelier's Principle** If a system in equilibrium is upset by a stress, the system shifts in whichever direction most directly absorbs the stress and restores equilibrium.

Provided that the applied stress hasn't been overwhelming, equilibrium will be restored. The rate of the unfavored reaction will eventually catch up to the opposing reaction. Both rates again become equal, and once more there is equilibrium. It's not the identical *system* that existed before the stress because the actual *quantities* of materials represented as reactants and products have changed, but there is once again equilibrium.

As we saw, the addition of heat to the water–water vapor system shifted the equilibrium to the right in favor of vapor. Le Châtelier's principle says that it *must* shift to the right because this is the only direction that absorbs the stress. The extra heat is used up as heat of vaporization and now exists as the molecular kinetic energy of the vapor molecules. Thus Le Châtelier's principle helps us predict the way a system at equilibrium *must* change under a given stress. Equilibria *always* change in whichever way absorbs an applied stress. It is as if the system "rolls with the punches."

A Liquid Boils When Its Vapor Pressure Equals the External Pressure Each liquid has a particular temperature at which its equilibrium vapor pressure exactly equals the external pressure of the atmosphere. At this point, the liquid's molecules can enter the vapor state, not just at the surface, *but everywhere throughout the liquid.* Bubbles of the vapor can now form *beneath* the surface, and they cause quite a commotion as they rise everywhere. This, of course, is the action called **boiling.** Each bubble is essentially pure vapor with a pressure equal to the opposing pressure of the atmosphere. Because the opposing pressure is no longer able to squeeze the bubbles back to liquid, and because the bubbles are much less dense than the liquid, they are free to rise to the surface.

The temperature at which a liquid's equilibrium vapor pressure equals 760 mm Hg (1 atm) is called the liquid's **normal boiling point.** Boiling, of course, can occur at other pressures. It happens at whatever temperature the vapor pressure equals the external pressure, but the associated temperature is not called the *normal* boiling point. For example, in Denver, Colorado, at an elevation of one mile where the atmospheric pressure is lower than it is at sea level, water boils at about 95 °C instead of 100 °C. The effect of this on cooking at high altitudes is described in Interaction 6.3. On the summit of Mount Everest, water boils just under 70 °C!

■ The liquid–vapor equilibrium of ether would be overwhelmed if so much heat were added that *all* of the liquid changed to vapor.

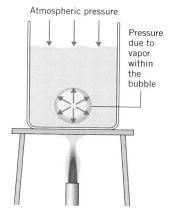

The vapor pressure inside the bubble overcomes the opposing pressure of the atmosphere at the boiling point.

■ On Mount Everest, the air pressure is roughly 250 mm Hg.

6.8 WATER AND THE HYDROGEN BOND

The physical properties of water are strongly affected by its polarity and the hydrogen bonds between its molecules.

Water is a particularly important liquid. Roughly 60% of the mass of an adult is water, and our bodies take in more water than all other materials combined. Water is the fluid in all cells, a heat-exchange agent, and the carrier in the bloodstream for the distribution of oxygen and all molecules from food, all hormones, minerals, and vitamins, and all disease-fighting agents. To understand many aspects of life at the

INTERACTION 6.3
COOKING AT HIGH ALTITUDE

As anyone who has tried to cook foods in boiling water at a higher altitude knows, it takes longer than at sea level. Where the atmospheric pressure is lower, water boils at a lower temperature. The chemical reactions caused by cooking are all endothermic. Therefore they do not occur as rapidly at a temperature of, say, 95 °C (roughly the boiling point of water in Denver, Colorado) as they do at 100 °C. No matter how high you turn up the stove setting as you prepare a soft-boiled egg using boiling water, you cannot raise the temperature of boiling water above its value at your altitude (Figure 1). Turning up the stove boils the water away *faster*, but it does not raise its temperature.

Of course, you could use a special pan with a tight lid, such as a pressure cooker. Now the steam cannot escape and its pressure can build up until the safety valve is activated. This higher pressure means that the temperature of the boiling water inside the pressure cooker is higher than in the open vessel, so the chemical reactions of cooking occur more rapidly.

These same principles are at work in steam sterilization equipment. To ensure that bacteria and viruses on surgical instruments are both quickly and completely destroyed, hospital workers can place the instruments in the equivalent of a pressure cooker where the water and steam temperature can be raised well above the normal boiling point of water.

FIGURE I Cooking with water at high altitude requires that you allow more time than at sea level.

molecular level, we must learn more about the high polarity of the water molecule and how this affects its properties.

■ Hydrides are compounds of hydrogen with another element.

Water's Boiling Point Is Unusually High The boiling points of *similar* substances generally increase with formula mass, a rule of thumb illustrated earlier by the noble gases. The group IVA hydrides—methane (CH_4), silane (SiH_4), and germane (GeH_4)—also follow the rule. Their boiling points increase regularly with increasing formula mass (see Figure 6.8). However, three simple hydrides with low formula masses, water, ammonia, and hydrogen fluoride, are each members of series displaying striking exceptions to the rule.

When the boiling points of the hydrides of the groups VA, VIA, and VIIA elements are plotted against their formula masses, the lowest members of a series do not fit a straight line plot (see Figure 6.8). Thus, among the group VA hydrides, ammonia (NH_3) boils far higher than "it should." It is badly off the straight line on which the other group VA hydrides fall, those of phosphorus (PH_3), arsenic (AsH_3), and antimony (SbH_3). Similarly, hydrogen fluoride (HF) does not fit the plot of the boiling points of the hydrides of the group VIIA elements, the halogens.

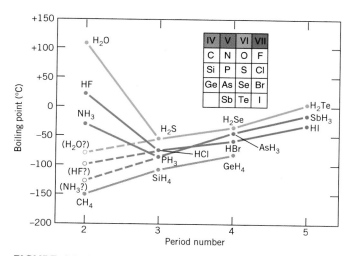

FIGURE 6.8 Boiling points versus formula masses for the binary, nonmetal hydrides of the elements in groups IVA, VA, VIA, and VIIA.

Water departs most of all from the normal trend in the boiling points of the hydrides of its group (VIA). If the boiling point of H_2O fell on the same line as the boiling points of H_2S, H_2Se, and H_2Te, water "should" boil at about -100 °C. But it actually boils at $+100$ °C, or 200 degrees higher.

The plots of Figure 6.8 suggest that forces of attraction are strong *between* molecules in liquid HF, H_2O, and NH_3, but not in methane (CH_4) These forces cannot be explained just as London forces, because the molecules of HF, H_2O, and NH_3 are too small. The molecules of these three hydrides, but not methane, must have unusually strong, *permanent* dipoles that allow relatively strong dipole–dipole attractions between their molecules.

Hydrogen Bonds Exist between Water Molecules   Oxygen is much more electronegative than hydrogen, so each of the two H—O bonds in H_2O are very polar with relatively large values of $\delta+$ on H and of $\delta-$ on O. The individual bond polarities in the H_2O molecule do not cancel each other because the molecule is bent. The water molecule as a whole is thus very polar, so polar that chemists consider the dipole–dipole attraction between water molecules strong enough to be called a *bond*. It's not a covalent bond or an ionic bond, but a dipole–dipole attraction having a special name—*hydrogen bond*. It occurs in other systems besides water, but the sizes of $\delta+$ and $\delta-$ are large enough for hydrogen bonds only when H is attached to an atom of one of the three most electronegative elements: O, N, or F. The **hydrogen bond** is the force of attraction between the $\delta+$ on H when held by F, O, or N and the $\delta-$ on some other O, N, or F atom. The hydrogen bond explains why HF, H_2O, and NH_3 have boiling points that are "out of line" for their molecular sizes.

Here are most of the several possibilities where hydrogen bonds can exist. Only partial structures are shown, and a dotted line is used to represent the hydrogen bond. (The solid lines, of course, are covalent bonds.)

$$\begin{matrix} \delta+ & \delta- & \delta+ & \delta- \\ H & \!-\!F & \cdots H & \!-\!F \end{matrix} \qquad \begin{matrix} \delta+ & \delta- & \delta+ & \delta- \\ H & \!-\!O & \cdots H & \!-\!O \end{matrix} \qquad \begin{matrix} \delta+ & \delta- & \delta+ & \delta- \\ H & \!-\!O & \cdots H & \!-\!N \end{matrix}$$

$$\begin{matrix} \delta+ & \delta- & \delta+ & \delta- \\ H & \!-\!N & \cdots H & \!-\!O \end{matrix} \qquad \begin{matrix} \delta+ & \delta- & \delta+ & \delta- \\ H & \!-\!N & \cdots H & \!-\!N \end{matrix}$$

■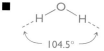

■ The direction of the polarity of the H—O bond is

$$\overrightarrow{H\!-\!O}$$

■ In large flexible molecules (like those of proteins), attractions between $\delta+$ and $\delta-$ sites can occur between parts of the same molecule.

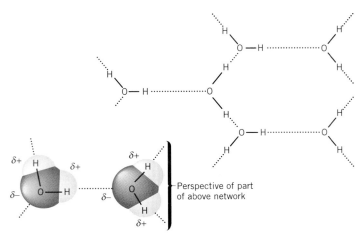

FIGURE 6.9   Hydrogen bonds in water (·····).

■ In a chain, the important link is the *weakest* link, not the strongest.

■ H_2O, bp 100 °C
   NH_3, bp −33.4 °C
   CH_4, bp −161.5 °C

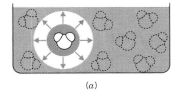

(a)

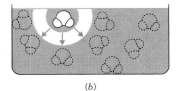

(b)

FIGURE 6.10   Surface tension. (a) In the interior of a sample of water, individual water molecules are attracted equally in all directions. (b) At the surface, nothing in the air counterbalances the downward pull that the surface molecules in water feel.

The hydrogen bond is a bridging bond between molecules, as illustrated for water in Figure 6.9. It is by no means as strong as a covalent bond, being roughly only 5% as strong. Nevertheless, the hydrogen bond is strong enough to make a difference not just in water but in such important yet different substances as muscle proteins, cotton fibers, and the chemicals of genes, DNA. In fact, among the biochemicals where they occur, hydrogen bonds are more important structurally than any other bond precisely because they are weak, not strong.

The hydrogen bond also helps us understand other properties of water such as the relatively high heat of vaporization of water noted earlier. An extra large input of heat per gram is necessary to boil water because energy is needed to overcome the hydrogen bonding force of attraction between water molecules.

The Hydrogen Bond Is Weaker in Liquid Ammonia than in Water   The H—N bonds in ammonia, NH_3, aren't as polar as the H—O bonds in water because N is less electronegative than O. Therefore, the hydrogen bonds between molecules in liquid ammonia aren't as strong as those that exist in water. This is chiefly why liquid ammonia doesn't have nearly as high a boiling point as that of water. Despite the lessened polarity of the H—N bond, and as suggested by the plots in Figure 6.8, hydrogen bonding nonetheless does occur in pure, liquid ammonia.

A Water Surface Acts Like a Skin Because of Hydrogen Bonding   All liquids possess a surface tension, but that of water is unusually high. **Surface tension** is a phenomenon in which the surface acts as though it were a thin, invisible, elastic membrane or skin. Water's surface tension explains, for example, why water forms tight droplets and doesn't spread out on a waxy surface but does spread out on clean glass. Surface tension is also the reason that a lung collapses under certain conditions.

Surface tension exists in water because its molecules so attract each other that they jam together where water meets air (see Figure 6.10). The dipole–dipole attractions (hydrogen bonds) that pull surface water molecules downward aren't counterbalanced by the almost zero forces from air molecules that pull the molecules upward. Thus, a net downward pull causes the surface jam-up of molecules responsible for surface tension.

On a greasy or waxed surface, water molecules gather together to form well-rounded beads (see Figure 6.11). The molecules in greases and waxes are relatively nonpolar so the inward-pulling forces of attraction in the water bead aren't matched by outward-pulling forces from the nonpolar molecules in wax or grease. Thus water does not spread out on a greasy or waxy surface.

Water does spread out on a clean, grease-free glass surface (Figure 6.12). Glass is rich in silicon–oxygen covalent bonds, so it is a very polar material. The $\delta+$ ends of H—O bonds in water are therefore attracted to the $\delta-$ charges on the oxygens in glass. Thus water molecules that are right on the glass find something to be more attracted to than they are to the molecules right behind them in the water. The water, therefore, *cannot help* but spread out. For the same reason, in a glass graduated cylinder or in a glass pipet, the boundary between the surface of an aqueous solution and the air, the *meniscus,* curves upward at the glass walls.

■ N$_2$ and O$_2$ molecules in air are nonpolar and so have no attraction for H$_2$O molecules.

FIGURE 6.11 Water forms beads on a waxed or greasy surface. Nothing in the surface has enough polarity to attract water molecules to make the water spread out. The net inward pull created by water molecules at the surface creates the bead, because this shape minimizes the total area of the droplet.

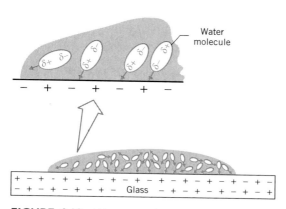

FIGURE 6.12 Water spreads out on a clean glass surface in response to strongly polar sites in the glass.

■ Glass consists mostly of silicon and oxygen. Si is less electronegative than H, so the S—O bond is more polar than the H—O bond.

Surface-Active Agents Reduce the Surface Tension of Water   There are many substances, called **surface-active agents** or **surfactants**, that lower the surface tension of water. All soaps and detergents are surfactants, for example. Soapy water won't bead on grease. A magician has to be careful that there is no soap or detergent whatsoever in the water used for the trick that sets a needle afloat on water. Before the 1980s, the leading cause of death among live-born preterm infants was a deficiency in their lungs of a natural surfactant, as discussed in Interaction 6.4.

6.9 SOLID STATE

Forces of attraction between the particles in a solid are greater than in a liquid or a gas.

The Particles in a Solid Do Not Move Around   Not only are atoms, molecules, or ions tightly packed in a solid, they also have fixed positions and fixed neighboring particles. This does not mean that they are completely at rest; they jiggle and vibrate about their fixed positions. But in the solid the forces of attraction between particles are just too strong to permit any of the movement that occurs in liquids or gases.

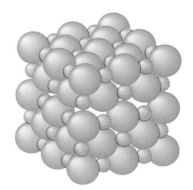

Solid (ionic)
Densely and orderly packed
Vibrations about fixed points

INTERACTION 6.4
SURFACTANT REPLACEMENT THERAPY FOR PRETERM BABIES

At the molecular level of life, a particularly important situation involving surfactants occurs. The moist membrane of an air sac (an alveolus) in the lungs carries a natural surfactant secreted by the membrane itself. Without it, water molecules would be attracted so strongly to the membrane that the air sac would collapse. If enough air sacs did this, breathing would be increasingly difficult and the entire lung could collapse. The membrane surfactant normally prevents this and so protects the lungs.

The lung's natural surfactant, however, does not fully mature until a baby is virtually full term and the cells lining the alveoli begin to make and secrete sufficient amounts. When born prematurely, a baby is at serious risk of respiratory distress syndrome—increased rate of respiration (tachypnea), chest retractions, and cyanosis (blood turning blue). As late as the early 1980s, respiratory distress syndrome was the leading cause of death among live-born preterm babies. Today, thanks largely to surfactant replacement therapy, this is no longer the case (Figure 1).

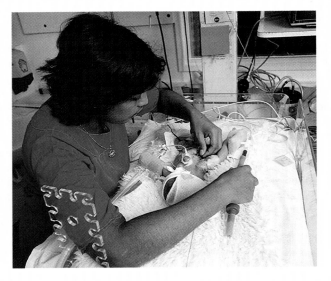

FIGURE 1 The survival of live-birth, preterm infants improved dramatically with the introduction of surfactant replacement therapy in the early 1980s.

A Solid–Liquid Equilibrium Exists at the Melting Point As the temperature of a solid is increased, the molecular kinetic energy and the vibrations of the individual particles become more and more intense. Eventually neighboring particles bump each other strongly enough to overcome the forces of attraction between them. Now the solid passes over into the liquid state; it *melts*. If the temperature of the system is carefully controlled, the rate at which particles leave the solid state and move around as a liquid can be made equal to the rate at which they return and take up fixed positions in the solid again. In other words, at the right temperature, the following equilibrium exists.

$$\text{solid} + \text{heat} \rightleftharpoons \text{liquid}$$

A mixture of ice pieces and water at 0 °C in a covered, well-insulated cup is an example of such an equilibrium system. The temperature at which an equilibrium exists between the solid and the liquid states of a substance is called its **melting point.**

The forward change in the above equilibrium is endothermic; it consumes heat. So if we put a stress on the equilibrium by adding heat—by raising the temperature—the equilibrium *must* shift to the right. Only such a shift uses up the stress, the added heat, and Le Châtelier's principle tells us that equilibria shift to *absorb* stresses. Of course, this shift means that some of the solid melts.

If we remove heat, if we cool the solid–liquid equilibrium, Le Châtelier's principle tells us that it must shift to the left; so some liquid will change to solid. Only a leftward shift can release some heat (the heat of fusion) to replace what is lost to the surroundings by cooling, and the equilibrium does what is necessary to replace heat being removed.

■ Heat of fusion is absorbed by a solid when it melts and the same heat is released when its liquid form solidifies again.

■ The heat of fusion of ethyl alcohol is 25 cal/g; that of gold is 15 cal/g.

The change of state from a solid to a liquid requires a characteristic quantity of heat, the **heat of fusion,** which is the heat needed to change 1 g of a solid to its liquid state at the melting point. Once again, water has an unusually high value; its heat of fusion is about 80 cal/g.

The icepack is a common application of the advantage given by the high heat of fusion of water. Ice is much more effective in an icepack than liquid water, even when the liquid form is at essentially the same temperature. The specific heat of liquid water is roughly 1.0 cal/g °C; its heat of fusion is about 80 cal/g. Thus, if *liquid* water at about 0 °C is to absorb 80 cal and yet increase in temperature only 1 °C, then 80 g of liquid are needed. But if *solid* water at roughly 0 °C is used to absorb 80 cal, then only 1 g of ice is needed.

Solids also exert a vapor pressure, but it usually is very weak. Some solids, like the naphthalene in mothballs, readily change directly from the solid to the vapor state, a change called **sublimation.** The solid form of carbon dioxide, "dry ice," also sublimes. To have solid CO_2 become a liquid, the system must be under a pressure of at least 5.2 atm.

Naphthalene sublimes; it passes directly to the vapor phase. Its vapors then condense on the cool surface of the flask filled with ice water where it forms beautiful crystals.

SUMMARY

Gas properties The four important variables for describing the physical properties of gases are number of moles (n), temperature T (in kelvins), volume (V), and pressure (P). We express pressure—force per unit area—in atmospheres (atm), mm Hg, torr, or pascals. Other important physical quantities in the study of gases are partial pressures, standard temperature and pressure (STP = 273 K and 760 mm Hg), and the molar volume at STP (22.4 L).

Gas laws All real gases obey, more or less, some important laws. Gas pressure is inversely proportional to volume (when n and T are fixed)—the pressure–volume law (Boyle's law). Gas volume is directly proportional to the Kelvin temperature (when n and P are fixed)—the temperature–volume law (Charles' law). The pressure of a gas is directly proportional to its Kelvin temperature (at fixed n and V)—Gay-Lussac's law. The general gas law incorporates Boyle's, Charles', and Gay-Lussac's laws.

Gas volume is directly proportional to the number of moles (when T and P are fixed). Equal volumes of gas (under identical T and P) have identical numbers of molecules (Avogadro's principle). Another gas law is that the total pressure of a mixture of gases equals the sum of their partial pressures (Dalton's law).

The ideal gas law incorporates all of the gas laws (except Dalton's): $PV = nRT$, where R, the universal gas constant, holds for all gases.

Kinetic theory If we imagine that an ideal gas consists of a huge number of very tiny, very hard particles in random, chaotic motion that do not attract or repel each other, the gas laws can be derived from the laws of motion. Out of this kinetic theory of gases came the insight that the Kelvin temperature of a gas is directly proportional to the average molecular kinetic energy of the gas particles.

Liquid state Liquids do not follow common laws as gases do, because essentially no space separates liquid particles from each other. In liquids, there are dipole–dipole attractions between particles originating from either permanent dipoles or temporary dipoles (London forces). Liquids can evaporate, and this "escaping tendency" gives rise to vapor pressure. Each liquid has a particular equilibrium vapor pressure that is a constant at each temperature, provided that care is taken to ensure a true dynamic equilibrium exists between the liquid and the vapor state. Evaporation is endothermic; condensation is exothermic. The heat needed to vaporize 1 g of a substance is its heat of vaporization. The heat of vaporization of water is unusually high.

When the liquid's vapor pressure equals the pressure of the atmosphere, the liquid boils. The normal boiling point is the temperature at which boiling occurs when the pressure is 760 mm Hg (1 atm).

Hydrogen bonds When hydrogen is covalently bonded to atoms of any of the three most electronegative elements (O, N, or F), its partial positive charge is large enough to be attracted to the partial negative charge on an atom of O, N, or F on a nearby molecule. This, the hydrogen bond, is a (relatively weak) force of attraction between the $\delta+$ on H in the polar bonds of the H—O, H—N, or H—F systems to the $\delta-$ of another O, N, or F.

Water   The higher electronegativity of oxygen over hydrogen and the angularity of the water molecule make it very polar, so polar that hydrogen bonds exist between its molecules. Hydrogen bonding helps to explain many of water's unusual thermal properties, such as its relatively high boiling point, its high heats of fusion and vaporization, and its high surface tension. Surfactants decrease the surface tension of water.

Dynamic equilibria   When the rates of two opposing changes, whether physical or chemical, are equal, the system is in dynamic equilibrium. If a stress, such as the addition or removal of heat, upsets an equilibrium, the equilibrium shifts in whichever direction absorbs the stress. If a change is exothermic, the reverse reaction is favored when more heat is added. If a change is endothermic, the forward reaction is favored by the addition of heat.

Solid state   The particles in a solid vibrate about fixed equilibrium points,but if the solid is heated, these vibrations eventually become so violent that the particles enter the liquid state. The temperature at which liquid and solid are in equilibrium is the melting point. Melting is endothermic and solidification is exothermic. The heat needed to melt 1 g of a solid at its melting point is its heat of fusion, and that of water is unusually high.

REVIEW EXERCISES

The answers to Review Exercises whose numbers are in color are found in Appendix E. The answers to the other Review Exercises are found in the Study Guide that accompanies this book. The more challenging questions are marked with asterisks.

General Properties of Gases

6.1   When the word *gas* is used in chemistry, does it refer to a *kind* of matter or a *state* of matter? What are the three *kinds* of matter? What are the three *states* of matter?

6.2   Gases generally share some common features, unlike liquids and solids. What are they?

6.3   To describe the physical state of a gas we provide what information about it, besides its chemical identity?

6.4   The property of *gas diffusion* ensures that a given sample of a gas will behave in what way?

Pressure

6.5   How is *pressure* defined?

6.6   What causes the atmosphere to have a pressure on the earth?

6.7   Why is the air pressure on top of a high mountain less than that at sea level?

6.8   What is the approximate weight in pounds of a column of air pressing on a 1-in.^{2} area of the earth's surface at sea level?

6.9   Referring to Review Exercise 6.8, if the area on which the air column rests is doubled to 2 in.^{2}, what else doubles, the *pressure* or the *weight* of the column of air? Explain.

6.10   What defines the *standard atmosphere*, the atmospheric pressure at sea level or something else? Explain.

6.11   What is the relationship between the *torr* and the *mm Hg?*

6.12   What is the name of the SI unit of pressure? Is it larger or smaller than the mm Hg?

6.13   Which number is closer to the number of pascals needed to equal one standard atmosphere: 10 Pa, 100 Pa, 1×10^3 Pa, 1×10^5 Pa, or 1×10^8 Pa?

6.14   Which exerts the higher pressure, a force of 100 kg acting on 25 cm^2 or a force of 25 kg acting on 5 cm^2?

6.15   Carry out the following unit conversions.
(a) 735 mm Hg to torr   (b) 740 mm Hg to atm
(c) 738 torr to mm Hg   (d) 1.45×10^3 Pa to mm Hg

6.16   What is the pressure in mm Hg of each of the following?
(a) 0.329 atm (summit of Mount Everest, world's highest mountain)
(b) 0.460 atm (summit of Mount Denali, highest mountain in the United States)

6.17   What is the pressure in kPa of each of the following? (These are the values of the pressures exerted individually by N$_2$, O$_2$, and CO$_2$, respectively, in typical inhaled air.)
(a) 595 mm Hg   (b) 160 mm Hg   (c) 0.300 mm Hg

6.18   In an unpressurized aircraft, a pilot's upper limit on altitude (assuming the availability of oxygen-enriched air) is about 40,000 ft (13 km, or 8 mi). The air pressure up there is roughly 0.20 atm. What is this in mm Hg? In torr? In kilopascals?

Specific Gas Laws and the Combined Gas Law

6.19   State the following laws in words. By what other name is each law known?
(a) temperature–volume law
(b) temperature–pressure law
(c) pressure–volume law
(d) law of partial pressures

6.20 Which of the four important variables in the study of the physical properties of gases are assumed to be held constant in each of the following laws?
(a) Boyle's law   (b) Charles' law
(c) Gay-Lussac's law   (d) Dalton's law
(e) Avogadro's principle   (f) general gas law

6.21 To compress nitrogen at 755 mm Hg from 740 mL to 525 mL, what must the new pressure be if the temperature is held constant?

6.22 A sample of helium at a pressure of 740 mm Hg and in a volume of 2.58 L was heated from 24.0 to 75.0 °C. The volume of the container expanded to 2.81 L. What was the final pressure (in mm Hg) of the helium?

6.23 What must be the new volume of a sample of nitrogen (in L) if 2.68 L at 745 mm Hg and 24.0 °C is heated to 375.0 °C under conditions that let the pressure change to 760 mm Hg?

6.24 When 280 mL of oxygen at 741 mm Hg and 18.0 °C was warmed to 33.0 °C, the pressure became 760 mm Hg. What was the final volume (in mL)?

6.25 After a sample of xenon with a volume of 532 mL was heated from 22.0 to 86.0 °C, its volume changed to 587 mL and its pressure became 789 mm Hg. What must have been its initial pressure in mm Hg?

Ideal Gas Laws

6.26 What volume in liters does 1.00 mol of O_2 occupy at 20.0 °C and 760 mm Hg?

6.27 A sample of 4.18 mol of H_2 at 18.0 °C occupies a volume of 24.0 L. Under what pressure, in atmospheres, is this sample?

6.28 If a steel cylinder of oxygen for respiratory therapy has a volume of 1.60 L and contains 10.0 mol of oxygen, under what pressure (in atm) is the oxygen if the temperature is 25.0 °C?

6.29 When the pressure in an oxygen cylinder with a volume of 4.50 L reaches 500 atm, the cylinder is likely to explode. If the cylinder contains 42.0 mol of oxygen at 24.0 °C, is it on the verge of exploding? (Calculate the pressure in atm.)

6.30 A steel cylinder with a volume of 25.0 L contains nitrogen under a pressure of 148 atm and a temperature of 25.0 °C. How many moles of nitrogen does the cylinder contain?

6.31 How many millimoles of oxygen are needed to fill a container with a volume of 750 mL at a temperature of 25.0 °C under a pressure of 750 mm Hg (roughly, laboratory conditions)?

***6.32** Under suitable conditions, water can be broken down by an electric current into hydrogen and oxygen according to the following equation.

$$2H_2O \longrightarrow 2H_2 + O_2$$

In one experiment, 875 mL of a dry sample of one of the product gases was collected at 748 mm Hg and 23.0 °C.
(a) How many moles of the gas were in this sample?
(b) The sample was found to have a mass of 1.133 g. What is the formula mass of the gas in the sample? (Recall that formula mass is the ratio of grams to moles.)
(c) Which gas was it, hydrogen or oxygen?
(d) Based on stoichiometry and Avogadro's principle, how many moles of the other gas were also collected?
(e) How many grams of the other gas were collected?

***6.33** Calcium carbonate decomposes as follows when it is strongly heated.

$$CaCO_3(s) \xrightarrow{heat} CaO(s) + CO_2(g)$$

In one experiment, 246 mL of CO_2 was collected at 740 mm Hg and 24.0 °C.

(a) How many moles of CO_2 formed?
(b) How many moles of $CaCO_3$ decomposed?
(c) How many grams of $CaCO_3$ decomposed?
(d) How many grams of CaO formed?

Dalton's Law of Partial Pressures

6.34 To what kind of a gas sample is the law of partial pressures relevant, and why is this law important for an understanding of the molecular basis of life?

6.35 What was it that John Dalton discovered about a mixture of gases?

6.36 The partial pressure of nitrogen in clean, dry air at 0 °C is 601 mm Hg when the total pressure is 760 mm Hg. The only other gas present to any appreciable extent is oxygen. If a sample of this air were sealed into a container at a pressure of 760 mm Hg and then all of the oxygen were removed, what would be the pressure exerted by the residual gas?

***6.37** A student was asked to prepare a sample of hydrogen gas by collecting it over water as it was produced by the reaction of zinc metal with hydrochloric acid.

$$Zn(s) + 2HCl(aq) \longrightarrow ZnCl_2(aq) + H_2(g)$$

The assignment was eventually to produce 250 mL of dry hydrogen as measured at STP. What are the minimum numbers of moles and grams of Zn needed to make this much H_2?

Kinetic Theory of Gases

6.38 Scientists asked, "What must gases be like for the gas laws to be true?" What was their answer?

6.39 What is true about an ideal gas that is not strictly true about any real gas?

6.40 Dalton's law of partial pressures implies that gas molecules from different gases actually leave each other alone in the mixture, both physically and chemically (except at moments of collisions, when they push each other around). Which one of the four postulates in the model of an ideal gas is based on Dalton's law?

6.41   How does the kinetic theory of gases explain the phenomenon of gas pressure?

6.42   How does the kinetic theory of gases account for Boyle's law (in general terms)?

6.43   What happens to the motions of gaseous molecules at 0 K?

6.44   How does the kinetic theory explain (in general terms) the volume–temperature law?

6.45   The pressure–temperature law (Gay-Lussac's law) can be explained in terms of the kinetic theory in what way (in general terms)?

The Liquid State, Vapor Pressure, and Dynamic Equilibria

6.46   Why aren't there universal laws for the physical behavior of liquids (or solids) as there are for gases?

6.47   Explain how there can be forces of attraction between atoms like those of argon (a monatomic noble gas), or between molecules of N_2, when these particles have no permanent dipoles, as in H—Cl.

6.48   Which has the higher boiling point, octane (C_8H_{18}) or butane (C_4H_{10})? (How can we tell without a table of boiling points?)

6.49   How does the kinetic theory explain
(a) How vapor pressure arises?
(b) Why vapor pressure rises with increasing liquid temperature?

6.50   Dimethyl sulfoxide (DMSO) is a controversial pain-killing drug the use of which is permitted by only a few states. Its boiling point is 189 °C.
(a) Is it more volatile or less volatile than water? Explain.
(b) The structure of DMSO has a double bond between S and O. Which atom of the S=O system carries a permanent $\delta+$ and which a permanent $\delta-$? Explain.
(c) What is the name of the chief force of attraction between DMSO molecules?

*6.51   Compare the following expressions.

A   Water + heat $\longrightarrow$ water vapor
B   Water + heat $\longleftarrow$ water vapor
C   Water + heat $\rightleftharpoons$ water vapor

Answer parts a–e by using the letter A, B, or C to indicate which expression best represents the answer.
(a) Which expression describes heat being liberated from the system?
(b) Which expression describes the net formation of liquid water from water vapor?
(c) Which expression describes a net endothermic change?
(d) Which expression shows opposing changes?
(e) Which expression, A or B, represents the forward change in expression C? Which represents the reverse change?

(f) Expression C can be the correct description for the water–water vapor system at 1 atm only if the temperature is what?
(g) What is true about the opposing changes in C?
(h) What special term applies to expression C? (It does not represent a *reaction* or a permanent *change*, but what?)

6.52   When a liquid and its vapor are in dynamic equilibrium at a given temperature, the rates of what two changes are equal?

6.53   Explain why a liquid's boiling point decreases as the external pressure decreases.

6.54   What is Le Châtelier's principle?

*6.55   Consider the following equilibrium existing in a sealed container.

$$\text{alcohol}(l) + \text{heat} \rightleftharpoons \text{alcohol}(g)$$

(a) What takes place when the reverse reaction occurs?
(b) Which reaction, the forward or the reverse, is endothermic?
(c) In which direction will the equilibrium shift if the system is in a sealed container and the pressure on the system is increased by some outside means not involving a change in the system temperature? (Hint: Which change, forward or reverse, is a volume-reducing change?)
(d) In which direction would the equilibrium shift if alcohol vapor could be removed from the container? Explain.
(e) If the system is cooled, will the equilibrium be affected? If so, in what way?

*6.56   Consider the following chemical equilibrium in which all the substances are gases.

$$N_2 + O_2 + \text{heat} \rightleftharpoons 2NO$$
$$\text{nitric oxide}$$

(a) Describe the chemical change for the forward reaction.
(b) What is the chemical change when the reverse reaction occurs?
(c) Which reaction, the forward or the reverse, is endothermic?
(d) Which reaction, the forward or the reverse, will be favored by adding heat?

Water and the Hydrogen Bond

6.57   The methane molecule, CH_4, does not become involved in hydrogen bonding. Why not?

6.58   Solid sodium hydroxide, NaOH, includes two kinds of chemical bonds. Which ones are they, and how do they differ?

6.59   Draw the structures of two water molecules. Write in $\delta+$ and $\delta-$ symbols where they belong. Then draw a correctly positioned dotted line between two molecules to symbolize a hydrogen bond.

*6.60   The hydrogen bond between two molecules of ammonia must be much weaker than the hydrogen bond between two molecules of water.

(a) How do boiling point data suggest this?

(b) What does this suggest about the relative sizes of the δ+ and the δ− sites in molecules of water and ammonia?

(c) Why are the δ+ and the δ− sites different in their relative amounts of fractional electric charge when we compare molecules of ammonia and water?

6.61 If it takes roughly 100 kcal/mol to break the covalent bond between O and H in H_2O, about how many kilocalories per mole are needed to break the hydrogen bonds in a sample of liquid water?

6.62 Explain in your own words how hydrogen bonding helps us understand each of the following.
(a) The high heats of fusion and vaporization of water
(b) The high surface tension of water

6.63 Explain in your own words why water forms tight beads on a waxy surface but spreads out on a clean glass surface.

6.64 What does a surfactant do to water's surface tension?

6.65 What common household materials are surfactants?

The Solid State

6.66 Describe the motions made by particles (e.g., ions or molecules) in a solid crystal.

6.67 What is the mechanism whereby heat causes a solid to melt?

6.68 What do we call the temperature at which a solid is in equilibrium with its liquid form?

6.69 At room temperature, nitrogen is a gas, water is a liquid, and sodium chloride is a solid. What do these facts tell us about the relative strengths of electrical forces of attraction in these substances?

6.70 Write an equilibrium expression for the sublimation of solid CO_2. Include "heat" in the appropriate place.

6.71 The following are some common observations. Using the kinetic theory, explain how each occurs in terms of what molecules are doing.
(a) Moisture evaporates faster in a breeze than in still air.
(b) Ice melts much faster if it is crushed than if it is left in one large block.
(c) Even if hung out to dry in below-freezing weather, wet clothes will become completely dry even though they freeze first.

Air Bags (Interaction 6.1)

6.72 Some air bags contain 130 g of sodium azide. Calculate the minimum volume of an air bag in liters needed to contain the nitrogen that is available from this mass of sodium azide, assuming that the nitrogen is collected under standard conditions of temperature and pressure.

Breathing at High Altitude (Interaction 6.2)

6.73 What must be the value of PO_2 for most people if they are to be able to take oxygen from the air?

6.74 Why is the partial pressure of oxygen at, say, 18,000 ft less than at sea level despite the fact that the volume percentage of oxygen is the same at both locations?

The Effect of Altitude on Cooking Times (Interaction 6.3)

6.75 Cooking an egg involves heat-induced chemical reactions as well as some physical changes. Why does it take longer to prepare a soft-boiled egg in Denver than in New York City?

Surfactant Replacement Therapy (Interaction 6.4)

6.76 Why are live-born preterm infants so much more at risk for respiratory distress syndrome than one-day-old, full-term infants?

6.77 What precaution against respiratory distress syndrome is now taken on behalf of preterm babies?

Additional Exercises

6.78 The diameter of the glass tube used to make a Torricelli barometer varies from barometer to barometer, yet the measured pressures are the same at the same time and location. How can this be?

*6.79 Imagine a Torricelli barometer constructed to use water instead of mercury.
(a) How high in *feet* will the column of water stand under standard atmosphere? (At 0 °C, the density of mercury is 13.5955 g/mL and that of water is 0.99987 g/mL).
(b) Why is the high density of mercury an advantage in a Torricelli barometer?

*6.80 When expressed in *inches* of mercury instead of mm Hg, the standard atmosphere is 29.92 in. Hg, which is the standard used by TV weather reporters in the United States. The lowest pressure ever recorded in the western hemisphere is 26.22 in. Hg (hurricane Gilbert, September 1988). What is this pressure in mm Hg?

6.81 When a sample of neon with a volume of 648 mL and a pressure of 0.985 atm was heated from 16.0 to 63.0 °C, its volume became 689 mL. What was its final pressure (in atm)?

6.82 A sample of argon with a volume of 6.18 L, a pressure of 761 mm Hg, and a temperature of 20.0 °C expanded to a volume of 9.45 L and a pressure of 373 mm Hg. What was its final temperature in kelvins and in °C?

6.83 At STP how many molecules of H_2 are in 22.4 L?

6.84 A sample of 1.00 mol of N_2 at 50.0 °C and 745 mm Hg occupies what volume, in liters?

6.85   A respiratory care unit had a steel cylinder containing 79.8 mol of oxygen with a volume of 10.0 L. At a temperature of 24 °C, what is the oxygen pressure (in atm)?

6.86   A container with a capacity of 750 mL held helium at a pressure of 740 mm Hg at a temperature of 25.0 °C.
(a) How many moles of helium were in the container?
(b) How many moles of H_2 could the container hold under identical conditions of pressure and temperature? (How can you tell without doing an additional calculation?)

6.87   What size container (in mL) is needed to hold 12.5 mmol of O_2 at 25.0 °C and a pressure of 0.965 atm?

*6.88   The following equation shows how ammonia can be made from nitrogen and hydrogen.

$$N_2(g) + 3H_2(g) \xrightarrow[\text{heat}]{\text{high pressure}} 2NH_3(g)$$

(a) If 6.00 mol of H_2 are consumed, how many moles of NH_3 are produced?
(b) If 250 L of nitrogen at 745 mm Hg and 25.0 °C are consumed, how many liters of ammonia can be made when the measurements of V and T are made at 745 mm Hg and 25.0 °C?
(c) If 75.0 g of N_2 are consumed, how many grams of H_2 are required and how many grams of NH_3 can be made?

6.89   A gas mixture at 745 mm Hg consists of at least two gases, helium and argon. Their partial pressures are 350 mm Hg for helium and 375 mm Hg for argon. Is it likely that any other gas is also present? How can you tell?

6.90   Those working out the kinetic theory have found that for 1 mol of an ideal gas, the product of pressure and volume is proportional to the average molecular kinetic energy of the ideal gas particles.
(a) To which of the four physical quantities used to describe a gas is the product of pressure and volume for 1 mol of a gas also proportional, according to the universal gas law (which makes no mention of kinetic energy)?
(b) If the product of P and V is proportional both to the average kinetic energy of the ideal gas particles and to the Kelvin temperature, what does this say about the relationship between the average kinetic energy and this temperature?

*6.91   Consider the following equilibrium that involves a common air pollutant, NO_2. Both substances are gases at room temperature.

$$2NO_2 \rightleftharpoons N_2O_4$$

When the temperature is reduced, this equilibrium shifts to the right.
(a) Which reaction, the forward or the reverse, is exothermic? Rewrite the equilibrium with "heat" positioned correctly.
(b) This reaction can be shifted in what direction by increasing the pressure on the mixture? Explain in terms of properties of gases and Le Châtelier's principle.

6.92   The hydrogen molecule, H—H, does not become involved in hydrogen bonding.
(a) What kind of bond occurs in a hydrogen molecule?
(b) Why can't this molecule become involved in hydrogen bonding?

6.93   Hydrogen bonds exist between two molecules of ammonia, NH_3. Draw the structures of two ammonia molecules. (You don't have to try to duplicate their tetrahedral geometry.) Put $\delta+$ and $\delta-$ signs where they should be located. Then draw a dotted line that correctly connects two points to represent a hydrogen bond.

6.94   If 456 mL of oxygen at 912 mm Hg is allowed to expand at constant temperature until its pressure is 760 mm Hg, what will be the volume of the oxygen sample?

*6.95   After the contents of a steel cylinder of pressurized oxygen with a volume of 18.5 L have been depleted until the pressure inside the cylinder is 1.00 atm, the cylinder is called "empty." How many moles of oxygen remain, however, if the temperature is 24.0 °C?

*6.96   Carbon dioxide can be made in the lab by the reaction of hydrochloric acid (HCl) with calcium carbonate according to the following equation.

$$CaCO_3(s) + 2HCl(aq) \longrightarrow CaCl_2(aq) + H_2O + CO_2(g)$$

How many grams of calcium carbonate and how many milliliters of 8.00 M HCl are needed to prepare 465 mL of dry CO_2 if it is collected at 20.0 °C and 745 mm Hg?

*6.97   The highest atmospheric pressure ever recorded in New York City was 31.06 in. Hg (December 25, 1949). What is this pressure in mm Hg and in kilopascals?

SOLUTIONS AND COLLOIDS

7

Light seems to stream sometimes because ultra tiny or colloidal sized particles of dust and moisture reflect light. It's the *Tyndall effect*, just one topic in our study of solutions and colloids in this chapter.

THIS CHAPTER IN CONTEXT

This chapter is about the general properties of homogeneous mixtures, so we'll build on our background in pure substances as we look at solutions and colloidal dispersions. One reason why we must study these special mixtures is that the bloodstream is simultaneously both. We're also interested in why some things dissolve in water and others don't. We'll look at one factor affecting the solubilities in blood of the gases of respiration, oxygen and carbon dioxide. And our study of dialysis will give us an insight into how the blood releases some things and absorbs others.

7.1 TYPES OF HOMOGENEOUS MIXTURES

The *sizes* of the particles intimately mixed with a solvent determine some physical properties of the mixtures.

■ Unlike pure substances (elements and compounds), mixtures do not obey the law of definite proportions.

Mixtures are either *homogeneous* or *heterogeneous*, but in either case they are always made of two or more nonreacting, pure substances occurring in *variable* proportions. **Homogeneous mixtures** are those in which the tiniest samples are everywhere identical in composition and properties. **Heterogeneous mixtures** are any that are not homogeneous. Most beverages are homogeneous mixtures, but orange juice with pulp is heterogeneous because a piece of the pulp itself does not have the same composition as the clear fluid. You can see, however, that the distinctions between homogeneous and heterogeneous depend altogether on what is meant by "tiniest sample." By this we normally mean a sample that can be manipulated and that may be as small as a fraction of a microgram or microliter. Even a microgram quantity of a pure substance will have a billion billion formula units.

■ "Normally" is a weasel word. What we really want to do is retain a flexibility in how to apply the terms "homogeneous" and "heterogeneous."

There are two kinds of relatively stable homogeneous mixtures: *solutions* and *colloidal dispersions*. A third kind, the *suspension*, is unstable, being homogeneous only when constantly stirred. The kinds of homogeneous mixtures differ fundamentally in the sizes of the particles involved, and the differences in size can alone cause interesting and important changes in properties.

In Solutions, the Dispersed Particles Are the Smallest A **solution** is a homogeneous mixture in which the particles of the solvent and all solutes have sizes of atoms, or ordinary ions and molecules. Their formula masses are no more than a few hundred; their diameters are 0.1 to 1 nm.

■ 1 nm = 10^{-9} m = 1 nanometer

We usually think of solutions as being liquids, but, in principle, the solvent or the solutes can be in any physical state—solid, liquid, or gas. Thus, there are several combinations that produce solutions (see Table 7.1).

Solutions are generally transparent—you can see through them—but they often are colored. Solutes do not settle out of solutions under the influence of gravity, and solutes cannot be separated from solutions by the use of filter paper. Among the many solutes carried in solution in the blood are the ions of NaCl and the molecules of glucose ($C_6H_{12}O_6$).

In Colloidal Dispersions, the Dispersed Particles Are Larger than Those in a Solution A **colloidal dispersion** is a homogeneous mixture in which the dispersed particles have diameters in the range of 1 nm to 1000 nm and consist of very large clusters of ions or molecules. Table 7.2 gives several examples of colloidal dispersions, and they include many familiar substances such as whipped cream, milk, dusty air, jellies, and pearls.

TABLE 7.1   Solutions

Kinds	Common Examples
Gaseous solutions	
Gas in a gas	Air
Liquid in a gas	(If droplets are present, a colloidal system)
Solid in a gas	(If particles are present, a colloidal system)
Liquid solutions	
Gas in a liquid	Carbonated beverages (carbon dioxide in water)
Liquid in a liquid	Vinegar (acetic acid in water), gasoline
Solid in a liquid	Sugar in water, seawater
Solid solutions	
Gas in a solid	Alloy of palladium and hydrogen[a]
Liquid in a solid	Toluene in rubber (e.g., rubber cement)
Solid in a solid	Zinc in copper (brass)

The blood also carries many substances in colloidal dispersions, including a variety of proteins. The molecules of many proteins are so huge that their formula masses run into the hundreds of thousands and so are called **macromolecules**. Some macromolecules are large enough to be of colloidal size and so form colloidal dispersions rather than solutions.

When colloidal dispersions are in a fluid state—liquid or gas—the dispersed particles, although large, are not large enough to be trapped by ordinary filter paper during filtration. They are large enough, however, to reflect and scatter light (Figure 7.1). Light scattering by a colloidal dispersion is called the **Tyndall effect**, after British scientist John Tyndall (1820–1893). The Tyndall effect is responsible for the milky, partly obscuring character of smog, and the way sunlight sometimes seems to stream through a forest canopy.

When a colloidal dispersion is a fluid, the large colloidally dispersed particles eventually settle out under the influence of gravity. Settling isn't always instantaneous, however; it can take a few seconds to many decades, depending on the system. A constant buffeting by solvent molecules helps to keep colloidal particles dispersed, causing them to move about erratically and unevenly, a behavior called the

■ The prefix *macro-* signifies something enormous in size relative to the suffix, such as "molecule" in *macromolecule*.

■ Solutions do not exhibit the Tyndall effect because the solute particles are too small.

FIGURE 7.1   Tyndall effect. The tube on the left contains a colloidal starch dispersion, and the tube on the right has a colloidal dispersion of Fe_2O_3 in water. The middle tube has a solution of Na_2CrO_4, a colored solute. The thin red laser light is partly scattered in the two colloidal dispersions, so it can be seen, but it passes through the middle solution unchanged.

TABLE 7.2   Colloidal Systems

Type	Dispersed Phase[a]	Dispersing Medium[b]	Common Examples
Foam	Gas	Liquid	Suds, whipped cream
Solid foam	Gas	Solid	Pumice, marshmallow
Liquid aerosol	Liquid	Gas	Mist, fog, clouds, certain air pollutants
Emulsion	Liquid	Liquid	Cream, mayonnaise, milk
Solid emulsion	Liquid	Solid	Butter, cheese
Smoke	Solid	Gas	Dust in smog
Sol	Solid	Liquid	Starch in water, jellies[c], paints
Solid sol	Solid	Solid	Black diamonds, pearls, opals, alloys

[a] The colloidal particles constitute the dispersed phase.
[b] The continuous matter into which the colloidal particles are scattered is called the dispersing medium.
[c] Sols that adopt a semisolid, semirigid form (e.g., gelatin desserts, fruit jellies) are called **gels**.

■ Robert Brown (1773–1858), a Scottish botanist, first observed this phenomenon when he saw the trembling of particles inside pollen grains viewed with a microscope.

Brownian movement. Light scintillating from colloidal particles as they engage in Brownian movement can be viewed with the aid of a good microscope.

In the most stable colloidal systems the dispersed particles carry like electrical charges. This also stabilizes colloidal dispersions because like-charged colloidal particles naturally repel each other. Such repulsions prevent the particles from gathering together and sticking to each other to make still larger particles that would be heavy enough to settle out. Protein molecules in living systems, for example, commonly carry net positive or negative charges that keep the molecules from "growing" into large clumps. (Other dissolved species of opposite charge, such as small ions, are present in solution to balance the charges on the colloidal particles.)

The Charge Status of Blood Colloids Is a Matter of Life and Death We've just had our first hint at how vitally important it is that the electrical charges on the particles in blood be normal. ("Normal" means that there is a proper *electrolyte balance*, both ions and charges, a concept briefly introduced in Interaction 4.1 and that will be expanded as we go.) Departures from normal could cause protein molecules in blood to aggregate and impede blood flow. So many factors are involved from several fields of chemistry, however, that we have to spread the study of the blood's electrolyte balance over parts of several chapters.

Emulsions Can Be Stabilized by Emulsifying Agents Colloidal dispersions of two liquids in each other, like oil and vinegar in salad dressing, are common and are called **emulsions**. They usually are not stable, and once made by shaking the mixture, the oil soon separates from the water. Sometimes, however, an emulsion is stabilized by a third component called an *emulsifying agent*. For example, mayonnaise is stabilized by egg yolk, whose electrically charged protein molecules coat the microdroplets of olive oil or corn oil and prevent them from merging into drops large enough to rise to the surface.

Fluid Suspensions Must Be Stirred to Remain Homogeneous In **suspensions**, the dispersed or suspended particles are over 1000 nm in average diameter, they separate under the influence of gravity, and they are large enough to be trapped by filter paper. A fluid suspension such as clay in water has to be stirred constantly to keep it from separating. A suspension is thus always on the borderline between being homogeneous and heterogeneous. The blood, while it is moving, is a suspension, besides being a solution and a colloidal dispersion. Suspended in circulating blood are its red and white cells and its platelets.

Table 7.3 gives a summary of the chief features of solutions, colloidal dispersions, and suspensions.

7.2 AQUEOUS SOLUTIONS AND HOW THEY FORM

Water dissolves best those substances whose ions or molecules can strongly attract water molecules.

■ We introduced the vocabulary of solutions in Section 5.4. You may want to review the meanings of *solute, solvent, unsaturated,* and *saturated* before continuing.

When sodium chloride crystals are dropped into water, their surfaces are instantly bombarded by water molecules, an action that works to dislodge the ions (Figure 7.2). But we might ask why Na^+ and Cl^- ions can leave their exceptionally stable environment in the crystal, the ions being surrounded as they are by oppositely charged neighbors. Yet water does dissolve salt, and solvents like gasoline or alcohol do not. How can water do this?

TABLE 7.3 **Characteristics of Three Mixtures: Solutions, Colloidal Dispersions, and Suspensions**

Particle Sizes Become Larger →		
Solutions	**Colloidal Dispersions**	**Suspensions**
All particles are on the order of atoms, ions, or small molecules (0.1–1 nm)	Particles of at least one component are large clusters of atoms, ions, or small molecules, or are very large ions or molecules (1–1000 nm)	Particles of at least one component may be individually seen with a low-power microscope (over 1000 nm)
Most stable to gravity	Less stable to gravity	Unstable to gravity
Most homogeneous	Also homogeneous but borderline	Homogeneous only if well stirred
Transparent (but often colored)	Often translucent or opaque, but may be transparent	Often opaque, but may appear translucent
No Tyndall effect	Tyndall effect	Not applicable (suspensions cannot be transparent)
No Brownian movement	Brownian movement	Particles separate unless system is stirred
Cannot be separated by filtration	Cannot be separated by filtration	Can be separated by filtration

Homogeneous to Heterogeneous →		

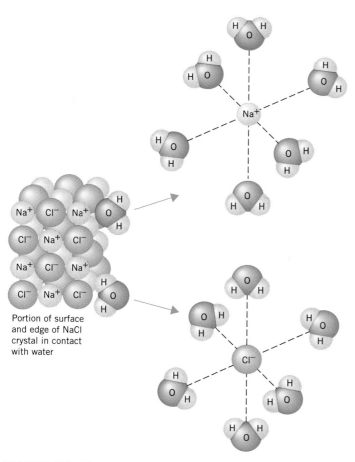

FIGURE 7.2 The hydration of ions helps ionic substances dissolve in water.

Portion of surface and edge of NaCl crystal in contact with water

Water Molecules Form Solvent Cages around Ions   Water molecules are very polar and have sizable partial charges. The $\delta-$ sites on their oxygen atoms attract Na^+ ions from the crystal, so once these ions are dislodged, they are surrounded by water molecules whose $\delta-$ ends point toward the positively charged ion (Figure 7.2). Similarly, water molecules also attract chloride ions and surround them (Figure 7.2). The $\delta+$ sites of water molecules point toward the negatively charged Cl^- ions. Both of the ions of NaCl can thus become surrounded by "cages" of water molecules—*solvent cages*. These cages are what stabilize the ions in water and permit the crystals to dissolve.

In its solvent cage, a chloride ion no longer has Na^+ ions as nearest neighbors, but several water molecules perform the same service, namely, they give a Cl^- ion an environment of opposite charge. Similarly, in their solvent cages, Na^+ ions no longer have Cl^- ions as nearest neighbors; water molecules take their place.

The phenomenon whereby water molecules are attracted to solute ions is called **hydration**, and the resulting clusters are called **hydrated ions**. The force of attraction between an ion and one end of the dipole of a polar molecule is called an **ion–dipole attraction**.

■ *Hydr- is from the Greek hydor, water.*

Water Is Unusual as a Solvent   Of all of the common solvents, only water has molecules both polar enough and small enough to form effective solvent cages around ions. Of course, in order for water to do this, its molecules must give up some of their attractions for each other. Only ionic substances or compounds made of very polar molecules can break up the hydrogen-bonded network between water molecules. Thus the formation of a solution isn't simply the separation of the solute particles from each other. It is also, to some extent, the separation of solvent molecules from each other.

When an ionic compound dissolves in water, we say that **dissociation** occurs because the oppositely charged ions separate—dissociate—from each other. Dissociation can be represented by an equation in which we take note of the physical states of the species.

$$NaCl(s) \xrightarrow{\text{dissociation}} Na^+(aq) + Cl^-(aq)$$

We should note here that chemists do not always include "(aq)" with the symbol of an ion dissolved in water. Often this designation is "understood." Usually, the context of the discussion makes very clear what is intended; whenever an ionic compound dissolves in water, its ions are *always* hydrated.

We should also note here that the effect of forming a solvent cage about an ion is not only to aid in forming a solution. The cage also increases the effective size of the ion. The channels built into cell membranes must thus be made so as to accommodate effective ionic sizes.

Crystals of ionic compounds made of doubly or triply charged ions, like $MgSO_4$ or Al_2O_3, are generally less soluble in water than crystals involving singly charged ions, as in KCl, $NaNO_3$, or even Na_2SO_4. Attractions between multiply charged ions in a crystal lattice are powerful and keep the crystals more or less intact in water. We mention this to show that whether or not an ionic compound dissolves depends on the outcome of competing forces, those within the crystal and those within a solvent cage.

Polar Molecular Compounds Also Dissolve in Water   Water is able to form solvent cages about polar molecules (Figure 7.3), particularly when they have oxygen or nitrogen atoms with $\delta-$ charges on them (the more such atoms the better). Hydrogen bonds can then form from their $\delta-$ sites to the $\delta+$ sites on water molecules.

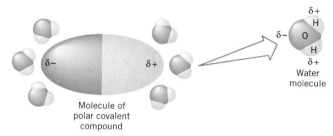

FIGURE 7.3 The hydration of a polar molecule helps polar molecular substances to dissolve in water.

A Review of Forces of Attraction

We'll pause here to go back over the forces involved in matter. *All of them are electrical*, so basic to the discussion are the general rules.

Unlike charges attract.

Like charges repel.

Within atoms, ions, and molecules there is the force of attraction of negatively charged electrons for positively charged atomic nuclei. This force holds an atom together until reactions occur that rearrange where electrons are relative to nuclei. Sometimes electrons transfer to give ions that turn out to be far more stable arrangements than existed in the atoms (assuming that ions of opposite charge or cages of water molecules are nearby). **Ion–ion attractions** are responsible for the **ionic bond** and are particularly powerful, as we can surmise from the fact that all ionic compounds are solids at room temperature. Ions in crystals, in other words, have almost no mobility; they can't wander around.

Sometimes electrons become shared between two nuclei giving rise to the **covalent bond**, which can be as strong as an ion–ion attraction. Nevertheless, reactions of molecules are possible that lead to even stronger covalent bonds.

Polar molecules have $\delta+$ and $\delta-$ sites. When an ion is nearby, an **ion–dipole attraction** arises and a solvent cage forms. Otherwise, polar molecules experience weaker **dipole–dipole attractions**. The **hydrogen bond** is a special dipole–dipole attraction involving the $\delta+$ charge on H when it's attached covalently to F, O, or N, and the $\delta-$ charge on another F, O, or N.

Nonpolar atoms or molecules develop **London forces**, which arise from the temporary dipoles induced in the particles as their electron clouds move near each other. The larger the molecule is, the larger is the potential London force.

■ Molecular substances with polar molecules and even ionic compounds also develop London forces on top of dipole–dipole or ion–ion attractions.

Dynamic Equilibrium Exists in a Saturated Solution A **saturated solution** is one in which there is a dynamic equilibrium between the undissolved and the dissolved solute, which we represent as follows.

$$\text{solute}_{\text{undissolved}} \rightleftharpoons \text{solute}_{\text{dissolved}} \tag{7.1}$$

In a saturated solution there is coming and going as solute particles leave the undissolved state and go into solution (the forward change) and other particles, at the same rate, leave the dissolved state and return to the undissolved condition (the reverse change). The *system*, here, is not homogeneous because both a solid and a liquid (the solution) are present. Hence, this equilibrium is called a *heterogeneous equilibrium*.

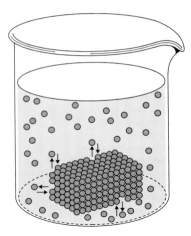

Dynamic equilibrium in a saturated solution

Each solid and most liquids have a limited solubility in water at a specific temperature. Hundreds of ionic compounds are only slightly soluble, and hundreds seemingly do not dissolve at all. In Section 5.4 we introduced such limits.

The Solubilities of Most Solids Increase with Temperature   Imagine a saturated aqueous solution of a solid at room temperature in which some of the undissolved solid rests on the bottom of the container. What will happen if we increase the temperature? Will more solid dissolve, or will some of the solute come out of solution and form crystals? What has been discovered for most solids is that *more* will dissolve as the temperature is increased. The reason is that *most solids dissolve endothermically* when the solution into which they are dissolving is at or very near the point of being saturated. In other words, such systems require heat to get more of the solid to dissolve into a solution already saturated. Thus, for solutes that dissolve endothermically, we can rewrite Equilibrium 7.1 by introducing an energy term, the *heat of solution*.

$$\text{solid}_{\text{undissolved}} + \text{solution} + \text{heat of solution} \rightleftharpoons \text{more concentrated solution}$$

Le Châtelier's principle applies to this equilibrium, as it does to all. When we heat a saturated solution of anything that dissolves endothermically, the stress is absorbed by a shift of the equilibrium to the right. Only a shift in this direction absorbs the additional heat and so "absorbs" the stress. The rate of the forward change increases, so more solute dissolves. The solution becomes even more concentrated. However, as more and more solute particles move out into the solution, the rate of the reverse change also increases, namely the return of solute to its undissolved state. Eventually, the rates of leaving and returning again become equal, and equilibrium is restored (assuming that undissolved solute is still present, of course).

The solubilities of some ionic compounds *decrease* with increasing temperature. Most common examples are salts of the sulfate ion, SO_4^{2-}, combined with metal ions carrying 2+ and 3+ charges, but a few are metal hydroxides like calcium hydroxide, $Ca(OH)_2$.

The Ions in Some Crystals Are Hydrated   If we let water evaporate from an aqueous solution of any one of several substances, the dry-appearing crystalline residue contains intact water molecules. They are held within the crystals in *definite* proportions, and for this reason these substances are true compounds; they obey the law of definite proportions. Such water-containing solids are called **hydrates**.

We write the formulas of hydrates in a special way to emphasize that intact water molecules are present. Thus the formula of the pentahydrate of copper(II) sulfate is written as $CuSO_4 \cdot 5H_2O$, where a centered dot separates the two parts of the formula. Another hydrate is photographer's "hypo," sodium thiosulfate pentahydrate, $Na_2S_2O_3 \cdot 5H_2O$. Still another is washing soda or sodium carbonate decahydrate, $Na_2CO_3 \cdot 10H_2O$.

The water present in a hydrate is called **water of hydration**. It usually can be driven out by heat, and when this is completely done the residue is called the **anhydrous form** of the compound. For example,

$$CuSO_4 \cdot 5H_2O(s) \xrightarrow{\text{heat}} CuSO_4(s) \quad + \quad 5H_2O(g)$$

copper(II) sulfate   copper(II) sulfate   (as steam)
pentahydrate (deep   (anhydrous form
blue crystals)   is white)

When a sample of blue $CuSO_4 \cdot 5H_2O$ is heated, water is expelled (see beads on the upper test tube wall) and anhydrous $CuSO_4$, which is white, forms.

Many anhydrous forms, even as solids, readily take up water and reform their hydrates. Plaster of paris, for example, although not completely anhydrous, contains relatively less water than gypsum. When we mix plaster of paris with water, it soon sets into a hard, crystalline mass according to the following equation.

$$(CaSO_4)_2 \cdot H_2O + 3H_2O \rightarrow 2CaSO_4 \cdot 2H_2O$$

plaster of paris   gypsum

■ The coefficient of 2 for gypsum applies to the entire formula, so a total four H_2O molecules are represented in $2CaSO_4 \cdot 2H_2O$.

Some compounds in their anhydrous forms are used as drying agents or desiccants. A **desiccant** is a substance that removes moisture from air by forming a hydrate. Any substance that can do this is said to be **hygroscopic**. Anhydrous calcium chloride, $CaCl_2$, is a common desiccant, and in humid air it draws enough water to form a liquid solution. Any substance this active as a desiccant is also said to be **deliquescent**. Thus, calcium chloride, which you can buy in hardware and building supply stores, is often used to dehumidify damp basements.

Our attention in this chapter is focused almost exclusively on water as the solvent. If you study organic chemistry, you will learn about others, such as various alcohols and hydrocarbons. One of the novel solvents is liquid carbon dioxide. In the last chapter we saw that carbon dioxide cannot be liquefied except under pressure. Large industrial systems can cope with this problem, however. Interaction 7.1 explains why doing so is worth the trouble.

7.3 SOLUBILITIES OF GASES

Pressure, temperature, and sometimes reaction with water affect the solubility of a gas in an aqueous solution.

By some means or another, all living things must exchange gases with the environment. Our bodies, for example, take in oxygen from the air and expel carbon dioxide. The processes intimately involve aqueous systems, so to understand these vital matters at the molecular level of life we have to look at the physical and chemical factors that affect gas solubilities. The chief physical influences are heat and pressure.

All Gases Are Less Soluble in Water at Higher Temperatures The solubilities of gases in water always decrease with increasing temperature. This is because the dissolving of gases in liquids, represented by the following equilibrium, is always an exothermic process.

$$gas_{undissolved} + solution \rightleftharpoons more\ concentrated\ solution + heat\ of\ solution$$

When heat is added, this equilibrium must shift to the left in favor of undissolved gas, in accordance with Le Châtelier's principle. It's the only way that the stress, heat, can be absorbed by the system. A shift to the left, of course, means that some of the gas leaves the solution. You have seen this often. Small bubbles appear in water soon after you start to heat it on the stove because dissolved air is leaving the water. (Don't confuse this with the boiling of the water.)

Gases Are More Soluble under Higher Partial Pressures Pressure is a factor that affects solubility when the solute is a gas. As seen in Figure 7.4, the solubilities of oxygen and nitrogen, two typical gases, are directly proportional to the applied pressure. The equilibrium expression is

$$gas_{undissolved} + solvent \rightleftharpoons solution \tag{7.2}$$

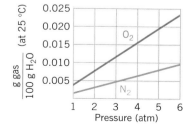

FIGURE 7.4 The solubilities of oxygen and nitrogen in water versus pressure.

INTERACTION 7.1
THE ENVIRONMENT FRIENDLY SOLVENT, LIQUID CO_2

Dry Ice Solid carbon dioxide looks like white ice; it's very cold, but it doesn't melt. Little wonder, then, that it's called *dry* ice (see Figure 1). You can buy it by the pound in large grocery stores, particularly in regions that attract people who love to fish and then quick freeze their catch.

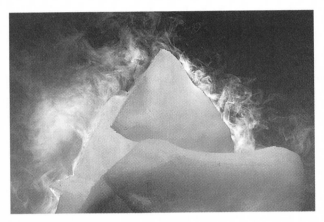

FIGURE 1 Dry ice does not melt under ordinary pressures.

Dry ice sublimes at −78 °C; the solid changes directly to its gaseous state without passing through a liquid phase. Thus, dry ice is not only very cold but it also leaves no puddles (unless it is loaded with ordinary water-ice that freezes out from a humid atmosphere). CO_2 has no odor, it's nonflammable, and it's nontoxic, which are major advantages. CO_2 does not poison the atmosphere or threaten the ozone layer (see Interaction 4.2, "Ozone in the Stratosphere and the Ozone Hole"). You can see why dry ice is a popular coolant for tough jobs. What you may not be aware of is the use of *liquid* CO_2 as an industrial solvent.

Liquid and Supercritical Carbon Dioxide The trick to getting dry ice to melt to a liquid rather than sublime is to increase the pressure on it to 5.2 atm; under this pressure dry ice melts and changes over to a liquid at −56.6 °C. It can exist as a liquid at higher temperatures, but the pressure must be increased. In fact, CO_2 can exist as a liquid at as high a temperature as 31.3 °C, but the pressure on it must now be a whopping 72.9 atm. Above 31.3 °C, CO_2 cannot exist as a liquid no matter what the pressure.[1] If the temperature of liquid CO_2 is raised above 31.3 °C, even under pressures above 73 atm, the visible boundary dividing the liquid from the gas disappears, and the system, now described only as a fluid, is said to be *supercritical* (see Figure 2).

Both liquid and supercritical CO_2 are excellent solvents. It's easy to get rid of solvent residues, but even if traces of CO_2 remain, they're harmless. The first significant industrial use of liquid CO_2 was to decaffeinate coffee. Liquid CO_2 dissolves the caffeine from coffee without removing all of the flavorful components. Other solvents, like certain chlorinated hydrocarbons (e.g., dichloromethane), that can be used to decaffeinate coffee release vapors that are not as friendly either to the atmosphere or to humans. Chlorinated hydrocarbons at sufficient levels are cancer causers in rats. And some chlorinated hydrocarbons, upon migrating into the stratosphere, are harmful to the ozone layer.

Wastewater Treatment Supercritical CO_2 is also being used to remove organic pollutants from the wastewater of pharmaceutical and chemical manufacturing plants. For example, an operation in Baltimore, put in place in 1989, treats over a million gallons per day. The wastewater is pumped into the top of a tall, pressurized column while the CO_2 is fed into the bottom. The CO_2 moves upward, dissolving the pollutants as the cleaner water settles to the bottom. From here the water is eventually sent into a regular sewage treatment plant. No other solvent does this job as well while leaving no harmful residues.

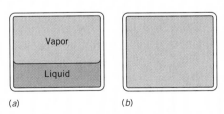

(a)   (b)

FIGURE 2 Liquid and supercritical systems. (*a*) The more dense liquid at the bottom can be detected by the interface between the phases. (*b*) In a supercritical fluid, the densities of the "liquid" and "vapor" phases are the same, so there is only one phase and no interface.

---

[1] The temperature above which a gas cannot be liquefied no matter how high we make the pressure is called the substance's *critical temperature* and the pressure needed at the critical temperature is the substance's *critical pressure*.

The equilibrium shifts to the right with increasing pressure because only such a change can absorb the volume-squeezing stress of extra pressure—yet another illustration of Le Châtelier's principle.

Similarly, if we reduce the pressure above a solution of a gas in water, we create a volume-expanding stress, and so equilibrium 7.2 must now shift to the left. Dissolved gas now leaves the solution, something you have often observed when you have opened a can or bottle of a soft drink. Under the suddenly lowered pressure, carbon dioxide fizzes out of solution.

William Henry (1775–1836) was the first to notice that gas solubility is directly proportional to gas pressure, so we now call this relationship **Henry's law** or the **pressure–solubility law**.

> **Henry's Law (Pressure–Solubility Law)** The concentration of a gas in a liquid at any given temperature is directly proportional to the partial pressure of the gas on the solution.

Stated in the form of an equation, Henry's law says

$$C_g = k_g P_g$$

where C_g is the concentration of the gas, k_g is a constant of proportionality, and P_g is the partial pressure of the gas above the solution. The reference is to *partial* pressure because each gas in a mixture of gases, like air, dissolves individually according to its own partial pressure and its own value of k_g.

The value of k_g for a given gas at a given temperature is a constant, and it does not change with the partial pressure. This lets us change the Henry's law equation to a form easier to use in calculations, because it lets us avoid actually having to know the constant k_g. For a given gas and using subscripts 1 and 2 to refer to different partial pressures we have an alternative equation for Henry's law.

$$\frac{C_1}{P_1} = \frac{C_2}{P_2} \qquad \text{(at constant temperature)} \qquad (7.3)$$

■ The two sides of Equation 7.3 equal each other because they each equal the same Henry's law constant.

■ **PRACTICE EXERCISE 1** How many milligrams of nitrogen dissolve in 100 g of water when the water is saturated with air and is in equilibrium with air that is saturated with water vapor? The partial pressure of nitrogen in air that is itself saturated with water vapor is 586 mm Hg. The solubility of pure nitrogen in water at 760 mm Hg is 1.90 mg/100 g H_2O.

The pressure–solubility relationship for solutions of gases in water is particularly important to people exposed to possible decompression sickness (the bends), as discussed in Interaction 7.2.

Water Reacts with Some Gases to Aid in Dissolving Them The chemical factor that affects the solubilities of some gases, like carbon dioxide, sulfur dioxide, and ammonia, is their ability to react with water. This makes a large difference. For example, at 20 °C, only 4.30 mg of oxygen dissolves in 100 g of water, but carbon dioxide is nearly 40 times as soluble, sulfur dioxide is nearly 2500 times as soluble, and ammonia is a whopping 12,000 times as soluble.

There are actually two reasons for this, and one is physical. The molecules of CO_2, SO_2, and NH_3 have $\delta+$ and $\delta-$ sites that can attract water molecules and be attracted by them, forming hydrogen bonds (Figure 7.5). The chemical fact is that a

■

Solubilities of Some Gases in Water in mg/100 g H_2O at 20 °C

O_2	4.3
CO_2	169
SO_2	10,600
NH_3	51,800

INTERACTION 7.2
DECOMPRESSION SICKNESS (THE BENDS)

People who work where the air pressure is high must return to normal atmospheric pressure slowly and carefully. Otherwise, they could experience the bends, or decompression sickness—severe pains in muscles and joints, fainting, and even deafness, paralysis, or death. At risk are deep-sea divers and those who work in deep tunnels where air pressures are increased to help keep out water.

Under high pressure, the blood dissolves more nitrogen and oxygen than at normal pressure, as Henry's law (and Figure 7.4) tells us. If blood thus enriched in nitrogen and oxygen is too quickly exposed to lower pressures, these gases suddenly come out of solution in blood. Their microbubbles block the tiny blood capillaries, close off the flow of blood, and lead to the symptoms we described.

If the return to normal pressure is made slowly, the gases leave the blood more slowly, and they can be removed as they emerge. The excess oxygen can be used by normal metabolism, and the excess nitrogen has a chance to be gathered by the lungs and removed by normal breathing. For each atmosphere of pressure above normal that the person was exposed to, about 20 minutes of careful decompression is usually recommended.

fraction of the molecules of CO_2, SO_2, and NH_3 actually reacts with water in solution. When water contains NH_3, for example, the following two equilibria exist, the first being physical and the second chemical.

$$NH_3(g) \rightleftharpoons NH_3(aq)$$
$$NH_3(aq) + H_2O \rightleftharpoons NH_4^+(aq) + OH^-(aq)$$

ammonium   hydroxide
ion   ion

■ In 1 M $NH_3(aq)$, only about 0.5% of the NH_3 molecules have reacted to form $NH_4^+(aq)$ + $OH^-(aq)$.

The forward reaction of the first equilibrium is aided by hydrogen bonding between water and ammonia. Then it is the *chemical* nature of ammonia to set up the second equilibrium; in the forward reaction, NH_3 tends to react with water to produce the ions shown. It isn't that the forward reaction occurs to much of an extent. The molar concentrations of the ions at equilibrium are only a small fraction of all unreacted molecules of $NH_3(aq)$. The left side of the equilibrium is definitely *favored*. However, the fact that the forward reaction occurs at all helps to draw $NH_3(aq)$ *out of the product side of the first equilibrium* and into solution by way of the second equilibrium's forward reaction.

The two equilibria involving the formation of aqueous CO_2 are similar to those of ammonia. In both equilibria, the *reactants* are favored.

$$CO_2(g) \rightleftharpoons CO_2(aq)$$
$$CO_2(aq) + H_2O \rightleftharpoons H_2CO_3(aq)$$

carbonic acid

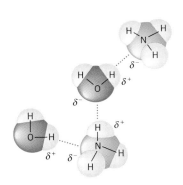

FIGURE 7.5 Hydrogen bonds (·····) between molecules of ammonia and water keep ammonia in solution.

Thus, to the (small) extent that $CO_2(aq)$ is removed by the second equilibrium and changed to H_2CO_3, the first equilibrium shifts to the right and so causes more $CO_2(g)$ to go into solution.

Similarly, SO_2 is involved in two equilibria in water.

$$SO_2(g) \rightleftharpoons SO_2(aq)$$
$$SO_2(aq) + H_2O \rightleftharpoons H_2SO_3(aq)$$

sulfurous acid

■ *Respiration* means all the activities that bring in *and use oxygen* and get rid of carbon dioxide.

Associated with Any Solution of a Gas in Water Is a *Gas Tension* Sometimes in discussions of human or animal respiration, the term *gas tension* is used to describe the availability of a gas from some body fluid. **Gas tension** is the partial pressure of a gas over a solution with which it is in equilibrium. It is an indirect measure of how much gas is in solution, because the more there is in solution the more there will be of the gas above the solution exerting a partial pressure. Thus, a high gas tension means a high availability of the gas from the solution.

In the blood returning to the lungs, the venous blood, the oxygen tension is only about 40 mm Hg. In the air freshly taken into the lungs, however, it is much higher, about 100 mm Hg. The differences in gas tension between two regions of the same system is called a *pressure gradient*. **Gradient** is a general term used to describe the occurrence of a quantity, like pressure, that varies in size from one location to another within the system. Because gases always tend to diffuse along a pressure gradient from a higher to a lower pressure, oxygen naturally tends to move from the lungs into the returning venous blood. This, of course, is the direction we want the oxygen to move. The blood becomes oxygen rich and it soon leaves the lungs in the arteries.

Carbon dioxide, on the other hand, has a gas tension in venous blood of about 46 mm Hg, but in the air inside the lungs its gas tension is a bit less, about 40 mm Hg. Thus, for CO_2, a waste product, the pressure gradient is from venous blood and into the lungs, which we want. CO_2 naturally migrates from venous blood into the alveoli of the lungs, from which it can be discharged as part of the exhaled air.

■ If you've ever heard of a mountain road or path having a "steep grade," you've encountered the concept, because "grade" means "gradient," here a change in altitude over distance.

■ Venous blood is the blood returning to the lungs in veins. Arterial blood is blood leaving the lungs.

7.4 PERCENTAGE CONCENTRATION EXPRESSIONS

The number of grams of solute in 100 g of solution is the percentage concentration of the solution.

For working with the stoichiometry of reactions in solution, no better concentration expression exists than that of molarity, the concentration in moles per liter. Many chemical situations arise, however, in which there is little need to know exact stoichiometry, yet some idea of concentration is useful. To serve this need, other expressions for concentrations have been developed. The most important of these are percentage concentrations.

Weight/Weight Percent   The **weight/weight percent (w/w%) concentration** of a solution is the number of grams of solute in 100 g of the solution. For example, a 10.0% (w/w) glucose solution has a concentration of 10.0 g of glucose in 100 g of solution. To make 100 g of this solution, you would mix 10.0 g of glucose with 90.0 g of the solvent for a total of 100 g.

■ We now use "percent" instead of "percentage" to conform to common usage.

EXAMPLE 7.1   Using Weight/Weight Percents

How many grams of 0.900% (w/w) NaCl solution contain 0.250 g of NaCl?

ANALYSIS   The concentration term, 0.900% (w/w), translates into the units of 0.900 g NaCl/100 g NaCl solution, so it gives us the following two conversion factors.

$$\frac{0.900 \text{ g NaCl}}{100 \text{ g NaCl soln}} \quad \text{and} \quad \frac{100 \text{ g NaCl soln}}{0.900 \text{ g NaCl}}$$

These two ratios are just equivalent ways of understanding the concentration, and we should remind ourselves that any expression of a concentration in any units can be expressed as either of two ratios, as we have done here. If we multiply the given, 0.250 g NaCl, by the second factor, the final units will be g NaCl soln.

SOLUTION

$$0.250 \text{ g NaCl} \times \frac{100 \text{ g NaCl soln}}{0.900 \text{ g NaCl}} = 27.8 \text{ g NaCl soln}$$

Thus 27.8 g of 0.900% (w/w) NaCl contains 0.250 g of NaCl.

EXAMPLE 7.2   Preparing Weight/Weight Percent Solutions

A special kind of saline solution, called isotonic saline, is sometimes used in medicine. Its concentration is 0.90% NaCl (w/w). How would you prepare 750 g of such a solution?

ANALYSIS   Once again, we have to translate the label on the bottle, 0.90% (w/w) NaCl, into conversion factors.

$$\frac{0.90 \text{ g NaCl}}{100 \text{ g NaCl soln}} \qquad \text{and} \qquad \frac{100 \text{ g NaCl soln}}{0.90 \text{ g NaCl}}$$

What we basically have to determine is the number of grams of NaCl that we must weigh out and dissolve in water to make the final mass equal to 750 g.

SOLUTION   To find the number of grams of NaCl, we multiply the given mass of the NaCl solution by the first conversion factor.

$$750 \text{ g NaCl soln} \times \frac{0.90 \text{ g NaCl}}{100 \text{ g NaCl soln}} = 6.8 \text{ g NaCl} \qquad \text{(rounded from 6.75)}$$

Thus if we dissolve 6.8 g NaCl in water and add enough water to make the final mass equal to 750 g, we can write the label to read 0.90% (w/w) NaCl.

■   PRACTICE EXERCISE 2   Sulfuric acid can be purchased from a chemical supply house as a solution that is 96.0% (w/w) H_2SO_4. How many grams of this solution contain 9.80 g of H_2SO_4 (or 0.100 mol)?

■   PRACTICE EXERCISE 3   How many grams of glucose and how many grams of water are needed to prepare 500 g of 0.250% (w/w) glucose?

Sometimes weight/weight percent solutions are prepared by diluting a more concentrated solution. The equation used to make the necessary calculation is similar to the one we derived in Section 5.6 for dilutions involving molar concentrations. We will simply give the equation here.

$$\text{g}_{\text{concd soln}} \times \text{percent (w/w)}_{\text{concd soln}} = \text{g}_{\text{dil soln}} \times \text{percent (w/w)}_{\text{dil soln}}$$

Volume/Volume Percent   A concentration expressed as a **volume/volume percent (v/v%)** gives us the number of volumes of one substance dissolved in 100 volumes of the mixture. This is often used for solutions of gases in gases or for solutions of liquids in liquids. For example, the concentration of oxygen in air is 21% (v/v), which means that there are 21 volumes of oxygen in 100 volumes of air. The unit used for volume can be any unit, as long as we use the same unit for the solute as for the solution. (Units cancel when we deal with true percentages.) Calculations involving volume/volume percents involve the same kinds of steps we used for problems of weight/weight percents.

Sometimes a weight/volume "percent" is used to express a concentration. **Weight/volume percent (w/v%)** means the number of grams of solute in 100 mL of the solution. It isn't a true percent because the units don't cancel. A concentration given as 0.90% (w/v) NaCl solution thus means

$$0.90 \text{ g of NaCl}/100 \text{ mL NaCl solution}$$

Weight/volume percent problems are handled through conversion factors just as we did for weight/weight percent problems.

The Trend Is Away from Using Percentages To avoid confusion over percentages and what they mean, explicit units should be given instead. Thus, instead of referring to a concentration of, say, 10.0% (w/v) KCl, the label or the report should read 10.0 g KCl/100 mL solution (or 10.0 g KCl/dL). When no units are supplied, only a percent sign, assume it to be a weight/weight percent. Thus "0.90% NaCl" means 0.90 g NaCl/100 g solution.

■ Often the units of g/100 mL will be given as g/dL because 100 mL = 1 dL.

Parts per Million and Parts per Billion For a very dilute solution, the concentration is sometimes given in **parts per million (ppm)**, which means the number of parts (in any unit) in a million parts (the same unit) of the solution. Parts per million might be interpreted as grams per million grams or pounds per million pounds. One ppm is analogous to one penny in a million pennies ($10,000) or 1 minute in a million minutes (about 2 years).

Parts per billion (ppb) similarly means parts per billion parts, such as grams per billion grams. This expression is used for extremely dilute systems. One ppb is like one penny in a billion pennies ($10 million), or like two drops of a liquid in a full, 33,000-gallon tank car.

7.5 OSMOSIS AND DIALYSIS

The selective migration of ions and molecules through cell membranes is an important mechanism for getting nutrients inside cells and waste products out.

Solutions generally have slightly lower melting points and slightly higher boiling points than their pure solvents. Aqueous solutions, for example, freeze not at 0 °C but a little below. They also boil not at 100 °C but a little higher. Interestingly, such effects depend not on what the solutes are but only on the ratio of solute to solvent *particles*. Properties of solutions or colloidal dispersions that depend only on the *number* of solute particles per unit quantity of solvent, not on the chemical identities of the solutes, are called **colligative properties**. The depression of the freezing point and the elevation of the boiling point are two examples. Two other examples are osmosis and dialysis, the abilities of components of a solution to migrate through certain kinds of membranes.

■ From the Greek *kolligativ*, depending on number and not on nature.

The effects of colligative properties, as we said, are related to concentrations, not to chemical identities. Consider, for example, two different solutes, NaCl and KBr. If we prepare one solution that has 1.0 mole of NaCl in 1000 g of water and another solution with 1.0 mole of KBr in 1000 g of water, they both freeze at the same temperature, at −3.4 °C (and not at 0 °C). And they both boil at 101 °C (at 760 mm Hg). Both the freezing and the boiling points are the same, despite the difference in chemical identity, because the ratios of the moles of ions to the moles of water in both solutions are identical.

The effects of solutes on freezing and boiling points are large only when the concentrations are very large. The whole basis for the use of an antifreeze mixture in a vehicle radiator, for example, is the large depression of the freezing point of the radiator fluid caused by the presence of the antifreeze in high concentration. A 50 : 50 mixture (v/v) of almost any commercially available vehicle antifreeze in water gives protection to about −40 °F.

■ The temperature of a mixture made from 33 g NaCl and 100 g ice is about −22 °C (−8 °F).

Osmosis Is the Diffusion of Solvent Molecules through Membranes   Cells in living systems are enclosed by cell membranes. On both sides of these are aqueous systems holding substances both in solution and in colloidal dispersion. Materials and water have to be able to move through cell membranes in either direction so that nutrients can enter cells and wastes can leave. Two factors control the movements through membranes, *active transport* and *dialysis*.

Active transport is the active involvement of specialized protein molecules embedded in a cell membrane to propel ions and molecules through it. Membrane-bound protein molecules establish "gates" that accept and move ions and molecules through membranes *by endothermic chemical reactions*. We can do no more here than mention this because we've not studied proteins.

■ The movements of Na⁺ and K⁺ ions into and out of cells is controlled by active transport mechanisms.

The other factor controlling movements through membranes, dialysis, depends on the semipermeable nature of a cell membrane. A **semipermeable membrane** is one that is able to let some *but not all* kinds of molecules and ions pass through it. Cellophane is an example of a synthetic semipermeable membrane. When cellophane separates water from a dilute solution that also contains colloidal sized particles, like starch molecules, for example, only water molecules and other small molecules and ions can migrate through it. Molecules of colloidal size are stopped. Membranes with this selectivity are called **dialyzing membranes**. It's as if cellophane has ultrafine pores just large enough to let the small particles through but too small for larger particles.

A semipermeable membrane that is so selective that only the *solvent* molecules can get through it is called an *osmotic membrane*. The effective sizes of ions, which are expanded by solvent cages as we noted, are apparently too large for the pores. No other molecules can pass either.

■ Essentially no purely osmotic membranes occur naturally.

Now let's see what difference a semipermeable membrane can make. We'll do this first by considering how an osmotic membrane makes possible a phenomenon called *osmosis*. **Osmosis** is the net migration of water (only) through an osmotic membrane from a solution with a lower concentration of solute (or from pure water) into the solution with a higher concentration of solute.

With the aid of Figure 7.6 let's see why there is a net flow in just one direction in osmosis. The figure describes the special case in which pure water is on one side of the membrane and a solution is on the other. Water molecules, of course, can move in *both* directions through the membrane, *but less readily from the solute side where the solute particles get in the way*. They prevent water molecules from leaving as frequently from the solute side as they are able to enter from the pure water, where no solute particles interfere. Thus water molecules move more frequently into the solution than leave it. So the solution becomes more dilute as well as larger in volume, and, in the right device, this can force a column of solution upward (Figure 7.6). Eventually, the weight of the rising column of water will exert a high enough back pressure to prevent any further rise, and then osmosis stops.

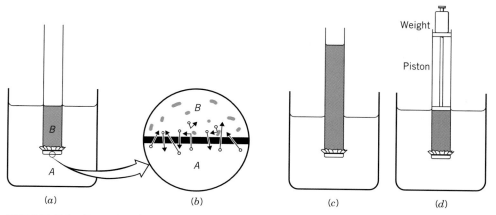

FIGURE 7.6 Osmosis and osmotic pressure. (*a*) In the beaker, *A*, there is pure water and in the tube, *B*, there is a solution. An osmotic membrane closes the bottom of the tube. (*b*) A microscopic view at the osmotic membrane shows how solute particles interfere with the movements of water molecules from *B* to *A*, but not from *A* to *B*. (*c*) The level in *A* has fallen and that in *B* has risen because of osmosis. A back pressure would be needed (*d*) to prevent osmosis, and the exact amount of pressure is the osmotic pressure of the solution in *B*.

Sometimes students have a problem with remembering the *direction* of osmosis. All that you have to recall is that *the net flow in osmosis always makes the more concentrated solution become more dilute*.

Osmotic Pressure Is a Measure of Concentration   The exact back pressure necessary to prevent osmosis is called the **osmotic pressure** of the solution, and its symbol is Π.

■ Π is the Greek capital letter pi.

The value of osmotic pressure is directly proportional to the molar concentration of the particles in the solution (at least at relatively low molarities), and the equation for osmotic pressure is almost identical to the ideal gas equation, $PV = nRT$. For osmotic pressure, Π,

$$\Pi V = nRT$$

If we rearrange terms to solve for Π, we get

$$\Pi = \frac{n}{V} \times RT$$

But n/V or moles per liter equals molarity, M, so

$$\Pi = MRT \tag{7.4}$$

You can see that osmotic pressure is directly proportional to the molar concentration.

The osmotic pressure of a solution has to be understood not as something that the solution is actually exerting, such as a hand pushing on a surface. Instead, osmotic pressure is a *potential* pressure, one that is directly related to concentration, that can be realized only when an osmotic membrane separates the solution from pure water. Even in relatively dilute solutions, the osmotic pressure can be very high, as the next worked example illustrates.

EXAMPLE 7.3   Calculating Osmotic Pressure

A dilute solution, 0.100 M sugar in water, is separated from pure water by an osmotic membrane. What is its osmotic pressure in mm Hg at a temperature of 25 °C or 298 K?

ANALYSIS   The calculation of osmotic pressure from molarity and temperature data requires Equation 7.4. To obtain the pressure in units of mm Hg, we must use 6.24×10^4 mL mm Hg/mol K for the value of R, and we must remember that M is in units of moles solute/1000 mL soln.

SOLUTION

$$\Pi = MRT$$
$$= \frac{0.100 \text{ mol}}{1000 \text{ mL}} \times 6.24 \times 10^4 \frac{\text{mL mm Hg}}{\text{mol K}} \times 298 \text{ K}$$
$$= 1.86 \times 10^3 \text{ mm Hg}$$

The solution has an osmotic pressure of 1.86×10^3 mm Hg. This means that a solution with this concentration can support a column of mercury 1.86×10^3 mm high, over 6 feet.

■   PRACTICE EXERCISE 4   What is the osmotic pressure of a 0.900 M glucose solution at 25 °C?

■   Water's density (1.00 g/mL) is much less than mercury's (13.6 g/mL), so the water column is 13.6 times higher.

If, instead of mercury, the column in Example 7.3 had been water (or the dilute solution), the column supported would have been 25.3 m (83.0 ft) high. Thus a relatively dilute solution, when separated from pure water, can be driven to a column height of several dozen feet. This phenomenon is one of the factors in the rise of sap in tall trees.

The Ions of an Ionic Compound Individually Affect Osmotic Pressure   Solute particles that cause osmotic pressure can be ions, molecules, or macromolecules. Just remember that osmotic pressure is a *colligative* property, so it depends only on the concentrations of the particles. Thus when the solute is an ionic compound, like sodium chloride, the concentration of *particles*—ions in this case—is twice the molar concentration of the salt itself (as given on the label of the bottle). For example, 0.10 M NaCl has a concentration of $2 \times (0.10)$ or 0.20 mol of all ions per liter, because NaCl breaks up into two ions for each formula unit that goes into solution. The osmotic pressure of 0.10 M NaCl is therefore twice as large as that of 0.10 M glucose, which does not break up into ions.

For an ionic compound like Na_2SO_4, for which three ions are released for each formula unit that dissolves—two Na^+ and one SO_4^{2-}—the concentration of particles in a 0.10 M solution is $3 \times (0.10) = 0.30$ mol of all ions per liter.

The labeled molarity of a solution thus does not reveal enough about a solution when we think about its osmotic pressure. To express the concentration of all osmotically active particles in the solution, scientists involved in the chemistry of health sometimes use a related concentration expression, called the solution's **osmolarity**, symbolized as **Osm**, the molar concentration of all solute particles active in osmosis or dialysis. Thus 0.10 M NaCl has a molarity of 0.10 mol/L of NaCl but an osmolarity of 0.20 mol/L. The osmolarity of 0.10 M Na_2SO_4 is 0.30 mol/L. The concentration term in Equation 7.4, M, must refer to the osmolarity of the solution.

■ **PRACTICE EXERCISE 5**   Assuming that any *ionic* solutes in this exercise break up completely into their constituent ions when they dissolve in water, what is the osmolarity of each solution?

(a) 0.010 M NH_4Cl (which ionizes as NH_4^+ and Cl^-)

(b) 0.005 M Na_2CO_3 (which ionizes as $2Na^+$ and CO_3^{2-})

(c) 0.100 M fructose (a sugar and a molecular substance)

(d) A solution that contains both fructose and NaCl with concentrations of 0.050 M fructose and 0.050 M NaCl

---

Small Solute Particles Pass through Dialyzing Membranes   **Dialysis** is like osmosis, only the membrane is more permeable. In dialysis, not only water molecules but also ordinary-sized ions and molecules move through the membrane. A *dialyzing membrane* can be thought of as having larger pores than an osmotic membrane. Cell membranes are, in part, dialyzing membranes.

Dialysis produces a net migration of water only if the fluid on one side of the dialyzing membrane has a higher concentration in colloidal substances than the other. Colloidal-sized particles are blocked by dialyzing membranes, so they get in the way of the movements of smaller particles through the membrane. The net flow of fluid in dialysis, as in osmosis, is from the side that has the lower concentration of colloidal substances to the side with the higher concentration. The effect is to make the concentrated solution more dilute.

The imbalance in concentration that is related to colloidally dispersed materials causes a **colloidal osmotic pressure**, which is similar to osmotic pressure in meaning. In the next section we present some situations involving life at the molecular level where osmotic pressure relationships are very critical and depend on the colloidal osmotic pressure of blood.

7.6 DIALYSIS AND THE BLOODSTREAM

When the osmotic pressure of blood varies too much, the result can be shock or damage to red blood cells.

The body tries to maintain the concentrations of all the substances that circulate in blood within fairly narrow limits. Quite complicated mechanisms exist to excrete or retain solutes or to excrete or retain water. When they fail, the consequences can be life threatening. In this section we will look briefly at two situations that arise when the osmotic pressure of blood changes too much.

The Brain Loses Blood Flowage in Shock   One feature of the shock syndrome is a dramatic increase in the permeability of the blood capillaries to colloidal-sized particles, particularly protein molecules. When these leave the blood, the colloidal osmotic pressure of blood decreases, another way of saying that the concentration of colloidal substances in blood decreases. The blood, in effect, becomes less concentrated. It is less able, therefore, to take up water from the spaces that surround the blood capillaries.

A sufficient decrease in the colloidal osmotic pressure of blood makes a net loss of water from blood possible. A loss of water means a loss of total blood volume. This makes it more difficult to bring nutrients to brain cells and carry wastes away. The brain functions much less well, and the result to the nervous system is called shock. When a person goes into shock, one of the many problems, but one that lies close to the central cause, is that blood capillaries become temporarily more permeable to the loss of macromolecules from blood.

Red Blood Cells Hemolyze in Water Millions of red blood cells circulate in the bloodstream, and their membranes behave as dialyzing membranes. Within each red cell is an aqueous fluid with dissolved and colloidally dispersed substances (Figure 7.7a). Although the colloidal particles are too large to dialyze, they contribute to the colloidal osmotic pressure. They help, therefore, to determine the direction of dialysis through the red cell membrane.

When red cells are placed in pure water, the fluid inside the red cell is more concentrated than the surrounding liquid. Dialysis now occurs to bring fluid *into the red cell*. Enough fluid moves in to make the cell burst open, as seen in Figure 7.7b. The rupturing of red cells is called **hemolysis**, and we say that the cells hemolyze.

On the other hand, when red cells are put into a solution with an osmolarity greater than their own fluid, dialysis occurs in the opposite direction—out of the cell and into the solution. Now the cells lose fluid volume, and shrivel and shrink. This process is called **crenation** (Figure 7.7c).

In some medical situations, body fluids need replacement or nutrients have to be given by intravenous drip. The osmolarity of the solution being added should match that of the fluid inside the red cells. Otherwise, hemolysis or crenation will occur.

Two solutions of equal osmolarity are called **isotonic solutions**. If one has a lower osmotic pressure than the other, the first is said to be *hypotonic* with respect to the second. A **hypotonic solution** has a lower osmolarity than the one to which it is compared. Red cells hemolyze if placed in a hypotonic environment, including pure water.

A **hypertonic solution** is one with a higher osmotic pressure than another. Thus 0.14 *M* NaCl is hypertonic with respect to 0.10 *M* NaCl. Red cells undergo crenation when they are in a hypertonic environment.

A 0.9% (w/w) NaCl solution, called **physiological saline solution**, is isotonic with respect to the fluid inside a red cell. Any solution to be added in any large quantity into the bloodstream has to be isotonic in this way.

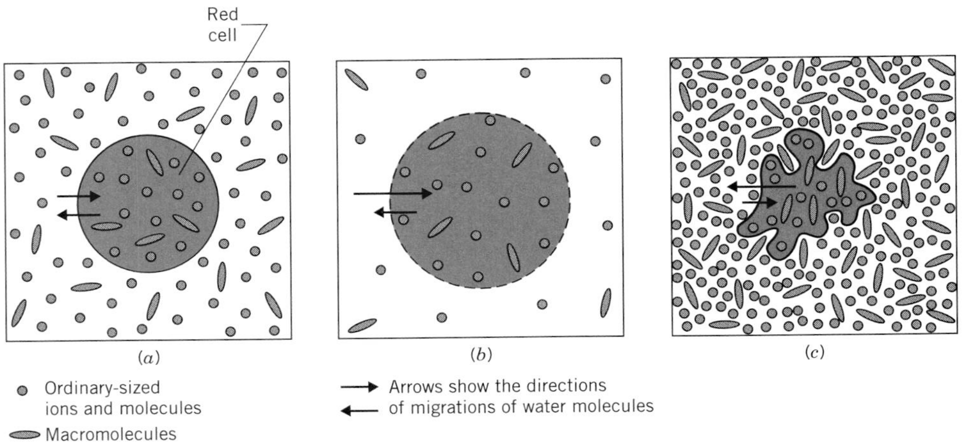

Red cell

FIGURE 7.7 Dialysis. (*a*) The red cell is in an isotonic environment. (*b*) Hemolysis is about to occur because the red cell is swollen by extra fluids brought into it from its hypotonic environment. (*c*) The red cell experiences crenation when it is in a hypertonic environment.

(*a*)   (*b*)   (*c*)

● Ordinary-sized ions and molecules

◯ Macromolecules

→ Arrows show the directions
← of migrations of water molecules

All the topics we have studied in this and the preceding section are important factors in the operation of artificial kidney machines, which are discussed in Interaction 7.3.

INTERACTION 7.3
HEMODIALYSIS

The kidneys cleanse the bloodstream of nitrogen waste products such as urea and other wastes. If the kidneys stop working efficiently or are removed, these wastes build up in the blood and threaten the life of the patient. The artificial kidney is one remedy.

The overall procedure is called hemodialysis—the dialysis of blood—and Figure 1 shows how it works. The bloodstream is diverted from the body and pumped through a long, coiled cellophane tube that serves as the dialyzing membrane. (The blood is kept from clotting by an anticlotting agent such as heparin.) A solution, called the dialysate, circulates outside of the cellophane tube. This dialysate is very carefully prepared not only to be isotonic with blood but also to have the same concentrations of all the essential substances that should be left in solution in the blood. When these concentrations match, the rate at which such solutes migrate out of the blood equals the rate at which they return. In this way several key equilibria are maintained, and there is no net removal of essential components. Figure 2 shows how this works. The dialysate, however, is kept very low in the concentrations of the wastes, so the rate at which they leave the blood is greater than the rate at which they can get back in. In this manner, hemodialysis slowly removes the wastes from the blood.

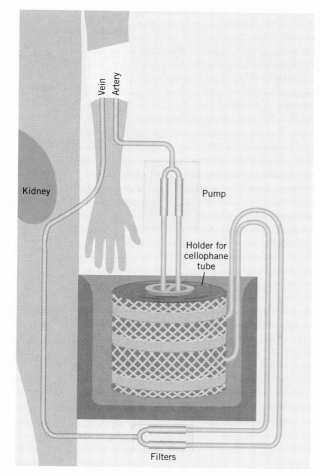

FIGURE 1

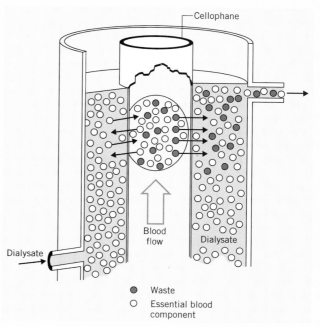

FIGURE 2

SUMMARY

Solutions Ions and molecules of ordinary size, if soluble in water at all, form solutions. These are homogeneous mixtures that neither gravity nor filtration can separate. The solubilities of most solids increase with temperature, because their dissolving is usually endothermic. (More energy is needed to break up the crystal than is recovered as the solvent cages form about the ions or molecules.)

Hydration The attraction of water molecules to ions or to polar molecules leads to a loose solvent cage that shields the ions or molecules from each other. This phenomenon is called hydration, and it helps to explain why some substances dissolve in water. Sometimes water of hydration is present in a crystalline material in a definite proportion to the rest of the formula unit, and such a substance is a hydrate. Heat converts most hydrates to their anhydrous forms. And some anhydrous forms serve as drying agents— desiccants.

Gas solubilities Gases dissolve in water exothermically, so the addition of heat to an aqueous solution of a gas drives the gas out of solution. The solubility of a gas is directly proportional to its partial pressure in the space above the solution (Henry's law). Some gases do more than mechanically dissolve in water; part of what dissolves forms hydrogen bonds with water and part reacts with water to form soluble species.

Percent concentration A variety of concentration expressions have been developed to provide ways to describe a concentration without going into molar concentrations. These include weight/weight percents, volume/volume percents, and hybrid descriptions that aren't true percentages.

Colloidal dispersions Large clusters of ions or molecules or macromolecules form not true solutions, but colloidal dispersions. These can reflect and scatter light (Tyndall effect), experience the Brownian movement, and (in time) succumb to the force of gravity (if the medium is fluid). Protective colloids, such as emulsifying agents, sometimes stabilize these systems. If the dispersed particles grow to an average diameter of about 1000 nm, they slip over into the category of suspended matter, and such systems must be stirred to maintain the suspension.

Osmosis and dialysis When a semipermeable membrane separates two solutions or dispersions of unequal osmolarities, a net flow occurs in the direction that, if continued, would produce solutions of identical osmolarities. When the membrane is osmotic, only the solvent can migrate, and the phenomenon is osmosis. The back pressure needed to prevent osmosis is called the osmotic pressure, and it's directly proportional to the concentration of all particles of solute that are osmotically active—ions, molecules, and macromolecules.

When macromolecules are present, their particular contribution to the osmotic pressure is called the colloidal osmotic pressure of a solution. It is this factor that operates when the membrane is a dialyzing membrane.

The permeability of blood capillaries changes temporarily when a person experiences shock, and macromolecules leave the blood. Their departure results in the loss of water, too, and the blood volume decreases.

Solutions of matched osmolarity are isotonic. Otherwise, one is hypertonic (more concentrated) with respect to the other, and the other is hypotonic (less concentrated) with respect to the first. Only isotonic solutions, or those that are nearly so, should be administered in large quantities intravenously.

REVIEW EXERCISES

The answers to Review Exercises whose numbers are in color are found in Appendix E. The answers to the other Review Exercises are found in the Study Guide that accompanies this book. The more challenging questions are marked with asterisks.

Homogeneous Mixtures

7.1 The definition of *homogeneous* depends on how we define "smallest sample." Explain.

7.2 Chlorine gas is a mixture of the following molecules, where the left superscripts denote the mass numbers of individual chlorine isotopes.

$$^{35}Cl—^{35}Cl \qquad ^{35}Cl—^{37}Cl \qquad ^{37}Cl—^{37}Cl$$

Under what circumstances can we call this mixture *homogeneous*? Explain.

7.3 Particle *size* is one basis for distinguishing among solutions, colloidal dispersions, and (stirred) suspensions. Explain why size works for this purpose.

7.4 Why are suspensions usually not considered homogeneous?

7.5 Which of the three kinds of homogeneous mixtures
(a) Can be separated into its components by filtration?
(b) Exhibits the Tyndall effect?
(c) Shows observable Brownian movement?
(d) Has the smallest particles of all kinds?
(e) Is likeliest to be the least stable at rest over time?

7.6 What kinds of particles make the most stable colloidal dispersions? Explain.

7.7 The blood is simultaneously a solution, a colloidal dispersion, and a suspension. Explain.

7.8 A colloidal dispersion gives the Tyndall effect but a solution doesn't. Explain.

7.9 What simple test could be used to tell whether a clear, colorless solution contained substances in colloidal dispersion?

7.10 What is an emulsion? Give some examples.

7.11 What is a sol? Give some examples.

7.12 What is a gel? Give an example.

Aqueous Solutions

7.13 In a crystal of sodium chloride the chloride ions are surrounded by oppositely charged ions (Na^+) as nearest neighbors. What replaces this kind of electrical environment for chloride ions when sodium chloride dissolves in water?

7.14 When we say that a sodium ion in water is *hydrated*, what does this mean? (Make a drawing as part of your answer.)

7.15 We have to distinguish between *how fast* something dissolves in water and *how much* can dissolve to make a saturated solution. The speed with which we can dissolve a solid in water increases if we (a) crush the solid to a powder, (b) stir the mixture, or (c) heat the mixture. Use the kinetic theory as well as the concept of forward and reverse processes to explain these facts.

7.16 Assuming that solid potassium bromide is dissolved in water:
(a) Write the equation for the dissociation of this compound as its solution forms.
(b) If a saturated solution is prepared with excess, undissolved KBr present, the rates of what two changes are equal in this saturated solution? Write an equilibrium expression.

7.17 Ammonium chloride dissolves in water endothermically. Suppose that you have a saturated solution of this compound, that its temperature is 30 °C, and that undissolved solute is present. Write the equilibrium expression for this saturated solution, and use Le Châtelier's principle to predict what will happen if you cool the system to 20 °C.

Hydrates

7.18 Write the equation for the decomposition of sodium sulfate decahydrate to its anhydrous form. (*Deca-* denotes ten.)

7.19 Why are hydrates classified as compounds and not as wet mixtures?

7.20 When water is added to anhydrous magnesium sulfate, the heptahydrate, called Epsom salts, forms. Write the equation. (*Hepta-* denotes seven.)

7.21 Anhydrous calcium chloride is hygroscopic. What does this mean? Does this property make it useful as a desiccant?

7.22 Sodium hydroxide is sold in the form of small pellets about the size and shape of split peas. It is a very deliquescent substance. What can happen if you leave the cover off of a bottle of sodium hydroxide pellets?

***7.23** When 6.45 g of the hydrate of compound X was strongly heated to drive off all of the water of hydration, the residue, the anhydrous form of X, had a mass of 3.40 g. What number should y be in the formula of the hydrate, $X \cdot yH_2O$? The formula mass of X is 201.27.

Gas Solubilities

7.24 The solubility of methane, the chief component in bunsen burner gas, in water at 20 °C and 1.0 atm is 0.025 g/L. What will be its solubility at 1.6 atm?

7.25 At 20 °C the solubility of nitrogen in water is 0.0150 g/L when the partial pressure of the nitrogen is 580 mm Hg. What is its solubility when the partial pressure is raised to 740 mm Hg?

7.26 Explain why carbon dioxide is more soluble in water than is nitrogen.

7.27 Using Le Châtelier's principle, explain why the solubility of a gas in water should decrease with decreasing partial pressure of the gas.

7.28 If the gas tension of O_2 in blood is described as 80 mm Hg, what specifically does this mean?

7.29 If in one region of the body the gas tension of oxygen over blood is 79 mm Hg and in a second region it is 60 mm Hg, which region (the first or the second) has a higher concentration of oxygen in the blood itself?

Percent Concentrations

7.30 If a solution has a concentration of 0.915% (w/w) NaOH, what two conversion factors can we write based on this value?

7.31 A solution bears the label 1.42% (w/v) KCl. What two conversion factors can be written for this value?

7.32 A solution of rubbing alcohol in water is described as 30% (v/v). What two conversion factors are possible from this value?

7.33 How many grams of solute are needed to prepare each of the following solutions?
(a) 500 g of 0.900% (w/w) NaCl
(b) 250 g of 3.20% (w/w) $KC_2H_3O_2$
(c) 125 g of 6.25% (w/w) NH_4Br
(d) 300 g of 1.50% (w/w) Na_2CO_3

7.34 How many grams of solute have to be weighed out to make each of the following solutions?
(a) 500 mL of 6.30% (w/v) NaBr
(b) 250 mL of 3.50% (w/v) NH_4Cl

(c) 50 mL of 2.00% (w/v) $Mg(NO_3)_2$
(d) 250 mL of 0.900% (w/v) NaCl

7.35 How many milliliters of ethyl alcohol have to be used to make 500 mL of 10.0% (v/v) aqueous ethyl alcohol solution?

*7.36 A chemical supply room has supplies of the following solutions: 3.00% (w/w) NaCl, 5.00% (w/w) $NaHCO_3$, and 3.50% (w/v) glucose. If the densities of these solutions can be taken to be 1.00 g/mL, how many milliliters of the appropriate solution would you have to measure out to obtain the following quantities?
(a) 3.00 g of NaCl   (b) 0.325 g of $NaHCO_3$
(c) 0.250 g of glucose   (d) 0.115 mol of NaCl
(e) 0.100 mol of glucose ($C_6H_{12}O_6$)

*7.37 A student needs 125.0 mL of 10.0% (w/w) aqueous sodium acetate, $NaC_2H_3O_2$. (The density of this solution is 1.05 g/mL.) Only the trihydrate of this compound, $NaC_2H_3O_2 \cdot 3H_2O$, is available, and the student knows that the water of hydration would just become part of the solvent once the solution was made. How many grams of the trihydrate must be weighed out to prepare the needed solution?

7.38 A student has to prepare 500 g of 2.50% (w/w) NaOH. The stock supply of NaOH is in the form of 10.0% (w/w) NaOH. How many grams of the stock solution have to be diluted to make the desired solution?

7.39 The stockroom has a 10.0% (w/w) HCl solution. How many grams of this solution have to be weighed out to prepare, by dilution, 500 g of 0.250% (w/w) HCl? If the density of the 10.0% solution is 1.05 g/mL, how many milliliters would provide the grams of the concentrated solution that are called for?

*7.40 Concentrated hydrochloric acid is available as 11.6 M HCl. The density of this solution is 1.18 g/mL.
(a) Calculate the percent (w/w) of HCl in this solution.
(b) How many milliliters of this concentrated acid have to be taken to prepare 250 g of a solution that is 15.0% (w/w) HCl?

*7.41 Commercial nitric acid comes in a concentration of 16.0 mol/L. The density of this solution is 1.42 g/mL.
(a) Calculate the percent (w/w) of nitric acid, HNO_3, in this solution.
(b) How many milliliters of the concentrated acid have to be taken to prepare 500 g of a solution that is 6.00% (w/w) HNO_3?

Osmosis and Dialysis

7.42 If a solution that contains 1.00 mol of sucrose in 1000 g of water freezes at −1.86 °C, what is the freezing point of a solution that contains 1.00 mol of glucose in 1000 g of water? (Both are compounds that do not break up into ions when they dissolve.)

7.43 A solution that contains 1.00 mol of glucose in 1000 g of water has a normal boiling point of 100.5 °C. Another solution that contains 1.00 mol of an unknown compound in 1000 g of water has a normal boiling point of 101.0 °C. What is the likeliest explanation for the higher boiling point of the second solution?

7.44 Explain in your own words and drawings how osmosis gives a net flow of water from pure water into a solution on the other side of an osmotic membrane.

7.45 In general terms, how does an osmotic membrane differ from a dialyzing membrane?

7.46 Explain in your own words why the osmotic pressure of a solution should depend only on the concentration of its solute particles and not on their chemical properties.

7.47 The equation for osmotic pressure (Equation 7.4) shows that this pressure is directly proportional to the Kelvin temperature. Use the kinetic model of molecules and ions in motion and other aspects of the general kinetic theory to explain why the osmotic pressure should increase with an increase in temperature.

7.48 Why is the osmolarity of 1.0 M NaCl not the same as its molarity?

*7.49 Which has the higher osmolarity, 0.10 M NaCl or 0.080 M Na_2SO_4? Explain.

*7.50 Solution A consists of 0.60 mol of NaCl, 0.12 mol of $C_6H_{12}O_6$ (glucose, a molecular substance), and 0.55 mol of starch (a colloidal, macromolecular substance), all in 1000 g of water. Solution B is made of 0.60 mol of NaBr, 0.12 mol of $C_6H_{12}O_6$ (fructose, a molecular substance related to glucose), and 0.005 mol of starch all in 1000 g of water. Which solution, if either, has the higher osmotic pressure? Explain.

7.51 What is the osmotic pressure (in mm Hg) of a 0.0100 M solution in water of a molecular substance at 25 °C?

7.52 Calculate the osmotic pressure in mm Hg of a 0.0125 M solution in water at 20.0 °C of a compound that breaks up into two ions per formula unit when it dissolves.

7.53 What happens to red blood cells in crenation?

7.54 Physiological saline solution has a concentration of 0.90% (w/w) NaCl.
(a) Is a solution that is 0.80% (w/w) NaCl described as hypertonic or hypotonic with respect to physiological saline solution?
(b) What would happen, crenation or hemolysis, if a red blood cell were placed (1) in 1.5% (w/w) NaCl? (2) In 0.5% (w/w) NaCl?

7.55 Explain how the loss of macromolecules from the blood can lead to the increased loss of water from blood and a reduction in blood volume.

Liquid Carbon Dioxide as a Solvent (Interaction 7.1)

7.56 If you were able to observe samples of liquid CO_2 and supercritical CO_2, what difference would you notice?

7.57 What is perhaps the most attractive advantage of using either liquid CO_2 or supercritical CO_2 as a solvent?

Decompression Sickness (Interaction 7.2)

7.58 The solubilities of which gases increase in blood to cause decompression sickness? Why do they increase?

7.59 How does an increased solubility of a gas in blood cause a problem when the individual comes back to normal pressure?

7.60 How does a slow decompression reduce the possibility of decompression sickness?

7.61 What is the "rule of thumb" about the rate of decompression needed to avoid decompression sickness?

Hemodialysis (Interaction 7.3)

7.62 What does *hemodialysis* mean?

7.63 During hemodialysis, what is the *dialysate*?

7.64 With respect to the following solutes in blood, what should be the concentration of the dialysate for effective hemodialysis, more or less concentrated or the same concentration?
(a) Na^+ (b) Cl^- (c) urea

Additional Exercises

7.65 The observation of the Brownian movement was important historically in the development of the kinetic theory of gases. Suggest a reason.

7.66 Carbon tetrafluoride, CF_4, has the unusually polar C—F bonds, and its molecules are tetrahedral, like those of methane, CH_4. This substance does not dissolve in water. Why won't water let CF_4 molecules in?

7.67 Write the equation for the dissociation of $Ca(NO_3)_2$ in water (using appropriate symbols for the *states* of the various species).

7.68 Suppose that you do not know and do not have access to a reference in which to look up the solubility of sodium nitrate, $NaNO_3$, in water at room temperature. Yet you need a solution that you know beyond doubt is saturated. How can you make such a saturated solution and know that it is saturated?

*7.69 When all of the water of hydration was driven off of 4.32 g of a hydrate of compound Z, the residue, the anhydrous form, Z, had a mass of 3.09 g. What is the formula of the hydrate (using the symbol Z as part of it)? The formula mass of Z is 90.0.

7.70 The solubility of a gas in water at 20 °C is 0.0176 g/L at 681 mm Hg. What is its solubility at 0.989 atm at 20 °C?

7.71 Explain why sulfur dioxide is much more soluble in water than oxygen.

7.72 Calculate the number of grams of solute needed to make each of the following solutions.
(a) 250 g of 0.600% (w/w) NaCl
(b) 125 g of 0.460% (w/w) NaI
(c) 100 g of 1.00% (w/w) $C_6H_{12}O_6$ (glucose)
(d) 50.0 g of 7.50% (w/w) H_2SO_4

7.73 A sample of 750 mL of 5.00% (v/v) aqueous methyl alcohol contains how many milliliters of pure methyl alcohol?

*7.74 The stockroom has the following solutions: 2.50% (w/w) NaBr, 0.750% (w/w) NH_4Cl, and 0.900% (w/v) NaCl. Assuming that the densities of these solutions are all 1.00 g/mL, how many milliliters of the appropriate solution have to be measured out to obtain the following quantities of solutes?
(a) 0.100 g of NaBr   (b) 0.200 g of NH_4Cl
(c) 0.100 mol of NaCl   (d) 0.125 mol of NaBr

*7.75 The stockroom has a solution labeled 50.0% H_2SO_4. Its density at 20 °C is 1.40 g/mL. What is the molarity of the solution?

*7.76 How many grams of $Na_2SO_4 \cdot 10H_2O$ have to be weighed out to prepare 125 mL of 5.00% (w/w) Na_2SO_4 in water? (The density of this solution is 1.09 g/mL.)

7.77 What is the osmotic pressure of a 0.0125 M solution of sugar in water at 20.0 °C?

*7.78 Which solution has the higher osmotic pressure, 5.0% (w/w) NaCl or 5.0% (w/w) KI? Both NaCl and KI break up in water in the same way—two ions per formula unit.

*7.79 For rehydration therapy for cholera patients, the World Health Organization (WHO) uses an aqueous solution with the following concentrations: 3.5 g NaCl/L, 2.5 g $NaHCO_3$/L, 1.5 g KCl/L, and 20 g of glucose per liter. Assuming that the ionic compounds break up fully into their ions (Na^+, Cl^-, HCO_3^-, and K^+), calculate the osmolarity of the solution. (Glucose is $C_6H_{12}O_6$.)

*7.80 Magnesium carbonate, a constituent of a limestone-like rock called dolomite, reacts with hydrochloric acid to give magnesium chloride, carbon dioxide, and water.
(a) Write the balanced equation for the reaction.
(b) How many milliliters of 15.0% hydrochloric acid (density = 1.073 g/mL) are necessary to react completely and exactly with 5.00 g of magnesium carbonate according to the balanced equation?
(c) How many milliliters of carbon dioxide would be produced if measured at 20.0 °C and 745 mm Hg?

8

ACIDS, BASES, AND IONIC COMPOUNDS

Lime is being sifted onto this lake in southern Norway to help combat the acids dumped into it from industrial Europe's northering winds. We'll learn about acid neutralization in this chapter.

THIS CHAPTER IN CONTEXT

We've mostly been developing background on pure substances, mixtures, and equations. Now we can get down to what many would call the more practical business of useful substances. One help will be the fact that the millions of compounds so far identified fall into a relatively small number of families, each with a set of similar properties. *Electrolytes* (they conduct electricity) constitute a very broad family of which the most important kinds are *acids, bases,* and *salts.*

What makes solutions acidic or basic are factors that, far more than most people realize, dominate the environment, commerce, and human well-being. You've heard, for example, of acid rain, and you know that corrosion, sped by acids, destroys bridges and vehicles. But you may not know that your blood is ever so slightly basic, and if illness or accident makes it even drift toward becoming acidic, you'll die unless countermeasures—natural or medical—kick in. We're talking here about major medical emergencies and about limitations on what you can do in strenuous sports and recreation. Much of our discussion of these matters is spread over this and the next chapter. One emphasis here will be on the most common compounds that can be used to neutralize acids and bases.

8.1 ELECTROLYTES

Solutions of ionic compounds in water conduct electricity.

All aqueous fluids of living systems, plants or animals, contain dissolved ions, so to understand these fluids we must study the chemical properties of their major ions. One of the great differences made when ions are present in water is the ability of the solutions to conduct electricity. If you ever need an electrocardiogram, for example, you will particularly value this property of body fluids.

■ Pure water conducts electricity very poorly.

Solutes Release Ions in Water by Dissociation or by Ionization We learned in the previous chapter that when ionic compounds dissolve in water their ions *dissociate,* meaning that they separate from each other as the crystals break up. We also learned that cages of water molecules surround the ions to stabilize them and allow them to exist apart from each other.

Many molecular compounds also generate ions as they dissolve in water, but the process is now called *ionization,* not dissociation. **Ionization** is the formation of ions *by a chemical reaction.* Hydrogen chloride, for example, *ionizes* as it dissolves in water. Using Lewis structures, we can diagram the reaction as follows.

■ The verb is *to ionize.*

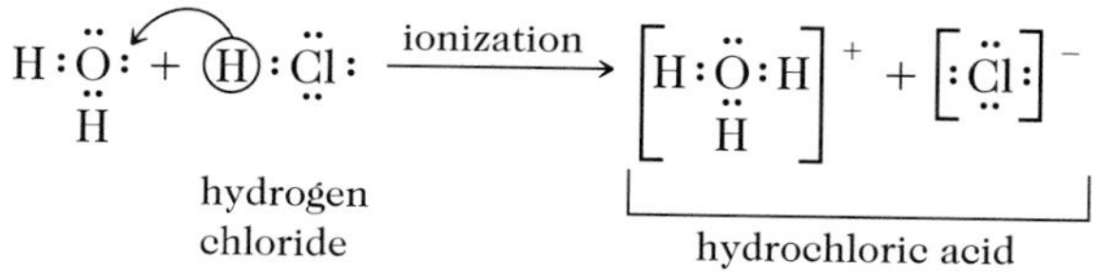

The H of HCl transfers from Cl to H_2O. Notice that H transfers without taking along any electrons. What really transfers, therefore, is not H but H^+. Although pure hydrogen chloride contains no ions, when HCl(g) dissolves in water essentially 100% of its molecules react with water to give hydronium and chloride ions. This *solution* is what is called *hydrochloric acid.* HBr(g) and HI(g) react the same way with water to give hydrobromic acid and hydroiodic acid, respectively.

■ For all practical purposes, no *molecules* of HCl remain un-ionized in dilute hydrochloric acid.

Ammonia is another compound that produces ions by reacting with water, but the extent of its ionization is quite small (about 0.5% in $1 \, M \, NH_3$). Lewis structures help us see what happens. Again, H^+ is what transfers.

$$H:\overset{..}{\underset{..}{O}}: + H:\overset{H}{\underset{\overset{|}{H}}{N}}:H \xrightarrow[\text{(aqueous solution)}]{\substack{\text{small percentage} \\ \text{ionization}}} \left[H:\overset{H}{\underset{\overset{|}{H}}{N}}:H\right]^+ + \left[H:\overset{..}{\underset{..}{O}}:\right]^-$$

ammonia   ammonium   hydroxide
                             ion   ion

As we learned in Section 7.3, this is really the forward reaction of an equilibrium:

$$NH_3(aq) + H_2O \rightleftharpoons NH_4^+(aq) + OH^-(aq)$$

Ions in Water Can Carry Electricity   Electricity in metals is a flow of electrons. A complete circuit includes a battery or generator that forces the electrons to move. (The energy of the flow is called *electrical energy*.) If the circuit is broken, the flow stops. The break might be just air at the gap of an open switch, or it might be a space filled with some inert insulating fluid, like sulfur hexafluoride. Pure water is also a relatively good insulator. If we add ions to the water, however, the solution conducts electricity much more readily.

The passage of electricity through a solution holding dissolved ions is called **electrolysis.** A solute that enables a solution to conduct electricity is called an **electrolyte.** (Sometimes the solution itself is called the electrolyte.) The passage of electricity causes chemical changes to the solutes or to the solvent, so electrical energy is forcing chemical reactions. The question now is, "How do electrolytes make electrolysis possible?"

Electrons do not move directly through a solution of electrolytes in the same way that they move through metals. Instead, the dissolved ions move (Figure 8.1). The plates or wires that dip into the solution are called **electrodes.** The battery forces electrons to one electrode, which makes it electron rich and negatively charged. In electrolysis, the negative electrode is called the **cathode.** The positive ions in solution—the *cations*—naturally are attracted to the cathode, because opposite charges attract. ("Cathodes attract cations.")

The electrons that make the cathode electron rich are "pumped" by the battery or generator from the other electrode, called the **anode,** which becomes electron poor and positively charged. Negative ions or *anions* are naturally attracted to the anode. ("Anodes attract anions.")

Electrolysis Causes a Redox Reaction   A cation removes electrons at the cathode at the same instant that an anion delivers them to the anode, as we'll show by the electrolysis of aqueous copper(II) bromide. This compound dissociates in water as follows.

$$CuBr_2(s) \xrightarrow{\text{dissociation}} Cu^{2+}(aq) + 2Br^-(aq)$$

When an electric current is passed through the solution, the following overall reaction is forced by electrolysis.

$$Cu^{2+}(aq) + 2Br^-(aq) \xrightarrow{\text{electrolysis}} Cu(s) + Br_2(l)$$

Of the products, copper, $Cu(s)$, forms at the cathode and bromine, $Br_2(l)$, at the anode (Figure 8.2). Let's see how this happens.

■ Electrical insulators prevent electricity from flowing.

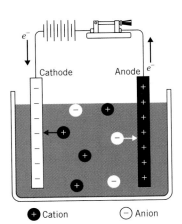

FIGURE 8.1 Electrolysis. Cations, positive ions, migrate to the cathode and remove electrons. Anions, negative ions, migrate to the anode and deposit electrons. The effect is a closed circuit.

Cu^{2+} cations are attracted to the cathode, where they pick electrons from the cathode's electron-rich surface and so are changed to Cu atoms. We can represent this by an equation in which electrons are shown as actual reactants.

$$Cu^{2+}(aq) + 2e^- \xrightarrow{\text{reduction}} Cu(s)$$

The oxidation number of copper thus changes from +2 to 0 and so becomes less positive. *Reduction,* therefore, is what happens to the Cu^{2+} ion. (*Reduction* here is clearly also a *gain of electrons,* the older definition of reduction.) Thus in electrolysis, reduction occurs at the cathode.

Br$^-$ anions are attracted to the anode where they deposit electrons at the anode's electron-poor surface. Two Br$^-$ ions give up electrons and one molecule of Br_2 forms. The oxidation number of bromine thus changes from -1 to 0 and so becomes more positive (less negative). *Oxidation* is happening to bromide ions. (And, by the older definition, there is clearly a loss of electrons as two Br$^-$ change to Br_2.) The anode reaction is

$$2Br^-(aq) \xrightarrow{\text{oxidation}} Br_2(l) + 2e^-$$

Thus, oxidation occurs at an anode during electrolysis.

During electrolysis we have something taking electrons from one electrode and something else putting them *simultaneously* on the other electrode. The effect is the same as if the electrons themselves were actually moving through the solution, but they move only through the wire.

Notice how the sum of the cathode and anode reactions gives the overall equation for the electrolysis.

At the cathode: $\quad Cu^{2+}(aq) + 2e^- \xrightarrow{\text{reduction}} Cu(s)$

At the anode: $\qquad\qquad 2Br^-(aq) \xrightarrow{\text{oxidation}} Br_2(l) + 2e^-$

Sum: $\qquad Cu^{2+}(aq) + 2Br^-(aq) \xrightarrow{\text{electrolysis}} Cu(s) + Br_2(l)$

The electrons cancel as this summation is made. They must cancel, of course, because we cannot have free electrons as actual reactants or products. But electrons can transfer. In electrolysis, they transfer from one dissolved species to the other through the wiring of the external circuit.

Molten Ionic Compounds Are Also Electrolytes For electrolysis to happen, ions must be mobile. When ions are immobilized in the solid state, no electrolysis occurs. If a crystalline ionic compound is heated until it melts, however, then its ions become mobile, and molten salts conduct electricity. Thus the term *electrolyte* refers either to a solution of ions or to the pure, solid ionic compound. (The term *electrolyte* does not apply to metals. Metals that conduct electricity are simply called *conductors.*)

Strong Electrolytes Give High Concentrations of Ions in Water Electrolytes are not equally good at enabling the flow of electricity. A "good" electrolyte is a substance that even in low concentrations makes possible a strong electrical current. A good or **strong electrolyte** is a solute that can readily supply ions; essentially 100% of its formula units dissociate (or ionize). Sodium hydroxide and sodium chloride are strong electrolytes because they *dissociate* 100% in water. Hydrochloric acid is a strong electrolyte because its initial solute, hydrogen chloride, has *ionized* 100% in water.

When a **weak electrolyte** is dissolved in water only a small percentage of its molar concentration changes to ions. Aqueous ammonia is a typical example. Thus

■ The ions of some electrolytes, like KNO_3, give more complex reactions at the electrodes. The solute does not change but water breaks down to H_2 and O_2, instead.

■ Br_2 is somewhat soluble in water, so $Br_2(aq)$ could just as well be used in these equations as $Br_2(l)$.

■ The electrolysis of anhydrous molten NaCl is the industrial synthesis of both sodium and chlorine.

$$2NaCl(l) \xrightarrow{\text{electrolysis}} 2Na(l) + Cl_2(g)$$

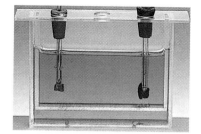

FIGURE 8.2 Electrolysis of $CuBr_2(aq)$. The solution is blue because of the copper(II) ion. The cathode, on the right, has a deposit of copper metal, and some has flaked off and fallen below it. The color around and below the anode, on the left, is brownish because Br_2 is forming.

■ In 1 *M* ammonia and 1 *M* acetic acid, the percentage ionization is less than 0.5%.

1 M $NH_3(aq)$ is a poor conductor and a weak electrolyte because only a small percentage of dissolved NH_3 molecules reacts with water to give ions. Acetic acid, the acid that gives vinegar its tart taste, and ascorbic acid (vitamin C) are other weak electrolytes and weak conductors.

Many substances are **nonelectrolytes,** whether they are in the liquid state or in solution. They do not conduct ordinary currents of electricity (e.g., household currents) at all. Although pure water will conduct a current under unusually high voltages, it is regarded as an example of a nonelectrolyte because under ordinary voltages it is a nonconductor. Ethyl alcohol and gasoline are others.

We can summarize the relationships we have just studied as follows. Be sure to notice the emphasis on *percentage* ionization as the feature dominating these definitions.

■ Weak electrolytes are generally molecular compounds that give ions by ionization, not dissociation.

> **Strong Electrolyte** One that is strongly dissociated or ionized in water—a high percentage ionization.
>
> **Weak Electrolyte** One that is weakly ionized in water—a low percentage ionization.
>
> **Nonelectrolyte** One that does not dissociate or ionize in water—essentially zero percentage ionization.

8.2 ACIDS AND BASES AS ELECTROLYTES

Acids supply hydrogen ions, and bases neutralize hydrogen ions.

The three principal ion producers in water are *acids*, *bases*, and *salts*. Acids and bases are more common than most people realize (Figure 8.3). In this section we will learn about the major acids and bases and what it means for an aqueous solution to be *acidic*, *basic*, or *neutral*.

Traces of H_3O^+ and OH^- Ions Form from the Self-Ionization of Water We begin with another look at water, because some of our definitions are related to its self-ionization. We have said in the previous section that pure water is a nonconductor.

(a)   (b)

FIGURE 8.3 (*a*) Some common acids. Vinegar contains acetic acid; vitamin C is ascorbic acid; and lemon juice has citric acid. (*b*) Some common bases. Both Drānō and Red Devil Lye contain sodium hydroxide, and milk of magnesia is a slurry of magnesium hydroxide in water.

Traces of ions, H_3O^+ and OH^-, are present, however, but not at concentrations high enough to conduct electricity at ordinary voltages. These ions come from the *self-ionization of water,* which is actually the forward reaction in the following chemical equilibrium.

$$2H_2O \rightleftharpoons H_3O^+(aq) + OH^-(aq)$$

A transfer of H^+ occurs from one molecule of water to another to give two ions in a 1:1 mole ratio, the **hydronium ion, H_3O^+**, and the **hydroxide ion, OH^-** (Figure 8.4). The forward reaction is certainly not favored; at 25 °C the concentration of each product ion is only 1.0×10^{-7} mol/L. This means that out of a little over 1 billion water molecules, only 2 molecules have changed into these ions at any one moment.

It is customary to use brackets, [ ], around a formula of a solute when we mean its concentration in the specific units of moles per liter. Thus, in pure water at 25 °C,

$$[H_3O^+] = [OH^-] = 1.0 \times 10^{-7} \text{ mol/L}$$

Although concentrations this low may seem too unimportant to mention, life itself hinges on holding the molar concentrations of H_3O^+ and OH^- ions in body fluids at about this level.

We have already learned that equilibria can be shifted, and shifts in water's self-ionization equilibrium are life threatening. *Almost all of what we will be studying about acids and bases is essential to the study of how the body controls the equilibrium for the self-ionization of water and the acid–base status of body fluids.*

The Hydronium Ion Is Often Referred To as the Hydrogen Ion The existence of ions in aqueous solutions of electrolytes was first proposed by Svante Arrhenius (1859–1927), a Swedish scientist. In the *Arrhenius theory of acids and bases,* all acids produce hydrogen ions, H^+, in water. Of course, H^+ is actually the hydrogen *atom,* $H\cdot$, minus its electron, so H^+ is a bare proton, a subatomic particle, and the nucleus of a hydrogen *atom.* Arrhenius had no way of knowing this. We now know that bare protons are *always* piggybacked on something else in solution; they have no independent existence as separate entities in solution any more than do electrons. Protons, however, can be *transferred* from one place to another, and H^+ is always held by an electron-pair bond to a water molecule or to something else. The "hydrogen ion" species in water is actually the hydronium ion, H_3O^+. It is the ability of an acid to transfer H^+ that makes it an acid.

In a practical sense, Arrhenius' supposition about H^+ wasn't too wide of the mark. H^+ is so easily available from H_3O^+ that scientists today commonly use the terms *proton, hydrogen ion,* and *hydronium ion* interchangeably. We will use *hydrogen ion* as a convenient nickname for *hydronium ion* ourselves, and we will often employ the symbol $H^+(aq)$ as a simpler way of writing $H_3O^+(aq)$.

Acids Make the H^+ Level Exceed the OH^- Level and Bases Do the Opposite When the molar concentrations of aqueous hydrogen ions and hydroxide ions are exactly

■ The *volt* is a unit of electrical force. Electricity at high voltage in a wire involves a powerful impelling force acting to move electrons.

■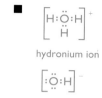
hydronium ion

hydroxide ion

■ Minute changes in the concentrations of acids or bases can switch enzymes on or off, and enzymes are essential to almost all reactions in living systems.

■ The phrase *acid–base balance* is sometimes used for *acid–base status.*

■ In 1884, Arrhenius nearly lost his bid for a doctoral degree for proposing ions, so rash was the idea considered. But in 1903, the idea earned him a Nobel Prize.

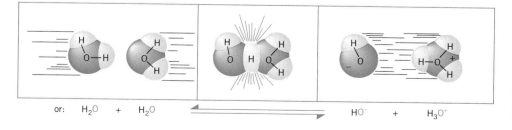

or: H_2O + H_2O HO^- + H_3O^+

FIGURE 8.4 In the self-ionization of water, H^+ transfers from one water molecule to another.

equal, as they are in pure water at any temperature, the solution is called a **neutral solution.**

Acids make the molar concentration of hydrogen ion higher than that of hydroxide ion. It's because acidic solutions all have an excess of hydrogen ion that they have many common properties. Acidic solutions, for example, turn blue litmus to a red color. Litmus is an example of an **acid–base indicator,** a compound whose color is different in acid than in base, so it can be used to tell if an aqueous solution is acidic or basic. Acidic solutions also have tart tastes, like solutions of acetic acid, citric acid, lactic acid, oxalic acid, and hydrochloric acid. But don't make a taste test without great care. Even dilute acids can corrode teeth.

Bases make the molar concentration of hydroxide ion greater than that of hydronium ion. Such solutions usually have a bitter taste and a soapy "feel," and they turn red litmus blue.

We can summarize the important conditions that define acidic, basic, and neutral solutions as follows.

Acidic solutions:	$[H^+] > [OH^-]$
Neutral solutions:	$[H^+] = [OH^-]$
Basic solutions:	$[H^+] < [OH^-]$

■ Paper impregnated with litmus dye is called litmus paper.

■

Acid	A Natural Source
Acetic acid	Vinegar
Citric acid	Lemons
Lactic acid	Sour milk
Oxalic acid	Rhubarb
Hydrochloric acid	Gastric juice

■ >, greater than
<, less than

Brønsted Broadened the Concept of Acids and Bases   We said earlier that it is the ability of something to transfer H^+ that makes it an acid. We can also say that it is the ability of something to accept H^+ that makes it a base. Johannes Brønsted (1879–1947), a Danish chemist, is generally credited with the development of these definitions.

■ Thomas Martin Lowry (1847–1936), an English chemist, proposed the same idea independently.

Brønsted Definitions of Acids and Bases

Acids are proton donors.
Bases are proton acceptors.

These definitions apply regardless of the solvent and even in the absence of any liquid solvent. Hydrogen chloride gas, for example, reacts with ammonia gas in a proton transfer reaction.

$$HCl(g) + NH_3(g) \longrightarrow NH_4Cl(s)$$

The product is a crystalline solid that forms in a cloud of microcrystals when fumes of ammonia and hydrogen chloride intermingle (Figure 8.5). The HCl molecules are proton donors; NH_3 molecules are proton acceptors. When H^+ transfers from the acid, HCl, to the base, NH_3, NH_4^+ ions and Cl^- ions form, and no solvent is involved. Yet an acid–base neutralization has occurred.

FIGURE 8.5 The reaction of $NH_3(g)$, from the bottle on the left, with $HCl(g)$, from the bottle on the right, produces a cloud of microcrystals of $NH_4Cl(s)$ by a neutralization reaction. These gaseous reactants are always present in the air spaces above concentrated solutions of aqueous ammonia and hydrochloric acid.

Acids and Bases Vary Widely in Strength   Acids and bases are quite different in their abilities to function as proton donors or proton acceptors. Water, for example, is extremely weak as both a donor and an acceptor, as we have just learned. Hydrogen chloride, $HCl(g)$, on the other hand, so readily donates H^+ that even such a weak acceptor as H_2O is able to take H^+ from $HCl(g)$. Essentially 100% of all hydrogen chloride molecules that dissolve in water react as follows.

$$HCl(g) + H_2O \longrightarrow H_3O^+(aq) + Cl^-(aq)$$

Because water is the solvent nearly always used in acid–base chemistry, *we normally define all strong acids with reference to their reactions with water.* If we use the general symbol HA for any acid, whether it is a gas, liquid, or solid, then all strong acids react as follows essentially 100%.

$$HA + H_2O \longrightarrow H_3O^+(aq) + A^-(aq)$$

We define a **strong acid** as one that is 100% ionized in this proton-donating reaction. The hydronium ion, itself, so readily donates a proton that we also apply the term *strong acid* to any aqueous solution of a strong acid.

A **weak acid** is one in which only a small percentage of its molecules ionize by reacting with water. You may have already noticed that the terms "strong" and "weak" are used alike with both acids and electrolytes. They refer to *percentage* ionization. All strong acids are strong electrolytes. All weak acids are weak electrolytes. Let's now look at a few of the most common strong and weak acids.

■ Think of *A* in H*A* or *A*⁻ as standing for the anion of the acid.

Hydrochloric Acid and Nitric Acid Are Strong Monoprotic Acids Hydrochloric acid is a **monoprotic acid** because the ratio of H_3O^+ ions to Cl^- ions in an aqueous solution of HCl is 1 to 1. Other monoprotic acids are those that can be made by dissolving the other hydrogen halide gases in water, for example, hydrofluoric acid, $HF(aq)$, hydrobromic acid, $HBr(aq)$, and hydroiodic acid, $HI(aq)$. All but hydrofluoric acid are strong acids.[1]

■ Because F^- ion can tie up Ca^{2+} ion at nerve endings—ions essential for sending nerve signals—and because $HF(aq)$ can carry F^- through the skin, HF causes excruciating pains and is a dangerous chemical.

■ **PRACTICE EXERCISE 1** Depict the ionization of HBr and HI in water using Lewis structures such as we did for HCl in the previous section.

Nitric acid, $HNO_3(aq)$, is also a strong, monoprotic acid. We can represent its ionization as follows:

$$H_2O + HNO_3(aq) \longrightarrow H_3O^+(aq) + NO_3^-(aq)$$

nitric acid (molecular formula)   hydronium ion   nitrate ion

nitric acid in water

nitric acid

Acetic Acid Is a Weak Monoprotic Acid Acetic acid is a typical organic acid, and like virtually all organic acids, it is a weak acid. **Organic compounds** are the compounds of carbon other than its oxides, the cyanides, or those related to earthlike substances such as limestone rocks and other carbonates. Compounds that are not organic compounds are called **inorganic compounds.**

The acetic acid molecule, $HC_2H_3O_2$, has four hydrogen atoms, but only one is attached to an oxygen atom. Only this one can transfer to a water molecule. The H—O bond in acetic acid is considerably stronger than the H—Cl bond in $HCl(g)$, so acetic acid does not as readily transfer H^+ to H_2O as does H—Cl. The chemical equilibrium in aqueous acetic acid is

$$H_2O + HC_2H_3O_2 \rightleftharpoons H_3O^+(aq) + C_2H_3O_2^-(aq)$$

acetic acid[2]   acetate ion

acetic acid
(The H in color is the proton available in acid–base reactions.)

acetate ion

[1] Hydrofluoric acid, $HF(aq)$, is unusual because its ions, H_3O^+ and F^-, attract each other so strongly in solution that they behave as if they were not very free of each other. $HF(aq)$ is thus classified as a *weak acid.*

[2] We will usually use $HC_2H_3O_2$ instead of the full structure as our symbol for acetic acid in this chapter and the next. Just remember that only one hydrogen is active in acid–base reactions, and that acetic acid is monoprotic.

In 0.1 M acetic acid, only 0.5% of all acetic acid molecules are at any moment ionized, so the forward reaction is definitely not favored. Yet the coming and going characteristic of all dynamic equilibria exists, and small percentages of hydronium and acetate ions are present.

The C=O Group Makes Acetic Acid More Acidic than Water   We might pause to ask here why acetic acid is an acid at all, why the proton of its H—O group more easily transfers than the proton of the H—O group in water. The difference is caused by the group that directly holds H—O in the acetic acid molecule, namely, the carbon–oxygen double bond, C=O. It's an electronegative group made so by its electronegative oxygen atom. Therefore, the C=O is able to pull some electron density away from the oxygen atom of the H—O group in acetic acid. This weakens the H—O bond, so the hydrogen more easily transfers to a water molecule than the hydrogen of a water molecule itself.

■ Sulfuric acid is the most widely used acid in industrial applications. Over 70 billion pounds (over 320 billion moles) are annually used in the United States.

■

sulfuric acid

■ A "lone oxygen" is one joined only to one other atom.

■

hydrogen sulfate ion

■

sulfate ion

■

phosphoric acid

Sulfuric Acid   Sulfuric acid, H_2SO_4, is the only common, stable inorganic **diprotic acid,** one that is able to give up two hydrogen ions per formula unit. The ionization of the first hydrogen ion is so easy that we do not write equilibrium arrows for the reaction; sulfuric acid is a strong acid.

$$H_2O + H_2SO_4 \longrightarrow H_3O^+(aq) + HSO_4^-(aq)$$

sulfuric   hydrogen
acid   sulfate ion

The two extra "lone oxygens" on sulfur in sulfuric acid help to make sulfuric acid an acid in a way similar to the effect of the lone oxygen in acetic acid.

The Hydrogen Sulfate Ion Is Also an Acid   An ion can be an acid as well as a molecule. For example, the hydrogen sulfate ion is a monoprotic acid. The transfer of H^+ from HSO_4^-, however, requires a positively charged particle (H^+) to pull away from one already oppositely charged (HSO_4^-). Although this is harder than the transfer of the first H^+ from H_2SO_4, it still happens. The following equilibrium exists, but the *reactants* are favored.

$$H_2O + HSO_4^-(aq) \rightleftharpoons H_3O^+(aq) + SO_4^{2-}(aq)$$

hydrogen   sulfate
sulfate   ion
ion

In 0.10 M HSO_4^- (present, for example, in 0.10 M $NaHSO_4$), the percentage ionization of the HSO_4^- ion is slightly less than 40%.[3]

Phosphoric Acid Is a Moderately Strong, Triprotic Acid   A **triprotic acid** is one that can potentially release three H^+ ions to proton acceptors. Phosphoric acid, H_3PO_4, is the only common example of an inorganic triprotic acid. Like the ionization of sulfuric acid, that of phosphoric acid occurs in steps, each one more difficult than the previous. Even the first step does not occur to 100% of the phosphoric acid molecules. In 1 M H_3PO_4, less than a third are ionized, so we have to write equilibrium expressions for all steps. In all equilibria, the *reactants* are favored.

---

[3] The double bonds from S to O and from P to O shown in the structures of the sulfuric acid and phosphoric acid systems violate the octet rule as we learned it. However, the valence shells of S and P, which are higher than level 2, can and often do hold more than 8 electrons. The double bonds shown in the margin structures reflect both this flexibility and the extra stabilization that additional bonds confer to the systems.

$$H_2O + H_3PO_4(aq) \rightleftharpoons H_3O^+(aq) + H_2PO_4^-(aq)$$

phosphoric
acid

dihydrogen
phosphate ion

$$H_2O + H_2PO_4^-(aq) \rightleftharpoons H_3O^+(aq) + HPO_4^{2-}(aq)$$

monohydrogen
phosphate ion

■ $H_2PO_4^-$ is a weak acid.

$$H_2O + HPO_4^{2-}(aq) \rightleftharpoons H_3O^+(aq) + PO_4^{3-}(aq)$$

phosphate ion

■ HPO_4^{2-} is a very weak acid.

The percentage ionization of H_3PO_4 is roughly 27% in a 0.10 M solution of phosphoric acid, a percentage too low to let us call it a strong acid. Although phosphoric acid is a good conductor of electricity in water, it's classified as a *moderate acid.*

It is important to learn the names and formulas of the three ions available from phosphoric acid, because the phosphate ion system occurs widely in the body. Relatives of phosphoric acid—diphosphoric acid and triphosphoric acid—are important systems in metabolism.

■

diphosphoric acid

Carbonic Acid Is Involved in Respiration   Carbonic acid, H_2CO_3, is a weak diprotic acid that is unusual because it is unstable. But its instability is actually vital to the body's ability to manage one of the respiratory gases, carbon dioxide. When carbon dioxide dissolves in water, a trace reacts with water to form carbonic acid, a weak acid.

$$CO_2(aq) + H_2O \rightleftharpoons H_2CO_3(aq)$$

carbonic acid

A small fraction of the carbonic acid molecules ionizes.

$$H_2CO_3(aq) + H_2O \rightleftharpoons H_3O^+ + HCO_3^-(aq)$$

bicarbonate ion

The bicarbonate ion is itself a (very) weak acid. A strong base, like OH^-, can remove its proton, however.

$$HCO_3^-(aq) + OH^- \rightleftharpoons CO_3^{2-}(aq) + H_2O$$

carbonate ion

The bicarbonate ion is the chief form in which waste carbon dioxide is carried from body tissues to the lungs. We'll give a brief overview of this process in Section 9.9, but we need to know something about the general reactions of acids and bases first.

■

triphosphoric acid

■

carbonic acid

■

bicarbonate ion

■

carbonate ion

■ **PRACTICE EXERCISE 2**   When sulfur dioxide dissolves in water, some of it reacts with the water to give a solution called sulfurous acid, traditionally written as H_2SO_3. Write the equations for the successive ionization equilibria of this weak acid.

■ Sulfurous acid is $SO_2 \cdot H_2O(aq)$, not $H_2SO_3(aq)$, but the latter formula is the more commonly used.

Table 8.1 summarizes the common aqueous acids, and you should now memorize the names and formulas of all the listed strong and moderate acids. We'll need this knowledge as we go along.

Sodium Hydroxide Is the Most Common Strong Base   Table 8.2 lists the common bases. The **strong bases** are those that dissociate nearly 100% in water, and they furnish OH^-, a strong proton-binding or proton-accepting species. Sodium hydroxide,

■ Solid NaOH is very hygroscopic, so its container must be promptly and tightly reclosed each time some is taken.

TABLE 8.1 Common Acids[a]

Acid	Formula	Percentage Ionization
Strong Acids		
Hydrochloric acid	HCI	Very high
Hydrobromic acid	HBr	Very high
Hydroiodic acid	HI	Very high
Nitric acid	HNO_3	Very high
Sulfuric acid[b]	H_2SO_4	Very high
Moderate Acids		
Phosphoric acid	H_3PO_4	27
Sulfurious acid[c]	H_2SO_3	20
Weak Acids		
Nitrous acid[c]	HNO_2	1.5
Acetic acid	$HC_2H_3O_2$	1.3
Carbonic acid[c]	H_2CO_3	0.2

[a] Data are for 0.1 M solutions of the acids in water at room temperature.

[b] *Concentrated* sulfuric acid (99%) is particularly dangerous not only because it is a strong acid but also because it is a powerful dehydrating agent. This action generates considerable heat at the reaction site, and at higher temperatures sulfuric acid becomes even more dangerous. Moreover, concentrated sulfuric acid is a thick, viscous liquid that does not wash away from skin or fabric very quickly.

[c] An unstable acid.

TABLE 8.2 Common Bases

Base	Formula	Solubility[a]	Percentage Ionization
Strong Bases			
Sodium hydroxide	NaOH	109	>90 (0.1 M solution)
Potassium hydroxide	KOH	112	>90 (0.1 M solution)
Calcium hydroxide	$Ca(OH)_2$	0.165	100 (saturated solution)
Magnesium hydroxide	$Mg(OH)_2$	0.0009	100 (saturated solution)
Weak Base			
Ammonia, aqueous	NH_3	89.9	1.3 (18 °C)[b]

[a] Solubilities are in grams of solute per 100 g of water at 20 °C.

[b] The ionization referred to here is the equilibrium:

$$NH_3(aq) + H_2O \rightleftharpoons NH_4^+(aq) + OH^-(aq)$$

NaOH, and potassium hydroxide, KOH, are examples of strong bases because they break up in water essentially 100% into the metal ion and OH^-, the actual base or proton acceptor.

$$NaOH(s) \xrightarrow{\text{water}} Na^+(aq) + OH^-(aq)$$

$$KOH(s) \xrightarrow{\text{water}} K^+(aq) + OH^-(aq)$$

Two other strong bases are the hydroxides of group IIA metals, magnesium hydroxide, $Mg(OH)_2$, and calcium hydroxide, $Ca(OH)_2$. They also ionize essentially

100% in water, but, as the data in Table 8.2 show, they are so insoluble in water that even saturated solutions provide only very dilute solutions of hydroxide ions.

$$Ca(OH)_2(s) \longrightarrow Ca^{2+}(aq) + 2OH^-(aq)$$
$$Mg(OH)_2(s) \longrightarrow Mg^{2+}(aq) + 2OH^-(aq)$$

Both sodium and potassium hydroxides can be prepared in solutions concentrated enough to be dangerous chemicals. In sufficient concentration, the hydroxide ion causes a severe chemical burn, certainly enough to be very hazardous to the eyes. Calcium and magnesium hydroxides, however, are so insoluble that they not only pose no grave danger to the skin, they are used internally in home remedies. Calcium hydroxide is a component of one commercial antacid tablet. A slurry of magnesium hydroxide in water, called "milk of magnesia," is used as an antacid and a laxative.

A **weak base** is a poor proton acceptor, one that is unable to take protons from water molecules to any appreciable extent. Ammonia is the most common example. A solution of ammonia in water, called *aqueous ammonia,* does have some excess hydroxide ion, but only a small percentage of ammonia molecules reacts to produce it, as we earlier learned. The following equilibrium is present, and the reactants are strongly favored:

$$\underset{\text{ammonia}}{NH_3(aq)} + H_2O \rightleftharpoons \underset{\substack{\text{ammonium}\\\text{ion}}}{NH_4^+(aq)} + OH^-(aq)$$

■ You'll sometimes see aqueous ammonia called "ammonium hydroxide," but this is misleading. NH_4OH is unknown as a pure compound.

A dilute solution (about 5%) of ammonia in water is sold as household ammonia in supermarkets. It's a good cleaning agent, but watch out for its fumes.

The names and formulas of the bases in Table 8.2 should also be memorized.

All Salts Are Strong Electrolytes   The third major family of ion-producing substances are the salts. We have to say something about them here—they'll be treated more fully in a later section—because salts are products of the reactions of acids and bases.

Salts are ionic compounds in which the positive ion is a metal ion or any other positive ion except H^+, and the negative ion is any except OH^- or O^{2-}. All salts are crystalline ionic solids at room temperature. All are strong electrolytes. When salts dissolve in water, their ions dissociate essentially 100%. Even for very insoluble salts, what little of them that does dissolve becomes 100% dissociated, and many can be made in aqueous solution by the reaction of an acid with a base.

■ Some chemists classify metal oxides, compounds with the oxide ion (O^{2-}), as *salts*. However, those that dissolve in water *react* with it completely, and $OH^-(aq)$ forms, not $O^{2-}(aq)$.

8.3 CHEMICAL PROPERTIES OF AQUEOUS ACIDS AND BASES

Acids react with hydroxides, bicarbonates, carbonates, ammonia, and active metals.

In this section we will principally study the reactions of the hydronium ion, H_3O^+, and the species that neutralize it. We'll use the symbol $H^+(aq)$ as shorthand for H_3O^+ in most of the equations. A special kind of equation, the *net ionic equation,* is particularly helpful when we concentrate on the chemical properties of this ion (or any others). We have occasionally been using such equations, and now we will learn how to write them.

■ We say *molecular* equation even though some of the chemicals in the equation might be ionic compounds.

A Net Ionic Equation Omits "Spectator" Species   The conventional equation for a reaction is called a **molecular equation** because it shows all of the substances in the molecular or empirical formulas that we would need for planning an actual experiment. The molecular equation for the reaction of sodium carbonate decahydrate with hydrochloric acid, for example, is

$$Na_2CO_3 \cdot {}_{10}H_2O(s) + 2HCl(aq) \longrightarrow 2NaCl(aq) + CO_2(g) + 11H_2O$$

Similarly, the molecular equation for the reaction of hydrochloric acid with aqueous sodium hydroxide is

$$HCl(aq) + NaOH(aq) \longrightarrow NaCl(aq) + H_2O$$

■ An insoluble salt would have had (s) after its formula.

However, as we now know, $HCl(aq)$ is really $H^+(aq)$ and $Cl^-(aq)$, and $NaOH(aq)$ is actually $Na^+(aq)$ and $OH^-(aq)$. We know this because we know the acid is a *strong* acid and the base is a *strong* base, so both must be essentially fully ionized in solution. We also know that $NaCl(aq)$ is fully ionized because all salts are strong electrolytes, and the (aq) by its formula tells us that the NaCl is in solution. The formula of water is the fourth formula in the equation, and water is a nonelectrolyte. Therefore, we cannot write its molecules as being separated into ions. (We ignore, of course, the self-ionization of water, because it occurs to an exceedingly low percentage.) We'll always assume that H_2O means $H_2O(l)$.

Using these facts we can expand the molecular equation into what is called the **ionic equation,** one that shows all of the dissolved species, whether ionic or molecular. We do this by replacing anything that we know is present as ions by the actual formulas of these ions. Thus the ionic equation for our example is

$$H^+(aq) + Cl^-(aq) + Na^+(aq) + OH^-(aq) \longrightarrow Na^+(aq) + Cl^-(aq) + H_2O$$

| These came from HCl(aq) | These came from NaOH(aq) | These came from NaCl(aq) | Not ionized |

The preparation of an ionic equation is actually just a scratch paper operation, because we next cancel all of the formulas that appear *identically* on opposites sides of the arrow. In the foregoing ionic equation, for example, we see that nothing happens either to $Na^+(aq)$ or to $Cl^-(aq)$. There is no reason, therefore, to let them remain in the equation when we just want to give full attention to the species that react or form. [$Na^+(aq)$ and $Cl^-(aq)$, of course, do serve one function; they give electrical neutrality to their respective compounds. Otherwise, they are nothing more than *spectator particles* in this reaction.] Formulas must be of the same physical state before they can be canceled. We could not, for example, cancel $HCl(g)$ by $HCl(aq)$, because their states are different.

■ Sometimes we don't cancel, but only reduce in number. If an ionic equation, for example, has

$$\ldots + 4H_2O \longrightarrow \ldots + 2H_2O$$

we can simplify it to

$$\ldots + 2H_2O \longrightarrow \ldots$$

We can cancel the spectator particles, whether they are ions or molecules, from the ionic equation. This leaves the **net ionic equation,** one that shows only the reacting species and the products they form. As you can see, this equation is a simple description of what happens when hydrochloric acid, *or any strong acid,* neutralizes sodium hydroxide, *or any hydroxide base,* in solution.

$$H^+(aq) + OH^-(aq) \longrightarrow H_2O$$

■ The material balance is what we need in any kind of balanced equation.

Net Ionic Equations Must Balance Both Electrically and Materially   For a net ionic equation to be balanced, two conditions must be met: a material balance and an electrical balance. We have **material balance** when the numbers of atoms of each element, regardless of how they are chemically present, are the same on both sides of the arrow. We have **electrical balance** when the algebraic sum of the charges to the left of the arrow equals the sum of the charges to the right.

EXAMPLE 8.1 Writing a Net Ionic Equation

Sulfuric acid is the most important acid in industrial use, and sometimes it has to be neutralized by sodium hydroxide. The reaction can be carried out to produce sodium sulfate, $Na_2SO_4(aq)$, and water. Write the molecular, ionic, and net ionic equations.

■ Sulfuric acid must be handled very carefully. See Table 8.1.

ANALYSIS 1 The question calls for three answers, so we'll proceed in steps. First we deal with the molecular equation. The complete formulas of the reactants and products must first be assembled in the pattern of an equation, which is then balanced to give the molecular equation.

SOLUTION 1, THE MOLECULAR EQUATION The balanced molecular equation is

$$H_2SO_4(aq) + 2NaOH(aq) \longrightarrow Na_2SO_4(aq) + 2H_2O$$

ANALYSIS 2 Using our knowledge about which acids and bases are strong and which salts are fully ionized, we disassemble the molecular equation to display all of the ions present or provided. The following facts about the reactants and products are relevant to this exercise. We show H_2SO_4 as breaking up entirely into $2H^+$ and SO_4^{2-} because the given reaction is with a strong base, not with water, and SO_4^{2-}, not HSO_4^-, forms.

$H_2SO_4(aq)$ means $2H^+(aq) + SO_4^{2-}(aq)$ (H_2SO_4 gives up *two* H^+ ions to the other reactant.)

$2NaOH(aq)$ means $2Na^+(aq) + 2OH^-(aq)$ (This is a strong, fully ionized metal hydroxide.)

$Na_2SO_4(aq)$ means $2Na^+(aq) + SO_4^{2-}(aq)$ (This is a salt, and *aq* tells us that it is in solution; hence, it is fully ionized.)

$2H_2O$ means $2H_2O$ (No breaking up occurs with this nonelectrolyte.)

SOLUTION 2, THE IONIC EQUATION The result of the analysis is the ionic equation.

$$2H^+(aq) + SO_4^{2-}(aq) + 2Na^+(aq) + 2OH^-(aq) \longrightarrow$$
$$2Na^+(aq) + SO_4^{2-}(aq) + 2H_2O$$

ANALYSIS 3 Finally, identical species in identical states on opposite sides of the arrow in the ionic equation are canceled.

$$2H^+(aq) + \cancel{SO_4^{2-}(aq)} + \cancel{2Na^+(aq)} + 2OH^-(aq) \longrightarrow$$
$$\cancel{2Na^+(aq)} + \cancel{SO_4^{2-}(aq)} + 2H_2O$$

SOLUTION 3, THE NET IONIC EQUATION This leaves us with a net ionic equation.

$$2H^+(aq) + 2OH^-(aq) \longrightarrow 2H_2O$$

Note that we can divide all the coefficients by 2 and convert them to smaller whole numbers. Thus the final net ionic equation is simply

$$H^+(aq) + OH^-(aq) \longrightarrow H_2O$$

CHECK We have both a material and an electrical balance for each equation.

■ PRACTICE EXERCISE 3   Write the molecular, the ionic, and the net ionic equations for the neutralization of nitric acid by potassium hydroxide. A water-soluble salt, $KNO_3(aq)$, and water form.

Strong Acids React with Metal Hydroxides To Give Water and a Salt   Example 8.1 and Practice Exercise 3 illustrate reactions of strong acids with metal hydroxides, and we saw in the net ionic equations that they are the reaction of a Brønsted acid, H^+, with a Brønsted base, OH^-. If we let M stand for any group IA metal, we can write the reactions of aqueous solutions of their hydroxides with a strong acid such as hydrochloric acid by the following general equation.

■ The group IA hydroxides are LiOH, NaOH, KOH, RbOH, and CsOH.

$$MOH(aq) + HCl(aq) \longrightarrow MCl(aq) + H_2O$$

The net ionic equation is

$$OH^-(aq) + H^+(aq) \longrightarrow H_2O$$

Only the group IA hydroxides are very soluble in water. Most of the others are relatively insoluble, but their solid forms can still neutralize strong acids. The net ionic equations for the reactions of the water-insoluble metal hydroxides with acid, therefore, reflect the insolubility of such hydroxides. The equations cannot show OH^- as a dissociated species.

$$M(OH)_2(s) + 2HCl(aq) \longrightarrow MCl_2(aq) + 2H_2O$$

The net ionic equation is

$$M(OH)_2(s) + 2H^+(aq) \longrightarrow M^{2+}(aq) + 2H_2O$$

Here, M stands for any metal in group IIA (except beryllium).

■ PRACTICE EXERCISE 4   When milk of magnesia is used to neutralize hydrochloric acid (stomach acid), solid magnesium hydroxide in the suspension reacts with the acid. Write the molecular and net ionic equations for this reaction.

Strong Acids React with Metal Bicarbonates to Give CO_2, H_2O, and a Salt   All metal bicarbonates react the same way with strong, aqueous acids. They react to give carbon dioxide, water, and a salt. For example, sodium bicarbonate and hydrochloric acid react as follows.

$$HCl(aq) + NaHCO_3(aq) \longrightarrow CO_2(g) + H_2O + NaCl(aq)$$

Potassium bicarbonate and hydrobromic acid give a similar reaction.

$$HBr(aq) + KHCO_3(aq) \longrightarrow CO_2(g) + H_2O + KBr(aq)$$

What is believed to form initially is not CO_2 and H_2O but H_2CO_3, carbonic acid. However, almost all of it promptly decomposes to CO_2 and H_2O, so the solution fizzes strongly as the reaction proceeds and CO_2 evolves.

Notice that the salt whose formula appears as a product in the molecular equation is always a combination of the cation of the bicarbonate (Na^+ or K^+ in our examples) and the anion of the acid (Cl^- or Br^- in our examples). Let's be sure we can write the formula of the salt that forms in these reactions before we continue.

NaHCO$_3$(aq) and HCl(aq) acid react to give NaCl(aq), H$_2$O, and CO$_2$(g), which can be seen bubbling out of this tube.

EXAMPLE 8.2   Writing the Formula of the Salt That Forms When a Bicarbonate Reacts with an Acid

What salt forms when lithium bicarbonate reacts with nitric acid?

ANALYSIS   The name, lithium bicarbonate, gives us the names of the ions involved, the lithium ion (Li^+) and the bicarbonate ion (HCO_3^-). The anion of nitric acid is NO_3^-, so this is the anion that must be paired with Li^+ if a salt is to form. The electrical charges, 1+ and 1−, on these ions tell us that they *must* combine in 1:1 ratio.

SOLUTION   The salt's formula is $LiNO_3$.

■ Pure bicarbonate salts generally involve only the group IA metal ions.

■   PRACTICE EXERCISE 5   What is the formula of the salt that forms when potassium bicarbonate reacts with sulfuric acid? Assume the salt is a sulfate and not a hydrogen sulfate.

In writing net ionic equations of reactions between strong acids and metal bicarbonates, we'll treat all metal bicarbonates as dissociated in water to the metal ion and the bicarbonate ion. As we will see, the reaction of bicarbonates with acids is one in which the bicarbonate ion is a base, a proton acceptor.

EXAMPLE 8.3   Writing Equations for the Reactions of Bicarbonates with Strong Acids

What are the molecular and the net ionic equations for the reaction of potassium bicarbonate with hydroiodic acid?

ANALYSIS   The products are CO_2, H_2O, and a salt. Using what we learned in Example 8.2, the salt must be a combination of the potassium ion, K^+, and the iodide ion, I^-. The salt is KI.

SOLUTION   The balanced molecular equation is

$$KHCO_3(aq) + HI(aq) \longrightarrow CO_2(g) + H_2O + KI(aq)$$

To prepare the ionic equation, we analyze each of the formulas in the molecular equation to see which can be disassembled into ions.

$KHCO_3(aq)$ means $K^+(aq)$ and $HCO_3^-(aq)$ (As we were told.)

$HI(aq)$ means $H^+(aq) + I^-(aq)$ (Because this is a fully ionized acid.)

$KI(aq)$ means $K^+(aq) + I^-(aq)$ (Because we treat all water-soluble salts as fully ionized.)

$CO_2(g)$ and H_2O stay the same   (Neither is ionized.)

With these facts the molecular equation expands into the ionic equation.

$$[K^+(aq) + HCO_3^-(aq)] + [H^+(aq) + I^-(aq)] \longrightarrow$$
$$CO_2(g) + H_2O + [K^+(aq) + I^-(aq)]$$

The $K^+(aq)$ and the $I^-(aq)$ cancel from each side of the arrow to leave the following net ionic equation.

$$H^+(aq) + HCO_3^-(aq) \longrightarrow CO_2(g) + H_2O$$

CHECK   The equation is balanced both materially and electrically.

The equation produced by Example 8.3 is the same net ionic equation for the reactions of all metal bicarbonates with all strong, aqueous acids. Had we wanted to be a bit more exact and used $H_3O^+(aq)$ instead of $H^+(aq)$, the net ionic equation would have been

$$H_3O^+(aq) + HCO_3^-(aq) \longrightarrow CO_2(g) + 2H_2O$$

The only difference is in how H_2O becomes balanced. The essential chemistry has not changed.

Because this reaction destroys the hydrogen ions of the acid, it must also be called an acid neutralization. In fact, the familiar "bicarb" used as a home remedy for acid stomach is nothing more than sodium bicarbonate. Stomach acid is roughly 0.1 M HCl, and bicarbonate ion neutralizes this acid by the reaction we have just studied. An overdose of "bicarb" must be avoided because it can cause a medical emergency involving the respiratory gases. Another use of sodium bicarbonate is as an isotonic solution given intravenously to neutralize acid in the blood. For still another use, see Interaction 8.1.

■ PRACTICE EXERCISE 6   Write the molecular, ionic, and net ionic equations for the reaction of sodium bicarbonate with sulfuric acid in which sodium sulfate, $Na_2SO_4(aq)$, is one of the products.

Strong Acids React with Carbonates to Give CO_2, H_2O, and a Salt   Carbonates react with hydrogen ions to give the same products as bicarbonates. Only the stoichiometry changes. The CO_3^{2-} ion is thus a base and, mole for mole, it neutralizes twice as much H^+ as the HCO_3^- ion. For example, sodium carbonate reacts with nitric acid as follows.

$$2HNO_3(aq) + Na_2CO_3(aq) \longrightarrow CO_2(g) + H_2O + 2NaNO_3(aq)$$

The net ionic equation, which can be obtained by the same kind of procedure shown in Example 8.3, is

$$2H^+(aq) + CO_3^{2-}(aq) \longrightarrow CO_2(g) + H_2O$$

Notice that there is both material and electrical charge balance. This net ionic equation describes the reactions of all of the carbonates of the group IA metals with all strong, aqueous acids.

■ PRACTICE EXERCISE 7   Write the molecular, ionic, and net ionic equations for the reaction of aqueous potassium carbonate, $K_2CO_3(aq)$, with sulfuric acid to give potassium sulfate, $K_2SO_4(aq)$, a water-soluble salt, and the other usual products.

INTERACTION 8.1
INSTANT CARBONATED BEVERAGES AND MEDICATIONS

As you no doubt know, you can buy fruit-flavored tablets that dissolve in water to give a fizzy drink. Alka-Seltzer and similar tablets contain a solid acid, citric acid, and solid sodium bicarbonate, besides aspirin (see Figure 1). The way that these tablets respond when they're dropped into water illustrates the importance of water as a solvent in the reactions of ions. In the crystalline materials, ions are not free to move, but as soon as these tablets hit the water and the ions become mobile, the ions start to react.

Citric acid is a triprotic acid, and we can represent it as H_3Cit, where Cit stands for the citrate ion, an ion with a charge of $3-$. The hydrogen ions liberated by citric acid when it is in solution react with the bicarbonate ions that become free to move around when sodium bicarbonate dissolves. This reaction gives the CO_2 that fizzes out of solution as it forms.

FIGURE 1 Alka-Seltzer.

$$H^+(aq) + HCO_3^-(aq) \longrightarrow CO_2(g) + H_2O$$

Only the carbonates of group IA metal ions (as well as ammonium carbonate) are very soluble in water. Most other carbonates are water-insoluble compounds. Calcium carbonate, $CaCO_3$, for example, is the chief substance in limestone and marble. Despite its insolubility in water, calcium carbonate reacts readily with strong, aqueous acids (with their hydrogen ions, of course). The products are soluble in water, so the insoluble carbonates dissolve by this reaction.

■ Stalactites and stalagmites in limestone caverns are chiefly deposits of limestone.

$$CaCO_3(s) + 2HCl(aq) \longrightarrow CO_2(g) + H_2O + CaCl_2(aq)$$

For water-insoluble carbonates, we have to write their entire formulas in net ionic equations, so the net ionic equation for the above reaction is

$$CaCO_3(s) + 2H^+(aq) \longrightarrow CO_2(g) + H_2O + Ca^{2+}(aq)$$

■ **PRACTICE EXERCISE 8** Dolomite, a limestone-like rock, contains both calcium and magnesium carbonates. Magnesium carbonate is attacked by nitric acid. The salt that forms is water soluble. Write the molecular, ionic, and net ionic equations for this reaction.

■ The addition of a few drops of hydrochloric acid to a rock sample is a field test for carbonate rocks. A positive test is the fizzing of an odorless gas.

Ammonia Neutralizes Strong Aqueous Acids An aqueous solution of ammonia is an excellent reagent for neutralizing acids. We learned in Section 4.6 how an unshared pair of electrons on nitrogen in ammonia can form a coordinate covalent bond to H^+ furnished by an acid. This makes ammonia an effective Brønsted base. For example,

$$NH_3(aq) + HCl(aq) \longrightarrow NH_4Cl(aq)$$

The net ionic equation is

$$NH_3(aq) + H^+(aq) \longrightarrow NH_4^+(aq)$$

FIGURE 8.6 The reaction of metallic sodium with water is violent. It produces hydrogen gas and enough heat to cause ignition and a shower of sparks.

All ammonium salts are soluble in water, so they liberate NH_4^+ ions in aqueous solutions. Many biochemicals have ammonia-like molecules that also neutralize hydrogen ions.

■ PRACTICE EXERCISE 9   Write the molecular and net ionic equations for the reaction of aqueous ammonia with (a) $HBr(aq)$ and (b) $H_2SO_4(aq)$.

Active Metals React with Strong Acids to Give Hydrogen Gas and a Salt   Many metals are attacked more or less readily by the hydrogen ion in solution. The products are generally hydrogen gas and a salt made of the cation from the metal and the anion from the acid. Zinc, for example, reacts with hydrochloric acid as follows:

$$Zn(s) + 2HCl(aq) \longrightarrow H_2(g) + ZnCl_2(aq)$$

The net ionic equation is

$$Zn(s) + 2H^+(aq) \longrightarrow H_2(g) + Zn^{2+}(aq)$$

Aluminum is also attacked by acids. Its reaction with nitric acid, for example, can be written as follows:

$$2Al(s) + 6HNO_3(aq) \longrightarrow 2Al(NO_3)_3(aq) + 3H_2(g)$$

The net ionic equation is

$$2Al(s) + 6H^+(aq) \longrightarrow 2Al^{3+}(aq) + 3H_2(g)$$

■ PRACTICE EXERCISE 10   Write the molecular and the net ionic equations for the reaction of magnesium with hydrochloric acid.

■ In oxidation, an oxidation number becomes more positive. In reduction, an oxidation number becomes more negative. Oxidation numbers were introduced in Section 4.3.

Metals Form an Activity Series in Their Reactions with Acids   Metals differ greatly in their tendencies to react with aqueous hydrogen ions. When they do, atoms of the metal are oxidized because they lose electrons and become metal ions. The oxidizing agent is H^+. The electrons are transferred to H^+, taken from H_3O^+ ions (sometimes from H_2O), and these protons are reduced and made electrically neutral. Two H atoms combine and emerge as a molecule of hydrogen gas, H_2. Thus the metal is oxidized by H^+ to a cation and H^+ is reduced by the metal to H_2.

The group IA metals such as sodium and potassium include the most reactive metals of all. They not only reduce H^+ taken from hydronium ions, they also reduce H^+ taken from water molecules. No acid need be present. The following reaction of sodium metal with water is extremely violent. It should be attempted only by an experienced chemist working with safety equipment, including a fire extinguisher (Figure 8.6).

$$2Na(s) + 2H_2O \longrightarrow 2NaOH(aq) + H_2(g)$$

Gold, silver, and platinum, in contrast, are stable not only toward water but also toward hydronium ions. Lead and tin react very slowly with acids. Figure 8.7 shows how the reactivities of iron, zinc, and magnesium differ toward 1 M HCl.

The vast differences in the reactivities of metals toward acids make it possible to arrange the metals in an order of reactivity. The result is the **activity series** of the metals, given in Table 8.3. Atoms of any metal above hydrogen in the series can transfer electrons to H^+, either from H_2O or from H_3O^+, to form hydrogen gas, and the metal atoms change to ions. The farther a metal is above hydrogen, the more reactive it is toward acids. The metals below hydrogen in the activity series do not transfer electrons to H^+.

FIGURE 8.7 Metals vary widely in their ease of oxidation. Iron is in the first tube, zinc in the second, and magnesium in the third, and all are exposed to HCl(*aq*) at the same molarity. All these metals can be oxidized to their metal ion states by H^+, and H^+ is reduced to hydrogen gas. Iron, the least readily oxidized of these metals, produces hardly any visible fizzing of hydrogen gas. Zinc, the next most easily oxidized of the three, reacts rather well. Magnesium reacts very vigorously and is the most easily oxidized of the three.

TABLE 8.3   The Activity Series of the Common Metals

Greatest tendency to become ionic	Potassium Sodium	React violently with water
	Calcium	Reacts slowly with water
React with hydrogen ions to liberate H_2	Magnesium Aluminum Zinc Chromium	React very slowly with water
	Iron Nickel Tin Lead	
	HYDROGEN	
Do not react with hydrogen ions	Copper Mercury Silver Platinum	
Least tendency to become ionic	Gold	

Strong, Moderate, and Weak Acids React at Different Rates with the Same Metal
The rate of the reaction of an acid with a metal depends on the acid as well as the metal. When compared at the same molar concentrations, strong acids react far more rapidly than weak acids (Figure 8.8). These differences reflect the differences in percentage ionizations, because the actual reaction, as we have said, is with the hydrogen ion. When the concentration of hydrogen ion is high, as it can be when the acid is strong, the reaction is vigorous. In 1 M HCl, the concentration of $H^+(aq)$ is also 1 M, because for each HCl one $H^+(aq)$ is released. However, in 1 M $HC_2H_3O_2$, acetic acid (a weak acid), the actual concentration of $H^+(aq)$ is closer to 0.004 M, which is about 1/250 as much. No wonder the liveliness of the reaction pictured in Figure 8.8, right tube, the reaction of zinc with 1 M acetic acid, is much less than in Figure 8.8, left tube, the reaction with 1 M HCl. In the center tube of Figure 8.8, the reaction is with 1 M H_3PO_4, a moderate acid, and the vigor of the reaction is somewhere in between that of the other two acids.

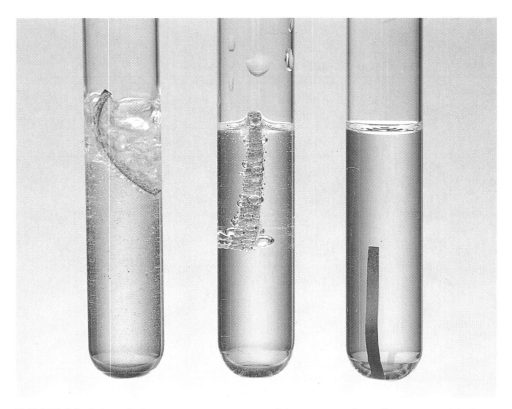

FIGURE 8.8 Relative hydrogen ion concentrations and the reactivity of zinc. Zinc reacts with hydrogen ion to give zinc ion and hydrogen gas, which fizzes out of the test tubes. Three different acids, ranging from strong to moderate to weak, are used here in identical molar concentrations. (*Left*) The acid is HCl(*aq*), a strong, fully ionized acid. (*Center*) The acid is H_3PO_4(*aq*), a moderate, partly ionized acid. (*Right*) The acid is acetic acid, $HC_2H_3O_2$(*aq*), a weak, poorly ionized acid. Although the molarities of the acids are the same, the actual molar concentrations of their hydrogen ions are greatly different, being highest in HCl(*aq*), where the bubbles of hydrogen are evolving the most vigorously, next highest in H_3PO_4(*aq*), and lowest in $HC_2H_3O_2$(*aq*).

8.4 STRENGTHS OF BRØNSTED ACIDS AND BASES— A QUALITATIVE VIEW

Acid–base equilibria involve *two* acids and *two* bases.

Our first goal in this section is to develop the ability to look at the formula of a chemical species and tell whether it has the capacity to alter the acid–base status of a body fluid or of an item of food or drink. We want to be able to do this while keeping what must be memorized to an absolute minimum, to the short list of strong acids. For this, we must learn to view the equilibria in *weak* acids the way Brønsted suggested.

■ Some additives are put into food products solely to control their acidity and thereby prolong their shelf lives.

Every Acid Has a Conjugate Base and Every Base Has a Conjugate Acid We wrote the following equilibrium for the ionization of acetic acid.

$$HC_2H_3O_2(aq) + H_2O \rightleftharpoons H_3O^+(aq) + C_2H_3O_2^-(aq)$$

In the forward proton transfer reaction, the proton donor (acid) is the acetic acid molecule, and the proton acceptor (base) is the water molecule. So the forward reaction is the reaction of a weak acid with a (very) weak base. But notice that the reverse reaction is also a proton transfer; it is also the reaction of an acid and a base. In the reaction from right to left, the acid is the hydronium ion, because it donates a proton. The base is the acetate ion, $C_2H_3O_2^-$, because it accepts a proton.

■ Because the discussion is *qualitative* (i.e., without specific physical quantities), we must remember that "strong" and "weak" are relative terms.

As Brønsted pointed out, in the acetic acid equilibrium we can identify *two* acids and *two* bases. Notice particularly that each base is related to one of the two acids in the equation. The acetate ion, one of the bases, is related to and comes from acetic acid, one of the acids. The other base, H_2O, is related to the other acid, H_3O^+. We can, therefore, label the species in our equilibrium expression as follows. (Remember that in this equilibrium the substances on the left are favored.)

$$HC_2H_3O_2(aq) + H_2O \rightleftharpoons H_3O^+(aq) + C_2H_3O_2^-(aq)$$

| weaker acid | weaker base | stronger acid | stronger base |

the favored species in this equilibrium

Pairs of particles like H_2O and H_3O^+ or $C_2H_3O_2^-$ and $HC_2H_3O_2$, whose formulas differ by just one H^+, are called **conjugate acid–base pairs.** Thus H_2O is the conjugate base of H_3O^+, and H_3O^+ is the conjugate acid of H_2O. Similarly, $C_2H_3O_2^-$ is the conjugate base of $HC_2H_3O_2$, and $HC_2H_3O_2$ is the conjugate acid of $C_2H_3O_2^-$.

conjugate pair

$$HC_2H_3O_2(aq) + H_2O \rightleftharpoons H_3O^+(aq) + C_2H_3O_2^-(aq)$$

conjugate pair

Before we go further, it will be useful, given the name or formula of one member of a conjugate acid–base pair, to be able to write the formula of the other, so let's study some examples.

EXAMPLE 8.4   Writing the Formula of a Conjugate Acid

What are the conjugate acids of NH_3 and PO_4^{3-}?

ANALYSIS All we have to do is change each formula by one H^+. When we do this we have to add not just the H but also the + charge. We add the charge algebraically.

SOLUTION $NH_3 + H^+ = NH_4^+$, the conjugate acid of NH_3.
$PO_4^{3-} + H^+ = HPO_4^{2-}$, the conjugate acid of PO_4^{3-}.

EXAMPLE 8.5   Writing the Formula of a Conjugate Base

What are the conjugate bases of nitrous acid, HNO_2, and the dihydrogen phosphate ion, $H_2PO_4^-$?

ANALYSIS We remove one H^+ from each formula, adjusting the charge to reflect one less positive charge.

SOLUTION When we take H^+ away from HNO_2, we're left with NO_2^-, the conjugate base of HNO_2. (When we take a charge of 1+ from a particle with a charge of 0, the remaining charge is 1−.) When we take H^+ away from $H_2PO_4^-$, both the H and a net of one + charge, we are left with HPO_4^{2-}. (When we take 1+ away from 1−, the result is 2−.)

■   PRACTICE EXERCISE 11   Write the formulas of the conjugate acids of the following particles.

(a) NO_3^- (b) SO_3^{2-} (c) CO_3^{2-} (d) SO_4^{2-} (e) Cl^- (f) H_2O (g) OH^-

■   PRACTICE EXERCISE 12   Write the formulas of the conjugate bases of the following particles.

(a) HCO_3^- (b) HPO_4^{2-} (c) H_2SO_4 (d) HSO_4^- (e) HBr   (f) H_3O^+ (g) H_2O

Strong Acids Have *Very* Weak Conjugate Bases; Weak Acids Have Moderately Strong Conjugate Bases When an acid is relatively *strong*, it gives up a proton more readily than a weaker acid. Some unit of its structure—the unit that becomes the conjugate base—is not holding the proton well. In other words, *every strong acid has a very weak conjugate base.* HCl(*g*) is a strong acid, for example, so Cl^-, its conjugate base, is a very weak base. Conversely, in a *weak* acid, the unit holding the proton—the unit that becomes the conjugate base—is a moderately strong base. This unit is holding the proton very well within the molecule of the acid. In other words, *every weak acid has a moderately strong conjugate base.* If the acid is *very weak,* like H_2O, its conjugate base, OH^-, is a very strong base.

We can summarize these observations as rules of thumb for conjugate acid–base relationships. The last two are just "opposite sides of the same coin" of the first two.

Conjugate Acid–Base Relationships, Brønsted Concept
The stronger an acid is, the weaker is its conjugate base.
The weaker an acid is, the stronger is its conjugate base.

The stronger a base is, the weaker is its conjugate acid.
The weaker a base is, the stronger is its conjugate acid.

These rules will enable us to judge when to expect a base to be relatively strong or weak just by using our knowledge of the list of strong acids. Let's work an example to show how this list enables us to figure out if a particular species is strong or weak. We'll first review how to tell if an acid not studied before is strong or weak.

EXAMPLE 8.6   Deducing Whether an Ion or Molecule Is a Weak Brønsted Acid

Lactic acid is the acid responsible for the tart taste of sour milk. Is lactic acid a strong or a weak acid?

SOLUTION   The list of strong acids does not include lactic acid. Hence, it is a weak acid. It's as simple as that (and we'd err very few times).

EXAMPLE 8.7   Deducing Whether an Ion or Molecule Is a Strong or a Weak Brønsted Base

Is the bromide ion a strong or a weak Brønsted base?

ANALYSIS   When the question deals with a potential *base,* we have to find the answer in a roundabout fashion. We accept this because the alternative would be to memorize a rather extensive list of the stronger Brønsted bases. Here's how to go about it.

Pretend that the potential base actually functions as a base, so write the formula of its conjugate acid. If Br^- were to be a base, its conjugate acid would be HBr. In water, HBr is hydrobromic acid. Now comes the crucial question. Is hydrobromic acid a strong acid? *We have to know the list of strong acids,* and HBr is on this list. So we know that HBr easily gives up a proton. Therefore we know that what remains when the proton so readily leaves HBr, Br^-, has to be a poor proton binder.

SOLUTION   Our answer is that Br^- is a weak Brønsted base.

EXAMPLE 8.8   Deducing Whether an Ion or a Molecule Is a Strong or a Weak Brønsted Base

Is the phosphate ion, PO_4^{3-}, a strong or a weak Brønsted base?

ANALYSIS   Using the strategy described in Example 8.7, we pretend that this ion actually is a base, a proton acceptor. So we give it a proton, and write the result, the conjugate acid. The conjugate acid of PO_4^{3-} is HPO_4^{2-}. This Brønsted acid isn't on our list of strong acids, so we conclude it's a weak acid. This means that PO_4^{3-} is a good proton binder (holding the proton as HPO_4^{2-}).

SOLUTION   A good proton binder is a relatively *strong* base, so the answer is that PO_4^{3-} is such a base.

■ PRACTICE EXERCISE 13   Classify the following particles as strong or as weak Brønsted acids.

(a) HSO_3^- (b) HCO_3^- (c) $H_2PO_4^-$

■ PRACTICE EXERCISE 14   Classify the following ions as strong or as weak Brønsted bases.

(a) I^- (b) NO_3^- (c) CN^- (d) NH_2^-

The Strongest Base We Can Have in Water Is OH^-   If we try to dissolve a base stronger than the OH^- ion in water, it reacts with water, takes a proton, and changes to the conjugate acid. For example, the oxide ion, O^{2-}, which is the conjugate *base* of OH^-, is a much stronger base than OH^-. If we add it to water in the form of sodium oxide, the following reaction occurs (very exothermically), and none of the oxide ions supplied by Na_2O is in the solution. They have all changed to hydroxide ions by the following reaction.

■ OH^- is the conjugate acid of O^{2-}.

$$Na_2O(s) + H_2O \longrightarrow 2NaOH(aq)$$

sodium oxide   sodium hydroxide

■ Solid Na_2O or any other group IA oxide cannot be stored exposed to humid air.

Thus, although sodium oxide is very soluble in water, its solution contains no oxide ions. It dissolves by reacting with water, and its oxide ions change to hydroxide ions. Most metal oxides that dissolve in water do so by reacting in this way. The ionic equation can be written:

$$O^{2-}(s) + H_2O \longrightarrow 2OH^-(aq)$$

To repeat, the strongest base that can exist in water is the hydroxide ion.

The Strongest Acid We Can Have in Water Is H_3O^+   If we try to dissolve in water any acid stronger than H_3O^+, it reacts with water to give hydronium ion. Hydrogen chloride, for example, is a stronger proton donor than H_3O^+. As we already know, when we bubble $HCl(g)$ into water, the following reaction occurs, a typical Brønsted acid–base reaction. We'll write it as an equilibrium, although the forward reaction occurs 100%.

$$HCl(g) + H_2O \rightleftharpoons H_3O^+(aq) + Cl^-(aq)$$

stronger   stronger   weaker   weaker
acid   base   acid   base

Evidently, the hydronium ion holds the proton better than it is held by the Cl atom in $HCl(g)$.

In All Brønsted Acid–Base Equilibria, the Weaker Acid and Weaker Base Are Favored   Now that we can make reasonable predictions of relative acid or base strengths, let's see how we can use this skill in predicting reactions. This will enable us to judge which side of an acid–base equilibrium is favored.

A logical consequence of our rules of thumb about acid–base strengths is that in all proton transfer equilibria the side having the weaker acid and base will always be favored over the side with the stronger acid and base. We can condense this to another rule of thumb concerning what to expect when we mix chemicals that can participate in acid–base equilibria: *The stronger always give way to the weaker in acid–base reactions.*

**EXAMPLE 8.9 Predicting Which Substances Are Favored
in an Acid–Base Equilibrium**

If we add hydrochloric acid to an aqueous solution of sodium cyanide, NaCN, will HCN and NaCl form to any significant extent? (If they do, the evolving HCN, hydrogen cyanide, may kill anyone who mixes these substances. Hydrogen cyanide is a very dangerous poison.)

ANALYSIS Because we are dealing with HCl(aq), the reagent actually consists of $H_3O^+(aq)$ and $Cl^-(aq)$. Because NaCN is a salt in solution, we are dealing with $Na^+(aq)$ and $CN^-(aq)$. $Na^+(aq)$ and $Cl^-(aq)$ would be spectator ions. The question, therefore, is, "Does the following reaction occur?"

$$H_3O^+(aq) + CN^-(aq) \longrightarrow H_2O + HCN(aq)$$

To apply the rule "the stronger gives way to the weaker" in proton transfer reactions, we first need to identify the acids and bases and then infer what is stronger and what is weaker. When we look for the conjugate pairs, we can see them as follows:

$$CN^-(aq) + H_3O^+(aq) \rightleftharpoons HCN(aq) + H_2O$$

conjugate pair (top, linking CN^- and HCN)

conjugate pair (bottom, linking H_3O^+ and H_2O)

SOLUTION Each pair must have an acid and each must have a base, so let's write in these labels (and, to reduce clutter, omit the lines that have served to connect conjugate pairs).

$$CN^-(aq) + H_3O^+(aq) \longrightarrow HCN(aq) + H_2O$$

  base   acid   acid   base

Now we decide which of the two acids is stronger. We know that H_3O^+ is the strongest acid species we can have in water. (We also know that because HCN isn't on the list of strong acids, it must be relatively weak.) So we modify our labels with this new information.

$$CN^-(aq) + H_3O^+(aq) \longrightarrow HCN(aq) + H_2O$$

  base   stronger   weaker   base
              acid   acid

Because the conjugate of a stronger acid must be a weaker base, and the conjugate of a weaker acid must be a stronger base, we can modify the remaining labels as follows.

$$CN^-(aq) + H_3O^+(aq) \longrightarrow HCN(aq) + H_2O$$

  stronger   stronger   weaker   weaker
  base   acid   acid   base

Finally, we can tell that the reaction must occur as written, because the stronger conjugates are always replaced by the weaker in acid–base reactions. If we write the equation as an equilibrium (as we should),

$$CN^-(aq) + H_3O^+(aq) \rightleftharpoons HCN(aq) + H_2O$$

| stronger base | stronger acid | weaker acid | weaker base |

then we must note that the products of the forward reaction are favored (the double arrows give no indication of this). Thus, HCN does form to a significant extent.

■ **PRACTICE EXERCISE 15**   When the meat preservative sodium nitrite, $NaNO_2$, enters the stomach and encounters the hydrochloric acid in gastric juice, can nitrous acid, HNO_2, be produced? Write the equilibrium expression for any net ionic interactions. State which are favored, the reactants or the products. (Nitrous acid is suspected of being a cause of cancer, but no evidence presently exists that it actually causes cancer in humans.)

■ Perchloric acid in Table 8.4 was not on our earlier list of strong acids because it is less common.

Acids and Bases Can Be Organized in Their Order of Strengths   Table 8.4 lists several acids and bases in their orders of increasing strength. Carbonic acid, for example, is a weaker acid than acetic acid so it stands below acetic acid in the table.

All acids above H_3O^+ in the column of acids in the table—those that we have learned are the strong acids—are essentially 100% ionized into the hydronium ion and the conjugate base in aqueous solutions. We also treat $HSO_4^-(aq)$ as a strong acid, although it is less strong than H_2SO_4. (Why?) $H_3PO_4(aq)$ is a moderate acid.

Moving down the column, the next acids—acetic acid, carbonic acid, the dihydrogen phosphate ion, the ammonium ion, the bicarbonate ion, and the monohydrogen phosphate ion—are all weak acids (becoming progressively weaker as we move down the column).

TABLE 8.4   **Relative Strengths of Some Brønsted Acids and Bases**

| | Brønsted Acid | | Brønsted Bases | |
	Name	Formula	Name	Formula
↑ Increasing acid strength	Perchloric acid	$HClO_4$	Perchlorate ion	ClO_4^-	↓ Increasing base strength
	Hydrogen iodide	HI	Iodide ion	I^-	
	Hydrogen bromide	HBr	Bromide ion	Br^-	
	Sulfuric acid	H_2SO_4	Hydrogen sulfate ion	HSO_4^-	
	Hydrogen chloride	HCl	Chloride ion	Cl^-	
	Nitric acid	HNO_3	Nitrate ion	NO_3^-	
	HYDRONIUM ION	H_3O^+	WATER	H_2O	
	Hydrogen sulfate ion	HSO_4^-	Sulfate ion	SO_4^{2-}	
	Phosphoric acid	H_3PO_4	Dihydrogen phosphate ion	$H_2PO_4^-$	
	Acetic acid	$HC_2H_3O_2$	Acetate ion	$C_2H_3O_2^-$	
	Carbonic acid	H_2CO_3	Bicarbonate ion	HCO_3^-	
	Dihydrogen phosphate ion	$H_2PO_4^-$	Monohydrogen phosphate ion	HPO_4^{2-}	
	Ammonium ion	NH_4^+	Ammonia	NH_3	
	Bicarbonate ion	HCO_3^-	Carbonate ion	CO_3^{2-}	
	Monohydrogen phosphate ion	HPO_4^{2-}	Phosphate ion	PO_4^{3-}	
	WATER	H_2O	HYDROXIDE ION	OH^-	
	Methyl alcohol	CH_3OH	Methoxide ion	CH_3O^-	
	Ammonia	NH_3	Amide ion	NH_2^-	
	Hydroxide ion	OH^-	Oxide ion	O^{2-}	
	Hydrogen	H_2	Hydride ion	H^-	

The acids from water and below in Table 8.4 ionize in water to such a low percentage that, except in discussions of the Brønsted concept, we do not refer to them as proton donors.

Moving over to the column of bases, the conjugate bases of all strong acids in Table 8.4 are such weak bases that we almost never refer to them as bases. Neither water nor the sulfate ion are routinely called bases either (except in discussions of the Brønsted concept). The dihydrogen phosphate ion is a weak base, and as we move down the list through the acetate ion, the bicarbonate ion, the monohydrogen phosphate ion, ammonia, the carbonate ion, and the phosphate ion, the bases get stronger. This means that as we move down through this series, the products in the following equilibrium expression become more and more favored. (We let B^- represent any base except NH_3.)

$$B^-(aq) + H_2O \rightleftharpoons BH(aq) + OH^-(aq)$$

The ions below OH^- in the table—CH_3O^-, NH_2^-, O^{2-}, and H^-—react quantitatively with water to give their conjugate acids. As we have said, no base stronger than OH^- can exist in water. In the reactions, for example, of NH_2^- and H^- with water, we don't normally even use equilibrium arrows; the reactions go to completion for all practical purposes.

$$NH_2^- + H_2O \longrightarrow NH_3(aq) + OH^-(aq)$$
$$H^- + H_2O \longrightarrow H_2(g) + OH^-(aq)$$

■ Sodium salts of all these strongly basic anions are known— $NaOCH_3$, $NaNH_2$, and NaH.

Our discussion of relative strengths of acids and bases has been qualitative because that is all that is needed for many uses. Sometimes, however, it is necessary to have numbers to describe these relative strengths, and we will describe such numbers in the next chapter.

The Ammonium Ion Is a Brønsted Acid   The ammonium ion occupies a special place in our study because many biochemicals, like proteins, have a molecular part that is very much like this ion. Although the ammonium ion is a weak acid (Table 8.4), it still is a Brønsted acid, and it can neutralize the hydroxide ion. When we add sodium hydroxide to a solution of ammonium chloride, the following reaction occurs.

■ The *protonated amino group,* —NH_3^+ is covalently bound to carbon in a large number of organic compounds, including proteins.

$$NaOH(aq) + NH_4Cl(aq) \longrightarrow NH_3(aq) + H_2O + NaCl(aq)$$

The net ionic equation is

$$OH^-(aq) + NH_4^+(aq) \longrightarrow NH_3(aq) + H_2O$$

This reaction neutralizes the hydroxide ion, and it leaves a solution of the weaker base, NH_3. (If the initial solution is concentrated enough, the final solution has a strong odor of ammonia.)

■ NH_3 is a much weaker base than OH^- so, when we compare their conjugate acids, we know that NH_4^+ is a stronger acid than H_2O.

In some medical emergencies, when the blood has become too alkaline or too basic, an isotonic solution of ammonium chloride is administered by intravenous drip. Its ammonium ions can neutralize some of the base in the blood and bring the acid–base status back to normal.

8.5 SALTS

A very large number of ionic reactions can be predicted from a knowledge of the solubility rules of salts.

Salts, as we have said, are ionic compounds whose cations are any except H^+ and whose anions are any except OH^- or O^{2-}. All are crystalline solids at room temperature, because forces of attraction between ions in crystals are very strong.

A **simple salt** is one that is made of just *two* kinds of oppositely charged ions. Examples are $NaCl$, $MgBr_2$, and $CuSO_4$. *Mixed salts* are those that have three or more different ions. Alum, used in water purification, is an example: $K_2SO_4 \cdot Al_2(SO_4)_3 \cdot 24H_2O$. As the formula of alum illustrates, the salt family includes hydrates. Some salts of practical value are given in Table 8.5.

Formation of Salts   In the laboratory, salts are obtained whenever an acid is used in any of the following ways. We summarize and review these methods here and show their similarities.

$$\text{Acid} + \text{metal hydroxide} \longrightarrow \text{a salt} + H_2O$$
$$\text{Acid} + \text{metal bicarbonate} \longrightarrow \text{a salt} + H_2O + CO_2$$
$$\text{Acid} + \text{metal carbonate} \longrightarrow \text{a salt} + H_2O + CO_2$$
$$\text{Acid} + \text{metal} \longrightarrow \text{a salt} + H_2$$

■ Notice that all of these reactions that produce salts also neutralize acid.

If the salt is soluble in water, we have to evaporate the solution to dryness to isolate it.

Many Salts Are Insoluble in Water   Sometimes a salt precipitates as it forms instead of remaining in solution. To predict when to expect this, we use a small number of solubility rules for ionic compounds. We say that a compound is *soluble* in water if it can form a solution with a concentration of at least 3 to 5% (w/w). When a *counter ion* is referred to in the following rules, it means the unnamed ion of the ionic compound. For example, in the lithium salt, $LiCl$, the counter ion is the chloride ion. In the hydroxide, $Ca(OH)_2$, the counter ion is Ca^{2+}.

TABLE 8.5   Some Salts and Their Uses

Formula and Name	Uses
BaSO4, barium sulfate	Used in the "barium cocktail" given before taking X-ray films of the gastrointestinal tract
$MgSO_4 \cdot 7H_2O$, magnesium sulfate heptahydrate (epsom salt)	Purgative
$(CaSO_4)_2 \cdot H_2O$, calcium sulfate hemihydrate (plaster of paris)	Plaster casts; wall stucco; wall plaster
$AgNO_3$, silver nitrate	Antiseptic and germicide; used in eyes of infants to prevent gonorrheal conjunctivitis; photographic film
$NaHCO_3$, sodium bicarbonate (baking soda)	Baking powders; effervescent salts; stomach antacid; fire extinguishers
$Na_2CO_3 \cdot 10H_2O$, sodium carbonate decahydrate (soda ash, sal soda, washing soda)	Water softener; soap and glass manufacture
NaCl, sodium chloride	Manufacture of chlorine, sodium hydroxide; preparation of food
$NaNO_2$, sodium nitrite	Meat preservative

Solubility Rules for Ionic Compounds in Water

1. All lithium, sodium, potassium, and ammonium salts are soluble regardless of the counter ion.

2. All nitrates and acetates are soluble, regardless of the counter ion.

3. All chlorides, bromides, and iodides are soluble, *except* when the counter ion is lead, silver, or mercury(I).

4. All sulfates are soluble *except* those of lead, calcium, strontium, mercury(I), and barium.

5. All hydroxides and metal oxides are insoluble *except* those of the group IA cations and those of calcium, strontium, and barium.

6. All phosphates, carbonates, sulfites, and sulfides are insoluble *except* those of the group IA cations and NH_4^+.

There are exceptions to the solubility rules, but we will seldom be wrong in applying them. One of their many applications is to predict possible reactions involving ionic compounds.

When we're able to *predict* reactions from a few facts, we sharply reduce the number of facts that should be memorized. Solubility rule 6 tells us, for example, that all phosphates (except those of group IA cations and NH_4^+) are insoluble in water. We'd predict, therefore, that calcium phosphate, $Ca_3(PO_4)_2$, is insoluble, and we would be right. Calcium phosphate, in fact, is extremely insoluble in water. Many cells of the body, however, must have calcium ions to work, but the same cells also contain phosphate ions. Interaction 8.2 describes how the body manages to supply calcium ions to cells without letting them become instantly precipitated as calcium phosphate.

Salts Can Form by *Double Replacement* ("Exchange of Partners") Reactions   In addition to the reactions we have already studied to make salts by acid–base neutralization, we can also make salts by a "change of partners" reaction called **double replacement.** For example, sodium carbonate and calcium chloride are both soluble in water. But if we mix aqueous solutions of the two, the following reaction occurs because CO_3^{2-} ion and Ca^{2+} ions cannot remain in solution in each other's presence beyond extremely trace concentrations. Their combination, $CaCO_3$, is too insoluble.

■ Some references use the term **metathesis reaction** for double replacement reaction. (The term *double displacement* is used by some, but it is not the preferred term.)

$$Na_2CO_3(aq) + CaCl_2(aq) \longrightarrow CaCO_3(s) + 2NaCl(aq)$$

combining CO_3^{2-} with Ca^{2+}

combining Na^+ with Cl^-

An ion from each salt combines with an ion from the other salt, which gives the informal name, "exchange of partners," to the reaction or, more formally, *double replacement.* The ionic equation for this reaction more clearly shows this exchange.

$$[2Na^+(aq) + CO_3^{2-}(aq)] + [Ca^{2+}(aq) + 2Cl^-(aq)] \longrightarrow$$
$$CaCO_3(s) + [2Na^+(aq) + 2Cl^-(aq)]$$

The net ionic equation is

$$Ca^{2+}(aq) + CO_3^{2-}(aq) \longrightarrow CaCO_3(s)$$

INTERACTION 8.2
CALCIUM CHANNELS IN CELL MEMBRANES

You have no doubt often noticed that your heartbeat increases in moments of stress, whether they occur before an exam, when you're late for meeting someone, or during a strenuous activity (Figure 1). Unseen, silently, and automatically, intricate events at the molecular level of life respond to the stress, and calcium ions are deeply involved. They participate in controlling muscle contractions, including heart muscle.

Each heartbeat requires the presence of calcium ion, but the fluid inside cells contains phosphate ion (or phosphate ion donors) at a concentration of about 10^{-3} mol/L. This turns out to be much too high to allow enough Ca^{2+} to be present inside the cell in a form that can be used. Yet, calcium ion *must* be available. The body solves this problem by keeping Ca^{2+} largely *outside* the cell in the surrounding fluid and away from PO_4^{3-} (or its donors).

There are mechanisms that permit the flow of Ca^{2+} ions through *calcium channels* embedded in the cell membrane only when these ions are needed inside the cell. Once in, Ca^{2+} combines with a *receptor protein* so that Ca^{2+} is not precipitated as $Ca_3(PO_4)_2$. The combination of Ca^{2+} with this protein changes the shape of the protein molecule, which converts it into a form needed by the cell to launch a contraction. After the muscle contraction occurs, the

FIGURE 1 Strenuous activity increase the heart rate, which calls on an accelerated use of calcium channels in heart muscle.

calcium ion is pumped back outside the cell! This goes on rapidly, repeatedly, and without our having to give thought to it. Pause for a moment and consider the exquisite beauty of this.

When the heart muscle contractions must be reduced in order to ease the burden on a stressed heart, medications called "calcium channel blockers" can be administered. They reduce the rate of flow of Ca^{2+} ions through the calcium channels and so retard the rate of contraction.

The precipitate in the beaker is being separated by *filtration*. It collects in a cone of filter paper in the funnel, and the *filtrate*—the clear solution—flows through into the beaker below.

The other product, NaCl, stays in solution in its dissociated form. You would have to filter off the precipitate of calcium carbonate and then evaporate the filtrate to dryness to obtain crystalline NaCl.

Not all combinations of solutes give double replacement reactions, as we can predict (and so not memorize), using the solubility rules. If you mixed solutions of NaI and KCl, for example, no combination of oppositely charged ions is insoluble, so the solution would just contain separated (and hydrated) ions of Na^+, K^+, I^-, and Cl^-. ("All sodium and potassium salts are soluble regardless of the counter ion.")

EXAMPLE 8.10   **Predicting Double Replacement Reactions of Salts**

What happens if we mix aqueous solutions of sodium sulfate and barium nitrate?

ANALYSIS By the solubility rules, we know that both sodium sulfate and barium nitrate are soluble in water, so their solutions contain their separated ions. When we pour the two solutions together, four ions experience attractions and repulsions. Hence, we must examine each possible combination of oppositely charged ions to see which, if any, makes a water-insoluble salt. When we find one, then we can write an equation for the reaction that produces this salt. Here are the possible combinations when Ba^{2+}, NO_3^-, Na^+, and SO_4^{2-} ions intermingle in water.

$Ba^{2+} + 2NO_3^- \xrightarrow{?} Ba(NO_3)_2(s)$ This possibility is obviously out, because barium ions and nitrate ions do not precipitate together from water. ("All nitrates are soluble.")

$2Na^+ + SO_4^{2-} \xrightarrow{?} Na_2SO_4(s)$ This possibility is also out. ("All sodium salts are soluble.")

$Na^+ + NO_3^- \xrightarrow{?} NaNO_3(s)$ No. Again, "All sodium salts are soluble."

$Ba^{2+} + SO_4^{2-} \xrightarrow{?} BaSO_4(s)$ Yes. Barium sulfate, $BaSO_4$, is not in any of the categories of water-soluble salts, so we conclude that it is an insoluble salt (rule 4.)

SOLUTION Because we predicted that $BaSO_4$ can form a precipitate, we can write a molecular equation, and we'll use some connector lines to show how partners exchange—how *double* replacement occurs.

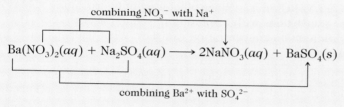

combining NO_3^- with Na^+

$$Ba(NO_3)_2(aq) + Na_2SO_4(aq) \longrightarrow 2NaNO_3(aq) + BaSO_4(s)$$

combining Ba^{2+} with SO_4^{2-}

The net ionic equation, however, is a better way to describe what happens.

$$Ba^{2+}(aq) + SO_4^{2-}(aq) \longrightarrow BaSO_4(s)$$

The sodium and nitrate ions are only spectators. To obtain the solid barium sulfate, we would filter the mixture and collect this compound on the filter. If we also wanted the sodium nitrate, we would evaporate the clear filtrate to dryness.

■ $BaSO_4$ is the white, insoluble substance in a flavored slurry given to patients as a "barium cocktail" before an X-ray film is taken of the intestinal tract. The barium ion stops X rays, so the tract is outlined on the film.

■ **PRACTICE EXERCISE 16** If solutions of sodium sulfide, Na_2S, and copper(II) nitrate, $Cu(NO_3)_2$, are mixed, what if anything will happen chemically? Write a molecular and a net ionic equation for any reaction.

One of the many uses of the solubility rules is to understand what it means for water to be called *hard water* and what it means to *soften* such water. These are discussed in Interaction 8.3.

Our study of the reaction of sodium sulfate and barium nitrate in Example 8.10 illustrates the power of knowing just a few facts for the sake of predicting an enormous number of others with a high probability of success. To predict the reactions of ions, the following should be well learned.

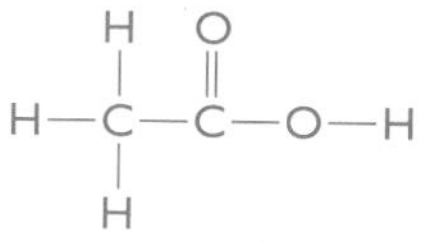

1. The solubility rules of ionic compounds (because then we can assume that all the other ionic compounds are insoluble).

2. The five strong acids in Table 8.1 (because then we can assume that all the other acids, including organic acids, are weak).

■ The chief organic acids all have the group: $-\overset{\overset{\displaystyle O}{\|}}{C}-O-H$

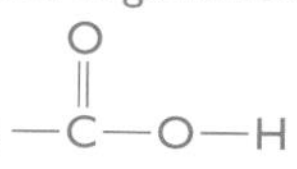

as in acetic acid:

$H-\overset{\overset{\displaystyle H}{|}}{\underset{\underset{\displaystyle H}{|}}{C}}-\overset{\overset{\displaystyle O}{\|}}{C}-O-H$

3. The first two strong bases of Table 8.2 (because then we can assume that all the other bases are either weak or are too insoluble in water to matter much).

A Summary

The Reactions of Ions in Which Gases, Molecules, or Precipitates Form We predict ions to react with each other in any of the following situations.

1. A gas can form that (mostly) leaves the solution. It could be
   (a) Hydrogen—from the action of acids on metals, or
   (b) Carbon dioxide—from acids reacting with carbonates or bicarbonates
2. An un-ionized, molecular compound forms that remains in solution. It could be
   (a) Water—from acid–base neutralizations or
   (b) A weak acid—by the action of H^+ on a strong Brønsted base, the conjugate base of any weak acid, or
   (c) Ammonia—by the reaction of OH^- with NH_4^+.
3. A precipitate forms—a water-insoluble salt or one of the water-insoluble hydroxides.

■ PRACTICE EXERCISE 17   When a solution of hydrochloric acid is mixed in the correct molar proportions with a solution of sodium acetate, $NaC_2H_3O_2$, essentially all of the hydronium ion concentration vanishes. What happens and why? Write the net ionic equation.

■ PRACTICE EXERCISE 18   What, if anything, happens chemically when each pair of solutions is mixed? Write net ionic equations for any reactions that occur.

(a) NaCl and $AgNO_3$ (b) $CaCO_3$ and HNO_3 (c) KBr and NaCl

---

The Common Ion Effect Shifts Solubility Equilibria to Favor Insoluble or Unionized Species The solubility rules predict the solubility of an individual salt when it is the lone solute in solution. In nature and in living systems, however, solutions this simple seldom occur. Usually, two or more electrolytes are present, so it is important to learn when the solubility of one compound might be changed by the presence of another.

   If the other solutes provide ions that are *entirely different* from those of the solute in question, then the solubility rules work, at least in solutions that are relatively dilute in all dissolved species. But if some other solute contributes one ion that is *common* to the salt in question, then the solubility of the latter is *reduced.* The reduction in solubility of a salt by the addition of another solute bearing an ion common to both is called the **common ion effect.** Let's see how it works.

   Suppose that we have a *saturated* solution of sodium chloride. The following equilibrium exists.

$$NaCl(s) \rightleftharpoons Na^+(aq) + Cl^-(aq)$$

What happens if we now pour into this solution some concentrated hydrochloric acid, a fully ionized acid? The ion common to both NaCl and HCl, of course, is the chloride ion. We can quickly increase its concentration by using *concentrated* $HCl(aq)$. We thus place a stress on the equilibrium. In accordance with Le Châtelier's principle, the equilibrium must shift to absorb this stress, and it *must* shift to the left. Only by this shift of the equilibrium can the system absorb the added Cl^- and so reduce the stress. But a shift to the left means the precipitation of some solid

HARD WATER

Groundwater that contains magnesium, calcium, or iron ions at a high enough level to interact with ordinary soap to form scum is called **hard water**. In **soft water** these "hardness ions"—Ca^{2+}, Mg^{2+}, Fe^{2+}, or Fe^{3+}—are either absent or are present in extremely low concentrations. (The anions that most frequently accompany the hardness ions are SO_4^{2-}, Cl^-, and HCO_3^-.)

Hard water in which the principal anion is the bicarbonate ion is called **temporary hard water**. Hard water in which the chief negative ions are anything else is called **permanent hard water**. When temporary hard water is heated near its boiling point, as in hot boilers, steam pipes, and instrument sterilizers, the bicarbonate ion breaks down to the carbonate ion. And this ion forms insoluble precipitates with the hardness ions. Their carbonate salts form, come out of solution, and deposit as scaly material that can even clog the equipment (see Figure 1). The equations for these changes are as follows.

The breakdown of the bicarbonate ion:

$$2HCO_3^-(aq) \longrightarrow CO_3^{2-}(aq) + CO_2(g) + H_2O$$

The formation of the scaly precipitate (using the calcium ion to illustrate):

$$CO_3^{2-}(aq) + Ca^{2+}(aq) \longrightarrow CaCO_3(s)$$

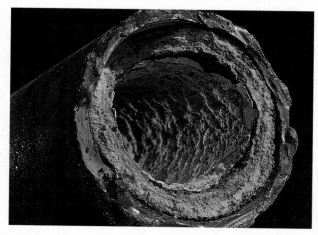

FIGURE I Boiler scale, a deposit of mostly calcium carbonate, has built up inside this hot water pipe.

Water Softening Removes the Hardness Ions Chemically Hard water can be softened in various ways. Most commonly, excess soap is used. Some scum does form, but then the extra soap does the cleaning work. To avoid the scum altogether, softening agents are added before the soap is used. One common water-softening chemical is sodium carbonate decahydrate, known as washing soda. Its carbonate ions take out the hardness ions as insoluble carbonates by the kind of reaction for which we wrote the previous net ionic equation.

Another home water-softening agent is household ammonia—5% (w/w) NH_3. We've already learned about the following equilibrium in such a solution.

$$NH_3(aq) + H_2O \rightleftharpoons NH_4^+(aq) + OH^-(aq)$$

In other words, aqueous ammonia has some OH^- ions, and the hydroxides of the hardness ions are not soluble in water. Therefore, when aqueous ammonia is added to hard water, the following kind of reaction occurs (illustrated using the magnesium ion this time).

$$Mg^{2+}(aq) + 2OH^-(aq) \longrightarrow Mg(OH)_2(s)$$

As hydroxide ions are removed by this reaction, more are made available from the ammonia–water equilibrium. (A loss of OH^- ion from this equilibrium is a stress, and the equilibrium shifts to the right in response, as we'd predict using Le Châtelier's principle.)

Still another water-softening technique is to let the hard water trickle through zeolite, a naturally occurring porous substance that is rich in sodium ions. When the hard water is in contact with the zeolite, sodium ions go into the water and the hardness ions leave the solution and attach themselves to the zeolite. Later, the hardness ions are flushed out by letting water that is very concentrated in sodium chloride trickle through the spent zeolite, and this restores the zeolite for reuse. Synthetic ion-exchange materials are also used to soften water by roughly the same principle.

Perhaps the most common strategy in areas where the water is quite hard is to use synthetic detergents instead of soap. Synthetic detergents do not form scums or precipitates with the hardness ions.

sodium chloride (Figure 8.9). After the activity has quieted, we still have a saturated solution, we once again have equilibrium, but we also have more solid NaCl, and we have a lower concentration of dissolved Na^+ ions.

FIGURE 8.9 The common ion effect. At the start there is a saturated solution of sodium chloride (first frame). When concentrated hydrochloric acid is added (middle frame), a white precipitate of sodium chloride appears, grows in quantity, and begins to settle (last frame).

The saturated sodium chloride system is a very small system. The principles of equilibria also apply to global systems. Thus, in Interaction 8.4, you can learn how the earth manages atmospheric CO_2 levels and how this level is an important factor in global warming.

Concentrations of Individual Ions in Solutions of Several Substances Are Often Given in Equivalents or Milliequivalents per Liter   Before we leave this introduction to salts, we must look at a concentration expression often used for their individual ions. It is based not on the moles of an ion per liter but on a quantity called an *equivalent* per liter. We'll see why soon.

One **equivalent** of an ion, abbreviated **eq**, is the number of grams of the ion that corresponds to Avogadro's number, one mole, of electrical charges. For example, when the charge is unity, either 1+ or 1−, it takes Avogadro's number of ions to have Avogadro's number of electrical charges. Thus 1 eq for ions such as Na^+, K^+, Cl^-, or Br^- is the same as the molar mass of each ion. The molar mass of Na^+ is 23.0 g Na^+/mol, so 1 eq of Na^+ equals 23.0 g of Na^+. This much sodium ion, 23.0 g of Na^+, contributes Avogadro's number of positive charges. Thus the equivalent mass of the sodium ion is 23.0 g Na^+/eq.

When an ion has a double charge, either 2+ or 2−, then the mass of one equivalent equals the molar mass divided by 2. For example, 1 mol of CO_3^{2-} ion equals 60.0 g of CO_3^{2-}, so 1 eq of CO_3^{2-} ion equals 30.0 g of CO_3^{2-} ion. This much carbonate ion carries Avogadro's number of negative charges. The extension of this to ions of higher charges should now be obvious.

The Equivalent Mass of an Ion Is Its Formula Mass Divided by Its Charge   You can see this in Table 8.6, which gives equivalent masses for a number of ions.

TABLE 8.6   Equivalents of Ions

Ion	g/mol	g/eq
Na^+	23.0	23.0
K^+	39.1	39.1
Ca^{2+}	40.1	20.1
Mg^{2+}	24.3	12.2
Al^{3+}	27.0	9.00
Cl^-	35.5	35.5
HCO_3^-	61.0	61.0
CO_3^{2-}	60.0	30.0
SO_4^{2-}	96.1	48.1

INTERACTION 8.4
EARTH'S MANAGEMENT OF CO_2 AND THE GREENHOUSE EFFECT

During the 1990s the hottest issue in climatology, the science of the climate, has been the possible *human* impact on the earth's average temperature, which stands at about 15 °C (59 °F). Such wide fluctuations in this value occur annually that it hasn't been clear to all specialists whether there is even a general *sustained* upward trend, to say nothing of how human activities might be factors. Climatologists do agree, however, that average temperature trends are significant only over a long period, at least a decade, not just over a year. Questions that we must consider before we can discuss the issue further include "Why is Earth warm at all?" "Is it becoming warmer?" "What natural events completely out of our control are factors in warming or cooling Earth?" "What human activities are factors?" "What can be done, if needed, about our activities?"

Why Earth Stays Relatively Warm Earth constantly receives energy, mostly from the sun but some from Earth itself, from the energy released by radioactive processes. Earth also radiates energy to outer space. It *must* do so at exactly the same average rate as it receives energy. Otherwise, Earth's average temperature would either increase or decrease drastically.

What modulates Earth's temperature is the atmosphere. It acts somewhat as the glass of a greenhouse, an analogy first proposed in 1822 by a French mathematician, Jean Fourier.[1] Without the atmosphere, Earth would be as cold and lifeless as Mars appears to be today. With too much of certain constituents in the atmosphere, Earth would heat up to a point at which life would also be impossible.

The insulating effect of the atmosphere is now dubbed the **greenhouse effect.** Trace substances, now called the *greenhouse gases*—water vapor, carbon dioxide, methane, dinitrogen monoxide (N_2O), and synthetic chemicals like the chlorofluorocarbons—rather than air's chief components (oxygen and nitrogen) absorb outgoing energy and reradiate some of it back to Earth.[2] It is this reradiation by the greenhouse gases that causes the insulating effect of

Earth's atmosphere (see Figure 1). The effect of the gases is that Earth's surface is an estimated 33 °C warmer than it would otherwise be. The concern today is not about the greenhouse gases themselves—they are largely of natural origin and perform an essential service—but about their concentration becoming steadily higher.

The Greenhouse Effect and Climate Small changes in the *average* Earth temperature could have major changes in world climate patterns. Tropical storms and hurricanes would intensify and perhaps destroy the coral reefs that protect coastlines. Huge ecosystems, like the Arctic Ocean and nearby Arctic tundra, as well as mangrove swamps and other coastal wetlands, would be under severe pressure.

It isn't that Earth has never before undergone large changes in overall climate with successful adaptations by most living species, but an accelerated greenhouse warming would entail an unprecedented *rate* of change, with which vulnerable species could not cope.

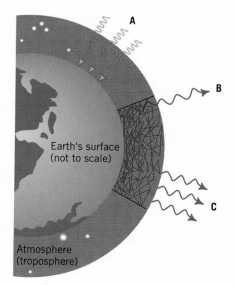

FIGURE 1 Greenhouse gases insulate Earth. Energy from the sun (A) enters Earth's atmosphere, but nearly a third is scattered back into space. Most of the remaining visible light energy is absorbed by Earth's surface and becomes heat. Part of the heat radiates directly to space (B), but much is randomly scattered about and absorbed by the atmosphere, some coming back to Earth, thus causing the greenhouse warming. Eventually this energy is also radiated to outer space (C). (Adapted with permission from D. B. Botkin and E. A. Keller, *Environmental Science,* 1995, John Wiley & Sons, Inc., New York. Developed by M. S. Manalis and E. A. Keller.)

---

[1] The analogy is actually not too good. The chief contribution of the glass of a real greenhouse is to prevent the wind from moving cold air in and warm air out. Still, the glass does absorb some outward-bound heat and reradiate it back inside.

[2] Both methane and nitrogen dioxide arise by the action of bacteria on materials in the soil, including fertilizers. Together they make up an estimated 20% of the greenhouse gases. The chlorofluorocarbons, although greenhouse gases made by humans, pose a more serious threat to the stratospheric ozone shield than to greenhouse warming (see Interaction 4.2, *Ozone in the Stratosphere and the Ozone "Hole"*).

Carbon Dioxide as the Chief Greenhouse Gas
The atmosphere receives each year an estimated 9×10^{15} mol of CO_2, largely from three sources: the respiration of plants and animals (over 93%), forest fires and the burning of other vegetable matter (2%), and the burning of fossil fuels—oil, coal, and natural gas (5%).

Two natural "sinks" take away over 95% of the released CO_2, each contributing about equally. One is *photosynthesis,* the conversion of CO_2 and H_2O in plants with the aid of chlorophyll (the green pigment in plants) and energy from the sun. The other sink is the uptake of CO_2 by the oceans as it reacts with calcium and magnesium ions to form carbonate deposits. The total uptake of CO_2 leaves roughly 5% of it in the air, a small but important net amount.

The increased rate of burning of fossil fuels throughout the last several decades is believed to be chiefly responsible for this extra amount of atmospheric CO_2. The additional carbon dioxide annually adds more to the increase in the concentration of the greenhouse gases than all other sources of greenhouse gases combined, about 55% of the total additions. Studies of ocean sediment cores and tree rings indicate that the carbon dioxide level of the atmosphere fluctuated between 200 and 300 ppm until the end of the 19th century. (One ppm, parts per million, is the same as 1 mg/L.)

Current Status as Estimated by Climatologists
What concerns scientists and government officials is the steady increase in the concentration of CO_2 in the atmosphere since the time when the burning of coal and oil became a major source of energy. From about 1900, the average CO_2 level in the atmosphere has risen from 280 to 345 ppm, and some scientists see the level going to 600 ppm in another 50 to 75 years, maybe in less time. This change alone could increase the average global temperature by 1.5 °C.

A serious problem for climatologists is in developing a reliable, testable computer program for making forecasts. To be on the safe side, a worldwide, phased reduction in CO_2 emissions was the subject of international negotiations in the early 1990s. Under a 1992 Climate Convention, industrial nations made a commitment to reduce their emissions of carbon dioxide to 1990 levels by the year 2000.

Has the greenhouse effect arrived? Is accelerated global warming now a certainty? Seeking answers to these and other questions is a multinational group of scientists called the United Nations Intergovernmental Panel on Climate Change (IPCC). In its 1995 report the panel concluded that there is a discernible human influence on global climate and that Earth's temperature could increase between 1 °C and 3.5 °C by the year 2100. If the warming were to reach the high end of this range, major ecological impacts throughout the world would occur, including serious impacts on human health. Tropical and subtropical regions would expand, opening larger populations to malaria, yellow fever, encephalitis, and dengue fever.

Atmospheric Haze and the Greenhouse Effect The problems of making long-range forecasts of changes in climate are huge, given our uncertain knowledge about how climate works. In its 1995 report, the IPCC said that some of the possible greenhouse warming is being canceled by haze particles that blanket large areas of the world. These colloidal aerosol particles consist largely of microscopic droplets of moisture containing sulfuric acid formed from sulfur oxides made by the combustion of sulfur-bearing fuels and vegetation. Some haze particles are also contributed by the several huge volcanic eruptions of recent years. Anybody who looks out the window of the airplane during a flight on a sunny day can see the whitish veil of aerosols over the ground below.

Haze particles reflect incoming solar radiation back to space and so work to cool the Earth. The most numerous of the haze particles are also the smallest and the most effective in reflecting solar radiation. The extent of haze cooling is uncertain, but it is not expected to compensate fully for the greenhouse warming. Over a period of decades, Earth's temperature is expected to change upward. The rate of this increase, however, is not known.

A Reference S. W. Matthews and J. A. Sugar, "Under the Sun," *National Geographic,* October 1990, page 66. See also R. A. Kerr, "Greenhouse Report Foresees Growing Global Stress," *Science,* November 3, 1995, page 731.

The advantage of the concept of the equivalent is the simplicity of a 1 to 1 ratio. Regardless of the amounts of charges on the individual ions, when cations and anions are present either in an ionic crystal or in a solution, we can always be sure that for every equivalent of positive charge there has to be one equivalent of negative charge. The condition of electrical neutrality in an ionic compound or a solution of ions is that

$$eq \text{ of cations} = eq \text{ of anions}$$

Or,

$$meq \text{ of cations} = meq \text{ of anions}$$

where the **milliequivalent**, or **meq**, is related to the equivalent by the relationship, 1000 meq = 1 eq.

The normal ranges of values of the concentrations of several components of blood are listed on the inside back cover, where you will see that many are given in units of meq/L.

SUMMARY

Electrolytes Solutes that are dissociated or ionized in water are electrolytes. Their solutions conduct electricity between a positively charged electrode, or anode, and a negatively charged electrode, or cathode. Cations that accept electrons from cathodes are reduced. Anions that deliver electrons to anodes are oxidized.

Ionization of water Trace concentrations of hydronium ions, H_3O^+, and hydroxide ions, OH^-, are always present in water. In neutral water, their molar concentrations are equal (and very low). In writing equations, we usually write H_3O^+ as H^+, calling the latter either the hydrogen ion or the proton. In explaining these reactions, however, we usually find it necessary to use the correct formula, H_3O^+.

Aqueous solutions of ions form either by the dissociation of ionic compounds as they dissolve or by the ionization of molecular substances as they react with water.

Chief ion producers Acids, bases, and salts are the common electrolytes. In the Brønsted concept, acids are chemical species that can donate hydrogen ions. Their aqueous solutions are also called acids. The five most common strong acids are hydrochloric, hydrobromic, hydroiodic, sulfuric, and nitric acid. All are monoprotic except sulfuric acid, which is diprotic. The chief acidic species in all is H_3O^+.

Bases are substances that accept (neutralize) hydrogen ions. Among the common bases are those that directly supply OH^- ion in water, like the hydroxides and oxides of sodium and potassium. Other common proton acceptors that readily take H^+ from H_3O^+ (but not from H_2O) are the carbonate and bicarbonate ions, the monohydrogen phosphate and phosphate ions, and ammonia.

Salts are ionic compounds involving any other ions but H^+, OH^-, or O^{2-}.

Strong and weak electrolytes; strong and weak acids and bases A strong electrolyte is one fully ionized or dissociated in solution, and all strong acids and strong bases are strong electrolytes. Salts in their molten states or in aqueous solutions are fully dissociated and are therefore all strong electrolytes. Remember that *strong* refers to percentage dissociation or ionization. Many salts are quite insoluble in water and so cannot supply a high concentration of ions. But what does dissolve of such salts is 100% dissociated.

Many molecular acids and bases ionize to a small percentage in water and so are weak electrolytes. Acetic acid and ammonia are examples. Many other molecular substances can be present in an aqueous system without being ionized and so are called nonelectrolytes. Pure water is a nonelectrolyte.

H_3O^+ is the strongest acid that can be present in water, and OH^- is the strongest base.

Conjugate acids and bases in the Brønsted concept All proton transfer reactions can be expressed in terms of equilibria in which two acids and two bases appear. An acid and a base whose formulas differ only by one H^+ are a conjugate acid–base pair.

In the Brønsted concept, the terms *strong* and *weak* are enlarged. A strong acid is one that is a good proton donor. It readily gives up H^+. A strong acid has a weak conjugate base. A weak acid has a strong conjugate base. A strong base is one that strongly binds a proton, and a weak base is one that cannot hold H^+ very well. A strong base has a weak conjugate acid, and a weak base has a strong conjugate acid.

Reactions of aqueous acids The hydronium ion in strong aqueous acids reacts with:

Metal hydroxides, to give a salt and water
Metal carbonates, to give a salt, carbon dioxide, and water
Metal bicarbonates, to give a salt, carbon dioxide, and water
Metals, to give the salt of the metal and hydrogen gas

A solution of an acid is neutralized when any sufficiently strong proton-binding species is added in the correct mole proportion to make the concentration of hydrogen ion and hydroxide ion equal (and very small).

Carbonic acid and carbonates Carbonic acid, H_2CO_3, is both a weak acid and an unstable acid. When it is generated in water by the reaction of any stronger acid with a bicarbonate or a carbonate salt, virtually all of the carbonic acid decomposes to carbon dioxide and water, and most of the carbon dioxide fizzes out. The carbonate ion and the bicarbonate ion are both Brønsted bases, and the bicarbonate ion is involved in carrying waste carbon dioxide from cells, where it is made, to the lungs.

Ammonia and the ammonium ion   Ammonia is a strong base toward H_3O^+ but a weak base toward H_2O. The ammonium ion is a strong acid toward OH^- but a weak acid toward H_2O. Ammonia can neutralize strong acids and the ammonium ion can neutralize strong bases.

Salts   The chemical properties of salts in water are the properties of their individual ions. If the anion of the salt is the conjugate base of a weak acid, as HCO_3^- is the conjugate base of H_2CO_3, then the salt can neutralize strong acids. Thus bicarbonates, carbonates, acetates, and the salts of other organic acids supply Brønsted bases, namely their anions.

If the cation of the salt is the conjugate acid of a weak base, in the way that NH_4^+ is the conjugate acid of NH_3, then the salt supplies a Brønsted acid in water. (NH_3 is a weak base so its conjugate acid, NH_4^+, is a Brønsted acid.)

Salts can be produced by any of the reactions of strong acids that have been studied (and summarized, above) as well as by double replacement reactions. The solubility rules are guides for the prediction of their reactions. If a combination of oppositely charged ions can lead to an insoluble salt, an un-ionized species that stays in solution, or a gas, then the ions react.

If a different compound that can furnish an ion that is identical to an ion of a salt already in solution is added to this solution, the solubility of the salt might be reduced enough to force it out of solution (common ion effect).

Equivalents of ions   An equivalent (eq) of an ion is the number of grams of the ion that carry Avogadro's number of positive or negative charges. It is calculated by dividing the molar mass of the ion by the size of the charge it carries. The concentration of an ion in a dilute solution is often given in meq/L, where 1000 meq = 1 eq, and meq means milliequivalent.

REVIEW EXERCISES

The answers to Review Exercises whose numbers are in color are found in Appendix E. The answers to the other Review Exercises are found in the Study Guide that accompanies this book. The more challenging questions are marked with asterisks.

Electrolytes

8.1   Magnesium nitrate is a crystalline solid and an electrolyte.
(a) What is its formula?
(b) It consists of which ions? (Give their names and formulas.)
(c) What does "electrolyte" mean in connection with this compound?
(d) By what process, dissociation or ionization, does this compound form its solution in water?

8.2   Hydrogen chloride gas is very soluble in water.
(a) Does it dissolve by dissociation or by ionization?
(b) Write the equation for the reaction when $HCl(g)$ dissolves in water.
(c) How does the *formation* of this solution differ from the way in which a solution of NaCl forms when crystals of NaCl are added to water?

8.3   Sulfur trioxide, SO_3, is a colorless gas that dissolves in water to give a solution that conducts electricity. By what *process*, dissociation or ionization, does it form this solution? How can you tell?

8.4   The word *electrolyte* can be understood in two ways. What are they? Give examples.

8.5   To which electrode do cations migrate?

8.6   The anode has what electrical charge, positive or negative?

8.7   An electrode that is negatively charged attracts what kinds of ions, cations or anions?

8.8   Explain in your own words how the presence of cations and anions in water enables the system to conduct electricity.

8.9   When NaOH(s) is dissolved in water, the solution is an excellent conductor of electricity, but when methyl alcohol is dissolved in water, the solution won't conduct electricity at all. What does this behavior suggest about the structural natures of NaOH and methyl alcohol? (Notice that both appear to have OH groups in their formulas.)

$$H-\overset{\displaystyle H}{\underset{\displaystyle H}{\vert}}\overset{\vert}{\underset{\vert}{C}}-OH$$

methyl alcohol

8.10 When lithium nitrate dissolves in water, its crystals break up entirely into Li^+ and NO_3^- ions. Do we call this compound a *weak* or a *strong* electrolyte?

8.11 In the liquid state, tin(IV) chloride, $SnCl_4$, is a nonconductor. What does this suggest about the structural nature of this compound?

***8.12** Molten sodium chloride conducts electricity. At the cathode one of its ions is reduced, and at the anode the other ion is oxidized.
(a) Write an equation for the reaction at the cathode (using electrons as species in the equation).
(b) Write an equation for the reaction at the anode (again, using electrons as species in the equation).
(c) Write the overall reaction for the electrolysis.

8.13 What families of compounds are the principal sources of ions in aqueous solutions?

8.14 Review the differences between atoms and ions by answering the following questions.
(a) Are there any atoms that have more than one nucleus? If so, give an example.
(b) Are there any ions with more than one nucleus? If so, give an example.
(c) Are there any ions that are electrically neutral? If so, give an example.
(d) Are there any atoms that are electrically charged? If so, give an example.

Acids and Bases as Electrolytes

8.15 Write the equilibrium equation for the self-ionization of water, and label the ions that are present.

8.16 Tell whether each of the following solutions is acidic, basic, or neutral.
(a) $[H^+] = 8.6 \times 10^{-7}$ mol/L and $[OH^-] = 6.8 \times 10^{-8}$ mol/L
(b) $[H^+] = 1.0 \times 10^{-7}$ mol/L and $[OH^-] = 1.0 \times 10^{-7}$ mol/L
(c) $[H^+] = 8.6 \times 10^{-8}$ mol/L and $[OH^-] = 1.2 \times 10^{-7}$ mol/L

8.17 Salts are all crystalline *solids* at room temperature. Why do you suppose they are solids and not liquids or gases?

8.18 How did Arrhenius define an acid? A base?

8.19 What features do the common aqueous acids share?

8.20 In the context of acid–base discussions, what are two other names that we can use for "hydrogen ion"? In reality, H^+ represents what particle?

8.21 Acids have a set of common reactions, and so do bases, but not salts. Explain.

8.22 How does litmus paper help us to tell whether a solution is acidic, basic, or neutral?

8.23 How did Brønsted define an acid? A base?

8.24 What are the names and the formulas of the aqueous solutions of the four hydrohalogen acids?

8.25 What is the difference between hydrochloric acid and hydrogen chloride?

8.26 Write the equation for the ionization of nitric acid in water.

8.27 Why is sulfuric acid with only 2 H in its formula called a diprotic acid but acetic acid, which has 4 H in its formula, is a monoprotic acid?

8.28 If we represent all diprotic acids by the symbol H_2A, write the equilibrium expressions for the two separate ionization steps.

8.29 Would the ionization of the second proton from a diprotic acid occur with greater ease or with greater difficulty than the ionization of the first proton? Explain.

8.30 Write the equations for the progressive ionizations of sulfuric acid. Include the names of the ions.

8.31 Write the equations for the progressive ionizations of phosphoric acid, including the names of the ions.

8.32 Which is the stronger acid in water, sulfurous acid or sulfuric acid? How can you tell?

sulfurous acid   sulfuric acid

8.33 Compare the structures of chlorous acid, $HClO_2$, and chloric acid, $HClO_3$.

chlorous acid   chloric acid

Which is the stronger acid? How can you tell?

8.34 Write the equilibrium expression for the solution of carbon dioxide in water that produces some carbonic acid.

8.35 Write the equilibrium expressions for the successive steps in the ionization of carbonic acid.

8.36 KOH is a strong base and a strong electrolyte. What do these terms mean in connection with this compound?

8.37 Magnesium hydroxide is only slightly soluble in water, and yet it is classified as a *strong* base. Explain.

8.38 Ammonia is very soluble in water, and yet it is called a weak base. Explain.

8.39   What are the names and formulas of two bases that are both strong and are capable of forming relatively concentrated solutions in water?

8.40   What is meant by aqueous ammonia? Why don't we call it "ammonium hydroxide?"

8.41   Write the names and the formulas of the five strong acids that we have studied.

8.42   What are the four strong bases—both the names and the formulas? Which are quite soluble in water?

Net Ionic Equations

8.43   Consider the following net ionic equation.

$$6H^+(aq) + Cu(s) + 3NO_3^-(aq) \longrightarrow$$
$$Cu^{2+}(aq) + 3NO_2(g) + 3H_2O$$

(a) Does it have material balance?
(b) Does it have electrical balance?
(c) Is it an *equation*?

*8.44   Complete and balance the following molecular equations, and then write the net ionic equations.
(a) $NaOH(aq) + H_2SO_4(aq) \rightarrow$
(b) $HCl(aq) + NaHCO_3(aq) \rightarrow$
(c) $NaHCO_3(aq) + HBr(aq) \rightarrow$
(d) $HNO_3(aq) + KHCO_3(aq) \rightarrow$
(e) $HBr(aq) + NH_3(aq) \rightarrow$
(f) $HNO_3(aq) + Ca(OH)_2(s) \rightarrow$
(g) $HCl(aq) + Mg(s) \rightarrow$

8.45   What are the net ionic equations for the following reactions of strong, aqueous acids? (Assume that all reactants and products are soluble in water.)
(a) With metal hydroxides
(b) With metal bicarbonates
(c) With metal carbonates
(d) With aqueous ammonia

8.46   Write net ionic equations for the reactions of all the water-insoluble group IIA carbonates, where you use $MCO_3(s)$ as their general formula, with nitric acid (chosen so that all the products are soluble in water).

8.47   If we let $M(OH)_2(s)$ represent the water-insoluble group IIA metal hydroxides, what is the general net ionic equation for all of their reactions with hydrochloric acid (chosen so that all the products are soluble in water)?

8.48   If we let $M(s)$ represent either calcium or magnesium metal, what net ionic equation represents the reaction of either with hydrochloric acid?

8.49   Zinc metal reacts more rapidly with which acid, 0.10 M hydrochloric acid or 0.10 M phosphoric acid? Explain.

8.50   How many moles of sodium bicarbonate can react quantitatively with 0.457 mol of HCl?

8.51   How many moles of sodium hydroxide can react quantitatively with 0.112 mol of H_2SO_4 (assuming that both H^+ in H_2SO_4 are neutralized)?

8.52   How many grams of sodium hydroxide will neutralize 6.24 g of HCl?

8.53   How many grams of sodium bicarbonate does it take to neutralize all of the acid in 24.6 mL of 0.755 M HCl?

8.54   How many grams of potassium carbonate will neutralize all of the acid in 32.9 mL of 0.435 M HCl?

*8.55   How many milliliters of 0.165 M KOH are needed to neutralize the acid in 28.6 mL of 0.212 M HBr?

*8.56   How many milliliters of 0.115 M NaOH are needed to neutralize the acid in 14.6 mL of 0.161 M H_2SO_4?

Strengths of Conjugate Brønsted Acids and Bases

8.57   What is the reason that OH^- is the strongest base we can have in water?

8.58   Why is H_3O^+ the strongest acid we can have in water?

8.59   What are the formulas of the conjugate acids of the following?
(a) HSO_4^- (b) Br^- (c) H_2O (d) $C_2H_3O_2^-$

8.60   Write the formulas of the conjugate bases of the following
(a) NH_3 (b) HNO_2 (c) HSO_3^- (d) H_2SO_3

8.61   Which member of each pair is the stronger Brønsted base?
(a) NH_3 or NH_2^- (b) OH^- or H_2O (c) HS^- or S^{2-}

8.62   Which member of each pair is the stronger Brønsted acid?
(a) H_2CO_3 or HCl   (b) H_2O or OH^-
(c) HSO_4^- or HSO_3^-

*8.63   If sodium phosphate and sodium hydrogen sulfate solutions are mixed in equimolar amounts of their solutes, the following ionic equilibrium is established.

$$HPO_4^{2-}(aq) + SO_4^{2-} \rightleftharpoons PO_4^{3-}(aq) + HSO_4^-(aq)$$

Which side is favored, the reactants or the products? How can you tell?

*8.64   Aspirin is a weak acid. We can represent it as H(*Asp*), and it has a sodium salt that we can symbolize as Na(*Asp*). When the sodium salt of aspirin is given as a medication and it encounters gastric juice, which contains $HCl(aq)$, the following ionic equilibrium is established (at least temporarily). Which side is favored, the reactants or products? How can you tell?

$$(Asp)^-(aq) + H_3O^+(aq) \rightleftharpoons H(Asp)(aq) + H_2O$$

*8.65   Suppose you are handed a test tube and told that it contains a concentrated solution of either ammonium chloride or potassium chloride. An aqueous solution of one of the substances that we have studied in this chapter could be added to the unknown solution as a test for deciding which of the two solutes is present. What is this test reagent, and what would you observe as a result of the test if the unknown contained ammonium chloride?

8.66 Complete and balance the following molecular equations.
(a) $K_2O(s) + H_2O \rightarrow$
(b) $LiNH_2(s) + H_2O \rightarrow$
(c) $NaH(s) + H_2O \rightarrow$

Salts

*8.67 Write the names and formulas of three compounds that, by reacting with hydrochloric acid, give a solution of potassium chloride. Write the molecular equations for these reactions.

*8.68 Write the names and formulas of three compounds that will give a solution of sodium bromide when they react with hydrobromic acid. Write the molecular equations for these reactions.

8.69 Which of the following compounds are insoluble in water (as we have defined solubility)?
(a) LiOH   (b) NH_4Br (c) Hg_2Cl_2
(d) $Mg_3(PO_4)_2$ (e) NaBr   (f) $PbSO_4$

8.70 Which of the following compounds are insoluble in water?
(a) KNO_3 (b) $AgNO_3$ (c) LiBr
(d) AgI   (e) $Ca_3(PO_4)_2$ (f) $(NH_4)_2SO_4$

*8.71 Assume you have separate solutions of each compound in the pairs below. Predict what happens chemically when the two solutions of a pair are poured together. If no reaction occurs, state so. If there is a reaction, write its net ionic equation.
(a) $MgCl_2$ and KOH   (b) $NaNO_3$ and $CaCl_2$
(c) KOH and H_2SO_4 (d) $Pb(NO_3)_2$ and NaCl
(e) NH_4Cl and K_2SO_4 (f) Na_2S and $Ni(NO_3)_2$
(g) Na_2SO_4 and $Ba(NO_3)_2$ (h) NaOH and HI
(i) Na_2S and $CuCl_2$ (j) $AgNO_3$ and NaBr
(k) $LiHCO_3$ and HI   (l) KCl and $AgNO_3$

*8.72 Soap is a mixture of the sodium salts of certain organic acids. One is sodium stearate, which we can represent as Na(Ste).
(a) Write the equilibrium expression for a saturated solution of this salt in water.
(b) What would happen to this equilibrium if a concentrated solution of sodium chloride were added to it?
(c) The NaCl solution need not be concentrated. Seawater is about 3% (w/w) NaCl, and soap doesn't work well when seawater is used. Suggest a reason.

Equivalents and Milliequivalents of Ions

8.73 The concentration of potassium ion in blood serum is normally in the range of 0.0035 to 0.0050 mol K^+/L. Express this range in units of milliequivalents of K^+ per liter.

8.74 The concentration of calcium ion in blood serum is normally in the range of 0.0042 to 0.0052 eq Ca^{2+}/L. Express this range in units of milliequivalents of Ca^{2+} per liter.

8.75 The potassium ion level of blood serum normally does not exceed 0.196 g of K^+ per liter. How many milliequivalents of K^+ ion are in 0.196 g of K^+?

Ion Mobility and Ionic Reactions (Interaction 8.1)

8.76 How can an acid, like citric acid, and a bicarbonate salt be stable in each other's presence since we know that acids and bicarbonates react to give an unstable acid?

8.77 What is the net ionic equation for the reaction between citric acid and sodium bicarbonate when something like an Alka-Seltzer tablet is dropped into water?

Calcium Channels (Interaction 8.2)

8.78 Cells must have both phosphate ions (or phosphate ion donors) and calcium ions to function, yet calcium phosphate is extremely insoluble in cell fluid (mostly water). How does heart muscle tissue solve this problem?

Hard Water (Interaction 8.3)

8.79 What is hard water?

8.80 What are the formulas of the "hardness ions"?

8.81 What chemical property of these ions and of ordinary soap makes it hard to use such soap in hard water?

8.82 What is temporary hard water? Why is it designated temporary?

8.83 What is permanent hard water?

8.84 What is meant by water softening?

8.85 Concerning washing soda as a water-softening agent,
(a) What is its molecular formula?
(b) What part of its formula is the active softening agent?
(c) What is the net ionic equation for its work in water where the hardness is caused by Ca^{2+}? By Mg^{2+}?

Greenhouse Effect (Interaction 8.4)

8.86 Explain briefly how the greenhouse effect works. What happens that tends to promote global warming?

8.87 What has largely been responsible for the steady increase in the atmospheric concentration of CO_2 during recent decades?

8.88 What are the two principal ways by which CO_2 is removed from the atmosphere?

8.89 What causes the haze that climatologists see as partially reversing the greenhouse effect? How does haze cause this reversal?

Additional Exercises

8.90 In which species is the covalent bond to hydrogen stronger, in HBr(g) or in $H_3O^+(aq)$? How do we know?

8.91 $HClO_4$ (perchloric acid) is one of the less common, strong acids. Represent its ionization in water by an equation.

8.92   When carbon dioxide is bubbled into pure water to form a solution, it takes only time and the help of a little warming to drive essentially all of it out of solution again. When this gas is bubbled into aqueous sodium hydroxide, however, it is completely trapped by a chemical reaction. If we assume that the reaction involves CO_2 and $NaOH$ in a mole ratio of one to one, what is the molecular equation for this trapping reaction?

*8.93   Complete and balance the following molecular equations, and then write the net ionic equations.
(a) $HNO_3(aq) + KOH(aq) \rightarrow$
(b) $K_2CO_3(aq) + HNO_3(aq) \rightarrow$
(c) $HBr(aq) + MgCO_3(s) \rightarrow$
(d) $CaCO_3(s) + HI(aq) \rightarrow$
(e) $NH_3(aq) + HI(aq) \rightarrow$
(f) $Mg(OH)_2(s) + HBr(aq) \rightarrow$
(g) $Al(s) + HCl(aq) \rightarrow$

*8.94   Sodium and potassium in group IA are higher in the activity series than calcium and magnesium in group IIA.
(a) What does it mean to be higher in the activity series?
(b) If you check back to Figure 3.1b, you will see that sodium and potassium have lower ionization energies than calcium and magnesium. In what way does this fact correlate with the *higher* positions of sodium and potassium in the activity series of the metals?

8.95   How many grams of magnesium carbonate react quantitatively with 4.68 g of HNO_3?

*8.96   How many grams of sodium bicarbonate does it take to neutralize all the acid in 28.9 mL of 1.05 M H_2SO_4?

*8.97   For an experiment that required 13.5 L of dry CO_2 gas (as measured at 745 mm Hg and 24 °C), a student let 4.62 M HCl react with marble chips, $CaCO_3$.
(a) Write the molecular and net ionic equations for this reaction.
(b) How many grams of $CaCO_3$ and how many milliliters of the acid are needed?

*8.98   How many liters of dry CO_2 gas are generated (at 740 mm Hg and 25 °C) by the reaction of $Na_2CO_3(s)$ with 325 mL of 5.85 M HCl? Write the molecular and the net ionic equations for the reaction, and calculate how many grams of Na_2CO_3 are needed.

8.99   Write the formulas of the conjugate acids of the following.
(a) HSO_3^- (b) HCO_3^- (c) I^- (d) NO_2^-

8.100   What are the conjugate bases of the following? Write their formulas.
(a) H_2CO_3 (b) $H_2PO_4^-$ (c) NH_4^+ (d) OH^-

8.101   Which member of each pair is the stronger Brønsted base?
(a) Br^- or HCO_3^- (b) $H_2PO_4^-$ or HSO_4^-
(c) NO_2^- or NO_3^-

8.102   Study each pair and decide which is the stronger Brønsted acid.
(a) $H_2PO_4^-$ or HPO_4^{2-} (b) H_2SO_3 or HSO_3^-
(c) NH_4^+ or NH_3

8.103   Identify the compounds that do not dissolve in water.
(a) $CaCO_3$ (b) NH_4NO_3 (c) Li_2CO_3
(d) $PbCl_2$ (e) Na_2SO_4 (f) Al_2O_3

8.104   Which of the following compounds do not dissolve in water?
(a) K_2CrO_4 (b) $AgBr$ (c) $FeCO_3$
(d) $Na_2Cr_2O_7$ (e) Li_2CO_3 (f) NH_4I

*8.105   Consider the following compounds:

$$Ca_3(PO_4)_2 \qquad CaHPO_4 \qquad Ca(H_2PO_4)_2$$

(a) How can the *relative* molar solubilities of these compounds be *predicted* rather than looked up in a table?
(b) Which would have the highest molar solubility in water? (Write its formula.)
(c) Which has the lowest molar solubility in water? (Write its formula.)

8.106   Suppose that the letter Z stands for some nonmetal that can form the following two acids.

(a) Which acid, **A** or **B**, would be the weaker acid? Explain.
(b) Which acid would be the stronger electrolyte? Explain.

8.107   Hydrazine, NH_2NH_2, dissolves in water to form a solution that turns red litmus blue. The solution is a poor conductor of electricity. Classify hydrazine as a weak or strong acid or a weak or strong base.

8.108   Barium hydroxide, $Ba(OH)_2$, is quite insoluble in water, but all of its formula units that do dissolve dissociate fully. Is this compound a strong or a weak electrolyte? Explain.

8.109   Methyl chloride, CH_3Cl, is a gas and does not dissolve in water. What do these facts tell us about its chemical structure? Explain.

8.110   When sodium bicarbonate and acetic acid are combined in a 1:1 mole ratio in water, a reaction occurs that produces sodium acetate and carbonic acid (which promptly decomposes to carbon dioxide and water, producing a fizz).
(a) Write the molecular and the net ionic equations for the reaction.
(b) Which is the weaker acid, carbonic or acetic acid? Explain.

*8.111   Compounds A and B are both white solids that dissolve in water. One is an ionic compound and the other is molecular. Discuss an experiment that could be conducted to find out which is molecular and mention any possible drawbacks to the kind of experiment you select. (How might the experiment give ambiguous results?)

*8.112   If you have separate solutions of each of the compounds given below and then mix the two of each pair together, what (if anything) happens chemically? If no reaction occurs, state so, but if there is a reaction write its net ionic equation.

(a) H_2S and $Cu(NO_3)_2$ (b) LiOH and HBr
(c) Na_2SO_4 and $Ba(NO_3)_2$ (d) $Pb(C_2H_3O_2)_2$ and Na_2SO_4
(e) $Ba(NO_3)_2$ and NaCl   (f) $KHCO_3$ and H_2SO_4
(g) Na_2S and $Cd(NO_3)_2$ (h) NaOH and HBr
(i) $Hg(NO_3)_2$ and KCl   (j) $NaHCO_3$ and HI
(k) KBr and NaCl   (l) $Pb(NO_3)_2$ and Na_2CrO_4

*8.113   The magnesium ion level in plasma normally does not exceed 0.0243 g of Mg^{2+}/L. How many milliequivalents of Mg^{2+} are in 0.0243 g of Mg^{2+}?

*8.114   A white solid is either KNO_3 or K_2O. Describe an experiment that you could carry out using only test tubes and aqueous solutions that would tell which compound is present. Assume that the lab has whatever other chemicals you need.

*8.115   A white solid is either Na_2CO_3 or $NaHCO_3$. A sample of the solid with a mass of 0.144 g requires 15.0 mL of 0.114 M HCl to react with it fully until the exact point is reached when no more CO_2 forms. Which compound is it?

*8.116   A white solid was a mixture of K_2CO_3 and KNO_3. A 0.624-g sample of the mixture consumed 21.5 mL of 0.156 M HCl before CO_2 stopped evolving. How many grams of K_2CO_3 were in the mixture?

8.117   The level of chloride ion in blood serum is normally quoted as 100 to 106 meq/L. How many grams and how many milligrams constitute 106 meq of Cl^-?

8.118   The sodium ion level in the blood is normally 135 to 145 meq/L. How many grams and how many milligrams of sodium ion constitute 135 meq of Na^+?

8.119   Give the formula of an acid and of something that can neutralize the acid that react with each other to give a solution of lithium iodide without liberating a gas.

9

REACTION KINETICS AND CHEMICAL EQUILIBRIA. ACID–BASE EQUILIBRIA

To reverse ecological damage in the Grand Canyon caused by low water flow, Glen Canyon dam was "opened" for a few days in 1996 to produce an artificial flood. If the acidity of the water had been that of human blood, it would have taken 3.5 minutes of full-bore flow for 1 mole of hydrogen ions (1 gram H^+) to pass through.

THIS CHAPTER IN CONTEXT

At the molecular level of life some reactions go very rapidly and others take forever. In this chapter we'll learn what we mean by "rate of reaction," and we'll develop the background we need to understand how reactions can vary so widely in their rates. Here will be our introduction to enzymes, the family of rate-controlling agents in living systems. In an example of background paying off, most of what we've learned about molecules in motion—the kinetic theory—will be brought to bear here.

The acidity of a body fluid has such a huge impact on the rates of what happens to us that a vocabulary of concepts and terms has sprung up to discuss it, concepts like pH and buffers. If you're headed into almost any aspect of medicine, therapy, and sports, you'll want to know how the acidity of blood affects your ability to breathe. We'll here take a giant stride forward into developing this area. It impinges on most medical emergencies as well as on limitations to sports accomplishments.

9.1 FACTORS THAT AFFECT REACTION RATES

The rate of a chemical reaction is affected by the nature of the reactants, their concentrations, the temperature, and a catalyst.

The field of chemistry that deals with the rates of chemical reactions, **kinetics,** gives us the factors that affect rates and how they work.

The Chemical Nature of the Reactants Is the Most Important Factor If visible action is what you want, you'd rather watch wood burning at a campfire than look at iron rust. The substances in wood react much more rapidly and entertainingly with oxygen than does iron. Clearly, *what* reacts is the first factor to determine reaction rate.

The Physical States of the Reactants Affect Rates *Homogeneous reactions* are those in which all of the reactants are in the dissolved state either in a liquid solvent or in the gaseous state. *Heterogeneous reactions* involve at least one reactant that is not intimately mixed with the others. Reactions depend on the natural kinetic motions of the reactant particles, which are all atoms, molecules, or ions in homogeneous reactions. But only in a fluid state, liquid or gas, can they get at each other efficiently. With heterogeneous reactions, the particles of at least one reactant consist of huge clumps of atoms, ions, or molecules, and the sizes of these clumps are critical, as described in Interaction 9.1.

Concentration Also Affects Reaction Rates Steel wool doesn't burn too well in air, even when heated first to red hotness, but when red hot steel wool is thrust into pure oxygen, a wondrous flame results (Figure 9.1). Air is only 21% oxygen. Using the more concentrated reactant, 100% oxygen, clearly permits a faster rate. Someone has estimated that if the air we breathe were 30% oxygen instead of 21%, no forest fire could ever be put out, and eventually all of the world's forests would disappear by uncontrollable fire. The higher concentration of oxygen would accelerate combustion too much. To continue our study of rates and concentrations we'll return to the simpler realm of homogeneous reactions.

"Rate" is like "speed," and speed or rate is always expressed as a ratio. You're familiar with the speed of travel, for example, and its most familiar units, namely, miles per hour in the United States, or kilometers per hour elsewhere.

FIGURE 9.1 Steel wool, after being heated to redness in a flame, burns spectacularly when dropped into pure oxygen.

INTERACTION 9.1
WHAT COULD DUST EXPLOSIONS AND FAT DIGESTION POSSIBLY HAVE IN COMMON?

If you have ever tried to set fire to a log of wood or a whole charcoal briquette, using only a match and no fire starting material like kindling or paper, you've experienced the difference that particle size makes to reaction rate. You've learned by experience that to launch a *heterogeneous reaction,* one in which the reactants are not in the same phase (all dissolved together), you should use something very easily ignited to start the main reaction.

In *homogeneous reactions,* the reactant particles are as small as they can get, being molecules (or atoms or ions), and they are intimately mixed when they are in the same phase. In heterogeneous reactions, however, the particles of at least one reactant are larger and consist of huge clumps of molecules. Only those molecules can react that are at the surfaces of the clumps. This is why we want to make as many of these molecules be parts of surfaces as we can before trying to ignite the system, why we use paper, or make small kindling, for example.

Grain Dust Explosions On a solid cube, 1 cm on a side, the surface area is 6 cm² (see Figure 1). Subdividing the cube into smaller cubes only 0.01 cm on a side gives a pile with a total surface area of 600 cm², a 100-fold increase. An egg-sized lump of coal, when pulverized even more, to the fineness of soot, picks up a total surface area of two to three football fields.

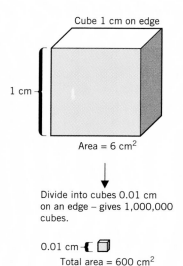

Cube 1 cm on edge

1 cm

Area = 6 cm²

Divide into cubes 0.01 cm on an edge – gives 1,000,000 cubes.

0.01 cm

Total area = 600 cm²

FIGURE 1   The total surface area increases as a solid is crushed to a powder.

Power companies that burn coal grind the coal first and then blow it into the combustion chamber, where it reacts almost explosively. Unplanned explosions occur from time to time in grain elevators when the concentration of dry grain dust in air reaches the right proportion (see Figure 2).

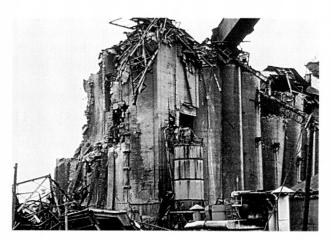

FIGURE 2   This grain elevator in New Orleans was destroyed by a dust explosion in 1977, killing 35 people.

Fat Digestion   One place where particle size impinges at the molecular level of life occurs with the fats and oils in our diets, like butter or salad oil. They are insoluble in water, but the digestive juices are dilute aqueous solutions. As you know, oil drops and water do not form a homogenous mixture, which would be a problem for digesting fats and oils because the globules can be chemically attacked only at their surfaces. The "trick" is to break the fat globules up into innumerable microglobules and create an emulsion. Fortunately, we have powerful surface-active agents, secreted in the bile juice from the gall bladder, which enable the emulsification of dietary fats and oils. Each resulting tiny globule is in the colloidal state and so has a very small surface area, but the total surface area of all of the globules is immense. Digestive reactions affecting the dietary fats and oils can therefore occur quite rapidly. People who produce no bile, perhaps as a result of the removal of the gall bladder, have to be watchful of the kind of fats and oils they use in their diets.

$$\text{Speed of travel} = \text{rate of vehicle motion} = \frac{\text{change in position}}{\text{time}} = \frac{\text{miles}}{\text{hour}}$$

Similarly, the rate of a chemical reaction is a ratio but, of course, the units are different.

$$\text{Rate of reaction} = \frac{\text{change in concentration}}{\text{time}} = \frac{\text{mol/L}}{\text{s}}$$

The concentration referred to here is the concentration of one of the species in the reaction, usually one of the reactants. The rate of the reaction then is the rate at which the molarity of the reactant decreases.

The rates of chemical reactions are not constant as the reaction proceeds. Let's discuss this by supposing the simplest possible reaction, the hypothetical conversion of substance X to Y.

$$X \longrightarrow Y$$

At the instant the reaction starts $[X]$ is at its highest and $[Y]$ is zero. As the reaction proceeds, however, the concentrations change; the value of $[X]$ decreases and that of $[Y]$ increases (Figure 9.2). Notice that early in the reaction the rates of the disappearance of X and the appearance of Y are rapid, so the curves are steepest. Late into the reaction, where the curves flatten out, further changes in concentrations take longer and longer. Thus, the *rate* of a reaction decreases with time.

In our hypothetical reaction, the rate at which $[X]$ decreases is identical to the rate at which $[Y]$ increases. In more complex reactions, a relationship this simple is almost never observed. What is found is that the rate is proportional to some mathematical combination of the molarities of two (or more) reactants, each molarity raised to some exponential power. In a slightly more complex but still hypothetical reaction of the type

$$X + Y \longrightarrow Z$$

the rate of the reaction may turn out to be related to concentrations in any one of the ways suggested by varying values of the exponents in the following equation.

$$\text{rate} \propto [X]^x[Y]^y$$

The values of x and y, which are exponents, must be discovered by doing experiments. Suppose it's found, for example, that simply doubling the concentration of X while holding the value of $[Y]$ constant doubles the reaction rate; then the exponent of $[X]$ must be 1. We won't need more details of how the exponents are determined,

■ Remember, the brackets, [ ], denote *moles per liter* concentrations.

■ If x were 2, then doubling the value of $[X]$ to $[2X]$ would quadruple the rate because 2^2 equals 4.

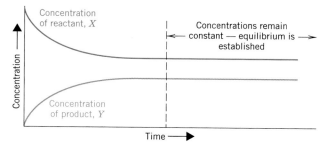

FIGURE 9.2 Changes in the concentrations of X and Y over time in the hypothetical reaction: $X \rightarrow Y$.

however, because the chief point that we want to carry forward is simply that *one factor that affects reaction rates is concentration.*

Temperature Affects Reaction Rates As a rough rule of thumb, an increase in the temperature of a reaction mixture by only 10 °C can be expected to double or triple a reaction rate. You have certainly observed, for example, that food spoils faster at room temperature than when cold in the refrigerator.

The reactions of metabolism also accelerate during a high fever, and these require oxygen. An increase in body temperature of about 0.5 °C increases the metabolic rate so much that the oxygen requirement of the body increases by at least 7%. To respond to the increased demand for oxygen the heartbeat must speed up, which puts a strain on the heart. Although a higher metabolic rate during a fever is part of the body's disease-fighting mechanism, a prolonged fever must be avoided. Thus, a patient in an intensive care unit who contracts a fever from pneumonia may be placed in cold wrappings to bring down the body temperature as quickly as possible.

■ Recall that *metabolism* refers to the sum total of all of the reactions of the body.

■ With an increase in the heartbeat, blood moves through the lungs faster, where it is oxygenated and where waste CO_2 is discharged.

Catalysts **Also Accelerate Reactions** Many reactions go at higher rates in the presence of tiny concentrations of what seem to be nonreactants. At least the substance causing the enhanced rate can be recovered unchanged at the end of the reaction. Outside agents that, in small concentrations, accelerate reactions without themselves being changed are called **catalysts,** and the phenomenon of such rate acceleration is called **catalysis.**

Virtually all reactions in plants and animals require special catalysts, only they're called **enzymes** and nearly all of them belong to a large family of biochemicals called *proteins.* An organism's mainline mechanism for governing metabolism is by controlling its enzymes.

If you have ever used dilute hydrogen peroxide to disinfect a wound you have seen the difference an enzyme makes. Blood contains *catalase,* an enzyme that accelerates the decomposition of hydrogen peroxide, causing a frothing of gaseous oxygen that helps to disinfect the wound.

$$2H_2O_2 \xrightarrow{\text{catalase}} 2H_2O + O_2$$

The enzyme is not itself permanently changed. The same decomposition does occur without the enzyme, but it is so slow at room temperature that if you look at a drugstore sample of hydrogen peroxide, you won't notice any bubbling action at all. The reason that catalase is present in the system in the first place is that hydrogen peroxide forms naturally in certain reactions. But it is toxic, so catalase protects us by destroying H_2O_2 as it forms.

A catalyst generally works in one of two ways. It can make a reaction occur much faster at the same temperature, or it can cause a reaction to take place at the same rate at a much lower temperature. A classic illustration is the decomposition of potassium chlorate into potassium chloride and oxygen. Notice in the following equations how the temperature at which the reaction can occur varies with the presence of manganese dioxide, a catalyst for the decomposition.

■ Sometimes the special conditions for a reaction are written above or below the arrow in the equation.

Without MnO_2, the temperature has to be 420 °C

$$2KClO_3 + \text{heat} \xrightarrow{420\ °C} 2KCl + 3O_2$$

With MnO_2, the temperature need be only 270 °C

$$2KClO_3 + \text{heat} \xrightarrow[MnO_2]{270\ °C} 2KCl + 3O_2$$

The rates of the evolution of oxygen are approximately equal under the sets of conditions given here, but the catalyst permits the decomposition to happen at a much lower temperature.

9.2 KINETIC THEORY AND CHEMICAL REACTIONS

Collision theory helps explain the factors that affect reaction rates.

Collision Theory Is One of the Major Theories in Kinetics   In chemical reactions, electrons and nuclei become reorganized relative to each other, but for this to happen, reactant particles have to collide. Only by a collision can the electrons and nuclei of the reactant particles be forced into the new arrangements of the products. This is the heart of **collision theory,** a theory concerning how reactions happen.

Central to collision theory is the idea that a given reaction will be faster the more frequently the reactant particles collide. *Collision frequency* is defined as the total number of collisions occurring between the reactant particles per unit of volume per second. The only way to increase the collision frequency *without changing the temperature* of the reacting mixture is to increase the concentrations of the reactant particles. It's like going from a stroll down a lonely country lane to an aisle of a very crowded store. An increase in the concentration of people in motion causes an increase in the "excuse-me" kind of bumps and collisions. If the molar concentration of one reactant is doubled, the frequency of all collisions must double because there are twice as many of its particles *in the same volume.* This is why reaction rate is a function of reactant concentrations.

An Increase in Temperature Also Increases Collision Frequency   The kinetic theory showed us that the temperature of a gas is proportional to the average molecular kinetic energy of the gas molecules. For a given gas, we saw that the increase in energy came as a result of an increase in molecular speeds, not molecular mass. The increase in speed must result in more frequent collisions, so this helps us understand why temperature is a factor in reaction rate.

Kinetic Energy Becomes Chemical Energy during Collisions between Reactant Particles   Nature operates under the law of conservation of energy. When two moving particles are about to collide, each has a certain kinetic energy. One can imagine a collision in which *both* particles stop. (This happens all the time in highway accidents.) If they stop, their kinetic energies go to zero, because KE equals $(1/2)mv^2$ and the value of v is now zero. Where did the kinetic energy go? Is it lost? If so, what of the law of conservation of energy?

Actually, the energy that existed as kinetic energy is not lost; it's transformed both into increased molecular kinetic energy (heat) and into the potential energy of distorted electron clouds. Relatively stable electron–nuclei arrangements are twisted temporarily into less stable arrangements that cannot last. They may, of course, twist back to the original arrangements of the reactants. When this happens, and it often does, the effect is that the colliding particles simply bounce off each other. The potential energy in the temporary and unstable arrangement at the instant of collision reconverts to kinetic energy much as a bouncing ball can hit a sidewalk, briefly stop with a distorted shape, then bounce away—still as a ball and not as some other substance. In other words, some collisions lead to no permanent change.

Following other, perhaps more violent collisions, reactant particles, during the brief moment of deformation, go through a rearrangement of their electrons and nuclei. As the system relaxes into a more permanent state, product particles form and we say that a *successful collision* has occurred. Thus the conversion of the kinetic energy of collision into the potential energy of distorted, unstable configurations makes a chemical reaction possible.

■ Many substances react solely because their molecules have received energy not by a collision but by absorbing it as heat, ultraviolet radiation, or some other form of energy.

■ The original kinetic theory concerned only an *ideal gas,* but the idea of particles in motion applies to all fluids.

■ If you have ever driven a bumper car at a carnival, you've sensed a relationship between speed and collision frequency.

■ A *successful* collision is simply one that produces products.

■ The *collision energy* is the sum of the kinetic energies of the colliding particles.

The Minimum Collision Energy for a Reaction Is Called the *Energy of Activation*

Generally, very light taplike collisions do not result in a reaction between the colliding particles. For each chemical reaction, there is a certain minimum collision energy that must develop before the new chemical bonds in the products can form. This minimum energy is called the reaction's **energy of activation** symbolized as E_{act}. Figure 9.3a illustrates what this means.

The vertical axis in both parts of Figure 9.3 represents changes in the *fraction* of all of the collisions taking place. The horizontal axis corresponds to increasing values that the collision energy can have, ranging from zero on the left to very large values—approaching infinity—on the right. The reactant particles have a large range of speeds, ranging from very low values (even a zero value for some, for a moment) to very high speeds. Therefore some collisions will be mere taps, whereas others will be extremely violent.

Figure 9.3a shows that in a sample of reactant particles, some collisions will be such slight taps that virtually no distortions of electron clouds occur. Thus, little if any kinetic energy changes into potential energy during such taps, and the fraction of all collisions that have zero collision energy is essentially zero. This why the curves begin at the "zero point," the intersection of the axes. Following the curves to the right we see that as the collision energy increases, the fractions of collisions having particular energies also increase. We eventually reach a maximum value of the fraction. Beyond it, collisions with increasingly higher energies become less and less likely, and the fractions with very high energies decline. The curve moves back down to the baseline.

For each reaction there is a value of collision energy that provides the exact amount of energy to make possible the electron–nuclei rearrangement of the chemical reaction. This particular collision energy is the energy of activation, E_{act}. When collisions change this much or more kinetic energy into potential energy, the reaction has enough energy to occur. Collisions with energies less than the energy of activation cannot lead to a chemical change. The colliding particles simply bounce apart.

It isn't enough, of course, to have sufficient energy. The colliding particles must hit each other just right, much as the runners in a relay race have to pass the baton correctly regardless of how fast or slowly they are moving at this critical moment in the race.

The **rate of a reaction** is the number of *successful* collisions that occur each second in each unit of volume of the reacting mixture. Generally, only a small fraction of the collisions have enough energy to be successful. The ratio of the shaded area marked A in Figure 9.3 to the total area under the curve, the area equal to $(A + B)$, represents this small fraction. We could write it as $A/(A + B)$.

■ All of the fractions represented by points on the curves in Figure 9.3 add up to unity, the total number of all collisions.

■ The *total* energy—kinetic plus potential—stays constant throughout the change, but it becomes apportioned differently.

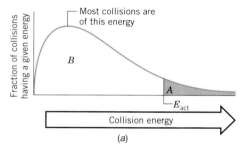 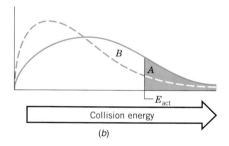

FIGURE 9.3   Energy of activation. (*a*) Only a small fraction of all collisions, represented by the ratio of areas, $A/(A + B)$, has enough energy for reaction. (*b*) This fraction greatly increases when the temperature of the reacting mixture is increased.

You can see from Figure 9.3 that if the energy of activation were very high, the shaded area to its right would be even smaller, so the fraction $A/(A + B)$ would be much smaller. In other words, *a high energy of activation means a small fraction of successful collisions and a slow rate of reaction.*

On the other hand, a reaction with a very small energy of activation would have a large fraction of successful collisions and a high rate of reaction. In the extreme, if A were to equal $(A + B)$, the fraction would equal 1, meaning that every collision would be successful no matter how low the energy of the collision. In practical terms, such a reaction would be extremely rapid—an explosion, essentially—because it would mean that simply mixing the reactants would cause instantaneous change.

At the other extreme, the energy of activation could be so high that A would equal zero, so the fraction $A/(A + B)$ would also be zero. Now no reaction ever occurs, and the "reactants" are eternally stable in each other's presence.

This analysis tells us that the rate of a reaction depends greatly on its energy of activation. A high energy of activation means a slow rate, and a low energy of activation means a fast rate.

A High Temperature Makes Collisions More Likely to Be Successful Figure 9.3*b* shows a second way in which temperature affects reaction rate. An increase in temperature not only increases the frequency of *all* collisions, it hugely increases the fraction of *successful* collisions. The curve of Figure 9.3*a* flattens as the temperature increases, and the curve's maximum shifts to the right, Figure 9.3*b*. However, *the energy of activation stays put.* Consequently, the area under the curve to the right of E_{act}, namely area *A,* grows as the temperature increases and the curve flattens, and you can see that $A/(A + B)$, therefore, increases. The increase in this ratio, of course, means a more rapid rate. We cannot go into the mathematical details, but they show that the influence of temperature on reaction rate is felt far more through its effect on the ratio $A/(A + B)$ than through its enhancement of the total collision frequency.

◼ The energy of activation sometimes changes as the temperature changes, but usually by relatively little. It's generally safe to say as a rule of thumb that E_{act} is not affected by temperature under ordinary conditions.

A Slow Reaction Can Still Be Very Exothermic An important distinction exists between a reaction's energy of activation and its *heat of reaction.* We'll use a *progress of reaction diagram* to do this (Figure 9.4). In such a diagram, the vertical axis gives *relative* values of the potential energies of the substances, either the reactants or the products, depending on which part of the plot gets our attention. The horizontal axis simply shows the direction of the chemical change.

We'll use the combustion of carbon and oxygen to illustrate how to follow a progress of reaction diagram. It's an exothermic reaction so we may write heat as a product.

$$C + O_2 \longrightarrow CO_2 + \text{heat of reaction}$$

Begin in Figure 9.4 on the left at site *A* with the unchanged reactants, carbon and oxygen. We know that these are quite stable together at or near room temperature. Coal (mostly carbon), after all, can be stored in air (with its 21% oxygen). To initiate a reaction between carbon and oxygen, we have to heat them (ignite the system). Heat gives their particles higher kinetic energies, and more and more collisions become closer to being successful. In the diagram, we are moving up the curve from *A,* because the potential energy of the system is increasing. We are climbing an "energy hill."

If a collision provides sufficient energy of activation, we are at the top of the energy barrier at location *B* in Figure 9.4. The electrons and nuclei of the reactants can now rearrange to give molecules of carbon dioxide. Some of the potential energy in

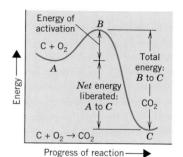

FIGURE 9.4 Progress of reaction diagram for the exothermic reaction of carbon with oxygen that produces carbon dioxide. The *heat of reaction* is the net energy liberated.

the unstable complex at the top of the energy hill now changes into the kinetic energy of newly forming carbon dioxide molecules.

Quite a drop in potential energy now occurs as the unstable complex breaks up into the more stable product C. Part of this potential energy goes to repay the cost of climbing the energy hill. A net excess, however, is liberated as heat, that is, as the molecular kinetic energy of CO_2 molecules. This net energy difference between the reactants and the products is called the **heat of reaction.** The heat of reaction comes from the conversion of some of the chemical energy in the electron–nuclei arrangements of carbon and oxygen into the molecular kinetic energy of CO_2 molecules.

As we know, once the reaction of carbon and oxygen starts, it continues spontaneously. The reaction is exothermic, and some of the energy represented in Figure 9.4 in the change from B to C activates (ignites) still unchanged particles of the reactants.

One might think that only a *rapid* exothermic reaction would produce a high heat of reaction, but this is not necessarily so. The oxidation of 1 mol of iron, for example, is a very slow reaction.

$$4Fe + 3O_2 \longrightarrow 2Fe_2O_3$$

Yet, it liberates over twice as much heat per mole of iron as the oxidation of one mole of carbon. Iron oxidizes very slowly, and carbon oxidizes so rapidly that we call it combustion. The difference lies in the energy barriers, the energies of activation. The energy barrier in the oxidation of iron is considerably higher than for the oxidation of carbon, so the rate of iron oxidation is slower.

Figure 9.5 explains this in terms of two hypothetical reactions shown in progress of reaction diagrams. The reaction on the right has the higher heat of reaction but also a higher energy of activation. It is the slower reaction. The one on the left has the much lower energy of activation, so its rate will be much faster. Yet it gives off less heat. Thus there is really no simple relationship between how rapidly an exothermic reaction occurs and how large its heat of reaction is.

No Net Release of Heat Occurs in Endothermic Reactions   Not all reactions liberate energy. Many won't occur without a continuous input of energy, as in the conversion of potassium chlorate into potassium chloride and oxygen, mentioned earlier. Now the heat of reaction has to be shown as if it were a reactant, because the reaction is endothermic.

■   The *quantity* of heat released is not a function of how long the release takes.

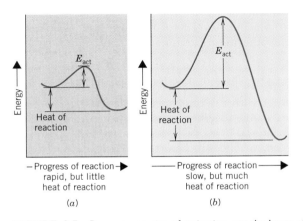

FIGURE 9.5   Because energies of activation, not the heats of reaction, dominate reaction rates, it is possible to have (*a*) a fast reaction with a small heat of reaction or (*b*) a slow reaction with a large heat of reaction. We can tell that the rate in part (*a*) is faster than that in part (*b*) because of its smaller E_{act}.

$$2KClO_3 + \text{heat of reaction} \longrightarrow 2KCl + 3O_2$$

As seen in a progress of reaction diagram, Figure 9.6, a good share of the energy of activation (A to B) is permanently retained by the product as potential energy. This share, the net energy retained, is represented by the vertical distance between A and C.

In an endothermic reaction there is a net conversion of kinetic energy (supplied by the steady input of heat) into the potential or chemical energy of the products. Both exothermic and endothermic reactions have energies of activation. But in the exothermic reaction there is still a net release of energy, whereas in the endothermic reaction there is a net absorption of energy.

Catalysts Decrease Energies of Activation by Devising Lower Energy Paths   With or without a catalyst, the decomposition of potassium chlorate is endothermic, as we saw in Figure 9.6. Figure 9.7 shows the progress of reaction diagram for the same reaction except that the catalyst MnO_2 is present. It illustrates one of the major facts about the entire phenomenon of catalysis, namely, *a catalyst does not change the heat of reaction.* What the catalyst does is enable the reactants to change to products by a different pathway, one that allows lower energy collisions to be successful. Thus, the reaction happens faster because the overall energy barrier is reduced, making the fraction of all collisions having enough energy to be successful larger with the catalyst than without it.

Enzymes generally work by reacting temporarily with a reactant, which then changes to product on the surface of the enzyme. Then the product molecules are released and the enzyme reverts to its original state, ready for more action. Let's illustrate this by a hypothetical reaction.

$$X + Y \longrightarrow Z$$

We suppose that the *direct* reaction of X and Y has a virtually insurmountable energy of activation. However, a molecule of X might react readily with an enzyme molecule to form an unstable package with *an altered electron cloud around X.* A collision with a molecule of Y might now occur more easily (that is, with a lower E_{act}) and bring about the rearrangement of electrons and nuclei necessary to form a molecule of Z. The series of steps would be

$$X + \text{catalyst} \longrightarrow \text{catalyst}{-}X$$
$$\text{catalyst}{-}X + Y \longrightarrow \text{catalyst}{-}X{-}Y$$
$$\text{catalyst}{-}X{-}Y \longrightarrow \text{catalyst} + Z$$

Despite having three steps in the pathway from reactants to products, no step has a high E_{act}.

9.3 CHEMICAL EQUILIBRIA REVISITED

An *equilibrium law* exists for every chemical equilibrium.

In a chemical equilibrium, the opposing reactions take place at identical rates, so now let's return to our study of equilibria by considering one kind, namely, acid–base equilibria. As we said, not all weak acids are equally weak, so what we'll seek is a number for each weak acid that will let us compare acid strengths. The number is called an *acid ionization constant,* but it's derived from another number, an *equilibrium constant,* which we'll have to treat first.

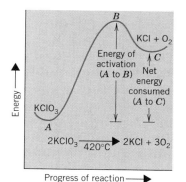

FIGURE 9.6   Progress of reaction diagram for the endothermic conversion of potassium chlorate into potassium chloride and oxygen.

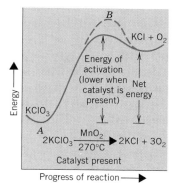

FIGURE 9.7   Progress of reaction diagram for the endothermic, catalyzed conversion of potassium chlorate into potassium chloride and oxygen. The dashed-line curve shows where the energy barrier was in the uncatalyzed reaction sketched in Figure 9.6. Notice that the net energy consumed, the heat of reaction, is identical to that of the uncatalyzed reaction, but the energy of activation is lower.

An Equilibrium Law and an Equilibrium Constant Exist for Each Equilibrium   In 1867, C. M. Guldburg and Peter Waage, two Norwegian scientists, discovered how the molar concentrations of species at equilibrium interact. Suppose that reactants A and B are in equilibrium with products C and D. The *equilibrium equation* is the following, where a, b, c, and d are the coefficients.

$$aA + bB \rightleftharpoons cC + dD$$

Recall that at equilibrium the *rates* of the forward and reverse reactions are equal. What Guldburg and Waage found is that the molar concentrations at equilibrium are related by an **equilibrium law,** which is actually an equation of the following form.

■ Equation 9.1 is often called the **law of mass action** (for historical reasons of no importance here). The ratio is the *mass action expression.* Its numerical value is called the *reaction quotient.*

$$K_{eq} = \frac{[C]^c [D]^d}{[A]^a [B]^b} \tag{9.1}$$

A unique equilibrium law exists for each chemical equilibrium. K_{eq} is called the **equilibrium constant** and it is calculated from the molar concentrations found to exist at equilibrium. K_{eq} generally has a different value for each equilibrium.

As we've learned, equilibria can be shifted by increasing or decreasing the temperature, so the value of K_{eq} at 25 °C, for example, is not the same as it is at 30 °C. The reason is that a temperature change usually does not affect the rates of the forward and the reverse reactions equally.

The Size of K_{eq} Indicates the *Position of Equilibrium*   One use we make of an equilibrium constant is to tell whether products or reactants are favored. So, to avoid confusion, it's necessary that everyone agrees that the value of K_{eq} always corresponds to the arrangement of terms in Equation 9.1, namely, *the products appear in the numerator and the reactants in the denominator.* Then the size of K_{eq} will always *directly* correspond with the *position* of the equilibrium.

When the value of K_{eq} is small, less than 1, the denominator in Equation 9.1 must be larger than the numerator. A larger denominator means that reactant concentrations are greater than those of the products. *A small value of K_{eq} means that the reactants are favored at equilibrium.*

Conversely, a value of K_{eq} greater than 1 means that the *products* are favored, because their molarities appear in the numerator of Equation 9.1. We can summarize the relationships of K_{eq} to positions of equilibria as follows.

$$\boxed{\begin{array}{l} K_{eq} < 1, \text{ reactants are favored at equilibrium} \\ K_{eq} > 1, \text{ products are favored at equilibrium} \end{array}}$$

In nearly all of the equilibria that we will study, those of weak acids and bases or of the self-ionization of water, K_{eq} will be less than 1. In aqueous acetic acid, for example, we have the following equilibrium,

$$HC_2H_3O_2(aq) + H_2O \rightleftharpoons H_3O^+(aq) + C_2H_3O_2{}^-(aq)$$

Putting the molar concentrations of the products in the numerator (and noting that all exponents are 1 because all chemical coefficients are 1), we find the equilibrium law with the known value of K_{eq} to be

■ When K_{eq} is greater than 10^2, we almost never express the equation as an equilibrium but use just a single arrow.

$$K_{eq} = \frac{[H_3O^+][C_2H_3O_2{}^-]}{[H_2O][HC_2H_3O_2]} = 3.2 \times 10^{-7} \; (25 \text{ °C})$$

We now have a number, K_{eq}, that can be used to *quantitatively* indicate how weak acetic acid is as an acid. K_{eq} is small for acetic acid, 3.2×10^{-7}, so the molarities of the product ions must be smaller than those of the un-ionized reactants. This is a distinct improvement from saying only that acetic acid is a weak acid (which begs the question, "How weak?").

K_{eq} **Remains Constant Even When the Equilibrium Shifts** If we add sodium acetate to the equilibrium in aqueous acetic acid, we add the acetate ion. The common ion effect now operates—a special case of Le Châtelier's principle—and the equilibrium shifts to the left to use up as much of the added acetate ion as possible. This shift changes the values of every term in the equilibrium law. However, an increase in $[C_2H_3O_2^-]$ is offset by a decrease in $[H_3O^+]$ and comparable changes in the denominator as the equilibrium adjusts to the stress of added acetate ion. When the new values of concentrations are measured and then inserted into the equilibrium law equation, the calculated K_{eq} is the same as before. This is really the heart of Gulburg and Waage's discovery.

It's important to realize that K_{eq} is a constant in the midst of other changes. Scientists have always paid attention to anything in nature that is a constant in the midst of change. No matter how we try to change the concentrations of individual species in the equilibrium and thereby make the equilibrium shift, the value of K_{eq} remains constant. This is why the word *law* is used in *equilibrium law*; it is the way nature behaves consistently.

9.4 ION PRODUCT CONSTANT OF WATER

The ion product constant of water is a modified equilibrium law.

Certain kinds of ionic equilibria have equilibrium laws that can be simplified without any loss in meaning. On our way to one of them (the acid ionization constant), we'll find another one, the *ion product constant of water,* very useful. It deals with one of nature's most important chemical equilibria, the self-ionization of water.

$$H_2O \rightleftharpoons H^+(aq) + OH^-(aq)$$

Its equilibrium law is

$$K_{eq} = \frac{[H^+][OH^-]}{[H_2O]} \tag{9.2}$$

■ The coefficients of the substances at equilibrium are all 1, so all of the exponents in Equation 9.2 are 1.

Let's see how Equation 9.2 can be simplified with no loss of precision. Notice first that the values of $[H^+]$ and $[OH^-]$ *in pure water* must be equal (because one of each ion forms). Experimentally, each has a value of 1.0×10^{-7} mol/L at 25 °C. These concentrations are so low that the formation of H^+ and OH^- ions from water molecules has essentially no effect on the value of $[H_2O]$ in Equation 9.2, even if we round to seven significant figures. In other words, the value of $[H_2O]$ is a constant, for all practical purposes.

In mathematics we learn that if we multiply one constant by another, we simply get a new constant. So if we multiply Equation 9.2 on both sides by the constant value of $[H_2O]$, then do an obvious cancellation, we obtain a new expression and a new constant.

$$K_{eq} \times [H_2O] = \frac{[H^+][OH^-]}{[H_2O]} \times [H_2O] = \text{a new constant}$$

■ $K_{eq} \times [H_2O]$ is one constant multiplied by another.

The new constant is the **ion product constant of water,** and its symbol is K_w.

$$K_w = [H^+][OH^-] \qquad (9.3)$$

■ If we add H^+, we shift

$$H_2O \rightleftharpoons H^+ + OH^-$$

■ K_w at Various Temperatures

Temperature (°C)	K_w
0	1.5×10^{-15}
10	3.0×10^{-15}
20	6.8×10^{-15}
25	1.0×10^{-14}
30	1.5×10^{-14}
40	3.0×10^{-14}
50	5.5×10^{-14}
60	9.5×10^{-14}

Equation 9.3 is not a true equilibrium law, because it omits the reactant's term in the denominator, but it still behaves exactly like such a law. No matter how we change $[H^+]$ by adding acid or base to a solution, the value of $[OH^-]$ in Equation 9.3 adjusts, and the *product* of the two terms remains equal to the constant, K_w. In fact, the only way to change K_w is to change the temperature, as data in the margin show. In all our work, we will assume a temperature of 25 °C. At 25 °C (and to two significant figures)

$$[H^+] = [OH^-] = 1.0 \times 10^{-7} \text{ mol/L}$$

Therefore,

$$\begin{aligned} K_w &= (1.0 \times 10^{-7})(1.0 \times 10^{-7}) \\ &= 1.0 \times 10^{-14} \qquad \text{(at 25 °C)} \end{aligned}$$

Knowing that K_w equals 1.0×10^{-14}, we can calculate the value of one of the two concentration terms, $[H^+]$ or $[OH^-]$, if we know the other.

EXAMPLE 9.1   Using the Ion Product Constant of Water

The value of $[H^+]$ of blood when measured at 25 °C is 4.5×10^{-8} mol/L. What is the value of $[OH^-]$, and is the blood acidic, basic, or neutral?

ANALYSIS The values of $[H^+]$ and $[OH^-]$ in any aqueous solution are always related by Equation 9.3 for K_w, so we substitute the value of $[H^+]$ in this equation and solve for $[OH^-]$.

SOLUTION

$$K_w = 1.0 \times 10^{-14} = (4.5 \times 10^{-8}) \times [OH^-]$$

$$[OH^-] = \frac{1.0 \times 10^{-14}}{4.5 \times 10^{-8}}$$

$$= 2.2 \times 10^{-7} \text{ mol/L}$$

■ A review of exponents is in Appendix A.

Because the value of $[H^+]$, 4.5×10^{-8} mol/L, is less than the value of $[OH^-]$, 2.2×10^{-7} mol/L, the blood is basic.

■ PRACTICE EXERCISE 1   For each of the following values of $[H^+]$, calculate the value of $[OH^-]$ and state whether the solution is acidic, basic, or neutral.

(a) $[H^+] = 4.0 \times 10^{-9}$ mol/L   (b) $[H^+] = 1.1 \times 10^{-7}$ mol/L
(c) $[H^+] = 9.4 \times 10^{-8}$ mol/L

9.5 pH CONCEPT

Very low levels of H^+ are more easily described and compared in terms of pH values than as molar concentrations.

Our interest in acid–base balance at the molecular level of life is usually with *weak* acids and bases and with very low concentrations of H^+ or OH^-. We therefore encounter very small numbers quite frequently, numbers usually expressed as negative exponentials, like 10^{-7}. When we're repeatedly faced with comparing two such numbers to see which is larger, like those in Example 9.1 (4.5×10^{-8} versus 2.2×10^{-7}), we usually must look in *two* places in each number, first, the exponents of the 10s and then the numbers before the 10s. To make the comparisons of such very small quantities easier, the Danish biochemist S. P. L. Sørenson (1868–1939), devised the concept of pH.

The pH of a Solution Is the Negative Logarithm of Its $[H^+]$ There are two completely equivalent ways of defining pH.

$$[H^+] = 1 \times 10^{-pH} \qquad (9.4)$$
$$pH = -\log [H^+] \qquad (9.5)$$

Equation 9.4 tells us that the **pH** of a solution is the negative power (the "p" in pH) to which the number 10 must be raised to express the molar concentration of a solution's hydrogen ions (hence, the "H" in pH). Equation 9.5 is the result of taking the logarithms of both sides of Equation 9.4 and relocating the minus sign.[1]

In pure water at 25 °C, $[H^+]$ equals 1.0×10^{-7} mol/L, a value in the identical exponential form as pH-defining Equation 9.4. Therefore, at a glance, we can see that the pH of pure water at 25 °C is 7.00. A pH of 7.00 thus corresponds to a neutral solution at 25 °C.[2]

There are analogous equations for expressing low concentrations of OH^- in terms of the **pOH** of a solution.

$$[OH^-] = 1 \times 10^{-pOH} \qquad (9.6)$$
$$pOH = -\log [OH^-] \qquad (9.7)$$

Values of pOH are seldom used but when they are, a simple relationship between pH and pOH exists.

$$pH + pOH = 14.00 \qquad (at\ 25\ °C) \qquad (9.8)$$

We can prove this equation by inserting the pH and the pOH expressions for $[H^+]$ and $[OH^-]$ into Equation 9.3.

$$K_w = (1.0 \times 10^{-pH})(1.0 \times 10^{-pOH}) = 1.0 \times 10^{-14} \qquad (at\ 25°C)$$

Mathematically, when we multiply numbers that involve exponents we *add* the exponents, so this equation means that

$$-pH + (-pOH) = -14$$

When we multiply both sides by -1, we get Equation 9.8.

■ In mass, 1 mol of H^+ has a mass only 1.0 g, so 1×10^{-7} mol of H^+ weighs 0.1 microgram (0.1 μg).

---

[1] Appendix A has a unit on logarithms as well as directions for using a scientific calculator to work with equations like either 9.4 or 9.5 in solving pH/$[H^+]$ problems.

[2] A word about significant figures in logarithms. The 7 in 7.00 comes from the *exponent* in 1.0×10^{-7}, so it actually does nothing more than set off a decimal point when we rewrite the number as 0.00000010. Hence the 7 in the pH value of 7.00 can't be counted as a significant figure. A pH value of 7.00 therefore has just *two* significant figures, those that *follow* the decimal point, just as there are but two significant figures in the value of the molar concentration of H^+, 1.0×10^{-7} mol/L. To repeat, the number of significant figures in any value of pH is the number of figures that *follow* the decimal point.

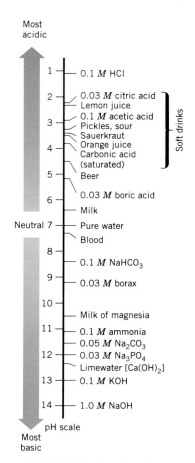

FIGURE 9.8 The pH scale and the pH values of several common substances.

Acidic Solutions Have pH Values Less than 7   Because pH occurs as a *negative* exponent in Equation 9.4, it takes a pH value less than 7.00 for a solution to be acidic and a value more than 7.00 for it to be basic. In pH terms, then, we have the following definitions of acidic, basic, and neutral solutions when their temperatures are 25 °C.

At 25 °C		
Acidic solution	pH < 7.00	
Neutral solution	pH = 7.00	(9.9)
Basic solution	pH > 7.00	

The pH values of several common substances are shown in Figure 9.8. Soft drinks, beer, and even milk are slightly acidic, as you can see, and sour pickles are sour for a now obvious reason.

Table 9.1 gives the correlations of pH, $[H^+]$, $[OH^-]$, and pOH values for the entire useful range of pH, 0 to 14. When the value of $[H^+]$ is 1 mol/L or higher, the pH concept is almost never used.

Seemingly Small pH Changes Can Mean Large $[H^+]$ Changes   One of the very deceptive features of the pH concept is that the actual hydrogen ion concentration changes greatly—by a factor of 10—for each change of only one unit of pH. For example, if the pH of a solution is zero (meaning that $[H^+] = 1 \times 10^0$ mol/L = 1 mol/L), only 1 L of water is needed to contain 1 mol of H^+. When the pH is 1, however, then 10 L of water (about the size of an average wastebasket) is needed to hold 1 mol of H^+. At a pH of 5, it takes a large railroad tank car full of water to include just 1 mol of H^+. If the pH of the water flowing over Niagara Falls, New York, were 10

TABLE 9.1   pH, $[H^+]$, $[OH^-]$, and pOH^a

pH	$[H^+]$	$[OH^-]$	pOH
Acidic Solutions			
0	1	1×10^{-14}	14.0
1.0	1×10^{-1}	1×10^{-13}	13.0
2.0	1×10^{-2}	1×10^{-12}	12.0
3.0	1×10^{-3}	1×10^{-11}	11.0
4.0	1×10^{-4}	1×10^{-10}	10.0
5.0	1×10^{-5}	1×10^{-9}	9.0
6.0	1×10^{-6}	1×10^{-8}	8.0
Neutral Solution			
7.0	1×10^{-7}	1×10^{-7}	7.0
Basic Solutions			
8.0	1×10^{-8}	1×10^{-6}	6.0
9.0	1×10^{-9}	1×10^{-5}	5.0
10.0	1×10^{-10}	1×10^{-4}	4.0
11.0	1×10^{-11}	1×10^{-3}	3.0
12.0	1×10^{-12}	1×10^{-2}	2.0
13.0	1×10^{-13}	1×10^{-1}	1.0
14.0	1×10^{-14}	1	0

^{a} Concentrations are in mol/L at 25 °C.

(which, of course, it isn't), an entire 1-hour flowage would be needed for 1 mol of H^+ to pass by. And at a pH of 14, the volume that would hold 1 mol of H^+ is about a quarter of the volume of Lake Erie, one of the Great Lakes. Seemingly small changes in pH numbers thus signify enormous changes in real concentrations of hydrogen ions. This is why small changes in pH can have dire consequences at the molecular level of life.

■ A solution at pH 4.56 has 10 times the concentration of H^+ as one at a pH of 5.56.

pH Refers to $[H^+]$, not to Un-ionized Acid Concentration   Another point about pH to be emphasized is that it refers to the molar concentration of *hydrogen ions,* not necessarily to the molar concentration of the solute contributing these ions. When the solute is a *weak* acid, only a small percentage of its molecules are ionized at equilibrium. Hence, the pH of a solution of a weak acid indicates the value of $[H^+]$, not the value of $[HA]$, the molarity of the weak acid solute.

With dilute solutions of strong, 100% ionized acids a simple correlation between pH and the acid's molarity does exist. For example, each molecule of HCl that goes into solution ionizes to give one H^+ ion and one Cl^- ion, because HCl is a strong acid. In 0.010 M HCl, therefore, $[H^+]$ equals 0.010 mol H^+/L or 1.0×10^{-2} mol H^+/L, so the pH is simply 2.00. Similarly, in 0.00010 M HNO_3, another strong monoprotic acid, $[H^+]$ equals 0.00010 mol/L or 1.0×10^{-4} mol/L. So the pH of this solution is 4.00.

The correlation between pOH and the concentration of a strong base, like NaOH, is also simple. In 0.0010 M NaOH, for example, $[OH^-]$ equals 0.0010 mol/L or 1.0×10^{-3} mol/L, so the pOH is simply 3.00. Because pH plus pOH equals 14.00 at 25 °C, a pOH of 3.00 means a pH of 11.00.

In all these simple correlations, the numbers were picked to let 1.0 stand before the 10 in the exponential expression. We will work one example involving a strong acid, for which the numbers do not have this relation, just to get used to using a scientific calculator for pH calculations.

EXAMPLE 9.2   Calculating pH from $[H^+]$

Lakes in upper New York State, and some New England areas as well as in the Boundary Waters Canoe Area of northern Minnesota, receive air pollutants, such as the oxides of sulfur and nitrogen, dissolved in rainfall, and they make the lake waters more acidic than normal. The water in one lake was found to have $[H^+]$ equal to 3.1×10^{-5} mol/L. Calculate the pH and the pOH of the lake water.

■ Rain made acidic by air pollutants is called **acid rain.**

ANALYSIS   The defining equation for pH, Equation 9.5, gives the relationship between $[H^+]$ and pH.

SOLUTION

$$\text{pH} = -\log [H^+]$$
$$= -\log (3.1 \times 10^{-5})$$

Enter 3.1×10^{-5} into your calculator. If your calculator has the function keys $\boxed{10^x}$ and $\boxed{\log}$, it almost certainly has the keys $\boxed{\text{EXP}}$ and $\boxed{+/-}$. (Your $\boxed{\text{EXP}}$ key might be labeled $\boxed{\text{EE}}$. Check your manual. If your calculator does not have these functions, you should buy one that does.) Remember, EXP means "times ten to the" as in "3.1 *times 10 to the* minus 5 power." And be doubly sure to remember to use the $\boxed{+/-}$ key to get a negative exponent from an entered positive number. To enter 3.1×10^{-5}, therefore, press the following keys.

$$\boxed{3} \quad \boxed{.} \quad \boxed{1} \quad \boxed{EXP} \quad \boxed{+/-} \quad \boxed{5}$$

The display screen should now look something like 3.1 −05. Now all you do is press the $\boxed{log}$ key. The display should now read −4.508638306. The pH is the negative of this, so change the sign. Also, round off to two significant figures, the number allowed by the value of $[H^+]$, 3.1×10^{-5}. The answer, therefore, is pH = 4.51.

The pOH is found from Equation 9.8.

$$pH + pOH = 14.00$$
$$4.51 + pOH = 14.00$$
$$pOH = 9.49$$

■   PRACTICE EXERCISE 2   Calculate the pH and the pOH in each of the following solutions. (a) 0.025 M HCl (b) 0.00025 M NaOH (hint: calculate pOH first, then the pH using Equation 9.8) (c) 0.00025 M $Ba(OH)_2$ (consider this to be 100% dissociated.)

■   PRACTICE EXERCISE 3   A blood specimen was found to have $[H^+] = 7.3 \times 10^{-8}$ mol/L. Calculate its pH. Is it acidic, basic, or neutral?

Another calculation that sometimes has to be made is to find $[H^+]$ from the pH. We'll work an example to show how your calculator can handle this.

EXAMPLE 9.3   Calculating $[H^+]$ from pH

■ Acid rain also harms forests and accelerates the corrosion of exposed objects made of metal, limestone, or marble.

Because of acid rain, the game fish in thousands of lakes in southern Norway are threatened. The pH of the lake waters is below 5.50. What value of $[H^+]$ corresponds to a pH of 5.50?

ANALYSIS   Equation 9.4 now becomes the best equation to use.

SOLUTION

$$[H^+] = 1 \times 10^{-pH}$$

We have to get 5.50 into the exponent as a negative number, so enter 5.50 into your scientific calculator *and then press the* $\boxed{+/-}$ *key.* You now have actually entered x for the $\boxed{10^x}$ key, so now press this key and the display will read 3.16227766^{-06}. This means 3.2×10^{-6}, after we round to the two significant figures allowed in the pH value of 5.50. Thus a pH of 5.50 means $[H^+]$ equals 3.2×10^{-6} mol/L.

■   PRACTICE EXERCISE 4   Calculate the values of $[H^+]$ for solutions of the following pH and state if the solutions are acidic or basic. (a) 6.34 (b) 7.89

■ PRACTICE EXERCISE 5   A blood sample had a pH of 7.28. What is [H⁺] for this sample?

■ PRACTICE EXERCISE 6   The pOH of a solution was 4.56. Calculate [H⁺].

The question we next consider is, "How are very small values of [H⁺] or their associated pH values measured in the lab?"

Litmus Is One of Many Acid–Base Indicators   The most common way to get a very rough idea of the pH of a solution is to use an acid–base indicator or a combination of them. A number of organic dyes are available for this purpose. Litmus, which we mentioned in the previous chapter, is one example; litmus is blue above a pH of about 8.5 and red below a pH of 4.5. Each indicator has its own pH range and set of colors, and Figure 9.9 gives a few examples. Phenolphthalein, for example, has a bright pink color at a pH above 10.0 and is colorless below a pH of 8.2. In the range of 8.2 to 10.0 phenolphthalein undergoes a gradual change from colorless to pale pink to deep pink. Bromothymol blue is blue above a pH of 7.6 and yellow below 6.0. Thus if you find that a solution turns phenolphthalein colorless (so the solution's pH is no higher than 8.2) but it makes bromothymol blue become blue (so the solution's pH is above 7.6), you would know that the pH of the solution is between 7.6 and 8.2.

■ Phenolphthalein (fee-noll-THAY-lean).

Commercial test papers are available that are impregnated with several indicator dyes. Their containers carry a color code, so you can match the color produced by a drop of solution to this code and so learn the pH of the solution (Figure 9.10).

When the solutions to be tested for pH are themselves highly colored, we can't use indicators. Moreover, we often need more than a rough idea of pH. For such situ-

pH 8.2   pH 10.0
Phenolphthalein

pH 6.0   pH 7.6
Bromothymol blue

pH 3.2   pH 4.4
Methyl orange

pH 9.4   pH 10.6
Thymolphthalein

FIGURE 9.9   The colors of some common acid–base indicators.

FIGURE 9.10 A pH test paper. A drop of the lemon juice in the beaker has been removed and put on the pH test paper causing the orange-red color. The pH of the lemon juice is therefore between 2 and 3, according to the color code.

FIGURE 9.11 A pH meter.

ations a variety of commercial pH meters are available (Figure 9.11) that come equipped with specially designed electrodes that can be dipped into the solution to be tested. With a good pH meter, pH values can be read to the second decimal place. In addition, pH meters with microelectrodes are available, which enable operating room personnel, for example, to measure the pH of a tiny sample of a body fluid.

9.6 ACID IONIZATION CONSTANTS

The strengths of weak acids are described quantitatively by their acid ionization constants, K_a.

■ The weak acids of greatest importance to our study are acetic acid, carbonic acid, and the organic acids produced as intermediates in metabolism.

We can now work on the problem of comparing the weaknesses of weak acids and bases. For this purpose, an *acid ionization constant*, K_a, is derived from the equilibrium constant for the weak acid. Let's represent any weak acid by HA, where A denotes the species that separates from H$^+$ when HA ionizes. Although given without an electrical charge, HA is not limited to electrically neutral species, like acetic acid. It can also represent a positively charged ion like the ammonium ion, or a negatively charged ion like the bicarbonate ion. All these are weak proton donors toward water. For a diprotic or triprotic acid, we'll use HA to consider the ionization of just one proton.

The equilibrium equation for the ionization of HA in water is as follows. Notice that one product is always H_3O^+ (or, later, H^+) and the other is always the conjugate base of the acid.

$$HA + H_2O \rightleftharpoons H_3O^+ + A^-$$
weak
Brønsted acid
conjugate base

To illustrate, the specific equilibria for the Brønsted acids already mentioned are

$$HC_2H_3O_2 + H_2O \rightleftharpoons H_3O^+ + C_2H_3O_2^-$$
$$NH_4^+ + H_2O \rightleftharpoons H_3O^+ + NH_3$$
$$HCO_3^- + H_2O \rightleftharpoons H_3O^+ + CO_3^{2-}$$

The general form for the equilibrium law for these equilibria is

$$K_{eq} = \frac{[H_3O^+][A^-]}{[HA][H_2O]} \tag{9.10}$$

■ The concentrations in the brackets are the concentrations *after equilibrium has been established.*

We can now simplify Equation 9.10. Precisely because weak acids are *weak*, little ionization occurs. So in their solutions *the molar concentration of water* has not changed, even to several significant figures. The value of [H_2O] in Equation 9.10 is thus essentially a constant. So let's multiply both sides of Equation 9.10 by this constant and cancel terms.

$$K_{eq} \times [H_2O] = \frac{[H_3O^+][A^-]}{[HA]\cancel{[H_2O]}} \times \cancel{[H_2O]} = \text{a new constant}$$

The new constant is called the **acid ionization constant,** and its symbol is K_a. If we now switch from H_3O^+ to H^+, the equation for K_a is

■ Some references call it the *acid dissociation constant.*

$$K_a = \frac{[H^+][A^-]}{[HA]} \tag{9.11}$$

Equation 9.11 is the (modified) equilibrium law for all monoprotic Brønsted acids. It's as if the equilibrium equation for the ionization of a weak acid, *HA,* is one in which we ignore the participation of water as a reactant and simply write the following as the equilibrium equation.

$$HA(aq) \rightleftharpoons H^+(aq) + A^-(aq)$$

We're not actually ignoring water as a reactant. We're simply remembering that the value of [H_2O] is unchanged because the percentage ionization is so small. The actual K_a equation for a specific acid can be figured out whenever it's needed; but let's do a worked example to make sure we can do this.

EXAMPLE 9.4   Writing Equations for K_a for Acids

What is the equation for K_a for formic acid, $HCHO_2$, the acid present in the stinging juices of ants?

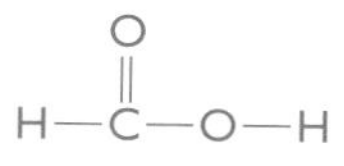

ANALYSIS The modified equilibrium equation for the ionization of formic acid is first written, and the K_a equation is then prepared from it.

SOLUTION The ionization of formic acid establishes an equilibrium, which, in our modified form, is

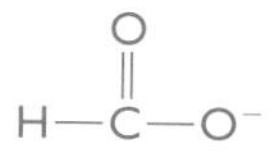

$$HCHO_2(aq) \rightleftharpoons H^+(aq) + CHO_2^-(aq)$$

The equation for K_a puts the products in the numerator and the reactants in the denominator.

$$K_a = \frac{[H^+][CHO_2^-]}{[HCHO_2]}$$

■
formic acid, $HCHO_2$

■
formate ion, CHO_2^-

■ **PRACTICE EXERCISE 7** Write the modified equilibrium equation and the K_a equation for acetic acid, $HC_2H_3O_2$.

■ **PRACTICE EXERCISE 8** Write the modified equilibrium equation and the K_a equation for the bicarbonate ion.

■ **PRACTICE EXERCISE 9** Write the modified equilibrium equation and the K_a equation for the ammonium ion.

K_a **Values Let Us Better Define Weak, Moderate, and Strong Acids**   The K_a values of several acids are given in Table 9.2. Because the products, including H^+, are in the *numerator* of the general equation for K_a (Equation 9.11), the value of K_a is small when the products are *not* favored. *The weaker the acid, the smaller is its K_a.*

A strong acid, on the other hand, generates a high percentage of H^+, so the K_a values of strong acids are high. *The stronger the acid, the greater is its K_a.* Truly strong acids, like hydrochloric acid or nitric acid, have K_a values so high they are ignored. (When the percentage ionization is 100%, the value of the denominator in Equation 9.11 is zero, and in mathematics the result of dividing a number by zero is infinity, that is, infinitely high.)

■ **PRACTICE EXERCISE 10** The K_a for HCN, hydrogen cyanide, is 6.2×10^{-10} and for ascorbic acid (vitamin C) it is 7.9×10^{-5}. Which is the stronger acid?

TABLE 9.2 K_a and pK_a Values for Acids at 25 °C[a]

Name of acid	Formula	K_a	pK_a
Perchloric acid	$HClO_4$	Large	
Hydroiodic acid	HI	Large	
Hydrobromic acid	HBr	Large	
Sulfuric acid	H_2SO_4	Large	
Hydrochloric acid	HCl	Large	
Nitric acid	HNO_3	Large	
HYDRONIUM ION	H_3O^+	55	
Hydrogen sulfate ion	HSO_4^-	1.0×10^{-2}	2.00
Phosphoric acid	H_3PO_4	7.1×10^{-3}	2.15
Citric acid	$H_3C_6H_5O_7$	7.1×10^{-4}	3.15
Ascorbic acid (vitamin C)	$H_2C_6H_6O_6$	7.9×10^{-5}	4.10
Acetic acid	$HC_2H_3O_2$	1.8×10^{-5}	4.74
Carbonic acid	H_2CO_3	4.5×10^{-7}	6.35
Dihydrogen phosphate ion	$H_2PO_4^-$	6.3×10^{-8}	7.20
Ammonium ion	NH_4^+	5.7×10^{-10}	9.24
Bicarbonate ion	HCO_3^-	4.7×10^{-11}	10.33
Monohydrogen phosphate ion	HPO_4^{2-}	4.5×10^{-13}	12.35
WATER	H_2O	1.8×10^{-16}	15.74
Hydroxide ion	OH^-	$.1 \times 10^{-36}$ (est.)	36

[a] For diprotic and triprotic acids, K_a is for the ionization of the first proton only. Data are rounded to two significant figures from the values given in E. H. Martell and R. M. Smith, *Critical Stability Constants,* Plenum Press, New York, 1974. For water, see R. Starkey, J. Norman, and M. Hintze, *J Chem. Ed. 63* (1986), p. 473. For OH^- see R. J. Myers, *J. Chem. Ed. 63* (1986), p. 687, and references cited therein.

Acid ionization constants are used to classify acids as weak, moderate, or strong according to the following criteria.

$K_a < 10^{-3}$	Weak acid
$K_a = 10^{-3}$ to 1	Moderate acid
$K_a > 1$	Strong acid

Most Transition Metal Cations and the Ammonium Ion Hydrolyze to Generate H_3O^+ Ions and Lower the pH of the Solution An aqueous solution of ammonium chloride, NH_4Cl, turns blue litmus red, so $[H^+]$ is greater than $[OH^-]$ in the solution. The extra hydrogen ions come from the forward reaction of the following equilibrium involving the NH_4^+ ion.

$$NH_4^+ + H_2O \rightleftharpoons NH_3(aq) + H_3O^+(aq)$$

This, of course, is nothing more than the ammonium ion acting as a weak acid toward water. (Its K_a is 5.7×10^{-10}.)

The reaction of a cation with water to generate hydronium ion is called the **hydrolysis of the cation.** The lesson to be carried forward here is that even certain *salts,* like ammonium chloride, can make a solution acidic even though their names do not have the word *acid* in them. It's a lesson we need to know, because of our interest in the acid–base status of a fluid at the molecular level of life. We have to be aware of any solute that gives a solution a pH other than 7.00. Ammonium salts are examples.

Even the hydrated cations of most metals can make a solution test acidic to litmus. The aluminum ion in water, for example, exists largely as $[Al(H_2O)_6]^{3+}$. The high positive charge on the central metal ion in this hydrated ion attracts electron density from the H—O bonds of the H_2O molecules it holds. These bonds are thus weakened so that one H_2O molecule in $[Al(H_2O)_6]^{3+}$ can donate H^+ to a solvent H_2O molecule.

$$[Al(H_2O)_6]^{3+}(aq) + H_2O \longrightarrow [Al(H_2O)_5(OH)]^{2+} + H_3O^+(aq)$$

Because H_3O^+ forms, the solution is acidic. A solution of 0.1 M $AlCl_3$ has a pH of about 3, for example, the same as 0.1 M acetic acid.

In sufficient concentration, all cations with charges of 3+ and most with charges of 2+ can generate enough hydronium ions in water to turn blue litmus red. The reason is that metal ions with 3+ and 2+ charges are very small. So they have high *densities* of charge, that is, high positive charges *per unit volume,* like the Al^{3+} unit within $[Al(H_2O)_6]^{3+}(aq)$. A site of high positive charge density is strongly electronegative. It attracts electron density, thus weakening H—O bonds in surrounding H_2O molecules. When the H—O bond is weak, H^+ can transfer away to generate H_3O^+.

The only common metal ions that do *not* hydrolyze enough to give solutions acidic to litmus are those of groups IA and IIA (except Be^{2+} of IIA), namely, Li^+, Na^+, and K^+, Mg^{2+}, Ca^{2+}, and Ba^{2+}. Evidently, except for Be^{2+}, the cations of groups IA and IIA do not have sufficiently high positive charge densities.

- "Hydrolysis" is from the Greek *hydro*, water, and *lysis*, loosening or breaking—breaking or loosening by water.

- Left behind when one H_2O in $(H_2O)_6$ gives up H^+ is one OH^- ion, so the charge on the complex drops from 3+ to 2+.

- A high charge density means a high charge in a small volume.

9.7 BASE IONIZATION CONSTANTS

Base ionization constants let us compare the strengths of weak bases.

Strong bases, like sodium hydroxide, dissociate completely in water to release OH^- ions.

$$NaOH(s) \xrightarrow{\text{dissociation}} Na^+(aq) + OH^-(aq)$$

Other, even stronger bases, like the oxide ion in sodium oxide, react completely with water to generate OH^- ions.

$$Na_2O(s) + H_2O \longrightarrow 2Na^+(aq) + 2OH^-(aq)$$

■   The weak bases of greatest importance in our study are NH_3, HCO_3^-, CO_3^{2-}, HPO_4^{2-}, $H_2PO_4^-$, and any of the conjugate bases of the weak organic acids that we will encounter.

Weak bases, like ammonia or the bicarbonate ion, react incompletely with water, usually to a small percentage, to make some OH^-. An equilibrium is established in which the unchanged base is favored. Ammonia and the bicarbonate ion, for example, generate OH^- ions in water in the following equilibria.

$$NH_3(aq) + H_2O \rightleftharpoons NH_4^+(aq) + OH^-(aq)$$
$$HCO_3^-(aq) + H_2O \rightleftharpoons H_2CO_3(aq) + OH^-(aq)$$

Enough forward reaction occurs to make $[OH^-]$ greater than $[H^+]$ in the resulting solutions, so they test basic to litmus.

To compare the strengths of weak bases, we again use a special equilibrium constant, the *base ionization constant*, K_b. The associated equilibrium is always of the following type, where we represent any base by B, regardless of its electrical charge. In this equilibrium, one product is always OH^- and the other is always the conjugate acid of the base.

$$B(aq) + H_2O \rightleftharpoons BH^+(aq) + OH^-(aq)$$

weak base   conjugate acid

Because the base is *weak*, the value of $[H_2O]$, for all practical purposes, does not change, and so a term for $[H_2O]$ is left out of the expression for K_b. Thus, the **base ionization constant** for this equilibrium, K_b, is defined by the following equation.

■ $[H_2O]$ is incorporated into K_b just as it was into K_a.

$$K_b = \frac{[BH^+][OH^-]}{[B]} \tag{9.12}$$

The K_b values for several bases are given in Table 9.3.

The Smaller the K_b, the Weaker the Base   Equation 9.12 has the products in the numerator, so when their concentrations are small, as they are when the base is weak, K_b has a small value. *The smaller the K_b, the weaker the base.* When the base is strong, then the products are in relatively high concentration, the numerator in Equation 9.12 now is larger, and the K_b value is higher. *The larger the K_b, the stronger the base.*

■   PRACTICE EXERCISE 11   The base ionization constant for CN^-, the cyanide ion, is 1.6×10^{-5} and for the bicarbonate ion is 2.6×10^{-8}. Which is the stronger base?

The interpretation of K_b values depends always on the ability to translate just the formula of the base both into its chemical equilibrium in water and into the specific equation for its K_b. The next example shows how this is done.

TABLE 9.3 K_b and pK_b Values for Bases at 25 °C[a]

Name of Base	Formula	$K_b{}^a$	pK_b
Oxide ion	O^{2-}	$.1 \times 10^{22}$	
HYDROXIDE ION	OH^-	55	
Phosphate ion	$PO_4{}^{3-}$	2.2×10^{-2}	1.66
Carbonate ion	$CO_3{}^{2-}$	2.1×10^{-4}	3.68
Ammonia	NH_3	1.8×10^{-5}	4.74
Monohydrogen phosphate ion	$HPO_4{}^{2-}$	1.6×10^{-7}	6.80
Bicarbonate ion	$HCO_3{}^-$	2.2×10^{-8}	7.65
Acetate ion	$C_2H_3O_2{}^-$	5.6×10^{-10}	9.26
Ascorbate ion	$HC_6H_6O_6{}^-$	1.3×10^{-10}	9.89
Citrate ion	$H_2C_6H_5O_7{}^-$	1.4×10^{-11}	10.85
Dihydrogen phosphate ion	$H_2PO_4{}^-$	1.4×10^{-12}	11.85
Sulfate ion	$SO_4{}^{2-}$	9.8×10^{-13}	12.01
WATER	H_2O	1.8×10^{-16}	15.74
Nitrate ion	$NO_3{}^-$	Very small	
Chloride ion	Cl^-	Very small	
Hydrogen sulfate ion	$HSO_4{}^-$	Very small	
Bromide ion	Br^-	Very small	
Iodide ion	I^-	Very small	
Perchlorate ion	$ClO_4{}^-$	Very small	

[a] K_b values (except for ammonia) were calculated from the K_a values obtained from the references cited for Table 9.2 and then rounded to two significant figures.

EXAMPLE 9.5 Writing Expressions for K_b for Brønsted Bases

The monohydrogen phosphate ion is a base. Write the equilibrium expression on which its K_b is based and then write the equation for K_b.

ANALYSIS To write the equilibrium expression for $HPO_4{}^{2-}$ in water, we put $HPO_4{}^{2-}$ and H_2O as the reactants, and OH^- and the conjugate acid of $HPO_4{}^{2-}$ as the products.

SOLUTION We figure out the formula of the conjugate acid of any base by adding one H^+ to the formula of the base, remembering to adjust the charge correctly. So the conjugate acid of $HPO_4{}^{2-}$ is $H_2PO_4{}^-$. Our equilibrium equation is

$$HPO_4{}^{2-}(aq) + H_2O \rightleftharpoons H_2PO_4{}^-(aq) + OH^-(aq)$$

Now we can write the equation for K_b, omitting H_2O and remembering that the products are always in the numerator. The answer, then, is

$$K_b = \frac{[H_2PO_4{}^-][OH^-]}{[HPO_4{}^{2-}]}$$

■ **PRACTICE EXERCISE 12** Write the equilibrium equations and the equations for K_b for each of the following Brønsted bases.

(a) $CO_3{}^{2-}$ (b) $C_2H_3O_2{}^-$ (acetate ion) (c) NH_3

The Anions of Many Common Salts Are Brønsted Bases   Most of the Brønsted bases are anions, and when their K_b values are greater than 10^{-13} they are available as sodium salts, such as Na_3PO_4, Na_2CO_3, $NaCN$, Na_2HPO_4, $NaHCO_3$, $NaC_2H_3O_2$, and NaH_2PO_4. None shows a hydroxide ion in its formula. Yet, dilute aqueous solutions of these salts, say about one-tenth molar, all test basic to litmus because their anions have reacted with water—it's called the **hydrolysis of anions**—to generate an excess of OH^- over H^+. The bicarbonate ion in aqueous $NaHCO_3$, for example, hydrolyzes to establish the following equilibrium in which OH^- ions are present in excess over H^+ ions.

■ One reason we are interested in any solute that can affect the pH of a solution is that enzyme action is very sensitive to pH.

$$HCO_3^-(aq) + H_2O \rightleftharpoons H_2CO_3(aq) + OH^-(aq)$$

Anions like Cl^-, Br^-, I^-, NO_3^-, and SO_4^{2-}, which are conjugate bases of strong acids, do not hydrolyze in this way. To summarize,

> Anions whose conjugate acids are weak acids hydrolyze in water and tend to make the solution basic.
>
> Anions of strong acids do not hydrolyze enough to affect litmus.

In the previous section we developed similar rules of thumb about cations, which we'll repeat here.

> Metal ions from group IA or IIA (except Be^{2+}) do not hydrolyze enough to affect litmus.
>
> Expect other metal ions as well as NH_4^+ to hydrolyze and make the solution acidic.

With these rules we achieve our goal, to predict correctly how a given salt will affect the pH of an aqueous solution. The exceptions are salts where both cation and anion hydrolyze, like ammonium acetate, $NH_4C_2H_3O_2$. Such salts have to be taken on a case by case basis, with the result hinging on the relative strengths of the cation as an proton producer and the base as a proton neutralizer. We will not work with such salts. But let's see how we can predict the hydrolysis of salts that respond to a simpler analysis.

EXAMPLE 9.6   Predicting How a Salt Affects the pH of Its Solution

Sodium phosphate, Na_3PO_4 ("trisodium phosphate"), is a strong cleaning agent for walls and floors. Is its aqueous solution acidic, basic, or neutral?

ANALYSIS   We must consider the *ions* that the salt releases in solution and decide what to expect of each.

SOLUTION Na_3PO_4 involves Na^+ and PO_4^{3-}. Na^+ does not hydrolyze, but PO_4^{3-} does. Its conjugate acid, HPO_4^{2-}, is not on our list of strong acids, so we can infer that it is a weak acid. We therefore expect PO_4^{3-} to be a relatively strong base and we expect it to hydrolyze as follows.

$$PO_4^{3-}(aq) + H_2O \rightleftharpoons HPO_4^{2-}(aq) + OH^-(aq)$$

This equilibrium generates some OH^- ions, so the solution will be basic.

In working Example 9.6 we did not need a table of Brønsted bases to predict that PO_4^{3-} would hydrolyze. We used our knowledge of just a few facts—which acids are strong acids in water and which cations do not hydrolyze—to figure out what we needed.

■ PRACTICE EXERCISE 13   Determine without the use of tables whether each ion can hydrolyze. If so, state whether it tends to make a solution acidic or basic.

(a) CO_3^{2-} (b) S^{2-} (c) HPO_4^{2-} (d) Fe^{3+} (e) NO_2^{-} (f) F^{-}

■ PRACTICE EXERCISE 14   Is a solution of potassium acetate, $KC_2H_3O_2$, acidic, basic, or neutral to litmus?

■ PRACTICE EXERCISE 15   Is a solution of copper(II) nitrate, $Cu(NO_3)_2$, acidic, basic, or neutral?

■ PRACTICE EXERCISE 16   Ammonium sulfate, $(NH_4)_2SO_4$, is a nitrogen fertilizer. Could the application of an aqueous solution of this fertilizer affect the pH of the soil? If so, will it increase or decrease the pH?

9.8 pK$_a$ AND pK$_b$ CONCEPTS

The negative logarithms of K_a and K_b, pK$_a$ and pK$_b$, are useful in the same way as pH.

For the same reason that the pH concept was devised, analogous pK$_a$ and pK$_b$ expressions, based on K_a and K_b, have been defined. The **pK$_a$** is the negative logarithm of K_a, and the **pK$_b$** is the negative logarithm of K_b.

■ This section is background for the quantitative treatment of buffers in Section 9.10.

$$pK_a = -\log K_a \qquad (9.13)$$
$$pK_b = -\log K_b \qquad (9.14)$$

Note carefully that pK$_a$ and pK$_b$ are *negative* logarithms. Therefore, the generalizations we want to carry forward from Equations 9.13 and 9.14 are the following.

The larger that the value of pK$_a$ *is, the weaker is the acid.*
The larger that the value of pK$_b$ *is, the weaker is the base.*

For example, pK$_a$ is 10.6 for carbonic acid, H_2CO_3, and 4.76 for acetic acid, $HC_2H_3O_2$. Therefore, carbonic acid is a weaker acid than acetic acid. We'll next study how to do some calculations, but the chief purpose of them is simply to get you used to the concepts of K_a, K_b, pK$_a$, and pK$_b$ and the typical sizes of the values involved. They're needed in the discussion of buffers to come.

EXAMPLE 9.7   Calculating pK$_a$ from K_a

The K_a for ascorbic acid (vitamin C) is 7.9×10^{-5} (at 25 °C). What is its pK$_a$ at this temperature?

ANALYSIS We use the defining equation for pK$_a$.

SOLUTION

$$pK_a = -\log K_a$$
$$= -\log (7.9 \times 10^{-5})$$
$$pK_a = 4.10 \text{ (rounded as per footnote 2)}$$

■ **PRACTICE EXERCISE 17**   An acid has a K_a value of 4.7×10^{-11}. Calculate its pK_a value.

The same procedure given in Example 9.7 is used to calculate values of pK_b for weak bases from K_b values.

For a Conjugate Acid–Base Pair, the Product of K_a and K_b Is K_w   A very simple relationship exists between K_a and K_b when we work with a conjugate acid–base pair.

$$K_a K_b = K_w \qquad \text{(for a conjugate acid–base pair)} \qquad (9.15)$$

To prove Equation 9.15, we substitute the expressions for K_a, K_b, and K_w into it and cancel what terms we can. In the equilibrium in a solution of a weak acid we have

$$HA \rightleftharpoons H^+ + A^- \qquad \text{and} \qquad K_a = \frac{[H^+][A^-]}{[HA]}$$

For a solution of the conjugate base of HA, which is A^- (put into solution as some salt, like NaA), we have

$$A^- + H_2O \rightleftharpoons HA + OH^- \qquad \text{and} \qquad K_b = \frac{[HA][OH^-]}{[A^-]}$$

We now multiply the expressions for K_a and K_b and cancel what we can.

$$K_a \times K_b = \frac{[H^+]\cancel{[A^-]}}{\cancel{[HA]}} \times \frac{\cancel{[HA]}[OH^-]}{\cancel{[A^-]}} = [H^+][OH^-]$$
$$= K_w \qquad \text{(proving Equation 9.15)}$$

■ Because $K_a \times K_b = K_w$,

$$K_a = \frac{K_w}{K_b} \qquad \text{and} \qquad K_b = \frac{K_w}{K_a}$$

When We Know Either K_a or K_b for a Conjugate Acid–Base Pair, We Can Calculate the Other   Equation 9.15 lets us calculate K_b when we know K_a for its conjugate acid or it lets us calculate K_a for an acid when we know K_b for its conjugate base. All the K_b values in Table 9.3, for example, were calculated from the values of K_a in Table 9.2. Remember, what we'll learn next works only for conjugate acid–base pairs.

EXAMPLE 9.8   Calculating K_b from K_a in a Conjugate Acid–Base Pair

What is the value of K_b at 25 °C for the conjugate base of hypochlorous acid, HOCl, whose K_a is 3.0×10^{-8}?

ANALYSIS   At 25 °C, $K_w = 1.0 \times 10^{-14}$, so we simply substitute this value and the given value of K_a, 3.0×10^{-8}, into Equation 9.15.

SOLUTION

$$(3.0 \times 10^{-8}) \times K_b = 1.0 \times 10^{-14}$$

Solving for K_b gives

$$K_b = \frac{1.0 \times 10^{-14}}{3.0 \times 10^{-8}}$$

$$= 3.3 \times 10^{-7}$$

Thus the K_b for OCl^-, the conjugate base of $HOCl$, is 3.3×10^{-7}.

■ **PRACTICE EXERCISE 18** For NH_4^+, $K_a = 5.7 \times 10^{-10}$. What is the conjugate base of this weak acid, and what is its K_b?

For a Conjugate Acid–Base Pair, pK_a Plus pK_b Equals 14.00 (25 °C) The relationship among K_a, K_b, and K_w leads to a simple relationship between pK_a and pK_b *for a conjugate acid–base pair.* If we take the logarithms of both sides of Equation 9.15, we obtain

$$\log (K_a \times K_b) = \log K_w$$

Or

$$\log K_a + \log K_b = \log K_w$$

After multiplying both sides by -1, we get

$$(-\log K_a) + (-\log K_b) = (-\log K_w)$$

But the first two terms define pK_a and pK_b, respectively, so this equation is equivalent to writing

$$pK_a + pK_b = -(\log 1.0 \times 10^{-14}) \qquad \text{(at 25 °C)}$$

Because $\log 1.0 \times 10^{-14} = 14.00$, we have the following relationship between the pK_a and pK_b values for an acid and its conjugate base at 25 °C.

$$pK_a + pK_b = 14.00 \qquad \text{(25 °C)} \qquad (9.16)$$

■ **PRACTICE EXERCISE 19** Hydrocyanic acid, HCN, is a very weak acid with a pK_a of 9.2 at 25 °C. Calculate the pK_b of its conjugate base, CN^-. Write the equation for the chemical equilibrium in which CN^- acts as a Brønsted base in water, and then write the equation for K_b.

9.9 BUFFERS

The pH of a solution can be held relatively constant if it contains a buffer—a weak base and its conjugate acid.

The addition of only one drop of concentrated hydrochloric acid to a liter of pure water drops the pH of the system from 7 to 4, a change of three units in pH but a 10^3 or 1000-fold change in acidity. If a change of this size occurred to your bloodstream, you would die. The pH of blood can't be allowed to change by more than 0.2 to 0.3 pH units from its normal pH of 7.35. All higher animal species are similarly sensitive. Interaction 9.2 tells of the threat posed by *acid rain* to the pH of lake water and how this affects fish.

■ One of the rules of logarithms is

$\log (a \times b) = \log a + \log b$

■ We could also say, $pK_w = -\log K_w$, so at 25 °C, $pK_w = 14.00$.

■ Hydrocyanic acid, HCN(*aq*), is a solution of hydrogen cyanide, HCN(*g*), in water.

■ $\dfrac{10^{-4}}{10^{-7}} = 1000$

INTERACTION 9.2
ACID RAIN

Enormous quantities of sulfur dioxide are generated worldwide from the combustion of coal and oil, which contain relatively small quantities of sulfur compounds. This sulfur becomes oxidized to gaseous SO_2 as the fuels burn. Although the sulfur content of a fuel seldom exceeds 3%, often much less, the vast tonnage of fuels consumed worldwide annually release hundreds of millions of tons of SO_2 per year into the atmosphere. It is a major contributor to "acid rain."

Sulfur dioxide dissolves in water by forming hydrates, $SO_2 \cdot nH_2O$, where n varies with concentration, temperature, and pH. The hydrates are in equilibrium with some hydronium ion and hydrogen sulfite ion, HSO_3^-, whose presence has long been explained simply in terms of $H_2SO_3(aq)$, sulfurous acid. Actual molecules of this species—H_2SO_3—have never been detected in or out of water, however. Nevertheless, for convenience in writing chemical equations, the formula H_2SO_3 is widely used for the solute in aqueous sulfur dioxide. It is the first ionization of sulfurous acid that generates virtually all of the hydrogen ion that this acid can produce in water.

$$H_2SO_3(aq) \rightleftharpoons H^+(aq) + HSO_3^-(aq)$$
$$pK_a = 1.92 \ (25\ °C)$$

The relatively low value of pK_a makes sulfurous acid a moderate acid. Thus when rain washes gaseous SO_2 from the atmosphere, the rainwater is acidic. Moreover, both oxygen and ozone (O_3) in smog convert some SO_2 to SO_3, particularly in sunlight when fine dust is present. When SO_3 reacts with water, sulfuric acid, a strong acid, forms. It also contributes to the acidity of rain where air pollution occurs.

Nitrogen Dioxide Is Another Major Air Pollutant That Contributes to Acid Rain As you know, nitrogen and oxygen are very stable toward each other *at ordinary temperatures and pressures*. When fuels are burned in vehicles, however, high temperatures and higher pressures cause some reaction to occur between nitrogen and oxygen to give nitrogen monoxide.

$$N_2(g) + O_2(g) \longrightarrow 2NO(g)$$

When the vehicle exhaust leaves and encounters a much cooler atmosphere, nitrogen monoxide reacts with oxygen to give nitrogen dioxide.

$$2NO(g) + O_2(g) \longrightarrow 2NO_2(g)$$

Nitrogen dioxide is responsible for the reddishness of smog. In water, NO_2 reacts to give two acids—HNO_3, a strong acid, and HNO_2, a weak acid.

$$2NO_2(g) + H_2O \longrightarrow HNO_3(aq) + HNO_2(aq)$$

Thus, oxides of both nitrogen and sulfur are chiefly responsible for **acid rain.** Rain as acidic as lemon juice (pH 2.1) was observed in 1964 in the northeastern section of the United States, and rain as acidic as vinegar (pH 2.4) fell at Pitlochry, Scotland, in 1974.

A Better Term for Acid Rain Is *Acid Deposition* Dry dust particles settling on buildings, metals, and soil also carry acidic materials adhering to their surfaces. Thus *acid deposition* refers to all means whereby acids enter the earth's system. Acid deposition is an acute problem in regions downwind from major users of sulfur-containing fuels. Southern parts of the Scandinavian peninsula receive acid deposition from Germany's Ruhr Valley, the English Midlands, and countries of eastern Europe. Parts of southern Canada and the northern United States get acid deposition from the great industrial belt curving from Boston to Chicago. Industrial areas of eastern Europe have released enormous loads of sulfur and nitrogen oxides into their atmospheres.

Acid deposition affects lakes, soil, vegetation, and building stone. It makes bodies of water too acidic for much aquatic life. Because CO_2 from the air is naturally present in water, the pH of fresh water in equilibrium with air is about 5.7. Below a pH of 5.5, the hatchlings of most game fish are unable to live.

■ Emergency room personnel at hospitals must be extremely well versed in recognizing signs of acidosis or alkalosis.

Acidosis and Alkalosis Are Life Threatening So critical is the maintenance of the pH of the blood that a special vocabulary exists to describe small shifts away from it. If the pH becomes lower, which means that the acidity of the blood is increasing, the condition is called **acidosis.** It's a feature of untreated diabetes and emphysema, for example.

If the pH of the blood increases, which means that the blood is tending to become more basic or alkaline, the condition is called **alkalosis.** An overdose of bicar-

Dissolved acids leach calcium and magnesium ions from the soil, which adversely affects vegetation and forests, because these ions are essential nutrients. Their loss also diminishes the ability of the soil to neutralize further acid deposition. Severe damage has occurred to many mature forests in or near heavily industrialized nations. In the Hubbard Brook Experimental Forest of New Hampshire the organic part of the soil lost 50% of its calcium content in the 20-year period ending in 1996. Since 1987, the forest has stopped growing, and calcium depletion may be the cause.

Acid deposition is corrosive to exposed metals such as railroad rails, vehicles, and machinery, as well as to stone building materials (Figure 1). Limestone and marble are particularly sensitive, because they are chiefly calcium carbonate, and carbonates are dissolved by acids.

$$CaCO_3(s) + 2H^+(aq) \longrightarrow Ca^{2+}(aq) + CO_2(g) + H_2O$$

Several major cathedrals in Europe need constant repair because of the attack of deposited acids on their limestone and marble.

No Easy Alternatives Exist to Sulfur-Containing Fuels Less reliance on sulfur-containing coal and oil might be thought to be a solution to the acid deposition problem. No doubt it would help considerably. The alternatives to coal and oil are either a drastic cutback in energy consumption made possible by a simpler, less consuming life style, or a switch to a heavier reliance on nuclear power, or new technology not yet tested on a huge scale. Nuclear power bears its own pollution ills and it presently is costlier in every way than power obtained from the burning of coal or oil. Meanwhile, as coal and oil are used, the removal of most of the SO_2 from smokestack gases is possible. SO_2, for example, is absorbed by wet calcium hydroxide by the following reaction.

$$SO_2(g) + Ca(OH)_2(s) \longrightarrow CaSO_3(s) + H_2O$$

FIGURE 1 Marble is calcium carbonate, which is readily attacked by acids. These marble columns in New York City have suffered extensive damage from acid deposition.

Not all SO_2 is removed, however, and given the enormous quantities of coal and oil burned worldwide, emissions of SO_2 still occur. In a *technological* sense, the problem is controllable. The remaining issues concerning emissions of the sulfur oxides are mostly personal, political, economic, and diplomatic.

Some References

1. V. A. Mohnen, "The Challenge of Acid Rain," *Scientific American,* August 1988, page 30.

2. L. O. Hedin and G. E. Likens, "Atmospheric Dust and Acid Rain," *Scientific American,* December 1996, page 88.

bonate, exposure to the low partial pressure of oxygen at high altitudes, or prolonged hysteria can cause alkalosis.

In their more advanced stages, acidosis and alkalosis are medical emergencies because they interfere with the smooth working of *respiration,* the physical and chemical apparatus that brings oxygen in, uses it, and removes waste carbon dioxide. We will learn in general terms in this section how the body controls the pH of its fluids.

■ Every form of life is very sensitive to slight changes of pH in internal fluids.

Buffers Prevent Serious Changes in pH   Certain combinations of solutes, called **buffers,** keep changes in pH to a minimum when strong acids or bases are added to an aqueous solution. One part of the buffer system is a base and neutralizes H^+, and the other part is the base's conjugate acid that neutralizes OH^-. Fluids that contain buffers are said to be *buffered* against the changes in pH that H^+ or OH^- ions otherwise cause.

The blood and other body fluids include buffers, and much of the body's work in maintaining its acid–base status depends on them. Both acidosis and alkalosis are greatly restrained and sometimes totally prevented by buffers. Let's first see what they are and how buffers work, and then we can study them quantitatively in a later section.

The Phosphate Buffer Is Important within Cells   The principal buffer at work inside cells is called the **phosphate buffer.** It consists of the pair of ions, HPO_4^{2-} and $H_2PO_4^-$, the monohydrogen and the dihydrogen phosphate ions. Notice that $H_2PO_4^-$ is the conjugate acid of HPO_4^{2-}, so $H_2PO_4^-$ is the member of this pair that is better able to neutralize base. Any added OH^- is neutralized by $H_2PO_4^-$, and this keeps the pH from increasing.

$$H_2PO_4^-(aq) + OH^-(aq) \longrightarrow HPO_4^{2-}(aq) + H_2O$$

The proton acceptor or base in the phosphate buffer is the conjugate base of $H_2PO_4^-$, the HPO_4^{2-} ion. It can neutralize H^+ and so keep the pH from decreasing.

$$HPO_4^{2-}(aq) + H^+(aq) \longrightarrow H_2PO_4^-(aq)$$

The Carbonate Buffer Is Important in the Blood   The principal buffer in blood is called the **carbonate buffer.** Traditionally, it has been described as the conjugate pair, H_2CO_3 and HCO_3^-, carbonic acid and the bicarbonate ion. Actually, the carbonic acid in blood is almost entirely in the form of $CO_2(aq)$. For every molecule of $H_2CO_3(aq)$, there are 400 molecules of $CO_2(aq)$. $CO_2(aq)$, however, is able to neutralize hydroxide ion *directly* by the following reaction.

$$CO_2(aq) + OH^-(aq) \longrightarrow HCO_3^-(aq) \tag{9.17}$$

Actually (and importantly), this is really the forward step in an equilibrium.

$$CO_2(aq) + OH^-(aq) \rightleftharpoons HCO_3^-(aq) \tag{9.18}$$

The equilibration of $CO_2(aq)$ and $OH^-(aq)$ with $HCO_3^-(aq)$ is catalyzed by what is one of the most rapidly acting enzymes in the body, *carbonic anhydrase.* Because of this enzyme's work, we can use $CO_2(aq)$ as a stand-in for $H_2CO_3(aq)$ in discussing the carbonate buffer, even though $CO_2(aq)$ is not an acid in the Brønsted sense.

The $CO_2(aq)$ of the blood's carbonate buffer can neutralize OH^-, as we just said, and thus prevent an increase in pH. If there should be some metabolic or respiratory problem that increases the blood's OH^- level, this OH^- is neutralized by $CO_2(aq)$ (Equation 9.17) and alkalosis is prevented.

The bicarbonate ion is the base of the blood's carbonate buffer. If in a particular metabolic or respiratory situation the blood's level of H^+ ion increases, the H^+ is neutralized by HCO_3^- and acidosis is prevented.

$$HCO_3^-(aq) + H^+(aq) \longrightarrow CO_2(aq) + H_2O \tag{9.19}$$

The Ability to Breathe Out CO_2 Is Essential to the Control of Acidosis   When acid is neutralized by the carbonate buffer (Equation 9.19), the blood's level of $CO_2(aq)$ increases. However, when the blood moves through the capillaries in the lungs,

■ In blood, $[HCO_3^-]$ is normally about 24 mmol/L (about 1.5 g/L).

■ The concentration of CO_2 in blood is normally about 1.2 mmol/L (about 0.05 g/L).

■ The "total CO_2" concentration of blood is $[HCO_3^-] + [CO_2]$ and is normally about 26 mmol/L.

gaseous CO_2 is released from dissolved CO_2 and breathed out. Because the gas leaves, *we cannot write this change as an equilibrium.*

$$CO_2(aq) \longrightarrow CO_2(g) \qquad (9.20)$$

The loss of one molecule of $CO_2(g)$ by this change means that the H^+ ion neutralized by the buffer (Equation 9.19) is now *permanently* neutralized. All these steps are summarized in Figure 9.12, where you can see that the H^+ to be neutralized ends up in molecule of water. The ability of this water molecule to form finally depends on the loss of the CO_2 molecule from the body. Everything is beautifully coordinated.

The blood thus brings to bear *two* mechanisms that rapidly handle an influx of H^+. It neutralizes H^+ by the work of the carbonate buffer, and it uses a physical process, *ventilation,* to make this neutralization permanent. **Ventilation** is the circulation of air into and out of the lungs. One control over ventilation is a site in the brain called the *respiratory center,* which monitors the $CO_2(aq)$ in the blood. When the CO_2 level increases, the respiratory center instructs the breathing apparatus to breathe more rapidly and deeply, a response called **hyperventilation.** This response increases the rate at which CO_2 is exhaled, and the resulting permanent loss of CO_2 shifts all the carbonate equilibria in their acid neutralizing directions.

■ A third mechanism involves the kidneys, which remove H^+ from blood and resupply HCO_3^-. But this work takes hours to days.

One of many lessons we can draw from our discussion is that anything that interferes with the loss of CO_2 from the body inhibits the neutralization of H^+ ion and tends to cause acidosis. **Hypoventilation,** or slow and shallow breathing, is one such interference. People with emphysema, for example, hypoventilate because they cannot help it. They are unable to breathe deeply enough. As a result they struggle with acidosis as well, because they have a problem with getting rid of CO_2 and so cannot neutralize H^+ as well as they should. Any other cause of involuntary hypoventilation that renders the body unable to breathe out CO_2, like asthma, pneumonia, or overdoses of narcotics or barbiturates, also threatens the system with acidosis.

■ This retention of CO_2 because of hypoventilation is called "the retention of acid" because CO_2, one way or another, neutralizes OH^-.

9.10 SOME QUANTITATIVE ASPECTS OF BUFFERS

Buffers help to keep a pH constant, but not necessarily at pH 7.

In this section we move our study of buffers to a quantitative level. Only by doing this will we be able to appreciate how significant the loss of CO_2 is to respiration.

■ Our backgrounds in acid ionization constants and the concept of pK_a will now serve us.

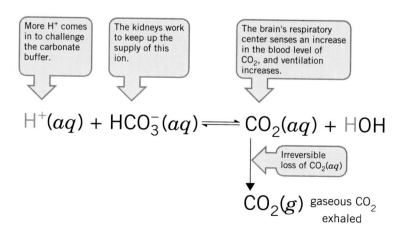

More H^+ comes in to challenge the carbonate buffer.

The kidneys work to keep up the supply of this ion.

The brain's respiratory center senses an increase in the blood level of CO_2, and ventilation increases.

$$H^+(aq) + HCO_3^-(aq) \rightleftharpoons CO_2(aq) + HOH$$

Irreversible loss of $CO_2(aq)$

$$CO_2(g) \quad \text{gaseous } CO_2 \text{ exhaled}$$

FIGURE 9.12 The irreversible neutralization of H^+ through the loss of CO_2 is one way the carbonate buffer system handles acidosis. It is the last step, the change of dissolved CO_2 into gaseous CO_2, which is exhaled, that draws all the equilibria to the right and makes H^+ "disappear" into H_2O.

We'll first have to explore the relationships among the pH of a buffered solution, the solute concentrations, and the relative acid–base strengths of the buffer components.

The pH Values of Buffered Solutions Can Be Estimated from Ionization Constants and Buffer Concentrations   To make the discussion general, we assume that the buffer is made up of a weak acid, H*A*, together with one its group IA salts, like Na*A*. A group IA salt, like the sodium salt, is preferred because we want a very soluble salt, one that is 100% ionized, and we want the cation to be strictly a spectator ion.

The H*A*/*A*$^-$ type of buffer system could involve any one of a number of weak acids of widely varying K_a values. So we cannot expect just one weak acid to work for the buffering of all ranges of pH. Our next goal, then, is an equation to tell us at what pH a specific H*A*/*A*$^-$ buffer system will operate. We work with Equation 9.11, which defines K_a for a weak acid, H*A*.

$$K_a = \frac{[H^+][A^-]}{[HA]} \qquad \text{(This was Equation 9.11.)}$$

Because we want to know $[H^+]$ and then pH, we rearrange this equation to give us $[H^+]$. (We can find pH after we find $[H^+]$.)

$$[H^+] = K_a \times \frac{[HA]}{[A^-]} \qquad (9.21)$$

All molar concentrations in Equations 9.11 or 9.21, remember, are the concentrations *at equilibrium* after the solution has been prepared. Strictly speaking, the terms [H*A*] and [*A*$^-$] do not mean the *initial* concentrations of solutes used to prepare the solution. To keep this point before us, we should write [H*A*] and [*A*$^-$] as [H*A*]$_{eq}$ and [*A*$^-$]$_{eq}$ instead. What makes life actually simpler when we deal with buffers, however, is that we can ignore the "strictly speaking" and safely make some neat simplifications. We can actually substitute data that are easily known, namely, the *initial* values of molarities of H*A* and *A*$^-$ for the equilibrium values. Initial values are easier to obtain because we get them directly from the moles of solutes that we use to make the buffer. We do not have to carry out more sophisticated chemical analyses to measure equilibrium values of [H*A*] and [*A*$^-$]. Here is why the simplifications work.

Because H*A* is a weak acid, little is ionized at equilibrium, even if H*A* were the only solute. But when its anion, *A*$^-$, is also present, the ionization of H*A* is suppressed even more. It's the common ion effect. The presence of *A*$^-$ from the salt puts a stress on the following equilibrium and keeps it shifted to the left, in favor of unionized H*A*.

$$HA(aq)_{eq} \rightleftharpoons H^+(aq)_{eq} + A^-(aq)_{eq}$$

The result is that the value of $[HA(aq)]_{eq}$ is essentially identical to that of $[HA(aq)]_{initial}$. Thus our first simplification is

$$[HA(aq)]_{eq} = [HA(aq)]_{\substack{\text{from initial} \\ \text{concentration} \\ \text{of the acid}}} = [\text{acid}]$$

Thus, we'll now let the symbol [acid] represent the *initial* molar concentration of the weak acid in the buffer.

To simplify $[A^-(aq)]_{eq}$, we note that not much $A^-(aq)_{eq}$ is supplied by the ionization of the weak acid. ("Weak" implies this.) Nearly 100% of the $A^-(aq)_{eq}$ is supplied, instead, by the salt, because the ionization of the weak acid is suppressed. Little happens to change the value of $[A^-(aq)]$ *as initially supplied by the salt*. We

■ Our simplifications will be

$$[HA]_{eq} = [HA]_{init}$$
$$[A^-]_{eq} = [A^-]_{init}$$

■ When [*A*$^-$] is high, then this equilibrium shifts to form H*A* in accordance with Le Châtelier's principle.

■ For all practical purposes, the *only* source of *A*$^-$ is the salt, Na*A*, used to prepare the buffer.

conclude that the value of $[A^-(aq)]_{eq}$ is essentially the same as $[A^-(aq)]_{initial}$. So our second simplification is

$$[A^-(aq)]_{eq} = [A^-(aq)]_{\substack{\text{from initial} \\ \text{concentration} \\ \text{of the salt}}} = [\text{anion}]$$

Thus, we may now let the symbol [anion] stand for the *initial* molar concentration of the other component of the buffer, the one supplied by the salt.

We can now substitute the new terms, [acid] and [anion], into Equation 9.21 for [HA] and [A^-], respectively. For a buffer system made of the HA/A^- pair, we have the following equation for [H^+].

For HA/A^- buffers:

$$[H^+] = K_a \times \frac{[\text{acid}]}{[\text{anion}]} \qquad (9.22)$$

Let's see how we can use what we have learned to find [H^+] and, from it, the pH of a buffer solution.

EXAMPLE 9.9   Calculating the pH of a Buffered Solution

To study the effect of a weakly acidic medium on the rate of growth of a species of bacteria, a biochemist prepared an "acetate buffer," a solution containing both acetic acid and sodium acetate. The respective concentrations were $0.11\,M$ $NaC_2H_3O_2$ and $0.090\,M$ $HC_2H_3O_2$. What is the pH of this solution?

ANALYSIS Equation 9.11 defines K_a, an equation you should now know. It is the only equation you really need to remember to work buffer problems. The rearranged form, Equation 9.22, is derived from it. But we'll use Equation 9.11 to emphasize how the simplifications work.

SOLUTION We note first, from Table 9.2, that K_a for acetic acid is 1.8×10^{-5}. We have

$$K_a = \frac{[H^+][A^-]}{[HA]} \qquad \text{(This is Equation 9.11.)}$$

Here is where we remember the substitutions that we may make when working with a buffer system, namely, [anion] for [A^-] and [acid] for [HA].

$$K_a = \frac{[H^+][A^-]}{[HA]} = \frac{[H^+][\text{anion}]}{[\text{acid}]}$$

We can use this equation to calculate [H^+].

$$[H^+] = K_a \times \frac{[\text{acid}]}{[\text{anion}]} = 1.8 \times 10^{-5} \times \frac{0.090}{0.11} \text{ mol/L}$$

$$= 1.5 \times 10^{-5} \text{ mol/L}$$

Finally we calculate the pH.

$$pH = -\log [H^+] = -\log (1.5 \times 10^{-5})$$

$$pH = 4.82$$

Thus the pH of this acetate buffer solution is 4.82.

■ The "given" are

$[\text{anion}] = 0.11 \text{ mol/L}$

$[\text{acid}] = 0.090 \text{ mol/L}$

$K_a = 1.8 \times 10^{-5}$

■ **PRACTICE EXERCISE 20**   A buffer solution was prepared using 0.085 M formic acid, $HCHO_2$, and sodium formate, $NaCHO_2$, dissolved in the same solution at a concentration of 0.12 mol/L. Calculate the pH of this solution. For formic acid, $K_a = 1.8 \times 10^{-4}$.

The pH of a Buffer Solution Can Also Be Found by the Henderson-Hasselbalch Equation   We can convert Equation 9.22 into a form that involves pH directly instead of $[H^+]$. Thus, if we take the logarithms of both sides of Equation 9.22, and then multiply every resulting term by -1, we get

$$-\log [H^+] = -\log K_a - \log \frac{[acid]}{[anion]}$$

■ $-\log [H^+] = pH$

$-\log K_a = pK_a$

We can recognize expressions for pH and pK_a in this, so we can write

$$pH = pK_a - \log \frac{[acid]}{[anion]}$$

If we note that

■ Another rule of logarithms:

$-\log a/b = +\log b/a$

$$-\log \frac{[acid]}{[anion]} = +\log \frac{[anion]}{[acid]}$$

We obtain

$$pH = pK_a + \log \frac{[anion]}{[acid]} \qquad (9.23)$$

Equation 9.23 is the **Henderson-Hasselbalch equation.**[3] Let's rework Example 9.9 using it.

EXAMPLE 9.10   Calculating the pH of a Buffered Solution

Calculate the pH of the buffer solution described in Example 9.9.

ANALYSIS   This is a problem for which the Henderson-Hasselbalch equation may be used.

■ $pK_a = -\log K_a$

SOLUTION   We note first that for acetic acid $K_a = 1.8 \times 10^{-5}$, so $pK_a = 4.74$. In the buffer of Example 9.9,

$$[anion] = 0.11 \text{ mol/L}$$

$$[acid] = 0.090 \text{ mol/L}.$$

Substituting these values into the Henderson-Hasselbalch equation, we get

---

[3] Some references give the Henderson-Hasselbalch equation as

$$pH = pK_a + \log \frac{[salt]}{[acid]}$$

In other words, [salt] is used instead of [anion], as though the two were always identical in value. They are identical only when the cation of the salt is of the form M^+ (e.g., Na^+ or K^+) so that each formula unit of the salt furnishes only *one* anion. But when the cation is of the form M^{2+} (e.g., Ca^{2+}), then *two* anions are released by the dissociation of only one formula unit of the salt. With such salts the value of [anion] is twice the value of [salt]. It is safest to stick with the form of Equation 9.23.

$$pH = pK_a + \log \frac{(0.11)}{(0.090)}$$

$$= 4.74 + \log 1.2$$
$$= 4.74 + 0.079$$

$$pH = 4.82 \qquad \text{(the same as calculated in Example 9.9)}$$

■ PRACTICE EXERCISE 21   Calculate the pH of a buffered solution made up to be 0.016 M sodium acetate and 0.12 M acetic acid.

Buffers Hold a pH Steady, but Not Necessarily at pH 7   One important point about buffers is the distinction between keeping a solution at a particular pH and keeping it neutral—at a pH of 7. Although it is certainly possible to prepare a buffer to work at pH 7, buffers can be made that will work at any pH value throughout the pH scale.

The Ratio [Anion]/[Acid] Dominates the pH Only after the Buffer Pair Is Chosen   The Henderson-Hasselbalch equation shows that two factors govern the pH of a buffered solution, namely, the pK_a of the weak acid in the buffer pair and the ratio of molarities, [anion]/[acid]. When this ratio is 1, the term drops out because the logarithm of 1 is 0. Now the pH of the buffer solution is simply the pK_a of the acid. Therefore, to decide what weak acid and salt should be used to prepare a buffer, we look for a weak acid with a pK_a as close to the desired pH as possible. Then we experimentally select values of [anion] and [acid] so as to adjust the ratio of molarities, [anion]/[acid], and nudge the buffer's working pH to the value we want.

■ When [anion] = [acid],

$$\log \frac{[\text{anion}]}{[\text{acid}]} = \log \frac{1}{1} = 0$$

So then pH = pK_a + 0.

Of course, when we work with the buffered solutions found in a living system, we have to take what nature gives us. And nature gives us carbonic acid [or the stand-in, $CO_2(aq)$] as the acid component of the chief blood buffer. So now the second factor in the Henderson-Hasselbalch equation, the ratio of molarities, becomes decisive for determining the exact pH of the buffer.

The second factor, as we said, is the *ratio* [anion]/[acid]. The pH of a buffer solution depends entirely on this once the acid component with its pK_a has been picked. Notice particularly that it isn't the absolute values of [anion] and of [acid] *but the ratio of these values* that determines the pH of the buffer. You could get a 1:1 ratio, for example, with [anion] and [acid] both equal to 0.50 mol/L or both equal to 0.10 mol/L. The pH of the buffer would be the same. One implication of this involves giving blood at a blood bank (see Interaction 9.3).

■ PRACTICE EXERCISE 22   How is the pH of a buffered solution related by an equation to the pK_a of the weak acid in the buffer when

(a)   The ratio of [anion] to [acid] is 10 to 1?

(b)   The ratio of [anion] to [acid] is 1 to 10?

Buffers Minimize but Do Not Completely Prevent pH Changes   If we add a small amount of a strong base, say 0.010 mol of pure NaOH, to 1.0 L of pure water, the concentration of OH^- ion changes from 1.0×10^{-7} to 1.0×10^{-2} mol/L, a whopping 100,000-fold increase in $[OH^-]$. The pH changes from 7 to 12, a change of 5 pH units. Now let's suppose that we add 0.010 mol of NaOH to 1.0 L of a hypothetical buffer, one in which the weak acid component has a pK_a of 7.00 and in which the molar concentrations of both anion and acid are 0.10 mol/L. Because [anion] equals

ON DRINKING FLUIDS AFTER DONATING BLOOD

If you have ever donated blood (Figure 1) you know that afterwards you are asked to drink more fluids than usual to replace the lost volume. Fortunately, and for good reason, replacing the lost blood volume by *water* rather than a buffer solution does not seriously affect the pH of your blood. The "good reason" is the relationship of the Henderson–Hasselbalch equation.

$$pH = pK_a + \log \frac{[\text{anion}]}{[\text{acid}]}$$

The value of [anion] is itself a *ratio,* the number of moles of anion per liter of solution. Likewise, the value of [acid] is also a ratio, the number of moles of acid per liter of solution. If we change the volume of the solution *identically* for both anion and acid simply by adding water, the *ratio* of [anion] to [acid] in the equation must remain the same as before. The actual values of the concentrations, [anion] and [acid], do decrease because the blood is being diluted, but both concentrations decrease *by the same factor.* Thus the ratio of the two, [anion] to [acid], is unchanged by the dilution and so the pH is unchanged. There is a limit to how much dilution of the blood one's system can stand, of course, but your thirst mechanism normally switches off any desire to exceed the limit.

FIGURE 1   After giving blood at an American Red Cross blood bank, this donor will drink extra fluids without endangering her blood pH.

[acid], the pH of the buffer must equal the specified pK_a or 7.00. A Henderson-Hasselbalch calculation would show that the new pH is 7.09. When the buffer is present, in other words, the addition of 0.010 mol of NaOH changes the pH by only 0.09 unit, from 7.00 to 7.09. It can be calculated that this amounts to a 1.25-fold increase in OH^- concentration, a vastly smaller change than the 100,000-fold increase. The lesson is that although buffers cannot completely prevent changes in pH, they keep the changes very small.

Dissolved CO_2 Must Be Factored into a Henderson-Hasselbalch Treatment of the Blood's Carbonate Buffer   In dealing with the carbonate buffer in real life, the Henderson-Hasselbalch equation must be slightly modified. We have to use the formula of carbon dioxide, not the formula of its associated weak acid, H_2CO_3, and we must also change one symbol, namely, use pK' instead of pK_a. The modified equation is

$$pH = pK' + \log \frac{[\text{HCO}_3^-(aq)]}{[\text{CO}_2(aq)]} \tag{9.24}$$

We can't use the usual symbol, pK_a, for two reasons. First, the bloodstream is not at 25 °C but at body temperature, 37 °C. Second, the buffer's acid component is not H_2CO_3 (with a pK_a of 6.35 at 25 °C) but CO_2. So an *apparent acid ionization constant,* symbolized as K', and a corresponding apparent pK' are used for aqueous CO_2. The value of pK' that works best is 6.1, so when we are dealing with the carbonate buffer in the body, Equation 9.24 becomes

■ K_a values are usually assumed to be for 25 °C unless another temperature is specified, like body temperature.

$$pH = 6.1 + \log \frac{[\text{HCO}_3^-(aq)]}{[\text{CO}_2(aq)]} \tag{9.25}$$

Normally, in human arterial blood, $[HCO_3^-]$ equals 24 mmol/L, and $[CO_2]$ equals 1.2 mmol/L. These data lead to a calculated pH of human arterial blood of 7.4. Thus, with the use of Equation 9.25, we have

$$pH = 6.1 + \log \frac{(24 \text{ mmol/L})}{(1.2 \text{ mmol/L})}$$
$$= 7.4$$

Human arterial blood actually has an average pH of 7.35, so the calculated and the observed values agree well. Now let's see how wondrously well the carbonate buffer in blood can work.

Normal Exhaling of CO_2, Hyperventilation, and Resupply of HCO_3^- by the Kidneys All Work to Combat Acidosis   Let's suppose that the blood is challenged with a sudden influx of acid in the equivalent of 10 mmol of HCl per liter of blood. This would neutralize 10 mmol/L of HCO_3^- and so reduce its concentration from 24 mmol/L to 14 mmol/L. The HCO_3^-, of course, changes almost entirely to $CO_2(aq)$, so 10 mmol/L of new $CO_2(aq)$ appears in the blood. If this new CO_2 could not be removed by breathing, the level of $CO_2(aq)$ would increase by 10 mmol/L, from 1.2 mmol/L to 11.2 mmol/L. This would be fatal because, by Equation 9.25, the resulting pH of the blood would be 6.2, far, far too low to permit life to continue. Thus,

$$pH = 6.1 + \log \frac{(14 \text{ mmol/L})}{(11.2 \text{ mmol/L})}$$
$$= 6.2$$

However, essentially all the $CO_2(aq)$ leaves the blood in the lungs and is breathed out as $CO_2(g)$. Although the level of HCO_3^- remains 14 mmol/L, the level of $CO_2(aq)$ quickly drops back to 1.2 mmol/L. So the pH quickly bounces back up to 7.2.

$$pH = 6.1 + \log \frac{(14 \text{ mmol/L})}{(1.2 \text{ mmol/L})}$$
$$= 7.2$$

■ A pH of 7.2 corresponds to fairly severe acidosis.

Of course, a blood pH of 7.2 is still too low for health, but it doesn't cause death.

Fortunately, when needed, another mechanism provides further upward readjustment in the pH of the blood. The body, if healthy, responds quickly to a lowering of the blood pH by increasing the rate of breathing, by causing *hyperventilation.* This forces even more CO_2 out of the blood and into the exhaled air. It's quite common for hyperventilation to pull the level of $CO_2(aq)$ from 1.2 mmol/L down to 0.70 mmol/L, sometimes a bit lower. This would bring the pH of the blood in our example—the sudden influx of 10 mmol/L of acid—back to 7.4.

$$pH = 6.1 + \log \frac{(14 \text{ mmol/L})}{(0.7 \text{ mmol/L})}$$
$$= 7.4$$

Thus two mechanisms have protected the system against the otherwise lethal assault of an influx of 10 mmol/L of strong acid. The buffer system has neutralized the acid, and the respiratory system has, by removing CO_2, readjusted the ratio of $[HCO_3^-]/[CO_2]$. Neither of these two mechanisms could save the situation alone.

The Kidneys Replenish the HCO_3^- of the Blood   The situation just described is not actually back to normal in all respects. By normal metabolism, the system will continue to produce CO_2. This will increase the denominator in the ratio $[HCO_3^-]/[CO_2]$, and so the pH will gradually decline again. The system must, therefore, also raise the level of HCO_3^-, and this is done principally by the chemical work of the kidneys.

The kidneys can manufacture "new" HCO_3^- to replenish the base of the blood's carbonate buffer, and as they do this the kidneys can also export H^+. It's hard to live without healthy kidneys, so it's wise to avoid the overuse of drugs, including alcohol. We won't delve further into this, but you can see that the management of the pH of the blood for the sake of respiration involves an intricate interplay of chemical and physiological events at the molecular level of life. The smooth, harmonious working of these events in the healthy body is one of the grandest, most beautiful "vistas" in all of nature. Moreover, if your intended career is a field of medicine, particularly in a primary care area like nursing, or any aspect of sports medicine, you should master the concepts of this section. The author has spoken with many experienced nurses and physicians who agree.

■ In battling severe acidosis, the pH of the urine can go as low as 4.

9.11   ACID–BASE TITRATIONS

At the end point of an acid–base titration, the number of moles of H^+ should match the number of moles of H^+ acceptor.

One of the very common kinds of chemical analysis is to determine the concentration of an acid or a base. The purpose is to find the molarity of some whole solute, like moles of acetic acid per liter of solution, not just to measure the pH of the solution. Thus we have to make some distinctions between the kinds of acid species present.

The pH of a Solution Refers to $[H^+]$, Not to $[HA]$ The pH of a solution tells us indirectly the *acidity* of a solution, what the concentration of hydronium ions is. It does not, however, disclose the **neutralizing capacity** of the solution—its capacity to neutralize a strong base. The 1 mol of acetic acid in 1 L of $1\ M\ HC_2H_3O_2$ can neutralize 1 mol of NaOH, yet this solution has an actual quantity of H_3O^+ of only about 0.004 mol, the result of acetic acid being a weak acid.

Always remember that acids are classified as *weak* or *strong* according to their abilities to transfer a proton to one particular and very weak base, H_2O. When sodium hydroxide is added to an acid, however, we are adding a very strong base, OH^-. This base takes H^+ not only from H_3O^+ but also from $HC_2H_3O_2$. Thus the neutralizing capacity of $1\ M$ acetic acid is considerably greater than its concentration of hydronium ions.

The Titration of an Acid with a Base Gives Data from Which Concentrations Can Be Calculated   The procedure used to measure the total acid (or base) neutralizing capacity of a solution is called **titration**. It involves comparing the volume of a solution of unknown concentration to the volume of a *standard solution* that exactly neutralizes it. A **standard solution** is simply one whose concentration is accurately known.

The apparatus for titration is shown in Figure 9.13. A carefully measured volume of the solution of unknown acidity (or basicity) is placed in a beaker or a flask. A very small amount of an acid–base indicator, like phenolphthalein, is added. Then a *standard solution* of the neutralizing reagent is added through a stopcock, portion

FIGURE 9.13   The apparatus for titration. By manipulating the stopcock, the analyst controls the rate at which the solution in the buret is added to the flask below.

■ The careful measurement of the concentration of a standard solution is called **standardization.** We say that we standardize the solution.

by portion, from a special tube called a *buret*, marked in 1-mL and 0.1-mL divisions (Figure 9.13). The addition is continued until the **end point**—a change in color, caused by the acid–base indicator, which signals that the unknown has been exactly neutralized.

End Points Ideally Occur at Equivalence Points   With a carefully selected acid–base indicator, the color change in an acid–base titration occurs when all the available hydrogen ions have reacted with all the available proton acceptors. This point in a titration is called the **equivalence point.**

A well-chosen indicator is one whose color at the equivalence point is the same as it would be in a solution made up of the *salt* that forms in the titration (and in the same concentration). When this salt has an ion that hydrolyzes, the equivalence point cannot be at pH 7.00. For example, when one mole of acetic acid has been exactly neutralized by one mole of sodium hydroxide, exactly one mole of sodium acetate ($NaC_2H_3O_2$) has been made. Because the acetate ion hydrolyzes (but not the sodium ion), this salt gives a solution that is slightly basic to litmus, not a solution with a pH of 7.00. So it would be poor to pick an indicator that changes color over an acidic range. (Phenolphthalein works very well in this titration.) Whether or not the indicator has been well chosen, the analyst has little choice but to stop the titration when the indicator's color changes. This stopping point is called the **end point** of the titration. In a well-run titration, of course, the end point and the equivalence point coincide.

In Chapter 5 we studied how to do calculations involving volumes and concentrations of solutions with an emphasis on calculating volumes. Acid–base titrations, however, are usually done to determine concentrations, so we'll work through an example to see how molarities can be calculated from titration data.

■ To reach the equivalence point the moles of H^+ used must be the same as (must be *equivalent to*) the moles of proton acceptor present.

■ The pH of 1 M $NaC_2H_3O_2$ is about 9.4.

EXAMPLE 9.11   Calculating Molarities from Concentration Data

---

A student titrated 25.0 mL of sodium hydroxide solution with standard sulfuric acid. It took 13.4 mL of 0.0555 M H_2SO_4 to neutralize the sodium hydroxide in the solution. What was the molarity of the sodium hydroxide solution? The equation for the reaction is

$$2NaOH(aq) + H_2SO_4(aq) \longrightarrow Na_2SO_4(aq) + 2H_2O$$

ANALYSIS   Be sure to understand the goal first. We are to calculate the molarity of the NaOH, which means the ratio of the moles of NaOH to liters of NaOH solution. We were given (indirectly) the liters of the NaOH solution, because 25.0 mL equals 0.0250 L. To find the moles of NaOH we need two conversion factors, one involving the molarity of the acid, and the other involving the coefficients in the equation.

The molarity of the H_2SO_4 solution, 0.0555 M, gives us the following conversion factors.

$$\frac{0.0555 \text{ mol } H_2SO_4}{1000 \text{ mL } H_2SO_4 \text{ soln}} \quad \text{and} \quad \frac{1000 \text{ mL } H_2SO_4 \text{ soln}}{0.0555 \text{ mol } H_2SO_4}$$

The first of these will enable us to calculate the moles of H_2SO_4 in 13.4 mL of 0.0555 M H_2SO_4.

The balanced equation gives us the following conversion factors that relate moles of acid used to moles of NaOH required.

$$\frac{1 \text{ mol } H_2SO_4}{2 \text{ mol NaOH}} \quad \text{and} \quad \frac{2 \text{ mol NaOH}}{1 \text{ mol } H_2SO_4}$$

The second factor is the one we'll use to find the moles of NaOH that are equivalent to the moles of acid in 13.4 mL of 0.0555 M H_2SO_4.

SOLUTION We first take the given volume of the acid and convert it into the number of moles of H_2SO_4 that were taken.

$$13.4 \text{ mL } H_2SO_4 \text{ soln} \times \frac{0.0555 \text{ mol } H_2SO_4}{1000 \text{ mL } H_2SO_4 \text{ soln}} = 7.44 \times 10^{-4} \text{ mol } H_2SO_4$$

The number of moles of NaOH that are equivalent to 7.44×10^{-4} mol H_2SO_4 *in the given reaction* is found by

$$7.44 \times 10^{-4} \text{ mol } H_2SO_4 \times \frac{2 \text{ mol NaOH}}{1 \text{ mol } H_2SO_4} = 1.49 \times 10^{-3} \text{ mol NaOH}$$

Thus 0.00149 mol of NaOH was present in 25.0 mL or 0.0250 L of NaOH solution. To find the molarity of the NaOH solution, we take the following ratio of moles to liters:

$$\frac{0.00149 \text{ mol NaOH}}{0.0250 \text{ L NaOH soln}} = 0.0596 \ M \text{ NaOH}$$

Thus the concentration of the NaOH solution is 0.0596 M.

■ **PRACTICE EXERCISE 23**   If it takes 24.3 mL of 0.110 M HCl to neutralize 25.5 mL of freshly prepared sodium hydroxide solution, what is the molarity of the NaOH solution?

■ **PRACTICE EXERCISE 24**   If 20.0 mL of 0.125 M solution of NaOH exactly neutralized the sulfuric acid in 10.0 mL of H_2SO_4 solution, what was the molarity of the H_2SO_4 solution?

SUMMARY

Reaction kinetics Virtually all chemical reactions have an energy of activation. Reactant particles more frequently surmount this barrier—the reaction happens faster—the more concentrated they are and the more readily their collisions have the proper combined total collision energy. Raising the temperature of a reacting mixture increases the frequency of successful collisions (makes the rate of reaction faster). A catalyst, such as any enzyme, lowers the energy of activation without affecting the overall heat of reaction, so a catalyst also increases the rate of a reaction.

Equilibrium laws An equilibrium law exists for every chemical equilibrium. It is of the form

$$K_{eq} = \frac{[C]^c[D]^d}{[A]^a[B]^b}$$

when the equilibrium is of the type

$$aA + bB \rightleftharpoons cC + dD$$

The equilibrium constant, K_{eq}, which is different for each equilibrium, depends only on the temperature. If any stress other than a temperature change is placed on an equilibrium, such as the addition of a common ion, all concentrations adjust, but K_{eq} stays the same. A large value of K_{eq} means that the products are favored at equilibrium; a small value means that the reactants are favored.

Ion-product constant of water, K_w The ion product constant of water, K_w, equals $[H^+][OH^-]$. At 25 °C, $K_w = 1.0 \times 10^{-14}$. If acids or bases are added, the value of K_w stays the same but individual values of $[H^+]$ and $[OH^-]$ adjust.

The pH concept A simple way to express very low values of $[H^+]$ is by pH, where $[H^+] = 1 \times 10^{-pH}$, or pH $= -\log [H^+]$. When pH < 7, the solution is acidic. When pH > 7, the solution is basic. An analogous pOH concept exists: pOH $= -\log [OH^-]$. At 25 °C, pH + pOH = 14.00.

To measure the pH of a solution we use indicators, dyes whose colors change over a narrow range of pH, or we use a pH meter.

Acid ionization constants A modified equilibrium law defining the acid ionization constant, K_a, exists for weak acids. Letting HA represent the acid,

$$K_a = \frac{[H^+][A^-]}{[HA]}$$

for the equilibrium,

$$HA \rightleftharpoons H^+ + A^-$$

For weak acids, $K_a < 10^{-3}$. For moderate acids K_a is between 1 and 1×10^{-3}, and for strong acids $K_a > 1$.

Besides weak acids, like HA, most metal ions in water generate hydrogen ions by the ionization of a water molecule attracted to the metal ion (e.g., in the hydrated metal ion). The exceptions that do not make a solution acidic are the group IA and IIA cations (below beryllium). Ammonium salts also give acidic solutions.

Base ionization constants Bases weaker than OH^- or O^{2-} occur mostly as anions, the conjugate bases of weak acids. Ammonia is also a base weaker than OH^-. A base, B, ionizes according to the equilibrium,

$$B + H_2O \rightleftharpoons BH^+ + OH^-$$

The base ionization constant is then

$$K_b = \frac{[BH^+][OH^-]}{[B]}$$

Strong bases have large values of K_b. Weak bases have small values of K_b.

The reaction of a basic anion with water is called the hydrolysis of the anion, and a salt of such an anion with any group IA or IIA metal (except beryllium) gives a basic solution in water. The anions that do not hydrolyze are the conjugate bases of strong acids, like Cl^-, Br^-, I^-, NO_3^-, and SO_4^{2-}.

pK_a and pK_b concepts The negative logarithm of K_a is pK_a, and the negative logarithm of K_b is pK_b.

$$pK_a = -\log K_a$$
$$pK_b = -\log K_b$$

The weaker the acid the larger its pK_a. The weaker the base the larger its pK_b. $pK_a + pK_b = 14.00$ at 25 °C.

Buffers Solutions that contain something that can neutralize OH^- ion (such as a weak acid) and something else that can neutralize H^+ ion (such as the conjugate base of the same weak acid) are buffered against changes in pH when either additional base or acid is added.

The phosphate buffer, which is present in the fluids inside cells of the body, consists of HPO_4^{2-} (to neutralize H^+) and $H_2PO_4^-$ (to neutralize OH^-).

The normal pH of blood is 7.35. A decrease in the pH of blood is called acidosis and an increase is called alkalosis. Either condition interferes with respiration, and extreme cases (pH < 7 or pH > 8) are lethal.

The carbonate buffer, which is the chief buffer in blood, consists of HCO_3^- (to neutralize H^+) and dissolved CO_2 (to neutralize OH^-). Dissolved CO_2 or $CO_2(aq)$, reacts directly with OH^- to give bicarbonate ion. In the lungs, bicarbonate ion is changed back to (waste) CO_2, which is expelled. The ability to exhale CO_2 is essential to the prevention of acidosis.

When metabolism or some deficiency in respiration produces or retains H^+ at a rate faster than the blood buffer can neutralize it, the lungs try to remove CO_2 at a faster rate (hyperventilation). Overall, for each molecule of CO_2 exhaled, one proton is neutralized.

The pH of a buffer solution consisting of a weak acid, HA, and its conjugate base, A^-, can be calculated from the defining equation for K_a. In such a buffer at equilibrium, for all practical purposes, $[HA]_{eq} = [acid]_{init}$ and $[A^-]_{eq} = [anion]_{init}$. So the defining equation for K_a can be modified to express $[H^+]$ as

$$[H^+] = K_a \times \frac{[acid]}{[anion]}$$

The defining equation for pH lets us convert this equation to the Henderson-Hasselbalch equation:

$$pH = pK_a + \log \frac{[anion]}{[acid]}$$

The two factors that determine at what pH a buffer works are the pK_a of the weak acid component and the *ratio* [anion]/[acid]. As this ratio varies from $10:1$ to $1:1$ to $1:10$, the pH varies as $pK_a \pm 1$.

A buffer thus does not keep a solution necessarily at a pH of 7, but it holds a fairly constant pH when extra acid or base enter.

At body temperature the Henderson-Hasselbalch equation for the carbonate buffer in blood takes the following working form in which $[CO_2(aq)]$ appears in the place of $[H_2CO_3]$.

$$pH = 6.1 + \log \frac{[HCO_3^-]}{[CO_2(aq)]}$$

If an acid neutralizes some HCO_3^-, the ratio in the log term decreases too much unless the lungs simultaneously breathe out the extra CO_2 produced. If the lungs are able to work, some hyperventilation helps this process, and the pH of the blood stays quite close to 7.35. Over a longer period, the supply of HCO_3^- is replenished by the work of the kidneys.

Acid–Base titration The concentration of an acid or a base in water can be determined by titrating the unknown solution with a standard solution of what can neutralize it.

The indicator is selected to have its color change occur at whatever pH the final solution would have were it made from the salt that forms by the neutralization. The unknown concentration can be calculated from the volumes of the acid and base used, the molarity of the standard solution, and the coefficients in the equation for the specific neutralization reaction.

REVIEW EXERCISES

The answers to Review Exercises whose numbers are in color are found in Appendix E. The answers to the other Review Exercises are found in the Study Guide that accompanies this book. The more challenging questions are marked with asterisks.

Collision Theory and the Factors That Affect Reaction Rates

9.1 What is the name of the field of chemistry that deals with the rates of chemical reactions?

9.2 How is the *rate of reaction* defined in mathematical terms? Include the units.

9.3 What factor distinguishes a homogeneous from a heterogeneous reaction?

9.4 What are the main factors that govern the rate of a reaction?

9.5 The rate of a given reaction is generally most rapid at the beginning of the reaction, rather than later.
(a) What factor affecting reaction rates is at work to account for this?
(b) How do we use the kinetic theory to explain this?

9.6 When an increase in the rate of a reaction has been caused by an increase in the concentration of one of the reactants, which of the following factors has been changed? (Identify them by letter.)

A   The frequency of collisions
B   The heat of reaction
C   The energy of activation
D   The frequency of successful collisions

9.7 As a rule of thumb, how much of a temperature increase doubles or triples the rates of most reactions?

9.8 How do we explain the rate-increasing effect of an increase in temperature?

9.9 Explain how an increase in body temperature can lead to a strain on the heart.

9.10 How can we increase the frequency of all collisions in a reacting mixture without raising the temperature?

9.11 In what way, if any, does a catalyst affect the following factors of a chemical reaction?
(a) The frequency of collisions
(b) The heat of reaction
(c) The energy of activation
(d) The frequency of successful collisions

9.12 What is the general name for the catalysts found in living systems?

9.13 Name the two catalysts mentioned in this chapter that are involved in metabolism and describe what they do by means of equations.

9.14 A catalyst speeds up a reaction at a given temperature. In what other way can a catalyst affect the ease of the reaction?

9.15 Collision theory holds that something must happen for a reaction to occur between two particles. What must occur and why?

9.16 In the chemical industry many catalysts are prized because they save energy. How do they do that?

9.17 In what *two* ways does an increase in temperature affect a reaction rate?

9.18 Distinguish between *collision frequency* and *rate of reaction*.

9.19 Consider the concepts of E_{act} and heat of reaction.
(a) Which of the two is associated with *rate*?
(b) Which of the two is associated with whether the reaction is endothermic or exothermic?
(c) How can a very exothermic reaction be very *slow*?
(d) If an energy of activation for a reaction is zero, what can be said to describe the reaction rate?
(e) A catalyst affects which factor the more, E_{act} or heat of reaction?

9.20 In terms of what we visualize as happening when two molecules interact to form products, how do we explain the existence of an energy barrier to the reaction— an energy of activation?

9.21 Study the accompanying progress of reaction diagram for the conversion of carbon monoxide and oxygen to carbon dioxide, and then answer the questions. The equation for the reaction is

$$2CO(g) + O_2(g) \longrightarrow 2CO_2(g)$$

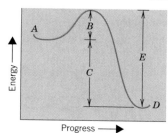

(a) What substance or substances occur at position A?
(b) What substance or substances occur at position D?
(c) Which letter labels the arrow that represents the heat of reaction?
(d) Which letter labels the arrow that stands for the energy of activation?
(e) Is this reaction endothermic or exothermic? How can you tell?
(f) Which letter labels the arrow that would correspond to the energy of activation if the reaction could go in reverse?

*9.22 Suppose that the following hypothetical reaction occurs.

$$A + B \longrightarrow C + D$$

Suppose further that this reaction is endothermic and that the energy of activation is numerically twice as large as the heat of reaction. Draw a progress of reaction diagram for this reaction, and draw and label arrows that correspond to the energy of activation and the heat of reaction.

9.23 The reaction of X and Y to form Z is exothermic. For every mole of Z produced, 10 kcal of heat are generated. The energy of activation is 3 kcal. Sketch the energy relationships on a progress of reaction diagram.

9.24 What did Guldburg and Waage discover about chemical equilibria?

*9.25 In setting up an equilibrium law for a system,
(a) What concentration units are assumed?
(b) How are the coefficients of the chemical equation used?
(c) The terms for which substances, reactants or products, appear in the denominator?
(d) Why must the temperature be specified?

9.26 Write the equilibrium laws for the following.
(a) $NH_3(aq) + H_2O \rightleftharpoons NH_4^+(aq) + OH^-(aq)$
(b) $2H_2(g) + O_2(g) \rightleftharpoons 2H_2O(g)$
(c) $N_2(g) + 3H_2(g) \rightleftharpoons 2NH_3(g)$

9.27 For the equilibrium in which ethylene reacts with water to give ethyl alcohol,

$$C_2H_4(g) + H_2O(g) \rightleftharpoons C_2H_5OH(g) \qquad K_{eq} = 8.3 \times 10^3$$

is the product favored or are the reactants? How can you tell?

Ion Product Constant of Water

9.28 Write the equation that defines K_w. How does it differ from the equilibrium law for water (written using H^+ and OH^-)?

9.29 What is the value of K_w at 25 °C?

9.30 At the temperature of the human body, 37 °C, the concentration of hydrogen ion in pure water is 1.56×10^{-7} mol/L. What is the value of K_w at 37 °C? Is this pure water acidic, basic, or neutral?

pH and pOH

9.31 What equation defines pH in exponential terms? In log terms?

9.32 The average pH of saliva is about 6.8. Is saliva acidic, neutral, or basic?

9.33 The pH of pancreatic juice, a digestive juice of the intestinal tract, is in the range of 7 to 8. Is pancreatic juice acidic, basic, or neutral?

9.34 What is the pH of $0.0001\ M$ $HCl(aq)$, assuming 100% ionization?

9.35 Explain why a pH of 7.00 corresponds to a neutral solution at 25 °C.

9.36 A certain brand of soft drink has a pH of 5.35. What is the concentration of hydrogen ion in moles per liter? Is the soft drink slightly acidic or basic?

*9.37 A solution of a monoprotic acid was prepared with a molar concentration of $0.010\ M$. Its pH was found to be 2.00. Is the acid a strong or a weak acid? Explain.

*9.38 The pH of a solution of a monoprotic acid was found to be 5.72, whereas its molar concentration was $0.0010\ M$. Is this acid a strong or a weak acid? Explain.

9.39 When a soil sample was stirred with pure water, the pOH of the water changed to 6.24. Did the soil produce an acidic or a basic reaction with the water?

Acid Ionization Constants

9.40 Write the equilibrium equation and the equation for K_a for the ionization of nitrous acid, HNO_2.

9.41 Write the equilibrium equation and the equation for K_a for the ionization of the hydrogen sulfite ion.

9.42 The K_a for the hydrogen sulfite ion, HSO_3^-, is 6.6×10^{-8} and for barbituric acid is 9.9×10^{-5}. Which is the stronger acid?

9.43 The K_a for the ammonium ion is 5.7×10^{-10} and for the hydrogen sulfide ion, HS^-, is 1×10^{-19}. Which is the stronger acid?

Base Ionization Constants

9.44   Write the equilibrium equation and the equation for K_b for the formate ion, HCO_2^-, acting as a base.

9.45   Write the equilibrium equation and the equation for K_b for the nitrite ion, NO_2^-, acting as a base.

9.46   Which is the stronger base, ammonia ($K_b = 1.8 \times 10^{-5}$) or the hypochlorite ion, OCl^- ($K_b = 3.3 \times 10^{-7}$)?

9.47   Explain why a solution of ammonium bromide tests slightly acidic.

9.48   Explain in your own words why a solution of sodium sulfide, Na_2S, is basic, not neutral.

*9.49   Predict whether each of the following solutions is acidic, basic, or neutral.
(a) KNO_3 (b) NH_4Br (c) $NaHCO_3$
(d) $FeCl_3$ (e) Li_2CO_3

9.50   Aspirin is a weak, monoprotic acid for which $K_a = 3.3 \times 10^{-4}$. Does a solution of the sodium salt of aspirin test acidic, basic, or neutral?

9.51   The K_b of the hydrogen sulfide ion, HS^-, is 1.1×10^{-7}. Does a solution of sodium hydrogen sulfide, NaSH, test acidic, basic, or neutral?

pK_a and pK_b

9.52   Calculate the pK_a values of the following acids.
(a) HF ($K_a = 6.8 \times 10^{-4}$), hydrofluoric acid
(b) HOCl ($K_a = 3.0 \times 10^{-8}$), hypochlorous acid

9.53   Calculate the pK_b values of the following bases.
(a) NO_2^- ($K_b = 1.4 \times 10^{-11}$), nitrite ion
(b) CHO_2^- ($K_b = 5.6 \times 10^{-11}$), formate ion

9.54   What are the K_b and pK_b values for the conjugate bases of the acids in Review Exercise 9.52?

9.55   What are the K_a and pK_a values for the conjugate acids of the bases given in Review Exercise 9.53?

9.56   Acid X has a pK_a of 5.68 and acid Y has a pK_a of 6.25. Which is the stronger acid? Which has the stronger conjugate base?

9.57   Base M has a pK_b of 8.02 and base N has a pK_b of 5.67. Which is the stronger base? Which has the weaker conjugate acid?

Buffers

9.58   In the study of the molecular basis of life, why is the study of buffers important?

9.59   What is acidosis? What is alkalosis?

9.60   In very general terms, why are both acidosis and alkalosis serious?

9.61   Following surgery, a patient experienced persistent vomiting and the pH of his blood became 7.49. (Normally it is 7.35.) Has the blood become more alkaline or more acidic? Is the patient experiencing acidosis or alkalosis?

9.62   A patient brought to the emergency room following an overdose of aspirin was found to have a pH of 7.18 for the blood. (Normally the pH of blood is 7.35.) Has the blood become more acidic or more basic? Is the condition acidosis or alkalosis?

9.63   What does it mean when we say that a solution of pH 7.45 is *buffered* at this pH?

9.64   What chemical species constitute the chief buffer inside cells?

9.65   Write the net ionic equation that shows how the phosphate buffer neutralizes OH^-.

9.66   How does the phosphate buffer neutralize acid? Write the net ionic equation.

9.67   What two chemical species make up the chief buffer in blood?

9.68   Write the net ionic equations that show how the chief buffer system in the blood works to neutralize hydroxide ion and hydrogen ion.

9.69   Explain in your own words, using equations as needed, how the loss of a molecule of CO_2 at the lungs permanently neutralizes a hydrogen ion.

9.70   What is meant by *ventilation* in connection with respiration? What is hyperventilation? Hypoventilation?

9.71   What does the respiratory center in the brain instruct the lungs to do when the level of CO_2 in blood increases? Why?

9.72   Why does the hypoventilation of someone with emphysema lead to acidosis?

9.73   A patient with untreated diabetes tends to hyperventilate. This is a natural response to what kind of change in the pH of the blood? Explain.

Buffer Calculations

9.74   When we use the defining equation for K_a, the concentrations in brackets refer to what condition, to the *initial* quantities of solutes used to prepare the solution or to the concentrations *after equilibrium* has been established?

9.75   Calculate the pH of a buffer solution consisting of 0.21 M acetic acid and 0.26 M sodium acetate.

9.76   Why is it better to use $[CO_2(aq)]$ instead of $[H_2CO_3(aq)]$ in the Henderson-Hasselbalch equation for the carbonate buffer in blood?

9.77   What is the apparent pK_a of the weak acid component of the carbonate buffer in blood?

9.78   Under normal pH conditions in human arterial blood, what values are usually assigned to the following? Be sure to include the correct units.
(a) $[HCO_3^-]$ (b) $[CO_2]$

*9.79   Suppose that normal human arterial blood is suddenly made to accept 11 mmol of HCl(aq) per liter of blood.

(a) What is the resulting pH of the blood if no CO_2 is allowed to escape?

(b) What is the resulting pH of the blood if 11 mmol/L of CO_2 can be quickly exhaled by normal processes?

(c) What further event will happen quickly to help bring the pH up still further toward normal?

(d) How does the body normally replace HCO_3^- lost by a battle with developing acidosis?

Acid–Base Titrations[4]

9.80 What does it mean to have a *standard* solution of, say, $HCl(aq)$?

9.81 When doing a titration, how does one know when the end point is reached?

9.82 What steps does an analyst take to ensure that the end point and the equivalence point in an acid–base titration occur together?

9.83 Give an example of a titration in which the equivalence point has a pH that is equal to 7.00. (Give a specific example of an acid and a base that, when titrated together, produce such a solution.)

9.84 Give a specific example of an acid and a base that, when titrated together, produce a solution with a pH greater than 7.00.

9.85 Individual aqueous solutions were prepared that contained the following substances. Calculate the molarity of each solution.

(a) 6.892 g of HCl in 500.0 mL of solution

(b) 8.090 g of HBr in 250.0 mL of solution

*9.86 If 20.00 g of a monoprotic acid in 100.0 mL of solution gives a concentration of 0.5000 M, what is the formula mass of the acid?

*9.87 A solution with a concentration of 0.2500 M could be made by dissolving 4.000 g of a base in 250.0 mL of solution. What is the formula mass of this base?

*9.88 How many grams of each solute are needed to prepare the following solutions?

(a) 1000 mL of 0.2000 M HCl

(b) 750.0 mL of 0.1025 M HNO_3

(c) 500.0 mL of 0.01125 M H_2SO_4

*9.89 A freshly prepared solution of sodium hydroxide was standardized with 0.1024 M H_2SO_4.

(a) If 19.46 mL of the base was neutralized by 21.28 mL of the acid, what was the molarity of the base?

(b) How many grams of NaOH were in each liter of this solution?

Explosions and Fat Digestion (Interaction 9.1)

9.90 Why does particle size matter in heterogeneous reactions?

[4] For all the calculations in the Review Exercises that follow, round atomic masses to their *second* decimal places before adding them to find formula masses.

Acid Rain (Interaction 9.2)

9.91 What are the chief gases that are responsible for acid rain and how do they get into the earth's atmosphere? (Write equations.) Why is *acid deposition* a better term?

9.92 Briefly describe some of the problems caused by acid rain in living systems and materials.

Post–Blood Donation Fluids (Interaction 9.3)

9.93 Consider two properties of any buffer, the pH it holds and the capacity of the buffer.

(a) We did not go into *buffer capacity* in the text. What do you suppose it refers to? Is it an extensive or an intensive property?

(b) Would diluting the blood by drinking much fluids (after donating blood, for example) challenge the pH of the blood or its buffer capacity?

Additional Exercises

9.94 Write the equilibrium law for the following reaction.
$$2CO(g) + O_2(g) \rightleftharpoons 2CO_2(g)$$

9.95 In the stratosphere, chlorine *atoms* destroy ozone molecules of the ozone shield by the following exothermic reaction.
$$Cl(g) + O_3(g) \longrightarrow ClO(g) + O_2(g)$$
It is generally true that the breaking of covalent bond *always* requires energy and that the formation of such a bond always releases energy.

(a) Which of the following progress of reaction diagrams, **A** or **B**, is more likely to be appropriate to describe this reaction? Explain.

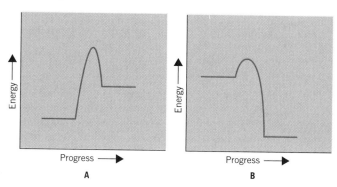

(b) Which covalent bond would cost *less* energy to break, the O—O bond in ozone or the Cl—O bond in ClO? Explain.

*9.96 Gaseous mononitrogen dioxide, which is the species that gives the reddish hue to smog, forms an equilibrium with gaseous dinitrogen tetroxide, which is colorless.

(a) Write the equilibrium equation, showing dinitrogen monoxide as the reactant.

(b) Write the equilibrium law for this reaction.

(c) At a certain temperature the equilibrium constant for this system is 10. Which species is favored at this temperature?

(d) As the temperature is raised, will the value of the equilibrium constant increase, decrease, or stay the same? Explain.

9.97 The higher the temperature, the higher the value of K_w. Why should there be this trend?

*9.98 "Heavy water" or deuterium oxide, D_2O, is used in nuclear power plants. It self-ionizes like water, and at 20 °C there is a concentration of D^+ ion of 3.0×10^{-8} mol/L. What is the value of K_w for heavy water at 20 °C?

*9.99 Calculate the pH of 0.115 M HCl.

9.100 What is the pOH of 0.01 M NaOH(aq), assuming 100% ionization? What is the pH of this solution?

9.101 The pH of a fruit juice was found to be 4.82. What is the value of $[H^+]$?

*9.102 A solution was prepared by dissolving 0.316 g of $Ba(OH)_2$ in a final volume of solution of 100 mL. What are the calculated values of pOH and pH for this solution, assuming that the $Ba(OH)_2$ is fully dissociated?

9.103 Write the equilibrium equation and the equation for K_a for the ionization of ammonia as an *acid*.

9.104 A solution of a monoprotic acid at a concentration of 3.9×10^{-4} M has a pH of 3.4. Is the acid a strong or a weak acid? Explain.

*9.105 Calculate the pH of 0.00045 M NaOH.

9.106 Write the equilibrium equation and the equation for K_b for the monohydrogen phosphate ion acting as a base.

*9.107 Predict whether each of the following solutions is acidic, neutral, or basic.
(a) Na_2SO_4 (b) K_2HPO_4 (c) K_3PO_4
(d) $Cr(NO_3)_3$ (e) $KC_2H_3O_2$

9.108 A patient entered the emergency room of a hospital after 3 weeks on a self-prescribed low-carbohydrate, high-fat diet and the regular use of the diuretic, acetazolamide. (A diuretic promotes the formation of urine and thus causes the loss of body fluid.) The pOH of the patient's blood was 6.82. What was the pH, and was the condition acidosis or alkalosis?

*9.109 In high-altitude sickness, the patient *involuntarily* overbreathes, and expels CO_2 from the body at a faster than normal rate. This results in an *increase* in the pH of the blood.
(a) Is this condition alkalosis or acidosis?
(b) Why should excessive loss of CO_2 result in an increase in the pH of the blood? (*Note:* Such a patient should be returned to lower elevations as soon as possible. It helps to rebreathe one's own air, as by breathing into a paper sack, because this helps the system retain CO_2.)

9.110 Why is it, in using the defining equation for pK_a in working buffer problems, we can substitute initial concentrations of acid and conjugate base for equilibrium concentrations?

*9.111 Calculate the pH of a buffer solution prepared as 0.015 M sodium acetate and 0.10 M acetic acid.

*9.112 What is the pH of a buffer solution made to be 0.19 M in dihydrogen phosphate ion and 0.22 M in its conjugate base, the monohydrogen phosphate ion?

*9.113 A 500 mL supply of a buffer solution is made of a weak acid, HA ($K_a = 6.21 \times 10^{-6}$), and its conjugate base, A^-, so that the solution contained 0.14 mol of the weak acid and 0.11 mol of the base.
(a) Calculate the pH of this solution.
(b) Calculate the pH of the solution after the addition of 0.020 mol of strong, monoprotic acid.
(c) Calculate the pH of a different sample of the original solution after the addition of 0.020 mol of NaOH.
(d) Calculate the pH that 500 mL of water would have had in parts (b) and (c) had no buffer been present.

*9.114 A buffer solution is prepared by dissolving 0.10 mol of solid sodium acetate in 1.0 L of water and as many moles of pure acetic acid as will make the final pH of the buffer equal to 5.00. Assuming that the amounts of solutes added to the 1.0 L of water do not affect the total volume in any significant way, how many moles of acetic acid must be used to make this buffer?

*9.115 A solution of NH_4Cl and NH_3 in water of roughly equal molarities is a buffer system.
(a) Which specific species in this buffer is able to neutralize OH^-? Write the net ionic equation.
(b) Which specific species in this buffer is able to neutralize H^+? Write the net ionic equation.
(c) The K_a of NH_4^+ is 5.7×10^{-10} (25 °C). Calculate the pH of a solution of this buffer in which $[NH_4Cl] = [NH_3]$.

9.116 At the equivalence point in a titration the pH is less than 7.00. Give a specific example of an acid and a base that, when titrated together, would produce this result.

*9.117 In standardizing a sodium carbonate solution, 22.48 mL of this solution was titrated to the end point with 19.82 mL of 0.1181 M HCl.
(a) Calculate the molarity of the Na_2CO_3 solution.
(b) How many grams of Na_2CO_3 does it contain per liter?

*9.118 A complex aqueous solution contains the following species, each at a concentration in excess of 0.01 M.

$$Na^+, K^+, NH_4^+, Cl^-, Br^-, \text{ and } NH_3$$

(a) Is the solution buffered?
(b) If so, what species constitute the buffer components?
(c) If the solution is buffered, write the net ionic equation for the reaction that occurs if a small amount of a strong base, like OH^-, is added.

Radioactivity and Nuclear Chemistry

Radiation streaming toward the Earth's magnetic poles causes spectacular displays of northern lights, seen here in Denali National Park, Alaska. Nearer home, radiation that we are about to study offers tools for diagnosing disease and halting cancer.

THIS CHAPTER IN CONTEXT

Almost anyone going into a health care career will eventually be in an environment where some radiation hazards exist. Moreover, any person wanting to be informed about radiation in the environment needs basic information. In this chapter we will study the various kinds of radiation, their beneficial uses, and their dangers.

10.1 ATOMIC RADIATION

Unstable atomic nuclei eject high-energy radiation as they change to more stable nuclei.

Some atomic nuclei are unstable, and isotopes having them are **radioactive,** meaning that they emit high-energy radiation. A radioactive isotope is called a **radionuclide.**

Unstable Nuclei Undergo Radioactive Decay   Radioactivity was discovered in 1896 when a French physicist, A. H. Becquerel (1852–1908), happened to store some well-wrapped photographic plates in a drawer that contained samples of uranium ore. The film became fogged, meaning that when developed the picture was like a photograph of fog.

Becquerel might have blamed the accident on faulty film, careless handling, or poor packaging had it not been that a mysterious radiation called X rays had recently been discovered by Wilhelm Roentgen (1845–1923), a German scientist. X rays were known to be able to penetrate the packaging of unexposed film and ruin it. What fogged Becquerel's film was a natural radiation that resembled X rays. It was soon found to be emitted by any compound of uranium as well as by uranium metal itself.

Several years later, two British scientists, Ernest Rutherford (1871–1937) and Frederick Soddy (1877–1956), explained radioactivity in terms of events inside unstable atomic nuclei. Such a nucleus undergoes a transformation, called a *disintegration,* and the overall process is called **radioactive decay.** A decaying nucleus may eject a tiny particle into space, or it may emit a powerful radiation like an X ray but called a gamma ray, or the nucleus may do both.

The nuclei that remain after decay are nearly always those of a different element, so decay usually converts one isotope into another, a phenomenon called **transmutation.** The natural sources of radiation on Earth emit one or more of three kinds: *alpha radiation, beta radiation,* or *gamma radiation.* We receive another called *cosmic radiation* from the sun and outer space. (See Interaction 10.1.)

Alpha Particles Are the Nuclei of Helium Atoms   One natural radiation is called **alpha radiation.** It consists of a stream of particles called *alpha particles* that move with a velocity almost one-tenth the velocity of light as they leave the atom. **Alpha particles** are clusters of two protons $(+)$ and two neutrons (n), so they are actually the nuclei of helium atoms (see Figure 10.1). They are the largest of the decay particles and have the most electrical charge, so when alpha particles travel in air, they soon collide with molecules of N_2 or O_2 and lose their energy (and charge). Alpha particles cannot penetrate even thin cardboard or the outer layer of dead cells on the skin, but they can cause a severe burn to the skin.

The most common isotope of uranium, uranium-238 or $^{238}_{92}U$, is an alpha emitter. Upon ejecting the alpha particle, a uranium-238 nucleus loses two protons, so the atomic number changes from 92 to 90. It also loses four units of mass number (two protons + two neutrons), so its mass number changes from 238 to 234. The result is that uranium-238 transmutes into an isotope of thorium, $^{234}_{90}Th$.

■ Becquerel shared the 1903 Nobel Prize in Physics with Pierre and Marie Curie.

■ Roentgen won the 1901 Nobel Prize in Physics, the first to be awarded.

■ Nobel Prizes in Chemistry were awarded to both Rutherford (1908) and Soddy (1921).

■ Isotopes have special symbols that give the mass number as a left superscript and the atomic number as a left subscript. Thus, in $^{238}_{92}U$, 238 is the mass number and 92 is the atomic number.

INTERACTION 10.1
COSMIC RAYS

Cosmic rays are streams of particles that enter our outer atmosphere from the sun and outer space. They consist mostly of high-energy protons plus some alpha and beta particles and the nuclei of the lower formula mass elements (up through number 26, iron).

Cosmic ray particles do not travel far when they enter the atmosphere because they quickly collide with the air's molecules and atoms. These collisions, however, generate all the subatomic particles, including neutrons and some others we haven't studied in this chapter. These secondary cosmic rays are what we are exposed to on the earth's surface. High-energy electromagnetic radiation is also produced, and this is the most dangerous component from a radiological health perspective.

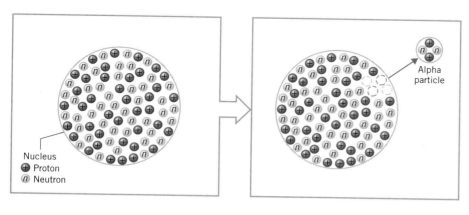

FIGURE 10.1 Emission of an alpha particle.

Beta Radiation Is a Stream of Electrons   Another natural radiation, **beta radiation,** consists of a stream of particles called **beta particles,** which are actually electrons. They are produced *within* the nucleus and then emitted (Figure 10.2). With less charge and a much smaller size, beta particles can penetrate matter, including air, more easily than alpha particles. Different sources emit beta particles with different energies, and those of the highest energy can reach internal organs from outside the body. Those of lower energy, however, are unable to penetrate the skin.

As a nucleus emits a beta particle, a neutron changes into a proton (see Figure 10.2). Thus there is no loss in mass number, but the atomic number *increases* by 1 unit because of the new proton. For example, thorium-234, $^{234}_{90}$Th, is a beta emitter, and when it ejects a beta particle it changes to an isotope of protactinium, $^{234}_{91}$Pa.

■ Losing one electron from the nucleus makes a proton out of a neutron, so no change in mass number occurs.

Gamma Radiation Often Accompanies Other Radiation   Atoms have different *nuclear* energy states just as they have different *electron* energy states. By ejecting small particles, unstable nuclei acquire a lower, more stable nuclear state. The en-

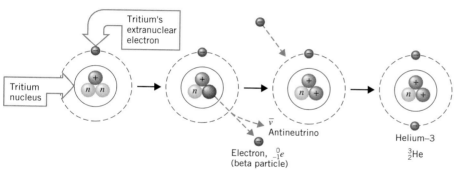

FIGURE 10.2 Emission of a beta particle.

TABLE 10.1   Radiation from Naturally Occurring Radionuclides

Radiation	Composition	Mass Number	Electrical Charge	Symbols
Alpha	Helium nuclei	4	2+	^4_2He, α
Beta	Electrons	0	1−	$^0_{-1}e$, β
Gamma	X-ray like	0	0	$^0_0\gamma$

ergy lost by the nucleus is usually all carried away by the moving particles, but often some photons of high-energy electromagnetic radiation are also emitted. This is called **gamma radiation,** and it is like X rays or ultraviolet rays, but with more energy. Like X rays, gamma radiation is quite penetrating and very dangerous. It easily travels through the entire body.

The composition and symbols of the three kinds of radiation studied so far are summarized in Table 10.1.

Nuclear Equations Must Display Balances of Both Mass Numbers and Atomic Numbers   In *chemical* reactions no changes in atomic nuclei occur. But nuclear reactions are nearly always accompanied by transmutations. Therefore **nuclear equations** are different from chemical equations in important ways. In particular, they must describe the changes in atomic numbers, mass numbers, and identities of radionuclides.

In nuclear equations the alpha particle is symbolized as ^4_2He, and although it is positively charged, the charge is omitted from the symbol. The particle soon picks up electrons, anyway, taking them from the matter through which it moves. The alpha particle thus becomes a neutral atom of helium.

The beta particle has the symbol $^0_{-1}e$ because its mass number is 0 and its charge is 1−. A photon of gamma radiation is symbolized simply by γ (or, sometimes, by $^0_0\gamma$). Because it is a photon, its mass number and charge are zero.

A nuclear equation is *balanced* when the sums of the mass numbers on either side of the arrow are equal and when the sums of the atomic numbers are equal. The alpha decay of uranium-238, for example, is represented by the following equation:

$$^{238}_{92}\text{U} \longrightarrow {}^{234}_{90}\text{Th} + {}^4_2\text{He}$$

You can see that the sums of the atomic numbers agree: 92 = 90 + 2. The sums of the mass numbers also agree: 238 = 234 + 4. The equation is balanced, but it's clearly a nuclear equation, not a chemical equation, because the atoms do not balance.

The beta decay of thorium-234, which also emits gamma radiation, is represented by the following nuclear equation.

$$^{234}_{90}\text{Th} \longrightarrow {}^{234}_{91}\text{Pa} + {}^0_{-1}e + \gamma$$

The sums of the atomic numbers agree: 90 = 91 + (−1). Likewise, the sums of the mass numbers agree, so the equation is balanced.

■ It's proper to think of the electron as having an atomic number of −1.

EXAMPLE 10.1   Balancing Nuclear Equations

Cesium-137, $^{137}_{55}\text{Cs}$, is one of the radioactive wastes that form in a nuclear power plant or an atomic bomb explosion. It decays by emitting both beta and gamma radiation. Write the nuclear equation for this decay.

ANALYSIS We start with an incomplete equation using the given information. Then we figure out any additional data we need.

SOLUTION The incomplete nuclear equation is

$$^{137}_{55}\text{Cs} \longrightarrow {}^{0}_{-1}e + {}^{0}_{0}\gamma + \underline{}$$

Mass number (A) goes here

Atomic symbol goes here

Atomic number (Z) goes here

We first have to figure out the atomic number, Z, so that we can find the atomic symbol for the product in the table inside the front cover. To calculate Z we remember that the atomic number (55) on the left side of the above equation must equal the sum of the atomic numbers on the right side.

$$55 = -1 + 0 + Z$$
$$Z = 56$$

From the periodic table element 56 is Ba (barium), but which isotope of Ba? Now we remember that the sums of the mass numbers on either side of the arrow must also be equal. Letting A represent the mass number of the barium isotope,

$$137 = 0 + 0 + A$$
$$A = 137$$

The balanced nuclear equation, therefore, is

$$^{137}_{55}\text{Cs} \longrightarrow {}^{0}_{-1}e + {}^{0}_{0}\gamma + {}^{137}_{56}\text{Ba}$$

EXAMPLE 10.2   Balancing Nuclear Equations

Until the 1950s, radium-226 was widely used as a source of radiation for cancer treatment. It emits both alpha and gamma rays. Write the equation for its decay.

ANALYSIS We have to look up the atomic number of radium, which turns out to be 88, so the symbol we'll use for this radionuclide is $^{226}_{88}\text{Ra}$. When one of its atoms loses an alpha particle, $^{4}_{2}\text{He}$, it loses 4 units in mass number, going from 226 to 222. And it loses 2 units in atomic number, going from 88 to 86. Thus the new radionuclide has a mass number of 222 and an atomic number of 86. We have to look up the atomic symbol for element number 86, which turns out to be Rn for radon.

SOLUTION Now we can assemble the nuclear equation.

$$^{226}_{88}\text{Ra} \longrightarrow {}^{222}_{86}\text{Rn} + {}^{4}_{2}\text{He} + \gamma$$

■ The radium used in cancer therapy was held in a thin, hollow gold or platinum needle to retain the alpha particles and all the decay products.

■ PRACTICE EXERCISE 1 Iodine-131 has long been used in treating cancer of the thyroid. This radionuclide emits beta and gamma rays. Write the nuclear equation for this decay.

■ PRACTICE EXERCISE 2 Plutonium-239 is a byproduct of the operation of nuclear power plants. It can be isolated from used uranium fuel and made into fuel itself or into atomic bombs. A powerful alpha and gamma emitter, it is one of the most dangerous of all known substances. Write the equation for its decay.

A Short Half-Life Means a Rapid Decay   Some radionuclides are much more stable than others, and we use the concept of a half-life to describe the differences. The **half-life** of a radionuclide, symbolized as $t_{1/2}$, is the time it takes for half of the atoms in a sample of a single, pure isotope to decay. (The atoms that decay don't just vanish, of course. They change into different isotopes.) Table 10.2 gives the half-lives of several radionuclides.

TABLE 10.2   Typical Half-Life Periods

Element	Isotope	Half-Life	Radiation or Mode of Decay
Naturally Occurring Radionuclides			
Potassium	$^{40}_{19}K$	1.3×10^9 y	Beta, gamma
Radon	$^{222}_{86}Rn$	3.82 d	Alpha
Radium	$^{226}_{88}Ra$	1590 y	Alpha, gamma
Thorium	$^{230}_{90}Th$	8×10^4 y	Alpha, gamma
Uranium	$^{238}_{92}U$	4.51×10^9 y	Alpha
Synthetic Radionuclides			
Tritium	$^{3}_{1}H$	12.26 y	Beta
Oxygen	$^{15}_{8}O$	124 s	Positron
Phosphorus	$^{32}_{15}P$	14.3 d	Beta
Technetium	$^{99m}_{43}Tc$	6.02 hr	Gamma
Iodine	$^{131}_{53}I$	8.07 d	Beta
Cesium	$^{137}_{55}Cs$	30 y	Beta
Strontium	$^{90}_{38}Sr$	28.1 y	Beta

The half-life of uranium-238 is 4.51×10^9 years, which means that an initial 100 g of this radionuclide would have only 50.0 g of uranium-238 left after 4.51×10^9 years. Strontium-90, a byproduct of nuclear power plants, is a beta emitter with a half-life of 28.1 years. Figure 10.3 shows graphically how an initial supply of 40 g is reduced successively by units of one-half for each half-life period. At the end of seven half-life periods (196.7 years, from 7×28.1), only a little over 0.3 g of strontium-90 remains in the sample.

The shorter the half-life, the larger the number of decay events per mole per second occurring in the isotope. Mole for mole, it's generally much safer to be near a sample that has a long half-life and thus decays very slowly than to be near one that has a short half-life and decays very rapidly. The potential danger of a radionuclide, however, is a function of more factors than its half-life, as we will soon learn.

A Succession of Decays Occurs in a Radioactive Disintegration Series   The decay of one radionuclide sometimes produces not a stable isotope but just another radionuclide. This may, in turn, decay to still another radionuclide, with the process repeating until a stable nuclide is finally reached. There are three such series still active in nature, called **radioactive disintegration series**, and uranium-238 is at the head of one (Figure 10.4). This series ends in a stable isotope of lead.

■ An extremely long half-life is typical of the radionuclides that head a radioactive disintegration series.

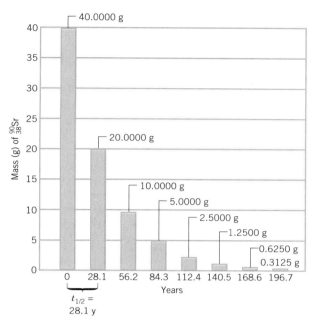

FIGURE 10.3 Each half-life period reduces the quantity of a radionuclide by a factor of two. Shown here is the pattern for strontium-90, a radioactive pollutant with a half-life of 28.1 years.

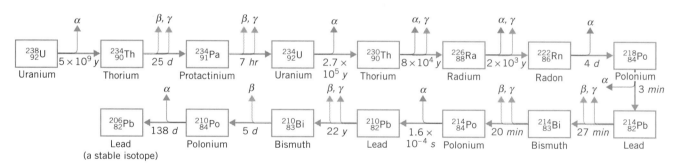

FIGURE 10.4 The uranium-238 radioactive disintegration series. The time given beneath the arrow is the half-life of the preceding isotope (y = year, d = day, hr = hour, min = minute, and s = second).

10.2 IONIZING RADIATION—DANGERS AND PRECAUTIONS

Atomic radiation creates unstable ions and radicals in tissue, which can lead to cancer, mutations, tumors, and birth defects.

The undesired effects of radiation are both *acute* and *latent*. Acute effects show up quickly and can be burns or any of the symptoms of radiation sickness, which we will study soon. Latent effects of radiation do not show themselves until some time after exposure. Cancer, particularly leukemia, is often a latent effect. Another is the alteration of a gene.

Policy Makers Consider That No Safe *Threshold Exposure* to Radiation Exists
Any kind of radiation that penetrates the skin or enters the body on food or through the lungs is considered harmful, and *the damage can accumulate* over a lifetime. Even the ultraviolet radiation in strong sunlight, which barely penetrates the skin, can alter the genetic molecules in skin cells so as to lead to skin cancer.

No "tiny bit of exposure," no **threshold exposure,** is considered to exist for ionizing radiation below which no harm is possible. Cells do have a significant capacity for the self-repair of radiation damage, and some exposures carry very low risks. The threshold exposure, surprisingly, appears to be high for transmittable genetic defects. The threshold is much lower for a fetus than it is for men and nonpregnant women.

The latent effects of radiation correlate well with the accumulated dose regardless of how it is received, so medical personnel who work with atomic radiation of any kind wear devices that automatically record their exposure. The exposure data are periodically logged into a permanent record book, and when the maximum permissible dose is attained, the worker must be transferred. Let's now see how the radiation has its effects and what are some steps for self-protection.

Unstable Ions and Radicals Are Produced in Tissue by Radiation The different kinds of atomic radiation are dangerous because they generate unstable, highly reactive particles in living tissue. Alpha and beta particles, therefore, as well as X rays and gamma rays, are called **ionizing radiations.** They can knock electrons from molecules as they strike them and so produce unstable polyatomic ions. Radiation, for example, can make ions from water molecules by the following reaction.

$$\text{H}-\overset{\cdot\cdot}{\underset{\cdot\cdot}{\text{O}}}-\text{H} \xrightarrow{\text{radiation}} [\text{H}-\overset{\cdot}{\underset{\cdot\cdot}{\text{O}}}-\text{H}]^+ + {}_{-1}^{0}e$$

The new cation, $[\text{H}-\overset{\cdot}{\underset{\cdot\cdot}{\text{O}}}-\text{H}]^+$, is unstable and one breakup path is

$$[\text{H}-\overset{\cdot}{\underset{\cdot\cdot}{\text{O}}}-\text{H}]^+ \longrightarrow \text{H}^+ \quad + \quad :\overset{\cdot}{\underset{\cdot\cdot}{\text{O}}}-\text{H}$$
$$\qquad\qquad\qquad\qquad\text{proton}\qquad\text{hydroxyl}$$
$$\qquad\qquad\qquad\qquad\qquad\qquad\text{radical}$$

A proton forms plus the *hydroxyl radical,* a kind of particle new to our study. It's a *neutral* particle but with an unpaired electron and without an octet for its oxygen atom. Any particle with an unpaired electron is called a **radical** and, with few exceptions, radicals are very reactive species.

■ Sometimes the term used is *free* radical.

■ Technical terms are
Carcinogen, a cancer causer
Tumorigen, a tumor causer
Mutagen, a mutation causer
Teratogen, a birth defect causer

The new ions and radicals react with stable substances around them, altering them in ways foreign to metabolism. When such chemical reactions happen in genes and chromosomes, the cell's genetic substances, subsequent reactions can lead to cancer, tumor growth, or a genetic mutation. If they happen in a sperm cell, an ovum, or a fetus, the result may be a birth defect. Relative to other tissues, the fetus is particularly sensitive to radiation, as we said.

Prolonged and repeated exposures to *low* levels of radiation are more likely to induce these problems than bursts of high-level radiation. It depends on whether the injured cell is still able to duplicate itself by cell division. High-energy radiation bursts usually kill a cell outright, or at least render it reproductively dead. For this reason, high doses of radiation are used in cancer treatment. But low-level radiation that leaves a cell reproductively viable can alter the cell contents in ways that affect the way in which the cell is reproduced.

■ No technology in any medical field is entirely risk-free. We take risks when we believe that the benefits outweigh them.

Cells do have a capacity for self-repair, as we have said, but no exposure is entirely risk-free. The widespread and routine use of X rays for public health screenings has long been curtailed.

The Collection of Symptoms Caused by Ionizing Radiation Is Called Radiation Sickness Molecules of hereditary materials in the cell's chromosomes are the primary site of the most serious radiation damage. Damage to these leads to all other problems. The first symptoms of radiation exposure, therefore, often occur in tissues whose cells divide most frequently, for example, the cells in bone marrow. These make white blood cells, so an early sign of radiation damage is a sharp decrease in the blood's white cell count. Cells in the intestinal tract also divide frequently, and even moderate exposure to X rays or gamma rays (as in radiation therapy for cancer) may produce intestinal disorders.

The set of symptoms caused by nonlethal exposures to radiation is called **radiation sickness.** Its symptoms include nausea, vomiting, a drop in the white cell count, diarrhea, dehydration, prostration, hemorrhaging, and the loss of hair. They often appear when sharp bursts of radiation are used to halt the spread of cancer.

Protection from Ionizing Radiation Is Achieved by Deploying Shields, Using Short Exposure Times, and Moving Away from the Source Shields have long been used for protection against ionizing radiation. (No doubt on a visit to the dentist you have had a lead apron placed over your chest before a dental X ray.) Alpha and beta rays are the easiest to stop, as the data in Table 10.3 show. Gamma radiation and X rays are stopped effectively only by particularly dense substances, like lead, a very dense metal but still fairly inexpensive. But notice in the data in Table 10.3 that even 30 mm (3.0 cm, a little over an inch) of lead reduces the intensity of gamma radiation by only 10%. A vacuum is not a radiation shield, of course, and air isn't much better. Low-density materials like cardboard, plastic, and aluminum are very poor shields, but concrete works well if it is thick (and it's much cheaper than lead).

TABLE 10.3   Penetrating Abilities of Some Common Kinds of Radiation[a]

Type of Radiation	Common Sources	Approximate Energy When from These Sources	Approximate Depth of Penetration of Radiation into		
			Dry Air	Tissue	Lead
Alpha rays	Radium-226 Radon-222 Polonium-210	5 MeV	4 cm	0.05 mm[b]	0
Beta rays	Tritium Strontium-90 Iodine-131 Carbon-14	0.01 to 0.2 MeV	6–300 cm	0.06–4 mm[b]	0.005–0.3 mm

			Thickness to Reduce Initial Intensity by 10%		
			Dry Air	Tissue	Lead
Gamma rays	Cobalt-60 Cesium-137 Radium-226 decay products	1 MeV	400 cm	50 cm	30 mm
X rays Diagnostic Therapeutic		Up to 90 keV Up to 250 keV	120 m 240 m	15 cm 30 cm	0.3 mm 1.5 mm

[a] Data from J. B. Little, *The New England Journal of Medicine*, vol. 275, pages 929–938, 1966.
[b] The protective layer of skin is about 0.07 mm thick. To penetrate it, alpha particles need about 7.5 MeV of energy and beta particles about 0.07 MeV.

Thus, by a careful choice of a shielding material, protection can be obtained.

Another strategy that helps to protect radiation technicians is the use of fast film. It cuts the *duration* of the exposure.

The least expensive self-protection step is to get as far from the radiation source as you can. Radiation, like light from a bulb, moves in straight lines spreading out in all the directions open to it from its source. From any point on the surface of the source, the radiation forms a cone of rays, so fewer rays can strike a unit of surface area the more distant the receiving surface is. The area of the base of such a cone increases with the square of the distance. Hence the radiation intensity I on a unit area diminishes with the square of the distance d from the source. This is the **inverse-square law** of radiation intensity.

Inverse-Square Law   The intensity of radiation is inversely proportional to the square of the distance from the source.

$$\text{radiation intensity, } I \propto \frac{1}{d^2} \tag{10.1}$$

This law is strictly true only in a vacuum, but it holds closely enough when the medium is air to make good estimates. If we move from one location, 1, to another location, 2, the variation of Equation 10.1 that we can use to compare the intensities, I_1 and I_2, at the two different places is given by the following equation.

$$\frac{I_1}{I_2} = \frac{d_2{}^2}{d_1{}^2} \tag{10.2}$$

■   PRACTICE EXERCISE 3   If the intensity of radiation is 25 units at a distance of 10 m, what does the intensity become if you move to a distance of 0.50 m?

■   PRACTICE EXERCISE 4   If you are receiving an intensity of 80 units of radiation at a distance of 6.0 m, to what distance would you have to move to reduce this intensity by half, to a value of 40 units?

No Escape from the Natural Background Radiation Is Possible   Shielding materials and distance can never completely eliminate our exposure to ionizing radiation. We are constantly exposed to **background radiation,** the radiation given off by naturally occurring radionuclides, by radioactive pollutants, cosmic rays, and medical X rays. About 50 of the roughly 350 isotopes of all elements in nature are radioactive. Different natural radionuclides (e.g., potassium-40 and carbon-13) are in the food we eat, the water we drink, and the air we breathe. Radiation enters our bodies with every X ray we receive. Radiation comes in showers of cosmic rays. The top 15 cm of soil on our planet has an average of 1 g of radium per square mile. Thus, radioactive materials are in the soils and rocks on which we walk and that we use to make building materials.

Radon, a *chemically* inert but radioactive gas and a product of the uranium-238 disintegration series, makes the largest contribution to our background radiation (see Interaction 10.2).

The amount of background radiation varies widely from place to place, and only estimates are possible. Table 10.4 gives the average radiation dose equivalents from various sources for the U.S. population. The table uses a unit of *dose equivalent* called the millirem (mrem), which we will define shortly, but, for purposes of com-

INTERACTION 10.2
RADON IN THE ENVIRONMENT

Radon-222 is a naturally occurring radionuclide in the family of noble gases produced by the uranium-238 disintegration series. Chemically, it is as inert as the other noble gases, but radiologically it is a dangerous air pollutant. It is an alpha emitter with a half-life of only 4 days. Produced in rocks and soil wherever uranium-238 is found, it migrates as a gas into the surrounding air. Basements not fully sealed act like fireplace chimneys to draw radon-222 into homes.

The first indication of how serious radon-222 pollution might be came when an engineer at a nuclear power plant in Pennsylvania set off radiation alarms just by his presence. The problem was traced to his home, where the radiation level in the basement was 2700 picocuries per liter of air. In the average home basement, the level is just 1 picocurie/L. (The picocurie is 10^{-12} curie.) The engineer had carried radon-222 and its radioactive decay products on his clothing into his workplace.

As a result of the incident, geologists went looking for unusual concentrations of uranium-238 nearby. They found that the Reading Prong, a formation of bedrock that cuts across Pennsylvania, New Jersey, New York State, and up into the New England states, is relatively rich in uranium-238.

Radon-222 enters the lungs with breathing, and some decays within the lungs. Several decay products in the series after radon-222 are not gases, like polonium-218 ($t_{1/2}$ 3 min, alpha emitter), lead-214 ($t_{1/2}$ 27 min, beta and gamma emitter) and polonium-214 ($t_{1/2}$ 1.6×10^{-4} s, alpha and gamma emitter). Left in the lungs, these can cause cancer, and hard rock miners have had a higher incidence of lung cancer than normal. Based on this experience, the U.S. Environmental Protection Agency (EPA), in the 1980s, tried to estimate how the radon in the air of normal homes might affect lung cancer cases. EPA scientists and their consultants estimated that roughly 10% of the country's annual deaths from lung cancer are caused by indoor radiation from radon-222. The EPA then recommended that the upper limit on radon-222 concentration in home air be set at 4 picocuries/L. As a result, an estimated 11 million home owners took action to lower radon levels in their basements at a cost of $1000 to $2000 per home.

The EPA estimates have since been challenged by several large studies both in the United States and elsewhere. A Finnish study, for example, reported in 1996 that residential exposure to radon "does not appear to be an important cause of lung cancer." Several additional, large-scale studies are under way in North America. The EPA continues to recommend (1996) that the home radon level in air be no higher than 4 picoCuries/L. Although there is no uncertainty that radon can cause lung cancer, the risks of low-level residential exposure are unknown.

TABLE 10.4 Average Radiation Doses Received Annually by the United States Population[a]

Type of Radiation	Dose (mrem)	Percent of Total	Type of Radiation	Dose (mrem)	Percent of Total
Natural radiation—295 mrem, 82%			*Artificial Radiation—65 mrem, 18%*		
Radon[b]	200	55	Medical X rays[d]	39	11
Cosmic rays[c]	27	8	Nuclear medicine	14	4
Rocks and soil	28	8	Consumer products	10	3
From inside the body	40	11	Others	2	<1
Subtotal	295		Grand total[e]	360	

[a] Data from "Ionizing Radiation Exposure of the Population of the United States." Report 93, 1987, National Council on Radiation Protection. These are averages. Individual exposures vary widely.

[b] See Interaction 10.2.

[c] Travelers in jet airplanes receive about 1 mrem per 1000 miles of travel.

[d] A normal chest X ray entails an exposure of 10 to 20 mrem.

[e] The federal standard for maximum safe occupational exposure in the United States is roughly 5000 mrem/y.

parison, a dose of 500 rem (500,000 mrem) given to all individuals in a large population would cause the deaths of half of them in 30 days. In relation to such a high dose, the intensity of natural background radiation is very small. At higher altitudes, the intensity of background radiation is greater because incoming cosmic rays, which contribute to the background, have had less opportunity to be absorbed and destroyed by the earth's atmosphere.

10.3 UNITS TO DESCRIBE AND MEASURE RADIATION

Units have been devised to describe the activity of a radioactive sample, radiation exposure, and radiation dose.

A number of units exist for a variety of measurements of radiation, and each unit exists to help answer one particular question. The SI units described here have been officially adopted in the United States, but the older units still appear often in the technical literature.

■ If you note carefully what these questions are, the units will be easier to learn.

The *Becquerel* Describes How Active a Sample Is   The **becquerel, Bq,** is the SI unit of activity and is used to answer the question, "How *active* is a sample of a radionuclide?" In terms of the number of disintegrations per second, or dps,

$$1 \text{ Bq} = 1 \text{ dps}$$

Mole for mole, a radionuclide has a higher activity the shorter its half-life is.

■ Marie Curie is one of two scientists to win two Nobel Prizes in a field of science, a share of the physics prize in 1903 and the chemistry prize in 1911.

The *curie*, the older unit of activity, is named after Marie Sklodowska Curie (1867–1934), a Polish scientist, who discovered radium. One **curie, Ci,** is equal to 3.7×10^{10} dps, which is the number of radioactive disintegrations that occur per second in a 1.0-g sample of radium.

$$1 \text{ Ci} = 3.7 \times 10^{10} \text{ dps} = 3.7 \times 10^{10} \text{ Bq}$$

This is an intensely active rate, so fractions of the Ci, such as the millicurie (mCi, 10^{-3} Ci), the microcurie (μCi, 10^{-6} Ci), and the picocurie (pCi, 10^{-12} Ci), are often used.

Exposure to X Rays or Gamma Rays Is Described in Terms of the Quantity of Ions They Produce in Dry Air   The older unit of X ray or gamma ray exposure is called the **roentgen.** It serves to answer the question, "How *intense* is the exposure to X-ray or gamma-ray radiation?" One roentgen creates a specific total number of charges (2.1×10^9 units) in 1 cm³ of dry air. If members of a large population are exposed to 650 roentgens, half will die in 1 to 4 weeks. (The rest will have radiation sickness.)

The SI has no special unit for exposure to X rays or gamma rays, only a symbol, *X*. *X* is the ratio of the total charge on the ions (of one sign) produced by the radiation per 1 kilogram of dry air.

■ *X* is a derived SI quantity, a ratio, and so is based on the SI units of the quantities in the ratio.

$$X = \frac{\text{coulombs of charge}}{\text{kg dry air}}$$

The *coulomb*, the SI unit of quantity of charge, equals the charge on 6.25 $\times 10^{18}$ electrons.

The *Gray* Describes the *Absorbed Dose*   The SI unit of *absorbed dose* is called the *gray*, named after a British radiologist, Harold Gray. It is used to answer the question, "How much *energy* is *absorbed* by a unit mass of tissue or other materials?" The **gray, Gy,** corresponds to the absorption of 1 joule (J) of energy per kilogram of tissue. (Recall that 1 cal equals 4.184 J.)

$$1 \text{ Gy} = 1 \text{ J/kg}$$

The older unit of absorbed dose, the **rad,** is 1/100th of a gray.

$$1 \text{ rad} = 10^{-2} \text{ Gy}$$

■ *Rad* comes from *r*adiation *ab*-sorbed *d*ose.

The roentgen and the rad are close enough in magnitude to be nearly equivalent from a health standpoint. Thus one roentgen of gamma radiation from cobalt-60 is equivalent to 0.96 rad in muscle tissue and 0.92 rad in compact bone.

About 6 Gy or 600 rad of gamma radiation would be lethal to most people despite the fact that this corresponds to a very small quantity of energy. But remember that it isn't the quantity of energy that matters, it's the formation of unstable radicals and ions. A 6-Gy dose delivered to water breaks up only one molecule in every 36 million, but the radicals produced begin a cascade of harmful reactions inside a cell.

The *Sievert,* the Unit of *Dose Equivalent,* Adjusts Absorbed Doses for Different Effects in Different Tissues A dose of 1 Gy of gamma radiation is not biologically the same as a dose of 1 Gy of beta radiation or of neutrons. Thus, the gray does not serve as a good basis for comparison when working with biological effects. The **sievert, Sv,** is the SI unit that satisfies the need for a way to express dose equivalent that is additive for different kinds of radiation and different target tissues. If we let D stand for absorbed dose in gray, Q for "quality factor" (meaning relative effectiveness for causing harm in tissue), and N for any other modifying factors, then the dose equivalent, H, in sieverts is defined as follows.

$$H = DQN$$

In other words, we multiply the absorbed dose, D, from some radiation by a factor Q that takes into account biologically significant properties of the radiation, and by any other factor N bearing on the net effect. The quality factor for alpha radiation is 20, but it is only 1.0 for beta and gamma radiation. The much larger size of the alpha particle accounts for its extra danger. Its size and charge enable it to strike molecules with considerable (although undesirable) efficiency.

The older unit of dose equivalent is called the **rem.** One rem of any given radiation is the dose that has, in a human being, the effect of one roentgen. The sievert is 100 times larger than the rem but is still a quantity small in terms of energy and yet significant in terms of danger. Even millirem quantities of radiation should be avoided, and when this isn't possible the workers must wear monitoring devices that allow the day-to-day exposures to be determined.

■ *Rem* comes from *r*oentgen *e*quivalent for *m*an.

The Electron-Volt Describes the Energy of X Rays or Gamma Rays The energies of X rays or gamma radiation are often described by the **electron-volt, eV,** defined as follows.

$$1 \text{ eV} = 1.602 \times 10^{-19} \text{ J}$$

■ The electron-volt is the energy an electron receives when it is accelerated by a voltage of 1 V.

Because the electron-volt is an extremely small amount of energy, roughly 10^{-19} joule, multiples of it are very common. They include the kiloelectron-volt (1 keV = 10^3 eV) and the megaelectron-volt (MeV = 10^6 eV). X rays used for diagnosis are typically 100 keV or less. The gamma radiation from cobalt-60 has energies of 1.2 and 1.3 MeV. Beta radiation of 70 keV or more can penetrate the skin, but alpha particles (which are much larger than beta particles) need energies of more than 7 MeV to do this. Alpha radiation from radium-226 has an energy of 5 MeV. The cosmic radiation that enters our outer atmosphere has energies ranging from 200 MeV to 200 GeV (1 GeV = 1 gigaelectron-volt = 10^9 eV).

■ Linear accelerators produce radiation for cancer treatment in the range of 6 to 12 MeV.

Film Dosimeters, Scintillation Counters, and Geiger Counters Measure Ionizing Radiation   A *dosimeter* is a device for measuring exposure. One common type is a film badge that contains photographic film, which becomes fogged by radiations. The degree of fogging, which is related to the exposure, can be measured.

Ionizing radiation also affects substances called phosphors, salts with traces of rare earth metal ions that scintillate when struck by radiation. The scintillations—brief, sparklike flashes of light—can be translated into doses of radiation received. Devices based on this technology are called *scintillation counters.*

Evacuated tubes that are fitted with two electrodes and hold a gas at very low pressure are used in devices such as the Geiger counter. This is a type of *ionization counter,* and Figure 10.5 shows how the tube itself is constructed. The tube, called a Geiger-Müller tube, is especially useful in measuring the beta and gamma radiation that has enough energy to penetrate the window. When the radiation enters the rarefied gaseous atmosphere in the tube, it creates ions that cause a brief pulse of electricity that the apparatus records as a "count."

■ Inside the screen of a color TV tube there is a coating that includes various phosphors; these glow with different colors when struck by the focused electron beam in the tube.

10.4 SYNTHETIC RADIONUCLIDES

Most radionuclides used in medicine are made by bombarding other atoms with high-energy particles.

Radioactive decay is nature's way of causing transmutations. They can also be caused artificially by bombarding atoms with high-energy particles. Several hundred isotopes that do not occur naturally have been made this way. Many have been used successfully in medicine, both in diagnosis and in treatment.

Various bombarding particles are used, including alpha particles, neutrons, and protons. The first artificial transmutation, observed by Rutherford, was the conversion of nitrogen-14 into oxygen-17 by alpha particle bombardment. Rutherford let alpha particles from a naturally radioactive source travel through a tube that contained nitrogen-14. He soon detected that another radiation was being generated, one far more penetrating than the alpha radiation he used. He showed that it consisted of high-energy protons and that atoms of oxygen-17 now existed in the tube.

To explain his observations, Rutherford reasoned that alpha particles had plowed right through the electron clouds of nitrogen-14 atoms and buried themselves in their nuclei. The strange new nuclei, called *compound nuclei,* evidently

■ These are gaseous protons, true subatomic particles, and not hydronium ions.

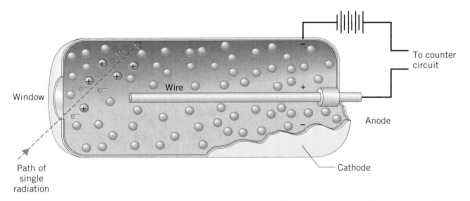

FIGURE 10.5   The basic features of a gas-ionization radiation detection tube such as used in a Geiger-Müller counter.

had too much energy to exist for long. They were excited nuclei of fluorine-18, and to rid themselves of excess energy each ejected a proton, leaving behind an atom of oxygen-17. The equation is

$$\underset{\substack{\text{alpha}\\\text{particle}}}{^4_2\text{He}} \;+\; \underset{\substack{\text{nitrogen}\\\text{nucleus}}}{^{14}_7\text{N}} \;\longrightarrow\; \underset{\substack{\text{fluorine}\\\text{nucleus}}}{^{18}_9\text{F}^*} \;\longrightarrow\; \underset{\substack{\text{oxygen}\\\text{nucleus}}}{^{17}_8\text{O}} \;+\; \underset{\text{proton}}{^1_1p}$$

■ The nucleus of $^{18}_9$F* is *compounded* of the nucleus of $^{14}_7$N and $^4_2$He, so is called a compound nucleus.

(The asterisk by the symbol for fluorine-18 signifies that the particle is a high-energy, compound nucleus.) Oxygen-17 is a rare but nonradioactive isotope of oxygen. Usually, however, transmutations caused by bombardments produce *radioactive* isotopes of other elements.

Electrically charged bombarding particles, like the alpha particle and the proton, can be given greater velocity and therefore greater energy when attracted by opposite charge. Particle accelerators, devices that do this, include some of the multi-million dollar hardware of atomic research. The interactions of their high-energy beams with selected targets have made possible the synthesis of dozens of new radionuclides.

■ Remember that the energy of a moving object increases with the *square* of its velocity. $KE = (1/2)mv^2$

Certain isotopes of uranium in atomic reactors eject neutrons, and although neutrons cannot be accelerated (they are electrically neutral), they have sufficient energy to serve as bombarding particles. Being neutral is an advantage, because the neutrons aren't repelled either by the electrons that surround an atom or by the nucleus. One of the very important applications of neutron bombardment is the synthesis of molybdenum-99 from molybdenum-98. With ^0_1n as the symbol for the neutron, the equation is

$$^{98}_{42}\text{Mo} + ^0_1n \longrightarrow ^{99}_{42}\text{Mo} + \gamma$$

As we will learn in Section 10.6, the decay of molybdenum-99 leads to one of the radionuclides often used in medicine.

10.5 RADIATION TECHNOLOGY IN THE FOOD INDUSTRY

Food irradiated with controlled doses of X rays, gamma rays, or electron beams is less likely to spoil.

X rays, gamma rays, and particle beams have long been used in medicine as technologies for diagnosis and for cancer treatments. In treatments for cancer the aim is to kill the cells in cancer tissue. This same kind of aim, to kill cells that can subdivide, such as those of disease-causing microorganisms, is behind food irradiation technology.

Food Irradiation Inhibits, Inactivates, or Kills Molds and Bacteria When food products are passed through a beam of gamma rays, X rays, or accelerated electrons, the effects depend on the energy of the beam. A low-dose beam—up to 100 kilorads—renders reproductively dead any insects that remain after harvest and inhibits the sprouting of potatoes and onions during storage. Such beams also inactivate trichinae (*Trichinella spiralis,* a nematode worm) in pork, the parasite that causes trichinosis.

Medium-dosage beams of radiation—100 to 1000 kilorads—significantly reduce the populations of *Salmonella* bacteria in poultry, fish, and other meats. Such radia-

■ The occurrence of trichinosis in the pork supply of the United States is small, and thorough cooking destroys it. (Pork is never served "rare.")

FIGURE 10.6 After 15 days of storage at 4 °C (38 °F), the unirradiated strawberries on the left became heavily covered by mold. Those on the right, however, had been protected by 200 kilorads of radiation.

tion also greatly extends the shelflives of strawberries (Figure 10.6) and certain other fruits, which otherwise form molds quickly.

High-dosage beams—1000 to 10,000 kilorads—sterilize poultry, fish, and other meats. They also kill microorganisms and insects on seasonings and spices.

Food Irradiation Does Not Appear To Produce Any Unique Radiolytic Products
Ionizing radiation causes chemical reactions in foods or any insects or microorganisms in them. Since water is generally the most abundant substance present, the primary products result from the splitting of water into radicals and ions. Some recombine to form water and others combine to give hydrogen peroxide. Otherwise, the primary products of irradiation react with food molecules to give secondary substances, called *radiolytic products.* Generally, however, these are the same as are produced by cooking or baking or by the subsequent digestion of foods.

■ In the body, hydrogen peroxide, H_2O_2, quickly breaks down to water and oxygen.

The irradiation of wheat and potatoes to control insects has been permitted in the United States for over 20 years, but no commercial operator is doing so. Some herbs and spices are now marketed after irradiation to reduce insects, bacteria, molds, and yeasts. The U.S. Food and Drug Administration (FDA) has approved the use of low-dosage radiation to control trichinae in pork and to inhibit the spoilage of fruit and vegetables. None of these operations is as yet widespread, mostly because they are expensive.

Besides cost, customer confidence is another barrier to further use of irradiation. Any technology with the word *nuclear* or *radiation* in it makes people nervous. The legitimate question is, "Does radiation produce radiolytic products that would not be present once foods have otherwise been processed, cooked, and digested?" If so, "Are the radiolytic products harmful at the levels at which they are present?"

Specialists in food irradiation claim that their research has yet to turn up any unique radiolytic products. Even benzene, for example, is not uniquely a product of radiation. Repeated environmental exposure to benzene is known to increase a person's likelihood of contracting leukemia, and the *high-level* irradiation of certain foods gives traces of benzene. But small traces of this substance are naturally present in some nonirradiated foods, like boiled eggs, at much higher levels than are produced by irradiation.

Still another safety issue concerns botulism, a particularly dangerous form of food poisoning caused by an odorless chemical, a toxin, produced by *Clostridium botulinum.* This bacterium is more resistant to radiation than the microorganisms that cause food spoilage. Organisms that spoil foods usually give them dreadful odors, which warn people away, but the botulinum toxin cannot be detected by odor. High-dose food irradiation could thus prevent the kind of food spoilage associated with odors, without necessarily destroying all of the botulinum bacillus. (Low-dose radiation kills too few of any kind of bacteria to pose this problem.)

■ A half a pound of the botulinus toxin would be enough to kill all of the people on Earth.

10.6 RADIATION TECHNOLOGY IN MEDICINE

Both in diagnosis and in cancer treatment, ionizing radiation is used when its benefits are judged to outweigh its harm.

For diagnostic or therapeutic medical uses, ionizing radiation is supplied either as X rays or electron beams. In diagnostic work, radiation is used to locate a cancer or tumor or to assess the function of some organ, like the thyroid gland. In therapeutic work, radiation is used to kill cancerous cells or to inhibit their growth. Generally, therapeutic doses are of much higher energy than those used for diagnostic purposes. We will look here at the radioactive chemicals used in diagnosis, and discuss beam radiation technologies in Interactions 10.3 to 10.5.

Chemical and Radiological Properties Are Factors in the Selection of Radionuclides for Medicine   The *chemical* properties of radionuclides are identical to those of the stable isotopes of the same element, so when used for their ionizing radiation they must be *chemically* compatible with the body. Moreover, their *chemistry,* not their radiation, is what guides them naturally to desired tissues. Iodine-127, for example, the only stable isotope of iodine, is used by the thyroid gland to make the hormone thyroxin. This chemical property will also guide iodine-131 to the thyroid gland. Once there, its beta radiation can be used to treat thyroid cancer.

Minimizing Harm and Maximizing Benefit Guide the Selection of Radionuclides in Medicine   Exposing anyone to any radiation entails some risks, because prolonged exposure can produce cancer. No such exposure is permitted unless the expected benefit from finding and treating a dangerous disease is thought to be greater than the risk. To minimize the risks, the radiologist uses radionuclides that, as much as possible, have the following properties.

■ *Radiology:* the science of radioactive substances and of X rays.

Radiologist: a specialist in radiology who also usually has a medical degree.

Radiobiology: the science of the effects of radiation on living things.

1. The radionuclide should have a half-life that is short. (Then it will decay *during* the diagnosis when the decay gives some benefit, and as little as possible of the radionuclide will decay later, when the radiation is of no benefit.)

2. The product of the decay of the radionuclide should have little if any radiation of its own. (Either the product should be a stable isotope or it should have a long half-life.) It should also be quickly eliminated.

3. The half-life of the radionuclide must be long enough for it to be prepared and administered to the patient before it all decays.

4. If the radionuclide is to be used for *diagnosis,* it should decay by penetrating radiation entirely, which means gamma radiation. (Nonpenetrating radiation, such as alpha and beta radiation, adds to the risk by causing internal damage without contributing to the detection of the radiation externally.

5. For uses in *therapy,* as in cancer therapy, nonpenetrating radiation is preferred because a radionuclide well placed in cancerous tissue *should* cause damage to such tissue.)

6. The diseased tissue should concentrate the radionuclide, giving a "hot spot" where the diseased area exists, or it should do the opposite and reject the radionuclide, making the diseased area a "cold spot" insofar as external detectors are concerned.

Let us now look briefly at a few of the more important radionuclides employed in medicine.

Technetium-99m Is a Widely Used Radionuclide in Medicine

Technetium-99m is made by the decay of molybdenum-99. You already know that gamma radiation often accompanies the emission of other types of radiation, but with molybdenum-99 the gamma radiation comes after a pause. Molybdenum-99 first emits a beta particle:

$$^{99}_{42}\text{Mo} \longrightarrow {}^{99m}_{43}\text{Tc} + {}^{0}_{-1}e$$

The other decay product is a metastable form of technetium-99, hence the m in 99m. Metastable means poised to move toward greater stability. Technetium-99m decays by emitting gamma radiation with an energy of 143 keV:

$$^{99m}_{43}\text{Tc} \longrightarrow {}^{99}_{43}\text{Tc} + \gamma$$

Technetium-99m almost ideally fits the criteria for a radionuclide intended for diagnostic work. Its half-life is short, 6.02 hr. Its decay product, technetium-99, has a very long half-life, 212,000 years, so it has too little activity to be of much concern. (Technetium-99 decays to a stable isotope of ruthenium, $^{99}_{44}\text{Ru}$.) The half-life of technetium-99m, although short, is still long enough to allow time to prepare it and administer it. It decays entirely by gamma radiation, which means that the maximum amount of the radiation gets to a detector to signal where the radionuclide is in the body. Finally, a variety of chemically combined forms of technetium-99m have been developed that permit either hot spots or cold spots to form.

One form of technetium-99m is the pertechnetate ion, TcO_4^-. It behaves in the body very much like a halide ion, so it tends to go where iodide ions, for example, go. It is eliminated by the kidneys, so it is used to assess kidney function. Other organs whose functions are also studied by technetium-99m are the liver, spleen, lungs, heart, brain, bones, and thyroid gland.

Technetium-99m technology has received competition from the CT scan (Interaction 10.3), the PET scan (Interaction 10.4), and the MRI (Interaction 10.5).

Iodine-131 and Iodine-123

The ability of the thyroid gland to concentrate iodide ion is so good that a small whole-body dose of radioactive iodine is concentrated nearly 1000-fold in this gland. When someone is suspected of having an underactive or an overactive thyroid, one technique is to let the patient drink a glass of flavored water that contains some iodine-123 (as I^- ion). This isotope is better than others because it has a short half-life (13.3 h) and it emits only gamma radiation (159 keV), a radiation powerful enough so that all can leave the thyroid and reach a detector. All of the radiation is therefore diagnostically useful. By placing the detector near the thyroid, the radiologist can tell how well the gland takes up iodide ion from circulation.

Iodine-131 also has a short half-life (8 days), but it emits both beta particles (600 keV) and gamma radiation (mostly 360 keV). The beta particles do not get out to the detector, but this makes iodine-131 useful for treating certain types of thyroid cancer.

The hydrated pertechnetate ion, TcO_4^-, has about the same radius and charge density as the hydrated iodide ion, and membranes of the thyroid gland do not distinguish between the two. When the purpose is to get any kind of radiation inside

INTERACTION 10.3
X RAYS AND THE CT SCAN

X rays are generated by bombarding a metal surface with high-energy electrons. These can penetrate the metal atom far enough to knock out one of its low-level orbital electrons, such as a 1s electron. This creates a "hole" in the electron configuration, and orbital electrons at higher levels begin to drop down. In other words, the creation of this "hole" leads to electrons changing their energy levels. The difference between two of the lower levels corresponds to the energy of an X ray, which is emitted.

The refinement of X-ray techniques and the development of powerful computers made possible the generation of a diagnostic technology called computerized tomography, or CT for short. A. M. Cormack (United States) and G. N. Hounsfield (England) shared the 1979 Nobel Prize in Physiology or Medicine for their work in the development of this technology. The instrument includes a large array of carefully positioned and focused X-ray generators. In the procedure called a CT scan, this array is rotated as a unit around the body or the head of the patient (see

Figure 1). Extremely brief pulses of X rays are sent in from all angles across one cross section of the patient (see Figure 2).

The changes in the X rays that are caused by internal organs or by tumors are sent to a computer, which then processes the data and delivers a picture of the cross section. It's like getting a picture of the inside of a cherry pit without cutting open the cherry. The CT scan is widely used for locating tumors and cancers.

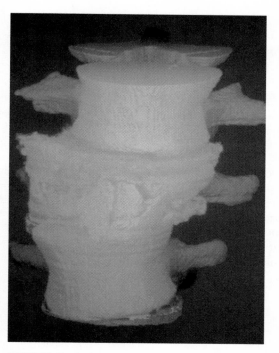

FIGURE 2 A three-dimensional image, based on 63 CT scans, of a section of the vertebrae of a man injured in a motorcycle accident. Both the compression and the twisting are clearly evident.

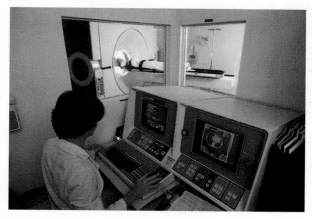

FIGURE 1 CT scan instrumentation.

the thyroid in order to detect a tumor or cancer, then Tc-99m, because of its short half life (6.02 h), is now preferred over any radionuclide of iodine.

Linear Accelerators Make High-Energy X Rays Betatrons are linear accelerators that generate X rays for therapeutic uses with energies in the range of 6 to 12 MeV. The machines are easily mounted so that they can rotate about a central point, which makes it easier to position the patient and plan the treatment.

INTERACTION 10.4
POSITRON EMISSION TOMOGRAPHY—THE PET SCAN

A number of synthetic radionuclides emit positrons. These are particles that have the same small mass as an electron but carry one unit of *positive* charge. (They're sometimes called positive electrons.) A positron forms by the conversion of a proton into a neutron, as follows:

$$\begin{array}{ccccc} {}^1_1p & \longrightarrow & {}^1_0n & + & {}^0_1e \\ \text{proton} & & \text{neutron} & & \text{positron} \\ \text{(in atom's} & & \text{(stays in} & & \text{(is emitted)} \\ \text{nucleus)} & & \text{nucleus)} & & \end{array}$$

Positrons, when emitted, last for only a brief interval before they collide with an electron. The two particles annihilate each other, and when this happens their masses convert entirely into energy in the form of two photons of gamma radiation (511 keV). The gamma radiation formed in this way is called annihilation radiation.

$$\begin{array}{ccccc} {}^{\;0}_{-1}e & + & {}^0_1e & \longrightarrow & {}^0_0\gamma \\ \text{electron} & & \text{positron} & & \text{gamma} \\ & & & & \text{radiation} \end{array}$$

The two photons leave the collision site in almost exactly opposite directions.

To make a medically useful technology out of this property of the positron, a positron-emitting nuclide must be part of a molecule with a chemistry that will carry it into the particular tissue to be studied. Once the molecule gets in, the tissue now has a gamma ra-

diator *on the inside.* Thus instead of X rays being sent through the body, as in a CT scan (Interaction 10.3), the radiation originates right within the site being monitored. The overall procedure is called positron emission tomography, or PET for short.

Three positron emitters are often used: oxygen-15, nitrogen-13, and carbon-10. Glucose, for example, can be made in which one carbon atom is carbon-11 instead of the usual carbon-12. Glucose can cross the blood–brain barrier and get inside brain cells. If some part of the brain is experiencing abnormal glucose metabolism, this will be reflected in the way in which positron-emitting glucose is handled,

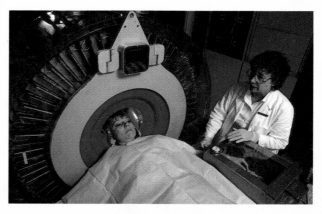

FIGURE 1 The patient is shown undergoing a brain scan performed by means of PET technology.

10.7 ATOMIC ENERGY AND RADIONUCLIDES

Nuclear power plants generate atomic wastes that must be kept from human contact for several centuries.

Large electrical power plants use the heat from the burning of some kind of fuel to convert water into high-pressure steam, which then forces turbines to spin. In nuclear power plants, *fission* is the source of heat, and it happens to a *nuclear fuel* inside a *reactor* (see Figure 10.7).

Fission is the disintegration of a large atomic nucleus into small fragments following neutron capture. It releases, besides heat, additional neutrons plus radioactive isotopes. So much heat is generated that unless the reactor is continuously cooled, usually by water, the whole system will melt very quickly. The coolant changes liquid water into steam at very high pressure, which drives turbines and thus changes some of the fission energy into electrical energy. Excess heat is released to the atmosphere or to a moving stream of cooler water from a river or lake.

and gamma radiation detectors on the outside can pick up the differences (see Figure 1). The use of the PET scan has led to the discovery that glucose metabolism in the brain is altered in schizophrenia, in manic depression, and by the presence of nicotine (see Figure 2). PET scanning technology is able to identify extremely small regions in the brain that are in early stages of breakdown and that CT and MRI techniques miss. A brain involved with Alzheimer's disease gives a PET scan different from a normal brain. The PET scan is particularly useful in detecting abnormal brain function in infants, as in early stages of epilepsy.

PET technology is being used to study a number of neuropsychiatric disorders, including Parkinson's disease. When a drug labeled with carbon-11 is used, its molecules go to the parts of the brain with nerve endings that release dopamine. A PET scan then discloses the dopamine-releasing potential of the patient. This potential becomes impaired in Parkinson's disease.

By labeling blood platelets with a positron emitter, scientists can follow the development of atherosclerosis in even the tiniest of human blood vessels. Blood flow in the heart can be monitored without having to insert a catheter.

FIGURE 2 *Left*: The control scan, a normal brain scan using PET technology. *Center*: PET scan of the brain of a volunteer injected with nicotine. *Right*: Color code, indicating rates of glucose metabolism. Note that the reduction in the rate of glucose metabolism is widespread in the presence of nicotine.

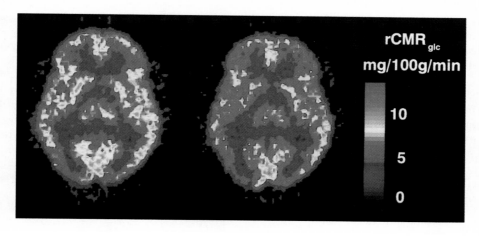

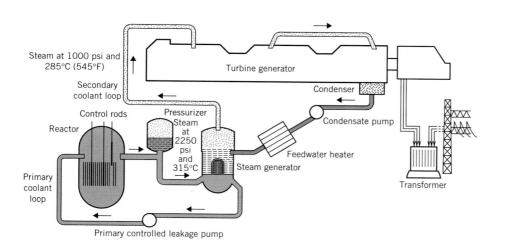

FIGURE 10.7 Pressurized water nuclear power station. Water in the primary coolant loop circulates around the reactor core and carries away the heat of fission. This water is sealed under pressure, so its temperature can rise well above its normal boiling point. The water delivers its heat to the water in the secondary loop, which then turns to high-pressure steam and drives the electrical turbines. Maximum efficiency is reached by having as large a temperature drop as possible between the inlet steam temperature in this loop and the outlet water temperature. (Drawing from WASH 1261, U.S. Atomic Energy Commission, 1973.)

INTERACTION 10.5
MAGNETIC RESONANCE IMAGING—THE MRI SCAN

The CT scan subjects a patient to large numbers of short bursts of X rays. The PET scan exposes the patient to gamma radiation that is generated on the inside. Thus both technologies carry the usual risks that attend ionizing radiation, and they are used when the potential benefits from correct diagnoses far outweigh such risks. MRI technology operates without these dangers. At least, none has been discovered so far. The principal developer of the first hardware for MRI imaging was Raymond Damadian.

MRI stands for magnetic resonance imaging. Atomic nuclei that have odd numbers of protons and neutrons behave as though they were tiny magnets, hence the *magnetic* part of MRI. The nucleus of ordinary hydrogen (which has no neutrons and one proton) constitutes the most abundant nuclear magnet in living systems, because hydrogen atoms are parts of water molecules and all biochemicals. Nuclear magnets spin about an axis much as the electron spins about its axis.

When molecules with spinning nuclear magnets are in a strong magnetic field and are simultaneously bathed with properly tuned radio frequency radiation (which is of very low energy), the nuclear magnets flip their spins. (This is the *resonance* part of MRI.) As they resonate, they emit electromagnetic energy that is biologically harmless, being of very low energy, and this energy is picked up by detectors. The data are fed into computers, which produce an image much like that of a CT or a PET scan, but without having subjected the patient to ultrahigh-energy electromagnetic radiation such as X rays or gamma rays. The MRI images are actually sophisticated plots of the distributions of the spinning nuclear magnets, of hydrogen atoms, for example.

MRI imaging has proved to be especially useful for studying soft tissue, the sort of tissue least well studied by X rays. Different soft tissues have different population densities of water molecules or of fat molecules (which are loaded with hydrogen atoms). And tumors and cancerous tissue have their own water inventories (see Figure 1). Calcium ions do not produce any signals to confuse MRI imaging, so bone, which is rich in calcium, is transparent to MRI.

MRI technology has developed into a method superior to the CT scan for diagnosing tumors at the rear and base of the skull and equal to the CT scan

for finding other brain tumors. MRI is now the preferred technology for assessing problems in joints (particularly the knees) and in the spinal cord, such as ruptured (herniated or "slipped") disks. Patients with heart pacemakers, embedded shrapnel, or surgical clips, however, present problems to the use of MRI because of the powerful magnets used.

CT scans are still better than MRI for the early detection of hemorrhages in the brain, so CT is the method of choice for finding them in potential stroke victims. CT scans are also still preferred for detecting tumors in the kidneys, lungs, pancreas, and the spleen.

A development in the offing is joining MRI and PET technologies so that the two scans can be taken simultaneously. The MRI, by itself, gives only an anatomical "map." The PET scan assesses actual tissue *function*. The marriage of the two has been made difficult by the interference that powerful magnetic fields (from MRI) give to the electronic detectors of PET instruments. Fiber optical lines and improved detectors, however, may make it possible to put the PET instruments far enough away from the magnets. This new technique would make it easier to study the effects of drugs on tissues, for example.

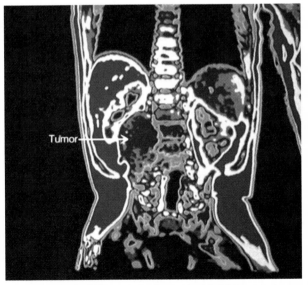

FIGURE 1 An MRI scan of a 7-month-old child revealed a malignant tumor pushing its way into the spinal canal. (The tumor was treated in time.)

One common nuclear fuel is the uranium-235 isotope, the only naturally occurring radionuclide that spontaneously undergoes fission when it captures slowly moving and relatively low-energy neutrons. After capturing such a neutron, the nucleus is made more unstable, and it spontaneously splits apart. It can split in a number of ways to give different products. We'll give the equation for just one of the many modes of splitting.

■ Neutrons moving too rapidly are poorly captured.

$$^{235}_{92}U + ^{1}_{0}n \xrightarrow[\text{capture}]{\text{neutron}} ^{236}_{92}U \xrightarrow{\text{fission}} ^{139}_{56}Ba + ^{94}_{36}Kr + 3^{1}_{0}n + \gamma + \text{heat}$$

As you can see, more neutrons are released than are used up. Hence, one fission event produces neutrons that can initiate more than one new fission. In other words, a **nuclear chain reaction** can take place (Figure 10.8). It would almost instantly envelop all uranium-235 atoms, except that both the reactor and the fuel are designed to permit control over the ratio of neutrons produced to neutrons captured. The best ratio is 1 : 1, the *critical* ratio. When the ratio goes higher, the system becomes *supercritical* and the reactor is in danger of destroying itself. In an atomic bomb, the fuel is relatively pure and the neutron ratio is intended to go higher so as to develop an atomic explosion. In a nuclear power plant, the fuel is usually uranium-235 at a

■ Plutonium-239, made in reactors, also can fission, and it is used as a nuclear fuel and in making atomic bombs.

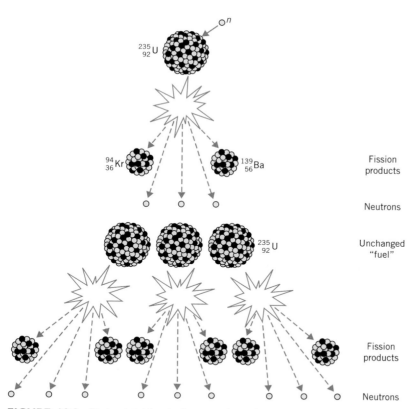

FIGURE 10.8 Fission is initiated when a nucleus of uranium-235 captures a neutron. The new nucleus splits apart, and more neutrons are released. These can initiate subsequent fission events, and unless this is prevented or at least tightly controlled, the whole mass of uranium-235 will detonate, as in the explosion of an atomic bomb. One method of control is to keep the mass of uranium-235 low, below what is called the critical mass. Now, enough neutrons escape before being captured by neighboring nuclei.

CHORNOBYL—AN EXPLOSION THAT SHOOK THE WORLD

On April 26, 1986, a large reactor blew up at the great atomic energy park located at Chornobyl, Ukraine.[1] It was then part of the Soviet Union, and the disaster together with the immense difficulties of dealing with it exposed so many flaws in the Soviet system of governance that some analysts believe that Chornobyl led to the collapse of the Soviet Union itself. How did the Chornobyl explosion happen? Probably as follows.

A chief engineer at Chornobyl ordered a "low power safety test," presumably an effort to find out how long the generators would continue to run without steam. So the steam flowage was blocked and the reactors were throttled back to low power. Unfortunately, the design of the reactor was flawed; it could not safely run at such low power. Equally unfortunately, important safety systems were disabled because they might have interfered with the test. When the operator realized that the reactor's power was soaring, it was too late, and, unlike the reactors at Three-Mile Island, the Chornobyl reactors had no containment vessels.

In a 5-second period, the reactor's power increased 1400-fold, and steam developed with enough pressure to blow off a cover plate weighing 2000 metric tons. The tops of all of the some 1600 cladding tubes were also torn off. Each erupted like a shotgun,

[1]The Russian spelling is Chernobyl, but the Ukrainians much prefer their own spelling, Chornobyl, a spelling now used by *Scientific American* and by *National Geographic.*

and several tons of radioactive, incandescent fuel and wastes were hurled into the night sky. No one knows how much radioactivity the explosion released; the official Soviet estimate was 90×10^6 Ci, which is probably a minimum estimate. Gamma radiation at the power plant rose to an intensity of 100 roentgens per hour, and it reached 100,000 roentgens per hour on the remaining roof of the structure. (A dose of 100 roentgens is hundreds of times the maximum *lifetime* dose that members of the general public may be permitted to receive, according to international guidelines.) Radiation sickness in the immediate aftermath of the explosion felled 31 workers, most of them extraordinarily brave firefighters working to contain the graphite fires. (Tons of graphite made up the neutron moderator for the reactor.) Eventually, the reactor building, still containing huge quantities of radioactive material, was encased in concrete (Figure 1).

The explosion sent a radioactive plume 3200 feet into the atmosphere to drift on the prevailing winds. The chief fallout problems stemmed from three radionuclides, cesium-137, strontium-90, and iodine-131. The human exposure to radioactive cesium-137 in the northern hemisphere from this accident is estimated to be 60% of the human exposure to this radionuclide from all previous atmospheric nuclear tests.

Iodine-131 enters food chains wherever it deposits. Milk from cows that have grazed on pastures contaminated by iodine-131 carries this isotope into the human diet. As we've said, the human thyroid gland concentrates iodine, and so radioactive iodine

concentration of 3% to 5%, too low to allow an *atomic* explosion to occur. However, if the power plant system inadvertently goes supercritical long enough, enough heat is generated to melt the reactor and to convert water into enough steam pressure to blow the reactor apart.

■ The fuel is a mixture of powdered oxides of uranium-235 and uranium-238 baked into ceramic-like pellets.

■ The circulating water is also a good neutron moderator.

Constant Cooling of the Reactor Is Vital A nuclear *reactor* is a device to enable the control of nuclear fission at the critical ratio so that the heat generated is removed rapidly enough to prevent a meltdown. The fuel is distributed among a large number of small-radius tubes (called cladding tubes) so that a significant fraction of neutrons can escape without causing fission. Control rods can be moved into or out of the spaces between these tubes. They are made of materials that can capture neutrons harmlessly. Circulating around and among the tubes is the coolant (water in nearly all reactors), which is heated by the atomic fission taking place.

In *pressurized water reactors,* the water is kept under pressure so that it can be heated above its normal boiling point. As seen in Figure 10.7, this water is in a closed, primary loop engineered so as to heat the water in a secondary loop. The steam thus generated in the secondary loop drives the turbines, and this operation cools the fluid back to liquid water for recirculation.

FIGURE I Wreckage of the power plant that blew up at Chornobyl.

can increase the incidence of thyroid cancer. Contaminated reindeer had to be destroyed in Finland, Sweden, and Norway. Both milk and crops were destroyed in the Ukraine, Poland, Germany, Austria, and Hungary. Another strategy to protect people from the ill effects of iodine-131 was to distribute nonradioactive sodium iodide to the populations with instructions about taking it in drinking water. By thus raising the level of nonradioactive iodide ion, the fraction of radioactive iodide ion taken up by the thyroid gland would be reduced.

Despite these measures, scientists of Belarus (the former Soviet state just north of Ukraine) reported in 1992 that a great increase in cancer of the thyroid had occurred among Belarus children since 1990 (114 new cases from 1990 to 1992 as compared to 15 cases from 1987 to 1989). In the Ukraine, nearly 600 new cases of thyroid cancer developed between 1986 and 1995. Estimates of the total human deaths that could be traceable to Chornobyl radiation releases are almost impossible to make.

Some References

1. Y. M. Shcherbak, "Ten Years of the Chornobyl Era," *Scientific American,* April 1996, page 44.

2. M. Freemantle, "Ten Years After Chernobyl," *Chemical and Engineering News,* April 29, 1996, page 18.

3. M. Edwards and G. Ludwig, "Living with the Chornobyl Monster," *National Geographic,* August, 1994, page 99.

Nuclear reactors are encased in huge chambers called containment vessels with thick walls made of steel-reinforced concrete. If a rupture occurs in the coolant lines that lead into and out of the reactor's core, and there is suddenly a loss of coolant around the core, an emergency backup coolant system is activated, and the reactor is shut down by means of the control rods. Between the time of the first loss of coolant and the successful operation of the backup and shutdown activities, the containment vessel is meant to retain any radioactivity.

A loss-of-coolant emergency occurred in 1979 at a nuclear power plant at Three-Mile Island in Pennsylvania. The emergency systems worked, but it took nearly 7 years of cleanup just to get the *undamaged* reactor back into operation. The worst technology-related environmental disaster in history occurred in 1986 at a nuclear power plant at Chornobyl in what is now the Ukraine (see Interaction 10.6).

Fission Products Are Potential Pollutants The new isotopes produced by fission are radioactive, and their decay leads to radioactive pollutants, which must be contained by the reactor and, later, by safe storage. They include strontium-90 (a bone-seeking element in the calcium family), iodine-131 (a thyroid-gland seeker), and cesium-137 (a group IA radionuclide that goes wherever Na^+ or K^+ can go). The U.S. government has set limits to the release of each radioactive isotope into the air and into the cooling water of nuclear power plants. Plants that operate in compliance with these standards expose people living near them to an extra dose of no more than 5% of the dose they normally receive from background radiation.

Atomic wastes that do escape enter food chains wherever they deposit. Milk from cows that have grazed on pastures contaminated by iodine-131 fallout, for example, carries this isotope into the human diet.

Wastes from Nuclear Power Plants Must Be Kept Apart from Human Contact for a Thousand Years   One of the most vexing problems of nuclear energy has been the permanent storage of long-lived radioactive wastes. Most are now in temporary storage at nuclear power plants. Because several waste radionuclides have very long half-lives, the wastes must be sequestered from all human contact for at least a thousand years, and scientists are seeking deep geologic formations out of all contact with mining operations or underground water supplies into which these wastes can be placed. The sites must be marked and continuously maintained so that archeologists several centuries hence will not unknowingly venture into them.

Alternatives to Nuclear Power Pose Several Problems   Power plants that use petroleum or coal are huge emitters of carbon dioxide, a greenhouse gas. When coal or petroleum also contains sulfur impurities, the power plants emit sulfur dioxide, a contributor to acid rain. Moreover, what few people know is that deaths from the mining of coal (so far) greatly exceed deaths related to nuclear power, including the mining of nuclear fuels. Many people call for drastic reductions in the use of all currently exploited forms of energy—by the conservation of energy, and by the use of wind energy, solar energy, and other technologies.

■   The entire reactor becomes a radioactive waste once the power plant's working lifetime is over.

SUMMARY

Atomic radiation   Radionuclides in nature emit alpha radiation (helium nuclei), beta radiation (electrons), and gamma radiation (high-energy X-ray-like radiation). This radioactive decay causes transmutation. The penetrating abilities of the different kinds of radiation are a function of the sizes of the particles, their charges, and the energies with which they are emitted. Gamma radiation, which has no associated mass or charge, is the most penetrating.

Each decay can be described by a nuclear equation in which mass numbers and atomic numbers on either side of the arrow must balance. To describe how stable a radionuclide is we use its half-life, and the shorter this is, the more radioactive is the radionuclide.

The decay of one radionuclide doesn't always produce a stable nuclide. Uranium-238 is at the head of a radioactive disintegration series that involves several intermediate radionuclides until a stable isotope of lead forms.

Ionizing radiation—dangers and precautions   When radiation travels in matter, it creates unstable ions and radicals that have chemical properties dangerous to health. Intermittent exposure can lead to cancer, tumors, mutations, and birth defects. Intense exposures cause radiation sickness and death. Intense exposures focused on cancer tissue are used in cancer therapy. The uses of distance, fast film, and dense shielding material are the best strategies to guard against the hazards of ionizing radiation. According to the inverse-square law, the intensity of radiation falls off with the square of the distance from the source. Complete protection, however, is not possible because of the natural background radiation, which now includes traces of radioactive pollutants.

Units of radiation measurement   To describe activity we use the SI unit of the becquerel, Bq, or the curie, Ci. To describe the intensity of exposure to X rays or gamma radiation, we use the SI unit X or the roentgen. The gray (Gy) or the rad is used to describe the absorbed dose, how much energy has been absorbed by a unit mass of tissue (or other matter). To put the damage that different kinds of radiation can cause when they have the same values of rads (or grays) on a comparable and additive basis, we use the sievert or the rem. Finally, to describe the energy possessed by a radiation, we use the electron-volt. Diagnostic X rays are on the order of 100 keV. The radiation used in cancer treatment is in the low MeV range. To measure radiation there are devices such as film badges, scintillation counters, and ionization counters (Geiger-Müller tubes).

Synthetic radionuclides A number of synthetic radionuclides have been made by bombarding various isotopes with alpha radiation, neutrons, or accelerated protons. The target nucleus first accepts the mass, charge, and energy of the bombarding particle, and then it ejects something else to give the new nuclide.

Radiation technology in the food industry X rays, gamma rays, and accelerated electrons can be used to inhibit or to kill insects, molds, and bacteria on seeds, potatoes, fruit, and meats. High doses fully sterilize the products. Low doses inhibit bacterial or mold growth. Radiolytic products, those formed by the chemical reactions induced by the radiation, are generally the same as the substances found naturally in food either before cooking and digestion or after. The resistance of the botulinum bacillus to radiation is one problem with the technology.

Radionuclides in medicine For diagnostic uses, the radionuclide should have a short half-life (but not so short that it decays before any benefit can be obtained). It should decay by gamma radiation only, and it should be chemi-

cally compatible with the organ or tissue so that either a hot spot or a cold spot appears. Its decay products should be as stable as possible, and capable of being eliminated from the body.

Technetium-99m is almost ideal for diagnostic work, particularly for assessing the ability of an organ or tissue to function. Radionuclides of iodine (iodine-123 and iodine-131) are used in diagnosing or treating thyroid conditions. Linear accelerators also provide high-energy (6 to 12 MeV) radiation for cancer therapy.

Atomic energy and radioactive pollutants The reactors of most nuclear power plants use uranium-235 as a fuel. When its atoms capture neutrons, they fission into smaller, usually radioactive atoms as neutrons are released that can cause additional fissions. The concentration of uranium-235 atoms is kept too small (3% to 5%) to make an atomic bomb type of explosion possible at a nuclear reactor. Circulating water keeps the reactor cool enough not to melt. The heat generated by fission converts water into high-pressure steam that drives electrical turbines. A containment vessel surrounding the reactor is intended to prevent the escape of radioactive materials should a loss-of-coolant emergency arise.

Radioactive byproducts of fission, such as iodine-131, strontium-80, and cesium-137, cannot be allowed to enter the food supply and must be contained. Some fission products have such long half-lives that radioactive wastes must be kept from human contact for a thousand years.

REVIEW EXERCISES

The answers to Review Exercises whose numbers are in color are found in Appendix E. The answers to the other Review Exercises are found in the Study Guide that accompanies this book. The more challenging questions are marked with asterisks.

Radioactivity and the Kinds of Radiation

10.1 Distinguish between *radioactive decay* and *transmutation*.

10.2 What are the names and symbols used in nuclear equations for the three types of naturally occurring atomic radiation?

10.3 The emission of which naturally occurring atomic radiation would *not* be accompanied by transmutation if it were the sole radiation from a radionuclide?

10.4 In balancing *chemical* equations we seek both a material balance and a charge balance.
(a) How do we check for material balance in a nuclear equation?
(b) Why do we ignore the question of charge balance by not showing electrical charges on reactants or products?

(c) In what other sense, however, do we consider charge balance?

10.5 The energy of an alpha particle is often higher than that of beta or gamma rays. Why, then, is it the least penetrating of the kinds of radiation?

10.6 The loss of an alpha particle changes the radionuclide's mass number by how many units? Its atomic number by how many units?

10.7 Why does the loss of a beta particle not change the radionuclide's mass number but *increases* its atomic number?

10.8 How many neutrons are in the nucleus of $^{232}_{92}U$?

10.9 If electrons do not exist in the nucleus, how can one originate in a nucleus in beta decay?

Nuclear Equations

****10.10** Write the symbols of the missing particles in the following nuclear equations.
(a) $^{211}_{82}Pb \rightarrow {}^{0}_{-1}e +$ _____
(b) $^{220}_{86}Rn \rightarrow {}^{4}_{2}He +$ _____ (c) $^{140}_{56}Ba \rightarrow {}^{0}_{-1}e +$ _____

*10.11   Write a balanced nuclear equation for each of the following changes.
(a) Alpha emission from neodymium-144
(b) Beta emission from potassium-40
(c) Beta emission from samarium-149
(d) Alpha and gamma emission from californium-251

Half-Lives

10.12   Lead-214 is in the uranium-238 disintegration series. Its half-life is 26.8 min. Explain in your own words what being in this series means and what *half-life* means.

10.13   Which would be more dangerous to be near, a radionuclide that has a short half-life and decays by alpha emission only, or a radionuclide that has the same half-life but decays by beta and gamma emission? Explain.

*10.14   A 12.00-ng sample of technetium-99*m* will still have how many nanograms of this radionuclide left after three half-life periods?

Dangers of Ionizing Radiation

10.15   We have ions in every fluid of the body. Why, then, is ionizing radiation dangerous?

10.16   What is a chemical *radical,* and why is it chemically reactive?

10.17   Ionizing radiations is a teratogenic agent. What does this mean?

10.18   Cesium-137 is a carcinogen. What does this mean?

10.19   What two properties of ionizing radiation are exploited in strategies for providing radiation protection?

10.20   Ionizing radiation is said to have no exposure threshold. What does this mean?

10.21   How is it that the same agent, radiation from a radionuclide, can be used both to cause cancer and to cure it?

10.22   The inverse-square law tells us that if we double the distance from a radioactive source, we will reduce the radiation intensity received by a factor of what number?

10.23   What general property of radiation is behind the inverse-square law?

10.24   List as many factors as you can that contribute to the background radiation.

10.25   Why does a trip in a high-altitude jet plane increase a person's exposure to background radiation?

10.26   A radiologist discovered that at a distance of 1.80 m from a radioactive source, the intensity of radiation was 140 millirad. How far should the radiologist move away to reduce the exposure to 2 millirad?

Units of Radiation Measurement

10.27   What SI unit is used to describe the activity of a radioactive sample?

10.28   What is the SI unit and the older unit that describe the intensity of an exposure to X rays?

10.29   What are the name and symbol of the SI unit used in describing how much energy a given mass of tissue receives from exposure to radiation? What are the name and symbol of the older, common unit?

10.30   How does the SI define the dose equivalent? What is meant by the *quality factor*? Which has the higher quality factor, alpha radiation or beta radiation? Explain.

10.31   Approximately how many rads would kill half of a large population within 4 weeks, assuming that each individual received this much? From a health protection standpoint, how do the roentgen and the rad compare in their potential danger?

10.32   We cannot add a 1-rad dose of gamma radiation to some organ to a 1-rad dose of neutron radiation to the same organ and say that the total biologically effective dose is 2 rads. Why not?

10.33   How is the problem implied by the previous review exercise resolved?

10.34   In units of mrem, what is the average natural background radiation received by the U.S. population, exclusive of medical sources, radioactive pollutants, and fallout?

10.35   What is the name of the energy unit used to describe the energy associated with an X ray or a gamma ray?

10.36   In the unit traditionally used (Review Exercise 10.35), how much energy is associated with diagnostic X rays?

10.37   In general terms, how does a film badge dosimeter work?

10.38   In your own words, how does a Geiger-Müller counter work? Why doesn't it detect alpha radiation?

Synthetic Radionuclides

*10.39   When manganese-55 is bombarded by protons, the neutron is one product. What else is produced? Write a nuclear equation.

*10.40   To make gallium-67 for diagnostic work, zinc-66 is bombarded with accelerated protons. When a nucleus of zinc-66 captures a proton, the nucleus of what isotope forms? Write the nuclear equation.

*10.41   When boron-10 is bombarded by alpha particles, nitrogen-13 forms and a neutron is released. Write the equation for this reaction.

*10.42   When nitrogen-14 is bombarded with deuterons, ^2_1H, oxygen-15 and a neutron form. Write the equation for this reaction.

*10.43   What bombarding particle can change sulfur-32 into phosphorus-32 and a proton? Write the equation.

Radiation Technology and the Food Industry

10.44 What is meant by the term *radiolytic product*?

10.45 What is meant by the term *unique* radiolytic product?

10.46 Are any radiolytic products known cancer-causing substances? Are they also found in nonirradiated foods? Give an example.

10.47 What nonradiation processes produce the same compounds as food irradiation?

10.48 What potential hazard in food is the most difficult to remove by radiation?

Medical Applications of Radiation

10.49 Why is it desirable to use a radionuclide of short half-life in diagnostic work, when we know that even small samples of such isotopes can be very active?

10.50 We know that gamma radiation is the most penetrating of all natural radiation. Why, then, is a diagnostic radionuclide that emits only gamma radiation preferable to one that gives, say, only alpha radiation?

10.51 Why is iodine-123 better for diagnostic work than iodine-131?

Atomic Energy and Radionuclides

10.52 What is *fission*?

10.53 In general terms, how does fission differ from radioactive decay?

10.54 What fundamental aspect of fission makes it possible for it to proceed as a chain reaction?

10.55 Which naturally occurring radionuclide is able to undergo fission?

10.56 What concentration is the fissionable isotope in atomic reactors? What concentration is it in an atomic bomb?

10.57 The heat generated by fission in a power plant reactor is carried away in what way? And for what purpose?

10.58 What makes a loss of coolant an emergency?

10.59 What events are supposed to happen in a loss-of-coolant emergency?

10.60 What is a "containment vessel" at a nuclear power plant and what is its purpose?

10.61 The reactors of nuclear power plants are fitted with many movable rods. (Some are made of carbon, for example.) These rods are effective in capturing neutrons. To turn down a reactor or to turn it off, these rods are pushed into the regions where fission occurs. How can their presence control the rate of fission?

10.62 Name three isotopes made in nuclear power plants that are particularly hazardous to health, and explain in what specific ways they endanger various parts of the body.

10.63 What fact about certain atomic wastes necessitates very long waste storage times?

Cosmic Rays (Interaction 10.1)

10.64 Where do cosmic rays originate?

10.65 What is the chief particle found in primary cosmic rays?

10.66 What happens to cosmic rays as they enter the earth's atmosphere?

Radon in the Environment (Interaction 10.2)

10.67 How is radon-222 produced in the environment?

10.68 What kinds of radiation does it emit?

10.69 Besides its own radiation, what other factors make radon-222 in the lungs particularly hazardous?

X Rays and CT Scans (Interaction 10.3)

10.70 In general terms, how are X rays prepared?

10.71 How does the CT scanner differ from an ordinary X-ray machine?

Positron Emission Tomography—The PET Scan (Interaction 10.4)

10.72 Compare the positron and electron in terms of mass and charge.

10.73 Describe how a positron forms in a positron-emitting radionuclide.

10.74 What property makes the lifetime of a positron extremely short?

10.75 When a radiologist uses the PET scan, what radiation is converted into an X-ray-like picture?

10.76 In general terms, how can a positron-emitting radionuclide be gotten inside a tissue, like the brain?

Magnetic Resonance Imaging—the MRI Scan (Interaction 10.5)

10.77 Why is the MRI less harmful a technique than the CT or PET scan?

10.78 How does MRI complement the use of X rays?

10.79 Why is bone transparent to the MRI?

Chornobyl (Interaction 10.6)

10.80 Was the explosion at Chornobyl like that of an atomic bomb or was it a steam explosion?

10.81 How can extra amounts of nonradioactive iodide ion in the diet provide some protection against radioactive iodide ion?

Additional Exercises

10.82   What is the *one* most appropriate term to apply to a *radionuclide*: element, isotope, compound, mixture, atom, ion, or molecule? Explain.

*10.83   Complete the following nuclear equations by writing the symbols of the missing particles.
(a) $^{149}_{62}\text{Sm} \rightarrow {}^{4}_{2}\text{He} + \underline{\hspace{1cm}}$
(b) $^{245}_{96}\text{Cm} \rightarrow {}^{4}_{2}\text{He} + \underline{\hspace{1cm}}$
(c) $^{22}_{9}\text{F} \rightarrow {}^{0}_{-1}e + \underline{\hspace{1cm}}$

*10.84   Give the nuclear equation for each of the following radioactive decays.
(a) Beta emission from rhenium-187
(b) Alpha and gamma emission from plutonium-242
(c) Beta emission from iodine-131
(d) Alpha emission from americium-243

*10.85   If a patient is given 12.00 ng of iodine-123 (half-life 13.3 h), how many nanograms of this radionuclide remain after 12 half-life periods (about a week)?

10.86   Using a Geiger-Müller counter, a radiologist found that in a 20-minute period the dose from a radioactive source would measure 80 mrad at a distance of 10.0 m. How much dose would be received in the same time by moving to a distance of 2.00 m?

10.87   A hospital purchased a sample of a radionuclide rated at 1.5 mCi. What does this rating mean?

10.88   Why should diagnostic radiation ideally be of much lower energy than radiation used in therapy, in cancer treatment, for example?

*10.89   To make indium-111 for diagnostic work, silver-109 is bombarded with alpha particles. What forms if the nucleus of silver-109 captures one alpha particle? Write the nuclear equation for this capture.

*10.90   The compound nucleus that forms when silver-109 captures an alpha particle (previous review exercise) decays directly to indium-111, plus *two* other identical particles. What are they? Write the nuclear equation for this decay.

*10.91   The isotope that forms when zinc-66 captures a proton is unstable in a novel way (novel at least to our study). This nucleus is able to capture one of its own electrons. When it does, what new nucleus forms? Write the equation for this kind of nuclear event, called *electron capture*.

*10.92   When fluorine-19 is bombarded by alpha particles, both a neutron and a nucleus of sodium-22 form. Write the nuclear equation, including the compound nucleus that is the intermediate.

*10.93   What bombarding particle could change aluminum-27 into phosphorus-32 and a proton? Write the equation.

10.94   The unit commonly used to describe how much radiation has been given to a food product is the *kilorad,* krad. Using conversion factors supplied in Section 10.3, calculate how much energy is in 1.00 krad in units of J/kg.

10.95   Why is radiation sickness called an *acute* effect to radiation exposure? What is a particularly common *latent* effect of radiation exposure?

10.96   The hydroxyl radical is an electrically neutral species, yet it is dangerous in tissue. Explain.

ORGANIC CHEMISTRY. SATURATED HYDROCARBONS

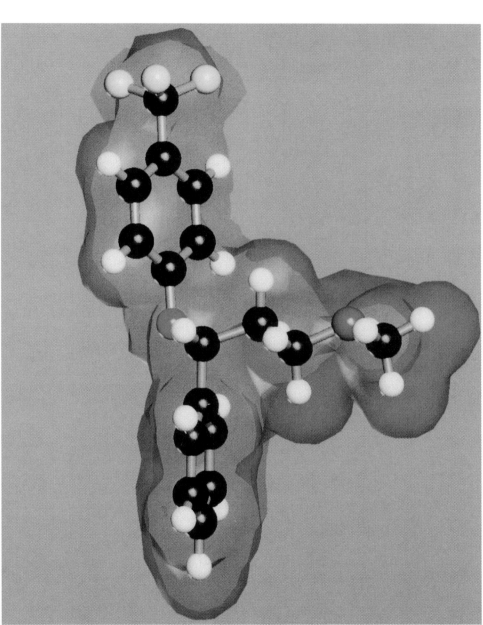

Prozac, seen here in a molecular model, is only one of several million organic compounds, but its calming effects on the mind have made it among the best known. What we'll learn about the architecture of organic molecules in this chapter will serve us for the rest of our study.

THIS CHAPTER IN CONTEXT

As we said in the Preface, our goal is to look at some of the major processes of life at the molecular level. Our questions are "How does nature work?" and, more specifically, "How does life work?" The answer to the latter is never as complete as we can make it without knowing the *chemistry* of life's processes. The surprising bonus is that once you know this, you'll even have an aesthetic experience, because the answers are beautiful.

Aside from water, the substances at the molecular level of life are almost entirely organic compounds. Most have complex molecules, however, with more than one *functional group,* so we're not going to tackle them right away. A functional group is a small part of a molecule that, well, *functions.* It is able to give the same set of chemical reactions wherever it's found. For the next few chapters we'll be surveying the functional groups that occur among biological chemicals, only we'll study the groups among much simpler compounds.

11.1 ORGANIC AND INORGANIC COMPOUNDS

Organic and inorganic compounds differ in composition, bond types, and molecular polarities.

Organic compounds are those made of *carbon* atoms covalently bonded to each other and to atoms of other nonmetals, like hydrogen, oxygen, nitrogen, sulfur, or the halogens. All others are **inorganic compounds,** but even they include a few that contain carbon, like the carbonates, bicarbonates, cyanides, and the oxides of carbon.

In the popular press, "organic" (as in "organic foods") has come to mean "produced without the use of pesticides or synthetic fertilizers or hormones." We will use the traditional meaning. **Organic chemistry** is the study of the structures, properties, and syntheses of organic compounds.

Wöhler's Experiment Opened the Door to the Laboratory Synthesis of Organic Compounds The word "organic" arose from an association with living organisms, because in the early days all organic compounds were isolated from living systems or their remains. Until 1828, all efforts to synthesize organic from inorganic compounds had failed, and out of such repeated failures the **vital force theory** emerged. It stated that it is actually impossible to make organic compounds in glass vessels, that a special *vital force* said to be found only in living systems was essential.

In 1828, presumably trying to make a sample of crystalline ammonium cyanate, NH_4NCO, Friedrich Wöhler (1800–1882) boiled the water from an aqueous solution containing the ammonium ion and the cyanate ion, NCO^-. The white solid he obtained, however, was an unexpected compound, urea. Ammonium cyanate, then as now, is regarded as an inorganic compound, but urea is clearly a product of metabolism. Wöhler had succeeded in making the first organic compound in a glass vessel. The heat used for boiling evidently caused the following reaction.

$$NH_4NCO \xrightarrow{\text{heat}} \underset{\text{urea}}{H-\overset{\overset{\displaystyle H}{|}}{N}-\overset{\overset{\displaystyle O}{\|}}{C}-\overset{\overset{\displaystyle H}{|}}{N}-H}$$

ammonium
cyanate

■ *Vita-* is from a Latin root meaning "life."

■ Urea is the chief nitrogen waste in the urine of animals. It is also manufactured from ammonia for use as a commercial fertilizer. Whether made by animals or machinery, urea is urea and is "organic." Plants accept no fertilizer other than what they are used to.

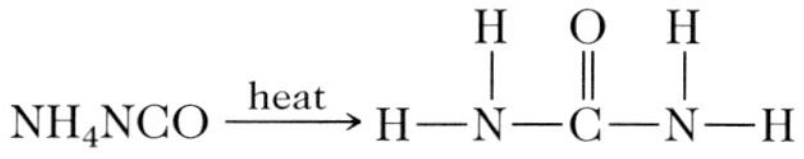

Following Wöhler's discovery, other organic compounds were made from inorganic substances, and more and more chemists began to question the vital force theory. Today, well over six million organic compounds are known, and all have been or could in principle be made from inorganic substances.

The significance to human well-being of the development of organic chemistry as a science is incalculable. Although large numbers of useful organic substances occur in nature and are still obtained from nature, one cannot imagine today's world of synthetic fabrics, dyes, and plastics, as well as most pharmaceuticals, without the ability to synthesize organic compounds in glass vessels. It's a huge human enterprise. Among all scientific specialties, there are more organic chemists than any other single kind of chemist. The education of those entering biochemistry, medicinal chemistry, pharmaceutical chemistry, polymer chemistry, molecular biology, and the primary health care fields of nursing and doctoring includes at its core the study of organic chemistry. One specialist at The Johns Hopkins School of Medicine insists that we cannot say we *know* what a disease is until we know its *chemistry,* and organic chemical principles are at the heart of this knowledge. The study of these principles begins with the kinds of *bonds* that hold organic molecules together.

Covalent Bonds, Not Ionic Bonds, Dominate Organic Molecules   The overwhelming prevalence of nonmetal atoms in organic compounds means that their molecular structures are dominated by *covalent* bonds. In contrast, most inorganic compounds are *ionic.* As we will see, carbon–carbon and carbon–hydrogen bonds are the most prevalent in organic molecules. These bonds are essentially nonpolar, so organic compounds tend to be relatively nonpolar except when atoms of such electronegative elements as oxygen and nitrogen are present. These structural facts are behind several major properties of organic compounds, like relatively low melting and boiling points and low solubility in water.

Most organic compounds have melting points and boiling points well below 400 °C, whereas most ionic compounds melt or boil far above this temperature. The reason is that the relatively nonpolar molecules in organic substances are unable to attract each other as strongly as can the oppositely charged ions in ionic compounds. But the *permanent* polarity of an organic molecule is only one factor that affects a boiling point or melting point. The *size* of the molecule is also a factor because the larger the size, the stronger are the London forces between molecules. And many organic molecules are huge, particularly those in starch, proteins, nucleic acids, polypropylene, and nylon.

Most organic compounds are relatively insoluble in water, whereas many ionic compounds are soluble. The reason is that weakly polar organic molecules are poorly hydrated, in contrast to ions. This simple fact is particularly relevant at the molecular level of life where the central fluid is water. We must, therefore, be alert during our study of organic chemistry to any molecular features that help molecules dissolve in water.

■ Organic *ions* tend to be very soluble in water.

11.2 Some Structural Features of Organic Compounds

Organic molecules can have straight or branched chains; they can be open chained or cyclic, saturated or unsaturated; and ring systems can be carbocyclic or heterocyclic.

The uniqueness of carbon among the elements is that its atoms can bond to each other successively many times and still form equally strong bonds to atoms of other nonmetals. A typical molecule in the familiar plastic polyethylene has hundreds of carbon atoms covalently joined in succession, and each carbon atom binds enough hydrogen atoms to fill out its full complement of four bonds.

■ Only a short segment of a typical molecule of polyethylene is shown here.

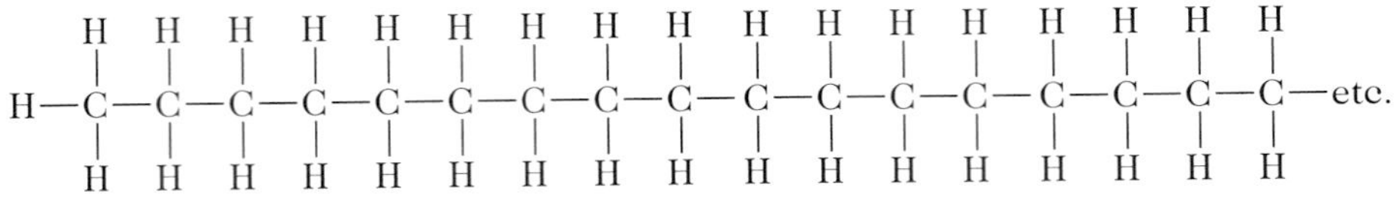

$$H-\overset{\displaystyle H}{\underset{\displaystyle H}{C}}-\overset{\displaystyle H}{\underset{\displaystyle H}{C}}-\overset{\displaystyle H}{\underset{\displaystyle H}{C}}-\overset{\displaystyle H}{\underset{\displaystyle H}{C}}-\overset{\displaystyle H}{\underset{\displaystyle H}{C}}-\overset{\displaystyle H}{\underset{\displaystyle H}{C}}-\overset{\displaystyle H}{\underset{\displaystyle H}{C}}-\cdots-etc.$$

polyethylene (small segment of one molecule)

The sequence of the heavier atoms, here the carbon atoms, is called the *skeleton* of the molecule, and it holds the hydrogen atoms. Many variations of heavier atom skeletons occur, and we will look at them next.

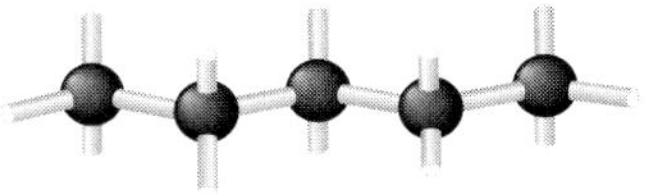

Carbon Skeletons Can Be in Straight Chains or Branched Chains   The carbon skeleton in the polyethylene molecule is described as a **straight chain.** *Straight* has a very limited and technical meaning here: the absence of carbon branches. This means that one carbon follows another, like the pearls in a single-strand necklace, with no additional carbons joined to the skeleton at intermediate points. Pentane illustrates a straight chain of five carbons. The 2-methylpentane molecule has a **branched chain,** a chain with at least one carbon atom joined to the skeleton between the ends of the main chain, like a charm hung on a bracelet.

Straight chain

Branched chain

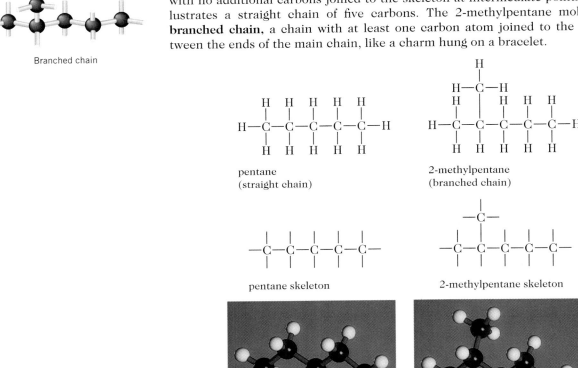

pentane
(straight chain)

2-methylpentane
(branched chain)

pentane skeleton

2-methylpentane skeleton

Some Features of Molecular Geometry Are Often "Understood" When Writing Structural Formulas   When you compare the ball-and-stick models of pentane and 2-methylpentane with their structural formulas, be sure to notice that the written (or printed) structures disregard the correct bond angles at carbon. A carbon that has four single bonds has a tetrahedral geometry with bond angles of 109.5°. The ball-and-stick models faithfully show the correct angles at each carbon, but the printed symbols do not. The point here is that it is perfectly all right to let bond angles be "understood" unless there is some important reason to the contrary.

It is definitely not "all right" to forget about the *geometry* of molecules, however. When we enter the study of biochemistry, particularly the way that enzymes work, we will quickly learn that the geometry of a molecule is just as important as any other feature of its structure. Molecules generally take up whatever shape is permitted by their *bonds* and the *electron clouds* surrounding the individual parts of the molecules. Bonds hold atoms together and whenever a carbon atom has four *single* bonds, the geometry at that point of the molecule will be tetrahedral. The electron clouds tend to push the parts of molecules away from each other and so influence the *overall* shape of the molecule. This "pushing" cannot succeed in splitting stable molecules apart, but it has a huge influence on how a molecule becomes twisted in shape, at least in molecules that have the flexibility permitted by "free rotation," studied next.

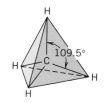
Tetrahedral carbon

Free Rotation at Single Bonds Is Also "Understood" in Structural Formulas   The molecules of pentane and 2-methylpentane are quite flexible, like a necklace. Figure 11.1 shows models of just a few of the many ways the carbon skeleton of the pentane molecule can be flexed. These twistings actually occur as pentane molecules collide with each other in the liquid or gaseous states, and they illustrate an important property of at least large segments of organic molecules, a property called **free rotation** about single bonds. Two clusters of atoms held by a single bond can rotate with respect to each other about the bond. It's as if you had two wheels held by a common axle and either wheel can spin if the other is stuck.

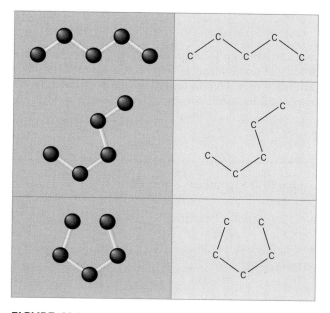

FIGURE 11.1   Free rotation at single bonds. Three of the innumerable conformations of the skeleton of the pentane molecule are shown here. (The hydrogens have been omitted.) Free rotation about single bonds easily converts one conformation into another.

■ The innumerable conformations of pentane cannot be separated from each other.

■ Because of free rotation, we have to be able to interpret zigzags. For example,

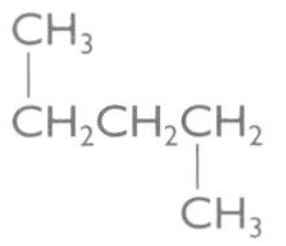

$$CH_3$$
$$|$$
$$CH_2CH_2CH_2$$
$$|$$
$$CH_3$$

is the same molecule as

$$CH_3CH_2CH_2CH_2CH_3$$

The differently twisted forms of pentane in Figure 11.1 are called *conformations,* and samples of liquid or gaseous pentane having molecules in only one conformation cannot be isolated. Free rotation prevents this. The physical and chemical properties of pentane, therefore, are the net results of the effects that the whole collection of conformations has on whatever physical agent or chemical reactant has been used to observe the property. Generally, however, one conformation is present in a relatively high concentration, the conformation that corresponds to the most stable arrangement of the electron clouds. The top conformation in Figure 11.1, for example, gets the electron clouds of the various parts of the pentane molecule as far apart as they can be within the molecule. With the forces of repulsion thus at a minimum in this conformation, the molecule has its lowest potential energy and so has its greatest stability. It's an example of *nature's preference for the lowest energy, most stable arrangements.* This preference is the basic principle behind all aspects of molecular shape within a given environment.

Condensed Structural Formulas Reduce Clutter with Little Sacrifice in Information
Just as we can leave some aspects of molecular geometry to the informed imagination, we can leave most of the bonds in a structural formula to it as well. Remembering that any carbon in a structural formula must have four bonds (or else carry some charge), we can group beside its symbol, C, all the hydrogen atoms held by it. If it holds three hydrogens, we can write CH_3 (or H_3C, but you don't see this as often). Just remember that these three hydrogen atoms are individually joined to the carbon. The structure of ethane illustrates this.

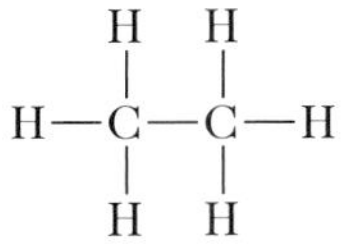

$$\begin{array}{cc} H & H \\ | & | \\ H-C-C-H \\ | & | \\ H & H \end{array} \qquad CH_3-CH_3 \quad \text{or} \quad H_3C-CH_3$$

ethane
(expanded
structure)

ethane (condensed structures)

A carbon holding two hydrogen atoms can be represented as CH_2 or (seen less often) as H_2C. When a carbon holds just one hydrogen, we can write it as CH or HC.

The result of these simplifications is called a **condensed structure,** or simply the **structure.** These kinds of structures will be used almost exclusively in our continuing study.

EXAMPLE 11.1   Condensing a Full Structural Formula

Condense the structural formula for 2-methylpentane.

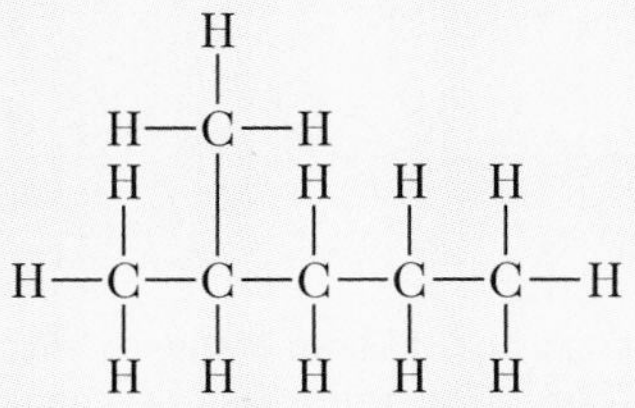

$$\begin{array}{c} H \\ | \\ H-C-H \\ H \quad | \quad H \quad H \quad H \\ | \quad | \quad | \quad | \quad | \\ H-C-C-C-C-C-H \\ | \quad | \quad | \quad | \quad | \\ H \quad H \quad H \quad H \quad H \end{array}$$

2-methylpentane

ANALYSIS Each unit of 3 H's attached to the same C becomes CH_3. Where 2 H's are joined to the same C, write CH_2. Where C holds only one H write CH.

SOLUTION

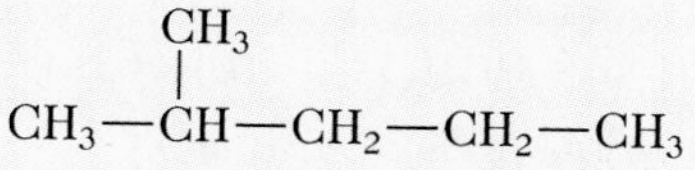

$$CH_3-\underset{\underset{CH_3}{|}}{CH}-CH_2-CH_2-CH_3$$

CHECK We formulate here a general check rule for all molecular structures. *Scan all bond connections to verify that the rules of covalence have been obeyed.* In the answer, each C has four bonds and each H one bond. When you find a violation, the answer cannot possibly be correct, so fix it.

■ PRACTICE EXERCISE 1 Condense the following expanded structural formulas.

(a) 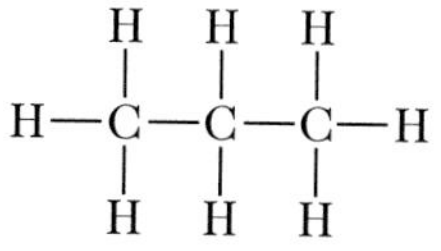
$$H-\underset{\underset{H}{|}}{\overset{\overset{H}{|}}{C}}-\underset{\underset{H}{|}}{\overset{\overset{H}{|}}{C}}-\underset{\underset{H}{|}}{\overset{\overset{H}{|}}{C}}-H$$

(b)
$$H-\underset{\underset{H}{|}}{\overset{\overset{H}{|}}{C}}-\underset{\underset{\underset{\underset{H}{|}}{H-C-H}}{|}}{\overset{\overset{H}{|}}{C}}-\underset{\underset{H}{|}}{\overset{\overset{H}{|}}{C}}-H$$

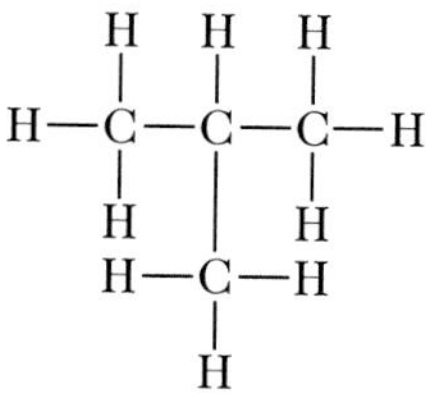

(c)
$$H-\underset{\underset{\underset{\underset{H}{|}}{H-C-H}}{|}}{\overset{\overset{\overset{\overset{H}{|}}{H-C-H}}{|}}{C}}-\underset{\underset{H}{|}}{\overset{\overset{\overset{\overset{H}{|}}{H-C-H}}{|}}{C}}\!\!-\!\!-\!\!\underset{\underset{H}{|}}{\overset{\overset{\overset{H}{|}}{H-C-H}}{C}}\!\!-\!\!-\!\!\underset{\underset{H}{|}}{\overset{\overset{H}{|}}{C}}-\underset{\underset{H}{|}}{\overset{\overset{H}{|}}{C}}-H$$

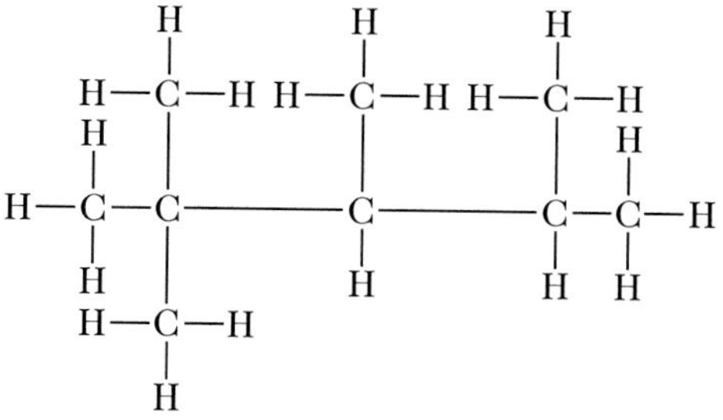

Even Most Single Bonds Can Be "Understood" When a *single* bond appears on a *horizontal* line, we need not write a straight line to represent it; we can leave such single bonds to the imagination. We do not do this, however, for bonds that are not on a horizontal line. Thus we can write the structure of 2-methylpentane, Example 11.1, as follows. Notice that the vertically oriented bond is shown by a line but that all other single bonds are understood.

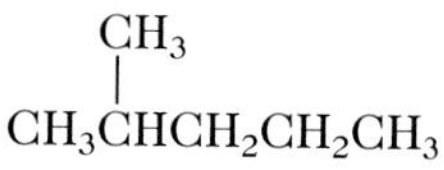

$$\underset{\text{2-methylpentane}}{CH_3\underset{\overset{|}{CH_3}}{CH}CH_2CH_2CH_3}$$

■ PRACTICE EXERCISE 2 Rewrite the condensed structures that you drew for the answers to Practice Exercise 1 and let the appropriate carbon–carbon single bonds be left to the imagination.

■ PRACTICE EXERCISE 3   Just to be certain that you are comfortable with condensed structures, expand each of the following to make them full, expanded structures with no bonds left to the imagination.

             CH₃   CH₃
               | |
(a) CH₃CH₃   (b) CH₃CHCHCH₃   (c) CH₃CH₂CCH₂CH₂CH₃
                | |
               CH₃   CH₃

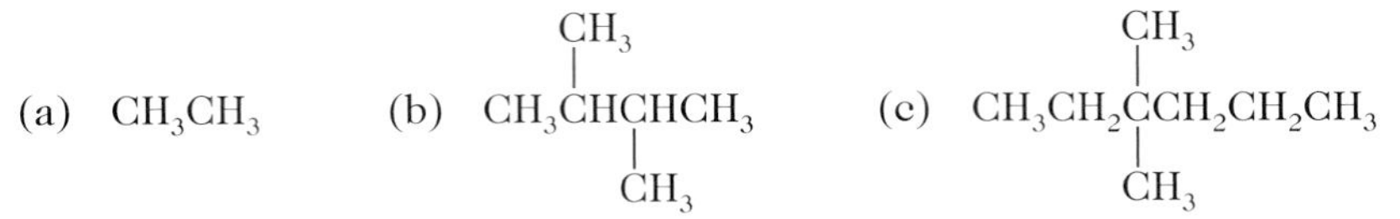

As indicated in the *check* part of Example 11.1, an important skill in using condensed structures is the ability to recognize errors. The most common is a violation of the rule that every carbon atom in a structure that carries no electrical charge must have exactly four bonds, no more and no fewer. Do the next Practice Exercise to test your skill in recognizing an error in structure.

■ PRACTICE EXERCISE 4   Which of the following structures cannot represent real compounds?

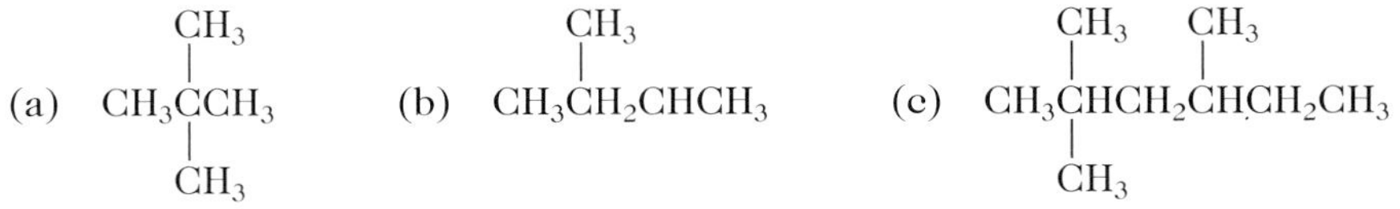

     CH₃   CH₃   CH₃   CH₃
      | | | |
(a) CH₃CCH₃ (b) CH₃CH₂CHCH₃ (c) CH₃CHCH₂CHCH₂CH₃
      | |
     CH₃   CH₃

When atoms other than carbon and hydrogen are present in a molecule, there is no major new problem in writing condensed structures. Remember that in molecules without double bonds, every oxygen or sulfur atom carrying no electrical charge must have two bonds, every nitrogen must have three, and every halogen atom must have just one.

Double and Triple Bonds Are Seldom Condensed   Another rule about condensed structures is that carbon–carbon double and triple bonds are never left to the imagination. Carbon–oxygen double bonds are sometimes condensed, as some of the following examples illustrate. Study them as illustrations of how to condense structures.

    H H
    | |
H—C—C—OH condenses to CH₃—CH₂—OH or to CH₃CH₂OH
    | |
    H H

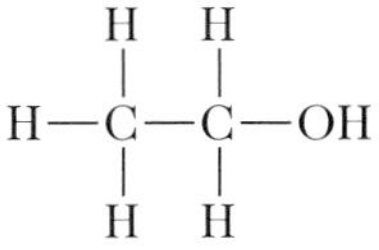

ethyl alcohol (in alcoholic drinks)

    H H
    | |
H—C=C—H condenses to CH₂=CH₂ or to H₂C=CH₂

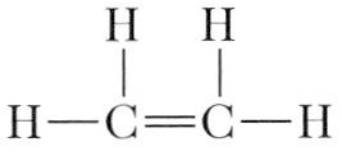

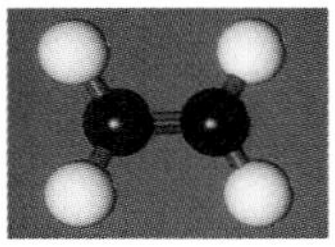

ethylene (raw material for making polyethylene)

    H O
    | ||
H—C—C—OH condenses to CH₃—C—OH or to CH₃COH and often to
    | CH₃CO₂H or to CH₃COOH
    H

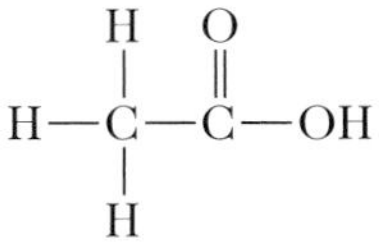

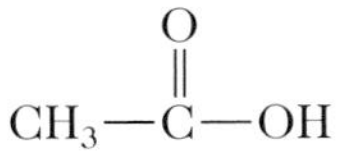

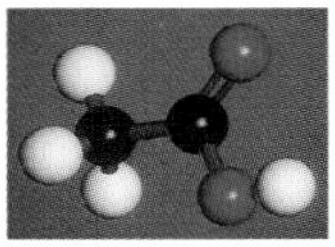

acetic acid (in vinegar)

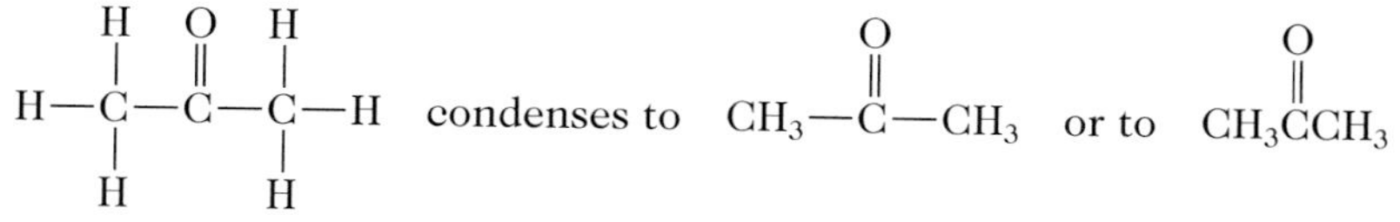

$$H{-}\underset{\underset{H}{|}}{\overset{\overset{H}{|}}{C}}{-}\overset{\overset{O}{\|}}{C}{-}\underset{\underset{H}{|}}{\overset{\overset{H}{|}}{C}}{-}H \quad \text{condenses to} \quad CH_3{-}\overset{\overset{O}{\|}}{C}{-}CH_3 \quad \text{or to} \quad CH_3\overset{\overset{O}{\|}}{C}CH_3$$

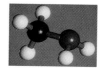

acetone (nail polish remover)

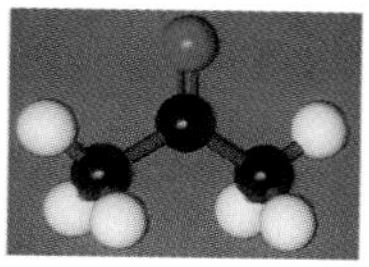

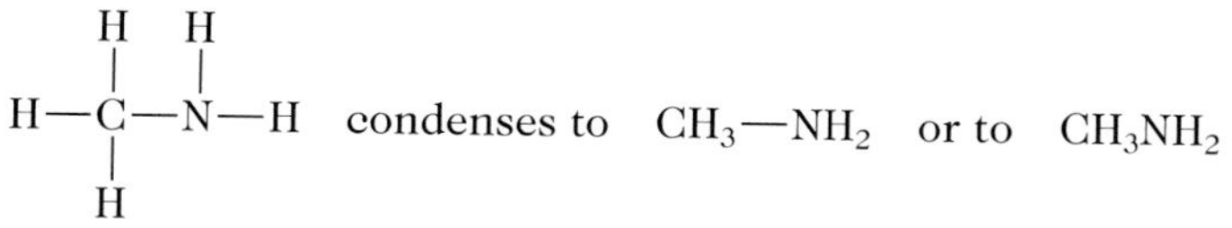

$$H{-}\underset{\underset{H}{|}}{\overset{\overset{H}{|}}{C}}{-}\underset{\underset{H}{|}}{\overset{\overset{H}{|}}{N}}{-}H \quad \text{condenses to} \quad CH_3{-}NH_2 \quad \text{or to} \quad CH_3NH_2$$

methylamine (in decaying fish)

Parentheses Are Sometimes Used to Condense Structures Further   Sometimes two or three identical groups that are attached to the same carbon are grouped inside a set of parentheses. For example,

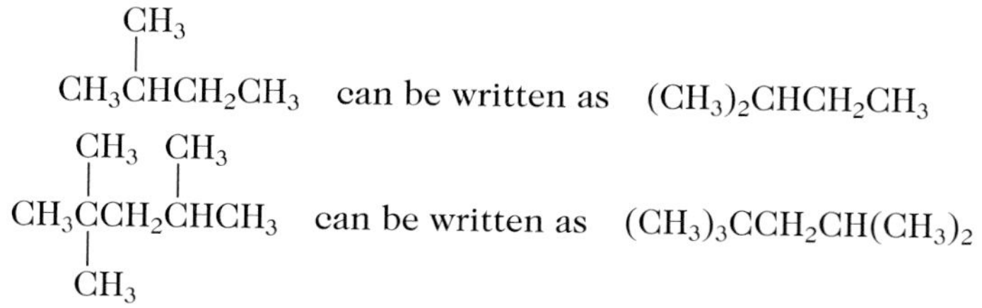

$$\underset{}{\overset{\overset{CH_3}{|}}{CH_3CHCH_2CH_3}} \quad \text{can be written as} \quad (CH_3)_2CHCH_2CH_3$$

$$\underset{\underset{CH_3}{|}}{\overset{\overset{CH_3 \quad CH_3}{| \quad \;\; |}}{CH_3CCH_2CHCH_3}} \quad \text{can be written as} \quad (CH_3)_3CCH_2CH(CH_3)_2$$

We will not do this often, but you will see it in many references and you should be aware of it.

Unsaturated Compounds Have Double or Triple Bonds   If its molecules have only single bonds, the compound is called a **saturated compound.** When one or more double or triple bonds are present, the substance is an **unsaturated compound.** Thus ethylene, acetic acid, and acetone, shown previously, are all unsaturated compounds, but ethyl alcohol and methylamine are saturated.

    Saturated describes any molecule all of whose atoms are directly holding as many other atoms as they can. Each carbon in ethylene, for example, is holding just three atoms, but in ethyl alcohol each is directly holding four. *Unsaturated* implies that something can be added, and we will see that unsaturated compounds can add certain substances, like hydrogen, to their double or triple bonds.

■ Molecules of edible oils, like olive oil or corn oil, have many double bonds and are described as "polyunsaturated."

Many Organic Molecules Contain Rings of Carbon Atoms   A carbon **ring** is an arrangement of three or more carbon atoms into a closed cycle. Molecules with this feature are called ring compounds or cyclic compounds. (Sometimes an all-carbon ring is described as *carbocyclic.*) Cyclohexane molecules, for example, have a ring of six carbon atoms. Cyclopropane, once an important anesthetic, is also a cyclic compound.

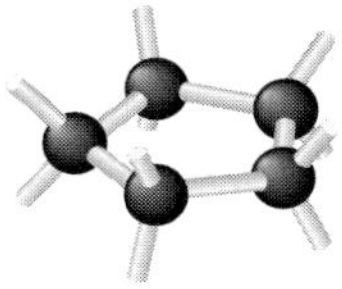

Carbon ring

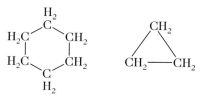

cyclohexane   cyclopropane

■ More than 2 billion pounds of cyclohexane are made annually in the United States, with over 90% being used to make nylon.

Cyclic compounds can have double bonds as in cyclohexene, but always remember that carbon–carbon double bonds are never left to the imagination. Rings, of course, can carry substituents, as in ethylcyclohexane. Not all the ring atoms have to be carbon atoms. They can be O, N, or S, too, and cyclic compounds with ring atoms other than C are called **heterocyclic compounds.** A simple example is tetrahydropyran.

■ The ring system in tetrahydropyran—5 C's plus 1 O in the ring—is widely present among molecules of carbohydrates.

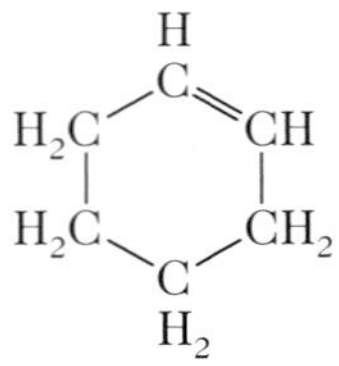

cyclohexene 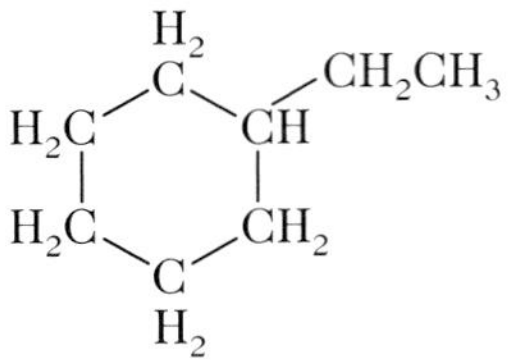   ethylcyclohexane 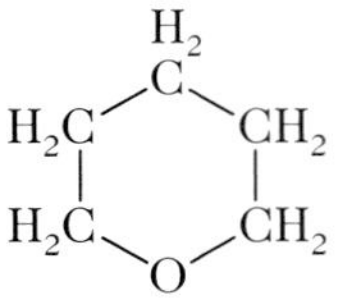   tetrahydropyran

The Rings of Cyclic Compounds Can Be Condensed to Simple Polygons   Since we can leave to our imaginations so many structural features, a polygon, a many-sided figure, becomes a handy way to condense rings. A square, for example, can represent cyclobutane. The photograph of the ball-and-stick model of cyclobutane and its progressively more condensed structures show what is left to the imagination when just a square is used. At each corner, we have to understand that there is a CH_2 group. Each line in the square is a carbon–carbon single bond.

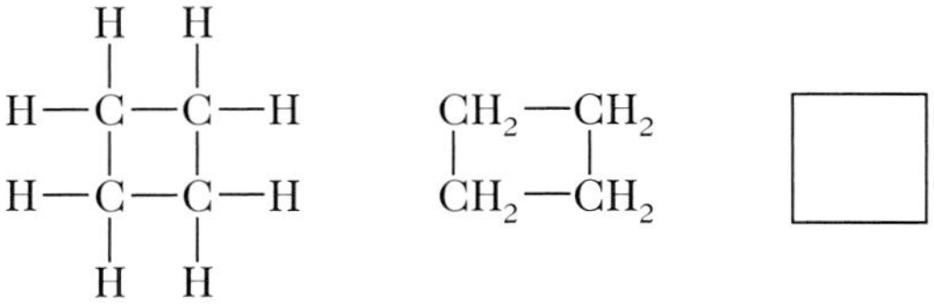

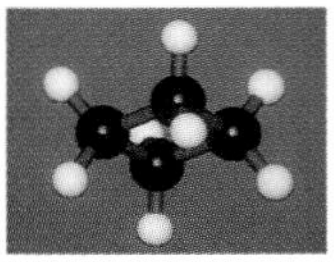

$$CH_2—CH_2$$
$$CH_2—CH_2$$

Three ways to represent the structure of cyclobutane

The model of methylcyclopentane and its progressively more condensed structures further show how a polygon can represent a ring.

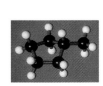

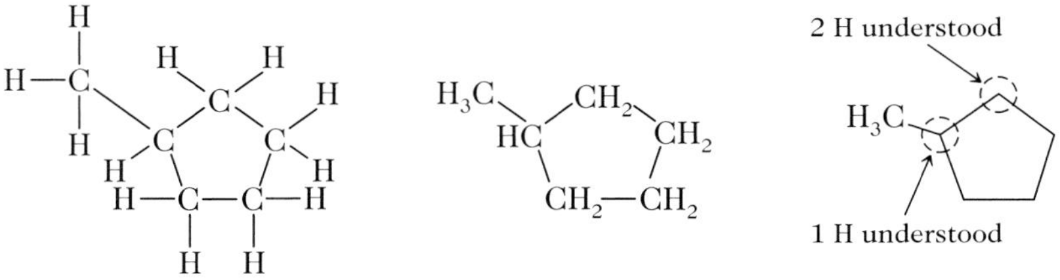

2 H understood

1 H understood

Three ways to represent the structure of methylcyclopentane

By convention, polygons like the hexagon for cyclohexane can be used to represent rings provided that we understand the following rules.

1. C occurs at each corner unless O or N (or another multivalent atom) is explicitly written at a corner.

2. A line connecting two corners is a covalent bond between adjacent ring atoms.

3. Remaining bonds, as required by the covalence of the atom at a corner, are understood to hold H atoms.

4. Double bonds are always explicitly shown.

We can illustrate these rules with the following cyclic compounds.

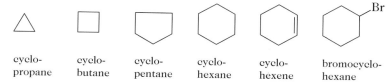

cyclo-   cyclo-   cyclo-   cyclo-   cyclo-   bromocyclo-
propane   butane   pentane   hexane   hexene   hexane

There is no theoretical upper limit to the size of a ring.

EXAMPLE 11.2   Understanding Condensed Structures of
Ring Compounds

To make sure that you can read a condensed structure when it includes a polygon for a ring system, expand this structure, including its side chains.

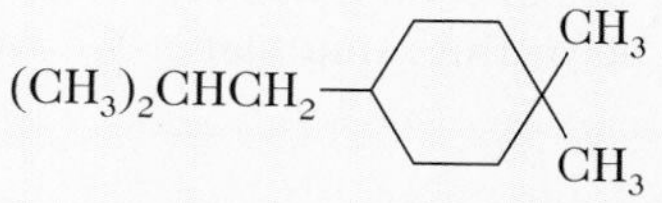

$$(CH_3)_2CHCH_2-\text{(ring)}\genfrac{}{}{0pt}{}{CH_3}{CH_3}$$

ANALYSIS Every carbon atom, every bond, and every H atom has to be shown explicitly.

SOLUTION

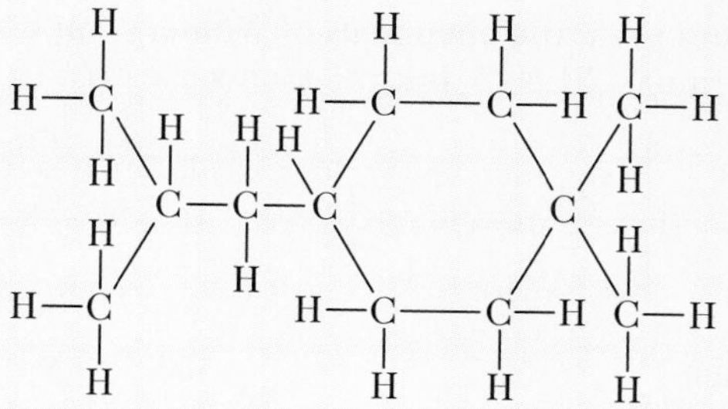

Note especially how we can tell the numbers of hydrogens that must be attached to a ring atom. We need as many as required to fill out a set of four bonds from each carbon. This example also shows a situation (on the right side of the ring) in which no bonding room is left for holding an H atom.

PRACTICE EXERCISE 5   Expand each of these two structures.

(a)   (b)

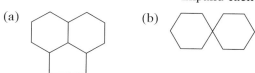

■ PRACTICE EXERCISE 6   Condense the following structure.

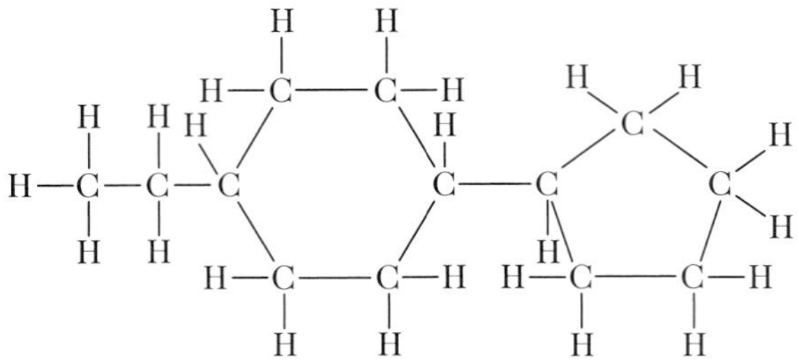

As you might expect, free rotation about the single bonds in a ring is not possible. Although there is a small amount of "flex" in rings, you'd have to break single bonds to get the kind of free rotation we observed with pentane (Figure 11.1).

When we use polygons to represent saturated rings that have six or more ring atoms, we gloss over one feature of such molecules. The ring atoms that make up the rings of this size do not all lie in the same plane. We will postpone a study of what this fact implies.

11.3 ISOMERISM

Compounds can have identical molecular formulas but different structures.

Ammonium cyanate and urea, the chemicals of Wöhler's important experiment, both have the molecular formula CH_4N_2O, but the atoms are organized differently.

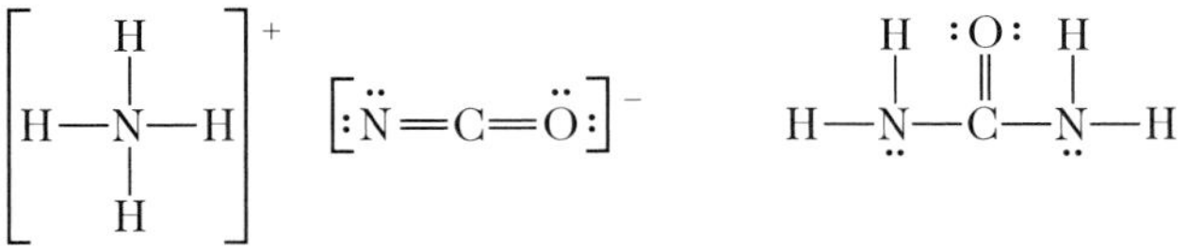

ammonium
cyanate, CH_4N_2O

urea, CH_4N_2O

■ "Isomer" has Greek roots— *isos,* the same, and *meros,* parts; that is, "equal parts" (but put together differently).

Compounds that have identical molecular formulas but different structures are called **isomers** of each other, and the existence of isomers is a phenomenon called **isomerism.** Isomerism is one reason why there are so many organic compounds.

There are several kinds of isomers, and we will consider just one type here, **constitutional isomers.** Constitutional isomers, once called *structural isomers,*[1] differ in the basic atom-to-atom connectivities. There are three constitutional isomers of C_5H_{12}, for example, pentane, 2-methylbutane, and 2,2-dimethylpropane.

■ The names in parentheses are the common names of these compounds. The letter *n* stands for *normal,* meaning the straight-chain isomer. *Neo* signifies *new,* as in a new isomer.

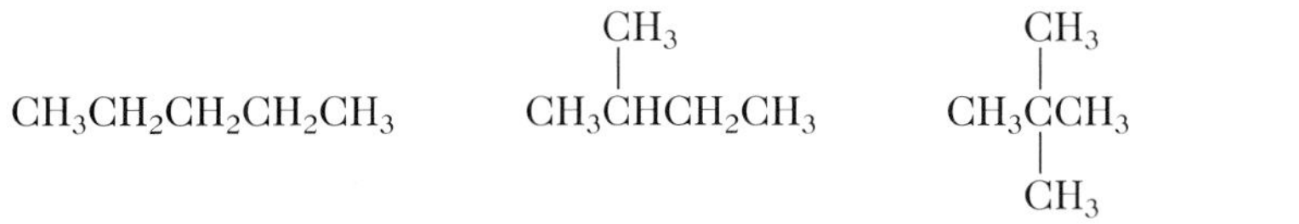

pentane
(*n*-pentane)

2-methylbutane
(isopentane)

2,2-dimethylpropane
(neopentane)

[1] The term "structural isomer" has fallen into disfavor because it is regarded as too broad, that it implies not only atom-to-atom connectivities but also geometrical differences. Another kind of isomerism, geometrical isomerism (next chapter), deals with the latter.

The larger the number of carbon atoms per molecule, the larger is the number of isomers. For example, C_8H_{18} has 18 isomers; $C_{10}H_{22}$ has 75, and $C_{20}H_{42}$ has 366,319. Someone has figured out that roughly 6.25×10^{13} isomers are possible for $C_{40}H_{82}$. (Only a tiny fraction have actually been prepared. It would take nearly 200 billion years to make each one at the rate of one per day.)

The isomers of pentane or of $C_{40}H_{82}$ have quite similar chemical properties because their molecules all have only C—C and C—H single bonds. Often, however, isomers have very different properties. There are two ways, for example, to organize the atoms in C_2H_6O into constitutional isomers, as seen near the top of Table 11.1. One isomer is ethyl alcohol and the other is dimethyl ether. They are radically different compounds, as the data in Table 11.1 show. For example, at room temperature, ethyl alcohol is a liquid and dimethyl ether is a gas. Ethyl alcohol reacts with sodium; dimethyl ether does not. It is quite common for isomers to have properties this different. Therefore, *we nearly always use structural rather than molecular formulas for organic compounds.* Only *structures* let us see at a glance how the atoms in the molecules are organized. The ability to recognize two structures as identical molecules, or as isomers or something else, is very important. We'll practice this next.

TABLE 11.1   Properties of Two Isomers: Ethyl Alcohol and Dimethyl Ether

Property	Ethyl Alcohol	Dimethyl Ether
Structure	CH_3CH_2OH	CH_3OCH_3
Boiling point	78.5 °C	−24 °C
Melting point	−117 °C	−138.5 °C
Density (25 °C)	0.79 g/mL (a liquid)	2.0 g/L (a gas)
Solubility in water	Soluble in all proportions	Slightly soluble

EXAMPLE 11.3   Recognizing Isomers

Which pair of structures represents a pair of isomers?

1. CH_3—O—CH_2CH_3 and CH_3CH_2—O—CH_3

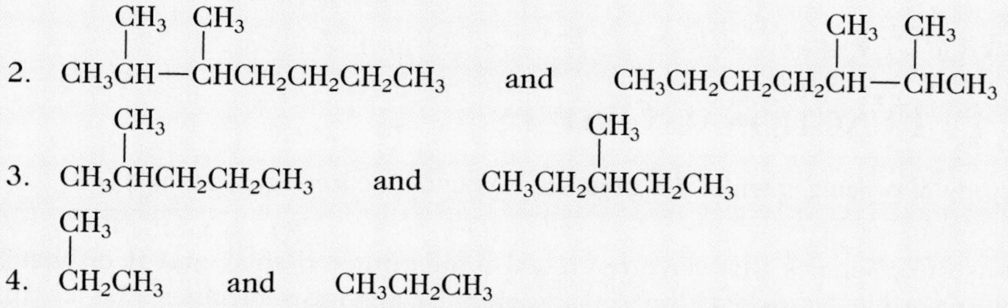

2. CH_3CH—$CHCH_2CH_2CH_2CH_3$ and $CH_3CH_2CH_2CH_2CH$—$CHCH_3$
    with CH_3 groups on the indicated carbons

3. $CH_3CHCH_2CH_2CH_3$ and $CH_3CH_2CHCH_2CH_3$
    with CH_3 groups on the indicated carbons

4. CH_2CH_3 and $CH_3CH_2CH_3$
    with CH_3 group on the indicated carbon

ANALYSIS   Unless you spot a difference that rules out isomerism immediately, the *first step* is to see whether the molecular formulas are the same. If they aren't, the two structures are *not* isomers. If the molecular formulas of the two structures are identical, they might be identical or they might be isomers.

In this problem, the members of each pair share the same molecular formula.

Pair 1: C_3H_8O Pair 2: C_9H_{20} Pair 3: C_6H_{14} Pair 4: C_3H_8

Next, to see whether a particular pair represents isomers we try to find at least one structural difference. If we can't, the two structures are identical; they are just oriented differently on the page, or their chains are twisted into different conformations. Don't be fooled by an "east-to-west" versus a "west-to-east" type of difference. The difference must be *internal* within the structure. (Whether you face east or west you're the same person!)

SOLUTION Pair 1 molecules are identical; they're only oriented differently. (Imagine using a pancake turner to flip the one on the left, left to right; it would then be the structure on the right.)

Pair 2 is also an example of an east versus west difference in orientation. These two structures are identical. Their internal sequences—their atom-to-atom connectivities—are the same.

Pair 3 molecules are isomers. In the first, a CH_3 group joins a five-carbon chain at the chain's second carbon, and in the second, this group is attached at the third carbon.

Pair 4 molecules are identical. The two structures differ only in the conformations of their chains. Recall that free rotation about bonds allows us to imagine the straightening out of a continuous, open chain.

■ PRACTICE EXERCISE 7 Examine each pair to see whether the members are identical, are isomers, or are different in some other way.

(a) $H-O-CH_3$ and CH_3-O-H

(b) $CH_3-NH-CH_3$ and $CH_3-CH_2-NH_2$

(c) 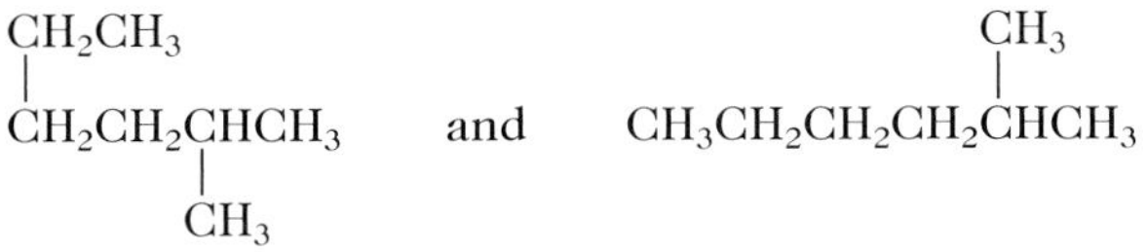 $\overset{\displaystyle CH_2CH_3}{\underset{\displaystyle CH_3}{\overset{|}{CH_2CH_2CHCH_3}}}$ and $\overset{\displaystyle CH_3}{\overset{|}{CH_3CH_2CH_2CH_2CHCH_3}}$

(d) $CH_2{=}CHCH_2CH_3$ and $CH_3CH{=}CHCH_3$

(e) 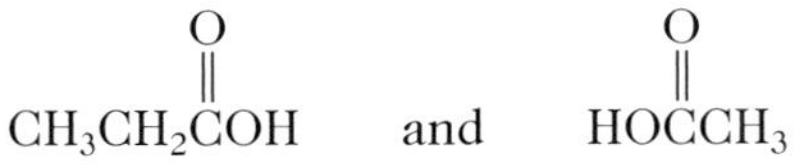 $CH_3CH_2\overset{\displaystyle O}{\overset{\|}{C}}OH$ and $HO\overset{\displaystyle O}{\overset{\|}{C}}CH_3$

11.4 FUNCTIONAL GROUPS

The study of organic chemistry is organized around functional groups.

Regions of molecules that have nonmetal atoms other than C and H or that have double or triple bonds are the specific sites in organic molecules that chemicals most often attack. These small structural units are called **functional groups**, because they are the chemically functioning parts of molecules. Sections of molecules consisting only of carbon and hydrogen and only single bonds are called the **nonfunctional groups**.

Each Functional Group Defines an Organic Family Although over six million organic compounds are known, there are only a handful of functional groups, and each one serves to define a family of organic compounds. Our study of organic chemistry will be organized around just a few of these families, those outlined in Table 11.2. Let's see how the idea of a family will greatly simplify our study.

TABLE 11.2   Some Important Families of Organic Compounds

Family	Characteristic Structural Features[a]	Examples
Hydrocarbons	Only C and H present	

Families of hydrocarbons

Alkanes: only single bonds		CH_3CH_3
Alkenes: C=C		CH_2=CH_2
Alkynes: C≡C		HC≡CH
Aromatics: benzene ring		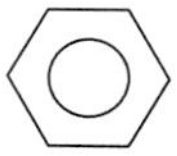

Family	Structural Features	Examples
Alcohols	ROH	CH_3CH_2OH
Ethers	ROR′	CH_3OCH_3
Thioalcohols	RSH	CH_3SH
Disulfides	RS—SR	CH_3S—SCH_3
Aldehydes	$\overset{\displaystyle O}{\overset{\|}{R}}CH$	$\overset{\displaystyle O}{\overset{\|}{CH_3}}CH$
Ketones	$\overset{\displaystyle O}{\overset{\|}{R}}CR'$	$\overset{\displaystyle O}{\overset{\|}{CH_3}}CCH_3$
Carboxylic acids	$\overset{\displaystyle O}{\overset{\|}{R}}COH$	$\overset{\displaystyle O}{\overset{\|}{CH_3}}COH$
Esters of carboxylic acids	$\overset{\displaystyle O}{\overset{\|}{R}}COR'$	$\overset{\displaystyle O}{\overset{\|}{CH_3}}COCH_3$
Esters of phosphoric acid	$\overset{\displaystyle O}{\overset{\|}{R}}O\overset{}{P}\underset{OH}{OH}$	$\overset{\displaystyle O}{\overset{\|}{CH_3}}O\underset{OH}{P}OH$
Esters of diphosphoric acid	ROPOPOH (O O above, HO OH below)	CH₃OPOPOH
Esters of triphosphoric acid	ROPOPO P(OH)₂ (O O O above, HO OH below)	CH₃OPOPOP(OH)₂
Amines	RNH_2, RNHR′, RNR′R″	CH_3NH_2 ; CH_3NHCH_3 ; $CH_3\overset{CH_3}{\overset{\|}{N}}CH_3$
Amides	$\overset{\displaystyle O\;\;\;R''(H)}{\overset{\|\;\;\;\;\;\;\;\;\;\;\;\|}{RC}}$—NR′(H)	$\overset{\displaystyle O}{\overset{\|}{CH_3}}CNH_2$

[a] R, R′, and R″ represent hydrocarbon groups—*alkyl groups*—defined in the text. R′(H) and R″(H) signify that the substituent can be either a hydrocarbon group or hydrogen.

The *alcohols* are an example of a major family of organic compounds. We have learned, for example, that ethyl alcohol is CH_3CH_2OH. Its molecules have the OH group attached to a chain of two carbons, but *chain length* is not what determines a compound's family. Chain length only bears on the *name* of that specific family member, as we'll soon see. The chain can be any length imaginable, and the substance will be in the alcohol family provided that somewhere on the chain there is an OH group attached to a carbon from which only single bonds extend. This functional group is called the *alcohol group*; it is very common in nature, being present in all carbohydrates and most proteins. Some examples of simple alcohols are the following.

■ Isopropyl alcohol is commonly used as a rubbing alcohol.

$$\overset{|}{\underset{|}{-C}}-O-H \qquad CH_3-OH \qquad CH_3CH_2-OH \qquad CH_3CH_2CH_2-OH \qquad \underset{\underset{OH}{|}}{CH_3CHCH_3}$$

| alcohol group | methyl alcohol | ethyl alcohol | propyl alcohol | isopropyl alcohol |

Because all these alcohols have the same functional group, they exhibit the same kinds of chemical reactions. When just one of these reactions is learned, it applies to all members of the family, literally to thousands of compounds. In fact, we will often summarize a particular reaction for an organic family by using a general *family formula*. All alcohols, for example, can be symbolized by R—OH, where R stands for a carbon chain (or ring), one of whatever length or branching (or ring size). All alcohols, for instance, react with sodium metal as follows.

■ R is from the German word *Radikal*, which we translate here to mean *group*, as in a group of atoms.

$$2R-OH + 2Na \longrightarrow 2R-ONa + H_2$$

If we wanted to write the specific example of this reaction that involves, say, ethyl alcohol, all we have to do is replace R by CH_3CH_2.

$$2CH_3CH_2-OH + 2Na \longrightarrow 2CH_3CH_2-ONa + H_2$$

Notice that this reaction changes only the OH group of the alcohol. Dimethyl ether, Table 11.1, which does not have the OH group, cannot give this reaction with sodium, as we have learned in the previous section. Thus, the functional group in dimethyl ether is not the same as that in ethyl alcohol.

Another important organic family is that of the *carboxylic acids*. All their molecules have the *carboxyl group*, which makes all its compounds weak acids. Acetic acid is one of the most common carboxylic acids.

■ The carboxyl group is present in all fatty acids, products of the digestion of the fats and oils in our diets.

$$\overset{O}{\underset{}{\overset{\|}{-C}}}-O-H \qquad RCO_2H \qquad \overset{O}{\underset{}{\overset{\|}{CH_3COH}}} \quad or \quad CH_3CO_2H \quad or \quad CH_3COOH$$

| carboxyl group | carboxylic acids | acetic acid |

We know that acetic acid neutralizes the hydroxide ion.

$$\overset{O}{\overset{\|}{CH_3COH}} + OH^- \longrightarrow \overset{O}{\overset{\|}{CH_3CO^-}} + H_2O$$

acetate ion

All molecules with the carboxyl group give the same reaction, so we can represent literally thousands of such reactions by just one simple equation.

$$\overset{O}{\overset{\|}{RCOH}} + OH^- \longrightarrow \overset{O}{\overset{\|}{RCO^-}} + H_2O$$

carboxylate ion

The groups characteristic of both carboxylic acids and the carboxylate ions are present in all proteins and their building blocks, the amino acids, so you can see why we must include them in our study.

The amino acids are examples of substances with more than one functional group in the same molecule. They have the carboxyl group as well as the *amino group*, a group that defines the family of the *amines*.

NH₂	CH₃NH₂	RNH₂	NH₂CH₂CO₂H
amino group	methyl- amine	simple amines	glycine, the simplest amino acid

The amines are ammonia-like compounds and, like ammonia, can neutralize hydrochloric acid or any other strong acid. The general equation is

$$RNH_2 + HCl \longrightarrow RNH_3Cl$$

Glycine, having both the amino group and the carboxyl group, is able to neutralize both strong acids (by the NH_2 group) and bases (by the CO_2H group).

You can see how powerful a learning tool the functional group is. In the next few chapters, we will study just a few of the reactions of the most important functional groups found at the molecular level of life. *Learning these reactions will be like mastering a set of map signs.* You can read thousands of maps intelligently once you know their common signs and symbols. Similarly, we'll be able to "read" some of the chemical and physical properties of astonishingly complex molecules with the knowledge of the properties of only a few functional groups.

The functional groups of a molecule generally occur in a setting of alkane-like groups. To be able to contrast the properties of functional groups with those of the alkane-like groups we turn our attention next to the alkanes themselves, the least reactive of the organic systems.

■ The amino group is a proton acceptor, like ammonia.

11.5 ALKANES AND CYCLOALKANES

Alkanes and cycloalkanes are saturated hydrocarbons.

Petroleum and natural gas are substances that consist almost entirely of a complex mixture of molecular compounds called hydrocarbons. Interaction 11.1 describes the nature of petroleum and other chemical fuels in greater detail. **Hydrocarbons** are made from the atoms of just two elements, carbon and hydrogen, and the covalent bonds between the carbon atoms can be single, double, or triple. The carbon skeletons can be chains or rings. These possibilities define the various kinds of hydrocarbons, which are outlined in Figure 11.2.

The **alkanes,** whether open chain or cyclic, are saturated hydrocarbons, those with only single bonds. Table 11.3 gives the first 10 straight-chain members of the family. The **alkenes** are hydrocarbons with one or more carbon–carbon double bonds, whether the skeletons are chains or rings. Hydrocarbons with one or more carbon–carbon triple bonds are **alkynes.** It is possible for one molecule to have both double and triple bonds, of course. We'll study the alkenes (and a little about the alkynes) in the next chapter.

One important distinction in Figure 11.2 is between aliphatic and aromatic hydrocarbons. **Aliphatic compounds** of whatever family have no benzene ring (or a similar system), and **aromatic compounds** are those with such rings. (We'll postpone the chemical implications of the benzene ring to the next chapter.)

■ Figure 11.2 and Table 11.3 are on page 342.

■ Because the bond angle at a triple bond is 180°, a ring has to be quite large to have a triple bond, and cycloalkynes are rare.

■ The first compounds found with benzene rings had pleasant odors and so were called *aromatic*. *Aliphatic* is from the Greek *aliphatos*, fat-like.

INTERACTION 11.1
ORGANIC FUELS

Fossil Fuels Long ago, nature used photosynthesis to transform solar energy into the chemical energy of ancient plants and then locked up this energy in the fossilized remains of these plants, the *fossil fuels*— principally petroleum, coal, and natural gas. These fuels are a legacy from the past now being consumed so rapidly that for the first time in history we are concerned about running out of them and about the impact on civilization if their disappearance occurs too suddenly for us to adapt.

Petroleum and Crude Oil The fossil fuels formed over a span of hundreds of thousands of years during the Carboniferous Period in geologic history, roughly 280 to 345 million years ago. Vast areas of the continents, little more than monotonous plains, then basked in sunlight near sea level. In the oceans, countless tiny, photosynthesizing plants like the diatoms—tons of them per acre of ocean surface during the early spring—soaked up solar energy to power their fugitive lives. Then they died. According to one theory, the death of each such plant released a tiny droplet of oily material that eventually settled into the bottom muds. The muds grew in thickness and compacted, sometimes into a rock called shale, sometimes into limestone and sandstone deposits. Under the pressure and heat of compacting, the oily matter changed into petroleum, a mixture of crude oil, water, and natural gas. (*Petroleum* is from *petra,* "rock" and *oleum,* oil.) In some parts of the world, the petroleum managed to move slowly through porous rock and collect into vast underground pools to form the great petroleum reserves of our planet. In other regions, this movement could not occur, and the oily substances remain to this day locked in enormous deposits of oil shales and oil sands.

The story is told of a gentleman in a western state who built a new home and made the fireplace out of a local shale. When he lit the first fire, both the fireplace and his house burned up! His shale was rich in oil; in the Green River region where Utah, Colorado, and Idaho meet, there is a shale formation estimated to contain 2000 billion barrels of *shale oil* (see Figure 1). (The United States uses roughly 5 billion barrels of oil per year.) When the oil shale is crushed and heated to about 260 °C, a substance essentially the same as petroleum is released. Rock qualifies as oil shale if it holds an average of 10 gallons of petroleum per ton of rock. The cost of wringing the petroleum from the rock is presently too high

FIGURE 1 Oil shale in the Mahogony zone near Parachute, Colorado.

for commercial exploitation, even if the management of the associated environmental problems is excluded from the cost.

Near Lake Athabasca in the province of Alberta, Canada, in a land area about the size of Lake Michigan, there are huge reserves of a material much like crude oil but intermixed with sand, not shale (see Figure 2). Each ton of this *oil sand* holds about two-thirds of a barrel of oil, and the total deposit of oil sand is estimated to contain over 600 billion barrels of oil. This is equivalent to about half of the entire petroleum reserves of the world.

Coal On the marshy lands bordering the ancient oceans, lush vegetation flourished and died in a moist

FIGURE 2 Syncrude's tar sands project in Alberta, Canada.

and sunny setting. The rate of decay of the remains of these plants, covered by stagnant, oxygen-poor, often acidic water, was slower than the rate at which the plants died. The slowly rotting mass accumulated to huge depths, became fibrous, and turned to *peat,* a woody material used as fuel in some regions of the world.

Where peat layers became thick enough or were compressed by later deposits of sedimentary rock like limestone and sandstone, the peat changed into lignite ("brown coal"). Although lignite is over 40% water, it is still an important fuel. Many lignite deposits became thick enough for further compaction to occur, and most of the water was squeezed out. Thus bituminous coal ("soft coal") formed, which has less than 5% water but contains considerable quantities of volatile matter. (*Volatile* means "easily evaporated.") When still further compaction took place and nearly all of the volatile matter was forced out, anthracite ("hard coal") formed, which is over 95% carbon.

The energy content in the coal reserves of the world exceeds that of the known petroleum reserves by a wide margin. In estimates that have the available supplies of petroleum lasting less than a century, the coal reserves are considered to have a lifetime of up to three centuries. Coal contains very small quantities of compounds of mercury, sulfur, radioactive elements and other potential pollutants. Their concentrations are very small, as we said, but because of the enormous quantities of coal annually burned, mostly at electric power plants, air pollution problems are caused, which often lead to water pollution as well.

The mercury pollution in many lakes north of the industrial regions of the United States stems in part from the mercury released by burning coal.

Natural Gas In both soft coal and petroleum, the most volatile hydrocarbons, methane and ethane, also accumulated. Natural gas is largely methane.

Useful Substances from Crude Oil by Refining Crude oil is a complex mixture of organic compounds, but nearly all are hydrocarbons. Small but vexing amounts of sulfur-containing compounds are also present. The object of refining crude oil is to separate this mixture into products of varying uses. Refinery operations yield mixtures of compounds called *fractions* that boil over certain ranges of temperatures (see the accompanying table). Roughly 500 compounds occur among the fractions boiling up to 200 °C; about a third are alkanes, a third are cycloalkanes, and a third are aromatic hydrocarbons. You can see in the table where the chief fuels for transportation—gasoline, diesel oil, and jet fuel—originate.

The gasoline fraction of crude oil does not provide nearly enough of the world's needs for gasoline. One of the strategies at petroleum refineries to make more gasoline is to subject high-boiling petroleum fractions to operations called *catalytic cracking* and *reforming.* In the presence of catalysts and when heated, large alkane molecules break up into smaller ones corresponding to lower boiling points that are in the range useful for gasoline engines.

Principal Fractions from Petroleum

Boiling Point Range (in °C)	Molecular Size	Principal Uses
Below 20	C_1 to C_4	Natural gas; heating and cooking fuel; raw materials for other chemicals
20–60	C_5 to C_6	Petroleum "ether," a nonpolar solvent and cleaning fluid
60–100	C_6 to C_7	Ligroin or light naphthas; nonpolar solvents and cleaning fluids
40–200	C_5 to C_{10}	Gasoline
175–325	C_{12} to C_{18}	Kerosene; jet fuel; tractor fuel
250–400	C_{12} and higher	Gas oil; fuel oil; diesel oil
Nonvolatile liquids	C_{20} and up	Refined mineral oil; lubricating oil; grease (a blend of soap in oil)
Nonvolatile solids	C_{20} and up	Paraffin wax; asphalt; road tar; roofing tar

TABLE 11.3   Straight-Chain Alkanes

IUPAC Names	Carbons	Molecular Formula	Structure	Boiling Point (°C)	Melting Point (°C)	Density (g/mL, 20 °C)
Methane	1	CH_4	CH_4	−161.5	−182.5	—
Ethane	2	C_2H_6	CH_3CH_3	−88.6	−183.3	—
Propane	3	C_3H_8	$CH_3CH_2CH_3$	−42.1	−189.7	—
Butane	4	C_4H_{10}	$CH_3(CH_2)_2CH_3$	−0.5	−138.4	—
Pentane	5	C_5H_{12}	$CH_3(CH_2)_3CH_3$	36.1	−129.7	0.626
Hexane	6	C_6H_{14}	$CH_3(CH_2)_4CH_3$	68.7	−95.3	0.659
Heptane	7	C_7H_{16}	$CH_3(CH_2)_5CH_3$	98.4	−90.6	0.684
Octane	8	C_8H_{18}	$CH_3(CH_2)_6CH_3$	125.7	−56.8	0.703
Nonane	9	C_9H_{20}	$CH_3(CH_2)_7CH_3$	150.8	−53.5	0.718
Decane	10	$C_{10}H_{22}$	$CH_3(CH_2)_8CH_3$	174.1	−29.7	0.730

[a] The molecular formulas of the open-chain alkanes fit the general formula C_nH_{2n+2}, where n = the number of carbon atoms per molecule.

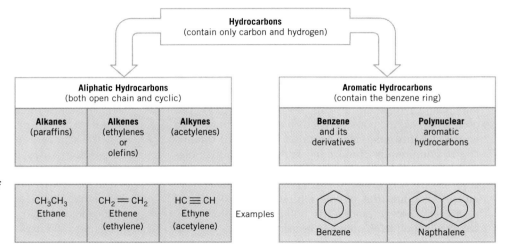

FIGURE 11.2 The several kinds of hydrocarbons. (The circles in the structures for benzene and naphthalene will be explained in the next chapter.)

Hydrocarbons of All Types, Saturated or Not, Have Common Physical Properties
Both the carbon–carbon and the carbon–hydrogen bonds are almost entirely nonpolar, so hydrocarbon molecules have very little if any overall polarity. Hydrocarbons of all types, for this reason, are insoluble in water, but they dissolve well in nonpolar solvents (like CCl_4). Indeed, many special mixtures of alkanes are themselves used as nonpolar solvents. Some people, for example, have used gasoline or lighter fluid—both are mixtures of alkanes—to remove tar spots or grease. (If you do, be sure to keep all flames away and work outside, never in a garage or other enclosed room.)

■ In the right proportion in air, hydrocarbon vapors explode when ignited.

   Hydrocarbons are not only insoluble in water but are generally less dense than water and so they will float on it. Thus, using water to put out a hydrocarbon fire, like flaming gasoline, will only float the flames over a wider area. Nonflammable foams or CO_2 extinguishers must be used instead.

Hydrocarbon Solvents Illustrate the Like-Dissolves-Like Rule   Grease and tar are relatively nonpolar materials, and their solubility in gasoline illustrates a very useful rule of thumb for predicting whether a solvent can dissolve some substance. It's called the **like-dissolves-like rule,** where "like" refers to a likeness in *polarity.* Polar solvents, such as water, are good for dissolving polar or ionic substances, like sugar or salt, because polar molecules or ions can attract water molecules around themselves, form solvent cages, and in this form freely intermingle with water molecules. Nonpolar solvents, like gasoline, do not dissolve sugar or salt, because nonpolar solvent molecules cannot be attracted to polar molecules or ions and form solvent cages.

Because of their low polarity and small size, the hydrocarbons that have 1 to 4 carbons per molecule are generally gases at or near room temperature. As the molecular size increases, however, London forces between molecules become stronger. Therefore, the boiling points of the straight-chain alkanes increase with chain length. Hydrocarbons that have from 5 to about 16 carbon atoms per molecule are usually liquids at room temperature (see Table 11.3). When alkanes have approximately 18 or more carbon atoms per molecule, the London forces are strong enough to make the alkanes (waxy) solids at room temperature. Paraffin wax, for example, is a mixture of alkanes whose molecules have 20 or more carbon atoms.

■ Most candles are made from paraffin.

Our First Molecular "Map Sign," Hydrocarbon-Like Portions of Molecules   Before moving on, let's pause to reflect on what we have done in relating physical properties to structural features. We have introduced the first molecular "map sign" in our study. *Substances whose molecules are entirely or even mostly hydrocarbon-like are likely to be insoluble in water but soluble in nonpolar solvents.* When we see an unfamiliar structure, we can tell at a glance whether it is mostly hydrocarbon-like. If it is, we can predict with considerable confidence that the compound is not soluble in water. The structure of cholesterol illustrates our point.

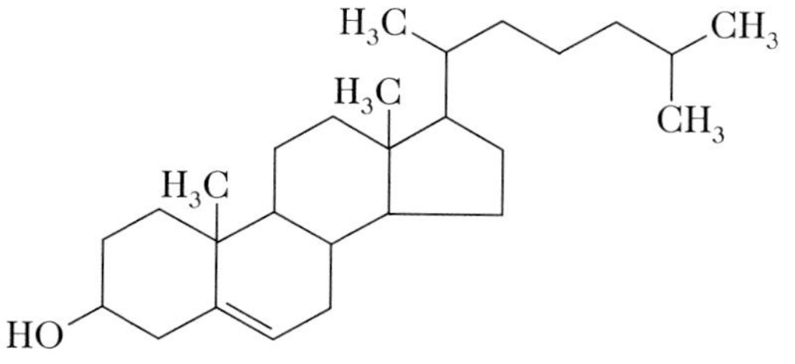

■ The cholesterol structure shown here extends to open-chain systems the formalism of rings. Thus we use a point *where two lines meet* to denote one C and as many H's as needed to fill out a covalence of 4 for the C atom. A line that terminates with nothing would denote CH_3. Thus, we could represent CH_3CH_2OH simply as

cholesterol

You probably know that cholesterol can form solid deposits in blood capillaries, and even close them. The heart must then work harder to sustain the flow of blood, and under this stress a heart attack can occur. Notice, now, that virtually the entire cholesterol molecule is hydrocarbon-like. It has only one polar group, the OH or alcohol group, and this takes up too small a portion of the molecule to make cholesterol sufficiently polar to dissolve either in water or in blood (which is mostly water).

What cholesterol illustrates is that by learning one very general fact, one "map sign," we don't have to memorize a long list of separate (but similar) facts about an equally long list of separate compounds that occur at the molecular level of life. With what we have just learned, we can look at the structures of hundreds of complicated compounds and confidently predict particular properties such as the likelihood of their being soluble in water. You have gained a powerful tool, one that immensely reduces what otherwise might have to be memorized.

EXAMPLE 11.4   Predicting Physical Properties from Structures

Study the following two structures and tell which is the structure of the more water-soluble compound.

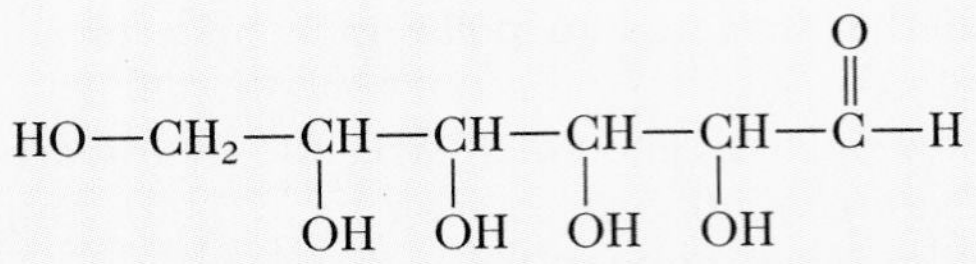

$$\text{HO}-\text{CH}_2-\underset{\underset{\text{OH}}{|}}{\text{CH}}-\underset{\underset{\text{OH}}{|}}{\text{CH}}-\underset{\underset{\text{OH}}{|}}{\text{CH}}-\underset{\underset{\text{OH}}{|}}{\text{CH}}-\overset{\overset{\text{O}}{\|}}{\text{C}}-\text{H}$$

$$\text{CH}_3\text{CH}_2\text{CH}_2\text{CH}_2\text{CH}_2\text{CH}_2\text{CH}_2\text{CH}_2\text{CH}_2\text{CH}_2\text{CH}_2\text{CH}_2\text{CH}_2\text{CH}_2\text{CH}_2\text{CH}_2\text{CO}_2\text{H}$$

ANALYSIS   The structure of the first compound has several polar OH groups but the second is almost entirely like an alkane.

SOLUTION   The first should be (and is) much more soluble in water. The first compound is glucose (in one of its forms), the chief sugar in the blood-stream. The second is stearic acid, which does not dissolve in water. It forms when we digest the fats and oils (lipids) in the diet.

■   PRACTICE EXERCISE 8   Which of the following is more soluble in gasoline?

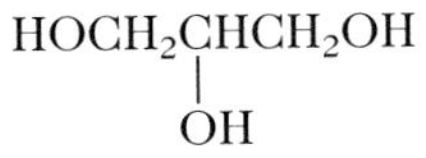

$$\underset{\text{glycerol}}{\text{HOCH}_2\underset{\underset{\text{OH}}{|}}{\text{CH}}\text{CH}_2\text{OH}} \qquad \underset{\text{2-methyl-1-butanol}}{\text{CH}_3\text{CH}_2\underset{\underset{\text{CH}_3}{|}}{\text{CH}}\text{CH}_2\text{OH}}$$

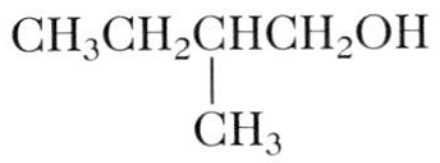

11.6   NAMING THE ALKANES AND CYCLOALKANES

An IUPAC name discloses the compound's family, the number of carbons in the parent chain, and the kinds and locations of substituents.

■   "Nomenclature" is from the Latin *nomen,* name, + *calare,* to call. Wealthy Romans had slaves called *nomenclators* whose duty it was to remind their owners of the names of important people who approached them on the street.

In chemistry, **nomenclature** means the rules for naming compounds. The formal ones are known as the **IUPAC rules,** after the International Union of Pure and Applied Chemistry, whose committees deal with such matters. All scientific societies in the world accept the IUPAC rules. They are so carefully constructed that only one name can be written for each compound, and only one structure can be drawn for each name.

The IUPAC names, unfortunately, are sometimes very long and difficult to write or pronounce. It's much easier to call table sugar *sucrose* than α-D-glucopyranosyl β-D-fructofuranoside, which illustrates why shorter names, referred to as *common names,* are still widely used. We will want to learn some common names, too, and you will see that even they usually have some system to them.

IUPAC Rules for Naming the Alkanes

1.   The name ending for all alkanes (and cycloalkanes) is *-ane.*

2.   The *parent chain* is the longest continuous chain of carbons in the structure. For example, the branched-chain alkane

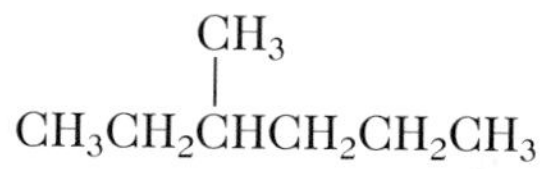

$$CH_3$$
$$|$$
$$CH_3CH_2CHCH_2CH_2CH_3$$

is regarded as being "made" from the following parent

$$CH_3CH_2CH_2CH_2CH_2CH_3$$

by replacing a hydrogen atom on the third carbon from the left with CH_3.

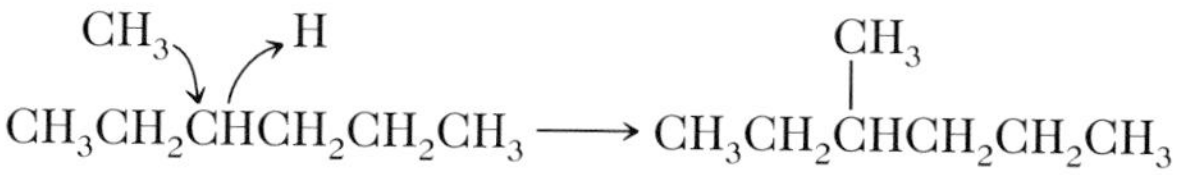

$$CH_3 \searrow \nearrow H \qquad\qquad CH_3$$
$$| \qquad\qquad\qquad\qquad |$$
$$CH_3CH_2CHCH_2CH_2CH_3 \longrightarrow CH_3CH_2CHCH_2CH_2CH_3$$

3. A prefix is attached to the name ending, *-ane*, that specifies the number of carbon atoms *in the parent chain.* The prefixes through parent chain lengths of 10 carbons are as follows *and should be learned.* The names in Table 11.3 show their use.

meth-	1 C	hex-	6 C
eth-	2 C	hept-	7 C
prop-	3 C	oct-	8 C
but-	4 C	non-	9 C
pent-	5 C	dec-	10 C

■ We won't need to know the prefixes for the higher alkanes.

Because the parent chain of our example has six carbons, the parent chain is named hexane—*hex-* for six carbons and *-ane* for being in the alkane family. The alkane whose name we are devising is regarded as a derivative of this parent, *hexane.*

4. The carbon atoms of the parent chain are numbered starting from whichever end of the chain gives the location of the first branch the lower of two possible numbers. Thus the correct direction for numbering our example is from left to right.

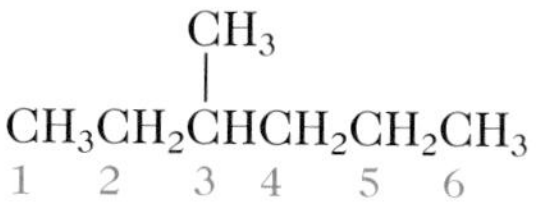

$$CH_3$$
$$|$$
$$CH_3CH_2CHCH_2CH_2CH_3$$
$$\;1\quad\; 2\quad\; 3\quad 4\quad\; 5\quad\; 6$$
(correct direction of numbering)

Had we numbered from right to left, the carbon holding the branch would have had a higher number, which is not allowed by the IUPAC rules for alkanes.

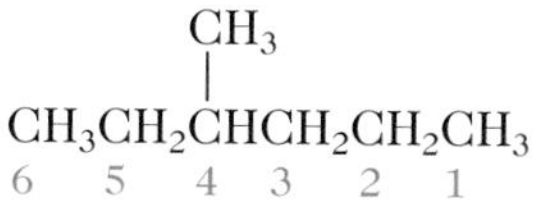

$$CH_3$$
$$|$$
$$CH_3CH_2CHCH_2CH_2CH_3$$
$$\;6\quad\; 5\quad\; 4\quad 3\quad\; 2\quad\; 1$$
(incorrect direction of numbering)

5. Name each branch attached to the parent chain. We must now pause and learn the names of some of the *alkyl groups,* groups with alkane-like branches.

The Alkyl Groups

Any branch that consists only of carbon and hydrogen and has only single bonds is called an **alkyl group,** and the names of all alkyl groups end in *-yl.* Think of an alkyl group as an alkane minus one H.

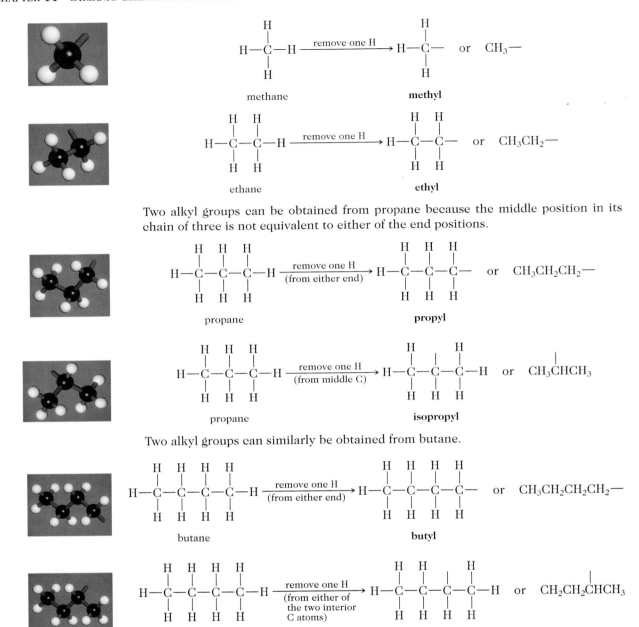

Two alkyl groups can be obtained from propane because the middle position in its chain of three is not equivalent to either of the end positions.

Two alkyl groups can similarly be obtained from butane.

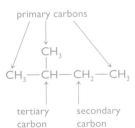

The last alkyl group is called the *secondary* butyl group (abbreviated *sec*-butyl) because the open bonding site is at a **secondary carbon,** a carbon that is directly attached to just two other carbons. A **primary carbon** is one to which just one other carbon is directly attached. The open bonding site in the butyl group, for example, is at a primary carbon atom. A **tertiary carbon** is one that holds directly three other carbons. We will encounter a tertiary carbon in a group that we will soon study.

Butane is the smallest alkane to have an isomer. The common name of the isomer is isobutane, and we can derive two more alkyl groups from it.

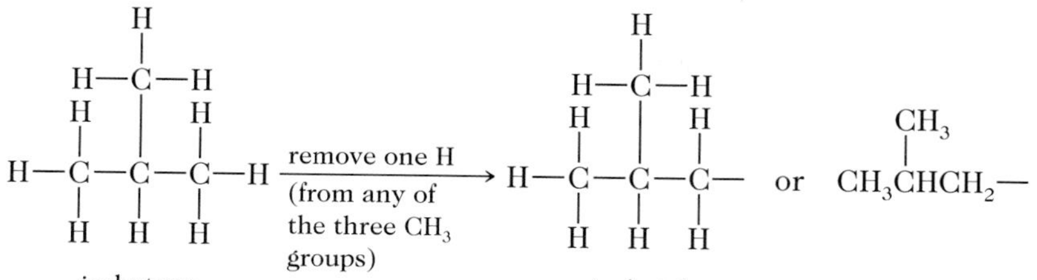

isobutane → isobutyl

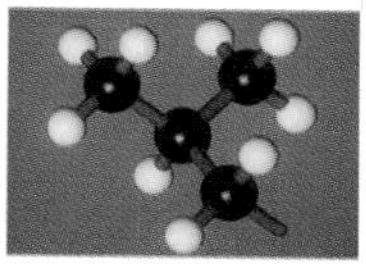

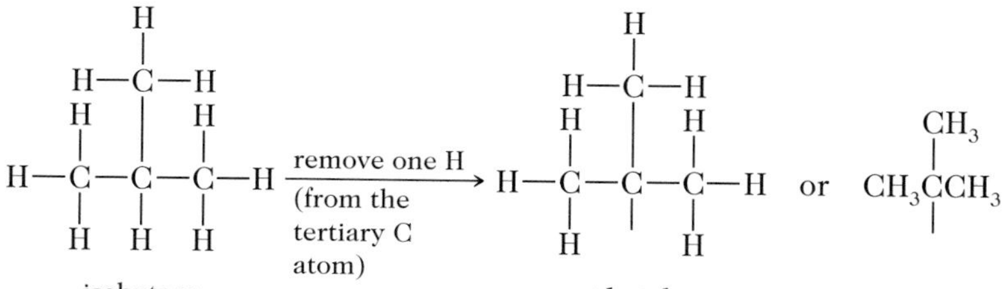

isobutane → *t*-butyl

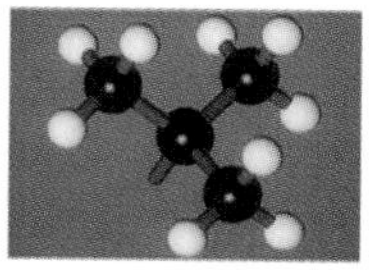

Notice that the open bonding site in the *tertiary*-butyl group (abbreviated *t*-butyl) occurs at a tertiary carbon.

The names and structures of these alkyl groups must now be learned. If you have access to ball-and-stick models, make models of each of the parent alkanes and then remove hydrogen atoms to generate the open bonding sites and the alkyl groups.

The prefix *iso-* in the name of an alkyl group, such as in isopropyl or isobutyl, has a special meaning. It can be used to name any alkyl group that has the following general features.

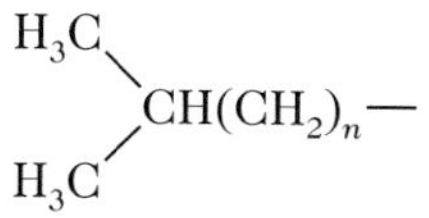

$$H_3C \diagdown \atop H_3C \diagup CH(CH_2)_n- \qquad (n = 0, 1, 2, 3)$$

$n = 0$, isopropyl group
$ = 1$, isobutyl group
$ = 2$, isopentyl group
$ = 3$, isohexyl group

■ The *Study Guide* accompanying this book has exercises that provide drills in recognizing alkyl groups when they are positioned in different ways on the page.

■ Here is another way to condense a structure. Thus $CH_3(CH_2)_3CH_3$ represents $CH_3CH_2CH_2CH_2CH_3$.

Notice that each of these names has a word fragment (*-prop-*, *-but-*, and so forth) associated with a number of carbon atoms. When these word fragments are attached to *iso*, they specify the *total* number of carbons in the alkyl group. Thus the isopropyl group has three carbons (indicated by *prop*) and the isobutyl group has four carbons (indicated by *but*).

We can now continue with the IUPAC rules for naming alkanes.

6. Attach the name of the alkyl group to the name of the parent as a prefix. Place the location number of the group in front of the resulting name and separate the number from the name by a hyphen. The name of our original example is thus 3-methylhexane.

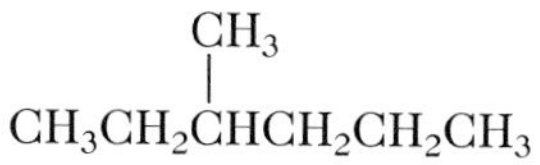

$$\overset{\displaystyle CH_3}{\underset{\displaystyle |}{CH_3CH_2CHCH_2CH_2CH_3}}$$

3-methylhexane

7. When two or more groups are attached to the parent, name each and locate each with a number. The names of alkyl substituents are assembled in their alphabetical order. Always use *hyphens* to separate numbers from words. Here is an application.

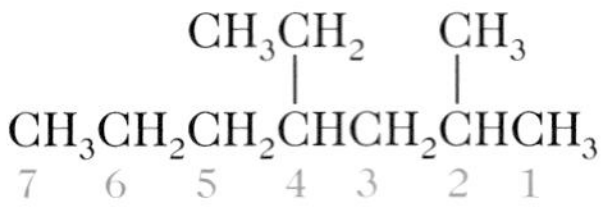

$$\text{CH}_3\text{CH}_2 \quad \text{CH}_3$$
$$| \qquad\quad |$$
$$\text{CH}_3\text{CH}_2\text{CH}_2\text{CHCH}_2\text{CHCH}_3$$
$$7 \quad 6 \quad 5 \quad 4 \quad 3 \quad 2 \quad 1$$

4-ethyl-2-methylheptane

8. When two or more substituents are identical, use such prefixes as di- (for 2), tri- (for 3), tetra- (for 4), and so forth; and specify the location number of every group. Always separate a number from another number in a name by a *comma*. For example,

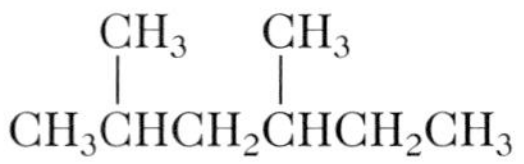

$$\text{CH}_3 \quad\; \text{CH}_3$$
$$| \qquad\quad |$$
$$\text{CH}_3\text{CHCH}_2\text{CHCH}_2\text{CH}_3$$

Correct name:	2,4-dimethylhexane
Incorrect names:	2,4-methylhexane
	3,5-dimethylhexane
	2-methyl-4-methylhexane

9. When identical groups are on the *same* carbon, repeat the number locating this carbon in the name. For example,

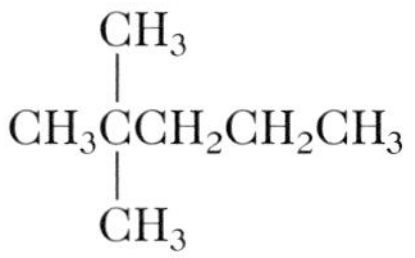

$$\text{CH}_3$$
$$|$$
$$\text{CH}_3\text{CCH}_2\text{CH}_2\text{CH}_3$$
$$|$$
$$\text{CH}_3$$

Correct name:	2,2-dimethylpentane
Incorrect names:	2-dimethylpentane
	2,2-methylpentane
	4,4-dimethylpentane

These are not all of the IUPAC rules for alkanes, but they will handle all of our needs. Study the following examples of correctly named compounds. Be sure to notice that in choosing the parent chain we sometimes have to go around a corner as the chain zigzags on the page.

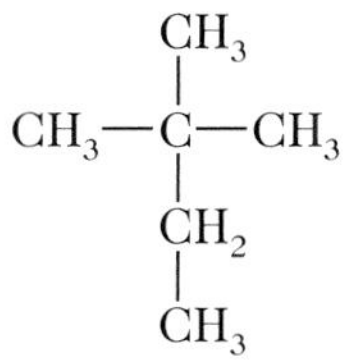

$$\text{CH}_3$$
$$|$$
$$\text{CH}_3—\text{C}—\text{CH}_3$$
$$|$$
$$\text{CH}_2$$
$$|$$
$$\text{CH}_3$$

2,2-dimethylbutane
not 2-ethyl-2-methylpropane

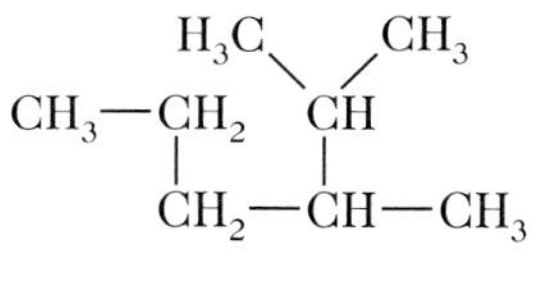

$$\text{H}_3\text{C} \quad\;\; \text{CH}_3$$
$$\diagdown \quad\; \diagup$$
$$\text{CH}_3—\text{CH}_2 \quad \text{CH}$$
$$| \qquad\quad |$$
$$\text{CH}_2—\text{CH}—\text{CH}_3$$

2,3-dimethylhexane
not 2-isopropylpentane

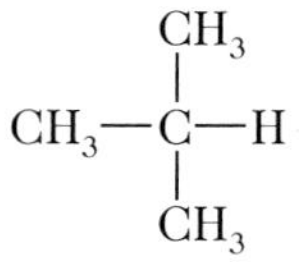

$$\text{CH}_3$$
$$|$$
$$\text{CH}_3—\text{C}—\text{H}$$
$$|$$
$$\text{CH}_3$$

2-methylpropane
not 1,1-dimethylethane

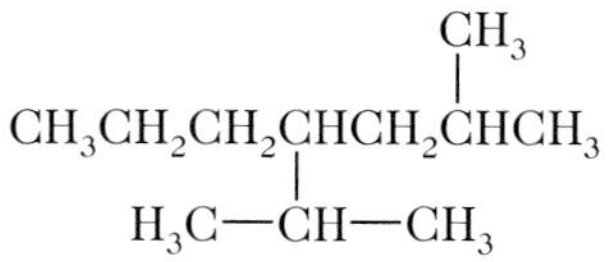

$$\text{CH}_3$$
$$|$$
$$\text{CH}_3\text{CH}_2\text{CH}_2\text{CHCH}_2\text{CHCH}_3$$
$$|$$
$$\text{H}_3\text{C}—\text{CH}—\text{CH}_3$$

4-isopropyl-2-methylheptane
not 4-isopropyl-6-methylheptane

EXAMPLE 11.5 Using the IUPAC Rules to Name an Alkane

What is the IUPAC name for the following compound?

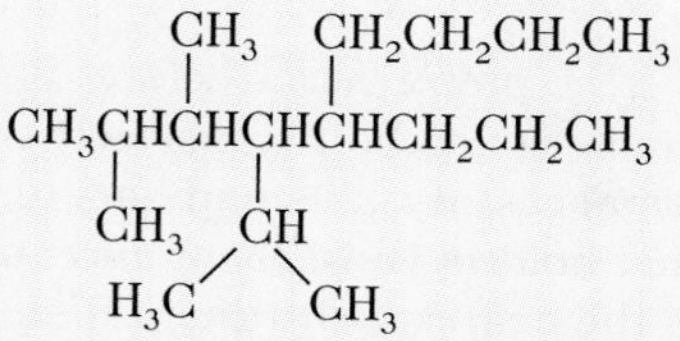

$$CH_3 \quad CH_2CH_2CH_2CH_3$$
$$CH_3CHCHCHCHCH_2CH_2CH_3$$
$$CH_3 \quad CH$$
$$H_3C \quad CH_3$$

ANALYSIS The compound is an alkane because it is a hydrocarbon with only single bonds. We must therefore use the IUPAC rules for alkanes.

SOLUTION The ending to the name must be *-ane*. The next step is to find the longest chain even if we have to go around corners. This chain is nine carbons long, so the name of the parent alkane is *nonane*. We have to number the chain from left to right, as follows, in order to reach the first branch with the lower number.

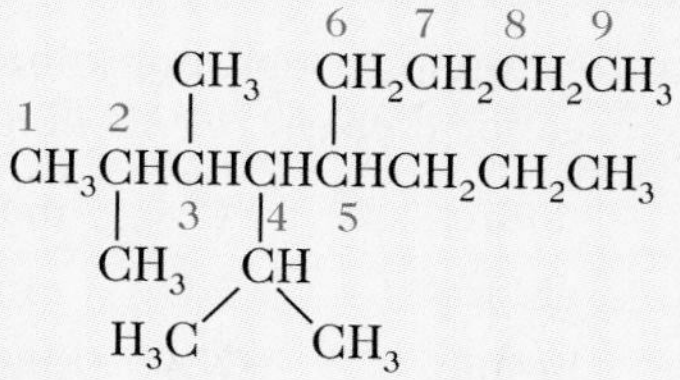

$$\overset{6}{C}H_3 \quad \overset{7}{C}H_2\overset{8}{C}H_2\overset{9}{C}H_2CH_3$$
$$\overset{1}{C}H_3\overset{2}{C}HCH\overset{4}{C}H\overset{5}{C}HCH_2CH_2CH_3$$
$$\overset{3}{C}H_3 \quad CH$$
$$H_3C \quad CH_3$$

At carbons 2 and 3 there are the one-carbon methyl groups. At carbon 4, there is a three-carbon isopropyl group (not the propyl group, because the bonding site is at the *middle* carbon of the three-carbon chain). At carbon 5, there is a three-carbon propyl group. (It has to be this particular propyl group because the bonding site is the *end* of the three-carbon chain in the group.) Alphabetically, *isopropyl* comes before *methyl*, which comes before *propyl*, so we must assemble these names as follows to make the final name. (Names of alkyl groups are alphabetized *before* any prefixes such as di- or tri- are affixed.)

4-Isopropyl-2,3-dimethyl-5-propylnonane

| hyphen separates a number from a word | comma separates two numbers | no hyphen, no comma, no space |

CHECK The most common mistake students make is in the discovery of the longest chain. Check your answer to make sure that you have not erred in this. Then be sure you have numbered from the correct end.

■ PRACTICE EXERCISE 9 Write the IUPAC names of the following compounds.

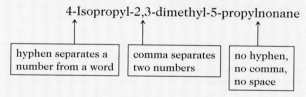

(a) 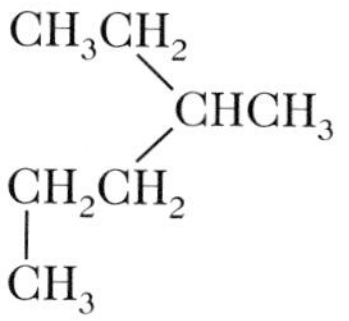
$$CH_3CH_2$$
$$CHCH_3$$
$$CH_2CH_2$$
$$CH_3$$

(b) 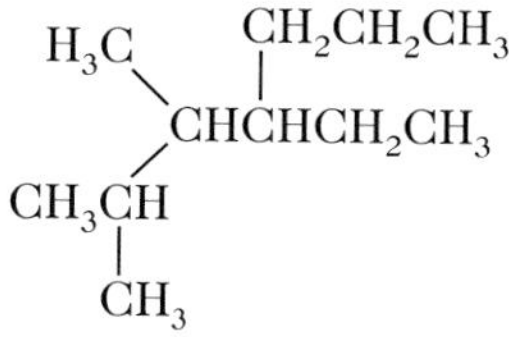
$$H_3C \quad CH_2CH_2CH_3$$
$$CHCHCH_2CH_3$$
$$CH_3CH$$
$$CH_3$$

(c) 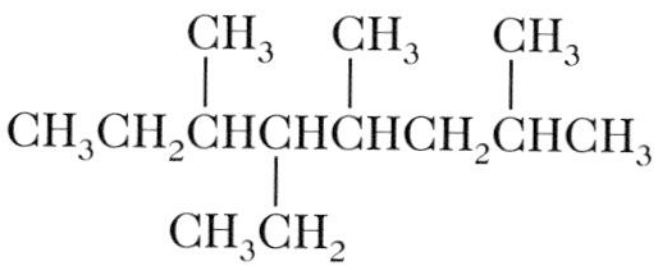
$$\qquad CH_3 \quad CH_3 \quad CH_3$$
$$CH_3CH_2CHCHCHCH_2CHCH_3$$
$$CH_3CH_2$$

IUPAC Rules for Naming Cycloalkanes

To name a cycloalkane, place the prefix *cyclo-* before the name of the straight-chain alkane that has the same number of carbon atoms as there are in the ring. This is illustrated in Table 11.4. When necessary, give numbers to the ring atoms by giving location 1 to a ring position that holds a substituent and numbering around the ring in whichever direction reaches the nearest substituent first. For example,

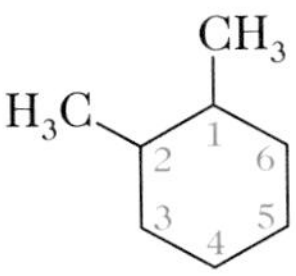

■ No number is needed when the ring has only one group. Thus,

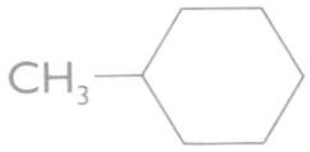

CH₃—⬡

is named methylcyclohexane, not 1-methylcyclohexane.

1,2-dimethylcyclohexane   1,2,4-trimethylcyclohexane

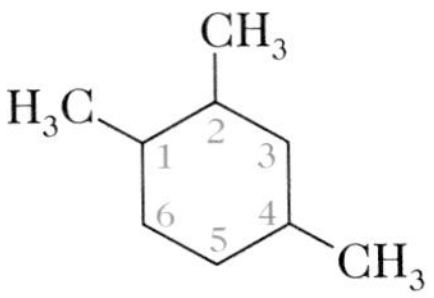

IUPAC Names of Substituents Other than Alkyl Groups

When halogen atoms, or nitro or amino groups, are joined to a carbon of a chain or ring, the following names are used for them in IUPAC nomenclature.

—F	fluoro	—I	iodo
—Cl	chloro	—NO₂	nitro
—Br	bromo	—NH₂	amino

■ No number is needed in *nitroethane* because the name is unambiguous as it stands.

For example, $CH_3CH_2CH_2CHCl_2$ is named 1,1-dichlorobutane in the IUPAC system. Nitroethane is $CH_3CH_2NO_2$.

■ PRACTICE EXERCISE 10   Write the condensed structures of the following compounds.

(a)   1-Bromo-2-nitropentane

(b)   5-Isopropyl-2,2,3,3,4,4-hexamethyloctane

(c)   5-*sec*-Butyl-6-*t*-butyl-2,2-diiodo-4-isopropyl-3-methylnonane

(d)   5,5-Di-*sec*-butyldecane

TABLE 11.4   Some Cycloalkanes

IUPAC Name	Structure	Boiling Point (°C)	Melting Point (°C)	Density (20 °C)
Cyclopropane	 △	−33	−127	1.809 g/L (0 °C)
Cyclobutane	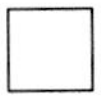 ☐	−13.1	−80	0.7038 g/L (0 °C)
Cyclopentane	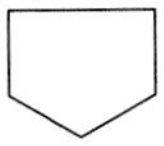 ⬠	49.3	−94.4	0.7460 g/mL
Cyclohexane	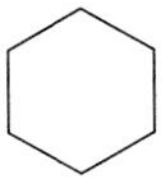 ⬡	80.7	6.47	0.7781 g/mL
Cycloheptane	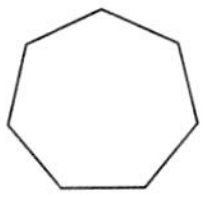 ⬡	118.5	−12	0.8098 g/mL

■ PRACTICE EXERCISE 11 Examine the structure of part (e) of Practice Exercise 10. Underline each primary carbon, place an asterisk above each secondary carbon, and point an arrow at each tertiary carbon.

Common Names

In some references you might see the names of straight-chain alkanes with the prefix *n*-, as in *n*-butane, the common name of butane. It stands for *normal,* which is a way of designating that the straight-chain isomer is regarded as the *normal* isomer, as in the common names, *n*-pentane, *n*-hexane, and so forth. It is used only when isomers are possible. (You would never see *n*-propane printed as a name, for example, because there are no isomers of propane.)

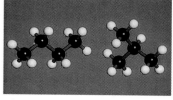

n-Butane and Isobutane.

Common Names of Alcohols, Amines, and Haloalkanes Employ the Names of the Alkyl Groups The following examples of some halogen derivatives of the alkanes, called *haloalkanes,* illustrate how common names are easily constructed. The IUPAC names are given for comparison.

■ The haloalkanes are examples of *organohalogen compounds*.

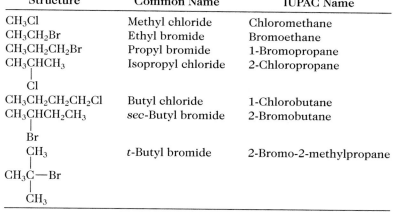

Structure	Common Name	IUPAC Name
CH_3Cl	Methyl chloride	Chloromethane
CH_3CH_2Br	Ethyl bromide	Bromoethane
$CH_3CH_2CH_2Br$	Propyl bromide	1-Bromopropane
$CH_3\underset{\underset{Cl}{\vert}}{C}HCH_3$	Isopropyl chloride	2-Chloropropane
$CH_3CH_2CH_2CH_2Cl$	Butyl chloride	1-Chlorobutane
$CH_3\underset{\underset{Br}{\vert}}{C}HCH_2CH_3$	*sec*-Butyl bromide	2-Bromobutane
$CH_3\overset{\overset{CH_3}{\vert}}{\underset{\underset{CH_3}{\vert}}{C}}—Br$	*t*-Butyl bromide	2-Bromo-2-methylpropane

■ PRACTICE EXERCISE 12 Give the common names of the following compounds.

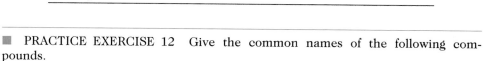

(a) $ClCH_2CH_3$ (b) $BrCH_2CH_2CH_2CH_3$

(c) $CH_3\overset{\overset{CH_3}{\vert}}{C}HCH_2Cl$ (d) $CH_3\overset{\overset{CH_3}{\vert}}{\underset{\underset{Br}{\vert}}{C}}CH_3$

11.7 CHEMICAL PROPERTIES OF ALKANES

Alkanes can burn and can give substitution reactions with the halogens, but they undergo almost no other reaction.

The chemistry of the alkanes and cycloalkanes is quite simple because very few chemicals react with them. This is why saturated hydrocarbons are nicknamed the

■ Mineral oil is a safe laxative (when used with care) because it is a mixture of high formula mass alkanes that undergo no chemical reactions in the intestinal tract.

paraffins, after the Latin *parum affinis,* meaning "little affinity" or "little reactivity." The alkanes are not chemically attacked by strong acids like sulfuric or hydrochloric acid, by strong bases like sodium hydroxide, by active metals such as sodium, by strong oxidizing agents such as the permanganate ion or the dichromate ion, or by any of the reducing agents. Among the few reactions of saturated hydrocarbons are combustion and halogenations, meaning reactions with the halogens, with F_2, Cl_2, and Br_2 (but not with I_2). In sunlight, oxygen attacks alkanes.

The Combustion of Alkanes or Any Hydrocarbon Gives CO_2 and H_2O Virtually all organic compounds burn, and the hydrocarbons are no exception. Thus, we burn mixtures of alkanes as fuel. Bunsen burner gas, for example, is mostly methane. Liquefied propane is used for fuel in areas that gas mains have not reached. Gasoline, diesel fuel, jet fuel, and heating oil are all mixtures of hydrocarbons, mostly alkanes.

If enough oxygen is available, the sole products of the complete combustion of *any hydrocarbon,* not just alkanes, are carbon dioxide and water plus heat. The (unbalanced) equation, regardless of the kind of hydrocarbon, is

$$\text{hydrocarbon} + O_2 \longrightarrow CO_2 + H_2O + \text{heat}$$

To illustrate, using propane,

$$CH_3CH_2CH_3 + 5O_2 \longrightarrow 3CO_2 + 4H_2O + 531 \text{ kcal/mol propane}$$

The same products, carbon dioxide and water, are also obtained by the complete combustion of any organic compound that consists only of carbon, hydrogen, and oxygen (for example, the alcohols). If insufficient oxygen is present, some carbon monoxide forms.

In the presence of sunlight, all hydrocarbons are eventually attacked by oxygen without combustion occurring. There are also bacteria with enzymes that are able to catalyze the oxidation of hydrocarbons. The products are mixtures of oxygenated hydrocarbons of various families that are much more soluble in water than the parents and are more readily degraded. When oil spills happen, both air oxidation and bacterial degradation occur, as further discussed in Interaction 11.2.

The Chlorination of Alkanes Is a *Substitution Reaction* In the presence of ultraviolet radiation or at a high temperature, alkanes react with chlorine to give organochlorine compounds and hydrogen chloride. An atom of hydrogen in the alkane is replaced by an atom of chlorine, and the replacement of one atom or group by another is called a **substitution reaction.** For example,

$$CH_4 + Cl_2 \xrightarrow[\text{or heat}]{\text{ultraviolet light}} CH_3Cl + HCl$$

$$\text{methyl chloride}$$

The reaction takes place by a chemical chain reaction as described in Interaction 11.3.

The hydrogen atoms in methyl chloride can also be replaced by chlorine, so as methyl chloride starts to form, its molecules compete with those of still unreacted methane for the remaining chlorine. This is how some methylene chloride (dichloromethane) forms when the chlorination of methane is carried out. In fact, when 1 mol of CH_4 and 1 mol of Cl_2 are mixed and allowed to react, several reactions eventually occur, and a mixture forms of four chlorinated methanes, hydrogen chloride, plus some unreacted methane. When it isn't possible to write a balanced equation, we represent the reaction by a flow of symbols.

INTERACTION 11.2
ENLISTING MICROBES AGAINST OIL SPILLS

Many microorganisms ("microbes") survive and prosper in environments completely unfriendly and even lethal to higher organisms. It stands to reason, therefore, that there will be important differences in the chemical makeups inside the cells of microbes and humans. Some of the differences occur with the kinds of enzymes, the catalysts of the reactions in living things. Like the members of all species, microbes have some unique enzymes and therefore can do some chemical reactions that humans cannot. Scientists over the last several decades have increasingly exploited this fact to develop chemical methods for doing jobs otherwise very hard to do. The cleanup of oil spills is only one of a growing list of examples.

What few people know is that the environment naturally contains over 30 different genera[1] of oil-degrading bacteria and fungi. They consume the hydrocarbons in oil, producing carbon dioxide, water, and other products that are mostly recyclable in the environment. This work is the chief reason that oil spills do not seem to result in *permanent* harm, although their short-term effects on birds, fish, the fishing industry, and tourism can be massive. The use of natural agents like oil-degrading microbes to consume substances that are alien to part of the environment and thus to fix or remedy the situation is called *bioremediation.* It's a useful supplement to traditional methods of oil spill cleanup (Figure 1).

In 1989, the *Exxon Valdez* went aground on Bligh Reef in Prince William Sound (Alaska), spilling over 35,000 metric tons of oil. An analysis of what happened to the oil, reported in 1994, estimated that about 20% of the oil evaporated and underwent oxidation, induced by sunlight, into products that then degraded further. (Solar energy initiates chemical chain

FIGURE I Containing oil spilled from tanker accidents by using trawlers and booms is only partly successful.

reactions between hydrocarbons and oxygen, like the chain reactions studied in Interaction 11.3.) Another 50% of the spilled oil degraded either in the water or on the beaches; 14% of the oil was recovered, 13% sank into bottom sediments where it has degraded to products indistinguishable from those from natural oil seeps; and the rest remains as highly weathered materials. Bioremediation of shoreline oil deposits was accelerated by the use of inorganic nitrogen and phosphorus fertilizers that helped to promote oil-eating microbes. The affected shore was largely made of porous cobbles, making for a good oxygen supply, which also helped the process.

[1] "Genera" is the plural of "genus," which is a classification of living things, a main subdivision of a *family,* and includes one or several *species.* Humans or *Homo sapiens,* for example, belong to family *Hominidae,* the genus *Homo,* and the species *sapiens.*

$$CH_4 + Cl_2 \xrightarrow[HCl]{Cl_2} CH_3Cl \xrightarrow[HCl]{Cl_2} CH_2Cl_2 \xrightarrow[HCl]{Cl_2} CHCl_3 \xrightarrow[HCl]{Cl_2} CCl_4$$

| methane (bp -162 °C) | methyl chloride (bp -24 °C) | methylene chloride (bp 40 °C) | chloroform (bp 61 °C) | carbon tetrachloride (bp 77 °C) |

■ Notice how the boiling points of the compounds (in parentheses) increase with formula mass.

INTERACTION 11.3
THE CHLORINATION OF METHANE—A FREE RADICAL CHAIN REACTION WITH COUNTERPARTS AT THE MOLECULAR LEVEL OF LIFE

Free Radicals Photons of the proper energy are absorbed by the electron pairs of covalent bonds and cause the bonds to break. Such an event produces not ions but electrically neutral particles in which an atom lacks an octet and has an unpaired electron. Particles with unpaired electrons are called *radicals,* sometimes *free radicals* because they so often form with the freedom to move about. The breaking of the bond in the chlorine molecule, Cl_2, for example, occurs as follows.

$$:\ddot{C}l:\ddot{C}l: \text{ + UV energy or heat} \longrightarrow :\ddot{C}l\cdot \text{ + } \cdot\ddot{C}l:$$

two chlorine atoms
(two free radicals)

Free Radical Chain Reaction in the Chlorination of Methane The reaction of methane with chlorine is a free radical reaction that begins by the breaking of the bond in Cl_2, as we just described. Chlorine is in group VIIA of the periodic table, so each chlorine atom has but seven valence shell electrons, not an octet. Each Cl atom is thus unstable and is able to launch the first of a two-step cycle of reactions called a *chemical chain reaction.* The first step is the reaction of the ·Cl atom with a methane molecule to generate a new radical, ·CH$_3$, a neutral particle called the methyl radical. (We'll often explicitly show only the unpaired electron of a radical or atom, not the full population of the valence shell.)

$$\cdot Cl + CH_4 \longrightarrow H-Cl + \cdot CH_3 \qquad (1)$$

Notice that the Cl atom recovers its outer octet within a molecule of H—Cl. If we use molecular models we can better imagine what happens when a chlorine atom strikes the H end of an H—C bond in methane hard enough.

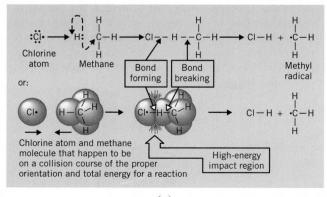

Chlorine atom and methane molecule that happen to be on a collision course of the proper orientation and total energy for a reaction

High-energy impact region

(a)

The carbon atom in the methyl radical also lacks an octet; it has seven valence shell electrons, one being unshared (and shown by the dot). When the unstable methyl radical collides hard enough with a still unreacted molecule of Cl—Cl, the Cl—Cl bond breaks and the new C—Cl bond in methyl chloride forms.

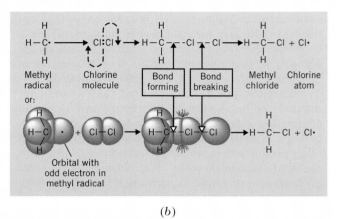

(b)

This is actually the second step of the chain reaction.

$$\cdot CH_3 + Cl_2 \longrightarrow CH_3Cl + \cdot Cl \qquad (2)$$

It is now very important to notice that one product of reaction 2, namely, ·Cl, is a necessary reactant for reaction 1. So reaction 1 occurs again but to a still unreacted methane molecule. And so a new methyl radical forms, which engages in reaction 2, again producing a methyl chloride molecule and yet another chlorine atom. You can see that a cycle of reactions 1 and 2 can continue until the methane and chlorine are used up, leaving methyl chloride and hydrogen chloride. Thus, once some chlorine atoms are initially generated by the action of heat or UV radiation on chlorine molecules, the chains are launched. This is why a relatively small amount of radiation can initiate chains leading to a huge number of product molecules.

Two chains are initiated by the breaking of the bond in one molecule of Cl_2 by heat or UV light because the chain-initiating event produces *two* ·Cl atoms. Upwards of 8000 molecules of methyl chloride can be generated by one chlorine atom formed from the initiating action of only one photon of UV radiation. The chains continue until two radicals happen to find each other and join. Some examples of *chain-termination* reactions are

$$2Cl \cdot \longrightarrow Cl_2$$
$$2CH_3 \cdot \longrightarrow CH_3 - CH_3$$
$$CH_3 \cdot + Cl \cdot \longrightarrow CH_3Cl$$

Other chains are launched, however, as some terminate, until one or both reactants are used up.

Multiple Chlorinations Because of the nature of the reaction, additional products are inevitable. Methyl chloride, CH_3Cl, forms while some unreacted Cl_2 remains, so a chlorine atom may collide with the H end of a H—C bond in CH_3Cl instead of at a H—C bond in another molecule of CH_4. This launches a new chain of reactions that converts CH_3Cl into CH_2Cl_2. The latter, of course, has its own H—C bonds, so still other chains can be started that convert CH_2Cl_2 into $CHCl_3$. You can see that the latter also has a H—C bond, so further chains can commence that lead to the formation of some CCl_4. These events occur more or less at random as the atoms and molecules whiz about in the gaseous state. Statistical probabilities largely govern what collisions occur but, as we mentioned in the text, a mixture of products is bound to form when chlorination is initiated in a mixture having a $1:1$ mol ratio of Cl_2 to CH_4.

Cancer, Aging, and Free Radicals Ultraviolet radiation and the generation of free radicals also have roles in the development of skin cancer. Remember that nearly all free radicals are inherently unstable because they lack outer octets. Free radicals are thus rogue species and trigger unwanted events in cells, some leading to cancer.

Atomic radiation is dangerous partly because some of the particles it generates in cells are radicals. The chemical changes that occur during aging, when muscles lose their suppleness and flexibility, involve the formation of free radicals that form not so much from exposure to radiation or sunlight but by natural processes involving peroxides, compounds with such general formulas as R—O—O—H and R—O—O—R′. Their O—O bonds break rather easily to give free radicals of the R—O· type that lead to the crosslinking of protein molecules. Heavy smoking and excessive exposure to sunlight (for example, by overtanning) also contribute, by means of free radical chemistry, to the deep wrinkling of the skin (see Figure 1). Vitamins A and C are known to scavenge and destroy free radicals and so they seem to provide some protection. Some skin aging is caused by the action of sunlight's ultraviolet component on genes in skin cells. Certain genes are activated to produce enzymes that break down the proteins in skin, collagens and elastins, that keep it supple. It's no accident that the bottoms of both babies and octogenarians have about the same smoothness.

FIGURE 1 Prolonged exposure to sunlight contributes to deep wrinkles.

■ Chloroform (bp 61 °C) was an anesthetic better suited for use in the tropics than diethyl ether (bp 35 °C) because of its higher boiling point.

■ The reaction with bromine is much slower than the one with chlorine.

Methylene chloride, chloroform, and carbon tetrachloride are examples of *organochlorine compounds,* and all are used as nonpolar solvents. Chloroform has been used as an anesthetic.

Bromine reacts with methane by the same kind of substitution as chlorine. Iodine does not react. Fluorine combines explosively with most organic compounds at room temperature, and complex mixtures form.

The higher alkanes can also be chlorinated. Ethyl chloride, plus more highly chlorinated products, form by the chlorination of ethane. Ethyl chloride (bp 11.5 °C) is used as a local anesthetic. When sprayed on the skin, ethyl chloride rapidly draws heat for its evaporation, which cools the area enough to prevent pain signals during minor surgery at the site.

When propane is chlorinated, both propyl chloride and isopropyl chloride form in roughly equal amounts (and again we have to write an unbalanced equation).

$$CH_3CH_2CH_3 + Cl_2 \xrightarrow{\text{ultraviolet light}} CH_3CH_2CH_2Cl + CH_3\overset{\underset{\displaystyle |}{\displaystyle Cl}}{C}HCH_3 + HCl$$

propane   propyl chloride   isopropyl chloride

Besides these products, some higher chlorinated compounds also form.

■ PRACTICE EXERCISE 13   (a) How many monochloro compounds of butane are possible? Give both their common and IUPAC names. (Consider only the monochloro compounds with the formula C_4H_9Cl.) (b) How many monochloro derivatives of isobutane are possible? Write both their common and their IUPAC names.(b) 2,4-Dichlorocyclopentane

---

SUMMARY

Organic and inorganic compounds   Most organic compounds are molecular and the majority of inorganic compounds are ionic. Molecular and ionic compounds differ in composition, in types of bonds, and in several physical properties.

Structural features of organic molecules   The ability of carbon atoms to join to each other many times in succession—in straight chains, in branched chains, as well as in rings—accounts in large measure for the existence of several million organic compounds. The skeletons of the rings can be made entirely of carbon atoms or there may be one or more other nonmetal atoms (heterocyclic compounds).

Full structural formulas of organic compounds are usually condensed by grouping the hydrogens attached to a carbon immediately next to this carbon; by letting single bonds on a horizontal line be understood; and by leaving bond angles and conformational possibilities to the informed imagination. Skeletons of rings are usually represented by simple polygons. Free rotation about single bonds is possible in open-chain compounds but not in rings.

Compounds without multiple bonds are saturated. Those with double or triple bonds are unsaturated. Carbon–carbon double or triple bonds are never "understood" in structures.

The families of organic compounds are organized around functional groups, parts of molecules at which most of the chemical reactions occur. Nonfunctional units can sometimes be given the general symbol R, as in ROH, the general symbol for all alcohols. These R groups are hydrocarbon-like groups.

Isomerism   Differences in the conformations of carbon chains do not create new compounds, but differences in the organizations of parts do. Isomers are compounds with identical molecular formulas but different structures. Constitutional isomers make up one kind of isomer, those whose molecules have different atom-to-atom connectivities. Sometimes constitutional isomers are in the same family, like butane and isobutane. Often they are not, like ethyl alcohol and dimethyl ether.

Hydrocarbons   Hydrocarbons are compounds in which the only elements are carbon and hydrogen. The alkanes are saturated hydrocarbons; the alkenes and alkynes are unsaturated. The alkenes have at least one double bond. The alkynes have at least one triple bond. The aromatic hydrocarbons have a benzene ring, and the aliphatic hydrocarbons do not. Being nonpolar compounds, the hydrocarbons are all insoluble in water, and many mixtures of alkanes are common, nonpolar solvents. The rule like-dissolves-like lets us predict solubilities.

Nomenclature of alkanes In the IUPAC system, a compound's family is always indicated by a name ending, like -*ane* for the alkanes. The number of carbons in the parent chain is indicated by a unique prefix for each number, like the prefix *but*- to denote four carbons in *butane*. The locations of side chains or other kinds of atoms or groups are specified in the final name by numbers assigned to the carbons to which they are attached. The numbering of the parent chain is done in the direction that locates the first branch at the lower of two possible numbers.

Alkane-like substituents are called alkyl groups, and the names and formulas of those having from one to four carbon atoms must be learned. Common names are still popular, particularly when the IUPAC names are long and cumbersome.

Chemical properties of alkanes Alkanes and cycloalkanes are generally unreactive at room temperature toward concentrated acids and bases, toward oxidizing and reducing agents, toward even the most reactive metals, and toward water. They burn, giving off carbon dioxide and water, and in the presence of ultraviolet light (or at a high temperature) they undergo useful substitution reactions with chlorine and bromine.

REVIEW EXERCISES

The answers to Review Exercises whose numbers are in color are found in Appendix E. The answers to the other Review Exercises are found in the Study Guide that accompanies this book. The more challenging questions are marked with asterisks.

11.1 Why are carbon compounds generally called *organic* compounds?

11.2 Consider the *bonding abilities* of the element carbon.
(a) What kind of chemical bond is nearly always displayed by a carbon atom, ionic, covalent, coordinate covalent, or hydrogen?
(b) What is unique about the element carbon in its bonding abilities?
(c) What bonding properties of carbon contribute to the huge *number* of possible organic compounds?

11.3 With respect to the *synthesis* of organic compounds, what specifically was the problem that organic chemists faced prior to 1828? What scientific theory had been devised to meet this problem?

11.4 What was Wöhler's goal when he evaporated an aqueous solution of ammonium cyanate to dryness? What happened instead? With respect to *scientific theory* at the time, what specifically did Wöhler accomplish?

11.5 How many single bonds are observed in neutral molecules at each of the following atoms?
(a) C   (b) O   (c) N   (d) H   (e) Cl

11.6 Which of the following compounds are inorganic?
(a) CH_3CH_2OH (b) CO_2 (c) $CHCl_3$
(d) $KHCO_3$ (e) Na_2CO_3

11.7 Are the majority of all compounds that dissolve in water ionic or molecular? Inorganic or organic?

11.8 Explain why very few organic compounds can conduct electricity either in an aqueous solution or as molten materials.

*11.9 Each compound described below is either ionic or molecular. State which it most likely is, and give one reason.
(a) A compound that melts at 281 °C, and burns in air.
(b) A compound that dissolves in water. When hydrochloric acid is added, the solution fizzes and an odorless, colorless gas is released, which can extinguish a burning flame.
(c) A compound that is a colorless gas at room temperature.
(d) A compound that melts at 824 °C and becomes white when heated.
(e) A compound that is a liquid and does not dissolve in water but does burn.
(f) A compound that is a liquid and does dissolve in water as well as burns.

Structural Features of Organic Molecules

11.10 One can write the structure of hexane as follows.

$$\begin{array}{ccc} CH_3 & CH_2 & \!\!\!-CH_2 \\ | & | & | \\ CH_2 & \!\!\!-CH_2 & CH_3 \end{array}$$

Are hexane molecules properly described as *straight chain* or as *branched chain,* in the sense in which we use these terms? Explain.

11.11 Which of the following structures are possible, given the numbers of bonds that various atoms can form?
(a) $CH_3CH_2CH_2OCH_3CH_3$
(b) $HOCH_2CH_2CH_2CH_3$
(c) $CH_2{=}CHCH_2NCH_3$
(d) $CH_3CH{=}CHCH_2OCH_3$
(e) $NH_2CH_2CH_2CH_3$

11.12 Write full (expanded) structures for each of the following *molecular* formulas. Remember how many covalent bonds nonmetals have in molecules: C, 4; N, 3; O, 2; H, Cl Br, 1 each. In some structures you will have to use double or triple bonds. (*Hint*: A trial-and-error approach will have to be used.)

(a) CH_4O (b) CH_2Cl_2 (c) N_2H_4 (d) C_2H_6
(e) CH_2O (f) CH_2O_2 (g) NH_3O (h) C_2H_4
(i) $CHCl_3$ (j) HCN   (k) C_2H_3N (l) CH_5N

11.13 Expand the structure of fluoxetine, marketed under the name of Prozac, the most widely sold antidepressant drug in the United States.

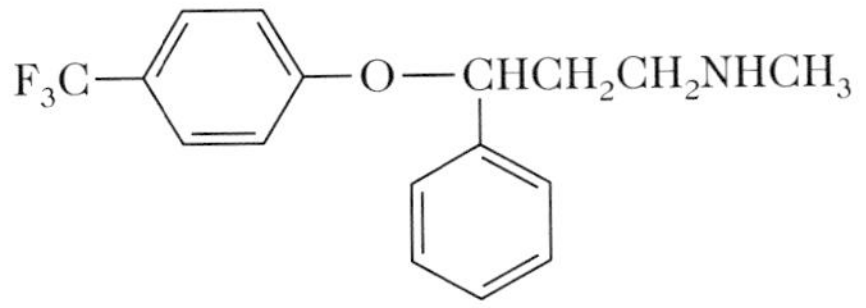

11.14 Expand the structure of melatonin, a hormone made in the pineal gland in the center of the brain, and sold in health food stores as a cure-all.

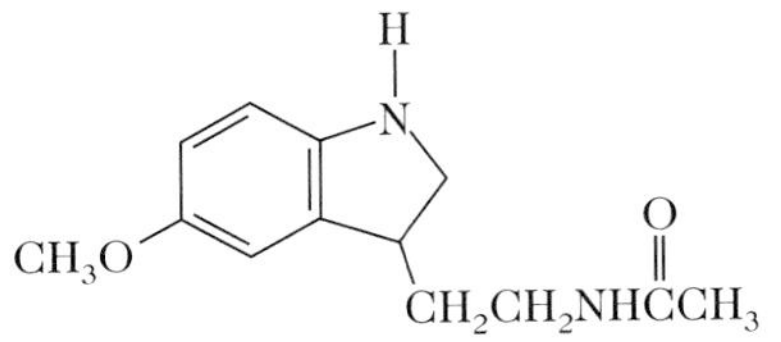

11.15 Write neat, condensed structures of the following.

(a)

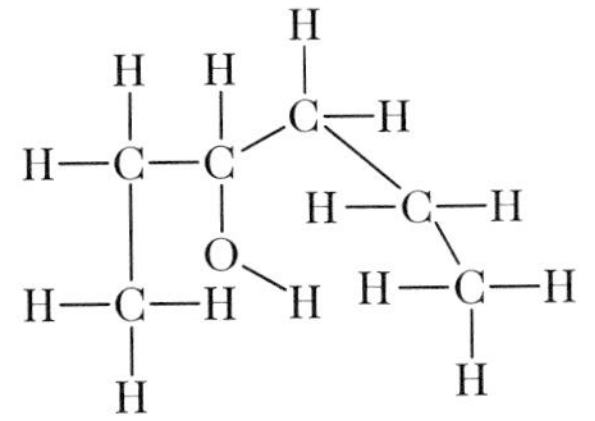

(b) Nicotinamide, a B vitamin.

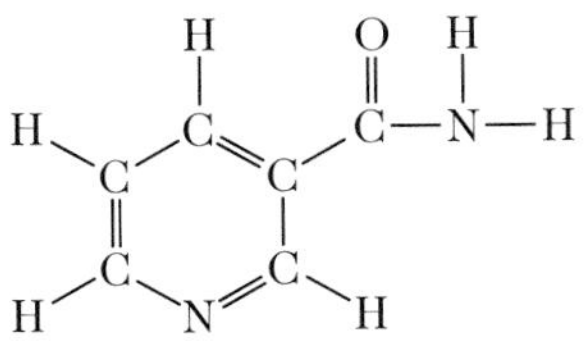

(c) Thiamine, vitamin B_1. Notice that one nitrogen atom has four bonds, so it has a positive charge.

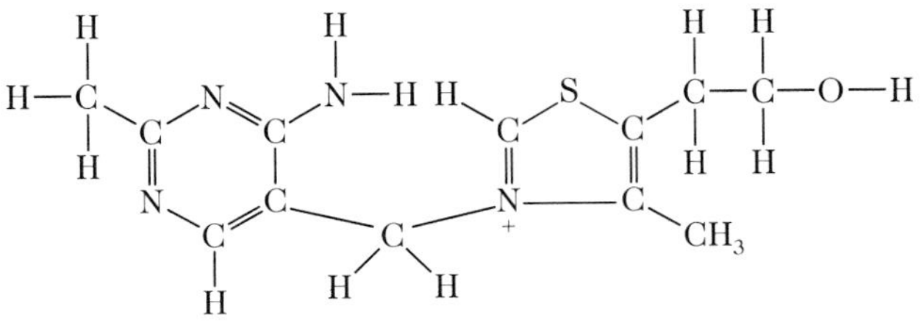

11.16 Why is the topic of molecular shape important at the molecular level of life?

11.17 A chemical bond is a force of attraction, so what is attracted to what in the C—C bond in ethane?

11.18 Free rotation can occur about single bonds (in open-chain structures) without breaking or weakening the bonds. Why is this possible?

11.19 Which of the following structures represent unsaturated compounds?

(a) 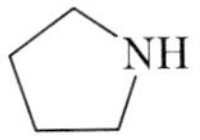   (b) $CH_3CH_2CCl_3$

(c) 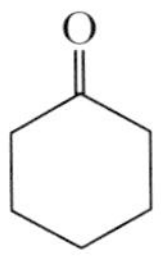   (d)

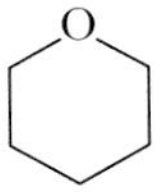

11.20 Which compounds are saturated?

(a) 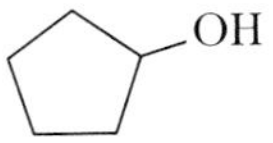   (b)

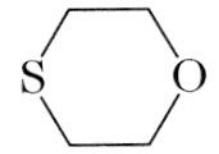

(c) 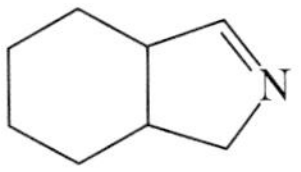   (d) C_2H_4

11.21 Examine each pair of compounds and decide whether the two are identical, are isomers, or are unrelated.

(a) $CH_3—CH_2—CH_3$ and $\overset{\displaystyle CH_3}{\underset{\displaystyle |}{CH_3CH_2}}$

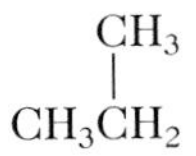

(b) 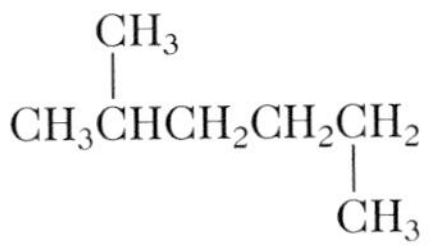 $CH_3\overset{\displaystyle |}{\underset{\displaystyle CH_3}{C}HCH_2CH_2CH_2}$ and $CH_3CH_2CH_2CH_2\overset{\displaystyle CH_3}{\underset{\displaystyle |}{C}HCH_3}$

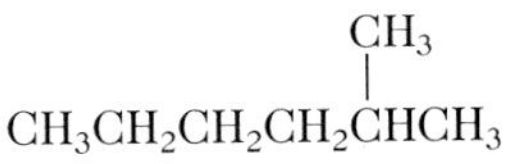

(c) $CH_3NHCH_2CH_3$ and $CH_3CH_2NHCH_3$

(d) 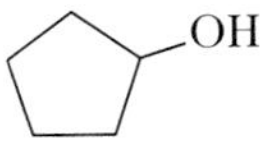   and

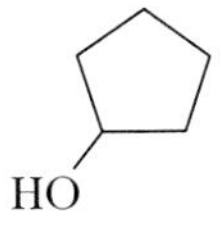

(e) 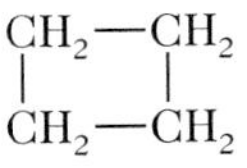 $CH_2—CH_2$
 $\,\,|\qquad\,\,|$
 $CH_2—CH_2$ and $CH_3CH=CHCH_3$

(f) 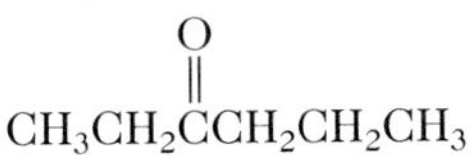 $CH_3CH_2\overset{\displaystyle O}{\overset{\displaystyle \|}{C}}CH_2CH_2CH_3$ and $CH_3CH_2CH_2\overset{\displaystyle O}{\overset{\displaystyle \|}{C}}CH_2CH_3$

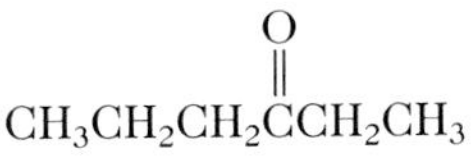

(g) 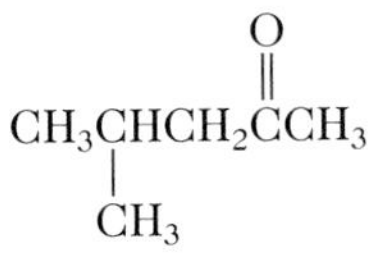 $CH_3\underset{\displaystyle CH_3}{\underset{\displaystyle |}{C}}HCH_2\overset{\displaystyle O}{\overset{\displaystyle \|}{C}}CH_3$ and $CH_3\underset{\displaystyle CH_3}{\underset{\displaystyle |}{C}}H\overset{\displaystyle O}{\overset{\displaystyle \|}{C}}CH_2CH_3$

(h) $HOCH_2\overset{\displaystyle O}{\overset{\displaystyle \|}{C}}CH_2CH_3$ and $CH_3CH_2CH_2\overset{\displaystyle O}{\overset{\displaystyle \|}{C}}OH$

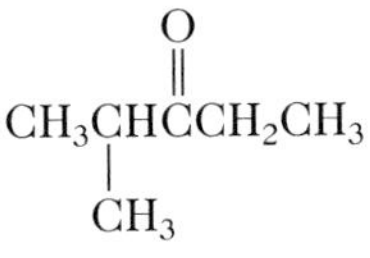

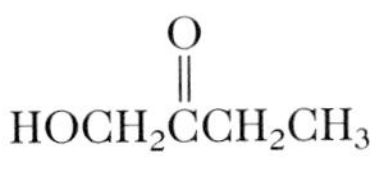

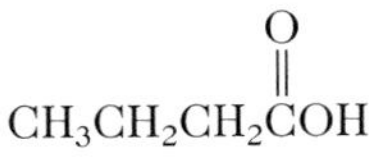

(i)  $\underset{\substack{O\\\parallel}}{HOCCH_2CHCH_3}$ and $\underset{\substack{\;\;\;\;\;\;\;\;\;O\\\;\;\;\;\;\;\;\;\;\parallel}}{CH_3CHCH_2COH}$
with CH_3 below on left and CH_3 above on right

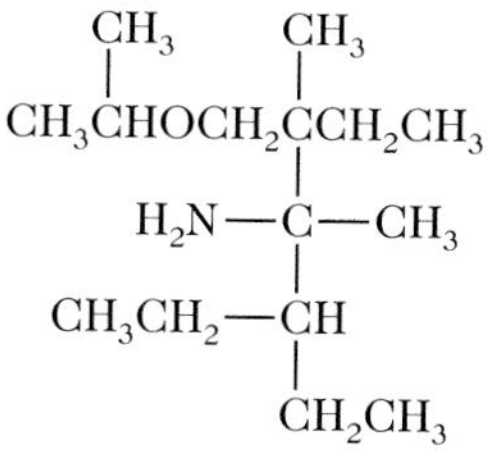

(j) 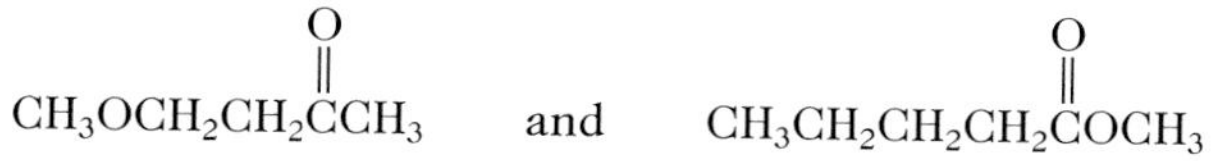 $\underset{\substack{O\\\parallel}}{CH_3OCH_2CH_2CCH_3}$ and $\underset{\substack{\;\;\;\;\;\;\;\;\;\;\;\;\;\;\;O\\\;\;\;\;\;\;\;\;\;\;\;\;\;\;\;\parallel}}{CH_3CH_2CH_2CH_2COCH_3}$

(k)  $\underset{\substack{O\\\parallel}}{HOCCH_2CHCH_3}$ and $\underset{\substack{O\\\parallel}}{HCOCH_2CHCH_3}$
with OH below on both

(l) $HOOCH_2CH_2OCH_3$ and $HOCH_2CH_2OCH_3$

(m) 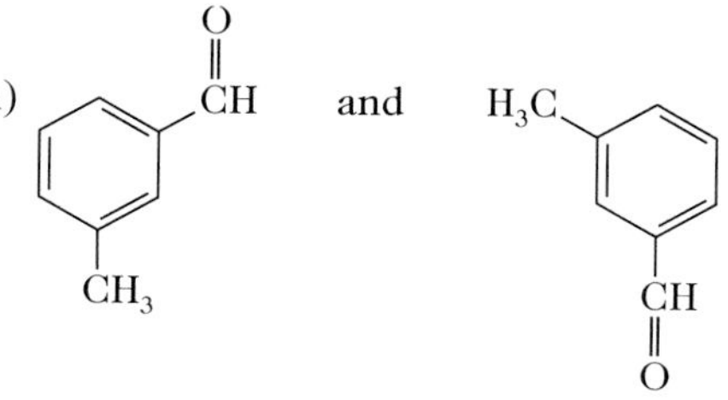 (benzene ring with $\overset{\substack{O\\\parallel}}{CH}$ group and CH_3 group) and (benzene ring with H_3C and $\underset{\substack{\parallel\\O}}{CH}$ group)

(n) $\underset{\substack{CH_3\;\;\;\;CH_3\\|\;\;\;\;\;\;\;\;\;|}}{CH_3CHOCH_2CCH_2CH_3}$ and

$$H_2N-\underset{\substack{|\\CH_3CH_2-CH\\|\\CH_2CH_3}}{\overset{\substack{|}}{C}}-CH_3$$

$$CH_3CH_2CH-\underset{\substack{|\\CH_3CH_2}}{\overset{\substack{NH_2\\|}}{C}}-\underset{\substack{|\\CH_3}}{\overset{\substack{CH_2CH_3\\|}}{C}}-\underset{\substack{|\\CH_3}}{\overset{\substack{CH_3\\|}}{CH_2OCHCH_3}}$$

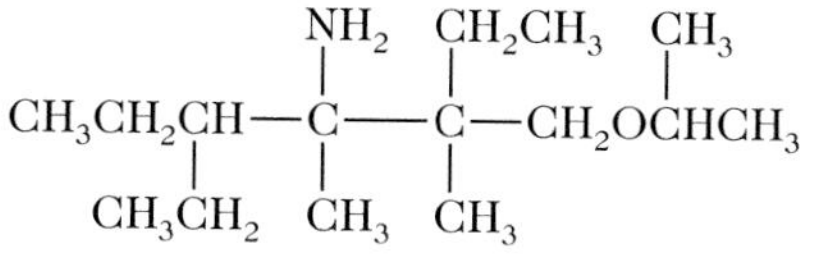

Families of Organic Compounds

11.22 Name the family to which each compound belongs.

(a) 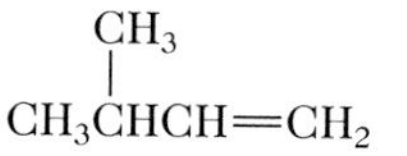 $\underset{\substack{|\\CH_3}}{\overset{\substack{CH_3\\|}}{CH_3CHCH}}=CH_2$ (b) 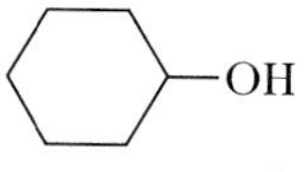 (cyclohexane ring)—OH

(c) $NH_2CH_2CH_2CH_3$ (d) 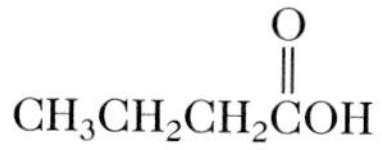 $\underset{\substack{O\\\parallel}}{CH_3CH_2CH_2COH}$

11.23 To which organic family does each compound belong?

(a) 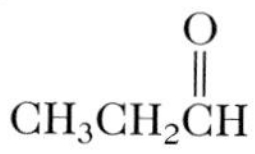 $\underset{\substack{O\\\parallel}}{CH_3CH_2CH}$ (b) 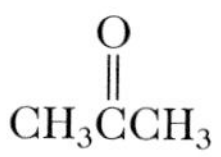 $\underset{\substack{O\\\parallel}}{CH_3CCH_3}$

(c) (square) (d) 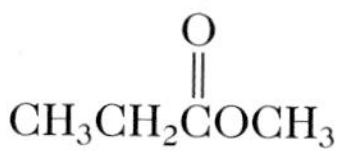 $\underset{\substack{O\\\parallel}}{CH_3CH_2COCH_3}$

11.24 Name the families to which the compounds in parts (a)–(k), (m), and (n) of Practice Exercise 11.21 belong. (A few belong to more than one family.)

11.25 Which compound must have the higher boiling point? Explain.

$CH_3CH_2CH_2CH_2OH$ CH_3OH

A   B

11.26 Which compound must be less soluble in gasoline? Explain.

$CH_3OCH_2CH_2CH_2CH_2CH_2CH_2CH_3$ $HOCH_2CH_2CH_2OH$

A   B

11.27 Suppose that you are handed two test tubes containing colorless liquids, and you are told that one contains pentane and the other holds hydrochloric acid. How can you use just water to tell which tube contains which compound without carrying out any chemical reaction?

11.28 Suppose that you are given two test tubes and are told that one holds methyl alcohol, CH_3OH, and the other hexane. How can water be used to tell these substances apart without carrying out any chemical reaction?

Nomenclature

11.29 There are five isomers of C_6H_{14}. Write their condensed structures and their IUPAC names.

11.30 Which of the isomers of hexane (Review Exercise 11.29) has the common name *n*-hexane? Write its structure.

11.31 Which of the hexane isomers (Review Exercise 11.29) has the common name isohexane? Write its structure.

11.32 Write condensed structures for the following compounds.
(a) Ethyl chloride   (b) Propyl bromide
(c) Isopropyl iodide   (d) *sec*-Butyl chloride

11.33 Write a condensed structure for each of the following compounds.
(a) Butyl chloride   (b) *t*-Butyl bromide
(c) Isobutyl iodide   (d) Isopropylcyclopentane

*11.34** The following are incorrect efforts at naming certain compounds. What are the most likely condensed structures and correct IUPAC names?
(a) 1,6-Dimethylcyclohexane
(b) 2,4,5-Trimethylhexane
(c) 1-Chloro-*n*-butane
(d) Isopropane

Reactions of Alkanes

11.35 Write the balanced equation for the complete combustion of heptane, a component of gasoline.

11.36 Gasohol is a solution of ethyl alcohol, CH_3CH_2OH, in gasoline. Write the equation for the complete combustion of ethyl alcohol.

11.37 What are the formulas and common names of all the compounds that can be made from methane and chlorine?

11.38 There are two isomers of $C_2H_4Cl_2$. What are their structures and IUPAC names?

Fossil Fuels (Interaction 11.1)

11.39 What are the three principal fossil fuels being used today?

11.40 What is the difference between petroleum, crude oil, and natural gas?

11.41 In general terms, describe what happened to change peat into lignite, lignite into soft coal, and soft coal into hard coal.

11.42 What does *fraction* mean in connection with oil refining?

11.43 What kinds of compounds predominate in the crude oil fractions that boil below 200 °C?

11.44 How do petroleum refineries increase the supply of gasoline?

Oil Spills (Interaction 11.2)

11.45 What is bioremediation and what does it have to do with oil spills?

Free Radical Reactions (Interaction 11.3)

11.46 Why is the chlorine atom called a free radical but the chlorine molecule is not?

11.47 The bromination of methane proceeds by a series of steps just like those of the chlorination of methane.
(a) Write the equation for the reaction that initiates the chain reactions.
(b) Write the two equations that constitute the chain reaction.
(c) Write two equations that illustrate how a chain reaction can be broken.

*11.48 The explosive gas-phase reaction of H_2 with Cl_2 that makes HCl is a free radical chain reaction.
(a) Write the balanced equation for the reaction.
(b) It is initiated by the breaking of the bond in Cl_2. Write the equation for this chain-initiating reaction as well as the equations for the chain reaction itself.

Additional Exercises

11.49 Write the condensed structures that would have the following molecular formulas.
(a) C_2H_7N (b) C_3H_8O (c) C_2H_6S

11.50 What are the identifying letters of the two structures that represent isomers among the following?

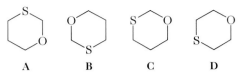

A   B   C   D

*11.51 There are nine isomers of C_7H_{16}. Write their condensed structures and their IUPAC names.

11.52 Write the condensed structures of the isomers of heptane (Review Exercise 11.51) that have the following names.
(a) *n*-Heptane   (b) Isoheptane

*11.53 Write the IUPAC name of the following.

*11.54 What is the IUPAC name of the following?

11.55 Write the structure of 1,1,3-trimethylcycloheptane.

11.56 The following names cannot be the correct names, but it is still possible to write structures from them. What are the correct IUPAC names and the condensed structures?
(a) 1-Chloroisobutane
(b) 2,4-Dichlorocyclopentane
(c) 2-Ethylbutane
(d) 1,3-6-Trimethylcyclohexane

*11.57 When propane reacts with chlorine, in addition to the formation of two isomeric monochloropropanes, some dichloropropanes also form. Write the structures and the IUPAC names for all the possible dichloropropanes.

*11.58 When cyclohexane is chlorinated, how many monochloro derivatives are possible? Write the name or names and structure or structures.

*11.59 What reaction, if any, will cyclohexane give with each reactant at room temperature?
(a) Aqueous NaOH
(b) Concentrated sulfuric acid
(c) Iodine

11.60 What is the name of the butyl group with nine equivalent hydrogen atoms?

*11.61 If 7.46 g of cyclopentane is chlorinated, what is the maximum number of grams of chlorocyclopentane that can be obtained?

UNSATURATED HYDROCARBONS

Different polymers do not mix well when recycled, so plastic items now bear symbols to indicate their composition. Those with a "1" inside the recycling logo are made of a polymer called PETE (for polyethylene terephthalate); those with a "2" consist of HDPE (high-density polyethylene); and those with a "3" are made of V (polyvinyl chloride). We have to study polymers because things like starch, proteins, enzymes, some hormones, and genes are natural polymers.

THIS CHAPTER IN CONTEXT

In this chapter we'll build on the introduction to names, symbols, and general physical properties of organic compounds studied in the previous chapter and move forward to our first organic functional group, the double bond. Our study will mostly be about the properties of this group wherever it occurs, not only in simple alkenes but anywhere. The complex, multifunctional molecules of vegetable oils, for example, or the molecules of a cell membrane are loaded with alkene groups.

12.1 OCCURRENCE

The carbon–carbon double bond occurs in nature largely in compounds having other functional groups.

■ The carbon–carbon double bond is sometimes called the *ene* function.

Unsaturation in hydrocarbons occurs as double bonds, triple bonds, or benzene rings. Hydrocarbons with double bonds are called **alkenes** and the carbon–carbon double bond is the **alkene group,** a group that is common at the molecular level of life (see Interaction 12.1).

Molecules of **aromatic hydrocarbons** contain the *benzene ring,* an unsaturated group that occurs in proteins and nucleic acids. We'll study the benzene ring and its interesting reactions near the close of the chapter. The carbon–carbon triple bond occurs in hydrocarbons called **alkynes,** but it is uncommon in nature.

The Physical Properties of Alkenes Resemble Those of Alkanes   The structures and some physical properties of several alkenes are given in Table 12.1. The first three alkenes, like the first four alkanes, are gases at room temperature. Like all hydrocarbons, alkenes are much less dense than water, are insoluble in water, and are soluble in nonpolar solvents.

TABLE 12.1   Properties of Some 1-Alkenes

Name (IUPAC)	Structure	BP (°C)	MP (°C)	Density (g/mL)[a]
Ethene	$CH_2\!=\!CH_2$	−104	−169	—
Propene	$CH_2\!=\!CHCH_3$	−48	−185	—
1-Butene	$CH_2\!=\!CHCH_2CH_3$	−6	−185	—
1-Pentene	$CH_2\!=\!CHCH_2CH_2CH_3$	30	−165	0.641
1-Hexene	$CH_2\!=\!CHCH_2CH_2CH_2CH_3$	64	−140	0.673
1-Heptene	$CH_2\!=\!CHCH_2CH_2CH_2CH_2CH_3$	94	−119	0.697
1-Octene	$CH_2\!=\!CHCH_2CH_2CH_2CH_2CH_2CH_3$	121	−102	0.715
1-Nonene	$CH_2\!=\!CHCH_2CH_2CH_2CH_2CH_2CH_2CH_3$	147	−81	0.729
1-Decene	$CH_2\!=\!CHCH_2CH_2CH_2CH_2CH_2CH_2CH_2CH_3$	171	−66	0.741
Cyclopentene	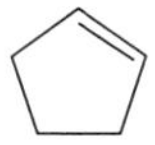	44	−135	0.722
Cyclohexene	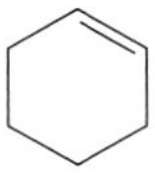	83	−104	0.811

[a] At 10 °C

INTERACTION 12.1
THE ALKENE DOUBLE BOND IN NATURE

Molecules with alkene groups are found among most of the major families of biochemically important substances. Cholecalciferol (vitamin D_3) and retinol (vitamin A_1), for example, are two of the vitamins that have several alkene groups. Their double bonds are slowly attacked by oxygen if foods containing these vitamins are boiled excessively.

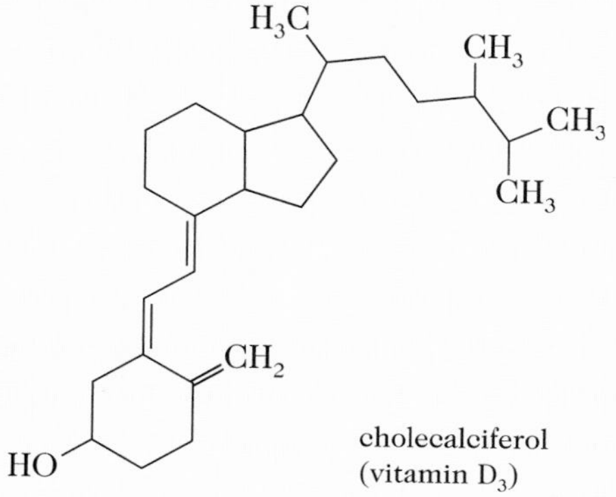

cholecalciferol
(vitamin D_3)

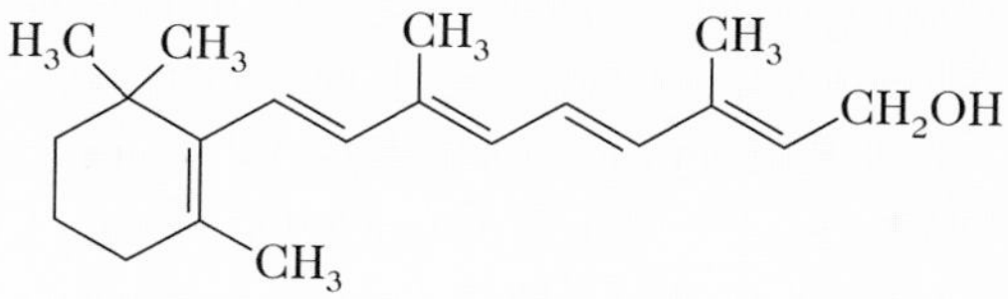

retinol
(vitamin A_1)

β-Carotene is one of the bright yellow-orange compounds in carrots and is called a *provitamin* because the body is able to convert it into vitamin A.

Several of the human sex hormones have alkene groups, including testosterone, the chief male sex hormone.

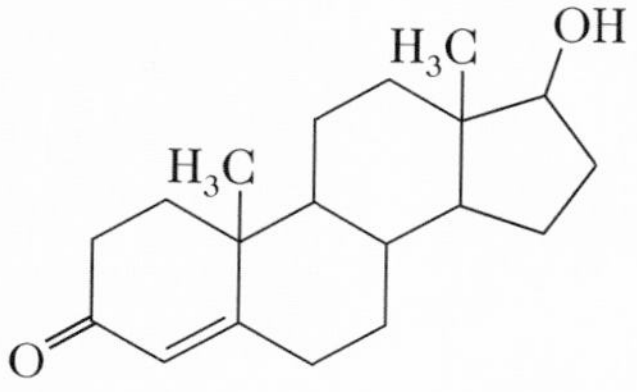

testosterone
(male sex hormone)

The edible fats and oils all have at least one alkene group per molecule, and the vegetable oils generally have two to four. Shown here is the structure of a molecule found in corn oil, a typical vegetable oil and a product widely used to make salad dressing. You can see why vegetable oils are described as *polyunsaturated*.

$$\begin{array}{l} \quad\quad\quad\quad O \\ \quad\quad\quad\quad \| \\ CH_2OC(CH_2)_7CH{=}CHCH_2CH{=}CH(CH_2)_4CH_3 \\ \quad\quad\quad | \quad O \\ \quad\quad\quad\quad \| \\ \quad\quad CHOC(CH_2)_{12}CH_3 \\ \quad\quad\quad | \quad O \\ \quad\quad\quad\quad \| \\ CH_2OC(CH_2)_7CH{=}CH(CH_2)_7CH_3 \end{array}$$

a molecule typical of those in vegetable oils

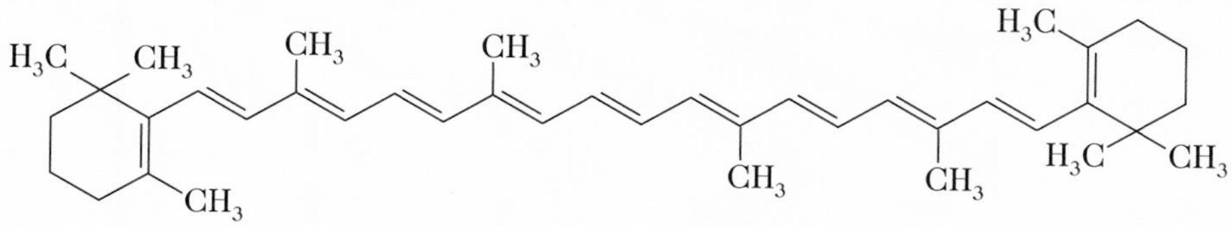

β-carotene

12.2 NAMING THE ALKENES

The IUPAC names of alkenes end in -*ene*, and the double bond takes precedence over substituents in numbering the parent chain.

As with the IUPAC rules for any family, the rules for the alkenes specify the *name ending*, how to pick out the *parent chain* or *parent ring*, how to *number the chain or ring,* and how to designate substituent groups. For alkenes and cycloalkenes, the IUPAC rules are as follows.

1. Use the ending *-ene* for all alkenes and cycloalkenes.

2. For open-chain alkenes, identify the parent chain as the longest sequence of carbons that *includes the double bond.* Name this chain as if it were that of an alkane and then change the *-ane* ending to *-ene.* This gives the basic *name* of the parent chain, except that the location of the double bond is yet to be specified. For example, the longest chain *that includes the double bond* in the first structure below has six carbons. A longer chain of seven carbons is present, but it does not include the double bond.

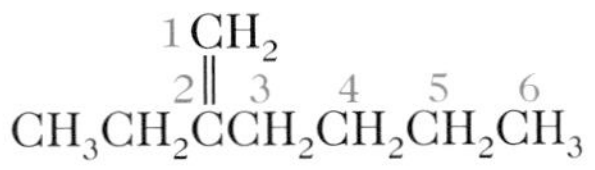

$$\overset{1}{C}H_2$$
$$\overset{2}{\parallel}\,\overset{3}{}\quad\overset{4}{}\quad\overset{5}{}\quad\overset{6}{}$$
$$CH_3CH_2CCH_2CH_2CH_2CH_3$$

The parent chain has six carbons, not seven. The (incomplete) name of the parent chain is hexene.

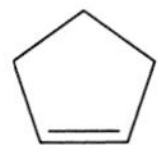

cyclopentene (complete name)

For cyclic alkenes, the ring is the parent in all situations we will encounter. In the second structure above, the parent cycloalkane is cyclopentane, so by changing the ending to *-ene* we have cyclopentene.

3. For open-chain alkenes, number the parent chain from whichever end gives the lower number to the first carbon of the double bond.

    This rule gives precedence to the location of the double bond over the location of the first substituent on the parent chain. For example,

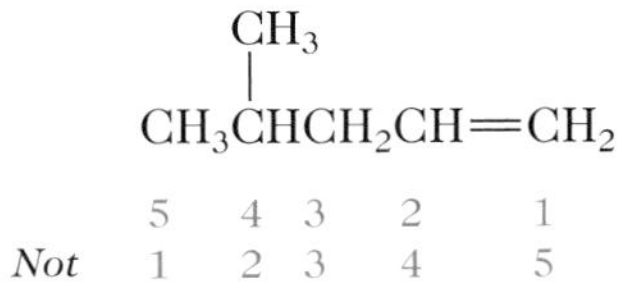

$$\overset{CH_3}{\underset{|}{}}$$
$$CH_3CHCH_2CH=CH_2$$

The double bond is at position 1, not 4.

Not

$$\begin{array}{ccccc} 5 & 4 & 3 & 2 & 1 \\ 1 & 2 & 3 & 4 & 5 \end{array}$$

4-methyl-1-pentene (complete name)
not 2-methyl-4-pentene

4. For cycloalkenes, always give position 1 to one of the two carbons at the double bond. To decide which carbon gets this number, number the ring atoms from carbon 1 *through the double bond* in whichever direction reaches a substituent first. For example, the numbers inside the ring represent the correct numbering, not the numbers outside the ring.

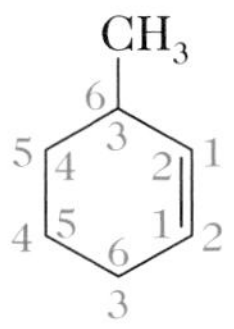

CH$_3$

3-methylcyclohexene (complete name)
not 6-methylcyclohexene

5. To the name begun with rules 1 and 2, place the number that locates the first carbon of the double bond as a prefix, and separate this number from the name by a hyphen.

6. If substituents are on the parent chain or ring, complete the name obtained by rule 5 by placing the names and location numbers of the substituents as prefixes.

Remember to separate numbers from numbers by commas, but use hyphens to connect a number to a word.

Several correctly named alkenes are shown next[1] with the common names of three given in parentheses. The ending *-ylene* characterizes the common names of open-chain alkenes.

$$CH_2\!\!=\!\!CH_2 \qquad CH_3CH\!\!=\!\!CH_2 \qquad CH_3\overset{\overset{\displaystyle CH_3}{|}}{C}\!\!=\!\!CH_2$$

ethene   propene   2-methylpropene
(ethylene)   (propylene)   (isobutylene)

■ The common name, ethylene, is also allowed by the IUPAC as the name for ethene.

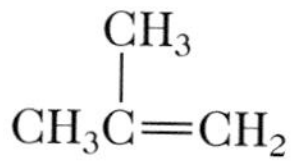

$$CH_3CH_2\overset{\overset{\displaystyle CH_3}{|}}{C}HCH_2CH\!\!=\!\!\overset{\overset{\displaystyle CH_3}{|}}{C}CH_3$$

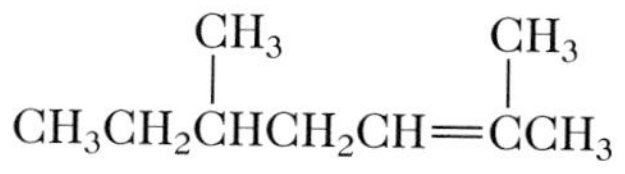

2,5-dimethyl-2-heptene

$$CH_3CH_2CH_2\overset{\overset{\displaystyle CH_3}{|}}{\underset{\underset{\displaystyle CHCH_2CH_2CH_3}{|}}{C}}\!\!=\!\!CH_2$$

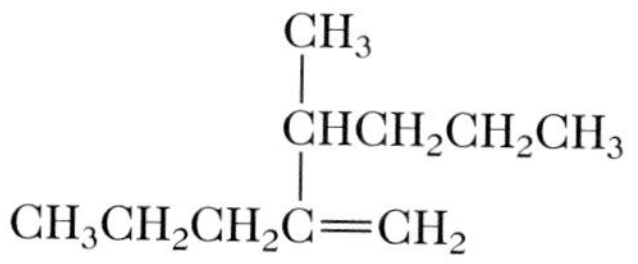

3-methyl-2-propyl-1-hexene

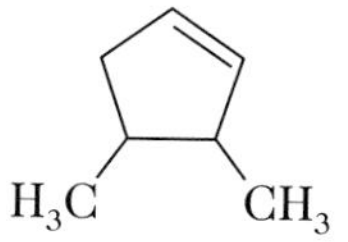

3,4-dimethylcyclopentene
not 4,5-dimethylcyclopentene

$$Cl(CH_2)_6CH\!\!=\!\!CH_2$$

8-chloro-1-octene

No number is used to locate the double bond in 3,4-dimethylcyclopentene because, by rule 4, the double bond can only be at position 1.

EXAMPLE 12.1   Naming an Alkene

Write the name of the following alkene.

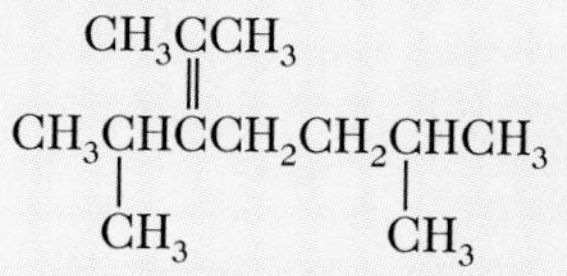

$$\begin{array}{c} CH_3\overset{\overset{\displaystyle CH_3}{\|}}{C}CH_3 \\ CH_3\overset{}{C}HCCH_2CH_2\overset{}{C}HCH_3 \\ {\scriptstyle |}\qquad\qquad{\scriptstyle |} \\ CH_3\qquad\quad CH_3 \end{array}$$

ANALYSIS The longest chain that includes the double bond must be identified and numbered from whichever end gives the first carbon of the double bond the lower of two possible numbers. The parent is an *alkene* with a chain of seven carbons, a heptene. The following numbering of the parent chain gives the double bond position 2. (The alternative numbering, right to left, would have given the double bond position 5.)

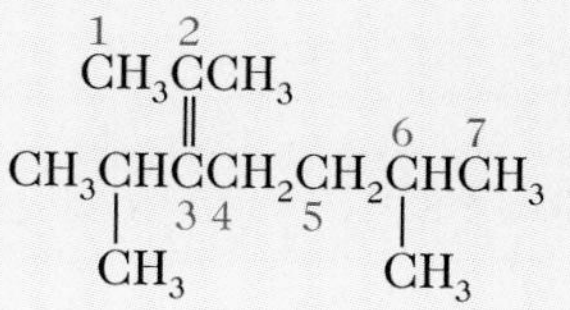

$$\begin{array}{c} \overset{1}{C}H_3\overset{2}{C}CH_3 \\ CH_3CHCCH_2CH_2CHCH_3 \\ {\scriptstyle 3\ 4}\quad{\scriptstyle 5}\quad{\scriptstyle 6\ 7} \\ CH_3\qquad\quad CH_3 \end{array}$$

---

[1] Propene is not named 1-propene because by rule 3 there is no 2-propene. 2-Methylpropene is not named 2-methyl-1-propene because the 1 is not needed.

The parent alkene is thus 2-heptene. It holds two methyl groups (positions 2 and 6) and one isopropyl group (position 3). The names and location numbers are next assembled into the name.

SOLUTION   The correct name is

3-Isopropyl-2,6-dimethyl-2-heptene

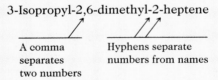

A comma
separates
two numbers

Hyphens separate
numbers from names

CHECK   The most common error that students make is to *fail to find the longest chain* that includes the double bond. The first check step, then, is to go back over the answer to see whether there is a chain holding the alkene group that is longer than 7 carbons. Use a colored pen to draw an enclosure for this chain so that all substituents are outside. Then move in from either end of the chain, counting carbons, to see which starting point yields the lower number for the beginning of the double bond. All the other numbers must then fall into place. Another common error is the failure to identify alkyl groups correctly, so double-check these.

■   PRACTICE EXERCISE 1   Write the IUPAC names for the following compounds.

(a) 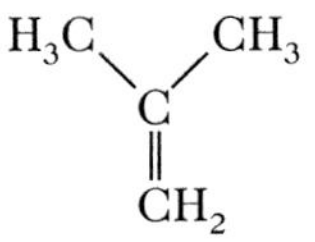
$$H_3C \diagdown \quad \diagup CH_3$$
$$C$$
$$\|$$
$$CH_2$$

(b) 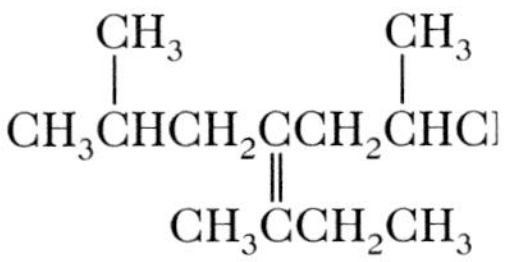 $CH_3CHCH_2CCH_2CHCl$ with CH_3, CH_3 above and $CH_3CCH_2CH_3$ below

(c) $CH_3CH{=}CHCl$

(d) $BrCH_2CH{=}CH_2$

(e) 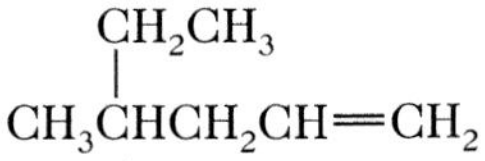 $CH_3CHCH_2CH{=}CH_2$ with CH_2CH_3

(f) 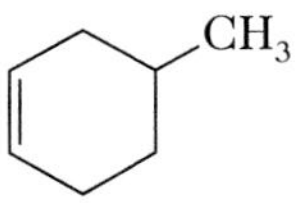   [cyclohexadiene ring with CH_3]

■   PRACTICE EXERCISE 2   Write condensed structures for each of the following.

(a)   4-Methyl-2-pentene
(b)   3-Propyl-1-heptene
(c)   4-Chloro-3,3-dimethyl-l-butene
(d)   2,3-Dimethyl-2-butene

When a compound has two double bonds, it is named as a *diene* with two numbers in the name to specify the locations of the double bonds. For example,

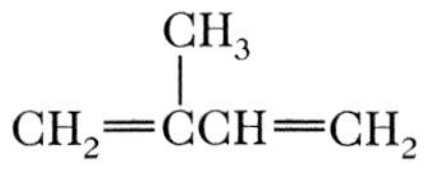

$$CH_2{=}CCH{=}CH_2$$ with CH_3

2-methyl-1,3-butadiene

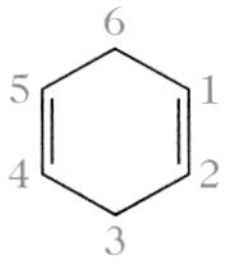

1,4-cyclohexadiene

This pattern can be easily extended to *trienes, tetraenes,* and so forth.

12.3 GEOMETRIC ISOMERS

The alkenes and cycloalkanes can exhibit geometric isomerism because there is no free rotation at the double bond or in a ring.

The six atoms at a double bond, the two carbons and the four atoms attached to them, all lie in the same plane, as illustrated in Figure 12.1, which shows the simplest alkene. The bond angles are 120°.

Some Alkene Isomers Have Identical Constitutions but Different Geometries Alkenes can exist as isomers in three ways. They can have different carbon skeletons, as shown by 1-butene and 2-methylpropene. These two compounds are *constitutional isomers*. Two alkenes can also have identical carbon skeletons but differ in the locations of their double bonds, as in 1-butene and 2-butene. These two are also constitutional isomers because their H atoms are attached differently to the skeleton.

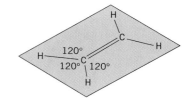

FIGURE 12.1 The geometry at a carbon–carbon double bond.

$$CH_2{=}CHCH_2CH_3 \qquad CH_2{=}\overset{\overset{\displaystyle CH_3}{|}}{C}CH_3 \qquad CH_3CH{=}CHCH_3$$

1-butene 2-methylpropene 2-butene

Some alkenes have identical constitutions *including the location of the double bond,* but differ only in the geometry at this bond, as seen in the two isomers with the constitution of 2-butene, $CH_3CH{=}CHCH_3$.

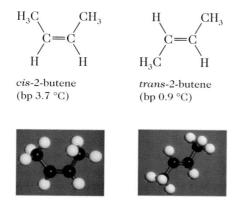

cis-2-butene
(bp 3.7 °C)

trans-2-butene
(bp 0.9 °C)

cis-2-Butene and *trans*-2-butene differ only in the *directions* taken by their end-of-chain methyl groups. Isomers that differ in geometry but have identical constitutions are called **geometric isomers,** and the phenomenon is called **geometric isomerism.** It is possible because there is no free rotation at the double bond. It's as if you pounded in two nails to hold a pair of boards together; you could not turn one board against the other. But you could if you had used only one nail (a single bond).

Two designated substituents on the same side of the double bond are said to be *cis* to each other. When they are on opposite sides, they are *trans* to each other. (Sometimes geometric isomerism is called *cis–trans isomerism.*) The designations of cis or trans can be made parts of the names of geometric isomers, as in the examples of *cis*- and *trans*-2-butene shown earlier.

■ The *side* of a double bond is not the same as the *end* of a double bond.

When There Are Two Identical Groups at One End of a Double Bond, Geometric Isomers Are Not Possible   If one end of a double bond has two *identical* groups, like two H's or two methyls, then there is nothing for a group at the other end to be *uniquely* cis or trans to. 1-Butene, for example, has two H atoms at one end of its double bond, so the ethyl group at the other end cannot be positioned to give geometric isomers. We can *write* structures that might appear to be isomers, but they are actually identical.

$$\begin{array}{c} H \\ \diagdown \\ C=C \\ \diagup \quad \diagdown \\ H \qquad H \end{array} \begin{array}{c} CH_2CH_3 \\ \diagup \end{array}$$

is the same as

$$\begin{array}{c} H \qquad H \\ \diagdown \quad \diagup \\ C=C \\ \diagup \quad \diagdown \\ H \qquad CH_2CH_3 \end{array}$$

1-butene   1-butene

Simply flop the *whole* first structure over, top to bottom, to get the second. Whole-molecule flopping, of course, does not reorganize bonds into any new structure or geometry. Thus, there are no geometric isomers of 1-butene.

Geometric isomerism also occurs when the atoms involved at the ends of the double bond are halogen atoms or other groups, for example,

$$\begin{array}{c} H_3C \qquad H \\ \diagdown \quad \diagup \\ C=C \\ \diagup \quad \diagdown \\ H \qquad Cl \end{array} \qquad\qquad \begin{array}{c} H_3C \qquad Cl \\ \diagdown \quad \diagup \\ C=C \\ \diagup \quad \diagdown \\ H \qquad H \end{array}$$

trans-1-chloro-1-propene   *cis*-1-chloro-1-propene

EXAMPLE 12.2   Writing the Structures of Cis and Trans Isomers

Write the structures of the cis and trans isomers, if any, of the following alkene.

$$CH_3CH{=}CHCH_2CH_3$$

2-pentene

ANALYSIS   Notice first that geometric isomerism is possible in 2-pentene. At *neither* end of the double bond are the two groups identical. At one end there are H and CH_3; at the other end, H and CH_2CH_3. Therefore, we must draw structures of the geometric isomers.

To show the geometry of each isomer correctly, we start by writing a carbon–carbon double bond without any attached groups, *spreading the single bonds at the carbon atoms at angles of about 120°.*

$$\begin{array}{c} \diagdown \qquad \diagup \\ C=C \\ \diagup \qquad \diagdown \end{array} \qquad\qquad \begin{array}{c} \diagdown \qquad \diagup \\ C=C \\ \diagup \qquad \diagdown \end{array}$$

Then we attach the two groups that are at one of the ends of the double bond. We attach them *identically* to make identical partial structures.

$$\begin{array}{c} H_3C \qquad\quad \diagup \\ \diagdown \qquad \diagup \\ C=C \\ \diagup \qquad \diagdown \\ H \end{array} \qquad\qquad \begin{array}{c} H_3C \qquad\quad \diagup \\ \diagdown \qquad \diagup \\ C=C \\ \diagup \qquad \diagdown \\ H \end{array}$$

Finally, at the other end of the double bond, we draw the other two groups, only this time be sure that they are switched in their relative positions.

SOLUTION   The geometric isomers of 2-pentene are

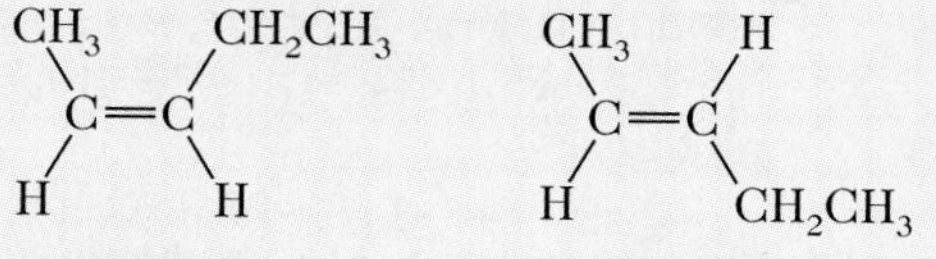

$$
\begin{array}{cc}
\underset{H}{\overset{CH_3}{\diagdown}}C=C\underset{H}{\overset{CH_2CH_3}{\diagup}} & \underset{H}{\overset{CH_3}{\diagdown}}C=C\underset{CH_2CH_3}{\overset{H}{\diagup}} \\
\textit{cis}\text{-2-pentene} & \textit{trans}\text{-2-pentene}
\end{array}
$$

CHECK   Be sure to check whether the two structures are geometric *isomers* and not two identical structures that are merely flip-flopped on the page.

■   PRACTICE EXERCISE 3   Write the structures of the cis and trans isomers, if any, of the following compounds.

(a) $CH_3CH_2\underset{\underset{CH_3}{|}}{C}=CHCH_3$ (b) $ClCH=CHCl$ (c) $CH_3\underset{\underset{CH_3}{|}}{C}=CH_2$ (d) $Cl\underset{\underset{Cl}{|}}{C}=CHBr$

Cyclic Compounds Can Also Have Geometric Isomers   The double bond is not the only source of restricted rotation; the ring is another. For example, two geometric isomers of 1,2-dimethylcyclopropane are known, and neither can be twisted into the other without breaking the ring open. This costs too much energy to occur spontaneously even at quite high temperatures.

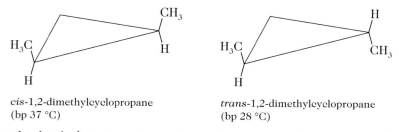

cis-1,2-dimethylcyclopropane   *trans*-1,2-dimethylcyclopropane
(bp 37 °C)   (bp 28 °C)

We'll see this kind of cis–trans isomerism in the many cyclic structures of carbohydrates; their *geometric* differences alone make most carbohydrates unusable in human nutrition. As we have said before, molecular geometry is as important in living processes as functional groups. For an interesting occurrence of cis–trans isomerism at the molecular level of vision, see Interaction 12.2.

12.4   ADDITION REACTIONS OF THE DOUBLE BOND

The carbon–carbon double bond adds H_2, Cl_2, Br_2, HX, H_2SO_4, and H_2O, and it is attacked by strong oxidizing agents, including ozone.

In an **addition reaction,** pieces of an adding molecule become attached to the carbon atoms at opposite ends of the double bond, which then becomes a single bond.

THE PRIMARY CHEMICAL EVENT THAT LETS US SEE

The retina in the eye of a human being has two kinds of cells, rods and cones, that can convert absorbed light into signals which the brain perceives as vision. Rods equip us to see shades of gray; cones give us color vision. We know more about how rods work than cones, so our brief discussion is limited to what happens in the rods when they absorb light.

Light behaves as a stream of tiny energy packets called *photons* having a wide range of energies. Photons in visible light have enough energy to break one of the two bonds of a double bond when it is a unit of a repeating pattern of alternating double and single bonds. Such a pattern occurs in 11-*cis*-retinal, the primary absorber of visible light in the rods of our eyes. *Cis* here means that the chain emerges on the same side of the double bond at carbons 11 and 12. In all-*trans*-retinal, the chain emerges trans at this location. Otherwise the structures are the same. The structure of vitamin A (shown in the previous Interaction) is remarkably similar to the retinals, and the body needs and uses this vitamin to make them.

Molecules of 11-*cis*-retinal must be joined to the protein *opsin* in order to couple the absorption of light photons to a nerve signal, and the coupling occurs at the carbon–oxygen double bond in 11-*cis*-retinal. If we represent everything except this double bond by R, then after the coupling the product can be represented by R—CH=N—{opsin}, which is called *rhodopsin.*

As we said, photons of visible light can break one of the two bonds of a double bond, while the remaining bond holds the molecule together. However, with only a single bond left between C-11 and C-12, *free rotation is possible.* This is exactly what happens when the vast carbon network of 11-*cis*-retinal in rhodopsin accepts a light photon. One side of the molecule flips to a trans configuration at the 11,12-linkage to form all-*trans*-retinal. This change in geometry affects the properties of the opsin portion of rhodopsin, and a series of chemical events now swiftly sends a signal to the brain. An equally swift set of chemical changes occurs in the cone of the eye to restore rhodopsin to its photon-accepting geometry. Thus cis–trans isomerization is at the heart of vision.

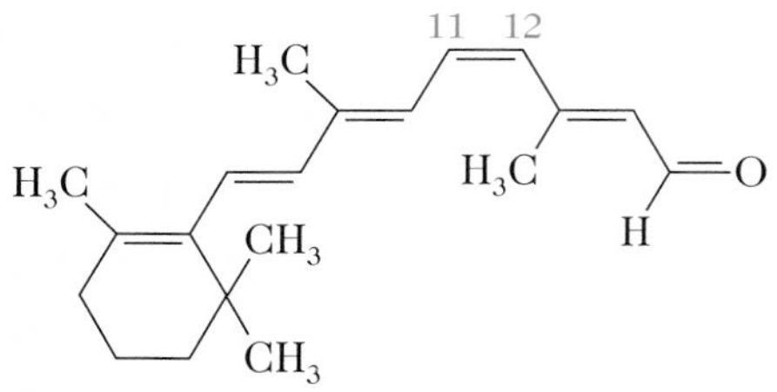

11-*cis*-retinal

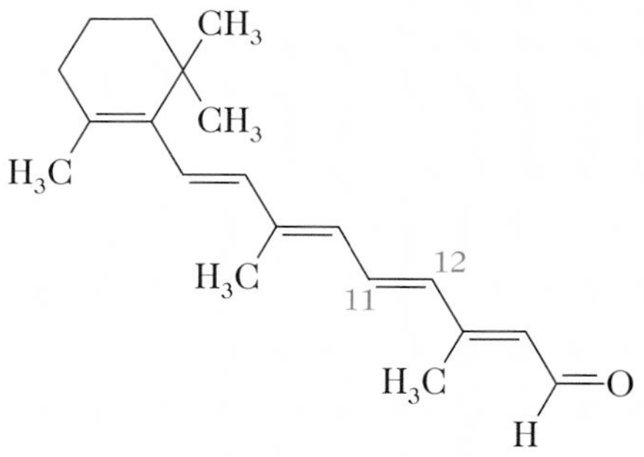

all-*trans*-retinal

Using X—Y as the adding molecule, all additions to an alkene group thus occur as follows.

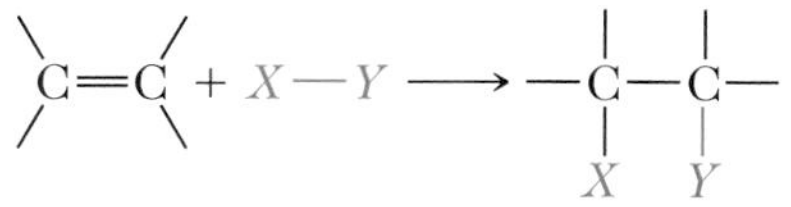

Hydrogen Adds to an Alkene Group and Saturates It   In the presence of a powdered metal catalyst, like powdered nickel or platinum, hydrogen adds to a double bond. The reaction, called *hydrogenation,* converts an alkene to an alkane.

■ The addition of hydrogen is often called the *reduction* of the double bond.

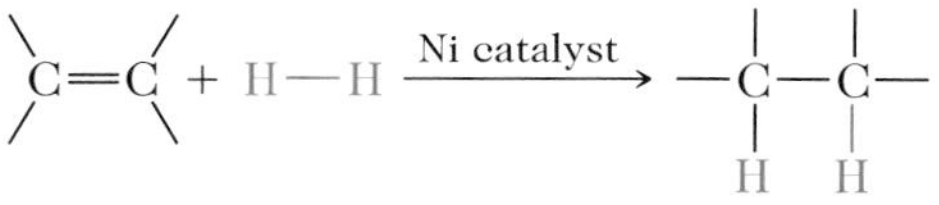

Specific examples involving alkenes are

$$CH_2{=}CH_2 + H{-}H \xrightarrow{\text{Ni catalyst}} \underset{\underset{H}{|}}{CH_2}{-}\underset{\underset{H}{|}}{CH_2} \quad \text{or} \quad CH_3CH_3$$

ethene   ethane

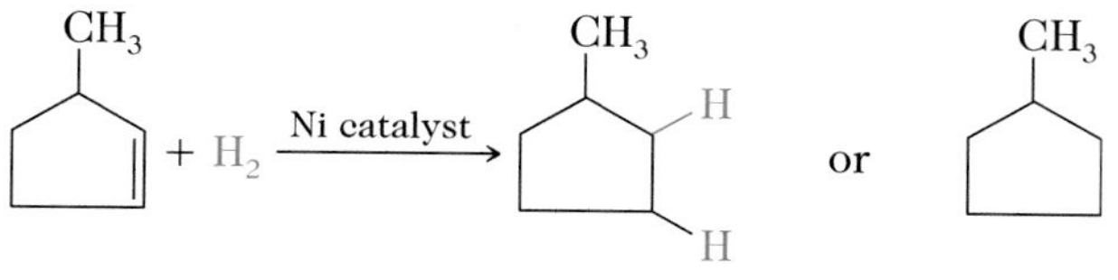

3-methylcyclopentene   methylcyclopentane

Notice that each product has the same carbon skeleton as the parent.

The Net Effect of Hydrogenation Occurs at the Molecular Level of Life Molecules of H_2 and powdered metal catalysts are unavailable in the body, of course. Yet, cells do have molecular carriers that deliver the pieces of H_2 to alkene groups. One piece of H_2 is the hydride ion, $H{:}^-$, and its carrier is an enzyme that transfers $H{:}^-$ to one end of the double bond. The other piece of H_2 is H^+, which is donated to the other carbon of the double bond by the same enzyme carrier or is plucked from a proton donor of the surrounding fluid. The *net* effect is the addition of H_2 because $H{:}^-$ and H^+ together add up to one $H{:}H$ molecule.

■ $H{:}^-$ (the hydride ion) must be donated by the carrier *directly* to the acceptor and not through the solution. When exposed directly to water, $H{:}^-$ reacts vigorously:

$$H{:}^- + H{-}OH \longrightarrow$$
$$H{-}H + OH^-$$

EXAMPLE 12.3   Writing the Structure of the Product of the Addition of Hydrogen to a Double Bond

Write the structure of the product of the following reaction.

$$\underset{\underset{}{}}{CH_3CH_2}\overset{\overset{CH_3}{|}}{C}{=}\overset{\overset{CH_3}{|}}{C}CH_2CH_2CH_3 + H_2 \xrightarrow{\text{Ni catalyst}} ?$$

ANALYSIS *The only change occurs at the double bond;* all the rest of the structure goes through the reaction unchanged. *This is true of all of the addition reactions that we will study.* Therefore, copy the structure of the alkene just as it is, except leave only a single bond where the double bond was. Then increase by one the number of H's at each C of the original double bond.

SOLUTION The structure of the product can be written as

$$\underset{\underset{H}{|}}{CH_3CH_2}\overset{\overset{CH_3}{|}}{C}{-}\underset{\underset{H}{|}}{\overset{\overset{CH_3}{|}}{C}}CH_2CH_2CH_3 \quad \text{or} \quad CH_3CH_2\overset{\overset{CH_3}{|}}{CH}{-}\overset{\overset{CH_3}{|}}{CH}CH_2CH_2CH_3$$

CHECK Make sure that the product has the identical carbon skeleton as the starting material. Then see that *each* carbon of the original alkene group has one more H.

One major goal in these chapters is to learn some chemical properties of functional groups. We have just learned a chemical "map sign" for the carbon–carbon double bond.

The alkene group adds hydrogen and changes to a single bond as each of its carbon atoms picks up one H atom.

This chemical fact about the alkene group has to be learned. But learn it by working specific examples using specific alkenes. Its pointless (and dreadfully tedious) to memorize individual reactions. Use your memory work to learn only the *kinds* of reactions each functional group gives and what *general changes* happen to a molecule. Now try the following *specific* examples. As you do, say to yourself the *general* fact being illustrated each time, the chemical map sign about hydrogenation.

■ PRACTICE EXERCISE 4 Write the structures of the products, if any, of the following.

(a) $CH_3CH{=}CH_2 + H_2 \xrightarrow{\text{Ni catalyst}}$

(b) $CH_3CH_2CH_3 + H_2 \xrightarrow{\text{Ni catalyst}}$

(c) 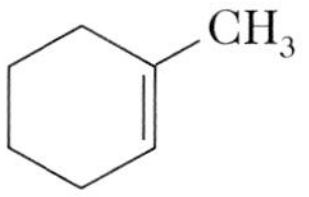 $+ H_2 \xrightarrow{\text{Ni catalyst}}$

(d) $CH_3(CH_2)_7CH{=}CH(CH_2)_7CO_2H + H_2 \xrightarrow{\text{Ni catalyst}}$

Chlorine and Bromine Also Add to Double Bonds The foregoing stated another alkene "map sign." Without any need for a special catalyst, both Cl_2 and Br_2 *rapidly* add to the carbon–carbon double bond. Iodine (I_2) does not add, and fluorine (F_2) reacts explosively with almost any organic compound to give a mixture of products. If you notice the similarity of this kind of reaction to hydrogenation, remembering the "map sign" will be so much easier.

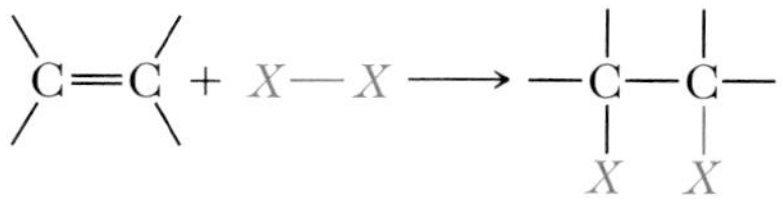

$$X = Cl \text{ or } Br$$

Specific examples are

■ Use a good fume hood and protective gloves when dispensing bromine.

$$CH_3CH{=}CH_2 + Br_2 \longrightarrow CH_3CH{-}CH_2$$

$$\underset{Br}{\qquad} \underset{Br}{\qquad}$$

propene   1,2-dibromopropane

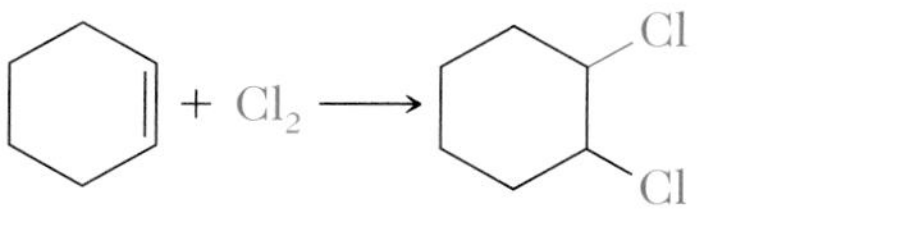

$+ Cl_2 \longrightarrow$

cyclohexene   1,2-dichlorocyclohexane

EXAMPLE 12.4 Writing the Structure of the Product of the Addition of Chlorine or Bromine to a Carbon–Carbon Double Bond

What compound forms in the following situation?

$$CH_3CH{=}CHCH_3 + Br_2 \longrightarrow ?$$

ANALYSIS As in hydrogenation problems, we rewrite the alkene except that a single bond is left where the double bond was.

$$CH_3CH{-}CHCH_3 \quad \text{(incomplete)}$$

Then we attach a bromine atom to each carbon of the former double bond.

SOLUTION

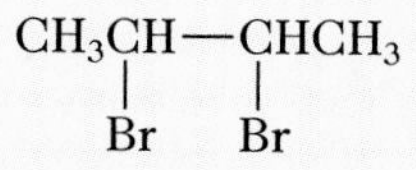

$$\underset{\underset{Br\quad Br}{|\qquad|}}{CH_3CH{-}CHCH_3}$$

2,3-dibromobutane

CHECK Make sure that the carbon skeleton in the product is identical to that of the original alkene, that the double bond has been replaced by a single bond, and that each carbon of the double bond carries one of the pieces (Br in this example) of the adding molecule.

■ Bromine is dark brown, but dibromoalkanes are colorless. So an organic compound that rapidly decolorizes bromine quite likely has a double bond.

■ **PRACTICE EXERCISE 5** Complete the following equations by writing the structures of the products. If no reaction occurs under the conditions shown, write "no reaction."

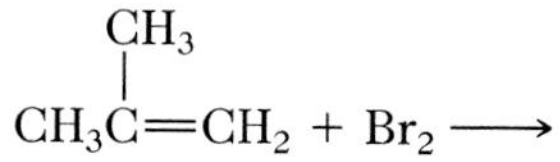

(a) $\underset{\underset{CH_3}{|}}{CH_3C}{=}CH_2 + Br_2 \longrightarrow$

(b) $CH_3CH_2CH_2CH_3 + Cl_2 \longrightarrow$

(c) $CH_2{=}CHCH_2CH_3 + Cl_2 \longrightarrow$

(d) $CH_3CH{=}CHCH_3 + H_2 \xrightarrow{\text{Ni catalyst}}$

Hydrogen Chloride, Hydrogen Bromide, and Sulfuric Acid Add Easily to Double Bonds We just gave another alkene "map sign." If gaseous hydrogen chloride or hydrogen bromide is bubbled into an alkene or if concentrated sulfuric acid is mixed with it, the following kind of reaction takes place. Notice that the pattern is the same in all these additions. We'll let H—G represent any of these reactants, where G stands for any electron-rich group that we will study, like Cl, Br, or OSO_3H.

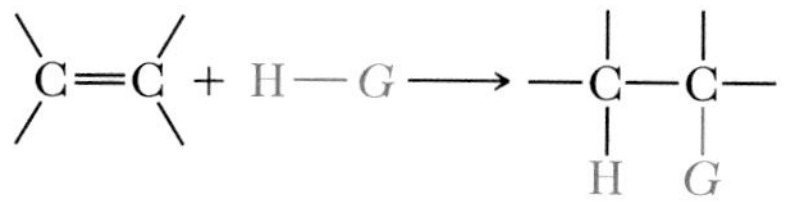

$$\underset{/}{\overset{\backslash}{C}}{=}\underset{\backslash}{\overset{/}{C}} + H{-}G \longrightarrow \underset{\underset{H\quad G}{|\quad|}}{-\overset{|}{C}-\overset{|}{C}-} \qquad (G = Cl, Br, \text{ or } OSO_3H)$$

An example:

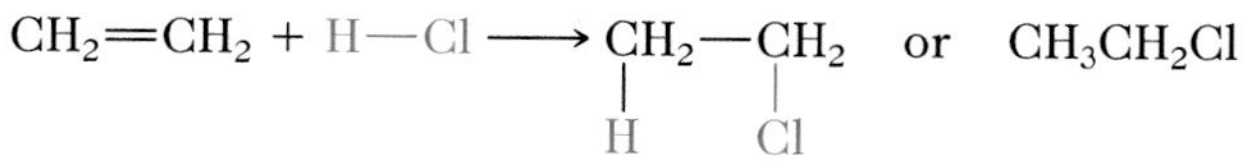

$$CH_2{=}CH_2 + H{-}Cl \longrightarrow \underset{\underset{H\qquad Cl}{|\qquad|}}{CH_2{-}CH_2} \quad \text{or} \quad CH_3CH_2Cl$$

Unsymmetrical Reactants Add Selectively to Unsymmetrical Double Bonds   We now have a small complication. H—G is not a symmetrical molecule, like H—H, Br—Br, or Cl—Cl. When *symmetrical* molecules add to a double bond, it doesn't matter which end of the double bond gets which half of the adding molecule. But it matters when H—G adds, *if the double bond is itself unsymmetrical.* An *unsymmetrical double bond* is one whose two carbon atoms hold unequal numbers of hydrogen atoms. For example, 1-butene, CH_2=$CHCH_2CH_3$, and propene, CH_3CH=CH_2, both have unsymmetrical double bonds. Each double bond has one carbon with two H's, whereas the other carbon has only one H, an unequal number. The double bond in 2-butene, CH_3CH=$CHCH_3$, however, is symmetrical; one H is at each carbon.

When H—Cl(g) adds to propene we could imagine obtaining 1-chloropropane, 2-chloropropane, or a mixture of the two, perhaps 50:50. Let's see what actually happens. We have

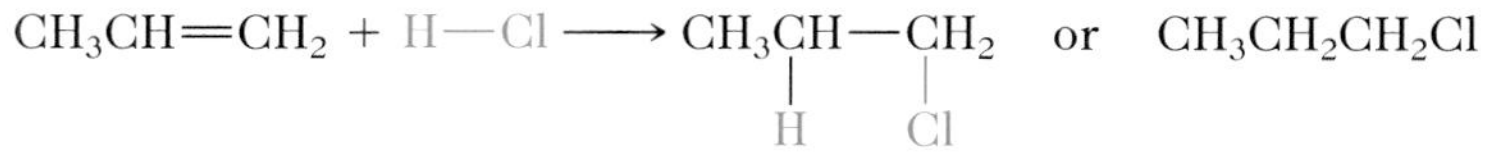

$$CH_3CH\text{=}CH_2 + H\text{—}Cl \longrightarrow \underset{\underset{H \qquad Cl}{|\qquad\ |}}{CH_3CH\text{—}CH_2} \quad \text{or} \quad CH_3CH_2CH_2Cl$$

1-chloropropane

or, if the addition occurs in the opposite sense, we have

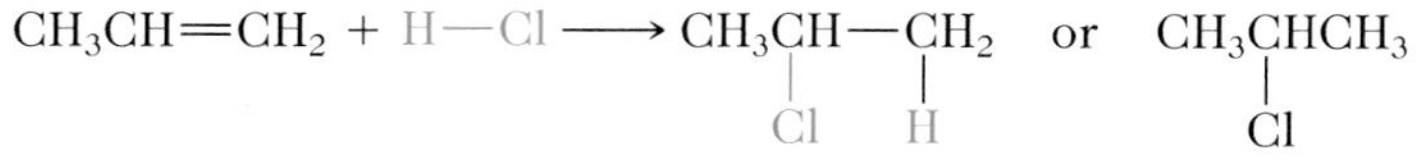

$$CH_3CH\text{=}CH_2 + H\text{—}Cl \longrightarrow \underset{\underset{Cl \qquad H}{|\qquad\ |}}{CH_3CH\text{—}CH_2} \quad \text{or} \quad \underset{\underset{Cl}{|}}{CH_3CHCH_3}$$

2-chloropropane
(the major product)

The actual product is largely 2-chloropropane; very little of its isomer, 1-chloropropane, forms. In other words, the reactant, H—Cl, adds to the unsymmetrical double bond selectively.

Markovnikov's Rule Predicts Directions in Unsymmetrical Additions   Vladimer Markovnikov (1838–1904), a Russian chemist, was the first to notice that unsymmetrical alkenes add unsymmetrical reactants mostly in one direction. Which is the preferred direction can be predicted by **Markovnikov's rule.**[2]

■ "Them that has, gits" applies here, too.

> **Markovnikov's Rule**   When an unsymmetrical reactant of the type H—G adds to an unsymmetrical alkene, the carbon with the greater number of hydrogens gets one more H.

The following examples illustrate Markovnikov's rule in action.

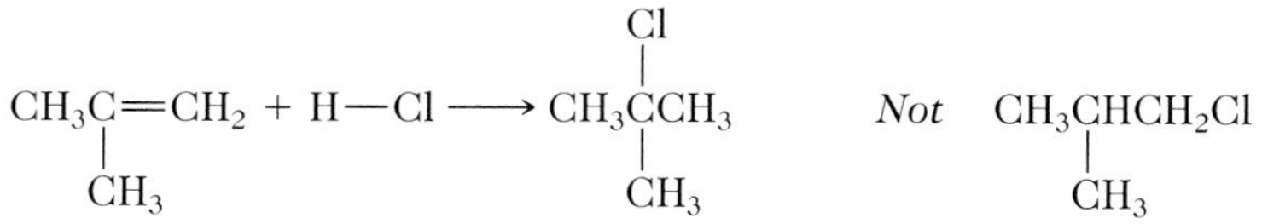

$$\underset{\underset{\underset{CH_3}{|}}{|}}{CH_3C\text{=}CH_2} + H\text{—}Cl \longrightarrow \overset{\overset{Cl}{|}}{\underset{\underset{CH_3}{|}}{CH_3CCH_3}} \quad\quad Not \quad \underset{\underset{CH_3}{|}}{CH_3CHCH_2Cl}$$

2-methylpropene   t-butyl chloride

---

[2] When hydrogen bromide is used, it is important that no peroxides, compounds of the type R—O—O—H or R—O—O—R, be present. Traces of peroxides commonly form in organic liquids that are stored in contact with air for long periods. When peroxides are present, the addition of H—Br occurs in the direction opposite to that predicted by Markovnikov's rule. Peroxides catalyze the anti-Markovnikov addition *only* of HBr, not of HCl or H_2SO_4.

1-methylcyclohexene + H—Cl ⟶ 1-chloro-1-methyl cyclohexane   (*Not* ...)

Concentrated Sulfuric Acid Actually Dissolves Alkenes   When an alkene is mixed with concentrated sulfuric acid, an addition reaction occurs, the hydrocarbon dissolves, and heat evolves. An alkane does not behave this way at all but merely forms a separate layer that floats on the sulfuric acid (see Figure 12.2).

The alkene dissolves because it reacts by an addition reaction to form an alkyl hydrogen sulfate. For example,

$$CH_3CH{=}CH_2 + H{-}O{-}\underset{\underset{O}{\|}}{\overset{\overset{O}{\|}}{S}}{-}O{-}H \longrightarrow CH_3\underset{CH_3}{\overset{\ \ }{\underset{|}{CH}}}{-}O{-}\underset{\underset{O}{\|}}{\overset{\overset{O}{\|}}{S}}{-}O{-}H$$

propene   sulfuric acid (H₂SO₄)   isopropyl hydrogen sulfate

Molecules of the alkyl hydrogen sulfates are very polar, and as soon as they form, they move smoothly into the polar sulfuric acid layer and out of the nonpolar alkene layer. The heat generated by the reaction and the strongly acidic nature of the mixture causes side reactions that generate black by-products.

■ The sodium salts of long-chain alkyl hydrogen sulfates are detergents, for example, CH₃(CH₂)₁₁OSO₃Na.

A Symmetrical Double Bond Does Not Add H—G Selectively   Markovnikov's rule does not apply when the double bond is symmetrical. Reactants like H—G add in

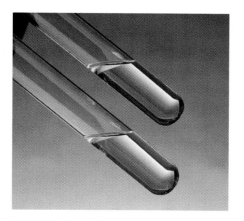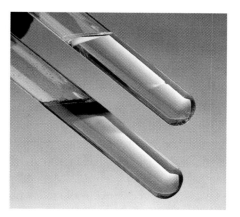

FIGURE 12.2   The effect of concentrated sulfuric acid on an alkane (cyclohexane) and an alkene (cyclohexene). In the photo on the left, the alkane (top tube) and the alkene (bottom tube) are seen as clear, colorless liquids. The photo on the right shows the two systems soon after concentrated sulfuric acid has been added to each. The alkane floats unaffected on the acid (top tube), but the alkene (bottom tube) has already begun to react and form an alkyl hydrogen sulfate (together with darkly colored matter produced by side reactions). The alkyl hydrogen sulfate is soluble in the remaining concentrated sulfuric acid, and so all of the alkene will appear to dissolve.

both of the two possible directions, and a mixture of isomers can form. For example,

$$CH_3CH_2CH\!=\!CHCH_3 + H\!-\!Br \longrightarrow CH_3CH_2\underset{\underset{Br}{|}}{C}HCH_2CH_3 + CH_3CH_2CH_2\underset{\underset{Br}{|}}{C}HCH_3$$

2-pentene   3-bromopentane   2-bromopentane

Actually, we should write two equations because two reactions go on simultaneously. Some of the 2-pentene molecules react to give 3-bromopentane; others react to give 2-bromopentane. In situations like this, however, chemists generally combine the two equations into one (and don't attempt to represent a *balanced* equation).

Water Adds to Double Bonds to Give Alcohols Water adds to the carbon–carbon double bond to give an alcohol, provided that an acid catalyst (or the appropriate enzyme) is present. Water alone or aqueous bases have no effect on alkenes. This kind of reaction is common at life's molecular level.

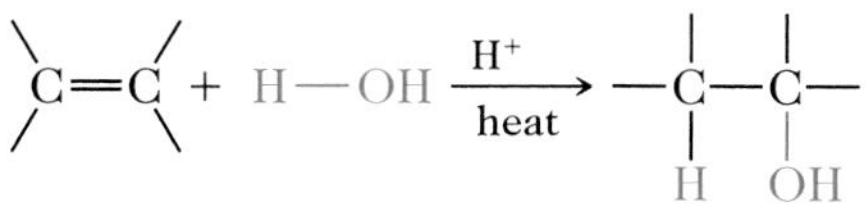

$$\underset{\text{alkene}}{\overset{\diagdown}{/}C\!=\!C\overset{\diagup}{\diagdown}} + H\!-\!OH \xrightarrow[\text{heat}]{H^+} \underset{\underset{H \quad OH}{|\quad|}}{\overset{|\quad|}{-C\!-\!C-}}$$

alkene   alcohol

Specific examples are

$$CH_2\!=\!CH_2 + H\!-\!OH \xrightarrow[240\,°C]{10\%\ H_2SO_4} CH_3CH_2OH$$

ethene   (closed vessel)   ethyl alcohol

$$\underset{\underset{CH_3}{|}}{CH_3C}\!=\!CH_2 + H\!-\!OH \xrightarrow[25\,°C]{10\%\ H_2SO_4} \underset{\underset{CH_3}{|}}{CH_3\overset{\overset{OH}{|}}{C}CH_3} \left(Not \quad \underset{\underset{CH_3}{|}}{CH_3CH}CH_2OH \right)$$

2-methylpropene   *t*-butyl alcohol

As you can see, Markovnikov's rule applies here, too. One H of H_2O goes to the carbon with the greater number of hydrogens, and the OH goes to the other carbon of the double bond.

Notice in the last example, and in all previous examples of addition reactions, that the carbon skeleton does not change. Although this is not always true, it will be in all the examples we will use as well as in all the Practice and Review Exercises.

EXAMPLE 12.5 Using Markovnikov's Rule

What product forms in the following situation?

$$CH_3CH\!=\!CH_2 + H\!-\!OH \xrightarrow[\text{heat}]{H^+}$$

ANALYSIS As in all of the addition reactions that we are studying, the carbon skeleton of the alkene can be copied over intact, except that a single bond is shown where the double bond was.

$$CH_3CH\!-\!CH_2 \qquad \text{(Incomplete)}$$

To decide which carbon of the original double bond gets the H atom from the water molecule, we use Markovnikov's rule. The H atom has to go to the CH_2 end because it has the greater number of hydrogens. The OH unit from H—OH goes to the other carbon.

SOLUTION The product is

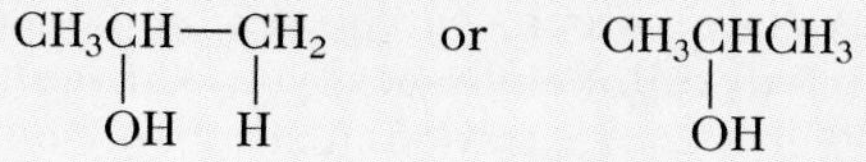

$$CH_3CH\!-\!CH_2 \quad \text{or} \quad CH_3CHCH_3$$
$$\underset{\displaystyle OH \quad H}{|\quad\ |} \qquad\qquad\quad \underset{\displaystyle OH}{|}$$

isopropyl
alcohol

CHECK At this stage be sure to see whether each carbon in the product has four bonds. If not, you can be certain that some mistake has been made. This is always a useful way to avoid at least some of the common mistakes made in solving a problem such as this. Then double-check that the carbon of the double bond initially having the greater number of H's has been given one more.

■ PRACTICE EXERCISE 6 Write structures for the product(s), if any, that would form under the conditions shown. If no reaction occurs, write "no reaction."

(a) $CH_2\!\!=\!\!CHCH_2CH_3 + HCl \longrightarrow$

(b) 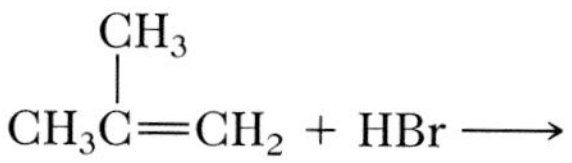 $\underset{\displaystyle\quad}{CH_3\overset{\displaystyle \overset{CH_3}{|}}{C}\!\!=\!\!CH_2} + HBr \longrightarrow$

(c) $CH_3CH\!\!=\!\!\overset{\displaystyle \overset{CH_3}{|}}{C}\!\!-\!\!\bigcirc + H_2O \xrightarrow[\text{heat}]{H^+}$

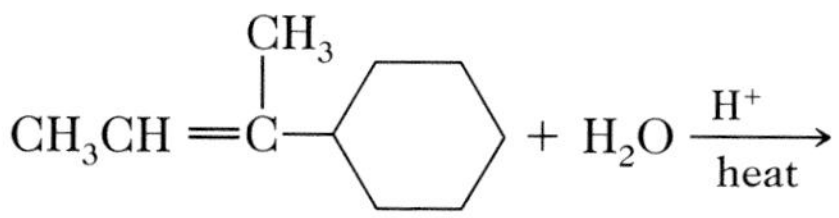

(d) 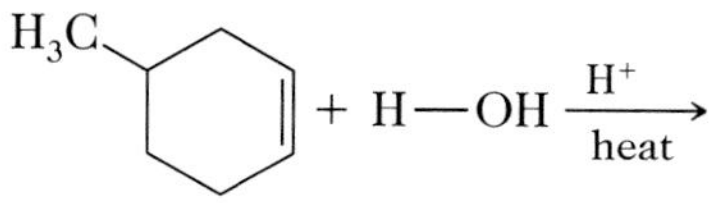 $\overset{H_3C}{\diagdown}\!\!\bigcirc + H\!-\!OH \xrightarrow[\text{heat}]{H^+}$

(e) 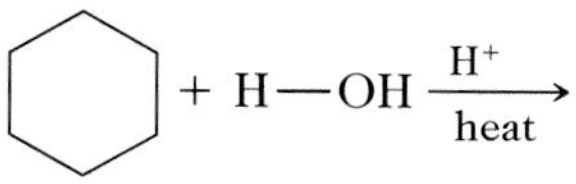 $\bigcirc + H\!-\!OH \xrightarrow[\text{heat}]{H^+}$

Double Bonds Are Attacked by Oxidizing Agents With two pairs of electrons, the double bond is more electron-rich than a single bond, so electron-seeking reagents attack it. These include oxidizing agents, which, by definition, are electron-accepting species. A hot solution of potassium permanganate ($KMnO_4$), for example, vigorously oxidizes molecules at carbon–carbon double bonds.

Oxidations begin as addition reactions but continue beyond to the point where alkene molecules are split apart at the double bond. We will not study any of the details or learn how to predict products. But it is important to our later study to know that a carbon–carbon double bond makes a molecule susceptible to attack by strong oxidizing agents. The products can be ketones, carboxylic acids, carbon dioxide, or mixtures of these. Alkanes, in sharp contrast, are inert toward these oxidizing agents.

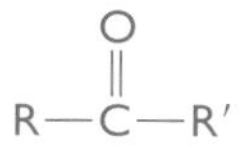

■

$$\underset{\text{ketones}}{R\!-\!\overset{\displaystyle \overset{O}{\|}}{C}\!-\!R'}$$

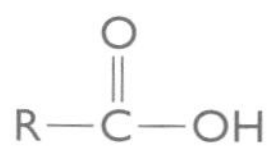

$$\underset{\text{carboxylic acid}}{R\!-\!\overset{\displaystyle \overset{O}{\|}}{C}\!-\!OH}$$

The permanganate ion is intensely purple in water, and as it oxidizes double bonds it changes to manganese dioxide, MnO_2, a brownish, sludgelike, insoluble solid (Figure 12.3).

■ Ozone destroys any vegetation that has the green pigment chlorophyll, because chlorophyll contains double bonds.

Ozone Is One of the Most Powerful Oxidizing Agents   Ozone, O_3, a pollutant in smog, is dangerous because it is a very powerful oxidizing agent that attacks biochemicals wherever they have carbon–carbon double bonds. Because such bonds occur in the molecules of all cell membranes, you can see that exposure to ozone must be kept very low. Even a concentration in air of only one part per million parts (1 ppm) warrants the declaration of a smog emergency condition. Interaction 12.3 describes how oxides of nitrogen and incompletely burned hydrocarbons released in the exhaust from vehicles contribute to the generation of ozone in smog.

12.5 HOW ADDITION REACTIONS OCCUR

The carbon–carbon double bond can accept a proton from an acid and change into a carbocation.

Because the double bond is somewhat electron-rich, it functions as a base, that is, as a proton acceptor, particularly when the proton donor is strong, like HCl, HBr, and $HOSO_3H$. When alkenes react with water, the *initial* step is the donation of a proton to the double bond *from the acid catalyst,* not from the water molecule. The water molecule is too weak a proton donor when the acceptor is a double bond.

■ In H—Cl, G is Cl. In H_2SO_4, G is OSO_3H. In dilute acid, the proton-donating species is H_3O^+, so G here is H_2O.

When an Alkene Accepts a Proton, a Reactive Carbocation Forms   Using the symbol H—G to represent any proton donor, the reaction we will now explain is the following addition of HG to propene.

$$CH_3CH{=}CH_2 + H{-}G \longrightarrow CH_3\underset{\underset{G}{|}}{C}HCH_3 \qquad (not \quad CH_3CH_2CH_2{-}G)$$

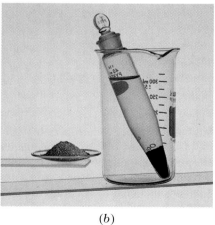

| (a) | (b) |

FIGURE 12.3 Permanganate oxidation. (*a*) Crystals of potassium permanganate are so deeply purple that they appear almost black, but a dilute aqueous solution is purple. (*b*) After an oxidizable compound has been added to the $KMnO_4$ solution and the mixture has been heated to complete the oxidation of the compound, the solution has no color but now contains a precipitate of MnO_2.

INTERACTION 12.3
OZONE IN SMOG

Because of their carbon–carbon double bonds, most materials of living tissue are attacked by ozone. The U.S. National Ambient[1] Air Quality Standard for ozone is a daily maximum 1 hour average ozone concentration of only 0.12 ppm (120 ppb). It's estimated that over 44 million people in the United States live in counties that do not meet this standard. How does ozone originate in the air where we live?

The *direct* and only source of ozone in the lower atmosphere is the combination of oxygen atoms with oxygen molecules.

$$O + O_2 + M \longrightarrow O_3 + M \qquad (1)$$

When the collision of O and O_2 occurs on the surface of *M,* which can be any molecule, like N_2 or O_2, some of the kinetic energy of the collision is absorbed by *M.* (Collisions that occur by themselves give O_3 molecules that break up instantly because they carry too much energy.) Thus the earlier question, "How does ozone originate?" becomes a new question, "How are oxygen *atoms* generated?" Once oxygen atoms are made, ozone will form. The chief source of oxygen atoms is the breakup of molecules of nitrogen dioxide, an air pollutant, a breakup made possible by solar energy.

$$NO_2 + \text{solar energy} \longrightarrow NO + O \qquad (2)$$

Thus the most efficient ozone production occurs in strong sunlight.

Oxides of Nitrogen The NO_2 for Equation 2 is made from NO, nitrogen monoxide, and NO is produced inside vehicle engine cylinders and power plant furnaces. In such situations the direct combination of nitrogen and oxygen is possible because of the high temperature and, in vehicles, high pressure.

$$N_2 + O_2 \xrightarrow{\frac{\text{high temperature}}{\text{pressure}}} 2NO \qquad (3)$$

As soon as newly made NO, now in the exhaust gas, hits the cooler outside air, it reacts further with oxygen to give nitrogen dioxide, NO_2, which gives smog its reddish-brown color (see Figure 1).

$$2NO + O_2 \longrightarrow 2NO_2 \qquad (4)$$

Thus an ample supply of NO_2 exists in air made smoggy by heavy traffic and power plants. To reduce

[1] *Ambient* means "all surrounding, all encompassing"; ppm means parts per million parts; ppb stands for parts per billion parts. One part per million parts translates into 1 mg/L.

(a)

(b)

FIGURE 1 The effect of smog on visibility. (a) A clear day. (b) A day of heavy smog. The reddish color is caused chiefly by NO_2.

NO_2 we must reduce the production of NO. In environmental literature, the two nitrogen oxides are lumped together by the formula NO_x.

The significance of reaction 2 is that solar energy "cracks" NO_2 to nitrogen monoxide and oxygen atoms. Thus, as the morning rush hour traffic increases and the sun moves higher, oxygen atoms are generated to create ozone by reaction 1.

Interestingly, nitrogen monoxide is able to destroy ozone.

$$NO + O_3 \longrightarrow NO_2 + O_2 \qquad (5)$$

So if reaction 2 produces both an ozone generator (O atoms) and an ozone destroyer (NO), what's the problem? Try it yourself; if we add reactions 1, 2 and 5 together, there is *no net chemical effect; no net ozone production.* Yet, the range of ozone concentration even in clean air, between 20 and 50 ppb, although very low, is still real. In more polluted urban areas, levels as high as 400 ppb commonly occur for brief periods of time each year. How then does a definite level of ozone develop in our atmosphere at all?

The answer is that reactions 1–5 proceed at their own rates, and these are unequal. In the atmosphere, the reactions actually do not cancel each other exactly. So there is a net effect after all; a small, steady-state concentration of ozone develops even in clean air. In smog, the heightened ozone level is tied to the level of NO_x. *No major, long-term solution to the problem of ozone in smog will happen without addressing the NO_x problem.* Smaller engines with lower horsepower are part of the solution because they generate lower levels of NO_x.

Unburned Hydrocarbons and the Ozone in Smog
Another culprit consists of unburned and partially oxidized hydrocarbons that enter the exhaust from the inefficient combustion of vehicle fuel. Reaction 5 is the reaction that can make most ozone disappear, but unburned and partially oxidized hydrocarbons are able to remove NO from reaction 5 *before the NO is used to destroy ozone.* Any removal of NO other than by reaction 5 enables a more rapid buildup in the ozone level in smog.

In sunlight and oxygen, unburned hydrocarbons are changed into organic derivatives of hydrogen peroxide called *peroxy radicals,* usually symbolized as R—O—O·, or simply RO_2. Peroxy radicals originate chiefly by the reaction of unburned hydrocarbons with hydroxyl radicals, HO. These arise by a number of mechanisms in polluted air, but we don't need the details. Suffice it to say, HO radicals and hydrocarbons lead to ROO radicals, and these destroy NO molecules.

$$ROO + NO \longrightarrow RO + NO_2$$

This reaction reduces the supply of ozone-destroying NO and so it allows for a more rapid buildup of the ozone level.

Because hydrocarbons lead to peroxy radicals, specialists generally agree that emissions of hydrocarbons must also be significantly reduced before the ozone problem can be solved. Engine manufacturers have tried to design vehicle cylinders to favor complete fuel combustion. The catalysts in mufflers also help to complete the oxidation of partially oxidized hydrocarbons that manage to leave in the exhaust. Because these catalysts are deactivated ("poisoned") by organic lead compounds, unleaded gasoline must be used in vehicles. Lead compounds, like tetraethyl lead, were once widely used to reduce engine noise ("knocking").

One strategy being tested by the U.S. Environmental Protection Agency (EPA) in the mid 1990s is to require that the gasoline sold in major cities with smog problems contain 2% oxygen by weight. This is not molecular oxygen but the oxygen in organic oxygen compounds, like ethanol, methyl *t*-butyl ether, and ethyl *t*-butyl ether. These substances, called *oxygenates,* favor the more complete combustion of fuels. (A 2% oxygen requirement translates into 11% methyl *t*-butyl ether by volume.)

Dimethyl ether is being developed as an alternative to diesel fuel. Its use results in lower NO_x emissions, almost soot-free combustion, and less engine noise.

We can write the first step in this addition reaction as follows.

What once were two electrons in a double bond are now the electrons of this bond.

$$CH_3CH{=}CH_2 + H{-}\ddot{G}{:} \longrightarrow CH_3\overset{+}{CH}{-}CH_2{-}H + {:}\ddot{G}{:}^-$$

propene   isopropyl carbocation

The curved arrows show how one of the two pairs of electrons of the double bond swings away from one carbon to form a bond to the proton.

One Carbon in a Carbocation Lacks an Octet The isopropyl cation that forms is an example of a **carbocation,** a positive ion in which carbon has a sextet, not an octet, of electrons. But notice that it is the *isopropyl* carbocation that forms, not the propyl carbocation, $CH_3CH_2CH_2^+$. We'll see why shortly, and when we do we'll understand Markovnikov's rule.

Carbocations are particularly unstable cations. All their reactions are geared to the recovery of an outer octet of electrons for carbon. The instant a carbocation forms, it strongly attracts any electron-rich species in the neighborhood, and just such particles are produced when H—Cl, H—Br, or H—OSO$_3$H donate a proton to the double bond. When the proton leaves any one of these acids, a conjugate base is released, which is Cl$^-$, Br$^-$, or $^-$OSO$_3$H, depending on the acid used. Each anion is an electron-rich species, just what a carbocation attracts. We'll continue to generalize by using the symbol G^- for any of these conjugate bases.

The newly formed carbocation reacts with G^- in the next step of the addition.

$$CH_3\overset{+}{C}HCH_3 + :\overset{..}{\underset{..}{G}}:{}^- \longrightarrow CH_3\overset{\overset{\displaystyle :\overset{..}{G}:}{\displaystyle |}}{C}HCH_3$$

isopropyl
carbocation

This restores the octet to carbon at the same time as it gives the product. As we noted earlier, the product is *not* $CH_3CH_2CH_2G$. Let us now see why.

The More Stable Carbocation Preferentially Forms   The stability of a carbocation, although always very low, does vary with the number of neighboring electron clouds that hunker around its positively charged center. Alkyl groups provide larger overall electron clouds than H atoms. Therefore, packing alkyl groups instead of H atoms around the positively charged carbon helps to stabilize the carbocation. Notice now that when the positive charge is on the secondary carbon of the isopropyl carbocation, the electron clouds of *two* alkyl groups (two methyl groups) crowd around it. When the charge is on the primary carbon of the propyl carbocation, the electron cloud of only one alkyl group (an ethyl group) is nearby.

$$CH_3\overset{+}{C}HCH_3 \qquad\qquad CH_3CH_2CH_2^+$$

isopropyl carbocation   propyl carbocation

As a result, the isopropyl carbocation is more stable than the propyl carbocation. Or, to generalize, *a secondary carbocation is more stable than a primary.* By extension, *a tertiary carbocation is more stable than a secondary.* The order of stability of carbocations is thus

$$CH_3^+ \quad < \quad R—CH_2^+ \quad < \quad R—\overset{\overset{\displaystyle R}{\displaystyle |}}{C}H^+ \quad < \quad R—\overset{\overset{\displaystyle R}{\displaystyle |}}{\underset{\underset{\displaystyle R}{\displaystyle |}}{C}}{}^+$$

| methyl carbocation | primary carbocation | secondary carbocation | tertiary carbocation |

increasing stability
of carbocations

least stable ——————————————————→ most stable

■ The R groups are alkyl groups, and they need not be identical.

This order of stability explains Markovnikov's rule. When the option of two different *kinds* of carbocations exists, as in the addition of H—G to propene, *the most stable carbocation always forms preferentially.* The double bond accepts H$^+$ from H—G so as to produce the more stable carbocation.

When the two possible carbocations are of the *same* type, both secondary, for example, then there is no preference. Both possible carbocations actually form, both accept G^-, and a mixture of isomeric products is produced.

The Water Molecule Is Too Weak a Proton Donor To Make a Carbocation from an Alkene   Water does not react with an alkene at all in the absence of an acid catalyst. The H_2O molecule is too weak a donor of H^+, as we said; H_2O holds its protons much too strongly. Thus when water does add to an alkene under acid catalysis, the first step in the mechanism of the reaction is a proton transfer *from the acid catalyst,* not from a water molecule. The water molecule has to wait for this to happen before it engages in the overall reaction. The job of the acid catalyst is to convert the alkene into a carbocation and thus temporarily make one carbon of the double bond into an electron-poor site.

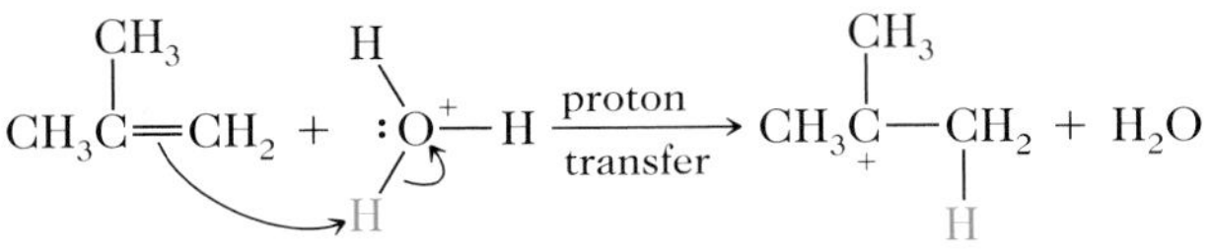

$$CH_3C{=}CH_2 + \ :\overset{+}{O}{-}H \xrightarrow[\text{transfer}]{\text{proton}} CH_3\overset{+}{C}{-}CH_2 + H_2O$$

t-butyl carbocation
(three alkyl groups on C^+)

The other possible but less stable carbocation, which does *not* form, is

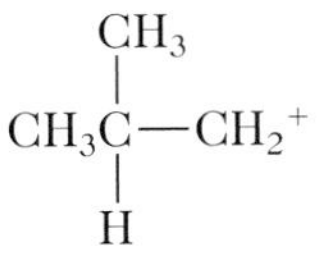

$$CH_3\overset{CH_3}{\underset{H}{\overset{|}{C}}}{-}CH_2^+$$

isobutyl carbocation
(one alkyl group on CH_2^+)

Once the *t*-butyl carbocation forms, what can it attract? What electron-rich particle is most abundantly available in water that contains only a *trace* concentration (a *catalytic* amount) of acid? It's a water molecule with its electron-rich oxygen atom. The *t*-butyl carbocation, therefore, combines with a water molecule as follows to give what is essentially a hydronium ion in which one H has been replaced by a *t*-butyl group.

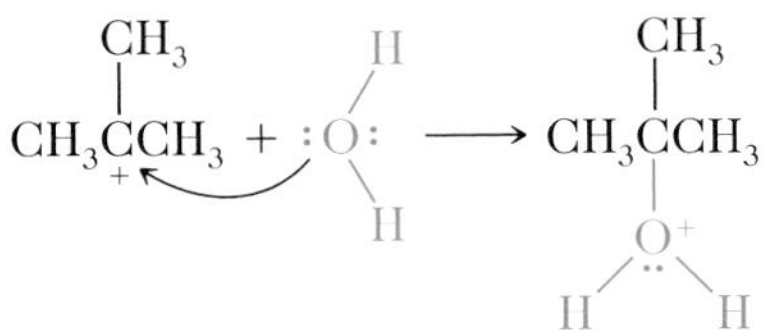

$$CH_3\overset{CH_3}{\overset{|}{\underset{+}{C}}}CH_3 + :\ddot{O}: \longrightarrow CH_3\overset{CH_3}{\overset{|}{C}}CH_3$$

Now the positive charge is on oxygen, but this is acceptable here because oxygen still has an outer octet.

In the last step, the catalyst, H_3O^+, is recovered by another proton transfer, this time from the *t*-butyl-substituted hydronium ion.

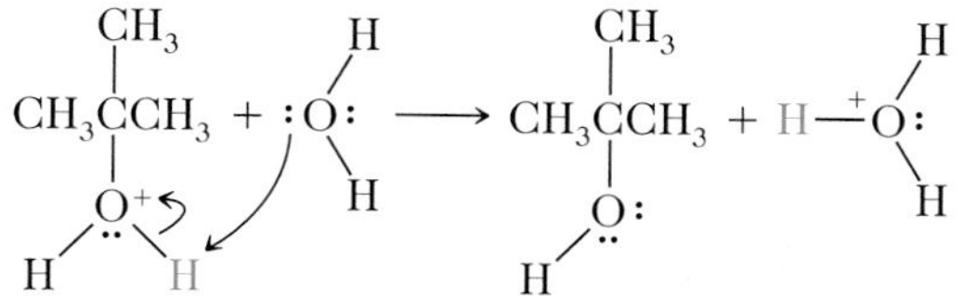

$$CH_3CCH_3 + :O: \longrightarrow CH_3CCH_3 + H{-}^+O:$$

| an alkyl- substituted hydronium ion | t-butyl alcohol | recovered catalyst |

■ **PRACTICE EXERCISE 7**   Write the condensed structures for the two carbo-cations that could conceivably form if a proton became attached to each of the following alkenes. State which carbocation is preferred. If both are reasonable, state that they are. Then write the structures of the alkyl chlorides that would form by the addition of hydrogen chloride to each alkene.

(a) $CH_3CH_2CH{=}CH_2$ (b) $CH_3\overset{\overset{\displaystyle CH_3}{|}}{C}{=}CH_2$ (c) 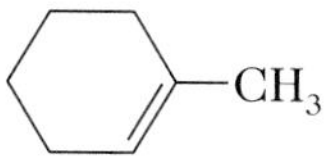   〈 〉—CH_3

(d) $CH_3CH{=}CHCH_3$ (e) $CH_3CH{=}CHCH_2CH_3$

(f)   The addition of water to 2-pentene (part e) gives a mixture of alcohols. What are their structures? Why is the formation of a mixture to be expected here but not when water adds to propene?

12.6 ADDITION POLYMERS

Hundreds to thousands of alkene molecules can join together to make one large molecule of a polymer.

Macromolecules Abound in Nature   A **macromolecule** is simply a molecule with a formula mass in the thousands. Some macromolecular substances, the *polymers,* have a unique structural regularity. A **polymer** is a substance consisting of macromolecules *all of which have repeating structural units,* up to many thousands. Two of the carbohydrates that we will study—cellulose and one component of starch— are *polysaccharides* characterized by the following system, where Gl stands for a molecular unit made from a glucose molecule (the details of which we'll leave to a later chapter).

etc.—Gl—O—Gl—O—Gl—O—Gl—O—Gl—O—Gl—O—Gl—O—etc.

section of a polysaccharide

Another component of starch as well as the starchlike polymer glycogen have similar sections but with many long branches also made of repeating Gl—O units.

Proteins consist almost entirely (and often completely) of polymers called *polypeptides* whose molecules have the following features. (We show only the molecular "backbone" and omit the substituents or *side chains* appended to it.) Notice the regularly repeating structural unit.

■ *Macro* signifies "huge" or "large scale." *Polymer* has Greek roots: *poly,* "many," and *meros,* "parts."

■ Starch molecules are a storage form for glucose in plants.

■ We store glucose units in the form of glycogen molecules principally in the liver, the kidneys, and muscles.

$$\text{etc.}-\underset{|}{\overset{\overset{\displaystyle O}{\|}}{NHCHC}}-\underset{|}{\overset{\overset{\displaystyle O}{\|}}{NHCHC}}-\underset{|}{\overset{\overset{\displaystyle O}{\|}}{NHCHC}}-\underset{|}{\overset{\overset{\displaystyle O}{\|}}{NHCHC}}-\underset{|}{\overset{\overset{\displaystyle O}{\|}}{NHCHC}}-\text{etc.}$$

Section of a polypeptide (The vertical single bonds are sites where
side chains are attached, about 20 different kinds being the options.)

The backbone is thus like the chain of a charm bracelet. Such chains have *identical*
links, so the bracelet acquires its uniqueness in the kinds of its charms (side chains)
and the order in which they are hung. Like charm bracelet chains, polypeptide backbones can be of varying lengths.

The molecules of DNA, one of the kinds of nucleic acids, the chemicals of heredity, similarly have identical backbones, but they can be of different lengths and have
different side chains (which we again omit). Notice, again, the regularly repeating
structural unit.

$$\text{etc.}-\underset{|}{\overset{\overset{\displaystyle O}{\|}}{OPO}}-\text{pentose}-\underset{|}{\overset{\overset{\displaystyle O}{\|}}{OPO}}-\text{pentose}-\underset{|}{\overset{\overset{\displaystyle O}{\|}}{OPO}}-\text{pentose}-\underset{|}{\overset{\overset{\displaystyle O}{\|}}{OPO}}-\text{pentose}-\text{etc.}$$

Section of a DNA molecule with the vertical bonds being sites for the attachment of
side chains, four kinds being used. ("Pentose" is a unit made from a sugar called
deoxyribose, which gives the D to DNA.)

These examples illustrate just one feature of all polymers; they have *repeating
structural units*. To study polymers, therefore, we focus on the origins of these units
and how they are joined together from small molecules. But let's learn the rudiments
of this by studying easier systems.

Polyethylene Is a Simple but Commercially Important Polymer   Under a variety of
conditions, many hundreds of ethylene molecules can reorganize their bonds, join
together, and change into one large molecule.

■ Certain acids also catalyze this
polymerization.

$$n CH_2{=}CH_2 \xrightarrow[\text{trace of } O_2]{\text{heat, pressure}} {-}(CH_2{-}CH_2)_n^{-} \qquad (n = \text{a large number})$$

ethylene   polyethylene
(repeating unit)

The *repeating unit* in polyethylene is $CH_2{-}CH_2$, and by using parentheses to enclose it we fashion a convenient way to represent a polymer structure. The parentheses mean that one repeating unit after another is joined together into an extremely long chain.

Chain-branching reactions also occur during the formation of polyethylene, so
the final product includes both straight- and branched-chain molecules. The molecules in a sample of a commercial polymer like polyethylene are never exact copies
of each other. Their chain lengths vary, and the extent of branching varies, but it is
still convenient to represent the polymer by showing its most characteristic repeating unit.

The starting material for making a polymer is called a **monomer,** and the reaction that changes a monomer into a polymer is called **polymerization.** Because
alkenes are nicknamed *olefins,* the polymers of alkenes are usually called *polyolefins*
(pol-y-**ol**-uh-fins).

Carbocations Are Intermediates in Some Polymerizations The polymerization catalysts used industrially vary widely. Some, for example, work by generating free radicals; others, by producing carbocation intermediates. For example, when the catalyst is a proton donor, like our generalized acid H*G*, it can convert ethylene into the ethyl carbocation.

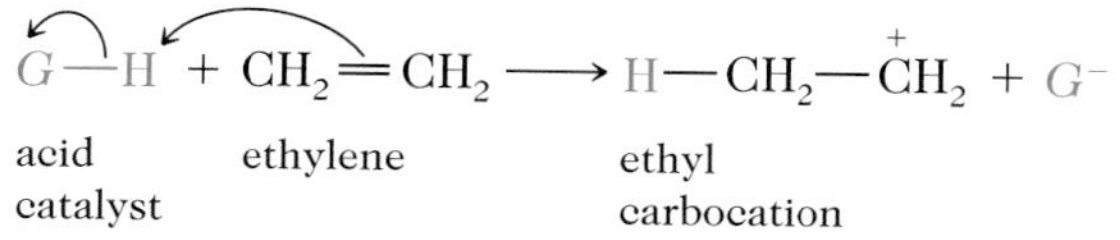

$$G{-}H \ + \ CH_2{=}CH_2 \longrightarrow H{-}CH_2{-}\overset{+}{C}H_2 \ + \ G^-$$

acid catalyst   ethylene   ethyl carbocation

■ The negative ions, *G*⁻, are also electron-rich but are present in minute traces compared to molecules of unreactive alkene.

When the only *abundant* electron-rich species around is *an alkene molecule* with its double bond, the new carbocation attracts the alkene. The new carbocation, therefore, restores an octet to its positively charged site by accepting an electron pair from an unreacted alkene molecule.

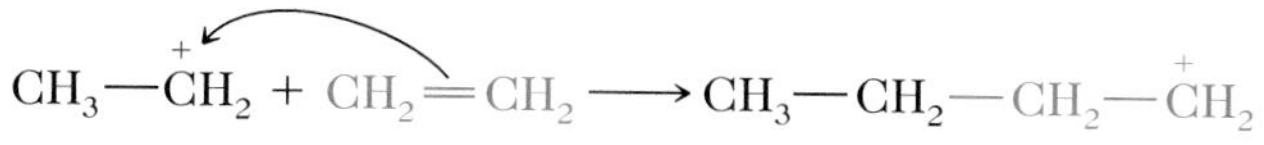

$$CH_3{-}\overset{+}{C}H_2 \ + \ CH_2{=}CH_2 \longrightarrow CH_3{-}CH_2{-}CH_2{-}\overset{+}{C}H_2$$

This reaction, of course, only creates a new and longer carbocation, one still surrounded mostly by unreacted molecules of alkene. So the new carbocation attracts still another molecule of alkene.

$$CH_3{-}CH_2{-}CH_2{-}\overset{+}{C}H_2 \ + \ CH_2{=}CH_2 \longrightarrow CH_3{-}CH_2{-}CH_2{-}CH_2{-}CH_2{-}\overset{+}{C}H_2$$

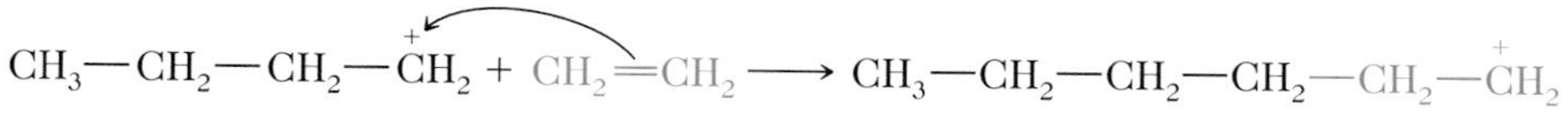

You can begin to see how this works. Yet another carbocation is produced, and it attracts another molecule of alkene. In this repetitive manner, the chain grows step by step, by one repeating unit of $CH_2{-}CH_2$ after another, until the positively charged site on a growing chain happens to pick up a stray anion, such as an anion left over from the catalyst. In this way, chains stop growing, some at different lengths than others. Actually, the catalyst should not be called a *catalyst,* because it is finally consumed and not regenerated. Here, the term *promoter* is better than *catalyst.*

■ Polymerization is an example of a *chemical chain reaction* because the product of one step initiates the next step.

The Methyl Side Chains Occur Regularly in Polypropylene When propene (common name, propylene) polymerizes, the methyl groups appear regularly, *on alternate carbons* of the main chain.

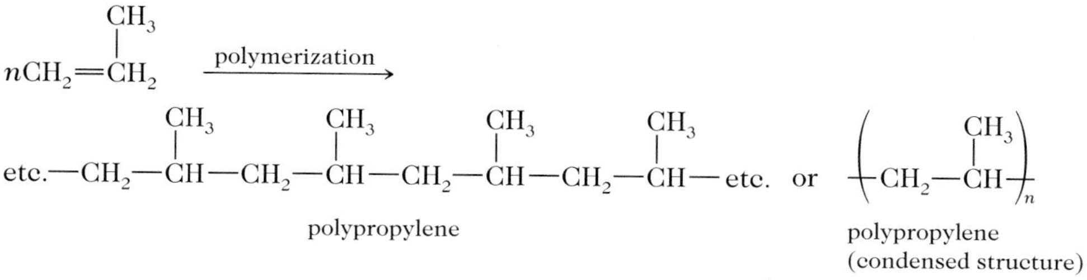

$$nCH_2{=}\overset{\overset{\displaystyle CH_3}{|}}{C}H_2 \ \xrightarrow{\text{polymerization}}$$

etc.$-CH_2-\overset{\overset{CH_3}{|}}{C}H-CH_2-\overset{\overset{CH_3}{|}}{C}H-CH_2-\overset{\overset{CH_3}{|}}{C}H-CH_2-\overset{\overset{CH_3}{|}}{C}H-$ etc.   or $\left(\!CH_2-\overset{\overset{CH_3}{|}}{C}H\!\right)_{\!n}$

polypropylene   polypropylene (condensed structure)

Polymerizations generally are orderly like this because the reactive intermediates have the same rules of stability that apply to simple carbocations so the same kind of intermediate always forms. (A secondary carbocation, for example, would always form in preference to primary carbocation.) A unique feature of propylene polymerization, however, is that by selecting a different promoter the methyl groups in polypropylene will line up all on the same side of the chain, or will alternate from side to side, or will project at random.

■ Each kind of polypropylene has its own commercial uses.

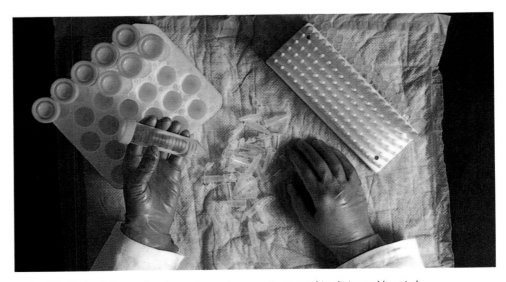

FIGURE 12.4   Polypropylene is used to make many items used in clinics and hospitals.

The Polyolefins Are Chemically Very Stable   Because polyolefins, like polypropylene and polyethylene, are fundamentally alkanes, they have all the chemical inertness of alkanes. Polyolefins, therefore, are popular raw materials for making many items, including containers that must be inert to food juices and to fluids used in medicine (Figure 12.4). Refrigerator boxes and bottles, containers for chemicals, sutures, catheters, various drains, and wrappings for aneurysms are commonly made of polyolefins. Polypropylene fibers are used to make indoor–outdoor carpeting and artificial turf because polypropylene is inert, won't mold, and is wear resistant.

Substituted Alkenes Are Monomers for Important Polymers   Monomers with carbon–carbon double bonds often carry other functional groups or halogen atoms. The resulting polymers are extremely important commercial substances, which we encounter often in our daily lives. Table 12.2 contains a few examples (see also Figure 12.5). Many dienes are used as monomers, too. Natural rubber is a polymer of a diene called isoprene (Table 12.2), and rubber is now made industrially.

FIGURE 12.5   Lucite and Plexiglas are two trade names for the polymer of methyl methacrylate.

12.7   BENZENE RING AND AROMATIC PROPERTIES

The benzene ring undergoes substitution reactions instead of addition reactions despite a high degree of unsaturation.

The molecular formula of benzene is C_6H_6, which indicates considerable unsaturation.[3] Its ratio of hydrogen to carbon is much lower than in two simple, saturated hydrocarbons with six carbons, hexane (C_6H_{14}) and cyclohexane (C_6H_{12}). We should expect benzene, therefore, to be some kind of alkene, or alkyne, or a combination. Alkynes give addition reactions very similar to those of alkenes (Interaction 12.4),

---

[3] The molecular formulas of all saturated, open-chain alkanes fit the general formula, C_nH_{2n+2}. For cyclic alkanes and open-chain alkenes, the general formula is C_nH_{2n}. For open-chain alkynes, it's C_nH_{2n-2}. Thus benzene, C_6H_6, or C_nH_{2n-6}, is highly unsaturated.

TABLE 12.2   Some Polymers of Substituted Alkenes[a]

Polymer	Monomer	Uses
Polyvinyl chloride (PVC)	$CH_2{=}CHCl$ vinyl chloride	Insulation, credit cards, bottles, plastic pipe
Saran	$CH_2{=}CCl_2$ vinylidene chloride and $CH_2{=}CHCl$ vinyl chloride	Packaging film, fibers, tubing
Teflon	$F_2C{=}CF_2$ tetrafluoroethylene	Nonstick surfaces, valves
Orlon	$CH_2{=}CHC{\equiv}N$ acrylonitrile	Fabrics
Polystyrene	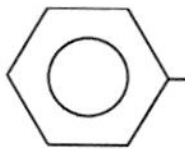 ⬡—$CH{=}CH_2$ styrene	Foamed items, insulation
Lucite	$\overset{\displaystyle CH_3}{\overset{\displaystyle \vert}{CH_2{=}CCO_2CH_3}}$ methyl methacrylate	Windows, coatings, molded items
Natural polymer		
Rubber	$\overset{\displaystyle CH_3}{\overset{\displaystyle \vert}{CH_2{=}C{-}CH{=}CH_2}}$ isoprene	Tires, hoses, boots

[a] The common names rather than the IUPAC names of the monomers are given.

and we might expect benzene to give addition reactions just as readily. There is one addition reaction that benzene does give; it adds hydrogen to give cyclohexane. Unlike the addition of hydrogen to an alkene, however, rigorous conditions of pressure and temperature are necessary to make benzene add hydrogen.

$$C_6H_6 + 3H_2 \xrightarrow[\substack{\text{high pressure} \\ \text{and temperature}}]{\text{catalyst}} \text{⬡}$$

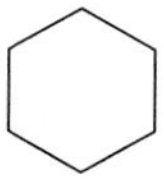

benzene   cyclohexane, C_6H_{12}

Benzene's Typical Reactions Are Substitutions, Not Additions   Alkenes (and alkynes) readily *add* chlorine and bromine without a catalyst, but benzene needs a catalyst, and the reaction is not simple addition but substitution. The catalyst is generally an iron halide (or iron itself).

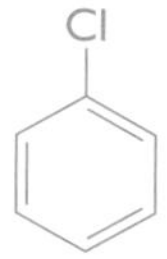

$$C_6H_6 + Cl_2 \xrightarrow[\text{FeCl}_3]{\text{Fe or}} C_6H_5Cl + HCl$$

benzene   chlorobenzene

chlorobenzene

$$C_6H_6 + Br_2 \xrightarrow[\text{FeBr}_3]{\text{Fe or}} C_6H_5Br + HBr$$

bromobenzene

Benzene also reacts, *by substitution,* with sulfur trioxide dissolved in concentrated sulfuric acid. (Recall that alkenes react exothermically with concentrated sulfuric acid by *addition.*)

INTERACTION 12.4
REACTIONS OF ALKYNES

Alkynes give the same kinds of addition reactions as alkenes, only one triple bond can add *two* molecules of a reactant. Usually, however, it is possible to con-trol the reaction so that only one molecule adds. Typical addition reactions are

$$CH_3C\equiv CH + H_2 \xrightarrow[\text{pressure}]{\text{Ni, heat,}} CH_3CH=CH_2 \xrightarrow{\text{more } H_2} CH_3CH_2CH_3$$

propyne   propene   propane

$$CH_3C\equiv CH + HCl \longrightarrow CH_3\overset{\overset{\displaystyle Cl}{|}}{C}=CH_2 \xrightarrow{+HCl} CH_3\overset{\overset{\displaystyle Cl}{|}}{\underset{\underset{\displaystyle Cl}{|}}{C}}CH_3$$

2-chloropropene   2,2-dichloropropane

■ Benzenesulfonic acid is about as strong an acid as hydrochloric acid. It is a raw material for the synthesis of aspirin.

$$C_6H_6 + SO_3 \xrightarrow[\text{room temperature}]{H_2SO_4 \text{ (concd)}} C_6H_5-\overset{\overset{\displaystyle O}{||}}{\underset{\underset{\displaystyle O}{||}}{S}}-O-H$$

benzenesulfonic acid

Benzene reacts with warm, concentrated nitric acid when it is dissolved in con-centrated sulfuric acid. We will now represent nitric acid as $HO-NO_2$, instead of HNO_3, because it loses the HO group during the reaction.

$$C_6H_6 + HO-NO_2 \xrightarrow[50-55\,°C]{H_2SO_4 \text{ (concd)}} C_6H_5NO_2 + H_2O$$

nitric acid   nitrobenzene

We have learned that alkenes (and alkynes) are readily oxidized by perman-ganate or dichromate ion, but benzene is utterly unaffected by these strong oxidizing agents even when boiled with them. (Ozone does attack benzene.)

In light of all these chemical properties, whatever benzene is, it isn't an alkene or alkyne. Yet it surely is unsaturated. The problem of what benzene is wasn't satis-factorily solved in organic chemistry until the early 1930s, roughly a century after its molecular formula was known and half a century after its skeleton structure had been determined. Many of the reactions referred to above helped to establish the structure of benzene. Let's see how this was done.

The Six H Atoms in C_6H_6 Are Chemically Equivalent to Each Other   When ben-zene is used to make chlorobenzene (or any of the other products shown above), only *one* monosubstituted compound forms. Only one C_6H_5-Cl exists. This re-minds us of what happens when ethane, CH_3CH_3, is chlorinated. Only one *mono*sub-stituted compound forms; only one CH_3CH_2Cl exists. It doesn't matter which H in CH_3CH_3 is replaced by Cl. The same is true for benzene. All six H atoms in C_6H_6 are equivalent to each other.

The Benzene Skeleton Has a Six-Membered Ring with a Hydrogen on Each Carbon
Although the reaction requires very strong conditions, cyclohexane forms when ben-zene is hydrogenated. Therefore, the six carbons of a benzene molecule must also form a six-membered ring. Because benzene's six H's are chemically equivalent, it

seems reasonable to put one H on each of the ring carbons. This gives a very symmetrical structure.

1
(incomplete)

2
(older structure
for benzene)

■ The older structure, **2,** is still widely used to represent benzene, although it is usually abbreviated further:

The trouble with the incomplete structure identified as **1** is that each carbon has only three bonds, not four. To solve this, chemists for several decades simply wrote in three double bonds, as seen in structure **2**. As they well knew, the difficulty with **2** is that it says that benzene is a triene, a substance with three alkene groups per molecule, and so should give addition reactions.

Because the three double bonds indicated in structure **2** are misleading in a chemical sense, scientists today often represent benzene simply by a hexagon with a circle inside, structure **3**.

■ Open-chain trienes, like 1,3,5-hexatriene, give addition reactions and are easily oxidized, like any ordinary alkene.

$$CH_2{=}CHCH{=}CHCH{=}CH_2$$

1,3,5-hexatriene

3
benzene

The ring system in **3**, the *benzene ring,* is planar. All its atoms lie in the same plane, and all the bond angles are 120°, as seen in the scale model of Figure 12.6. The circle in **3** also conveys something uncovered by advanced bonding theory, namely that three pairs of the benzene ring's electrons, the three pairs represented by the second bonds of the double bonds in structure **2**, move within a large ring-encompassing space. These six electrons are said to be *delocalized.* By existing in a relatively large space, the six repel each other less than if they were confined to individual double bonds (as in **2**). With less repulsions between electrons, the benzene ring is much more stable than structure **2** would suggest.

When a hydrogen atom held by a ring carbon is replaced by another group, the six-electron network is not broken up. But if an addition were to occur, the ring system would no longer be that of benzene but, instead, that of a cyclic diene. For example,

■ Electrons repel each other, so they are managed with greater overall stability in systems that allow them more room.

$$\text{+ Cl}_2 \xrightarrow[\text{(does not happen)}]{\times}$$

It costs the system far more energy to react by this addition reaction than to react by a substitution reaction, so the benzene ring strongly resists addition reactions.

Aromatic Compounds Have Benzene Rings That Give Substitution Reactions Any substances whose molecules have benzene rings and whose rings give substitution reactions instead of addition reactions are called **aromatic compounds.** The term is

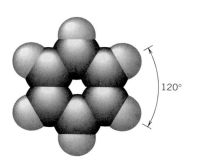

FIGURE 12.6 Scale model of a molecule of benzene. It shows the relative volumes of space occupied by the electron clouds of the atoms.

a holdover from the days when most of the known compounds of benzene actually had aromatic fragrances, but now the term does not refer to odor. Although oil of wintergreen and vanillin do have pleasant fragrances, aspirin has no odor. Yet all three have the benzene ring and all are classified as aromatic compounds.

oil of wintergreen   vanillin   aspirin

1-Phenylpropane is an example of an aromatic compound with an aliphatic side chain. (In IUPAC nomenclature, the group C_6H_5, derived from benzene by removing one H atom, is called the **phenyl** group, which rhymes with "kennel.") So stable is the benzene ring toward oxidizing agents that alkylbenzenes like 1-phenylpropane are attacked by hot permanganate at *the side chain* and not at the ring. Benzoic acid can be made by the oxidation of 1-phenylpropane (to use an unbalanced equation).

$$C_6H_5CH_2CH_2CH_3 \xrightarrow{\text{hot KMnO}_4(aq)} C_6H_5\overset{\overset{\text{O}}{\|}}{\text{C}}\text{OH} \quad (+ \text{MnO}_2)$$

1-phenylpropane   benzoic acid

Most of the side chain is destroyed, but the ring is not attacked. Sodium dichromate also oxidizes side chains.

The dichromate ion is bright orange in water, and when it acts as an oxidizing agent, it changes to the bright green, hydrated chromium(III) ion, $Cr^{3+}(aq)$, as seen in Figure 12.7.

Not all benzene derivatives have rings that resist oxidation. Rings that hold either the OH or the NH_2 groups, for example, are *very* readily oxidized, and such sys-

■ = C_6H_5

phenyl group

■ The liver has enzymes that put OH groups on benzene rings, making the products easier to break down into manageable wastes or to be made into substances needed by the body.

(a)

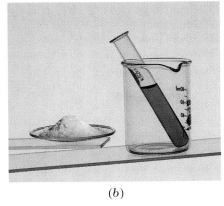
(b)

FIGURE 12.7 Dichromate oxidation. (a) Crystals of sodium dichromate are bright orange and so is a dilute aqueous solution. (b) After the dichromate solution has been mixed with some oxidizable compound and heated, the solution takes on the green color of the hydrated Cr^{3+} ion.

tems are present in proteins. We won't take this any further here because our goal has been to learn *general* properties of the benzene ring. Benzene-like aromatic rings also occur in the heterocyclic rings of all nucleic acids.

12.8 NAMING COMPOUNDS OF BENZENE

Common names dominate the nomenclature of simple derivatives of benzene.

The names of several monosubstituted benzenes are straightforward. The substituent is indicated by a prefix to the word *benzene*. For example,

nitrobenzene   fluorobenzene   chlorobenzene   bromobenzene   iodobenzene

■ All these compounds are oily liquids.

Other aromatic compounds have common names that are always used.

  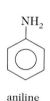

toluene   phenol   aniline   benzoic acid   benzaldehyde   benzene-sulfonic acid

■ Phenol was the first antiseptic used by British surgeon Joseph Lister (1827–1912), the discoverer of antiseptic surgery.

Ortho, Meta, and *Para* Are Terms for 1,2-, 1,3-, and 1,4-Relationships   When two or more groups are attached to the benzene ring, both what they are and where they are must be specified. One common way to indicate the relative locations of two groups in disubstituted benzenes is by the prefixes *ortho-, meta-,* and *para-,* which usually are abbreviated *o-, m-,* and *p-,* respectively. Two groups that are in a 1,2-relationship are *ortho* to each other, as in 1,2-dichlorobenzene, commonly called *o-*dichlorobenzene. A 1,3-relationship is designated *meta,* as in *m-*dichlorobenzene. In *p-*dichlorobenzene, the substituents are 1,4- or *para* relative to each other.

ortho or 1,2
*o-*dichlorobenzene   meta or 1,3
*m-*dichlorobenzene   para or 1,4
*p-*dichlorobenzene

A disubstituted benzene is usually named as a derivative not of benzene but of a monosubstituted benzene when the latter has a common name, like toluene or aniline. Then the *o-, m-,* or *p-* designations are used to specify relative positions of the two groups. For example,

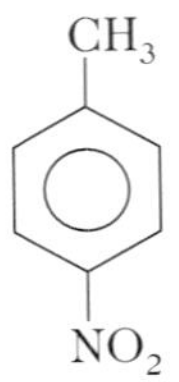

p-nitrotoluene

(*not* 4-nitro-1-
methylbenzene)

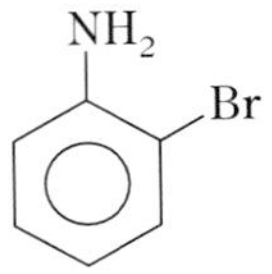

o-bromoaniline

(*not* 2-bromo-1-
aminobenzene)

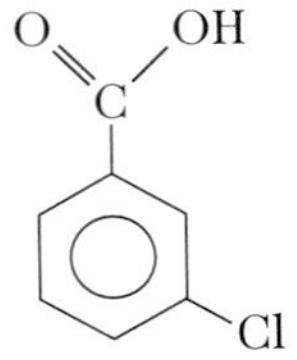

m-chloro-
benzoic acid

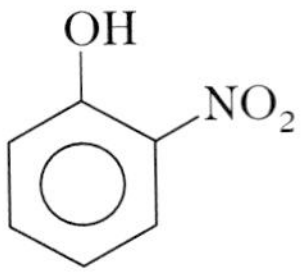

o-nitrophenol

When we have trisubstituted benzenes (or higher), we cannot use the *ortho,
meta,* or *para* designations with sufficient precision to make an unambiguous name.
We must now use numbers assigned to ring positions in such a way as to use the lowest numbers possible. For example,

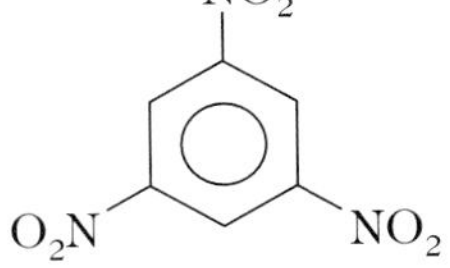

1,3,5-trinitrobenzene,
TNB

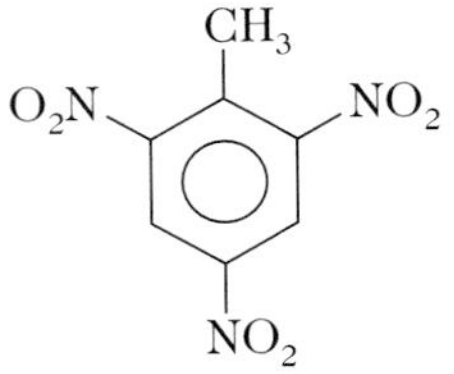

2,4,6-trinitrotoluene,
TNT

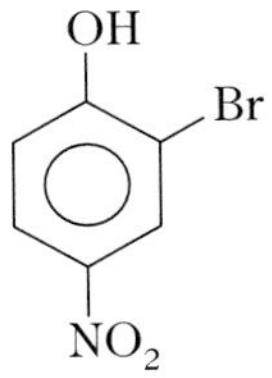

2-bromo-4-nitrophenol

■ TNT is an important explosive.
TNB is an even better explosive
than TNT, but it is more expensive
to make.

SUMMARY

Alkenes The lack of free rotation at a double bond makes
geometric (cis–trans) isomers possible, but they exist only
when the two groups are not identical at *either* end of
the double bond. Cyclic compounds also exhibit cis–trans
isomerism.

Alkenes and cycloalkenes are given IUPAC names by a
set of rules very similar to those used to name their corresponding saturated forms. However, the double bond takes
precedence both in selecting and in numbering the main
chain (or ring). The first unsaturated carbon encountered
in moving down the chain or around the ring through the
double bond must have the lower number.

Addition reactions Several compounds add to the carbon–carbon double bond—H_2, Cl_2, Br_2, HCl, HBr,
$HOSO_3H$, and H_2O (in the presence of an acid catalyst).
The kinds of products that can be made are outlined in the
accompanying chart of the reactions of alkenes. When both
the alkene and the reactant are unsymmetrical, the addition proceeds according to Markovnikov's rule—the end of
the double bond that already has the greater number of hydrogens gets one more. The double bond is vigorously attacked by strong oxidizing agents, like the permanganate
ion (MnO_4^-) and ozone (O_3).

How additions occur The carbon–carbon double bond is
a proton acceptor toward strong proton donors. The proton
becomes attached to one carbon at the double bond, using
one pair of electrons in the double bond to make the single
bond to this hydrogen. The other carbon of the original
double bond becomes positively charged. The result is a
carbocation, which then accepts an electron-rich particle
to complete the formation of the product.

Polymerization of alkenes The polymerization of an
alkene is like an addition reaction. The alkene serves as the
monomer, and one alkene molecule adds to another, and so
on, until a long chain with a repeating unit forms—the
polymer molecule.

Aromatic compounds Aromatic compounds contain rings
in which six electrons exist in a ring-encompassing space, a
stabilizing feature that makes the ring resistant to oxidation except when it holds OH or NH_2 as a substituent.
Moreover, when aromatic compounds undergo reactions at
the benzene ring, substitutions rather than additions occur.
In this way, the closed-circuit six-electron network of the
ring remains unbroken.

Chemical Properties

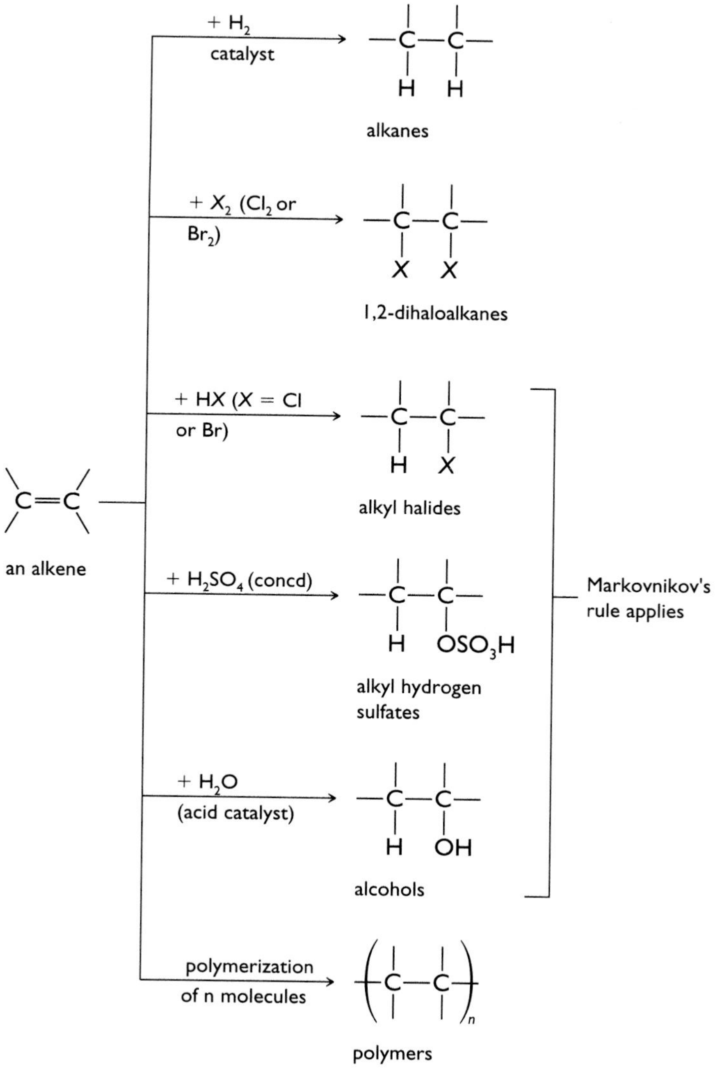

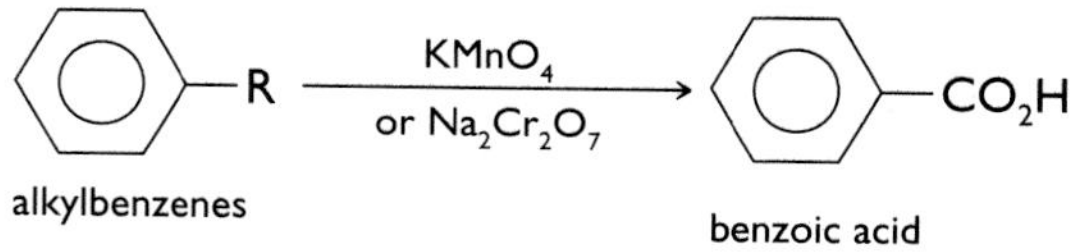

an alkene

+ H₂ / catalyst → alkanes

+ X₂ (Cl₂ or Br₂) → 1,2-dihaloalkanes

+ HX (X = Cl or Br) → alkyl halides

+ H₂SO₄ (concd) → alkyl hydrogen sulfates

+ H₂O (acid catalyst) → alcohols

Markovnikov's rule applies

polymerization of n molecules → polymers

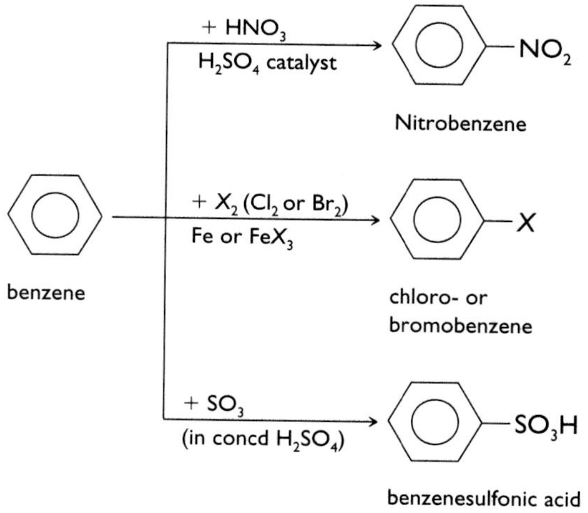

benzene

+ HNO₃ / H₂SO₄ catalyst → Nitrobenzene (NO₂)

+ X₂ (Cl₂ or Br₂) / Fe or FeX₃ → chloro- or bromobenzene (X)

+ SO₃ (in concd H₂SO₄) → benzenesulfonic acid (SO₃H)

alkylbenzenes (R) — KMnO₄ or Na₂Cr₂O₇ → benzoic acid (CO₂H)

REVIEW EXERCISES

The answers to Review Exercises whose numbers are in color are found in Appendix E. The answers to the other Review Exercises are found in the Study Guide that accompanies this book. The more challenging questions are marked with asterisks.

Occurrence and Physical Properties

12.1   Examine the following structural formulas and *use their identifying letters* to answer the questions about them.

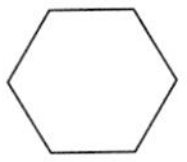

A (cyclohexane hexagon)

$HC{\equiv}CCH_2CH_3$ **B**

$CH_3CH{=}CHCH_3$ **C**

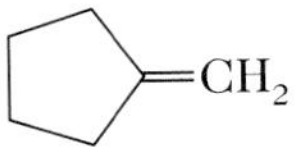

${=}CH_2$ (cyclopentane)   **D**

$CH_2{=}CH{-}CH{=}CH_2$ **E**

(a) What are the letters of the compounds that are members of the *alkene* family?
(b) Which is a *saturated* compound?
(c) The prefix *but-* would appear in the names of which compounds?
(d) Which compounds are relatively insoluble in water?

12.2   In view of the theme of this book, the molecular basis of life, what has justified our including the alkene group in our study?

Nomenclature

12.3   Write the condensed structures of the following compounds.
(a) Propylene
(b) *cis*-3,4-Dimethyl-3-hexene
(c) Isobutylene
(d) 1,2-Dichlorocyclopentene
(e) 2,4-Dimethylcyclohexene
(f) 1,3-Dimethylcyclohexene

12.4   What are the IUPAC names of the following compounds?

(a) $CH_3(CH_2)_2CH{=}CHCH_2CH_3$

(b) 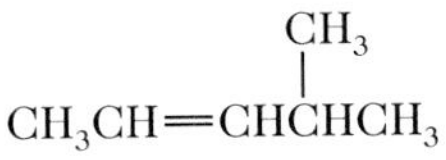 $CH_3CH{=}CHCHCH_3$ with CH_3 substituent

(c) 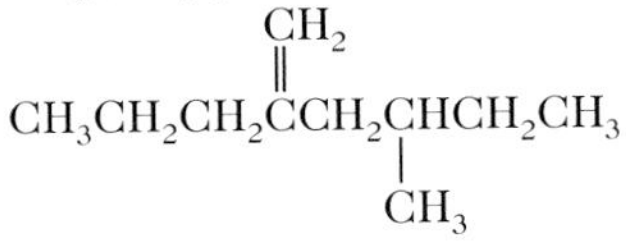 $CH_3CH_2CH_2\overset{\underset{\|}{CH_2}}{C}CH_2CHCH_2CH_3$ with CH_3

(d) 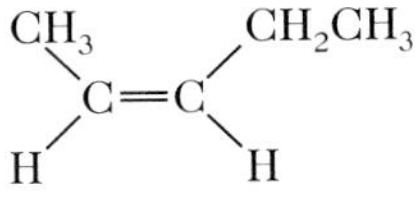 $\underset{H}{\overset{CH_3}{}}C{=}C\underset{H}{\overset{CH_2CH_3}{}}$
(include *cis* or *trans* designation)

(e) 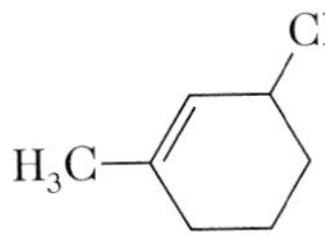 H₃C— (cyclohexene with CH₃)

(f) 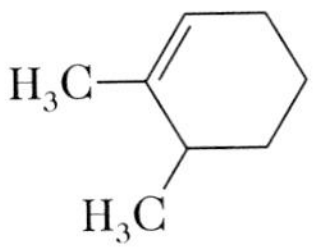 H₃C— (cyclohexene with H₃C)

*12.5   Write the condensed structures and the IUPAC names for all the isomeric pentenes, C_5H_{10}. Include cis and trans isomers.

12.6   Write the condensed structures and the IUPAC names for all the isomeric methylcyclopentenes.

*12.7   Write the condensed structures and the IUPAC names for all the isomeric dimethylcyclopentenes. Include the cis and trans isomers. Remember that all six atoms directly involved with an alkene group are in the same plane.

Cis–Trans Isomerism

12.8   Which of the following pairs of structures represent identical compounds or isomers?

(a) 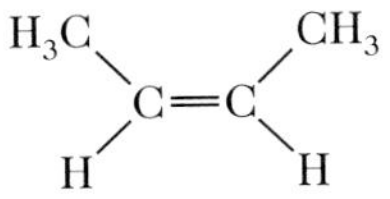 $\underset{H}{\overset{H_3C}{}}C{=}C\underset{H}{\overset{CH_3}{}}$ and $\underset{H_3C}{\overset{H}{}}C{=}C\underset{CH_3}{\overset{H}{}}$

(b) $\underset{H}{\overset{Br}{}}C{=}C\underset{H}{\overset{Cl}{}}$ and $\underset{Br}{\overset{H}{}}C{=}C\underset{Cl}{\overset{H}{}}$

(c) (cyclopentene with Br)   and   (cyclopentene with Br)

(d) $CH_3CH_2C{=}CHCH_3$ with CH_3 and $CH_3CH{=}CCH_2CH_3$ with CH_3

(e) $CH_3C{=}CHCHCH_3$ with CH_3 and $\overset{CH_3}{\underset{CH_3CCH_3}{CHCHCH_3}}$

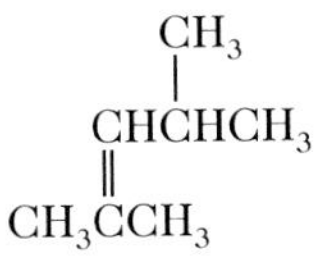

12.9   Study the following structures to discover which are able to exhibit cis–trans isomerism. For those that do, write the structures of the cis and trans isomers.
(a) $CH_3CH_2CH{=}CHCH_3$
(b) 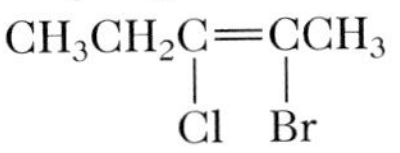 $CH_3CH_2C{=}CCH_3$ with Cl and Br
(c) 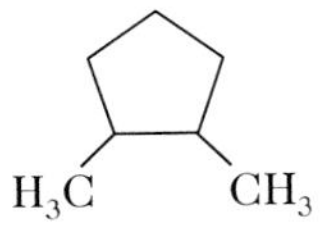 (cyclopentane with H₃C and CH₃)
(d) 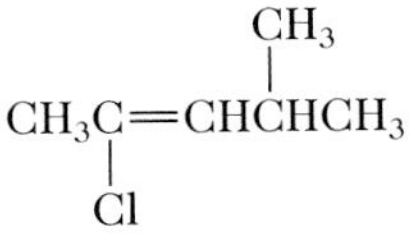 $CH_3C{=}CHCHCH_3$ with CH_3 and Cl

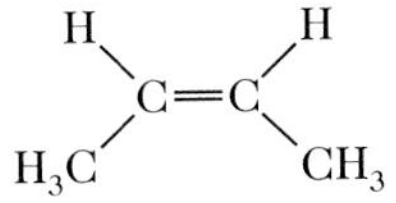

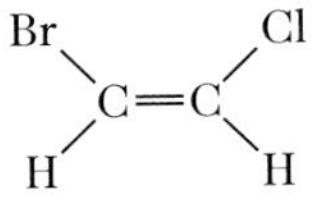

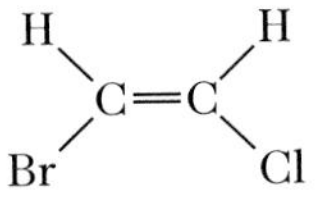

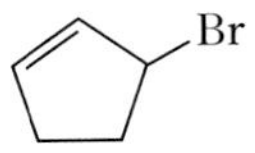

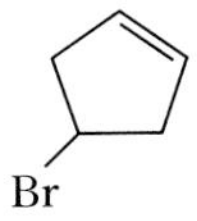

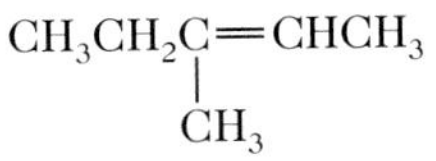

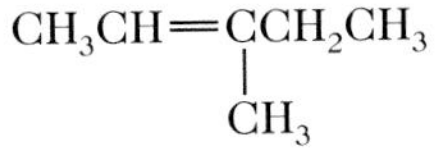

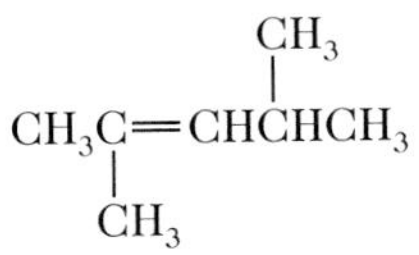

12.10 Identify which of the following compounds can exist as cis and trans isomers and write the structures of these isomers.
(a) $FClC=CHBr$
(b) H₃C   CH₃

(c) $CH_3CH=CHCH=CH_2$

Reactions of the Carbon–Carbon Double Bond

12.11 Write equations for the reactions of 2-methyl-propene with the following reactants.
(a) Cold, concentrated sulfuric acid
(b) Hydrogen in the presence of a nickel catalyst
(c) Water, in the presence of an acid catalyst
(d) Hydrogen chloride
(e) Hydrogen bromide
(f) Bromine

12.12 Write equations for the reactions of 1-methylcyclopentene with the reactants listed in Review Exercise 12.11.

12.13 Write equations for the reactions of 2-methyl-2-butene with the compounds given in Review Exercise 12.11.

12.14 Ethane is insoluble in concentrated sulfuric acid, but ethene dissolves readily. Write an equation to show how ethene is changed into a substance polar enough to dissolve in concentrated sulfuric acid, which, of course, is highly polar.

*12.15** Pentene, C_5H_{10}, has several isomers, and most but not all of them add bromine, dissolve in concentrated sulfuric acid, and react with potassium permanganate. Give the structure of at least one isomer of C_5H_{10} that gives *none* of these reactions.

*12.16** Cyclohexene can be oxidized to adipic acid, one of the raw materials for the synthesis of nylon. When potassium permanganate is the oxidizing agent, the first step produces the potassium salt of adipic acid, $K_2C_6H_8O_4$, according to the following incomplete equation.

$3C_6H_{10} + 8KMnO_4 \longrightarrow$

$3K_2C_6H_8O_4 + 8MnO_2 + \underline{\quad} + 2H_2O$

(a) Using only what you know about balanced equations in general, fill in the blank in this equation.
(b) How many grams of potassium permanganate are needed for the oxidation of 12.3 g of cyclohexene?
(c) How many grams of the potassium salt, $K_2C_6H_8O_4$, can form?

How Addition Reactions Occur

12.17 1-Butene reacts with hydrogen iodide in the same way it reacts with hydrogen chloride, by an addition reaction. The chief product is 2-iodobutane, not 1-iodobutane. Explain.

Polymerization

12.18 Polystyrene is a polymer of phenylethene, $C_6H_5CH=CH_2$. (In *foamed polystyrene,* the molten polymer has microbubbles blown into it as it hardens.) The structure of polystyrene is quite regular, like that of polypropylene. Write the structure of polystyrene in two ways.
(a) One that shows three repeating units, one after the other
(b) The condensed form of this structure

12.19 Write the condensed structure of polyvinyl acetate, the polymer of vinyl acetate.

vinyl acetate

Polyvinyl acetate is converted to Butvar, another polymer, by reactions to be studied in a later chapter. Butvar is the glue that binds two glass sheets together to form safety glass.

12.20 Gasoline is mostly a mixture of alkanes. However, when a sample of gasoline is shaken with aqueous potassium permanganate, a brown precipitate of MnO_2 appears, and the purple color of the permanganate ion disappears. What kind of hydrocarbon is evidently also present?

12.21 If gasoline rests for months in the fuel line of some engine, the line slowly accumulates a sticky material, which can clog the fuel line and also make the carburetor work poorly or not at all. In view of your answer to Review Exercise 12.20, what is likely to be happening, chemically?

Aromatic Properties

12.22 Sulfanilamide, one of the sulfa drugs, has no odor at all, but it is still classified as an aromatic compound. Explain.

sulfanilamide

12.23 Write equations for the reactions, if any, of benzene with the following compounds.
(a) Sulfur trioxide (in concentrated sulfuric acid)
(b) Concentrated nitric acid (in concentrated sulfuric acid)
(c) Hot sodium hydroxide solution
(d) Hydrochloric acid
(e) Chlorine (alone)
(f) Hot potassium permanganate
(g) Bromine in the presence of $FeBr_3$

12.24 Write the structure of any compound that would react with hot potassium permanganate to give benzoic acid.

12.25 Explain why benzene strongly resists addition reactions and gives substitution reactions instead.

Names of Aromatic Compounds

12.26 Write the condensed structure of each compound.
(a) Toluene   (b) Benzaldehyde
(c) Phenol   (d) Benzoic acid
(e) Aniline   (f) Nitrobenzene

12.27 Give the condensed structures of the following compounds.
(a) *p*-Nitrobenzoic acid
(b) *m*-Bromotoluene
(c) *o*-Chloronitrobenzene
(d) 2,4-Dinitrophenol

Alkene Group in Nature (Interaction 12.1)

12.28 What structural feature occurs in vitamins D and A to explain why these vitamins deteriorate somewhat when heated in air?

12.29 What structural difference exists between vegetable oils and animal fats?

Chemistry of Vision (Interaction 12.2)

12.30 When cis–trans isomerization takes place, do groups at the double bond break off from the chain and exchange places?

12.31 How does a photon of the proper energy enable a cis isomer to change into a trans isomer?

12.32 The retina in the eye of an owl has only rods, not cones. What connection is there between this circumstance and the ability of an owl to see in very dim light? Can an owl see in a situation of total blackness?

Ozone in Smog (Interaction 12.3)

12.33 What event in a vehicle engine launches the production of ozone in smog? (Write an equation.)

12.34 How is NO_2 formed in smog? (Write an equation.)

12.35 How is NO_2 involved in the production of ozone in smog? Write equations.

12.36 Why is ozone dangerous to living things?

12.37 In areas with severe smog problems, air quality authorities seek to reduce emissions of hydrocarbons, even those arising from the use of power lawn mowers and similar machines. What is the connection between these uses of fuels and the ozone in smog?

Reactions of Alkynes (Interaction 12.4)

12.38 Write the structures of the products that form when one mole of 2-butyne reacts with one mole of each of the following.
(a) H_2 (b) Cl_2 (c) Br_2

12.39 Write the structures of the products that form when one mole of 2-butyne reacts with two moles of each of the following.
(a) H_2 (b) Cl_2 (c) Br_2

Additional Exercises

12.40 Consider the use of general formulas like C_nH_{2n+2} to define hydrocarbons.
(a) Name a group of hydrocarbons besides alkenes that fit the formula C_nH_{2n}.
(b) Name a group of hydrocarbons besides alkynes that fit the formula C_nH_{2n-2}.

12.41 Give the IUPAC name of the following compound.

$$\begin{array}{ccc} CH_3 & CH_3 & \\ | & | & \\ CH_2CHCHCHCH_3 & & \\ | & | & \\ CH_3CH_2 & CH=CH_2 & \end{array}$$

12.42 If the structure of the previous review exercise exhibits cis–trans isomerism, write the structure of the cis isomer.

12.43 Consider structures **A** and **B**. Structure **A** can exist as cis–trans isomers but **B** cannot. Explain.

A   B

*12.44 Write the condensed structures and IUPAC names for all the isomeric butynes. The IUPAC rules for naming alkynes are identical with the rules for naming alkenes, except that the name ending is *-yne,* not *-ene.*

*12.45 Write the condensed structures and the names for all the open-chain dienes with the molecular formula C_5H_8. (Note that there can be two double bonds from the same carbon.)

12.46 Write equations for the reactions of 1-methylcyclohexene with the following reactants. (Do not attempt to predict whether cis or trans isomers form.)

(a) Cold, concentrated sulfuric acid
(b) Hydrogen in the presence of a nickel catalyst
(c) Water, in the presence of an acid catalyst
(d) Hydrogen chloride
(e) Hydrogen bromide
(f) Bromine

12.47 The molecules of a hydrocarbon, *A,* with the molecular formula C_7H_{12} have the following structure, which is complete except for the location of one double bond. Compound *A* is *not* capable of existing as cis and trans isomers. However, the *product* of the hydrogenation of compound *A* is capable of existing as cis and trans isomers. Write the structure of *A*. (If more than one structure is possible, write them all.)

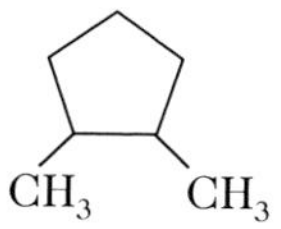

incomplete
structure of *A*

12.48 When 2-pentene reacts with hydrogen chloride, the product consists of a mixture of roughly equal parts of 2-chloropentane and 3-chloropentane. Explain why substantial proportions of *both* isomers form.

12.49 PVC (polyvinyl chloride) is made by the polymerization of vinyl chloride (chloroethene). Write the condensed structure of PVC.

12.50 Dipentene has a very pleasant, lemonlike fragrance, but it is not classified as an aromatic compound. Why?

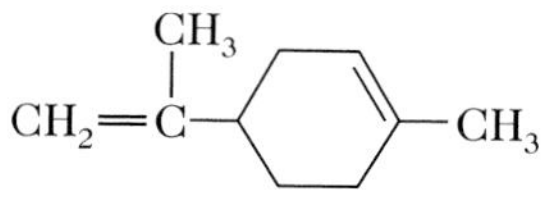

dipentene

12.51 Phthalic acid is one of the raw materials for making a polymer that is used in automobile finishes.

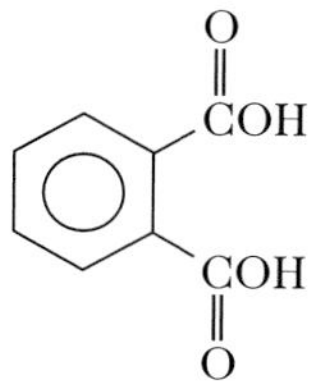

phthalic acid

What hydrocarbon with the formula C_8H_{10} could be changed to phthalic acid by the action of hot potassium permanganate? (Write its structure.)

12.52 Write the structures of the products to be expected in the following situations. If no reaction is to be expected, write "no reaction." To work this kind of exercise, you have to be able to do three things.

(1) *Classify* a specific organic reactant into its proper family. Do this first.
(2) *Recall* the short list of chemical facts about the family. (If there is no matchup between this list and the reactants and conditions specified by a given problem, assume that there is no reaction.)
(3) *Apply* the recalled fact, which might be some "map sign" associated with a functional group, to the specific situation.

Study the next two examples before continuing.

EXAMPLE 12.6   Predicting Reactions

What is the product, if any, of the following?
$$CH_3CH_2CH_2CH_3 + H_2SO_4 \longrightarrow ?$$

ANALYSIS We note first that the organic reactant is an alkane, so we next turn to the list of chemical properties about all alkanes that we learned. With this family, of course, the list is very short. Except for combustion and halogenation, we have learned no reactions for alkanes, and we assume, therefore, that there aren't any others, not even with sulfuric acid. Hence, the answer is "no reaction."

EXAMPLE 12.7   Predicting Reactions

What is the product, if any, in the following situation?

$$CH_3CH{=}CHCH_3 + H_2O \xrightarrow{\text{acid catalyst}} ?$$

ANALYSIS We first note that the organic reactant is an alkene, so we review our mental "file" of reactions of the carbon–carbon double bond.

1. Alkenes add hydrogen (in the presence of a metal catalyst) to form alkanes.
2. They add chlorine and bromine to give 1,2-dihaloalkanes.
3. They add hydrogen chloride and hydrogen bromide to give alkyl halides.
4. They add sulfuric acid to give alkyl hydrogen sulfates.
5. They add water in the presence of an acid catalyst to give alcohols.
6. They are attacked by strong oxidizing agents.
7. They polymerize.

These are the chief chemical facts, the principal "map signs," about the carbon–carbon double bond that we have studied, and we see that the list includes a reaction with water in the presence of an acid catalyst. We remember that in all addition reactions the double bond changes to a single bond and the pieces of the adding molecule end up on the carbons at ends of the double bond. We also have to remember Markovnikov's rule to tell us which pieces of the water molecule go to each carbon. However, in this specific example, Markovnikov's rule does not apply because the alkene is symmetrical.

SOLUTION

$$CH_3CH_2\underset{\underset{\displaystyle OH}{|}}{C}HCH_3$$

Now work the following parts. (Remember that C_6H_6 stands for benzene and that C_6H_5 is the phenyl group.)

(a) $CH_2{=}CHCH_3 + H_2O \xrightarrow{\text{acid catalyst}}$

(b) $CH_2{=}CH\underset{\underset{\displaystyle CH_3}{|}}{C}HCH_3 + H_2 \xrightarrow{\text{Ni catalyst}}$

(c) $C_6H_6 + Br_2 \xrightarrow{\text{FeBr}_3}$

(d) $CH_3CH_2CH_2CH_2CH_2CH_3 + H_2SO_4(\text{concd}) \longrightarrow$

(e) 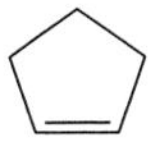 $+ H_2O \xrightarrow{\text{acid catalyst}}$

(f) $CH_3CH_2CH_2CH{=}CH_2 + H_2 \xrightarrow{\text{Ni catalyst}}$

(g) 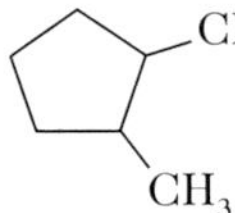 $CH_3 + H_2SO_4(\text{concd}) \longrightarrow$

(h) $C_6H_5CH{=}CHC_6H_5 + H_2O \xrightarrow{\text{acid catalyst}}$

(i) 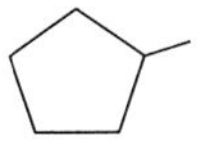 $CH_3 + H_2 \xrightarrow{\text{Ni catalyst}}$

(j) $CH_3CH_2CH(CH_3)_2 + O_2 \xrightarrow[\substack{\text{(Balance the}\\\text{equation.)}}]{\substack{\text{complete}\\\text{combustion}}}$

(k) $C_6H_6 + H_2O \xrightarrow{\text{acid catalyst}}$

12.53   Write the structures of the products in the following situations. If no reaction is to be expected, write "no reaction."

(a) $CH_3CH{=}CHCH_2CH{=}CH_2 + 2H_2 \xrightarrow{\text{Ni catalyst}}$

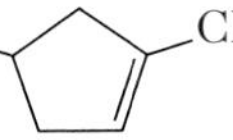

(b) $+ H_2O \xrightarrow{\text{acid catalyst}}$

(c) $CH_3\underset{\underset{\displaystyle CH_3}{|}}{\overset{\overset{\displaystyle CH_3}{|}}{C}}{=}CCH_3 + HBr \longrightarrow$

(d) 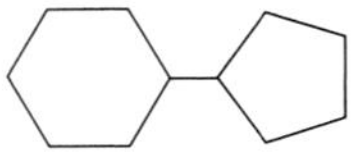 $+ H_2SO_4(\text{concd}) \longrightarrow$

(e) 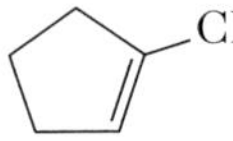 $CH_3 + Cl_2 \longrightarrow$

(f) 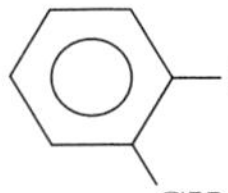 $CH_3 \xrightarrow[\substack{\text{(Balance the}\\\text{equation.)}}]{\substack{\text{complete}\\\text{combustion}}}$

(g) $C_6H_6 + Cl_2 \xrightarrow{\text{FeCl}_3}$

(h) $C_6H_6 + NaOH(aq) \longrightarrow$

(i) $CH_3CH_2CH{=}\underset{\underset{\displaystyle CH_3}{|}}{C}CH_3 + HCl \longrightarrow$

(j) 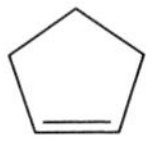 $+ NaOH(aq) \longrightarrow$

(k) $CH_2{=}CHCH_2CH{=}CHCH_2CH{=}CH_2 + 3Cl_2 \longrightarrow$

(l)  $+ 3Cl_2 \longrightarrow$

ALCOHOLS, PHENOLS, ETHERS, AND THIOALCOHOLS

13

The glucose in the juice of these grapes in Napa Valley, California, changes into ethyl alcohol and carbon dioxide during wine making. Glucose is also the chief sugar in our blood, only we use it for its chemical energy. To understand glucose wherever it's found we must learn about alcohols, the principal substances studied in this chapter.

THIS CHAPTER IN CONTEXT

We must study the alcohol group because it occurs in most of the compounds in living things. The chemistry of carbohydrates, for example, is little more than that of alcohols (this chapter) and of aldehydes and ketones (next chapter). Studying ethers here will also help us with carbohydrates. The phenol and thioalcohol groups occur among proteins. In other words, we're continuing to study functional groups vital at the molecular level of life, but first, according to plan, doing so among simple systems.

13.1   OCCURRENCE, TYPES, AND NAMES OF ALCOHOLS

In the alcohol family, molecules have the OH group attached to a saturated carbon atom.

In **alcohols,** the OH group is covalently held by a *saturated* carbon atom, one having only *single* bonds. Only then is the OH group called the **alcohol group.** The *simple alcohols* or **monohydric alcohols** are those with one OH per molecule and no other functional group (Table 13.1). Beverage alcohol is ethyl alcohol (see Interaction 13.1). A molecule of a **dihydric alcohol** or **glycol,** like 1,2-ethanediol (ethylene gly-

TABLE 13.1   Some Common Alcohols

Name[a]	Structure	BP (°C)
Methanol	CH_3OH	65
Ethanol	CH_3CH_2OH	78.5
1-Propanol	$CH_3CH_2CH_2OH$	97
2-Propanol	CH_3CHCH_3 | OH	82
1-Butanol	$CH_3CH_2CH_2CH_2OH$	117
2-Butanol (*sec*-butyl alcohol)	$CH_3CH_2CHCH_3$ | OH	100
2-Methyl-1-propanol (isobutyl alcohol)	CH_3 | CH_3CHCH_2OH	108
2-Methyl-2-propanol (*t*-butyl alcohol)	CH_3 | CH_3COH | CH_3	83
1,2-Ethanediol (ethylene glycol)	$CH_2{-}CH_2$ | | OH OH	197

1,2-Propanediol (propylene glycol)	$CH_3{-}CH{-}CH_2$
	      | |
	      OH OH
		189

1,2,3-Propanetriol (glycerol)	$CH_2{-}CH{-}CH_2$ | | | OH OH OH	290

[a] The IUPAC names are given with the common names in parentheses.

INTERACTION 13.1
ETHYL ALCOHOL AND ALCOHOLISM

Throughout all of human history people have drunk alcoholic beverages both for their pleasing tastes and their ability to alter moods. Because of the consequences of alcohol abuse, opposition to the use of such beverages is also widespread.

With respect to alcohol abuse as an issue of public policy in the United States, there are roughly two camps. One camp sees alcohol itself as the culprit and so encourages abstinence. The other camp sees alcohol abusers at the heart of the problem and so espouses treatment centers and drug abuse education for them and temperance for the rest. We can do no more in this discussion than concentrate on facts to which both camps would agree.

Alcoholic Beverages Pure ethyl alcohol is a colorless liquid with a taste so burning that no one drinks it pure. Users always drink dilute forms. Beer and ale are 3% to 6% alcohol; wine is 7% to 14%. (When the concentration reaches this point, the enzymes that catalyze the fermentation of fruit sugar, mostly glucose, are deactivated.) Fortified wines result from adding alcohol to wine to make the concentration about 20%. Stronger liquors are made by distilling wines or beers—brandy from wine and whiskey from a beerlike mixture. Such distilled beverages or "spirits" are 40% to 55% ethyl alcohol.

The "proof" of a beverage is always twice the percentage of alcohol. Pure alcohol is thus 200 proof. Historically, early distillers or their customers poured a small amount of the product onto a mound of gunpowder. If the powder would still burn, it was "proof" that the product had not been watered down. The lower limit on a positive test was called "100 proof" material, which we now know to be 50% alcohol.

Effects of Alcohol What is termed the *typical drink* contains 0.75 ounce of ethyl alcohol, which may be obtained by a pint of beer (16 oz of 4.5% alcohol); a 5-oz glass of wine (14% alcohol); a 3.5-oz glass of fortified wine, like sherry (20%); or a 1.5-oz "shot" of whiskey (50%). Observant friends of an adult, nonalcoholic, "standard," 150-lb male will notice changes in his behavior and mood at a blood alcohol level of 0.050% (39 mg/dL, 8.5 mmol/L)—the result of two typical drinks in succession. The mood changes of this state are largely what encourage moderate drinking. The effects of drinking, however, depend on too many factors to warrant easy generalizations.

Body mass is only one factor in how quickly the blood alcohol level increases during drinking. The "standard" 120-lb adult, nonalcoholic female generally reaches a blood level of 0.050% with less drinking than the standard adult male. Food is another factor. The presence of food in the system slows down the absorption of the alcohol. Ethyl alcohol is rather quickly metabolized; the rate of complete metabolism for the "standard" male is about one drink per hour.

As the blood alcohol level increases above 0.050%, the depressant effect of alcohol on the brain kicks in, and at a level of 0.10% ethyl alcohol (79 mg/dL, 17 mmol/L), voluntary motor actions, such as walking, hand and arm movements, and speech, become clumsy. Most states of the United States define "intoxication" in terms of a blood alcohol level of 0.10%. The driving laws of many states, however, specifically allow lower values—even as low as 0.05%—to be evidence for DWI, "driving while under the influence." At a level of 0.20% (157 mg/dL, 34 mmol/L) the typical nonalcoholic male's control over the entire motor area of the brain is impaired and the brain region controlling emotional behavior is affected. Staggering, boisterous talking, or anger (or weeping) appear; the person is a certified "drunk" to those who observe him. At a level of 0.30%, the user is very confused and may lapse into a stupor. With 0.40 to 0.50% blood alcohol, the individual will be in a coma and those parts of the brain that control the heartbeat and breathing are so affected that death results.

Ethyl alcohol is absorbed directly into the bloodstream along any part of the intestinal tract. Enzymes in both the intestinal tract and the liver work to detoxify it, but the liver does most of such work. One enzyme takes ethyl alcohol to the aldehyde stage, producing acetaldehyde (CH_3CHO), which is chiefly responsible for cirrhosis. In time, the acetaldehyde is oxidized to the acetate ion, $CH_3CO_2^-$, which can be metabolized normally.

The Breathalyzer Test Law enforcement officers use a breath-analyzing device called a *breathalyzer* to determine the blood alcohol level of drivers suspected of being under the influence (Figure 1). It has been shown that the partial pressure of ethyl alcohol vapor in exhaled breath is proportional to the concentration of alcohol in the blood. The suspect, therefore, is asked to exhale through a tube that leads the air into

FIGURE I The administration of the breathalyzer test along a highway.

aqueous sodium dichromate. The yellow dichromate ion changes to the green chromium(III) ion, Cr^{3+}, as it oxidizes ethyl alcohol to acetic acid.

$$3CH_3CH_2OH + 2Cr_2O_7^{2-} + 16H^+ \longrightarrow$$

ethyl alcohol   dichromate
            (yellow)

$$3CH_3CO_2H + 4Cr^{3+} + 11H_2O$$

       acetic acid   (green)

The instrument measures the amount of color change and translates the information directly into blood alcohol concentration.

Alcoholism Ethyl alcohol is the chemical cause of *alcoholism.* Although "alcoholism" is not easily defined, most specialists agree that the definition must focus on the frequency of those symptoms that relate to drinking itself rather than something else. The essential features of alcoholism are a powerful craving for and a dependency on ethyl alcohol to the point that drinking or thinking about the next drink becomes the major preoccupation of life and withdrawal is very painful.

Alcoholism is accompanied by disorders of the nervous system and muscles. In its early stages, it causes a fatty liver and in its later stages *cirrhosis* of the liver, an incurable and fatal disorganization of liver structure brought about by the development of nodules surrounded by fibrous tissue. One-third of heavy drinkers suffer heart damage and nearly half experience an increasing weakness of skeletal muscles. Alcoholics also have impaired immune systems and lowered resistance to pneumonia and other infectious diseases. You can see how all of the by-products of an alcoholic lifestyle would contribute immensely to a nation's medical costs.

Malnutrition, which is an insufficiency of proper nutrients (and not necessarily a lack of enough calories), complicates but does not cause the medical disorders of alcohol abuse. Two "fifths" of whiskey a day would supply enough calories for metabolism but, unless food is eaten, neither vitamins nor proteins are furnished. (A "fifth" equals one-fifth of a gallon, or a

■ We'll soon learn about systems with the OH group that are not alcohols.

■ The following system, the 1,1-diol, is unstable; only rare examples are known.

        OH
        |
    R—C—OH
        |
        R′

     1,1-diols

■ Pronounce 1° as *primary,* 2° as *secondary,* and 3° as *tertiary.*

col), has two OH groups. The molecules of a **trihydric alcohol,** like 1,2,3-propanetriol (glycerol), have three alcohol groups. All types are commercially important substances (see Interaction 13.2).

$$\begin{array}{ccc} CH_2—CH_2 \\ | \qquad | \\ OH \quad OH \end{array}$$

1,2-ethanediol
(ethylene glycol,
a dihydric alcohol)

$$\begin{array}{ccc} CH_2—CH—CH_2 \\ | \qquad | \qquad | \\ OH \quad OH \quad OH \end{array}$$

1,2,3-propanetriol
(glycerol, a trihydric
alcohol)

$$CH_2—CH—CH—CH—CH—\overset{\displaystyle O}{\overset{\displaystyle \|}{CH}}$$
$$\;\; | \qquad | \qquad | \qquad | \qquad |$$
$$OH \quad OH \quad OH \quad OH \quad OH$$

glucose, a sugar
(open form of molecule)

Carbohydrates, like glucose, have several OH groups per molecule. To be stable, however, one carbon atom generally can hold only one OH group. Exceptions are rare. If such should form during a reaction, it breaks up.

The Alcohol Group Occurs as a 1°, 2°, or 3° System An alcohol is classified as primary (1°), secondary (2°), or tertiary (3°) according to the condition of the carbon that holds the OH group. When held by a 1° carbon, one that has only one R group directly joined to it, the alcohol is a **primary alcohol.** In a **secondary alcohol,** the

bottle containing this amount. It's thus four-fifths of a quart or a little more than 750 mL.)

When taken in their totality over an entire population, individual human decisions concerning personal uses of alcoholic beverages are responsible for widespread misery and economic burdens on individuals, families, and society. In the late 1980s in urban areas, the third most frequent cause of death among those 25 to 64 years of age was cirrhosis of the liver. The National Institute on Alcohol Abuse and Alcoholism estimates that the combined costs of health expenses and lost productivity are roughly $120 billion annually in the United States.

Alcoholism Treatment The conventional wisdom at alcoholism treatment centers is that alcoholics have a disease that arises from a strong genetic predisposition and is characterized by physical dependency. The conventional treatment is to get the alcoholic to stop drinking and henceforth to practice total abstinence. Less conventional views note that the long-term success of the orthodox treatment requires a step beyond abstinence. Success depends on changes in both outlook and values that confer a satisfying lifestyle not dependent on heavy drinking (or longing for it). Your genes may or may not dispose you to becoming addicted to ethyl alcohol. But you choose to take this or any other drug as an option in the quest for some desirable end, such as acceptance into a social group, release from stress, a change in mood, or some other goal.

Alcohol and Pregnancy *Fetal alcohol syndrome* appears to be caused directly by excessive ethyl alcohol itself, not by its oxidation products. The problem is that the fetus cannot detoxify ethyl alcohol, and the alcohol can stop cell division. The syndrome leads to babies with impaired mental abilities and many other disorders. The conventional wisdom today favors total abstinence for those who are pregnant, advice that can come too late for those unaware of their condition.

Some Additional Readings

1. D. Antai-Otong, "Helping the Alcoholic Patient Recover," *The American Journal of Nursing,* August 1995, page 22.

2. S. Schenker and K. V. Speeg, "The Risk of Alcohol Intake in Men and Women," *The New England Journal of Medicine,* January 11, 1990, page 127.

3. M. E. Charness, R. P. Simon, and D. A. Greenberg, "Ethanol and the Nervous System," *The New England Journal of Medicine,* August 17, 1989, page 442.

4. G. E. Vaillant, *The Natural History of Alcoholism.* Cambridge, MA, Harvard University Press, 1983.

5. H. Fingarette, *Heavy Drinking: The Myth of Alcoholism as a Disease.* University of California Press, 1988.

OH group is held by a 2° carbon. When the OH group is joined to a 3° carbon, the alcohol is a **tertiary alcohol.** The value of these subclasses will surface when we learn that the different classes of alcohols respond differently to oxidizing agents.

$$R{-}CH_2{-}OH \qquad R{-}\underset{\underset{R'}{|}}{CH}{-}OH \qquad R{-}\underset{\underset{R''}{|}}{\overset{\overset{R'}{|}}{C}}{-}OH$$

primary alcohol   secondary alcohol   tertiary alcohol

■ The R groups in 2° and 3° alcohols don't have to be the same.

■ PRACTICE EXERCISE 1 Classify each of the following as monohydric or dihydric. For each found to be monohydric, classify it further as 1°, 2°, or 3°. If the structure is too unstable to exist, state so.

(a) $CH_3\underset{\underset{OH}{|}}{CH}CH_3$
(b) a cyclohexene ring with —OH
(c) $CH_3\underset{\underset{OH}{|}}{\overset{\overset{OH}{|}}{C}}CH_3$

INTERACTION 13.2
ALCOHOLS IN OUR DAILY LIVES

Methanol (methyl alcohol, wood alcohol) When taken internally in enough quantity, methanol causes either blindness or death. In industry, it is used as the raw material for making formaldehyde (which is used to make polymers), as a solvent, and as a denaturant (poison) for ethanol. It is also used as the fuel in canned heat (e.g., Sterno), as well as in burners for fondue pots.

Most methanol is made by the reaction of carbon monoxide with hydrogen under high pressure and temperature:

$$2H_2 + CO \xrightarrow[\substack{350-400\ °C \\ ZnO/Cr_2O_3\ \text{catalyst}}]{3000\ \text{lb/in.}^2} CH_3OH$$

Ethanol (ethyl alcohol, grain alcohol) Some ethanol is made by the fermentation of sugars, but most is synthesized by the addition of water to ethene in the presence of a catalyst. A 70% (v/v) solution of ethanol in water is used as a disinfectant.

In industry, ethanol is used as a solvent and to prepare pharmaceuticals, perfumes, lotions, and rubbing compounds. For these purposes, the ethanol is adulterated by poisons that are very difficult to remove so that the alcohol cannot be sold or used as a beverage. (Nearly all countries derive revenue by taxing potable, i.e., drinkable, alcohol.)

2-Propanol (isopropyl alcohol) 2-Propanol is a common substitute for ethanol for giving back rubs. In solutions with concentrations from 50% to 99% (v/v), 2-propanol is used as a disinfectant. It is twice as toxic as ethanol.

1,2-Ethanediol (ethylene glycol) and 1,2-Propanediol (propylene glycol) Ethylene and propylene glycols are the chief components in permanent-type antifreezes. Their great solubility in water and their very high boiling points make them ideal for this purpose. An aqueous solution that is roughly 50% (v/v) in either glycol does not freeze until about −40 °C (−40 °F). 1,2-Ethanediol has a sweet taste, but it is highly toxic. The lethal dose for adults is about 100 mL. 1,2-Propanediol, on the other hand, is far less toxic (possibly because the anions of its oxidation products, pyruvic acid and acetic acid, are normally formed in metabolism).

1,2,3-Propanetriol (glycerol, glycerin) Glycerol, a colorless, syrupy liquid with a sweet taste, is freely soluble in water and insoluble in nonpolar solvents. It is one product of the digestion of the fats and oils in our diets. Because it has three OH groups per molecule, each capable of hydrogen bonding to water molecules, glycerol can draw moisture from humid air. It is sometimes used as a food additive to help keep foods moist.

An oily compound made from glycerol and nitric acid, **nitroglycerin,** is a powerful explosive. When pure, it detonates from concussion. Interestingly, although nitroglycerin is toxic, it has a place in medicine.

$$O_2NOCH_2CHCH_2ONO_2$$
$$|$$
$$ONO_2$$

nitroglycerin
(1,2,3-glyceryl trinitrate)

People who have periodic attacks of intense pain (angina pectoris), centered in heart muscle because of vasoconstriction (constriction of the blood vessels), are able to administer to themselves carefully controlled amounts of a vasodilator (a dilator of blood vessels). Nitroglycerin is commonly used for this purpose.

(d) [structure: cyclohexane with two OH groups]

(e) $HOCH_2\overset{\displaystyle CH_3}{\underset{\displaystyle CH_3}{C}}CH_3$

(f) [benzene ring]—CH_2OH

(g) $CH_3\overset{\displaystyle CH_3}{\underset{\displaystyle CH_3}{C}}OH$

(h) $CH_3CH_2\overset{\displaystyle CH_3}{C}HOH$

(i) $CH_3\overset{\displaystyle OH}{\underset{\displaystyle OH}{C}}OH$

Not All Compounds with the OH Group Are Alcohols   As you can see by the following structures, the OH group also occurs in the family of the *phenols,* where it is
attached to a benzene ring, and in the family of the *carboxylic acids,* where it is attached to a carbon with a double bond to oxygen.

■ Phenol is only the simplest
member of the family of *phenols,*
compounds with OH groups attached to benzene rings that may
carry any number and kind of other
ring substituents.

phenol	carboxylic acid	enol system (unstable)

When the OH group is directly attached to an alkene group, the system, called an
enol (ene + ol), is unstable. Most of our attention in this chapter will be given to
simple alcohols but with some study of phenols. We'll study the carboxylic acids in a
later chapter.

■ **PRACTICE EXERCISE 2**   Classify the following as alcohols, phenols, or carboxylic acids.

(a) 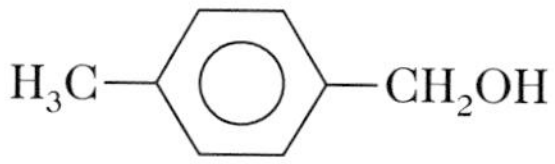 H_3C—⟨ ⟩—CH_2OH (b) 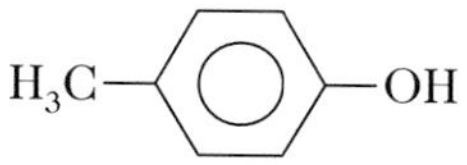 H_3C—⟨ ⟩—OH

(c) 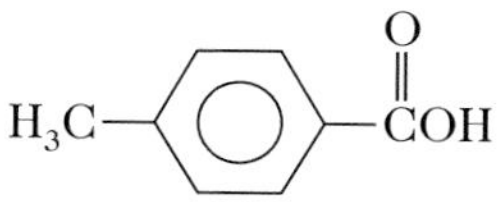 H_3C—⟨ ⟩—$\overset{\overset{\displaystyle O}{\|}}{C}OH$ (d) CH_2=$CHOH$

(e) $CH_3CH_2CH_2CH_2OH$ (f) 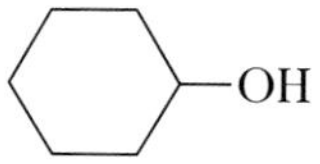   ⟨ ⟩—OH

**Common Names of Alcohols Are Popular When Their Alkyl Groups Are Easily
Named**   Simple alcohols have common names devised by writing the word *alcohol*
after the name of the alkyl group holding the OH group. For example,

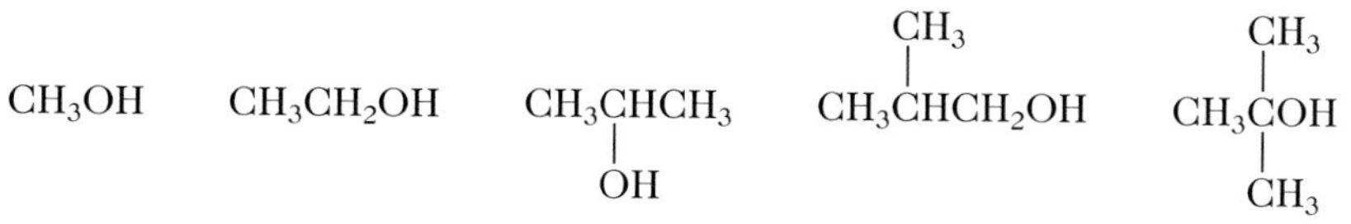

$$CH_3OH \qquad CH_3CH_2OH \qquad CH_3\underset{\underset{\displaystyle OH}{|}}{C}HCH_3 \qquad CH_3\underset{\underset{\displaystyle CH_3}{|}}{C}HCH_2OH \qquad CH_3\underset{\underset{\displaystyle CH_3}{|}}{\overset{\overset{\displaystyle CH_3}{|}}{C}}OH$$

methyl alcohol	ethyl alcohol	isopropyl alcohol	isobutyl alcohol	*t*-butyl alcohol

IUPAC Names of Alcohols End in *-ol*   The IUPAC rules for naming alcohols are similar to those for naming alkanes in that the name is based on the idea of a *parent alcohol* with substituents on its chain.

1.   Determine the parent alcohol by selecting the longest chain of carbons *that includes the carbon atom to which the OH group is attached.* Name the parent alcohol by changing the name ending of the alkane that corresponds to this chain
from *-e* to *-ol.* Examples are

$$CH_3OH$$

methanol
(complete name;
parent alkane
is methane)

$$CH_3CH_2CH_2OH$$

propanol
(incomplete name—
see rules 2 and 3;
parent alkane
is propane)

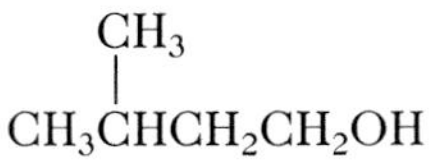

$$\overset{\displaystyle CH_3}{\underset{\displaystyle |}{CH_3CHCH_2CH_2OH}}$$

a substituted butanol
(incomplete name—
see rules 2 and 3;
parent alkane
is butane)

■ No number is needed to specify the location of the OH group in the names "methanol" and "ethanol."

2. Number the parent chain from whichever end gives the lower number to the carbon holding the OH group. The location of the OH group thus takes precedence over the location of alkyl groups (or halogen atoms). For example,

$$\underset{4321}{\overset{\displaystyle CH_3}{\underset{\displaystyle |}{CH_3CHCH_2CH_2OH}}}$$

3-methyl-1-butanol
not 2-methyl-4-butanol

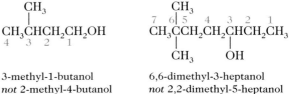

6,6-dimethyl-3-heptanol
not 2,2-dimethyl-5-heptanol

3. Write the number that locates the OH group in front of the name of the parent, and separate this number from the parent's name by a hyphen. Next, alphabetically prefix the names of the substituents and their location numbers. Use commas and hyphens in the usual way. For example,

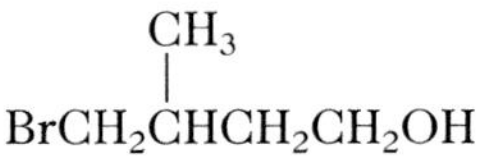

$$\overset{\displaystyle CH_3}{\underset{\displaystyle |}{BrCH_2CHCH_2CH_2OH}}$$

4-bromo-3-methyl-1-butanol
not 3-methyl-4-bromo-1-butanol

4. When two or more OH groups are present, use name endings such as *-diol* (for two OH groups), *-triol,* and so forth. Immediately in front of the name of the *parent portion* of the alcohol, write the two, three, or more numbers that show the locations of the OH groups. Thus,

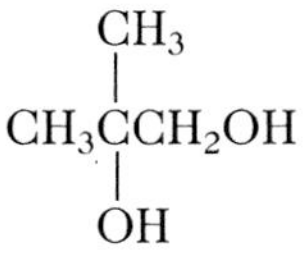

$$\overset{\displaystyle CH_3}{\underset{\displaystyle \underset{\displaystyle OH}{|}}{\overset{\displaystyle |}{CH_3CCH_2OH}}}$$

2-methyl-1,2-propanediol

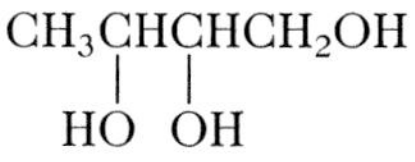

$$\underset{\displaystyle HOOH}{\overset{\displaystyle ||}{CH_3CHCHCH_2OH}}$$

1,2,3-butanetriol

5. If no parent alcohol name is possible or convenient, then the OH group can be treated as substituent named *hydroxy.* For example,

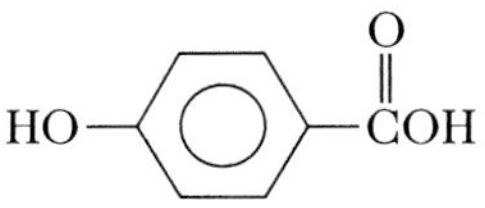

4-hydroxybenzoic acid

EXAMPLE 13.1   Using the IUPAC Rules to Name an Alcohol

What is the IUPAC name of the following compound?

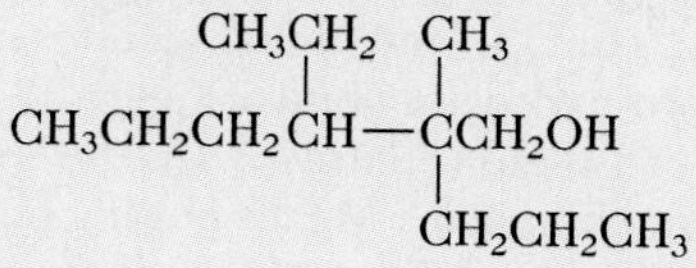

$$CH_3CH_2CH_2CH—CCH_2OH$$

with CH_3CH_2 and CH_3 on the upper branches and $CH_2CH_2CH_3$ below.

ANALYSIS The compound is in the alcohol family, so the ending to its IUPAC name is *-ol*. The IUPAC rules for alcohols require the parent chain to be the longest carbon chain *that includes the carbon atom to which the OH group is attached.* In the given structure, this is a six-carbon chain. We must now number the chain from whichever of its ends lets the carbon bearing the OH group have the lower number. Thus,

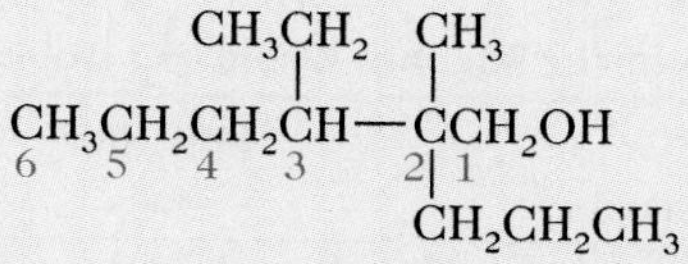

$$\underset{6}{CH_3}\underset{5}{CH_2}\underset{4}{CH_2}\underset{3}{CH}—\underset{2}{C}\underset{1}{CH_2OH}$$

with CH_3CH_2 and CH_3 on the upper branches and $CH_2CH_2CH_3$ below.

■ The longest carbon chain in the molecule has eight carbon atoms, but it doesn't include the OH group.

Therefore, the final name must end in -1-hexanol. Next, note that carbon 2 holds both a methyl and a propyl group, and carbon 3 has an ethyl group.

SOLUTION The final name, arranging the alkyl groups alphabetically, is

3-ethyl-2-methyl-2-propyl-1-hexanol

■ **PRACTICE EXERCISE 3**   Give the IUPAC names of the following.

(a) 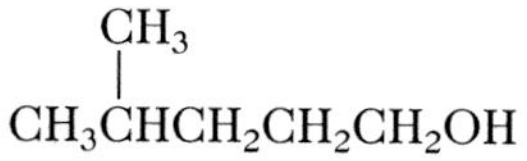 $CH_3CHCH_2CH_2CH_2OH$ with CH_3 branch
(b) 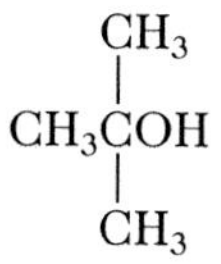 CH_3COH with CH_3 above and CH_3 below
(c) 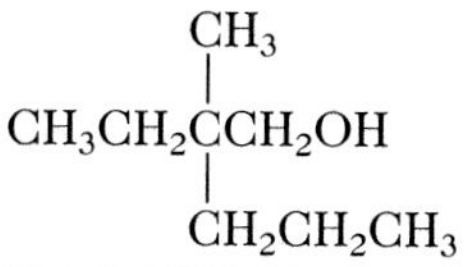 $CH_3CH_2CCH_2OH$ with CH_3 above and $CH_2CH_2CH_3$ below
(d) 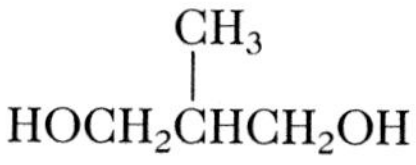 $HOCH_2CHCH_2OH$ with CH_3 branch

13.2 PHYSICAL PROPERTIES OF ALCOHOLS

Hydrogen bonding dominates the physical properties of alcohols.

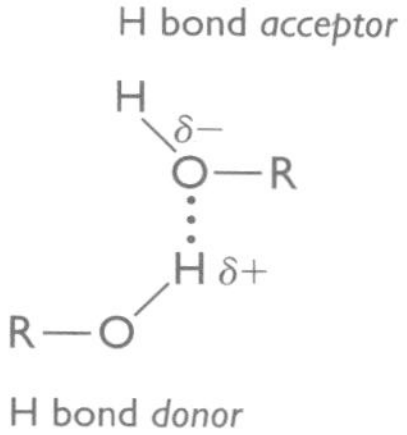

Alcohols have much higher boiling points and much greater solubilities in water than alkanes of the same formula mass. The difference is the OH group. It is quite polar, and it can both donate and accept hydrogen bonds.

Hydrogen Bonds between Molecules Raise Boiling Points To get a sense of how significantly the OH group provides forces of attraction between molecules, compare

■ The *donor* OH group has the H from which the H bond (····) extends to the $\delta-$ site on the *acceptor*.

H bond *acceptor*

$$\underset{R—O}{\overset{H}{\underset{\delta+}{\overset{\delta-}{O}—R}}}$$

H bond *donor*

the boiling points of some alcohols to those of alkanes with comparable formula masses (Table 13.2). The differences are caused by hydrogen bonding. The hydrogen bond, remember, is a force of attraction between opposite *partial* charges, namely, the $\delta+$ charge on H in the OH group and the $\delta-$ charge on the O of another group. No such partial charges exist in molecules of alkanes because C and H have nearly identical electronegativities. Thus, an individual alcohol molecule is attracted to two neighboring molecules by two hydrogen bonds (Figure 13.1*a*). Water molecules have three hydrogen bonds between them (Figure 13.1*b*). So water, despite having a lower formula mass than methyl alcohol, has a higher boiling point.

Ethylene glycol (Table 13.2), a dihydric alcohol, can have as many as four hydrogen bonds between molecules, two from each OH group, and it boils nearly 200 °C higher than butane and 100 °C higher than water.

The OH Group Makes Compounds More Soluble in Water We're always interested in any structural feature that affects how a biochemical dissolves in water, the chief fluid in living systems. Methane, like all hydrocarbons, is insoluble in water.[1] When

■ Only three elements, F, O, and N, have atoms that are electronegative enough to participate significantly in hydrogen bonds.

TABLE 13.2   The Influence of the Alcohol Group on Boiling Points

Name	Structure	Formula Mass	BP (°C)	Difference in BP
Ethane	CH_3CH_3	30	−89	154
Methyl alcohol	CH_3OH	32	65	
Propane	$CH_3CH_2CH_3$	44	−42	120
Ethyl alcohol	CH_3CH_2OH	46	78	
Butane	$CH_3CH_2CH_2CH_3$	58	0	197
Ethylene glycol	$HOCH_2CH_2OH$	62	197	

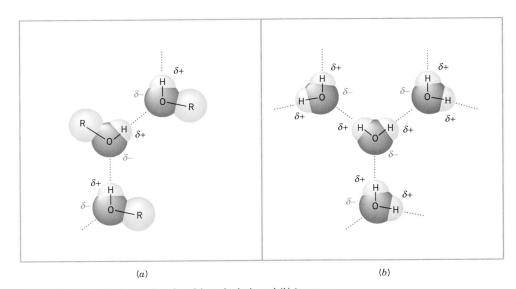

(a) (b)

FIGURE 13.1 Hydrogen bonding (*a*) in alcohols and (*b*) in water.

[1] Nothing is *totally* insoluble in water. The size of Avogadro's number (6.02×10^{23}) is so great that even if something can form a solution with a molarity of only, say, 1×10^{-14} mol/L—immeasurably small—there are still roughly a little over a billion (10^9) molecules of the solute in each liter of solution!

we put an OH group on methane to make methyl alcohol, however, water solubility increases hugely. The difference is that hydrogen bonds form between molecules of methyl alcohol and those of water. Such bonds enable alcohol molecules to slip into the hydrogen bonding network between water molecules (Figure 13.2). The CH_3 group in CH_3OH is too small to interfere.

As the size of the R group in a monohydric alcohol molecule increases, however, alcohol molecules become more and more alkane-like. A long R group interferes with the alcohol's ability to dissolve in water. In 1-decanol, for example,

$$CH_3CH_2CH_2CH_2CH_2CH_2CH_2CH_2CH_2CH_2OH$$

1-decanol

the small, waterlike OH group is overwhelmed by the long, flexing hydrocarbon chain, for which water molecules have no attraction. Water molecules will not move apart to let those of 1-decanol into solution. This alcohol and most with five or more carbons are thus insoluble in water. But they do dissolve in nonpolar solvents like diethyl ether and gasoline.

■ PRACTICE EXERCISE 4   1,2-Propanediol and 1-butanol have similar formula masses. In which of these is hydrogen bonding between molecules more extensive and stronger? How do the data in Table 13.1 support the answer? Which would be more soluble in water?

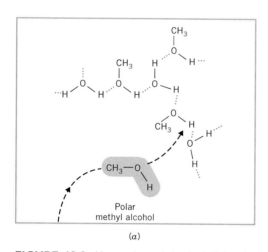

(a)

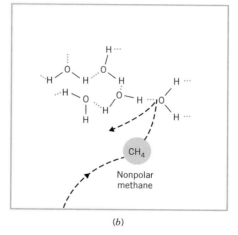

(b)

FIGURE 13.2 How a short-chain alcohol dissolves in water. (a) The alcohol molecule can take the place of a water molecule in the hydrogen-bonding network of water. (b) An alkane molecule cannot break into the hydrogen-bonding network in water, so the alkane cannot dissolve.

13.3 CHEMICAL PROPERTIES OF ALCOHOLS

The loss of water (dehydration) and the loss of hydrogen (oxidation) are two important reactions of alcohols.

Alcohols react with both inorganic and organic compounds, but we will study only the inorganic reactants in this chapter. Before continuing, however, we need to point out some properties that the alcohols do *not* have, but that may be thought possible because of their OH groups.

Despite the OH Group, Alcohols Are Not Strong Bases   Alcohols are monoalkyl derivatives of water, and, like water, they are extremely weak proton donors (acids) and proton acceptors (bases). Alcohols do not ionize to give either OH^- ions or H^+ ions. A solution of methyl alcohol (or any alcohol) in water thus has a neutral pH.

■ When an alcohol dissolves in water, it doesn't raise or lower the pH.

Alcohols Can Be Dehydrated to Alkenes   In laboratory vessels, the action of heat and a strong acid catalyst causes the dehydration of an alcohol to an alkene. A water molecule splits out and a carbon–carbon double bond emerges. The pieces of the water molecule, one H and one OH, come from *adjacent* carbons.

$$-\overset{|}{\underset{|}{C}}-\overset{|}{\underset{|}{C}}- \xrightarrow[\text{heat}]{\text{H}^+ \text{ catalyst}} \overset{\backslash}{\underset{/}{C}}=\overset{/}{\underset{\backslash}{C} } + \text{H---OH}$$
$$\quad\quad\underset{\text{H}\quad\text{OH}}{}$$

alcohol   alkene

Specific examples are as follows.

$$CH_3CH_2OH \xrightarrow[\text{heat, 170 °C}]{\text{H}^+ \text{ catalyst}} CH_2{=}CH_2 + H_2O$$

ethanol   ethene

$$\underset{\text{2-butanol}}{CH_3CH_2\overset{\overset{\displaystyle OH}{|}}{C}HCH_3} \xrightarrow[\text{heat}]{\text{H}^+ \text{ catalyst}} \underset{\substack{\text{2-butene}\\\text{(chief product)}}}{CH_3CH{=}CHCH_3} + \underset{\substack{\text{1-butene}\\\text{(minor product)}}}{CH_3CH_2CH{=}CH_2} + H_2O$$

How the acid catalyst works is explained in Appendix D.

When It Is Possible for Two Alkenes to Be Made, the More Highly Branched Alkene Forms   As seen in the last example above, water can split out in two ways from 2-butanol. 2-Butanol molecules have H atoms on *two* carbons adjacent to the carbon holding the departing OH group. Because either one of these H atoms could combine with the OH group to form H_2O, two alkenes are possible, 2-butene and 1-butene, and some of each forms. When such options exist, however, one alkene product predominates, generally *the more highly branched alkene*. This is the alkene with the greater number of alkyl groups attached to the double bond. 2-Butene is more branched than 1-butene because it has two alkyl groups at the double bond (two methyl groups) and 1-butene has only one (an ethyl group).

■ $CH_3{-}CH{=}CH{-}CH_3$

2-butene
(two alkyl groups)

$CH_2{=}CH{-}CH_2CH_3$

1-butene
(one alkyl group)

In acid-catalyzed dehydrations of alcohols, the more branched alkene predominates because it is the more stable, as combustion experiments have shown. For example, both 1-butene and 2-butene are C_4H_8, and the combustion of either occurs by the same equation.

$$C_4H_8(g) + 6O_2(g) \longrightarrow 4CO_2(g) + 4H_2O + \text{heat of combustion}$$

■ When some alcohols undergo dehydration, their carbon skeletons rearrange, but we won't study any examples of this.

However, slightly less heat per mole of C_4H_8 is released when 2-butene is burned than 1-butene. The only way by which 2-butene can release less energy than 1-butene is *to have less energy initially*. Having less energy always means being more stable, so 2-butene must be slightly more stable than 1-butene. (None of this explains *why* 2-butene is more stable. The explanation is complex.)

The complication of two possible products of alcohol dehydration is not a problem when the reaction is enzyme catalyzed. Enzymes direct reactions in specific ways and not always toward the more stable products.

■ Enzymes are exceedingly selective in what they do and how they control reactions.

EXAMPLE 13.2 Writing the Structure of the Alkene That Forms When an Alcohol Undergoes Dehydration

What is the product of the dehydration of isobutyl alcohol?

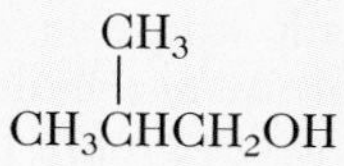

$$CH_3CHCH_2OH$$
$$|$$
$$CH_3$$

isobutyl alcohol

ANALYSIS To predict a product of dehydration, we rewrite the structure of the alcohol but leave off the OH group and one H from a carbon adjacent to the one holding the OH group. We then write a double bond between these two carbons.

SOLUTION

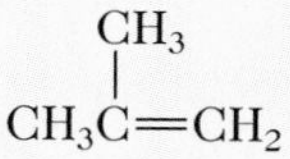

$$CH_3C=CH_2$$
$$|$$
$$CH_3$$

2-methylpropene

CHECK Ask the following questions of the answer. Is it an *alkene?* (Alcohol dehydration give alkenes.) Is its carbon skeleton the same as in the starting material? (Changes in the carbon skeleton do not occur in all of the examples that we will study.) Does each carbon have four bonds from it? (The rules of covalence must be obeyed.) By answering "yes" to these questions you have made a thorough check.

■ PRACTICE EXERCISE 5 Write the structures of the alkenes that can be made by the dehydration of the following alcohols.

(a) $CH_3CH_2CH_2OH$
(b) 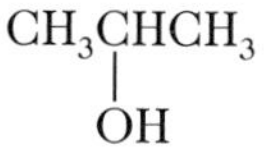 CH_3CHCH_3 with OH
(c) 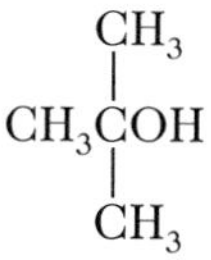 CH_3COH with CH_3 and CH_3
(d) 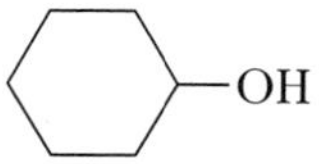 ⬡—OH

1° and 2° Alcohols Are Dehydrogenated by Strong Oxidizing Agents Organic chemists often use the following rules of thumb to tell whether an oxidation or a reduction has occurred. These rules are quicker and give the same results as are obtained by the use of oxidation numbers. With organic compounds,

1. An *oxidation* is the loss of 2 H or the gain of O by a molecule.
2. A *reduction* is the loss of O or the gain of 2 H by a molecule.

The oxidation of an alcohol is an example of the loss of 2 H, which is why the reaction is often called *dehydrogenation* rather than oxidation. Enzymes that catalyze such reactions are called *dehydrogenases*.

■ In the body, the enzyme *alcohol dehydrogenase* (ADH) oxidizes methyl alcohol (wood alcohol) to formaldehyde, $CH_2=O$, which has toxic effects and can cause blindness and death.

In studying dehydrogenation or oxidation, we are more interested in what happens to the organic molecule than in developing balanced equations. We only want to know what the oxidized organic molecule changes into. To serve our limited needs, we therefore simplify in two ways. We'll largely use unbalanced "reaction sequences," not balanced equations, and we'll use the symbol (O) *for any oxidizing agent that can bring about the oxidation given by a reaction sequence.* Two oxidizing agents commonly used in the lab are potassium permanganate, $KMnO_4$, and sodium dichromate, $Na_2Cr_2O_7$, which we introduced in the previous chapter.

■ The term *in vitro* means "in a glass vessel," that is, carried out in laboratory glassware. The term *in vivo* means in a living cell.

When an oxidizing agent causes an alcohol molecule to lose hydrogen, molecular hydrogen (H_2) does not itself form either in vitro or in vivo. Instead, the hydrogen atoms end up in a water molecule whose oxygen atom comes from the oxidizing agent. *One H comes from the OH group of the alcohol, and the other H comes from the carbon that has been holding the OH group.* Left behind in the organic molecule is a carbon–oxygen double bond, a carbonyl group. Study the following schematic carefully to learn where precisely the two H atoms originate.

$$-\overset{|}{\underset{H}{C}}-O\; + (O)\; \longrightarrow\; \overset{\diagdown}{\diagup}C=O + H-O\diagdown_H$$

$$\left[H\!:^- +\; H^-\right] \xrightarrow{\;(O)\;}$$

What makes this reaction an *oxidation* is specifically the *loss of the electron pair* of the C—H bond in the alcohol system. In the schematic above, we see this pair first on $H:^-$ and then incorporated into the water molecule. The $H:^-$ ion (the hydride ion) never becomes free in the aqueous medium, however. We show it in brackets above only to help you track the oxidation.

In the body, when an alcohol is oxidized by a series of in vivo reactions called the *respiratory chain,* the electron pair of the C—H bond is transferred from enzyme to enzyme until it finally does lodge in a water molecule, as indicated by our schematic.

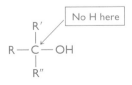

3° alcohol system

3° Alcohols Cannot Be Dehydrogenated   Only 1° and 2° alcohols can be oxidized by the loss of 2 H atoms. Molecules of 3° alcohols do not have an H atom on the carbon that holds the OH group, so 3° alcohols cannot be oxidized by dehydrogenation.

Strong Oxidizing Agents Change a 1° Alcohol Group to a Carboxyl Group   The organic product of the oxidation of either a 1° or a 2° alcohol has a carbon–oxygen double bond. Each subclass of alcohol, however, is oxidized to a different organic family. A 1° alcohol is oxidized first to an aldehyde, but because *aldehydes are more easily oxidized than 1° alcohols,* particularly in vitro, it is seldom practical to use $MnO_4^-(aq)$ or $Cr_2O_7^{2-}(aq)$ to prepare aldehydes. As soon as aldehyde groups appear in the presence of these ions, the aldehyde groups begin to use up oxidizing agent and change to carboxyl groups. In the presence of sufficient oxidizing agent, the successive stages in the oxidation of a 1° alcohol group are

$$RCH_2OH \xrightarrow[\substack{MnO_4^-(aq)\ or\\ Cr_2O_7^{2-}(aq)}]{(O)} \overset{O}{\overset{\|}{R}CH} \xrightarrow{more\ (O)} \overset{O}{\overset{\|}{R}COH}$$

1° alcohol   aldehyde   carboxylic acid

In the lab, therefore, when either aqueous permanganate or dichromate is used, the oxidation of a 1° alcohol is generally carried out with enough oxidizing agent to take the oxidation of the 1° alcohol all the way to its corresponding carboxylic acid. No carbon atoms are lost. Carboxylic acids strongly resist further oxidation. The dichromate oxidation of 1-propanol to propanoic acid, for example, occurs by the following equation.

$$3CH_3CH_2CH_2OH + 2Cr_2O_7^{2-} + 16H^+ \longrightarrow 3CH_3CH_2CO_2H + 4Cr^{3+} + 11H_2O$$

1-propanol   propanoic acid

Enzyme-Catalyzed Oxidations Convert a 1° Alcohol Group to an Aldehyde Group
The problem of halting the oxidation of a 1° alcohol at the aldehyde stage does not occur in body cells, because *different enzymes are required for each oxidation step.* One enzyme handles the oxidation of a 1° alcohol to the corresponding aldehyde. A different enzyme is needed to take an aldehyde group to the next oxidation stage, and this enzyme is generally not in the same place where the aldehyde is made. In general, remembering what (O) stands for, the reaction is

■ Certain vitamins in the diet, like riboflavin and niacin, provide molecular acceptor units for $H:^-$ in dehydrogenase enzymes.

$$RCH_2OH \xrightarrow[\text{(enzyme catalyzed)}]{\text{(O)}} \overset{\displaystyle O}{\overset{\|}{R C H}} + H_2O$$

1° alcohol   aldehyde

An oxidizing enzyme, a dehydrogenase, accepts $H:^-$ from the CH unit that holds the OH group, and a proton, H^+, slips away from the O atom of the OH group. With some enzymes, H^+ is simply neutralized by the buffer system of the cell fluid. Other enzymes accept both $H:^-$ and H^+. As we said, eventually $H:^-$ is changed into H_2O.

■ A dehydrogenase changes (temporarily) to its reduced form when it accepts $H:^-$.

Because our chief interest lies in what can occur in living systems, we will not be further concerned about in vitro oxidations using strong, aqueous oxidizing agents except as they may be used in lab tests. What interests us almost entirely is what specific aldehyde is made from a 1° alcohol in vivo and what carboxylic acid can eventually form from a 1° alcohol.

EXAMPLE 13.3   **Writing the Structure of the Product of the Oxidation of a Primary Alcohol**

What aldehyde and what carboxylic acid could be made by the oxidation of ethyl alcohol?

ANALYSIS The CH_2OH group is changed to $CH=O$ when a 1° alcohol is oxidized to an aldehyde. Anything attached to the carbon of the CH_2OH group, like the CH_3 group in our example, is retained in the final structure of the aldehyde. When the aldehyde is further oxidized to a carboxylic acid, the $CH=O$ group is changed to CO_2H.

SOLUTION The aldehyde corresponding to CH_3CH_2OH is $CH_3CH=O$, ethanal. The carboxylic acid is CH_3CO_2H, acetic acid.

CHECK Is the first product an *aldehyde?* (Does it have the $CH=O$ group?) Is the second product a *carboxylic acid?* (Does it have the CO_2H group, sometimes written COOH?) Are the carbon skeletons of the two products identical with that of the 1° alcohol? Does each carbon have four bonds and each oxygen two? By answering "yes" to these questions you have made a thorough check.

■   PRACTICE EXERCISE 6   Write the structures of the aldehydes and carboxylic acids that can be made by the oxidation of the following alcohols.

$$\overset{\displaystyle CH_3}{\underset{\displaystyle |}{}}$$

(a) CH_3CHCH_2OH (b) 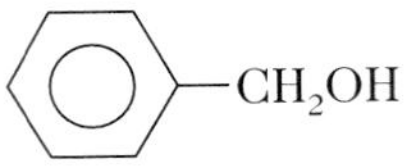   —CH_2OH

Secondary Alcohols Are Oxidized to Ketones   Ketones strongly resist further oxidation, so they are easily made by the oxidation of 2° alcohols using strong oxidizing agents like MnO_4^- or $Cr_2O_7^{2-}$. In vivo, dehydrogenases accomplish the identical overall reaction. In general, for 2° alcohols,

$$\overset{\displaystyle OH}{\underset{\displaystyle |}{RCHR'}} + (O) \longrightarrow \overset{\displaystyle O}{\overset{\displaystyle ||}{RCR'}} + H_2O$$

2° alcohol   ketone

Specific in vitro examples are:

$$\overset{\displaystyle OH}{\underset{\displaystyle |}{CH_3CHCH_2CH_3}} \xrightarrow{Cr_2O_7^{2-},\ H^+} \overset{\displaystyle O}{\overset{\displaystyle ||}{CH_3CCH_2CH_3}}$$

2-butanol   2-butanone

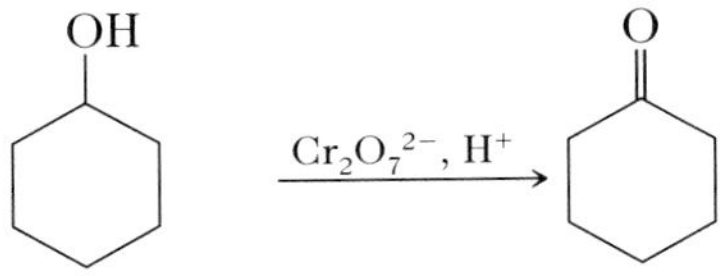

cyclohexanol   cyclohexanone

EXAMPLE 13.4   Writing the Structure of the Product of the Oxidation of a Secondary Alcohol

What ketone forms when 2-propanol is oxidized?

ANALYSIS   2-Propanol is a 2° alcohol, and every 2° alcohol has a CHOH group. The oxidation strips both H atoms from CHOH, creates a double bond between C and O, and so changes CHOH to C=O. (The two H atoms emerge in a molecule of water.) *The skeleton of all of the heavy atoms in the 2° alcohol, like C and O, remains intact.*

SOLUTION

■   The name chemists commonly use for propanone is *acetone*.

$$\overset{\displaystyle OH}{\underset{\displaystyle |}{CH_3CHCH_3}} \quad \text{is oxidized to} \quad \overset{\displaystyle O}{\overset{\displaystyle ||}{CH_3CCH_3}} \quad \text{plus } H_2O$$

2-propanol   propanone

CHECK   Is the product a *ketone?* (Secondary alcohols are oxidized to ketones.) Does the product have the same carbon skeleton as the starting material? Does each carbon have four bonds and each oxygen atom two?

■ PRACTICE EXERCISE 7   Write the structures of the ketones that can be made by the oxidation of the following alcohols.

(a) 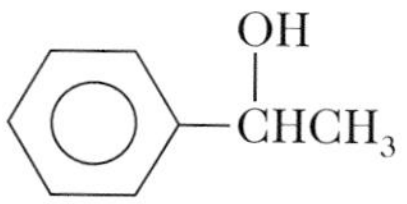   (b)

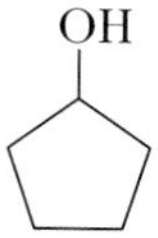

■ PRACTICE EXERCISE 8   What are the products of the oxidation of the following alcohols? If the alcohol is a 1° alcohol, show the structures of both the aldehyde and the carboxylic acid that could be made, depending on the conditions. If the alcohol cannot be oxidized, write "no reaction."

(a) 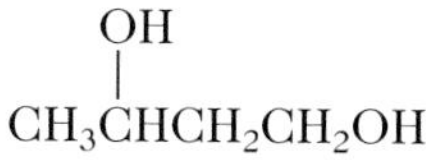   CH₃CHCH₂CH₂OH   (with OH above first CH)

(b) 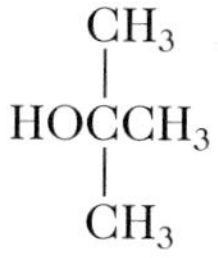   HOCCH₃ (with CH₃ above and CH₃ below)

(c) 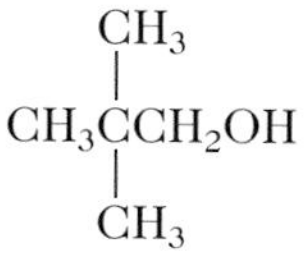   CH₃CCH₂OH (with CH₃ above and CH₃ below)

(d) 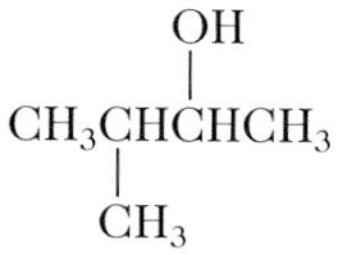   CH₃CHCHCH₃ (with OH above second C and CH₃ below third C)

The foregoing discussion of the chemical properties of alcohols leaves us with the following structural "map signs."

1.   Alcohols can be dehydrated to alkenes, with the most branched alkene generally forming in vitro.

2.   1° Alcohols are oxidized in vivo to aldehydes; in vitro to carboxylic acids.

3.   2° Alcohols are oxidized both in vivo and in vitro to ketones.

13.4 PHENOLS

Phenols are weak acids that can neutralize sodium hydroxide, and their benzene rings are easily oxidized.

For a compound to be classified as a *phenol,* its molecules must have at least one OH group directly attached to a benzene ring. The simplest member of this family also carries the name phenol, and it is a raw material for making aspirin. Phenols are widespread in nature and in commerce (see Interaction 13.3).

Phenols Are Weak Acids   In sharp contrast to alcohols, phenols are acidic, but they are *weak* acids, as the K_a values in the margin show. Phenol is itself a much weaker acid than acetic acid. Yet phenols in general are strong enough acids to neutralize OH⁻. For example,

■ For phenol, $K_a = 1.0 \times 10^{-10}$ (25 °C), which can be compared to $K_a = 1.8 \times 10^{-5}$ for acetic acid.

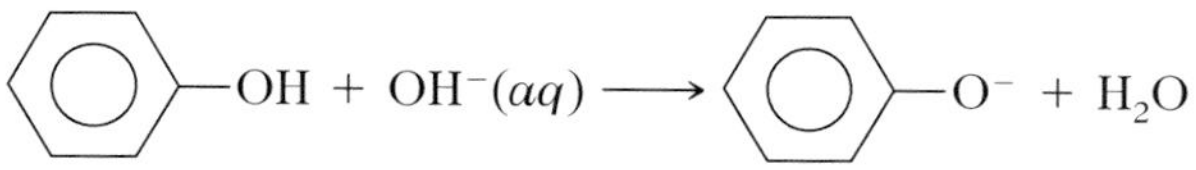

phenol   phenoxide ion

INTERACTION 13.3
PHENOLS IN OUR DAILY LIVES

Phenol (carbolic acid) Phenol, the first antiseptic to be used in surgery, kills bacteria by denaturing their proteins, that is, by destroying the abilities of their proteins to function normally. Phenol, however, is dangerous to healthy tissue, because it is a general protoplasmic poison, so other antiseptics have been developed.

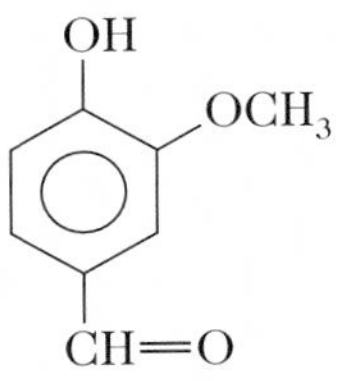

vanillin 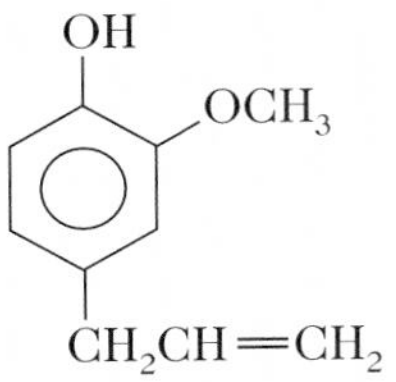   eugenol 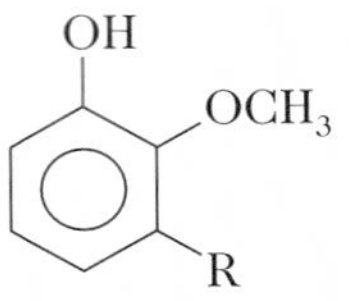   the urushiols

Vanillin The vanilla bean is the source of vanillin, a valued flavoring agent with three functional groups—phenol, ether, and aldehyde. Most vanillin is manufactured today from another phenol, eugenol.

Eugenol The odor of cloves is caused chiefly by eugenol, which is used in making perfumes and, as we have said, for the manufacture of vanillin.

Urushiols The active irritant in poison ivy or poison oak is a mixture of similar compounds called the urushiols. The R group on the ring is a straight 15-carbon chain, but different urushiols have from zero to three alkene groups in this chain.

BHA and BHT Butylated hydroxyanisol (BHA) and butylated hydroxytoluene (BHT) are widely used as antioxidants in gasoline, lubricating oils, rubber, edible fats and oils, and materials used for packaging foods that might turn rancid. Being phenols, and therefore easily oxidized themselves, they act by competing with the oxidations of the materials they are designed to protect.

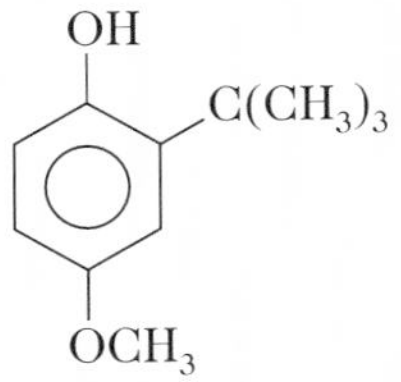

BHA (butylated hydroxyanisole) 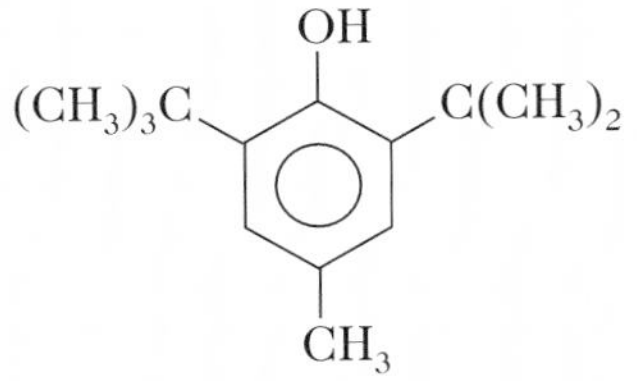   BHT (butylated hydroxytoluene)

Phenol but not an alcohol is able to donate H^+ to OH^- because the *anion* of phenol, $C_6H_5O^-$, is more stable than the anion from an alcohol, RO^-. The reason is that the extra electron responsible for the negative charge on $C_6H_5O^-$ is somewhat spread out into the same ring-encircling space occupied by six of the benzene ring electrons. *The surest way to stabilize negative charge is to spread it out* ("like charges repel"), but the negative charge on the RO^- ion cannot spread out. The alcohol's anion, RO^-, therefore, is less stable and it forms to a lesser extent from ROH than does the phenoxide ion from a phenol. Therefore, ROH is a poorer proton donor than a phenol.

The Ring in a Phenol Is Easily Oxidized In vivo, special enzymes are able to place an additional OH group on certain phenolic rings. Such a reaction, of course, is an oxidation because the ring gains oxygen. In vitro, oxidations with traditional oxidizing agents, like permanganate or dichromate, also take place very readily. But they give complex mixtures of highly colored products and so can seldom be usefully employed to make specific substances. Even phenol crystals left exposed to air slowly turn dark because the oxygen in the air attacks phenol.

Why is the ring in phenol so susceptible to oxidation? Remember that oxidizing agents are electron seekers. So the ease of oxidation of the benzene ring in phenol suggests that this ring is more electron-rich than it is in benzene. Advanced bonding theory indeed shows that some of the electron density associated with the two unshared electron pairs on the oxygen in phenol spreads out into the space occupied by six of the electrons of the benzene ring. The increased electron density at the ring, therefore, exposes it to oxidative attack.

Phenols Are Not Dehydrated   Unlike alcohols, phenols cannot be dehydrated. Dehydration would put a *triple* bond into a six-membered ring, and this ring is too small to accommodate the linear geometry at a triple bond.

13.5 ETHERS

The ethers are almost as chemically unreactive as the alkanes.

Ethers are compounds whose molecules have two organic groups joined to the same oxygen atom, R—O—R′ (Table 13.3). A carbon atom joined to the bridging oxygen atom cannot be the carbon of a carbonyl group, C=O. Thus, the first three examples below are ethers, but methyl acetate is an ester, not an ether, because the bridging oxygen is attached to a carbonyl carbon. We'll study esters in a later chapter.

$$CH_3CH_2—O—CH_2CH_3 \qquad CH_3—O—\bigcirc \qquad CH_3\overset{\overset{O}{\|}}{C}—O—CH_3$$

diethyl ether   methyl phenyl ether   methyl acetate
                                                                  (an *ester*, not
                                                                  an ether)

In the common names of ethers, the names of the groups attached to O are placed before the word *ether*.

Ethers are nearly as chemically stable as the alkanes. They do not react with strong acids, with bases, or with strong oxidizing or reducing agents. Of course, like all organic compounds, ethers burn. When diethyl ether was widely used as an anesthetic precautions against fires were stringent. Two groups of ethers are discussed in Interaction 13.4.

Ethers Cannot Donate Hydrogen Bonds   Because the ether group cannot donate hydrogen bonds, the boiling points of simple ethers are more like those of alkanes of comparable formula masses than those of the alcohols. The oxygen of the ether group can accept hydrogen bonds, however, so ethers are more soluble in water than alkanes. Both 1-butanol and its isomer, diethyl ether, for example, dissolve in water to the extent of about 8 g/100 mL, but pentane is insoluble in water.

Ethers Can Be Prepared from Alcohols   We learned earlier in this chapter that alcohols can be dehydrated by the action of heat and an acid catalyst to give *alkenes*. If a lower a temperature is used for this reaction, however, water splits out *between*

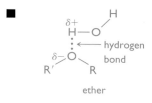

Compound	BP (°C)	Solubility in Water
Pentane	36	0.036 g/dL[a]
Diethyl ether	35	8.4 g/dL[a]
1-Butanol	118	11 g/dL[b]

[a] At 15 °C.
[b] At 25 °C.

TABLE 13.3   Some Ethers

Common Name	Structure	BP (°C)
Dimethyl ether	CH_3OCH_3	−23
Methyl ethyl ether	$CH_3OCH_2CH_3$	11
Methyl *t*-butyl ether	$CH_3OC(CH_3)_3$	55.2
Diethyl ether	$CH_3CH_2OCH_2CH_3$	34.5
Dipropyl ether	$CH_3CH_2CH_2OCH_2CH_2CH_3$	91
Methyl phenyl ether	$CH_3OC_6H_5$	155
Diphenyl ether	$C_6H_5OC_6H_5$	259
Divinyl ether	$CH_2{=}CHOCH{=}CH_2$	29

INTERACTION 13.4
ETHERS IN COMMERCE AND THE ENVIRONMENT

Gasoline Additives or Alternative Fuels Both methyl *t*-butyl ether (MTBE) and ethyl *t*-butyl ether (ETBE), as well as ethanol, are being mixed with gasoline for sale in a number of smog-ridden urban areas. When the gasoline already contains some oxygen (in the form of these "oxygenates") the combustion is more complete and so generates less carbon monoxide and less partly oxidized hydrocarbons. The latter contribute to a more rapid buildup of the ozone level in smog, as we discussed in Interaction 12.3.

Dioxins *Dioxin* is the name of a family of compounds now produced mostly by complex reactions during industrial processes and the incineration of materials that contain chlorine compounds, both inorganic and organic (see Figure 1). 2,3,7,8-TCDD is the flagship molecule in the dioxin toxic waste fleet.

TCDD
(2,3,7,8-tetrachlorodibenzo-*p*-dioxin)

FIGURE I Traces of dioxins sometimes form from other organic chlorine compounds in the high temperature regions of municipal incinerators, like the one pictured here.

The 2,3,7,8-TCDD molecule has two ether groups that bridge two parallel benzene rings, each holding two chlorine atoms. There are actually 75 chlorine-containing dioxins of widely varying toxicity, some having up to eight Cl atoms per molecule, but 2,3,7,8-TCDD is regarded as the most dangerous. None of the dioxins has any known value to humans and they offer nothing but risks.

In commercial incinerators, nearly all of the dioxins that form are caught in the solids, called the fly ash, which are produced but then trapped and put into landfills. The U.S. Environmental Protection Agency (EPA) estimated in 1994 that, country-wide, 600 to 700 kg of dioxins enter the environment by incineration. When EPA scientists evaluated this according to the varying toxicity of the dioxins, however, the amount was deemed equivalent to a little under 15 kg of 2,3,7,8-TCDD. Spread over a nation, this may seem trivial, but the health hazards are potentially so great that the EPA regards even this amount as unacceptable.

In the air, the dioxins are rapidly degraded by sunlight when moisture is present, being destroyed in days. 2,3,7,8-TCDD is only slowly broken down in soil, however, and when it enters the ground water it is recycled. TCDD enters the food chain when taken up by plants and animals, so inhalation is not the only way humans are exposed.

In male guinea pigs, 2,3,7,8-TCDD is extremely toxic, more so than strychnine, the nerve gases, and cyanide. In male or female hamsters, however, the lethal dose is nearly 2000 times greater than for male guinea pigs, demonstrating that toxicity can be very species sensitive. In humans, 2,3,7,8-TCDD appears to be even less toxic, but it does cause a form of acne (chloracne) in humans as well as digestive distress, pains in the joints, and psychiatric effects.

2,3,7,8-TCDD is teratogenic (causer of birth defects) and carcinogenic (causer of cancer) in experimental animals. It is probably carcinogenic in humans, but how much so is not known yet. Cancer, however, is not the only threat. The EPA report said that noncancer effects of dioxins in humans likely include disruptions of the reproductive and immune systems as well as our mechanisms for handling hormones. Traces of dioxin may affect a developing fetus. Extrapolating from experimental animals to humans is an uncertain enterprise, so, not surprisingly, the EPA report launched an instant controversy among scientists.

two alcohol molecules rather than from within one molecule. The product then is an ether, not an alkene. In general,

$$R—O—H + H—O—R \xrightarrow{\text{acid catalyst}} R—O—R + H_2O$$

two alcohol molecules   ether

A specific example is

$$2CH_3CH_2OH \xrightarrow[140\ °C]{H_2SO_4} CH_3CH_2—O—CH_2CH_3 + H_2O$$

ethyl alcohol   diethyl ether

Our interest in this is that a reaction very much like making an ether from two molecules bearing OH groups is important to the study of carbohydrates.

■ Earlier we learned that concentrated H_2SO_4 acts on ethanol to give ethene when the temperature is higher (170 °C).

EXAMPLE 13.5   **Writing the Structure of an Ether That Can Form from an Alcohol**

If the conditions are right, 1-butanol can be converted to an ether. What is the structure of this ether?

ANALYSIS What this question asks is to complete the following equation.

$$CH_3CH_2CH_2CH_2OH \xrightarrow[\text{heat}]{\text{acid catalyst}} \text{what ether?}$$

As usual, the structure of the starting material gives us most of the answer. We know that the ether must get its organic groups from the alcohol, so we start an ether molecule by writing one O atom with two bonds from it.

$$—O—$$

Then, taking alkyl groups from the alcohol, we attach two, one to each bond.

SOLUTION The structure of the ether is

$$CH_3CH_2CH_2CH_2—O—CH_2CH_2CH_2CH_3$$

Remembering that water is the other product, the balanced equation is

$$2CH_3CH_2CH_2CH_2OH \xrightarrow[\text{heat}]{\text{acid catalyst}}$$

1-butanol

$$CH_3CH_2CH_2CH_2—O—CH_2CH_2CH_2CH_3 + H_2O$$

dibutyl ether

We need *two* molecules of the alcohol to make one molecule of the ether, but always remember that *we balance an equation after we have written the correct formulas for reactants and products.*

CHECK Is the product truly an *ether*? (Is it of the form R—O—R?) Are the R groups identical to the R group of the parent alcohol? (In this example are they *butyl* groups?) Does each carbon atom have four bonds and each oxygen two?

■   PRACTICE EXERCISE 9   Write the structures of the ethers to which the following alcohols can be converted.

(a) CH_3OH (b) $CH_3CH_2CH_2OH$ (c) 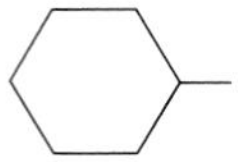   ⬡—OH

13.6   THIOALCOHOLS AND DISULFIDES

Both the SH group and the S—S system are important in proteins.

■   *Mercaptan* is a contraction of *mercury capturer.* Compounds with SH groups form precipitates with mercury ions.

Alcohols, R—O—H, can be viewed as alkyl derivatives of water, H—O—H. Similar derivatives of hydrogen sulfide, H—S—H, are also known and are in the family called the **thioalcohols** or the **mercaptans** (Table 13.4).

$$R—S—H \qquad R—S—R' \qquad R—S—S—R'$$

thioalcohols   thioethers   disulfides
(mercaptans)

Dialkyl derivatives of water, the ethers (R—O—R′), have their sulfur counterparts, too, the *thioethers.* (You can see that the prefix *thio-* indicates the replacement of an oxygen atom by a sulfur atom.)

■   Lower formula mass thioalcohols are responsible for the considerable respect usually given to skunks.

The SH group is variously called the *thiol group,* the *mercaptan group,* or the *sulfhydryl group.* The IUPAC nomenclature of thioalcohols is easily inferred from the names in Table 13.4. One of the building blocks of proteins, an amino acid called cysteine (Table 13.4), bears the SH group, so some very important properties of proteins depend on this group.

Thioalcohols Are Oxidized to Disulfides   Only one reaction of thioalcohols is important in our study of proteins, namely oxidation. Thioalcohols can be oxidized to **disulfides,** compounds whose molecules have two sulfur atoms joined by a covalent bond, R—S—S—R′. In general,

$$R—S—H + H—S—R + (O) \longrightarrow R—S—S—R + H_2O$$

two molecules of   one molecule
a thioalcohol   of a disulfide

A specific example is

$$2CH_3SH + (O) \longrightarrow CH_3—S—S—CH_3 + H_2O$$

methanethiol   dimethyl disulfide

TABLE 13.4   Some Thioalcohols

Name	Structure	BP (°C)
Methanethiol	CH_3SH	6
Ethanethiol	CH_3CH_2SH	36
1-Propanethiol	$CH_3CH_2CH_2SH$	68
1-Butanethiol	$CH_3CH_2CH_2CH_2SH$	98
Cysteine	$^+NH_3CHCO_2^-$	(Solid)
(a monomer for proteins)	$\mid$	
	CH_2SH	

EXAMPLE 13.6 Writing the Product of the Oxidation of a Thioalcohol

What is the product of the oxidation of ethanethiol?

ANALYSIS Because the oxidation of the SH group generates the —S—S— group, begin simply by writing this group down.

$$-S-S-$$

Then we attach two alkyl groups (from the thioalcohol), one on each S atom.

SOLUTION Ethanethiol furnishes ethyl groups, so attach one of each of these groups to the S atoms.

$$CH_3CH_2-S-S-CH_2CH_3$$

diethyl disulfide

If the problem had called for an equation, we would have had to use the coefficient of 2 for the ethanethiol.

$$2CH_3CH_2SH + (O) \longrightarrow CH_3CH_2-S-S-CH_2CH_3 + H_2O$$

However, always remember that balancing an equation comes *after* you have written the correct formulas for the reactants and products. And the problem called only for the product.

CHECK Is the product a *disulfide*? (Thiols are oxidized to disulfides.) Are the groups attached to the S atoms the same as present in the reactant? Does each carbon have four bonds and each sulfur two?

Disulfides Are Reduced to Thioalcohols The sulfur–sulfur bond occurs widely among proteins, and what we need to know is that it is very easily reduced. The products are two separate SH groups. Disulfides are reduced to thiols. Let's look at this using a simple system. We'll use (H) to represent any reducing agent that can do the task, just as we used (O) for an oxidizing agent. In general,

$$R-S-S-R + 2(H) \longrightarrow R-S-H + H-S-R$$

one molecule   two molecules of
of disulfide   a thioalcohol

A specific example is

$$CH_3CH_2-S-S-CH_2CH_2CH_3 + 2(H) \longrightarrow CH_3CH_2-S-H + H-S-CH_2CH_2CH_3$$

■ PRACTICE EXERCISE 10 Complete the following equations by writing the structures of the products that form. If no reaction occurs, write "no reaction."

(a) $CH_3SSCH_3 + (H) \longrightarrow$?   (b) $CH_3CHCH_3 + (O) \longrightarrow$?
                                                 |
                                                 SH

(c)  + (H) ⟶ ?   (d) 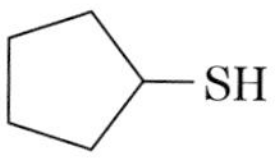 —SH + (O) ⟶ ?

SUMMARY

Alcohols The alcohol system has an OH group attached to a saturated carbon. The IUPAC names of simple alcohols end in *-ol,* and their chains are numbered to give precedence to the location of the OH group. The common names have the word *alcohol* following the name of the alkyl group.

Alcohol molecules hydrogen bond to each other and to water molecules. By the action of heat and an acid catalyst alcohols can be dehydrated internally to give carbon–carbon double bonds or externally to give ethers. Primary alcohols can be oxidized to aldehydes and ultimately to carboxylic acids. Secondary alcohols can be oxidized to ketones. Tertiary alcohols cannot be oxidized (without breaking up the carbon chain). The OH group in alcohols does not function well as either an acid or a base.

Phenols When the OH group is attached to a benzene ring, the system is the phenol system, and it is now acidic enough to neutralize strong bases. In addition, the ring is vulnerable to oxidizing agents.

Ethers The ether system, R—O—R′, does not react at room temperature or body temperature with strong acids, bases, oxidizing agents, or reducing agents. It can accept hydrogen bonds but cannot donate them.

Thioalcohols The thioalcohols or mercaptans, R—S—H, are easily oxidized to disulfides, R—S—S—R. Disulfides are reduced to the original thioalcohols.

Reactions studied Without attempting to present balanced equations, or even all of the inorganic products, we can summarize the reactions studied in this chapter as follows.

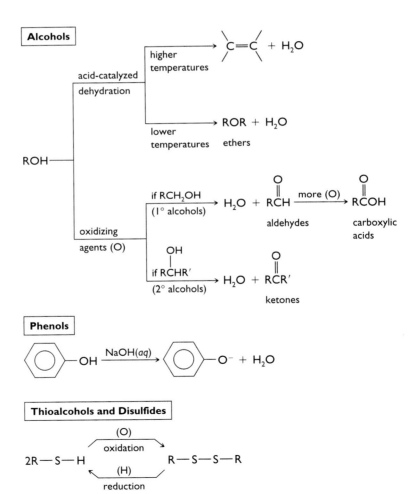

REVIEW EXERCISES

The answers to Review Exercises whose numbers are in color are found in Appendix E. The answes to the other Review Exercises are found in the Study Guide that accompanies this book. The more challenging questions are marked with asterisks.

Functional Groups

13.1 Name the functional groups identified by the numbers in the structure of estradiol, a human estrogenic hormone. If a group is an alcohol, state whether it is a 1°, 2°, or 3° alcohol. Distinguish between an alcohol group and a phenol.

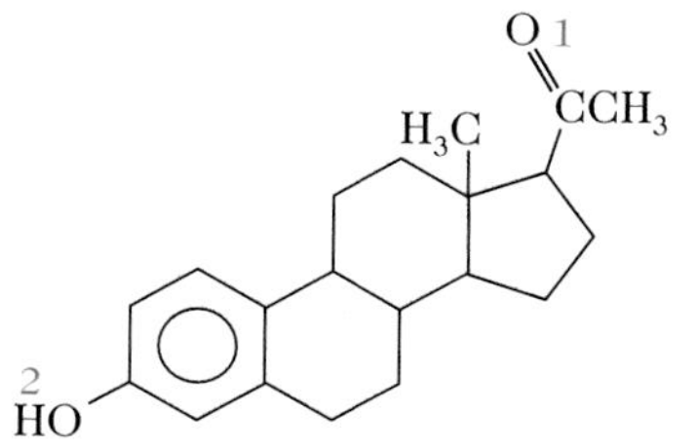

estradiol

13.2 Give the names of the functional groups identified by the numbers in the structure of cortisol, one of 28 hormones secreted by the cortex of the adrenal gland. If a group is an alcohol, state whether it is a 1°, 2°, or 3° alcohol. Distinguish between an alcohol group and a phenol.

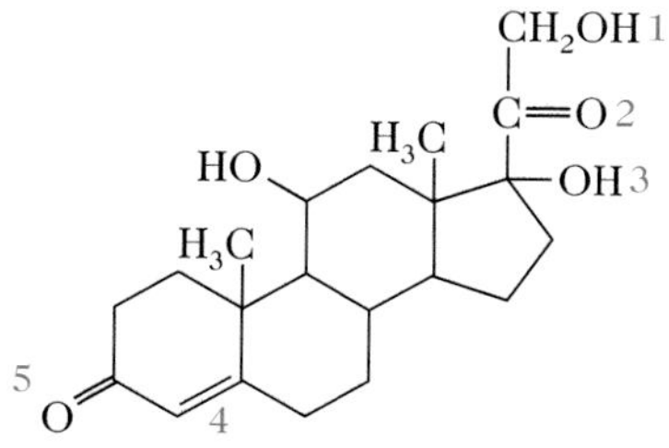

cortisol

Structures and Names

13.3 Write the structure of each compound.
(a) Ethyl alcohol   (b) Methyl alcohol
(c) Butyl alcohol   (d) Propyl alcohol

13.4 What is the structure of each of the following compounds?
(a) Isobutyl alcohol   (b) Isopropyl alcohol
(c) Glycerol   (d) t-Butyl alcohol

13.5 What are the common names of the following compounds?

(a) HOCH with CH₃ groups (b) HOCH₂—CH₃

(c) CH₃CHCH₂CH₃ with OH (d) CH₃CCH₃ with OH and CH₃

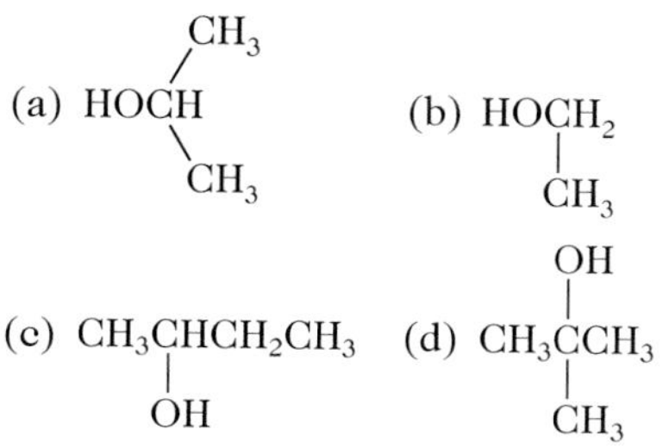

13.6 What is the structure and the IUPAC name of the simplest, *stable* dihydric alcohol?

13.7 Give the structure and the IUPAC name of the simplest, *stable* trihydric alcohol.

13.8 Give the IUPAC names for the compounds listed in Review Exercise 13.3.

13.9 Give the IUPAC names for the compounds listed in Review Exercise 13.4.

13.10 Give the IUPAC name for the following compound.

$$CH_3CH_2CHCH_2CH_2CH_3 \text{ with } CH_2CH_2OH$$

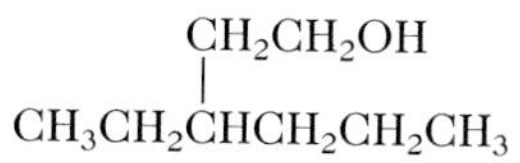

13.11 Write the IUPAC name of the following compound.

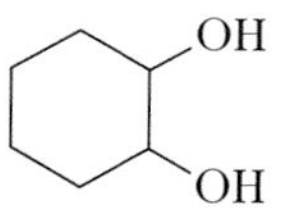

13.12 When ethyl alcohol dissolves in water, its molecules slip into the hydrogen-bonding network in water. Draw a figure that illustrates this. Use dotted lines for hydrogen bonds, and place the δ+ and δ− symbols where they belong.

*****13.13** Arrange the following compounds in their order of increasing boiling points. Place the letter symbol of the compound that has the lowest boiling point on the left end of the series, and arrange the remaining letters in the correct order.

$CH_3CH_2CH_2OH$ CH_3CH_3 $HOCH_2CH_2OH$ $CH_3CH_2OCH_3$
A   B   C   D
< < <
Lowest bp   Highest bp

Chemical Properties of Alcohols

13.14 Write the structures of the alkenes that form when the following alcohols undergo acid-catalyzed dehydration. Where more than one alkene is possible, identify which one most likely forms in the greatest relative amount.

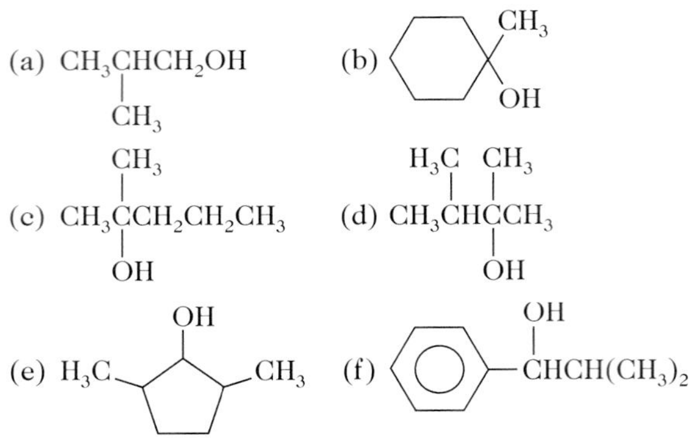

(a) CH_3CHCH_2OH
       $|$
       CH_3

(b) cyclohexane with CH₃ and OH

(c) $CH_3CCH_2CH_2CH_3$
       $|$
       OH
   (with CH₃ above)

(d) $CH_3CHCHCH_3$
   (H_3C CH_3 above, OH below)

(e) H_3C —⬠— CH_3 (cyclopentane with OH)

(f) —$CHCH(CH_3)_2$ (with OH above)

13.15 Write the structures of the products of the oxidation of the alcohols given in Review Exercise 13.14. If the alcohol is a 1° alcohol, give the structures of both the aldehyde and the carboxylic acid that could be made by varying the quantities of the oxidizing agent.

*13.16 Write the structures of any alcohols that could be dehydrated to give each of the following alkenes. In some instances, more than one alcohol would work.

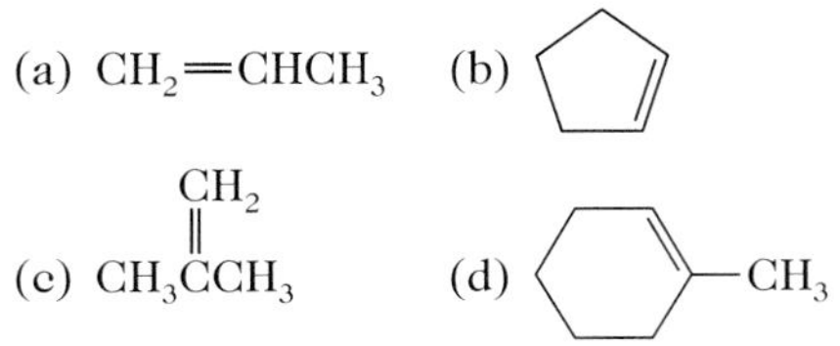

(a) $CH_2{=}CHCH_3$ (b) ⬠ (cyclopentene)

(c) CH_3CCH_3
   (with CH_2 double bond above)

(d) ⬡—CH_3 (methylcyclohexene)

*13.17 Write the structure of any alcohol that could be used to prepare each of the following compounds by an oxidation.

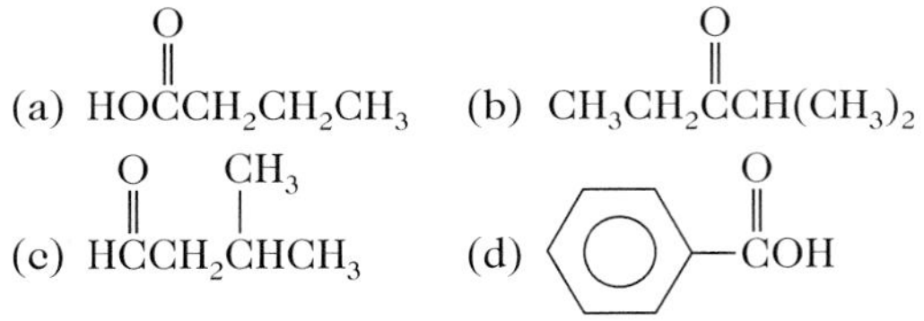

(a) $HOCCH_2CH_2CH_3$ (with O double bond)

(b) $CH_3CH_2CCH(CH_3)_2$ (with O double bond)

(c) $HCCH_2CHCH_3$ (with O double bond and CH_3)

(d) ⬡—COH (with O double bond)

Phenols

13.18 Write the structures of the three isomeric monochlorophenols and give their names (using the o-, m-, and p-designations).

*13.19 What is one difference in *chemical* properties between **A** and **B**?

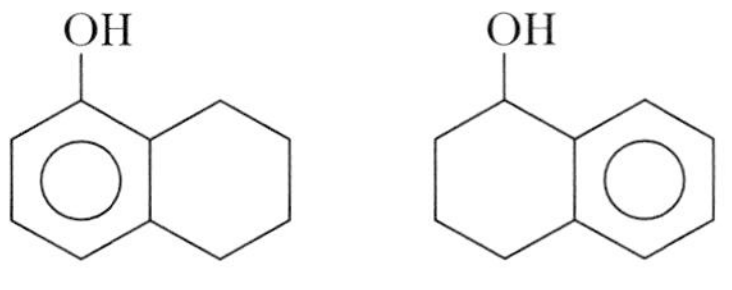

A   B

13.20 A compound was either **A** or **B**.

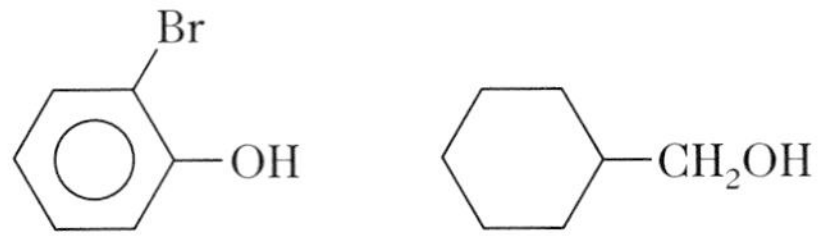

A   B

The compound dissolved in aqueous sodium hydroxide but not in water. Which compound was it? How can you tell? (Write an equation.)

Ethers

13.21 Write the structures of the ethers that can be made from the following alcohols.

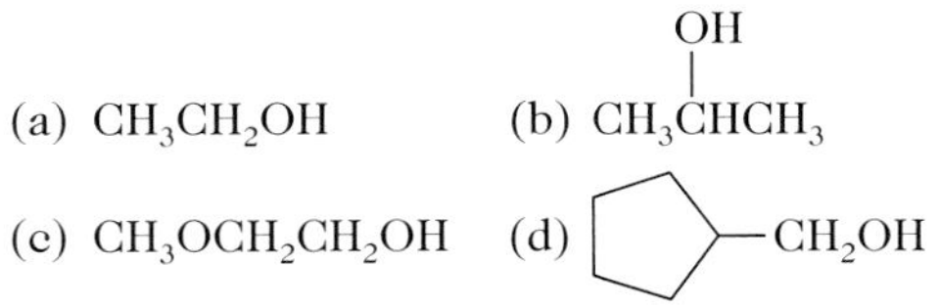

(a) CH_3CH_2OH

(b) CH_3CHCH_3 (with OH above)

(c) $CH_3OCH_2CH_2OH$

(d) ⬠—CH_2OH

*13.22 Write the structures of the alcohols that could serve as the starting materials to prepare each of the following ethers.

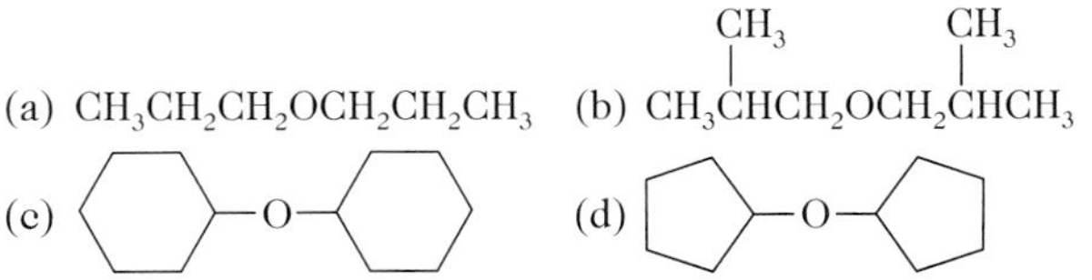

(a) $CH_3CH_2CH_2OCH_2CH_2CH_3$

(b) $CH_3CHCH_2OCH_2CHCH_3$ (with CH_3 above each CH)

(c) ⬡—O—⬡

(d) ⬠—O—⬠

*13.23 Suppose that a mixture of 0.50 mol of ethanol, 0.50 mol of methanol and a catalytic amount of sulfuric acid is heated under conditions that favor only ether formation. What organic products will be obtained? Write their structures.

13.24 What happens chemically when the following compound is heated with aqueous sodium hydroxide?

$$CH_3CH_2CH_2OCH_2CH_2CH_3$$

Thioalcohols and Disulfides

13.25 We did not study rules for naming thioalcohols, but the *patterns* of the names given in this chapter make these rules obvious. Write the structures of the following compounds.
(a) Diethyl disulfide   (b) 1,2-Propanedithiol
(c) Isopropyl mercaptan   (d) 1-Propanethiol

13.26 Complete the following reaction sequences by writing the structures of the organic products that form.
(a) $CH_3CH_2CH_2SH + (O) \longrightarrow$
(b) $(CH_3)_2CHCH_2SSCH_2CH(CH_3)_2 + (H) \longrightarrow$

(c)  ⬠ (with S—S) $+ (H) \longrightarrow$

(d) $HSCH(CH_3)_2 + (O) \longrightarrow$

13.27 Ethanol, methanethiol, and propane have nearly the same formula masses, but ethanol boils at 78 °C, methanethiol at 6 °C, and propane at −42 °C. What do the boiling points suggest about the possibility of hydrogen bonding in the thioalcohol family? Does hydrogen bonding occur at all? Are the hydrogen bonds as strong as those in the alcohol family?

Ethyl Alcohol and Alcoholism (Interaction 13.1)

13.28 What ranges of percentages of ethyl alcohol occur in beer, wine, and "spirits"?

13.29 A vodka rated as 110 proof has what percentage of ethyl alcohol?

13.30 At what level of blood alcohol do the depressant effects become noticeable in most nonalcoholic drinkers?

13.31 Most states define intoxication as corresponding to what level of blood alcohol?

13.32 Describe in detail the connection between the concentration of ethyl alcohol in the blood and what happens (chemically and physically) in the breathalyzer test.

13.33 If the "standard," nonalcoholic adult male consumes two "typical drinks" over a 2-hour period, what will his blood alcohol level most likely be 3 hours from the start of the drinking?

13.34 Name some of the reasons that alcoholism contributes so much to a nation's health bill.

Alcohols in Our Daily Lives (Interaction 13.2)

13.35 Give the name of the alcohol used in each of the following ways.
(a) As the alcohol in beverages
(b) As a rubbing alcohol (name two)
(c) As a moisturizer in some food products
(d) As a fuel in "canned heat"
(e) As a permanent antifreeze (name two)
(f) To manufacture a vasodilator

13.36 Give the name of the alcohol made by
(a) The digestion of fats or oils in the diet
(b) The fermentation of sugars
(c) The hydrogenation of carbon monoxide

Phenols in Our Daily Lives (Interaction 13.3)

13.37 What is the name of a member of the phenol family that is used in or involved in each of the following ways?
(a) A flavoring agent
(b) A antiseptic used once used in surgery
(c) A substitute for cloves
(d) An irritant in poison ivy

13.38 Both BHA and BHT interfere with the air oxidation of food materials. What chemical property do these food additives have that accounts for this?

13.39 Phenol is no longer used as an antiseptic. Why?

Ethers In Medicine (Interaction 13.4)

13.40 TCDD is a member of what family of pollutants?

13.41 What is currently an entry for trace quantities of TCDD into the environment?

13.42 Contrast the fates of TCDD in air and in the ground.

Additional Exercises

13.43 Give the common names of the following compounds.
(a) CH_3CH_2OH (b) $HOCH_2CH_2OH$

(c) $HOCH_2\overset{\overset{\displaystyle OH}{|}}{C}HCH_2OH$ (d) $CH_3\overset{\overset{\displaystyle OH}{|}}{C}HCH_3$

13.44 Give the IUPAC names for the compounds listed in Review Exercise 13.43.

*__13.45__ Write the IUPAC name of the following compound.

13.46 Draw a figure that illustrates a hydrogen bond between two molecules of ethyl alcohol. Use a dotted line to represent this bond, and write in the $\delta+$ and the $\delta-$ symbols where they belong.

*__13.47__ Arrange the following compounds in their order of increasing solubility in water. Place the letter symbol of the compound that has the least solubility on the left end of the series, and arrange the remaining letters in the correct order.

$CH_3CH_2CH_2\overset{\overset{\displaystyle }{|}}{C}HCH_2CH_3$
$\overset{}{|}$
OH

A

$HOCH_2CH_2\overset{\overset{\displaystyle }{|}}{C}HCH_2OH$
$\overset{}{|}$
OH

B

$CH_3CH_2CH_2CH_2CH_3$

C

$$\underset{\substack{\text{lowest} \\ \text{solubility}}}{} < \underset{\substack{\text{highest} \\ \text{solubility}}}{} < \underline{\hspace{3cm}}$$

13.48 An unknown alcohol is either **A** or **B**.

When the unknown is shaken with a few drops of aqueous potassium permanganate, the purple color disappears and a brown precipitate forms. Which alcohol, **A** or **B**, is it? Explain how the given information led you to your conclusion.

13.49   An unknown liquid is either **A** or **B**.

When it was shaken with aqueous $Na_2Cr_2O_7$ in acid, the orange color of the mixture did not change. What is the unknown? Explain. What would have been *seen* if the other compound had been the liquid?

13.50   Which compound, **A** or **B**, more readily begins to discolor upon being left exposed to the air? Explain.

13.51   Which of the compounds, **A** or **B**, of Review Exercise 13.50 is more readily dehydrated by the action of aqueous sulfuric acid? Explain.

13.52   Which of the following compounds contain the ether function?
(a) $CH_3CH_2OCH_2CH_2CH_2OCH_3$

(b) $CH_3\overset{\overset{\displaystyle O}{\|}}{C}OCH_3$

(c) $CH_3\overset{\overset{\displaystyle O}{\|}}{OC}OCH_3$

(d) —OCH_3

(e) $CH_3CH_2OOCH_2CH_3$

(f)

(g) $CH_3OCH_2CH_2\overset{\overset{\displaystyle O}{\|}}{C}CH_3$

*__13.53__   Examine each of the following sets of reactants and conditions and decide whether a reaction occurs. If one does, write the structures of the organic products. If no reaction occurs, write "no reaction."

Some of the parts involve *alcohols* and their reactions. If the reaction is an oxidation of a 1° alcohol, write the structure of the *aldehyde* that can form, not the carboxylic acid.

When an alcohol is in the presence of an acid catalyst, we have learned that the alcohol might be dehydrated to an *alkene* or to an *ether*. To differentiate between these, use the following guide. When the alcohol structure has no coefficient, then write the structure of the alkene that can form. When the alcohol structure has a coefficient of 2, then give the structure of the ether that is possible. (This violates our rule that balancing an equation is the *last* step in writing an equation, but we need a signal here to tell what kind of reaction is intended.)

(a) —OH $\xrightarrow[\text{heat}]{H_2SO_4}$

(b) $2HOCH_2CH_3$ $\xrightarrow[\text{heat}]{H_2SO_4}$

(c) $CH_3\overset{\overset{\displaystyle OH}{|}}{CH}CH_2CH_3 + (O) \longrightarrow$

(d) —$CH_3 + H_2$ $\xrightarrow{\text{Ni catalyst}}$

(e) $CH_3\overset{\overset{\displaystyle H_3C}{|}}{CH}\overset{\overset{\displaystyle CH_3}{|}}{\underset{\underset{\displaystyle OH}{|}}{C}}CH_3 + (O) \longrightarrow$

(f) $CH_3\overset{\overset{\displaystyle }{}}{\underset{\underset{\displaystyle OH}{|}}{CH}}CH_3 + NaOH(aq) \longrightarrow$

(g) $CH_3CH_2CH{=}CHCH_2CH_3 + H_2O$ $\xrightarrow[\text{heat}]{H_2SO_4}$

(h) $\xrightarrow{H_2SO_4}$

(i) $CH_3CH{=}CH$— $+ H_2$ $\xrightarrow{\text{Ni catalyst}}$

(j) —$\overset{\overset{\displaystyle OH}{|}}{CH}CH_2CH_3 + (O) \longrightarrow$

*__13.54__   Write the structure of the principal organic product that would be expected in the following situations. Follow the directions given for Review Exercise 13.53. If no reaction occurs, write "no reaction."
(a) $CH_3\overset{\overset{\displaystyle }{}}{\underset{\underset{\displaystyle OH}{|}}{CH}}CH_3 + (O) \longrightarrow$

(b) $CH_3CH{=}\overset{\overset{\displaystyle CH_3}{|}}{C}CH_3 + HCl(g) \longrightarrow$

(c) H_3C— $+ (O) \longrightarrow$

(d) $2CH_3CHOH \xrightarrow[\text{heat}]{H_2SO_4}$
  $\overset{|}{CH_3}$

(e) $HOCH_2CH_2\overset{\overset{\displaystyle CH_3}{|}}{C}HCH_3 + (O) \longrightarrow$

(f) 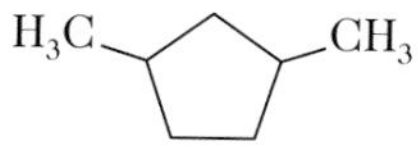 $H_3C\underset{}{\diagdown}\hspace{1.5cm}CH_3 + H_2O \xrightarrow[\text{heat}]{H_2SO_4}$

(g) $CH_3\overset{\overset{\displaystyle CH_3}{|}}{\underset{\underset{\displaystyle CH_3}{|}}{C}}OH \xrightarrow[\text{heat}]{H_2SO_4}$

(h) 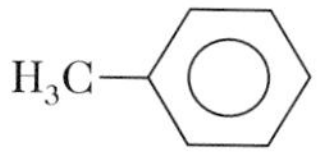 $H_3C\!-\!\!\langle\bigcirc\rangle\!+ NaOH(aq) \longrightarrow$

(i) 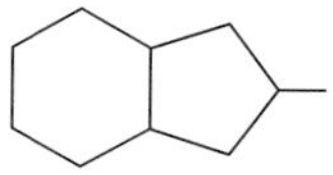 $\langle\text{bicyclic}\rangle\!-\!OH + (O) \longrightarrow$

(j) $CH_3\overset{\overset{\displaystyle CH_2OH}{|}}{C}HCH_2CH_3 + (O) \longrightarrow$

*13.55 2-Propanol (C_3H_8O) can be oxidized to acetone (C_3H_6O) by potassium permanganate according to the following equation.

$$3C_3H_8O + 2KMnO_4 \rightarrow 3C_3H_6O + 2MnO_2 + 2KOH + 2H_2O$$

(a) How many moles of acetone can be prepared from 3.00 mol of 2-propanol?
(b) How many moles of potassium permanganate are needed to oxidize 0.114 mol of 2-propanol?
(c) A student began with 15.6 g of 2-propanol. What is the minimum number of grams of potassium permanganate needed for this oxidation?
(d) Referring to part (c), what is the maximum number of grams of acetone that could be made? How many grams of MnO_2 are also produced?

*13.56 1-Propanol, C_3H_8O, can be oxidized to propanoic acid, $C_3H_6O_2$, according to the following net ionic equation.

$$3C_3H_8O + 2Cr_2O_7^{2-} + 16H^+ \rightarrow 3C_3H_6O_2 + 4Cr^{3+} + 11H_2O$$

(a) Transform the net ionic equation into a *molecular equation* assuming that the sodium salt of the dichromate ion is used and that the aqueous acid is HCl(aq).
(b) For each mole of 1-propanol used, how many moles (in theory) of sodium dichromate are needed according to the molecular equation?
(c) The stockroom carries sodium dichromate only as its dihydrate. Write the formula of the dihydrate. How many moles of the dihydrate are needed for each mole of 1-propanol used, according to the equation? Write the molecular equation for the oxidation when sodium dichromate dihydrate is used.
(d) What is the maximum number of grams of propanoic acid that can be obtained if the reaction is carried out starting with 10.6 g of 1-propanol?
(e) What is the minimum number of grams of sodium dichromate dihydrate that are needed to use up 10.6 g of 1-propanol according to the balanced equation?

*13.57 The oxidation of cyclopentanol to its corresponding ketone using sodium dichromate in aqueous sulfuric acid proceeds according to the following net ionic equation.

$$3C_5H_{10}O + Cr_2O_7^{2-} + 8H^+ \rightarrow 3C_5H_8O + 2Cr^{3+} + 7H_2O$$

Assume that because of side reactions and losses during the purification of the product ketone, only 50.0% of the theoretically possible ketone can actually be isolated. If you are assigned the task of preparing 10.0 g of the ketone under these limitations, what is the minimum number of grams of cyclopentanol that you must use at the beginning? What is the minimum number of grams of sodium dichromate that you must also use?

14

ALDEHYDES AND KETONES

Honey contains a 1:1 mixture of glucose and fructose. The former has an aldehyde group and the latter, which is sweeter than cane sugar, comes with a ketone group. What we study in this chapter is essential to an understanding of all members of the sugar or carbohydrate family.

THIS CHAPTER IN CONTEXT

A knowledge of some of the properties of aldehydes and ketones is essential to understanding carbohydrates. So our study of the major functional groups involved at the molecular level of life continues, but among simpler systems.

14.1 STRUCTURES AND PHYSICAL PROPERTIES OF ALDEHYDES AND KETONES

Both aldehydes and ketones contain the carbonyl group.

All *simple sugars* are either polyhydroxy aldehydes or polyhydroxy ketones, as illustrated in the structures of glucose and fructose.

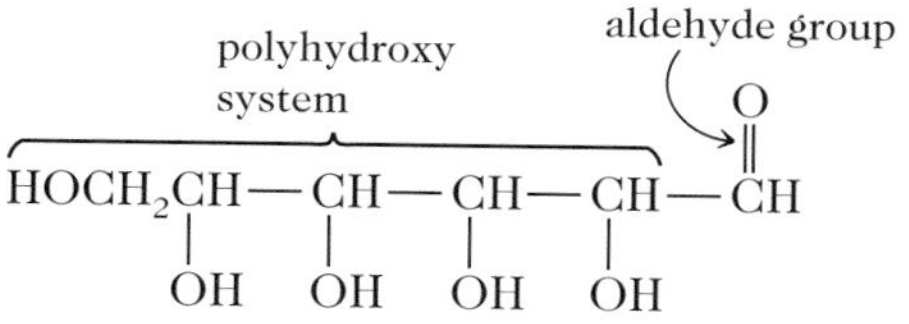

glucose (open-chain form)

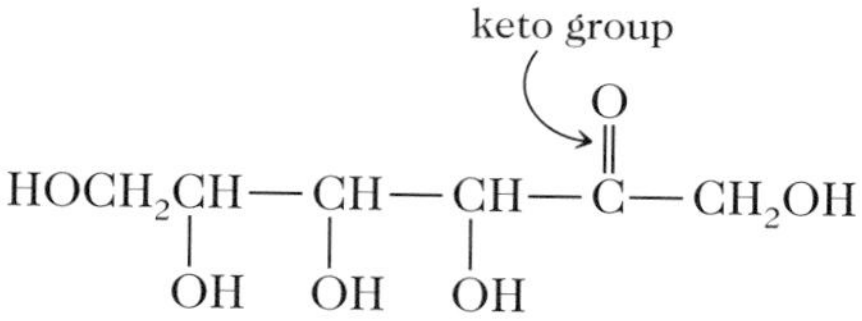

fructose (open-chain form)

■ We show here only one form of each of the glucose and fructose molecules.

Many other naturally occurring substances, including some vitamins and hormones, are also aldehydes or ketones (see Interaction 14.1).

Aldehydes and Ketones Have the *Carbonyl Group* Both aldehydes and ketones contain the carbon–oxygen double bond, called the **carbonyl group** (pronounced car-bon-EEL).

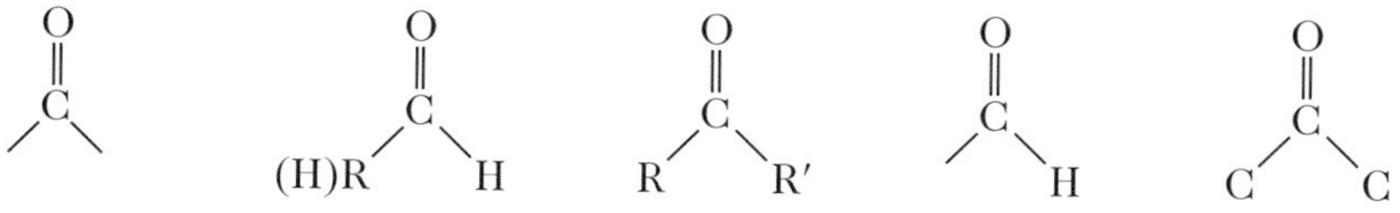

carbonyl group | aldehydes | ketones | aldehyde group | ketone system

This group also occurs in carboxylic acids and their salts, esters, anhydrides, and amides, subjects of following chapters.

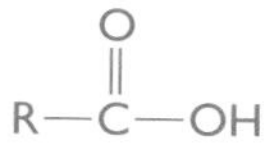

Aldehydes Have the CH=O Group For a compound with a carbonyl group to be an **aldehyde,** the carbon of C=O *must bond to at least one H.* The other single bond *must* be to C (or to another H), but not to O, N, or S. Thus all aldehydes have the CH=O group, called the **aldehyde group.** You'll often see it written as CHO as in the general formula for aldehydes, RCHO.

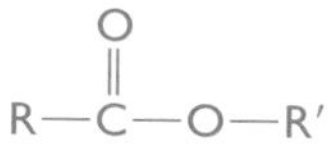

Simple aldehydes (Table 14.1) have unpleasant odors. This is partly because, by air oxidation, liquid aldehydes contain traces of their corresponding carboxylic acids, which also have some of the worst odors in chemistry.

In *Ketones*, the Carbonyl Carbon Is Joined Directly to Two Carbons For a compound to be a **ketone,** its carbonyl group *must* be flanked on *both* sides by bonds to carbons (see Table 14.2). Only then is the carbonyl group actually a **keto group.** The structure of a ketone is sometimes condensed to RCOR′. Ketones also have distinctive but generally not disagreeable odors.

■

carboxylic acids

esters

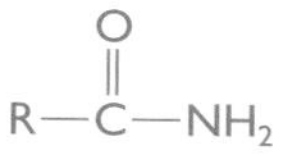

amides

SOME IMPORTANT ALDEHYDES AND KETONES

Formaldehyde Pure formaldehyde is a gas at room temperature, and it has a very irritating and distinctive odor. It is quite soluble in water, so it is commonly marketed as a solution called formalin (37%) to which some methanol has been added. In this and more dilute forms, formaldehyde was once commonly used as a disinfectant and as a preservative for biological specimens. (Concern over formaldehyde's potential hazard to health has caused these uses to decline.) Most formaldehyde today is used to make various plastics, such as Bakelite.

Acetone Acetone is valued as a solvent. Not only does it dissolve a wide variety of organic compounds, but it is also miscible with water in all proportions. Nail polish remover is generally acetone. Should you ever use "superglue," it would be a good idea to have some acetone (nail polish remover) handy because superglue can stick your fingers together so tightly that it takes a solvent such as acetone to get them unstuck.

Acetone is a minor by-product of metabolism, but in some situations (e.g., untreated diabetes) enough is produced to give the breath the odor of acetone.

Some Aldehydes and Ketones in Metabolism The aldehyde group and the keto group occur in many compounds at the molecular level of life, like pyridoxal (a B vitamin) and estrone (a female sex hormone).

pyridoxal, one of the vitamins (B_6)

estrone, a female sex hormone

TABLE 14.1   Aldehydes

Name	Structure	Formula Mass	BP (°C)	Solubility in Water
Methanal	$CH_2{=}O$	30.0	−21	Very soluble
Ethanal	$CH_3CH{=}O$	44.0	21	Very soluble
Propanal	$CH_3CH_2CH{=}O$	58.1	49	16 g/dL (25 °C)
Butanal	$CH_3CH_2CH_2CH{=}O$	72.1	76	4 g/dL
Benzaldehyde	$C_6H_5CH{=}O$	106.1	178	0.3 g/dL

polarization of the carbonyl group

hydrogen bond (··) between a water molecule and a carbonyl group

Aldehyde and Keto Groups Are Planar, Unsaturated, and Moderately Polar Because oxygen is more electronegative than carbon, the carbonyl group makes aldehydes and ketones (moderately) polar compounds. Their boiling points and solubilities in water reflect this. Aldehydes and ketones boil higher than alkanes, but lower than alcohols, when compared with substances of nearly the same formula masses (Table 14.3).

The lack of the OH group means that aldehydes and ketones cannot donate hydrogen bonds, but only accept them. This is sufficient to make the low formula mass aldehydes and ketones relatively soluble in water. As the total carbon content increases, of course, their solubility decreases (see Tables 14.1 and 14.2).

14.2 NAMING ALDEHYDES AND KETONES

The IUPAC name ending for aldehydes is *-al* and for ketones is *-one*.

The Common Names of Simple Aldehydes Are Derived from Those of Carboxylic Acids The common names of the simple aldehydes are actually more used than

TABLE 14.2   Ketones

Name	Structure	Formula Mass	BP (°C)	Solubility in Water
Acetone	$\overset{\text{O}}{\overset{\|}{\text{CH}_3\text{CCH}_3}}$	58.1	56	Very soluble
Butanone	$\overset{\text{O}}{\overset{\|}{\text{CH}_3\text{CCH}_2\text{CH}_3}}$	72.1	80	33 g/dL (25 °C)
2-Pentanone	$\overset{\text{O}}{\overset{\|}{\text{CH}_3\text{CCH}_2\text{CH}_2\text{CH}_3}}$	86.1	102	6 g/dL
3-Pentanone	$\overset{\text{O}}{\overset{\|}{\text{CH}_3\text{CH}_2\text{CCH}_2\text{CH}_3}}$	86.1	102	5 g/dL
Cyclopentanone	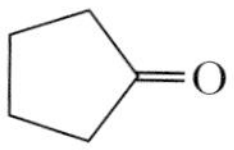 (cyclopentane ring)=O	84.1	129	Slightly soluble
Cyclohexanone	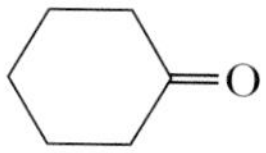 (cyclohexane ring)=O	98.1	156	Slightly soluble

TABLE 14.3   Boiling Point versus Structure

Name	Structure	Formula Mass	BP (°C)
Butane	$\text{CH}_3\text{CH}_2\text{CH}_2\text{CH}_3$	58.1	0
Propanal	$\text{CH}_3\text{CH}_2\text{CH}=\text{O}$	58.1	49
Acetone	$\overset{\text{O}}{\overset{\|}{\text{CH}_3\text{CCH}_3}}$	58.1	56
1-Propanal	$\text{CH}_3\text{CH}_2\text{CH}_2\text{OH}$	60.1	98
1, 2-Ethanediol (ethylene glycol)	$\text{HOCH}_2\text{CH}_2\text{OH}$	62.1	198

their IUPAC names. What is easy about their common names is that they all (well, nearly all) end in *-aldehyde* with prefixes taken from the common names of the corresponding carboxylic acids. So we have to study the common names of both families here.

The simple carboxylic acids have been known for centuries, and their common names are based on their natural sources. Formic acid, for example, is present in the stinging fluid of ants, and the Latin root for ants is *formica*. So this one-carbon acid is called formic acid. The prefix in formic acid is *form-*, so the one-carbon aldehyde is called *formaldehyde*. Here are the four simplest carboxylic acids and their common names together with the structures and names of their corresponding aldehydes.

■ Formic acid also appears to have an aldehyde group, but its second bond from C is to another O, not to H (or C), and it is classified as a carboxylic acid.

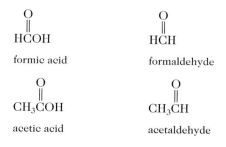

formic acid   formaldehyde

acetic acid   acetaldehyde

■ L. *acetum*, vinegar

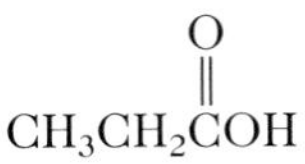

O
‖
CH₃CH₂COH

propionic acid

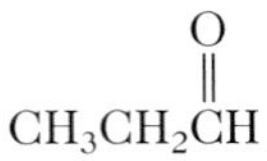

O
‖
CH₃CH₂CH

propionaldehyde

■ Gr. *proto*, first, and *pion*, fat

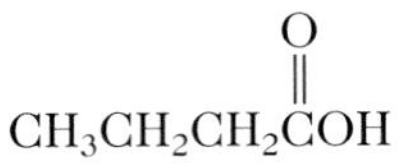

O
‖
CH₃CH₂CH₂COH

butyric acid

O
‖
CH₃CH₂CH₂CH

butyraldehyde

■ L. *butyrum*, butter

In the aromatic series we have the following examples.

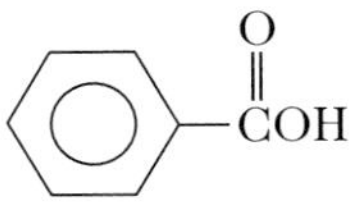

benzoic acid 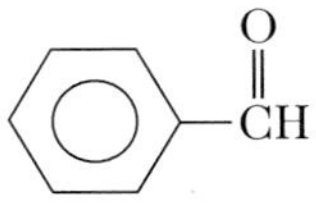   benzaldehyde

The IUPAC Names of Aldehydes End in -al   As with the IUPAC names of the alcohols, those of the aldehydes are based on the idea of a *parent aldehyde.* Here are the rules.

1. Select as the parent aldehyde the longest chain *that includes the carbon atom of the aldehyde group.*

   The parent aldehyde in the following structure is a five-carbon aldehyde.

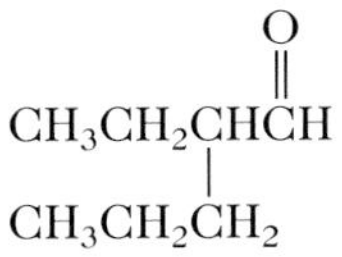

O
‖
CH₃CH₂CHCH
|
CH₃CH₂CH₂

   There is a longer chain in the structure, one of six carbons, but it doesn't include the carbon of the aldehyde group, so this longer chain may not be used in selecting and naming the parent.

2. Name the parent by changing the *-e* ending of the corresponding alkane to *-al.*

   In the example shown with rule 1, the alkane that corresponds to the correct chain is pentane, so the name of the parent aldehyde in this structure is *pentanal.*

3. Number the chain to give the carbon atom of the carbonyl group number 1.

   Precedence is accorded the aldehyde group. Regardless of where other substituents occur on the parent chain, they have to take whatever numbers they receive following the assignment of 1 to the carbonyl carbon atom.

4. Assemble the rest of the name in the same way that is used in naming alcohols, except do not include "1" to specify the location of the aldehyde group.

   The carbonyl carbon cannot have any other number but 1, so we do not include this number. Thus, using the example begun under rule 1, we have

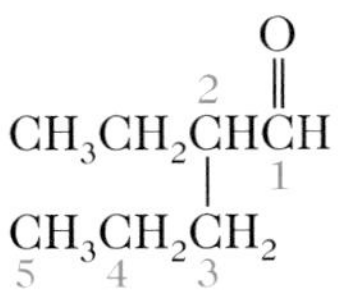

2-ethylpentanal
Not: 2-ethyl-1-pentanal

EXAMPLE 14.1 Writing the IUPAC Name of an Aldehyde

What is the IUPAC name of the following compound?

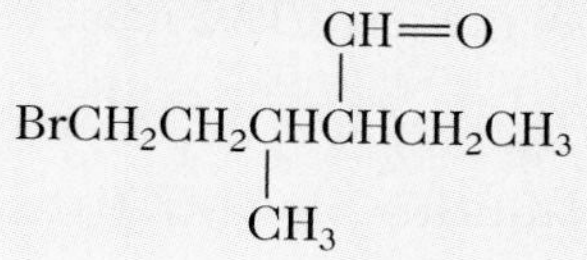

$$\begin{array}{c} \text{CH}{=}\text{O} \\ | \\ \text{BrCH}_2\text{CH}_2\text{CHCHCH}_2\text{CH}_3 \\ | \\ \text{CH}_3 \end{array}$$

ANALYSIS First, we identify the parent aldehyde. The longest chain that includes the carbon atom of the aldehyde group has five carbons, so the name of the parent aldehyde is *pentanal*. Next, we number the chain beginning with the carbon atom of the aldehyde group.

$$\begin{array}{c} \overset{1}{\text{CH}}{=}\text{O} \\ {}^{5}\quad {}^{4}\quad {}^{3}| \\ \text{BrCH}_2\text{CH}_2\text{CHCHCH}_2\text{CH}_3 \\ \quad\quad |\ {}^{2} \\ \quad \text{CH}_3 \end{array}$$

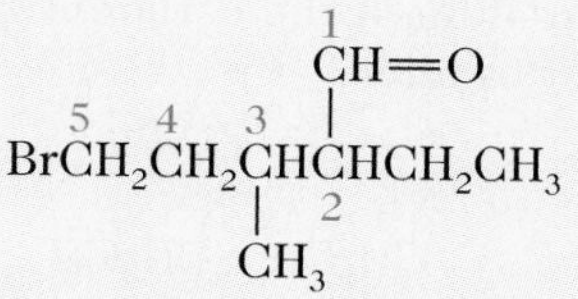

At position 2 there is an ethyl group; at 3, a methyl group; and at 5, a bromo group.

SOLUTION We organize the names of the groups alphabetically. The name is

5-bromo-2-ethyl-3-methylpentanal

CHECK Remember that the most common error is in failing to find the *longest chain that includes the carbonyl group.* Whenever you check your work in writing a name from a structure, test every conceivable way of finding the parent chain.

■ PRACTICE EXERCISE 1 Write the IUPAC names of the following aldehydes.

(a) 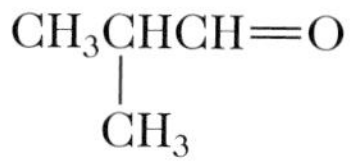 $\begin{array}{c} \text{CH}_3\text{CHCH}{=}\text{O} \\ | \\ \text{CH}_3 \end{array}$ (b) $\begin{array}{c} \text{CH}_3\text{CHCH}_2\text{CH}{=}\text{O} \\ | \\ \text{Br} \end{array}$ (c) $\begin{array}{c} \text{CH}_3 \quad\ \text{CH}_3 \\ |\qquad\ | \\ \text{CH}_3\text{CHCH}_2\text{CCH}_2\text{CHCH}{=}\text{O} \\ |\qquad\ | \\ \text{CH}_3\text{CH}_2\ \ \text{CH}_3 \end{array}$

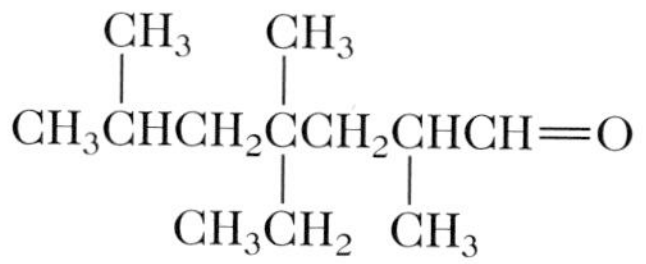

■ PRACTICE EXERCISE 2 What is wrong with the name *2-isopropylpropanal?*

IUPAC Names of Ketones End in *-one* The IUPAC rules for naming ketones are identical to those for the aldehydes, except for two obvious changes. The name of the parent ketone must end in *-one* (not *-al*) and the keto group must be located by a number. In numbering the chain, the location of the keto group takes precedence, not the locations of substituents. Thus, the chain of the following ketone that must be selected for naming purposes is numbered as shown.

■ Pronounce *-one* as *own.*

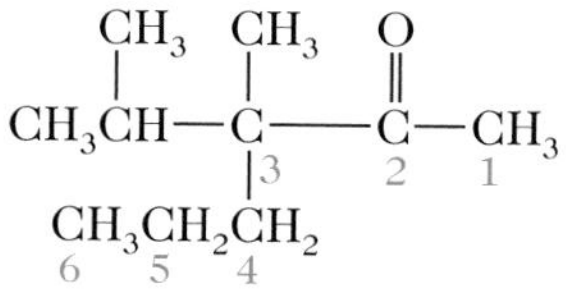

$$\begin{array}{c} \text{CH}_3 \quad\ \text{CH}_3 \quad \text{O} \\ |\qquad\ |\qquad\ \| \\ \text{CH}_3\text{CH}{-}\underset{|\,3}{\text{C}}{-}\overset{}{\underset{2}{\text{C}}}{-}\underset{1}{\text{C}}{-}\text{CH}_3 \\ \text{CH}_3\text{CH}_2\text{CH}_2 \\ {}_{6}\quad\ {}_{5}\quad {}_{4} \end{array}$$

The parent ketone is therefore 2-hexanone, not 2-pentanone. At carbon 3 the chain has a methyl group plus an isopropyl group, so the complete name of the ketone is 3-isopropyl-3-methyl-2-hexanone.

■ PRACTICE EXERCISE 3   Write the IUPAC names of the following ketones.

$$\text{(a)} \quad \underset{\displaystyle \text{O}}{\overset{\displaystyle \|}{\text{CH}_3\text{CH}_2\text{CCH}_3}} \qquad \text{(b)} \quad \underset{\displaystyle \text{CH}_3}{\overset{\displaystyle |}{\text{CH}_3\text{CHCH}_2\text{CH}_2\text{CH}_2\text{CCH}_3}}\ \overset{\displaystyle \text{O}}{\overset{\displaystyle \|}{}} \qquad \text{(c)}$$

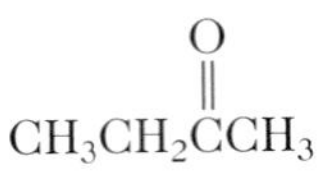

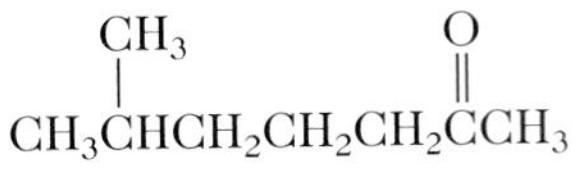

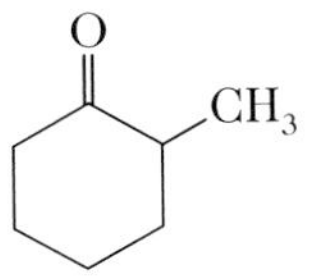

The Simplest Ketone Is Usually Called Acetone, Not Propanone   Quite often the simpler ketones are given common names that are made by naming the two alkyl groups attached to the carbon atom of the carbonyl group and then following these names by the word *ketone.* For example,

$$\underset{\text{methyl ethyl ketone}}{\overset{\displaystyle \text{O}}{\overset{\displaystyle \|}{\text{CH}_3\text{CH}_2\text{CCH}_3}}} \qquad \underset{\text{diethyl ketone}}{\overset{\displaystyle \text{O}}{\overset{\displaystyle \|}{\text{CH}_3\text{CH}_2\text{CCH}_2\text{CH}_3}}} \qquad \underset{\substack{\text{(dimethyl ketone)} \\ \text{acetone}}}{\overset{\displaystyle \text{O}}{\overset{\displaystyle \|}{\text{CH}_3\text{CCH}_3}}}$$

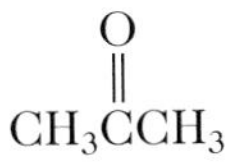

■ The name *acetone* stems from the fact that this *ket*one can be made by heating the calcium salt of *ace*tic acid.

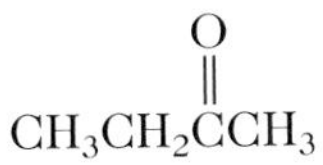

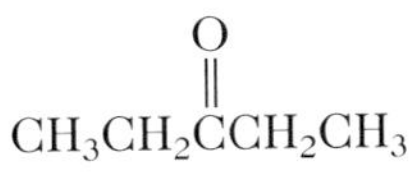

As we noted, the name *acetone* is almost always used for dimethyl ketone or propanone.

■ PRACTICE EXERCISE 4   Write the structures of the following ketones.

(a) Ethyl isopropyl ketone   (b) Methyl phenyl ketone

(c) Dipropyl ketone   (d) Di-*t*-butyl ketone

14.3   OXIDATION OF ALDEHYDES AND KETONES

The aldehyde group is easily oxidized to the carboxylic acid group, but the keto group is difficult to oxidize.

The oxidation of a 1° alcohol to an aldehyde requires special reagents, because aldehydes are themselves easily oxidized by even mild reactants. This has led to some simple test-tube tests for aldehydes.

■ The Tollens' test is sometimes called the *silver mirror test.* Glucose gives this test.

The *Tollens' Test* Produces a Silver Mirror   One very mild oxidizing agent, called **Tollens' reagent,** consists of an alkaline solution of the silver ion in combination with two ammonia molecules, $[\text{Ag}(\text{NH}_3)_2]^+$. It oxidizes the aldehyde group to a carboxyl group (rather, to its anion form), and the silver ion is reduced to metallic silver.

■ Tollens' reagent must be freshly made, because it deteriorates on standing.

$$\text{RCH}{=}\text{O}(aq) + 2[\text{Ag}(\text{NH}_3)_2]^+(aq) + 3\text{OH}^-(aq) \longrightarrow$$
$$\text{RCO}_2{}^-(aq) + 2\text{Ag}(s) + 2\text{H}_2\text{O} + 4\text{NH}_3(aq)$$

When this reaction occurs in a thoroughly clean, grease-free test tube, the silver plates to the glass as a beautiful mirror. In fact, this is one technique used to manufacture mirrors. However, a positive Tollens' test is the formation of metallic silver *in any form,* either as a mirror in a clean test tube or as a grayish precipitate. The appearance of silver is dramatic evidence that a reaction occurs, so Tollens' reagent provides a simple test, called **Tollens' test,** to tell whether an unknown compound is an aldehyde or a ketone.

Benedict's Test Produces a Brick-Red Precipitate   **Benedict's reagent,** another mild oxidizing agent, consists of a basic solution of the copper(II) ion and the citrate ion, the anion of citric acid, which causes the tart taste of citrus fruits. The medium has to be slightly basic in order for an aldehyde group to be oxidized by the reagent. However, Cu^{2+} normally forms an extremely insoluble precipitate of CuO in a basic environment. The citrate ion prevents this by a mechanism to be studied shortly.

When an easily oxidized compound like glucose is added to a test tube that contains some Benedict's reagent, and the solution is warmed, Cu^{2+} ions are reduced to Cu^+ ions. The citrate ion is unable to keep the Cu^+ ion in the dissolved state in the basic medium. The newly formed Cu^+ ions are instantly changed by the base (OH^-) into a precipitate of copper(I) oxide, Cu_2O. We'll write the equation using the general structure of an aldehyde, $RCH{=}O$, but *simple* aldehydes (implied by this structure) give complex results.[1]

$$RCH{=}O(aq) + 2Cu[citrate]^{2+}(aq) + 5OH^-(aq) \longrightarrow$$
$$RCO_2^-(aq) + Cu_2O(s) + 3H_2O + 2[citrate]$$

The Benedict's reagent has a brilliant blue color caused by the copper(II) ion, but Cu_2O has a brick-red color (Figure 14.1). Therefore the visible evidence of a positive **Benedict's test** is the disappearance of a blue color and the appearance of a reddish precipitate.

Simple aldehydes do not give the test as well as do aldehydes with neighboring oxygens. Three systems, one not even an aldehyde, and all of which occur among various carbohydrates, give positive Benedict's tests:

O ‖ RCHCH | OH	O O ‖ ‖ RC—CH	O ‖ RCHCR′ | OH
α-hydroxy aldehyde (present in glucose)	α-keto aldehyde	α-hydroxy ketone (present in fructose)

Benedict's Test Has Been a Common Method for Detecting Glucose in Urine   In certain conditions, like diabetes, the body cannot prevent some of the excess glucose in the blood from being present in the urine, so testing the urine for its glucose concentration has long been used in medical diagnosis. Clinitest tablets, a convenient solid form of Benedict's reagent, contain all the needed reactants in their solid forms. To test for glucose, a few drops of urine are mixed with a tablet, and as the tablet dissolves, the heat needed for the test is generated. The color that develops is compared with a color code on a chart provided with the tablets. Specialists in the control of diabetes, however, prefer to monitor the carbohydrate status of a diabetic

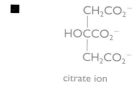

citrate ion

FIGURE 14.1 Benedict's reagent is a slightly alkaline solution of citrate ion and Cu^{2+}, which makes the solution at the rear intensely blue. In a positive Benedict's test, the blue color disappears as a brick-red slurry of Cu_2O forms (foreground test tube).

■   A carbon atom immediately adjacent to a carbonyl group is often called an alpha (α) carbon:

alpha carbon

■   Other tests for glucose are based on enzyme-catalyzed reactions, which we will learn more about later.

---

[1] Although simple aldehydes do give a reaction with the Benedict's reagent that changes the blue color, the gummy solid that forms is not Cu_2O. Even the equation that we write here for carbohydrates is an oversimplification because in a *warm* alkaline medium, which is involved in the Benedict's test, carbohydrates undergo complex reactions. Yet, Cu_2O does form when carbohydrates are tested with Benedict's reagent. See R. Daniels, C. C. Rush, and L. Bauer, *Journal of Chemical Education,* Vol. 37, 1960, page 205.

When aqueous sodium hydroxide is added to aqueous silver nitrate, a tan, mudlike precipitate of silver oxide forms.

person by determining the glucose in the *blood* instead of in the urine. Not all patients, however, can or are willing to manage blood tests, particularly when they are needed frequently.

Complex Ions Are Present in Tollens' and Benedict's Reagents   The reagents for the Tollens' and Benedict's tests involve a species new to our study, one of great importance at the molecular level of life, the *complex ion.* Tollens' reagent is prepared by adding sodium hydroxide to dilute silver nitrate. This causes the very insoluble silver oxide, Ag_2O, to precipitate. Undissolved Ag_2O would be unable to give the Tollens' test. When dilute ammonia is added next, however, its molecules are able to pull silver ions out of the solid silver oxide by forming a soluble complex ion, called the *silver diammine ion,* $[Ag(NH_3)_2]^+$. Thus the silver oxide dissolves.

A **complex ion**—often simply called a **complex**—consists of a metal ion that has strongly attracted a definite number of **ligands,** species that are either negative ions or neutral but electron-rich molecules, like ammonia. Examples of ligands include any of the halide ions (F^-, Cl^-, Br^-, and I^-), the cyanide ion (CN^-), the hydroxide ion, and many anions of organic acids (like the citrate ion). Among the common, electrically neutral ligands are water and ammonia. Other ligands are organic molecules with *amino groups,* NH_2. The most common metal ions that can form complex ions are those of the transition metal elements in the periodic table, like Ag^+ and Cu^{2+}. Two complex ions of Cu^{2+} and neutral ligands are $Cu(H_2O)_4^{2+}$ and $Cu(NH_3)_4^{2+}$, which are both blue but with strikingly different intensities of color (Figure 14.2). The citrate ion is a negatively charged ligand that forms the (blue) complex with Cu^{2+} in Benedict's reagent.

Many Complex Ions Are Important at the Molecular Level of Life   Uncomplexed transition metal ions are, in general, insoluble when the pH is greater than 7; they precipitate as their hydroxides or oxides. Thus virtually all of the trace transition metal ions required in nutrition, like Cu^{2+}, Co^{2+}, Fe^{2+}, and several others, can exist in the slightly alkaline fluids of the cell or in blood only as complex ions. Many electron-rich organic ligands, however, are able to form water-soluble complexes with such metal ions and so allow them to be in solution even at pHs greater than 7. The iron(II) ion, for example, is insoluble in base, but in blood (pH 7.35) it occurs in a complex ion called *heme,* the red species in hemoglobin and the oxygen carrier in blood.

Inside cells, the phosphate ion level is sufficiently high to form insoluble phosphates with calcium ion, but this must be prevented for many reasons, not least of which is that cells would mineralize and die. Cells prevent this by forming soluble complex ions between Ca^{2+} and a variety of electron-rich molecular units on protein molecules. Complex ions thus have absolutely vital functions at the molecular level of life.

■ Hemoglobin exists inside erythrocytes, red blood cells.

FIGURE 14.2 The hydrated copper(II) ion, $Cu(H_2O)_4^{2+}$ (left), gives a pale blue color to its aqueous solution. At the same molar concentration the ammoniated copper(II) ion, $Cu(NH_3)_4^{2+}$ (right), causes a deeper blue.

14.4 REDUCTION OF ALDEHYDES AND KETONES

Aldehydes and ketones are reduced to alcohols when hydrogen adds to their carbonyl groups.

Aldehyde and Keto Groups Can Be Reduced by Hydrogen or by Hydride Ion
Aldehydes are reduced to 1° alcohols and ketones to 2° alcohols by the direct addi-

tion of hydrogen or by hydride ion transfer. Either hydrogenation is the *reduction of* an aldehyde or ketone because the molecules gain hydrogen. For example, under heat and pressure and in the presence of a finely divided metal catalyst, the carbonyl groups of aldehydes and ketones add hydrogen according to the following equations.

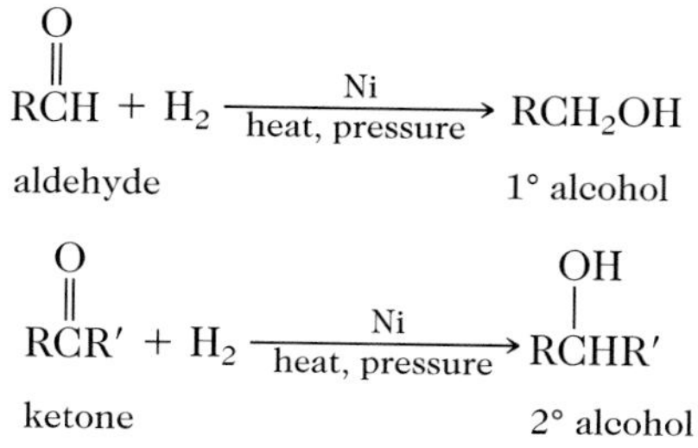

$$\underset{\text{aldehyde}}{\overset{\overset{\displaystyle O}{\|}}{RCH}} + H_2 \xrightarrow[\text{heat, pressure}]{Ni} \underset{1° \text{ alcohol}}{RCH_2OH}$$

$$\underset{\text{ketone}}{\overset{\overset{\displaystyle O}{\|}}{RCR'}} + H_2 \xrightarrow[\text{heat, pressure}]{Ni} \underset{2° \text{ alcohol}}{\overset{\overset{\displaystyle OH}{|}}{RCHR'}}$$

The experimental conditions for catalytic hydrogenations cannot exist in living systems, of course. However, the *net effects* are the same in living cells, where hydride ion transfers bring about reductions.

The Aldehyde or Keto Group Is Reduced by Acceptance of the Hydride Ion The only way $H:^-$ can be supplied in living systems is by a donor that transfers it *directly* to the acceptor. Although the hydride ion is a powerful reducing agent in its free form, it reacts vigorously with water to give hydrogen gas, leaving the *relatively* much weaker base, OH^-.

One hydride acceptor is the carbonyl group. When an aldehyde or keto group accepts $H:^-$, the following reaction occurs and the anion of an alcohol forms. We'll represent a donor of $H:^-$ by the symbol *Mtb*:H, where *Mtb* refers to a *metabolite*, a chemical intermediate in metabolism.

■ *Mtb*:H in the body is often made from a B vitamin unit incorporated into an enzyme.

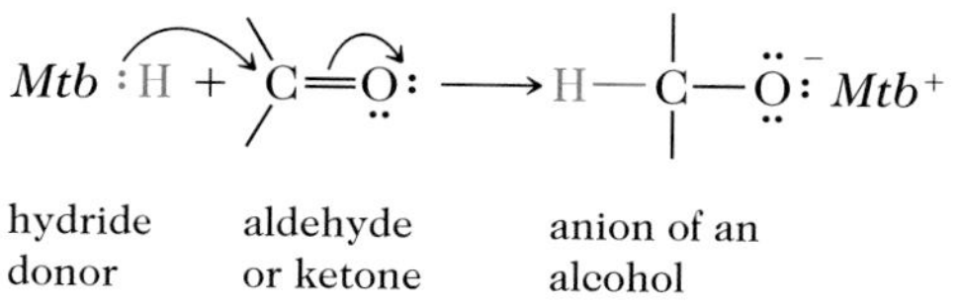

$$Mtb:H + \overset{\diagdown}{\underset{\diagup}{C}} = \ddot{O}: \longrightarrow H - \overset{|}{\underset{|}{C}} - \ddot{\underset{..}{O}}:^- Mtb^+$$

hydride donor   aldehyde or ketone   anion of an alcohol

The alcohol anion is a strong proton acceptor, stronger even than a hydroxide ion. So in the instant when the newly formed anion emerges, it takes a proton either from a water molecule or from some other proton donor in the surrounding buffer system. Thus the final organic product is an alcohol.

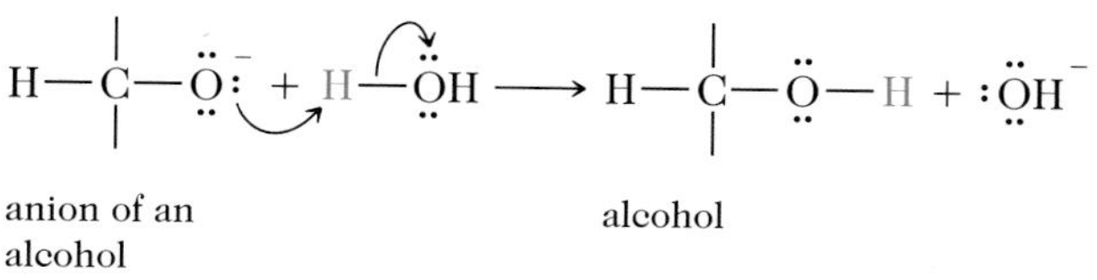

$$H - \overset{|}{\underset{|}{C}} - \ddot{\underset{..}{O}}:^- + H - \ddot{\underset{..}{O}}H \longrightarrow H - \overset{|}{\underset{|}{C}} - \ddot{\underset{..}{O}} - H + :\ddot{\underset{..}{O}}H^-$$

anion of an alcohol   alcohol

As we've already noted, another name for *hydrogenation* is *reduction,* and when the carbonyl carbon atom accepts the pair of electrons carried by $H:^-$, it *gains* this pair and so is reduced.

One of the many examples of hydride ion reduction in cells is one of the steps in the metabolism of glucose, namely, the reduction of the keto group in the pyruvate ion to give the lactate ion.

■ Remember, a gain of electrons is reduction.

■ NAD$^+$ is a structural unit in several enzymes and is made from a B vitamin. NAD∶H, usually written NADH, is the reduced form of NAD$^+$. We explicitly use NAD∶H here to emphasize that the species is a donor of H∶$^-$. Later, we'll generally write NADH.

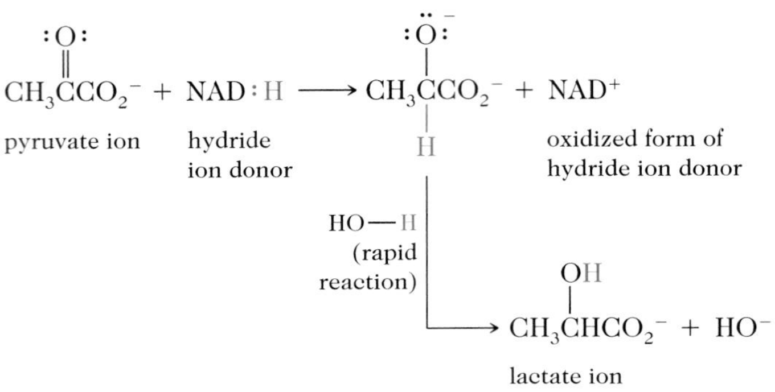

$$CH_3CCO_2^- + NAD∶H \longrightarrow CH_3CCO_2^- + NAD^+$$

pyruvate ion   hydride
                 ion donor

H

oxidized form of
hydride ion donor

HO—H
(rapid
reaction)

OH

$$CH_3CHCO_2^- + HO^-$$

lactate ion

NAD$^+$ stands for *nicotinamide adenine dinucleotide,* but right now, all we need to know about NAD$^+$ is that its reduced form, NAD∶H, is a good donor of the hydride ion. What we must next learn is how to write the products of reduction when we know the reactants. We're interested only in net, overall effects.

EXAMPLE 14.2   Writing the Structure of the Product
of the Reduction of an Aldehyde or Ketone

What is the product of the reduction of propanal?

ANALYSIS   All the action is at the carbonyl group. It changes to an alcohol group. So we copy over the structure of the given compound, change the double bond to a single bond, and supply the two H atoms, one to the oxygen of the original carbonyl group and one to the carbon.

SOLUTION   The product of the reduction of propanal is 1-propanol.

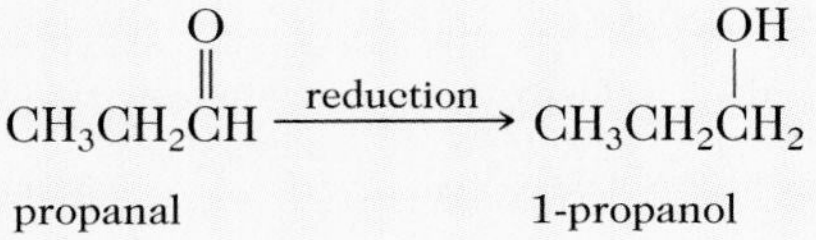

$$CH_3CH_2CH \xrightarrow{\text{reduction}} CH_3CH_2CH_2$$

propanal   1-propanol

CHECK   One common error is to change the structural skeleton, so be sure to check that the *sequence* of all of the atoms heavier than H, like C and O, *has not changed.* Another error is to write a structure which includes a violation of the rules of bonding for the heavy atoms—four bonds for C and two for O in neutral species. So check the answer, atom by atom, to see that each C has four bonds and each O has two. If you find an error, fix it.

■ PRACTICE EXERCISE 5   Write the structures of the products that form when the following aldehydes and ketones are reduced.

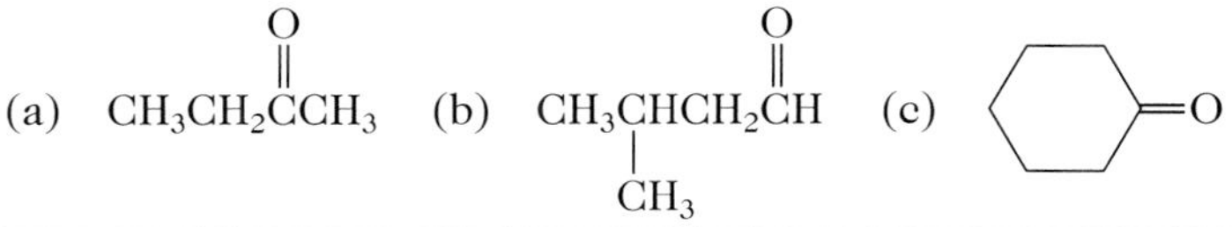

(a) $CH_3CH_2CCH_3$ (b) CH_3CHCH_2CH (c)

CH_3

14.5 REACTIONS OF ALDEHYDES AND KETONES WITH ALCOHOLS

1,1-Diethers—acetals or ketals—form when aldehydes or ketones react with alcohols.

What we study in this section is vital to understanding carbohydrate structure.

Alcohols Add to the Carbonyl Groups of Aldehydes and Ketones When a solution of an aldehyde in an alcohol is prepared, molecules of the alcohol add to those of the aldehyde, and the following equilibrium mixture forms.

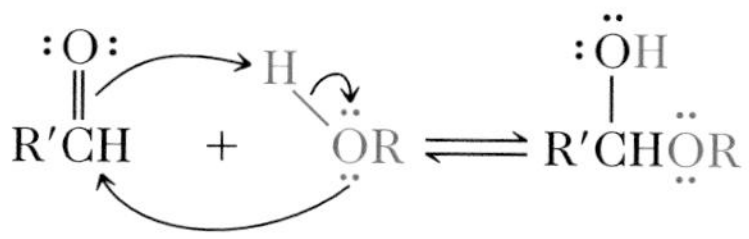

aldehyde   alcohol   hemiacetal

The reaction resembles the addition of water to alkenes with one difference, namely, the *direction* of the addition of the *unsymmetrical* reactant, ROH, to C=O is dominated by the permanent polarity of C=O. In the ROH molecule, the H is the more positive unit than the RO. Therefore, the H always goes to the O of C=O where there is a permanent (and opposite) $\delta-$ charge. ("Unlike charges attract.") The alcohol's RO unit, which also has a $\delta-$ on O, always goes to the C of the carbonyl group where there is a permanent $\delta+$ charge.

The product, a **hemiacetal**, has molecules that *always* have a carbon atom holding both an OH group and an OR group. We create a new organic family for the product because when the two groups, OH and OR, are this close to each other, they so modify each other's properties that we have neither an ordinary alcohol nor an ordinary ether group. For example, unlike either alcohols or ethers, hemiacetals are not very stable. They readily break back down to aldehydes and alcohols. Ordinary ethers, we learned, strongly resist reactions that break their molecules. We study hemiacetals only because the system is stable when it occurs among carbohydrates. Glucose molecules, for example, exist as cyclic hemiacetals.

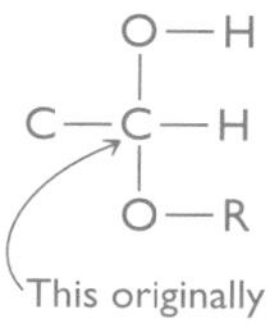

When a ketone is dissolved in an alcohol, a similar reaction occurs to give an equilibrium in which the product is called a **hemiketal** to signify its origin from a ketone.

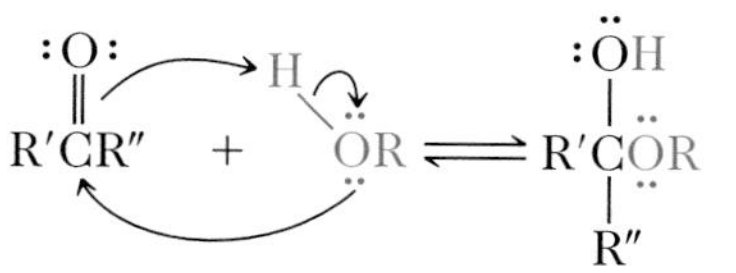

ketone   alcohol   hemiketal

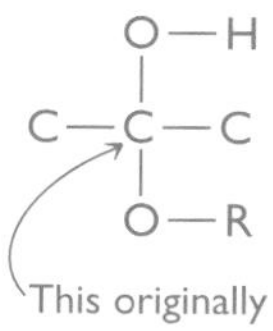

The position of equilibrium overwhelmingly favors the reactants, the ketone and alcohol, so hemiketals are even less stable than hemiacetals. The hemiketal system, however, does occur among carbohydrates. One form of fructose, for example, is a cyclic hemiketal. The continuation of our study of these systems will deal almost entirely with hemiacetals because the extension of the principles to hemiketals is straightforward.

The relative ease with which hemiacetals break back down means that the hemiacetal system is a site of structural weakness, even among carbohydrates. For this reason, we have to learn how to recognize the system when it occurs in a structure.

■ The hemiacetal system:

This originally was the carbon atom of an aldehyde group.

■ The hemiketal system:

This originally was the carbon atom of a keto group.

■ The ring system of **3** also occurs in glucose.

EXAMPLE 14.3 Identifying the Hemiacetal System

Which of the following structures has the hemiacetal system? Place an asterisk by any carbon atoms that were initially the carbon atoms of aldehyde groups.

$$CH_3-O-CH_2-CH_2-OH \qquad CH_3-O-CH_2-OH \qquad \begin{array}{c} CH_2-O \\ H_2C \qquad\qquad CH-OH \\ CH_2-CH_2 \end{array}$$

$$\qquad\qquad\qquad 1 \qquad\qquad\qquad\qquad\qquad 2 \qquad\qquad\qquad\qquad 3$$

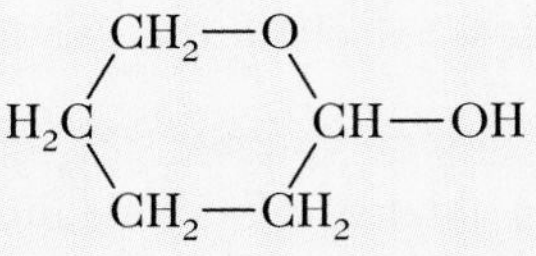

ANALYSIS To have the hemiacetal system, the molecule must have a carbon to which are attached one OH group and one —O—C unit.

SOLUTION In structure **1** there is an OH group and an —O—C unit, but they are not joined to the *same* carbon. Therefore **1** is not a hemiacetal. It has only an ordinary ether group plus an alcohol group.

In structure **2**, the OH and the —O—C are joined to the same carbon, so **2** is a hemiacetal. Similarly, in structure **3**, the carbon on the far right corner of the ring holds both an OH group and —O—C unit, and **3** is also a hemiacetal, a cyclic hemiacetal. Structure **3** shows the way in which the hemiacetal system occurs in many carbohydrates, as a cyclic hemiacetal. The asterisks (*) in the following structures identify carbon atoms that were originally carbonyl carbons.

$$CH_3-O-\overset{*}{C}H_2-OH \qquad\qquad \begin{array}{c} CH_2-O_{\;*} \\ H_2C \qquad\qquad \overset{*}{C}H-OH \\ CH_2-CH_2 \end{array}$$

$$\qquad\qquad\qquad 2 \qquad\qquad\qquad\qquad\qquad\qquad 3$$

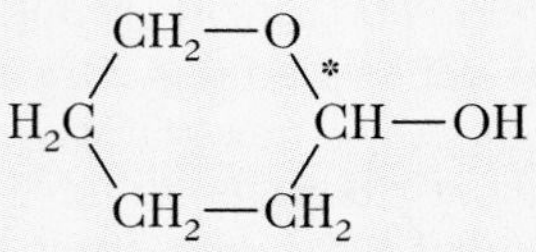

CHECK Make sure that any structure identified as a hemiacetal has at least one carbon attached to *two* O atoms by single bonds, that *one* of the O atoms is part of the OH group, and that the other is joined by its second single bond to C.

■ PRACTICE EXERCISE 6 Identify the hemiacetals or hemiketals among the following structures, and place an asterisk by the carbon atoms that initially were part of the carbonyl groups of parent aldehydes or ketones.

(a)

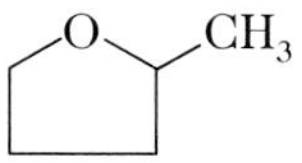

(b) 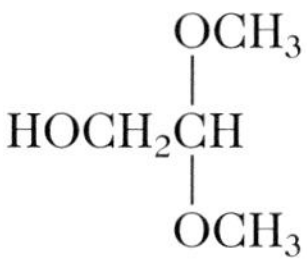 $\begin{array}{c} OCH_3 \\ | \\ HOCH_2CH \\ | \\ OCH_3 \end{array}$

(c) $HOCH_2OCH_2CH_3$

(d)

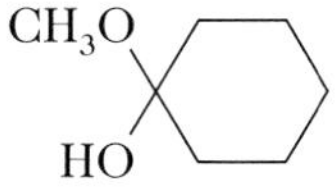

Another skill needed for our study of carbohydrates is the ability to write the structure of a hemiacetal that could be made from a given aldehyde and alcohol.

EXAMPLE 14.4 **Writing the Structure of a Hemiacetal Given Its Parent Aldehyde and Alcohol**

Write the structure of the hemiacetal that is present at equilibrium in a solution of propanal in ethanol.

ANALYSIS A hemiacetal must have a carbon atom to which both an OH group and an OR group are attached. This carbon atom *is provided by the aldehyde.* The structure of the alcohol, CH_3CH_2OH, tells us that the R group in OR of the hemiacetal is CH_2CH_3.

SOLUTION The structure of the hemiacetal formed from propanal and ethyl alcohol is

$$\underset{\substack{\text{from the}\\\text{aldehyde}}}{\underbrace{CH_3CH_2\overset{\displaystyle\overset{\textstyle OH}{|}}{C}H}}\underset{\substack{\text{from the}\\\text{alcohol}}}{\underbrace{OCH_2CH_3}}$$

CHECK Find the carbon holding *two* O atoms and check the other two atoms or groups that it also holds *against the original aldehyde* (or ketone). These two atoms or groups—here, H and CH_3CH_2—must match those of the aldehyde (or ketone). Finally check what else the two O atoms are holding; one must hold H (to make it an OH group) and the other must hold the alkyl group from the original alcohol.

■ **PRACTICE EXERCISE 7** Write the structures of the hemiacetals that are present in the equilibria that involve the following pairs of compounds.

(a) Ethanal and methanol   (b) Butanal and ethanol

(c) Benzaldehyde and 1-propanol   (d) Methanal and methanol

Still another skill that will be useful in studying carbohydrates is the ability to write the structures of the aldehyde and alcohol that are liberated by the breakdown of a hemiacetal.

EXAMPLE 14.5 **Writing the Breakdown Products of a Hemiacetal**

What aldehyde and alcohol form when the following hemiacetal breaks down?

$$CH_3CH_2CH_2\overset{\displaystyle\overset{\textstyle OH}{|}}{C}H-O-CH_2CH_3$$

ANALYSIS The key step to solving this problem lies in analyzing the given structure. First, pick out the carbon atom of the original carbonyl group; it's the one holding *both* an OH group and an OR unit. Anything else this carbon holds—usually one H and one hydrocarbon group—completes what we need in order to write the structure of the aldehyde. The R group of OR is the hydro-carbon group of the original alcohol. Our analysis thus gives us:

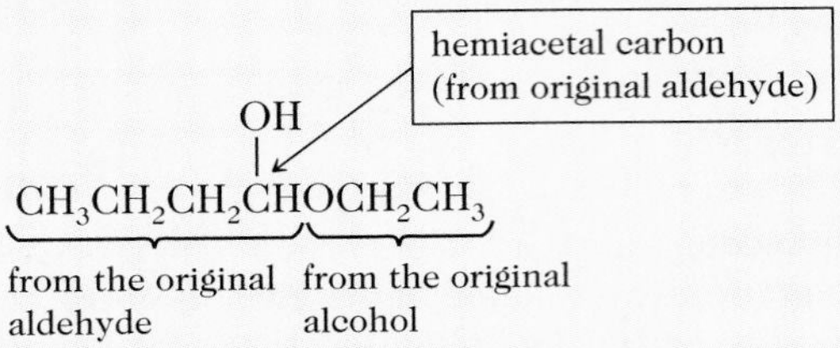

$$\underbrace{CH_3CH_2CH_2\overset{\displaystyle OH}{\overset{|}{C}H}}_{\text{from the original aldehyde}}\underbrace{OCH_2CH_3}_{\text{from the original alcohol}}$$

hemiacetal carbon (from original aldehyde)

The original aldehyde is thus the unbranched four-carbon aldehyde, butanal, and the original alcohol is seen to be the two-carbon alcohol, ethanol.

SOLUTION   The products of the breakdown of the given hemiacetal are

$$\underset{\text{butanal}}{CH_3CH_2CH_2\overset{\displaystyle O}{\overset{||}{C}}H} + \underset{\text{ethanol}}{HOCH_2CH_3}$$

CHECK   It is essential to notice that *only one bond* is affected when we disassemble the initial hemiacetal, the C—O bond of the hemiacetal carbon, not any other C—O bond, not a C—C bond, and not a C—H bond. Failure to learn this is the most common error that students make in working problems like this. Examine closely, therefore, your answer to see that the original molecule is ruptured *only* at the C—O bond at the hemiacetal carbon. *Break no other bond.*

■ PRACTICE EXERCISE 8   Write the structures of the breakdown products of the following hemiacetals.

$$\text{(a)}\quad CH_3CH_2\overset{\displaystyle OH}{\overset{|}{C}H}OCH_3 \qquad \text{(b)}\quad CH_3CH_2O\overset{\displaystyle OH}{\overset{|}{C}H}CH_2CH_3$$

■ The acetal system:

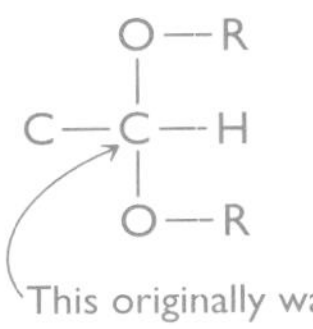

$$\overset{\displaystyle O-R}{\underset{\displaystyle O-R}{\overset{|}{C}{-}\overset{|}{C}{-}H}}$$

This originally was the carbon atom of an aldehyde group.

■ The ketal system:

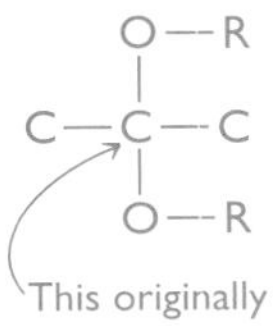

$$\overset{\displaystyle O-R}{\underset{\displaystyle O-R}{\overset{|}{C}{-}\overset{|}{C}{-}C}}$$

This originally was the carbon atom of a keto group.

Acetals and Ketals Form When Alcohols React Further with Hemiacetals and Hemiketals   Hemiacetals and hemiketals do resemble alcohols in one important property. They can undergo a reaction that looks like the formation of an ether. An ordinary ether does not form, however, but a special kind, a 1,1-diether called an **acetal** or a **ketal**. The overall change that leads to an acetal is as follows.

$$\underset{\text{hemiacetal}}{R'\overset{\displaystyle OH}{\overset{|}{C}H}OR} + H{-}OR \xrightarrow{\text{acid catalyst}} \underset{\text{acetal}}{R'\overset{\displaystyle OR}{\overset{|}{C}H}OR} + H_2O$$

Hemiketals give the identical kind of reaction, but the products are called *ketals*. Unlike hemiacetals and hemiketals, both acetals and ketals are stable compounds that can be isolated and stored.

The difference between the formation of an acetal and an ordinary ether is that *acetals form and break more readily*. As a rule, when two functional groups are very close to each other in a molecule, each modifies the properties of the other in some way. Here, the OR group makes the OH group attached to the same carbon much more reactive toward the splitting out of water with an alcohol.

In a structural sense, an acetal is a 1,1-diether, but "1,1" does not refer to the numbering of the chain and the "ether" part of 1,1-diether does not connote "resistance to breaking up." The "1,1" means only that the two OR groups come to the *same* carbon. In this sense, ketals are also 1,1-diethers. The molecules of many carbohydrates, like sucrose (table sugar), lactose (milk sugar), and starch, also have the 1,1-diether system, which is why we study acetals and ketals.

Ordinary ethers, R—O—R, do not break up in dilute acid or base, but *acetals and ketals are stable only if they are kept out of contact with aqueous acids.* In water, acids (or digestive enzymes) catalyze the hydrolysis of acetals and ketals to their parent alcohols and aldehydes (or ketones). In aqueous *base,* however, the acetal (ketal) system is stable.

Hydrolysis is the only chemical reaction of acetals and ketals that we need to study; it is the reaction by which carbohydrates are digested. Before we study this reaction further, we must be sure that we can recognize the acetal or ketal system when it occurs in a structure.

EXAMPLE 14.6 Recognizing the Acetal or Ketal Systems and Identifying Which Carbon Atoms Came from Parent Carbonyl Carbons

Examine each structure to see whether it is an acetal or a ketal. If it is either, identify the carbon atom furnished by the carbonyl carbon of the parent aldehyde or ketone.

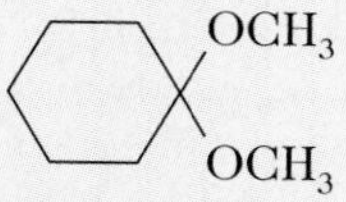

ANALYSIS We have to find one carbon that holds two OR types of groups. The two *must* join to the *same* carbon.

SOLUTION Structure **4** has such a carbon in its central CH_2 unit. This carbon was initially the carbonyl carbon atom of an aldehyde (methanal), because it also holds at least one H atom. Structure **5** similarly has such a carbon, in the CH unit, a carbon that also came from an aldehyde group (because it has at least one H atom). In structure **6**, we can also find a carbon that holds two O—C bonds. This carbon lacks an H atom, however, so it must have come from the carbonyl group of a ketone system.

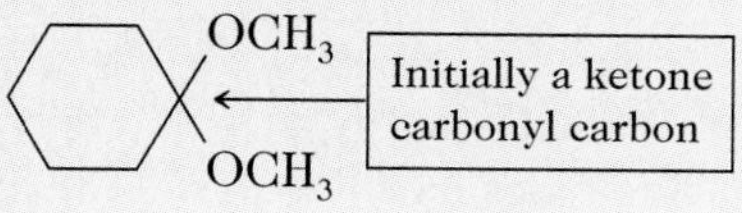

■ PRACTICE EXERCISE 9 Which of the following two compounds, if either, is an acetal or a ketal? Identify the carbon atom that came originally from the carbonyl group of a parent aldehyde or ketone.

(a) $CH_3OCH_2CH_2OCH_3$ (b) $CH_3CH_2OCOCH_2CH_3$ with CH_3 above and CH_3 below

Because acetals and ketals can be hydrolyzed, we have to be able to examine the structures of such compounds and write the structures of their parent alcohols and aldehydes (or ketones), the hydrolysis products. A worked example shows how this can be done.

EXAMPLE 14.7 Writing the Structures of the Products of the Hydrolysis of Acetals or Ketals

What are the products of the following reaction?

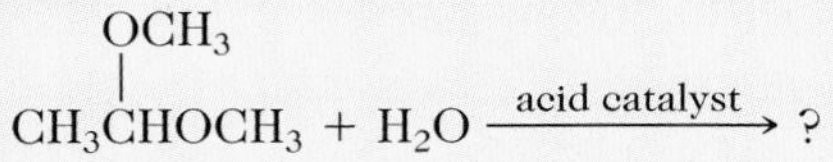

$$\underset{\text{CH}_3\text{CHOCH}_3}{\overset{\text{OCH}_3}{|}} + \text{H}_2\text{O} \xrightarrow{\text{acid catalyst}} ?$$

ANALYSIS The best way to proceed is to find the carbon atom in the structure that holds *two* oxygen atoms. This carbon is the carbonyl carbon atom of the parent aldehyde (or ketone). Break both of its bonds to these oxygen atoms. *Do not break any other bonds.* Then at the carbon that once held two oxygens, make a carbonyl group. The other groups, those of the OR type (here, OCH$_3$) become alcohols

SOLUTION The final products of the hydrolysis of the given acetal are ethanal and methanol.

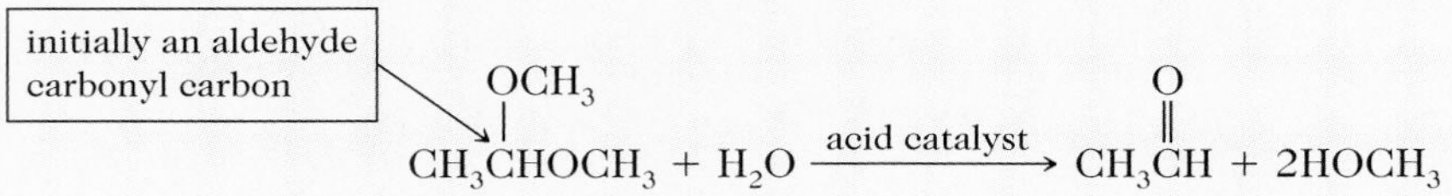

initially an aldehyde carbonyl carbon

$$\underset{\text{CH}_3\text{CHOCH}_3}{\overset{\text{OCH}_3}{|}} + \text{H}_2\text{O} \xrightarrow{\text{acid catalyst}} \overset{\text{O}}{\underset{}{\text{CH}_3\text{CH}}} + 2\text{HOCH}_3$$

■ PRACTICE EXERCISE 10 Write the structures of the aldehydes (or ketones) and the alcohols that are obtained by hydrolyzing the following compounds. If they do not hydrolyze like acetals or ketals, write "no reaction."

(a) CH$_3$OCH$_2$OCH$_3$ (b) CH$_3$OCH$_2$CH$_2$OCH$_2$CH$_3$ (c) $\underset{\text{CH}_3}{\underset{|}{\overset{\text{H}_3\text{C OCH}_3}{\overset{|\ \ |}{\text{CH}_3\text{CHCOCH}_3}}}}$

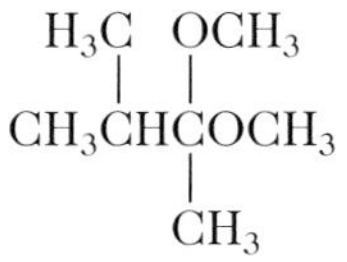

Other Reactions Involving Aldehydes and Ketones Aldehydes and ketones are able to enter into reactions that create larger molecules from smaller molecules by forming new carbon–carbon bonds. The *aldol condensation* is one such reaction. The simplest illustration is the reaction of ethanal with itself in the presence of dilute sodium hydroxide or an enzyme (called *aldolase*).

■ The mechanism of the aldol condensation is given in Appendix D.

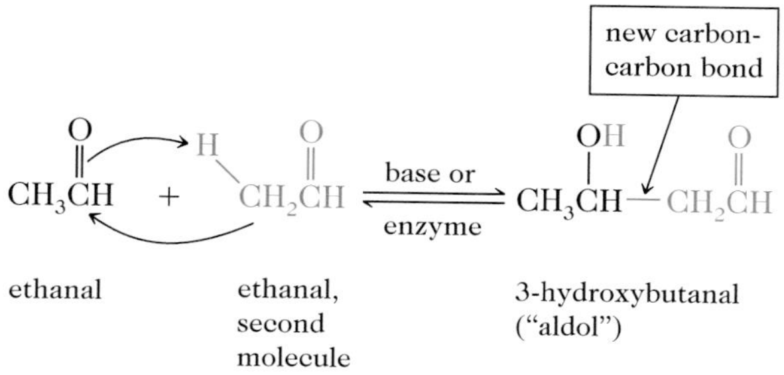

It is little more than the addition of one aldehyde molecule to the carbonyl double bond of another, but it is a *reversible* reaction, as the equilibrium arrows indicate. The forward reaction makes a new carbon–carbon single bond. The reverse reaction, called the *reverse aldol,* breaks this same bond.

An aldol condensation occurs in those cells of ours where glucose molecules are made from smaller molecules. The breakdown of glucose occurs by a different series of reactions called *glycolysis,* and one step is a reverse aldol condensation.

SUMMARY

Naming aldehydes and ketones The IUPAC names of aldehydes and ketones are based on a parent compound, one with the longest chain that includes the carbonyl group. The names of aldehydes end in *-al* and of ketones in *-one,* and the chains are numbered so as to give the carbonyl carbons the lower of two possible numbers.

Physical properties of aldehydes and ketones The carbonyl group confers moderate polarity, which gives aldehydes and ketones higher boiling points and solubilities in water than hydrocarbons but lower boiling points and solubilities in water than alcohols (that have comparable formula masses).

Ease of oxidation of aldehydes Aldehydes are easily oxidized to carboxylic acids, but ketones resist oxidation. Aldehydes give a positive Tollens' test and ketones do not. α-Hydroxy aldehydes and ketones give the Benedict's test.

Complex ions The reagents for the Tollens' and Benedict's tests contain complex ions, the silver diammine complex in Tollens' reagent and the copper(II) citrate complex ion in Benedict's test. In the first, Ag^+, the central metal ion, holds two molecules of the ligand, NH_3. The ligand for the Cu^{2+} ion in Benedict's reagent is the negative ion of citric acid. Complex ions help to keep transition metal ions in solution, even in base.

Hemiacetals and hemiketals When an aldehyde or a ketone is dissolved in an alcohol, some of the alcohol adds to the carbonyl group of the aldehyde or ketone. The equilibrium thus formed includes molecules of a hemiacetal (or hemiketal). The chart at the end of this summary outlines the chemical properties of the aldehydes and ketones we have studied.

Hemiacetals and hemiketals are usually unstable compounds that exist only in an equilibrium involving the parent carbonyl compound and the parent alcohol (which generally is the solvent). Hemiacetals and hemiketals readily break back down to their parent carbonyl compounds and alcohols. When an acid catalyst is added to the equilibrium, a hemiacetal or hemiketal reacts with more alcohol to form an acetal or ketal.

Acetals and ketals Acetals and ketals are 1,1-diethers that are stable in aqueous base or in water but not in aqueous acid. Acids catalyze the hydrolysis of acetals and ketals, and the final products are the parent aldehydes (or ketones) and alcohols.

Summary of reactions (We omit the aldol condensation here.)

Aldehydes

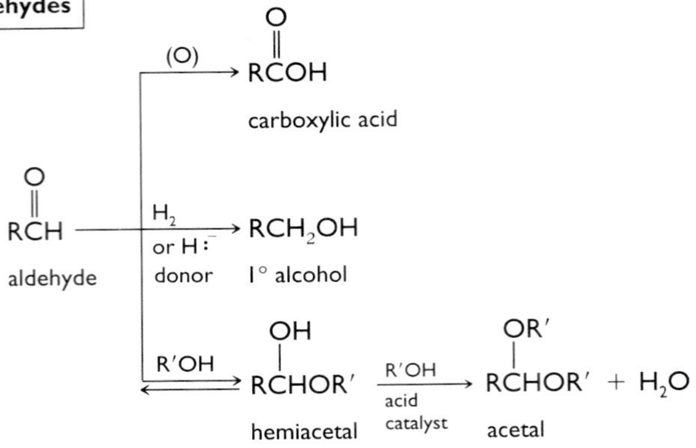

Ketones

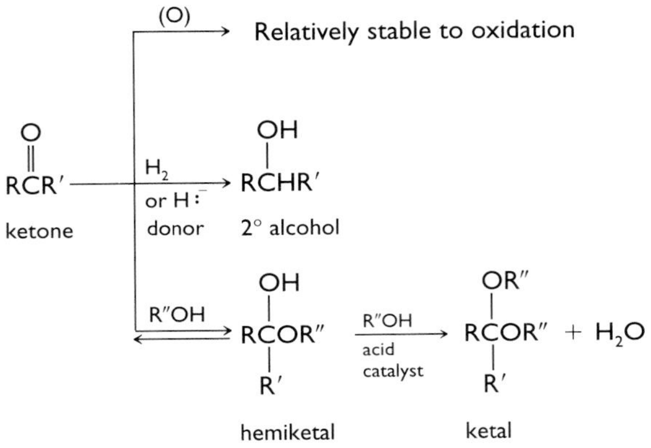

Acetals and Ketals

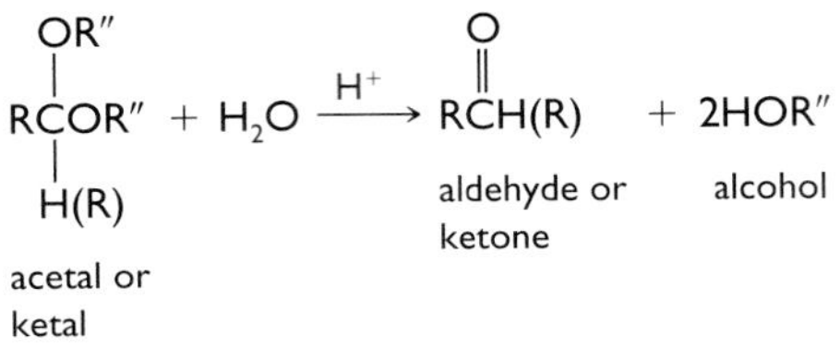

REVIEW EXERCISES

The answers to Review Exercises whose numbers are in color are found in Appendix E. The answers to the other Review Exercises are found in the Study Guide that accompanies this book. The more challenging questions are marked with asterisks.

Names and Structures

14.1 Give the names of the functional groups present in the following structural formulas.

(a) CH_3CCH_3 (with =O above central C)

(b) CH_3CH_2CH (with =O above terminal C)

(c) $HOCCH_2CH_2CH_3$ (with =O above C)

(d) CH_3CHO

(e) 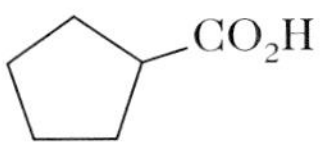 (cyclopentyl)CO_2H

(f) 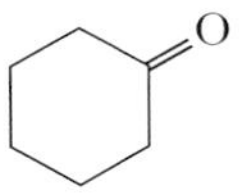 (cyclohexanone structure)

14.2 To display the *structural* differences among aldehydes, ketones, carboxylic acids, and esters, write the structure of one example of each using three carbons per molecule.

14.3 Write the structure of each of the following compounds.
(a) 3-Methylpentanal
(b) 3-Bromocyclopentanone
(c) 1-Cyclopentyl-2-butanone
(d) Acetaldehyde
(e) Acetone
(f) Butyraldehyde

14.4 What are the structures of the following compounds?
(a) Benzoic acid
(b) 2-Ethylhexanal
(c) 2,2-Dimethylcyclohexanone
(d) 1,3-Diphenyl-2-propanone

(e) Propionaldehyde

(f) Benzaldehyde

14.5 Although we can write structures that correspond to the following names, when we do, we find that the names aren't proper. How should these compounds be named in the IUPAC system?

(a) 6-Methylcyclohexanone

(b) 1-Hydroxy-1-propanone (Give the common name.)

(c) 1-Methylbutanal

(d) 2-Methylethanal

(e) 2-Propylpropanal

14.6 The following names can be used to write structures, but the names turn out to be improper. What should be their IUPAC names?

(a) 2-sec-Butylbutanal

(b) 1-Phenylethanal

(c) 4,5-Dimethylcyclopentanone

(d) 1-Hydroxyethanal (Give the common name.)

(e) 1-Butanone

14.7 If the IUPAC name of **A** is 2-ketopropanal, then what is the IUPAC name of compound **B**?

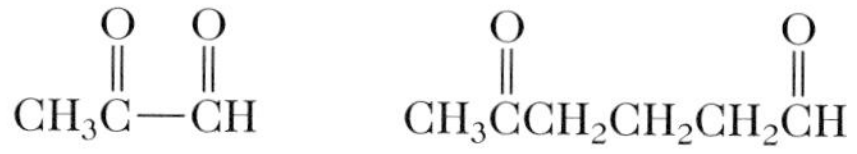

$$CH_3C-CH \qquad CH_3CCH_2CH_2CH_2CH$$

A **B**

14.8 Write the IUPAC names of the following compounds.

(a) $CH_3CCH_2CH_2CHCH_3$
     $CH_3CH_2CH_2$

(b) $CH_3CH_2CHCHCH_3$
     CH_3
     CH

(c)

(d) $HCCH_2CCH_3$
     $CH_2CH_2CH_3$
     CH_3

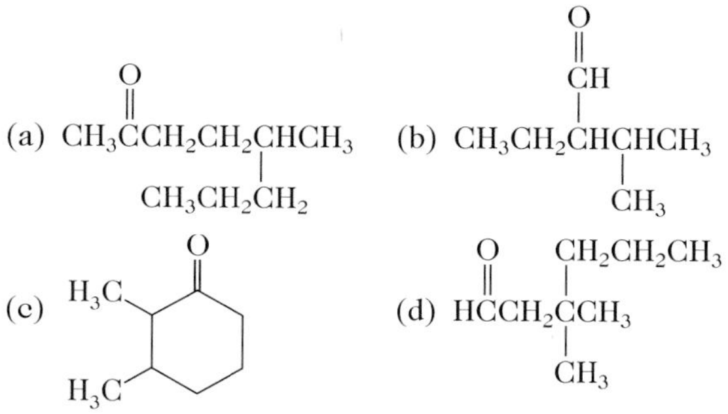

14.9 If the common name of $CH_3CH_2CH_2CH_2CO_2H$ is valeric acid, then what is the most likely common name of the following compound?

$$CH_3CH_2CH_2CH_2CHO$$

14.10 If the common name of **E** is glyceraldehyde, what is the most the likely common name of **F**?

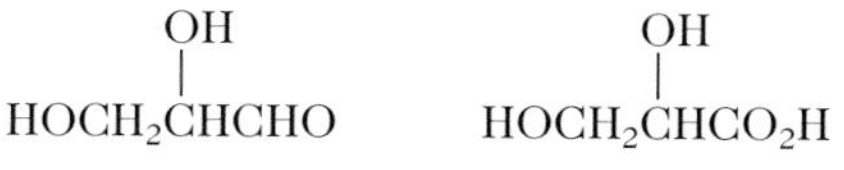

$$HOCH_2CHCHO \qquad HOCH_2CHCO_2H$$
$$OH \qquad\qquad OH$$

E **F**

Physical Properties of Aldehydes and Ketones

14.11 Arrange the following compounds in their order of increasing boiling points. Do this by placing the letters that identify them in the correct order, starting with the lowest

boiling compound and moving in order to the highest boiling compound. (They have about the same formula masses.)

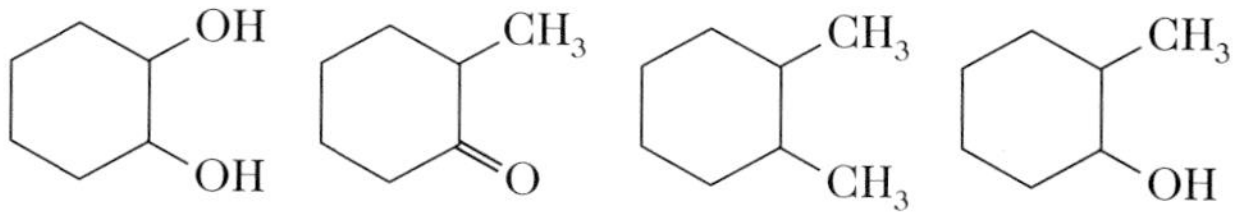

A   B   C   D

14.12 Arrange the following compounds in their order of increasing boiling points. Do this by placing the letters that identify them in the correct order, starting with the lowest boiling compound on the left in the series and moving to the highest boiling compound. (They all have about the same formula mass.)

$$CH_3CH_2CH_2CH_2CHCH_3 \qquad HOCH_2CH_2CH_2CHCH_3$$
$$CH_3 \qquad\qquad\qquad CH_3$$

A **B**

$$HOCH_2CH_2CH_2CHOH \qquad CH_3CH_2CH_2CH_2CCH_3$$
$$CH_3 \qquad\qquad\qquad O$$

C **D**

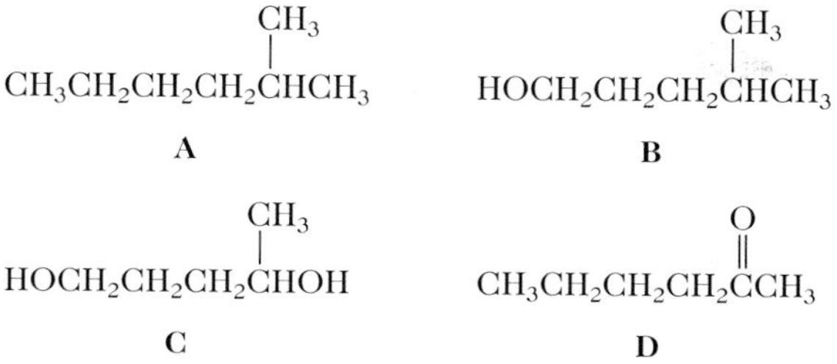

14.13 Reexamine the compounds of Review Exercise 14.11, and arrange them in their order of increasing solubility in water.

14.14 Arrange the compounds of Review Exercise 14.12 in their order of increasing solubility in water.

14.15 Draw the structure of a water molecule and a molecule of propanal and align them on the page to show how the propanal molecule can accept a hydrogen bond from the water molecule. Use a dotted line to represent this hydrogen bond and place $\delta+$ and $\delta-$ symbols where they are appropriate.

Oxidation of Alcohols and Aldehydes

14.16 What are the structures and the IUPAC names of the aldehydes and ketones to which the following compounds can be oxidized?

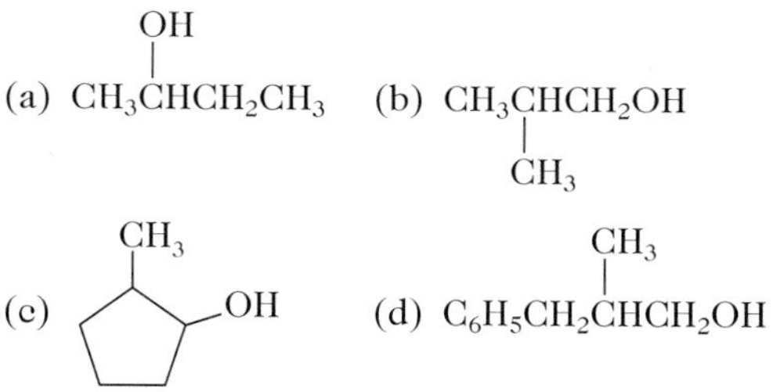

(a) $CH_3CHCH_2CH_3$
     OH

(b) CH_3CHCH_2OH
     CH_3

(c)

(d) $C_6H_5CH_2CHCH_2OH$
     CH_3

14.17   Examine each of the following compounds to see whether it can be oxidized to an aldehyde or to a ketone. If it can, write the structure of the aldehyde or ketone.

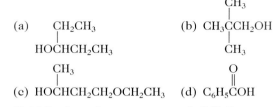

(a) CH_2CH_3
           |
        $HOCHCH_2CH_3$

(b) CH_3
        |
     CH_3CCH_2OH
        |
     CH_3

(c) CH_3
        |
     $HOCHCH_2CH_2OCH_2CH_3$

(d)   O
       ‖
     C_6H_5COH

*__14.18__   An unknown compound, C_3H_6O, reacted with permanganate ion to give $C_3H_6O_2$, and the same unknown also gave a positive Tollens' test. Write the structures of C_3H_6O and $C_3H_6O_2$.

*__14.19__   An unknown compound, $C_3H_6O_2$, could be oxidized easily by permanganate ion to $C_3H_4O_3$, and it gave a positive Benedict's test. Write Structures for $C_3H_6O_2$ and $C_3H_4O_3$.

*__14.20__   Which of the following compounds can be expected to give a positive Benedict's test? All are intermediates in metabolism.

(a) OH
       |
    $HOCH_2CHCHO$

(b)   O
         ‖
    $HOCH_2CCH_2OH$

(c) OH
       |
    $HOCH_2CHCHCHO$
              |
             OH

(d)   O
         ‖
    $HOCH_2CH_2CCH_3$

14.21   Concerning complex ions,
(a) The cations of what kinds of elements are usually involved?
(b) What *kinds* of particles commonly are ligands?
(c) Give the *formulas* of three examples of negatively charged ligands that come from the same family in the periodic table.
(d) Give the *formulas* of two electrically neutral ligands that form complex ions with Cu^{2+}. Write the formulas of these complex ions.

14.22   Concerning complex ions in test reagents,
(a) What is the *formula* of the complex ion in Tollens' reagent and why is it important that Ag^+ be so complexed?
(b) What is the *function* of the citrate ion in Benedict's reagent?

14.23   What is the *formula* of the precipitate that forms in a positive Benedict's test?

14.24   Clinitest tablets are used for what?

14.25   What is one practical commercial application of the Tollens' reagent system?

14.26   One of the steps in the metabolism of fats and oils in the diet is the oxidation of the following compound:

OH
|
$CH_3CHCH_2CO_2^-$

Write the structure of the product of this oxidation.

*__14.27__   One of the important series of reactions in metabolism is called the *citric acid cycle*. Structures **A** and **B** are of compounds (actually, anions) that participate in this cycle. One of them, isocitric acid, is oxidized to a ketone. The other is not. Which one is isocitric acid, **A** or **B**? Write the structure of its corresponding ketone.

$CH_2CO_2^-$
|
$HOCCO_2^-$
|
$CH_2CO_2^-$

A

$HOCHCO_2^-$
|
$CHCO_2^-$
|
$CH_2CO_2^-$

B

Reduction of Aldehydes and Ketones

14.28   The hydride ion, reacts as follows with water:
$$H{:}^- + H{-}OH \longrightarrow H{-}H + OH^-$$
The hydride ion reacts in a similar way with CH_3CH_2OH. Write the net ionic equation for this reaction.

14.29   Consider the reaction that occurs when a hydride ion transfers from its donor (which we can write *Mtb*:H) to ethanal.
(a) Write the structure of the organic anion that forms when the hydride ion is transferred to ethanal.
(b) What is the net ionic equation of the reaction of the anion formed in part (a) with water?
(c) What is the IUPAC name of the organic product of the reaction with water of the product of part (b)?

*__14.30__   The metabolism of aspartic acid, an amino acid, occurs by a series of steps. A portion of this series is indicated below, where NAD:H is a reducing agent that becomes NAD^+ as it transfers hydride ion.

$^+NH_3CHCO_2^-$ $\xrightarrow{\text{two steps}}$ $^+NH_3CHCO_2^-$ $\xrightarrow{NAD:H}$
| |
$CH_2CO_2^-$ $CH_2CH{=}O$

aspartate ion

$NAD^+ + {}^+NH_3CHCO_2^-$
                       |
                   (?) $\xrightarrow{H_2O}$ **B** $+ OH^-$

A

Complete the structure of **A**, and write the structure of **B**.

14.31 Write the structures of the aldehydes or ketones that could be used to make the following compounds by reduction (hydrogenation).

(a) 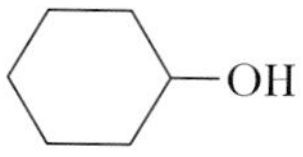 [cyclohexyl]—OH

(b) $CH_3\overset{\displaystyle OH}{\underset{\displaystyle |}{CH}}CH_2CH_3$

(c) $HOCH_2\overset{\displaystyle CH_3}{\underset{\displaystyle |}{CH}}CH_2CH_3$

(d) 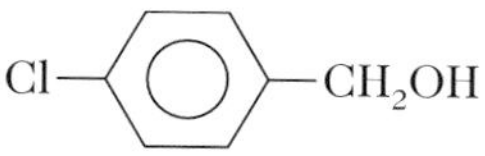 Cl—[benzene ring]—CH_2OH

Hemiacetals and Acetals. Hemiketals and Ketals

14.32 Examine each structure and decide whether it represents a hemiacetal, a hemiketal, an acetal, a ketal, or something else.

(a) $CH_3O\overset{\displaystyle CH_3}{\underset{\displaystyle |}{CH}}OH$

(b) $CH_3\overset{\displaystyle OCH_3}{\underset{\displaystyle |}{CH}}OCH_3$

(c) $CH_3O\overset{\displaystyle CH_3}{\underset{\displaystyle |}{\underset{\displaystyle CH_3}{\overset{\displaystyle |}{C}}}}OCH_3$

(d) $CH_3O\overset{\displaystyle CH_3}{\underset{\displaystyle |}{CH}}CH_2OCH_3$

14.33 Examine each structure and decide whether it represents a hemiacetal, a hemiketal, an acetal, a ketal, or something else.

(a) $CH_3CH_2OCH_2OH$

(b) $HOCH_2CH_2OCH_2CH_2CH_3$

(c) 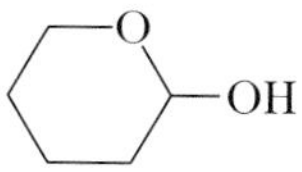 [ring with O]—OH

(d) $HOCH_2CH_2\overset{\displaystyle OCH_2CH_3}{\underset{\displaystyle |}{CH}}OCH_3$

14.34 Write the structures of the hemiacetals and the acetals that can form between butanal and the following two alcohols.
(a) Methanol   (b) Ethanol

14.35 What are the structures of the hemiketals and the ketals that can form between acetone and these two alcohols?
(a) Methanol   (b) Ethanol

*14.36 Write the structure of the hydroxyaldehyde (a compound having both the alcohol group and the aldehyde group in the same molecule) from which the following hemiacetal forms in a ring-closing reaction. (You may leave the chain of the open-chain compound somewhat coiled.)

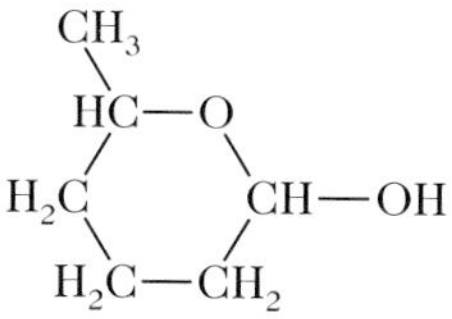

[ring structure with CH_3, HC—O, H$_2$C, CH—OH, H$_2$C—CH$_2$]

*14.37 One form in which a glucose molecule exists is given by the following structure. (*Note:* The atoms and groups that are attached to the carbon atoms of the six-membered ring must be seen as projecting *above* or *below* the ring.)

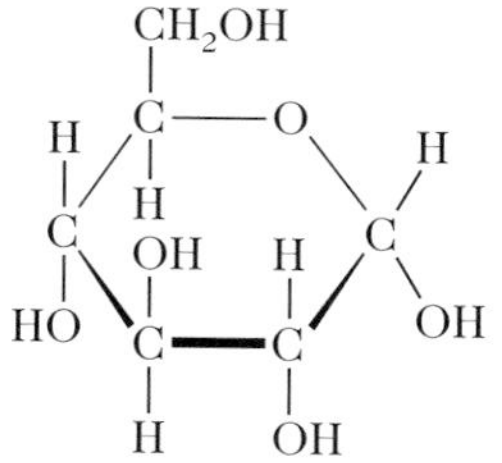

[glucose ring structure with CH$_2$OH, H, C, O, H, C, H, OH, H, C, HO, C, OH, H, OH]

(a) Place an asterisk by the hemiacetal carbon.
(b) Write the structure of the open-chain form that has a free aldehyde group. (You may leave the chain coiled.)

14.38 The digestion of some carbohydrates is simply their hydrolysis catalyzed by enzymes. Acids catalyze the same kind of hydrolysis of acetals and ketals. Write the structures of the products, if any, that form by the action of water and an acid catalyst on the following compounds.

(a) $CH_3O\overset{\displaystyle CH_3}{\underset{\displaystyle |}{CH}}OCH_3$

(b) $CH_3CH_2O\overset{\displaystyle CH_3}{\underset{\displaystyle |}{\underset{\displaystyle CH_3}{\overset{\displaystyle |}{C}}}}OCH_2CH_3$

(c) 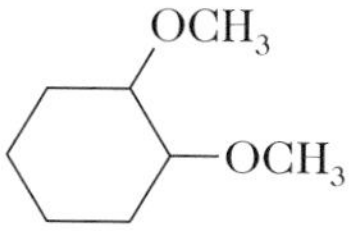 [cyclohexane ring with OCH$_3$ and OCH$_3$]

(d) 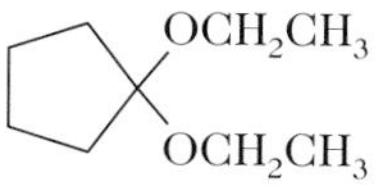 [cyclopentane ring with OCH$_2$CH$_3$ and OCH$_2$CH$_3$]

14.39 What are the structures of the products, if any, that form by the acid catalyzed reaction of water with the following compounds?

(a) $CH_3CH_2\overset{\displaystyle OCH_2CH_3}{\underset{\displaystyle |}{CH}}OCH_3$

(b) $CH_3\overset{\displaystyle CH_3}{\underset{\displaystyle |}{CH}}O\overset{\displaystyle CH_3}{\underset{\displaystyle |}{CH}}OCH_3$

(c) $CH_3O\overset{\displaystyle OCH_3}{\underset{\displaystyle |}{CH}}CH_2OCH_3$

(d) 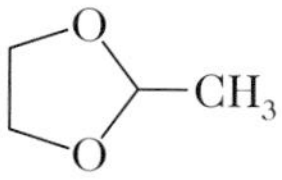 [dioxolane ring with O, O, CH$_3$]

Important Aldehydes and Ketones (Interaction 14.1)

14.40 Give the name of a specific aldehyde or ketone described in Interaction 14.1 that is
(a) A female sex hormone
(b) A good nail polish remover
(c) A preservative

Additional Exercises

14.41 Write the IUPAC names of the following compounds.

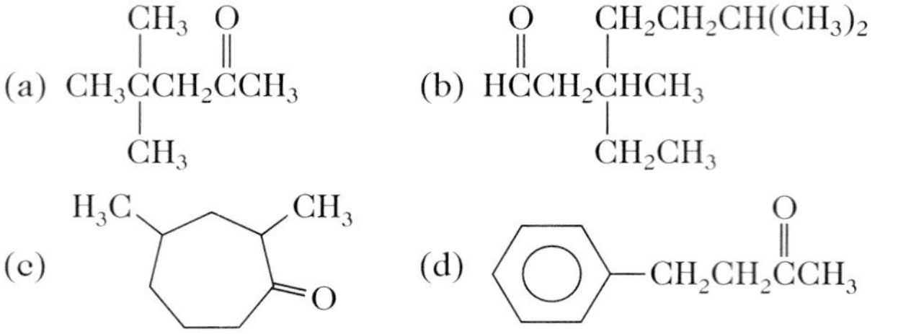

(a) $CH_3CCH_2CCH_3$ (with CH_3 groups and O)

(b) $HCCH_2CHCH_3$ (with O, $CH_2CH_2CH(CH_3)_2$, and CH_2CH_3)

(c) [structure: seven-membered ring with H_3C, CH_3 substituents and O]

(d) [structure: benzene ring]—$CH_2CH_2CCH_3$ (with O)

14.42 Draw the structures of molecules of methanol and acetone, and align them on the page to show how a hydrogen bond (which you are to indicate by a dotted line) can exist between the two. Place $\delta+$ and $\delta-$ symbols where they are appropriate.

*14.43 Neither the lactate ion nor the pyruvate ion gives a positive Tollens' test. When the body metabolizes the lactate ion, it oxidizes it to the pyruvate ion, $C_3H_3O_3^-$. Using only these facts, write the structure of the lactate ion.

14.44 If a donor of a hydride ion ($Mtb\!:\!H$) transfers it to a molecule of acetone,
(a) What is the structure of the organic ion that forms?
(b) What happens to the organic ion formed in part (a) in the presence of water? (Write a net ionic equation.)
(c) What is the IUPAC name of the organic product of the reaction of part (b)?

*14.45 One of the steps the body uses to make long-chain carboxylic acids is a reaction similar to the following reaction.

$$CH_3CCH_2CS\text{—}\boxed{enzyme} + NAD\!:\!H \longrightarrow$$

$$\boxed{?}\text{—}CH_2CS\text{—}\boxed{enzyme} + NAD^+$$

$$\mathbf{A} \quad \Big\downarrow^{H_2O} \longrightarrow \mathbf{B} + OH^-$$

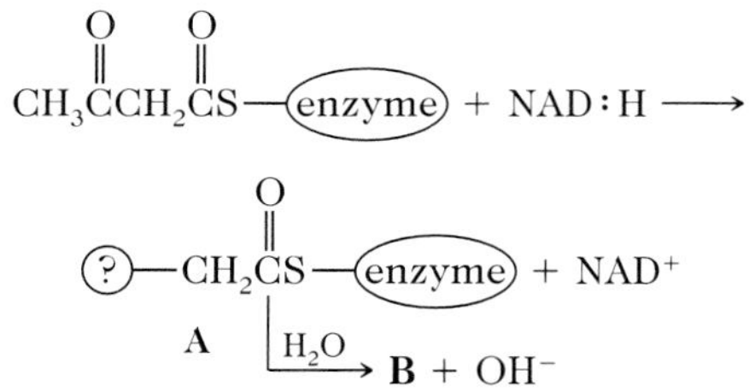

Complete the structure of **A**, and write the structure of **B**.

*14.46 Write the structure of a hydroxy ketone (a molecule that has both the OH group and the keto group) from which the following hemiketal forms in a ring-closing reaction. (You may leave the chain of the open-chain compound somewhat coiled.)

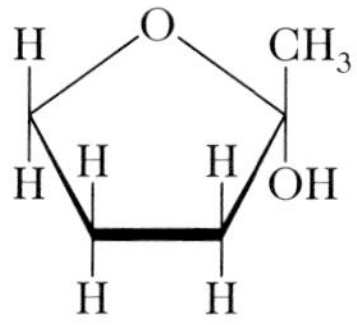

[structure: five-membered ring with O, H, CH₃, OH, and H substituents]

*14.47 Fructose occurs together with glucose in honey, and it is sweeter to the taste than table sugar. One form in

which a fructose molecule can exist is given by the following structure.

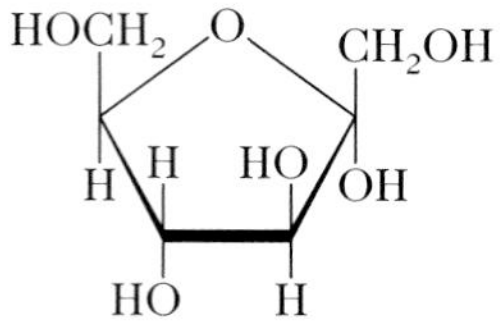

[structure: five-membered ring with HOCH₂, O, CH₂OH, H, HO, OH, HO, H]

(a) Place an asterisk by the carbon of the hemiketal system that came initially from the carbon atom of a keto group.
(b) In water, fructose exists in an equilibrium with an open-chain form of the given structure. This form has a keto group in the same molecule as five OH groups. Draw the structure of this open-chain form (leaving the chain coiled somewhat as it was in the structure that was given).

*14.48 Complete the following reaction sequences by writing the structures of the organic products that form. If no reaction occurs, write "no reaction." (Reviewed here too are some reactions of earlier chapters.)

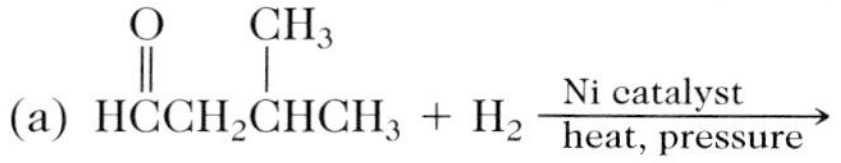

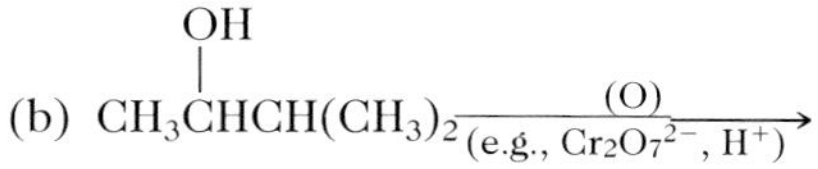

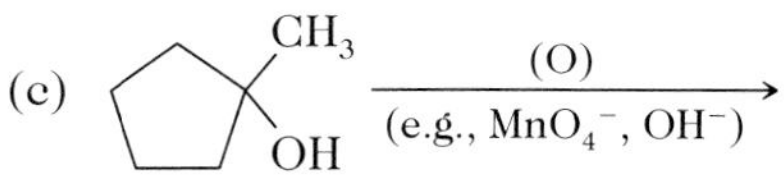

(a) $HCCH_2CHCH_3 + H_2 \xrightarrow[\text{heat, pressure}]{\text{Ni catalyst}}$ (with O and CH_3)

(b) $CH_3CHCH(CH_3)_2 \xrightarrow[\text{(e.g., Cr}_2O_7^{2-},\ H^+)]{\text{(O)}}$ (with OH)

(c) [cyclopentane ring with CH_3 and OH] $\xrightarrow[\text{(e.g., MnO}_4^-,\ OH^-)]{\text{(O)}}$

(d) $CH_3CH_2CH\!=\!CH_2 + H_2 \xrightarrow[\text{heat, pressure}]{\text{Ni catalyst}}$

(e) $CH_3OH + CH_3CH_2CHO \rightleftharpoons$

(f) $CH_3CHO + Mtb\!:\!H \xrightarrow[\text{by H}^+]{\text{followed}}$

   (where $Mtb\!:\!H$ is a metabolite able to donate hydride ion)

(g) $CH_3CHO + 2CH_3CH_2OH \xrightarrow[\text{catalyst}]{\text{acid}}$

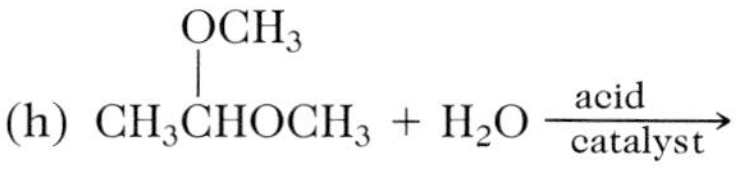

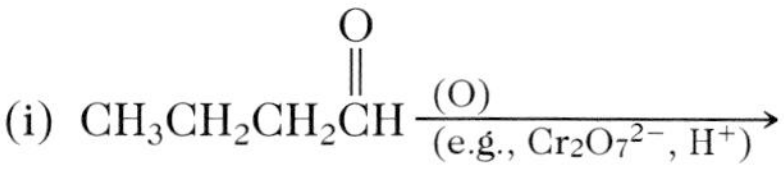

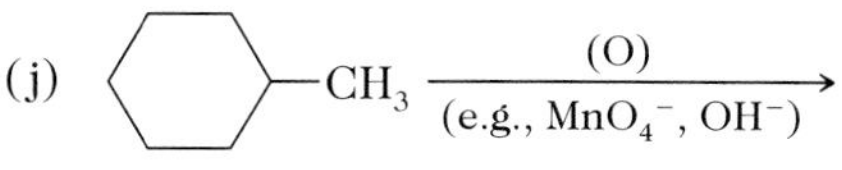

(h) $CH_3CHOCH_3 + H_2O \xrightarrow[\text{catalyst}]{\text{acid}}$ (with OCH_3)

(i) $CH_3CH_2CH_2CH \xrightarrow[\text{(e.g., Cr}_2O_7^{2-},\ H^+)]{\text{(O)}}$ (with O)

(j) [cyclohexane ring]—$CH_3 \xrightarrow[\text{(e.g., MnO}_4^-,\ OH^-)]{\text{(O)}}$

*14.49 Write the structures of the organic products that form in each of the following situations. If no reaction occurs, write "No reaction." (Some of the situations constitute a review of reactions in earlier chapters.)

(a) $O=$⬠ $\xrightarrow[\text{(e.g., MnO}_4^-,\ \text{OH}^-)]{\text{(O)}}$

(b) $CH_3CH_2\overset{\overset{\displaystyle OH}{|}}{C}CH_2CH_3 \xrightarrow[\text{(e.g., MnO}_4^-,\ \text{OH}^-)]{\text{(O)}}$
　　　　　$\underset{\displaystyle CH_3}{|}$

(c) $CH_3CHO + CH_3CH_2CH_2OH \rightleftharpoons$

(d) $CH_3CH_2\overset{\overset{\displaystyle CH_3}{|}}{O}CHCH_2OCH_3 + H_2O \longrightarrow$

(e) $CH_3\overset{\overset{\displaystyle CH_3}{|}}{C}OCH_2CH_3 \rightleftharpoons$
　　　　$\underset{\displaystyle OH}{|}$

(f) ⬠$-\overset{\overset{\displaystyle O}{\|}}{C}H + Mtb\!:\!H \xrightarrow[\text{by } H^+]{\text{(followed}}$

(where $Mtb\!:\!H$ is a metabolite able to donate hydride ion)

(g) $CH_3CH_2\overset{\overset{\displaystyle CH_3CH_2O}{|}}{C}CH_3 + H_2O \xrightarrow[\text{catalyst}]{\text{acid}}$
　　　　　$\underset{\displaystyle CH_3CH_2O}{|}$

(h) $CH_3\overset{\overset{\displaystyle OH}{|}}{C}H-$⬠$\xrightarrow[\text{(e.g., Cr}_2O_7^{2-},\ H^+)]{\text{(O)}}$

(i) (furan) $+ H_2 \xrightarrow{\text{Ni catalyst}}$

(j) (naphthalene)$-\overset{\overset{\displaystyle O}{\|}}{C}H + 2CH_3OH \xrightarrow[\text{catalyst}]{\text{acid}}$

*14.50 Catalytic hydrogenation of compound **A** (C_3H_6O) gave **B** (C_3H_8O). When **B** was heated strongly in the presence of sulfuric acid, it changed to compound **C** (C_3H_6). The acid-catalyzed addition of water to **C** gave compound **D** (C_3H_8O); and when **D** was oxidized, it changed to **E** (C_3H_6O). Compounds **A** and **E** are isomers, and compounds **B** and **D** are isomers. Write the structures of compounds **A** through **E**.

*14.51 When compound **F** ($C_4H_{10}O$) was gently oxidized, it changed to compound **G** (C_4H_8O), but vigorous oxidation changed **F** (or **G**) to compound **H** ($C_4H_8O_2$). Action of hot sulfuric acid on **F** changed it to compound **I** (C_4H_8). The addition of water to **I** (in the presence of an acid catalyst) gave compound **J** ($C_4H_{10}O$), a compound that could not be oxidized. Compounds **F** and **J** are isomers. Write the structures of compounds **F** through **J**.

*14.52 A student was assigned the preparation of the dimethyl acetal of butanal for which the equation is

$$CH_3CH_2CH_2CH{=}O + 2CH_3OH \xrightarrow[\text{catalyst}]{\text{acid}}$$
$$CH_3CH_2CH_2CH(OCH_3)_2 + H_2O$$

A solution of 12.5 g of butanal in 50.0 mL of methanol was prepared for accomplishing this reaction. The density of methanol is 0.787 g/mL.
(a) How many moles of butanal were taken?
(b) How many moles of methanol were used?
(c) Was sufficient methanol taken? (Calculate the minimum number of grams of methanol that would be required.)
(d) How many grams of water would be obtained?
(e) Offer a reason for using an excess quantity of methanol.

*14.53 In Review Exercise 12.19, polyvinyl acetate was described as a polymer from which Butvar is made and that Butvar is used to make safety glass. You may wish to refer to your answer to Review Exercise 12.19 for the parts of this Review Exercise.
(a) Prepare the structure of a segment of the polyvinyl acetate molecule consisting of *two* repeating units.
(b) Polyvinyl acetate can be converted into the corresponding *polyvinyl alcohol* by the replacement of all of the CH_3CO groups attached to the oxygens of polyvinyl acetate's chain by hydrogen atoms. This leaves a very long alkane chain with OH groups on every other carbon atom. Write the structure of a segment of polyvinyl alcohol consisting of *two* repeating units.
(c) Write the structure of the *monomer* of polyvinyl alcohol. This monomer does not exist (explaining why its polymer must be made indirectly). Is the monomer properly called an *alcohol* by our definition? What kind of an "alcohol" is it? Is this kind of "alcohol" stable? Does polyvinyl alcohol have OH groups that are properly called alcohol groups? Explain.
(d) Butvar is made by the combination of butanal with polyvinyl alcohol. *Cyclic* acetal systems form in which every other carbon of the main alkane chain is part of the ring system of these cyclic acetals. The generic name of this new polymer is polyvinyl butyral. Write the structure of a segment of this polymer that includes one cyclic acetal system.

15

CARBOXYLIC ACIDS AND ESTERS

Those who race in yachts want sails made of strong, lightweight, quick-drying fabrics, like Dacron. Chemically, Dacron is a synthetic polyester. We'll learn about esters in this chapter, but mostly to be able to understand edible fats and oils or cell membranes.

This Chapter in Context

To understand cell membranes, as well as the fats and oils of our diets, we need to learn about carboxylic acids and their salts and esters. For fundamental insights into the molecular powerhouse of most cells, we must study phosphoric acid esters and anhydrides. Our study of the major functional groups at the molecular level of life thus continues.

15.1 Occurrence, Names, and Physical Properties of Carboxylic Acids

The carboxylic acids are polar compounds whose molecules form strong hydrogen bonds to each other.

Carboxylic acids and *sulfonic acids* are the two main types of organic acids. The former occur widely in nature (Table 15.1); the latter are less common.

$$\overset{\displaystyle O}{\overset{\displaystyle \|}{-C}}-OH \quad \text{or} \quad -CO_2H \quad \text{or} \quad -COOH \qquad \overset{\displaystyle O}{\underset{\displaystyle O}{\overset{\displaystyle \|}{\underset{\displaystyle \|}{-S}}}}-OH \quad \text{or} \quad -SO_3H$$

carboxyl group

sulfonic acid group

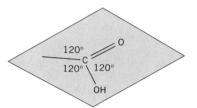

The carboxyl group has a planar geometry at the carbonyl group.

In **carboxylic acids,** the carbonyl carbon holds an OH group and either another C or an H atom. Those with long, straight, alkane-like chains are called **fatty acids,** because they're products of the digestion of butterfat, olive oil, and similar fatty substances in the diet.

Formic acid, the simplest acid, has a sharp, irritating odor and is responsible for the sting of nettle plants and certain ants. Acetic acid, the second acid in Table 15.1, gives tartness to vinegar, where its concentration is 4% to 5%. Butyric acid, a C_4 acid, causes the odor of rancid butter. Valeric acid is named after one of its sources, the valerian plant. Valeric acid has a very obnoxious odor. Other acids with vile odors are caproic, caprylic, and capric acids, which get their names from the Latin *caper,* meaning "goat," also a reference to their odor.

Dicarboxylic acids have two carboxyl groups per molecule. Oxalic acid, the simplest example, gives the sour taste to rhubarb. Citric acid, a *tricarboxylic acid,* causes the tartness of citrus fruits. Lactic acid, which has both a carboxyl and a 2° alcohol group, gives the tartness to sour milk.

■ "Carboxyl" comes from *carbonyl + hydroxyl.*

$$HO-\overset{\displaystyle O}{\overset{\displaystyle \|}{C}}-\overset{\displaystyle O}{\overset{\displaystyle \|}{C}}-OH \qquad\qquad \underset{\displaystyle CH_2CO_2H}{\overset{\displaystyle CH_2CO_2H}{HOCCO_2H}} \qquad\qquad \overset{\displaystyle OH}{\underset{\displaystyle CH_3CHCO_2H}{|}}$$

oxalic acid
(in rhubarb)

citric acid
(in lemon juice)

lactic acid
(in sour milk)

■ The lactate ion is made in muscles during strenuous exercise.

All carboxylic acids are weak acids, and they exist in the form of their anions both in basic solutions and in their salts. *It is largely as their anions that carboxylic acids occur in living cells and body fluids.*

TABLE 15.1 Carboxylic Acids

n	Structure	Name[a]	Origin of Name	MP (°C)	BP (°C)	Solubility in Water[b]	K_a (25 °C)
1	HCO_2H	Formic acid (methanoic acid)	L. *formica*, ant	8	101	∞	1.8×10^{-4} (20 °C)
2	CH_3CO_2H	Acetic acid (ethanoic acid)	L. *acetum*, vinegar	17	118	∞	1.8×10^{-5}
3	$CH_3CH_2CO_2H$	Propionic acid (propanoic acid	L. *proto, pion*, first, fat	−21	141	∞	1.3×10^{-5}
4	$CH_3(CH_2)_2CO_2H$	Butyric acid (butanoic acid)	L. *butyrum*, butter	−6	164	∞	1.5×10^{-5}
5	$CH_3(CH_2)_3CO_2H$	Valeric acid (pentanoic acid)	L. *valere*, valerian root	−35	186	4.97	1.5×10^{-5}
6	$CH_3(CH_2)_4CO_2H$	Caproic acid (hexanoic acid)	L. *caper*, goat	−3	205	1.08	1.3×10^{-5}
7	$CH_3(CH_2)_5CO_2H$	Enanthic acid (heptanoic acid)	Gr. *oenanthe*, vine blossom	−9	223	0.26	1.3×10^{-5}
8	$CH_3(CH_2)_6CO_2H$	Caprylic acid (octanoic acid)	L. *caper*, goat	16	238	0.07	1.3×10^{-5}
9	$CH_3(CH_2)_7CO_2H$	Pelargonic acid (nonanoic acid)	Pelargonium, geranium	15	254	0.03	1.1×10^{-5}
10	$CH_3(CH_2)_8CO_2H$	Capric acid (decanoic acid)	L. *caper*, goat	32	270	0.015	1.4×10^{-5}
12	$CH_3(CH_2)_{10}CO_2H$	Lauric acid (dodecanoic acid)	Laurel	44	—	0.006	—
14	$CH_3(CH_2)_{12}CO_2H$	Myristic acid (tetradecanoic acid)	*Myristica*, nutmeg	54	—	0.002	—
16	$CH_3(CH_2)_{14}CO_2H$	Palmitic acid (hexadecanoic acid)	Palm oil	63	—	0.0007	—
18	$CH_3(CH_2)_{16}CO_2H$	Stearic acid (octadecanoic acid)	Gr. *stear*, solid	70	—	0.0003	—

Miscellaneous

	Structure	Name[a]	Origin of Name	MP (°C)	BP (°C)	Solubility in Water[b]	K_a (25 °C)
	$C_6H_5CO_2H$	Benzoic acid	Gum benzoin	122	249	0.34 (25 °C)	6.5×10^{-5}
	$C_6H_5CH{=}CHCO_2H$	Cinnamic acid (trans isomer)	Cinnamon	132	—	0.04	3.7×10^{-5}
	$CH_2{=}CHCO_2H$	Acrylic acid	L. *acer*, sharp	13	141	Soluble	5.6×10^{-5}
	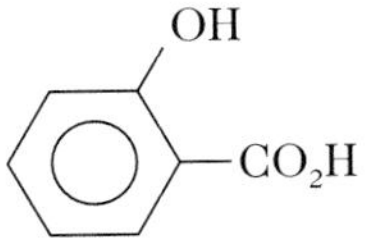 (Salicylic acid structure: benzene ring with OH and —CO₂H)	Salicylic acid	L. *salix*, willow	159	211	0.22 (25 °C)	1.1×10^{-3} (19 °C)

[a] In parentheses below each common name is the IUPAC name.
[b] In grams of acid per 100 g of water at 20 °C except where noted otherwise.

$$R-\overset{\overset{\displaystyle O}{\|}}{C}-O^- \quad \text{or} \quad RCO_2^- \quad \text{or} \quad RCOO^-$$

Symbols of anions of carboxylic acids

IUPAC Names of Carboxylic Acids End in -*oic acid* In the IUPAC rules for naming carboxylic acids, the *parent acid* is that of the longest chain that includes the carboxyl group. To name the parent acid, change the ending of the name of the alkane having the same number of carbons (the parent alkane) from *-e* to *-oic acid*. To number the chain of the parent acid, *always* give position 1 to the carbon atom of the carbonyl group. Examples of IUPAC names are given in parentheses in Table 15.1.

To name anions of carboxylic acids, the *carboxylate ions*, change the ending of the name of the parent *acid* from *-ic* to *-ate*, and omit the word *acid*. This rule applies to both IUPAC and common names. For example,

■ Always use the name of the *acid*, not the name of the alkane, to devise the name of the anion of the acid.

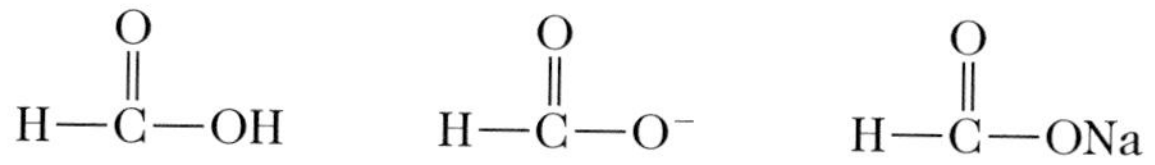

$$
\begin{array}{ccc}
\overset{\displaystyle O}{\underset{\displaystyle \|}{\;}} & \overset{\displaystyle O}{\underset{\displaystyle \|}{\;}} & \overset{\displaystyle O}{\underset{\displaystyle \|}{\;}} \\
H\!-\!C\!-\!OH & H\!-\!C\!-\!O^- & H\!-\!C\!-\!ONa
\end{array}
$$

| methanoic acid | methanoate ion | sodium methanoate |
| (formic acid) | (formate ion) | (sodium formate) |

■ Pronounce "oate" as "oh-ate."

EXAMPLE 15.1 Naming a Carboxylic Acid and Its Anion

The common name of the following carboxylic acid is isovaleric acid. What is the IUPAC name of this acid and its sodium salt?

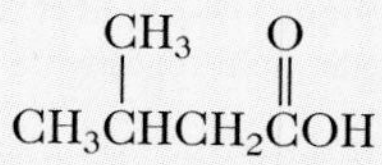

$$\underset{\text{CH}_3\text{CHCH}_2\text{COH}}{\overset{\displaystyle \text{CH}_3 \quad\;\; \text{O}}{}}$$

ANALYSIS The longest chain that includes the carboxyl group has four carbon atoms, so the parent acid is named by changing the name *butane*, the parent alkane, to *butanoic acid*. To number the chain, we start with the carboxyl group's carbon.

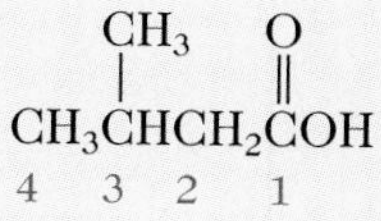

$$\underset{4 \quad 3 \quad 2 \quad 1}{\underset{\text{CH}_3\text{CHCH}_2\text{COH}}{\overset{\displaystyle \text{CH}_3 \quad\;\; \text{O}}{}}}$$

SOLUTION The methyl group is at position 3, so the IUPAC name of this acid is 3-methylbutanoic acid.

To name its anion, we drop *-ic acid* from the name of the acid and add *-ate*. Therefore the anion is 3-methylbutanoate, and the sodium salt is sodium 3-methylbutanoate. (Its common name is sodium isovalerate.)

■ **PRACTICE EXERCISE 1** What are the IUPAC names of the following compounds?

(a) 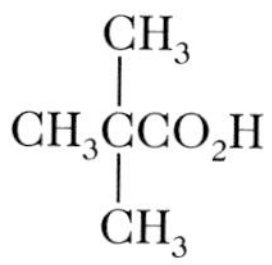 $\underset{\underset{\text{CH}_3}{|}}{\overset{\overset{\text{CH}_3}{|}}{\text{CH}_3\text{CCO}_2\text{H}}}$

(b) 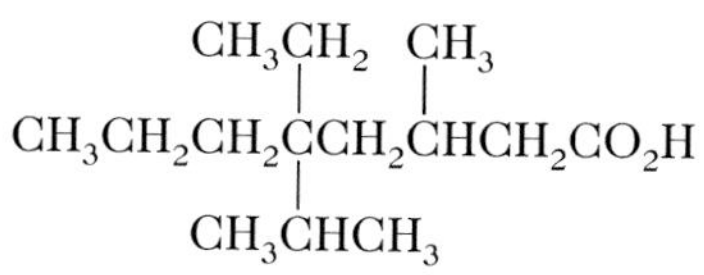 $\underset{\underset{\text{CH}_3\text{CHCH}_3}{|}}{\overset{\overset{\text{CH}_3\text{CH}_2 \quad\; \text{CH}_3}{| \qquad\;\; |}}{\text{CH}_3\text{CH}_2\text{CH}_2\text{CCH}_2\text{CHCH}_2\text{CO}_2\text{H}}}$

(c) $\text{CH}_3\text{CO}_2\text{Na}$

(d) 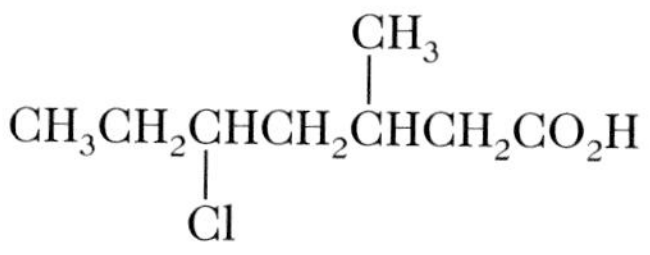 $\underset{\underset{\text{Cl}}{|}}{\overset{\overset{\quad\quad\quad\; \text{CH}_3}{\quad\quad\quad\; |}}{\text{CH}_3\text{CH}_2\text{CHCH}_2\text{CHCH}_2\text{CO}_2\text{H}}}$

■ **PRACTICE EXERCISE 2** If the IUPAC name of $HO_2CCH_2CO_2H$ is propanedioic acid (and not 1,3-propanedioic acid), what must be the IUPAC name for $HO_2CCH_2CH_2CH_2CO_2H$?

■ Oleic acid is actually the *cis* isomer. The name of the *trans* isomer is elaidic acid.

■ **PRACTICE EXERCISE 3** If the IUPAC name of $CH_3CH{=}CHCH_2CH_2CO_2H$ is 4-hexenoic acid, what must be the IUPAC name of the following acid? Its common name is *oleic acid*, and it is one of the products of the hydrolysis of almost any edible vegetable oil or animal fat. (The name of the straight-chain alkane with 18 carbon atoms is octadecane.)

$$CH_3(CH_2)_7CH{=}CH(CH_2)_7CO_2H$$

oleic acid (common name)

Carboxylic Acid Molecules Hydrogen Bond to Each Other Carboxylic acids have higher boiling points than alcohols of comparable formula masses because molecules of carboxylic acids form hydrogen-bonded pairs.

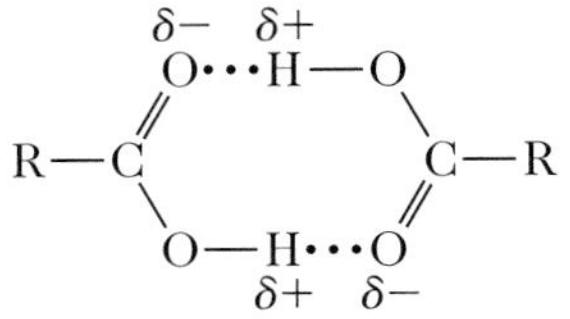

Hydrogen bonds ($\cdots$) hold two molecules of a carboxylic acid together.

This makes the effective formula mass of a carboxylic acid much higher than its calculated formula mass, and therefore the boiling point is higher.

The lower formula mass carboxylic acids (C_1–C_4) are soluble in water largely because the carboxyl group has *two* oxygen atoms that can accept hydrogen bonds from water molecules. In addition, the carboxyl group has the OH group that can donate hydrogen bonds.

■ Remember, we're interested in how structure affects solubility in water because water is the fluid medium in the body.

15.2 ACIDITY OF CARBOXYLIC ACIDS

The carboxylic acids are weak acids toward water but strong acids toward the hydroxide ion.

Aqueous solutions of carboxylic acids contain the following species in equilibrium.

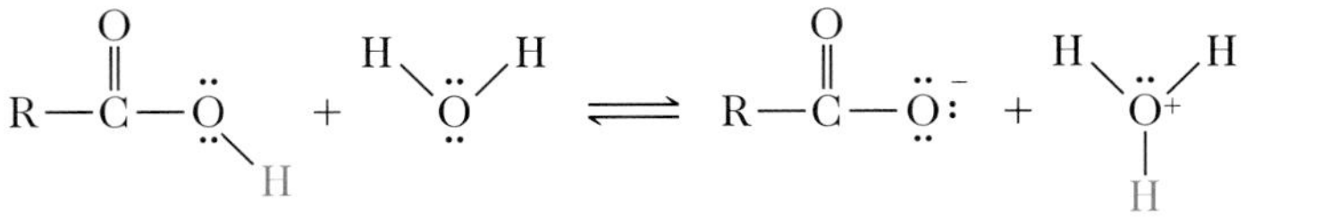

$$K_a = \frac{[RCO_2^-][H^+]}{[RCO_2H]}$$

| carboxylic acid | water | carboxylate ion | hydronium ion |
| (weaker acid) | (weaker base) | (stronger base) | (stronger acid) |

The K_a values of several carboxylic acids are given in Table 15.1, and you can see that most are on the order of 10^{-5}. Thus the carboxylic acids are weak acids (poor proton donors) toward water, and their percentage ionizations are low. For example, at 25 °C in a 1 *M* solution, the percentage ionization of acetic acid is only about 0.5%. In other words, only 1 molecule in 200 is ionized.

Carboxylic Acids Are Stronger Acids than Phenols or Alcohols Carboxylic acids ($K_a \approx 10^{-5}$) are several billion times stronger acids than alcohols ($K_a \approx 10^{-16}$), and roughly 100,000 times stronger acids than phenols ($K_a \approx 10^{-10}$). The greater acidity of carboxylic acids reflects the greater stability of carboxylate anions relative to that of the anions of alcohols or phenols. In the carboxylate ion, the negative charge is adjacent to the strongly electronegative carbonyl group, which withdraws some electron density of the charge. This helps to stabilize the anion of the acid by helping to spread the charge throughout a larger space than is possible in the anions of alcohols or phenols. A more stable ion means one easier to form.

Carboxylic Acids Are Neutralized by Strong Bases Toward relatively strong bases, like the hydroxide, carbonate, and bicarbonate ions, carboxylic acids are good proton donors. The bicarbonate ion in body fluids thus helps to neutralize the carboxylic acids we produce by normal metabolism. Otherwise, the pH of body fluids, like blood, would fall too low (become too acidic) to sustain life.

With hydroxide ion, the reaction is as follows.

$$RCO_2H + OH^- \longrightarrow RCO_2^- + H_2O$$

stronger acid · stronger base · weaker base · weaker acid

With bicarbonate ion, the chief base in the buffer system of the blood, the following reaction occurs.

$$RCO_2H + HCO_3^- \longrightarrow RCO_2^- + H_2O + CO_2$$

stronger acid · stronger base · weaker base · weaker acid

Some specific examples are as follows.

$$CH_3CO_2H + OH^- \longrightarrow CH_3CO_2^- + H_2O$$

acetic acid · acetate ion

$$CH_3(CH_2)_{16}CO_2H + OH^- \longrightarrow CH_3(CH_2)_{16}CO_2^- + H_2O$$

stearic acid (insoluble in water) · stearate ion (soluble in water)

$$C_6H_5CO_2H + HCO_3^- \longrightarrow C_6H_5CO_2^- + H_2O + CO_2$$

benzoic acid (insoluble in water) · benzoate ion (soluble in water)

■ **PRACTICE EXERCISE 4** Write the structures of the carboxylate anions that form when the following carboxylic acids are neutralized.

(a) $CH_3CH_2CO_2H$ (b) 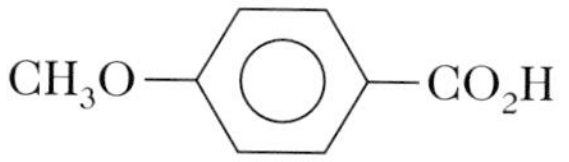 CH_3O—⬡—CO_2H (c) $CH_3CH=CHCO_2H$

Carboxylate Ions Are More Soluble in Water than Their Parent Acids The purified salts consisting of carboxylate ions and metal ions are genuine salts, that is, assemblies of oppositely charged *ions*, so all are solids at room temperature (Table 15.2). The sodium salts, for example, are soluble in water but completely insoluble in nonpolar solvents, like ether or gasoline. Traces of several carboxylate salts are added to foods or food packaging to retard molds (see Interaction 15.1).

■ Phenols are strong enough acids to neutralize OH^-, a strong base, but are not strong enough acids to neutralize HCO_3^-, a weak base.

■ The negative charges on the anions are balanced by the presence of some cation, the Na^+ or K^+ ion, for example. However, we'll work largely with net ionic equations.

■ The stearate ion is one of several organic ions in soap.

INTERACTION 15.1
FIBERS AND FOOD ADDITIVES.
SOME IMPORTANT CARBOXYLIC ACIDS AND SALTS

Acetic Acid Most people experience acetic acid directly in the form of its dilute solution in water, which is called vinegar. Acetic acid is an important industrial chemical, and nearly 5 billion pounds (38 billion moles) are manufactured each year in the United States. Some goes into the manufacture of acetate rayon, which is done by converting alcohol groups in cellulose into acetate ester groups (see Figure 1).

The acetate ion is one of the major intermediates in the metabolism of carbohydrates, lipids, and proteins.

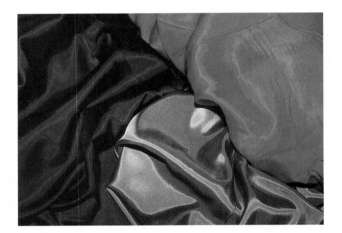

FIGURE 1 Acetate rayon is a lustrous fabric made from cellulose and acetic acid.

Propanoic Acid and Its Salts Propanoic acid occurs naturally in Swiss cheese in a concentration that can be as high as 1%. Its sodium and calcium salts are food additives used in baked goods and processed cheese to retard the formation of molds or the growth of bacteria. (On ingredient labels, these salts are listed under their common names, sodium or calcium propionate.)

Sorbic Acid and the Sorbates Sorbic acid, or 2,4-hexanedienoic acid, $CH_3CH=CHCH=CHCO_2H$, and its sodium or potassium salts are added in trace concentrations to a variety of foods to inhibit the growth of molds and yeasts. The sorbates often appear on ingredient lists for fruit juices, fresh fruits, wines, soft drinks, sauerkraut and other pickled products, and some meat and fish products. For food products that usually are wrapped, such as cheese and dried fruits, solutions of sorbate salts are sometimes sprayed onto the wrappers.

Sodium Benzoate Traces of sodium benzoate inhibit molds and yeasts in products that normally have pH values below 4.5 or 4.0. (The sorbates work better at slightly higher pH values—up to 6.5.) You'll see sodium benzoate on ingredient lists for beverages, syrups, jams and jellies, pickles, salted margarine, fruit salads, and pie fillings. Its concentration is low—0.05% to 0.10%—and neither benzoic acid nor the benzoate ion accumulates in the body.

TABLE 15.2 Some Sodium Salts of Carboxylic Acids

Common Name	Structure	MP (°C)	Solubility Water	Solubility Ether
Sodium formate (sodium methanoate)	HCO_2Na	253	Soluble	Insoluble
Sodium acetate (sodium ethanoate)	CH_3CO_2Na	323	Soluble	Insoluble
Sodium propionate (sodium propanoate)	$CH_3CH_2CO_2Na$	—	Soluble	Insoluble
Sodium benzoate	$C_6H_5CO_2Na$	—	66 g/100 mL	Insoluble
Sodium salicylate	(structure: benzene ring with OH and CO_2Na)	—	111 g/100 mL	Insoluble

Carboxylate Ions Are Good Proton Acceptors or Bases Because carboxylate ions are the anions of *weak* acids, we can infer that they themselves must be good bases (good proton acceptors), especially toward H_3O^+, a strong proton donor. At room temperature, the following neutralization of H_3O^+ by a carboxylate ion occurs virtually instantaneously. *It is the most important reaction of the carboxylate ion that we will study* because it makes the CO_2^- group a neutralizer of acid at the molecular level of life.

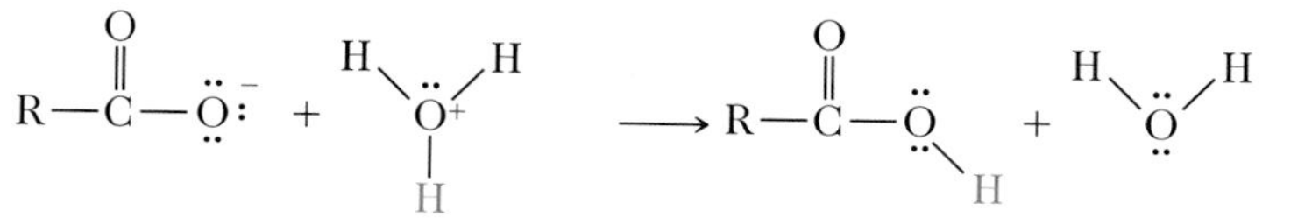

| carboxylate ion | hydronium ion | carboxylic acid | water |
| (stronger base) | (stronger acid) | (weaker acid) | (weaker base) |

For example,

$$C_6H_5CO_2^- + H_3O^+ \longrightarrow C_6H_5CO_2H + H_2O$$

benzoate ion   benzoic acid
(soluble in water)   (insoluble in water)

$$CH_3(CH_2)_{16}CO_2^- + H_3O^+ \longrightarrow CH_3(CH_2)_{16}CO_2H + H_2O$$

stearate ion   stearic acid
(soluble in water)   (insoluble in water)

The Carboxylic Acid Group Is a Solubility "Switch" Be sure to notice in the equations above the huge difference that H^+ makes in the solubilities of several species. The CO_2H group is, therefore, a "solubility switch" for an entire molecule, because a water-insoluble carboxylic acid almost instantly dissolves when we add a base and RCO_2H changes to RCO_2^-. Similarly, a water-soluble carboxylate anion, RCO_2^-, instantly changes to its much less soluble RCO_2H form when we add a strong acid. In other words, by suitably adjusting the pH of an aqueous solution, we can make a substance with a CO_2H group more soluble or less soluble in water.

■ **PRACTICE EXERCISE 5** Write the structures of the organic products of the reactions of the following compounds with dilute hydrochloric acid at room temperature.

(a) 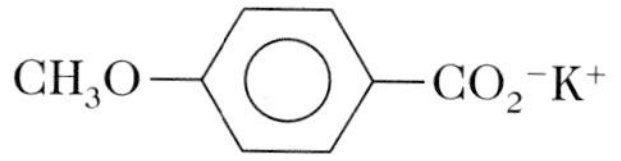 CH_3O—⬡—$CO_2^-K^+$ (b) $CH_3CH_2CO_2^-Li^+$

(c) $(CH_3CH{=}CHCO_2^-)_2Ca^{2+}$

Carboxylic Acids Strongly Resist Oxidation Simple carboxylic acids or their anions (those with no other functional groups) are the stable end products of the oxidations of 1° alcohols and aldehydes, as we have learned. Not even hot solutions of permanganate or dichromate ion break down the carboxyl group. The carboxylic acids will burn, of course, to give carbon dioxide and water.

15.3 CONVERSION OF CARBOXYLIC ACIDS TO ESTERS

Carboxylic acids can be used directly or indirectly to make esters from alcohols.

Carboxylic acids are the parents for several families that collectively are called **acid derivatives**. They include **acid chlorides, anhydrides, esters,** and **amides.** They're called acid *derivatives* because they can be made from the acids and they can be hydrolyzed back to the acids.

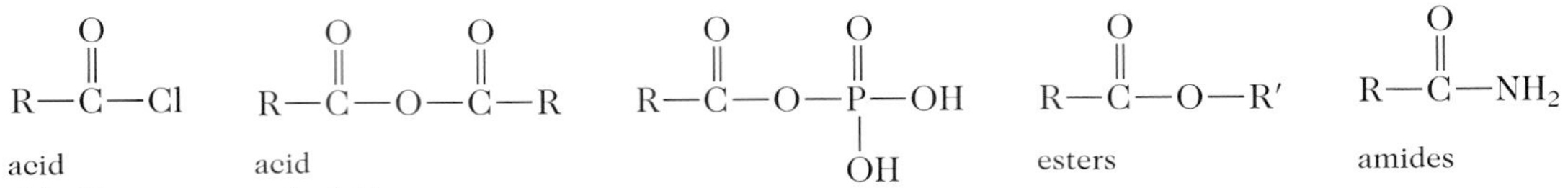

$$R-\overset{\displaystyle O}{\overset{\|}{C}}-Cl \qquad R-\overset{\displaystyle O}{\overset{\|}{C}}-O-\overset{\displaystyle O}{\overset{\|}{C}}-R \qquad R-\overset{\displaystyle O}{\overset{\|}{C}}-O-\overset{\displaystyle O}{\overset{\|}{\underset{\underset{\displaystyle OH}{|}}{P}}}-OH \qquad R-\overset{\displaystyle O}{\overset{\|}{C}}-O-R' \qquad R-\overset{\displaystyle O}{\overset{\|}{C}}-NH_2$$

acid   acid   esters   amides
chlorides   anhydrides

mixed anhydrides
with phosphoric acid

The **acyl group,** a carboxyl group minus OH, occurs in all of these.

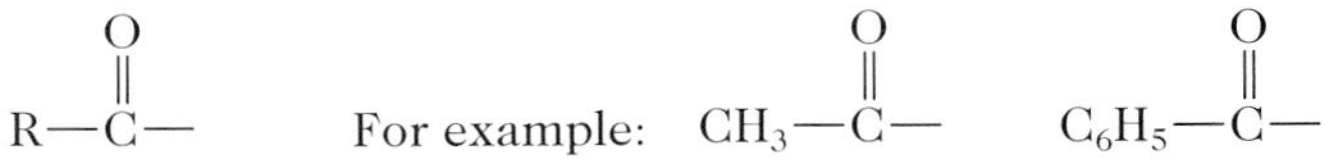

$$R-\overset{\displaystyle O}{\overset{\|}{C}}- \qquad \text{For example:} \quad CH_3-\overset{\displaystyle O}{\overset{\|}{C}}- \qquad C_6H_5-\overset{\displaystyle O}{\overset{\|}{C}}-$$

acyl group   acetyl group   benzoyl group

The reactions by which the acid derivatives are made as well as the reactions of the derivatives themselves, either in vitro or in vivo, generally occur as **acyl group transfer reactions.** The ability to transfer an acyl group, however, varies widely among the acid derivatives, as we will see.

Acid Chlorides Are the Most Reactive of Acid Derivatives Acid chlorides are very reactive acyl group transfer agents. They react exothermically with water to give the parent acids and hydrochloric acid. An acyl group transfers from Cl of the acid chloride to OH. We call the chloride ion the *leaving group.*

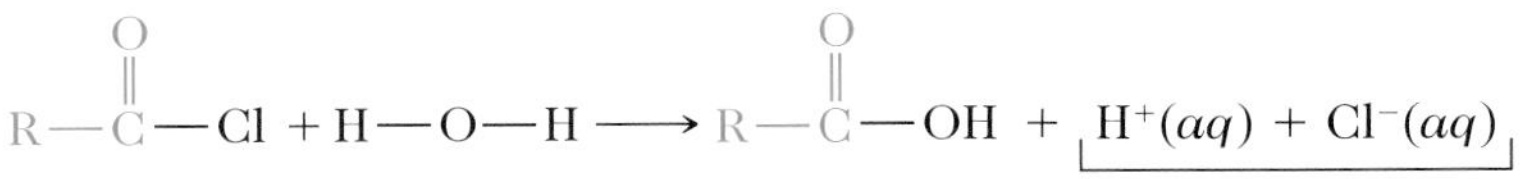

$$R-\overset{\displaystyle O}{\overset{\|}{C}}-Cl + H-O-H \longrightarrow R-\overset{\displaystyle O}{\overset{\|}{C}}-OH + \underline{H^+(aq) + Cl^-(aq)}$$

acid chloride   carboxylic acid   hydrochloric acid

Acid chlorides also readily transfer acyl groups to alcohols to give esters.

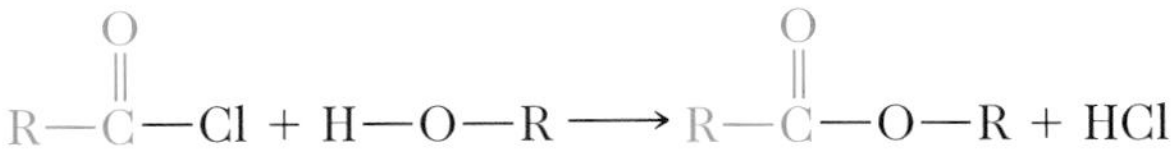

$$R-\overset{\displaystyle O}{\overset{\|}{C}}-Cl + H-O-R \longrightarrow R-\overset{\displaystyle O}{\overset{\|}{C}}-O-R + HCl$$

acid chloride   alcohol   ester

For example,

■ We'll learn how to name esters soon, but we will not develop the rules for naming acid chlorides or acid anhydrides.

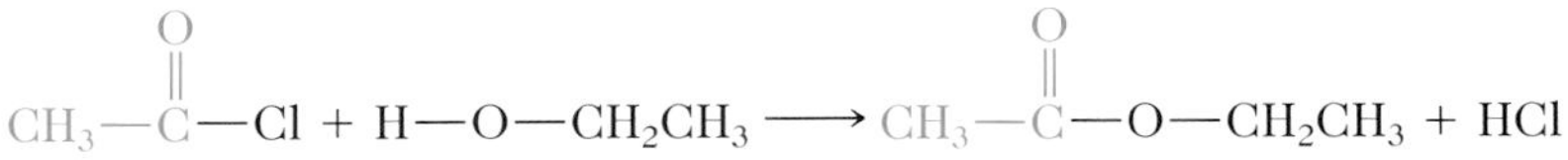

$$CH_3-\overset{\displaystyle O}{\overset{\|}{C}}-Cl + H-O-CH_2CH_3 \longrightarrow CH_3-\overset{\displaystyle O}{\overset{\|}{C}}-O-CH_2CH_3 + HCl$$

acetyl chloride   ethyl alcohol   ethyl acetate

This is an example of **esterification,** a reaction that makes an ester. We say that both the alcohol and carboxylic acid are *esterified.*

Acid Anhydrides Are Also Good Acyl Transfer Reactants When a carboxylic acid anhydride reacts with an alcohol, esterification also occurs. An acyl group transfers to OR. An ester and a carboxylic acid form, the latter coming from the leaving group in the anhydride. In general,

■ Either of the two acyl groups of the acid anhydride could be transferred. We've picked one.

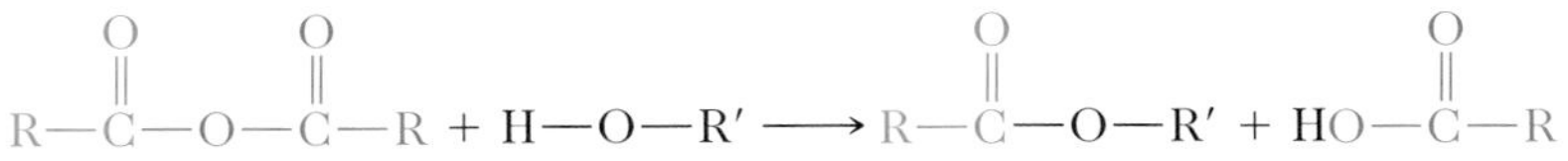

$$R-\overset{\displaystyle O}{\overset{\|}{C}}-O-\overset{\displaystyle O}{\overset{\|}{C}}-R + H-O-R' \longrightarrow R-\overset{\displaystyle O}{\overset{\|}{C}}-O-R' + HO-\overset{\displaystyle O}{\overset{\|}{C}}-R$$

acid anhydride   alcohol   ester   carboxylic acid

Phenols, like alcohols, can be also esterified. Aspirin, for example, is prepared by the reaction between acetic anhydride and the phenol group in salicylic acid.

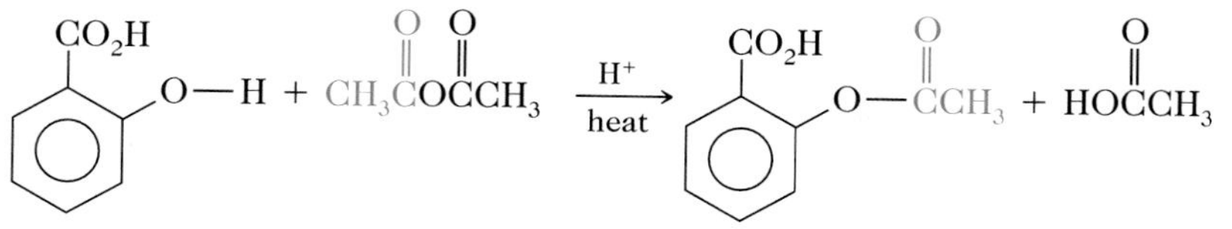

salicylic acid — acetic anhydride — acetylsalicylic acid (aspirin) — acetic acid

■ Both salicylic acid and acetic anhydride are common, readily available organic chemicals.

Direct Esterification of Acids Is Another Synthesis of Esters When a solution of a carboxylic acid in an alcohol is heated in the presence of a strong acid catalyst, the following species become involved in an equilibrium.

$$\underset{\text{carboxylic acid}}{R{-}\overset{\overset{\displaystyle O}{\|}}{C}{-}OH} + \underset{\text{alcohol}}{H{-}O{-}R'} \underset{}{\overset{H^+}{\rightleftharpoons}} \underset{\text{ester}}{R{-}\overset{\overset{\displaystyle O}{\|}}{C}{-}O{-}R'} + H_2O$$

When the alcohol is in *excess,* the equilibrium shifts so much to the right (in accordance with Le Châtelier's principle) that the reaction, called *direct esterification,* is a good method for making an ester. For example,

$$\underset{\text{acetic acid}}{CH_3{-}\overset{\overset{\displaystyle O}{\|}}{C}{-}OH} + \underset{\substack{\text{ethyl alcohol}\\ \text{(large excess)}}}{H{-}O{-}CH_2CH_3} \xrightarrow{H^+} \underset{\text{ethyl acetate}}{CH_3{-}\overset{\overset{\displaystyle O}{\|}}{C}{-}O{-}CH_2CH_3} + H_2O$$

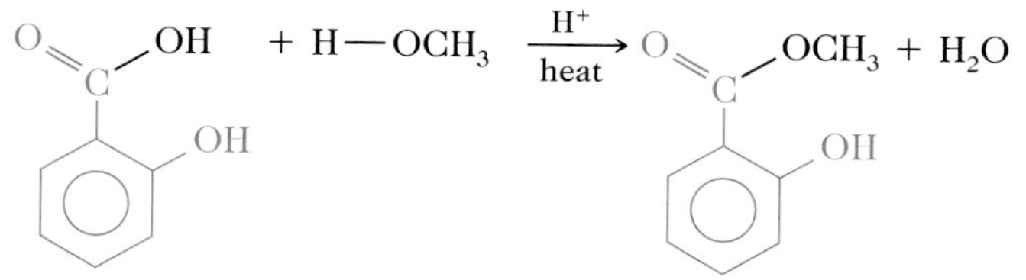

salicylic acid — methyl alcohol (large excess) — methyl salicylate (oil of wintergreen)

Notice that salicylic acid has *two* groups that can form an ester. Esterification of its phenolic OH group by acetic anhydride gives aspirin; esterification of its carboxyl group by methyl alcohol gives methyl salicylate.

EXAMPLE 15.2 Writing the Structure of a Product of Direct Esterification

What ester can be made from benzoic acid and methyl alcohol?

ANALYSIS We need the *structures* of the starting materials, and in an esterification it is sometimes helpful to let the OH groups "face" each other.

$$\underset{\text{benzoic acid}}{C_6H_5\overset{\overset{\displaystyle O}{\|}}{C}{-}OH} + \underset{\text{methyl alcohol}}{H{-}OCH_3}$$

■ Although it is not important in predicting correct structures of products, always erase the OH group from the carboxylic acid, not the alcohol. This will make it easier to learn a reaction coming up in the next chapter.

We know that a molecule of water splits out between the acid and the alcohol during esterification. For the pieces of H_2O, we take the OH from the acid and the H from the alcohol. This operation leaves the following fragments.

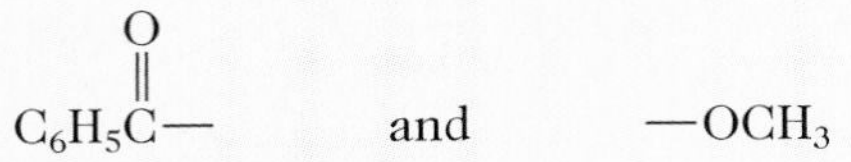

$$\underset{\displaystyle C_6H_5\overset{\textstyle O}{\overset{\|}{C}}-}{} \qquad \text{and} \qquad -OCH_3$$

All that remains is to join these fragments.

SOLUTION   The ester that forms is methyl benzoate.

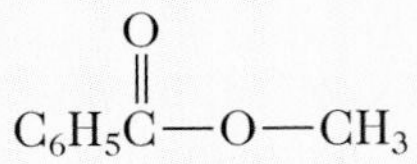

$$C_6H_5\overset{\textstyle O}{\overset{\|}{C}}-O-CH_3$$

■ **PRACTICE EXERCISE 6**   Write the structures of the esters that form by the direct esterification of acetic acid by the following alcohols.

(a)   Methyl alcohol   (b)   Propyl alcohol   (c)   Isopropyl alcohol

---

■ **PRACTICE EXERCISE 7**   Write the structures of the esters that can be made by the direct esterification of ethyl alcohol by the following acids.

(a)   Formic acid   (b)   Propionic acid   (c)   Benzoic acid

---

The Acid Catalyst Helps Direct Esterification   Without an acid catalyst, direct esterification proceeds very slowly. The catalyst, by donating H^+ to the O atom of the carbonyl group of the carboxylic acid, makes the C atom of the carbonyl group much more positive in charge.

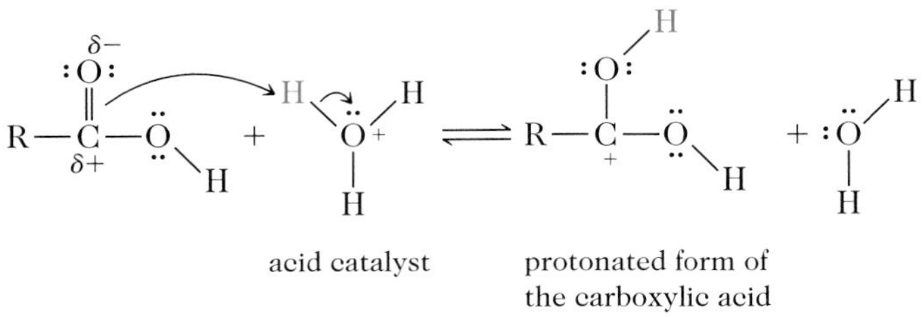

acid catalyst   protonated form of
                       the carboxylic acid

In the protonated form of the carboxyl group, the original carbonyl carbon is now much more attractive to the O of the alcohol. The alcohol molecule can now more readily attack. Following this attack, a shift of a proton from one O atom to another occurs.

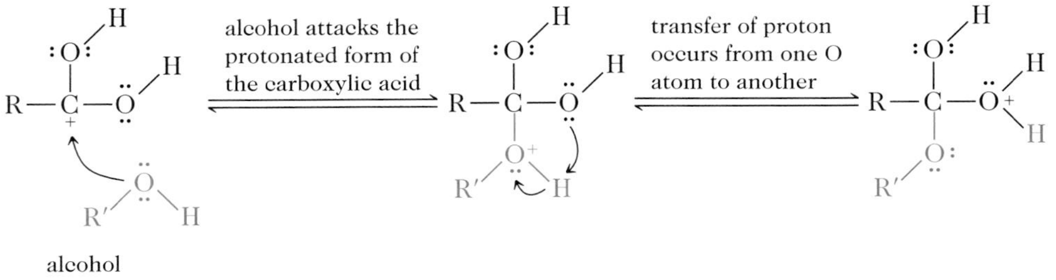

With this proton transfer, the system now suddenly has a very stable built-in leaving group, a water molecule, which now breaks off.

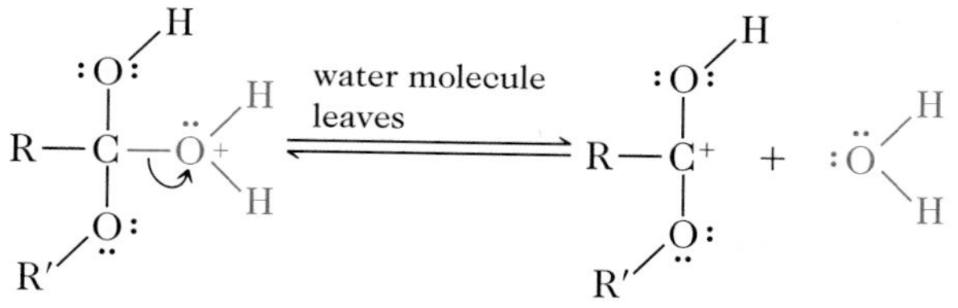

Finally, H^+ transfers to an acceptor in the medium, thus freeing the proton catalyst for more chemical work, and the ester molecule emerges.

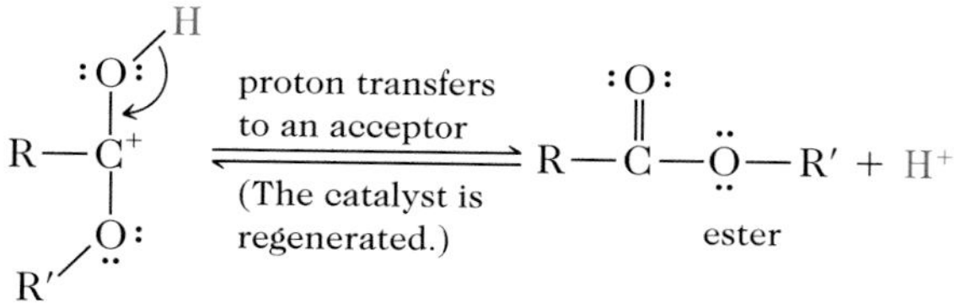

Every step involves an equilibrium. Each step involves simple principles of proton transfers and the attractions of unlike charges. What drives *all* of the equilibria to the right in vitro in favor of the ester is simply an excess of the alcohol. The shifts are all in accordance with Le Châtelier's principle.

15.4 OCCURRENCE, NAMES, AND PHYSICAL PROPERTIES OF ESTERS

The ester group has bonds from a carbonyl group to oxygen and then to a hydrocarbon group.

The functional group of an ester is the central structural feature of all of the edible fats and oils as well as a number of constituents of body cells (see also Table 15.3). Be sure that you can recognize the ester group and can specifically pick out the *ester linkage,* because it's where an ester breaks apart when it reacts with water. It's the single bond between the carbonyl C and the O that holds the ester's alkyl group.

■ The molecules of fats and oils — triacylglycerols — have three ester groups per molecule.

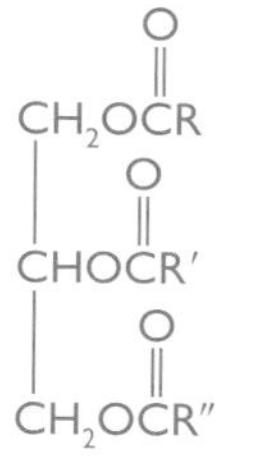

triacylglycerol
(general structural
features)

two general formulas
for esters

ester group (carbonyl–
oxygen–carbon system)

One interesting feature about acids and their esters is that the low formula mass acids have vile odors, but their esters have some of the most pleasant fragrances in all of nature (see Table 15.4). Interaction 15.2 describes some important esters in more detail.

The Acid Portions of Esters and Carboxylate Ions Have Identical Names Both the common and IUPAC names of esters are devised in the same way. For the moment,

TABLE 15.3   Esters of Carboxylic Acids

Name[a]	Structure	MP (°C)	BP (°C)	Solubility in Water[b]
Ethyl esters of straight-chain carboxylic acids, $RCO_2C_2H_5$				
Ethyl formate (ethyl methanoate)	$HCO_2C_2H_5$	−79	54	Soluble
Ethyl acetate (ethyl ethanoate)	$CH_3CO_2C_2H_5$	−82	77	7.35[c]
Ethyl propionate (ethyl propanoate)	$CH_3CH_2CO_2C_2H_5$	−73	99	1.75
Ethyl butyrate (ethyl butanoate)	$CH_3(CH_2)_2CO_2C_2H_5$	−93	120	0.51
Ethyl valerate (ethyl pentanoate)	$CH_3(CH_2)_3CO_2C_2H_5$	−91	145	0.22
Ethyl caproate (ethyl hexanoate)	$CH_3(CH_2)_4CO_2C_2H_5$	−68	168	0.063
Ethyl enanthate (ethyl heptanoate)	$CH_3(CH_2)_5CO_2C_2H_5$	−66	189	0.030
Ethyl caprylate (ethyl octanoate)	$CH_3(CH_2)_6CO_2C_2H_5$	−43	208	0.007
Ethyl pelargonate (ethyl nonanoate)	$CH_3(CH_2)_7CO_2C_2H_5$	−45	222	0.003
Ethyl caprate (ethyl decanoate)	$CH_3(CH_2)_8CO_2C_2H_5$	−20	245	0.0015
Esters of acetic acid, CH_3CO_2R				
Methyl acetate	$CH_3CO_2CH_3$	−99	57	24.4
Ethyl acetate	$CH_3CO_2CH_2CH_3$	−82	77	7.39[c]
Propyl acetate	$CH_3CO_2CH_2CH_2CH_3$	−93	102	1.89
Butyl acetate	$CH_3CO_2CH_2CH_2CH_2CH_3$	−78	125	1.0[d]
Miscellaneous esters				
Methyl acrylate	$CH_2{=}CHCO_2CH_3$	—	80	5.2
Methyl benzoate	$C_6H_5CO_2CH_3$	−12	199	Insoluble
Natural waxes	$CH_3(CH_2)_nCO_2(CH_2)_nCH_3$ $n = 23-33$, carnauba wax $n = 25-27$, beeswax $n = 14-15$, spermaceti			

[a] Common names; IUPAC names are in parentheses.
[b] In grams of ester per 100 g H_2O at 20 °C (unless otherwise specified).
[c] At 25 °C.
[d] At 22 °C.

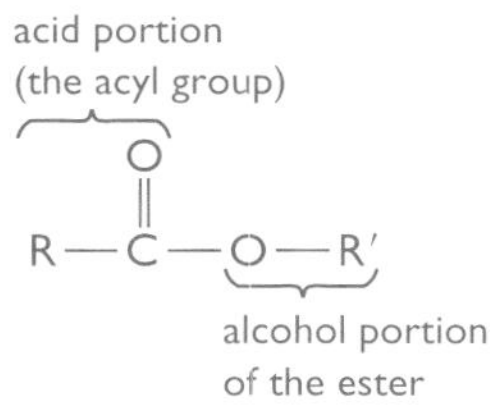

■ acid portion
(the acyl group)

R—C—O—R′
  ‖
  O

alcohol portion
of the ester

simply ignore the alcohol portion of an ester and focus on its acid portion. Pretend that you are naming the *anion* of the acid. Remember that in both the common and the IUPAC names for the anion, the *-ic* ending of the name of the parent acid is changed to *-ate.* Thus salts of acetic acid (ethanoic acid) are called acetate salts (common name) or ethanoate salts (IUPAC). Similarly, esters of acetic acid are acetate esters (common) or ethanoate esters (IUPAC).

Once you have the name of the acid portion, simply write the name of the alkyl group of the ester's alcohol portion in front of it (as a separate word). Here are some examples; the IUPAC names are in parentheses.

Ester	Name of Parent Acid	Name of Acid Portion	Alkyl Group	Name of Ester
$\overset{\displaystyle O}{\underset{\displaystyle \|}{CH_3COCH_3}}$	Acetic acid (ethanoic acid)	Acetate (ethanoate)	Methyl	Methyl acetate (methyl ethanoate)
$\overset{\displaystyle O}{\underset{\displaystyle \|}{CH_3CH_2OCH}}$	Formic acid (methanoic acid)	Formate (methanoate)	Ethyl	Ethyl formate (ethyl methanoate)

The IUPAC name of the following ester, for example is isopropyl hexanoate.

$$CH_3CH_2CH_2CH_2CH_2\overset{O}{\overset{\|}{C}}-O-\overset{CH_3}{\overset{|}{C}}HCH_3$$

The alkyl group on O is isopropyl, and the acyl group is from the straight-chain, six-carbon hexanoic acid. Changing "-ic acid" to "-ate" gives "hexanoate." Try the following exercises. Always remember to identify and name the acyl portion of an ester as if it were a carboxylic acid; then change "-ic acid" to "-ate" and place the alkyl group's name in front of it.

■ PRACTICE EXERCISE 8   Write the IUPAC names of the following esters.

(a) $CH_3CH_2\overset{O}{\overset{\|}{C}}OCH_3$ (b) $CH_3CH_2\overset{CH_3}{\overset{|}{C}}HCH_2\overset{O}{\overset{\|}{C}}OCH_2CH_2CH_3$

■ PRACTICE EXERCISE 9   Using the patterns developed, write the common names of the following esters.

(a) $CH_3CO_2\overset{CH_3}{\underset{CH_3}{\overset{|}{\underset{|}{C}}}}CH_3$ (b) $CH_3CH_2CH_2CO_2CH_2CH_3$

TABLE 15.4   Fragrances of Some Esters

Name	Structure	Fragrance
Ethyl formate	$HCO_2CH_2CH_3$	Rum
Isobutyl formate	$HCO_2CH_2CH(CH_3)_2$	Raspberries
Pentyl acetate	$CH_3CO_2CH_2CH_2CH_2CH_2CH_3$	Bananas
Isopentyl acetate	$CH_3CO_2CH_2CH_2CH(CH_3)_2$	Pears
Octyl acetate	$CH_3CO_2(CH_2)_7CH_3$	Oranges
Ethyl butyrate	$CH_3CH_2CH_2CO_2CH_2CH_3$	Pineapples
Pentyl butyrate	$CH_3CH_2CH_2CO_2(CH_2)_4CH_3$	Apricots
Methyl salicylate	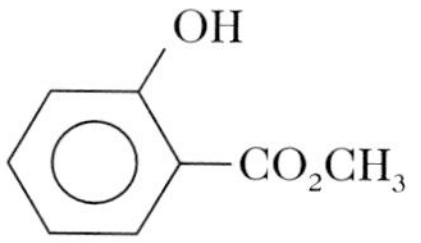	Oil of wintergreen

INTERACTION 15.2
FOOD ADDITIVES, MEDICINALS AND FIBERS. SOME IMPORTANT ESTERS

Esters of *p*-Hydroxybenzoic Acid—The Parabens Several alkyl esters of *p*-hydroxybenzoic acid— referred to as *parabens* on ingredients labels— are used to inhibit molds and yeasts in cosmetics, pharmaceuticals, and food.

Salicylates Certain esters and salts of salicylic acid are analgesics (pain suppressants) and antipyretics (fever reducers). The parent acid, salicylic acid, is itself too irritating to the stomach for these uses, but sodium salicylate and acetylsalicylic acid (aspirin) are commonly used. Methyl salicylate, a pleasant-smelling oil, is used in liniments, because it readily migrates through the skin.

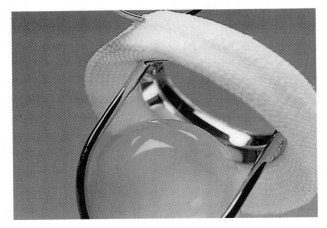

FIGURE 1 The knitted tubing for this aortic heart valve is made of Dacron fiber.

sodium salicylate · acetylsalicylic acid (aspirin) · methyl salicylate (oil of wintergreen)

Dacron Dacron, a polyester of exceptional strength, is widely used to make fabrics and film backing for recording tapes. The name *Dacron* actually applies just to the fiber form of this polyester. When it is cast as a thin film, its name is *Mylar*. Dacron fabrics have been used to repair or replace segments of blood vessels and heart valves (see Figure 1).

The formation of Dacron and many other polyfunctional polymers starts with two difunctional monomers, *aAa* and *bBb*. Their functional groups are able to react with each other to split out a small molecule, *ab*. The monomer fragments, *A* and *B*, join end to end to make a very long, polymer molecule. In principle, the polymerization can be represented as follows.

$$aAa + bBb + aAa + bBb + aAa + bBb + \ldots \text{etc.} \rightarrow$$
$$\text{—A—B—A—B—A—B— etc.} + n(ab)$$

a copolymer

Because two monomers are used, the reaction is called *copolymerization*.

One monomer used to make Dacron is ethylene glycol, which has two alcohol OH groups. The other monomer is dimethyl terephthalate, which has two methyl ester groups. The copolymerization of these two monomers depends on a reaction of esters with alcohols that we will not study. The *ab* molecule that splits out is methyl alcohol. The copolymerization proceeds as follows.

etc. C—OCH₃ + H—OCH₂CH₂O—H + CH₃O—C⬡C—OCH₃ + H—OCH₂—etc.

ethylene glycol · dimethyl terephthalate

etc. C—OCH₂CH₂O—C⬡C—OCH₂—etc. + nCH₃OH

repeating unit in Dacron/Mylar

466

The Ester Group Is Polar but It Cannot Donate Hydrogen Bonds The inability of the ester group to donate hydrogen bonds makes the boiling points of esters lower than those of the parent acids. For example, although methyl acetate has a higher formula mass than acetic acid, it boils at 57 °C, whereas acetic acid boils at 118 °C.

Because it has oxygen atoms, an ester group can *accept* hydrogen bonds, which allows at least the lower formula mass esters to be relatively soluble in water. Although the ester group is itself polar, the *overall* polarity of ester molecules is relatively low, so esters are generally soluble in nonpolar solvents.

15.5 SOME REACTIONS OF ESTERS

Ester molecules are broken apart by water in the presence of either acids or bases.

Esters react with water very slowly unless a catalyst or promoter is present. In vitro, strong acids are catalysts; in vivo, enzymes work, as in the digestion of dietary fats and oils. Strong bases promote the breakup of esters, too, but are consumed (neutralized) by the reaction.

Esters Hydrolyze to Their Parent Acids and Alcohols Esters react with water to give carboxylic acids and alcohols, a reaction called the *hydrolysis of an ester.* Generally an excess of water is used to shift what is actually an equilibrium in favor of the products. In general, using an acid catalyst to speed up the reaction,

$$\mathrm{R\overset{\displaystyle O}{\overset{\|}{C}}{-}OR' + H{-}OH \xrightarrow[\text{heat}]{H^+} R\overset{\displaystyle O}{\overset{\|}{C}}{-}OH + H{-}OR'}$$

ester — carboxylic acid — alcohol

For example,

$$\mathrm{CH_3\overset{\displaystyle O}{\overset{\|}{C}}{-}OCH_2CH_3 + H_2O \xrightarrow{H^+} CH_3\overset{\displaystyle O}{\overset{\|}{C}}{-}OH + H{-}OCH_2CH_3}$$

ethyl acetate — acetic acid — ethyl alcohol

methyl benzoate — methyl alcohol — benzoic acid

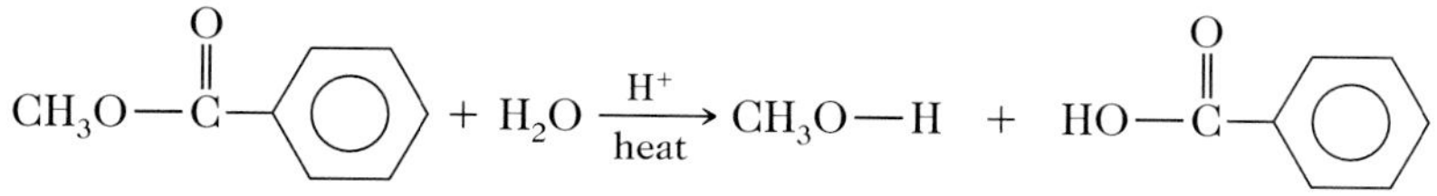

Notice that the *only* bond to break in hydrolysis is the "ester bond," the one that joins the carbonyl group to the O atom. (The most common student mistake is to break the bond on the other side of the carbonyl group.) Notice also that the products are always the "parents" of the ester and that the names of the parents are "given away" in the name of the ester. Methyl benzoate, for example, hydrolyzes to *methyl* alcohol and *benzoic* acid.

EXAMPLE 15.3 **Predicting the Products of an Ester Hydrolysis**

What are the products of the hydrolysis of the following ester?

$$\text{CH}_3\overset{\overset{\displaystyle O}{\|}}{\text{C}}\text{—O—CH}_2\text{CH}_2\text{CH}_3$$

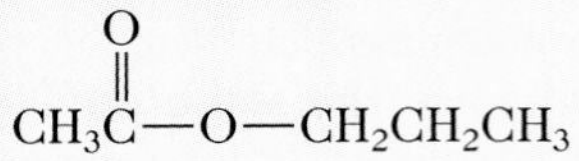

ANALYSIS Finding the ester bond, the carbonyl-to-oxygen bond, is the crucial step because this is the bond that is broken when an ester hydrolyzes. It doesn't matter in which direction this bond happens to point on the page; it is the *only* bond that breaks.

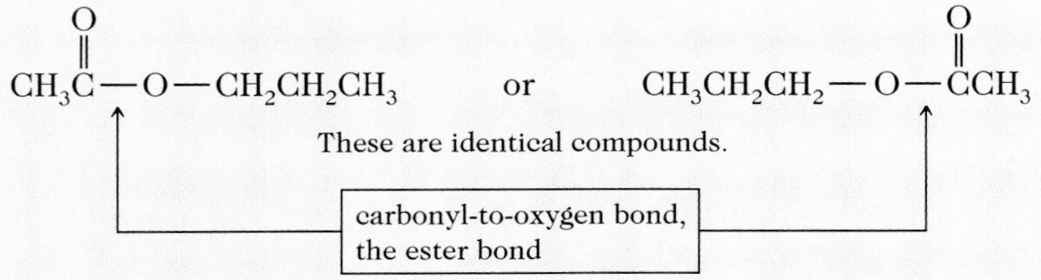

$$\text{CH}_3\overset{\overset{\displaystyle O}{\|}}{\text{C}}\text{—O—CH}_2\text{CH}_2\text{CH}_3 \qquad \text{or} \qquad \text{CH}_3\text{CH}_2\text{CH}_2\text{—O—}\overset{\overset{\displaystyle O}{\|}}{\text{C}}\text{CH}_3$$

These are identical compounds.

carbonyl-to-oxygen bond,
the ester bond

Break the carbonyl–oxygen bond. Erase it and separate the fragments. If the ester were written as follows, we would have

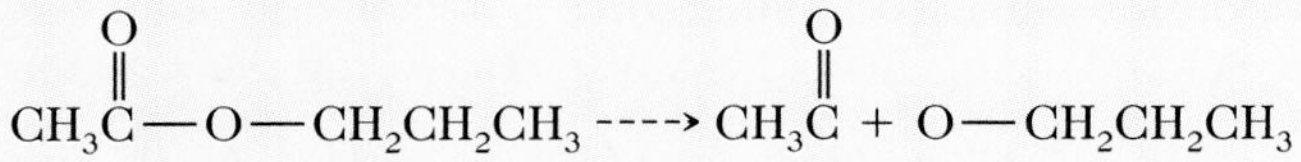

$$\text{CH}_3\overset{\overset{\displaystyle O}{\|}}{\text{C}}\text{—O—CH}_2\text{CH}_2\text{CH}_3 \dashrightarrow \text{CH}_3\overset{\overset{\displaystyle O}{\|}}{\text{C}} + \text{O—CH}_2\text{CH}_2\text{CH}_3$$

On the other hand, if the ester's structure were written in the opposite direction, we'd have

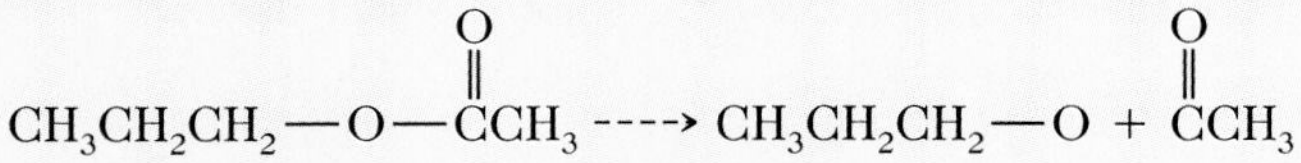

$$\text{CH}_3\text{CH}_2\text{CH}_2\text{—O—}\overset{\overset{\displaystyle O}{\|}}{\text{C}}\text{CH}_3 \dashrightarrow \text{CH}_3\text{CH}_2\text{CH}_2\text{—O} + \overset{\overset{\displaystyle O}{\|}}{\text{C}}\text{CH}_3$$

Either way gives the same answer. Next we attach the pieces of the water molecule to make the "parents" of the ester, attaching OH to the carbonyl carbon and H on the oxygen of the other fragment.

SOLUTION
The products, therefore, are propyl alcohol and acetic acid.

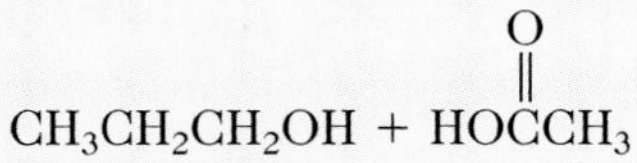

$$\text{CH}_3\text{CH}_2\text{CH}_2\text{OH} + \text{HO}\overset{\overset{\displaystyle O}{\|}}{\text{C}}\text{CH}_3$$

CHECK Reexamine the structure of the ester. Its alcohol portion, the R group on O, has three carbons in a straight chain, so the alcohol must, too. The acid portion has two carbons, so the acid we wrote is correct. Double-check each C and O for the correct number of bonds.

■ PRACTICE EXERCISE 10 Write the structures of the products of the hydrolysis of the following esters.

(a) 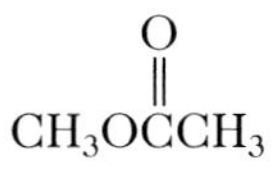 $\text{CH}_3\text{O}\overset{\overset{\displaystyle O}{\|}}{\text{C}}\text{CH}_3$
(b) $\text{CH}_3\text{CH}_2\overset{\overset{\displaystyle O}{\|}}{\text{C}}\text{O}\overset{\overset{\displaystyle CH_3}{|}}{\text{CH}}\text{CH}_3$
(c) $\text{CH}_3\overset{\underset{\displaystyle CH_3}{|}}{\text{CH}}\overset{\overset{\displaystyle O}{\|}}{\text{C}}\text{OCH}_2\text{CH}_2\text{CH}_3$

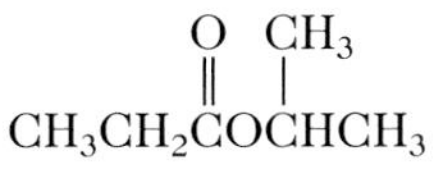

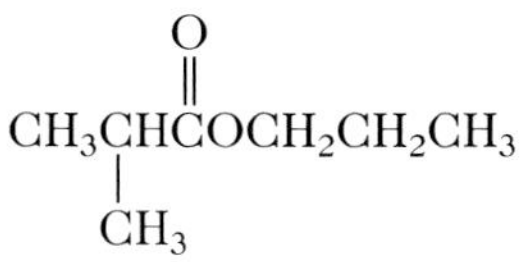

Ester hydrolysis is the reverse of direct esterification. Both involve the identical species in the identical chemical equilibrium that we studied previously. So if we were to take an ester and water in a 1:1 mole ratio, not all of the ester molecules would change into molecules of the parent acid and alcohol. Some of the ester molecules and some of the water molecules would still be unchanged. When we want to make sure that all of the ester is hydrolyzed, we use a large excess of water, as we said. In accordance with Le Châtelier's principle, this excess of one reactant shifts the equilibrium in favor of making the products of hydrolysis.

Esters Are Saponified by Bases   If an aqueous base instead of a strong acid is used to promote the breakup of an ester, the products are the *salt* of the parent acid and the parent alcohol. The reaction is called **saponification,** and it requires a full mole of base (not just a catalytic trace) for each mole of ester bonds. The base *promotes* the reaction, but, unlike a true catalyst, it is permanently changed (neutralized). No equilibrium forms, because one product, the *anion* of the parent acid, cannot be converted into an ester by a direct reaction with alcohols. In the lab, the base used is often sodium hydroxide. We'll write net ionic equations, but remember that wherever you see an anion, there is always some cation around (the Na^+ ion when NaOH is used). In general,

■ L. *sapo,* soap, and *onis,* to make. Ordinary soap is made by the saponification of the ester groups in fats and oils.

$$RC\overset{O}{\overset{\|}{-}}OR' + OH^- \xrightarrow{\text{heat}} RC\overset{O}{\overset{\|}{-}}O^- + H-OR'$$

ester   carboxylate   alcohol
                         anion

Specific examples are as follows. (Again, assume that OH^- comes from an aqueous solution of NaOH.)

$$CH_3C\overset{O}{\overset{\|}{-}}OCH_2CH_3 + OH^- \xrightarrow{\text{heat}} CH_3C\overset{O}{\overset{\|}{-}}O^- + H-OCH_2CH_3$$

ethyl acetate   acetate ion   ethyl alcohol

$$C_6H_5C\overset{O}{\overset{\|}{-}}OCH_3 + OH^- \xrightarrow{\text{heat}} C_6H_5C\overset{O}{\overset{\|}{-}}O^- + H-OCH_3$$

methyl benzoate   benzoate ion   methyl alcohol

Saponification occurs approximately as follows.

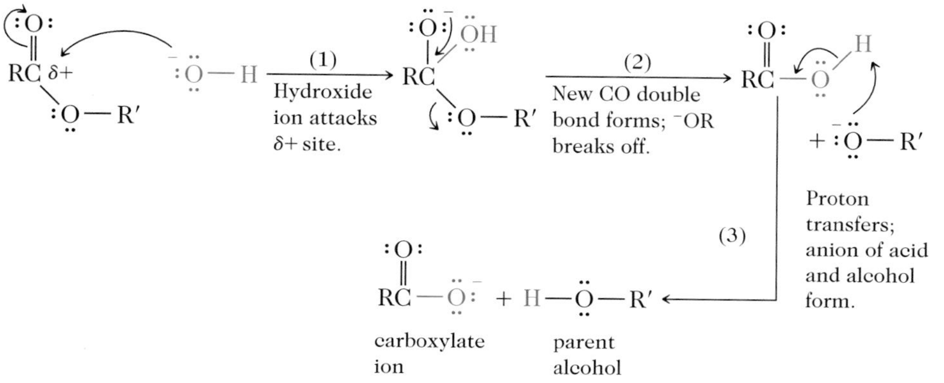

In step (1), the hydroxide ion is attracted to the $\delta+$ site on the carbonyl group's carbon atom, and the double bond breaks open to a single bond. In step (2), the double

■ The steps occur much more smoothly than implied by the way the mechanism has been written.

bond reforms as $^-$OR′ is expelled. It's an even stronger base than OH$^-$, so in step (3) it takes a proton where indicated. The end products are the anion of the parent acid and the parent alcohol. The steps are not reversible because the $\delta-$ site on the O atom of the alcohol has no attraction for the carboxylate ion; like charges repel.

EXAMPLE 15.4   Writing the Structures of the Products of Saponification

---

What are the products of the saponification of the following ester?

$$\underset{\text{CH}_3\text{CH}_2\overset{\displaystyle O}{\overset{\|}{\text{C}}}-\text{OCH}_3}{}$$

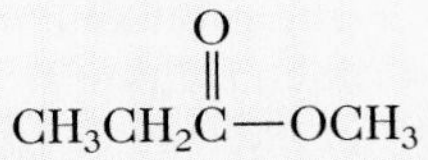

ANALYSIS   Saponification is very similar to ester hydrolysis. The ester bond is broken. *Break only this bond.* Separate the fragments.

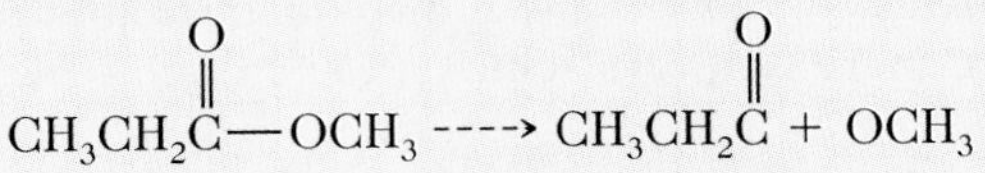

$$\text{CH}_3\text{CH}_2\overset{\displaystyle O}{\overset{\|}{\text{C}}}-\text{OCH}_3 \dashrightarrow \text{CH}_3\text{CH}_2\overset{\displaystyle O}{\overset{\|}{\text{C}}} + \text{OCH}_3$$

Now change the fragment that has the carbonyl group into the *anion* of a carboxylic acid. Do this by attaching O$^-$ to the carbonyl carbon atom. Then attach an H atom to O of the other fragment to make the alcohol.

SOLUTION   The products are

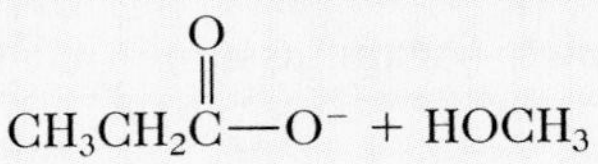

$$\text{CH}_3\text{CH}_2\overset{\displaystyle O}{\overset{\|}{\text{C}}}-\text{O}^- + \text{HOCH}_3$$

■   PRACTICE EXERCISE 11   Write the structures of the products of the saponification of the following esters.

(a) 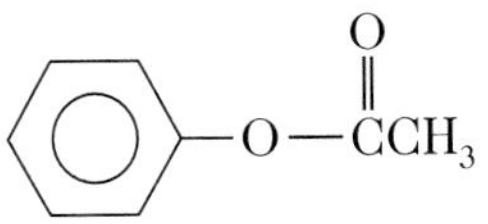   (b)

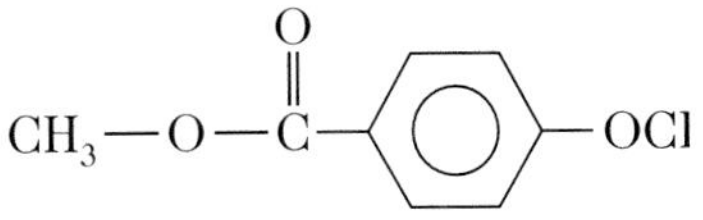

15.6 ORGANOPHOSPHATE ESTERS AND ANHYDRIDES

Anions of esters of phosphoric acid, diphosphoric acid, and triphosphoric acid are some of the most widely distributed kinds of substances in living organisms.

Phosphoric acid appears in several forms in the body, but the three fundamental parents of all are phosphoric acid, diphosphoric acid, and triphosphoric acid.

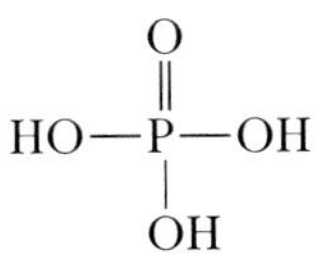

phosphoric acid 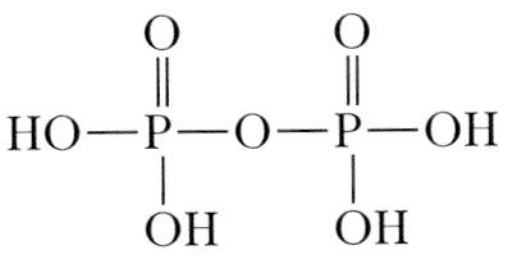   diphosphoric acid 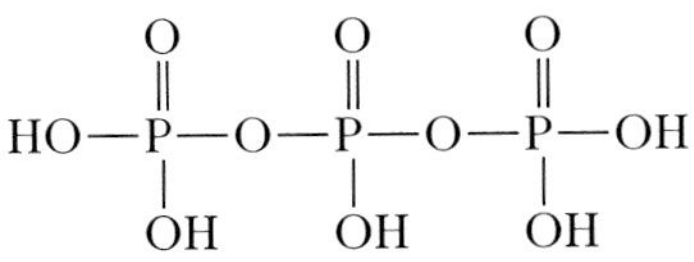   triphosphoric acid

They are all polyprotic acids, but at the slightly alkaline pH of body fluids, they occur, instead, as mixtures of anions. The net charge on each anion and their relative amounts are functions of pH.

Esters of Alcohols and Phosphoric Acid Are Monophosphate Esters If you look closely at the structure of phosphoric acid, you can see that part of it resembles a carboxyl group.

$$
\begin{array}{cc}
\text{O} & \text{O} \\
\parallel & \parallel \\
\text{HO---P---} & \text{HO---C---} \\
\vert & \\
\end{array}
$$

Part of a phosphoric   Part of a carboxylic
acid molecule   acid molecule

It isn't surprising therefore that esters of phosphoric acid exist and that they are structurally similar to esters of carboxylic acids.

$$
\begin{array}{cc}
\text{O} & \text{O} \\
\parallel & \parallel \\
\text{R'O---P---} & \text{R'O---C---} \\
\vert & \\
\end{array}
$$

Part of a phosphate ester   Part of a carboxylate ester

One large difference between the two kinds of esters is that a phosphate ester is still a diprotic acid. Its molecules still carry two proton-donating OH groups. Therefore, depending on the pH, a phosphate ester can exist in any one of three forms, and there usually is an equilibrium mixture of all three.

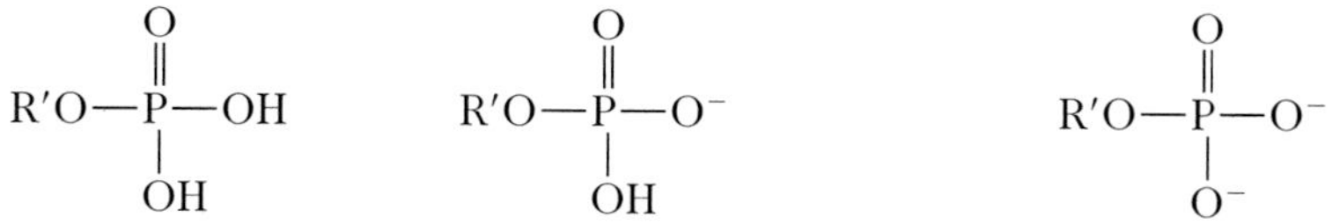

$$
\begin{array}{ccc}
\text{O} & \text{O} & \text{O} \\
\parallel & \parallel & \parallel \\
\text{R'O---P---OH} & \text{R'O---P---O}^- & \text{R'O---P---O}^- \\
\vert & \vert & \vert \\
\text{OH} & \text{OH} & \text{O}^-
\end{array}
$$

phosphate ester   phosphate ester   phosphate ester
(as a diprotic acid)   (as a singly ionized species) (as a doubly ionized species)

favored at low pH   favored at pH   favored at pH
                             values just below 7   values above 7

At the pH of most body fluids (slightly more than 7), phosphate esters exist mostly as the doubly ionized species, a dinegative ion. All forms, however, are generally soluble in water, and one reason that the body converts many substances into phosphate esters may be to improve their solubilities.

Diphosphoric Acid Forms Diphosphate Esters with Alcohols A diphosphate ester actually has three kinds of functional groups, namely, an ester group, three proton-donating OH groups, and a phosphoric anhydride system.

■ Each of the two OH groups in a phosphate ester can be converted into an ester. Nucleic acid molecules, for example, have the phosphate diester system.

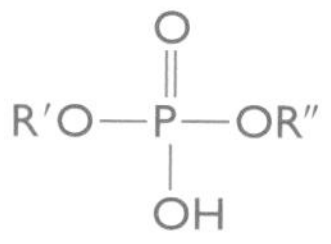

$$
\begin{array}{c}
\text{O} \\
\parallel \\
\text{R'O---P---OR''} \\
\vert \\
\text{OH}
\end{array}
$$

a phosphate diester

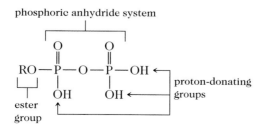

phosphoric anhydride system

Notice the similarity of part of the structure of this diphosphate ester to that of an anhydride of a carboxylic acid.

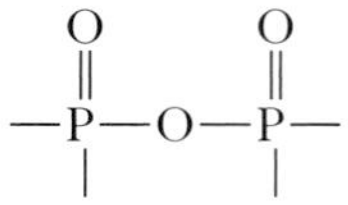

$$\begin{array}{ccc} \overset{\displaystyle O}{\overset{\|}{-P}}-O-\overset{\displaystyle O}{\overset{\|}{P}}- & \qquad & \overset{\displaystyle O}{\overset{\|}{-C}}-O-\overset{\displaystyle O}{\overset{\|}{C}}- \\ | & & \\ \end{array}$$

part of the diphosphate   part of a carboxylic
system   acid anhydride system

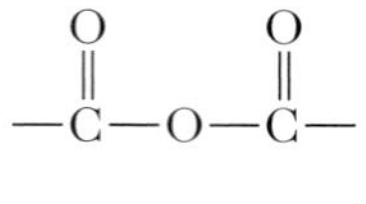

the phosphoric   the carboxylic
anhydride group   anhydride group

Adenosine diphosphate, or ADP, is one of many diphosphate esters in the body, existing largely as a triply charged anion.

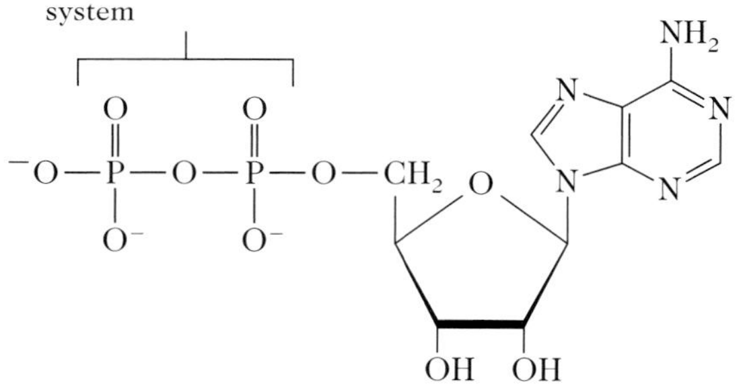

adenosine diphosphate, ADP
(fully ionized form)

The Phosphoric Anhydride Group Is a Major Storehouse of Chemical Energy in Living Systems ADP reacts with water either to give adenosine and two phosphate ions, or to give adenosine monophosphate and one phosphate ion. Either hydrolysis is *very* slow in the absence of an enzyme, but when it occurs, considerable energy per mole is released.

ADP also reacts with alcohols in a reaction that resembles hydrolysis because it breaks up the phosphoric anhydride group.

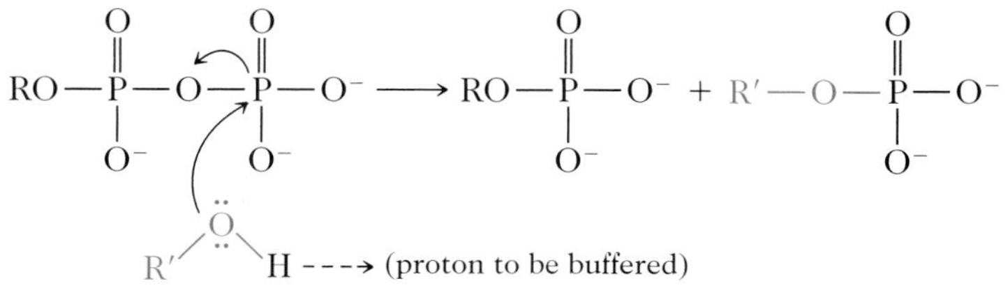

The phosphoric anhydride group turns out to be the chief means for storing chemical energy in cells. The group is like a coiled spring because, as you can see in the above structure, the central chain bears three oxygen atoms with full negative charges. The charges repel each other (as like charges always do), and this internal repulsion sets the molecule all aquiver, primed to break apart, given the right reactant.

Molecules with alcohol groups are such reactants, but the reaction in the body still requires an enzyme. Without it, a negatively charged ADP anion actually *repels* an electron-rich species, like the O atom at the business end of ROH. Yet, if ROH is

FIGURE 15.1 The negatively charged oxygen atoms in the phosphoric anhydride system of ADP screen the phosphorus atoms. The charges deflect incoming, electron-rich particles, like molecules of an alcohol or water, whose O atoms must reach the anhydride's phosphorus atoms. This is why phosphoric anhydride systems react very slowly in cells with ROH or H_2O, unless a specific enzyme is present. When the enzyme is available, the large chemical energy of the anhydride becomes available.

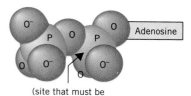

(site that must be
attacked by H_2O)

to attack the phosphoric anhydride system, this O atom must strike a P atom, as shown in the equation above. The P atoms, however, are buried within a clutch of negatively charged O atoms (see the molecular model in Figure 15.1). Thus the internal tension in the anion of the phosphoric anhydride system cannot be relieved *unless an enzyme for the reaction is present.*

We have here an important principle about how the body controls its chemistry, namely, by controlling enzymes, either their availability or their activity. Thus, *the body controls energy-releasing reactions of diphosphates by its control of the enzymes for these reactions.* Water could make the same kind of exothermic attack on a diphosphate ester as an alcohol, but *the body has no enzymes inside cells that catalyze this reaction.* Hence, energy-rich diphosphates can exist in cells despite the abundance of water.

Alcohols and Triphosphoric Acid Form Triphosphate Esters   Adenosine triphosphate, or ATP, is the most common and widely occurring member of a small family of energy-rich triphosphate esters. Because the triphosphates have two phosphoric anhydride systems in each molecule, on a mole-for-mole basis the triphosphates are among the most energy-rich substances in the body. They're called the "powerhouses of cells."

adenosine triphosphate, ATP
(fully ionized form)

Triphosphates are actually much more widely used in cells as sources of energy than the diphosphates. The overall, multistep reaction for the contraction of a muscle, for example, can be summarized very simply as follows.

$$\text{Relaxed muscle} + \text{ATP} \xrightarrow{\text{enzyme}} \text{contracted muscle} + \text{ADP} + P_i$$

We introduce the symbol P_i in this equation to stand for the *set* of inorganic phosphate ions, mostly $H_2PO_4^-$ and HPO_4^{2-}, produced in the breakup of ATP and present at equilibrium at body pH.

As the above equation says, muscular work requires ATP, and if the body's supply of ATP were used up *with no way to remake it,* we'd rapidly become helpless. *The resynthesis of ATP from ADP and P_i is one of the major uses of the chemical energy in the carbohydrates and the fats and oils of our diets.*

SUMMARY

Acids and their salts The carboxyl group, CO_2H, is a polar group that confers moderate water solubility to a molecule without preventing its solubility in nonpolar solvents. This group is very resistant to oxidation. Toward OH^-, CO_3^{2-}, and HCO_3^-, carboxylic acids are strong proton donors, whereas alcohols are not. Toward H_2O, however, carboxylic acids are weak acids. Therefore their conjugate bases, the carboxylate anions, are good proton acceptors toward the hydronium ions of strong acids.

Salts of carboxylic acids are ionic compounds, and the potassium or sodium salts are very soluble in water. Hence, the carboxyl group is one of nature's important "solubility switches." An insoluble acid becomes soluble in base, but it is thrown out of solution again by the addition of acid.

The derivatives of acids studied in this chapter—acid chlorides, anhydrides, and esters—can be made from the acids and are converted back to the acids by reacting with water. We can organize the reactions we have studied for the carboxylic acids and esters as follows.

Reactions of carboxylic acids

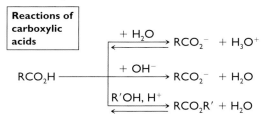

Formation of esters

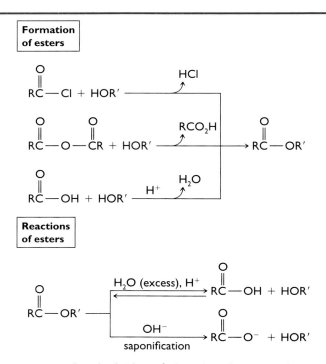

Reactions of esters

Esters and anhydrides of the phosphoric acid system
Esters of phosphoric acid, diphosphoric acid, and triphosphoric acid occur in living systems largely as anions, because these esters are also polyprotic acids. In addition, esters of diphosphoric and triphosphoric acid are phosphoric anhydrides and so are energy-rich compounds. Their reactions with water or alcohols are very exothermic, but the reactions are also very slow unless a catalyst (an enzyme) is present.

REVIEW EXERCISES

The answers to Review Exercises whose numbers are in color are found in Appendix E. The answers to the other Review Exercises are found in the Study Guide that accompanies this book. The more challenging questions are marked with asterisks.

Structures and Names of Carboxylic Acids and Their Salts

15.1 Of the following structures, which has the following functional group?
(a) Alcohol   (b) Carboxyl group   (c) Ketone
(d) Enol

$$CH_3CH_2COH \quad CH_3CCH_2OH \quad CH_3CO_2H$$

  A   B   C

$$HO_2CCH_2CH_3 \quad HOCH_2CH{=}CH_2$$

  D   E

15.2 The carboxylic acids obtained by the hydrolysis of fats and oils in the diet have what general name?

15.3 What is the common name of the acid in vinegar? In sour milk?

15.4 Write the structures of the following substances.
(a) Formic acid   (b) Butyric acid
(c) Propionic acid   (d) Benzoic acid

15.5 What are the structures of the following?
(a) Acetate ion
(b) 2-Chloro-4-methylheptanoic acid
(c) Pentanedioic acid
(d) 2,3-Dimethylbutanoate ion

15.6 Write the IUPAC names of the following compounds.

(a) HO_2CCH_3 (b) 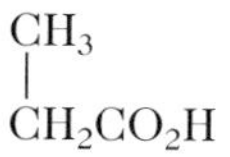 $\overset{\overset{\displaystyle CH_3}{|}}{CH_2CO_2H}$

(c) $CH_3CH_2CO_2Na$ (d)  $CH_3\overset{\overset{\displaystyle CH_3}{|}}{C}HCH_2CH_2CO_2H$

15.7 What are the IUPAC names of the following compounds?

(a) 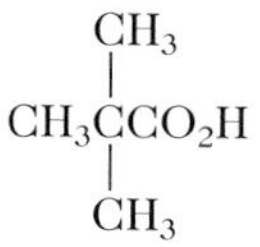 $CH_3\overset{\overset{\displaystyle CH_3}{|}}{\underset{\underset{\displaystyle CH_3}{|}}{C}}CO_2H$ (b) 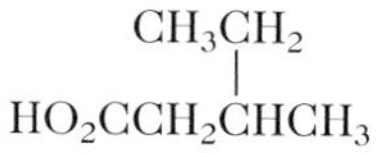 $HO_2CCH_2\overset{\overset{\displaystyle CH_3CH_2}{|}}{C}HCH_3$

(c) $CH_3CH_2CH_2CO_2K$ (d) $C_6H_5CO_2Na$

15.8 One of the compounds whose concentration in blood increases in unchecked diabetes has the following structure.

$$CH_3\overset{\overset{\displaystyle OH}{|}}{C}HCH_2CO_2H$$

If its IUPAC name is 3-hydroxybutanoic acid and its common name is β-hydroxybutyric acid, what are the IUPAC and common names for its sodium salt?

*15.9 The citric acid cycle is one of the major metabolic sequences in the body. One acid in this series of reactions is commonly called fumaric acid, which has the following structure.

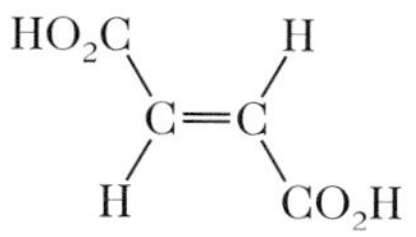

Which of the following is its correct IUPAC name?
(a) *trans*-Dibutenoic acid
(b) *trans*-Butenedioic acid
(c) *trans*-Dibutenedioic acid
(d) *cis*-Ethenedicarboxylic acid

Physical Properties of Carboxylic Acids

15.10 Draw a figure that shows how two acetic acid molecules can pair in a hydrogen-bonded form.

*15.11 Explain why acetic acid is much more soluble in water than pentane.

15.12 Give the following compounds in their order of increasing solubility in water. Do this by arranging their identifying letters in a row in the correct order, placing the letter of the least soluble on the left.

$$\underset{A}{HCO_2H} \qquad \underset{B}{CH_3CH_2CH_2CH_2CH_2CH_3} \qquad \underset{C}{CH_3CH_2CH_2CO_2H}$$

*15.13 Give the following compounds in their order of increasing boiling points by arranging their identifying letters in a row in the correct order. Place the letter of the lowest boiling compound on the left.

$$\underset{A}{HO_2CCH_2CH_2CO_2H} \qquad \underset{B}{CH_3CH_2CH_2CH_3}$$

$$\underset{C}{CH_3CH_2OH} \qquad \underset{D}{CH_3CO_2H}$$

Carboxylic Acids as Weak Acids

15.14 Concerning an aqueous solution of acetic acid:
(a) Write the equation for the equilibrium that is present.
(b) In what direction will this equilibrium shift, toward acetic acid or toward the acetate ion, if hydrochloric acid is added? Explain.
(c) Write the K_a equation for acetic acid.
(d) Using data in a table in this chapter, is acetic acid a stronger or a weaker acid than salicylic acid?

15.15 Arrange the following compounds in their order of increasing acidity by placing their identifying letters in a row, placing the letter of the least acidic compound on the left.)

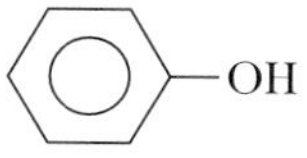

$\underset{B}{H_2SO_4}$ $\underset{C}{CH_3CH_2OH}$ $\underset{D}{CH_3CO_2H}$

15.16 Write the net ionic equation for the complete reaction, *if any,* of aqueous sodium hydroxide with each of the compounds in Review Exercise 15.15 at room temperature.

15.17 What are the net ionic equations for the reactions of the following compounds with aqueous sodium hydroxide at room temperature?
(a) $HOCH_2CH_2CH_2CO_2H$ (b) $HO_2CCH_2CH_2CO_2H$

Salts of Carboxylic Acids

15.18 Which compound, **A** or **B**, is more soluble in water? Explain.

$$\underset{A}{CH_3(CH_2)_6CO_2Na} \qquad \underset{B}{CH_3(CH_2)_6CO_2H}$$

*15.19   Suppose that you add 0.1 mol of hydrochloric acid to an aqueous solution that contains 0.1 mol of the compound given in each of the following parts. If any reaction occurs rapidly at room temperature, write its net ionic equation.
(a) $CH_3CH_2CO_2^-$
(b) $^-O_2CCH_2CH_2CH_2CO_2^-$
(c) NH_3

Esterification and Reactivity

15.20   What are the structures of the reactants that are needed to make ethyl propanoate from ethanol and each of the following kinds of starting materials?
(a) An acid chloride
(b) A carboxylic acid anhydride
(c) By direct esterification

15.21   In order to prepare methyl benzoate, what are the structures of the reactants needed for each kind of approach?
(a) By direct esterification
(b) From an acid chloride
(c) From an acid anhydride

*15.22   The reaction of ethyl alcohol with acetyl chloride is rapid.
(a) What is the structure of the organic product?
(b) How might the relatively high speed of the reaction be explained?

15.23   What are the structures of the products of the esterification by methyl alcohol of each compound?
(a) Propionic acid
(b) 2-Methylbutanoic acid
(c) p-Bromobenzoic acid
(d) Phthalic acid. (Show the esterification of both of the carboxyl groups.)

phthalic acid

15.24   When ethanoic acid is esterified by each of the following compounds, what are the structures of the esters that form?
(a) Methanol
(b) 2-Methyl-1-propanol
(c) p-Nitrophenol
(d) $HOCH_2CH_2CH_2OH$ (1,3-propanediol) (Show the esterification of both alcohol groups.)

*15.25   Explain by means of equations how H^+ works as a catalyst in the direct esterification of acetic acid by ethyl alcohol.

*15.26   How do we explain the fact that esters react much more slowly with water than acid chlorides do?

Structures and Physical Properties of Esters

15.27   Write the structures of the following compounds.
(a) Ethyl formate   (b) Ethyl p-chlorobenzoate

15.28   What are the structures of the following compounds?
(a) t-Butyl propionate
(b) Isopropyl 2-methylbutanoate

*15.29   Arrange the following compounds in their order of increasing boiling points. Do this by placing their identifying letters in a row, starting with the lowest boiling compound on the left.

$CH_3CO_2CH_2CH_2CH_3$ $CH_3CH_2CH_2CH_2CO_2H$

        A   B

$CH_3CH_2CH_2CH_2CH_3$ $HO_2CCH_2CH_2CH_2CH_2CH_3$

        C   D

Reactions of Esters

15.30   Write the equation for the acid-catalyzed hydrolysis of each compound. If no reaction occurs, write "No reaction."

(a) $CH_3CH_2\overset{\displaystyle O}{\overset{\|}{C}}OCH(CH_3)_2$ (b) $CH_3CH_2\overset{\displaystyle O}{\overset{\|}{C}}O\!-\!\bigcirc$

(c) $CH_3\overset{\displaystyle O}{\overset{\|}{C}}CH_2OCH_3$ (d) $CH_3CH_2OCH_2CH_2CH_3$

15.31   What are the equations for the acid-catalyzed hydrolyses of the following compounds? If no reaction occurs, write "No reaction."

(a) $C_6H_5\overset{\displaystyle O}{\overset{\|}{C}}OCHCH_3$ (b) $C_6H_5O\overset{\displaystyle O}{\overset{\|}{C}}CHCH_3$
     $\quad\quad\quad\quad\underset{\textstyle CH_3}{|}$ $\quad\quad\quad\quad\underset{\textstyle CH_3}{|}$

(c) $CH_3O\overset{\displaystyle O}{\overset{\|}{C}}CH_2\overset{\displaystyle O}{\overset{\|}{C}}OCH_3$ (d) $C_6H_5\overset{\displaystyle O}{\overset{\|}{C}}CH_2OCH_2CH_3$

15.32   The metabolism of fats and oils involves the complete hydrolysis of molecules such as the following. What are the structures of its hydrolysis products?

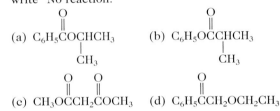

15.33   What are the structures of the organic products of the saponification of the compounds in Review Exercise 15.30 by aqueous NaOH?

15.34 What forms, if anything, when the compounds of Review Exercise 15.31 are subjected to saponification by aqueous KOH? Write the structures of the organic products.

15.35 What are the products of the saponification of the compound given in Review Exercise 15.32? (Assume that aqueous NaOH is used.)

*15.36 A pharmaceutical chemist needed to prepare the ethyl ester of an extremely expensive and rare carboxylic acid in order to test this form of the drug for its side effects. Direct esterification had to be used. How could the conversion of all the acid to its ethyl ester be maximized? Use RCO_2H as a symbol for the acid in any equations you write.

Phosphate Esters and Anhydrides

15.37 Write the structures of the following compounds.
(a) Monomethyl phosphate
(b) Monoethyl diphosphate
(c) Monopropyl triphosphate

15.38 State one apparent advantage to the body of its converting many compounds into phosphate esters.

15.39 What part of the structure of ATP is particularly responsible for its being described as an *energy-rich* compound? Explain.

15.40 Why is ATP more difficult to hydrolyze than acetyl chloride?

Common Acids and Salts (Interaction 15.1)

15.41 Name a compound that is
(a) Used to manufacture a kind of rayon
(b) Present in vinegar
(c) A food additive put into wrappers of cheese

Common Esters (Interaction 15.2)

15.42 Esters of *p*-hydroxybenzoic acid are referred to by what common name? How are these esters used in commerce?

15.43 What is meant by a *copolymer*?

15.44 What copolymer has been used in surgical grafts?

15.45 Salicylates are described as analgesics and antipyretics. What do these terms mean?

15.46 Why is salicylic acid, the parent of the salicylates, structurally modified for medicinal uses?

15.47 Concerning salicylic acid,
(a) What two functional groups does it have?
(b) Which functional group is esterified in acetylsalicylic acid?
(c) Which group is esterified in methyl salicylate?

Additional Exercises

15.48 Consider compounds **A** and **B**.

$$C_6H_5CO_2H \qquad C_6H_5OH$$

A   B

(a) Which is the stronger acid?
(b) Write the structure of the conjugate base of each compound.
(c) Which is the stronger conjugate base of the two you wrote for part (b)?

15.49 Consider a solution of formic acid in water.
(a) Write the equation for the equilibrium that is present.
(b) In what direction will this equilibrium shift, toward formic acid or toward the formate ion, if NaOH is added? Explain.
(c) Write the K_a equation for formic acid.
(d) Using data in a table in this chapter, is formic acid a stronger or a weaker acid than benzoic acid?

*15.50 Which solution has the lower pH, 0.1 M CH_3CO_2H or 0.1 M C_6H_5OH? Explain.

15.51 Which compound, **A** or **B**, is more soluble in ether? Explain.

A   B

*15.52 Suppose that you have each of the following compounds in a solution in water. What reaction, if any, will occur rapidly at room temperature if an equimolar quantity of hydrochloric acid is added? Write net ionic equations.
(a) $HOCH_2CH_2CO_2^-$
(b) $HOCH_2CH_2CO_2H$

(c)

*15.53 Methyl alcohol reacts rapidly with acetic anhydride to give methyl acetate and acetic acid. How might the very rapid rate of the reaction be explained?

*15.54 Water does not react readily with methyl acetate to form methyl alcohol and acetic acid. The hydroxide ion, on the other hand, more readily attacks methyl acetate (to form methyl alcohol and the acetate ion). What might be the reason for the higher reactivity of the OH^- ion over H_2O toward methyl acetate?

15.55 Suppose that a way could be found to remove H_2O as rapidly as it is produced in direct esterification. What would this do to the equilibrium in this reaction, shift it to the right (favoring the ester) or to the left (favoring the carboxylic acid and the alcohol)? Explain.

*15.56   Arrange the following compounds in their order of increasing solubilities in water by placing their identifying letters in the correct sequence, beginning with the least soluble on the left.

CH$_3$CH$_2$CH$_2$CH$_2$CO$_2$Na   CH$_3$CH$_2$CH$_2$CH$_2$CO$_2$CH$_2$CH$_3$

A   B

CH$_3$CH$_2$CH$_2$CH$_2$CO$_2$H   CH$_3$CH$_2$CH$_2$CH$_2$CH$_3$

C   D

*15.57   Cyclic esters are known compounds. What is the structure of the product when the following compound is hydrolyzed?

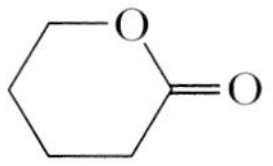

15.58   Write structure of the organic ion that forms when the compound of Review Exercise 15.57 is saponified.

*15.59   Write the steps in the mechanism of the acid-catalyzed hydrolysis of methyl acetate. (Remember, this is the exact reverse of the acid-catalyzed, direct esterification of acetic acid by methyl alcohol.)

*15.60   To prepare a sample of oil of Niobe, methyl benzoate, a student heated a solution of 5.64 g of benzoic acid in 25.0 mL of methyl alcohol in the presence of a small amount of sulfuric acid as a catalyst.
(a) What is the maximum number of grams of methyl benzoate obtainable from the mass of benzoic acid used?
(b) What is the minimum number of grams and milliliters of methyl alcohol needed for the complete conversion of benzoic acid to methyl benzoate? (The density of methyl alcohol is 0.787 g/mL.)
(c) What advantage is there in using an excess of methyl alcohol? (What principle is involved?)

*15.61   Complete the following reaction sequences by writing the structures of the products. If no reaction occurs, state so. (These reactions constitute a review of this and earlier chapters on organic chemistry.)

(a) CH$_3$OCCH$_2$CH$_3$ + H$_2$O $\xrightleftharpoons[\text{catalyst}]{\text{acid}}$

(b) CH$_3$COH + NaOH(aq) $\longrightarrow$

(c) C$_6$H$_5$CHO + (O) $\xrightarrow[\text{e.g., K}_2\text{Cr}_2\text{O}_7(aq)]{}$

(d) 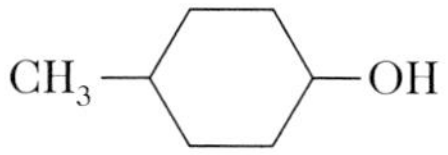 CH$_3$—⟨ ⟩—OH $\xrightarrow[\text{heat}]{\text{H}_2\text{SO}_4}$

(e) CH$_3$CHCH$_2$COCH$_3$ + NaOH(aq) $\longrightarrow$
          |
          OH

(f) CH$_3$CHCH$_2$CH$_3$ + (O) $\xrightarrow[\text{e.g., MnO}_4^-(aq)]{}$

(g) CH$_3$CCl + CH$_3$CH$_2$CH$_2$OH $\longrightarrow$

          OCH$_3$
          |
(h) CH$_3$CHOCH$_3$ + H$_2$O $\xrightleftharpoons[\text{catalyst}]{\text{acid}}$

(i) 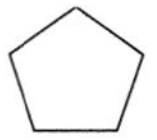 ⬠ + H$_2$SO$_4$ $\longrightarrow$

(j) CH$_3$CH$_2$OH + C$_6$H$_5$CO$_2$H $\xrightleftharpoons[\text{catalyst}]{\text{acid}}$

(k) CH$_3$CH$_2$CH═CH$_2$ + HCl(g) $\longrightarrow$

(l) CH$_3$CH$_2$OCH$_2$CH$_2$CH$_2$CH$_3$ + H$_2$O $\longrightarrow$

*15.62   Write the structures of the products, if any, that form in the following situations. If no reaction occurs, state so. (Some of these constitute a review of the reactions of earlier chapters.)

(a) 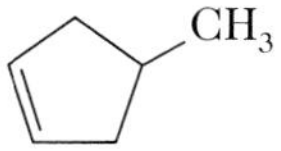 ⬠ with CH$_3$ + H$_2$ $\xrightarrow{\text{Ni or Pt}}$

(b) CH$_3$COCCH$_3$ + CH$_3$OH $\longrightarrow$

          CH$_3$ O
          | ||
(c) CH$_3$CHCH$_2$CO$^-$ + HCl(aq) $\longrightarrow$

          OH
          |
(d) CH$_3$CH$_2$CCH$_3$ + (O) $\xrightarrow[\text{e.g., MnO}_4^-(aq)]{}$
          |
          CH$_3$

(e) CH$_3$CH$_2$OCCH$_2$CH$_2$COCH$_3$ + NaOH(aq) $\longrightarrow$
          (excess)

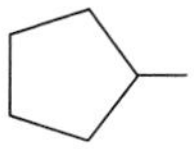

(f) ⬠—CH$_2$CO$_2$H + CH$_3$OH $\xrightleftharpoons[\text{catalyst}]{\text{acid}}$

(g) CH$_3$CH$_2$OCCH$_2$CH$_2$OCCH$_2$CH$_3$ + NaOH(aq) $\longrightarrow$
          (excess)

          OCH$_2$CH$_3$
          |
(h) C$_6$H$_5$CHOCH$_2$CH$_3$ + H$_2$O $\xrightleftharpoons[\text{catalyst}]{\text{acid}}$

(i) HO$_2$CCH$_2$CH$_2$CH$_2$CH$_3$ + NaOH(aq) $\longrightarrow$

(j) CH$_3$CH$_2$CH$_2$OCCH$_2$CH$_2$OCCH$_3$ + H$_2$O $\xrightleftharpoons[\text{catalyst}]{\text{acid}}$
          (excess)

(k) HOCH$_2$CCH$_2$CH$_2$CO$_2$H + CH$_3$OH $\xrightleftharpoons[\text{catalyst}]{\text{acid}}$
          (excess)

(l) CH$_3$OCH$_2$CH$_2$CH$_2$CO$_2^-$ + HCl(aq) $\longrightarrow$

AMINES AND AMIDES

16

Mountaineers trust their lives to ropes and lines that give a little and that do not easily fray or snap. Fibers made of nylon, a synthetic polyamide, meet these criteria superbly. Proteins are also polyamides but are made from different monomers.

THIS CHAPTER IN CONTEXT

To understand proteins as well as nucleic acids we need to know a small number of the chemical properties of amines and amides. With this chapter we'll complete our survey of the functional groups that are important at the molecular level of life.

16.1 OCCURRENCE, NAMES, AND PHYSICAL PROPERTIES OF AMINES

The amino group, NH_2, has some of the properties of ammonia.

Both the amino group and its protonated form occur in proteins, enzymes, and nucleic acids (the chemicals that carry our genes). When a carbonyl group is attached to nitrogen, the properties change sufficiently from those of an amine to make it convenient to create a different family, the *amides.*

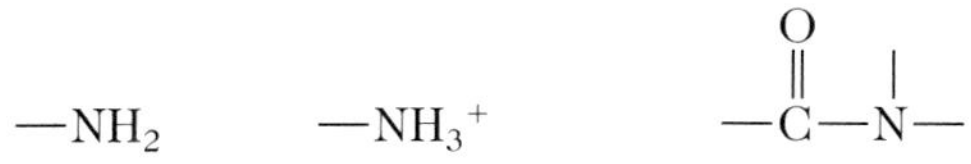

$$\begin{array}{ccc} -NH_2 & -NH_3{}^+ & \overset{\displaystyle O}{\underset{}{\overset{\|}{-C}}}\!-\!\overset{|}{N}\!- \\ \text{amino} & \text{protonated} & \text{amide} \\ \text{group} & \text{amino group} & \text{system} \end{array}$$

Amines Are Ammonia-Like Compounds The **amines** are organic relatives of ammonia in which one, two, or all three of the hydrogen atoms on an ammonia molecule have been replaced by a hydrocarbon group. Some examples are

- These are the common names, not the IUPAC names.

$$\begin{array}{cccc} & & \overset{\displaystyle CH_3}{\overset{|}{}} & \\ CH_3NH_2 & CH_3NHCH_3 & CH_3NCH_3 & CH_3NHCH_2CH_3 \\ \text{methylamine} & \text{dimethylamine} & \text{trimethylamine} & \text{ethylmethylamine} \end{array}$$

Several amines are listed in Table 16.1. All are basic, like ammonia.

It's quite important to realize that for a compound to be an amine, not only must its molecules have a nitrogen with three bonds, but also none of these bonds can be to a carbonyl group. If such a system is present—a carbonyl–nitrogen bond—the substance is an **amide.** Thus the structure given by **1** is an amide, not an amine. However, the structure given by **2** is not an amide, because there is no carbonyl–nitrogen bond. Instead, **2** has two functional groups, a keto group and an amino group.

- The carbonyl–nitrogen bond occurs in protein molecules, where it is called the **peptide bond.**

$$\begin{array}{cc} \overset{\displaystyle O}{\overset{\|}{RCNH_2}}\ \text{amide group} & \overset{\displaystyle O}{\overset{\|}{RCCH_2NH_2}}\ \text{keto group} \\ & \qquad\qquad \text{amino group} \\ \mathbf{1} & \mathbf{2} \end{array}$$

The chemical difference is that amines are basic and amides are not. Another difference is that the carbon–nitrogen bond in amines can't be broken by water, but the carbonyl–nitrogen bond in amides can, provided an appropriate catalyst is present.

If one or more of the groups attached directly to nitrogen in an amine is a benzene ring, then the amine is an *aromatic* amine. Otherwise, it is classified as an

TABLE 16.1   Amines

Common Name	Structure	BP (°C)	Solubility in Water	K_b (25 °C)
Methylamine	CH_3NH_2	−6	Very soluble	4.4×10^{-4}
Dimethylamine	$(CH_3)_2NH$	8	Very soluble	5.3×10^{-4}
Trimethylamine	$(CH_3)_3N$	3	Very soluble	0.5×10^{-4}
Ethylamine	$CH_3CH_2NH_2$	17	Very soluble	5.6×10^{-4}
Diethylamine	$(CH_3CH_2)_2NH$	55	Very soluble	9.6×10^{-4}
Triethylamine	$(CH_3CH_2)_3N$	89	14 g/dL	5.7×10^{-4}
Propylamine	$CH_3CH_2CH_2NH_2$	49	Very soluble	4.7×10^{-4}
Aniline	$C_6H_5NH_2$	184	4 g/dL	3.8×10^{-10}

aliphatic amine. Thus benzylamine is an aliphatic amine and aniline, *N*-methylaniline, and *N,N*-dimethylaniline are all aromatic amines.

aniline   *N*-methylaniline   *N,N*-dimethylaniline   benzylamine

The Common Names of Amines Usually End in *-amine*   The common names of the simple, aliphatic amines are made by writing the names of the alkyl groups attached to nitrogen in front of the word *amine* (and leaving no space). We have already seen how this works. Here are three more examples.

$$\underset{\text{isobutylamine}}{\overset{\displaystyle CH_3}{CH_3CHCH_2NH_2}} \qquad \underset{\text{ethylisopropylamine}}{\overset{\displaystyle CH_3}{CH_3CHNHCH_2CH_3}} \qquad \underset{\text{ethylmethylpropylamine}}{\overset{\displaystyle CH_3}{CH_3CH_2NCH_2CH_2CH_3}}$$

In complex systems, the names of some amines use the name *amino* as a substituent name for the NH_2 group. Thus isobutylamine can be named 1-amino-2-methylpropane. We will not develop IUPAC names for amines.

Nitrogen-containing, heterocyclic amine rings occur in both proteins and nucleic acids. Those in nucleic acids involve either the pyrimidine ring or the purine ring system.

pyrimidine

purine

■ PRACTICE EXERCISE 1   Give common names for the following compounds.

(a) $(CH_3)_2NCH(CH_3)_2$ (b)   ⬡—NH_2 (c) $(CH_3)_2CHCH_2NHC(CH_3)_3$

■ PRACTICE EXERCISE 2   Write the structures of the following compounds.

(a)   *t*-Butyl-*sec*-butylamine   (b)   *p*-Nitroaniline   (c)   *p*-Aminobenzoic acid

N—H Groups in Amines Are Involved in Hydrogen Bonding   We have sometimes used boiling point data to tell us something about forces between molecules. The data in the margin tell us, for example, that when compounds of similar formula mass are compared, the boiling points of amines are higher than those of alkanes but lower than those of alcohols. This suggests that the forces of attraction between molecules are stronger in amines than in alkanes, but they are weaker in amines than in alcohols.

Compound	Formula Mass	BP (°C)
CH_3CH_3	30	−89
H_3NH_2	31	−6
H_3OH	32	65

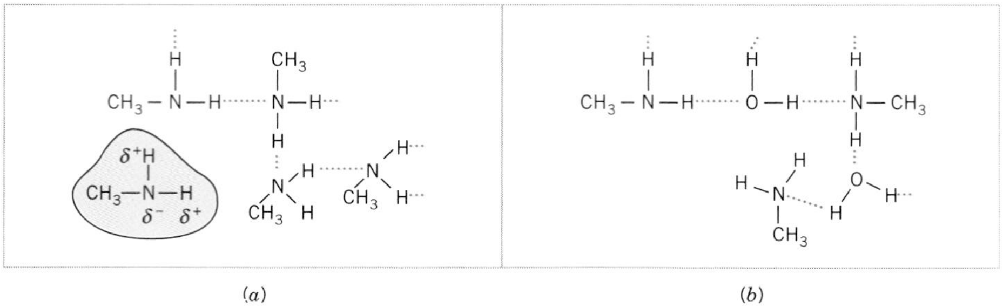

FIGURE 16.1   Hydrogen bonds (*a*) in amines and (*b*) in aqueous solutions of amines.

We can understand these trends in terms of hydrogen bonds. When a hydrogen atom is bound to oxygen or nitrogen but not to carbon, the system can donate and accept hydrogen bonds. Nitrogen, however, has a lower electronegativity than oxygen, so the polarity of the N—H bond in amines is weaker than the polarity of the O—H bond in alcohols. *The N—H system, therefore, develops weaker hydrogen bonds than the O—H system.* As a result, amine molecules can't attract each other as strongly as alcohol molecules, so amines boil lower than alcohols (of comparable formula masses). But amine molecules, nevertheless, do develop some hydrogen bonds (Figure 16.1), which alkane molecules cannot do, so amines boil higher than alkanes (of comparable formula masses). Hydrogen bonding also helps amines to be much more soluble in water than alkanes, as Figure 16.1 also depicts. Hydrogen bonds within the molecules of proteins and nucleic acids stabilize their special shapes, without which the substances lose their abilities to carry out their biological functions.

■ The processes in the body that lead to the sensations of odor or taste begin with *chemical* reactions.

Odor isn't actually a *physical* property, but we should note that the amines with lower formula masses smell very much like ammonia. The odors of amines become very "fishy" at slightly higher formula masses.

16.2 CHEMICAL PROPERTIES OF AMINES

The amino group is a proton acceptor, and the protonated amino group is a proton donor.

We will examine two chemical properties of amines that will be particularly important to our study of biochemicals, namely, the basicity of amines, in this section, and their conversion to amides, in the next.

Aliphatic Amines Are About as Basic as Ammonia   When ammonia dissolves in water, the following equilibrium becomes established and a small concentration of hydroxide ion forms together with the ammonium ion.

$$NH_3 \ + \ H_2O \rightleftharpoons NH_4^+ \ + \ OH^-$$

ammonia   ammonium
ion

A very similar equilibrium forms when an amine dissolves in water. All that is different is that an alkyl group has replaced one of the H atoms of NH_3 (or NH_4^+).

$$R-NH_2 + H_2O \rightleftharpoons R-NH_3^+ + OH^-$$

amine   protonated
                         amine

How much the products are favored in this equilibrium is expressed by the value of the **base ionization constant, K_b**.

$$K_b = \frac{[RNH_3^+][OH^-]}{[RNH_2]}$$

■ Remember, the brackets [ ] denote the moles-per-liter concentration of the compound or ion that the brackets enclose.

Table 16.1 gives the K_b values for several amines. The K_b of ammonia is 1.8×10^{-5}, so you can see that most amines have K_b values slightly greater than that of ammonia. Thus the aliphatic amines are generally slightly stronger bases than ammonia. Like ammonia, water-soluble amines cause the hydroxide ion concentration of the medium to become greater than the hydrogen ion concentration, so the aqueous solutions of amines are basic (and turn red litmus blue).

Like ammonia, compounds with amino groups can also neutralize hydronium ions. The following acid–base neutralization occurs rapidly and essentially completely at room temperature.

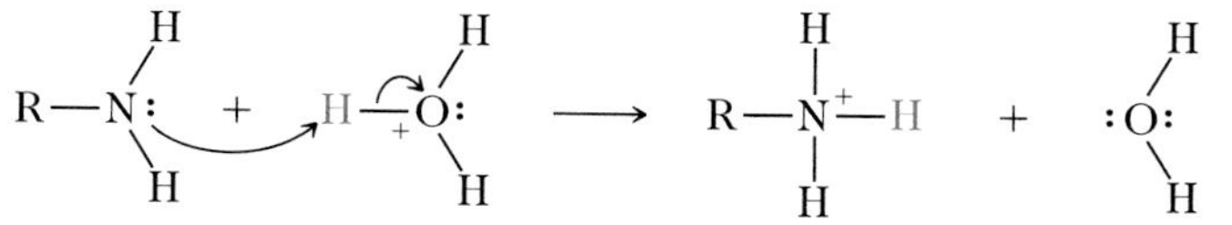

amine   hydronium ion   protonated amine
(or ammonia,   (or the ammonium
when R = H)   ion when R = H)

For example, hydrochloric acid is neutralized by methylamine as follows.

$$CH_3NH_2 + H_3O^+ + Cl^- \longrightarrow CH_3NH_3^+ + Cl^- + H_2O$$

methylamine   hydrochloric acid   methylammonium
                                        chloride

It doesn't matter if the nitrogen atom in an amine bears one, two, or three hydrocarbon groups. The amine can still neutralize strong acids, because the reaction involves just the unshared pair of electrons on the nitrogen, not any of the bonds to the other groups.

■ Tetraalkylammonium ions, R_4N^+, are also known, but they can't be basic because they have no unshared electron pair on nitrogen.

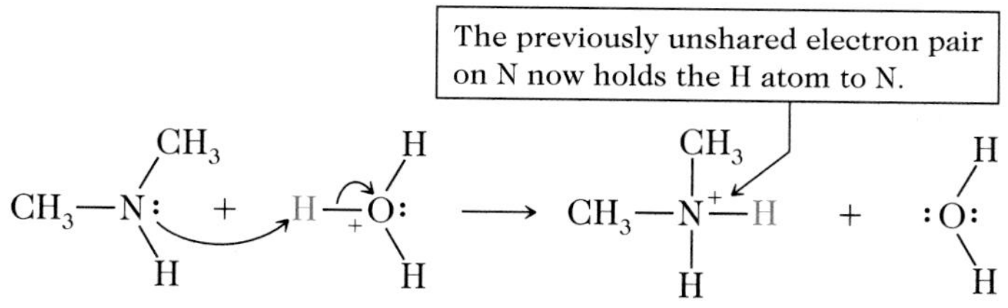

The previously unshared electron pair on N now holds the H atom to N.

dimethylamine   hydronium ion   dimethylammonium ion

Notice that the structure of the protonated form of an amine is drawn by attaching H to N and placing a plus sign by the N.

■ **PRACTICE EXERCISE 3** What are the structures of the cations that form when the following amines react completely with hydrochloric acid?

(a) Aniline   (b) Trimethylamine   (c) $NH_2CH_2CH_2NH_2$

■ Some amine salts are internal salts, like all of the amino acids, the building blocks of proteins.

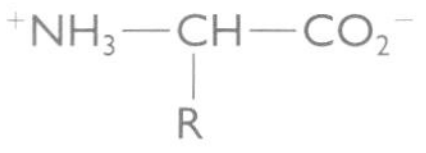

$$^+NH_3—CH—CO_2^-$$
$$|$$
$$R$$

General formula of all amino acids. R = an organic group but not always a simple alkyl group

Protonated Amines Can Neutralize Strong Bases   A combination of a protonated amine and an anion make up an organic salt called an **amine salt.** Table 16.2 gives some examples and, like all salts, amine salts are crystalline solids at room temperature. In addition, like the salts of the ammonium ion, nearly all amine salts of strong acids are soluble in water *even when the parent amine is not.* Amine salts are much more soluble in water than amines because the *full* charges carried by the ions of an amine salt can be much better hydrated by water molecules than the amine itself, where only the small, partial charges of polar bonds occur.

Protonated amine cations also neutralize the hydroxide ion and revert to amines in the following manner (where we show only skeletal structures).

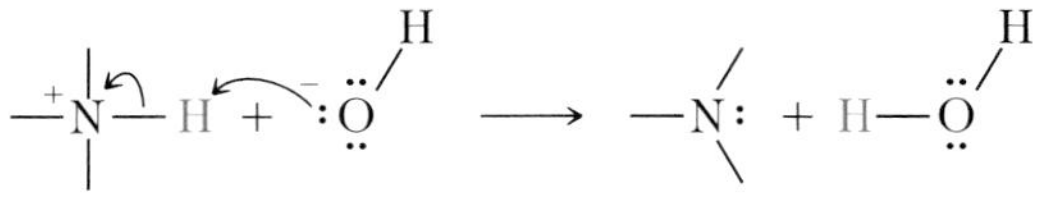

| protonated amine | hydroxide ion | amine |

For example,

$$CH_3NH_3^+ + OH^- \longrightarrow CH_3NH_2 + H_2O$$
$$CH_3CH_2\overset{+}{N}H_2CH_3 + OH^- \longrightarrow CH_3CH_2NHCH_3 + H_2O$$

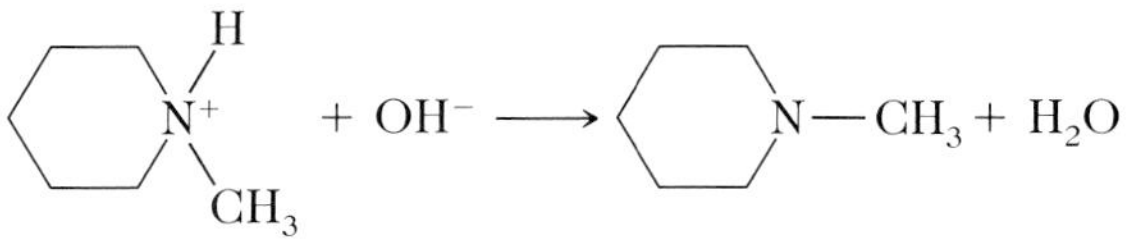

■   PRACTICE EXERCISE 4   Write the structures of the products after the following protonated amines have reacted with OH⁻ in a 1:1 mole ratio.

(a)

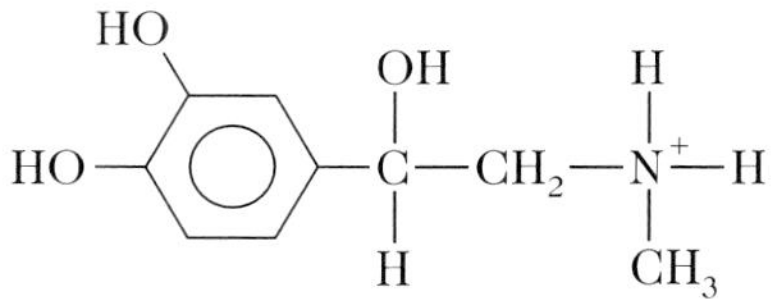

Epinephrine (adrenaline) is a hormone given here in its protonated form. As the chloride salt in a 0.1% solution, it is injected in some cardiac failure emergencies. (See also Interaction 16.1.)

■   *Hallucinogens* are drugs that cause illusions of time and place, make unreal experiences or things seem real, and distort the qualities of things.

(b)

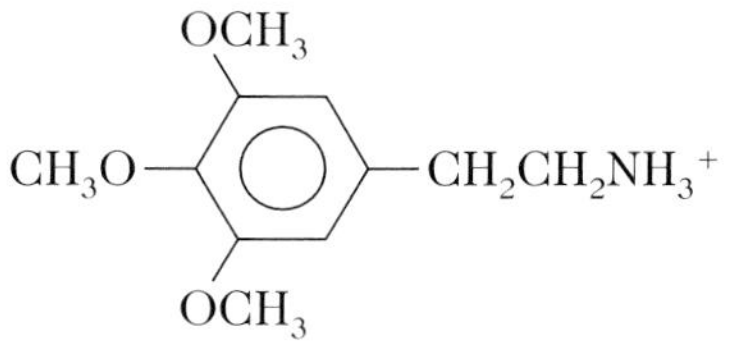

Mescaline is a mind-altering hallucinogen shown here in its protonated form. It is isolated from the mescal button, a growth on top of the peyote cactus. Indians in the southwestern United States have used it in religious ceremonies.

TABLE 16.2   Amine Salts

Name	Structure	MP (°C)
Methylammonium chloride	$CH_3NH_3^+Cl^-$	232
Dimethylammonium chloride	$(CH_3)_2NH_2^+Cl^-$	171
Dimethylammonium bromide	$(CH_3)_2NH_2^+Br^-$	134
Dimethylammonium iodide	$(CH_3)_2NH_2^+I^-$	155
Tetramethylammonium hydroxide[a]	$(CH_3)_4N^+OH^-$	130–135 (decomposes)

[a] This compound is as strong a base as NaOH because its OH⁻ ion fully dissociates in water.

SOME PHYSIOLOGICALLY ACTIVE AMINES

Maybe you have heard the expression, "I need to get my adrenalin flowing." Adrenalin—or *epinephrine,* its technical name—is a hormone made by the adrenal gland. When you have a sudden fright, a trace amount of epinephrine immediately leaves this gland, enters circulation, and is picked up by heart muscle cells. As a result your heartbeat strengthens, your blood pressure rises, and glucose in storage in the cells moves into circulation, all of which activity gets the body rapidly ready to respond to the threat.

Epinephrine and its simpler cousin, norepinephrine, are two of the many *hormones* used by our bodies, compounds made in special glands to serve as chemical messengers between parts of the body. Hormones are released in response to particular stimuli, such as fright, food odor, or sugar ingestion. When secreted, the hormone moves to the organ or tissue it's meant for and there it activates a metabolic series of reactions that are the biochemical response to the initial stimulus.

Norepinephrine has effects similar to those of epinephrine, and because these two hormones are secreted by the adrenal gland, they are called **adrenergic agents.**

HO—, HO— ⟨ring⟩ —CHCH$_2$NHCH$_3$ (OH)

epinephrine

HO—, HO— ⟨ring⟩ —CHCH$_2$NH$_2$ (OH)

norepinephrine

Synthetic epinephrine is the active ingredient in Primatine Mist.

Several useful drugs mimic epinephrine and norepinephrine, and all are classified as *adrenergic drugs.* Most of them, like epinephrine and norepinephrine, are related structurally to β-phenylethanolamine. In nearly all uses, the compounds are prepared as dilute solutions of their acid salts.

⟨ring⟩ —CHCH$_2$NH$_2$ (OH)

β-phenylethanolamine

Several β-phenylethanolamines carry two phenol groups and are called the **catecholamines,** after *catechol,* the common name of 1,2-dihydroxybenzene. Ethylnorepinephrine and isoproterenol are examples.

HO—, HO— ⟨ring⟩ —CHCHNH$_2$ (OH), CH$_2$CH$_3$

ethylnorepinephrine
(used against asthma in children)

HO—, HO— ⟨ring⟩ —CHCH$_2$NHCH(CH$_3$)$_2$ (OH)

isoproterenol
(used in treating emphysema and asthma)

The β-phenylethylamines, named after 1-amino-2-phenylethane, are another family of physiologically active amines. You may have heard of dopamine, for example, which is also a catecholamine. In schizophrenia, parts of the brain receive too much dopamine. Synthetic dopamine is used to treat shock associated with severe congestive heart failure.

The *amphetamines,* which are also β-phenylethylamines, include Dexedrine ("speed") and Methedrine ("crystal," or "meth"). Millions of these "pep pills" or "uppers" are sold illegally, constituting a serious drug abuse problem. The dangers of overuse include suicide, belligerence and hostility, paranoia, and hallucinations. Amphetamines, however, can be legally prescribed as stimulants and antidepressants, and sometimes they are prescribed for weight-control programs.

HO—, HO— ⟨ring⟩ —CH$_2$CH$_2$NH$_2$

dopamine

⟨ring⟩ —CH$_2$CHNH$_2$ (CH$_3$)

Dexedrine

⟨ring⟩ —CH$_2$CHNHC (CH$_3$)

Methedrine

■ The other important solubility switch that we have studied involves the carboxylic acid group.

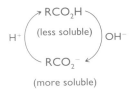

The Amino Group Is a Solubility Switch   We have just learned that putting a proton on an amino group and taking if off are easily done at room temperature simply by adding an acid and then a base. We have also learned that the protonated amine is more soluble in water than the amine. This makes the amino group an excellent "solubility switch."

The solubility of an amine can be switched on simply by adding enough strong acid to protonate it. Triethylamine, for example, is insoluble in water, but we can switch on its solubility by adding a strong acid, like hydrochloric acid. The amine dissolves as its protonated ionic form is produced.

$$(CH_3CH_2)_3N\colon + HCl(aq) \longrightarrow (CH_3CH_2)_3\overset{+}{N}HCl^-(aq)$$

triethylamine   hydrochloric   triethylammonium
(water insoluble)   acid   chloride (water soluble)

We can just as quickly and easily bring the amine back out of solution by adding a strong base, like the hydroxide ion. It takes the proton off the protonated amine and gives the less soluble form.

$$(CH_3CH_2)_3\overset{+}{N}HCl^-(aq) + OH^- \longrightarrow (CH_3CH_2)_3N\colon + H_2O$$

triethylammonium   (e.g., supplied   triethylamine
chloride (water soluble)   by NaOH)   (water insoluble)

The significance of this "switching" relationship is that the solubilities of complex compounds that have the amine function can be changed almost instantly simply by adjusting the pH of the medium.

One application of this property involves medicinals. A number of amines obtained from the bark, roots, leaves, flowers, or fruits of various plants are useful drugs. These naturally occurring, acid-neutralizing, physiologically active amines are called **alkaloids,** and morphine, codeine, and quinine are just three examples.

morphine   codeine   quinine

To make it easier to administer alkaloidal drugs in the dissolved state, we often prepare them as their water-soluble amine salts. Morphine, for example, a potent sedative and painkiller, is often given as morphine sulfate, the salt of morphine and sulfuric acid. Quinine, an antimalarial drug, is available as quinine sulfate. Codeine, sometimes used in cough medicines, is often present as codeine phosphate. Interaction 16.1 tells about a few other physiologically active amines, most of which are also prepared as their amine salts.

16.3 AMIDES OF CARBOXYLIC ACIDS

Amides are neutral nitrogen compounds that can be hydrolyzed to carboxylic acids and ammonia (or amines).

The carbonyl–nitrogen bond is sometimes called the **amide bond,** because it is the bond that forms when amides are made, and it is the bond that breaks when amides are hydrolyzed. As the following general structures show, an amide can be derived either from ammonia or from amines. Those derived from ammonia itself are often referred to as *simple* amides.

O ‖ RCNH₂	O ‖ RCNHR′	O R″ ‖ │ RC—NR′	O ‖ —C—N—
amides of ammonia (simple amides)	amides of amines	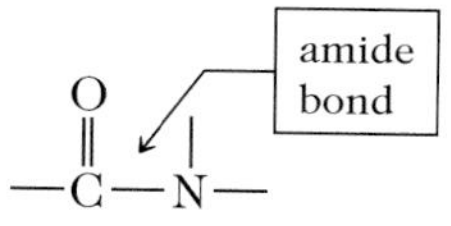	amide group

We study the amide system because all proteins are essentially polyamides, polymers whose molecules have regularly spaced amide bonds. Nylon is a synthetic polymer with repeating amide groups (see Interaction 16.2).

Table 16.3 lists several low formula mass amides. Their molecules are quite polar, and when they have an H atom bonded to N, they can both donate and accept hydrogen bonds. These forces add up so much in simple amides that all except methanamide are solids at room temperature. Simple amides have considerably higher boiling points than alkanes, alcohols, or even carboxylic acids of comparable formula mass, as the data in the margin show. When we study proteins, we'll see how hydrogen bonding is involved in stabilizing the shapes of protein molecules, shapes that are as important to the functions of proteins as anything else about their structures.

Compound	Formula mass	BP (°C)
$CH_3CH_2CH_2CH_3$	58	−42
$CH_3CH_2CH_2OH$	60	97
CH_3CO_2H	60	118
CH_3CONH_2	59	222

The Names of Simple Amides End in *-amide*   The common names of simple amides are made by replacing *-ic acid* by *-amide,* and their IUPAC names are devised by replacing *-oic acid* by *-amide.* In the following examples, notice how we can condense the structure of the amide group.

TABLE 16.3   Amides of Carboxylic Acids

IUPAC Name	Structure	MP (°C)
Methanamide	$HCONH_2$	3
N-Methylmethanamide	$HCONHCH_3$	−5
N,N-Dimethylmethanamide	$HCON(CH_3)_2$	−61
Ethanamide	CH_3CONH_2	82
N-Methylethanamide	$CH_3CONHCH_3$	28
N,N-Dimethylethanamide	$CH_3CON(CH_3)_2$	−20
Propanamide	$CH_3CH_2CONH_2$	79
Butanamide	$CH_3CH_2CH_2CONH_2$	115
Pentanamide	$CH_3CH_2CH_2CH_2CONH_2$	106
Hexanamide	$CH_3CH_2CH_2CH_2CH_2CONH_2$	100
Benzamide	$C_6H_5CONH_2$	133

INTERACTION 16.2
NYLON, A POLYAMIDE

The term *nylon* is a coined name that applies to any synthetic, long-chain, fiber-forming polymer with repeating amide linkages. One of the most common members of the nylon family, nylon-66, is made from 1,6-hexanediamine and hexanedioic acid.

$$NH_2CH_2CH_2CH_2CH_2CH_2CH_2NH_2$$

1,6-hexanediamine

$$\underset{\text{hexanedioic acid}}{HOCCH_2CH_2CH_2CH_2COH}$$

etc. $\left[C(CH_2)_4CNH(CH_2)_6NH \right]_n$ etc.

repeating unit in nylon-66

(The "66" means that each monomer has six carbon atoms.) To be useful as a fiber-forming polymer, each nylon-66 molecule should contain from 50 to 90 of each of the monomer units. Shorter molecules form weak or brittle fibers.

When molten nylon resin is being drawn into fibers, newly emerging strands are caught up on drums and stretched as they cool. Under this tension, the long polymer molecules within the fiber line up side by side, overlapping each other, to give a finished fiber of unusual strength and beauty (see Figure 1). Part of nylon's strength comes from the innumerable hydrogen bonds that extend between the polymer molecules and that involve their many regularly spaced amide groups.

Nylon is more resistant to combustion than wool, rayon, cotton, or silk, and it is as immune to insect attack as fiberglass. Molds and fungi do not attack nylon molecules either. In medicine, nylon is used in specialized tubing, and as velour for blood contact surfaces. Nylon sutures were the first synthetic sutures and are still commonly used.

FIGURE 1 This woman's life depends on the strength of nylon when she is parasailing.

$$\underset{\substack{\text{acetamide (common name)}\\\text{ethanamide (IUPAC name)}}}{CH_3CNH_2 \quad \text{or} \quad CH_3CONH_2}$$

$$\underset{\substack{\text{butyramide (common name)}\\\text{butanamide (IUPAC name)}}}{CH_3CH_2CH_2CNH_2 \quad \text{or} \quad CH_3CH_2CH_2CONH_2}$$

The simplest aromatic amide is called benzamide, $C_6H_5CONH_2$, where C_6H_5 signifies the phenyl group. We'll not need to know the rules for naming other kinds of amides.

■

■ PRACTICE EXERCISE 5 Write the IUPAC names of the following amides.

$$\text{(a)} \quad \underset{\substack{|\\CH_3}}{CH_3CH_2CHCH_2CH_2CNH_2} \quad \text{(b)} \quad \underset{\substack{|\\CH_3CH_2}}{CH_3CH_2CHCNH_2}$$

Unlike the Amines, Amides Are Not Proton Acceptors   One reason for creating a separate family for the amides apart from the amines is that, unlike amines, amides are not proton acceptors or bases. They're not proton donors or acids either. *Amides are neutral in an acid–base sense.* The amide group, in other words, does not affect the pH of an aqueous system.

The electronegative carbonyl group on the nitrogen atom causes the acid–base neutrality of amides. Although both an amide and an amine have an unshared pair of electrons on nitrogen, in the amide this pair is drawn back so tightly by the electron withdrawing ability of the carbonyl group that the electron pair shown on the nitrogen atom of the amide cannot actually accept and hold a proton.

■ The oxygen atom of the carbonyl group is what makes the whole group electronegative.

Amides Are Made from Amines by Acyl Group Transfer Reactions   Amides can be made from amines just as esters can be made from alcohols. Either acid chlorides or acid anhydrides react smoothly with ammonia or amines to give amides. (The amine, of course, must have at least one hydrogen atom on nitrogen, because one hydrogen has to be replaced as the amide forms.) We can illustrate these reactions using ammonia.

$$
\underset{\text{acid chloride}}{R-\overset{\overset{\displaystyle O}{\|}}{C}-Cl} + 2NH_3 \longrightarrow \underset{\text{amide}}{R-\overset{\overset{\displaystyle O}{\|}}{C}-NH_2} + \underset{\text{ammonium chloride}}{NH_4Cl}
$$

$$
\underset{\text{acid anhydride}}{R-\overset{\overset{\displaystyle O}{\|}}{C}-O-\overset{\overset{\displaystyle O}{\|}}{C}-R} + 2NH_3 \longrightarrow \underset{\text{amide}}{R-\overset{\overset{\displaystyle O}{\|}}{C}-NH_2} + \underset{\substack{\text{ammonium salt of}\\\text{the carboxylic acid}}}{NH_4O-\overset{\overset{\displaystyle O}{\|}}{C}-R}
$$

These reactions are further examples of *acyl group transfer reactions.* The acyl group in the acid chloride, for example, transfers from the Cl atom to the N atom of the amine (or ammonia). An acyl group can transfer from an acid anhydride to N, also. In both of these reactions, a stable, weakly basic leaving group, Cl^- or RCO_2^-, is released by the transferring acyl group. Direct acyl transfer from a carboxylic acid, however, is more difficult because the leaving group is the less stable strong base, OH^-.

■ When heated strongly, the ammonium salts of carboxylic acids change to the corresponding simple amides as water is expelled.

In the body, other kinds of acyl carrier molecules serve instead of ordinary acid chlorides and anhydrides as sources of the acyl group. When proteins are made from amino acids, for example, the acyl portions of amino acids—they are called *aminoacyl units*—are held by carrier molecules.

$$
\underset{\underset{\displaystyle R}{|}}{NH_2-CH-\overset{\overset{\displaystyle O}{\|}}{C}}-\boxed{\substack{\text{carrier}\\\text{molecule}}}
$$

aminoacyl unit

When a cell makes an amide bond, it transfers an aminoacyl group from its carrier molecule to the nitrogen atom of the amino group. The carrier molecule is released to be reused.

■ R is some organic group, but not necessarily an alkyl group.

$$
\underset{\underset{\displaystyle R}{|}}{NH_2-CH-\overset{\overset{\displaystyle O}{\|}}{C}}-\boxed{\substack{\text{carrier}\\\text{molecule}}} + NH_2-R' \xrightarrow{\substack{\text{aminoacyl}\\\text{group transfer}}} \underset{\underset{\displaystyle R}{|}}{NH_2-CH-\overset{\overset{\displaystyle O}{\|}}{C}}-NH-R' + \boxed{\substack{\text{carrier}\\\text{molecule}}} + H^+
$$

This is the aspect of making amides—aminoacyl transfers—that is of greatest interest as we prepare for our upcoming study of biochemistry.

EXAMPLE 16.1   Writing the Structure of an Amide That Can Be Made from the Acyl Group of an Acid and an Amine

---

What amide can be made from the following two substances, assuming that a suitable acyl group transfer process is available?

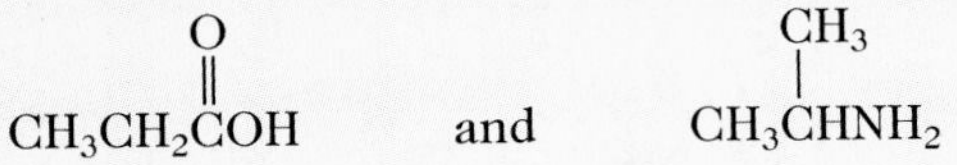

$$\underset{CH_3CH_2\overset{\displaystyle O}{\overset{\displaystyle \|}{C}}OH}{} \qquad and \qquad \underset{CH_3\overset{\displaystyle CH_3}{\overset{\displaystyle |}{C}}HNH_2}{}$$

ANALYSIS The amide system must be part of the structure we seek, so the best way to proceed is to write the skeleton of the amide system and then build on it. It doesn't matter how we orient this skeleton, left to right or right to left, as we'll demonstrate by showing both approaches.

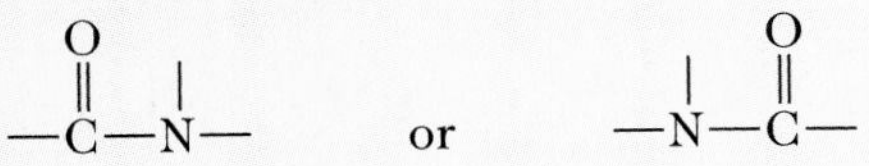

$$-\overset{O}{\overset{\|}{C}}-\overset{|}{N}- \qquad or \qquad -\overset{|}{N}-\overset{O}{\overset{\|}{C}}- \quad (Incomplete)$$

Then we look at the acid to see what else must be on the carbon atom of this skeleton. It's an ethyl group, so we write it in.

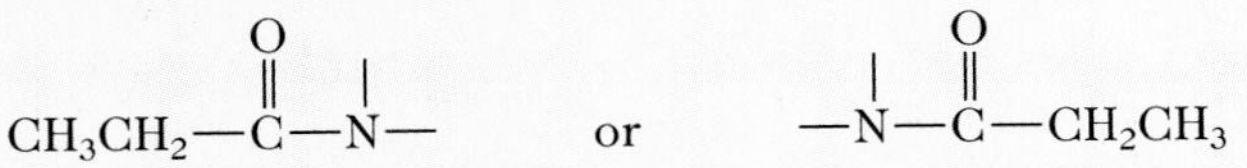

$$CH_3CH_2-\overset{O}{\overset{\|}{C}}-\overset{|}{N}- \qquad or \qquad -\overset{|}{N}-\overset{O}{\overset{\|}{C}}-CH_2CH_3 \quad (Incomplete)$$

Then we look at the amine to see what group(s) it carries. It has an isopropyl group, so we attach it to the N atom. (If there *had been two* organic groups on N in the amine, we would attach both, of course.)

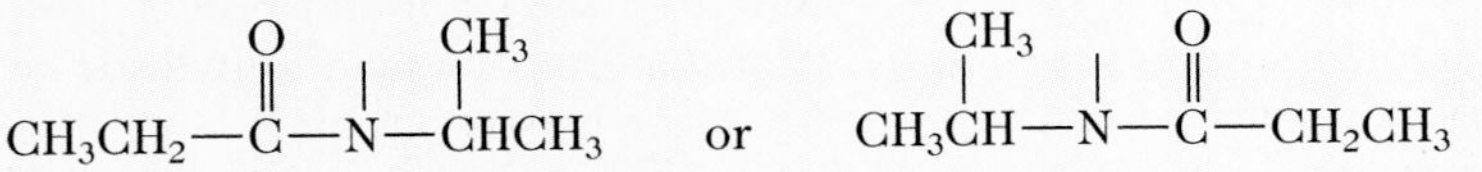

$$CH_3CH_2-\overset{O}{\overset{\|}{C}}-\overset{|}{N}-\overset{CH_3}{\overset{|}{C}}HCH_3 \qquad or \qquad CH_3\overset{CH_3}{\overset{|}{C}}H-\overset{|}{N}-\overset{O}{\overset{\|}{C}}-CH_2CH_3 \quad (Incomplete)$$

Finally, of the two H atoms on N in the amine, one survives, and our last step is to write it in. (Recall that N needs three bonds in a neutral species.)

SOLUTION The final answer is

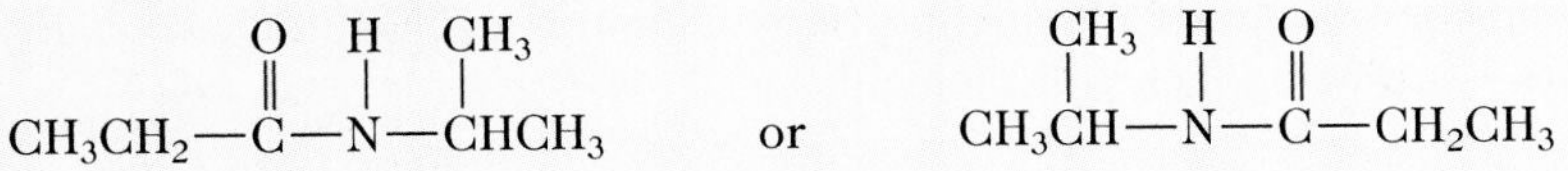

$$CH_3CH_2-\overset{O}{\overset{\|}{C}}-\overset{H}{\overset{|}{N}}-\overset{CH_3}{\overset{|}{C}}HCH_3 \qquad or \qquad CH_3\overset{CH_3}{\overset{|}{C}}H-\overset{H}{\overset{|}{N}}-\overset{O}{\overset{\|}{C}}-CH_2CH_3$$

These structures, of course, are identical.

■ PRACTICE EXERCISE 6 What amides, if any, could be made by suitable acyl group transfer reactions from the following pairs of compounds?

(a) CH_3NH_2 and $CH_3\overset{CH_3}{\overset{|}{C}}HCO_2H$ (b) $NH_2C_6H_5$ and CH_3CO_2H

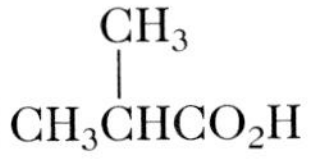

(c) 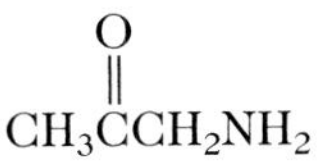 $CH_3\overset{O}{\overset{\|}{C}}CH_2NH_2$ and CH_3NH_2 (d) CH_3CO_2H and $CH_3\overset{CH_3}{\overset{|}{N}}CH_3$

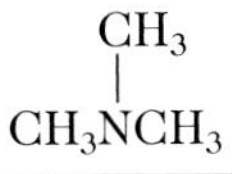

Amides Are Hydrolyzed to Their Parent Amines and Acids   The only reaction of amides that we will study is their hydrolysis, a reaction in which the amide bond breaks and we obtain the amide's parent acid and amine (or ammonia). The hydrolysis of an amide does not occur easily. In vitro, either acids or bases are needed to promote the reaction. In vivo, enzymes catalyze amide hydrolysis, and this reaction is all that is involved in the overall chemistry of the digestion of proteins.

■   The digestive tract provides several protein-digesting enzymes called *proteases*.

In vitro, when an acid promotes the hydrolysis of an amide, one of the products, the amine, neutralizes the acid. (This is why we don't say that the acid *catalyzes* the hydrolysis. Catalysts, by definition, are reaction promoters that are not used up.) Thus instead of obtaining the amine itself, we get the salt of the amine. For example,

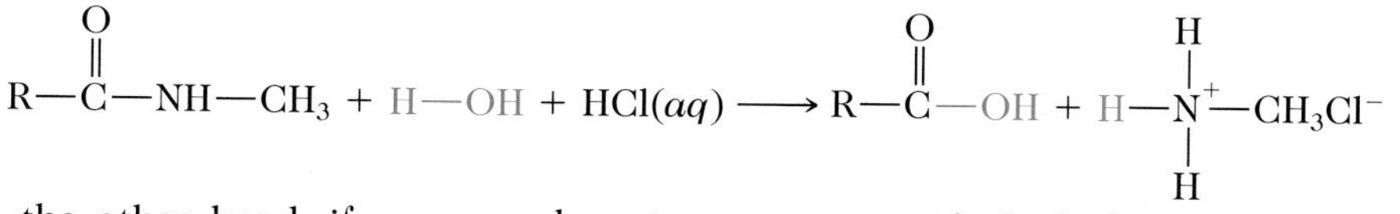

$$R-\overset{\overset{\displaystyle O}{\|}}{C}-NH-CH_3 + H-OH + HCl(aq) \longrightarrow R-\overset{\overset{\displaystyle O}{\|}}{C}-OH + H-\overset{\overset{\displaystyle H}{|}}{\underset{\underset{\displaystyle H}{|}}{N^+}}-CH_3Cl^-$$

On the other hand, if we use a base to promote amide hydrolysis, then the carboxylic acid that forms neutralizes the base, and we get the salt of the carboxylic acid. For example,

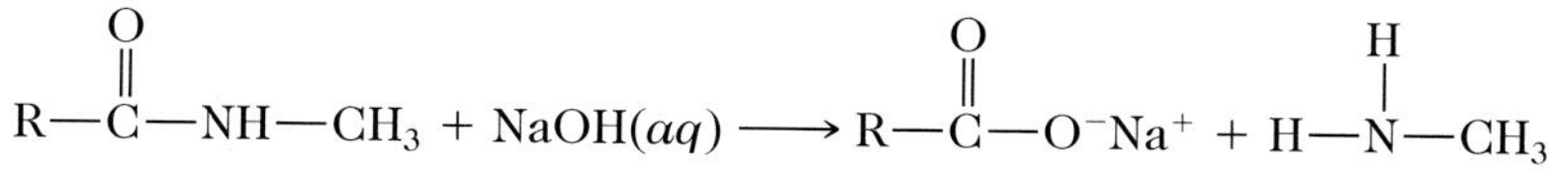

$$R-\overset{\overset{\displaystyle O}{\|}}{C}-NH-CH_3 + NaOH(aq) \longrightarrow R-\overset{\overset{\displaystyle O}{\|}}{C}-O^-Na^+ + H-\overset{\overset{\displaystyle H}{|}}{N}-CH_3$$

When enzymes catalyze this hydrolysis, they are not used up by the reaction. Because our applications of this reaction concern in vivo situations at the molecular level of life, we'll write amide hydrolysis as a simple reaction with water to give the free carboxylic acid and the free amine. Here are some examples. How the acid and the amine actually emerge depends on the pH of the medium and the buffers present.

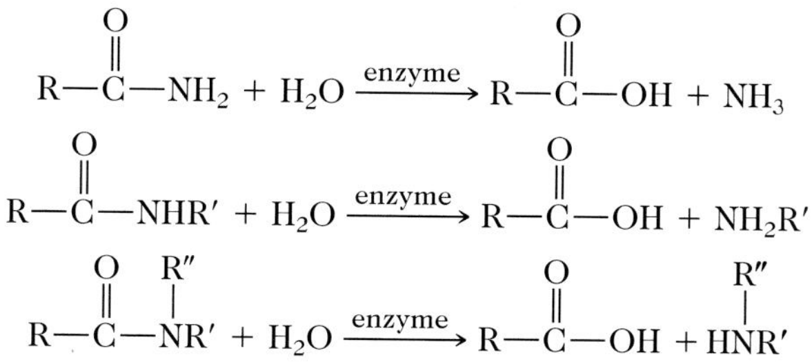

$$R-\overset{\overset{\displaystyle O}{\|}}{C}-NH_2 + H_2O \xrightarrow{\text{enzyme}} R-\overset{\overset{\displaystyle O}{\|}}{C}-OH + NH_3$$

$$R-\overset{\overset{\displaystyle O}{\|}}{C}-NHR' + H_2O \xrightarrow{\text{enzyme}} R-\overset{\overset{\displaystyle O}{\|}}{C}-OH + NH_2R'$$

$$R-\overset{\overset{\displaystyle O}{\|}}{C}-\overset{\overset{\displaystyle R''}{|}}{N}R' + H_2O \xrightarrow{\text{enzyme}} R-\overset{\overset{\displaystyle O}{\|}}{C}-OH + H\overset{\overset{\displaystyle R''}{|}}{N}R'$$

EXAMPLE 16.2   Writing the Products of the Hydrolysis of an Amide

Acetophenetidin (phenacetin) was used in some brands of headache remedies.

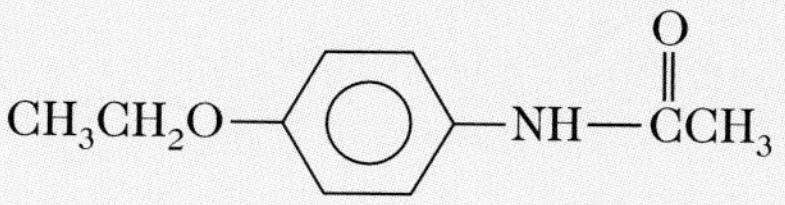

$$CH_3CH_2O-\!\!\left\langle\!\!\bigcirc\!\!\right\rangle\!\!-NH-\overset{\overset{\displaystyle O}{\|}}{C}CH_3$$

acetophenetidin

If this compound is an amide, what are the products of its hydrolysis?

ANALYSIS   Acetophenetidin does have the amide bond, NH to carbonyl, so it can be hydrolyzed. (The functional group on the left side of this structure is an *ether,* and *ethers do not react with water.*) Because the amide bond breaks when an amide is hydrolyzed, simply erase this bond from the structure and separate the parts. *Do not break any other bond.*

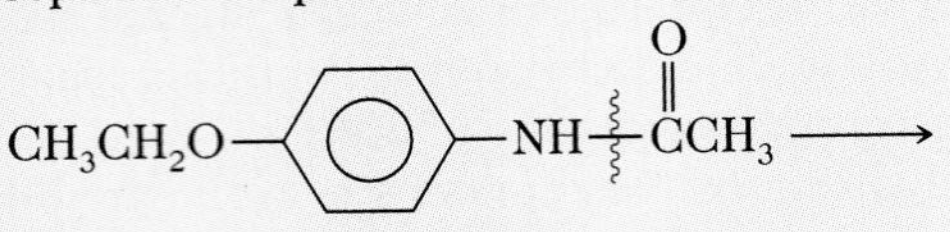

$$CH_3CH_2O-\!\!\!\left\langle\bigcirc\right\rangle\!\!\!-NH\overset{\underset{\displaystyle\vdots}{}}{}\overset{\displaystyle O}{\overset{\|}{C}}CH_3 \longrightarrow$$

acetophenetidin

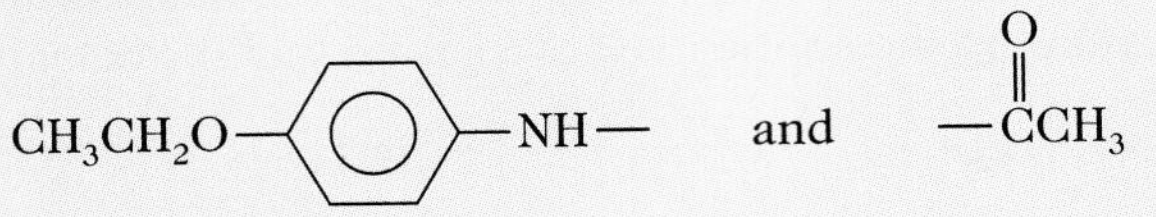

$$CH_3CH_2O-\!\!\!\left\langle\bigcirc\right\rangle\!\!\!-NH- \qquad \text{and} \qquad -\overset{\displaystyle O}{\overset{\|}{C}}CH_3 \quad \text{(Incomplete)}$$

We know that the hydrolysis uses HO—H to give a *carboxylic acid* and an *amine,* so we put an HO group on the carbonyl group and we put H on the nitrogen of the other fragment.

SOLUTION   The products of the hydrolysis of acetophenetidin are

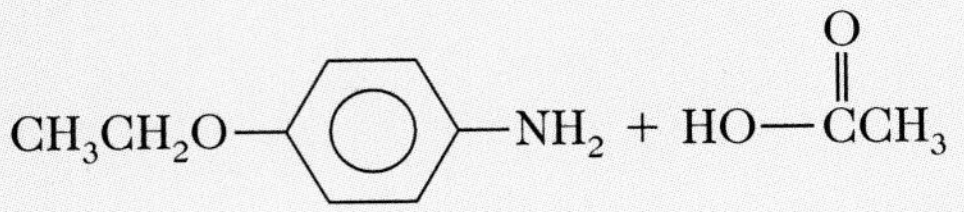

$$CH_3CH_2O-\!\!\!\left\langle\bigcirc\right\rangle\!\!\!-NH_2 + HO-\overset{\displaystyle O}{\overset{\|}{C}}CH_3$$

■   PRACTICE EXERCISE 7   For all compounds in the following list that are amides, write the products of their hydrolysis.

(a) 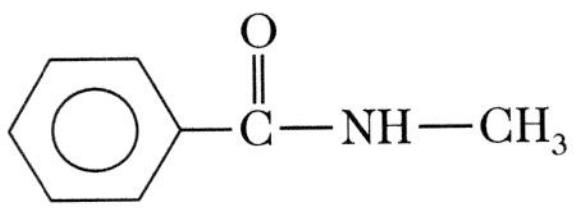 $\left\langle\bigcirc\right\rangle\!\!-\overset{\displaystyle O}{\overset{\|}{C}}-NH-CH_3$ (b) $\left\langle\bigcirc\right\rangle\!\!-\overset{\displaystyle O}{\overset{\|}{C}}-CH_2-NH_2$

(c) $\left\langle\bigcirc\right\rangle\!\!-NH-\overset{\displaystyle O}{\overset{\|}{C}}-CH_3$ (d) $CH_3-\overset{\displaystyle O}{\overset{\|}{C}}-NH-CH_2CH_2-NH-\overset{\displaystyle O}{\overset{\|}{C}}-CH_3$

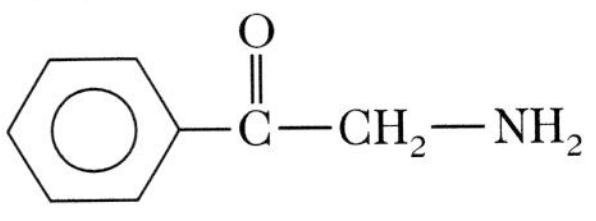

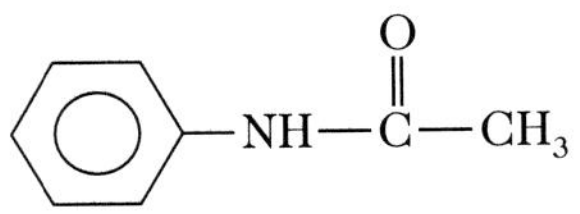

■   PRACTICE EXERCISE 8   The following structure illustrates some of the features of protein molecules. What are the products of the complete, enzyme-catalyzed hydrolysis (the digestion) of this substance? (A typical protein would hydrolyze to give several hundred and up to several thousand of the kinds of small molecules produced by hydrolysis in this example.)

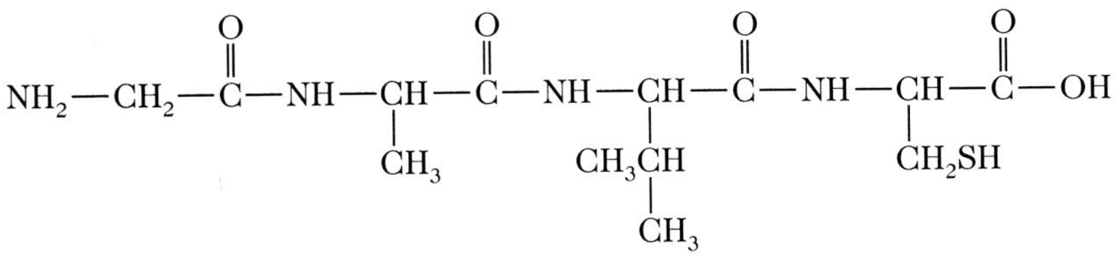

$$NH_2-CH_2-\overset{\displaystyle O}{\overset{\|}{C}}-NH-\underset{\underset{\displaystyle CH_3}{|}}{CH}-\overset{\displaystyle O}{\overset{\|}{C}}-NH-\underset{\underset{\underset{\displaystyle CH_3}{|}}{\underset{\displaystyle CH_3CH}{|}}}{CH}-\overset{\displaystyle O}{\overset{\|}{C}}-NH-\underset{\underset{\displaystyle CH_2SH}{|}}{CH}-\overset{\displaystyle O}{\overset{\|}{C}}-OH$$

SUMMARY

Amines and protonated amines When one, two, or three of the hydrogen atoms in ammonia are replaced by an organic group (other than a carbonyl group), the resulting compound is an amine. The nitrogen atom can be part of a ring, as in heterocyclic amines. Like ammonia, the amines are weak bases, and all can form salts with strong acids. The cations in these salts are protonated amines.

Amine salts are far more soluble in water than their parent amines. Protonated amines are easily deprotonated by any strong base to give back the original and usually far less soluble amine. Thus any compound with the amine function has a "solubility switch," because its solubility in an aqueous system can be turned on by adding acid (to form the amine salt) and turned off again by adding base (to recover the amine).

Amides The carbonyl–nitrogen bond, the amide bond, can be formed by letting an amine or ammonia react with anything that can transfer an acyl group (e.g., an acid chloride or an acid anhydride). Amides are neither basic nor acidic, but are neutral compounds. Amides can be made to react with water to give back their parent acids and amines. The accompanying chart summarizes the reactions studied in this chapter.

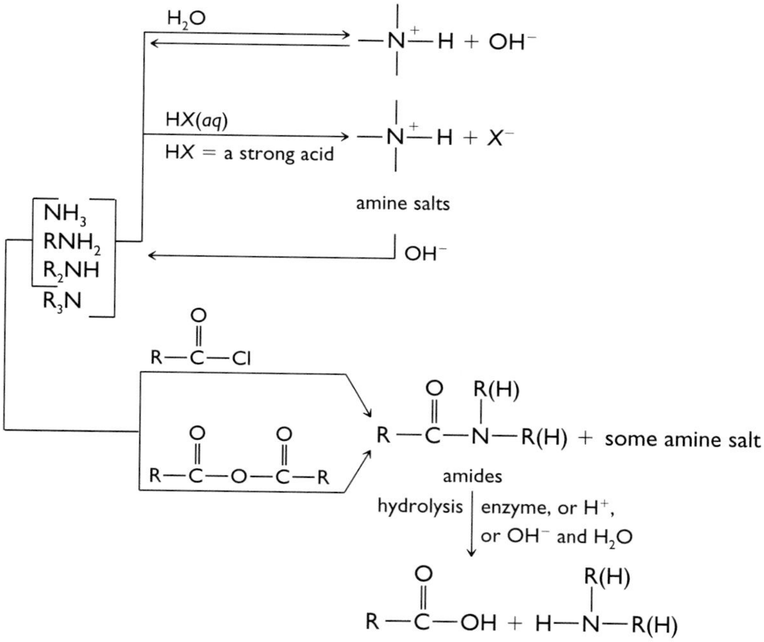

REVIEW EXERCISES

The answers to Review Exercises whose numbers are in color are found in Appendix E. The answers to the other Review Exercises are found in the Study Guide that accompanies this book. The more challenging questions are marked with asterisks.

Structures of Amines and Amides—Review of Functional Groups

16.1 Identify the functional groups in the following.

(a) $CH_3CCH_2CNH_2$

(b) $HOCCH_2CCH_2NH_2$

(c) $CH_3OCCH_2CNH_2$

(d)

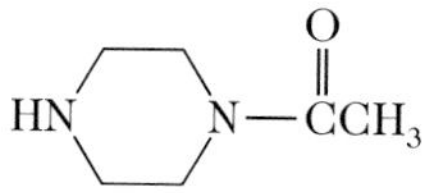

16.2 Classify each of the following as aliphatic, aromatic, or heterocyclic amines or amides. Name any other functional groups that are present.

(a) 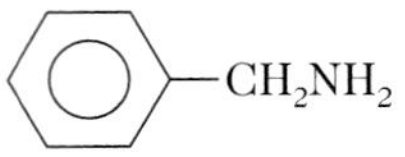 ⬡—CH_2NH_2

(b) 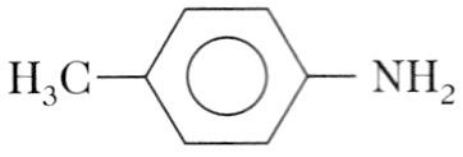 H_3C—⬡—NH_2

(c) 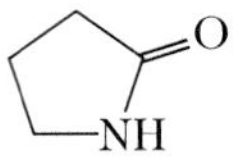 [pyrrolidinone structure with O and NH]

(d) 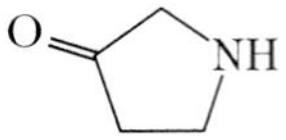 [pyrrolidinone structure with O and NH]

*16.3 The following compounds are all very active physiological agents. Name the numbered functional groups that are present in each.

(a) 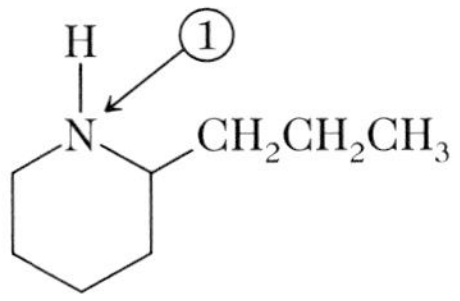 [piperidine ring with H and ① pointing to N, $CH_2CH_2CH_3$]

Coniine, the poison in the extract of hemlock that was used to execute the Greek philosopher Socrates

(b) 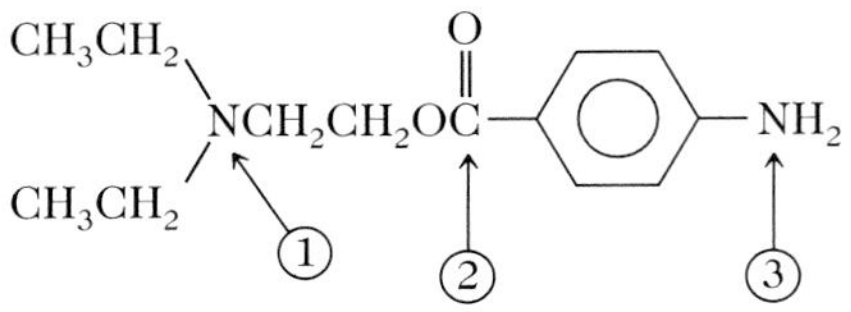 CH_3CH_2 \ $NCH_2CH_2O\overset{\overset{O}{\|}}{C}$—⬡—$NH_2$ / CH_3CH_2 with ① ② ③

Novocaine, a local anesthetic

(c) 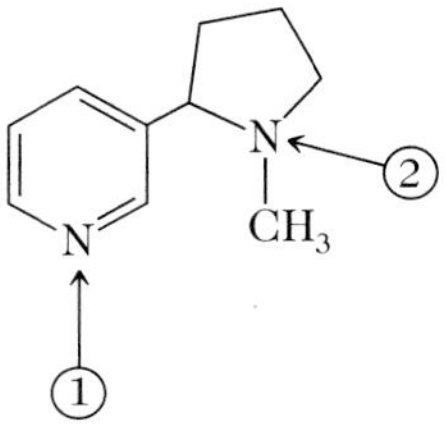 [pyridine ring connected to pyrrolidine ring with N—CH_3, numbered ① and ②]

Nicotine, a poison in tobacco leaves

(d) 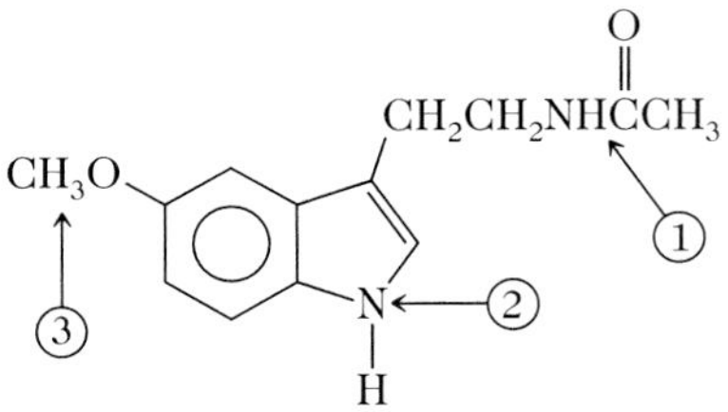 CH_3O—[indole ring structure]—$CH_2CH_2NH\overset{\overset{O}{\|}}{C}CH_3$ with ①, ② at N—H, ③

Melatonin, a human hormone made in the pineal gland

(e) 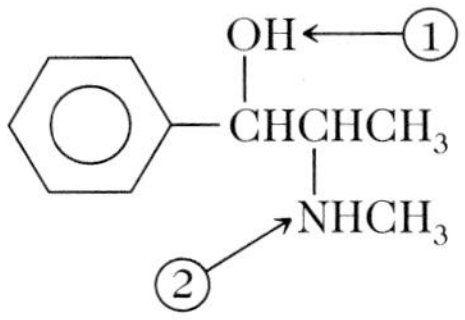 ⬡—$\overset{OH\leftarrow①}{\underset{NHCH_3}{\overset{|}{C}HCHCH_3}}$ with ②

Ephedrine, a bronchodilator

*16.4 Some extremely potent, physiologically active compounds are in the following list. Name the functional groups that they have.

(a) 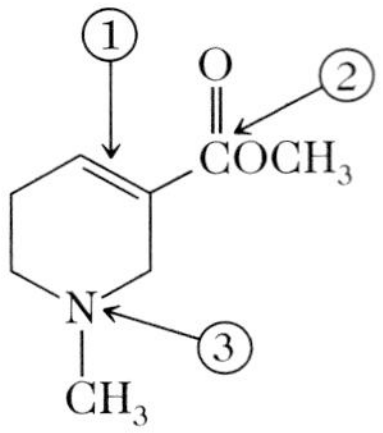 [cyclohexene ring with ① pointing to top, $\overset{\overset{O}{\|}}{C}OCH_3$ with ②, N—CH_3 with ③]

Arecoline, the most active component in the nut of the betel palm. This nut is chewed daily as a narcotic by millions of inhabitants of parts of Asia and the Pacific islands.

(b) 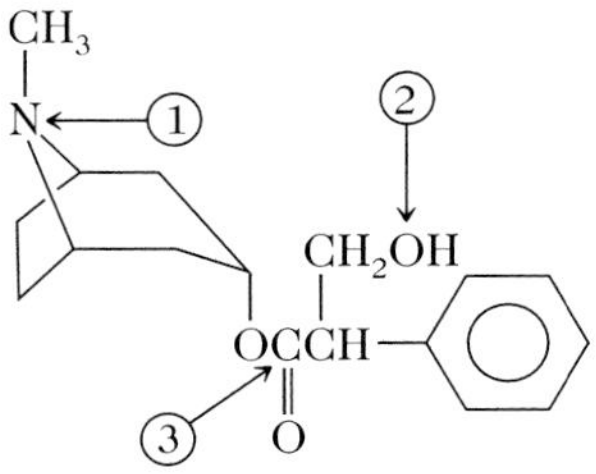 [bicyclic tropane structure with N—CH_3, ①, CH_2OH with ②, $O\overset{\overset{}{}}{C}CH$—⬡, ③ and O]

Hyoscyamine, a constituent of the seeds and leaves of henbane, and a smooth muscle relaxant. (A similar form is called atropine, a drug used to counteract nerve poisons.)

(c) 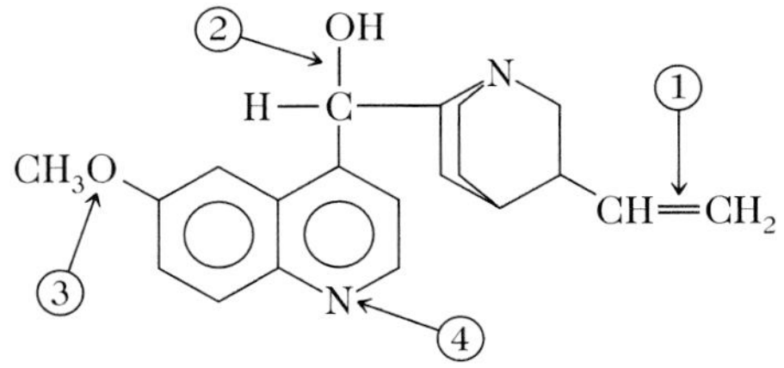 CH_3O—[quinoline ring structure] with H—$\overset{OH}{\underset{}{C}}$, ②, [quinuclidine ring with N], $CH=CH_2$ with ①, ③, ④

Quinine, a constituent of the bark of the chinchona tree in South America and used to treat malaria.

(d) 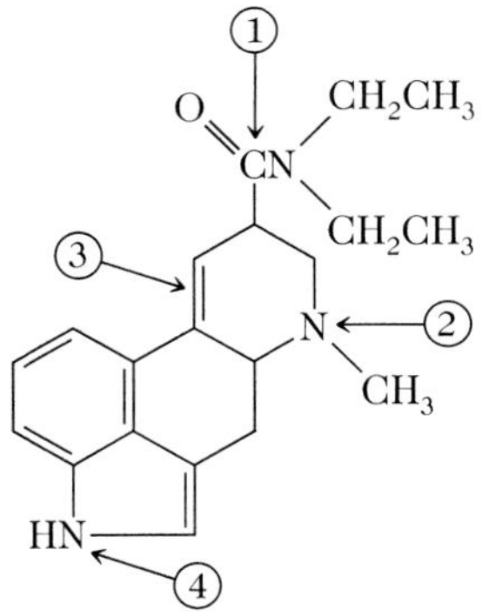 [lysergic acid diethylamide complex ring structure] $O=\overset{}{C}N\overset{CH_2CH_3}{\underset{CH_2CH_3}{}}$ with ①, N—CH_3 with ②, ③, HN— with ④

Lysergic acid diethylamide (LSD), a constituent of diseased rye grain and a notorious hallucinogen.

16.5 Give the common names of the following compounds or ions.

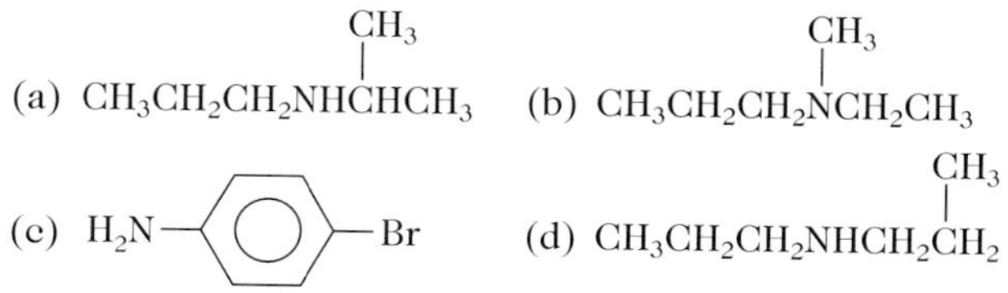

(a) CH$_3$CH$_2$CH$_2$NHCHCH$_3$
$\quad$ with CH$_3$ on the CH

(b) CH$_3$CH$_2$CH$_2$NCH$_2$CH$_3$
$\quad$ with CH$_3$ on the N

(c) H$_2$N—⬡—Br

(d) CH$_3$CH$_2$CH$_2$NHCH$_2$CH$_2$
$\quad$ with CH$_3$ branch

16.6 What are the common names of the following compounds?

(a) (CH$_3$CH$_2$)$_3$N$^+$Cl$^-$

(b) ⬡—NHCHCH$_3$ (with CH$_3$ on the CH)

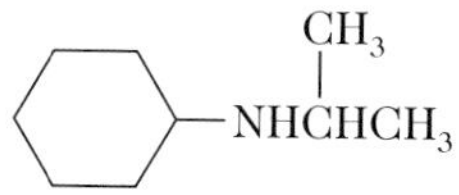

(c) (CH$_3$)$_3$CNH$_2$

(d) [(CH$_3$)$_2$CH]$_3$N

Chemical Properties of Amines and Amine Salts

16.7 Complete the following reaction sequences by writing the structures of the organic products. If no reaction occurs, write "No reaction."
(a) CH$_3$CH$_2$CH$_2$NH$_2$ + HCl(aq) →
(b) CH$_3$CH$_2$CH$_2$NH$_3$$^+Cl^-$ + NaOH(aq) →
(c) CH$_3$CH$_2$CH$_2$NH$_3$$^+Cl^-$ + HCl(aq) →
(d) CH$_3$CH$_2$CH$_2$NH$_2$ + NaOH(aq) →

16.8 Write the structures of the organic products that form in each situation. Assume that all reactions occur at room temperature. (Some of the named compounds are described in Review Exercises 16.3 and 16.4.) If no reaction occurs, write "No reaction."

(a) ⬡NH + HCl(aq) ⟶

(b) structure with COCH$_3$ + OH$^-$(aq) ⟶

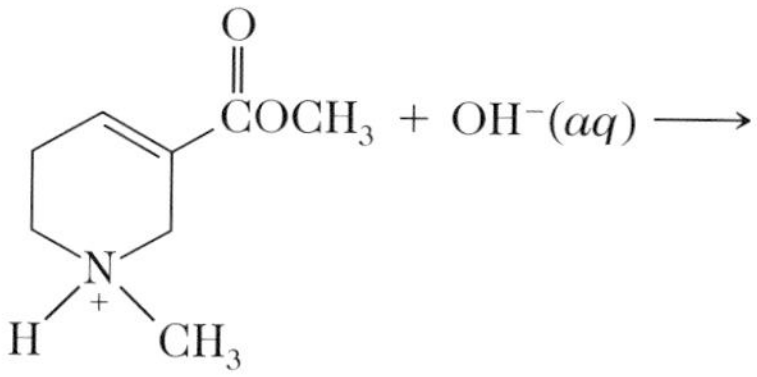

protonated form
of arecoline

(c) structure + OH$^-$(aq) ⟶

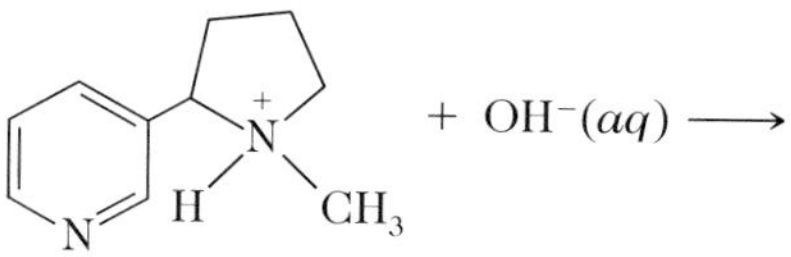

protonated form
of nicotine

(d) ⬡—CHCHCH$_3$ + HCl(aq) ⟶
$\quad$ with OH and NHCH$_3$

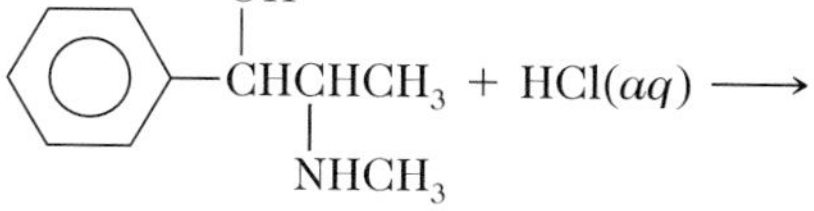

ephedrine

(e) structure with CH$_3$, N, CH$_2$OH, OCCH—⬡, =O + HCl(aq) $\xrightarrow{\text{at room temperature}}$

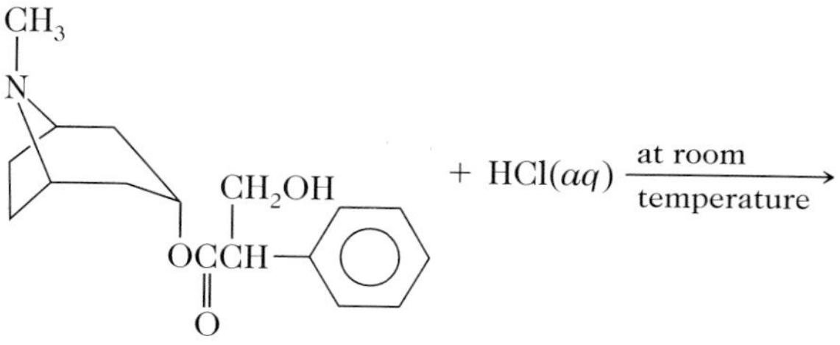

hyoscyamine

16.9 Which is the stronger base, **A** or **B**? Explain.

$$NH_2CH_2\overset{\overset{\displaystyle O}{\|}}{C}CH_3 \qquad CH_3\overset{\overset{\displaystyle O}{\|}}{C}NHCH_2CH_3$$
$\qquad\quad$ **A** $\qquad\qquad\qquad$ **B**

16.10 Which is the stronger proton acceptor, **A** or **B**? Explain.

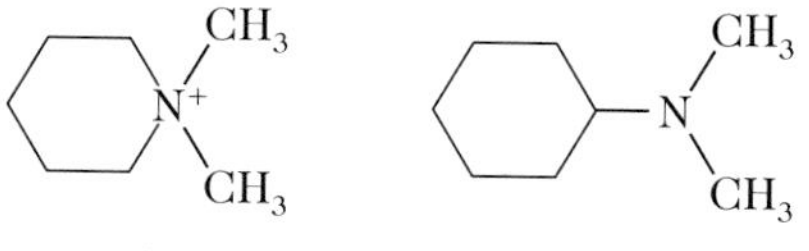

structure **A** (piperidinium with two CH$_3$)   structure **B** (cyclohexyl-N with two CH$_3$)

$\qquad$ **A** $\qquad\qquad\qquad$ **B**

Names and Structures of Amides

16.11 What are the IUPAC names of the following compounds?

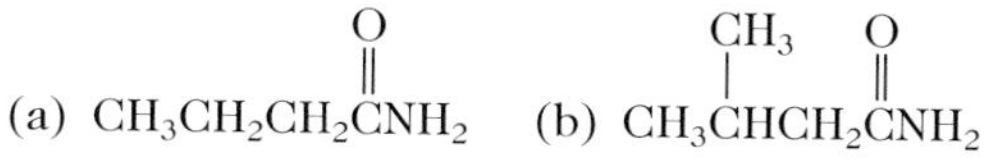

(a) CH$_3$CH$_2$CH$_2$$\overset{\overset{\displaystyle O}{\|}}{C}NH_2$

(b) CH$_3$$\overset{\overset{\displaystyle CH_3}{|}}{CH}CH_2$$\overset{\overset{\displaystyle O}{\|}}{C}NH_2$

16.12 If the common name of hexanoic acid is caproic acid, what is the common name of its simple amide?

16.13 If C$_6$H$_5$CONHCH$_3$ is the structure of N-methylbenzamide, what is the structure of N,N-dimethylbenzamide?

16.14 If ethanediamide has the structure shown, what is the structure of butanediamide?

$$NH_2\overset{\overset{\displaystyle O}{\|}}{C}—\overset{\overset{\displaystyle O}{\|}}{C}NH_2$$
ethanediamide

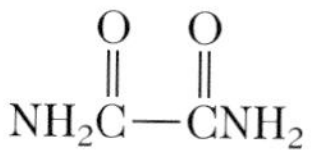

Synthesis of Amides

16.15 Write the equations for two ways to make acetamide using ammonia as one reactant.

16.16 What are two ways to make N-methylacetamide if methylamine is one reactant? Write the equations.

*16.17   Examine the following acyl group transfer reaction.

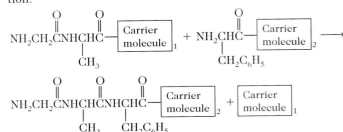

(a) Which specific acyl group transferred? (Write its structure.)
(b) How many amide bonds are showing (or implied) in the product?

Reactions of Amides

16.18   What are the products of the hydrolysis of the following compounds? (If no hydrolysis occurs, state so.)

(a) $CH_3CH_2NHCCH_3$ (with C=O)

(b) $CH_3CHNHCCH_2CH_3$ (with CH_3 and C=O)

(c) $CH_3NHC—CHCH_3$ (with O and CH_3)

(d) $CH_3CCH_2NHCH_3$ (with C=O)

16.19   Write the structures of the products of the hydrolysis of the following compounds. If no reaction occurs, state so. If more than one bond is subject to hydrolysis, be sure to hydrolyze all of them.

(a) $NH_2CHCH_2CNHCH_2COH$ (with CH_3 and two C=O)

(b) $NH_2CCH_2CHCNH_2$ (with two C=O and H_3C)

(c) [six-membered ring with NH, C=O, and CH_3 substituent]

(d) NH_2CNH_2 (with C=O)

Physiologically Active Amines (Interaction 16.1)

16.20   What are hormones and, in very broad terms, what is their function?

16.21   Hormones secreted by the adrenal gland are called what kinds of agents?

16.22   Name two hormones secreted by the adrenal gland.

16.23   Drugs that tend to mimic the two hormones secreted by the adrenal gland are called what kinds of drugs?

16.24   To be a *catecholamine* as well as a β-phenylethanolamine, a compound must have what structural features?

16.25   Is dopamine, a β-phenylethylamine, also a catecholamine?

16.26   In what general family of the physiologically active amines are the amphetamines found?

16.27   What is the chemical name of each?
(a) "Speed"   (b) "Uppers"

Nylon (Interaction 16.2)

16.28   What functional group is present in nylon-66?

16.29   The strength of a nylon fiber is attributed in part to what relatively weak bond?

Additional Exercises

*16.30   What is the structure of lysergic acid? The structure of its *N,N*-diethylamide is given in an earlier Review Exercise.

16.31   What is the structure of the amide that can form between acetic acid and ephedrine? The structure of ephedrine is given in an earlier Review Exercise.

*16.32   If the following anhydride is mixed with ammonia, what possible organic products that are not salts can form? Write their structures.

$$CH_3CH_2COCCH_3$$ (with two C=O)

*16.33   What are all the functional groups we have studied that can be changed by each of the following reactants? Write the equations for the reactions, using general symbols such as ROH or RCO_2H and so forth to illustrate these reactions, and name the organic families to which the reactants and products belong.
(a) Water, either with an acid or an enzyme catalyst.
(b) Hydrogen (or a hydride ion donor) and any needed catalysts and special conditions.
(c) An oxidizing agent represented by (O), such as $Cr_2O_7^{2-}$ or MnO_4^-, but not ozone and not oxygen as used in combustion.

16.34   We have described three functional groups that typify those involved in the chemistry of the digestion of carbohydrates, fats and oils, and proteins. What are the names of these groups and to which type of food does each belong?

*16.35   A student performed an experiment that hydrolyzed 1.65 g of benzamide.
(a) What is the maximum number of grams of benzoic acid that could be obtained?
(b) How many milliliters of 0.482 *M* HCl would be needed to convert all of the ammonia that can form from the hydrolysis of the sample of benzamide into ammonium chloride?

*16.36   Write the structures of the organic products that would form in the following situations. If no reaction occurs, state so. These exercises constitute a review of nearly all the organic reactions we have studied.

(a) $CH_3COCH_3 + H_2O \xrightarrow[\text{catalyst}]{\text{acid}}$ (with C=O)

(b) $CH_3\overset{O}{\overset{\|}{C}}CH_2CH_3 + H_2 \xrightarrow{\text{Ni or Pt}}^{\text{pressure}}$

(c) $CH_3CH_2CH_2CH_2CH_3 + MnO_4^-(aq) \longrightarrow$

(d) $CH_3CH_2\overset{O}{\overset{\|}{C}}Cl + NH_3 \longrightarrow$
(excess)

(e) ⬡$-CH_2CH_3 + Cr_2O_7^{2-}(aq) \longrightarrow$

(f) $CH_3\overset{CH_3}{\overset{|}{C}H}CH_2CH_3 + NaOH(aq) \longrightarrow$

(g) $CH_3\overset{OCH_2CH_3}{\overset{|}{C}H}OCH_2CH_3 + H_2O \xrightarrow{\text{acid}}^{\text{catalyst}}$

(h) $CH_3CH_2\overset{O}{\overset{\|}{C}}H + Cr_2O_7^{2-}(aq) \longrightarrow$

(i) $CH_3CH_2\overset{O}{\overset{\|}{C}}CH_3 + Cr_2O_7^{2-}(aq) \longrightarrow$

(j) $CH_3OH + CH_3CH_2\overset{O}{\overset{\|}{C}}OH \rightleftharpoons^{H^+}$

(k) $CH_3CH_2\overset{O}{\overset{\|}{C}}NH_2 + NaOH(aq) \xrightarrow{\text{heat}}$

(l) $CH_3CH_2\overset{O}{\overset{\|}{C}}H + 2CH_3OH \xrightarrow{\text{acid}}^{\text{catalyst}}$

(m) $CH_3CH_2SH + (O) \longrightarrow$

(n) $C_6H_5\overset{O}{\overset{\|}{C}}OCH_2\overset{CH_3}{\overset{|}{C}H}CH_3 + NaOH(aq) \longrightarrow$

(o) $NH_2CH_2CH_2\overset{CH_3}{\overset{|}{C}H}CH_3 + HCl(aq)$

(p) $CH_3CH{=}CHCH_2OCH_3 + H_2 \xrightarrow{\text{Ni or Pt}}$

*16.37 What are the structures of the organic products that form in the following situations? (If there is no reaction, state so.) These reactions review most of the chemical properties of functional groups we have studied.

(a) $H\overset{O}{\overset{\|}{C}}CH_2CH_2OCH_3 + MnO_4^-(aq)$

(b) ⬠$-CO_2H + NaOH(aq) \longrightarrow$

(c) ⬡ $+ H_2O \xrightarrow{H^+}$

(d) $NH_2CH_2CH_2NH_2 + HCl(aq) \longrightarrow$
(excess)

(e) $CH_3\overset{O}{\overset{\|}{C}}CH_2\overset{CH_3}{\overset{|}{C}H}CH_3 + H_2 \xrightarrow{\text{Ni or Pt}}^{\text{pressure}}$

(f) $CH_3(CH_2)_5CH_3 + Cr_2O_7^{2-}(aq) \longrightarrow$

(g) $CH_3CH_2\overset{O}{\overset{\|}{C}}H + 2CH_3OH \xrightarrow{\text{acid}}^{\text{catalyst}}$

(h) $CH_3OCH_2\overset{O}{\overset{\|}{C}}OCH_2CH_3 + NaOH(aq) \longrightarrow$

(i) $CH_3CH_2\overset{OH}{\overset{|}{C}H}\overset{}{C}HCH_3 + MnO_4^-(aq) \longrightarrow$ (with CH_3)

(j) $CH_3\overset{OCH_2CH_3}{\overset{|}{C}}OCH_2CH_3 + H_2O \xrightarrow{\text{acid}}^{\text{catalyst}}$ (with CH_3)

(k) $CH_3CH_2O\overset{O}{\overset{\|}{C}}(CH_2)_3\overset{O}{\overset{\|}{C}}OCH_3 + H_2O \xrightarrow{\text{acid}}^{\text{catalyst}}$

(l) $CH_3O\overset{CH_2CH(CH_3)_2}{\overset{|}{C}H}OCH_3 + H_2O \xrightarrow{\text{acid}}^{\text{catalyst}}$

(m) $CH_3\overset{O}{\overset{\|}{C}}\overset{O}{\overset{\|}{C}}CH_3 + NH_3 \longrightarrow$
(excess)

(n) $CH_3SSCH_3 \xrightarrow{\text{reduction (2H)}}$

(o) $CH_3(CH_2)_4CO_2H + CH_3CH_2OH \rightleftharpoons^{H^+}$

(p) $CH_3NH\overset{O}{\overset{\|}{C}}CH_2\overset{CH_3}{\overset{|}{C}H}NH\overset{O}{\overset{\|}{C}}CH_2CH_3 + H_2O \xrightarrow{\text{enzyme}}$
(excess)

16.38 Write the IUPAC names of the following.
(a) $CH_3CH_2CH_2CO_2CH_3$
(b) $(CH_3)_2CHCH_2Br$
(c) $O{=}CHCH_2CH(CH_3)_2$
(d) $CH_3CH_2CH{=}CHCH_3$
(e) $(CH_3)_3CCH_2CH(CH_3)_2$
(f) $CH_3CH_2CH_2COCH(CH_3)_3$
(g) $(CH_3)_3COH$
(h) $HO_2C(CH_2)_3CH_3$
(i) CH_3CO_2Na

16.39 Identify by letter which of the following compounds would be more soluble in water at a pH of 12 than at a pH of 7. Explain.

$CH_3(CH_2)_6CO_2CH_3 \qquad CH_3(CH_2)_6CO_2H \qquad CH_3(CH_2)_6CH_2NH_2$

A \qquad\qquad **B** \qquad\qquad **C**

17

STEREOISOMERISM

To reduce the glare caused by polarized light glancing from this stream in California, the photographer used a polarizing filter. Solutions of most substances at the molecular level of life affect another property of polarized light. It's because their molecules have "handedness," which we'll learn about in this chapter.

THIS CHAPTER IN CONTEXT

Your two hands are so alike that you'd be excused if you said they're identical. (We're ignoring wrinkles and fingerprints.) But if you tried to put your right hand into a left hand glove, you'd realize the error. The molecules of nearly all organic compounds at the molecular level of life are also capable of existing in "left hand" and "right hand" forms. *The difference this makes is as important as anything in biochemistry.* It affects how all enzymes and hormones work, for example, because their work requires that their molecules snugly fit others. This chapter is meant to tell you about this, the most subtle way by which isomers can be different. You may be surprised by the beauty of it, too.

17.1 TYPES OF ISOMERISM

Constitutional isomers and stereoisomers are the two broad classes of isomers.

Compounds that have identical molecular formulas can be different in two general ways, as *constitutional isomers* or as *stereoisomers*.

■ Constitutional isomers are often called *structural isomers,* an older name now being replaced.

Constitutional Isomers Differ in Molecular Frameworks   Butane and isobutane, both C_4H_{10}, have different chains. Ethanol and dimethyl ether, both C_2H_6O, have different functional groups. Each pair illustrates **constitutional isomerism** because their members differ in the ways by which the atoms are joined to each other. The molecules of **constitutional isomers** have different atom-to-atom sequences.

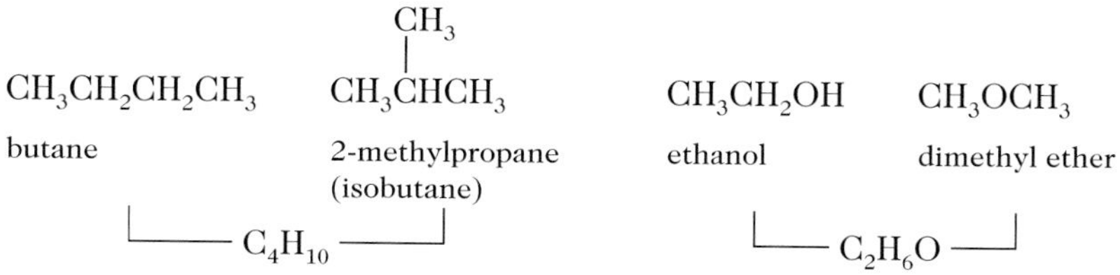

$$CH_3CH_2CH_2CH_3 \qquad CH_3\overset{\overset{\displaystyle CH_3}{|}}{C}HCH_3 \qquad CH_3CH_2OH \qquad CH_3OCH_3$$

butane   2-methylpropane   ethanol   dimethyl ether
                  (isobutane)

$$\underbrace{\qquad\qquad}_{C_4H_{10}} \qquad\qquad \underbrace{\qquad\qquad}_{C_2H_6O}$$

Stereoisomers Differ Only in Geometry   *cis*-2-Butene and *trans*-2-butene are **stereoisomers,** meaning that they have identical constitutions but different geometries.

■ STER-ee-oh-EYE-som-ers

$$CH_3CH=CHCH_3$$

2-butene

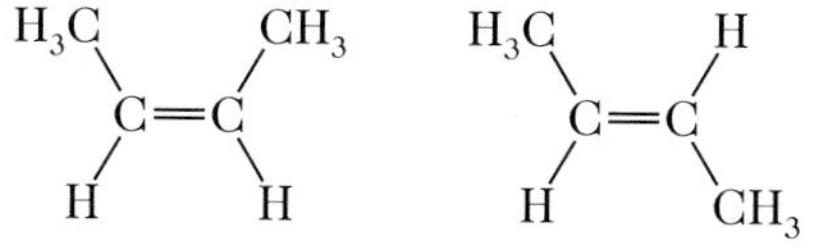

cis-2-butene   *trans*-2-butene

Stereoisomers of 2-butene

Stereoisomerism depends on the possibility that compounds can be geometrically different despite having identical molecular formulas, functional groups, and heavy-atom skeletons. In the case of the 2-butene stereoisomers, the lack of free rotation at the double bond is what ensures that one cannot easily switch to the other.

■ Recall that the four atoms attached directly to the carbon atoms of the double bond all lie in the same plane.

■ dye-a-STER-ee-o-mers

en-AN-tee-o-mers

There are two broad families of stereoisomers, *diastereomers* (illustrated by *cis*- and *trans*-2-butene) and *enantiomers*. This chapter is mostly about enantiomers, but to define them requires more background. They come about as close to being identical as your left and right hands, yet they display dramatic differences in chemistry at the molecular level of life. One such difference is illustrated by the bitter-sweet story of asparagine.

Asparagine is a white solid first isolated in 1806 from the juice of asparagus. The asparagine obtained from this source has a bitter taste. Its molecular formula is $C_4H_8N_2O_3$, and its structure is given by **1**.

■ Asparagine is a building block for making proteins in the body.

$$\overset{\displaystyle O}{\overset{\displaystyle \|}{NH_2CCH_2\overset{\displaystyle |}{\underset{\displaystyle NH_2}{C}}HCO_2H}}$$

1

■ Vetch is a member of a genus of herbs some of which are useful as fodder for cattle.

In 1886, a chemist isolated from sprouting vetch a white substance with the same molecular formula *and constitution,* but it had a sweet taste. To have names for these, the one isolated from asparagus is now called L-asparagine, and the one from vetch sprouts is D-asparagine.

Here were two different substances seemingly answering to the same structure despite what is almost a dogma in chemistry, namely, the principle of "*one substance—one structure*": If two samples of matter have identical physical and chemical properties, then they must be identical at the level of their individual formula units. If two samples differ in even one way in their fundamental properties, then there has to be at least one difference in the way that the atoms are put together into their molecules.

Taste is a chemical sense, so the two samples of asparagine do have one difference in chemical property. Under the one substance—one structure rule, therefore, the molecules of D- and L-asparagine must be structurally different in at least one way. This difference arises from a peculiar lack of symmetry in their molecules that makes possible two different relative configurations of their molecular parts. The molecules of asparagine display "handedness." We'll see how in the next section.

■ The letters D and L will acquire more specific meaning in the next chapter. Consider them to be only labels now.

17.2 MOLECULAR CHIRALITY

The molecules of many substances have a handedness like that of the left and the right hands.

Two molecular models of asparagine are shown in Figure 17.1*a*. Study each to see that it, or its simplified version in Figure 17.1*b*, represents asparagine, structure **1**. Notice again how like each other the two molecular structures are, particularly that the same four groups are attached to a central carbon atom. Yet one represents a bitter-tasting compound, and the other, a sweet-tasting compound. In some way these two seemingly identical structures must be different, and yet the difference isn't removed by rotating groups around single bonds.

Two Materials Whose Molecules Can Be Superimposed Are Samples of the Same Compound To understand how the two asparagine structures are different, we first have to learn the ultimate test for deciding whether two structures are the same. *For two structures to be identical, it must be possible to superimpose their three-di-*

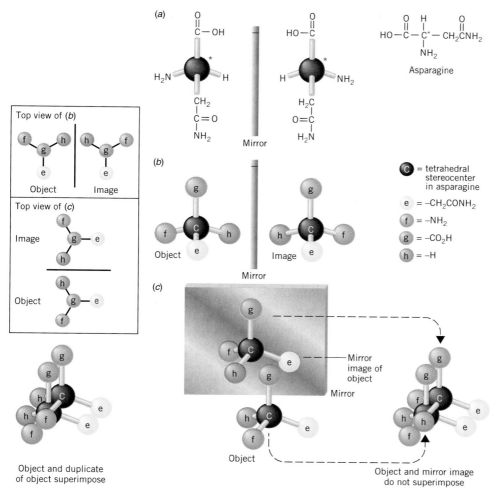

FIGURE 17.1 The two stereoisomers of asparagine. (*a*) Two ways of joining the four groups to the carbon marked by the asterisk. (*b*) Simplified representations of the models in part (*a*). (*c*) What is in front of the mirror is identical with the model on the left in part (*b*). What is seen in the mirror as the image is identical with the model on the right in part (*b*). You'll mentally have to spin the mirror image 120° counterclockwise about the bond from C to **g** to see that they are the same. The object and its mirror image do not superimpose, so they can't be identical. Instead, they are enantiomers.

mensional models. This means to do a manipulation that you can carry to completion only in your mind, but the use of molecular models helps. **Superimposition** is the mental blending of one molecular model with another so that the two would coincide in every atom and bond *simultaneously* if the operation could actually be completed. When working with molecular models, it's fair to twist parts about single bonds to find conformations that can be superimposed, but it's not legal to break any bonds. Superimposition is illustrated in the lower left hand side of Figure 17.1*c*, where two *identical* models of *one* of the asparagines are used.

The test of superimposition is failed by the two *different* asparagines of Figure 17.1*b*. To show this we've taken the model on the left in this part, turned it counterclockwise by 120° around the vertical bond from C to the group **g**, and then placed it as the object in front of the mirror shown in Figure 17.1*c*. Now look at the reflection

■ In some references you'll see the word "superposition" used for "superimposition."

■ If you have access to molecular models, by all means use them now.

of this model in the mirror. If you made an exact molecular model of its reflection, the new model would be identical to that of the *other* asparagine, the one on the right in Figure 17.1*b*. This is how nearly identical the two asparagines are. In the lower right of part of Figure 17.1*c*, you can see that these two models, the one in front of the mirror as the object and the model of its reflection, do not superimpose. The two asparagine molecules do not superimpose; *they must be different.*

The Two Asparagines Are Enantiomers   Pairs of stereoisomers whose molecules are related as object to mirror image that cannot be superimposed are called **enantiomers.**

It's now very important to remember two general facts about isomers of any kind. *They are truly different substances;* they are different compounds that share the same molecular formula while differing in the arrangements of their atoms. Enantiomers are just special kinds of isomers.

All isomers other than enantiomers that qualify as *stereoisomers* are called *diastereomers.* **Diastereomers** are stereoisomers that are not enantiomers. Thus all purely geometric (cis–trans) isomers are diastereomers but other examples also exist, as we'll see in Interaction 17.1 later in this chapter. Figure 17.2 sorts out the kinds of isomers we have studied.

■ **Constitutional isomers:** Identical molecular formulas but different arrangements of atoms.

Stereoisomers: Identical *constitutions* but different geometries.

Enantiomers: Pairs of *stereoisomers* whose molecules are related as object to mirror image but cannot be superimposed.

Diastereomers: *Stereoisomers* that are not enantiomers.

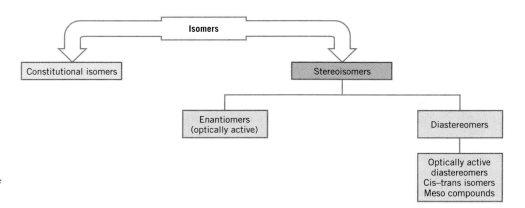

FIGURE 17.2 The relationships of various kinds of isomers.

Molecules of a Pair of Enantiomers Have Opposite Configurations   Lack of free rotation is not the cause of the asparagine enantiomers. Their molecules, instead, have *opposite configurations*. To show what this means, we have repositioned their abbreviated molecular models in Figure 17.3. Imagine a mirror standing between the two and perpendicular to the page to see that they are related as object to mirror image.

You're now going to let your eyes make a special scanning trip around each molecule. Imagine that the bond from C to H in each model is the column of the steering wheel of a car. Then imagine that the remaining three groups—**e**, **f**, and **g**—are distributed around the steering wheel itself. Now move your eyes from **g** to **e** and then to **f**. When you do this with the asparagine model on the left in Figure 17.3, your eyes move clockwise. But to make the identical trip—**g** to **e** to **f**—in the model on the right, your eyes move counterclockwise. These clockwise versus counterclockwise arrangements of identical parts around the same central axis are what having *opposite configuration* means. The four groups on the central carbon in one asparagine are the same four groups as in the other, but they are configured oppositely in space.

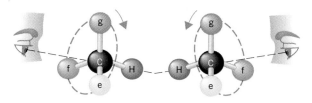

FIGURE 17.3 When the two stereoisomers of asparagine are viewed down the same axis, the C—H bond axis, the remaining three groups at the central carbon atom have opposite configurations.

Molecules of a Pair of Enantiomers Have Opposite Chirality We have to remind ourselves now that any object has a mirror image. Spheres, cubes, broom handles, water glasses, and so on, can all be reflected in a mirror. It's only when the model of the object and the model of the mirror image cannot be superimposed that we call the two enantiomers.

Your two hands are like a pair of enantiomers, if you disregard small differences such as wrinkles, scars, rings, and fingerprints. Place your left hand in front of a mirror near its edge. Place your other hand just off the edge of the mirror and notice that the reflection of your left hand in the mirror is the same as your right hand (disregarding, as we said, the small differences). Your right hand is the mirror image reflection of your left hand.

Next, try to superimpose the two hands. Because the mirror image of your left hand is your real right hand, use your two hands to see whether they superimpose. If you put them palm to palm it seems as though all the fingers do superimpose. But remember, you have to carry this blending through to completion (in your mind), and when you do, the palm of one hand comes out on the back side of the other. The palms won't superimpose when the fingers seem to. And if you try to get the palms to come out right, then the fingers come out all wrong. Left and right hands, although related as object to mirror image, don't superimpose. They are related as enantiomers.

Notice, now, how the two hands have opposite configurations. Look at them down the same axis, as we did with the asparagine models in Figure 17.3, say, down the axis from palm to backside. This means that both palms will face you. To make the trip from thumb to little finger, touching every other finger on the way, you have to scan in one direction for one hand and in exactly the opposite direction for the other. Thus the two hands have opposite configurations.

The little experiment with the hands has been used for decades to teach about the configurational differences of enantiomers, and this is why we say that molecules of different enantiomers have different *handedness*. The technical term for handedness is **chirality** (from the Greek word for "hand," which is *cheir*). We say that *the molecules of enantiomers have opposite chirality,* meaning opposite handedness. We also say that the molecules of any given enantiomer are **chiral**—they possess handedness or chirality.

The opposite of chiral is **achiral**. An achiral molecule is one that is symmetrical enough so that its model and the model of its mirror image do superimpose. The methane molecule is achiral, for example. Some examples of larger achiral objects include a cube, a sphere, a broom handle, and a water glass.

Chirality Can Make Enormous Differences at the Molecular Level of Life One asparagine enantiomer tastes sweet and the other tastes bitter—a large (although a somewhat trivial) difference that chirality can make. The details are not fully

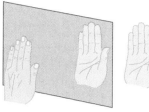

The image of the left hand is just like the right hand in the relative orientations of the fingers, thumb, and palm.

The two hands are related as an object to its mirror image, but they can't be superimposed.

A left hand glove does not fit the right hand.

known, but the taste mechanism probably begins with a chemical reaction that is catalyzed by an enzyme. The enzymes involved, *like all enzymes,* themselves consist of very large, chiral molecules with molecular surfaces that are different for each enzyme.

The substance whose reaction an enzyme catalyzes is called the **substrate** for that enzyme. An enzyme works by letting a molecule of the substrate come and temporarily fit into the contours on the enzyme's surface. This idea of fitting can be illustrated by a return to our hands, only now we'll add gloves. Gloves, like hands, are chiral, and a glove fits well only to its matching hand. The left hand fits well into the left glove, not the right glove. Now suppose that an enzyme responsible ultimately for the sensation of a sweet taste is like a glove for the right hand. This means that only the substrate molecules that have the matching handedness can interact with this enzyme. Substrate molecules of the opposite handedness can't fit to this enzyme.

We can now shift back from this analogy of hands and gloves to chiral molecules with the aid of Figure 17.4. To make it easier, we have used simple geometric forms to create two enantiomers, and indentations that match these forms are part of the enzyme surface. One enantiomer can fit to the enzyme surface, but no matter how you turn the model of the other enantiomer you can't get it to fit to the same enzyme surface.

There is, evidently, a different enzyme whose surface chirality matches the other asparagine enantiomer that lets us know that this other enantiomer has a different taste. The phenomenon of chirality is absolutely central to this difference. And the different chemical properties that relate to the taste of asparagine are illustrated in countless ways at the molecular level of life. We'll see time after time that *differences in geometry and configuration are as important as functional groups to the chemical reactions of life.*

Chirality Does Not Affect Reactions with Achiral Compounds   If the enzyme molecule were not chiral, it could not discriminate between enantiomers. In fact, *enan-*

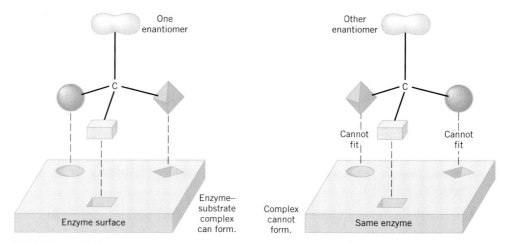

FIGURE 17.4 Because an enzyme is chiral, it can accept substrate molecules of only one of a pair of enantiomers. To illustrate this difference, we have used simple geometric forms. On the left, the enzyme can accept as a substrate the molecule of one enantiomer. On the right, the same enzyme can't accept a molecule of the other enantiomer, because the shapes don't match.

tiomers *have identical chemical properties toward all reactants whose molecules are achiral*—reactants such as H_2O, NaOH, HCl, Cl_2, NH_3, and H_2. A broom handle, which isn't chiral, is an analogy. It fits just as easily to the left hand as to the right. It can't discriminate between the hands. In like manner, *the molecules of an achiral reactant cannot tell the difference between molecules of enantiomers.* To summarize, enantiomers react differently toward reactants whose molecules are chiral but identically toward achiral reactants.

Two Enantiomers Have Identical Physical Properties, with One Exception   Two enantiomers have identical melting points, boiling points, densities, and solubilities in common (i.e., achiral) solvents like water, diethyl ether, or ethanol. This *must* be so, because the molecules of two enantiomers *must* have identical molecular polarities. They have, after all, identical intranuclear distances and bond angles.

Returning once again to our analogy with the hands, you can see that the distance, say, from the tip of the thumb to the tip of the little finger is the same in both hands. You can pick any other such intrahand distance that you please, and it is the same in both hands. Similarly, the angle between any two fingers is the same in both hands when the hands are spread identically. In like manner, if we pick any distance between atoms or any bond angle in one asparagine enantiomer, we will find it to be the same in the other. When all these distances and angles are identical in the two enantiomers, their molecules as a whole must have identical polarities. This is responsible for the identical physical properties that we mentioned. There is one difference in physical property, however, which we will study in the next section.

Molecules with One Carbon Holding Four Different Groups Are Chiral   It is important that we be able to recognize when a potential substrate is chiral, because molecules of opposite chirality have such different chemical properties at the molecular level of life. How, then, can we look at a structure and tell whether its molecules are chiral without making molecular models of both object and mirror image?

In all examples of chiral molecules that we'll encounter, their molecules always have at least one carbon to which *four different groups* are attached. A carbon that holds four different atoms or groups is called a **tetrahedral stereocenter.** A *stereocenter* is an atom in a molecule whose attached groups can be arranged in different configurations to give different stereoisomers. A *tetrahedral* stereocenter is simply a stereocenter having four single bonds arranged tetrahedrally.[1]

The asparagine molecule has one (and just one) tetrahedral stereocenter, and when a molecule has only one such center, we can be absolutely certain that the molecule as a whole is itself chiral. Here, then, is one way to predict whether a substance consists of chiral molecules; we look for a tetrahedral stereocenter. If we find *one,* we can be certain that the molecules are chiral. If we find more than one tetrahedral stereocenter, we have to be careful.[2] The complication is treated in Interaction 17.1.

■   An older (but still widely used) name for *tetrahedral stereocenter* is **chiral atom.** When the atom is C, the term is *chiral carbon atom.*

---

[1] Alkenes that are cis and trans isomers of each other (e.g., the geometric isomers of 2-butene) have two stereocenters, two carbon atoms each holding *three* atoms or groups that, when differently arranged, give the two stereoisomers—the cis and the trans. The two carbon atoms at the double bond are thus not tetrahedral stereocenters but *trigonal* stereocenters.

[2] Probably 99.99% of all examples of substances that have two or more tetrahedral stereocenters also have chiral molecules. A few exceptions, however, exist where the molecule is achiral despite having tetrahedral stereocenters. These are the *meso compounds* discussed in Interaction 17.1.

INTERACTION 17.1
MESO COMPOUNDS AND DIASTEREOMERS

Tartaric acid is a normal constituent of grapes. During the fermentation of grape juices into wine, the monopotassium salt separates as an insoluble substance called "tartar."

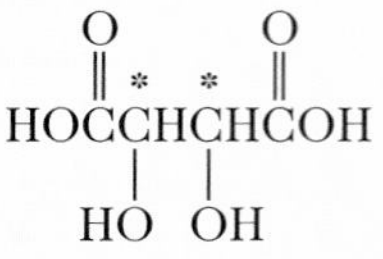

$$\underset{\text{tartaric acid}}{\underset{\displaystyle HO \quad OH}{\overset{\displaystyle O \qquad O}{HOCCHCHCOH}}}$$

Tartaric acid molecules have two tetrahedral stereocenters. They are *identical* because the sets of four atoms or groups attached to each are identical; both centers hold HO, H, CO_2H, and $CH(OH)CO_2H$. When we prepare perspective drawings of the structures of all of the possible stereoisomers of tartaric acid, we obtain those shown below. Solid wedges denote bonds coming forward and dashed-line wedges mean bonds going rearward. The first two, D- and L-tartaric acid, are related as nonsuperimposable object and mirror image and are thus *enantiomers*. Notice that they have identical melting points and identical degrees, but opposite signs of specific rotation.

D-tartaric acid
$[\alpha]_D^{20}$ −11.98°
mp 170 °C

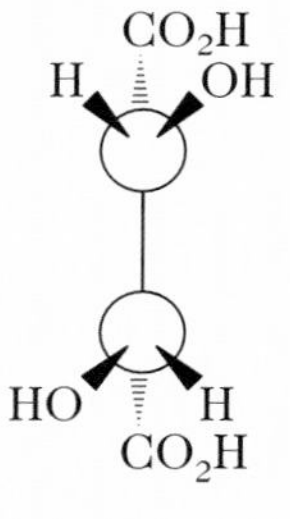

L-tartaric acid
$[\alpha]_D^{20}$ +11.98°
mp 170 °C

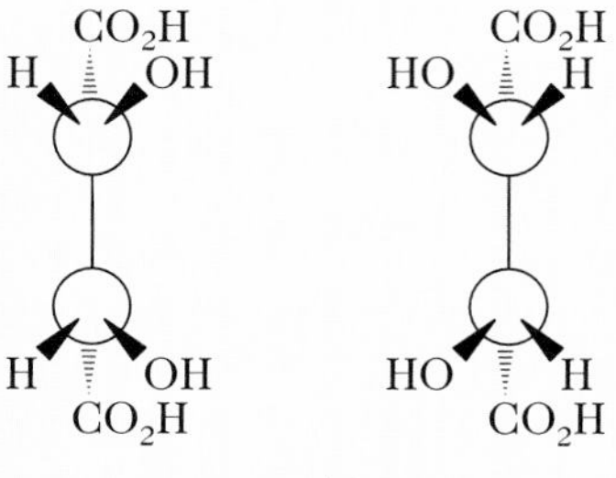

meso-tartaric acid
$[\alpha]_D^{20}$ 0°
mp 140 °C

Only one of the remaining two perspective structures has been given a name, *meso*-tartaric acid. Al-

though its mirror image structure has been drawn to make a point, the object and mirror image are actually superimposable. If you rotate the mirror image of *meso*-tartaric acid by 180° in the plane of the paper and around an axis perpendicular to the plane (piercing the paper at the bond *between* the two carbons), you soon realize that the two structures are identical (and so superimposable). Despite having two tetrahedral stereocenters, the *meso*-tartaric acid molecule is achiral. Therefore *meso*-tartaric acid is optically inactive; it has zero specific rotation. Optical activity *requires* chirality, and *meso*-tartaric acid doesn't have it. According to our definitions, because *meso*-tartaric acid is a nonenantiomeric isomer of a pair of enantiomers, it is a *diastereomer* of each enantiomer.

meso-Tartaric acid is one example of a common occurrence and, historically, it gave part of its name to the kind of diastereomer it exemplifies. An achiral diastereomer of a set of stereoisomers that includes some that are chiral is called the **meso** isomer. There are thus only three stereoisomers of tartaric acid. The equation, 2^n = number of stereoisomers, cannot be applied to tartaric acid because it works only when the *n* tetrahedral stereocenters are *different*, when the sets of four atoms or groups at the center differ in at least one way. No general equation exists for calculating the number of stereoisomers when two (or more) stereocenters are identical.

In Example 17.2, the structures of the four stereoisomers of threonine were shown. (Two bear the name "threonine" and two have the name "allothreonine.") Their molecules also have two tetrahedral stereocenters, but the centers are not identical. Hence the correct number of stereoisomers is predicted by the equation we have used. Each of the threonine enantiomers is a diastereomer of each of the allothreonine enantiomers.

Diastereomers do not have identical physical or chemical properties (although the chemical properties will be very similar with all achiral reactants). You can see in the threonine–allothreonine system how melting points and specific rotations for diastereomers are different. The set of intramolecular distances and bond angles in one diastereomer isn't exactly duplicated in any other diastereomer. Hence, molecules of one diastereomer should be expected to have at least slightly different polarities than those of any other diastereomer in the set. Such differences cannot help but cause differences in physical properties.

EXAMPLE 17.1   Identifying Tetrahedral Stereocenters

Amphetamine exists as a pair of enantiomers. One of them has its own name, Dexedrine. Find the tetrahedral stereocenter in amphetamine, and list the four groups attached to it.

$$CH_3$$
$$|$$
$$C_6H_5CH_2CHNH_2$$

amphetamine

ANALYSIS   A tetrahedral stereocenter has *four different* attached atoms or groups. Amphetamine has one such center.

SOLUTION   The tetrahedral stereocenter is labeled with an asterisk.

$$CH_3$$
$$|$$
$$C_6H_5CH_2CHNH_2$$
$$*$$

The four groups are $C_6H_5CH_2$, H, CH_3, and NH_2.

■ Amphetamine, a stimulant, is a controlled substance in the United States.

■   PRACTICE EXERCISE 1   Place an asterisk next to each tetrahedral stereocenter in the following structures.

(a)

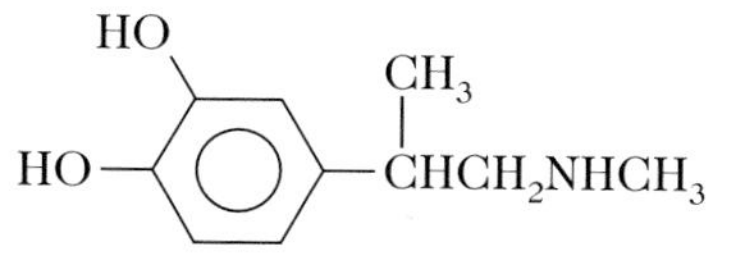

HO—, HO— ring with —CHCH$_2$NHCH$_3$ and CH$_3$ group

Epinephrine, a hormone (see Interaction 16.1)

(b) CH_3CHCO_2H
           $|$
           OH

Lactic acid, the sour constituent in sour milk

(c) $CH_3CHCHCO_2{}^-$
           $|$ $|$
           HO $NH_3{}^+$

Threonine, one of the amino acid building blocks of proteins

(d) $HOCH_2CH—CH—CHCH$ with O double bonded at right
                   $|$ $|$ $|$
                   OH OH OH

Ribose, a sugar unit in one of the two kinds of nucleic acids (ribonucleic acid or RNA).

A Molecule with *n Different* Tetrahedral Stereocenters Has 2^n Stereoisomers
When a molecule has two or more tetrahedral stereocenters, as in parts (c) and (d) of Practice Exercise 1, then it becomes useful to judge whether these centers are *different*. When used in this context, "different" means that the sets of four atoms or groups at the various tetrahedral stereocenters have at least one difference. Two tetrahedral stereocenters are said to be *different* if the set of four groups at one is not duplicated by the set at the other. Whenever the tetrahedral stereocenters in a molecule are different in this sense—as they are in parts (c) and (d) of Practice Exercise 1—then the substance can exist in the forms of 2^n stereoisomers, where n is the number of different tetrahedral stereocenters. These 2^n stereoisomers occur as half as many *pairs* of enantiomers. We'll see this illustrated in the next example.

EXAMPLE 17.2 Judging Whether Tetrahedral Stereocenters Are Different and Calculating the Number of Stereoisomers

The threonine molecule, part (c) of Practice Exercise 1, has two tetrahedral stereocenters, labeled here by asterisks.

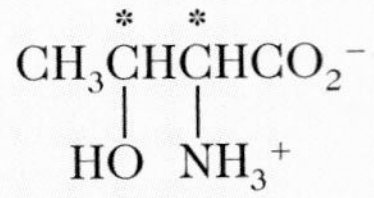

$$\overset{*\quad *}{CH_3CHCHCO_2^-}$$
$$|\quad|$$
$$HO\quad NH_3^+$$

threonine

Are these tetrahedral stereocenters different? If so, how many stereoisomers of threonine are there?

ANALYSIS To compare the groups attached at each tetrahedral stereocenter, we should make a list of the sets of four different groups and compare them. If the lists are not identical in every respect, then the two tetrahedral stereocenters are different.

At one tetrahedral stereocenter:	At the other tetrahedral stereocenter:
CH_3 H	CH_3CH H
	$\quad\quad\;\;|$
	$\quad\quad\;OH$
HO $CHCO_2^-$	
$\quad\quad\;\;\;|$	CO_2^- NH_3^+
$\quad\quad\;NH_3^+$	

SOLUTION The sets are obviously different, so $n = 2$, the number of different tetrahedral stereocenters in a threonine molecule. Therefore, $2^n = 2^2 = 4$, the number of stereoisomers of threonine. These occur as half of 4 or 2 pairs of enantiomers. The complete set has the following structures. One pair of enantiomers is on the left. Just imagine that the mirror is between them and is perpendicular to the page. The other pair is on the right. Solid wedges denote bonds that project forward. Dashed wedges are bonds that project backward. Solid lines are bonds in the plane of the page.

■ Only L-threonine works as a building block for making proteins in the body. There is no enzyme that can accept any of the other optical isomers as substrates.

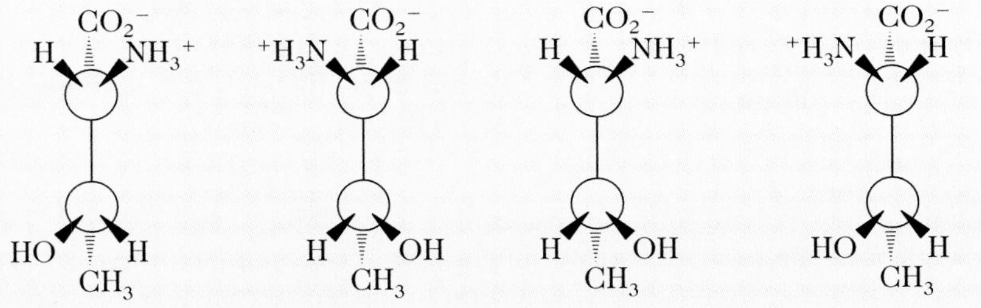

■ The labels D and L are explained in the next chapter, as we said. The meaning of the experimental values given for the symbol $[\alpha]_D^{26}$ is explained in the next section.

D-threonine	L-threonine	D-allothreonine	L-allothreonine
$[\alpha]_D^{26}$ +28.3°	$[\alpha]_D^{26}$ −28.3°	$[\alpha]_D^{26}$ −9.6°	$[\alpha]_D^{26}$ +9.6°

■ PRACTICE EXERCISE 2 Examine the structure of ribose that was given in part (d) of Practice Exercise 1. (a) How many different tetrahedral stereocenters does it have? (b) How many stereoisomers are there of this structure? (Only one is actually the ribose that can be used by the body.) (c) How many pairs of enantiomers correspond to this structure?

■ PRACTICE EXERCISE 3 Write the structure of 2,3-butanediol and place an asterisk by each carbon that is a tetrahedral stereocenter. Are they *different* tetrahedral stereocenters?

17.3 OPTICAL ACTIVITY

The members of a pair of enantiomers affect polarized light in equal and opposite ways.

We mentioned earlier that the two members of a pair of enantiomers differ in one physical property. To describe it we first have to learn something about polarized light.

The Electromagnetic Oscillations of Polarized Light Are All in the Same Plane
Light is electromagnetic radiation in which the intensities of the electric and magnetic fields set up by the light source oscillate in a regular way. In ordinary light, these oscillations occur equally in all directions about the line that defines the path of the light ray.

Certain materials, such as the polarizing film in the lenses of Polaroid sunglasses, affect ordinary light in a special way. Polarizing film interacts with the oscillating electrical field of any light passing through it to make this field oscillate *in just one plane.* The light that emerges is now **plane-polarized light** (see Figures 17.5 and 17.6a).

If we look at some object through polarizing film and then place a second film in front of the first, we can rotate one film until the object can no longer be seen (see Figures 17.6b and 17.7). If we now rotate one film by 90°, we'll see the object at maximum brightness again. The first film seems to act as a lattice fence, forcing any light that goes through it to vibrate only in the direction allowed by the long spaces between the slats. This light then moves on to the molecular slats of the second film. If the second film's slats are perpendicular to those of the first, the light has no freedom to oscillate, and it cannot get through the second film. At intermediate angles, fractional amounts of light can go through the second film. Only when the slats of both films are *parallel* to each other can the light leaving the first film slip easily through the second film with its maximum intensity.

■ You can try this out using two Polaroid sunglass lenses. The lenses of these glasses reduce glare by cutting out the plane-polarized light produced when sunlight reflects from a plane surface such as a road, a snowfield, or a lake.

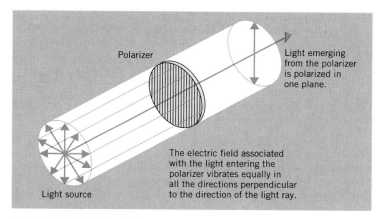

FIGURE 17.5 When light passes through polarizing film, it becomes polarized light.

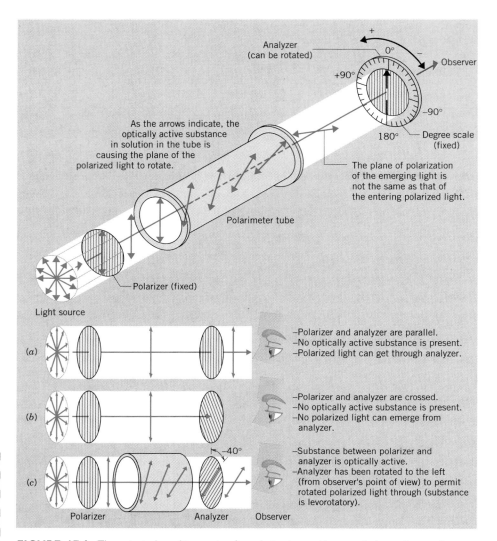

FIGURE 17.6 The principal working parts of a polarimeter and how optical rotation can be measured.

FIGURE 17.7 When two polarizing films are "crossed," no light can get through to the observer.

An Enantiomer Can Rotate the Plane of Plane-Polarized Light   When a solution of D-asparagine in water is placed in the path of plane-polarized light, the *plane* of polarization is twisted or rotated. Any substance that can rotate the plane of plane-polarized light is said to be **optically active.** We have not actually explained *how* this phenomenon happens—we are unable to do so. We have only reported that it does take place. Quite often the members of a set of stereoisomers in which any or all are optically active are called **optical isomers.** Thus the two asparagines are optical isomers of each other. All of the threonine stereoisomers of Example 17.2 are likewise optical isomers of each other.

The *Polarimeter* Measures the Degree of Optical Activity   The instrument used to detect and measure optical activity is called a **polarimeter** (see Figure 17.6). Its principal working parts are a *polarizer* for the light beam, a tube for holding solutions in the path of the polarized light, an *analyzer* (actually, just another polarizing device), and a circular scale for measuring the number of degrees of rotation. When the "slats" of the polarizer and the analyzer are parallel and the tube contains no optically active material, the polarized light emerges from the analyzer with maximum intensity. Let's assume that we start with this parallel orientation of polarizer and analyzer.

When a solution of one pure enantiomer is placed in the light path, the plane-polarized light encounters molecules of just one chirality. They cause the plane of oscillation of the polarized light to rotate. Hence, the plane of oscillation of the polarized light that *leaves* the solution is now no longer parallel with the analyzer (see Figure 17.6c). Consequently, not as much light gets through the analyzer, so the observed light intensity is now less than the original maximum. To restore the original intensity, the operator can rotate the analyzer to the right or to the left a definite number of degrees until the analyzer is once again parallel *with the light that leaves the tube.*[3]

The operator, looking *toward* the light source, might find that rotating the analyzer to the right (clockwise) restores the original light intensity with fewer degrees of rotation than rotating the analyzer to the left (counterclockwise). When such a rightward rotation works, the degrees are recorded as positive, and the optically active substance is said to be **dextrorotatory**. If the fewer degrees of rotation are found by a leftward rotation, then the degrees are recorded as negative and the substance is said to be **levorotatory**. In Figure 17.6c, the reading is $\alpha = -40°$, where α (including the plus or minus sign) stands for the observed **optical rotation**, the observed number of degrees of rotation caused by the solution.

■ Latin, *dextro*, right, and *levo*, left.

The value of α varies with the temperature of the solution, the wave length of the light used, and the solvent, but not in any simple way. Consequently, when α is recorded, the temperature, light wave length, and solvent used are also reported.

The *Specific Rotation* of an Optically Active Compound Is One of Its Physical Constants   The observed rotation, α, is related in simple ways to the concentration of the solution and the length of the tube.

$$\alpha \propto \text{concentration}$$

$$\alpha \propto \text{path length}$$

Both facts tell us that the degree of rotation of the plane-polarized light is a function of the *population* of the chiral molecules. Either by increasing the concentration or by making the tube longer, we can force the polarized light to be in greater contact with chiral molecules.

Because α is *directly* proportional both to concentration (c) and to path length (l), α is proportional to the products of these two.

$$\alpha \propto cl \tag{17.1}$$

The unit used for length, l, is the decimeter.

We can convert expression 17.1 to an equation by inserting a constant of proportionality, given the symbol $[\alpha]$ and the name **specific rotation**.

$$\alpha = [\alpha]cl \tag{17.2}$$

The unit traditionally used here for concentration is g/mL. By rearranging this equation, we obtain the usual form of the definition of specific rotation. The temperature (t) and the light wave length (λ) are placed by the closing bracket of $[\alpha]$. The solvent is reported in a separate statement.

$$[\alpha]_\lambda^t = \frac{\alpha}{cl} \tag{17.3}$$

■ This is a rare use of the decimeter unit in chemistry.

I decimeter (dm) =
10 centimeters (cm)

■ Be sure to notice that the units of c are g/mL, not g/100 mL. If an optically active pure liquid is in the polarimeter tube, then the units of its concentration equal those of its density.

[3] This description of the measurement emphasizes only the essential principle involved. The operator has other options that yield identical results.

Quite often polarimeters are used with the intense yellow light of a sodium vapor lamp like those that illuminate the streets in some cities. The symbol for this light is D, so when we write that $[\alpha]_D^{20} = +5.42°$ for D-asparagine, we mean that the temperature of the solution was 20 °C and that a sodium vapor lamp was used.

For L-asparagine in water, $[\alpha]_D^{20} = -5.42°$. Thus we see that the *only* physical difference between the two enantiomers of asparagine is the *direction* of the rotation of the plane of plane-polarized light. The numbers of degrees are identical; only the signs of rotation are opposite. Any pair of enantiomers is like this. All physical properties of a pair of enantiomers are the same—densities, boiling points, and melting points, for example—but the *signs* of the numerically identical degrees of rotation are opposite.

Specific Rotation Provides an Analytical Tool   The specific rotation of an optically active compound is an important physical constant, comparable to its melting point, boiling point, or density. If we know the value of $[\alpha]_\lambda^t$ for a compound, it's easy to see from Equation 17.3 that we have a way to determine the concentration of a solution. We measure the observed rotation, α, for the solution when it is in a tube of known path length, l, and use these data together with the specific rotation to calculate the concentration, c.

Sometimes the measurement of optical activity is used to identify a substance. A measurement of the observed rotation, α, of a solution of known concentration, c, in a tube of known path length, l, is made and the specific rotation is calculated. The calculated value is then compared to a table of specific rotations to see what matches.

A 1:1 Mixture of Enantiomers Is Optically Inactive   It is important to realize that *optical activity* and *optical isomerism* are not the same. *Optical activity* refers to a phenomenon observable with a special instrument, and what we see is the number of degrees of rotation of the solution. *Optical isomerism* is part of our explanation of optical activity. We *infer* optical isomerism from the observation of optical activity.

What happens if we mix two enantiomers together in a 1:1 ratio? Now as the plane-polarized light travels through a polarimeter tube it encounters some molecules forcing its plane to twist to the right but it meets an identical number forcing its plane to twist just as much to the left. *The result is no net change to the plane of vibration of the plane-polarized light.* The operator of the polarimeter would be bound by the definitions that we have introduced to report that the substance in the tube is *optically inactive*. Any 1:1 mixture of enantiomers is optically inactive and is called a **racemic mixture.** Thus a substance like a racemic mixture can be made entirely of chiral molecules and yet be optically inactive.

Enantiomers Normally Have Large Differences in Biological Properties   Asparagine is actually one of the amino acids that the body uses as a building block for proteins. However, only one enantiomer works. If a racemic mixture of asparagine enantiomers is given in the diet, the body can use only one. This is generally true about pairs of enantiomers. If the body uses one enantiomer for a specific purpose, it cannot use the other for the same purpose. Indeed the other may even be a very dangerous substance. Interaction 17.2 describes the tragic case of one such example.

". . . MIRROR, MIRROR ON THE WALL, WHAT'S THE SAFEST . . . ?"

Thalidomide can exist as one or another of a pair of enantiomers. (Can you spot the tetrahedral stereocenter?)

thalidomide

It was once widely prescribed in Europe as a sedative–tranquilizer, particularly for pregnant women. Tragically, the prescribed drug was the racemic mixture, and *only one enantiomer gives the desired effect*. The other enantiomer disrupts fetal development during the first 12 weeks of pregnancy causing phocomelia—seal or flipperlike arms—and often abnormalities of the digestive tract, the eyes, and the ears. Between 1957 and 1962, 10,000 babies were born in West Germany with such thalidomide-caused problems. The drug was withdrawn from the market but not before many tragic births had occurred. The episode furnishes a dramatic example of how great can be the differences between enantiomers, even though their molecules are so nearly identical as to be related as object and mirror image.

Some children born of wives of servicemen stationed in Germany, women being treated by German physicians, also suffered phocomelia. Thalidomide, however, was never approved for use in the United States because Dr. Frances Oldham Kelsey of the U.S. Food and Drug Administration insisted *before* birth defects appeared in Europe that the testing of thalidomide had not been thorough enough.

Despite the bad reputation of thalidomide, it has survived as an experimental medication for certain rare diseases. It is basically an immunosuppressive agent, and it has been used with some success to treat one form of leprosy as well as a graft-versus-host disease, in which white cells from transplanted bone marrow attack host tissues. Experiments are also being done using thalidomide to treat rheumatoid arthritis, some ulcerative diseases affecting the mouth and genital areas, and inflammatory skin diseases, like discoid lupus erythematosus.

Racemic Switches A number of medications that until recently were marketed as racemic mixtures are now undergoing "racemic switches," meaning that racemic drugs already approved have been redeveloped and submitted for approval as single enantiomers. The approval process is taking place for several in the United States, and other countries have already approved some. Austria, for example, allows over-the-counter selling of the dextrorotatory enantiomer of a popular antiinflammatory drug, ibuprofen. (Can you find its tetrahedral stereocenter?)

ibuprofen (Advil, Motrin)   fenfluoramine

phentermine

This enantiomer is claimed to reach therapeutic levels in the blood in 12 minutes, as compared to 30 minutes for the racemic mixture.

Fenfluoramine (as a racemic mixture) has long been sold in weight-loss programs, but only the dextrorotatory enantiomer is effective. In early 1996, the Food and Drug Administration (FDA) approved the use of this enantiomer as an antiobesity drug, sold as Dexfenfluoramine or as Redux. However, users have a higher risk than nonusers of primary pulmonary hypertension, which can be fatal. Hence, the drug is only for very obese people *under a physician's care.*

Some weight-loss programs have used "fen/phen," a combination of the (cheaper) racemic mixture of fenfluoramine enantiomers together with phentermine. The FDA never approved the *combination*, however. By mid-1997, the danger of using the combination became so alarming that a general warning was widely aired. Many users developed heart conditions requiring open-heart surgery! Drugs in combination often and unexpectedly interact to produce effects not seen by the use of any of the individual drugs alone. The "fen/phen" case was a dramatic example.

SUMMARY

Stereoisomerism Stereoisomers are isomers whose molecules have identical constitutions but different geometries. The two kinds of stereoisomers are enantiomers and diastereomers. Enantiomers are pairs of stereoisomers whose molecules are mirror images but they do not superimpose. Diastereomers are stereoisomers that are not related as enantiomers. Almost always, an enantiomer molecule has a tetrahedral stereocenter, which is usually a carbon atom holding four different atoms or groups. If a molecule has n *different* tetrahedral stereocenters, then the number of stereoisomers is 2^n and these occur as half as many pairs of enantiomers. A 1:1 mixture of enantiomers, a racemic mixture, is optically inactive.

Optical activity Optical activity is a natural phenomenon detected by means of a polarimeter. A substance is optically active if polarized light that passes through a substance or its solution undergoes a rotation in its plane of polarization.

Specific rotation The specific rotation of an optically active substance, $[\alpha]^t_\lambda$, is what its observed rotation is at one unit of concentration (1 g/mL) in one unit of path length (1 dm). It varies, but not in a simple way, with the temperature and the wavelength of the light used. Values of specific rotation can be used to determine concentrations of optically active substances.

Properties of enantiomers Enantiomers are identical in every physical property except the signs of their specific rotation. They are also identical in every chemical respect provided that the molecules or ions of the reactant are achiral. When the reactant particles are chiral, then one enantiomer reacts differently with the reactant than the other enantiomer, a phenomenon always observed when an enzyme is acting as a (temporary) reactant.

REVIEW EXERCISES

The answers to Review Exercises whose numbers are in color are found in Appendix E. The answers to the other Review Exercises are found in the Study Guide that accompanies this book. The more challenging questions are marked with asterisks.

Structural Isomers and Stereoisomers

17.1 What specifically must be true about two compounds before they can be called *constitutional isomers*?

17.2 What are the structures of the two *simplest* compounds of each of the following families that are constitutional isomers?
(a) Alcohols
(b) Ketones
(c) Alkenes

17.3 What must be true about two compounds before they can be called *stereoisomers*?

17.4 There are two kinds of stereoisomers. What are their names and how is each kind defined?

17.5 Classify the following pairs of structures as constitutional isomers, stereoisomers, identical, or unrelated.

(a)   and

(b)   and

(c)   and

(d)   and

17.6 What is the "one substance—one structure" principle?

Optical Isomers

17.7 Consider the structures of fructose and glucose. Both are present in equal concentrations in honey with fructose having the sweeter taste.

$$\text{HOCH}_2\text{CH}-\text{CH}-\text{CH}-\overset{\displaystyle O}{\overset{\displaystyle \|}{\text{C}}}\text{CH}_2\text{OH}$$
$$\qquad\quad | \qquad | \qquad |$$
$$\qquad\;\text{OH} \quad \text{OH} \quad \text{OH}$$

fructose (open form)

$$\text{HOCH}_2\text{CH}-\text{CH}-\text{CH}-\text{CH}-\overset{\displaystyle O}{\overset{\|}{\text{CH}}}$$
$$\qquad\quad\ \underset{\text{OH}}{|}\quad\underset{\text{OH}}{|}\quad\underset{\text{OH}}{|}\quad\underset{\text{OH}}{|}$$

glucose (open form)

(a) What kinds of isomers are fructose and glucose to each other?

(b) If within either or both of them you spot tetrahedral stereocenters, place an asterisk by each such center.

(c) How many of the tetrahedral stereocenters qualify as *different* tetrahedral stereocenters in each structure?

(d) How many stereoisomers of each compound are possible?

(e) How many *pairs* of enantiomers are possible for fructose? For glucose?

17.8 One of the approximately 20 building blocks of protein molecules is glycine: $^+\text{NH}_3\text{CH}_2\text{CO}_2^-$. Does glycine have stereoisomers? How can you tell?

17.9 One of the important intermediate substances in the body's energy-producing metabolism is an anion of the following acid, citric acid.

$$\begin{array}{c}\text{CH}_2\text{CO}_2\text{H}\\ |\\ \text{HOCCO}_2\text{H}\\ |\\ \text{CH}_2\text{CO}_2\text{H}\end{array}$$

(a) Does citric acid have stereoisomers? How can you tell?

(b) One of the possible monomethyl esters of citric acid has chiral molecules. Write its structure. Place an asterisk by its tetrahedral stereocenter.

Properties of Enantiomers

17.10 The melting point of $(-)$-cholesterol is 148.5 °C. What is the melting point of $(+)$-cholesterol? How can we know what this melting point *must* be without actually making the measurement?

17.11 Explain why enantiomers should have identical physical properties (except for the sign of specific rotation).

17.12 Explain why enantiomers cannot have different chemical properties toward reactants whose molecules or ions are achiral. (Use an analogy if you wish.)

17.13 Explain why enantiomers have different chemical properties toward reactants whose molecules or ions are chiral. (Use an analogy if you wish.)

Specific Rotation

*17.14 Pantothenic acid was once called vitamin B$_3$.

$$\underset{\underset{\text{H}_3\text{C}}{|}}{\overset{\overset{\text{H}_3\text{C}}{|}}{\text{HOCH}_2\text{C}}}-\underset{\underset{\text{OH}}{|}}{\text{CH}}\overset{\overset{O}{\|}}{\text{C}}\text{NHCH}_2\text{CH}_2\text{CH}_2\text{CO}_2\text{H}$$

pantothenic acid

Only its dextrorotatory form can be used by the body, and its specific rotation is $[\alpha]_D^{25} = +37.5°$. In a tube 1.00 dm long and at a concentration of 1.00 g/100 mL, what is the *observed* optical rotation of the levorotatory enantiomer?

*17.15 A solution of sucrose (table sugar) in water at 25 °C in a tube that is 10.0 cm long gives an observed rotation of $+2.00°$. The specific rotation of sucrose in water at this temperature and the same wavelength of light is $+66.4°$. What is the concentration of the sucrose solution in g/100 mL?

17.16 Quinine sulfate, an antimalarial drug, has a specific rotation in water at 17 °C of $-214°$. If a solution of quinine sulfate in this solvent in a 1.00-dm tube and under the same conditions of temperature and wavelength has an observed rotation of $-10.4°$, what is its concentration in g/100 mL?

*17.17 Strychnine and brucine are structurally similar compounds that have extremely bitter tastes, and both are very poisonous. The specific rotation in chloroform at 20 °C of brucine is $-127°$ and that of strychnine under identical conditions is $-139°$. If a solution of one of these in chloroform at a concentration of 1.68 g/100 mL in a tube 1.00 dm long gave an observed rotation of $-2.34°$, which of the two compounds was it? Do the calculation.

17.18 Corticosterone and cortisone are two substances used to treat arthritis. Under identical conditions of solvent and temperature, the specific rotation of corticosterone is $+223°$ and that of cortisone is $+209°$. If a solution of one of these at a concentration of 1.48 g/100 mL and in a tube 1 dm long has an observed rotation of $+3.10°$, which compound was in the solution? Do the calculation.

Other Optical Isomers (Interaction 17.1)

17.19 Examine the structures of D-threonine and D-allothreonine as given in Example 17.2.

(a) Why don't these qualify as enantiomers?

(b) Why are they described as being members of the *same set* of optical isomers?

(c) Why aren't they called geometric isomers?

(d) What kind of stereoisomers are they?

17.20 Two models of methane, related as object to mirror image, superimpose. Why isn't methane called a meso compound?

17.21 Although *meso*-tartaric acid is optically inactive, it is described as an *optical isomer* of D- or L-tartaric acid. Why?

". . . Mirror, Mirror . . ." (Interaction 17.2)

17.22 What is "racemic switching" and why is it done?

18 CARBOHYDRATES

When the corn is green, photosynthesis makes corn fields immense converters of solar to chemical energy. Much of the corn being harvested here will feed livestock, thus transferring energy to animals. The primary product of photosynthesis is glucose, the chief topic of this chapter, and glucose is the favorite source of chemical energy for our brains.

THIS CHAPTER IN CONTEXT

You have already studied more biochemistry than you may realize. Ions, molecules, acids, bases, salts, buffers, redox reactions, the chief organic functional groups, and chirality are all helpful topics for a "molecule's eye view" of what's going on in our cells. Yet this chapter marks our formal entry into our quest for the molecular basis of living processes. That's why we'll take a broad overview of the remainder of the text as our initiation. Most of our focus will be on the molecular basis of life in the human body.

18.1 BIOCHEMISTRY—AN OVERVIEW

Building materials, information, and energy are basic essentials for life.

Biochemistry is the systematic study of the chemicals of living systems, their organization, and the principles of their participation in the processes of life.

The Cell Is the Smallest Unit That Lives   The molecules of living systems are lifeless, yet life has a molecular basis. Whether studied in cells or when isolated from them, the chemicals at the foundation of life obey all of the known laws of chemistry and physics. Yet, *in isolation,* not one compound of a cell has life. The intricate *organization* of molecules in a cell is as important to life as the chemicals themselves. Thus the cell is the smallest unit of matter that lives and can make a new cell like itself.

The Life of a Cell Requires Materials, Information, and Energy   A cell in particular and a whole organism in general has three basic needs, namely, materials, information, and energy. Without the daily satisfaction of these, human life at any of the many loftier levels, like creativity, relationships, and love, would be severely constrained.

In the next three chapters we'll study the organic materials of life, starting with the three main classes of foodstuffs—carbohydrates, lipids, and proteins. We use their molecules to build and run our bodies and to try to stay in some state of repair. Plants also rely heavily on carbohydrates for cell walls, and animals obtain considerable energy from carbohydrates made by plants. Lipids (fats and oils) serve many purposes. They are used as materials to make cell membranes and as sources of chemical energy. Proteins are particularly important in both the structures and functions of cells, whether of plants or of animals. Because of the central catalytic role of proteins in regulating chemical events in cells, we will immediately follow our study of proteins with an examination of enzymes, which make up a particular family of proteins. We've mentioned enzymes often as the special catalysts in living systems.

■ Cornstarch, potato starch, table sugar, and cotton are all carbohydrates.

■ Butter, lard, margarine, and corn oil are all examples of lipids known as fats and oils.

The Circulatory System Delivers Needed Compounds and Carries Wastes Away   Few of the substances in the diet are in forms directly usable by our bodies. That's why digestive enzymes catalyze the breakdown of carbohydrates, lipids, and proteins into much smaller molecules. They are then delivered into circulation. All tissues depend on the bloodstream to deliver raw materials and oxygen and to remove wastes as well as to carry hormones and disease-fighting agents. One of the major topics to come, one already begun in our study, is the molecular basis for using oxygen and releasing carbon dioxide during metabolism.

■ The reactions of *digestion* process food molecules into (usually) smaller molecules which then enter circulation and later participate in *metabolism.*

■ Neurotransmitters carry chemical signals from one nerve cell to the next.

■ The field of *molecular biology* deals with the nucleic acids.

There is both contentment and chemical energy in a peanut butter sandwich.

■ The oxidized and reduced forms of polyhydroxy aldehydes and ketones, as well as certain amino derivatives, are also in the family of carbohydrates.

■ *Deoxy-* means lacking an oxygen where one normally is.

Every Cell Has an Information System   Enzymes, hormones, and neurotransmitters are components of the intricate information system in the body. Without information, the materials and energy delivered to the body could produce only rubbish, just as monkeys swinging hammers would only reduce a stack of lumber to splinters. But carpenters, using the same materials and expending no more raw energy, can build a building, because they possess information in the form of plans and experience.

Although enzymes are major players in the cell's information system, they do not *originate* the cellular script. They only help to carry out directions. These are encoded in the molecular structures of nucleic acids, compounds that are able to direct the synthesis of enzymes. Hormones and neurotransmitters, two other components of cellular information, depend on the presence of the right enzymes not only for their own existence but also for their final services. You can see that a study of the enzyme makers, nucleic acids, must be included in any study of the molecular basis of life. The study of nucleic acids will also help us to see how different species can take essentially the same raw materials and energy, synthesize their unique sets of enzymes, and thus turn out differently. Both a kitten and a human baby thrive and grow on the same raw materials found in milk.

Some Materials Are Used Mainly for Their Chemical Energy   The molecular basis of energy for life is another broad topic to come. One of the kinds of questions is "How can one get the energy for running, skipping, and laughing out of a sandwich?" Our study of biochemical energetics will also give us numerous occasions to peer deeply into the molecular basis of some significant disorders and diseases. The other side of the coin in a study of the molecular basis of life is a study of the molecular basis of disease.

To supply materials for any use—parts, information, or energy—each organism has basic nutritional needs. These include not just organic materials, but also minerals, water, and oxygen. Thus, after learning about the materials of life and how they are processed and used, we will close our study with a broad survey of nutritional needs.

18.2   INTRODUCTION TO MONOSACCHARIDES

The monosaccharides are carbohydrates that cannot be hydrolyzed.

Carbohydrates are aldehydes and ketones with many OH groups, or substances that form these when hydrolyzed. They include the simple sugars, like glucose, as well as table sugar, starch, and cellulose. Carbohydrates are the primary products of **photosynthesis,** the complex series of reactions in plants by which CO_2, H_2O, and minerals are converted to plant chemicals and oxygen using the solar energy absorbed by the green pigment, chlorophyll. Interaction 18.1 discusses photosynthesis further.

The Simple Sugars Do Not React with Water   The carbohydrates that cannot be hydrolyzed are called the **monosaccharides,** or **simple sugars,** having the empirical formula $(CH_2O)_n$. Those with aldehyde groups are called **aldoses** and those with keto groups are **ketoses.** The names of virtually all carbohydrates similarly end in *-ose*.

Whether aldoses or ketoses, monosaccharides with three carbons are trioses, those with four are tetroses, and this pattern continues with pentoses, hexoses, and higher sugars as well. Two trioses, glyceraldehyde and dihydroxyacetone, occur in metabolism, and two pentoses, ribose and 2-deoxyribose, are needed by nucleic acids.

PHOTOSYNTHESIS

The energy released when a piece of wood burns came originally from the sun, yet wood isn't just bottled sunlight. It's a complex, highly organized mixture of compounds, mostly organic. The solar energy needed to make them is temporarily stored in wood in the form of their distinctive, energy-rich arrangements of electrons and nuclei.

The complex compounds in wood are made from very simple, energy-poor arrangements of electrons and nuclei, namely, carbon dioxide, water, and soil minerals. Only plants, not animals, can use solar energy in this way, and the name of the overall process is **photosynthesis.** The simplest statement of photosynthesis in equation form is

$$nCO_2 + nH_2O \xrightarrow[\text{energy}]{\text{solar}} \overset{\text{chlorophyll}}{\underset{\text{plant enzymes}}{}} (CH_2O)_n + nO_2$$

The symbol (CH_2O) stands for a molecular unit in carbohydrates—to make glucose, n must equal 6—but plants use the energy of carbohydrates to make proteins, lipids, and many other compounds. In the final analysis, the synthesis of all the materials in our bodies consumes solar energy, and all our activities that use energy ultimately depend on a steady flow of solar energy through plants to the plant materials that we eat. The meat and dairy products in our diets also depend on the consumption of plants by animals.

Chlorophyll is the green pigment, usually found in plant leaves, that absorbs solar energy in a way that enables photosynthesis to happen. A large number of steps and several enzymes are involved, and the overall rate increases with air temperature and with the concentration of CO_2 in air.

Be sure to notice the second product of photosynthesis, namely, oxygen, because this is the way that our global oxygen supply is continuously regenerated. Roughly 400 billion tons of oxygen are set free by photosynthesis each year, and about 200 billion tons of carbon (as CO_2) are used up. Of all of this activity, only about 10% to 20% occurs in land plants. The rest is done by tiny phytoplankton and algae in Earth's oceans. In principle it would be possible to dump so much poison into the oceans that the cycle of photosynthesis would be gravely affected. It is quite clear that the nations of the world must see that this does not happen.

When plants die and decay, their carbon atoms end up eventually in carbon dioxide again, and the reactions of decay consume oxygen. The combustion of fuels such as petroleum, coal, and wood also uses oxygen. And animals consume oxygen during respiration. Thus there exists a grand cycle in nature in which atoms of carbon, hydrogen, and oxygen move from CO_2 and H_2O into complex forms plus molecular O_2. The latter then interact in various ways to regenerate CO_2 and H_2O.

Someone has estimated that all the oxygen in Earth's atmosphere is renewed by this cycle once in about 20 centuries, and that all the CO_2 in the atmosphere and Earth's waters goes through this cycle every three centuries.

glyceraldehyde   dihydroxyacetone

ribose   2-deoxyribose

The nutritionally important monosaccharides are glucose, galactose, and fructose. Glucose, $C_6H_{12}O_6$, is a hexose with an aldehyde group and so is also called an **aldohexose.** (We're interested only in terms here; structures will come soon.) Galactose, also an aldohexose, is a stereoisomer of glucose. Fructose has a keto group and so is a **ketohexose;** it's a constitutional isomer of glucose and galactose.

■ *Oligosaccharide* molecules yield from three to a few dozen monosaccharide molecules when they are hydrolyzed.

Disaccharides and Polysaccharides Make Up the Other Families of Carbohydrates

The monosaccharides are the monomer units of di- and polysaccharides. **Disaccharides,** like sucrose (table sugar), maltose, and lactose (milk sugar), are carbohydrates that can be hydrolyzed to two monosaccharides. Starch and cellulose are called **polysaccharides** because, when one of their molecules reacts with water, it gives hundreds of monosaccharide molecules.

All Monosaccharides Are Reducing Carbohydrates

Carbohydrates are sometimes described by their abilities to react with Tollens' and Benedict's reagents. Silver ion is reduced to silver metal in Tollens' test and the copper(II) ion is reduced to Cu^+ (in Cu_2O) in Benedict's test, so carbohydrates that give these reactions are called **reducing carbohydrates.** They include all monosaccharides and nearly all disaccharides. Sucrose is not a reducing carbohydrate, however, and neither are the polysaccharides. We'll see why shortly.

■ Glucose is also called corn sugar, because it can be made by the hydrolysis of cornstarch.

Glucose Is Nature's Most Widely Used Organic Monomer

If we count all of its combined forms, (+)-glucose is perhaps the most abundant organic species on Earth. It's the building block for molecules of cellulose, a polysaccharide that makes up about 10% of all the tree leaves of the world (on a dry mass basis), about 50% of the woody parts of plants, and nearly 100% of cotton. Glucose is also the monomer for starch, a polysaccharide in many of our foods, particularly grains and tubers. Glucose and fructose are the major components of honey. Glucose is also commonly found in plant juices. Because it is by far the most common carbohydrate in blood, glucose is often called **blood sugar,** although this term strictly applies to the mixture of all of the carbohydrates in blood.

■ Massachusetts General Hospital regards a concentration of glucose in the blood of 70 to 100 mg/100 mL (3.9 to 6.1 mmol/L) to be the "normal" range for a healthy adult who has not eaten for a few hours.

One Form of Glucose Is a Pentahydroxy Aldehyde

Simple alcohols, ROH, can form acetate esters, CH_3CO_2R. Glucose forms a pentaacetate, so five of the six oxygens in $C_6H_{12}O_6$ are in alcohol groups. The sixth oxygen is in an aldehyde group because glucose is easily oxidized to a C_6 monocarboxylic acid by reagents, like Tollens' reagent, that do not oxidize alcohol groups.

Under strong, forcing conditions, glucose can be reduced to a straight-chain derivative of hexane, which tells us that the six carbons in glucose are in a straight chain. The five OH groups must be strung out, one on each of five carbons, because 1,1-diols are not stable. Glucose is thus 2,3,4,5,6-pentahydroxyhexanal. In fact, all aldohexoses have the same basic skeleton; they are all optical isomers of each other.

■ A 1,1-diol consists of the following system.

$$\begin{array}{c} OH \\ | \\ -C-OH \\ | \end{array}$$

$$\underset{6}{HOCH_2}\underset{5}{CH}-\underset{4}{CH}-\underset{3}{CH}-\underset{2}{CH}-\underset{1}{\overset{\overset{\displaystyle O}{\|}}{C}}-H$$
$$\begin{array}{cccc} & | & | & | & | \\ & OH & OH & OH & OH \end{array}$$

Basic structure of all aldohexoses, including glucose

■ When we use the term "stereocenter" we'll always mean "tetrahedral stereocenter." (The older term for such a center at a C atom is **chiral carbon.**)

Carbons 2, 3, 4, and 5 in the glucose chain are all tetrahedral stereocenters. Each center has a *unique* set of four different groups, so the centers are all different. The number of stereoisomers of a compound whose molecules have n different tetrahedral stereocenters is 2^n. In 2,3,4,5,6-pentahydroxyhexanal, n equals 4, so there must be 2^4 or 16 stereoisomers, occurring as eight pairs of enantiomers. (+)-Glucose is one of these 16; galactose is another. None of the remaining 14 stereoisomers is nutritionally important. Clearly, to know what glucose really is, we must look more closely at the stereoisomers of glucose.

18.3 D- AND L-FAMILIES OF CARBOHYDRATES

The C-5 position in glucose has the same configuration as (+)-glyceraldehyde, putting glucose in the D-family of stereoisomers.

In this section we will study the actual configurations at the tetrahedral stereocenters in monosaccharides.

All Naturally Occurring Monosaccharides Belong to the Same Optical Family Let's retreat from the complexities of (+)-glucose and go back to the simplest aldose, glyceraldehyde. The structures of its two enantiomers are shown in Figure 18.1. Both are known, and the enantiomer labeled D-(+)-glyceraldehyde actually has the absolute configuration shown. **Absolute configuration** refers to the actual arrangement in space about each stereocenter in a molecule. When we know the absolute configuration of (+)-glyceraldehyde, we also know that of (−)-glyceraldehyde, because its molecules *must* be the mirror image of the molecules of (+)-glyceraldehyde (see also Figure 18.1).

Chemists have used the absolute configurations of the enantiomers of glyceraldehyde to devise configurational or optical families for the rest of the carbohydrates. Any compound that has a configuration like that of (+)-glyceraldehyde and can be related to it by known reactions is said to be in the **D-family.** For example, (−)-glyceric acid is in the D-family because it can be made from D-(+)-glyceraldehyde by an oxidation that doesn't disturb any of the four bonds to the stereocenter, as the following equation shows.

■ As long as no bond to the stereocenter is disturbed in the reaction, no change in configuration can possibly occur.

When the molecules of a compound are the mirror images of an enantiomer in the D-family, the compound is in the **L-family.**

The letters D and L are only family names. *They have nothing to do with actual signs of the values of their specific rotations,* [α]. No way exists, in fact, to tell from the *sign* of the specific rotation whether a compound is in the D- or the L-family.

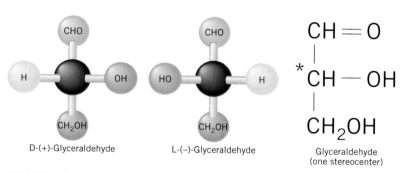

FIGURE 18.1 The absolute configurations of the enantiomers of glyceraldehyde.

These letters signify something about absolute configuration only. Later we'll see that while D-glucose is dextrorotatory, D-fructose is levorotatory. Yet both are in the D-family.

Fischer Projection Formulas Simplify Absolute Configurations   When a molecule has several stereocenters, it becomes quite difficult for most people to make a perspective, three-dimensional drawing of an absolute configuration. Emil Fischer, a chemist who unraveled most of the carbohydrate structures, devised a way around this, and his structural representations are called *Fischer projection formulas.* To make them, we follow a set of rules that let us project onto a plane surface the three-dimensional configuration of each stereocenter in a molecule.

■ Emil Fischer (1852–1919), a German chemist, won the second Nobel Prize in Chemistry in 1902.

RULES FOR WRITING FISCHER PROJECTION FORMULAS

1.   Visualize the molecule with its main carbon chain vertical and with the bonds that hold the chain together projecting to the rear at each stereocenter. *Carbon-1 is at the top.*

2.   Mentally flatten the structure, stereocenter by stereocenter, onto a plane surface. See Figures 18.2 and 18.3.

3.   In the projected structure, represent each stereocenter either as the intersection of two lines or conventionally as C.

4.   The horizontal lines at a stereocenter actually represent bonds that project *forward,* out of the plane of the paper.

5.   The vertical lines at a stereocenter actually represent bonds that project *rearward,* behind the plane.

A Fischer projection formula can have more than one intersection of lines, each representing a stereocenter, as seen in Figure 18.3. Always remember that at each stereocenter, a horizontal line is a bond coming toward you and a vertical line is a bond going away from you.

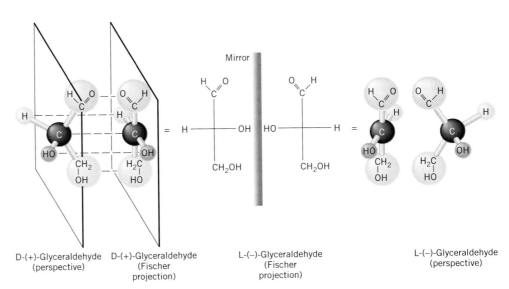

D-(+)-Glyceraldehyde   D-(+)-Glyceraldehyde   L-(–)-Glyceraldehyde   L-(–)-Glyceraldehyde
(perspective)   (Fischer   (Fischer   (perspective)
                          projection)   projection)

FIGURE 18.2   The relationships of the perspective (three-dimensional) drawings of D-(+)-glyceraldehyde and L-(–)-glyceraldehyde to their corresponding Fischer projection formulas.

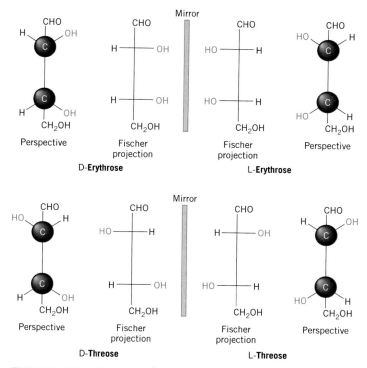

FIGURE 18.3 The four aldotetroses in their perspective and Fischer projection formulas. There are two different tetrahedral stereocenters, so there are $2^2 = 4$ stereoisomers that occur as two pairs of enantiomers, those of D- and L-erythrose and those of D- and L-threose.

Once we have one plane projection structure, it's easy to draw the mirror image, as we saw in Figures 18.2 and 18.3. We can easily test for superimposition, too, provided we strictly heed one important additional rule. We may never (mentally) lift a Fischer projection formula out of the plane of the paper. We may only slide it and rotate it within the plane, and a rotation must be by 180°, not 90°. This rule is necessary because if we turn a Fischer projection formula out of the plane and over or rotate it by only 90°, we actually make groups that project in one direction project oppositely, but the operation that we do on the paper will not show this reversal.

EXAMPLE 18.1 Writing Fischer Projection Formulas

Write the Fischer projection formulas for the stereoisomers of glyceric acid.

$$HOCH_2CHCO_2H$$
$$|$$
$$OH$$

glyceric acid

ANALYSIS There are three carbons in the chain, but only one is joined to four different atoms or groups. Only the center carbon is a stereocenter. Therefore in our equation for calculating the number of stereoisomers, $n = 1$, so

$2^n = 2$ and glyceric acid has two enantiomers. We represent its lone stereocenter by the intersection of two perpendicular lines, and we make two of these, one for each enantiomer.

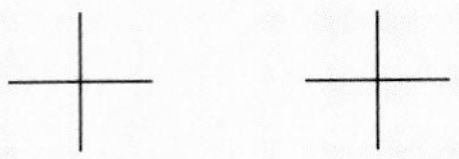

$$+ \qquad + \qquad \text{(Incomplete)}$$

Then we attach the other two carbons. According to the rules, we have to put C-1, the carbon with the carbonyl group in CO_2H, at the top. This then requires that we place CH_2OH at the lower end of the vertical line.

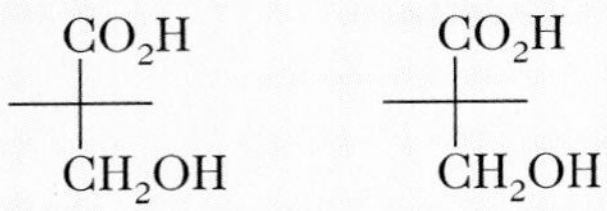

$$\begin{array}{c} CO_2H \\ + \\ CH_2OH \end{array} \qquad \begin{array}{c} CO_2H \\ + \\ CH_2OH \end{array} \qquad \text{(Incomplete)}$$

SOLUTION We know that the OH group at C-2 can be either on the right or the left, so we finish the Fischer projection formulas.

$$\begin{array}{c} CO_2H \\ H\!-\!\!-\!\!-OH \\ CH_2OH \end{array} \qquad \begin{array}{c} CO_2H \\ HO\!-\!\!-\!\!-H \\ CH_2OH \end{array} \qquad \text{(Complete)}$$

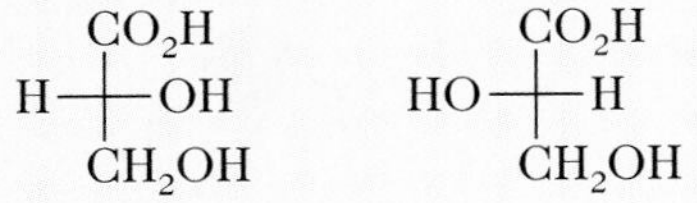

These are the two enantiomers of glyceric acid. The one on the left is D-glyceric acid, and the other is L-glyceric acid.

■ D-Glyceric acid is levorotatory but its salts are dextrorotatory, which illustrates again that the sign of a specific rotation cannot be deduced from the D- or L-family membership.

■ PRACTICE EXERCISE 1   Fischer projection formulas that correspond to the various optical isomers of tartaric acid are given below. (a) Which are identical? (b) Which are related as enantiomers? (c) One is a *meso* compound (Interaction 17.1). Which is it?

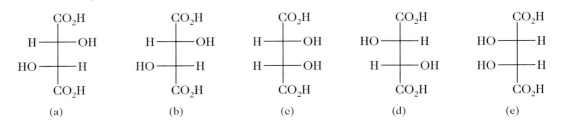

$$\begin{array}{c} CO_2H \\ H\!-\!\!-\!OH \\ HO\!-\!\!-\!H \\ CO_2H \end{array} \qquad \begin{array}{c} CO_2H \\ H\!-\!\!-\!OH \\ HO\!-\!\!-\!H \\ CO_2H \end{array} \qquad \begin{array}{c} CO_2H \\ H\!-\!\!-\!OH \\ H\!-\!\!-\!OH \\ CO_2H \end{array} \qquad \begin{array}{c} CO_2H \\ HO\!-\!\!-\!H \\ H\!-\!\!-\!OH \\ CO_2H \end{array} \qquad \begin{array}{c} CO_2H \\ HO\!-\!\!-\!H \\ HO\!-\!\!-\!H \\ CO_2H \end{array}$$

(a)   (b)   (c)   (d)   (e)

■ PRACTICE EXERCISE 2   Write Fischer projection formulas for the stereoisomers of the following compound.

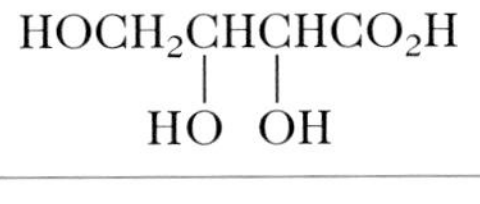

$$\begin{array}{c} HOCH_2CHCHCO_2H \\ \qquad | \quad | \\ \qquad HO \ OH \end{array}$$

All Naturally Occurring Carbohydrates Are in the D-Family Monosaccharides are assigned to the D-family or the L-family according to the projection of the OH group *at the stereocenter farthest from the carbonyl group.* The compound is in the D-family when this OH group projects to the right *in a Fischer projection formula* oriented so that the carbonyl group is at or near the top. When this OH group projects to the left, the substance is in the L-family. We have already illustrated these rules by

D- and L-glyceraldehyde (Figure 18.2), and by the enantiomers of threose and erythrose (Figure 18.3). It doesn't matter how the other OH groups at the other stereocenters project. Membership in the D- or L-family is determined solely by the projection of the OH on the stereocenter farthest from the carbonyl carbon in a properly drawn Fischer projection structure.

Because the nutritionally important carbohydrates are all in the D-family, throughout the rest of this book we'll assume that the D-family is meant whenever the family membership of a carbohydrate isn't given.

Figure 18.4 gives the Fischer projection formulas of all the aldoses in the D-family from the aldotriose through the aldohexoses. There are eight D-aldohexoses. The enantiomers of these constitute eight L-aldohexoses (not shown). In all, therefore, there are 16 optical isomers of the aldohexoses, as we calculated earlier. Figure 18.5 gives the Fischer projection formulas of several important ketoses. The ketotriose dihydroxyacetone, the two ketopentoses ribulose and xylulose, and the ketohexose fructose are the biologically important ketoses.

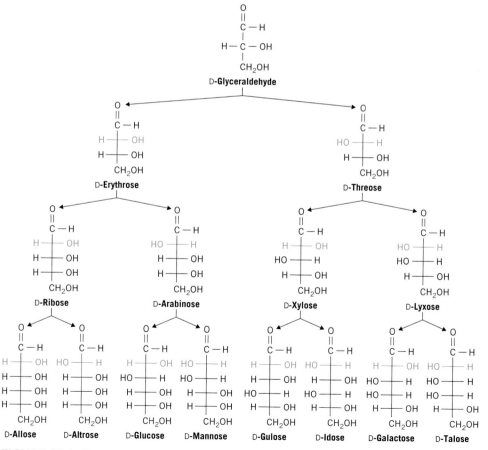

FIGURE 18.4 The D-family of the aldoses through the aldohexoses. Notice that in all of them, the OH group on the stereocenter that is farthest from the carbonyl group projects to the right. Each pair of arrows points to a pair of aldoses whose configurations are identical except at C-2.

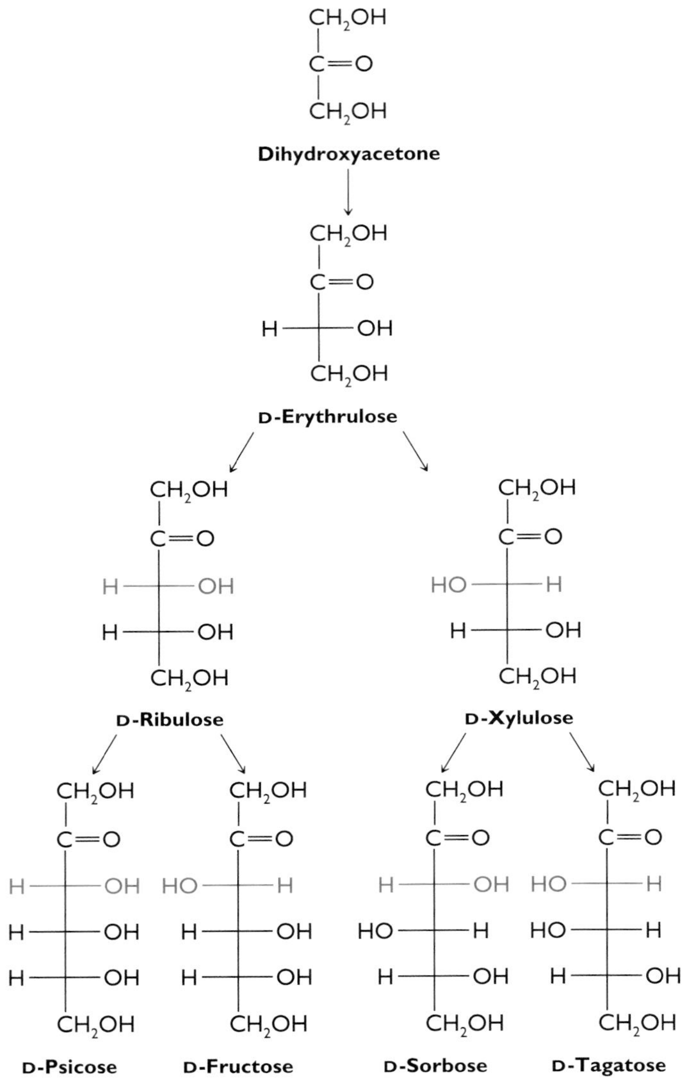

FIGURE 18.5 The D-ketoses having three to six carbon atoms. Notice that in all of them, the OH group of the stereocenter that is farthest from the carbonyl group projects to the right. Each pair of arrows points to a pair of ketoses whose configurations are identical except at C-3.

18.4 CYCLIC FORMS OF MONOSACCHARIDES

The internal addition of an OH group to an aldehyde or keto group can produce two cyclic forms of monosaccharides.

Fresh Glucose Solutions Gradually Change in Optical Rotation   Although glucose is optically active, when we try to measure its optical activity, it behaves in a very strange way. A *freshly prepared* solution of (+)-glucose has a specific rotation of

$[\alpha]_D^{20} = +113°$. As the solution ages, however, its specific rotation slowly changes until it stabilizes at a value of +52°. We'll call a glucose solution with this specific rotation an "aged glucose solution."

By a special method of recovery (which we'll not discuss), we can recover crystalline (+)-glucose from the aged solution, the same glucose in every respect as before. A *freshly prepared* solution of this recovered glucose again shows a specific rotation of +113°, but it ages in the identical way until the specific rotation stabilizes again at +52°. This cycle can be repeated as often as we please.

It is possible by another method of recovering glucose from an aged solution to obtain a slightly different crystalline compound. Its freshly prepared solution has a specific rotation of +19°, but it also changes with time to +52°, the same value that was observed for the other aged solution. Using this second method to recover glucose, we can repeat this cycle as often as we please, too.

To summarize, from the original aged solution we can use one method to recover the solute and get back glucose with a specific rotation of +113°. With the second recovery technique, we get a glucose with a specific rotation of +19°. We can interconvert these forms through the aged solution as often as we please. This change over time of the optical rotation of an optically active substance, one that can be recovered from an aged solution without any other apparent change, is called **mutarotation**. All the hexoses and most of the disaccharides mutarotate. Let us now see what is behind it.

■ The "aging" of a glucose solution occurs almost instantly when hydroxide ion is present.

Glucose Molecules Exist Mostly in Cyclic Forms   Built into the *same* molecule of 2,3,4,5,6-pentahydroxyhexanal are the two functional groups needed to make a hemiacetal, the OH group and the CH=O group. Recall that hemiacetals form by the addition of an alcohol to an aldehyde.

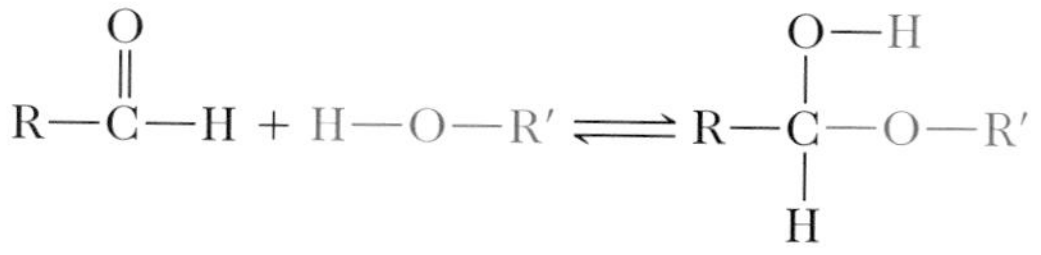

aldehyde   alcohol   hemiacetal

Suppose now that the OH group is on the *same chain* as the CH=O group. We would have something like

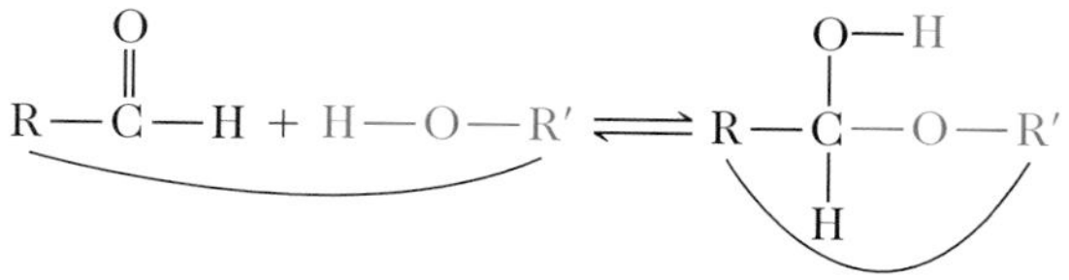

R and R′ would be joined   a cyclic hemiacetal
if CH=O and HO were in
the *same* molecule.

This is what happens to the open form of glucose. The C-5 OH group adds to the aldehyde group. The open structure has to coil for this to happen, as shown in structure **2**, in the following paragraph. It undergoes ring closure, and a new OH group, the hemiacetal OH, appears at C-1. This C-1 OH, however, can emerge on one side of the ring or the other. It depends on the way the O atom of the C=O group points just before ring closure.

If, at the moment of ring closure, the C-1 OH comes out on the side of the ring opposite to the CH$_2$OH unit (involving C-6), one cyclic form of glucose emerges, namely, the alpha form, **1**, called α-glucose. If the new C-1 OH group comes out on the same side of the ring as the CH$_2$OH unit, the beta form of glucose, called β-glucose, **3**, forms.

■   The H at C-2 and C-5 and the OH at C-3 do not stick *inside* the ring. They stick *above or below the plane* of the ring.

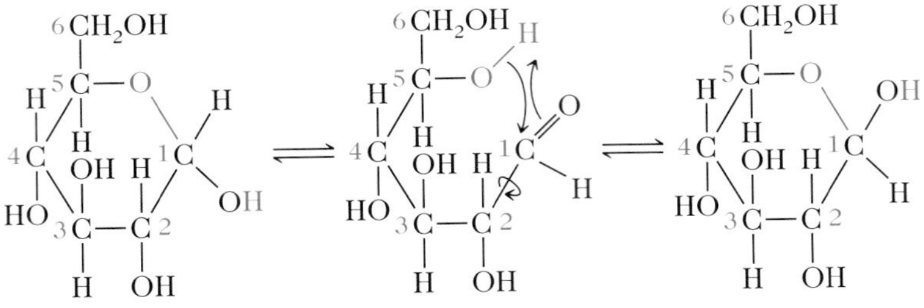

1	**2**	**3**
α-glucose	open form of glucose	β-glucose

The Six-Membered Rings of Glucose Are Actually Not Flat   The carbon atoms in glucose are all tetrahedral, so the bonds from them normally are at angles of 109.5°. The two bonds from the O atom in the ring are also close to this. A *flat* hexagon ring, however, would have internal angles of 120°. A *saturated* six-membered ring, therefore, cannot be flat as indicated by structures **1** or **3**. Instead, such rings are nonplanar so that normal bond angles are possible. We show next the three forms of glucose as they are known to exist in nonplanar conformations called *chair forms*.

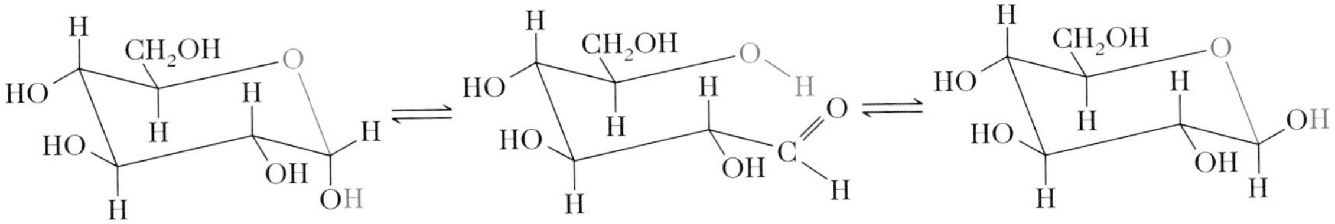

4 (= 1)	**5 (= 2)**	**6 (= 3)**
α-glucose	open form of glucose	β-glucose

The nonplanar forms of glucose

■   The isomeric, cyclic forms of any given carbohydrate that differ *only* in the configuration of the hemiacetal (or hemiketal) carbon are called **anomers.** Thus **4** (= **1**) and **6** (= **3**) are anomers.

Interaction 18.2 discusses chair conformations of six-membered rings in more detail and explains why they are preferred. Having called this to your attention, we will often use the planar designations anyway, because major references in biochemistry do so. Many times we will show both forms, however.

In Aged Glucose Solutions, All Three Forms of Glucose Exist in Equilibrium α-Glucose is the form of the glucose molecules when a freshly prepared aqueous solution has a specific rotation of +113°. Hemiacetals are unstable, however, and glucose in its cyclic forms is a hemiacetal. First one molecule of **1** and then another opens up to give form **2**. As soon as molecules of **2** appear, they can and do reclose. Figure 18.6 shows how free rotation about the C-1 to C-2 bond can reposition the aldehyde group, so that either one side or the other side of the carbonyl system faces the C-5 OH group at the moment of ring closure.

■   This arrangement of CH$_2$OH relative to the ring O atom also ensures that the optical configurations at all stereocenters of natural glucose are correctly displayed in our structures.

To keep the two glucose forms straight, use the CH$_2$OH group and the ring O atom as points of reference. Notice that in both of the cyclic forms of glucose the CH$_2$OH unit sticks upward from the plane *when the ring is drawn with its oxygen in the upper right-hand corner.* When we use these specific orientations, we are certain to be drawing a member of the D-family of the aldohexoses. Now notice in structure **3** that the OH group at C-1 projects *upward* in β-glucose and so is on the same

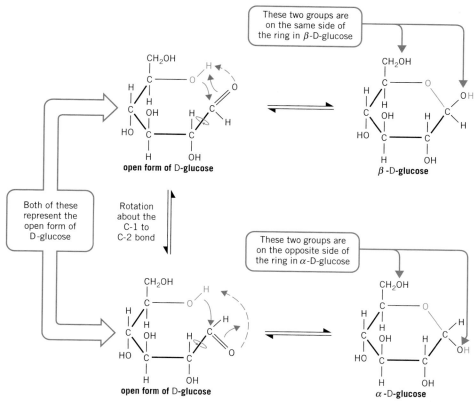

FIGURE 18.6 The α and β forms of D-glucose arise from the same intermediate, the open-chain form. Depending on how the aldehyde group, CH=O, is pointing when the ring closes, one ring form or the other results.

side of the ring as the CH_2OH group to the left of the ring O atom. In α-glucose, structure **1**, you can see that the OH at C-1 projects *downward* on the opposite side of the ring from the CH_2OH group. We should now use the names β-D-glucose and α-D-glucose for **3** and **1** but, as we have said, we'll always mean the D-family (unless something else is stated). The projections of all the other OH groups in both structures **1** and **3** are identical. If we changed any of their orientations, we would have the structure of a molecule that isn't any form of glucose, but one of its stereoisomers instead.

Ring Opening and Ring Closing Occur During Mutarotation   The structures of the two cyclic forms of glucose differ only in the orientation of the OH group at C-1. With this in mind, let's review what happens during mutarotation. As we said, when α-glucose is freshly dissolved in water, it has a specific rotation of +113°. But its molecules open and close, because the hemiacetal system easily breaks apart and re-forms. Some of the newly formed open-chain molecules reclose as α-glucose, and some as β-glucose. These events take place during mutarotation, whether we start with α- or β-glucose, until one grand, dynamic equilibrium involving all three forms of glucose is established. The identical equilibrium develops from either of the cyclic forms of glucose, because we obtain an aged solution that has the identical specific rotation, +52°. We can express the equilibrium in words as follows.

$$\alpha\text{-glucose} \rightleftharpoons \text{open-chain form of glucose} \rightleftharpoons \beta\text{-glucose}$$

■ All forms of glucose are used by living systems.

THE BOAT AND CHAIR FORMS OF SATURATED, SIX-MEMBERED RINGS

The inside angle of a regular hexagon is 120°, so the six atoms of a saturated six-membered ring cannot lie in the same plane and also have the tetrahedral bond angles of 109.5°. The cyclohexane ring resolves this by twisting into a nonplanar shape called the **chair form**.

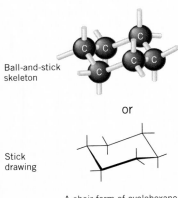

Ball-and-stick skeleton

or

Stick drawing

A chair form of cyclohexane

Even when one CH_2 unit of this ring is replaced by an oxygen atom, as in the rings of the glucose forms, the same kind of chair conformation predominates.

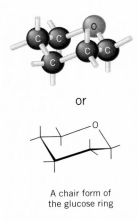

or

A chair form of the glucose ring

The **boat form** is another conformation of the ring that permits normal bond angles, and the ring has enough flexibility to be able to twist from the chair to the boat form. As the drawings show, if you twist one end of the chair form upward, you get the boat, and if you twist the opposite end downward you get an alternative chair form. Thus there are *two* chair forms and one boat. All three have normal bond angles.

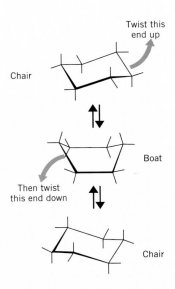

Chair

Twist this end up

Boat

Then twist this end down

Chair

The two chair forms of cyclohexane are equally stable, but the boat form is less stable than a chair. In the boat form, as you can see in the scale model, the electron clouds by the hydrogen atoms are closer to one another, particularly the two hydrogens at opposite ends, marked *a,* one at the "prow" and one at the "stern." They nudge each other in the boat form but are as far from each other as possible in the chair form.

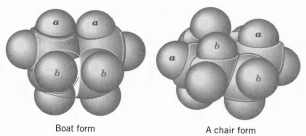

Boat form A chair form

Cyclohexane (scale model)

Similarly, the electron clouds marked *b* are closer to each other in the boat than in the chair form. Electron clouds repel each other, so the boat form is less stable than the chair form.

When the six-membered ring holds substituents, as it does among the carbohydrates, the alternative chair forms are no longer equivalent. We must, therefore, have labels to distinguish the two ways that bonds and substituents can be oriented. Positions around the perimeter are called *equatorial positions* because they are located roughly on the ring's equator. They are indicated by the black bonds and their attached H atoms in structure **A**. The positions that project above or below the average plane and are parallel to the axis through this plane are called *axial substituents,* indicated in color in **A**.

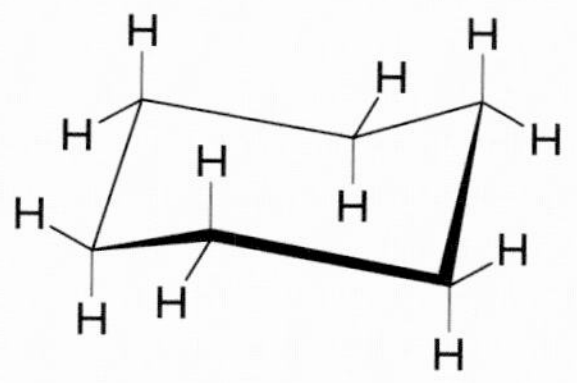

When chair form **A** twists into its alternative chair form, **B**, every equatorial position changes to axial and every axial to equatorial.

In an actual sample of cyclohexane, the two chair forms exist in equilibrium, and they constantly flip-flop back and forth. In a sense, the flat hexagonal structure that we usually draw for the ring of cyclohexane is an average of these two forms, and in most situations we can ignore the true bond angles of the six-membered ring.

The equivalency of the two chair forms vanishes, as we said, when the ring bears substituents. The electron clouds of axial substituents nudge one another more than do those of equatorial substituents. Thus equatorial orientations are more stable. As a rule, therefore, saturated six-membered rings take up whichever chair form puts the maximum number of bulky substituents in equatorial positions. This important fact dominates the conformations of the ring forms of the aldohexoses. The beta form of glucose, for example, is able to have every ring substituent oriented equatorially, *the only aldohexose in which this is possible.* In its alternative chair form, however, all substituents would be oriented axially, and this form does not occur.

A

⇅

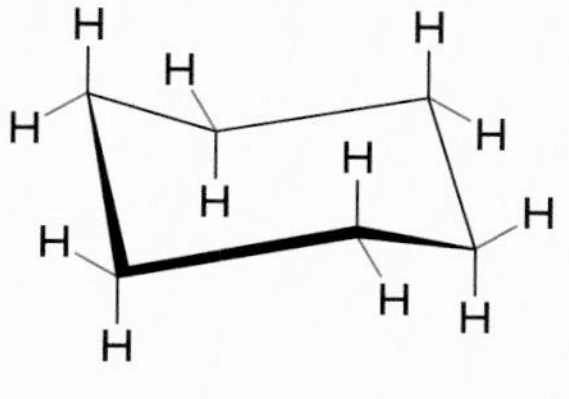

B

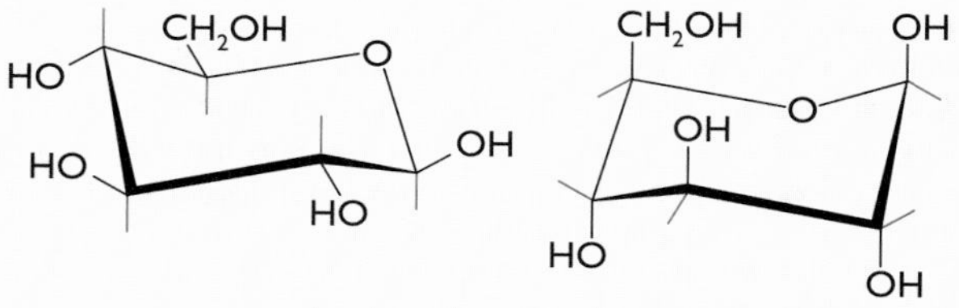

β-Glucose
(more stable;
all substituents
are equatorial)

β-Glucose
(less stable;
all substituents
are axial)

Since most people find it easier to draw flat rings for the aldohexoses, and most references in biochemistry use them widely, we will generally use them also. They correctly show *relative* projections of groups on a ring—the up or down orientations—but their bond angles are not correct.

At equilibrium, the solute is 36% α-glucose, 64% β-glucose, and scarcely a trace (<0.05%) of the open-chain form.

The method of recovering solid glucose from an aged solution determines which form of glucose nestles into crystals. One recovery method succeeds in getting just α-glucose molecules to start the formation of crystals. The molecules of β-glucose can't fit to this crystal and help it grow, so they remain in solution. But the loss of molecules of α-glucose from the equilibrium puts a stress on it, and the equilibrium shifts (in accordance with Le Châtelier's principle) to replace the lost molecules. In this way, all the β-glucose molecules eventually get changed to α-glucose molecules and nestle into the growing crystals.

The other method of crystallization succeeds in getting crystals started from just the β-glucose molecules, so eventually all the α-glucose molecules get switched over to the β form and join the growing crystals.

Molecules of the open-chain form never crystallize. They occur only in the solution. Of course, they have the aldehyde group and so are the ones attacked by Tollens' or Benedict's reagents. As open-chain molecules are removed from the equilibrium by oxidation, they are replaced. A steady shifting of the equilibrium replenishes open-form molecules from those of closed forms. This is why glucose gives the chemical properties of a pentahydroxy aldehyde despite the fact that glucose is in either one cyclic form or the other—entirely in the solid state and almost entirely in solution. This is also why it is acceptable to define a monosaccharide as an aldehyde (or ketone) with multiple OH groups rather than as a cyclic hemiacetal.

Before continuing, you should now pause to learn how to write the cyclic forms of glucose (using flat rings). Figure 18.7 outlines the steps in mastering this that have worked well for many students. Although six-membered rings are not *flat*, as we said, the projections of the groups on the ring, relative to the CH₂OH unit, are faithfully shown by these kinds of drawings.

Galactose Is a Stereoisomer of Glucose   Galactose is an aldohexose that occurs in nature mostly as a structural unit in larger molecules, like the disaccharide lactose. It is also a sugar in peas. Galactose differs from glucose only in the orientation of the C-4 OH group. Like glucose, it is a reducing sugar, it mutarotates, and it exists in solution in three forms, alpha, beta, and open.

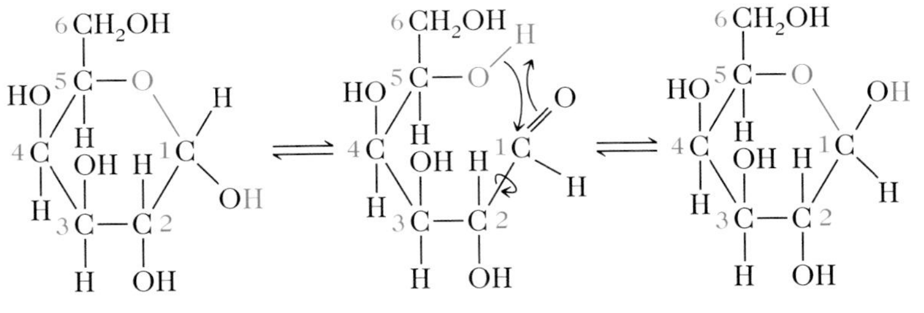

■ This is another example of Le Châtelier's principle in action.

■ Lactose is milk sugar.

■ Carbohydrates that differ *only* in the orientation of the OH at *only one* stereocenter other than the hemiacetal or hemiketal carbon are called **epimers** of each other.

■ D-Glucose and D-galactose are epimers at C-4.

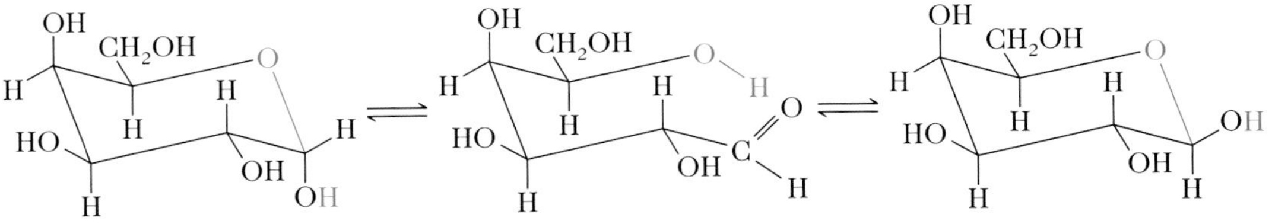

7
α-galactose

8
open form of galactose

9
β-galactose

10 (= 7)
α-galactose

11 (= 8)
open form of galactose

12 (= 9)
β-galactose

1. First write a six-membered ring with an oxygen in the upper right-hand corner.

2. Next "anchor" the terminal CH₂OH unit on the carbon to the left of the oxygen. (Let all the H's attached to ring carbons be "understood.")

3. Continue in a *counterclockwise* way around the ring, placing the OH's first down, then up, then down.

4. Finally, at the last site on the trip, how the last OH is positioned depends on whether the alpha or the beta form is to be written. The alpha is "down," the beta "up."

β-D-Glucose   **β-D-Glucose**   **α-D-Glucose**

If this detail is immaterial, or if the equilibrium mixture is intended, the structure may be written as

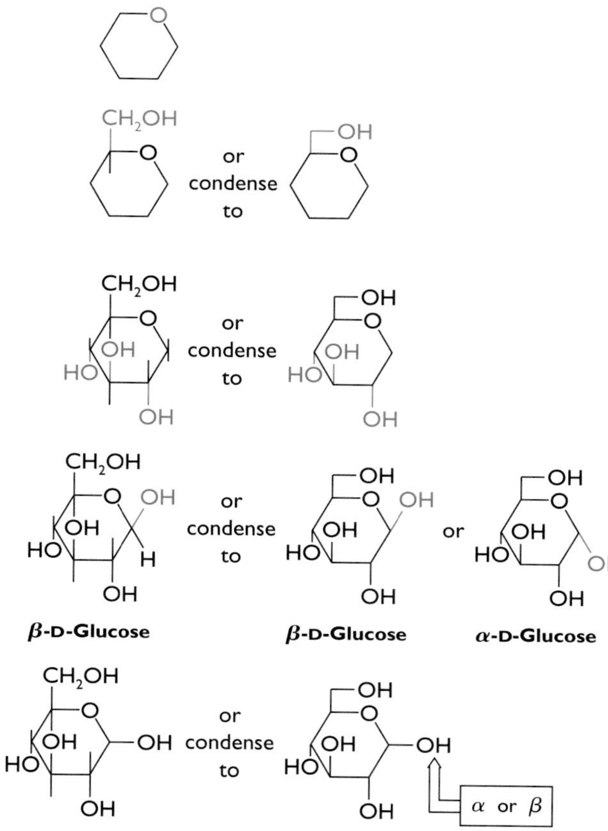

α or β

FIGURE 18.7   How to draw the cyclic forms of D-glucose in a highly condensed way.

Fructose Occurs in a Five-Membered, Cyclic Hemiketal Form   Fructose, the most important ketohexose, is found together with glucose and sucrose in honey and in fruit juices. It, too, can exist in more than one form, one open and two cyclic hemiketals. The hemiketal carbon is C-2.

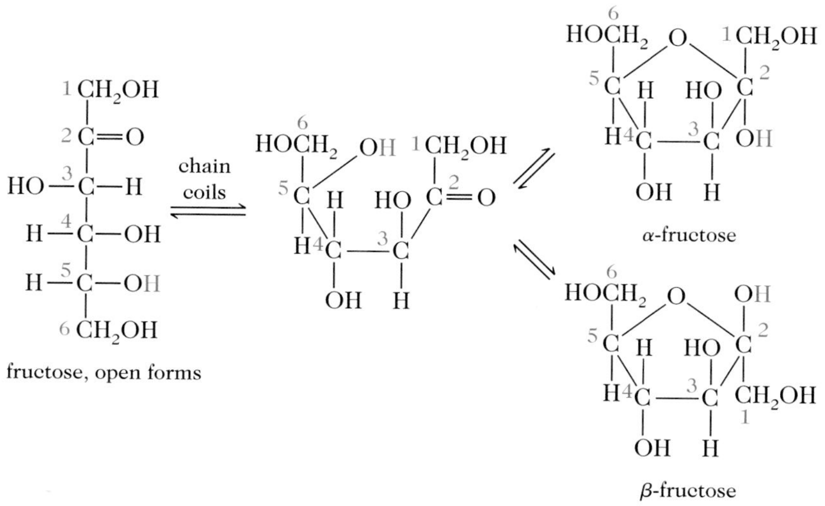

α-fructose

β-fructose

fructose, open forms

■ The internal angle of a pentagon (108°) is very close to the tetrahedral angle, so the five-membered ring is nearly flat.

■ An old name for fructose is *levulose*, after its levorotatory power.

■ Mono- and diphosphoric acid esters of fructose and glucose are important compounds in metabolism.

D-Fructose is strongly levorotatory with a specific rotation of $[\alpha]_D^{20} = -92.4°$. Fructose is a reducing sugar, not because it's a ketone (simple ketones do not easily oxidize) but because its molecules have the α-hydroxyketone system, a system that reduces Benedict's reagent.

Ribose and 2-Deoxyribose Are Aldopentoses Important to Nucleic Acid Structures Both ribose and 2-deoxyribose are building blocks of nucleic acids, and each can exist in three forms, two cyclic hemiacetals and an open form. We show just one cyclic form of each. A *deoxycarbohydrate* is one with a molecule that lacks an OH group where normally such a group is expected. Thus 2-deoxyribose is the same as ribose except that there is no OH group at C-2, just two H's instead.

■ Ribose gives the R to RNA, a nucleic acid. Deoxyribose gives the D to DNA, another nucleic acid and the chemical of genes.

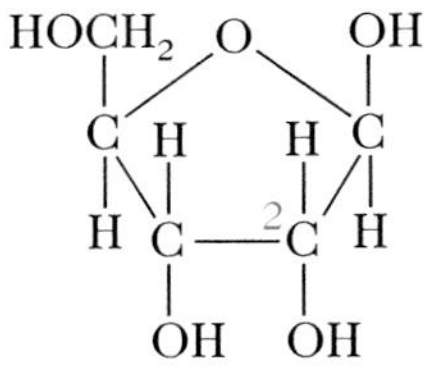

β-ribose

β-2-deoxyribose

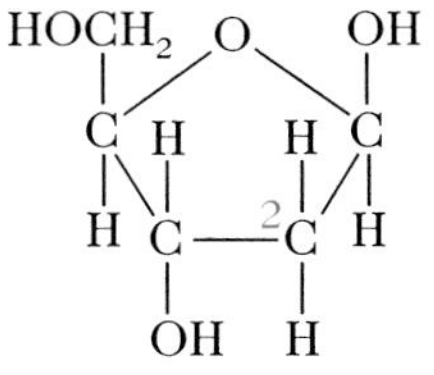

Ribose furnishes the sugar units to molecules of ribonucleic acid or RNA. Deoxyribose is used to make molecules of deoxyribonucleic acid, DNA.

18.5 DISACCHARIDES

The disaccharides are glycosides (sugar acetals) that can be hydrolyzed to monosaccharides.

The aldoses, as we have just seen, are hemiacetals. Like all hemiacetals, they react with alcohols in the presence of a catalyst to give acetals.

$$\underset{\text{hemiacetal}}{\text{R—}\overset{\text{O—R}'}{\text{CH}}\text{—OH}} + \underset{\text{alcohol}}{\text{H—O—R}'} \xrightarrow[\text{heat}]{H^+} \underset{\text{acetal}}{\text{R—}\overset{\text{O—R}'}{\text{CH}}\text{—O—R}'} + \text{H—OH}$$

The Sugar Acetals Are Called Glycosides Like all hemiacetals, those of the cyclic forms of the monosaccharides also form acetals. Methanol, for example, reacts with the cyclic hemiacetal system of glucose to give either an α- or a β-acetal, depending on how the new OCH_3 group becomes oriented at the ring. If it's on the same side of the ring as our reference CH_2OH group, then we have the β form. If it is on the opposite side, then we have the α form.

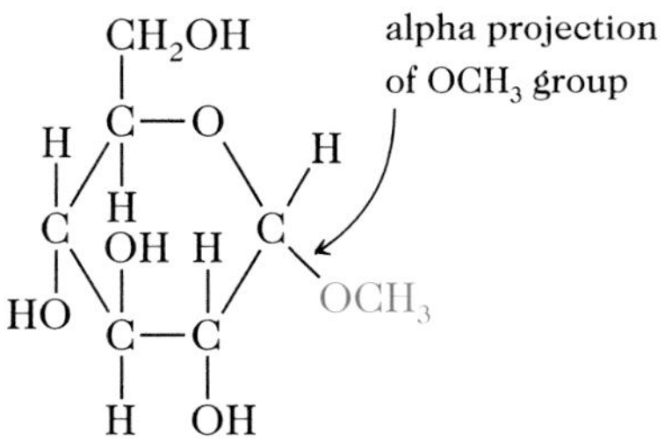

an alpha glucoside

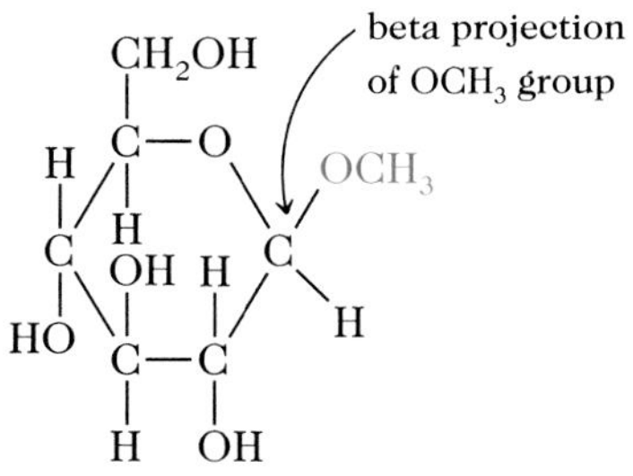

a beta glucoside

Alternatively,

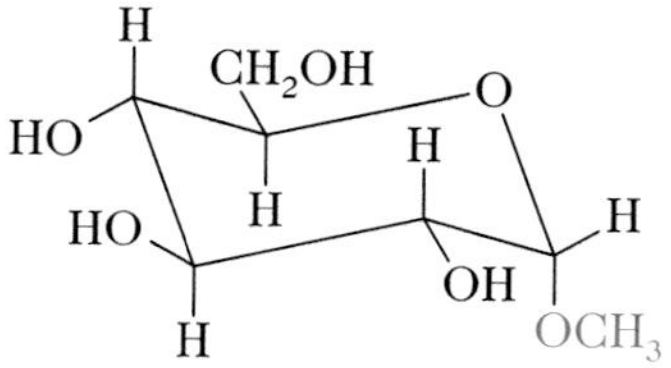

methyl α-D-glucoside
$[\alpha]_D^{25} = +158°$

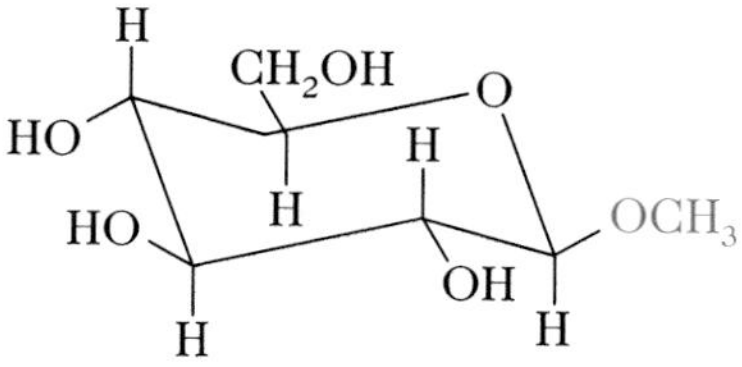

methyl β-D-glucoside
$[\alpha]_D^{25} = -33°$

Sugar acetals have the general name of **glycosides.** To name the glycoside of a specific sugar, the *-ose* in the name of the sugar is replaced by *-oside.* Thus a glycoside made from glucose is called a *glucoside.* One made from galactose is a *galactoside.*

The sugar acetals or glycosides made from simple alcohols are stable enough to isolate, but they are readily hydrolyzed when an acid catalyst (or the appropriate enzyme) is present. The glycosides do not mutarotate and do not give positive Benedict's tests because their rings cannot open to expose an aldehyde group.

The Disaccharides Are Glycosides That Use a Second Sugar as the Alcohol All the disaccharides are glycosides made from the cyclic hemiacetal unit of one sugar and one of the alcohol groups of another sugar. *An acetal oxygen "bridge" thus links two monosaccharide units in disaccharides.* This acetal unit, like all acetals, reacts readily with water in the presence of an acid or enzyme catalyst, and this hydrolysis frees the original monosaccharide molecules.

The three nutritionally important disaccharides are maltose, lactose, and sucrose. All are in the D-family. We'll first show their relationships to monosaccharides by word equations, and then we'll look more closely, but briefly, at their structures.

$$\text{Maltose} + H_2O \xrightarrow[\text{enzyme (maltase)}]{H^+ \text{ or}} \text{glucose} + \text{glucose}$$

$$\text{Lactose} + H_2O \xrightarrow[\text{enzyme (lactase)}]{H^+ \text{ or}} \text{glucose} + \text{galactose}$$

$$\text{Sucrose} + H_2O \xrightarrow[\text{enzyme (sucrase)}]{H^+ \text{ or}} \text{glucose} + \text{fructose}$$

■ You can see why glucose is of such central interest to carbohydrate chemists.

Maltose Is Made from Two Glucose Units Maltose or malt sugar does not occur widely as such in nature, although it is present in germinating grain. It occurs in corn syrup, which is made from cornstarch, and it forms from the partial hydrolysis of starch.

Maltose is made from two glucose units. They are joined by an acetal oxygen bridge that in carbohydrate chemistry is called a **glycosidic link.**

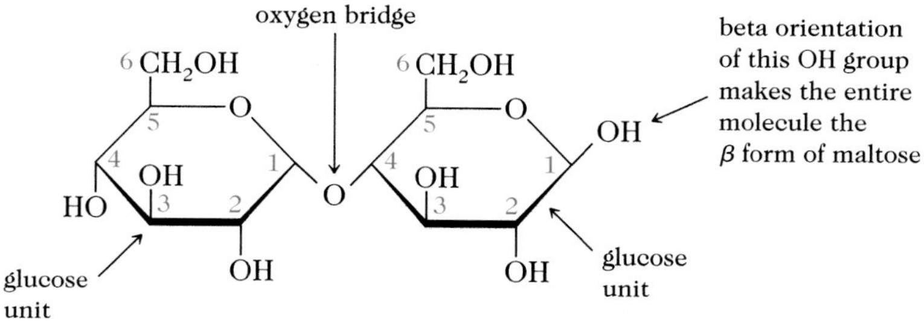

oxygen bridge

beta orientation of this OH group makes the entire molecule the β form of maltose

glucose unit

glucose unit

β-maltose

■ If the OH group at C-1 on the far-right glucose unit projected downward instead of upward, the structure would be that of α-D-maltose instead of β-D-maltose.

Or, in perspective, we have

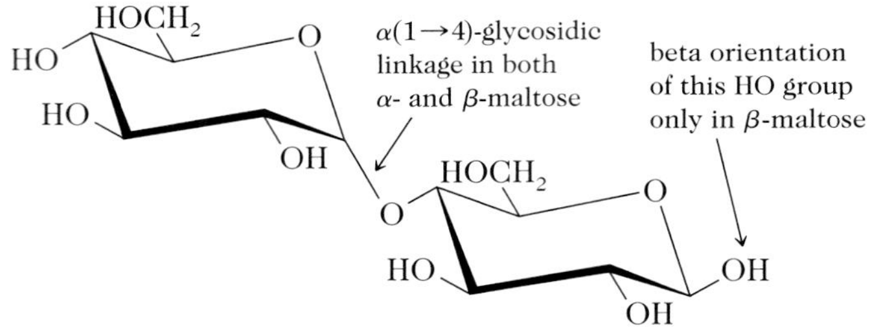

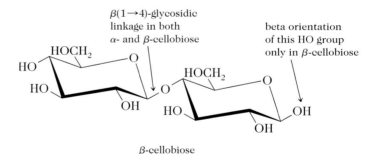

HOCH$_2$
HO
HO
OH
$\alpha(1\rightarrow4)$-glycosidic linkage in both α- and β-maltose
beta orientation of this HO group only in β-maltose
O
HOCH$_2$
HO
OH
OH

β-maltose

In maltose, the bridging oxygen of the glycosidic link joins the C-1 position of the glucose unit on the left to the C-4 of the glucose unit on the right. Such a glycosidic link is designated as $(1 \rightarrow 4)$.

The bond to the bridging oxygen from C-1 (on the left) points in the *alpha* direction, so the glycosidic link is more fully described as $\alpha(1 \rightarrow 4)$. Had this link pointed in the beta direction, it would have been described as $\beta(1 \rightarrow 4)$. But then the disaccharide would not have been maltose but cellobiose, a different disaccharide.

The purely geometric difference between $\alpha(1 \rightarrow 4)$ and $\beta(1 \rightarrow 4)$ that marks the difference between maltose and cellobiose may seem to be a trifle, but the difference to us is that we can digest maltose but not cellobiose. Just this difference in *geometry* bars humans from an enormous potential food source, for nature could supply much of it from the polysaccharide cellulose. We have maltase, an enzyme that catalyzes the digestion (the hydrolysis) of maltose. We have no enzyme for cellobiose (or, for that matter, cellulose), although some organisms do.

$\beta(1\rightarrow4)$-glycosidic linkage in both α- and β-cellobiose
beta orientation of this HO group only in β-cellobiose

HOCH$_2$
HO
HO
OH
O
HOCH$_2$
HO
OH
OH

β-cellobiose

Maltose Retains a Hemiacetal Unit, So It Is a Reducing Sugar and It Mutarotates
The glucose unit on the 4 side of an $\alpha(1 \rightarrow 4)$ glycosidic link in maltose still has a hemiacetal group. This part of maltose, therefore, can open and close, so maltose can exist in three forms, α-, β-, and the open form. Maltose therefore mutarotates and is a reducing sugar. The ring-opening action occurs only at the hemiacetal part, not at the oxygen bridge.

Lactose Links Galactose by a $\beta(1 \rightarrow 4)$ Bridge to Glucose   Lactose or milk sugar occurs in the milk of mammals—4% to 6% in cow's milk and 5% to 8% in human milk. It is also a by-product in the manufacture of cheese.

Lactose is a galactoside. From C-1 of its galactose unit there is a $\beta(1 \rightarrow 4)$-glycosidic link to C-4 of a glucose unit. The glucose unit therefore still has a free hemiacetal system, so lactose mutarotates and is a reducing sugar.

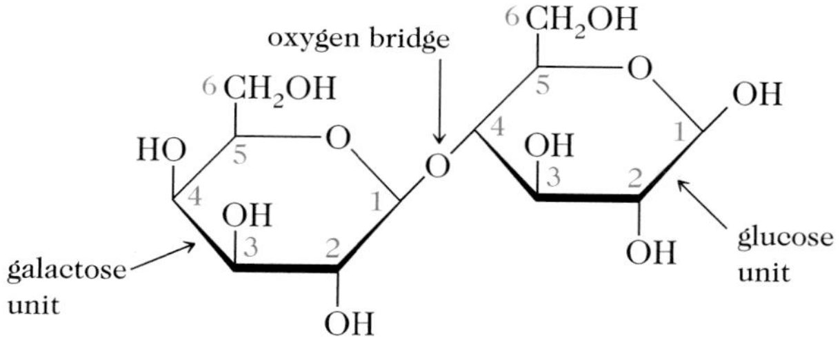

β-lactose

Alternatively,

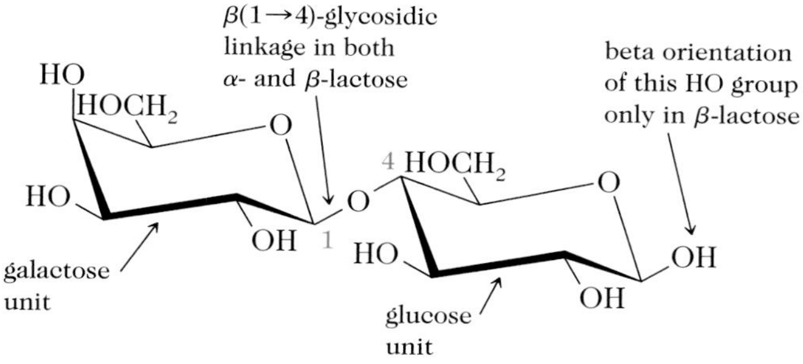

β-lactose

■ The adults in some ethnic human groups lack the digestive enzyme lactase that catalyzes hydrolysis of the β-glycosidic link between galactose and glucose units in lactose.

Sucrose Links a Glucose Unit to a Fructose Unit Sucrose, our familiar table sugar, is obtained from sugarcane or from sugar beets. Its structure links a glucose to a fructose unit by an oxygen bridge in such a way that *no hemiacetal or hemiketal group remains.* Neither ring in sucrose, therefore, can open and close spontaneously in water. Hence, sucrose neither mutarotates nor gives positive tests with Tollens' or Benedict's reagents. It's our only common nonreducing disaccharide.

■ Beet sugar and cane sugar are identical compounds, sucrose.

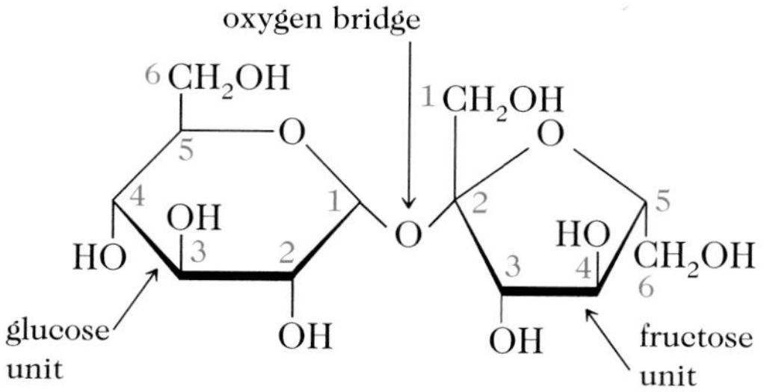

sucrose

The 1 : 1 mixture of glucose and fructose that forms when sucrose is hydrolyzed is called *invert sugar,* and it makes up the bulk of the carbohydrate in honey. (The sign of specific rotation inverts from plus to minus when sucrose, $[\alpha]_D^{20}$ +66.5°, changes to invert sugar, $[\alpha]_D^{20}$ −18.9°, and this inversion of the sign is the origin of the term *invert* sugar.)

18.6 POLYSACCHARIDES

Starch, glycogen, and cellulose are all polyglucosides.

In this section we will study the structures and some of the properties of three polymers of glucose — starch, glycogen, and cellulose.

Plants Store the Chemical Energy of Glucose in Starch Molecules When glucose is made by photosynthesis, solar energy becomes stored as chemical energy, which the plants can use for chemical work. Free glucose, however, is very soluble in water. A plant, therefore, would have to retain considerable water if its cells had to hold free glucose molecules in solution. Otherwise, the concentration of the cell fluid would be too high, causing osmotic pressure imbalances that would upset proper movements of plant fluids. The plant avoids the problem of storing glucose without too much water with it by converting glucose to its much less soluble polymer, starch. Starch is particularly abundant in plant seeds and tubers, where its energy is used for sprouting and growth. Animals that include plants in their diets also take advantage of the chemical energy in starch.

Starch is a mixture of two kinds of polymers of α-glucose, *amylose* and *amylopectin.* In amylose, the glucose units are joined by a linear succession of α(1 → 4)-glycosidic links (Figure 18.8). The lengths of the amylose "chains" vary within the same sample, but over 1000 glucose units occur per amylose molecule. Formula masses ranging from 150,000 to 600,000 have been measured. The long amylose molecules coil into spiral-like helices, which tuck a good fraction of the OH groups inside and away from contact with water. Thus amylose is only slightly soluble in water.

Amylopectin molecules have both α(1 → 4)- and α(1 → 6)-glycosidic links (Figure 18.9). The α(1 → 6) bridges link the C-1 ends of linear amylose units to C-6 positions of certain glucose units that are parts of other amylose chains (see again Figure 18.9). There are hundreds of such links per molecule, so amylopectin is heavily branched, and the branches prevent any coiling of the polymer. This leaves many more OH groups exposed to water than in amylose, so amylopectin tends to be somewhat more soluble in water than is amylose. However, neither dissolves well. The "solution" is actually a colloidal dispersion, so it scatters light and appears milky (Tyndall effect).

Natural starches are about 10% to 20% amylose and 80% to 90% amylopectin. Neither is a reducing carbohydrate and neither mutarotates. One unique test that starch does give is the **iodine test,** which can detect extremely minute traces of starch.[1] When a drop of iodine reagent is added to a starch dispersion, an intensely

■ Up to a million glucose units per molecule have been found in some amylose samples, making amylose one of nature's largest molecules.

■ So-called *soluble* starch is partially hydrolyzed starch, and its smaller molecules more easily dissolve in water.

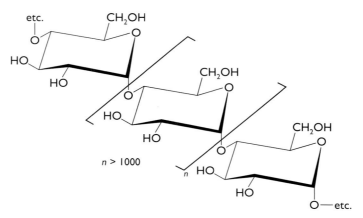

FIGURE 18.8 Amylose—partial structure.

---

[1] The starch–iodine reagent is made by dissolving iodine, I_2, in aqueous potassium iodide, KI. Iodine by itself is very insoluble in water, but iodine molecules combine with iodide ions to form the triiodide ion, I_3^-. Molecular iodine is readily available from this ion if some reactant can react with it.

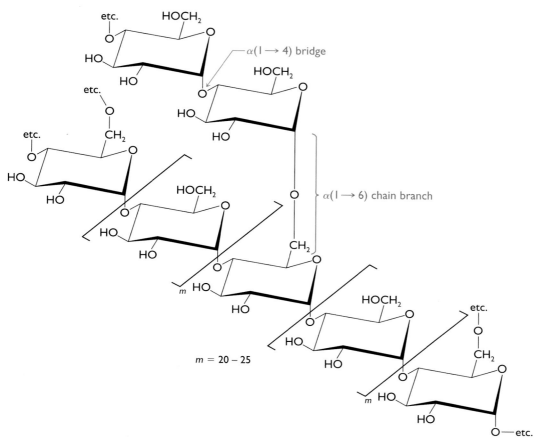

FIGURE 18.9 Amylopectin—partial structure. When m = 6 to 12, the structure would represent glycogen, a polysaccharide with branches occurring more frequently than in amylopectin.

purple color develops as iodine molecules become trapped within the vast network of starch molecules (Figure 18.10). In a starch sample undergoing hydrolysis, this network gradually breaks up so the system slowly loses its ability to give the test.

(a)   (b)

FIGURE 18.10 The iodine test for starch. (a) The starch dispersion is so dilute that it does not even appear cloudy. (b) A few drops of iodine reagent cause a purple color to develop, which becomes quite intense.

The glycosidic links in starch are easily hydrolyzed in the presence of acids or digestive enzymes, which we have. The complete digestion of starch gives only glucose. Partial hydrolysis produces *dextrin,* which consists of smaller polyglucose molecules and has been used to manufacture mucilage and paste.

Glycogen Is the Storage Form of Glucose in Animals   We and many animals use plant starch for food. Digestion hydrolyzes starch, and what glucose our bodies cannot use right away is changed into an amylopectin-like polymer called *glycogen.* In this form, we store the chemical energy of glucose units principally in the liver and in muscle tissue. Normally we don't excrete excess glucose. If our diet provides more than enough glucose to replenish our glycogen reserves, any extra is converted to fat (to the satisfaction of a huge weight-watcher industry).

■ Glycogen is sometimes called animal starch.

Glycogen molecules are essentially like those of amylopectin, perhaps even more branched. When the values of *m* in the structure of amylopectin (Figure 18.9) are in the range of 6 to 12, the structure would be that of glycogen. The formula masses of various samples of glycogen have been reported in the range of 300,000 to 100,000,000, which correspond roughly to 1700 to 600,000 glucose units per molecule.

Cellulose Is a Polymer of β-Glucose   Much of the glucose a plant makes by photosynthesis goes to make cellulose and other substances that it needs to build its cell walls and its rigid fibers. Cellulose is thus a major component of the food fiber in our diets. Cellulose, unlike starch or glycogen, has a geometry that allows its molecules to line up side by side, overlap each other, and twist into fibers.

■ Cellobiose is to cellulose what maltose is to starch.

The huge geometric difference that allows cellulose to form fibers but not amylose is the orientation of the oxygen bridge. Cellulose is a polymer of the beta form of glucose (Figure 18.11). All the oxygen bridges are $\beta(1 \rightarrow 4)$. Cellulose molecules, moreover, have no branches corresponding to the $\alpha(1 \rightarrow 6)$ branches in amylopectin. All the substituents in the rings in cellulose project in the most stable directions (the equatorial directions as discussed in Interaction 18.2). The cellulose molecule is thus quite ribbonlike, so it's easy for neighboring molecules to nestle to each other where hydrogen bonds between molecules stabilize the aggregations. With twistings of these collections, cellulose fibers of great strength are possible.

As we have noted, humans have no enzyme that can catalyze the hydrolysis of a beta-glycosidic link in cellulose, so none of the huge supply of cellulose in the world,

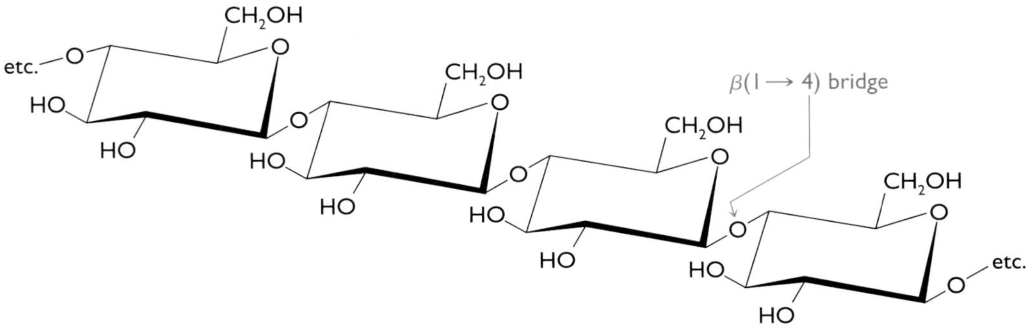

FIGURE 18.11   Cellulose—partial structure. In cotton, this polymer of β-D-glucose has from 2000 to 26,000 glucose units, depending on the variety. The strength of a cotton fiber comes in part from the thousands of hydrogen bonds that can exist between parallel and overlapping cellulose molecules.

or the cellobiose that could be made from it, is nutritionally useful to us. Many bacteria have this enzyme, however, and some strains dwell in the stomachs of cattle and other animals. Bacterial action converts cellulose in hay and other animal feed into small molecules that the larger animals can then use. Fungi and termites also can hydrolyze cellulose, enabling them to cause the decay of woody debris.

The oxygen bridges in the cellulose of cotton fabrics, being acetal systems, are hydrolyzed when a trace of acid catalyst is present. Perhaps you have discovered this the morning after spilling some acid on your jeans in lab.

■ If you spill *concentrated* acid on your clothes or skin, flush the area with water immediately.

SUMMARY

Carbohydrates   Carbohydrates are aldehydes or ketones with multiple OH groups or are glycosides of these. Those that can't be hydrolyzed are the monosaccharides, which in pure forms exist as cyclic hemiacetals or cyclic hemiketals that can mutarotate and that are reducing sugars.

D- and L- families of carbohydrates   Fischer projection structures of open-chain forms of monosaccharides are made according to a set of rules. The carbon chain is positioned vertically with any carbonyl group as close to the top as possible. At each stereocenter, this chain projects toward the back. Any groups on bonds that appear horizontal project forward. If the OH group of a carbohydrate that is farthest from the carbonyl group projects to the right in a Fischer projection structure, the carbohydrate is in the D-family. If this OH group projects to the left, the substance is in the L-family.

Monosaccharides   The three nutritionally important monosaccharides are glucose, galactose, and fructose—all in the D-family. Glucose is the chief carbohydrate in blood. Galactose, which differs from glucose only in the orientation of the OH at C-4, is obtained (together with glucose) from the hydrolysis of lactose. Fructose, a reducing ketohexose, differs from glucose only in the location of the carbonyl group. It's at C-2 in fructose and at C-1 in glucose.

Disaccharides   The disaccharides are glycosides whose molecules split up into two monosaccharide molecules when they react with water. Maltose is made of two glucose units joined by an $\alpha(1 \rightarrow 4)$-glycosidic link. In a molecule of lactose (milk sugar)—a galactoside—a galactose unit joins a glucose unit by a $\beta(1 \rightarrow 4)$ oxygen bridge. In sucrose (cane or beet sugar), there is an oxygen bridge from C-1 of a glucose unit to C-2 of a fructose unit. Both maltose and lactose retain hemiacetal systems, so both mutarotate and are reducing sugars. They also exist in α and β forms. Sucrose is a nonreducing disaccharide. The digestion of these disaccharides gives their monosaccharide units.

Polysaccharides   Three important polysaccharides of glucose are starch (a plant product), glycogen (an animal product), and cellulose (a plant fiber). In molecules of each, $(1 \rightarrow 4)$-glycosidic links occur. They're alpha bridges in starch and glycogen and beta bridges in cellulose. In the molecules of the amylopectin portion of starch as well as in glycogen, numerous $\alpha(1 \rightarrow 6)$ bridges also occur. No polysaccharide gives a positive test with Tollens' or Benedict's reagents. Starch gives the iodine test. As starch is hydrolyzed its molecules successively break down to dextrins, maltose, and finally glucose. Humans have enzymes that catalyze the hydrolysis of $\alpha(1 \rightarrow 4)$- and $\alpha(1 \rightarrow 6)$-glycosidic links, but not the $\beta(1 \rightarrow 4)$-glycosidic links of cellulose.

REVIEW EXERCISES

The answers to Review Exercises whose numbers are in color are found in Appendix E. The answers to the other Review Exercises are found in the Study Guide that accompanies this book. The more challenging questions are marked with asterisks.

Biochemistry

18.1   Substances in the diet must provide raw materials for what three essentials for life?

18.2   What are the three broad classes of foods?

18.3   What kind of compound carries the genetic "blueprints" of a cell?

18.4   What is as important to the life of a cell as the chemicals that make it up or that it receives?

Carbohydrate Terminology

18.5   Examine the following structures and identify by letter(s) which structure(s) fit each of the labels. If a particular label is not illustrated by any structure, state so.

$$
\begin{array}{cccc}
\text{CH=O} & & \text{CH}_2\text{OH} & \\
| & \text{CH=O} & | & \text{CH=O} \\
\text{CHOH} & | & \text{C=O} & | \\
| & \text{CHOH} & | & \text{CHOH} \\
\text{CHOH} & | & \text{CHOH} & | \\
| & \text{CHOH} & | & \text{CHOH} \\
\text{CHOH} & | & \text{CHOH} & | \\
| & \text{CHOH} & | & \text{CH}_2 \\
\text{CHOH} & | & \text{CHOH} & | \\
| & \text{CH}_2\text{OH} & | & \text{CH}_2\text{OH} \\
\text{CH}_2\text{OH} & & \text{CH}_2\text{OH} & \\
\textbf{A} & \textbf{B} & \textbf{C} & \textbf{D}
\end{array}
$$

(a) Ketose   (b) Deoxy sugar   (c) Aldohexose
(d) Aldopentose

18.6 Write the structure (open-chain form) that illustrates
(a) Any ketopentose
(b) Any aldotetrose

18.7 What is the structure and the common name of the simplest aldose?

18.8 What is the structure and common name of the simplest ketose?

18.9 A sample of 0.0001 mol of a carbohydrate reacted with water in the presence of a catalyst and 1 mol of glucose was produced. Classify this carbohydrate as a mono-, di-, or polysaccharide.

18.10 An unknown carbohydrate gives a positive Benedict's test. Classify it as a reducing or a nonreducing carbohydrate.

18.11 An unknown carbohydrate, **A,** gives the following reaction.

$$\textbf{A} + H_2O \xrightarrow{\text{H}^+ \text{catalyst}} \text{galactose} + \text{glucose} + \text{xylose}$$

(a) What should be the coefficient of H_2O to balance this equation?
(b) How is **A** classified, as a mono-, di-, or trisaccharide?

*18.12 An unknown carbohydrate could be reduced by a series of steps to butane. It gave a positive Benedict's test, and its molecules had just one stereocenter. What is a structure consistent with these facts?

18.13 A student proposed the following two structures as the likeliest candidate structures for a carbohydrate being studied.

$$
\text{HOCH}{-}\underset{|}{\text{CH}}{-}\underset{|}{\text{CH}}{-}\underset{|}{\text{CH}}{-}\underset{|}{\text{CH}}{-}\overset{\displaystyle O}{\overset{\|}{\text{CH}}}
$$
$$
\quad\;\; \text{OH} \quad \text{OH} \quad \text{OH} \quad \text{OH} \quad \text{OH}
$$
A

$$
\text{CH}_2{-}\underset{|}{\text{CH}}{-}\underset{|}{\text{CH}}{-}\underset{|}{\text{CH}}{-}\underset{|}{\text{CH}}{-}\overset{\displaystyle O}{\overset{\|}{\text{CH}}}
$$
$$
\;\; \text{OH} \quad \text{OH} \quad \text{OH} \quad \text{OH} \quad \text{OH}
$$
B

One of these structures is highly unlikely. Which one, and why?

18.14 What is the name of the most abundant carbohydrate in blood?

Absolute Configurations

18.15 Consider a monomethyl ether of glyceraldehyde.

$$
\text{CH}_3\text{OCH}_2\text{CHCH} \atop \text{OH}
$$

(with carbonyl $\overset{O}{\overset{\|}{}}$)

(a) How many stereoisomers are possible for this compound?
(b) Draw the Fischer projection structures of the stereoisomers according to the conventions used for carbohydrates.
(c) Correctly label each structure as D or L according to the conventions used for carbohydrates.

18.16 The simplest ketotriose is never drawn in the form of a Fischer projection structure. Why not?

18.17 Write the Fischer projection formula of L-glucose. (Refer to Figure 18.4).

18.18 What is the Fischer projection formula of D-2-deoxyribose?

18.19 Sorbitol, $C_6H_{14}O_6$, is found in the juices of many fruits and berries (e.g., pears, apples, cherries, and plums). It can be made by the addition of hydrogen to D-glucose.

$$
\underset{\text{D-glucose}}{C_6H_{12}O_6} + H_2 \xrightarrow[\text{heat and pressure}]{\text{Ni}} \underset{\text{D-sorbitol}}{C_6H_{14}O_6}
$$

Write the Fischer projection structure of D-sorbitol.

18.20 The magnesium salt of D-gluconic acid (Glucomag) is used as an antispasmodic and to treat dysmenorrhea. D-Gluconic acid, $C_6H_{12}O_7$, forms by the mild oxidation of D-glucose. What is the Fischer projection formula of D-gluconic acid?

Cyclic Forms of Carbohydrates

*18.21 Consider the following cyclic hemiacetal.

(a) What is the structure of its open form? (Write the open form with its chain coiled in the same way it is coiled in the closed form, above.)
(b) At which specific carbon (by number) does this com-

pound differ from naturally occurring glucose? (The hemiacetal carbon has position 1 in the ring, and the ring is numbered clockwise from it.)

(c) Is the compound in the D- or the L-family? (How can you tell without writing a Fischer projection structure?)

(d) With the aid of your answer to part (b) and Figure 18.4, what is the name of this compound?

18.22   Examine the following structure. If you judge that it is either a cyclic hemiketal or a cyclic hemiacetal, write the structure of the open-chain form (coiled in like manner as the chain of the ring).

18.23   Allose is identical with glucose except that in its cyclic forms the OH group at C-3 projects on the opposite side of the ring from the CH_2OH group. Allose mutarotates like glucose. Write the structures of the three forms of allose that are in equilibrium after mutarotation gives a final value of specific rotation. Which structures are α- and β-allose?

18.24   If fewer than 0.05% of all galactose molecules are in their open-chain form at equilibrium in water, how can galactose give a strong, positive Tollens' test?

18.25   At equilibrium, after mutarotation, a glucose solution consists of 36% α-glucose and 64% β-glucose (and just a trace of the open form). Suppose that in some enzyme-catalyzed process the beta form is removed from this equilibrium. What becomes of the other forms of glucose?

18.26   Study the cyclic form of β-ribose in Section 18.4 again. If its designation as β-ribose signifies a particular relationship between the CH_2OH group at C-4 and the OH group at C-1, what must the cyclic formula of α-ribose be?

18.27   Is the following structure that of α-fructose, β-fructose, or something else? Explain.

18.28   Could 4-deoxyribose exist as a cyclic hemiacetal with a five-membered ring (one of whose atoms is O)? Explain.

*18.29   With the aid of Figure 18.5 and the cyclic forms of D-fructose given in this chapter, write the structures of the cyclic forms of D-sorbose and correctly label your structures as α- or β-forms.

Glycosides

18.30   Using cyclic structures, write the structures of ethyl α-glucoside and ethyl β-glucoside. Are these two enantiomers or are they some other kind of stereoisomers?

18.31   What are the structures of methyl α-galactoside and methyl β-galactoside? Could these be described as cis–trans isomers? Explain.

Disaccharides

18.32   What are the names of the three nutritionally important disaccharides?

18.33   What is invert sugar?

18.34   Why isn't sucrose a reducing sugar?

*18.35   Consider the following structure.

(a) Does it have a hemiacetal system? Where? (Draw an arrow to it or circle it.)

(b) Does it have an acetal system? Where? (Circle it.)

(c) By what specific symbolism would the oxygen bridge be described? As an example of the kind of symbolism meant, recall that a bridge may be described as $α(1 \rightarrow 6)$.

(d) Does this substance give a positive Benedict's test? Explain.

(e) In what specific structural way does it differ from maltose?

(f) What are the names of the products of the acid-catalyzed hydrolysis of this compound?

*18.36   Maltose has a hemiacetal system. Write the structure of maltose in which this group has changed to the open form.

*18.37   When lactose undergoes mutarotation, one of its rings opens up. Write this open form of lactose.

Polysaccharides

18.38   Name the polysaccharides that give only D-glucose when they are completely hydrolyzed.

18.39   What is the main structural difference between amylose and cellulose?

18.40   How are amylose and amylopectin alike structurally?

18.41   How are amylose and amylopectin different structurally?

18.42   Why can't humans digest cellulose?

18.43   What is the iodine test? Describe the reagent and state what it is used to test for and what is seen in a positive test.

18.44   How do amylopectin and glycogen compare structurally?

18.45   How does the body use glycogen?

Photosynthesis (Interaction 18.1)

18.46   The energy available in glucose originates in the sun. Explain in general terms how this happens.

18.47   Write the overall equation for photosynthesis.

18.48   What is the name and color of the energy-absorbing pigment in plants?

18.49   In what general region of planet Earth is most of the photosynthesis carried out? By what organisms?

18.50   Describe in general terms the oxygen cycle of our planet, including the function of photosynthesis in it.

Boat and Chair Forms (Interaction 18.2)

18.51   Practice drawing the skeletons of the two chair forms of six-membered rings. Start by simply tracing the structures. Why are chair forms of saturated, six-membered rings more stable than boat forms?

18.52   Why are equatorial positions for groups attached to six-membered rings more stable than axial?

*18.53   Draw the structures of the following.
(a) A chair form of cyclohexane. Label the bonds that can hold substituents as being axial (*a*) or equatorial (*e*).
(b) The least stable structure of *trans*-1,2-dimethylcyclohexane.
(c) The most stable structure of *trans*-1,2-dimethylcyclohexane.
(d) The most stable form of D-glucose.
(e) The most stable form of D-allose. (Hint: Refer to the Fischer projection structure of D-allose in Figure 18.4 and note where it is the same as the Fischer projection structure of D-glucose and where it differs.)

Additional Exercises

*18.54   A student in an advanced lab was assigned the task of determining the structure of a carbohydrate. The empirical formula was found to be CH_2O and the compound had a molecular mass of 150. When it was allowed to react with as much acetic anhydride as it could, it was changed to $C_{13}H_{18}O_9$. Very gentle oxidation (Tollens' reagent) changed the carbohydrate into a monocarboxylic

acid, $C_5H_{10}O_6$. Vigorous reduction yielded pentane. Write a structure for the open-chain carbohydrate that is consistent with these observations.

18.55   Suppose that the aldehyde group of D-glyceraldehyde is oxidized to a carboxylic acid group, and that this is then converted to a methyl ester under conditions that do not touch any of the four bonds to C-2. What is the Fischer projection formula of this methyl ester? To what family, D or L, does it belong? Explain.

*18.56   Mannose mutarotates like glucose. Mannose is identical with glucose except that in the cyclic structures the OH at C-2 in mannose projects on the same side of the ring as the CH_2OH group. Write the structures of the three forms of mannose that are in equilibrium after mutarotation gives a steady value of specific rotation. Identify which corresponds to α-mannose and which to β-mannose.

18.57   Write the cyclic structure of α-3-deoxyribose and place an asterisk by its hemiacetal carbon.

18.58   Trehalose is a disaccharide found in young mushrooms and yeast, and it is the chief carbohydrate in the hemolymph of certain insects. On the basis of its structural features, answer the following questions.

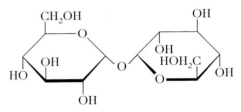

(a) Is trehalose a reducing sugar? Explain.
(b) Can trehalose mutarotate? Explain.
(c) Identify, by name only, the products of the hydrolysis of trehalose.

*18.59   A freshly prepared aqueous solution of sucrose gives a negative Tollens' test (as expected), but when the solution has stood at room temperature for about a week it gives this test. Explain.

*18.60   A freshly prepared solution (actually a dispersion) of starch in water gives a positive iodine test. If this solution is warmed with a trace of human saliva, however, the ability of the solution to give this test gradually disappears. How might this observation be explained?

*18.61   How many atoms (C and O) are there in the largest size ring possible for a cyclic hemiacetal form of an aldotetrose?

*18.62   Glyceraldehyde does not form a cyclic hemiacetal. Offer an explanation for this fact.

LIPIDS 19

The polyunsaturated oils squeezed from peanuts, olives, and corn are in the lipid family, which we'll study in this chapter. We use them for chemical energy and to build components for all of our cell membranes.

THIS CHAPTER IN CONTEXT

■ "Lipid" is from the Greek *lipos,* fat.

What nature tucks into our tummy cells during periods of abundant eating are large ester molecules with long hydrocarbon chains. They're rich in C—H bonds, these members of the *lipid* family, and the systems are good storehouses of chemical energy. The outer membranes of every cell of the body are also made of such molecules, plus membrane-bound proteins and cholesterol. The molecules that we'll study in this chapter are clearly critical to life.

19.1   WHAT LIPIDS ARE

The lipids include the edible fats and oils, all esters of long-chain fatty acids and glycerol.

■ Extraction means to shake or stir a mixture with a solvent that dissolves just part of the mixture.

When undecomposed plant or animal material is crushed and ground with a nonpolar solvent, like ether, whatever dissolves is classified as a **lipid.** Thus it's in terms of a lab *operation,* solvent extraction, that we define lipids, and the operation catches a large variety of relatively nonpolar substances, all in the lipid family (Figure 19.1). Among the many substances that won't dissolve in nonpolar solvents are carbohydrates, proteins, nucleic acids, other very polar organic substances, inorganic salts, and water.

Lipids Are Broadly Subdivided According to the Presence of Hydrolyzable Groups
One of the major subgroups of lipids, the **hydrolyzable lipids,** consists of compounds with one or more groups that react with water, in nearly all examples, *ester groups.* This subgroup includes neutral fats, waxes, phospholipids, and glycolipids.

■ What is *neutral* about the neutral fats is the absence of electrical charges on their molecules.

Neutral fats include such familiar food products as butterfat, lard (pork fat), tallow (beef fat), olive oil, corn oil, and peanut oil. Thus some neutral fats are solids and others are liquids at room temperature. The solid neutral fats are generally from animals and so are called the *animal fats.* The liquid neutral fats are from plants and so are called the *vegetable oils.* Both fats and oils, at 9.0 kcal per gram, make up the highest calorie components of the diet, which can be compared to 4.0 kcal/g for either proteins or carbohydrates. (Remember that the "calorie" used in food science is actually the kilocalorie.)

Another major subgroup are the **nonhydrolyzable lipids,** which lack groups that react with water. The *steroids,* which include cholesterol and many sex hormones,

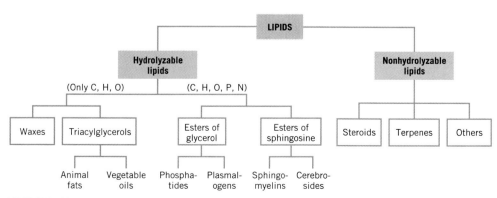

FIGURE 19.1   Lipid families

are examples (Section 19.4). Many plants produce *terpenes,* another family of non-hydrolyzable lipids, which are responsible for the pleasant odors of many plant oils. Oil of rose, for example, is 40% to 60% geraniol, a terpene alcohol with a sizable hydrocarbon unit and thus soluble in ether.

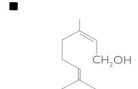

geraniol, a terpene and a component of rose oil

The *Fatty Acids* Are Mostly Long-Chain, Unbranched Monocarboxylic Acids
When hydrolyzable lipids react with water, among the products are carboxylic acids (or their anions) that are dubbed the **fatty acids.** Those obtained from the lipids of most plants and animals share the following features.

STRUCTURAL FEATURES OF THE COMMON FATTY ACIDS

1. They are usually *mono*carboxylic acids, RCO_2H.
2. The R group is usually a long *unbranched* chain.
3. The number of carbon atoms is almost always *even.*
4. The R group can be saturated, or it can have one or more double bonds, which are cis.

Thus just two functional groups are present in the fatty acids, the alkene double bond and the carboxyl group.

The most abundant *saturated* fatty acids are palmitic acid, $CH_3(CH_2)_{14}CO_2H$, and stearic acid, $CH_3(CH_2)_{16}CO_2H$, which have 16 and 18 carbons, respectively. Refer back to Table 15.1 for the other saturated fatty acids obtainable from lipids—the acids with more carbon atoms than acetic acid but with *even* numbers of carbons, like butanoic, hexanoic, octanoic, and decanoic acids. Fatty acids with fewer than 16 carbons, however, are relatively rare in nature. Interestingly, butterfat has many, namely, the C_4, C_6, C_8, and C_{10} acids. Butyric acid (C_4) is 3% to 4% of the total, enough to cause rancid butter to have its unusually offensive odor. (Butter turns rancid both by the hydrolyzing action of bacterial enzymes and by air oxidation.)

The *unsaturated* fatty acids most commonly obtained from lipids are listed in Table 19.1 and include palmitoleic acid (C_{16}) and the C_{18} acids, oleic, linoleic, and linolenic acids. All have cis double bonds. Oleic acid (C_{18}, one alkene group) is the most abundant and most widely distributed fatty acid in nature.

The cis geometry of the alkene groups of the unsaturated fatty acids causes kinks in their long hydrocarbon chains, which affects their melting points, as you can see in the last column of Table 19.1. The more kinks there are, the lower is the melting point, because kinks inhibit the closeness of the packing of molecules in crystals. Such closeness is required if there are to be strong forces of attraction between molecules, forces that are behind higher melting points. The relationship be-

Model of stearic acid, a saturated fatty acid.

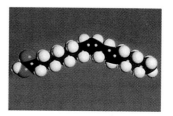

Model of oleic acid.

TABLE 19.1 Common Unsaturated Fatty Acids

Name	Double Bonds	Total Carbons	Structure	MP (°C)
Palmitoleic acid	1	16	$CH_3(CH_2)_5CH=CH(CH_2)_7CO_2H$	32
Oleic acid	1	18	$CH_3(CH_2)_7CH=CH(CH_2)_7CO_2H$	4
Linoleic acid	2	18	$CH_3(CH_2)_4CH=CHCH_2CH=CH(CH_2)_7CO_2H$	−5
Linolenic acid	3	18	$CH_3CH_2CH=CHCH_2CH=CHCH_2CH=CH(CH_2)_7CO_2H$	−11
Arachidonic acid	4	20	$CH_3(CH_2)_4CH=CHCH_2CH=CHCH_2CH=CHCH_2CH=CH(CH_2)_3CO_2H$	−50

INTERACTION 19.1
THE PROSTAGLANDINS AND HOW ASPIRIN WORKS

The prostaglandins, so named because they were initially believed to originate in the prostate gland, were discovered in the mid-1930s by a Swedish scientist, Ulf von Euler (Nobel Prize, 1970). They didn't arouse much interest in medical circles until the late 1960s, when it became increasingly evident that prostaglandins are involved in a large number of processes. Like hormones, the prostaglandins are chemical messengers that turn on cellular reactions, but, unlike hormones, they work right within the cell where they are made and do not move in circulation to their target cells. Nearly all cells in mammals, except red blood cells, make prostaglandins and similar compounds. All are C_{20} compounds and so are called *eicosanoids* (from the Greek *eikosi,* twenty). Two of the over 20 members of the prostaglandin family are PGE_2 and $PGF_{2\alpha}$. (A dashed line bond in a structure indicates that the attached group projects behind the plane of the five-membered ring; a solid wedge bond means that the group projects forward.)

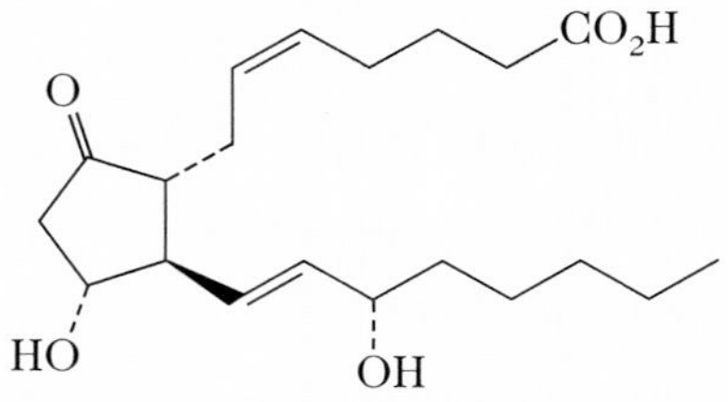

PGE₂

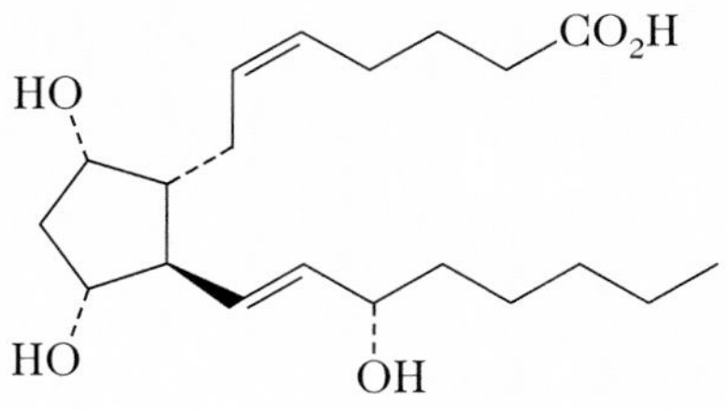

PGF₂α

Like hormones, the prostaglandins all work at extremely low concentrations to produce profound physiological effects, such as the following.

1. Activation of the inflammatory response in the joints (as in rheumatoid arthritis)

2. The production of fever

3. The production of pain

4. The induction of blood clotting

5. The induction of labor and other reproductive functions

6. The regulation of sleeping

You can easily see why pharmaceutical companies are intensely interested in discovering substances that can control or inhibit prostaglandin synthesis. Through such research might come medications for a number of humanity's common physical afflictions.

Aspirin, Pain, and Fever If an old remedy called aspirin comes to mind as you look at the first three effects listed above, you're right on track (see Figure 1). Aspirin is now known to combat both pain and fever by inhibiting prostaglandin synthesis. It does so by inactivating a key catalyst, an enzyme called PGH_2 synthase, which is essential to the formation of prostaglandins. John Vane, a British scientist, shared the 1982 Nobel Prize in Physiology or Medicine for this discovery.

Prostaglandins are made from arachidonic acid, one of the essential fatty acids. By coiling a molecule of arachidonic acid, as shown below, you can see how its structure needs only a ring closure on its left side and three more oxygen atoms to become $PGF_{2\alpha}$ (structure above). The oxygen atoms are all provided by molecular oxygen itself.

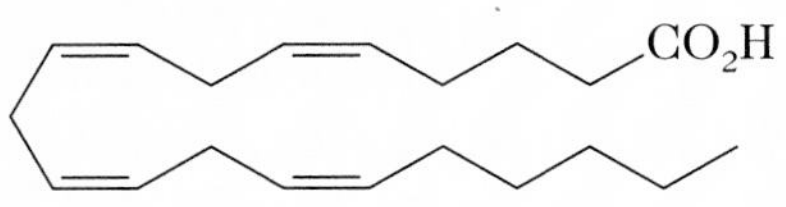

arachidonic acid

PGH_2 synthase ← aspirin inhibits

$PGH_2 \longrightarrow PGF_{2\alpha}$

FIGURE 1 Aspirin was first produced commercially by the Bayer company in 1899. Seen here is a unit of Bayer's modern production line. Aspirin is now the most widely used medicine in the world. In the United States alone, the annual production is over 15,000 tons.

When aspirin reduces the rate at which a prostaglandin is made, it reduces the ability of the system to transmit pain and to develop a higher body temperature. So aspirin fights the onsets of both pain and fever.

Aspirin and Heart Attacks The intermediate PGH₂ (structure not shown) has more than one fate, depending on the tissue. In white blood cells, for exam- ple, PGH₂ is used to make a substance (thromboxane A₂) that stimulates the aggregation of platelets, a step in the clotting of blood. Aspirin, therefore, acting in platelets to inhibit the formation of this aggregation factor, helps to prevent a myocardial infarction (heart attack) arising from a blood clot. Each year over 500,000 Americans experience their first symptoms of coronary heart disease, and roughly half die before getting to a hospital. Aspirin therapy can prevent up to 20% of all first-time or subsequent infarctions, making aspirin today an essential cardiovascular medication.

Because aspirin inhibits clotting, it does prolong internal bleeding. Thus, when physicians want to see whether a patient's stools contain blood, the patient is told not to use aspirin for a few days before the test.

Aspirin as an Acetylating Agent We'll be learning more about proteins and enzymes in later chapters, but enzymes generally have alcohol groups. The aspirin molecule has an acetyl group, and this group is transferred to an OH group of the enzyme molecule when the two interact, rendering the enzyme inactive.

$$\text{aspirin} + \left(\begin{array}{c}\text{PGH}_2\\\text{synthase}\end{array}\right)-\text{CH}_2\text{OH} \longrightarrow$$

$$\text{salicylic acid} + \left(\begin{array}{c}\text{PGH}_2\\\text{synthase}\end{array}\right)-\text{CH}_2\text{OCCH}_3$$

acetylated PGH₂ synthase (inactive enzyme)

A Reference C. C. Mann and M. L. Plummer. *The Aspirin Wars: Money, Medicine, and 100 Years of Rampant Competition.* New York: Alfred A. Knopf, 1991.

tween double bonds per molecule and melting point carries over to the neutral fats. Animal fats have fewer alkene groups per molecule than vegetable oils, so animal fats are likely to be solids at room temperature and vegetable oils tend to be liquids.

■   Omega (ω) designation

CH$_3$ ω-1
|
CH$_2$ ω-2
|
CH ω-3
‖
CH
|
(remainder of fatty acid)

■   PRACTICE EXERCISE 1   To visualize how a cis double bond introduces a kink into a molecule, write the structure of oleic acid in a way that correctly shows the cis geometry of the alkene group. (Without the double bond, the entire side chain can stretch out into a perfect zigzag conformation, as in stearic acid, a conformation that makes it easy for two side chains to nestle very close to each other.)

The *prostaglandins* are an unusual family of fatty acids with 20 carbons, five-membered rings, and a wide variety of effects in the body. See Interaction 19.1.

In the late 1980s, one small group of fatty acids, the omega-3 fatty acids, appeared in scientific debates about the value of fish or marine oils in the diet. Interaction 19.2 describes them further.

■   The name *triglycerides* is common in the older scientific literature on triacylglycerols.

The Triacylglycerols (or Triglycerides) are Triesters of Glycerol and Fatty Acids

The molecules of the most abundant lipids are **triacylglycerols** or **triglycerides,** esters between glycerol and three fatty acids.

$$\begin{array}{c}
\text{glycerol} \ \begin{cases} \boxed{\text{fatty acyl unit}} \ \ CH_2\!-\!O\!-\!\overset{\displaystyle O}{\overset{\|}{C}}\!-\!R \\[6pt] \boxed{\text{fatty acyl unit}} \ \ CH\!-\!O\!-\!\overset{\displaystyle O}{\overset{\|}{C}}\!-\!R' \\[6pt] \boxed{\text{fatty acyl unit}} \ \ CH_2\!-\!O\!-\!\overset{\displaystyle O}{\overset{\|}{C}}\!-\!R'' \end{cases}
\end{array}$$

As you can see, there are no (+) or (−) charges on triacylglycerol molecules and so they are unlike the more complex hydrolyzable lipids to be studied in Section 19.3. The triacylglycerols are thus the neutral fats mentioned earlier.

In a particular fat or oil, certain fatty acids predominate, others either are absent or are present in trace amounts, and virtually all of the molecules are triacylglycerols. Data on the fatty acid compositions of several fats and oils are listed in Table 19.2. Oleic acid (C_{18}; one alkene group), as we said, is the most abundant *monounsaturated* fatty acid, and its acyl group is very common among both the fats and oils. Notice particularly, however, that the vegetable oils deserve their nickname as *polyunsaturated* products. They incorporate more of the acyl groups of the polyunsaturated fatty acids, like those of linoleic and linolenic acid, than do the animal fats. The saturated fatty acyl units of palmitic and stearic acids are far more common in animal fats, which are thus called the *saturated fats.* Notice that coconut, palm kernel, and palm oils (plus cocoa fat in chocolate) are vegetable oils that are still high in saturated fatty acids. Among the animal fats, butterfat wins the distinction of being highest in saturated fatty acids.

The three acyl units in the triacylglycerol molecules of a given fat or oil are usually contributed by two or three *different* fatty acids. All that we can say about the structures of fats and oils are that they are mixtures of different molecules that share common structural features. Thus we cannot give *one* structure for all of cottonseed oil, but we can describe what is a fairly typical molecule; it's one like that of structure **1**.

■   An acyl group has the general structure

$$\overset{\displaystyle O}{\overset{\|}{R C}}\!-$$

INTERACTION 19.2
THE OMEGA-3 FATTY ACIDS AND HEART DISEASE

Omega-3 refers to the location of a double bond third in from the far end of a long chain fatty acid, particularly those with 18, 20, and 22 carbons. Just as omega, ω, is the last letter in the Greek alphabet, so the omega position in a fatty acid is the one farthest from the carboxyl group. Arachidonic acid (Table 19.1) is an omega-6 C_{20} fatty acid because it has a double bond at the sixth carbon from the omega end. Linolenic acid (Table 19.1) is an omega-3 fatty acid. Two other omega-3 acids are considered by some to be important in metabolism:

$$CH_3CH_2CH{=}CHCH_2(CH{=}CHCH_2)_4(CH_2)_4CO_2H$$

ω-3-eicosapentaenoic acid

$$CH_3CH_2CH{=}CHCH_2(CH{=}CHCH_2)_5CH_2CO_2H$$

ω-3-docosahexaenoic acid

The basis of the interest in these is that native Alaskans have low incidences of heart disease despite relatively high cholesterol levels in their diets (from fish oils and fish liver). Some scientists believe that the high level of the omega-3 fatty acids in marine oils provides protection against disease.

Human milk, but not infant formulas, contains ω-3-docosahexaenoic acid (DHA), which the infant's retina and brain uses. Studies using monkeys have found that a deficiency of DHA in a monkey's infancy can cause functional changes that may be permanent.

TABLE 19.2 Fatty Acid Composition (in Percentages) of Vegetable Oils and Animal Fats

Vegetable Oils and Shortening	Polyunsaturated Fatty Acids[a]	Monounsaturated Fatty Acids[a]	Total Unsaturated Fatty Acids[b]	Saturated Fatty Acids[a]
Safflower oil	75	12	87	9
Sunflower oil	66	20	86	10
Corn oil	59	24	83	13
Soybean oil	58	23	81	14
Cottonseed oil	52	18	70	26
Canola oil	33	55	88	7
Olive oil	8	74	82	13
Peanut oil	32	46	78	17
Margarine, soft tub[c]	31	47	78	18
Margarine, stick[c]	18	59	77	19
Shortening, vegetable[c]	14	51	65	31
Palm oil	9	37	46	49
Coconut oil	2	6	8	86
Palm kernel oil	2	11	13	81
Animal Fats				
Tuna fat[d]	37	26	63	27
Chicken fat	21	45	66	30
Hog fat (lard)	11	45	56	30
Mutton fat	8	41	49	47
Beef fat	4	42	46	50
Butter fat	4	29	33	62

[a] Percentages of total fat.
[b] Values in this column are the sum of the poly- and monounsaturated fatty acid percentages in the preceding columns. The values in this column plus those in the last column do not add up to 100% because each item has a small and varying amount of other fatty substances.
[c] Made by hydrogenating soybean plus cottonseed oil.
[d] White tuna, canned in water; data are for the drained solids.
Data from the National Heart, Lung, and Blood Institute, Public Health Service, National Institutes of Health, U.S. Department of Human Services. NIH Publication 88-2696 (1987).

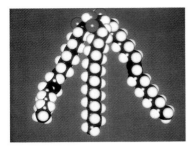

Model of structure **1**.

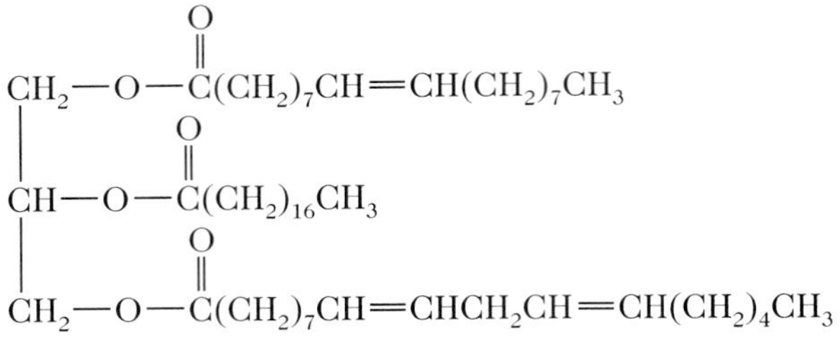

$$CH_2-O-\overset{\overset{\displaystyle O}{\|}}{C}(CH_2)_7CH=CH(CH_2)_7CH_3$$
$$CH-O-\overset{\overset{\displaystyle O}{\|}}{C}(CH_2)_{16}CH_3$$
$$CH_2-O-\overset{\overset{\displaystyle O}{\|}}{C}(CH_2)_7CH=CHCH_2CH=CH(CH_2)_4CH_3$$

1

Plant Waxes Are Simple Esters with Long Hydrocarbon Chains   The waxes occur as protective coatings on fruit and leaves as well as on fur, feathers, and skin. Nearly all **waxes** are esters of long-chain monohydric alcohols and long-chain monocarboxylic acids in both of which there is an *even* number of carbons. As many as 26 to 34 carbon atoms can be incorporated into *each* of the alcohol and the acid units, which makes the waxes almost totally hydrocarbon-like.

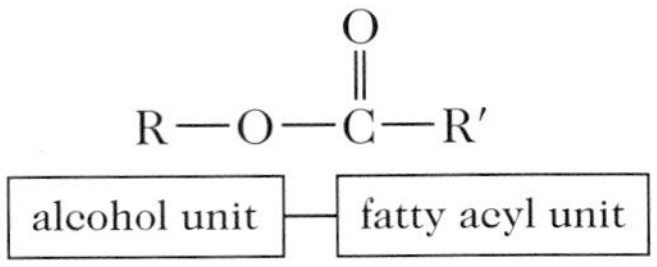

$$R-O-\overset{\overset{\displaystyle O}{\|}}{C}-R'$$

| alcohol unit | fatty acyl unit |

components of waxes

Any particular wax, like beeswax, consists of a mixture of similar compounds that share the kind of structure shown above. In molecules of lanolin (wool fat), however, the alcohol portion is contributed by steroid alcohols, which have large ring systems that we'll study in Section 19.4. Waxes exist in sebum, a secretion of human skin that helps to keep the skin supple.

■ Lanolin is used to make cosmetic skin lotions.

■   PRACTICE EXERCISE 2   One particular ester in beeswax can be hydrolyzed to give a straight-chain primary alcohol with 26 carbons and a straight-chain carboxylic acid with 28 carbons. Write the structure of this ester.

19.2   CHEMICAL PROPERTIES OF TRIACYLGLYCEROLS

Triacylglycerols can be hydrolyzed, saponified, and hydrogenated.

■   The solubility of free fatty acids in water is extremely low, only about 10^{-6} mol/L, so to be transported in blood fatty acids must be carried on protein molecules.

Triacylglycerols Can Be Hydrolyzed   When we need the chemical energy of the triacylglycerols stored in our fat tissue, a special enzyme (a lipase) catalyzes their complete hydrolysis. The fatty acid molecules, bound to proteins (albumins) in the blood, are then sent to the liver. In general, the equation for the hydrolysis of a triacylglycerol is

$$
\begin{array}{c}
\text{CH}_2\text{—O—}\overset{\displaystyle O}{\overset{\|}{\text{C}}}\text{—R} \\
\text{CH—O—}\overset{\displaystyle O}{\overset{\|}{\text{C}}}\text{—R}' \;+\; 3\text{H}_2\text{O} \xrightarrow[\substack{\text{(triacyl-}\\ \text{glycerol}\\ \text{lipase)}}]{\text{enzyme}} \\
\text{CH}_2\text{—O—}\overset{\displaystyle O}{\overset{\|}{\text{C}}}\text{—R}''
\end{array}
\qquad
\begin{array}{c}
\text{CH}_2\text{—OH } + \text{ HO—}\overset{\displaystyle O}{\overset{\|}{\text{C}}}\text{—R} \\
\text{CH—OH } + \text{ HO—}\overset{\displaystyle O}{\overset{\|}{\text{C}}}\text{—R}' \\
\text{CH}_2\text{—OH } + \text{ HO—}\overset{\displaystyle O}{\overset{\|}{\text{C}}}\text{—R}''
\end{array}
$$

triacylglycerol   glycerol   fatty acids

When we *digest* triacylglycerols, hydrolysis is usually not complete. The digestive enzyme (pancreatic lipase) takes the hydrolysis only to monoacylglycerols, fatty acids, and some diacylglycerols. A synthetic fatlike product, Olestra, while loaded with fatty acyl groups (esterified with the OH groups of sugar), cannot be digested at all (see Interaction 19.3).

Soaps Are Made by the Saponification of Triacylglycerols The saponification of the ester links in triacylglycerols by the action of a strong base (e.g., NaOH or KOH) gives glycerol and a mixture of the salts of fatty acids.

$$
\begin{array}{c}
\text{CH}_2\text{—O—}\overset{\displaystyle O}{\overset{\|}{\text{C}}}\text{—R} \\
\text{CH—O—}\overset{\displaystyle O}{\overset{\|}{\text{C}}}\text{—R}' \;+\; 3\text{NaOH}(aq) \xrightarrow{\text{heat}} \\
\text{CH}_2\text{—O—}\overset{\displaystyle O}{\overset{\|}{\text{C}}}\text{—R}''
\end{array}
\qquad
\begin{array}{c}
\text{CH}_2\text{—OH } + \text{ Na}^{+}\text{—O—}\overset{\displaystyle O}{\overset{\|}{\text{C}}}\text{—R} \\
\text{CH—OH } + \text{ Na}^{+}\text{—O—}\overset{\displaystyle O}{\overset{\|}{\text{C}}}\text{—R}' \\
\text{CH}_2\text{—OH } + \text{ Na}^{+}\text{—O—}\overset{\displaystyle O}{\overset{\|}{\text{C}}}\text{—R}''
\end{array}
$$

triacylglycerol   glycerol   mixture of sodium salts
of fatty acids (soap)

The salts make up "old-fashioned" soap (e.g., Ivory soap), and how soaps exert their detergent action is described in Interaction 19.4.

■ **PRACTICE EXERCISE 3** Write a balanced equation for the saponification of **1** (in Section 19.1) with sodium hydroxide.

The Hydrogenation of Vegetable Oils Gives Solid Shortenings and Margarine When hydrogen is made to add to some of the double bonds in vegetable oils, the oils become like animal fats, both physically and structurally. One very practical consequence of such partial hydrogenation is that the oils change from being liquids at room temperature to solids. Many people prefer a solid, lardlike shortening for cooking, instead of a liquid oil. So the manufacturers of such "hydrogenated vegetable oils" as Crisco and Spry take inexpensive, readily available vegetable oils, like corn oil, soybean oil, and cottonseed oil, and catalytically add hydrogen to some (not all) of the alkene groups in their molecules. We say that the double bonds become *saturated*. Unlike natural lard, the vegetable shortenings have no cholesterol.

■ Except for the absence of cholesterol, hydrogenated vegetable oils are chemically and nutritionally identical to animal fats.

■ **PRACTICE EXERCISE 4** Write the balanced equation for the complete hydrogenation of the alkene links in structure **1**, Section 19.1.

INTERACTION 19.3
FAKE FAT—HOW OLESTRA ESCAPES DIGESTION

Sucrose, as you know, is table sugar, and it's probably the most commonly used sweetener around the home.

sucrose

The sucrose molecule has eight OH groups. Chemists at Proctor & Gamble discovered over 30 years ago that these can be made into esters with six, seven, or eight fatty acids per sucrose molecule, and they named the mixture *Olestra*. The media dubs it "fake fat." It's an oily material, like a cooking oil, and it can be used as one. It tastes like a cooking oil, too, but probably not to everybody.

Olestra (with 6 of 8 OH groups esterified);
R groups are long chain and not all the same.

Giving Enzymes the Cold Shoulder You can well imagine how the addition of six to eight long, fatty acid hydrocarbon chains to one sucrose molecule affects properties. Physically, the transformation converts a hydrophilic molecule (sucrose) into one that is hydrophobic (Olestra).

Chemically, sucrose is changed into something that lipase, a fat-digesting (ester-hydrolyzing) enzyme, cannot touch. None of the lipase molecules "fit" the large molecular shoulders of Olestra. Moreover, Olestra molecules also cannot fit to those of *sucrase,* the digestive enzyme that catalyzes the hydrolysis of sucrose. Olestra, in other words, gives lipase and sucrase the proverbial cold shoulder. Because Olestra molecules cannot be digested, none of the fatty acids and none of the sucrose chemically embedded in Olestra can be sent out of the digestive tract into circulation for eventual metabolism or fat storage. Even if some slight digestion of Olestra occurs, what forms are normal products of digestion anyway and so are nontoxic. Olestra is a nutritional novelty, a product that tastes like a cooking oil and can be used as a cooking oil but that adds no calories to the diet!

Some Controversy Exists—A Modestly Warm Shoulder from the FDA In early 1996, after a 25 year review process, the U.S. Food and Drug Administration (FDA) gave its limited approval of the use of Olestra but only to prepare "savory snack foods," like potato chips, crackers, and tortilla chips (see Figure 1). Why the limits and why so long to decide? What's the controversy?

The *long-term* health effects have not been studied to everyone's satisfaction at the FDA. The FDA has found what the law says has to be determined, namely, "a reasonable certainty of no harm" from Olestra.

The chief lipid material in *margarine* is also produced from vegetable oils by partial hydrogenation, but it is done with special care so that the final product can melt on the tongue, one property that makes butterfat so pleasant. (If all the alkene groups in a vegetable oil were hydrogenated, instead of just some of them, the product would be like beef or mutton fat, relatively hard materials that would not melt on the tongue.)

The popular brands of peanut butter, those with peanut oils that do not separate, are made by the partial hydrogenation of the oil in real peanut butter. The peanut oil changes to a solid, hydrogenated form at room temperature, and therefore it does not separate.

Another problem is that when olestra is the chief dietary "fat," the body's ability to absorb the fat-soluble vitamins, vitamins A, D, E, and K, from the intestinal tract is inhibited. So the FDA requires that products made with Olestra have extra amounts of these vitamins to compensate. It's like shifting an equilibrium by increasing the concentration of "reactants," namely, the vitamins, in order to favor the "product," namely, sufficient vitamin absorption. Still another problem is that Olestra may cause abdominal cramping and loose stools in a small percentage of users.

In the view of the FDA, the problems with Olestra are deemed inconvenient but not dangerous, and they appear not to affect most users. Still, the FDA requires that Proctor & Gamble conduct studies on Olestra's long-term effects, and the FDA's Food Advisory Committee will review the findings. Moreover, the FDA requires that the labels of products made with Olestra carry the following warning.

> This product contains Olestra. Olestra may cause abdominal cramping and loose stools. Olestra inhibits the absorption of some vitamins and other nutrients. Vitamins A, D, E, and K have been added.

If it all works as hoped, people who want their snacks have something besides diet pills to use in the struggle to control obesity and heart disease. Test marketing of Olestra-based products in small markets occurred in 1996. National marketing is planned for the late 1990s.

References

1. Henry Blackburn, "Olestra and the FDA," *The New England Journal of Medicine*, April 11, 1996, page 984.

2. Ellin Doyle, "Olestra? The Jury's Still Out," *Journal of Chemical Education*, April 1997, page 370.

FIGURE 1 Potato chips made using Olestra as the cooking oil have fewer calories per gram than those made conventionally.

19.3 PHOSPHOLIPIDS

Phospholipid molecules have very polar or ionic sites in addition to long hydrocarbon chains.

Phospholipids are esters of either glycerol or sphingosine, which is a long-chain, dihydric amino alcohol with one double bond.

$$CH_3(CH_2)_{12}CH{=}CHCHCHCH_2OH$$
$$\underset{HO}{|}\quad\underset{NH_2}{|}$$

sphingosine

INTERACTION 19.4
HOW DETERGENTS WORK

Soap Water is a very poor cleaning agent because it can't penetrate greasy substances, the "glues" that bind soil to skin and fabrics. When just a little soap is present, however, water cleans very well, especially warm water. Soap is a simple chemical, a mixture of the sodium or potassium salts of the long-chain fatty acids obtained by the saponification of fats or oils.

Detergents Soap is just one kind of detergent. All detergents are surface-active agents that lower the surface tension of water. All consist of ions or molecules that have long hydrocarbon portions plus ionic or very polar sections at one end. The accompanying structures illustrate these features and show the varieties of detergents that are available.

Although soap is manufactured, it is not called a synthetic detergent. This term is limited to detergents that are not soap, that is, not the salts of naturally occurring fatty acids obtained by the saponification of lipids. Most synthetic detergents are salts of sulfonic acids, but others have different kinds of ionic or polar sites. The great advantage of synthetic detergents is that they work in hard water and are not precipitated by the hardness ions—Mg^{2+}, Ca^{2+}, and the two ions of iron. These ions form messy precipitates ("bathtub ring") with the anions of the fatty acids present in soap. The anions of synthetic detergents do not form such precipitates.

Figure 1 shows how detergents work. In Figure 1*a* we see the hydrocarbon tails of the detergent work their way into the hydrocarbon environment of the grease layer. ("Like dissolves like" is the principle at work here.) The ionic heads stay in the water phase, and the grease layer becomes pincushioned with electrically charged sites. In Figure 1*b* we see the grease layer breaking up, aided with some agitation or scrubbing. Figure 1c shows a magnified view of grease globules studded with ionic groups and, being like charged, these globules repel each other. They also tend to dissolve in water, so they are ready to be washed down the drain.

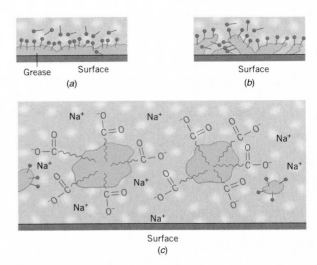

FIGURE 1

$$CH_3(CH_2)_{14}CO_2^-Na^+$$

soap—an anionic detergent

$$CH_3(CH_2)_{13}OSO_3^-Na^+$$

a sodium alkyl sulfate—
an anionic detergent

$$CH_3(CH_2)_8 \!-\!\! \bigcirc \!\!-\! SO_3^-Na^+$$

a sodium alkylbenzenesulfonate—
an anionic detergent

$$CH_3(CH_2)_{11}\overset{+}{N}(CH_3)_3Cl^-$$

a trimethylalkylammonium
ion—a cationic detergent

$$CH_3(CH_2)_8O(CH_2CH_2O)_nH$$

a nonionic detergent

The phospholipids all have small but very polar molecular parts that are extremely important in the formation of cell membranes. We will survey their structures largely to demonstrate how the phospholipids can be both polar and hydrocarbon-like.

The Glycerophospholipids (Phosphoglycerides) Have Phosphate Units Plus Two Acyl Units The **glycerophospholipids** occur in two broad types, the *phosphatides* and the *plasmalogens*. Both are esters of glycerol. Molecules of the **phosphatides** have two ester bonds to fatty acids plus one ester bond to phosphoric acid. The phosphoric acid unit, in turn, is joined by a phosphate ester link to a small alcohol molecule. Without this link, the compound is called *phosphatidic acid*.

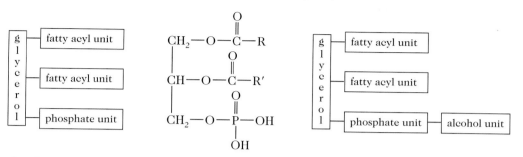

phosphatidic acid components   phosphatidic acid   phosphatide components

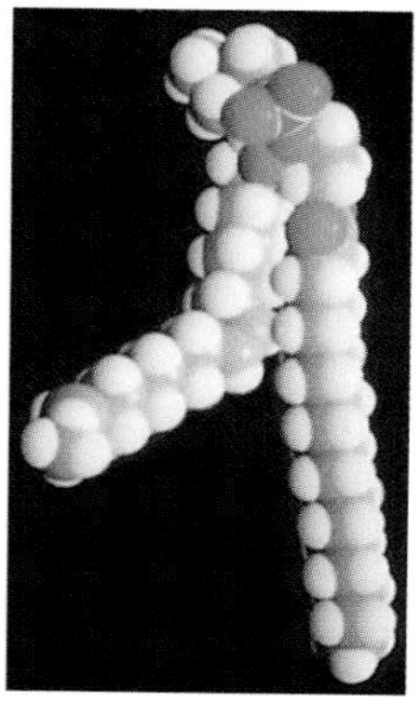

A typical phosphatide.

Three particularly important phosphatides are esters between phosphatidic acid and either choline, ethanolamine, or serine, forming, respectively, phosphatidylcholine (lecithin), **2**, phosphatidylethanolamine (cephalin), **3**, and phosphatidylserine, **4**.

$$HOCH_2CH_2\overset{+}{N}(CH_3)_3$$
choline

$$HOCH_2CH_2NH_2$$
ethanolamine

$$HOCH_2\underset{\underset{NH_3^+}{|}}{CH}CO_2^-$$
serine

As the structures of **2**, **3**, and **4** given below show, one part of each phosphatide molecule is very polar because it carries full electrical charges. These are partly responsible for the somewhat greater solubility in water of phosphatides compared to triacylglycerols. The remainder of a phosphatide molecule is nonpolar and hydrocarbon-like, so phosphatides can be extracted from animal matter by relatively nonpolar solvents.

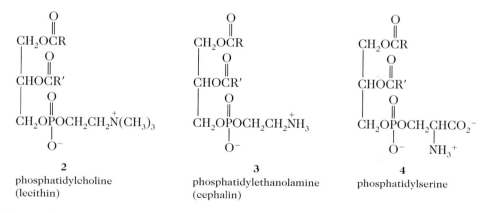

2
phosphatidylcholine
(lecithin)

3
phosphatidylethanolamine
(cephalin)

4
phosphatidylserine

■ *Cephalin* is from the Greek *kephale,* head. Cephalin is found in brain tissue.

These three are the most common hydrolyzable lipids used to make animal cell membranes.

Lecithin is a powerful emulsifying agent for triacylglycerols, and this is why egg yolks, which contain lecithin, are used to make the emulsions found in mayonnaise, ice cream, candies, and cake dough. When pure, lecithin is a clear, waxy, and very hygroscopic solid. In air, it is quickly attacked by oxygen, which makes it turn brown in a few minutes.

■ *Lecithin* is from the Greek *lekithos,* egg yolk—a rich source of this phospholipid.

The Plasmalogens Have Both Ether and Ester Groups   The **plasmalogens,** as we said, make up another family of glycerophospholipids, and they occur widely in the membranes of both nerve and muscle cells. They differ from the phosphatides by the presence of an unsaturated *ether* group instead of an acyl group at one end of the glycerol unit.

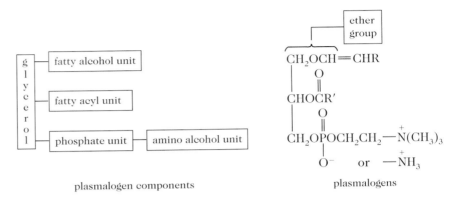

plasmalogen components   plasmalogens

Plasmalogen molecules, like phosphatides, also carry electrically charged positions as well as long hydrocarbon chains.

The Sphingolipids Are Based on Sphingosine, Not Glycerol   The two types of sphingosine-based lipids or **sphingolipids** are the *sphingomyelins* and the *cerebrosides,* and they are also important constituents of cell membranes, particularly those of nerve cells. The sphingomyelins are phosphate diesters of sphingosine. Their acyl units occur as acylamido parts, and they come from unusual fatty acids that are not found in neutral fats. Like the molecules of the phosphatides, those of the sphingolipids have two nonpolar tails.

The cerebrosides are not phospholipids. Instead they are **glycolipids,** lipids with a sugar unit (i.e., galactose or glucose) and not a phosphate ester system. The sugar unit, with its many OH groups, provides a strongly polar hydrophilic site, and it is usually a D-galactose or a D-glucose unit, or an amino derivative of these. The cerebrosides are prevalent in the membranes of brain cells.

A sphingomyelin.

sphingolipid components   sphingomyelins   cerebrosides

19.4 STEROIDS

Cholesterol and other steroids are nonhydrolyzable lipids.

Steroids are aliphatic compounds of high formula mass whose molecules include a characteristic four-ring feature called the *steroid nucleus*, structure **5**. It has three six-membered rings and one five-membered ring.

■ Steroid alcohols are called *sterols*.

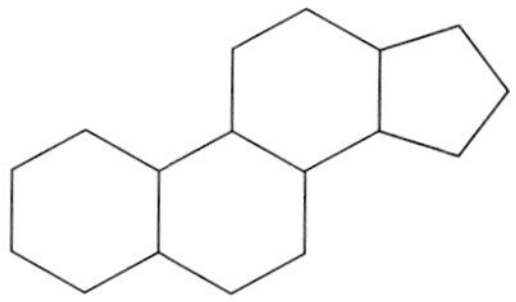

5
steroid nucleus

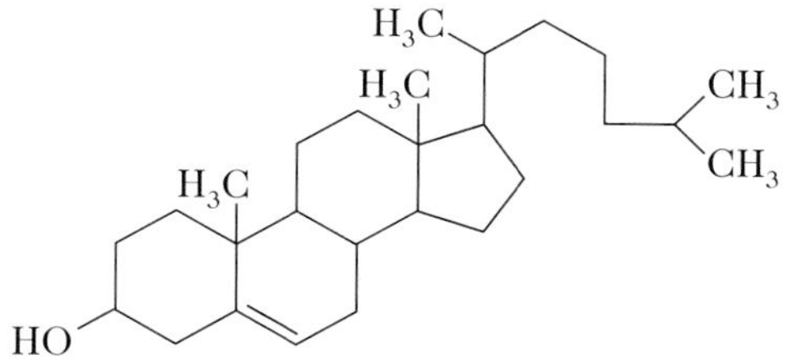

cholesterol

Cholesterol.

Several steroids and their functions are given in Table 19.3, where you can learn of the diverse roles they have in the body.

TABLE 19.3   Important Steroids

Vitamin D_3 Precursor

Irradiation of this derivative of cholesterol by ultraviolet light opens one of the rings to produce vitamin D_3. Meat products are sources of this compound.

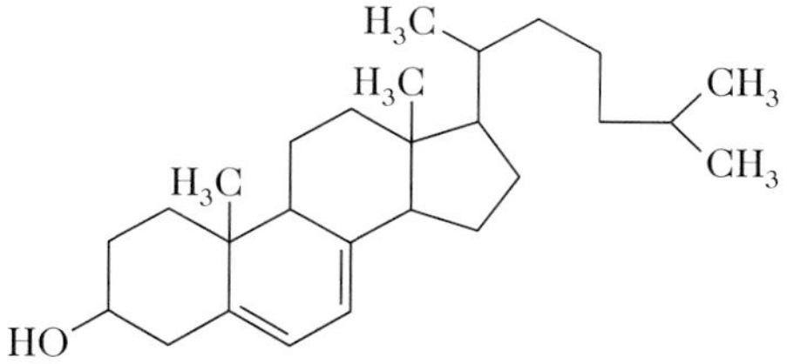

7-dehydrocholesterol

ultraviolet radiation

Vitamin D_3 is an antirachitic factor. Its absence leads to rickets, an infant and childhood disease characterized by faulty deposition of calcium phosphate and poor bone growth.

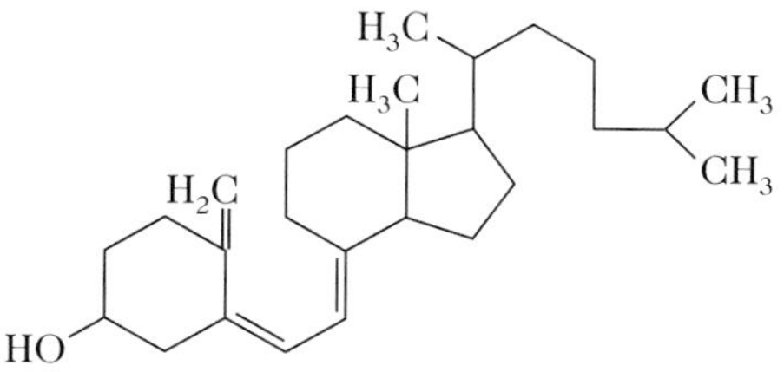

vitamin D_3

Adrenocortical Hormone

Cortisol is one of the 28 hormones secreted by the cortex of the adrenal gland. Cortisone, very similar to cortisol, is another such hormone. When cortisone is used to treat arthritis, the body changes much of it to cortisol by reducing a keto group to the 2° alcohol group that you see in the structure of cortisol.

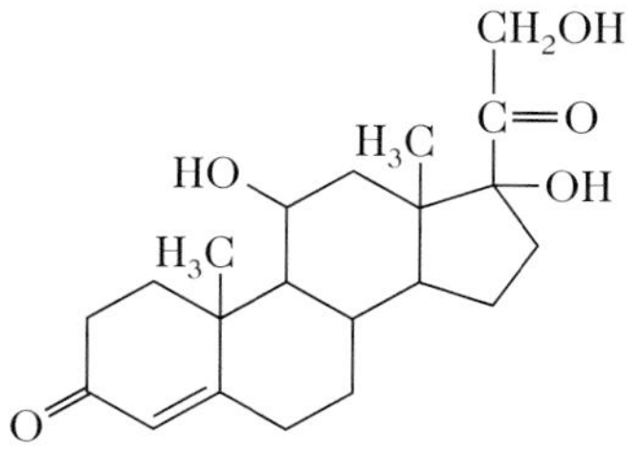

cortisol

Sex Hormones

Estradiol is a human estrogenic hormone.

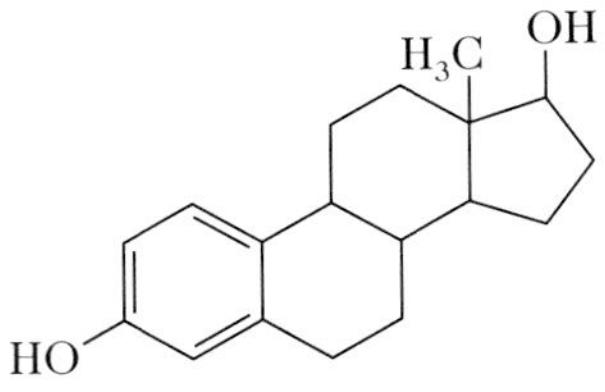

estradiol

Progesterone, a human pregnancy hormone, is secreted by the corpus luteum.

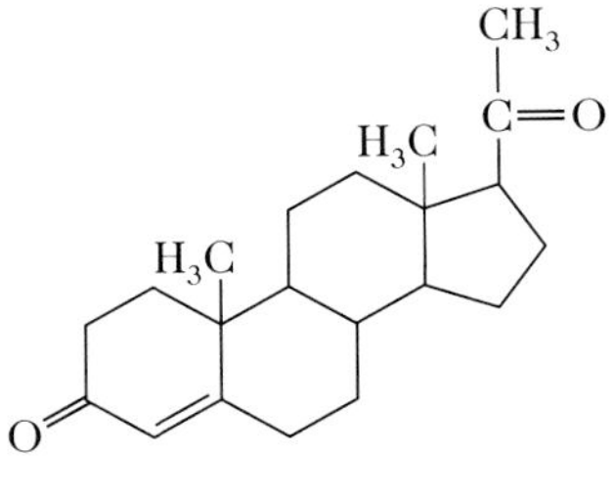

progesterone

Testosterone, a male sex hormone, regulates the development of reproductive organs and secondary sex characteristics.

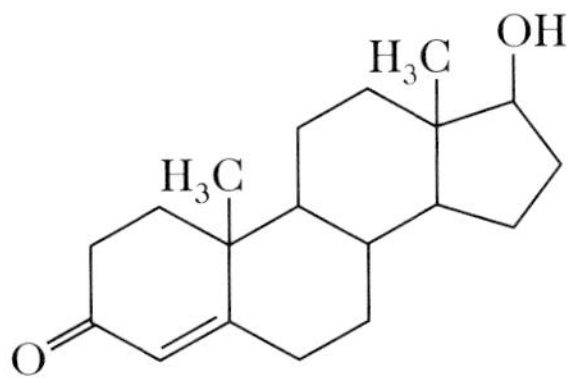

testosterone

Androsterone is another male sex hormone.

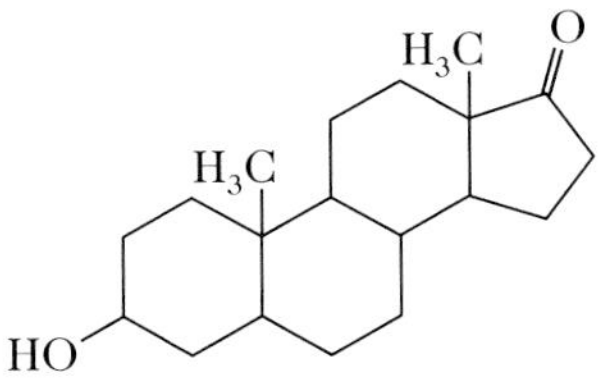

androsterone

Synthetic Hormones in Fertility Control

Most oral contraceptive pills contain one or two synthetic, hormone-like compounds. (Synthetics must be used because the real hormones are broken down in the body.)

Synthetic Estrogens

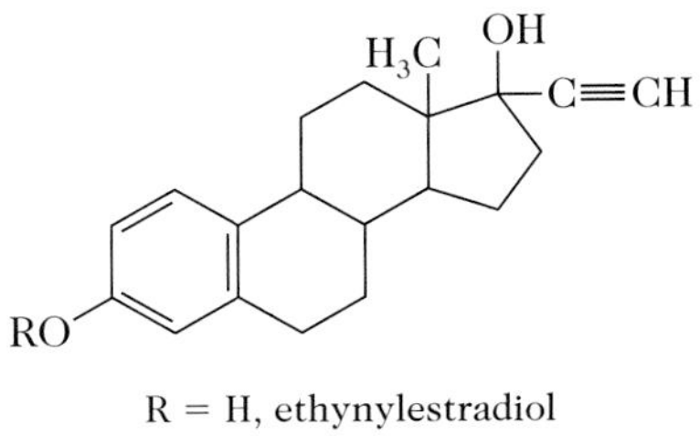

R = H, ethynylestradiol
R = CH$_3$, mestranol

Synthetic Progestin

The most widely used pills have a combination of an estrogen and a progestin. The triphasic pills (introduced in 1984) provide three levels of hormones over the course of one month.

An antiprogesterone called mifepristone, or RU 486, is a relatively new development in birth control technology. It acts to prevent the implantation of a fertilized ovum in the uterus (see Interaction 20.3).

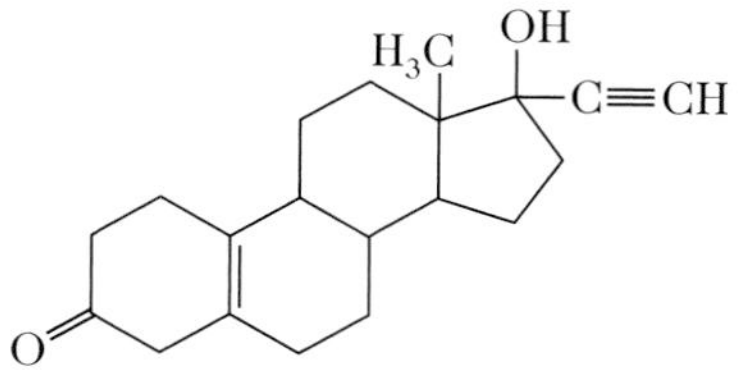

norethynodrel

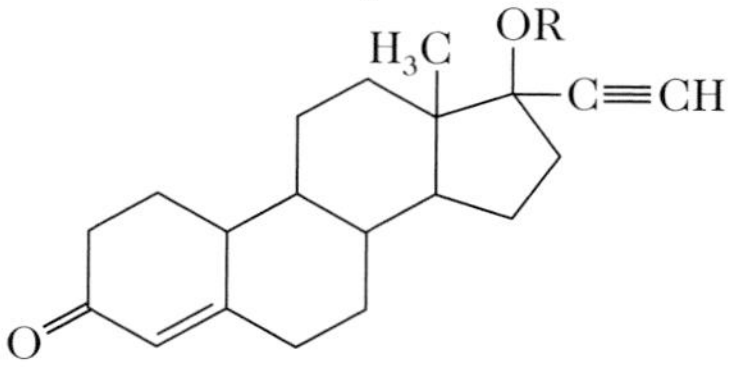

R = H, norethindrone

R = COCH$_3$, norethindrone acetate

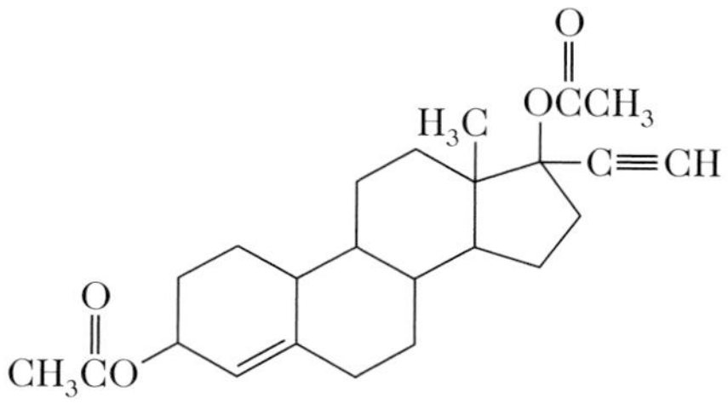

ethynodiol diacetate

Cholesterol Molecules Are Components of Cell Membranes   Cholesterol is an unsaturated steroid alcohol that makes up a significant part of the membranes of animal cells. The membrane of a human red blood cell (erythrocyte), for example, is about 25% cholesterol by mass. Cholesterol is thus not a demon molecule; it's essential to all cell membranes. *Too much cholesterol* is the demon, particularly when the mechanism for eliminating the excess is not working well (which we'll study in a later chapter). Cholesterol, moreover, is the body's raw material for making bile salts and steroid hormones, including the sex hormones listed in Table 19.3. Little cholesterol is found in plants, but they have compounds with similar structures.

Cholesterol enters the body via the diet, but up to 800 mg per day can be synthesized in the liver from two-carbon acetate units. Some cholesterol made in the liver is converted to esters of cholesterol and some is used to make the bile salts, like sodium cholate (see below).

Dietary cholesterol is put into circulation in the bloodstream as components of *very low density lipoproteins* (or VLDL). Esters of cholesterol occur in higher density complexes. The relationships between cholesterol, esters of cholesterol, the various kinds of lipoproteins, and the risk of heart disease will be discussed when we study the metabolism of lipids. We need to know more about proteins, genes, and enzymes first.

■   Cholesterol is the chief constituent in gallstones.

■   *Lipoprotein* molecules are combinations of lipids and proteins.

sodium cholate
(a bile salt)

cholesteryl ester
(R is long chain)

Bile Salts Are Detergents   Bile salts made in the liver are secreted into the intestinal tract where they function as powerful surface-active agents (detergents). They aid both in the digestion of dietary lipids and also the absorption of fat-soluble vitamins and fatty acids from the digestive tract into circulation.

19.5   CELL MEMBRANES—THEIR LIPID COMPONENTS

Cell membranes consist of a lipid bilayer containing molecules of proteins and cholesterol.

Cell membranes are made of both lipids and proteins. The lipid components are what keep a cell's insides in and its outsides out. The protein components perform services such as accepting hormone molecules and relaying them or their "messages" inside; providing passages—closable molecular channels—for small ions and molecules; and acting as "pumps" to move solutes across a cell membrane. We can only introduce the general features of animal cell membranes in this section. The services performed by the membrane proteins will be described in more detail after we have studied proteins and enzymes.

■ *Hydrophilic—from the Greek hydor,* water, and *philos,* loving. *Hydrophobic—from the Greek phobikos,* hating.

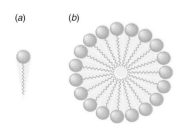

(a)   *(b)*

FIGURE 19.2   Detergent micelle. *(a)* Space-filling requirements of an amphipathic detergent or soap molecule with one hydrophobic tail (wavy line) and a hydrophilic head (blue sphere). *(b)* Micelle in water. The hydrophobic tails gather together as the hydrophilic heads have maximum exposure to the aqueous medium.

Both Hydrophilic and Hydrophobic Groups Are Necessary for Cell Membranes The principal lipids of animal cell membranes are not triacyglycerols but more complex lipids, like phospholipids and glycolipids, as well as cholesterol. The polar or ionic sites of phospholipids and glycolipids are called **hydrophilic groups,** because they are able to attract water molecules. Hydrophilic groups force their molecules to take up positions in membranes that let the groups be in maximum contact with the water of the fluids both inside and outside the cell.

The nonpolar, hydrocarbon sections of membrane lipids are called **hydrophobic groups** because they are water avoiding. They tend to force their molecules to let them be positioned *within* a cell membrane so that they are out of contact with water as much as possible.

Substances like the phospholipids or glycolipids, with both hydrophilic and hydrophobic groups, are called **amphipathic compounds.** (Soaps and detergents are also examples.) When an amphipathic compound is mixed in the right proportion with water, its molecules spontaneously group into *micelles.* A **micelle** is a globular collection of amphipathic molecules in which hydrophobic contacts are minimized and hydrophilic interactions are maximized. Figure 19.2 shows how a micelle forms when the amphipathic molecules have a *single* hydrocarbon "tail," like a detergent or soap molecule.

In the Lipid Bilayer of Cell Membranes, Hydrophobic Groups Intermingle between the Membrane Surfaces As we said, molecules of phospholipids and glycolipids have two hydrophobic hydrocarbon "tails." When made of such molecules, a micelle takes up an extended disklike shape (Figure 19.3). Further extension of the shape shown in the figure produces a **lipid bilayer,** which consists of two rows of lipid molecules aligned side by side in a sheetlike array. The lipid bilayer is the basic architecture of an animal cell membrane (Figure 19.4). The hydrophobic molecular "tails" intermingle in the center of the bilayer so that they can get as far away from water molecules as possible. In a sense, these "tails" dissolve in each other (following the "like-dissolves-like" rule). The hydrophilic "heads" stick out into the aqueous phase to be in contact with water as much as possible. *These water-avoiding and water-attracting properties, not covalent bonds, are thus the major "forces" that stabilize the cell membrane.*

Cholesterol Molecules Also Help to Stabilize Membranes Cholesterol molecules are somewhat long and flat. In the lipid bilayer, they occur with their long axes lined up side by side with the hydrocarbon chains of the other lipids. The cholesterol OH groups are hydrogen-bonded to O atoms of ester groups of the membrane lipid molecules. Because the cholesterol units are relatively rigid, much more so than the fatty acid chains, cholesterol molecules help to keep a membrane from being too fluidlike.

The Lipid Bilayer Is Self-Sealing If a pin were stuck through a cell membrane and then pulled out, the lipid layer would close back spontaneously. This flexibility is allowed because, as we said, no covalent bonds hold neighboring lipid molecules to each other. Only the net forces of attraction that we imply when we use the terms *hydrophobic* and *hydrophilic* are at work. Yet the bilayer is strong enough to hold a cell together, and it is flexible enough to let things in and out. Water molecules move back and forth easily through the membrane, but other molecules and ions are vastly less free to move. Their migrations depend on the protein components of the membrane, also indicated in Figure 19.4. We must, therefore, interrupt our discussion of membranes until we have learned more about proteins, particularly such families of proteins as enzymes and receptors.

(a)

(b)

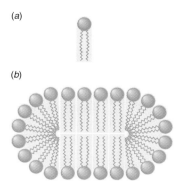

FIGURE 19.3 Phosphoglyceride micelle in water. (*a*) Space-filling requirements for an amphipathic phosphoglyceride, which has two hydrophobic tails (wavy lines) and a hydrophilic head (blue sphere). (*b*) A disklike micelle whose "wall" for the most part (top and bottom segments) is a lipid bilayer.

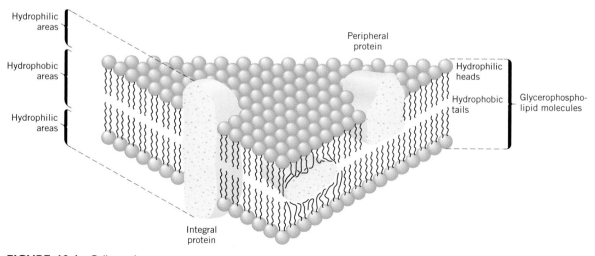

FIGURE 19.4 Cell membrane.

SUMMARY

Lipids   Lipids are ether-extractable substances in animals and plants, and they include hydrolyzable esters and non-hydrolyzable compounds. The esters are generally of glycerol or sphingosine with their acyl portions contributed by long-chain carboxylic acids called fatty acids. The fatty acids obtained from lipids by hydrolysis generally have long chains of even numbers of carbons, seldom are branched, and often have one or more alkene groups. The alkene groups are cis. Because molecules of all lipids are mostly hydrocarbon-like, lipids are soluble in nonpolar solvents but not in water.

Triacylglycerols   Molecules of neutral fats, those without electrically charged sites or sites that are similarly polar, are esters of glycerol and a variety of fatty acids, both saturated and unsaturated. Vegetable oils have more double bonds per molecule than animal fats. The triacylglycerols can be hydrogenated, hydrolyzed, and saponified.

Waxes   Molecules of the waxy coatings on leaves and fruit, or in beeswax or sebum, are simple esters between long-chain alcohols and fatty acids.

Glycerophospholipids   Molecules of the glycerophospholipids are esters both of glycerol and of phosphoric acid. A second ester bond from the phosphate unit goes to a small alcohol molecule that can also have a positively charged group. Thus this part of a glycerophospholipid is strongly hydrophilic. The two types of glycerophospholipids are the phosphatides and the plasmalogens. Both are vital to animal cell membranes. In phosphatide molecules there are two fatty acyl ester units besides the phosphate system. In plasmalogens, there is one fatty acyl unit and a long-chain, unsaturated ether unit in addition to the phosphate system.

Sphingomyelins   Sphingomyelins are esters of sphingosine, a dihydric amino alcohol. They also have a strongly hydrophilic phosphate system.

Glycolipids   Also sphingosine based, the glycolipids use a monosaccharide instead of the phosphate-to-small-alcohol unit to provide the hydrophilic section. Otherwise, they resemble the sphingomyelins.

Steroids   Steroids are nonhydrolyzable lipids with the steroid nucleus of four fused rings (three being C_6 rings and one a C_5 ring). Several steroids are sex hormones, and oral fertility control drugs mimic their structure and functions. Cholesterol, the raw material used by the body to make bile salts and other steroids, is manufactured in the liver. Cholesterol is carried in circulation as cholesteryl esters in lipoprotein complexes. Cholesterol molecules are essential components of animal cell membranes.

Animal cell membranes   A double layer of phospholipids or glycolipids plus cholesterol and assemblies of protein molecules make up the cell membrane. The hydrophobic tails of the amphipathic lipids intermingle within the bilayer, away from the aqueous phase. The hydrophilic heads are in contact with the aqueous medium. Cholesterol molecules help to stiffen the membrane.

REVIEW EXERCISES

The answers to Review Exercises whose numbers are in color are found in Appendix E. The answers to the other Review Exercises are found in the Study Guide that accompanies this book. The more challenging questions are marked with asterisks.

Lipids in General

19.1   Gasoline is soluble in nonpolar solvents, yet it isn't classified as a lipid. Explain.

19.2   Cholesterol has no ester group, yet we classify it as a lipid. Why?

19.3   Ethyl acetate has an ester group, but it isn't classified as a lipid. Explain.

19.4   What are the criteria for deciding whether a substance is a lipid?

Fatty Acids

19.5   What are the structures and the names of the two most abundant *saturated* fatty acids?

19.6   Write the structures and names of the *unsaturated* fatty acids that have 18 carbons each and that have no more than three double bonds. Show the correct geometry at each double bond.

19.7   Write the equations for the reactions of myristic acid with
(a) NaOH(*aq*)
(b) CH_3OH (when heated in the presence of acid).

19.8   What are the equations for the reactions of oleic acid with each substance?
(a) Br_2
(b) NaOH(*aq*)
(c) H_2 (in the presence of a catalyst)
(d) CH_3OH (heated with an acid)

19.9 Which of the following acids, **A** or **B**, is more likely to be obtained by the hydrolysis of a lipid? Explain.

$$CH_3(CH_2)_{12}CO_2H \qquad CH_3\overset{\overset{\displaystyle CH_3}{\displaystyle |}}{CH}(CH_2)_{11}CO_2H$$

$$\mathbf{A} \qquad\qquad\qquad \mathbf{B}$$

19.10 Without writing structures, state what kinds of chemicals the prostaglandins are.

Triacylglycerols

19.11 Write the structure of a triacylglycerol that involves linolenic acid, linoleic acid, and palmitic acid, besides glycerol.

19.12 What is the structure of a triacylglycerol made from glycerol, stearic acid, oleic acid, and palmitic acid?

19.13 Write the structures of all the products that would form from the complete hydrolysis of the following lipid. (Show the free carboxylic acids, not their anions.)

$$\begin{array}{l} \overset{\overset{\displaystyle O}{\displaystyle \|}}{CH_2OC}(CH_2)_7CH{=}CHCH_2CH{=}CH(CH_2)_4CH_3 \\[2pt] \overset{\overset{\displaystyle O}{\displaystyle \|}}{CHOC}(CH_2)_{14}CH_3 \\[2pt] \overset{\overset{\displaystyle O}{\displaystyle \|}}{CH_2OC}(CH_2)_7CH{=}CH(CH_2)_7CH_3 \end{array}$$

19.14 Write the structures of the products that are produced by the saponification (by NaOH) of the triacylglycerol whose structure was given in Review Exercise 19.13.

*19.15 The hydrolysis of a lipid produced glycerol, myristic acid, linoleic acid, and oleic acid in equimolar amounts. Write a structure that is consistent with these results. Is there more than one structure (constitution) that can be written? Explain.

19.16 The hydrolysis of 1 mol of a lipid gives 1 mol each of glycerol and oleic acid and 2 mol of stearic acid. The lipid molecule is chiral. Write its structure. Is more than one structure possible? Explain.

19.17 What is the structural difference between the triacylglycerols of the animal fats and the vegetable oils?

19.18 Products such as corn oil are advertised as being "polyunsaturated." What does this mean in terms of the structures of the molecules that are present? Corn oil is "more polyunsaturated" than what?

19.19 What chemical reaction is used to make oleomargarine?

19.20 Lard and butter are chemically almost the same substances, so what is it about butter that makes it so much more desirable a spread for bread than, say, lard or tallow?

Waxes

19.21 One component of beeswax has the formula $C_{36}H_{72}O_2$. When it is hydrolyzed, it gives $C_{18}H_{36}O_2$ and $C_{18}H_{38}O$. Write the most likely structure of this compound.

*19.22 When all the waxes from the leaves of a certain shrub are separated, one has the formula of $C_{60}H_{120}O_2$. Its structure is **A**, **B**, or **C**. Which is it most likely to be? Explain why the others can be ruled out.

$$CH_3(CH_2)_{56}CO_2CH_2CH_3 \qquad CH_3(CH_2)_{29}CO_2(CH_2)_{28}CH_3$$

$$\mathbf{A} \qquad\qquad\qquad\qquad \mathbf{B}$$

$$CH_3(CH_2)_{28}CO_2(CH_2)_{29}CH_3$$

$$\mathbf{C}$$

Phospholipids

19.23 Why are the phosphatides and plasmalogens both called glycerophospholipids?

19.24 What site in a glycerophospholipid carries a negative charge? What atom carries a positive charge?

19.25 In general terms, how are the sphingomyelins and cerebrosides structurally alike? How are they structurally different?

19.26 What structural unit provides the most polar groups in a molecule of a glycolipid? (Name it.)

19.27 Phospholipids are not classified as neutral fats. Explain.

19.28 Phospholipids are common in what part of a cell?

19.29 What are the names of the two types of sphingosine-based lipids?

19.30 Are the sugar units that are incorporated into the cerebrosides bound by glycosidic links or by ordinary ether links? How can one tell? Which kind of link is more easily hydrolyzed (assuming an acid catalyst)?

*19.31 The complete hydrolysis of 1 mol of a phospholipid gave 1 mol each of the following compounds: glycerol, linolenic acid, oleic acid, phosphoric acid, and the cation, $HOCH_2CH_2N(CH_3)_3{}^+$.
(a) Write a structure of this phospholipid that is consistent with the information given.
(b) Is the substance a glycerophospholipid or a sphingolipid? Explain.
(c) Are its molecules chiral or not? How can you tell?
(d) Is it lecithin or a cephalin? How can you tell?

19.32 When 1 mol of a certain phospholipid was hydrolyzed, there was obtained 1 mol each of lauric acid, oleic acid, phosphoric acid, glycerol, and $HOCH_2CH_2NH_2$.
(a) What is a possible structure for this phospholipid?
(b) Is it a sphingolipid or a phosphoglyceride? Explain.
(c) Can its molecules exist as enantiomers or not? Explain.
(d) Is it a cephalin or a lecithin? Explain.

Steroids

19.33 What is the steroid detergent in our bodies?

19.34 What is the name of a vitamin that is made in our bodies from a dietary steroid by the action of sunlight on the skin?

19.35 Give the names of three steroidal sex hormones.

19.36 What is the name of a steroid that is part of the cell membranes in animal tissues?

19.37 What is the raw material used by the body to make bile salts? How does the body use the bile salts?

19.38 How does the body carry cholesterol in circulation in the bloodstream?

Cell Membranes

19.39 Describe in your own words what is meant by the *lipid bilayer* structure of cell membranes.

19.40 How do the hydrophobic parts of phospholipid molecules avoid water in a lipid bilayer?

19.41 Besides lipids, what kinds of substances are present in a cell membrane?

19.42 What kinds of forces hold a cell membrane together?

19.43 What functions do the proteins of a cell membrane serve?

The Prostaglandins (Interaction 19.1)

19.44 Name the fatty acid that is used to make prostaglandins.

19.45 What effect is aspirin believed to have on prostaglandins, and how is this related to aspirin's value in working against (a) pain and fever and (b) heart attacks?

The Omega-3 Fatty Acids and Heart Disease (Interaction 19.2)

19.46 What is it about the structure of linolenic acid that lets us call it an omega-3 acid?

19.47 What source of the omega-3 acids is relatively rich in the C_{20} and C_{22} acids?

19.48 Why have the omega-3 acids aroused the interests of people in nutrition and in medicine?

Olestra (Interaction 19.3)

19.49 Although Olestra molecules are loaded with nutritionally useful fatty acids, they cannot be digested. Explain.

Detergent Action (Interaction 19.4)

19.50 What kind of chemical is soap?

19.51 For household laundry work, which product is generally preferred, a synthetic detergent or soap? Why?

19.52 Why are soap and sodium alkyl sulfates called *anionic* detergents?

19.53 Explain in your own words how a detergent can loosen oils and greases from fabrics.

Additional Exercises

19.54 Examine the following structure.

$$
\begin{array}{l}
\text{CH}_2\text{OC(CH}_2)_7\text{CH}\!=\!\text{CHCH}_2\text{CH}\!=\!\text{CH(CH}_2)_5\text{CH}_3 \\
\quad\ \ \overset{\text{O}}{\overset{\|}{}} \\
\text{CHOC(CH}_2)_{11}\text{CH}_3 \\
\quad\ \ \overset{\text{O}}{\overset{\|}{}} \\
\text{CH}_2\text{OC(CH}_2)_7\text{CH}\!=\!\text{CH(CH}_2)_8\text{CH}_3
\end{array}
$$

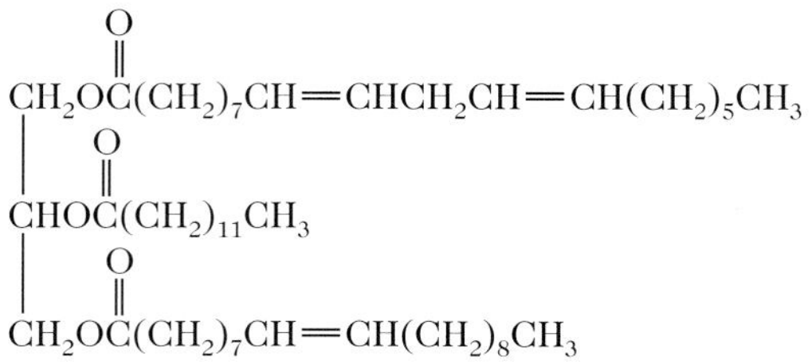

(a) Is it a triester of glycerol?
(b) Does it have hydrophobic groups?
(c) What are its hydrophilic functional groups?
(d) What would form if all of its alkene groups were hydrogenated?
(e) Is this molecule likely to be found among naturally occurring triacylglycerols? Explain.

*19.55 Examine the following structure.

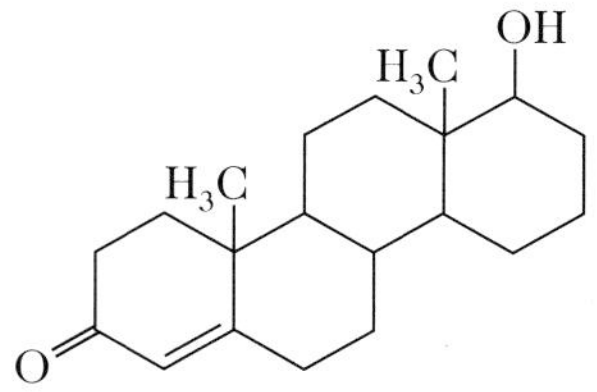

(a) Is it amphipathic? Explain.
(b) Is it a member of the steroid family? How can one tell?

PROTEINS

20

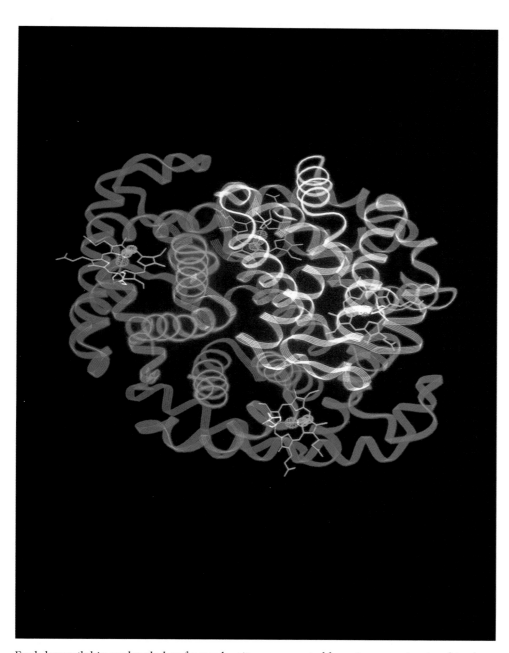

Each hemoglobin molecule has four subunits, represented here in computer graphics by twisted coils, two blue and two purple. Heme molecules are in red, one per coil, and each carries an oxygen-binding site. Hemoglobin is important in our study for two reasons. We couldn't live without it, and its molecules represent every subtlety in protein structure that we'll study, including the twisting of coiled molecules.

THIS CHAPTER IN CONTEXT

Proteins constitute about half of the body's dry mass, and they're found in virtually all parts of all cells. They give strength and elasticity to skin and blood vessels. Proteins of muscles and tendons act as cables to help us move our bones. Other proteins reinforce the bones themselves, as well as teeth, much as steel rods reinforce concrete. Antibody proteins protect us against disease. Hemoglobin, albumins, and globulins—all proteins in blood—are haulers of otherwise blood-insoluble oxygen and lipids. Nearly all enzymes, some hormones and neurotransmitters, and cell membrane receptors are proteins. No other class of compounds is involved in such a variety of functions, all essential to life. They deserve the name *protein,* taken from the Greek *proteios,* "of the first rank," so they merit our careful study. If you look for it, you'll see beauty here, too, because protein structure is elegant.

20.1 AMINO ACIDS. THE BUILDING BLOCKS OF PROTEINS

Living things select from about 20 α-amino acids to make the polypeptides in proteins.

■ The chief elements in proteins are C, H, O, N, and S.

The dominant structural units of **proteins** are polymers of very high formula mass called **polypeptides.** Metal ions and small organic molecules or ions are often present also. The relationship of these parts to whole proteins is shown in Figure 20.1. Many proteins, however, are made entirely of polypeptide molecules.

■ With few exceptions, the amino acids not in the standard set consist of modifications of standard amino acid molecules made *after* a polypeptide has been put together.

Polypeptides Are Made from α-Amino Acids   The monomers of polypeptides are α-amino acids, having the general structure **1**. Twenty make up the "standard" set (Table 20.1), and in any given polypeptide most are used many times. The same set of 20 standard amino acids is used by all species of plants and animals. A few others are present in certain tissues as well as in some bacteria.

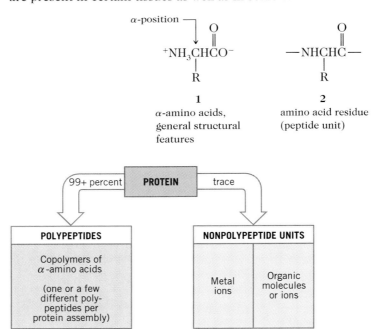

FIGURE **20.1**   Components of proteins. Some proteins consist exclusively of polypeptide molecules, but most also have nonpolypeptide units such as small organic molecules or metal ions, or both.

TABLE 20.1 Amino Acids: $^+NH_3CHCO_2^-$
$$|$$
$$R$$

Type	Side Chain, R	Name	Three-Letter	One-Letter	pI
Side chain is nonpolar	—H	Glycine	Gly	G	6.06
	—CH_3	Alanine	Ala	A	6.11
	—$CH(CH_3)_2$	Valine	Val	V	6.00
	—$CH_2CH(CH_3)_2$	Leucine	Leu	L	6.04
	—$CHCH_2CH_3$ $\mid$ CH_3	Isoleucine	Ile	I	6.04
	—$CH_2C_6H_5$	Phenylalanine	Phe	F	5.91
	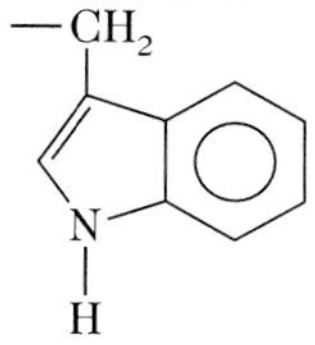 —CH_2 (indole)	Tryptophan	Trp	W	5.88
	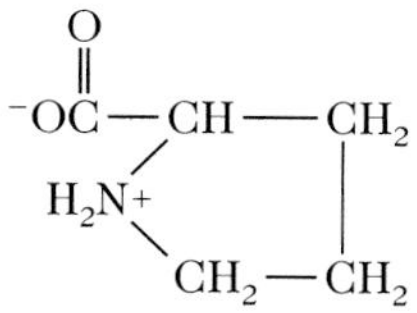 (complete structure, proline ring)	Proline	Pro	P	6.30
Side chain has a hydroxyl group	—CH_2OH	Serine	Ser	S	5.68
	—$CHOH$ $\mid$ CH_3	Threonine	Thr	T	5.64
	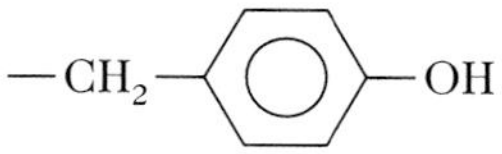 —CH_2—(phenyl)—OH	Tyrosine	Tyr	Y	5.63
Side chain has a carboxyl group or an amide group	—CH_2CO_2H	Aspartic acid	Asp	D	2.98
	—$CH_2CH_2CO_2H$	Glutamic acid	Glu	E	3.08
	—CH_2CONH_2	Asparagine	Asn	N	5.41
	—$CH_2CH_2CONH_2$	Glutamine	Gln	Q	5.65
Side chain has a basic amino group	—$CH_2CH_2CH_2CH_2NH_2$	Lysine	Lys	K	9.47
	—$CH_2CH_2CH_2NHCNH_2$ ($\overset{NH}{\parallel}$)	Arginine	Arg	R	10.76
	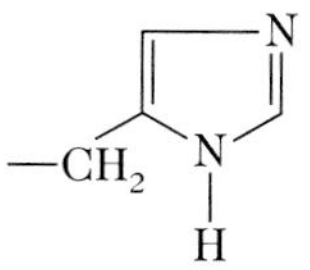 —CH_2—(imidazole)	Histidine	His	H	7.64
Side chain contains sulfur	—CH_2SH	Cysteine	Cys	C	5.07
	—$CH_2CH_2SCH_3$	Methionine	Met	M	5.74

Hundreds of **amino acid residues,** sometimes called **peptide units,** are joined in a single polypeptide molecule. Clearly, before studying how polypeptides are put together, we need to learn more about their monomers.

The Molecules of α-Amino Acids Exist in Dipolar Ionic Forms   As you can see in Table 20.1, the amino acids are all α-amino acetic acids with **side chains** or **R** groups at the alpha position. In the solid state, amino acids exist entirely in the form shown by **1,** called a **dipolar ion** or **zwitterion.** Although, overall, it is an electrically neutral particle, it has a positive and a negative charge on different sites. Thus, amino acids are exceedingly polar compounds and, like salts, they have melting points that are considerably higher than those of most molecular compounds. They also tend to be much more soluble in water than in nonpolar solvents.

Structure **1** is actually an *internally neutralized molecule.* We can imagine that **1** started out with a regular amino group, NH_2, plus an ordinary carboxyl group, CO_2H. But the amino group, a proton acceptor, took a proton from the carboxyl group, a proton donor, to give the dipolar ionic form, **1.** Of course, **1** has its own (weaker) proton donating group, NH_3^+, and its own (also weaker) proton accepting group, CO_2^-, so these dipolar ions can neutralize acids or bases of sufficient strength, like H_3O^+ and OH^-. Because polypeptides generally have at least one base-neutralizing NH_3^+ group and one acid-neutralizing CO_2^- group, *proteins are able to serve as buffers in body fluids.*

For amino acids to exist in water as dipolar ions, **1,** the pH has to be about 6 to 7. If we make the pH much lower (more acidic) or much higher (more basic), the form of the amino acid changes. For example, if we add enough strong acid like $HCl(aq)$ to a solution of an amino acid to lower the pH to about 1, the CO_2^- groups accept H^+ and change to CO_2H groups in structure **2.** As **2,** the molecules are positively charged (cations) and can migrate to a negative electrode (cathode) in an electrolysis experiment.

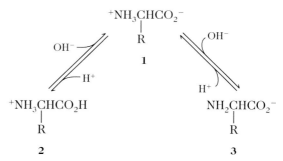

On the other hand, if we add enough strong base to an amino acid solution to raise the pH to about 11, then OH^- takes H^+ from NH_3^+ groups, making them NH_2 groups, and the amino acid molecules change to structure **3.** Because **3** is a negative ion (an anion), it can migrate to the positive electrode (anode) in an electrolysis apparatus.

A pH Exists for Each Amino Acid at Which No Net Migration in an Electric Field Occurs   In an aqueous solution of an amino acid, a dynamic equilibrium exists among **1, 2,** and **3.** Imagine that we now send an electric current through electrodes dipping into the solution. Cations of form **2** cannot help but migrate to the cathode. Anions of form **3** migrate to the anode. Neutral molecules, **1,** migrate nowhere. A molecule such as **1** with *equal* numbers of positive and negative charges is called an **isoelectric molecule,** and it cannot migrate in an electric field.

■   Hereafter, when we say "amino acid," we'll mean α-amino acid.

■   α-Amino acids melt around 300 °C but their simple esters, which cannot be dipolar ions, generally melt around 100 °C.

■   We can still call dipolar ions *molecules* because, overall, they are electrically neutral particles.

■   In an electrolysis experiment, the negatively charged electrode is called the *cathode,* so positively charged ions (*cations*) are attracted to it. Similarly, the positively charged electrode, the *anode,* attracts *anions,* ions bearing negative charge.

Because the equilibrium is *dynamic,* a migrating cation like **2** could, as it churns along, flip a proton to some acceptor, become **1** and isoelectric (neutral), and stop moving to the cathode. In another instant, it (now **1**) could shed still another proton, become **3**, and so be made to turn around and head for the anode! Similarly, an anion on its way to the anode might pick up a proton, become neutral, and also stop dead. Then it might take another proton, become **2**, and get turned around. In the meantime, an isoelectric molecule, **1**, might either donate or accept a proton, become electrically charged, and start its own migration. What a flurry of activity! The question is, what overall *net* migration occurs and how is this net effect influenced by the pH of the solution?

Although much coming and going occurs in an amino acid solution, the net molar concentrations of the species stay the same at equilibrium. If either **2** or **3** is in any molar excess because of the pH, then some *net* migration occurs toward one electrode or the other. When the net molar concentration of **2** (a cation), for example, is greater than that of **3** (an anion), some statistical net movement of amino acid units to the cathode occurs.

Remember, however, that these equilibria can be shifted by adding acid or base. By carefully adjusting the pH, in fact, we can so finely tune the concentrations at equilibrium that *no net migration occurs.* At the right pH, the rates of proton exchange are such that each unit that is not **1** spends an equal amount of time as **2** and as **3**. (And the concentrations of **2** and **3** are very low.) Thus, any net migration to one electrode is blocked.

The pH at which no net migration of an amino acid can occur in an electric field is called the **isoelectric point** of the amino acid, and the symbol of this pH value is **pI** (see Table 20.1, last column). But, you ask, what has all this got to do with proteins?

Proteins, Like Amino Acids, Have Isoelectric Points

As we will soon see, all proteins have NH_3^+ or CO_2^- groups or can acquire them by a change in the pH of the surrounding medium. Whole protein molecules, therefore, can also be isoelectric at the right pH. *Each protein thus has its own isolectric point.* Now think of what can happen if the pH of a solution of a protein is changed to have some value other than the pI value of the protein. The entire electrical condition of the huge protein molecules can be made either cationic or anionic almost instantly, at room temperature, by adding strong acid or base, that is, by changing the pH of the medium. *Such changes in the electrical charge of a protein have serious consequences at the molecular level of life.*

Being electrically charged can dramatically affect chemical reactions, for example, or greatly alter protein solubility. If proteins are to serve their biological purposes, some must not be allowed to go into solution and others must not be permitted to precipitate. We'll return to this concept in this and later chapters, *but the discussion focuses our attention again on how important it is that an organism control the pH values of its fluids.*

But let's get back to the amino acids. We will next survey the types of amino acid side chains and how they affect the properties of polypeptides and proteins. These include how a polypeptide molecule will spontaneously fold and twist into its distinctive and vital final shape. You should memorize the side chains of a minimum of five amino acids: glycine, alanine, cysteine, lysine, and glutamic acid are suggested because they represent *types.* The next example shows how to use this information to write a complete amino acid structure.

■ These shifts of H^+ ions illustrate Le Châtelier's principle at work.

■ Casein, the protein in milk, precipitates when milk turns sour because a change in pH causes the casein molecules to become isoelectric.

■ The amino acid proline is the only one of the standard 20 in which the α-amino group is itself joined to one end of the side chain.

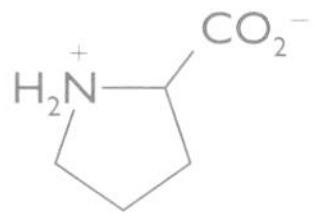

proline

EXAMPLE 20.1   Writing the Structure of an Amino Acid

What is the structure of cysteine?

ANALYSIS   Regard any α-amino acid as having two structural features, a unit common to *all* α-amino acids plus a unique side chain at the α-position. So write the common unit first and then add the side chain.

SOLUTION   The common unit for all but one amino acid (proline) is

$$^+NH_3CHCO_2{}^-$$

Next, either look up or recall from memory the side chain for the particular amino acid. For cysteine, this is CH_2SH, so simply attach this group to the α-carbon. Cysteine is

$$^+NH_3CHCO_2{}^-$$
$$CH_2SH$$

■   PRACTICE EXERCISE 1   Write the structures of the dipolar ionic forms of glycine, alanine, lysine, and glutamic acid. (Strive to do this from memory.)

Several Amino Acids Have Hydrophobic Side Chains   The first amino acids in Table 20.1, including alanine, have essentially nonpolar, hydrophobic side chains. When a long polypeptide molecule folds into its distinctive shape, these hydrophobic groups tend to be folded next to each other as much as possible rather than next to highly polar groups or to water molecules.

Water avoidance by nonpolar side chains is called the **hydrophobic interaction** of side chains, and two factors are at work. One is the strong tendency of water molecules to form hydrogen bonds between each other and so "reject" molecules that cannot themselves "offer" hydrogen bonds to water. Water molecules, in a sense, club together forcing nonpolar groups to stay away and be by themselves. The other factor in hydrophobic interactions is that of *London forces* of attraction. These are attractive forces between groups made possible by temporary dipoles, dipoles that increase and collapse as the molecular electron clouds squeeze each other during collisions (or near misses). The larger the side chain is, the stronger the London force is.

Some Amino Acids Have Hydrophilic OH Groups on Side Chains   The second set of amino acids in Table 20.1 are those whose side chains bear alcohol or phenol OH groups. They are polar and hydrophilic, and they can donate and accept hydrogen bonds. As a long polypeptide chain folds into its final shape, side chains with OH groups tend to stick out into the surrounding aqueous phase, to which they are attracted by hydrogen bonds.

Two Amino Acids Have Side-Chain Carboxyl Groups   The side chains of aspartic and glutamic acid carry proton donating CO_2H groups. Because body fluids are generally slightly basic, they actually occur mostly as $CO_2{}^-$ groups, their protons having been neutralized. The solution would have to be made quite acidic to prevent this, that is, to force protons back to make $CO_2{}^-$ groups stay as CO_2H. This is why the pI

■   Aspartame, a popular artificial sweetener, has an aspartic acid residue.

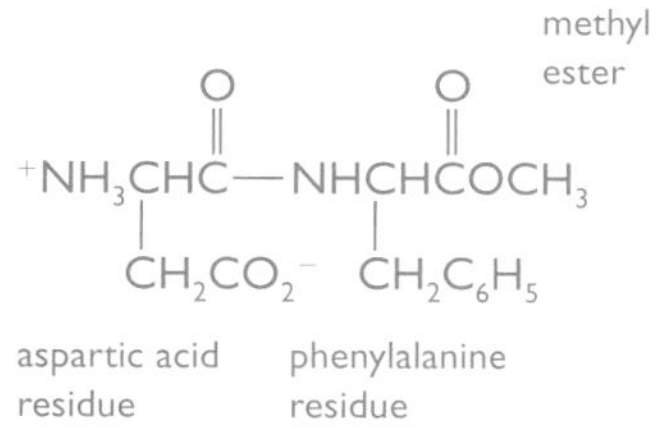

aspartic acid   phenylalanine
residue   residue

Aspartame
(NutraSweet)

values of aspartic and glutamic acid are much lower (more acidic) than the pI values of amino acids with nonpolar side chains. At a higher (more basic) pH, aspartic and glutamic acids would not be isoelectric but would bear net charges of 1−.

Aspartic acid and glutamic acid often occur as asparagine and glutamine, in which their side chain CO_2H groups have become amide groups, $CONH_2$, instead. These are also polar, hydrophilic groups, *but they are not electrically charged.* They are neither proton donors nor proton acceptors, so the pI values of asparagine and glutamine are higher than those of aspartic or glutamic acids.

■ PRACTICE EXERCISE 2   Write the structure of aspartic acid (in the manner of **1**) with the side-chain carboxyl (a) in its carboxylate form and (b) in its amide form.

Lysine, Arginine, and Histidine Have Basic Groups on Side Chains   The extra NH_2 group on lysine makes its side chain basic and hydrophilic. Remember that an amine in water makes the pH of the solution greater (more basic) than 7 because of the OH^- ion present in the following equilibrium.

$$RNH_2(aq) + H_2O \rightleftharpoons RNH_3^+(aq) + OH^-(aq)$$

The addition of OH^- to this equilibrium would shift the equilibrium to the left in accordance with Le Châtelier's principle. (The extra base pulls H^+ from the NH_3^+ group.) A solution of lysine has to be made *basic,* therefore, to prevent its side-chain NH_2 group from existing in the protonated form, NH_3^+, and so affect the net charge on the lysine molecule. This is why the pI value of lysine, 9.47, is relatively high. Arginine and histidine have similarly basic side chains.

■ PRACTICE EXERCISE 3   Write the structure of arginine in the manner of **1**, but with its side-chain amino group in its protonated form. (Put the extra proton on the $=NH$ unit, not the NH_2 unit of the side chain.)

■ PRACTICE EXERCISE 4   Is the side chain of the following amino acid hydrophilic or hydrophobic? Does it have an acidic, a basic, or a neutral group?

$$^+NH_3CHCO_2^-$$
$$|$$
$$CH_2CH_2CONH_2$$

Cysteine and Methionine Have Sulfur-Containing Side Chains   The side chain in cysteine has an SH group. Molecules with this group are easily oxidized to disulfides, and disulfides are easily reduced to SH groups.

$$2RSH \underset{\text{(reduction)}}{\overset{\text{(oxidation)}}{\rightleftharpoons}} RS{-}SR + H_2O$$

Cysteine and its oxidized form, cystine, are interconvertible by oxidation and reduction, a property of far-reaching importance in some proteins.

$$H—SCH_2CHCO_2^- \xrightarrow[H_2O]{\text{(O)}\atop\text{(oxidation)}} SCH_2CHCO_2^-$$

$$+ \quad NH_3^+ \qquad\qquad\qquad NH_3^+$$

$$H—SCH_2CHCO_2^- \xleftarrow[2(H)]{\text{(reduction)}} SCH_2CHCO_2^-$$

$$NH_3^+ \qquad\qquad\qquad\qquad NH_3^+$$

Cysteine   Cystine
(two molecules)

The **disulfide link** contributed by cystine is especially common in protective proteins, such as those of hair, fingernails, and the shells of certain crustaceans.

The Alpha Position in All Amino Acids Except Glycine is a Stereocenter   All the amino acids except glycine consist of chiral molecules and can exist as pairs of enantiomers. For each possible pair of enantiomers, however, nature supplies just one of the two (with a few rare exceptions). All the naturally occurring amino acids, moreover, belong to the same optical family, the L-family (Figure 20.2). Thus, all the proteins in our bodies, including all enzymes, are made from L-amino acids, and *all are chiral.*

20.2   OVERVIEW OF PROTEIN STRUCTURE

Protein molecules can have four levels of complexity.

Protein structures are more complicated by far than those of carbohydrates or lipids, and every aspect of their structures is vital at the molecular level of life. We'll ease our way into this field, therefore, with a broad overview of protein architecture.

Protein Structure Involves Four Features   There are four possible levels of complexity in the structures of proteins. Disarray at any level almost always renders the protein biologically useless. A protein having the structure and overall shape that it normally possesses in a living system and that permit it to function biologically is called a **native protein.** The same protein when made to lose its molecular shape even while retaining its original molecular constitution (covalent structure) is a **denatured protein.** (We'll return to denaturing agents later.)

The first and most fundamental level of protein structure, the **primary structure,** concerns only the *sequence of amino acid residues* in the protein's polypeptide(s). However, this sequence ultimately determines the three-dimensional shape of a protein and thus determines how a protein can function.

The next level of protein complexity, the **secondary structure,** also concerns only individual polypeptides. But it entails noncovalent forces, particularly the hydrogen bond. Secondary structure is the particular way in which polypeptide strands coil, intertwine, or line up side to side.

The **tertiary structure** of a polypeptide concerns the further coiling, bending, kinking, or twisting of secondary structures. If you've ever played with a coiled spring, you know that the coil (secondary structure) can be bent and twisted (tertiary structure). By and large, noncovalent forces such as hydrogen bonds and hydrophobic interactions stabilize tertiary shapes. In many polypeptides, attractions and repulsions between electrically charged sites on side chains are also involved in determining the overall shape.

■   Some D-amino acid residues occur in bacterial cell membranes, which helps these disease-causing agents to survive in higher animals, where the enzymes for attacking polypeptides work only with L-forms.

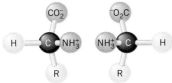

FIGURE 20.2   The two possible enantiomers of α-amino acids whose molecules have just one tetrahedral stereocenter. The absolute configuration on the right, which is in the L-family, represents virtually all the naturally occurring α-amino acids.

When polypeptides have disulfide bonds, they form from SH groups *after* the polypeptide has been made in the cell, so the S—S covalent bond is usually classified as a feature of tertiary not primary structure.

Finally, some proteins have **quaternary structure.** It develops by the coming together of two or more polypeptides, often with other relatively small molecules or ions, that aggregate in a precise manner to form one grand whole. Each polypeptide unit—now actually a *subunit* of the protein—has all previous levels of structure.

20.3 PRIMARY STRUCTURES OF PROTEINS

The backbones of all polypeptides of all plants and animals have a repeating series of N—C—C(=O) units.

The Peptide Bond Joins Amino Acid Residues in a Polypeptide The **peptide bond** is the covalent bond that forms when amino acids are put together to make a polypeptide. It's nothing more than an amide system, carbonyl-to-nitrogen. We'll illustrate this by putting together just two amino acids to form a *dipeptide.*

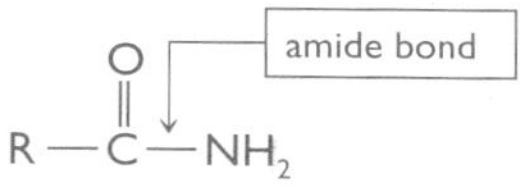

Suppose that glycine acts at its carboxyl end and alanine acts at its amino end so that by a series of steps (not given in detail) a molecule of water splits out and a carbonyl-to-nitrogen bond, a peptide bond, is created.

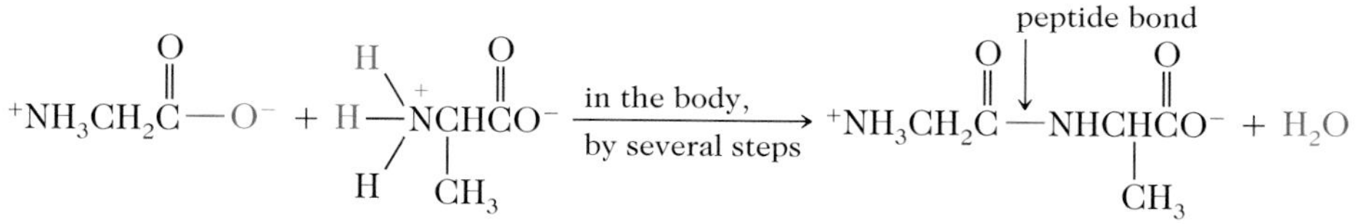

■ *Simple* amides have the following structure

■ How a cell causes a peptide bond to form involves a cell's nucleic acids.

glycine (Gly)   alanine (Ala)   glycylalanine (Gly-Ala)   **4**

Of course, there is no reason we could not picture the roles reversed so that alanine acts at its carboxyl end and glycine at its amino end. A different dipeptide how forms, but an isomer of the first.

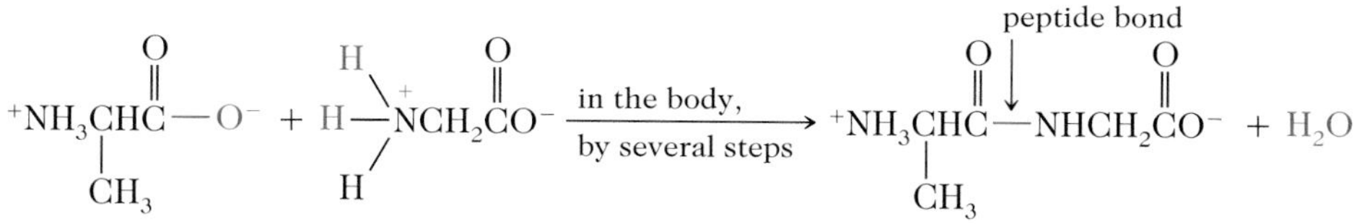

alanine (Ala)   glycine (Gly)   alanylglycine (Ala-Gly)   **5**

The product of the union of any two amino acid residues by a peptide bond is called a **dipeptide,** and all dipeptides have the following features.

■ The *di-* in *dipeptide* signifies that *two* amino acid residues are present and not the number of peptide bonds.

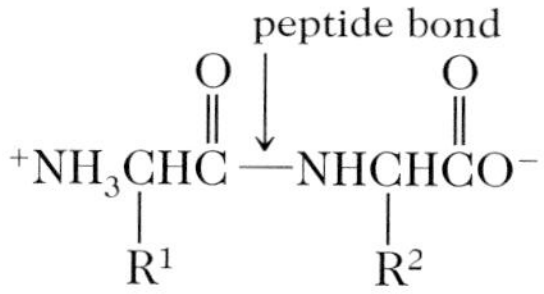

dipeptide

Structures **4** and **5** differ only in the sequence in which the side chains, H and CH$_3$, occur on the α-carbons. This is fundamentally how polypeptides also differ, in their sequences of side chains.

EXAMPLE 20.2   Writing the Structure of a Dipeptide

What are the two possible dipeptides that can be put together from alanine and cysteine?

ANALYSIS Both dipeptides must have the same backbone, so we write two of these first. We follow the convention that such backbones are always written in the N to C—left to right—direction, but this is only a convention.

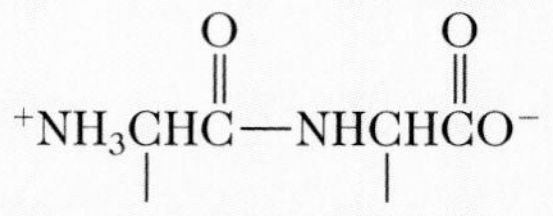

$$\overset{\displaystyle O}{\overset{\displaystyle \|}{^+NH_3CHC}}-\overset{\displaystyle O}{\overset{\displaystyle \|}{NHCHCO^-}} \qquad and \qquad \overset{\displaystyle O}{\overset{\displaystyle \|}{^+NH_3CHC}}-\overset{\displaystyle O}{\overset{\displaystyle \|}{NHCHCO^-}}$$

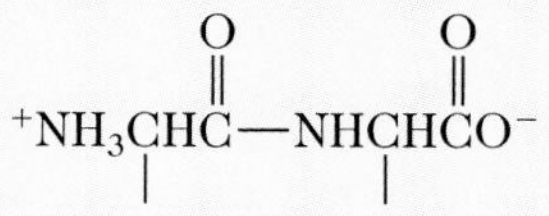

Then either from memory or by using a table, we know that the two side chains are CH$_3$ for alanine and CH$_2$SH for cysteine. We simply attach these in their two possible orders to make the finished structures.

SOLUTION

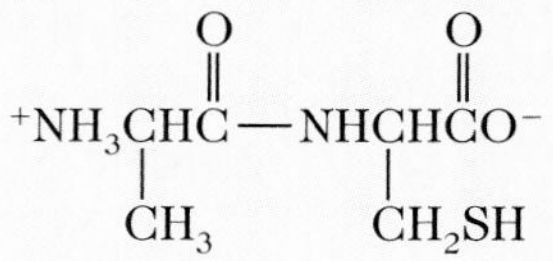

$$\overset{\displaystyle O}{\overset{\displaystyle \|}{^+NH_3CHC}}-\overset{\displaystyle O}{\overset{\displaystyle \|}{NHCHCO^-}} \qquad and \qquad \overset{\displaystyle O}{\overset{\displaystyle \|}{^+NH_3CHC}}-\overset{\displaystyle O}{\overset{\displaystyle \|}{NHCHCO^-}}$$
$$\quad\;\; CH_3 \qquad CH_2SH \qquad\qquad\qquad CH_2SH \qquad CH_3$$
$$\qquad\qquad Ala\text{-}Cys \qquad\qquad\qquad\qquad\qquad Cys\text{-}Ala$$

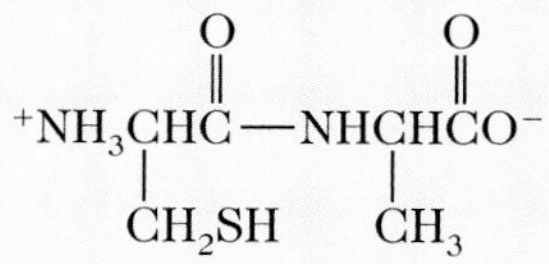

It would be worthwhile now simply to memorize the easy repeating sequence in a dipeptide, because it carries forward to higher peptides.

$$nitrogen-carbon-carbonyl-nitrogen-carbon-carbonyl$$

(Remember that the "carbon" in "nitrogen–carbon–carbonyl" is the *alpha* carbon. Remember also that the direction—left to right means nitrogen–carbon–carbonyl—is conventional, not a law of nature.)

■   PRACTICE EXERCISE 5   Write the structures of the two dipeptides that can be made from alanine and glutamic acid.

Three-Letter Symbols for Amino Acid Residues Simplify the Writing of Polypeptide Structures   Each amino acid has been assigned a three-letter symbol, given in the third from the last column in Table 20.1.[1] Three-letter symbols are used to write very condensed structures of polypeptides, but there are conventions to heed. A series of three-letter symbols, each separated by a hyphen, may represent a polypeptide structure, provided that the first symbol (reading left to right) is the free amino end, $^+NH_3$, and the last symbol is the free carboxylate end, CO_2^-. The structure

[1] In some applications, which we'll not encounter, it is more convenient to use single-letter symbols. These are given in the next to the last column of Table 20.1.

of the dipeptide **4**, for example, can be rewritten as Gly-Ala, and its isomer **5** as Ala-Gly. In both, the backbones are identical.

Dipeptides still have NH_3^+ and CO_2^- groups, so a third amino acid can react at either end. In general,

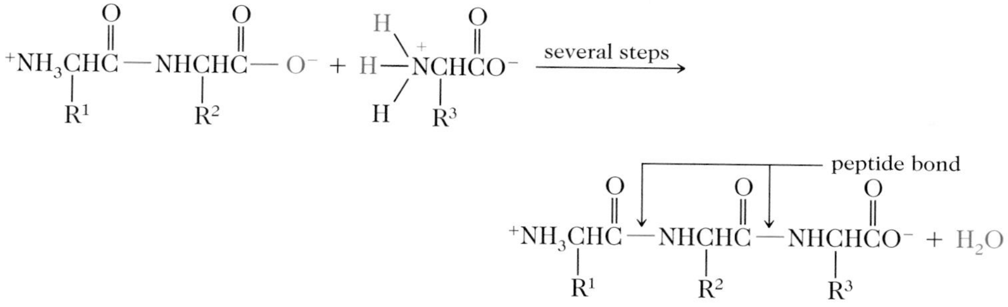

- A polypeptide is actually a *copolymer*, because the monomer units are not identical.

peptide bond

If we start with glycine, alanine, and phenylalanine, the tripeptide, Gly-Ala-Phe, would be only one of six possible tripeptides that involve these three different amino acids. The set of all possible sequences for a tripeptide made from Gly, Ala, and Phe is as follows.

Gly-Ala-Phe	Ala-Gly-Phe	Phe-Gly-Ala
Gly-Phe-Ala	Ala-Phe-Gly	Phe-Ala-Gly

Each tripeptide still has NH_3^+ and CO_2^- groups at each end that can interact with a fourth amino acid to make a tetrapeptide. And this would still have the end groups from which the chain could be extended even further. You can see how a repetition of this pattern many hundreds of times would produce a long polymer, namely, a polypeptide.

The Sequence of Side Chains on the Repeating N—C—C(=O) Backbone Is the *Primary* Structure of All Polypeptides All polypeptides have the same backbone "skeletons." They differ in length (n) and in the kinds and sequences of side chains.

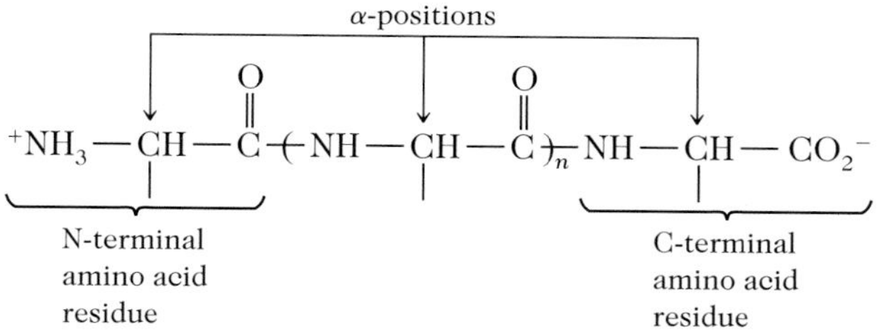

Polypeptide "backbone" (n can equal several thousand)

- We'll often use the designations *N-terminal unit* and *C-terminal unit* for the residues with the free α-NH_3^+ and the free α-CO_2^- groups, respectively.

As the number of amino acid residues increases, some used several times, the number of possible polypeptides increases rapidly. For example, if 20 different amino acids are incorporated, each used only once, there are 2.4×10^{18} possible isomeric polypeptides! (And a polypeptide with only 20 amino acid residues in each molecule is a very small polypeptide.)

The *Peptide Group* Is Trans-Planar *Rotation about the peptide bond is not free* despite having the appearance of a single bond. The unshared electron pair on the N atom of the NH group interacts with the C and O atoms of the carbonyl group to give the C—N peptide bond enough of the character of a double bond to prevent free rotation about it. Thus, in what we can call the *peptide group* of atoms, namely, from one α-carbon through the carbonyl–nitrogen unit to the next α-carbon, the four atoms lie in the same plane (Figure 20.3). The H on N of the NH group and the O on C of the carbonyl group are held trans to each other, on opposite sides of the backbone. The rigidity of the peptide group places some constraints on the flexibility of the polypeptide chain and so affects its overall shape.

20.4   SECONDARY STRUCTURES OF PROTEINS

The α-helix and the β-pleated sheet are two important kinds of secondary protein structures.

Once a cell puts together a polypeptide, noncovalent forces, like the hydrogen bond and hydrophobic interactions, determine how a polypeptide twists into a particular native shape. Hydrophobic interactions largely "drive" the formation of an overall shape, and hydrogen bonds stabilize it.

The α-Helix Is a Major Secondary Structure   One of the most common configurations is the *α-helix,* a coiled configuration of a polypeptide strand discovered by Linus Pauling and R. B. Corey (Figure 20.4). In the **α-helix,** the polypeptide backbone coils as a right-handed screw, which permits all of its side chains to stick to the outside of the coil.

Hydrogen Bonds Stabilize α-Helices   In α-helices, hydrogen bonds extend from the H atoms of polar NH units to oxygen atoms of carbonyl units situated four residues farther along the backbone. Individually, single hydrogen bonds are weak forces of attraction, but they add up much like the individual forces that hold a zipper strongly shut.

Generally, in very long polypeptides, only *segments* of molecules have an α-helix configuration. Coils 11 residues long are common, but 53 residues occur in one polypeptide's α-helix. Uncoiled portions of a molecule or a segment with a different feature (discussed below) occur between α-helix units.

■   Wood screws have right-handed helices. Its turns are in the same direction taken by your curling right hand fingers when your thumb points along the direction in which the helix advances.

■

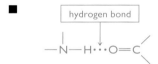

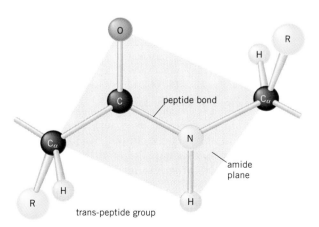

FIGURE 20.3   The trans nature of the *peptide group* showing how one C_α atom, the carbonyl carbon atom, the nitrogen atom, and the next C_α atom in the polypeptide chain are all in the same plane. The H in NH is trans to the O in C=O. (Adapted by permission from D. Voet and J. G. Voet, *Biochemistry,* 2nd edition, John Wiley & Sons, Inc., 1995.)

FIGURE 20.4 The α-helix. The polypeptide backbone follows the spiraling ribbon as a right-handed helix. The oxygen atoms of carbonyl groups are in red, nitrogen atoms of the peptide NH group are in dark blue, and the H atoms of NH are in white. Side-chain groups, R, are represented here only by simple spheres in purple. Note how they project to the *outside* of the ribbon. Dashed lines show how hydrogen bonds extend from each NH group's H atom to a carbonyl group's O atom four residues along the backbone. (Copyright by Irving Geis.)

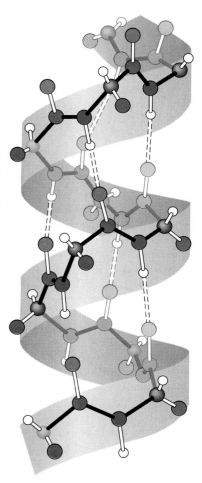

A Left-Handed Helix Characterizes Polypeptide Strands in Collagen   The *collagens*, the most abundant proteins in vertebrates, are a family of extracellular proteins that give strength to bone, teeth, cartilage, tendons, skin, blood vessels, and certain ligaments. Glycine contributes a third of the amino acid residues. Another 15% to 30% are those of proline and 4-hydroxyproline, an amino acid not on the list of the standard 20. Residues from 3-hydroxyproline and 5-hydroxylysine, also not on the list of 20, occur in collagen as well.

4-hydroxyproline   3-hydroxyproline   5-hydroxylysine

The ring systems in the molecules of proline and its hydroxylated relatives limit the flexibility of a strand, restricting the coiling of collagen polypeptides to *left-handed* helices. Moreover, there is a further level of structure to collagen, tertiary structure, which we'll consider shortly.

Vitamin C Is Essential to the Synthesis of Collagen   The reactions that put OH groups on proline rings (or lysine side chains) occur *after* the initial polypeptide is made. An enzyme is involved that requires ascorbic acid (vitamin C), and this is the molecular reason that vitamin C is essential to growing children and to their forming strong bones and teeth. When an adult's diet is deficient in ascorbic acid, wounds do not heal well, blood vessels become fragile, and a vitamin deficiency disease called *scurvy* results.

The β-Pleated Sheet Is a Side-by-Side Array of Polypeptide Units   Pauling and Corey also discovered another kind of secondary structure in which hydrogen bonds hold things together. It's one in which adjacent segments of polypeptide chains line up side by side to form a sheetlike array. Often it is slightly pleated and so is called the **β-pleated sheet** (Figure 20.5). Side chains project above and below the sheet's surface. As few as two segments of a polypeptide strand and as many as 15 have been found in the same pleated sheet, each strand ranging from 6 to 15 residues long.

When a hairpin turn carries the polypeptide chain from one segment of a pleated sheet to the next, the strands run in opposite direction. Other kinds of turns are possible but are less common (Figure 20.6). The pleated sheet is a feature in portions of the polypeptides of many proteins and is the dominant feature in fibroin, silk protein (seen in Figure 20.5).

■ Linus Pauling won the 1954 Nobel Prize in Chemistry for discovering the α-helix and β-sheet configurations.

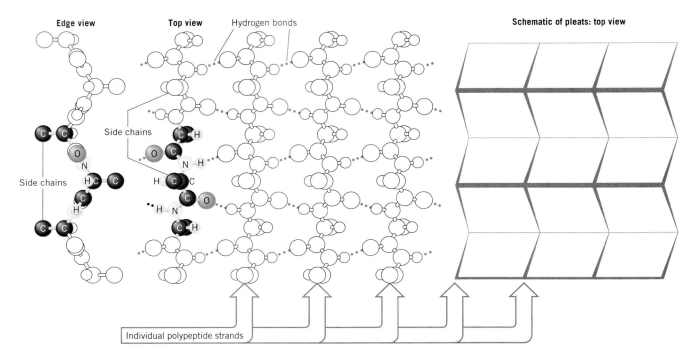

FIGURE 20.5 The β-pleated sheet.

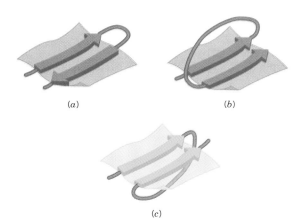

FIGURE 20.6   Segments of polypeptide strands can become adjacent to and parallel with each other in more than one way. (*a*) A hairpin loop brings the next segment into an antiparallel arrangement. (*b*) A back-and-over loop allows the next segment to have a parallel alignment. (*c*) A left-handed crossover loop also permits a parallel alignment. The loops are made of segments of the polypeptide chain. (Reproduced with permission from D. Voet and J. G. Voet, *Biochemistry,* 2nd edition, John Wiley & Sons, Inc., 1995.)

20.5 TERTIARY STRUCTURES OF PROTEINS

Tertiary structures are the results of folding, bending, and twisting of secondary structures.

Once primary and secondary structures are in place, the final shaping of a protein molecule occurs. All these activities happen spontaneously in cells, sometimes in a matter of seconds, sometimes several minutes, after the polypeptide molecule has been made. The "rules" followed by the polypeptide to give these shapes are not fully known.

Tertiary Protein Structure Involves the Folding and Bending of Secondary Structure When α-helices take shape, their side chains tend to project outward where, in an aqueous medium, they can be in contact with water. Even in water-soluble proteins, however, as many as 40% of the side chains are hydrophobic. Because such groups cannot break up the hydrogen-bonding networks among the water molecules, an entire α-helix or β-sheet undergoes further twisting and folding until the hydrophobic groups, as much as possible, are tucked to the inside, away from water, and the hydrophilic groups stay exposed to water. Thus the final shape of the polypeptide, its **tertiary structure,** emerges in response to simple molecular forces set up by the water-avoiding and the water-attracting properties of the side chains. In fact, hydrophobic interactions sometimes determine the best secondary structure for the polypeptide.

Disulfide Bonds Can Give Loops in Polypeptides or Join Two Strands Together Polypeptides with disulfide bonds receive them by the oxidation of SH groups during the development of tertiary structure. If the SH group on the side chain of cysteine appears on two neighboring polypeptide molecules, then mild oxidation is all it takes to link the two molecules by a disulfide bond. Such cross-linking, catalyzed by enzymes, can also occur between parts of the same polypeptide molecule to create a closed loop.

■ The symbols for cystine are

Cys   C
 | and |
Cys   C

Ionic Bonds Also Stabilize Tertiary Structures Another force that can stabilize a tertiary structure is the attraction between a full positive and a full negative charge, each occurring on a particular side chain. At the pH of body fluids, the side chains of both aspartic acid and glutamic acid carry CO_2^- groups. The side chains of lysine and arginine carry NH_3^+ groups. These oppositely charged groups attract each other, like the attraction of oppositely charged ions in an ionic crystal. The attraction is an *electrical attraction,* and is sometimes called a **salt bridge.**

Hydrophobic Interactions Significantly Affect Myoglobin Molecules In the tertiary structure of myoglobin, the oxygen-holding protein in muscle tissue, about 75% of the single polypeptide molecule consists of α-helix segments (Figure 20.7a). Virtually all of the hydrophobic groups are folded inside where they avoid water as much as possible, and the hydrophilic groups are on the outside. Heme, a nonprotein molecule, completes the native structure of myoglobin (Figure 20.7b).

Polypeptides Often Incorporate Prosthetic Groups into Their Tertiary Structures A nonprotein, organic compound that associates with a polypeptide, like heme in myoglobin, is called a **prosthetic group.** It is often the focus of the protein's biological purpose. Heme, for example, is the actual oxygen holder in myoglobin. Heme

■ "Prosthetic" is from the Greek *prosthesis,* an addition.

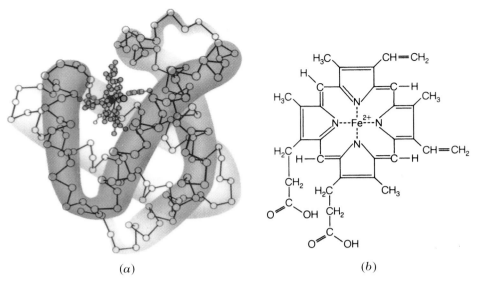

(a)   (b)

FIGURE 20.7 (a) Myoglobin (sperm whale), a polypeptide with 153 amino acid residues. The tubelike forms outline the eight segments that are in an α-helix. The flat, purple structure is the heme unit and the red circle is an oxygen molecule. Only the atoms that make up the backbone of the chain are indicated (by circles). The side chains have been omitted except for the two electrically charged groups that bind Fe^{2+} in the heme unit. (b) Heme. [Part (a) Copyright by Irving Geis.]

serves the same function in **hemoglobin,** the oxygen carrier in blood. The heme molecule is held in the folded globin molecule by electrical attractions between two electrically charged side chains and the Fe^{2+} ion in heme.

β-Sheets Are Often Twisted as Well as Pleated   Many polypeptides incorporate both helices and sheets within the same structure. In Figure 20.8, we see an artist's representation of two views of the molecule of a 247-residue enzyme, triose phosphate isomerase. Each broad, flat arrow is a segment of the chain in a β-sheet arrangement with a neighboring segment, also shown as a broad arrow. There are eight segments of the strand in β-sheet regions of this enzyme, and the entire sheet is itself twisted to form a barrel-like configuration (called a β-barrel). α-Helix units occur between the strands involved in the β-sheet.

20.6 QUATERNARY STRUCTURES OF PROTEINS

For many proteins, the native form emerges only as two or more polypeptides assemble into a quaternary structure.

Proteins like myoglobin have finished shapes at the tertiary level, being made of single polypeptide molecules, sometimes with prosthetic groups. Many proteins, however, are aggregations of two or more polypeptides, giving **quaternary structure** to the protein. One molecule of the enzyme phosphorylase, for example, consists of two tightly aggregated molecules of the same polypeptide. If the two become sepa-

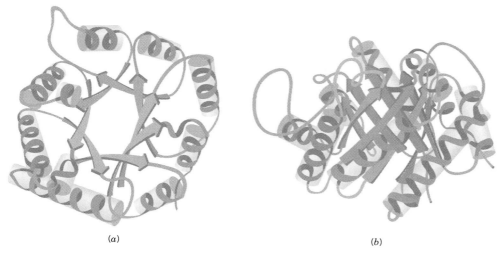

(a)   (b)

FIGURE 20.8 The molecule of the enzyme triose phosphate isomerase illustrates how both α-helix segments and β-sheet arrays can be incorporated together. The β-sheet consists of eight parallel segments, but as indicated by the twists of the arrows, the sheet is itself twisted. The top view (a) and the side view (b) both indicate how this twist creates a barrel-like cylindrical structure. (Reproduced with permission from D. Voet and J. G. Voet, *Biochemistry,* 2nd edition, John Wiley & Sons, Inc., 1995.)

rated, the enzyme can no longer function. Individual molecules of polypeptides that make up an intact protein are called the protein's *subunits.*

Hemoglobin has four subunits, two of one kind (designated α-subunits) and two of another (called the β-subunits), each subunit supporting a heme molecule (Figure 20.9). A combination of hydrophobic and electrostatic interactions as well as hydrogen bonds hold the subunits together. These forces do not work unless each subunit has the appropriate primary, secondary, and tertiary structural features. If even one amino acid residue is wrong, the results can be very serious, as in the example of sickle-cell anemia, described in Interaction 20.1.

Covalent Cross-Linkages Occur in Collagen The polypeptide units in collagen, each with about 1000 amino acid residues and each in a left-handed helix, assemble in units of three molecules each. The three left-handed helices wrap around each other in a relatively open *right*-handed helix of helices to form the **triple helix,** cablelike system called *tropocollagen* (Figure 20.10). Between the polypeptide strands of the triple helix *covalently bonded* molecular bridges are erected by a series of reactions that cause lysine side chains to link together. *Covalent* cross-links are better able than hydrogen bonds or hydrophobic interactions to resist forces that would undo the collagen's tertiary structure. A microfiber or *fibril* of collagen forms when individual tropocollagen cables overlap lengthwise. The mineral deposits in bones and teeth become tied into the protein at the gaps between the heads of tropocollagen molecules and the tails of others.

As people grow older, additional covalent cross-links develop between collagen strands leading to less muscular flexibility and agility. Free radicals, species with unshared electrons and generated by components of tobacco smoke as well as by overexposure to the ultraviolet portions of sunlight, accelerate the aging of skin.

■ The protein in the cornea of the eye is also a member of the collagen family.

■ A collagen fibril only 1 mm in diameter can hold a suspended mass as large as 10 kg (22 lb).

■ Meat from old animals is tougher because of their more highly cross-linked collagen.

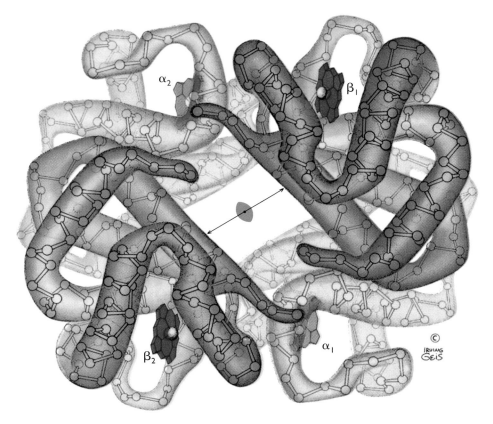

FIGURE 20.9 Hemoglobin. Four polypeptide chains, each with one heme molecule represented here by the colored, flat plates that contain spheres (Fe^{2+} ions), are nestled together. Only the atoms of the backbones are shown (by circles). The central cavity, indicated by the double-headed arrow, has enough room to hold an organic anion not shown here, 2,3-bisphosphoglycerate (BPG), until hemoglobin starts to load up with oxygen molecules. (Copyright Irving Geis.)

FIGURE 20.10 The triple helix of collagen. (*a*) Schematic drawing. (*b*) Electron micrograph of collagen fibrils from the skin. A fibril is an orderly aggregation of collagen molecules aligned side by side but overlapping each other in a regularly repeating manner that produces the banded appearance. (Schematic copyright Irving Geis.)

(*a*)

(*b*)

INTERACTION 20.1
SICKLE-CELL ANEMIA AND ALTERED HEMOGLOBIN

The decisive importance of the primary structure to all other structural features of a polypeptide or its associated protein is illustrated by the grim story of sickle-cell anemia. This inherited disease is widespread among those whose roots are in equatorial regions of central and western Africa.

In its mild form, where only one parent carries the genetic trait, the symptoms of sickle-cell anemia are seldom noticed except when the environment has a low partial pressure of oxygen, as at high altitudes. In the severe form, when both parents carry the trait, the infant usually dies by the age of 2 unless treatment is begun early. The problem is *an impairment in blood circulation traceable to the altered shape of hemoglobin in sickle-cell anemia.* The altered shape is particularly a problem after the hemoglobin has delivered oxygen and is on its way back to the heart and lungs for more.

The fault at the molecular level lies in a β-subunit of hemoglobin. One of the amino acid residues should be glutamic acid, but is valine instead. Thus instead of a side-chain CO_2^- group, which is electrically charged and hydrophilic, there is an isopropyl side chain, which is neutral and hydrophobic. Normal hemoglobin, symbolized as HbA, and sickle-cell hemoglobin, HbS, therefore have different patterns of electrical charges. Both have about the same solubility in well-oxygenated blood, but oxygen-free molecules of HbS clump together inside red cells and precipitate. This deforms the cells into a telltale sickle shape (see Figure 1). The distorted cells are harder to pump through capillaries, where the cells often create plugs. Sometimes the red cells split open. Any of these events places a strain on the heart. The error in one side chain seems insignificant, but it is far from small in human terms.

The sickle cell trait offers some resistance to malaria, which almost certainly explains why the trait survives largely where this tropical disease is most common. Normally, the mosquito-borne para-

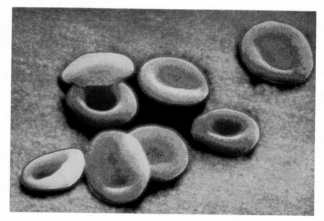

FIGURE 1 Electron micrographs of a normal red blood cell, above, and a sickle cell, below.

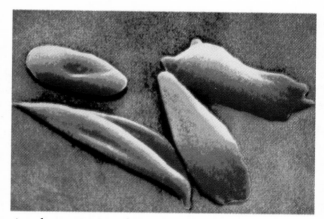

site that causes malaria resides within a red blood cell. However, the parasite cannot survive very long inside a sickled cell. The parasite has a high need for potassium ion, but the membrane of a sickled cell allows too much potassium ion to get through and escape. Thus people with the sickle-cell trait are statistically more likely than individuals unprotected from malaria to live long enough to bear children and so pass the trait to their offspring.

20.7 COMMON PROPERTIES OF PROTEINS

Even small changes in the pH of a solution can affect a protein's solubility and its physiological properties.

Although proteins come in many diverse biological types, they generally have similar chemical properties toward ordinary substances because they have similar functional groups.

Protein Digestion Is Hydrolysis The digestion of a protein is nothing more than the hydrolysis of its peptide bonds to give a mixture of amino acids. Different digestive enzymes, all in the family of *proteases,* handle the cleavage of peptide bonds, each enzyme working best when particular side chains are close by.

To illustrate digestion, we'll use the hydrolysis of a tripeptide.

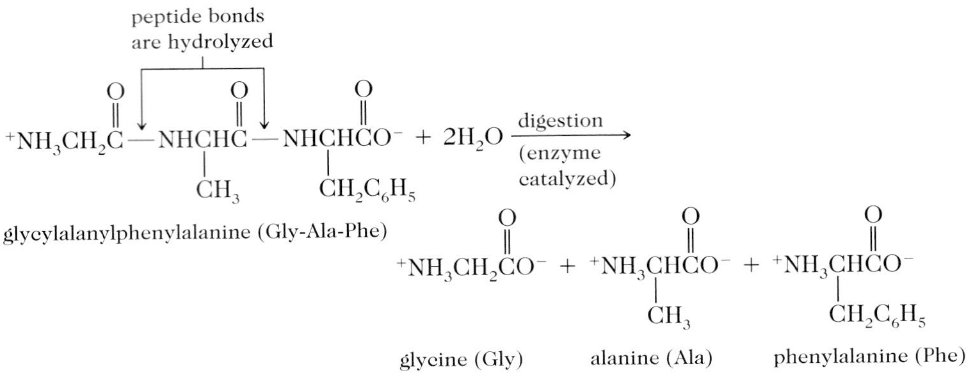

peptide bonds
are hydrolyzed

$$^+NH_3CH_2C \overset{O}{\overset{||}{}} -NHCHC \overset{O}{\overset{||}{}} -NHCHCO^- + 2H_2O \xrightarrow[\text{(enzyme catalyzed)}]{\text{digestion}}$$

glycylalanylphenylalanine (Gly-Ala-Phe)

$$^+NH_3CH_2CO^- + {}^+NH_3CHCO^- + {}^+NH_3CHCO^-$$

glycine (Gly)   alanine (Ala)   phenylalanine (Phe)

Protein *Denaturation* Is the Loss of Protein Shape Peptide bonds are not hydrolyzed when a protein is *denatured.* All that has to happen is some disruption of secondary or higher structural features, so all proteins are vulnerable. **Denaturation** is the disorganization of the overall molecular shape of a native protein. It can occur as an unfolding or uncoiling of helices, or as the separation of subunits. Because native proteins have their overall shapes in an aqueous environment where water molecules are intimately involved with hydrophobic interactions, even the removal of water can cause the denaturation of many proteins.

Usually, denaturation is accompanied by a major loss of solubility. When egg white is whipped or heated, for example, as when you cook an egg, its albumin molecules unfold and become entangled among themselves. The system no longer blends with water—it's insoluble—and it no longer allows light to pass through—it's opaque.

Table 20.2 describes several reagents or physical forces that cause denaturation, together with brief explanations of how they work. How effective a given denaturant is depends on the kind of protein. Those of hair and skin and of fur or feathers quite strongly resist denaturation because they are rich in disulfide links.

In recent years, several proteins have been discovered that, after denaturation, can be *renatured.* The examples are proteins denatured by the cleavage of disulfide links. When treated with an enzyme for oxidizing SH groups, the original S—S linkages of such denatured proteins are restored.

Changing the Shape of a Prion Renders It Lethal One dramatic example of the importance of polypeptide shape is given by the *prions* (proteinaceous infectious particles). Prions are proteins normally made in the cells of many vertebrates, but sometimes the prions of healthy animals, including humans, are induced to take up altered molecular shapes. According to perhaps the most widely held theory, the shape-inducing agents are not bacteria or viruses but rather the altered prions of diseased animals. Altered prions are observed, for example, in mad cow disease (bovine spongiform encephalopathy or BSE), which threw England and Europe into a panic in the mid 1990s. It's a neurological disorder in the same family as scrapie, kuru, and Creutzfeldt-Jacob disease (see Interaction 20.2).

TABLE 20.2 Denaturing Agents for Proteins

Denaturing Agent	How the Agent May Operate
Heat	Disrupts hydrophobic interactions and hydrogen bonds by making molecules vibrate too violently. Produces co-agulation, as in the frying of an egg.
Microwave radiation	Causes violent vibrations of molecules that disrupt hydrogen bonds and hydrophobic interactions.
Ultraviolet radiation	Probably operates much like the action of heat (e.g., sun burning)
Violent whipping or shaking	Causes molecules in globular shapes to extend to longer lengths and then entangle (e.g., beating egg white into meringue).
Soaps	Probably affect hydrogen bonds and salt bridges.
Organic solvents (e.g., ethanol, acetone, 2-propanol)	May interfere with hydrogen bonds because these solvents can also form hydrogen bonds or can disrupt hydrophobic interactions. Quickly denature proteins in bacteria, killing them (e.g., the disinfectant action of 70% ethanol).
Strong acids and bases	Disrupt hydrogen bonds and salt bridges. Prolonged action leads to actual hydrolysis of peptide bonds.
Salts of heavy metals (e.g., salts of Hg^{2+}, Ag^+, Pb^{2+})	Cations combine with SH groups and form precipitates. (These salts are all poisons.)

Protein Solubility Depends Greatly on pH Because some side chains as well as the end groups of polypeptides bear electrical charges, an entire polypeptide molecule can bear a net charge. Because these charged groups are either proton donors or acceptors, the net charge is easily changed by changing the pH. For example, CO_2^- groups become electrically neutral CO_2H groups as they pick up protons when a strong acid is added.

Suppose that the net charge on a polypeptide is $1-$, and that one extra CO_2^- is responsible for it. When all molecules have the same charge, they repel each other (and stay in solution). However, when acid is added, the following change could destroy the net charge (the oval represents the polypeptide molecule).

$$\text{(oval) } NH_3^+ \; CO_2^- \; CO_2^- + H_3O^+ \longrightarrow \text{(oval) } NH_3^+ \; CO_2^- \; CO_2H + H_2O$$

net charge: $1-$ → net charge: 0

The product is now isoelectric, *and the molecules no longer repel each other.* On the other hand, a polypeptide might have a net charge of $1+$, caused by one extra NH_3^+ group. Now addition of OH^- could cause a polypeptide to become isoelectric. Either way, the change is from an electrically charged system to one that is neutral.

Each protein, as we said, has a characteristic pH, its *isoelectric point,* at which its net charge is zero and at which it cannot migrate in an electric field. This matters because electrically *neutral* polypeptide molecules can clump together to become such enormous megaparticles that they simply drop out of solution (Figure 20.11). A

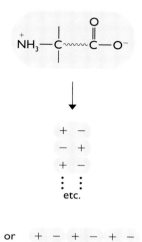

FIGURE 20.11 Several isoelectric protein molecules (top) can aggregate into very large clusters that no longer dissolve in water.

INTERACTION 20.2
MAD COWS, PRIONS, AND PROTEIN SHAPES

A Dogma Under Attack   Until the early 1980s one of the standard "dogmas" in biology was that agents that transmit infectious diseases—chiefly bacteria, viruses, and fungi—always contain some nucleic acid. As you probably already know, nucleic acids are the chemicals of heredity, like DNA and RNA, and we'll study them in a later chapter. During the 1980s, evidence mounted, however, that nucleic acids are not directly involved in a few transmissible and always fatal diseases, like *scrapie* and *kuru.* These diseases affect the central nervous systems, chiefly of mammals, turning their brains into pitted sponges and the victims into demented wrecks. The diseases have similar characteristics, one being that most develop very slowly, so slowly that for a long time scientists ascribed their origins to unknown "slow viruses."

Scrapie is a neurological disorder of goats and sheep, the later becoming so irritated by scrapie that they impulsively scrape off their wool (hence the name "scrapie").

Kuru is a degenerative and always fatal brain disease once found among the Fore people of Papua New Guinea, who would honor their dead by eating their brains. The victims lost their coordination (ataxia) and became demented. When the Fore people stopped the ritual, kuru virtually disappeared.

Creutzfeldt-Jacob disease is a neurological disease found elsewhere among humans that is very similar to kuru, possibly even identical with it. Some cases of Creutzfeldt-Jacob disease appear to result from inheritance, but the symptoms do not develop until the victims are about 60 years old.

A scrapielike agent also causes *mad cow disease* (bovine spongiform encephalopathy or BSE) in cattle, a disease characterized by severe disorientation, staggering, and an inability to stand. In early 1996, British health authorities suspected that the agent responsible for mad cow disease might have caused several new cases of a variation of Creutzfeldt-Jakob disease among the younger British population who, unknowingly, ate contaminated beef. The possibility that the agent, whatever it is, might be transmitted from mad cows to humans by the eating of beef threw all of Europe into a panic, and the sale of British beef (a three billion dollar a year industry) was banned in many countries.

Prions   The agents of neurological diseases such as scrapie lose no potency when subjected to conditions that destroy nucleic acids. Conditions that denature only proteins, however, destroy the scrapie agent's ability to cause the disease. It was evidence like this that led Stanley Prusiner (University of California School of Medicine, San Francisco) to propose that some infectious, transmittable diseases are caused by previously unknown agents that lack nucleic acid and consist solely of protein. Prions appear to consist of relatively small protein molecules (actually glycoproteins), one being found to have only 208 amino acid residues.

Fatal Changes in the *Shapes* of Prion Protein Molecules   The prion protein, symbolized as PrP (for prion protein), is a variation of a protein normally made in the cells of many vertebrates *and apparently necessary to the organisms.* We'll use the symbol PrP^C for the normal, cellular prion protein (the C is for cell) and the symbol PrP^{Sc} for the abnormal, scrapie prion (the Sc is for scrapie). Within a given prion disease, the amino acid sequences in the two prion proteins are identical in nearly all cases studied! The difference between a victim's PrP^C and PrP^{Sc} appears to be only in the *shapes* of the molecules. Molecular regions that are rich in α-helices in PrP^C have changed over to β-regions in PrP^{Sc}. The change in molecular shape renders the PrP^{Sc} protein resistant to protein-digesting enzymes, so when humans eat infected food, the PrP^{Sc} molecules are not all destroyed by digestion. It's not known with certainty, but scientists think that some PrP^{Sc} molecules or major fragments of them manage to get into the lymph system and are thus delivered by circulation to the target tissues. PrP^{Sc} molecules (or fragments) then somehow cause a cell's normal PrP^C molecules to become configurationally changed. With the help of Figure 1, we may consider a current theory about how this occurs.

A PrP^{Sc} molecule fits to a PrP^C molecule (Figure 1a), which induces a change in the PrP^C molecule itself. Then the original PrP^{Sc} molecule disengages (Figure 1b); the number of PrP^{Sc} molecules has now doubled. Each of the two PrP^{Sc} molecules can then attach to unchanged PrP^C molecules, changing them also (Figure 1c). Eventually, all of the original PrP^C molecules are changed (Figure 1d).

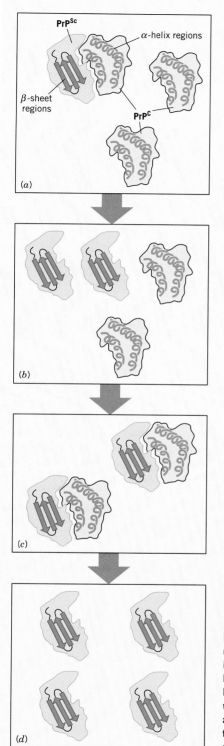

Protein-digesting processes in the cell can only partially degrade the new but alien PrP^{Sc} molecules, and their larger fragments aggregate in clusters of rodlike particles very much like the particles (amyloid plaques) seen in the brain cells of people who have died of Alzheimer's disease. Indeed, some scientists speculate that other proteins, behaving like prions, have a role in more common neurodegenerative diseases, like Alzheimer's disease, Lou Gehrig's disease (amyotrophic lateral sclerosis), and Parkinson's disease.

Not all specialists in prion diseases hold that the evidence compels belief that altered prions, instead of viruses, are the *causes* of the diseases. The changed prions might be among the *results*. These specialists argue that the prion theory is not established until prions have been isolated from diseased organisms, purified so as to convince all scientists that no viral particles (and so no nucleic acids) are present, and then used to cause a prion disease in an animal. No doubt evidence for one view or the other will emerge in the late 1990s.

A Reference

Stanley B. Prusiner, "The Prion Diseases," *Scientific American,* January 1995, page 48.

FIGURE I According to one theory, the change of normal PrP^C molecules to PrP^{Sc} molecules in infectious scrapielike diseases is initiated by an alien PrP^{Sc} molecule (or a major fragment). (*a*) The PrP^{Sc} unit binds to a molecule of PrP^C. (*b*) The binding induces the PrP^C molecule to undergo a change in shape, resulting in an additional molecule of PrP^{Sc}. (*c*) The two new PrP^{Sc} molecules similarly alter more two unchanged PrP^C molecules, resulting in a total of four new PrP^{Sc} molecules. (*d*) Eventually, all of the original PrP^C molecules have changed.

protein is least soluble in water when the pH equals the protein's isoelectric point. Therefore, when a protein must be in *solution* to work, as is true for many enzymes, the pH of the medium must be kept away from the protein's isoelectric point. Buffers in body fluids ensure this. We cannot overemphasize how much of life depends on the control of the pH of body fluids.

Casein or milk protein (pI 4.7) illustrates the effect of pH on solubility. As milk turns sour, its pH drops from a normal value of 6.3 to 6.6 toward 4.7. At 4.7, casein molecules become isoelectric, denature, clump together, and separate as curds. As long as the pH of milk is something *other than* the pI for casein, the protein remains colloidally dispersed.

20.8 CELL MEMBRANES REVISITED — GLYCOPROTEIN COMPONENTS

Glycoproteins provide "recognition sites" on the surfaces of cell membranes.

In the previous chapter we introduced the general features of cell membranes, giving particular attention to membrane lipids. With our background in proteins, we can now take a closer look at membranes.

Membrane Proteins Help to Maintain Concentration Gradients   If the cell membrane were an ordinary dialyzing membrane, any kind of small molecule or ion could move freely in and out of the cell. The way cells work, however, requires that many concentration *gradients* exist between the cell interior and whatever fluid is outside. A **gradient** is an unevenness in the value of some physical property throughout a system or between the inside and outside of a membrane. A *concentration gradient* exists in a solution, for example, when one region of a solution has a higher concentration of solute than another, such as in unstirred coffee just after you add sugar.

Gradients are generally unstable compared to thoroughly mixed up systems. Because of the random motions of ions and molecules in liquids and gases, the natural tendency is for gradients eventually to disappear and for solute concentrations to become uniform. Yet the membranes of living cells maintain a number of gradients. For example, both sodium ions and potassium ions have quite different concentrations inside and outside of cells (see table in margin). *Such gradients must be maintained against nature's spontaneous tendency to destroy them, or we die.*

Here is where some of the proteins in cell membranes carry out a vital function. One kind of assembly of membrane protein molecules can move sodium ions *against* their gradient, meaning that concentration differences are intensified rather than wiped out. When too many sodium ions move to the inside of a cell, they are "pumped" back out despite the existence of a higher sodium ion concentration in the external fluid. It's done by a special molecular machinery, made of membrane-bound polypeptides, called the *sodium–potassium pump.* The same pump can move potassium ions back inside a cell. This movement of any solute *against* its concentration gradient requires chemical energy and is an example of **active transport.** Other reactions in cells supply the needed chemical energy.

Gap Junctions Enable Substances to Move Directly from One Cell to Another   In the cells of most tissues of multicelled organisms, membrane proteins provide a route for the *direct* movements of ions and molecules from one cell to another.

	Concentration (mmol/L)	
Ion	Plasma	Cells
Na⁺	135–145	10
K⁺	3.5–5.0	125

These routes are through **gap junctions,** tubules fashioned from membrane proteins that "rivet" cells together (Figure 20.12). So many of such junctions occur in some tissues that the entire tissue is interconnected from within.

Gap junctions in bone tissue, for example, enable bone cells at some distance from capillaries to receive nourishment and to remove wastes. Heart muscle is able to contract *synchronously* because gap junctions allow ions to move easily between cells. The gaps are large enough to allow certain ions, like Ca^{2+}, and certain relatively small molecules (up to formula masses of about 1200) to pass but are not large enough for macromolecules like proteins and nucleic acids to get through.

Calcium ion appears to control the diameters of gap junctions, one of the crucial functions of this dietary mineral. When the concentration of Ca^{2+} is very low ($<10^{-7}\,M$), the channels are fully open. At higher levels, the gaps close down and become completely shut at about $5 \times 10^{-5}\,M\,Ca^{2+}$. One consequence of this control is that if part of an interconnected mass of cells is injured, the closure of gaps limits the damage.

Some Proteins in Cell Membranes Are Receptors for Hormones and Neurotransmitters A **receptor** is a membrane-bound protein whose unique molecular shape enables its molecules to fit only to those of a *substrate,* a compound with which it is supposed to interact. A receptor is thus able to "recognize" the molecules of just one compound from among the hundreds whose molecules bump against it. A receptor is like a dock on an orbiting space vehicle; it is shaped to be a perfect match to the docking module of an incoming vehicle. This is roughly how specific hormones are able to find and stop only at the cells where they are meant to stop, by being able to fit to unique molecular docks or receptors. Some medications mimic hormones by binding to receptors and so tricking the system. The antiprogesterone birth control agent, RU 486, is an example (see Interaction 20.3).

Docking to a membrane protein also helps neurotransmitters send a signal from one nerve cell to the next. Their molecules, released by one cell, move to the surface of the next across the very narrow gap between them. There they attach quickly to the right receptor, whereupon further biochemical changes are induced. All that we've learned about protein shapes is relevant to these docking maneuvers.

■ Neurotransmitters are organic molecules that help carry nerve signals from the end of one nerve cell to the beginning of the next.

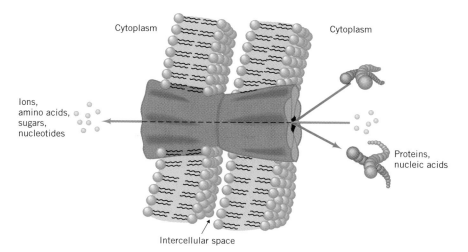

FIGURE 20.12 Gap junctions. A protein-fashioned channel (in reddish brown) between two cells enables some small particles to pass directly from one cell to another. The blue spheres each with two wavy tails are phospholipid molecules of the cell membranes. (Reproduced with permission from D. Voet and J. G. Voet, *Biochemistry,* 2nd edition, John Wiley & Sons, Inc., 1995.)

INTERACTION 20.3
MIFEPRISTONE (RU 486)—RECEPTOR BINDING OF A SYNTHETIC ANTIPREGNANCY COMPOUND

When released from an ovarian follicle, the natural female hormone, progesterone, acts to prepare the system for pregnancy both by inhibiting further production of ova (egg cells) and by preparing the uterus for the implantation of the fertilized ovum. The action of progesterone involves the binding of its molecules to protein receptors within cells of the lining of the uterus (the endometrium).

Antiprogesterones (Table 19.3) mimic the work of progesterone in that they cause a pseudopregnant state ("false pregnancy state") and so suppress the production of ova. Fertilization cannot occur without an ovum, of course, and so the synthetic progestin-containing medications are birth control pills.

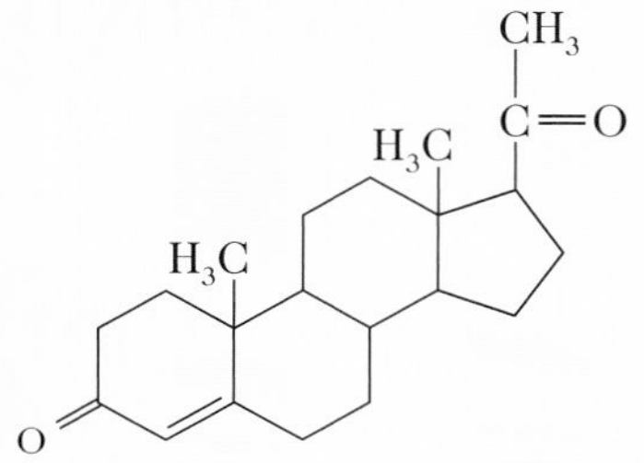

progesterone

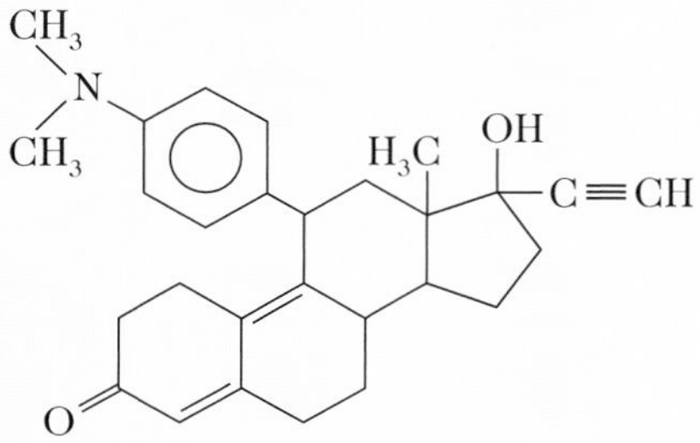

mifepristone (RU 486)

RU 486 (RU for Roussel-Uclaf, a French pharmaceutical company) was first prepared in 1980. It was soon discovered to be a strong binder to the progesterone binding sites on the receptor protein for progesterone. This action blocks the normal action of progesterone, and RU 486 is thus an *antiprogesterone*. If RU 486 is taken during the 5- to 6-day postcoital period (the period immediately following intercourse), its blocking action prevents pregnancy by suppressing the implantation of a fertilized ovum in the uterus. If used within 72 hours after unprotected intercourse, its failure rate is very low. It can thus be used as a "morning after" pill.

RU 486, followed by the use of two prostaglandin-like compounds, also induces abortion. Thus RU 486 is also an abortifacient (an abortion inducing agent). Under medical supervision, the use of RU 486 for this purpose has a 96% success record. The failures include continued pregnancy, only partial expulsion of the fetus, and the need for procedures to stem uterine bleeding.

That RU 486 "prevents pregnancy" is a controversial statement, because some view the onset of pregnancy as occurring at the moment of fertilization. Others regard pregnancy as not starting until the fertilized ovum has become implanted in the uterus. The controversy thus involves the questions, "When does *pregnancy* begin?" and the not identical question, "When does *human life* begin?" Around these questions have surged some of the stormiest waters of the prolife–prochoice controversy.

Membrane Proteins Are Glycoproteins   Carbohydrate molecules contribute one of the most significant features of membranes, so much so, in fact, that essentially all cells turn out to be "sugar coated." The sugars are mostly *oligosaccharides,* carbohydrates whose molecules can be hydrolyzed to three or more—up to a few dozen—monosaccharide molecules.

Some membrane carbohydrates are covalently joined to lipids, making up the membrane's **glycolipids.** Others are bound to proteins, forming the membrane's **glycoproteins** (Figure 20.13). Most proteins, in fact, are glycoproteins. Several thousand have been identified.

The oligosaccharides of glycoproteins generally contain sulfur or nitrogen, the

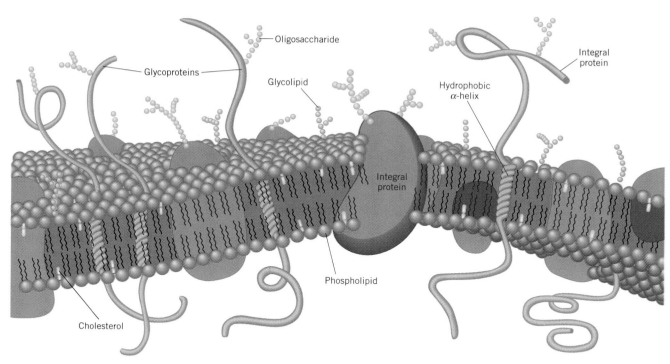

FIGURE 20.13 Glycoproteins as structural units in a cell membrane. The blue spheres each with two wavy tails are phospholipid molecules. Shown in yellow are cholesterol molecules. Chains of green beads represent glycolipids. Chains of yellow beads attached to polypeptides and proteins (in reddish brown) are oligosaccharide units. (Reproduced with permission from D. Voet and J. G. Voet, *Biochemistry,* 2nd edition, John Wiley & Sons, Inc., 1995.)

sulfur occurring in SO_3^- groups. Nitrogen is present as an amine or an amide group. *Amino sugars* are those in which an OH group is replaced by NH_2. The most common amino sugar is D-glucosamine, occurring as its *N*-acetyl derivative—*N*-acetyl-D-glucosamine. It and systems like it are usually joined by what is called an *N-link* to a polypeptide at an asparagine residue.

■ The *N* signifies that the acetyl group is attached to *nitrogen.*

β-D-glucosamine

N-acetyl-*β*-D-glucosamine

N-acetyl-*β*-D-glucosamine unit
N-linked to a polypeptide unit

One Kind of Glycoprotein Gives Resiliency to Cartilage   Table 20.3 gives the structures and chief uses of three of several carbohydrate polymers known as *glycosaminoglycans*[2] or *mucopolysaccharides*. Two glycosaminoglycans form the gel-like material called **ground substance** present in cartilage and in which fibers of collagen and another fibrous protein, *elastin,* are embedded. The fibers give *tensile strength* to cartilage, the strength to withstand stretching tension without breaking. Ground substance gives cartilage *flexibility* and shock absorbency, as we'll explain next.

Because their molecules have so many groups with hydrogen bonding ability, it isn't surprising that glycosaminoglycan molecules are "sticky," making such substances thick, slimy, viscous materials resembling mucus. It's also not surprising that glycosaminoglycan molecules attract and hold large numbers of water molecules. It's this that gives ground substance its spongy, resilient nature needed for the cartilage in bone joints. As cartilage is squeezed during bumps and jolts, water is forced out. When the pressure is released, water rushes back in. In fact, this "tidal flow" of water is what carries metabolic wastes away and brings in nutrients to cartilage tissue, a tissue that lacks blood vessels. The movement of water caused by flexing the joints makes it easy to understand why the cartilage in joints becomes somewhat fragile, even brittle, during long periods of no exercise.

Monosaccharide Units Have Many Ways to Combine into Oligosaccharides   Glycoproteins have a large subfamily of N-linked oligosaccharides whose structures include several monosaccharides in both "straight-chain" and "branched-chain" connections. Many of the linkages between monosaccharide units involve oxygen atoms at ring positions not used for linking by the nutritionally important disaccharides or polysaccharides. Table 20.3, in fact, illustrates $(1 \rightarrow 3)$ linkages.

■ In carbohydrate chemistry, amylose and cellulose are examples of "straight-chain" polysaccharides; amylopectin and glycogen are "branched-chain."

The possibilities for a number of $(1 \rightarrow n)$ linkages having varying geometries give living systems far more options for joining monosaccharide units than for joining amino acid units. Two identical amino acids, for example, can be joined in only one way by a peptide bond, but two identical monosaccharides can be linked to form 11 different disaccharides. Someone has calculated that only four *different* monosaccharide units can be linked in over 35,000 unique tetrasaccharides. *Thus an almost unlimited variety of structurally different oligosaccharides is available for making glycoproteins.* This momentous fact at the molecular level of life is behind the almost incredible spectrum of biological properties observed throughout the living world.

■ The varying glycoproteins that provide this uniqueness are called **glycoforms.**

Different oligosaccharides, N-linked to polypeptides in a cell membrane, make possible the unique abilities of the membranes not only of species but also of individuals within species to discriminate among substances cruising near their cells. The ABO blood groups; the inability of the sperm of one species to fertilize the egg of any other; the action of bacteria normally at just one tissue and not at others; the existence of bacterial infections in animals that do not affect humans (and vice versa); the ability of hormones to be snared only by their own "target" tissues; these properties and many others can be traced to the occurrence of so many different oligosaccharides residing on the surfaces of cell membranes.

■ The oligosaccharide units also are involved in the actions of toxins, viruses, and bacteria.

[2] In the term "glycosaminoglycan," *glycos-* is from "glycose," the generic name for monosaccharides; "glycan" is the generic name of all polysaccharides. The *gly-* part of both glycose and glycan is replaced by the prefix of a specific monosaccharide when naming specific polysaccharides. Thus starch is a *glucan* because it is a polymer of glucose.

TABLE 20.3   Glycosaminoglycans—Their Repeating Disaccharide Units[a]

Structure	Description
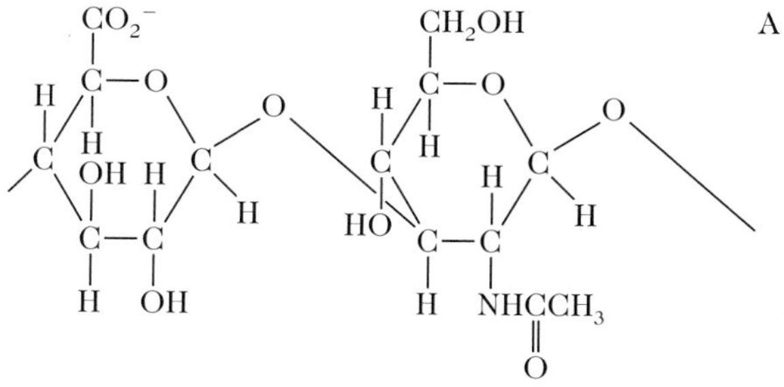 Hyaluronate monomer system	A component of ground substance particularly in connective tissue, in fluid that lubricates joints (synovial fluid), and the vitreous humor of the eye. Depending on the location, from 250 to 25,000 of these disaccharide units are joined into the polymer. (Note the N-acetyl-D-glucosamine unit on the right.)
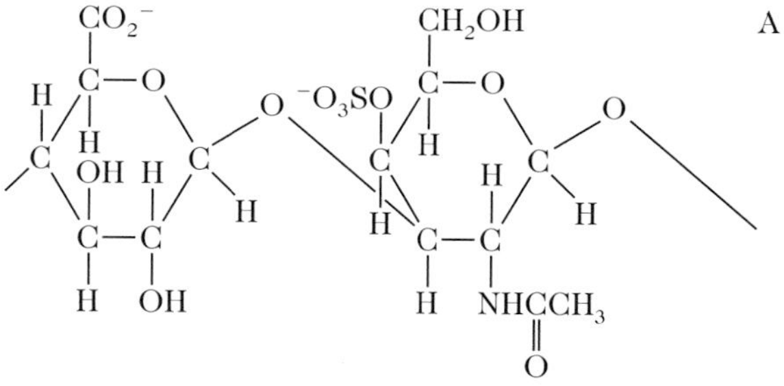 Chondroitin 4-sulfate monomer system	A major component of cartilage and other connective tissues (after the Greek *chondros*, cartilage). There is also a 6-sulfate relative of this system. (Note that the unit on the right is derived from D-galactose, not D-glucose.)
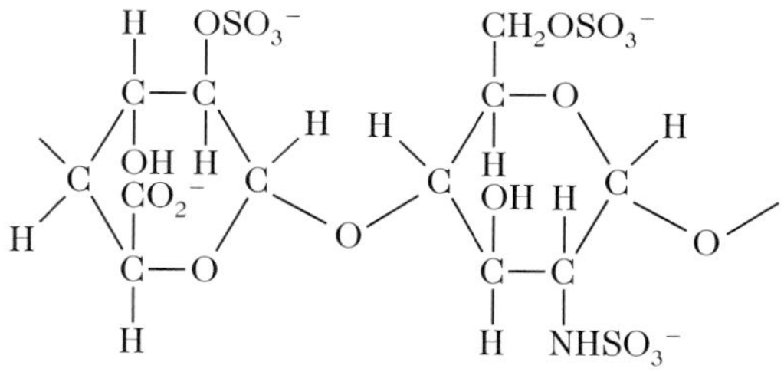 Heparin monomer system	Occurs not in connective tissue, like the above, but in mast cells, cells that line the walls of arteries, particularly in the lungs, liver, and skin. Heparin inhibits the formation of blood clots, and its release from mast cells when an injury occurs prevents clotting from going too far. Heparin is widely used postsurgically to control clotting.

[a] In order to show the ring structures of the individual sugar units in their conventional array, distortions of connecting bonds must sometimes be tolerated in such structures.

The Linkage of Oligosaccharides to Proteins Largely Occurs at Protein Surfaces, Not in Their Interiors   The N-links of oligosaccharides to polypeptides occur most often where the polypeptide strand is following a bend between segments of secondary structure, like β-sheets or α-helices. The oligosaccharide units, therefore, have little if any direct effect on tertiary structure but project, instead, from protein surfaces. This is why we could say near the start of this section that cell membranes are "sugar coated." The "sugar" consists of oligosaccharide units extending into the surrounding spaces from the glycoproteins that make up parts of cell membranes. Oligosaccharide units contribute to the adhesion between cells, and they have critical functions in all of the activities that depend on the "recognition" of hormones and neurotransmitters by a cell.

20.9 CLASSES OF PROTEINS

Three criteria for classifying proteins are solubility, composition, and biological function.

We began this chapter with hints about the wide diversities of the kinds and uses of proteins. Now that we know about their structures, we can better understand how so many types of proteins with so many functions are possible. The following major classifications of proteins and their several examples give substance to the chapter's introduction.

Proteins Can Be Classified According to Solubility When proteins are classified by their solubilities, two families are the **fibrous proteins** and the **globular proteins.**

Fibrous Proteins

1. **Collagens** occur in bone, teeth, tendons, skin, blood capillaries, cartilage, and some ligaments. When such tissue is boiled with water, the portion of its collagen that dissolves is called *gelatin.*

2. **Elastins,** which have elastic, rubberlike qualities, are also in cartilage and are found in stretchable ligaments, the walls of large blood vessels like the aorta, the lungs, and the necks of grazing animals. Elastin, like collagen, is rich in glycine residues and proline, but not in hydroxyproline. Elastin chains are cross-linked by covalently bonded units that are largely responsible for elastin's elasticity.

3. **Keratins** occur in hair, wool, animal hooves, horns, nails, porcupine quills, and feathers. The keratins are rich in disulfide links, which contribute to the unusual stabilities of these proteins to environmental stresses.

4. **Myosins** are the proteins in contractile muscle.

5. **Fibrin** is the protein of a blood clot. During clotting, fibrin forms from its precursor, fibrinogen, by an exceedingly complex series of reactions.

Globular Proteins

Globular proteins are soluble in water or in water that contains salts.

1. **Albumins** are present in egg white and in blood. In the blood, the albumins are buffers, transporters of water-insoluble molecules of lipids or fatty acids, and carriers of metal ions, like Cu^{2+} ions, that are insoluble in aqueous media at pH values higher than 7.

2. **Globulins** include antibodies, factors of the body's defenses against diseases. In addition, enzymes, many transport proteins, and receptor proteins are globulins.

Proteins Can Be Classified According to Biological Function Perhaps no other system more clearly dramatizes the importance of proteins than classifying them by their biological function.

1. Enzymes. The biological catalysts.

2. Contractile muscle. With stationary filaments, myosin, and moving filaments, actin.

3. Hormones. Such as growth hormone, insulin, and others.

4. Neurotransmitters. Such as the enkephalins and endorphins.

■ When meat is cooked, some of its collagen changes to gelatin, which makes the meat easier to digest.

■ Elastin is not changed to gelatin by hot water.

5. Storage proteins. Those that store nutrients that the organism will need such as seed proteins in grains, casein in milk, ovalbumin in egg white, and ferritin, the iron-storing protein in human spleen.

6. Transport proteins. Those that carry things from one place to another. Hemoglobin and the serum albumins are examples already mentioned. Ceruloplasmin is a copper-carrying protein.

7. Structural proteins. Proteins that hold a body structure together, such as collagen, elastin, keratin, and glycoproteins in cell membranes.

8. Protective proteins. Those that help the body to defend itself. Examples are the antibodies and fibrinogen.

9. Toxins. Poisonous proteins. Examples are snake venom, diphtheria toxin, and *Clostridium botulinum* toxin (a toxic substance that causes some types of food poisoning).

SUMMARY

Amino acids   About 20 α-amino acids supply the amino acid residues that make up a polypeptide. The molecules of all but one (glycine) are chiral and in the L-family. In the solid state or in water at a pH of roughly 6 to 7, amino acids exist as dipolar ions or zwitterions. Isoelectric points are the pH values of solutions in which amino acid (or protein) molecules are isoelectric. For amino acids without acidic or basic side chains, the pI values are in the range of 6 to 7. Amino acids with CO_2H groups on side chains have lower pI values. Those with basic side chains have higher pI values. Several amino acids have hydrophobic side chains, but the side chains in others are strongly hydrophilic. The SH group of cysteine opens the possibility of disulfide cross-links between or within polypeptide units.

Polypeptides   Amino acid residues are held together by peptide (amide) bonds, so the repeating unit in polypeptides is —NH—CH—CO—. Each amino acid residue has its own side chain. This repeating system with a unique sequence of side chains constitutes the primary structure of a polypeptide.

Once the primary structure is fashioned, the polypeptide coils and folds into higher features—secondary and tertiary—that are stabilized largely by hydrophobic interactions and hydrogen bonds. The most prominent secondary structures are the α-helix—a right-handed helix—and the β-pleated sheet. Individual polypeptides in collagen, which has an abundance of glycine, proline, and hydroxylated proline residues, are in a left-handed helix. Disulfide bonds form from SH groups on cysteine residues as many proteins assume their tertiary structure.

Proteins   Many proteins consist just of one kind of polypeptide. Many others have nonprotein, organic groups—prosthetic groups—or metal ions. And still other proteins—those with quaternary structure—involve two or more polypeptides whose molecules aggregate in definite ways, stabilized by hydrophobic interactions, hydrogen bonds, and salt bridges. Thus the terms *protein* and *polypeptide* are not synonyms, although for some specific proteins they turn out to be.

Because of their higher levels of structure, proteins can be denatured by agents that do nothing to peptide bonds. A few denatured proteins can be renatured, but this is uncommon. The acidic and basic side chains of polypeptides affect protein solubility, and when a protein is in a medium whose pH equals the protein's isoelectric point, the substance is least soluble. The amide bonds (peptide bonds) of proteins are hydrolyzed during digestion.

Membrane proteins—glycoproteins   Incorporated into the lipid bilayer membranes of cells are proteins (and lipids) with attached oligosaccharide units of widely varying structure. Some of the proteins of a cell membrane provide conduits by which active transport processes can maintain concentration gradients. Other proteins provide gap junctions for direct movements, cell to cell, of certain dissolved species. The oligosaccharides of the membrane proteins stick out away from the membrane surface. They do not appear to affect the overall shapes of their attached polypeptides, and they serve as cell-recognition features for molecules moving near the cell. Certain oligosaccharides, the glycosaminoglycans, make up ground substance, which gives elasticity and shock absorbency to cartilage.

REVIEW EXERCISES

The answers to Review Exercises whose numbers are in color are found in Appendix E. The answers to the other Review Exercises are found in the Study Guide that accompanies this book. The more challenging questions are marked with asterisks.

Amino Acids

20.1   One of the following structures is *not* of an amino acid on the list of standard 20. Which one is not on the list? How can you tell without looking at Table 20.1?

$$NH_2CH_2CH_2CH_2CH_2CHCO_2^-$$
$$| $$
$$NH_3^+$$

A

$$^+NH_3CH_2CHCO_2^- \qquad ^+NH_3CHCO_2^-$$
$$| \qquad\qquad\qquad |$$
$$CH_3 \qquad\qquad\qquad CH_2CO_2H$$

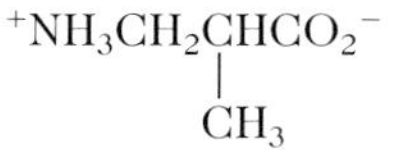

B 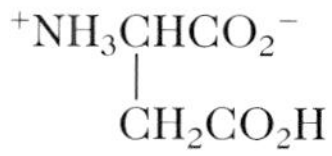   **C**

20.2   The following amino acid is on the standard list.

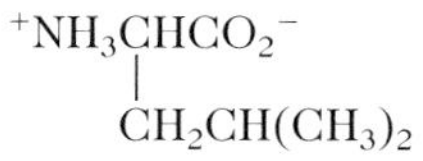

$$^+NH_3CHCO_2^-$$
$$|$$
$$CH_2CH(CH_3)_2$$

(a) What part of its structure would be its amino acid *residue* in the structure of a polypeptide? (Write the structure of this residue.)
(b) With the aid of Table 20.1, write the name and the three-letter symbol of this amino acid.
(c) Is its side chain hydrophobic or hydrophilic?

20.3   What structure will nearly all the molecules of glycine have at a pH of about 1?

20.4   What structure will most of the molecules of alanine have at a pH of about 12?

20.5   Pure alanine does not melt, but at 290 °C it begins to char and decompose. However, the ethyl ester of alanine, which has a free NH₂ group, has a low melting point, 87 °C. Write the structure of this ethyl ester, and explain this large difference in melting point.

*20.6   The ethyl ester of alanine is a much stronger base—more like ammonia—than alanine. Explain this.

20.7   Which of the following amino acids has the more hydrophilic side chain? Explain.

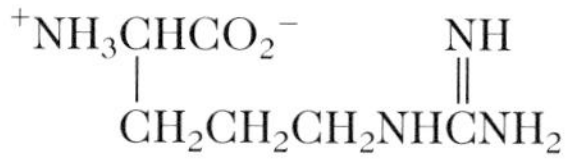

$$^+NH_3CHCO_2^- \qquad NH$$
$$| \qquad\qquad\qquad ||$$
$$CH_2CH_2CH_2NHCNH_2$$

$$^+NH_3CHCO_2^-$$
$$|$$
$$CH_3CHCH_2CH_3$$

A 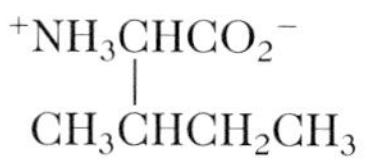   **B**

20.8   Which of the following amino acids has the more hydrophobic side chain? Explain.

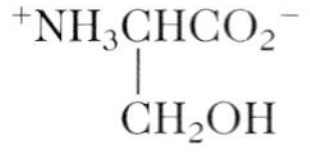

$$^+NH_3CHCO_2^- \qquad ^+NH_3CHCO_2^-$$
$$| \qquad\qquad\qquad |$$
$$CH_2OH \qquad\qquad\quad CH_2C_6H_5$$

A 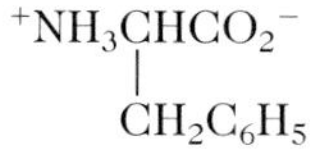   **B**

20.9   Glutamic acid can exist in the following form.

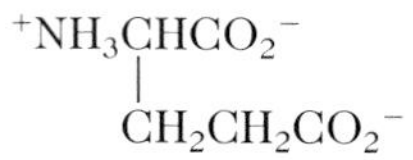

$$^+NH_3CHCO_2^-$$
$$|$$
$$CH_2CH_2CO_2^-$$

(a) Would this form predominate at a pH of 2 or a pH of 9? Explain.
(b) To which electrode, the anode or the cathode—or to neither—would aspartic acid in this form migrate in an electric field?

20.10   When it is said that a substance is poorly soluble in water because of a *hydrophobic interaction,* what does "hydrophobic interaction" mean?

20.11   What kind of a reactant is required to convert cysteine into cystine: an acid, a base, an oxidizing agent, or a reducing agent?

*20.12   Write two equilibrium equations that show how glycine, in its isoelectric form, can serve as a buffer.

20.13   Complete the following Fischer projection formula to show correctly the absolute configuration of L-serine.

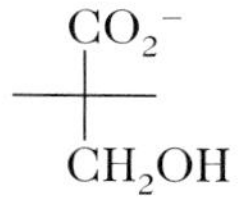

$$CO_2^-$$
$$\underline{\quad\quad|\quad\quad}$$
$$|$$
$$CH_2OH$$

Polypeptides

*20.14   Each of the following structures has an amide linkage. Each can be hydrolyzed to glycine and lysine. The amide linkage in one of the two structures, however, cannot properly be called a *peptide bond.* This is true of which structure? Why?

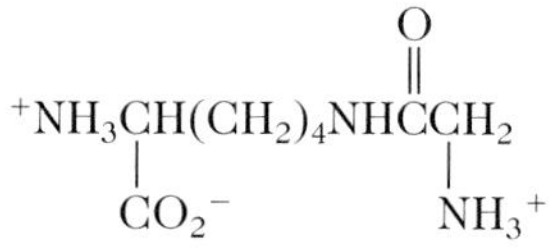

$$\qquad\qquad\qquad O$$
$$\qquad\qquad\qquad ||$$
$$^+NH_3CH(CH_2)_4NHCCH_2$$
$$| \qquad\qquad\qquad\qquad |$$
$$CO_2^- \qquad\qquad\qquad\quad NH_3^+$$

$$\qquad\qquad\qquad O$$
$$\qquad\qquad\qquad ||$$
$$^+NH_3CH_2CNHCHCO_2^-$$
$$\qquad\qquad\qquad\quad |$$
$$\qquad\qquad\qquad (CH_2)_4NH_2$$

A 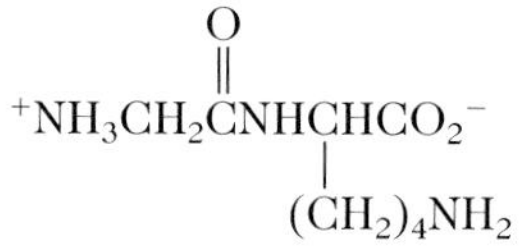   **B**

20.15   Write both the conventional and the condensed structures (three-letter symbols) of the dipeptides that can be made from lysine and cysteine.

20.16 What are the condensed structures of the dipeptides that can be made from glycine and glutamic acid? (Do not use the three-letter symbols.)

20.17 Using three-letter symbols, write the structures of all of the tripeptides that can be made from lysine, glutamic acid, and cysteine.

20.18 Write the structures in three-letter symbols of all of the tripeptides that can be made from glycine, cysteine, and alanine.

20.19 What is the conventional structure of Val-Ile-Phe?

****20.20** Write the conventional structure for Val-Phe-Ala-Gly-Leu.

****20.21** Write the conventional structure for Asp-Lys-Glu-Thr-Tyr.

****20.22** Compare the side chains in the pentapeptide of Review Exercise 20.20 (call it **A**) with those in the following, which we can call **B**.

Lys-Glu-Asp-Thr-Ser

(a) Which of the two, **A** or **B**, is the more hydrocarbon-like?
(b) Which is probably more soluble in water? Explain.

****20.23** Compare the side chains in the pentapeptide of Review Exercise 20.21, which we'll label **C**, with those in Phe-Leu-Gly-Ala-Val, which we can label **D**. Which of the two would tend to be less soluble in water? Explain.

****20.24** If the tripeptide Gly-Cys-Ala were subjected to mild oxidizing conditions, what would form? Write the structure of the product using three-letter symbols.

20.25 What is meant by a *peptide group*? Describe its geometry.

20.26 What atoms or groups are trans to each other in a trans-planar peptide group?

Higher Levels of Protein Structure

20.27 Which *level* of polypeptide complexity concerns the molecular "backbone" and the sequence of side chains?

20.28 What is meant by *native* protein?

20.29 To what level of protein complexity is the disulfide bond normally assigned?

20.30 The disulfide bond is a *covalent* bond. Why isn't it assigned to the primary level of polypeptide structure?

20.31 An enzyme consists of two polypeptide chains associated together in a unique manner. To what level of protein structure is this detail assigned?

20.32 Does the trans-planar nature of the peptide group enlarge or reduce the *range of geometrical options* available to a polypeptide?

20.33 Describe the specific geometrical features of an α-helix structure. What force of attraction stabilizes it? Between what two kinds of sites in the α-helix does this

force operate? How do the side chains become positioned in the α-helix?

20.34 Give a brief description of the secondary structure of an individual polypeptide strand in collagen.

20.35 What function does ascorbic acid (vitamin C) perform in the formation of strong bones?

20.36 Describe the structure and geometry of tropocollagen. How is tropocollagen made into a collagen fibril?

20.37 Bridges between the polypeptide strands in collagen have what principal feature: a hydrophobic interaction, an electrostatic attraction (salt bridge), a disulfide system, or some other kind of covalent linkage?

20.38 What specific force of attraction stabilizes a β-pleated sheet? Where do the side chains take up positions?

20.39 Does an α-helix or a β-sheet describe the *entire* secondary structure of a polypeptide? If not, how do these features occur?

20.40 What factors affect the bending and folding of α-helices in the presence of an aqueous medium?

20.41 What is meant by a salt bridge?

20.42 When is the disulfide bond normally put into place during the formation of a protein?

20.43 In what way does hemoglobin represent a protein with quaternary structure (in general terms only)?

20.44 How do myoglobin and hemoglobin compare (in general terms only)?
(a) Structurally—at the quaternary level
(b) Where they are found in the body
(c) In terms of their prosthetic group(s)
(d) In terms of their functions in the body

Properties of Proteins

****20.45** What products form when the following polypeptide is completely digested? Write the structures.

$$^+NH_3CHCONHCHCONHCHCONHCHCONHCH_2CO_2^-$$
$$\underset{CH_2OH}{|} \quad \underset{CH_3}{|} \quad \underset{\underset{H_3C\quad CH_3}{CH}}{|} \quad \underset{(CH_2)_4NH_2}{|}$$

20.46 Explain why a protein is least soluble in an aqueous medium that has a pH equal to the protein's pI value.

20.47 What is the difference between the *digestion* and the *denaturation* of a protein?

20.48 Some proteins can be denatured by a reducing agent but then completely renatured by a mild oxidizing agent. What functional groups are involved?

Cell Membranes

20.49 What is meant by "gradient" in the term "concentration gradient"?

20.50 Which has the higher level of Na^+, plasma or cell fluid?

20.51 Does cell fluid or plasma have the higher level of K^+?

20.52 In which fluid, plasma or cell fluid, would the level of sodium ion increase if the sodium ion gradient could not be maintained?

20.53 What does the sodium–potassium pump do?

20.54 What does "active" refer to in the term "active transport"?

20.55 What is a *gap junction* and what services does it perform?

20.56 The concentration of what species appears to control the size of the opening of a gap junction?

Glycoproteins and Cell Membranes

20.57 What does the prefix *glyco-* refer to in "glycoprotein"?

20.58 In general terms only, in what structural way does an oligosaccharide differ from a mono- or a disaccharide? From a polysaccharide?

20.59 The term "glycan" refers to what?

20.60 A "D-glucosaminoglucan" would be made of what monomer?

20.61 What is *ground substance*?

20.62 What kinds of substances provide tensile strength to cartilage and what substance gives cartilage its resiliency and shock-absorbing properties?

20.63 What role does the hydrogen bond play in the ability of cartilage tissue to carry out its functions?

20.64 What structural fact about monosaccharides (including the amino sugars) makes possible the huge variety of possible oligosaccharides?

Types of Proteins

20.65 What experimental criterion distinguishes between fibrous and globular proteins?

20.66 What is the relationship between collagen and gelatin?

20.67 How are collagen and elastin alike? How are they different?

20.68 What experimental criterion distinguishes between the albumins and the globulins?

20.69 What is fibrin and how is it related to fibrinogen?

20.70 What general name can be given to a protein that carries a carbohydrate molecule?

Sickle-Cell Anemia (Interaction 20.1)

20.71 What is the primary *structural* fault in the hemoglobin of sickle-cell anemia?

20.72 What happens in blood cells in sickle-cell anemia that causes their shapes to become distorted?

20.73 What problems are caused by the distorted shapes of the red cells?

Prions (Interaction 20.2)

20.74 Prions appear to be normal substances in vertebrates, so what goes wrong at the molecular level in a prion disease?

Mifepristone (RU 486) (Interaction 20.3)

20.75 What is meant by a "receptor protein"?

20.76 In general terms only, how does RU 486 work in the early postcoital period?

Additional Exercises

20.77 When an oligosaccharide unit is cleaved from its glycoprotein, the overall *shape* of the protein section is largely unchanged. Explain.

20.78 Write the structure of a pentapeptide that would hydrolyze to give only alanine.

*20.79 Consider the following structure.

$$\overset{O}{\overset{\|}{^-OCCHNHCCHNHCCHNH_3^+}}$$

$$\underset{CH_2SH\ \ CH_3\qquad CH_2C_6H_5}{}$$

(a) If a polypeptide were *partially* hydrolyzed, could a molecule of this structure possibly form in theory? Explain.
(b) What is the three-letter symbol of the *N*-terminal residue?
(c) How would the structure of this compound be represented using the three-letter symbols and following the rules for writing such a structure?
(d) Would a mild reducing agent have any affect on this compound? If so, write the structure of the product.

ENZYMES

Most people on viewing this sunset on the Oregon coast would stand in awe, oblivious to the enzymes involved in the quickening of the pulse. We have no idea of the molecular basis of our concept of beauty, but we do know that our senses respond rapidly to its presence. The awareness seems to be instantaneous. In some way, cellular chemicals are involved, and we do know something about the factors that affect the rates of cellular reactions.

THIS CHAPTER IN CONTEXT

With our background in proteins, we can now easily move to a study of enzymes. Virtually all are proteins. All are made under the supervision of genes (DNA), and even these syntheses are enzyme catalyzed. We've opened the door on enzymes in earlier chapters. Now we step inside.

21.1   THE NATURE OF ENZYMES

Enzymes are biological catalysts whose activities often depend on cofactors made from B vitamins or metal ions.

A few enzymes are made of nucleic acid, specifically RNA, but they are exceptions. In this chapter we will only study enzymes that are proteins. Basic requirements of any theory about how enzymes work are to explain the unusual specificity of enzymes, the dependence of their activity on pH and temperature, and their amazing catalytic speed.

■ All enzyme molecules and most substrate molecules are chiral; they have handedness.

Enzymes Are Specific for Substrates   *Enzyme specificity* means that a given enzyme acts in vivo on just one substrate or on one kind of bond. The enzyme that catalyzes the hydrolysis of $\alpha(1 \rightarrow 4)$ oxygen bridges in starch, for example, does not work on the $\alpha(1 \rightarrow 6)$ bridges. Some protein-digesting enzymes are even specific about which side chains may be near a peptide bond that they help to hydrolyze.

■ Many enzymes have been isolated for use in catalyzing chemical reactions *in vitro*.

Enzymes Work Best over Narrow Ranges of pH and Temperature   Enzymes are most active only in the pH range normal to their environment in the body. Fumarase, for example, an enzyme involved in the breakdown of sugar or fatty acids, works best at a pH of just below 7 (Figure 21.1). Pepsin, the stomach's protein-digesting enzyme, is most active at a pH of about 2, which is roughly how acidic the stomach juices are.

Enzyme activity is also at a peak only over a small range of temperatures. With ordinary reactions, rates decrease as the temperature decreases, and they increase with increasing temperature. Although enzyme activity is slowed at colder temperatures, it is not increased at increasingly higher temperatures. This is because enzymes generally lose their native molecular shapes (are denatured) by heat.[1]

Enzymes Display Remarkable Rate Enhancements   Enzymes, like all catalysts, affect *rates* by providing a way for a reaction to occur with a lower energy barrier (a lower energy of activation) than the uncatalyzed reaction. Even small reductions of energy barriers cause spectacular rate increases. The enzyme carbonic anhydrase (CA) is an example. It catalyzes the interconversion of bicarbonate ion and protons with carbon dioxide and water.

$$CO_2 + H_2O \xrightleftharpoons[]{\text{carbonic anhydrase}} HCO_3^- + H^+$$

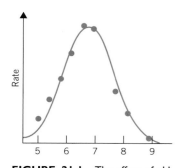

FIGURE 21.1   The effect of pH on the rate of a reaction catalyzed by an enzyme (fumarase). (Reproduced by permission from D. Voet and J. G. Voet, *Biochemistry,* 2nd edition, 1995, John Wiley & Sons, Inc.)

[1] The enzymes in extremophiles, microorganisms that live under extreme temperatures, are exceptions (and have even been nicknamed extremozymes). Some systems thrive in Arctic waters at a temperature of 0 °C. Others live in deep-sea thermal vents at temperatures above 100 °C. (See M. T. Madigan and B. L. Marrs, *Scientific American,* April 1997, page 82.)

In actively metabolizing cells, where CO_2 is being made, so its supply is relatively high, this equilibrium shifts to the right (and so uses up CO_2). In blood circulating through the lungs, where exhaling keeps the supply of CO_2 low, the equilibrium must shift to the left to release CO_2, *and the same enzyme participates in both changes.* Each molecule of carbonic anhydrase aids in the conversion of 600,000 molecules of CO_2 *each second!* This is 10 million times faster than the uncatalyzed reaction, which makes the speed of action of carbonic anhydrase among the highest of all known enzymes.

Enzymes Get Equilibria Established Extremely Rapidly In a chemical equilibrium, it is important to remember that a catalyst speeds up *equilibration.* It accelerates *both* the forward and the reverse reactions. Whether the equilibrium shifts to the right or to the left doesn't depend on the catalyst at all. It depends strictly on the inherent equilibrium constants; on the relative concentrations of reactants and products; on whether other reactions feed substances into the equilibrium or continuously remove them; and on the temperature. All the catalyst does is to speed up whatever shift in equilibrium is mandated by these conditions.

Most Enzymes Consist of Polypeptides plus Cofactors The molecules of most enzymes include a nonpolypeptide component called a **cofactor.** The polypeptide portion is then called the **apoenzyme,** but without the cofactor there is no enzymatic activity.

The cofactor of some enzymes is simply a metal ion, and most of the trace metal ions of nutrition are enzyme cofactors. Zn^{2+}, for example, is the metal ion in carbonic anhydrase. Fe^{2+} occurs in the cytochromes, a family of enzymes needed for biological oxidations.

In other enzymes the cofactor is an organic molecule or ion called a **coenzyme.** Some enzymes have both a coenzyme and a metal ion cofactor.

B Vitamins Are Used to Make Coenzymes We'll now look briefly at some monstrous structures simply to reinforce the point that B vitamin molecules are incorporated into coenzyme molecules. Thiamine diphosphate, for example, is a coenzyme with structure **1,** shown here in the fully ionized form that it has at the pH of body fluids. It's a diphosphate ester of thiamine, a B vitamin.

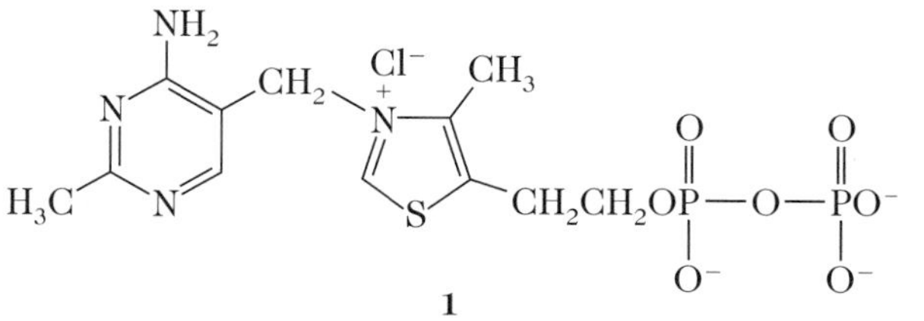

1
thiamine diphosphate

Nicotinamide, another B vitamin, is part of the structure of nicotinamide adenine dinucleotide, **2a,** another important coenzyme. Mercifully, its long name is usually shortened to NAD^+ (or, sometimes, just NAD). The bottom half of the NAD^+ molecule is from adenosine monophosphate, AMP. The upper half is almost like its lower portion except that a molecule of nicotinamide has replaced the two-ring heterocyclic unit.

■ The equilibrium *must* shift to the right to make HCO_3^- and H^+ when the supply of CO_2 is high—a consequence of Le Châtelier's principle.

■ When the diet is deficient in thiamine, often called vitamin B_1, a disease called *beriberi* develops.

■ Nicotinamide's other name is *niacin.* A deficiency of this vitamin leads to *pellagra.*

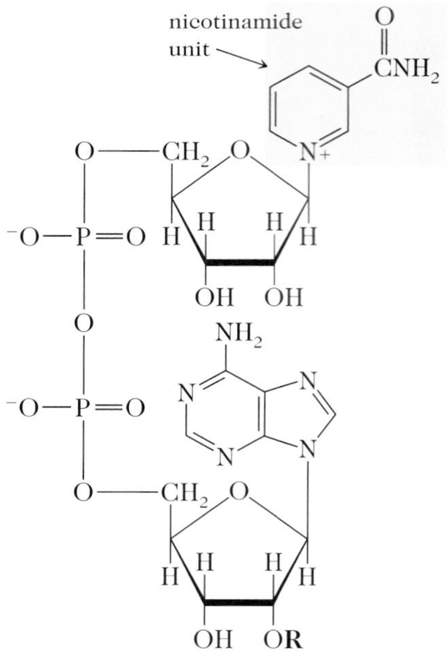

2
a NAD^+ **R** = H
b $NADP^+$ **R** = PO_3^{2-}

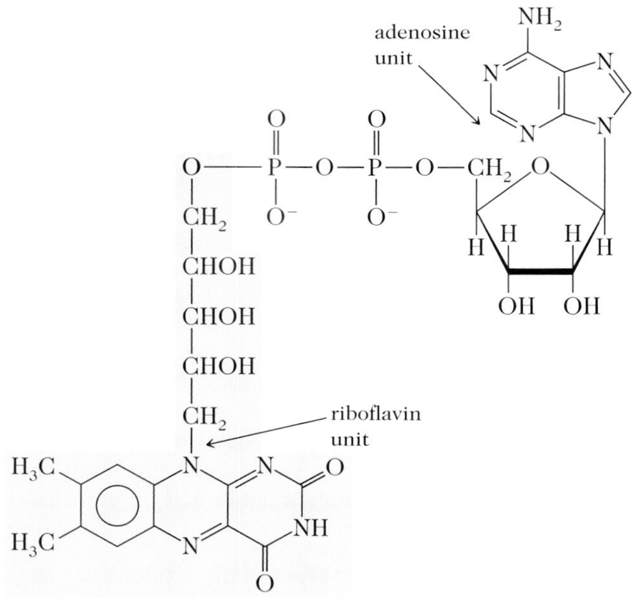

3
FAD

■ The P in $NADP^+$ refers to the extra phosphate ester unit.

Nicotinamide occurs in yet another major coenzyme, nicotinamide adenine dinucleotide phosphate, **2b,** a phosphate ester of NAD^+. Its name is usually shortened to $NADP^+$ (or, sometimes, just NADP). Both NAD^+ and $NADP^+$ are coenzymes in biological redox reactions.

Quite often an equation involving an enzyme with a recognized coenzyme is written with the symbol of the coenzyme itself standing for a whole enzyme. NAD^+, for example, is the cofactor for the enzyme that catalyzes the body's oxidation of ethyl alcohol to acetaldehyde. It serves, in fact, as the actual acceptor of the hydride ion, $H\!:\!^-$, given up by ethyl alcohol.

$$CH_3CH_2OH + NAD^+ \longrightarrow CH_3CH{=}O + NAD\!:\!H \quad + H^+$$

ethanol   ethanal   reduced   hydrogen
                                                        form of NAD^+ ion (buffered)

The NAD^+ unit in the enzyme accepts $H\!:\!^-$ from the alcohol, and we can write this part of the reaction by the following equation.

$$NAD^+ + H\!:\!^- \longrightarrow NAD\!:\!H$$

By accepting the *pair of electrons* in $H\!:\!^-$, NAD^+ is reduced, and $NAD\!:\!H$ (usually written as NADH) is called the *reduced form* of NAD^+. $NADP^+$ can also accept hydride ion, and its reduced form is written as NADPH.

We learned earlier that we may not call something a catalyst unless it undergoes no *permanent* change, which the foregoing examples seem to contradict. In the body, however, a reaction that alters an enzyme is followed by one that regenerates it. The NADH produced by the oxidation of ethyl alcohol, for example, is recovered in the next step in which one enzyme cofactor is FAD (for flavin adenine dinucleotide), **3.** FAD incorporates still another B vitamin, riboflavin. FAD can accept $H\!:\!^-$ from $NAD\!:\!H$, change to $FADH_2$ (the second H is H^+ from the buffer), and so regenerate NAD^+. The overall reaction is

■ Riboflavin is vitamin B_2.

$$NADH + FAD + H^+ \longrightarrow NAD^+ + FADH_2$$

The FAD-containing enzyme is, of course, now in its reduced form, $FADH_2$. $FADH_2$ passes on its load of hydrogen and electrons in yet another step and so is re-oxidized and restored to FAD. The steps continue, but we'll stop here. The main points are that B vitamins are key parts of coenzymes, and that the catalytic activities of the associated enzymes directly involve the parts of the molecules contributed by these vitamins.

Flavin mononucleotide or FMN is a near relative of FAD and also contains riboflavin. The reduced form of FMN is $FMNH_2$, and FMN is also involved in biological oxidations.

Enzymes Are Named after Their Substrate or Reaction Types Nearly all enzymes have names that end in -*ase*. The prefix is either from the name of the substrate or from the kind of reaction. For example a **hydrolase** catalyzes hydrolysis reactions. An *esterase* is a hydrolase that aids the hydrolysis of esters. A *lipase* works on the hydrolysis of lipids. A *peptidase* or a *protease* catalyzes the hydrolysis of peptide bonds.

An **oxidoreductase** handles a redox equilibrium. Sometimes an oxidoreductase is called an *oxidase* when the favored reaction is an oxidation and a *reductase* when the reaction is a reduction. A **transferase** catalyzes the transfer of a group from one molecule to another, and a *kinase* is a special transferase that handles phosphate groups. Other broad categories of enzymes are the **lyases,** which catalyze elimination reactions that produce double bonds; **isomerases,** which cause the conversion of a compound into an isomer; and **ligases,** which cause the formation of bonds at the expense of chemical energy in triphosphates, like ATP.

An International Enzyme Commission has developed a system of classifying and naming enzymes that places considerable chemical information into the enzyme's name. The names of the principal reactants, separated by a colon, are written first and then the name of the kind of reaction is written as a prefix to -*ase*. For example, in moving from left to right in the following equilibrium, an amino group transfers from the glutamate ion to the pyruvate ion.

$$^-O_2CCH_2CH_2\underset{\underset{NH_3^+}{|}}{C}HCO_2^- + CH_3\overset{\overset{O}{||}}{C}CO_2^- \rightleftharpoons \ ^-O_2CCH_2CH_2\overset{\overset{O}{||}}{C}CO_2^- + CH_3\underset{\underset{NH_3^+}{|}}{C}HCO_2^-$$

glutamate ion pyruvate ion α-ketoglutarate ion alanine

The systematic name for the enzyme is *glutamate:pyruvate aminotransferase*. In all but formal publications, such a cumbersome (but unambiguous) name is seldom used. This enzyme, for example, is often referred to simply as GPT (after the older name, glutamate:pyruvate transaminase) and sometimes as alanine transaminase or ALT. We'll largely stick with common names of enzymes.

■ PRACTICE EXERCISE 1 What is the most likely substrate for each of the following enzymes? (a) sucrase (b) glucosidase (c) protease (d) esterase

---

Enzymes Often Occur as a Family of Similar Compounds Called *Isoenzymes* with Identical Functions Identical reactions are often catalyzed by enzymes with identical cofactors but slightly different apoenzymes. These variations are called **isoenzymes** or **isozymes.**

■ Whenever we see -*ase* as a suffix in the name of any substance or type of reaction, the word is the name of an enzyme.

■ The digestive enzymes *trypsin, chymotrypsin,* and *pepsin,* all peptidases, have old (nonsystematic) names that do not end in -*ase.*

■ This reaction, incidentally, is an example of how the body can make an amino acid—here, alanine—from other substances.

■ Here "iso-" signifies the same catalytic function, not identical molecular formulas.

■ When supplies of ATP are low and those of ADP are therefore high, the forward reaction of this equilibrium becomes a major path for making more ATP in muscle cells.

Creatine kinase or CK, for example, consists of two polypeptide chains labeled M (for skeletal muscle) and B (for brain). It occurs as three isoenzymes. All catalyze the transfer of a phosphate group in the following equilibrium, which serves to resupply actively metabolizing tissues with the chemical energy of ATP.

$$\underset{\text{creatine phosphate}}{\overset{O}{\underset{O^-}{\overset{\|}{\underset{|}{-OPONHCNCH_2CO_2^-}}}}\underset{CH_3}{\overset{NH_2^+}{\overset{\|}{C}}}} + ADP \underset{\xrightarrow{\text{creatine}}}{\overset{\text{kinase}}{\rightleftharpoons}} \underset{\text{creatine}}{NH_2\overset{NH_2^+}{\overset{\|}{\underset{CH_3}{C}}}NCH_2CO_2^-} + ATP$$

One CK isoenzyme, called CK(MM), has two M units and occurs in skeletal muscle. Another, CK(BB), has two B units and occurs in brain tissue. The third, CK(MB), has one M and one B polypeptide, and it is present almost exclusively in heart muscle, where it accounts for 15% to 20% of the total CK activity. The rest is contributed by CK(MM).

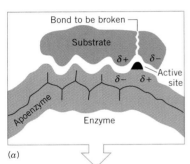
(a)

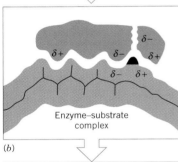
(b)

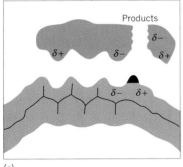
(c)

FIGURE 21.2 The lock-and-key model for enzyme action. (a) The enzyme and its substrate fit together to form an enzyme–substrate complex. (b) A reaction, such as the breaking of a chemical bond, occurs. (c) The product molecules separate from the enzyme.

21.2 ENZYME–SUBSTRATE COMPLEX

The shape of an enzyme molecule allows only its substrate molecules to fit to it and be activated for reaction.

When an enzyme catalyzes a reaction of a substrate, molecules of each must momentarily fit to each other. This temporary combination is called an **enzyme–substrate complex.** It is part of a series of chemical equilibria that carry the substrate through a number of changes until the products of the overall reaction form.

$$\underset{\text{enzyme}}{E} + \underset{\text{substrate}}{S} \rightleftharpoons \underset{\substack{\text{enzyme–}\\\text{substrate}\\\text{complex}}}{E-S} \rightleftharpoons$$

$$\underset{\substack{\text{substrate-}\\\text{activated}\\E-S\\\text{complex}}}{E-S^*} \rightleftharpoons \underset{\substack{\text{enzyme–}\\\text{product}\\\text{complex}}}{E-P} \rightleftharpoons \underset{\substack{\text{enzyme}\\\text{(recovered)}}}{E} + \underset{\text{product}}{P}$$

The first equilibrium is the binding of the enzyme to the substrate. It is like the fitting of a key (the substrate molecule) to a tumbler lock (the enzyme), so the theory is often called the *lock-and-key theory* of enzyme action. Shaped pieces that fit together are said to have *complementary shapes;* or we say that there is *complementarity* between the two shapes (Figure 21.2). *The need for complementarity between molecules of enzyme and substrate explains the specificity of an enzyme.*

For an enzyme–substrate complex to form, there must actually be two kinds of complementarity. The first is what we have already implied—*geometrical complementarity:* a square peg fits a square hole better than a round hole. The other is *physical complementarity,* which concerns factors other than shape—hydrophobic interactions, hydrogen bonds, and electrical charges of *opposite* nature nestling *nearest* each other as the complex forms.

We can now see why enzyme activity is affected by pH. Enzymes, being proteins, carry electrical charges. Their number and distribution are themselves factors in complementarity, and they change with pH as H^+ ions move among proton-accepting and -donating groups. Moreover, as the charges change, a protein molecule often takes on an altered shape, destroying complementarity.

We can now also understand why enzymes stop working when the temperature increases too much. Heat denatures proteins, meaning that the molecules become reshaped. A reshaped enzyme clearly cannot fit to its normal substrate.

Another factor in the high degree of specificity of enzymes comes from their chiral natures and the chiralities of substrates as well. To illustrate, HIV protease is an enzyme that helps to break up the HIV virus (human immunodeficiency virus). Like all enzymes, it is made from amino acids that are all in the L-family. The natural substrate for HIV protease is a protein likewise made of all L-family amino acids. However, D-amino acids have been successfully used to make D versions of both HIV protease and its substrate. The all-D enzyme works only with the all-D substrate; the all-L (natural) enzyme cleaves only the all-L substrate.

Still another example of the significance of chirality to complementarity is the ability of trypsin, a digestive protease, to affect only substrates made of L-amino acids. The enzymes involved in the metabolism of glucose are similarly effective only with D-glucose, not L-glucose units.

The Flexibility of a Protein Molecule Permits Induced Complementarity   To get the substrate to fit to the enzyme depends on some flexibility in the enzyme molecule much as a lock flexibly adapts as the key is inserted. As the substrate molecule nestles onto the enzyme, the molecular groups of the substrate induce the enzyme molecule to adjust its shape to achieve the best fit (Figure 21.3). The initial contact with substrate and enzyme may cause changes in tertiary structure in the polypeptide of the protein. Such changes, which induce stress in the polypeptide, force the enzyme to modify its shape further. The phenomenon is called **induced fit.**

Proteins consist of huge molecules, and not *all* parts are ever *directly* involved in catalysis. Just some of the enzyme's sequence of amino acid residues make up the enzyme's *binding sites,* having side chains with shapes and polar sites complementary to the substrate. Other groups on the enzyme, called *catalytic sites,* handle the actual catalytic work in the complex. Binding sites and catalytic sites are seldom the same. Catalytic sites are often supplied by coenzymes, as we said.

■ Hormone uptake only by specific cells also depends on a flexible lock-and-key kind of recognition.

■ Many enzymes consist of two or more polypeptide subunits *each of which has binding and catalytic sites and all of which become involved in the overall reaction.*

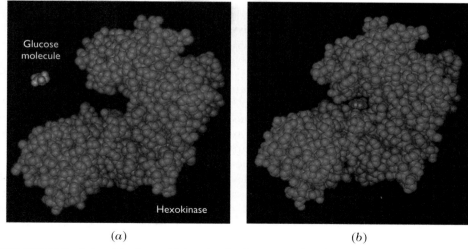

Glucose molecule

Hexokinase

(a)   (b)

FIGURE 21.3   Induced fit theory. (a) A molecule of an enzyme, hexokinase, has a gap into which a molecule of its substrate, glucose, can fit. (b) The entry of the glucose molecule induces a change in the shape of the enzyme molecule, which now surrounds the substrate entirely.

The Initiation of Changes to the Substrate Leads to the Transition State   The fit achieved by the enzyme and substrate in the E—S complex is not perfect. However, the intermolecular forces that cause the complex to *begin* its formation in the first place now continue to work. These forces distort and stretch chemical bonds in the substrate to improve the fit. The chemical energy for this distortion is generally provided by the *gain in overall stability achieved in the complex*. The result of such changes is the conversion of the initial enzyme–substrate complex, E—S, into a substrate activated complex, E—S^*. In E—S^*, the fit between enzyme and substrate is as good as it can be, and the substrate molecule has reached a unique condition of both shape and internal energy called its *transition state*. The perfecting of the enzyme–substrate fit as the transition state *forms*, rather than the initial, somewhat imperfect fitting of enzyme to substrate, largely accounts for the high catalytic power of an enzyme.

Whether E—S^* collapses to return to the reactants or to proceed to the products depends on how much reactant and product concentrations are building up or declining, all in accordance with Le Châtelier's principle. If product forms (see Figure 21.2b), we might suppose that its molecule has a different distribution of electrical charges than that of the reactant molecule (see Figure 21.2c). The enzyme–product complex, E—P, does not hold together, and the product molecule P slips off. The enzyme is then ready to receive another substrate molecule.

This has been a broad and simplified view of enzyme catalyses intended to fortify one point, namely, that theories of how enzymes work start with the idea of induced fitting based on both geometrical and physical complementarity. An increasing number of detailed mechanisms that explain how specific enzymes or teams of enzymes work is accumulating. Interaction 21.1, for example, describes how the principle of complementarity is at work among antibodies, antigens, and the ABO blood groups.

21.3   KINETICS OF SIMPLE ENZYME–SUBSTRATE INTERACTIONS

At high substrate concentrations, an enzyme's rate enhancement levels off.

We know that reaction rates are sensitive to the *concentrations* of reactants. In many reactions between two species, doubling the initial concentration of one species, holding the other's constant, doubles the rate. Many enzyme-catalyzed reactions are also like this if we treat their enzymes as actual (although temporary) reactants. At some fixed initial enzyme concentration, $[E_0]$, doubling the concentration of the substrate, $[S]$, doubles the reaction rate, V. What is significant about enzyme-catalyzed reactions is that this rate enhancement cannot be indefinitely extended to higher initial substrate concentrations. Eventually, at some higher initial substrate concentration, rate acceleration ceases and the rate levels off.

We can see what this means with the aid of Figure 21.4, a plot of initial rates versus initial values of $[S]$. Imagine a series of experiments in all of which the molar concentration of the enzyme, $[E_0]$, is the same. We assume a simple reaction, meaning one for which the enzyme is able to handle only *one* substrate molecule at a time. We will vary only the initial concentration of the substrate, $[S]$, from experiment to experiment. As we said, in most ordinary reactions, the initial rate would double each time we double the initial concentration of one reactant.

In our series of experiments, we do observe something like this, but only in those trials that have *low* initial values of $[S]$, as in part A of the plot in Figure 21.4.

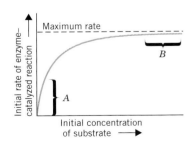

FIGURE 21.4   Initial rates of an enzyme-catalyzed reaction plotted versus the initial concentrations of substrates when the concentration of the enzyme is fixed in each experiment. The sections labeled A and B are discussed in the text.

Here, a small increase in [S] does cause a proportionate increase in initial rate. The curve rises steadily.

In succeeding experiments, however, at higher and higher initial values of [S], the initial rates respond less and less until, in part *B* of the plot, the initial rates are constant *regardless of the value of* [S]. *The reason for the leveling off is that we now have enough substrate molecules to saturate all of the active sites of all the enzyme molecules.* Any additional substrate molecules have to wait their turns, so to speak.

The relationship between *V*, [E_o], and [S] for the kind of reaction just described is given by the following equation.

$$V = \frac{k[E_o][S]}{K_M + [S]} \qquad (21.1)$$

The symbol *k* stands simply for a proportionality constant. The symbol K_M is another constant, called the *Michaelis constant,* and it has a particular value for a specific enzyme-catalyzed reaction. Notice what happens when [S] has very small values, approaching zero. At very low values of [S] the denominator becomes essentially identical with K_M, and Equation 21.1 becomes

$$V = \frac{k}{K_M}[E_o][S] \qquad (21.2)$$

■ In the early part of the 20th century, biochemists Leonor Michaelis and Maude Menten were pioneer scientists in the field of enzyme kinetics.

The ratio, k/K_M is a constant, being a ratio of other constants. Equation 21.2, in other words, says that the velocity of the reaction is directly proportional to [S] at *low* values of [S]. "Directly proportional" translates into a *straight line* or *linear* plot of initial rate versus [S], which is approximately what we see in Figure 21.4. At low initial values of [S], the initial rates lie nearly on a straight line.

The curve bends, as you can see, at higher values of [S]. As [S] becomes high enough, the K_M term in the denominator in Equation 21.1 is overwhelmed by the ever larger [S] term. Eventually, at sufficiently high values of [S], the [S] term in the *numerator* can be canceled by the entire denominator (which is now almost entirely contributed by [S] anyway), leaving the following simple expression as the equation for the velocity, where V_m means the *maximum* velocity.

$$V_m = k[E_o] \qquad (21.3)$$

But [E_o] is a *constant,* the concentration of the enzyme (in any form, free or combined in complex), and *k* is, of course, also a constant. So the maximum velocity, V_m, *must* be a constant. Thus at high values of [S], the rate of the enzyme-catalyzed reaction must level off at a maximum, as you can see it does in Figure 21.4. If we substitute the expression from Equation 21.3 into Equation 21.1, we obtain what is called the *Michaelis-Menten equation* for the rate of a simple enzyme-catalyzed reaction.

$$V = \frac{V_m[S]}{K_M + [S]} \qquad (21.4)$$

■ The *E*—S complex is sometimes called the Michaelis complex to honor the work of Leonor Michaelis.

The reason for a *constant* rate at high values of [S] is that all enzyme molecules are now saturated with substrate and are in the form of the complex *E*—*S*. Further catalysis depends on the freeing of enzyme from enzyme–product complexes. At one specific value of [S], namely, [S] equal to K_M, Equation 21.4 reduces to give

$$V = \frac{1}{2}V_m \qquad (21.5)$$

MOLECULAR COMPLEMENTARITY AND IMMUNITY, AIDS, AND THE ABO BLOOD GROUPS

The *immune system,* as large and complex as the nervous system, is the body's array of defenses against *pathogens*—disease-causing microorganisms and viruses. We cannot in a brief special topic do justice to the immune system, of course, but we can take note of some of the ways in which it shares basic operating principles with the enzyme–substrate reaction. The concept of the fitting of a substrate to an enzyme by means of geometrical and physical complementarity is also at the molecular base of the body's immune system as well as the existence of blood type groups.

When a pathogen has penetrated the first line of defense, the physical barriers of skin and mucous membranes, white blood cells known as *lymphocytes* go into action. They all begin life in bone marrow, but not all mature there. Two kinds of immunity involving two kinds of lymphocytes are recognized. One is *cellular immunity,* and it is handled by *T-lymphocytes* or *T-cells* (after *t*hymus tissue, where T-cells mature). Cellular immunity handles viruses that have gotten inside cells, as well as parasites, fungi, and foreign tissue.

AIDS
The human immunodeficiency virus (HIV) is able to destroy certain kinds of T-cells, the *helper T-cells.* This renders the immune system deficient in its ability to handle infections and results in AIDS, acquired immune deficiency syndrome. Relatively nonlethal diseases normally handled routinely by the body thus become lethal in AIDS victims.

Antigen–Antibody Reaction
The second kind of immunity is *humoral immunity* (after an old word for fluid, *humor*). Humoral immunity is the responsibility of the *B-lymphocytes* or *B-cells* (because they mature in *b*one marrow). B-Cells act mostly against bacterial infections but also against those workings of viral infections that occur outside of cells. We'll limit the continuing discussion to the work of B-cells.

B-cells carry and manufacture *antibodies,* glycoproteins that are able, by an interaction like that between substrate and enzyme, to attract and take antigens out of circulation and defeat the spread of the pathogen. An *antigen* is any molecular species or any pathogen that induces the immune system to make antibodies as well as gives the immune system a molecular–cellular memory for the antigen. Thus, at a later invasion of the same antigen, the immune system is poised for a far more rapid defensive response than it initially had. A *vaccine* is able to start the initial defensive response leading to the molecular memory for the antigen without causing the disease itself.

Figure 1 represents an antibody that is *dipolar;* it has two cross-linking molecular groups. The antigen in Figure 1 is represented by a unit that can become bound to at least three antibody binding sites. *The antibody protein is specific for just one antigen.* As

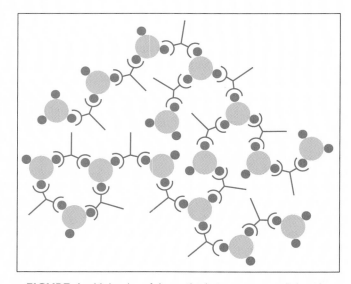

FIGURE 1 Molecules of the antibody, in green, cross-link with antigen particles, in red, to form a mass resembling a huge copolymer. (Reproduced by permission from D. Voet and J. G. Voet, *Biochemistry,* 2nd edition, 1995, John Wiley & Sons, Inc.)

you can see, the interaction of antibody with its antigen essentially "polymerizes" the entire system into one vast "copolymer." The product is now in a far less soluble form and it bears molecular markings that are recognized by other white cells (phagocytes) that engulf the "polymer" and destroy it. There are some antigen–antibody complexes that are destroyed by a series of interacting proteins called the *complement system*. By tying up the antigens, the antibodies prevent the spread of the infection and thus allow the system the time needed to destroy the antigens.

The ABO Blood Groups

The surfaces of red blood cells carry projecting oligosaccharide units of glycolipids. There are differences, however, among individuals in the structures of these sugar residues. One of the consequences of these differences is the existence of blood group systems, one being the *ABO system*. You might have type A, type B, or type O blood. Some have a combination type, AB blood.

If you are of type A and by a transfusion are given blood from a type B person, the red cells from the type B blood will clump together (agglutinate), likely causing a blockage of blood capillaries that could be fatal. Thus your type A blood is able to "see" something in type B blood as a foreign material. Type A blood contains in the serum portion (the liquid minus the cells) an antibody against type B blood. What specifically is the antigen in the type B blood is the molecular unit at the tip of an oligosaccharide joined to a glycolipid of the (type B) red cell. This is why

each kind of red cell is described as carrying an *antigen*, one of three types, A, B, and H.

If you are of type A, your serum includes anti-B antibodies. People with type B blood have anti-A antibodies. Those with AB blood have neither anti-A nor anti-B antibodies. (AB blood type people are able to accept transfusions from people of any blood type. However, in all but emergency situations, transfusions are normally done using type AB blood.) Type O people carry the H "antigen" on their red cells, and their blood has *both* anti-A and anti-B antibodies. Type O people, therefore, can *receive* transfusions only from individuals with type O blood. At the same time, type O people can *give* transfusions to all types—they are universal donors—because the antigen on the red cells in type O people actually has no "enemies" in other types of blood, no antibodies that can attack and agglutinate type O red cells when they are transfused into people of other blood types.

The H antigen in type O blood is given the name antigen because it is the precursor to the A and B antigens of other blood types. Type A individuals make type A antigen by adding an *N*-acetylgalactosamine residue to the tip of a glycolipid on the red cell. Type B people make type B antigen by adding a galactose residue to the same glycolipid. Type O individuals simply lack the enzymes needed for these transformations. The differences among these enzymes are thought to involve single amino acid residue substitutions in the enzymes' polypeptides. Table 1 summarizes donor–acceptor relationships for the blood types.

TABLE I Blood Type Acceptor–Donor Options

If Your Blood Type Is	You Can Accept Blood from One of This Type	You Can Donate Blood to One of This Type
O	O	O, A, B, AB
A	A or O	A or AB
B	B or O	B or AB
AB	AB, A, B, O	AB

This equation gives meaning to the Michaelis constant, K_M; K_M is the substrate concentration at which the reaction rate is one-half of the maximum rate.

Our discussion involving Equations 21.1 to 21.5 concerned the response of rate to concentrations of enzyme and substrate when the enzyme carries only one active catalytic site. Let's now turn our attention to enzymes with more than one site; they offer the system numerous pathways for the regulation of enzyme action.

21.4 REGULATION OF ENZYMES

Enzymes are switched on and off by initiators, effectors, inhibitors, genes, poisons, hormones, and neurotransmitters.

■ The prevention of enzyme synthesis often involves the regulation of genes and their work of directing the synthesis of polypeptides.

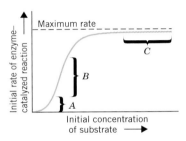

FIGURE 21.5 Initial rates of enzyme-catalyzed reactions plotted versus initial values of substrate concentrations, at fixed enzyme concentration, when an allosteric effect is observed. Sections labeled *A* and *B* of the curve—a sigmoid curve—are discussed in the text.

A cell cannot be allowed to do everything at once. Some of its possible reactions have to be shut down while others occur. One way to keep a reaction switched off is to prevent its enzyme from being made in the first place. Hormones and neurotransmitters are natural regulators of enzymes, and we'll learn about them in the next chapter. In this section we study several other means to control enzymes.

Some Enzymes Initially Resist the Formation of an Enzyme–Substrate Complex
One of the very significant features of many enzymes is that the initial sharp rise of the curve in Figure 21.4 does not occur when the concentration of the substrate, [*S*], is low. It's as if the enzyme is inactive at low substrate concentrations. For such enzymes, the plots of initial values of [*S*] versus initial rates look more like the curve in Figure 21.5. The plot has a lazy S shape, so it's called a *sigmoid plot* (after *sigma,* Greek for S). A sigmoid plot means that the rate increases very slowly with initial substrate concentration, then it takes off in the normal response of rate to concentration, and finally it levels off in the usual way.

Sigmoid rate plots are found among enzymes that remain inactive *until a sufficient concentration of substrate forces them into active forms.* The significance of a sigmoid curve is that it suggests something about *enzyme activation.*

The Catalytic Sites in Some Enzymes Are Activated by Reactions at Other Sites
Enzymes with sigmoid rate curves (see Figure 21.5) consist of two or more polypeptide units *each* with a catalytic site that normally is in an *inactive* configuration, even in the presence of some substrate. Inactive catalytic sites, however, are eventually activated by the substrate, but only when the initial concentration of the substrate is high enough.

Let's suppose, for simplicity, that our enzyme is made of just two polypeptide chains, each having a catalytic site. We'll represent each site by a geometric shape, as shown in Figure 21.6*a.* We have to suppose that the shape of each site is not quite complementary to the substrate, that the substrate must itself induce the correct fit, because the enzyme's response is sluggish at low substrate concentration. We're in region *A* of the sigmoid rate curve (Figure 21.5), the region of the slower than normal rate.

Battering by a substrate molecule induces a conformational change in the polypeptide unit, enabling the substrate to fit to one catalytic site. When this occurs, the other polypeptide simultaneously adopts a new shape *causing the second catalytic site to become active.* Now the same enzyme molecule can accept the second substrate molecule *much more easily than the first.* The enzyme is now being used at maximum efficiency, and the rate of reaction takes off, putting the system into part *B* of Figure 21.5.

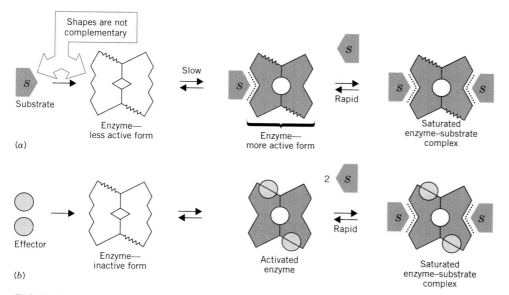

FIGURE 21.6 Allosteric activations. (*a*) Allosteric activation by substrate. (*b*) Allosteric activation by effector.

The phenomenon of one active site being activated by an event occurring *elsewhere* on the enzyme is called **allosteric activation**. The enzyme's subunits, in other words, cooperate with each other to cause full activation, but this doesn't happen until the level of substrate concentration has climbed high enough to start the process. *The activity of an enzyme subject to allosteric activation is thus regulated by how much its services are needed.* The "need," of course, increases as the concentration of substrate increases, because something has to be done about the increasing level of substrate.

We'll see in a later chapter how a similar allosteric effect is caused by oxygen when it interacts with hemoglobin, which is not an enzyme, and how this enables hemoglobin to operate at 100% efficiency, or very nearly so.

Allosteric Activation Can Be Caused by Effectors Instead of Substrates   The catalytic sites of some enzymes are activated by substances called **effectors** that are not substrates. When their molecules bind allosterically to the enzyme, meaning at a location distinct from the catalytic site, they force a configurational change that activates the enzyme (see Figure 21.6*b*). The effector may, for example, be a molecule whose own metabolism *needs the products* made by the enzyme it activates.

Nerve Signals Can Indirectly Tell an Effector to Work   Two of the important effectors are *calmodulin,* a protein found in most cells, and *troponin,* another protein present in muscle cells. Neither works as an effector, however, until it has itself been activated. The activator is calcium ion, and cells control their calcium ion levels by active transport mechanisms mediated by nerve signals.

Normally, the level of calcium ion that moves freely in solution in the cytosol is only about 10^{-7} mol/L. It must be kept extremely low, because the cytosol contains phosphate ion, which forms an insoluble salt with Ca^{2+}. The level of Ca^{2+} just outside the cell is about 10^{-3} mol/L, very considerably higher. Despite the concentration gradient, which nature would normally erase by diffusion, the calcium ions stay outside until something changes the cell's permeability to them. Nerve signals do this.

■ *Allo-,* other; -*steric,* space—the other space or the other site.

■ The cytosol is the *solution* in the cytoplasm and does not include the organelles in the cytoplasm.

The overall sequence is roughly as follows. A nerve signal opens protein channels in the cell membrane for Ca^{2+} ions. They enter the cell and bind to calmodulin or troponin. The effector is thus activated, and it then activates an enzyme to cause some chemical work or to cause muscle contraction. When the signal is over, the channels close, and Ca^{2+} ions are pumped back out through other portals. The effector is thus inactivated. This mechanism thus connects nerve signals to specific chemical activities in cells.

■ Other kinds of proproteins are known, like *proinsulin*, the precursor to the hormone *insulin,* a blood sugar regulator. The conversion of proinsulin to insulin entails the removal of a 33-residue polypeptide unit.

Some Enzymes Are Activated by the Removal of a Polypeptide Unit   Several digestive enzymes are first made in inactive forms called **zymogens** or **proenzymes.** Their polypeptide strands have several more amino acid residues than the enzyme, and these extra units cover over the active site. Then, when the active enzyme is needed, a complex process is launched that clips off the extra units and, by exposing the active site, activates the enzyme.

One of the functions of *enteropeptidase,* a compound released in the upper intestine when food moves in from the stomach, is to convert the zymogen trypsinogen into the enzyme trypsin, which helps to digest proteins. Trypsin is activated by the deletion of a small polypeptide unit in trypsinogen. When no food is present, there is no need for trypsin, but when food enters, enteropeptidase comes in as well, and then trypsin is activated. It's all a rather beautiful instance of coordination.

■ Kinases are the enzymes for this phosphorylation, and they must themselves be activated (usually by Ca^{2+}) only when needed.

Phosphorylation Activates Some Enzymes   Glycogen phosphorylase, the enzyme that catalyzes the hydrolysis of glycogen to glucose, is made in an inactive state. However, when a serine side chain, CH_2OH, is changed to its phosphate ester, $CH_2OPO_3^{2-}$, the enzyme is activated.

Inhibitors Can Keep Enzymes Switched Off until They Are Needed   Some substances, called **inhibitors,** bind reversibly to the enzyme and prevent it from working. In **allosteric inhibition,** molecules of the inhibitor bind to the enzyme somewhere other than the active site. This affects the shape of the active site or a binding site, and the enzyme–substrate complex cannot form (Figure 21.7a). Then if some reaction changes the inhibitor so that it no longer sticks to the enzyme, the catalyst becomes active.

■ Inhibitors are *negative effectors.*

In **competitive inhibition,** the inhibitor is a nonsubstrate molecule with a shape similar enough to that of the true substrate *that it can compete with the substrate for attachment to the active site.* When the inhibitor molecules lock to the enzyme's active sites, but do not undergo a reaction and leave, then the enzyme has become useless to the true substrate (see Figure 21.7b).

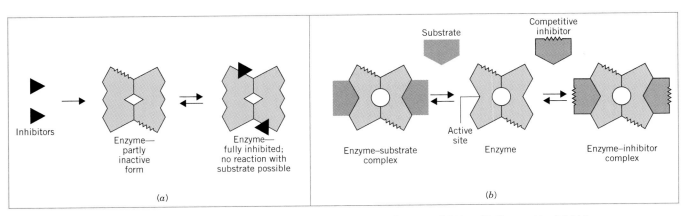

FIGURE 21.7   Enzyme inhibition. (*a*) Allosteric inhibition. (*b*) Competitive inhibition.

A competitive inhibitor doesn't have to be a product of the enzyme's own work. It can be something else the cell makes, or it could be the molecules of a medication. What its molecules must do is *resemble* those of the normal substrate enough to bind to the active site of the enzyme.

Sometimes an inhibitor is the *product* of the reaction being catalyzed or one of the products produced later in a series of connected reactions. Now we have **feedback inhibition.** As the level of such a product increases, its molecules "feed back" with increasing success as inhibitors to one of the enzymes in the series of reactions that helped to make the product. *A feedback inhibitor works as an allosteric, noncompetitive inhibitor.* The amino acid isoleucine, for example, is a feedback inhibitor. It is made from another amino acid, threonine, by a series of steps, each with its own enzyme. As more and more isoleucine is made, its molecules more and more inhibit the enzyme involved with threonine in the first step, E_1, of the series.

$$\underset{\text{threonine}}{\overset{+}{N}H_3\overset{|}{C}HCO_2^- \atop \overset{|}{C}HOH \atop \overset{|}{C}H_3} \xrightarrow{E_1} \xrightarrow{E_2} \xrightarrow[{\substack{\text{inhibition of } E_1 \\ \text{by molecules} \\ \text{of isoleucine}}}]{E_3} \xrightarrow{E_4} \xrightarrow{E_5} \underset{\text{isoleucine}}{\overset{+}{N}H_3\overset{|}{C}HCO_2^- \atop \overset{|}{C}HCH_3 \atop \overset{|}{C}H_2CH_3}$$

■ Both threonine and isoleucine have two tetrahedral stereocenters. Can you spot them, and can you tell how many stereoisomers there are of each?

The beautiful feature of feedback inhibition is that the system for making a product shuts down automatically when enough product is made. Then, as the cell consumes this product, it eventually uses even product molecules that have been serving as inhibitors. The result is that when the product concentration has dropped very low, the enzyme needed to make more is released from its bondage.

Feedback inhibition is very common in nature. It helps to maintain a condition of **homeostasis** in which disturbances to systems by stimuli are minimized, because the stimulus is able to start a series of events that restores the system to the original state. Body temperature is a condition maintained by homeostatic mechanisms. The body does this so well that even small changes in temperature tell us that something is wrong.

A familiar homeostatic mechanism in the home is the work of a furnace controlled by a thermostat. When the room becomes hot enough (the desired "product"), the thermostat trips and the furnace shuts off. In time, the temperature drops, the thermostat trips back, and the furnace restarts.

Antibiotics Inhibit Enzymes   A broad family of compounds called **antimetabolites** includes some made by bacteria and fungi and called **antibiotics.** Antibiotics inhibit or prevent the normal metabolism of a disease-causing bacterial system. Some antibiotics work by inhibiting an enzyme that the bacterium needs for its own growth. Both the sulfa drugs and penicillin work in this way.

■ An antimetabolite is called an *antibiotic* when it is the product of the growth of a fungus or a natural strain of bacteria.

Poisons Often Cause Irreversible Enzyme Inhibition   The most dangerous **poisons** are effective even at very low concentrations because they are powerful inhibitors of enzymes. By inhibiting an enzyme, "a little poison goes a long way." The cyanide ion, for example, forms a strong complex with one of the metal ion cofactors in an enzyme needed for our use of oxygen.

Enzymes that have SH groups are denatured and deactivated by such poisonous heavy metal ions as Hg^{2+}, Pb^{2+}, Cu^{2+}, and Ag^+.

Nerve gases and their weaker cousins, the organophosphate insecticides, inactivate enzymes of the nervous system. In the mid 1990s evidence appeared that the Gulf War syndrome suffered by many soldiers might have enzyme inactivation at its root.

■ PRACTICE EXERCISE 2 The following overall change is accomplished by a series of steps, each with its own enzyme.

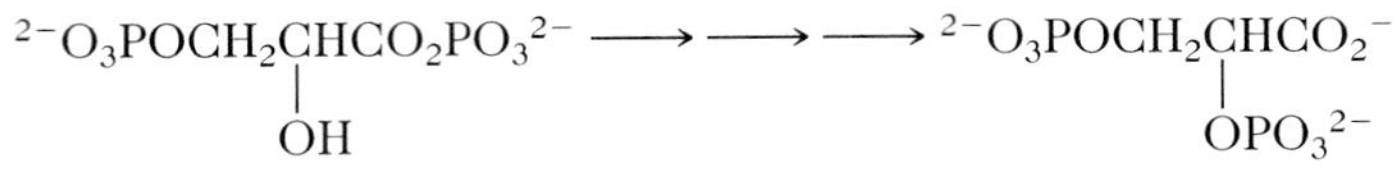

$$^{2-}O_3POCH_2CHCO_2PO_3{}^{2-} \longrightarrow \longrightarrow \longrightarrow {}^{2-}O_3POCH_2CHCO_2{}^{-}$$
$$\qquad\quad | \qquad\qquad\qquad\qquad\qquad\qquad\qquad\qquad\qquad\quad |$$
$$\qquad\quad OH \qquad\qquad\qquad\qquad\qquad\qquad\qquad\qquad\quad OPO_3{}^{2-}$$

1,3-bisphosphoglycerate   2,3-bisphosphoglycerate
(1,3-BPG)   (2,3-BPG)

One of the enzymes in this series is inhibited by 2,3-BPG. What kind of control is exerted by 2,3-BPG on this series? (Name it.)

21.5 ENZYMES IN MEDICINE

The specificity of the enzyme for its substrate provides several methods of medical diagnosis.

Enzymes that normally work only inside cells are not found in the blood except at extremely low concentrations. When cells are diseased or injured, however, their enzymes spill into the bloodstream. Much can be learned about the disease or injury by detecting such enzymes and measuring their levels.

Enzyme Assays of Blood Use Substrates as Chemical "Tweezers" Despite the enormous complexity of blood and the very low levels of enzymes in it, enzyme assays of blood are relatively easy to carry out. *The substrate for the enzyme is used to find its own enzyme,* and the specificity of the enzyme–substrate system ensures that it will find nothing else. If no enzyme is present to match the substrate, nothing happens. Otherwise, the extent of the reaction of the substrate measures the concentration of the enzyme. In this section, we learn about some examples of this medical technology.

Viral Hepatitis Is Detected by the Appearance of GPT and GOT in Blood Heart, muscle, kidney, and liver tissue all contain the enzyme GPT (for glutamate:pyruvate aminotransferase), which we introduced earlier. The liver, however, has about three times as much GPT as any other tissue, so the appearance of GPT in the blood generally indicates liver damage or a viral infection of the liver, such as viral hepatitis. The blood level of another enzyme, GOT (for glutamate:oxaloacetate aminotransferase), also increases in viral hepatitis.

Heart Attacks Cause Increased Levels of Three Enzymes in Blood Serum A *myocardial infarction* (MI) is the withering of a portion of the heart muscle following some blockage of the blood vessels that supply it with oxygen and nutrients. Such blockage can be caused by deposits, by hardening, or by a clot. If the patient survives, the withered muscle becomes scar tissue, and the outlook for a reasonably active life is generally good, particularly if treatment is started promptly. A diagnosis of an infarction of exceptionally high reliability can be made by the analysis of the serum for several enzymes and isoenzymes.

■ The popular term for this set of events is *heart attack.*

When a myocardial infarction occurs, the serum levels of three enzymes normally confined inside heart muscle cells begin to rise (Figure 21.8). These are CK (Section 21.1), GOT (just described) and LD. LD stands for lactate dehydrogenase, which catalyzes the formation of the oxidation–reduction equilibrium between lactate and pyruvate.

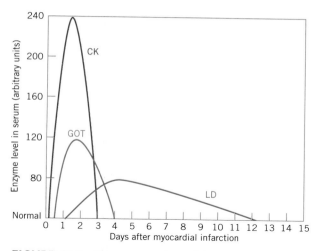

FIGURE 21.8 The concentrations of three enzymes in blood serum increase after a myocardial infarction. Here CK is creatine kinase; GOT is glutamate:oxaloacetate aminotransferase; and LD is lactate dehydrogenase.

$$\underset{\text{pyruvate}}{CH_3\overset{O}{\overset{\|}{C}}CO_2^- } + NAD\!:\!H + H^+ \overset{\text{lactate dehydrogenase (LD)}}{\underset{\text{(an NAD}^+\text{enzyme)}}{\rightleftharpoons}} \underset{\text{lactate}}{CH_3\overset{OH}{\overset{|}{C}}HCO_2^- } + NAD^+$$

As we learned earlier, the CK enzyme occurs as three isoenzymes, CK(*MM*), CK(*BB*), and CK(*MB*). To be sure that the increase in serum CK level is caused by injury to the *heart* tissue, the clinical chemist uses a special technique to analyze specifically for the CK(*MB*) isoenzyme common to heart muscle tissue.

As additional confirmation of an infarction, the serum LD fraction is further separated by a special technique into the five LD isoenzymes. The *relative* serum concentrations of these five differ distinctively between a healthy person (Figure 21.9*a*) and one who has suffered an infarction (Figure 21.9*b*). Of particular importance are the relative levels of the LD_1 and the LD_2 isoenzymes. Normally the LD_1 level is less than that of the LD_2, but after a myocardial infarction an "LD_1-LD_2 flip" occurs; their relative concentrations become reversed. The level of LD_1 rises higher than that of LD_2. When both the CK(*MB*) band and the LD_1-LD_2 flip occur, the diagnosis of a myocardial infarction is essentially 100% certain.

■ The numbered subscripts of the LD isoenzymes are simply the order in which their molecules are separated by a special method.

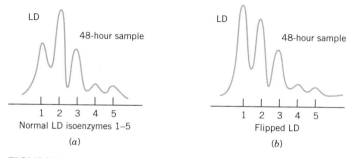

FIGURE 21.9 The five lactate dehydrogenase isoenzymes. (*a*) The normal pattern of the relative concentrations of the isoenzymes. (*b*) The pattern after a myocardial infarction. Notice the reversal in relative concentration between the first two, LD_1 and LD_2. This is the LD flip.

This technique works well for severe heart attacks, but in mild attacks, the CK(*MB*) level sometimes does not increase. In 1996 another blood assay was approved, this one being an analysis for one of the three subunits of troponin (Section 21.4), which also spill from damaged heart muscle cells. The troponin level is a good indicator not only if a heart attack has occurred (instead of just chest pains) but also of the likelihood that the patient will survive.

The Blood Glucose Level Can Be Determined Enzymatically   The regular determination of the level of glucose in blood is important to people with diabetes, because when this level is poorly managed, several complications can occur. One commercially available test uses a combination of chemicals, including enzymes, that react with blood glucose to generate a dye. The intensity of the resulting dye is proportional to the blood glucose level.

Enzymes Can Be Immobilized on Solid Supports   In some applications, enzymes are physically immobilized onto the surfaces of extremely tiny, inert plastic beads. *Heparinase,* for example, is immobilized onto beads that are then used to catalyze the breakdown of the heparin added to blood sent through a hemodialysis machine. The added heparin inhibits the clotting of the blood outside the body, but it has to be removed before the blood goes back into the body. The heparinase on the beads catalyzes this removal.

Several enzymes are immobilized on the tips of special electrodes where they act in the analyses of specific substances in a fluid, such the urea or cholesterol levels in blood.

A Natural Blood Clot–Dissolving Enzyme Can Be Activated by Other Enzymes
If you cut yourself, there is set in motion a huge cascade of enzyme-catalyzed reactions that bring about the formation of fibrin from a circulating polypeptide, fibrinogen. Fibrin's molecules are long and stringy, and they tangle, forming a brush-heap mat that entraps water and puts a seal, a blood clot, on the cut. After the wound heals, the clot must be dissolved. No part of a blood clot must break loose and circulate to the heart, because it will be stopped by tiny capillaries in heart muscle tissue. Such a blockage is one cause of a myocardial infarction. Clots in the lungs and the brain are obviously very serious as well.

To dissolve the fibrin of a clot, the body normally converts a zymogen, *plasminogen,* into the enzyme *plasmin.* (We described the chemistry of zymogen activation in the previous section.) Plasminogen actually is absorbed out of circulation by the fibrin *as the clot forms.* Its eventual activation, therefore, occurs exactly where its active form, plasmin, is needed. Plasmin, a protease, then catalyzes the hydrolysis of fibrin, and the clot "dissolves."

One of the interesting facts about plasminogen is that it becomes more susceptible to activation when bound to fibrin than when it is simply in circulation. In time, a circulating *tissue plasminogen activator* does what its name implies. It catalyzes the conversion of plasminogen to plasmin at the site of the blood clot. The plasmin then goes to work to help dissolve the clot.

Therapy for a myocardial infarction is intended to open blocked capillaries as rapidly as possible, before the oxygen starvation of surrounding heart muscle tissue spreads the damage too widely. *Plasminogen activation therapy* has become the most commonly used method for treating clot-related infarctions. When this therapy is applied as soon as possible following a myocardial infarction, the chances are better that long-term heart muscle damage will be slight.

Plasminogen activation therapy involves introducing into circulation one of the commercially available plasminogen activating enzymes, such as tissue plasminogen

In the Chemstrip MatchMaker device, the blood glucose level is measured by the intensity of a dye produced enzymatically and converted into milligrams of glucose per deciliter (100 mL) of blood.

■   This therapy is called *thrombolytic therapy* because it *lyses* (breaks down) *thrombi* (blood clots).

activator (TPA) or streptokinase. When therapy is started within the first 4 hours of a myocardial infarction, TPA appears to have a small edge in effectiveness.

TPA is now also used to treat those kinds of strokes caused by blood clots (ischemic strokes), which account for 80% of strokes. This is the first actual *treatment* of a stroke; before, all that could be done was to provide rehabilitative therapy to victims. Streptokinase has also been tried, but it tends to make matters worse by increasing the risk of bleeding in the brain.

■ Tissue plasminogen activator made by genetic engineering is referred to as *recombinant* tissue plasminogen activator and symbolized as rTPA.

SUMMARY

Enzymes Enzymes are the catalysts in cells. Some consist wholly of one or more polypeptides, and other enzymes include a cofactor besides the polypeptides. The cofactor can be an organic coenzyme, a metal ion, or both. Some coenzymes are phosphate esters of B vitamins and, in these examples, the vitamin unit usually furnishes the enzyme's active site. Because they are mostly polypeptide in nature, enzymes are vulnerable to all of the conditions that denature proteins. The name of an enzyme, which almost always ends in *-ase*, usually discloses either the identity of its substrate or the kind of reaction it catalyzes.

Some enzymes occur as small families called isoenzymes in which the polypeptide components vary slightly from tissue to tissue in the body. An enzyme is very specific both in the kind of reaction it catalyzes and in its substrate. Enzymes make possible reaction rates that are substantially higher than the rates of uncatalyzed reactions.

Induced fitting When an enzyme–substrate complex forms, the active site is brought together with the part of the substrate that is to react. Binding sites on the enzyme guide the substrate molecule in and induce a change in the conformation of the enzyme molecule to produce the best fit of the substrate. The recognition of the enzyme by the substrate occurs as a flexible, lock-and-key model that involves complementary shapes and electrical charges.

Enzyme kinetics An enzyme is a reactant in the initial phase of the reaction as it functions catalytically. At a sufficiently high initial substrate concentration, the active sites on all the enzyme molecules become saturated by substrate molecules, so at still higher initial substrate concentrations there is little further increase in the initial rate. The rate behavior of enzyme-catalyzed reactions for which the enzyme has one catalytic site is described by the Michaelis-Menten equation. At low substrate concentration, the equation produces a straight-line plot of rate versus substrate concentration. At high substrate concentration, the equation shows that the rate must reach a maximum value.

Some enzymes seem to respond sluggishly to small increases in substrate concentration when the latter is very low. These display a sigmoid rate curve, and they require activation by substrate molecules before they produce their dramatic rate enhancements.

Regulation of enzymes Enzymes that display a sigmoid rate curve can be activated allosterically either by their own substrates or by effectors. Some enzymes are activated by genes. Some enzymes that are parts of the membranes of cells (or small bodies within cells) are activated by the interaction between a hormone or a neurotransmitter and its receptor protein. The work of many of these is to cause changes in the calcium ion level in a cell.

Other enzymes, such as certain digestive enzymes, exist as zymogens (proenzymes) and are activated when some agent acts to remove a small part that blocks the active site. The kinase enzymes are activated by being phosphorylated.

Enzymes can be inhibited by a nonproduct inhibitor or by competitive feedback that involves a product of the enzyme's action. Some inhibitors act allosterically; they bind to the enzyme at some location other than the catalytic site. Some of the most dangerous poisons bind to active sites and irreversibly block the work of an enzyme, or they carry enzymes out of solution by a denaturant action. Many antibiotics and other antimetabolites work by inhibiting enzymes in pathogenic bacteria.

Medical uses of enzymes The serum levels of many enzymes rise when the tissues or organs that hold these enzymes are injured or diseased. By monitoring these serum levels, and by looking for certain isoenzymes, diseases such as viral hepatitis and myocardial infarctions are diagnosed.

When a blood clot threatens or causes a heart attack, streptokinase or recombinant tissue plasminogen activator (rTPA) can be used to initiate the hydrolysis of the fibrin of the clot. rTPA is used also to treat victims of ischemic strokes.

REVIEW EXERCISES

The answers to Review Exercises whose numbers are in color are found in Appendix E. The answers to the other Review Exercises are found in the Study Guide that accompanies this book. The more challenging questions are marked with asterisks.

Nature of Enzymes

21.1   What are (a) the function and (b) the composition, in general terms only, of an enzyme?

21.2   To what does *specificity* refer in enzyme chemistry?

21.3   Define and distinguish among the following terms.
(a) Apoenzyme   (b) Cofactor   (c) Coenzyme

21.4   Write the equation for the equilibrium catalyzed by carbonic anhydrase. What is particularly remarkable about the enzyme?

21.5   What in general does an enzyme do to an equilibrium?

Coenzymes

21.6   What B vitamin is involved in the $NAD^+/NADH$ system?

21.7   The active part of either FAD or FMN is furnished by which vitamin?

21.8   Complete and balance the following equation.

$$\underset{\substack{| \\ CH_3CHCH_3}}{OH} + NAD^+ \longrightarrow \underset{\substack{\| \\ CH_3CCH_3}}{O} + \underline{\hspace{1cm}} + \underline{\hspace{1cm}}$$

21.9   Complete and balance the following equation.

$$\underline{\hspace{1cm}} + NADH + FAD \longrightarrow NAD^+ + \underline{\hspace{1cm}}$$

21.10   In what structural way do NAD^+ and $NADP^+$ differ? What formula can be used for the reduced form of $NADP^+$?

Kinds of Enzymes

21.11   What *kind* of reaction does each of the following enzymes catalyze?
(a) An oxidase   (b) Transmethylase
(c) Hydrolase   (d) Oxidoreductase

21.12   What is the difference between lactose and lactase?

21.13   What is the difference between a hydrolase and hydrolysis?

21.14   What are isoenzymes (in general terms)?

21.15   What are the three isoenzymes of creatine kinase? Give their symbols and state where they are principally found.

Theory of How Enzymes Work

21.16   What name is given to the part of an enzyme where the catalytic work is carried out?

21.17   How is enzyme specificity explained?

21.18   What is the induced fit theory?

21.19   The Michaelis-Menten equation applies to an enzyme having how many catalytic sites?

*21.20   Which relationship, $V \propto [S]$ or $V \propto [E_o]$, applies under each of the following conditions?
(a) A relatively high concentration of substrate
(b) A relatively low concentration of substrate

21.21   When the velocity of the enzyme-catalyzed reaction obeying the Michaelis-Menten equation equals half of the maximum velocity, what does the Michaelis constant compare to?

Enzyme Activation and Inhibition

21.22   What does *allosteric* mean?

21.23   If the plot of initial reaction rate versus initial substrate concentration at constant $[E]$ has a sigmoid shape, what does this signify about the active site(s) on the enzyme?

21.24   How does a substrate molecule activate an enzyme whose rate curve is sigmoid?

21.25   How does an effector differ from a substrate in causing allosteric activation?

21.26   What are the names of two important effectors? Which one is used in muscle cells?

21.27   What are the approximate concentrations of calcium ion in the cytosol and the fluid just outside a cell? Why doesn't simple diffusion wipe out this concentration gradient?

21.28   Why must the concentration of Ca^{2+} be so low in the cytosol?

21.29   Does the concentration of Ca^{2+} in the cytosol measure the amount of Ca^{2+} in the whole cytoplasm? Explain.

21.30   What does Ca^{2+} do to calmodulin or troponin?

21.31   When Ca^{2+} combines with troponin, what happens with respect to other proteins in the cell? What then happens to Ca^{2+}?

21.32   What is the relationship of a zymogen to its corresponding enzyme? Give an example of an enzyme that has a zymogen.

21.33   What is enteropeptidase and what does it do?

21.34   What is plasmin and in what form does it normally circulate in the blood?

21.35   What role does a phosphorylation reaction have in connection with some enzymes?

21.36   How does competitive inhibition of an enzyme work?

21.37 Feedback inhibition of an enzyme works in what way?

21.38 Why is feedback inhibition an example of a homeostatic mechanism?

*21.39 How do competitive inhibition and allosteric inhibition differ?

21.40 How do the following poisons work?
(a) CN^- (b) Hg^{2+}
(c) Nerve gases or organophosphate insecticides

21.41 What are antimetabolites, and how are they related to antibiotics?

21.42 In broad terms, how does penicillin work?

Enzymes in Medicine

21.43 If an enzyme such as CK or LD is normally absent from blood, how can a *serum* analysis for either tell anything? (Answer in general terms.)

21.44 What is the significance of CK(*MB*) in serum in trying to find out whether a person has had a heart attack and not just some painful injury in the chest region?

21.45 What CK isoenzyme would increase if the injury in Review Exercise 21.44 were to skeletal muscle?

21.46 What is the LD flip, and how is it used in diagnosis?

21.47 How are subunits of troponin used in diagnosis?

21.48 How is immobilized heparinase used?

21.49 Name two enzymes that are available to help dissolve a blood clot. Which one occurs naturally in human blood?

21.50 What substance makes up most of a blood clot, and what happens to it when TPA works?

21.51 What property of TPA makes it useful in the treatment of ischemic stroke?

Complementarity and Immune Responses (Interaction 21.1)

21.52 What is a pathogen (in general terms)?

21.53 What two kinds of immunity are recognized and how do they differ?

21.54 The HIV particle attacks what kind of immunity and in what way?

21.55 What is an antibody? An antigen?

21.56 At the molecular level, what aspects of molecular structure explain the high specificity of the immune response?

21.57 In what kind of immunity are the B-cells operative, and what kind of substance do B-cells eventually make to counter an alien material?

21.58 Why is a glycolipid on a red blood cell of an A-type individual referred to as an antigen and not an antibody?

21.59 At the molecular level involving their red blood cells, in what specific ways do the A, B, and O types differ?

21.60 What is present in the blood of an A-type person that makes receiving blood from a B type dangerous?

21.61 Why is it that O types can donate blood to people of any type but can receive blood only from other O types?

Additional Exercises

21.62 How does the plot of initial rate versus initial [S] at constant [E] look when (a) an allosteric effect is occurring and (b) no allosteric effect is observed? (Draw pictures.)

*21.63 If you drink enough methanol, you will become blind or die. One strategy to counteract methanol poisoning is to give the victim a nearly intoxicating drink of dilute ethanol. As the ethanol floods the same enzyme that attacks the methanol, the methanol gets a lessened opportunity to react and it is slowly and relatively harmlessly excreted. Otherwise, it is oxidized to formaldehyde, the actual poison from an overdose of methanol:

$$CH_3OH \xrightarrow{\text{dehydrogenase}} CH_2O$$

methanol   formaldehyde

What kind of enzyme inhibition might ethanol be achieving here? (Name it.)

*21.64 Referring to the equation given on page 615 for the conversion of threonine to isoleucine, what is the maximum number of milligrams of isoleucine that could be made from 150.0 mg of threonine?

22 HORMONES AND NEUROTRANSMITTERS

If hormones could speak, and you can be sure they're pumping here, they might want to ask the brain if it has lost its mind! But hormones "know" better; they're busy here mobilizing everything her body owns to meet the upcoming challenge at the bottom.

THIS CHAPTER IN CONTEXT

Our bodies are like corporations. They're organizations of highly specialized units. Information must flow smoothly among the parts to maintain a well-run system. Although genes are the *sources* of information, its *flow* involves hormones and neurotransmitters, chemical messengers sent out in response to a variety of signals. We said some time ago that the rock bottom needs of a living system are materials, energy, and information. Our study of the molecular basis of information thus continues in this chapter.

22.1 CHEMICAL COMMUNICATION—HOW IT WORKS

Hormones and neurotransmitters are the chief methods by which cells communicate with one another.

Hormones are made in specialized organs, the *endocrine glands,* and secreted into the bloodstream. They usually travel some distance to particular **target tissues** or **target cells** where they launch chemical responses. They might move as near as neighboring tissue or as far away as 15 to 20 cm. The signal for releasing a hormone could be something conveyed by one of our senses, such as light or an odor, or it might be a stress, or a variation in the level of a particular substance in some fluid, like blood. Insulin, for example, is released when the level of glucose in blood increases. Epinephrine is sent out when we're suddenly frightened.

A **neurotransmitter** is a chemical made in one nerve cell, a neuron, and sent to the next neuron. The distinctions between hormones and neurotransmitters are thus not great, concerning mostly how far they travel. How they cause what they do bears close similarities, as we will see.

■ Greek *hormon,* arousing.

Cell Receptors Identify Messengers by a Lock-and-Key Mechanism   At a target cell, a hormone or neurotransmitter delivers its message by binding to a cell **receptor.** Each receptor is a unique glycoprotein whose molecules can accept just the messenger intended for it. A lock-and-key mechanism is at work. Sometimes molecules of toxic substances, virus particles, or even dangerous bacteria "recognize" a glycoprotein on the cells of one particular tissue and cause entirely unwanted changes.

Receptor-like proteins are also involved in cell-to-cell adhesion. Sperm cells carry receptor-like glycoproteins, making them able to lock to molecular units only on the ova of the same species, thanks again to a lock-and-key mechanism. Antibodies "recognize" antigens in the body's immune response by a lock-and-key mechanism.

■ We can imagine a particular oligosaccharide on a membrane's surface acting like a fishing line, dangling a unique lure at its tip that is attractive only to one kind of hormone or neurotransmitter.

Chemical Messengers Enter Cells by Four Major Mechanisms   The formation of a receptor−messenger complex changes the receptor structure, so that now it is activated to do something. It may be to activate a gene, or an enzyme, or to alter the permeability of a cell membrane so that certain ions or small molecules can move across it. In neurons, the activation of a receptor sends the nerve signal on. We'll consider specific examples later. We offer only a broad overview here.

Figure 22.1 outlines the principal ways by which signals enter cells. Some hormones, once "recognized" by the target cell, move directly through the cell membrane, enter the cytosol, and find a receptor inside the cell. Steroid hormones work in this way. See 1 in Figure 22.1. They bind to receptors close to or inside the cell nucleus, where they induce changes in the way the cell uses DNA.

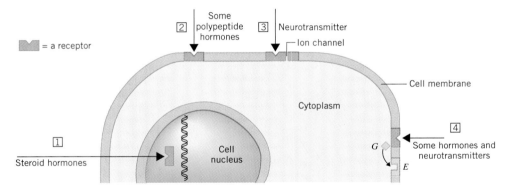

FIGURE 22.1   The ways by which hormones get chemical messages into cells.

Polypeptide hormones cannot migrate directly through cell membranes, so they bind to receptors that are integral parts of the membrane. See ② in Figure 22.1. Insulin and growth hormone work in this way.

Neurotransmitters also bind to membrane-bound receptors, and this opens channels through the membrane for metal ions, ③. We saw in the previous chapter how movements of calcium ions can affect calmodulin or troponin and so activate a series of enzyme-catalyzed reactions.

Some membrane receptors, in accepting neurotransmitters, hormones, or even light photons, activate a subunit of what is nicknamed *G-protein,* ④ in Figure 22.1. It has three subunits plus a cofactor and is bound on the cytosol side of the lipid bilayer. Activated G-protein then activates an enzyme (*E* in Figure 22.1). Remarkably, a large variety of cells share just a few mechanisms for taking advantage of the G-protein. We'll study two, the cyclic AMP and the inositol phosphate cascades.

The Formation and Hydrolysis of Cyclic AMP Is One Way by Which Many Cells Pass on Messages   Cyclic nucleotides, particularly 3′,5′-cyclic AMP, are important secondary chemical messengers, meaning that they form inside a cell following the action of a primary messenger at a cell receptor. How cyclic AMP works is sketched in Figure 22.2.

At the top of the figure we see a hormone—it could just as well be a neurotransmitter—that locks to a receptor molecule at the cell surface. The hormone–receptor complex activates a G-protein subunit whose cofactor is a molecule very much like ADP, only it's made using guanine, not adenine; hence the G.

■ American scientists A. G. Gilman and M. Rodbell shared the 1996 Nobel Prize in Physiology or Medicine for research on the G-protein.

■ *Cyclic* refers to the *extra* ring of the phosphate diester system.

ADP, adenosine diphosphate   GDP, guanine diphosphate

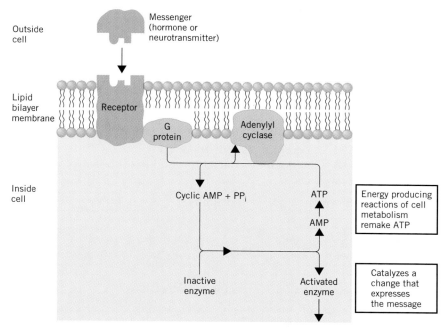

FIGURE 22.2 The activation of the enzyme adenylyl cyclase by a hormone (or a neurotransmitter). The hormone–receptor complex releases a unit of the G-protein, which activates this enzyme. It then catalyzes the formation of cyclic AMP, which, in turn, activates an enzyme inside the cell.

The activation of the G-protein subunit involves exchanging GDP for its triphosphate form, GTP, taken from the cytosol, and which, like ATP, has a third phosphate unit. The energy-rich, activated G-protein finds and activates a molecule of an inactive form of the enzyme adenylyl cyclase, which is also an integral part of the cell membrane.

Once adenylyl cyclase is activated, the "message" is on the inside of the cell membrane, because now the enzyme promptly catalyzes the conversion of ATP into cyclic AMP and diphosphate ion, PP_i.

■ E. W. Sutherland, Jr., an American scientist, won the 1971 Nobel Prize in Physiology or Medicine for his work on cyclic AMP.

The newly formed cyclic AMP now activates an enzyme, which, in turn, catalyzes a reaction, the original goal of the message.

Finally, an enzyme called phosphodiesterase catalyzes the hydrolysis of cyclic AMP to AMP, and this shuts off the cycle. Energy-producing reactions in the cell will now remake ATP from the AMP.

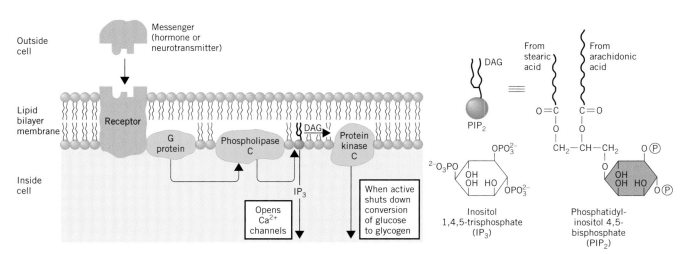

cyclic AMP   AMP

It all seems like such a convoluted process, but when there are several steps there are several opportunities for the system to control things. Let's summarize the steps in this remarkable chemical cascade.

■ Hormones that work through the cyclic AMP cascade that we will encounter later include epinephrine, glucagon, norepinephrine, and vasopressin.

1. A signal releases the hormone or neurotransmitter.

2. It travels to its target cell.

3. The primary messenger molecule finds its target cell by a lock-and-key mechanism and combines with the receptor, which results in the activation of a subunit of the G-protein.

4. The activated G-protein subunit activates the enzyme adenylyl cyclase.

5. Adenylyl cyclase catalyzes the conversion of ATP to cyclic AMP.

6. Cyclic AMP activates an enzyme inside the cell.

7. The enzyme catalyzes a reaction, one that corresponds to the primary message of the hormone or neurotransmitter.

8. Cyclic AMP is hydrolyzed to AMP, which is reconverted to ATP, and the system returns to the preexcited state.

■
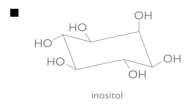

inositol

The Inositol Phosphate System Works in a Way Roughly Similar to the Cyclic AMP System Another of the G-protein systems involves phosphate and trisphosphate esters of inositol, one of the stereoisomers of hexahydroxycyclohexane. *The inositol phosphate system is at the center of the control of a cell's metabolism.* One of its beautiful features is that the same signal that sets off muscle contraction also mobilizes the needed chemical energy. We follow it with the aid of Figure 22.3.

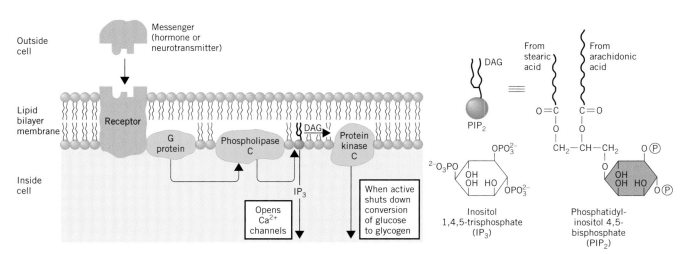

FIGURE 22.3 The inositol phosphate system takes a message from a hormone or neurotransmitter and activates two enzymes to work cooperatively.

Some primary messenger binds to the cell receptor, the G-protein is activated and, in turn, it activates an enzyme called phospholipase C. The pattern so far, as you can see, is quite similar to the cyclic AMP system. Phospholipase C now catalyzes the hydrolysis of a phosphate ester bond in a membrane-bound phospholipid, which we'll call PIP$_2$ (because its full name makes eyes glaze over). The products are two fragments *that are both messengers.* One we'll call IP$_3$ (an inositol trisphosphate) and the other DAG (a diacylglycerol).

DAG is mostly hydrocarbon-like, so it stays in the lipid bilayer of the cell membrane, where it activates a membrane-bound enzyme, protein kinase C. This enzyme helps the cell obtain glucose whose chemical energy the cell will soon need (because IP$_3$ is doing its own work).

The other fragment, IP$_3$, causes the opening of a channel for Ca^{2+} ions, and they flood into the cell. Recall that calcium ions trigger the contraction of a muscle. Of course, muscle work requires energy, so you can see that the two fragments that resulted from one signal, DAG and IP$_3$, work together. IP$_3$ tells the cell to do some work; DAG gets busy with the energy supply (glucose).

■ There are several variations of the G-protein.

■ The phosphoinositol cascade mediates the following activities:
Glycogenolysis in the liver
Insulin secretion
Smooth muscle contraction
Platelet aggregation

22.2 HORMONES AND NEUROTRANSMITTERS

Many drugs, both licit and illicit, intervene in the work of hormones and neurotransmitters.

Structurally, Hormones Come in Four Broad Types   The principal endocrine glands of the human body and the major hormones they secrete are too numerous to catalog in detail. It is impossible to do justice to a subject as vast as hormones in one section of one chapter, so what follows is a very broad sketch of a few chemical aspects of hormone action. In later chapters, where specific hormones are particularly relevant to a topic, we will take them up as needed. Let's here consider only general features of hormones. They come in four broad types.

Some hormones, the *steroid hormones,* are made from cholesterol and so have largely hydrocarbon-like molecules. Being hydrophobic (and lipophilic), they slip easily through the lipid bilayers of target cells. Inside they find their final receptors, and the hormone–receptor complexes move to DNA molecules where they bind and affect the transcriptions of genetic messages. The sex hormones like estradiol, progesterone, and testosterone work in this way.

Among the *polypeptide hormones* are growth factors as well as insulin, oxytocin, and thyroid-stimulating hormone. These are able to alter the permeabilities of their target cells to the migrations of small molecules. The growth factors, for example, help get amino acids inside cells where they are needed for growth. Insulin helps to get glucose inside its target cells. Either the absence of insulin or the absence (or inactivity) of insulin receptors results in the disease diabetes mellitus ("diabetes"). Several neurotransmitters are also polypeptides, and they alter the permeability of a neuron membrane to Ca^{2+} and Na$^+$. The cross-membrane movements of these ions are involved in the electrical signal that flows down a neuron.

The *prostaglandins* (Interaction 19.1) are now classified as hormones, as *local* hormones because they work where they are made.

Finally, a number of hormones are relatively simple *amino compounds* made from amino acids. These include epinephrine and thyroxine. Some are also neurotransmitters. One hormone, melatonin, has become the modern equivalent of snake oil, a "cure" for almost anything (see Interaction 22.1)

INTERACTION 22.1
MELATONIN—HOPE OR HYPE?

There has probably been no over-the-counter pill to receive more hype since Carter peddled "little liver pills," only there may be something to melatonin (see Figure 1), a hormone made naturally by one of our glands. Still, a huge amount of faith concerning the *human* benefits of melatonin has been placed on relatively few experiments with mice. Not only health food stores but also a few scientists suggest that melatonin is just the thing for grievances ranging from

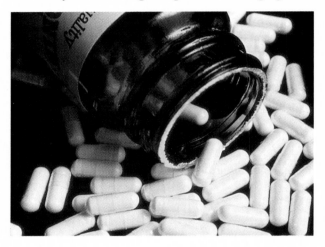

FIGURE 1 Melatonin is widely available without prescription.

autism to schizophrenia, from Alzheimer's disease to jet lag. But where scientists (well, most of them) are cautious about exaggerating melatonin's possibilities, health food stores weigh in with unblushing, declarative statements. It's a cure for AIDS, cancer, cataracts, depression, diabetes, epilepsy, influenza, and Parkinson's disease. It is a sleeping pill and a weight-control agent, and it sharply slows the aging process. So what's the truth about melatonin (at least as of 1997)?

melatonin

Melatonin and the Brain Melatonin is a natural substance (and so cannot be patented), an actual human hormone produced by a tiny, pea-sized gland called the pineal gland, which occurs in the center of the brain. Melatonin is also a hormone widely used in other animals, like protozoa, birds, mice, and rats, the latter being popular in animal studies of new drugs.

Melatonin and the Day/Night Cycle The release of melatonin by the pineal gland is a function of the time of day! When the retina of the eye is receiving daylight, the melatonin level in the blood is relatively low, but at night the level increases. Thus, the sensitivity of the melatonin level to the day/night cycle may be a factor in telling other parts of the body when it's night or day. (As the nights lengthen in the fall in the northern hemisphere, the melatonin level in migratory birds may help to signal that it's time to go where the sun shines more.)

Melatonin and Aging Melatonin levels decrease markedly as we get older. People who cannot resist drawing (and publishing) inferences have, therefore, suggested that all manner of age-associated illnesses as well as aging itself can be prevented or sharply retarded simply by taking this inexpensive pill. Some experiments with mice have indeed bolstered these claims, but no one yet has a clue as to the possible long-term side effects of taking excessive melatonin doses for years. No one knows how higher than normal levels of melatonin might produce a drug interaction with, say, Prozac (another "miracle" drug). A *drug interaction* occurs when the simultaneous presence of two drugs in the system causes effects that are not seen with either drug alone. In other words, the two drugs can interact with each other or with the metabolism or use of each other.

The U.S. Food and Drug Administration has received very few complaints from users of over-the-counter melatonin, and, besides, melatonin is not a synthetic substance under regulatory control. Much more research on melatonin must be done, but getting companies to conduct costly, long-term trials runs into the poor payback prospects of a substance that cannot be patented. Time will tell, as it always does.

Some References

1. R. J. Reiter and J. Robinson. *Melatonin.* New York: Bantam Books, 1995.

2. F. W. Turek, "Melatonin hype hard to swallow," *Nature,* January 25, 1996, page 295.

Neurotransmitters Cross the Synaptic Gap from One Neuron to the Next A partial list of neurotransmitters is given in Table 22.1. Some are nothing more than simple amino acids (like glycine). Others are β-phenylethylamines (like norepinephrine) or catecholamines (like dopamine), and many are polypeptides (like the enkephalins).

Each nerve cell has a fiberlike part called an *axon* that reaches to the face of the next neuron or to one of its filament-like extensions called *dendrites*. A nerve im-

TABLE 22.1   Neurotransmitters

Monoamines

Acetylcholine
$(CH_3)_3\overset{+}{N}CH_2CH_2O\overset{O}{\overset{\|}{C}}CH_3$

Dopamine

$CH_2CH_2NH_2$ — benzene ring with OH, OH

Norepinephrine

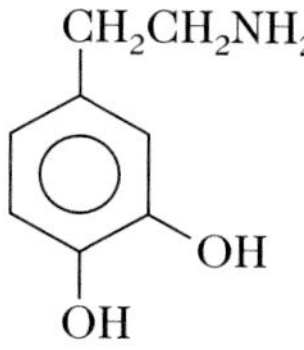

$CHCH_2NH_2$ — benzene ring with OH, OH

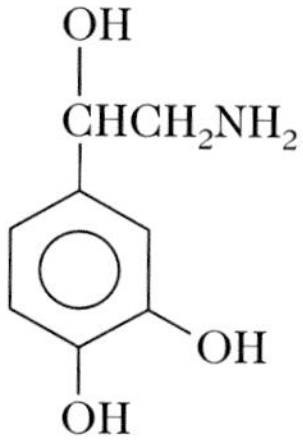

Serotonin

HO — indole ring — $CH_2CH_2NH_2$

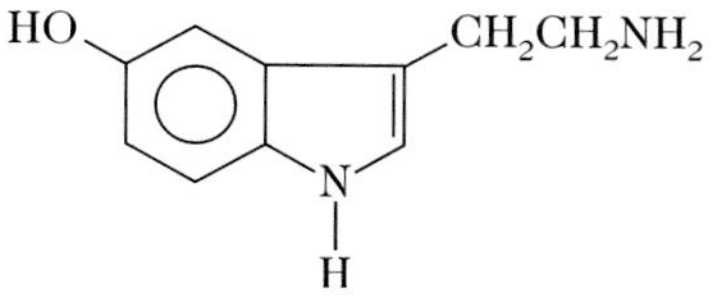

Amino Acids

Glycine
$^+NH_3CH_2CO_2^-$

γ-Aminobutyric acid, GABA
$^+NH_3CH_2CH_2CH_2CO_2^-$

Glutamic acid
$^+NH_3CHCO_2^-$
$\quad\quad | $
$\quad CH_2CH_2CO_2H$

Neuropeptides

Met-Enkephalin   Tyr-Gly-Gly-Phe-Met

Leu-Enkephalin   Tyr-Gly-Gly-Phe-Leu

β-Endorphin   Tyr-Gly-Gly-Phe-Met-Thr-Ser-Glu-Lys-Ser
                      |
                      Gln-Thr-Pro-Leu-Val-Thr-Leu-Phe-Lys-Asn
                      |
                      Ala-Ile-Val-Lys-Asn-Ala-His-Lys-Gly-Gln

Substance P   Arg-Pro-Lys-Pro-Gln-Gln-Phe-Phe-Gly-Leu-Met-NH$_2$

Angiotensin II   Asp-Arg-Val-Tyr-Ile-His-Pro-Phe-NH$_2$

Somatostatin   Ala-Gly-Cys-Lys-Asn-Phe
                      | \Phe
                      Cys |
                      | Trp
                      Ser-Thr-Phe-Thr ———————/

■ The traveling wave of electrical charge moves rapidly, but still not as rapidly as electricity moves in electrical wires.

pulse consists of a traveling wave of electrical charge that sweeps down the axon as small ions migrate at different rates between the inside and the outside of the neuron. To get the impulse launched into the next neuron requires neurotransmitters. They're made within the neuron from amino acids and stored in sacs called *vesicles* located near the ends of the axons.

A very narrow, fluid-filled gap called the *synapse* occurs between the terminal of an axon and the end of the next neuron. Neurotransmitters move across the synapse when the electrical wave triggers their release from vesicles.

When neurotransmitter molecules lock to their receptors on the other side of the synapse, adenylyl cyclase (or phospholipase C) is activated. We'll use the adenylyl cyclase system to illustrate the process (Figure 22.4). The formation of cyclic AMP is catalyzed, and newly formed cyclic AMP initiates whatever change is programmed by the chemicals in the target neuron. An enzyme then deactivates adenylyl cyclase by catalyzing the release of the neurotransmitter molecule.

If the newly released neurotransmitter were a hormone, it would be swept away in the bloodstream, but it's not. It's still a neurotransmitter in the synapse, so unless the system wants it to act again, it must be removed or deactivated. A number of options are open, depending on the neurotransmitter. One method is to break it up by a chemical reaction, the option used where acetylcholine functions.

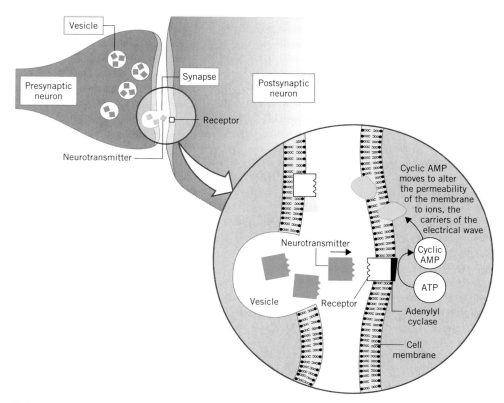

FIGURE 22.4 In neurotransmission, the neurotransmitter molecules, released from vesicles of the presynaptic neuron, travel across the synapse. At the postsynaptic neuron, they find their receptors, and the cyclic AMP system becomes activated.

The Neurotransmitter Acetylcholine Is Swiftly Hydrolyzed   Acetylcholine is a neurotransmitter in the autonomic nervous system or ANS. Local anesthetics like nupercaine, procaine, and tetracaine are blockers of the acetylcholine receptor. Nicotine molecules bind to the receptors in one part of the brain (see Interaction 22.2).

When acetylcholine molecules have done their work, they are catalytically hydrolyzed to choline and acetic acid, the enzyme being acetylcholinesterase.

$$(CH_3)_3\overset{+}{N}CH_2CH_2O\overset{\overset{\displaystyle O}{\|}}{C}CH_3 + H_2O \xrightleftharpoons{\text{acetylcholinesterase}} (CH_3)_3\overset{+}{N}CH_2CH_2OH + HO\overset{\overset{\displaystyle O}{\|}}{C}CH_3$$

acetylcholine   choline   acetic acid

Within two milliseconds (2×10^{-3} s) of the release of acetylcholine into the synapse, all its molecules are broken down. The synapse is now cleared for a fresh release of acetylcholine from the presynaptic neuron, if the signal for its release continues. If it doesn't, the action shuts down.

The Botulinum Toxin Prevents a Neuron from Releasing Acetylcholine   The botulinum toxin is an extremely powerful toxic agent made by the food-poisoning botulinum bacterium and works by preventing the release of acetylcholine into the synapse. Without this neurotransmitter, the cholinergic nerves of the ANS can't work.

Nerve Poisons Deactivate Acetylcholinesterase   Nerve gases are organophosphates that deactivate acetylcholinesterase. Some insecticides are mild versions of organophosphate nerve poisons, and they work in the same way. The deactivation of acetylcholinesterase means that the signal transmitted by acetylcholine can't be turned off. In the case of nerve poisons intended as poison gases in warfare, the signal continues unabated until the heart fails, usually in a minute or two.

Drugs that block the action of a neurotransmitter are called **antagonists** to the neurotransmitter. The neurotransmitter itself is sometimes referred to as an **agonist**.

Some Used Neurotransmitters Are Reabsorbed by the Presynaptic Neuron   Norepinephrine and serotonin are deactivated by being reabsorbed by the sending neuron, where they are then degraded by the action of enzymes called **monoamine oxidases** or MAO.

Drugs That Inactivate Monoamine Oxidases Are Used to Treat Depression   Among other places, norepinephrine works in the brainstem, where mood regulation is centered. If monoamine oxidases are inactivated, then an excess of norepinephrine builds up in brainstem cells, and some spills back into the synapse and sends signals on. In some mental states, like depression, an abnormally *low* level of norepinephrine develops, so now one would *want* to inactivate the monoamine oxidases. This would leave what norepinephrine there is to carry on its work. Thus, some of the antidepressant drugs, like iproniazid, work by inhibiting monoamine oxidases.

Other antidepressants, like amitriptyline (e.g., Elavil) and imipramine (e.g., Tofranil), inhibit the reabsorption of norepinephrine by the sending neuron. Without this reabsorption and then degradation by monoamine oxidases, the level of norepinephrine and its signal-sending work stay high.

Margin notes:

■ The ANS nerves handle the signals that run the organs that have to work autonomously (without conscious effort), such as the heart and the lungs.

■ The ANS nerves that use acetylcholine are called the *cholinergic nerves*.

■ The nerves that use norepinephrine are called the *adrenergic nerves* (after an earlier name for norepinephrine, noradrenaline).

■

iproniazid

INTERACTION 22.2
NICOTINE—HIJACKER OF ACETYLCHOLINE RECEPTORS

Poisons and Toxicity Tobacco smoke contains over 4000 compounds. About 400 are poisons, including nicotine and carbon monoxide, and another 40 are carcinogens (cancer causers). In this discussion we'll learn how poisonous nicotine is, whether deaths from smoking are caused directly by nicotine, and, if not, just what nicotine does do. First, we need to define some terms.

A *poison* is any substance that causes sickness or death when taken in sufficient amounts, but the phrase "in sufficient amounts" makes the definition slippery. Everybody knows that arsenic is a poison, yet we all ingest traces of it from natural food sources with no apparent harm. Nicotine is a poison, but we do not die simply from inhaling the traces of it in a smoke filled room. Salt, on the other hand, is not considered to be a poison. Yet, if you drank seawater (3% sodium chloride) for a day or two instead of regular water, you would so upset the electrolyte balance of your blood that you would die. So what is a poison, really?

In common talk, to be called a poison means that a substance can kill in small doses, some poisons being "deadlier" than others. To compare the deadliness of poisons, each must be evaluated for its *toxicity.* (Those who study toxicities are called *toxicologists*). Toxicity is a quantitative rating of a substance in terms of its *lethal dose,* which is not the amount per individual that would kill everybody receiving it but rather the quantity that would kill an agreed-on percentage. Commonly, this is taken to be 50%, and a poison's LD_{50} value is the quantity per kilogram of body mass that is the *lethal* dose to 50% of an animal population.

It is difficult to obtain LD_{50} data on the human toxicities of poisons, of course, so tests involving mice or rats are generally used. With few exceptions, the resulting data provide enough margin of error for assessing human risks. The avenue of administration must also be reported, whether orally (by mouth) or subcutaneously (beneath the skin), because toxicities vary according to different modes of entry. The oral LD_{50} value for ethyl alcohol (beverage alcohol) in rats, for example, is 7 g/kg. According to a classification used by toxicologists (see Table 1), ethyl alcohol is therefore rated as *slightly toxic.* But this is only one category in a whole range of toxicities. Less than 1 mg/kg of hydrogen cyanide, the gas chamber agent, will kill an adult human, and this poison is rated as *supertoxic.*

The way the term is used publicly, what people call a *toxic substance* or *poison* roughly turns out to be any substance that is lethal at 50 mg/kg or less. Nicotine, for example, when injected into the bloodstreams of mice, causes death at a level of 0.3 mg nicotine per kilogram of body mass, making pure nicotine a supertoxic agent.

nicotine

Although nicotine is supertoxic, the dose obtained by smoking a cigarette is obviously not lethal. Of the roughly 9 mg of nicotine in one cigarette, a 70-kg man smoking such a cigarette takes into his blood-

TABLE I   Toxicity Classes*

Toxicity Class	Probable Lethal Dose in Humans	
	mg/kg	For a 70-kg (150-lb) Man
Supertoxic	Less than 5	A taste (less than 7 drops)
Extremely toxic	5–50	Between 7 drops and 1 teaspoon
Very toxic	50–500	Between 1 teaspoon and 1 ounce
Moderately toxic	500–5,000	Between 1 ounce and 1 pound
Slightly toxic	5,000–15,000	Between 1 pound and 4 pounds
Practically nontoxic	Above 15 g/kg	More than 4 pounds

* Adapted by permission from Alphonse Poklis and Amadeo J. Pesce, "Toxicology," Chapter 47 in *Clinical Chemistry, Theory, Analysis, and Correlation,* Lawrence A. Kaplan and Amadeo J. Pesce, Editors. St. Louis: The C. V. Mosby Company, 1984.

stream, via his lungs, between 1 and 3 mg of nicotine. In terms of quantity of nicotine per kilogram of body mass, this amounts roughly to 0.01 to 0.04 mg/kg, well below a lethal dose. Yet, according to the U.S. Centers for Disease Control and Prevention, roughly 420,000 Americans die each year from the *effects* of cigarette smoking. The deaths are mostly from lung and heart diseases and cancer. Tobacco products are the only materials legally sold that, when used as intended by the manufacturers, eventually cause death. Cigarette smoking is the leading cause of avoidable, premature death in the United States. Why, then, does a quarter of the U.S. population smoke?

Nicotine as an Addictive Drug To label a drug *addictive,* medical organizations of both the United States and other countries have agreed on particular criteria. These include compulsive use even in the face of a desire to quit, psychoactive effects (those produced in the brain), behavioral changes, and withdrawal symptoms. The withdrawal symptoms for

nicotine include restlessness (anxiety, difficulty in concentrating or sleeping), reduced heart rate, depression, a craving sensation, and weight gain.

There is little doubt that beginning smokers who become hooked enjoy the reward of feeling good. Experienced smokers, on the other hand, smoke not only for the pleasure but also to avoid withdrawal symptoms, pains that begin in a matter of minutes after the last smoke to a few hours, and they last up to 2 weeks. "Feeling good" to the experienced smoker is actually related as much, if not more to the disappearance of the discomforts of withdrawal than to the arousal pleasure caused by nicotine. Thus, the nicotine in tobacco, whether smoked or smokeless, is the addicting agent, the chemical whose psychoactive effects have smokers asking for more. The substances that actually cause respiratory diseases and cancer are other materials, not nicotine, including compounds found in the tars of the tobacco smoke. Cigarettes sold with no nicotine were a commercial flop; smokers seek their nicotine "fix." They have given up

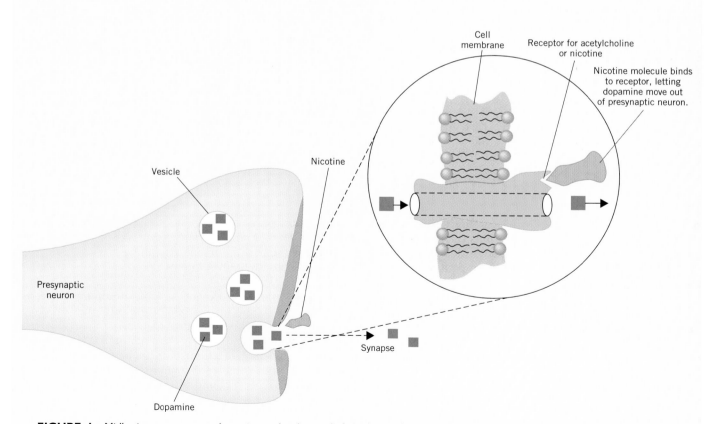

FIGURE I Midbrain neurons carry dopamine molecules ready for release when nicotine molecules bind to acetylcholine receptors.

their ability to control their behavior, which is at the heart of the meaning of addiction. To succeed in quitting smoking requires effort, will power, and motivation. What makes nicotine addictive?

Nicotine and Acetylcholine Receptors Nicotine molecules change the structure and function of key receptors in the midbrain, the brain's mesolimbic area (which includes the prefrontal cortex, the nucleus accumbens, and the ventral tegmental area). Nicotine molecules move onto cholinergic receptor sites in the midbrain normally targeted by the neurotransmitter acetylcholine (see Figure 1). The result in the midbrain is the release of dopamine, another neurotransmitter and one associated with feelings of arousal and pleasure. The amphetamine drugs as well as cocaine work in much the same way at their molecular levels of action.

The binding of nicotine to receptor proteins retards their normal turnover rate, the rate at which these proteins are destroyed. But the rate of synthesis of new receptor proteins is not slowed. The result is that increasing exposure to nicotine gives to nicotine addicts a greater number of such receptors than

they would otherwise have. With time, however, more and more receptors are desensitized or permanently inactivated, and this is why nicotine *tolerance* develops with time. Drug tolerance means that a given dose becomes less and less able to produce the desired effects, so more of the drug is sought. Nicotine tolerance drives the smoker to more and more smoking and deeper inhaling. Such behavior only accelerates the exposure to the agents in smoke that cause all of the medical problems.

References

1. Lois R. Ember, "The Nicotine Connection," *Chemical and Engineering News,* November 28, 1994, page 8.

2. C. E. Bartecchi, T. D. MacKenzie, and R. W. Schrier, "The Human Cost of Tobacco Use," *The New England Journal of Medicine,* March 31, 1994, page 907, and April 7, 1994, page 975.

3. D. A. Kessler, "Nicotine Addiction in Young People," *The New England Journal of Medicine,* July 20, 1995

amitriptyline (Elavil)

imipramine (Tofranil)

fluoxetine (Prozac)

chlorpromazine

haloperidol

The most widely prescribed antidepressant today is fluoxetine (Prozac). It inhibits the reabsorption of serotonin. Fluoxetine was approved in 1987 specifically for major depression, but pill poppers without this ailment are taking it for mood alteration, with what long-term consequences no one yet knows.

Dopamine Excesses Occur in Schizophrenia Dopamine, like norepinephrine, is also a monoamine neurotransmitter. It occurs in neurons of the midbrain involved with feelings of pleasure and arousal as well as with control of certain movements.

In schizophrenia the dopamine-using neurons are overstimulated, because either the releasing mechanism or the receptor mechanism is overactive. Drugs commonly used to treat schizophrenia, like chlorpromazine (e.g., Thorazine) and haloperidol (Haldol), bind to dopamine receptors and thus inhibit their signal-sending work.

Amphetamine Abuse Causes Schizophrenia-Like Symptoms Stimulants like the amphetamines work by triggering the release of dopamine into the arousal and pleasure centers of the brain. The effect is therefore a "high." But it's easy to abuse the amphetamines. When this occurs, the same kind of overstimulation associated with schizophrenia results, leading to delusions of persecution, hallucinations, and other disturbances of the thought processes.

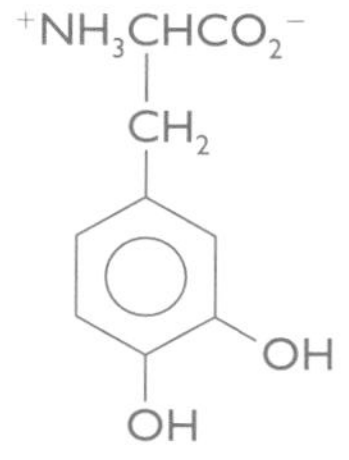

Dopamine Releasing Neurons Have Degenerated in Parkinson's Disease   When the dopamine using neurons in the brain degenerate, as in Parkinson's disease, an extra supply of dopamine itself is needed to compensate. This is why L-DOPA (*levorotatory dihydroxyphenylalanine*) is used to treat the disease. The neurons that still work use it to make extra dopamine.

■ L-DOPA structure

L-DOPA

GABA Inhibits Nerve Signals   The normal function of some neurotransmitters is to *inhibit* instead of initiate signals. Gamma-aminobutyric acid (GABA) is an example, and as many as a third of the synapses in the brain have GABA available.

The inhibiting work of GABA can be made even greater by mild tranquilizers such as diazepam (e.g., Valium) and chlordiazepoxide hydrochloride (Librium), as well as by ethanol. The augmented inhibition of signals reduces anxiety, affects judgment, and induces sleep. Widespread abuses of Valium, Librium, and ethanol also occur.

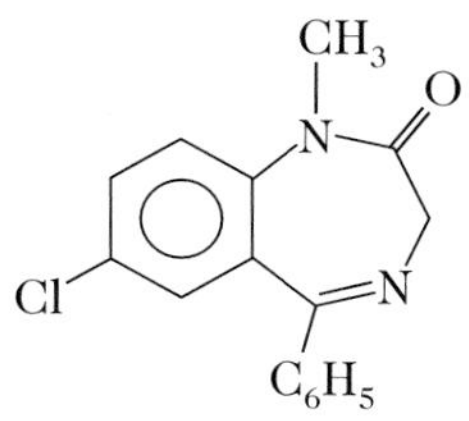

■ $^+NH_3CH_2CH_2CH_2CO_2^-$

GABA

diazepam (Valium)

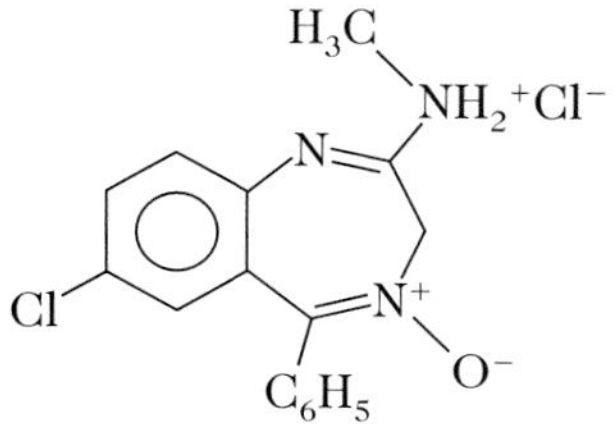

chlordiazepoxide
hydrochloride (Librium)

An oversupply of *receptors* for GABA may explain petit mal or "absence" seizures in children. (They are lapses in consciousness for a few seconds, and may occur as often as 100 times a day.) Receptor blockers, given to lethargic mice, decreased the number of seizures, but receptor-enhancing agents increased them.

GABA Is Deficient in Huntington's Chorea   The victims of Huntington's chorea, a hereditary neurological disorder, suffer from speech disturbances, irregular movements, and a steady mental deterioration, all related to a deficiency in GABA. Unhappily, GABA can't be administered in this disease, because it can't move out of circulation and into the regions of the brain where it works.

■ Greek, *choreia*, dance.

Several Polypeptides Act as Pain-Killing Neurotransmitters   As we said, some neurotransmitters are relatively small polypeptides (see Table 22.1). One type includes the *enkephalins*; another consists of the *endorphins*. Both types are powerful pain inhibitors. One of them, dynorphin, is the most potent painkiller yet discovered, being 200 times stronger than morphine, an opium alkaloid that is widely used to relieve severe pain. Sites in the brain that strongly bind molecules of morphine also bind those of the enkephalins, so these natural painkillers are now often referred to as the body's natural opiates.

■ *En-* or *end-*, within; *kephale*, brain; *-orph-*, from morphine.

Many Neurotransmitters Exert More than One Effect   Several neurotransmitters can be received by more than one kind of receptor. For example, at least three types of receptors for the opiates have been identified. Receptor multiplicity may explain how some neurotransmitters have multiple effects. Thus not only do opiates reduce pain, they also affect emotions, induce sleep, and affect the appetite, with each of the opiate receptors handling a different one of these functions.

Calcium Channel Blockers Reduce the Vigor of Heart Muscle Contractions   As we have often seen, calcium ions are major secondary chemical messengers, and neurotransmitters are able to open channels for calcium ions through cell membranes. Heart muscle tissue receives such signals at a rate that paces the heart as its muscles contract and relax during the heartbeat. Calcium ions are what finally deliver the message to contract. Then the cell pumps them back out and the muscle relaxes until another cycle starts.

Drugs like nifedipine (Procardia, Adalat), diltiazem (Cardizem), and verapamil (Isoptin) find calcium ion channels in heart muscle and block them. Not all are blocked, of course, so the effect is to reduce the migrations of Ca^{2+} through cell membranes. These calcium channel blockers (also called calcium antagonists or slow channel blocking agents) thus make each heart muscle contraction less vigorous. This reduces the risk of heart attacks in people known to be at risk, like those who experience angina pectoris and cardiac arrhythmias (heartbeat irregularities).

22.3 RETROGRADE MESSENGERS

■ "Retrograde" means moving or directed backward.

Some neurotransmitters feed back to the sending cell.

Nitric Oxide Is a Chemical Messenger   Not all of the work of chemical messengers is done by a common mechanism, namely, one involving lock-and-key fitting of substrates to receptors. A widespread system of nerves uses one of the simplest of all molecules, that of nitric oxide, NO, as a neurotransmitter or as an effector. This tiny molecule diffuses easily through cell membranes without any help from receptors or special channels. Thus its actions cannot be blocked by the usual type of receptor inhibitors or blockers. Its activity depends on its chemical properties far more than its molecular geometry. And what a range of activities it has.

Nitric oxide is involved in platelet aggregation, cell adhesion, the control of vasodilation (and so control of blood pressure), inhibition of tumor growth, penile erection, and enhancement of immunity to malaria, to name some of its work. On the negative side, the brain cell damage accompanying an ischemic stroke may be caused by an overstimulation of the enzyme that helps to make NO, thus flooding too much of it into the tissue. Nitric oxide, after all, is one of the poisons in smog (which is another example of the principle that it is the *size of the dose* that makes something a poison).

NO cannot be regulated by interventions with receptors; it has none. It can be regulated only by controlling its synthesis. NO synthase (NOS) is the enzyme that makes NO, and different variations of it occur in brain cells, endothelial tissue, and macrophages (a type of white blood cell that destroys foreign materials in the blood). With the aid of Figure 22.5, we can visualize how some of newly made NO does its work in a postsynaptic neuron and other NO diffuses back to the presynaptic neuron to condition it for additional release of the neurotransmitter, glutamate. The numbers in what follows refer to the boxed numbers in this figure.

1. A signal sends glutamate from its vesicles into the synapse.

2. Glutamate opens a receptor/gate for Ca^{2+} ions, which now enter the postsynaptic neuron.

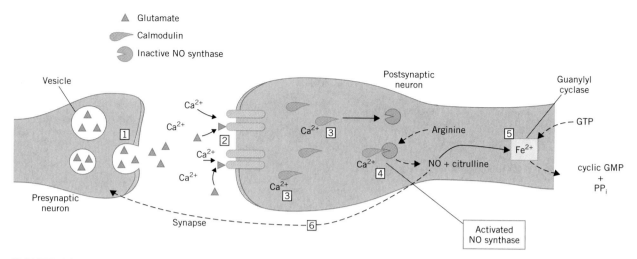

FIGURE 22.5 Nitric oxide and the NO synthase system. The boxed numbers refer to the text discussion.

3. Ca^{2+} binds to calmodulin.

4. The Ca^{2+}–calmodulin complex activates NO synthase and NO is now made. (Arginine is the amino acid raw material.)

5. Some newly made NO is drawn to Fe^{2+} ion in a heme unit of the enzyme guanylyl cyclase, which catalyzes the formation of cyclic GMP (cGMP) from GTP. cGMP is like cyclic AMP, a secondary chemical messenger that initiates a response in the cell.

6. Some NO molelcules diffuse back to the presynaptic neuron, where they condition it for even more releases of glutamate.

Nitric Oxide May Be Involved in the Storage of Memory   Learning involves putting experiences, information, and data into memory in such a way that they can be retrieved and applied in new situations. If life has a molecular basis, then there must be a molecular basis to memory, too. What this basis is has long intrigued neuroscientists. According to one broad model of how some types of learning take place, a nerve cell that receives a signal associated with something to be remembered manufactures NO. When some NO moves back to the signal-sending cell it strengthens the connection between the two cells. Each additional time that the same signal is sent, the signal is further strengthened. NO is therefore sometimes called a *retrograde messenger.* This theory is tentative, but it is known that when a binder of NO is present in the brain cells of experimental rats they are unable to learn certain spatial tasks. NO is evidently necessary to learning in this situation.

Carbon Monoxide Is Probably a Retrograde Messenger   Carbon monoxide, another noxious gas, might also be a brain messenger and possibly involved in a long-term learning-linked process. (One is certainly tempted to say, "Of all things!") In relatively large concentrations when carried into the body in inhaled air, CO binds so strongly to heme units in hemoglobin that the latter cannot carry oxygen, which causes death. Yet in small concentrations manufactured within the brain, CO activates guanylyl cyclase, the same enzyme activated by NO. CO thus has many of the same effects as NO.

■ GTP is a triphosphate like ATP, only it uses guanine instead of adenine.

■ Abnormal NO systems have been found in some people with hypertension and others with high blood cholesterol levels.

SUMMARY

Chemical communication with hormones and neuro-transmitters The carbohydrate tails of membrane glyco-proteins project away from the membrane surface and provide recognition sites for chemical messengers (and other particles). The messenger molecule and the receptor protein form a complex. One common response to the formation of the complex is the release of a G-protein. In the adenylyl cyclase cascade, this activates adenylate cyclase, which then triggers the formation of cyclic AMP. In turn, cyclic AMP sets off other events, such as the activation of an enzyme that catalyzes a reaction, one that is ultimately what the "signal" of the neurotransmitter was all about.

The activation of a G-protein is followed by the inositol phosphate cascade. Phospholipase C is activated, which breaks up a phospholipid in the membrane into two enzyme activators, PIP_2 and DAG. PIP_2 initiates the rapid release of Ca^{2+} from cytoplasm stores to cause muscle contraction, and DAG helps to keep the cell's glucose level high so that its metabolism can supply the energy.

Hormones Endocrine glands secrete hormones, and these primary chemical messengers travel to their target cells in the blood, where they activate a gene, or an enzyme, or affect the permeability of a cell membrane. They recognize their own target cells by binding to specific receptor proteins.

The steroid hormones can move into a cell to its nucleus and there find a receptor. Polypeptide hormones bind to membrane-bound receptors to initiate their action.

Neurotransmitters In response to an electrical signal, vesicles in an axon release a neurotransmitter that moves across the synapse. Its molecules bind to a receptor protein on the next neuron, and then the pattern is much like that of hormones. The result, however, is to open channels through the cell membrane for the migration of ions.

Neurotransmitters include amino acids, monoamines, and polypeptides. Some neurotransmitters *activate* some response in the next neuron, whereas others *deactivate* some activity. A number of medications work by interfering with neurotransmitters or with the opening of calcium ion channels.

Retrograde messengers Nitric oxide is a neurotransmitter that does not work through the usual receptor mechanisms. Its small size, nonpolar nature, and easy diffusability across membranes enable it to move in any direction, even in a retrograde manner. In brain neurons, NO can move back to the signal-sending cell.

REVIEW EXERCISES

The answers to Review Exercises whose numbers are in color are found in Appendix E. The answers to the other Review Exercises are found in the Study Guide that accompanies this book The more challenging questions are marked with asterisks.

Chemical Communication

22.1 What are the names of the sites of the synthesis of (a) hormones and (b) neurotransmitters?

22.2 What do the lock-and-key and induced fit concepts have to do with the work of hormones and neurotransmitters?

22.3 In what general ways do hormones and neurotransmitters resemble each other?

22.4 What general name is given to the substance on a target cell that recognizes a hormone or neurotransmitter?

22.5 Name the protein that both the cyclic AMP and the inositol phosphate systems use to activate something inside the cell.

22.6 What function does adenylyl cyclase have in the work of at least some hormones?

22.7 How is cyclic AMP involved in the work of some hormones and neurotransmitters?

22.8 After cyclic AMP has caused the activation of an enzyme inside a cell, what happens to the cyclic AMP to stop its action until more is made?

*22.9 In what ways does the inositol "cascade" resemble the cyclic AMP cascade?

22.10 When the G-protein of the inositol cascade completes its work, it has produced *two* new messengers. In general terms, what does each one do next?

Hormones

22.11 What are the four broad types of hormones?

22.12 What structural fact about the steroid hormones makes it easy for them to get through a cell membrane?

22.13 In each case, what substance (or kind of substance) can enter a target cell more readily following the action of the hormone?
(a) Insulin   (b) Growth hormone   (c) A neurotransmitter

22.14 Why are the prostaglandins called *local hormones*?

Neurotransmitters

22.15 What happens to acetylcholine after it has worked as a neurotransmitter? What is the name of the enzyme that catalyzes this change? In chemical terms, what does a nerve gas poison do?

22.16 How does a local anesthetic such as procaine affect the functioning of acetylcholine as a neurotransmitter?

22.17 How does the botulinum toxin work?

22.18 What, in general terms, are the monoamine oxidases, and in what way are they important?

22.19 What does iproniazid do chemically in the neuron signaling that is carried out by norepinephrine?

22.20 In general terms, how do antidepressants such as amitriptyline or imipramine work?

22.21 How does fluoxetine (Prozac) interact with serotonin?

22.22 The overactivity of which neurotransmitter is thought to be one biochemical problem in schizophrenia?

22.23 How do the schizophrenia-control drugs chlorpromazine and haloperidol work?

22.24 How can the amphetamines, when abused, give schizophrenia-like symptoms?

22.25 How does L-DOPA work in treating Parkinson's disease?

22.26 Which common neurotransmitter in the brain is a signal inhibitor?

22.27 How do such tranquilizers as Valium and Librium affect the neurotransmitter of Review Exercise 22.26?

22.28 Why is enkephalin called one of the body's own opiates? How does it appear to work?

22.29 How do the calcium channel blockers reduce the risk of a heart attack?

Retrograde Messengers

22.30 What is meant by the term *retrograde messenger*?

22.31 What physical property of nitric oxide enables it to act in a retrograde manner?

22.32 How is NO synthase activated?

22.33 How does NO activate guanylyl cyclase?

22.34 In some tissues, carbon monoxide appears to act as a messenger. It has many of the same effects as NO. What reaction of CO might explain this?

Melatonin (Interaction 22.1)

22.35 Melatonin belongs to what category of substances; enzymes, hormones, neurotransmitters, or secondary messengers?

22.36 The pineal gland releases melatonin more during the day or the night?

22.37 What is meant by *drug interaction*?

Nicotine (Interaction 22.2)

22.38 Define *poison*.

22.39 Pure nicotine is *supertoxic*. What does this mean? And if it's supertoxic, why aren't smokers dying simply from smoking a cigarette?

22.40 For something to be labeled addictive, what must be true about it?

22.41 How do nicotine molecules affect the release of dopamine?

22.42 How does nicotine *tolerance* develop?

Additional Exercise

22.43 Truffles are edible, potato-shaped fungi that grow underground in certain parts of France, and are highly prized by gourmet cooks and gourmands. Pigs are used to locate truffles buried as much as one meter below the surface because truffles carry traces of a steroid, androsten-16-en-3-ol, which is a powerful sex attractant for pigs. A sex attractant is a species-specific chemical compound made and released in trace amounts by a female member of a species toward which a male member experiences a powerful sexual response. A male pig, of course, does not initially know that it is a truffle emitting the attractant. (It appears that all sex attractants for humans are *non*chemical, being better understood, perhaps, as public relations activities.)

androsten-16-en-3-ol   androsterone

Notice the structural similarity between androsten-16-en-3-ol and a human sex hormone, androsterone.

(a) In terms of what general theory would we explain how androsten-16-en-3-ol has a particular specificity in pigs and not in humans but androsterone has a specificity in humans and not in pigs?

(b) In order to convert androsten-16-en-3-ol into androsterone *in vitro*, what specific *series* of changes in functional groups would have to be carried out. Your answer would begin with something like "First, we have to change such and such a group into . . . . Then this new group would have to be changed . . . ."(All needed changes involve one-step reactions we have studied).

23 EXTRACELLULAR FLUIDS OF THE BODY

If too much carbon dioxide is lost from the blood and expelled from the lungs during high-altitude work, the pH of the blood may increase. We'll study many other life-threatening situations involving the blood's pH in this chapter.

This Chapter in Context

It is vitally important that the pH of the blood stay very close to 7.35. In this chapter we'll see why. Nothing less than our ability to breathe is at stake. Health care professionals, particularly those in emergency rooms or critical care units, say that there are no topics in chemistry quite as important as the acid–base balance of the blood and the chemistry of respiration. The necessity to breathe presupposes, of course, that oxygen will be used on compounds delivered from our diets. So we'll begin with the digestion and absorption of nutrients and their delivery by the bloodstream.

23.1 Digestive Juices

The end products of digestion include monosaccharides, fatty acids, monoacylglycerols, and amino acids.

Life engages two environments, the external environment that we commonly think of, and the *internal environment,* which we usually take for granted. When healthy, our bodies have nearly perfect control over the latter, and this enables us to handle large external changes more or less well, changes such as huge temperature fluctuations, chilling winds, stifling humidity, and a fluctuating atmospheric tide of dust and pollutants.

Cells Exist in Contact with Interstitial Fluids and Blood The **internal environment** consists of all of the **extracellular fluids,** those that aren't actually inside cells. About three-quarters of our extracellular fluids consist of **interstitial fluid,** which is in the spaces or interstices *between* cells. Blood makes up nearly all the rest. Lymph, cerebrospinal fluid, digestive juices, and synovial fluids (lubricants of joints) are also extracellular fluids.

■ The fluids of the internal environment make up about 20% of the mass of the body.

The Digestive Tract Has Access to Hormones and Hydrolases The **digestive juices** are dilute solutions of electrolytes and hydrolytic enzymes (or zymogens) either in the cells lining the intestinal tract or in solutions that enter the tract from various organs (Figure 23.1). Specialized cells manufacture and release hormones in response to various signals, such as arriving nutrients, the distension they cause to the wall of the tract, or a change in pH. Some hormones are secreted into the tract itself, and others enter the bloodstream.

Taken as a whole, the digestive tract is itself a vast endocrine gland, the largest we have. The lower part of the stomach releases *gastrin,* which helps to stimulate the release of gastric juice. Cells in the upper intestinal tract release the hormone *cholecystokinin.* It controls the release of material from the stomach into the upper intestinal tract so that digestion proceeds at a sufficiently slow rate to be complete. Cholecystokinin also tells the gallbladder to release bile, and it makes the pancreas release pancreatic zymogens. *Secretin,* another hormone, makes the pancreas release bicarbonate ion for neutralizing incoming gastric acid.

Saliva Provides a Starch Splitting Enzyme, α-Amylase The flow of **saliva** is stimulated by the sight, smell, taste, and even the thought of food. Besides water (99.5%), saliva includes a food lubricant called *mucin* (a glycoprotein) and an enzyme, *α-amylase.* It catalyzes the partial hydrolysis of starch to dextrins and maltose, and it works best at the pH of saliva, 5.8 to 7.1. Proteins and lipids pass through the mouth essentially unchanged.

■ Dextrins are partial breakdown products of amylopectin, a component of starch.

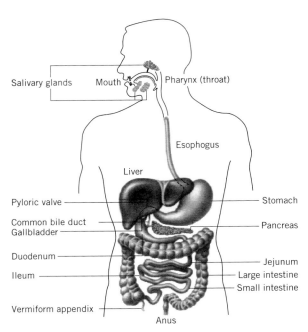

FIGURE 23.1   Organs of the digestive tract.

Gastric Juice Starts the Digestion of Proteins   When food is on the way to the stomach, *acetylcholine* (a neurotransmitter) is made in gastric cells and *histamine* is also manufactured. When food arrives and distends the stomach, cells in the gastric lining release *gastrin*. Gastrin, acetylcholine, and histamine stimulate the release of the fluids that combine to give **gastric juice.** One kind of gastric cell secretes mucin, which coats the stomach to protect it against its own digestive enzymes and its acid. Mucin is continuously produced and only slowly digested. If for any reason its protection of the stomach is hindered, part of the stomach itself could be digested, making or aggravating an ulcer.

Another kind of gastric cell (parietal cells) secretes hydrogen ion at a concentration of roughly 0.15 mol/L (pH 0.8 or over a million times more acidic than blood). A K^+–H^+ ion "pump," using the chemical energy of ATP, takes K^+ ion out of stomach fluids and puts H^+ into the fluids. The K^+ is then sent back out along with Cl^- ion. Because chloride ion is the chief anion in gastric juice, the overall effect is the secretion of hydrochloric acid. Histamine is the specific stimulator of the K^+–H^+ ion pump. In the medley of medications used to treat ulcers, some act to prevent the secretion of acid and some attack a bacterium that causes ulcers (see Interaction 23.1).

Gastric cells also secrete *pepsinogen,* a zymogen that gastric acid changes into *pepsin,* a protease. Pepsin catalyzes the only important digestive work in the stomach, the hydrolysis of some of the peptide bonds of proteins to make smaller polypeptides. Pepsin's optimum pH is about 2, reflecting how acidic stomach fluid becomes as it mixes with incoming food. Gastric acid coagulates proteins so they are retained in the stomach longer for pepsin's work.

Adult gastric juice also has a *gastric lipase,* but it does not start its work until it arrives in a medium of higher pH in the upper intestinal tract.

The gastric juice of infants is less acidic than an adult's. To compensate for the protein-coagulating work normally done by stomach acid, infant gastric juice con-

■   A *protease* is an enzyme that catalyzes the digestion of proteins.

INTERACTION 23.1
GASTRIC JUICE AND ULCER TREATMENT

Food Stimulates the Release of Gastric Juice When food begins its trip into the stomach, gastric cells in the stomach lining launch a synthesis of *histamine,* making it from the α-amino acid histidine by the loss of a molecule of carbon dioxide (by decarboxylation).

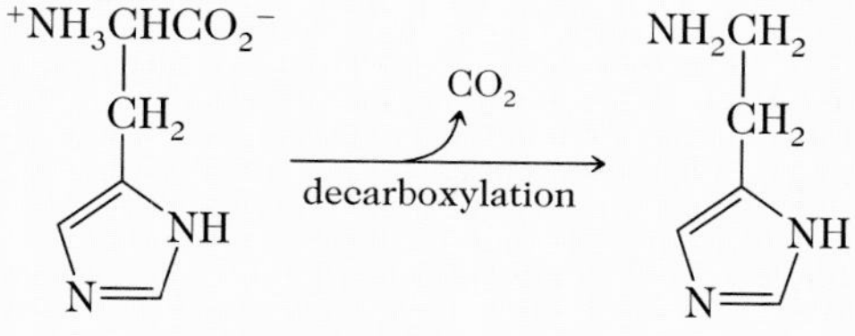

$$^+NH_3CHCO_2^- \xrightarrow[\text{decarboxylation}]{CO_2} NH_2CH_2$$

histidine → histamine

Histamine is the specific stimulator of the K^+–H^+ ion pump.

Molecular Mimics of Histamine Inhibit the Pump Cimetidine (Tagamet), ranitidine (Zantac), and omeprazole (Prilosec) are drugs used in the treatment of ulcers. They inhibit the secretion of gastric juice and so are classified as antisecretory agents. Their molecules are able to mimic histamine mole-

cules and bind to the receptor protein of the K^+–H^+ pump without stimulating the pump to work. By preventing histamine binding, the K^+–H^+ pump shuts down. Now the secretion of hydrochloric acid into the stomach is severely inhibited, giving time for the stomach to heal the ulcer.

The Ultimate Cause of Ulcers Is a Bacterium Until the mid-1980s, the conventional wisdom was that high stress and poor diets caused ulcers. However, evidence has accumulated that possibly most ulcers are caused by a bacterium, *Helicobacter pylori.* (It's believed to infect 60% of Americans by the age 60, and 10% of Americans will develop an ulcer.) In 1996, the U.S. Food and Drug Administration approved a two-drug attack, a combination of the antibiotic clarithromycin (Biaxin) and omeprazole (Prilosec). A three-drug approach is also in the works, one using a combination of ranitidine (Zantac), clarithromycin, and bismuth citrate.

A Reading M. J. Blaser, "The Bacteria behind Ulcers," *Scientific American,* February 1996, page 104.

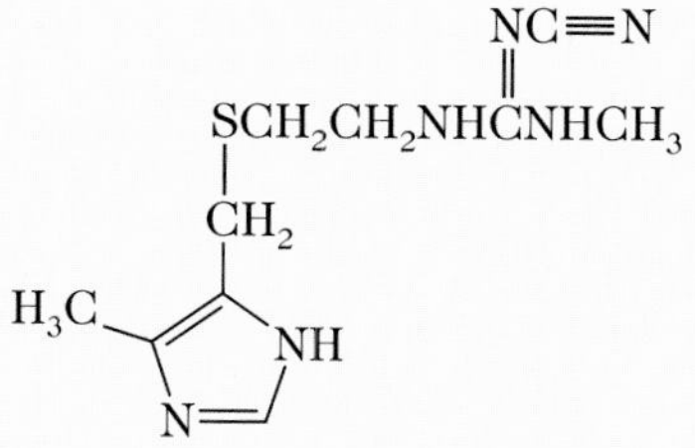

cimetidine (Tagamet)

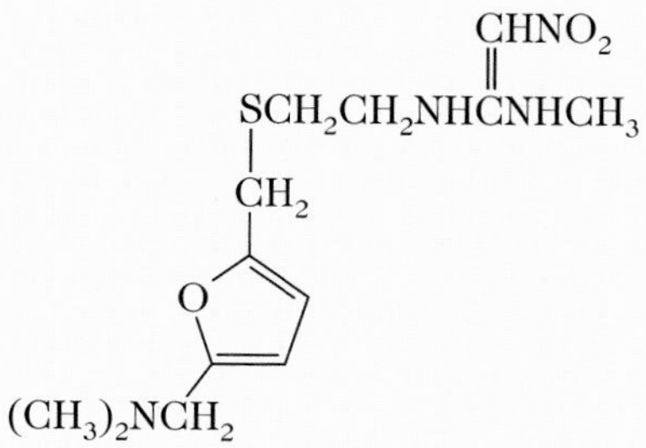

ranitidine (Zantac)

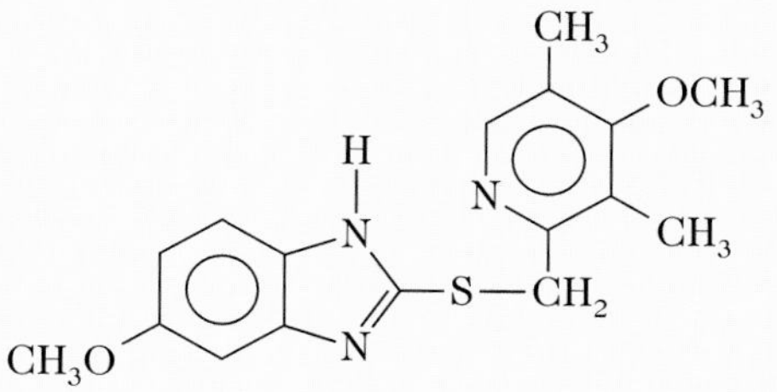

omeprazole (Prilosec)

tains *rennin,* a powerful protein coagulator. Because the pH of an infant's gastric juice is higher than that of an adult, an infant's gastric lipase gets an early start on lipid digestion.

The churning and digesting activities in the stomach produce a liquid mixture called *chyme.* This is released in portions through the pyloric valve into the duodenum, the first 12 inches of the upper intestinal tract.

Pancreatic Juice Furnishes Several Zymogens and Enzymes As soon as chyme enters the duodenum, two hormones, cholecystokinin and secretin, are released that circulate to the pancreas and induce it to release two juices. One is almost entirely dilute sodium bicarbonate, which neutralizes the acid in chyme. The other, **pancreatic juice,** carries enzymes or zymogens that become involved in the digestion of practically everything in food. It contributes an α-amylase similar to that in saliva, a *lipase, nucleases,* and zymogens for protein-digesting enzymes.

■ The nucleases include ribonuclease (RNase) and deoxyribonuclease (DNase).

■ Enteropeptidase used to be called enterokinase.

The conversion of the proteolytic zymogens in pancreatic juice to active enzymes begins with a "master switch" enzyme called *enteropeptidase*. It's released from cells that line the duodenum when chyme arrives, and it then catalyzes the formation of trypsin from its zymogen, trypsinogen.

$$\text{Trypsinogen} \xrightarrow{\text{enteropeptidase}} \text{trypsin}$$

Trypsin then catalyzes the change of the other zymogens into enzymes.

$$\text{Procarboxypeptidase} \xrightarrow{\text{trypsin}} \text{carboxypeptidase}$$

$$\text{Chymotrypsinogen} \xrightarrow{\text{trypsin}} \text{chymotrypsin}$$

$$\text{Proelastase} \xrightarrow{\text{trypsin}} \text{elastase}$$

■ These proteases must exist as zymogens first or they will catalyze the self-digestion of the pancreas, which does happen in acute pancreatitis.

Trypsin, chymotrypsin, and *elastase* catalyze the hydrolysis of large polypeptides to smaller ones. *Carboxypeptidase,* working in from C-terminal ends of small polypeptides, carries the action further to amino acids and di- or tripeptides.

■ The structure of a typical bile salt was given on page 562.

Bile Salts Are Surfactants Needed to Manage Triacylglycerols and Fat-Soluble Vitamins In order to digest triacylglycerols, the lipase in pancreatic juice needs the help of the powerful detergents in bile, called the *bile salts*. These emulsify water-insoluble fatty materials and so greatly increase the exposure of lipids to water and lipase. The digestion of triacylglycerols gives the anions of fatty acids and monoacylglycerols (plus some diacylglycerols).

Bile enters the duodenum from the gallbladder. Its secretion is stimulated by a hormone released when chyme contains fatty material. Bile is also an avenue of excretion, because it can carry cholesterol and breakdown products of hemoglobin into the developing feces. These hemoglobin products constitute the bile pigments, which give color to feces.

Bile salts also assist in the absorption of fat-soluble vitamins (A, D, E, and K) from the digestive tract into the blood. Some bile pigments are also reabsorbed, and they eventually leave the body via the urine. Thus bile pigments are responsible for the color of both feces and urine.

■ These intestinal cells last only about 2 days before they self-digest. They are constantly being replaced.

Cells of the Intestines Carry Several Digestive Enzymes The term **intestinal juice** embraces not only a secretion but also the enzyme-rich fluids found inside certain kinds of cells that line the duodenum and jejunum (see Figure 23.1). The secretions deliver both an amylase and enteropeptidase, which we have just described.

The other enzymes in this region work within their cells as digestible compounds are already being absorbed. An *aminopeptidase,* working inward from N-terminal ends of small polypeptides, hydrolyzes them to amino acids. Hydrolases, *sucrase, lactase,* and *maltase,* handle the digestion of disaccharides. Sucrose is hydrolyzed to fructose and glucose, lactose to galactose and glucose, and maltose to glucose. An intestinal lipase and enzymes for the hydrolysis of nucleic acids (*nucleases*) are also present.

As the anions of fatty acids and mono- and diacylglycerols migrate into the cells of the duodenal lining, they are reconstituted into triacylglycerols. These are taken up by the lymph system and delivered to the blood in complexes of lipids and proteins called *chylomicrons.* (We'll study what happens to them and to dietary cholesterol in Chapter 27.)

Some Vitamins and Essential Amino Acids Are Made in the Large Intestine No digestive functions are performed in the large intestine. Microorganisms in residence there, however, make vitamins K and B, plus some essential amino acids. These are absorbed by the body, but their contribution to overall nutrition in humans is not large. Water and sodium chloride are reabsorbed from the ileum, and undigested matter, microorganisms, and some water make up the feces.

23.2 BLOOD AND THE ABSORPTION OF NUTRIENTS BY CELLS

The balance between the blood's pumping pressure and its colloidal osmotic pressure tips at capillary loops.

The circulatory system is one of our two main lines of chemical communication between the external and internal environments (Figure 23.2). All of the veins and arteries together make up the **vascular compartment**. The **cardiovascular compartment** includes this plus the heart.

■ The nervous system with its neurotransmitters is the other line of communication.

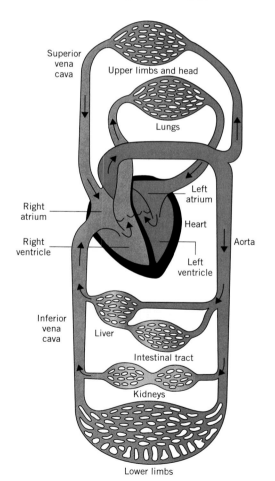

FIGURE 23.2 Human circulatory system. When oxygen-depleted venous blood (blue areas) returns to the heart, it is pumped into the capillary beds of the alveoli in the lungs to reload oxygen and get rid of carbon dioxide. Then the freshly oxygenated blood (red areas) is distributed by the arteries throughout the body, including the heart muscle.

■ About 8% of the body's mass is blood. In the adult, the blood volume is 5 to 6 L.

Blood Moves Nutrients, Oxygen, Messengers, Wastes, and Disease Fighters throughout the Body Blood in the pulmonary branches moves through the lungs where waste carbon dioxide is exchanged for oxygen. Oxygenated blood then moves to the rest of the system via the systemic branches.

At the intestinal tract, blood picks up the products of digestion. Most are immediately "examined" at the liver, and many alien chemicals are chemically modified to make it easier for the body to eliminate them. While flowing through the kidneys, blood replenishes its buffer supplies and eliminates waste nitrogen, mostly as urea. *The pH of blood and its electrolyte levels depend largely on the kidneys.* At endocrine glands, blood picks up hormones whose secretions are often in response to something present in the blood.

White cells in blood provide protection against infection. Red cells or **erythrocytes** carry oxygen and waste bicarbonate ion. *Platelets* are needed for blood clotting and other purposes. Blood also carries several zymogens used by the blood-clothing mechanism.

■ About a quarter of the plasma proteins are replaced each day.

Proteins in Blood Are Vital to Its Colloidal Osmotic Pressure The principal types of substances in whole blood are summarized in Figure 23.3. Among the proteins, the **albumins** help carry hydrophobic molecules, like fatty acids and steroid hormones, and they contribute 75% to 80% of the osmotic effect of the blood. Some **globulins** carry ions (e.g., Fe^{2+} and Cu^{2+}) that otherwise are insoluble when the pH is greater than 7. Other globulins are antibodies that help to protect against infectious disease. **Fibrinogen** is carried in blood to be converted to an insoluble form, **fibrin**, when a blood clot must form.

The levels of solutes, including proteins, in various components of the major body fluids are shown in Figure 23.4. The higher level of protein in blood is the principal reason that it has a higher osmotic pressure than interstitial fluid.[1] The *total* osmotic pressure of blood is caused, of course, by all the dissolved and colloidally dispersed solutes. Small ions and molecules, however, can move back and forth between the blood and the interstitial compartment more readily than large protein molecules. So it's proteins that give blood a higher effective osmotic pressure than interstitial fluid. The difference is the blood's *colloidal osmotic pressure.*

■ The osmolarity of plasma is about 290 mOsm/L.

Because of the higher osmotic pressure of blood, water tends to flow into the blood from the interstitial compartment. This cannot be allowed to happen everywhere and continually, however, or the interstitial spaces and then the cells would

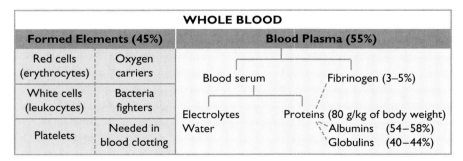

FIGURE 23.3 Major components of blood.

---

[1] As a reminder and a useful memory aid, high solute concentration means high osmotic pressure; and solvent flows in osmosis or dialysis from a region where the solute is dilute to the region where it is concentrated. The "goal" of this flow is to even out the concentrations everywhere.

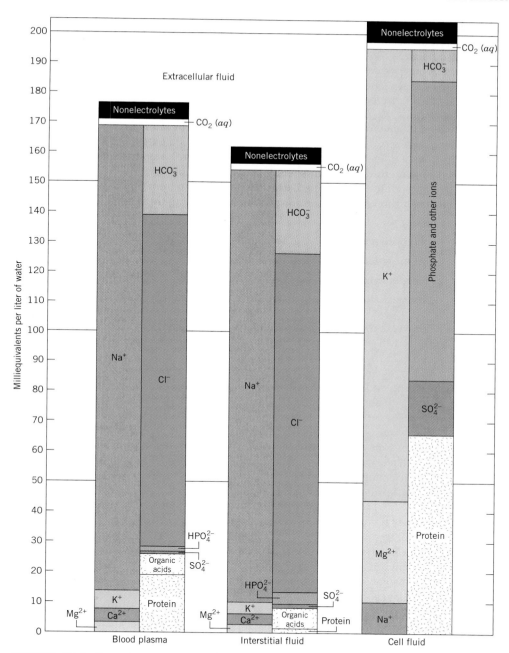

FIGURE 23.4 Electrolyte composition of body fluids. (Adapted by permission from J. L. Gamble, *Chemical Anatomy, Physiology and Pathology of Extracellular Fluids,* 6th ed. Cambridge, MA: Harvard University Press, 1954.)

eventually become too dehydrated to maintain life. Before we see how this problem is managed, we need to survey some of the **electrolytes** in blood.

The Chief Ions in Blood Are Na$^+$, Cl$^-$, and HCO$_3^-$ The sodium ion is the chief cation in both the blood and the interstitial fluid and the potassium ion is the major cation inside cells (see Figure 23.4). A sodium–potassium pump, a special ATP-run protein complex, maintains these gradients. Both ions are needed to maintain osmotic pressure relationships, and both are a part of the system for regulating acid–base balance. Both ions are also needed for the smooth working of the muscles and the nervous system.

Changes in the concentrations of sodium and potassium ion in blood can lead to serious medical emergencies, so a special vocabulary has been developed to describe various situations. We will see here how a technical vocabulary can be built on a few word parts, and we will use some of these word parts in later chapters, too. For example, we use *-emia* to signify "in the blood." *Hypo-* indicates a condition of a low concentration of something, and *hyper-* is the opposite, a condition of a high concentration of something. We can specify this "something" by a word part, too. Thus *-nat-* signifies sodium (from the Latin *natrium* for sodium), and *-kal-* designates potassium (from the Latin *kalium* for potassium). Putting these together gives us the following terms.

■ Another word part is *-uria,* of the urine. Thus *glucosuria* means glucose in the urine.

Hyponatremia:	low level of sodium ion in blood
Hypokalemia:	low level of potassium ion in blood
Hypernatremia:	high level of sodium ion in blood
Hyperkalemia:	high level of potassium ion in blood

■ The **milliequivalent,** abbreviated **meq,** is analogous to the *millimole* (10^{-3} mol). Think of the mass of 1 meq of an ion as equal to the number of milligrams that contains 1 millimole of electrical charge. Thus 1 mmol of Cl^- contains 1 meq of negative charge; 1 mmol of Ca^{2+} contains 2 meq of positive charge.

Na^+ and K^+ Levels Are Regulated in Tandem   The normal range for the sodium ion level in blood is 135 to 145 meq/L (Figure 23.5) and that for potassium ion is 3.5 to 5 meq/L (Figure 23.6). See also Table 23.1.

If our kidneys cannot make urine or if we drink water faster than the kidneys can handle it, the sodium level of the blood decreases, and signs of *hyponatremia* develop, namely, flushed skin, fever, a dry tongue, and a noticeable decrease in urine output.

Hypernatremia occurs from excessive losses of water under circumstances in which sodium ions are not lost, as in diarrhea, diabetes, and even in some high-protein diets. (The loss of *solvent,* water, without the loss of solute, Na^+, means that Na^+ has a higher concentration.)

Because the level of potassium ion in blood is so low, small changes are particularly dangerous. Severe *hyperkalemia* leads to death by heart failure, a danger with crushing injuries, severe burns, or heart attacks, anything that breaks cells open so that K^+ ions spill into circulation.

At the other extreme, severe *hypokalemia,* caused by any unusual losses of body fluids, including excessive sweating, can lead to death by heart failure. Both water *and electrolytes* must be replaced during severe exercise.

The serum levels of sodium ions and potassium ions are regulated in tandem by the kidneys. When the intake of Na^+ is high, the body loses K^+. When the intake of K^+ is high, the body loses Na^+. It's the total positive charge that must be maintained

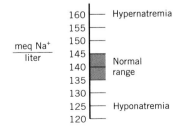

FIGURE 23.5   The sodium ion level in the blood.

TABLE 23.1   Sodium and Potassium Ions in the Human Body

Area	Na^+	K^+
Total body	2700–3000 meq	3200 meq
Plasma level	135–145 meq/L	3.5–5.0 meq/L
Intracellular level	10 meq/L	125 meq/L
Mass of 1 meq	23.0 mg	39.1 mg

so as to balance the total negative charge. One reason we can't tolerate seawater (3% NaCl), and will die if we drink it, is that it upsets the sodium–potassium balance in the body.

Mg^{2+} Is Second to K^+ as a Cation Inside Cells The normal ranges for the levels of calcium ion and magnesium ion in the body are given in Table 23.2. The level of magnesium ion in the blood is even lower than that of K^+ (see Figure 23.4), so small variations can mean large trouble. *Hypomagnesemia,* for example, is observed in alcoholism, in untreated diabetes, and when the kidneys aren't working properly. Some of its signs are muscle weakness, insomnia, and cramps in the legs or the feet. Injections of isotonic magnesium sulfate solution may be used to restore Mg^{2+} to serum.

On the opposite side, *hypermagnesemia* can lead to cardiac arrest, and it can be brought on by the overuse of magnesium-based antacids, like milk of magnesia, $Mg(OH)_2$.

Nearly All the Body's Ca^{2+} Is in Bones and Teeth The Ca^{2+} that is not in bones and teeth is absolutely vital because it's an important secondary chemical messenger for activating enzymes and for muscle contraction.

Hypocalcemia can be brought on by vitamin D deficiency, the overuse of laxatives, an impaired activity of the thyroid gland, and even by hypomagnesemia. *Hypercalcemia,* on the other hand, is caused by the opposite conditions, namely, an overdose of vitamin D, the overuse of antacids containing calcium ion, or an overactive thyroid. In severe hypercalcemia, the heart functions poorly.

Almost All the Negative Charge Needed to Balance Cationic Charges in Blood Are Provided by Cl^- and HCO_3^- The chloride in blood helps to maintain osmotic pressure relationships, the acid–base balance, and the distribution of water in the body. It has a function in oxygen transport that we'll study later. The bicarbonate ion is the chief acid-neutralizing buffer in blood and the principal form in which waste CO_2 is carried.

In *hypochloremia,* an excessive loss of Cl^- from the blood occurs. For every Cl^- lost, blood either must lose one $(+)$ charge, like Na^+, or retain one extra $(-)$ charge on some other anion. Electrical neutrality dictates these simple facts. The chief ion retained is HCO_3^-. Because HCO_3^- tends to raise the pH of a fluid, *a condition of hypochloremia can cause the pH of blood to increase* (which is alkalosis).

By the same token, *hyperchloremia,* an increase in the level of Cl^- in blood, could mean that HCO_3^- has to be dumped via the urine, which would mean a loss of base from the blood. Thus *hyperchloremia can cause a decrease in blood pH* (which is acidosis).

Fluids That Leave the Blood Must Return in Identical Volume Blood vessels undergo extensive branching until the narrowest tubes, called capillaries, are reached.

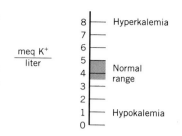

FIGURE 23.6 The potassium ion level in the blood.

■ In blood, the normal range of $[Cl^-]$ is 100 to 106 meq/L (1 meq $Cl^- = 35.5$ mg).

■ Sports beverages like Gatorade resupply the body with electrolytes.

TABLE 23.2 Calcium and Magnesium Ions in the Human Body

Area	Ca^{2+}	Mg^{2+}
Total body	6×10^4 meq (1.2 kg)	1000 meq (12 g)
Plasma level	4.2–5.2 meq/L	1.5–2.0 meq/L
Intracellular level	*	35 meq/L
Mass of 1 meq	20.0 mg	12.2 mg

* Free in solution in the cytosol, about 10^{-7} mol/L.

Blood enters a capillary loop as *arterial blood,* but it leaves on the other side of the loop as *venous blood* (Figure 23.7).

During the switch, fluids and nutrients leave the blood and move into the interstitial fluids and then into tissue cells. These fluids must return to the blood *in the same volume,* but now they must carry the wastes of metabolism. The rate of this diffusion of fluids throughout the body is high, about 25 to 30 L/s. Some fluids return to circulation by way of lymph ducts, which are thin-walled, closed-end capillaries that bed in soft tissue.

■ The lymph system makes antibodies and it has white cells that help defend the body against infectious diseases.

Blood Pressure Overcomes Osmotic Pressure on the Arterial Side of a Capillary Loop   Interstitial fluids, being more dilute than blood, have a lower osmotic pressure. The natural tendency, therefore, is for fluid to move from interstitial fluid into blood. On the arterial side of a loop, however, the blood pressure is sufficiently high to overcome this. Water and dissolved solutes are forced, instead, from the blood *into* the interstitial space. While there, exchanges of chemicals occur, nutrients move into cells, and cells get rid of wastes.

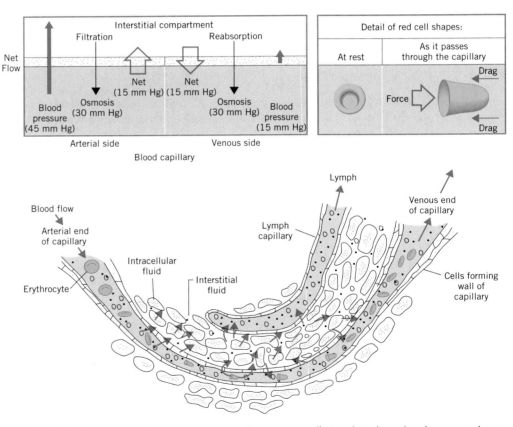

FIGURE 23.7   The exchange of nutrients and wastes at capillaries. As indicated at the top, on the arterial side of a capillary loop the blood pressure counteracts the pressure from dialysis and osmosis, and fluids are forced to leave the bloodstream. On the venous side of the loop, the blood pressure has decreased below that of dialysis and osmosis, so fluids flow back into the bloodstream. On the top right and the bottom is shown how a normal red cell distorts as it squeezes through a capillary loop. Red cells in sickle-cell anemia do not pass through as smoothly. The bottom drawing also shows how some fluids enter the lymph system.

Osmotic Pressure Overcomes Blood Pressure on the Venous Side of a Capillary Loop As blood emerges from the thin constriction of a capillary into the venous side, its pressure drops. Now it's too low to prevent the natural diffusion of fluids from the interstitial space into the bloodstream. By this time, of course, the fluids are carrying waste products.

Blood Loses Albumins in the Shock Syndrome When capillaries become more permeable to blood proteins, as they do in such trauma as sudden severe injuries, major surgery, and extensive burns, protein molecules leave the blood. Unfortunately, this means a loss of the blood's colloidal osmotic pressure, the pressure which normally helps fluids return from tissue to the bloodstream. As a result, the total volume of circulating blood drops quickly, which drastically reduces the blood's ability to carry oxygen and to remove carbon dioxide. The drop in blood volume and the resulting loss of oxygen to the brain send the victim into traumatic **shock.**

■ The prompt restoration of blood volume is mandatory in the treatment of shock.

Blood Also Loses Proteins in Kidney Disease and Starvation Sometimes the proteins in the blood are lost at malfunctioning kidneys. The effect, although gradual, is a slow but unremitting drop in the blood's colloidal osmotic pressure. Fluids accumulate in the interstitial regions. Because this takes place more slowly and water continues to be ingested, there is no sudden drop in blood volume as in shock. The victim appears puffy and waterlogged, a condition called **edema.**

■ Greek, *oidema*, swelling.

Edema can also appear at one stage of starvation, when the body has metabolized its circulating proteins to make up for the absence of dietary proteins.

Any obstruction in the veins can also cause edema, as in varicose veins and certain forms of cancer. Now it is the venous blood pressure that rises, creating a back pressure that reduces the rate at which fluids can return to circulation from the tissue areas. Localized swelling resulting from a blow is a temporary form of edema caused by injuries to the capillaries.

23.3 BLOOD AND THE EXCHANGE OF RESPIRATORY GASES

The binding of oxygen to hemoglobin is allosteric, and it is affected by the pH, the pCO_2, and the pO_2 of the blood.

The carrier of oxygen in blood is **hemoglobin,** a complex protein found inside erythrocytes. It consists of four subunits, each with one molecule of heme, an organic group that holds an iron(II) ion, the actual oxygen-binding unit. Two of the subunits are identical and have the symbol α. The other two are also identical and have the symbol β.

■ Each red cell carries about 2.8×10^8 molecules of hemoglobin.

When hemoglobin is oxygen free, it is called *deoxyhemoglobin*; in this state the molecule has a cavity in which a small organic anion nestles. This is the **2,3-bis-phosphoglycerate** ion, called simply **BPG,** and it has an important function in oxygen transport.[2]

■
$$^{2-}O_3POCH_2CHCO_2^-$$
$$|$$
$$OPO_3^{2-}$$

BPG (2,3-bisphosphoglycerate)

The Subunits of Hemoglobin Cooperate in Its Oxygenation The **oxygen affinity** of blood is the percentage to which blood has all of its hemoglobin molecules saturated

---

[2] The name 2,3-diphosphoglycerate (DPG) has been supplanted by 2,3-bisphosphoglycerate (BPG). "Diphospho-" signifies a diphosphate ester, that is, an ester of diphosphoric acid. "Bisphospho" signifies two (hence, "bis") monophosphate ester groups.

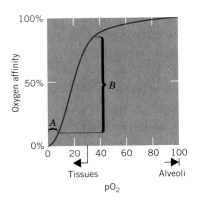

FIGURE 23.8 Hemoglobin–oxygen dissociation curve. Regions A and B are discussed in the text.

■ This natural resistance is needed in working cells where we want no restrictions on the deoxygenation of blood.

■ Carbon monoxide binds 150 to 200 times more strongly to hemoglobin than does oxygen and thus prevents the oxygenation of hemoglobin and causes internal suffocation.

■ About 20% of a smoker's hemoglobin is more or less permanently tied up by carbon monoxide.

with oxygen. A fully laden hemoglobin molecule carries four oxygen molecules and is called **oxyhemoglobin.** For maximum efficiency, all hemoglobin molecules should leave the lungs as fully loaded oxyhemoglobin. Let's see what factors ensure this.

First, the *partial pressure of oxygen is higher in the lungs* than anywhere else in the body; pO_2 is 100 mm Hg in freshly inhaled air in alveoli and only about 40 mm Hg in oxygen-depleted tissues. This partial pressure gradient helps oxygen to migrate from the lungs into the bloodstream. The higher partial pressure *pushes* oxygen into the blood.

Second, newly arrived *oxygen reacts with hemoglobin,* binding to its Fe^{2+} ions to form oxyhemoglobin. This reaction *pulls* oxygen into the blood. Notice that a plot of oxygen affinity versus pO_2 has a sigmoid shape (Figure 23.8), which we learned to associate with an allosteric effect among enzymes (Section 21.4). *An allosteric effect similarly helps to load hemoglobin with oxygen.* At low values of pO_2, in region A of the plot, the ability of the blood to take up oxygen (its oxygen affinity) increases only slowly with increases in pO_2. But eventually the oxygen affinity takes off, and rises very steeply with still more increases in pO_2, in region B of the plot. Eventually, the oxygen affinity starts to level off. It almost seems that at low partial pressures of oxygen there is a small "molecular dam" that thwarts the efforts of oxygen molecules to be joined to hemoglobin.

What is thought to be happening is as follows. We'll represent deoxyhemoglobin by structure **1**, below. Each circle in **1** is a polypeptide subunit with its heme unit but without oxygen. The oval figure centered within the structure represents one BPG anion. When the first O_2 molecule manages to bind ($\mathbf{1 \rightarrow 2}$), it induces a change in the shape of the affected subunit and a β subunit, which we have represented as a change from a circle to a square.

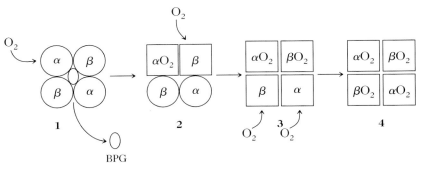

This change tends to squeeze the BPG unit out, breaking the "dam." The next oxygen molecule now enters more readily ($\mathbf{2 \rightarrow 3}$). As the subunit it enters changes its shape, *the remaining two subunits allosterically change shape* and become very receptive to the third and fourth molecules of oxygen. These two flood in ($\mathbf{3 \rightarrow 4}$), doing so far more readily than either could bind to completely deoxygenated hemoglobin. Thus, when a hemoglobin molecule accepts its first oxygen molecule, it's almost certain that it will readily accept three more and become fully oxygenated. Few if any partially oxygenated hemoglobin molecules leave the lungs.

For the sake of the remaining discussion, we'll simplify the above by letting the symbol HHb represent an entire hemoglobin molecule. The first H in HHb stands for a potential hydrogen ion, and we overlook the fact that more than one is actually present in hemoglobin. (We're now also overlooking the fact that *four* molecules of O_2 bind to one of hemoglobin.) With this in mind, we can represent the oxygenation of hemoglobin as the *forward* reaction in the following equilibrium where oxyhemoglobin is represented as the anion, HbO_2^-.

$$HHb + O_2 \rightleftharpoons HbO_2^- + H^+ \qquad (23.1)$$

hemoglobin   oxyhemoglobin

Two facts indicated by this equilibrium are that HHb is a weak acid and that it becomes a stronger acid as it becomes oxygenated. It's strong enough to be produced in its ionized state as HbO_2^- and H^+. *The presence of H^+ in Equilibrium 23.1 means that the equilibrium can be shifted one way or another simply by changing the pH,* a fact of enormous importance at the molecular level of life, as we'll see.

To understand the oxygenation of hemoglobin, we have to see how various stresses shift Equilibrium 23.1 to the *right* in the lungs. One stress, as we've already noted, is the relatively high value of pO_2 (100 mm Hg) in the alveoli. This stress acts on the left side of 23.1, so it helps to shift the equilibrium to the right. (This is the "push" mentioned earlier.)

$$HHb + O_2 \longrightarrow HbO_2^- + H^+$$

| In the red cell | From air | In the red cell | In the red cell |

Another stress, not evident from Equation 23.1, is the removal of H^+ *as it forms.* The red cell in the lungs is carrying waste HCO_3^-. The newly forming H^+ ions are therefore promptly equilibrated with HCO_3^- and CO_2 by the help of carbonic anhydrase. (We write only the forward reaction of this equilibrium here because this is how the equilibrium shifts when both H^+ and HCO_3^- are high.)

$$H^+ + HCO_3^- \xrightarrow{\text{carbonic anhydrase}} CO_2 + H_2O \qquad (23.2)$$

| In the red cell | In the red cell | | In the red cell |

■ The stimulation of HHb to bind O_2 caused by the removal of H^+ is called the **Bohr effect** after Christian Bohr, a Danish scientist (and the father of nuclear physicist Niels Bohr).

■ Carbonic anhydrase is one of the body's fastest working enzymes.

This switch from the appearance of H^+ as a product to its disappearance as a reactant is called the **isohydric shift.**

We see here one of the beautiful examples of coordinated activity in the body, the coupling of the uptake of oxygen to the release of the carbon dioxide to be exhaled. The neutralization of the H^+ ions produced from the uptake of O_2 produces CO_2. Then the loss of CO_2 from the red cell by exhaling pulls this and all previous equilibria to the *right* in the lungs.

$$CO_2 \xrightarrow{\text{exhaling}} CO_2$$

| In the red cell | In exhaled air |

Thus the uptake of O_2 as hemoglobin oxygenates, which *pushes* the equilibria to the right, simultaneously produces a chemical (CO_2) whose loss *pulls* the same equilibria to the right.

Let's now see how waste CO_2 is picked up at cells that have produced it and is carried to the lungs; and let's also see how this also works cooperatively with the *release* of oxygen at cells that need it.

Hemoglobin Subunits and BPG Cooperate in the Deoxygenation of Oxyhemoglobin in Metabolizing Tissues   Consider, now, a tissue that has done some chemical work, used up some oxygen, and made some waste carbon dioxide. When a fully oxygenated red cell arrives in such a tissue, some of the events we just described reverse themselves.

■ CO_2 molecules diffuse in body fluids 30 times more easily than O_2 molecules, so the partial pressure gradient for CO_2 need not be as steep as that for O_2.

We can think of this reversal as beginning with the diffusion of waste CO_2 from the tissue into the blood. An impetus for this diffusion is the higher pCO_2 (50 mm Hg) in active tissue versus its value in blood (40 mm Hg). Once the CO_2 arrives in the blood, it moves inside a red cell where it encounters carbonic anhydrase. Equation 23.2 is therefore run in reverse. It's part of the equilibrium managed by carbonic anhydrase, and it shifts to *make* HCO_3^- (thus consuming CO_2). From 60% to 90% of all waste CO_2 returns to the lungs as the bicarbonate ion.

$$H_2O + CO_2 \xrightarrow{\text{carbonic anhydrase}} HCO_3^- + H^+ \qquad (23.2, \text{reversed})$$

From the working tissue   In the red cell

Of course, this generates hydrogen ions. If you'll look back to Equation 23.1, which is also part of an equilibrium, you will see that an increase in the level of H^+ (caused by the influx of waste CO_2) can only make Equation 23.1 run in reverse.

$$H^+ \quad + \quad HbO_2^- \longrightarrow HHb + O_2 \qquad (23.1, \text{reversed})$$

Just made in the red cells at working tissue | In the red cell; just arrived at tissue | In the red cell | Will diffuse into the tissue needing oxygen

This reaction, another isohydric shift, not only neutralizes the acid generated by the arrival of waste CO_2, *but also helps to force oxyhemoglobin to give up its oxygen.* Notice the cooperation. The tissue that needs oxygen has made CO_2 and, hence, it has indirectly made the H^+ that is required to release its needed oxygen from newly arrived HbO_2^-. It's all rather beautiful.

■ The high negative charge on BPG keeps it from diffusing through the red cell's membrane.

The deoxygenation of HbO_2^- is also aided by the BPG anions that were pushed out when HbO_2^- formed. These anions are still inside the red cell, and as soon as an O_2 molecule leaves oxyhemoglobin, a BPG anion starts to move back into HHb. The changes in shapes of the hemoglobin subunits now operate in reverse, and all oxygen molecules smoothly leave. It's all or nothing again, and the efficiency of the unloading of oxygen is so high that if one O_2 molecule leaves, the other three follow essentially at once. BPG helps this to happen. Partially deoxygenated hemoglobin units do not slip through and go back to the lungs.

BPG and Hemoglobin Levels Are Higher in People Living at High Altitudes   It's interesting that those who live and work at high altitudes, such as the populations in Nepal in the Himalayan Mountains or the people in the Andes Mountains in Bolivia, have 20% higher levels of BPG in their blood and more red blood cells than those who live at sea level. The extra red cells give them more hemoglobin to help them carry more oxygen per milliliter of blood, and the extra BPG increases the efficiencies of both loading and unloading oxygen.

■ No conditioning at a low altitude can get the cardiovascular system ready for a low pO_2 at a high altitude.

When lowlanders take trips to high altitudes, their bodies start to build more red cells and to make more BPG so that they can function better where the partial pressure of atmospheric oxygen is lower. Those who patiently wait during the few days that it takes for these events to occur before they set off on strenuous backpacking expeditions are less likely to suffer high-altitude sickness, a condition that can cause death.

To summarize the chemical reactions we have just studied, we can write the following equations. The cancel lines show how we arrive at the overall net results, Equations 23.3 and 23.4.

Oxygenation These reactions occur in the lungs.

$$HHb \quad + \quad O_2 \quad \longrightarrow \quad HbO_2^- \quad + \quad H^+$$

In the red cell	From air	Will go in red cell to tissue	In the red cell

$$H^+ \quad + \quad HCO_3^- \quad \xrightarrow{\text{CA}} \quad CO_2 \quad + \quad H_2O$$

Just made	In red cell (but from tissues)	In red cell (but will be exhaled)	

■ CA is carbonic anhydrase.

$$\mathbf{HHb} \quad + \quad \mathbf{O_2} \quad + \quad \mathbf{HCO_3^-} \quad \longrightarrow \quad \mathbf{HbO_2^-} \quad + \quad \mathbf{CO_2} \quad + \quad \mathbf{H_2O} \qquad (23.3)$$

In the red cell	From air	In red cell (but from tissues)	Will go in red cell to tissue	Will leave the lungs in exhaled air	

■ Net effect of oxygenating hemoglobin

Deoxygenation These reactions occur wherever tissues are low in oxygen.

$$CO_2 \quad + \quad H_2O \quad \xrightarrow{\text{CA}} \quad HCO_3^- \qquad\qquad + \quad H^+$$

Waste from tissues		In red cell (in blood still within tissue but will go to the lungs)	In the red cell

$$H^+ \quad + \quad HbO_2^- \quad \longrightarrow \quad HHb \qquad + \quad O_2$$

Just made	In red cell (in blood within tissues)	In red cell (will return to the lungs)	Goes into tissue needing it

$$\mathbf{CO_2} \quad + \quad \mathbf{H_2O} \quad + \quad \mathbf{HbO_2^-} \quad \longrightarrow \quad \mathbf{HHb} \quad + \quad \mathbf{HCO_3^-} \quad + \quad \mathbf{O_2} \qquad (23.4)$$

Waste from tissues	In red cell (in blood within tissues)	In red cell (will go to the lungs)	Goes in blood to the lungs	Goes into tissue

■ Net effect of deoxygenating oxyhemoglobin

The summarizing equations do not show the importance of two factors in both the loading and the unloading of oxygen, that of BPG and that of the concentration of H$^+$. Concerning the concentration of H$^+$, Figure 23.9 shows the plots of oxygen affinity versus pO$_2$ at two different values of pH, one relatively low (pH 7.2) compared with the normal value of 7.35, and the other relatively high (pH 7.6). You may recall that the value of pO$_2$ in the vicinity of oxygen-starved cells is low, around 30 to 40 mm Hg. Notice in Figure 23.9 that in this range of low pO$_2$, the blood's ability to hold oxygen is much less at the lower pH (7.2) than it is at a higher value (7.6). And a lower pH does develop in actively metabolizing cells, caused chiefly by the presence of the CO$_2$ these cells have made. This drop in pH caused by CO$_2$ thus assists in the deoxygenation of HbO$_2^-$ exactly where it must happen.

Precisely where O$_2$ should be unloaded there is a chemical signal (a lower pH) that makes it occur. It's an altogether beautiful example of how a set of interrelated chemical equilibria shift in just the directions that are required for health and life. We'll later use Figure 23.9 to help us understand how both acidosis and alkalosis are serious threats.

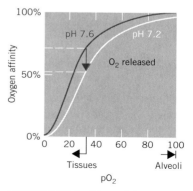

FIGURE 23.9 Hemoglobin–oxygen dissociation curves at two different values of the pH of blood.

Some CO$_2$ Is Carried to the Lungs on Hemoglobin   Another route exists for carrying waste carbon dioxide away. Not all CO$_2$ ends up as HCO$_3^-$. Some reacts with hemoglobin, which has just been released by deoxygenation of oxyhemoglobin (Equation 23.4).

$$CO_2 + HHb \longrightarrow HbCO_2^- + H^+ \qquad\qquad (23.5)$$

Waste from tissue	Just released by deoxygenation of HbO$_2^-$	Carbamino-hemoglobin (in red cells)

(This is actually one reaction of an equilibrium.)   **Carbaminohemoglobin,** HbCO$_2^-$, forms, and some waste CO$_2$ travels to the lungs in this form.

Notice in Equation 23.5 that H$^+$ also forms, just as it does when waste CO$_2$ reacts with water to make HCO$_3^-$. Thus, hydrogen ions are made whether waste CO$_2$ is changed to HCO$_3^-$ and H$^+$ or to HbCO$_2^-$ and H$^+$, and remember that H$^+$ is needed and is consumed by the unloading of O$_2$ from HbO$_2^-$. Either route of handling waste CO$_2$ helps to release O$_2$ *where it is needed.* Now let's follow the circulation back to the lungs.

When a red cell comes back to the lungs, fresh H$^+$ will now be *generated* by the combination of HHb with oxygen.

$$HHb + O_2 \rightleftharpoons HbO_2^- + H^+ \qquad\qquad (23.1, \text{again})$$

The H$^+$ now forces the reaction of Equation 23.5 to shift into reverse. Of course, this releases CO$_2$ *where it can be exhaled,* and it gets hemoglobin ready to take on more oxygen.

Chloride Ion Is Also Needed to Deoxygenate Hemoglobin   The chloride ion is still another factor, not mentioned so far, that helps to unload oxygen from oxyhemoglobin. Hemoglobin, HHb, binds chloride ion, and it binds it better than does oxyhemoglobin. The following equilibrium exists along with all the others, where HHb(Cl$^-$) represents chloride ion bound to hemoglobin.

$$HHb(Cl^-) + O_2 \rightleftharpoons HbO_2^- + Cl^- + H^+ \qquad\qquad (23.6)$$

Therefore to *unload* oxygen (that is, do the reverse of 23.6), chloride ion must also be available *inside* the red cell. The red cell obtains it by a mechanism called the *chloride shift.* Let's see how it occurs.

As chloride ions are drawn into a red cell to react with some of the HHb that is freed up in active tissue, negative ions must move out so that there is electrical charge balance. The negative ions that leave the red cell are newly formed bicarbonate ions. Although HCO$_3^-$ ions must be made *inside* red cells, because that's where carbonic anhydrase is, most leave to make the trip *outside* red cells.

For every chloride ion that enters the red cell, one HCO$_3^-$ ion leaves, and this switch is called the **chloride shift.** When the red cell gets back to the lungs, the various new chemical stresses make all the equilibria, including the chloride shift, run in reverse.

If you're bewildered by all these equilibria and how they are made to shift in the correct directions, you're almost certainly not alone among your classmates. This isn't easy material, but it is so much at the heart of so many aspects of health and one's ability to have an active life that the effort to master it is very worthwhile. As you make this effort, get the key equilibria down and memorized. For each one ask, "What are the stresses that can make it shift, and where in the body is each stress important, in the lungs or at actively metabolizing cells?" The key stresses are the following.

■   For those entering careers in nursing and respiratory therapy, there is no other single topic of such career-lasting importance as the chemistry of respiration and its associated electrolyte balance.

1. The relative partial pressures of O_2.
2. The relative partial pressures of CO_2.
3. The changes in the levels of H^+ caused by the influx of CO_2.
4. The permanent loss of CO_2 through exhaling.

After you have studied the various equilibria from the stress point of view, and you can write all the equilibria and discuss the influence of various stresses, then you might find Figure 23.10 a useful way to review. *Don't tackle it without this prior study.* Follow the direction of the large U-shaped arrow that curves around the legend, and use the boxed numbers to follow the events. Notice particularly that *the reactions that occur in the red cell when it is in metabolizing tissue are just the reverse of those that happen when the cell is in the lungs.* The reasons are the changes in the stresses to equilibria listed above.

■ For those who simply want to understand the respiratory demands (and limitations) of active sports, including skiing at altitude, the chemistry of respiration is the key.

Myoglobin Binds O_2 More Strongly than Hemoglobin   Myoglobin is a heme-containing protein in red muscle tissue such as heart muscle, where it binds and stores oxygen for the tissue's needs. Unlike hemoglobin, myoglobin has only one polypeptide unit and only one heme unit per molecule. Moreover, no allosteric effect operates when it binds O_2, as the shape of the myoglobin–oxygen dissociation curve indicates (Figure 23.11, page 659).

Myoglobin's oxygen affinity is greater than that of hemoglobin, especially in the range of pO_2 associated with actively metabolizing tissues. Consequently, *myoglobin is able to take oxygen from oxyhemoglobin.*

$$HbO_2^- + HMy \longrightarrow HHb + MyO_2^-$$

■ HMy = myoglobin; MyO_2^- = oxymyoglobin

This ability is vital to heart muscle which, as much as the brain, must have an assuredly continuous supply of oxygen. When oxymyoglobin, MyO_2^-, gives up its oxygen for the cell's needs, it can at once get a fresh supply from circulating blood. Not only does this cell now have CO_2 and H^+ available to deoxygenate HbO_2^-, it also has the superior oxygen affinity of its own myoglobin to draw more O_2 into the cell.

Fetal Hemoglobin Binds Oxygen More Strongly than Adult Hemoglobin   The hemoglobin in a fetus is slightly different from that of an adult, and it has a higher oxygen affinity than adult hemoglobin. This helps to ensure that the fetus successfully pulls oxygen from the mother's oxyhemoglobin to satisfy its own needs.

23.4   ACID–BASE BALANCE OF THE BLOOD

The proper treatment of acidosis or of alkalosis depends on knowing whether the underlying cause is a metabolic or a respiratory disorder.

Acid–base balance in the blood exists when the pH of blood is in the range of 7.35 to 7.45. A decrease in pH (acidosis) or an increase (alkalosis) is serious and requires prompt attention, because all of the equilibria that involve H^+ in oxygenation or deoxygenation of blood are sensitive to pH. If the pH falls below 6.8 or rises above 7.8, life is not possible.

■ Acidosis is sometimes called *acidemia*, and alkalosis is called *alkalemia*.

Disturbances in Either Metabolism or Respiration Can Upset the Blood's Acid–Base Balance   In general, acidosis results from either the retention of acid or the loss of base by the body, and these can be induced by disturbances in either me-

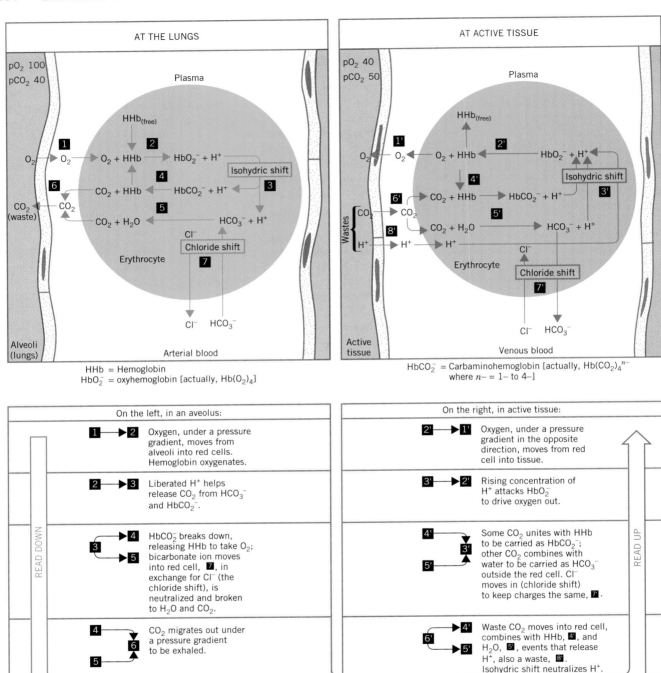

FIGURE 23.10 Oxygen and carbon dioxide exchange in the blood.

tabolism or respiration. Similarly, alkalosis results either from the loss of acid or from the retention of base, and some disorder in either metabolism or respiration can be the underlying cause.

A malfunction in respiration can be caused by any kind of injury or disease of the lungs or of the *respiratory centers*. These are units in the brain that sense changes in the pH and pCO_2 of the blood and instruct the lungs to breathe either more rapidly or more slowly.

In this section we'll study four situations, metabolic and respiratory acidosis as well as metabolic and respiratory alkalosis. We will learn how the values of pH, pCO_2, and the molar concentration of HCO_3^- in serum, $[HCO_3^-]$, change in each situation.

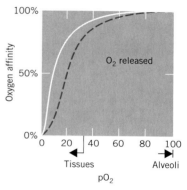

FIGURE 23.11 The myoglobin–oxygen dissociation curve is the solid line. The dashed line is the curve for hemoglobin. Over the entire range of pH in tissues that may need oxygen, the oxygen affinity of myoglobin is greater than that of hemoglobin.

Metabolic Acidosis Receives a Respiratory Compensation, Hyperventilation In **metabolic acidosis,** the lungs and the respiratory centers are working, and the problem is metabolic. Acids are being produced faster than they are neutralized, or they are being exported too slowly.

Excessive loss of base, as from severe diarrhea, can also result in metabolic acidosis. (In diarrhea, the alkaline fluids of the duodenum leave the body, and as base migrates to replace them, a depletion of base occurs somewhere else, such as in the blood, at least for a time.)

As the pH of the blood falls and the molar concentration of H^+ rises, there are parallel *but momentary* increases in the values of pCO_2 and $[HCO_3^-]$. The value of pCO_2 starts to increase because the carbonate buffer, working hard to neutralize the extra H^+, manufactures CO_2.

$$H^+ \; + \; \underset{\substack{\text{Produced} \\ \text{by acidosis}}}{HCO_3^-} \longrightarrow H_2O + CO_2$$

The kidneys work harder during this situation to try to keep up the supply of HCO_3^-.

The chief compensation for metabolic acidosis, however, involves the respiratory system. The respiratory centers, which are sensitive to changes in pCO_2, instruct the lungs to blow the CO_2 out of the body. The lungs, in other words, hyperventilate. As the equation above shows, the loss of each molecule of CO_2 means a net neutralization of one H^+ ion.

Hyperventilation, however, is usually overdone. So much CO_2 is blown out that pCO_2 actually decreases; and a low arterial pCO_2 is called **hypocapnia**. Thus, as the blood pH decreases, so too do the values of pCO_2 (from hyperventilation) and $[HCO_3^-]$ (from the reaction with H^+). We can summarize the range of a number of clinical situations that involve metabolic acidosis as follows.

- Normal values (arterial blood):

pH = 7.35–7.45

pCO_2 = 35–45 mm Hg

$[HCO_3^-]$ = 19–24 meq/L

(1.0 meq HCO_3^- = 61 mg HCO_3^-)

- (↓) means a decrease from normal and (↑) means an increase. Some typical values are in parentheses. Note that some changes do not necessarily bring values outside the normal ranges.

Clinical Situations of Metabolic Acidosis

Lab results	pH ↓ (7.20); pCO_2 ↓ (30 mm Hg); $[HCO_3^-]$ ↓ (12 meq/L)
Typical patient	An adult male comes to the clinic with a severe infection. He does not know that he has diabetes.
Range of causes	Diabetes mellitus; severe diarrhea (with loss of HCO_3^-); kidney failure (to export H^+ or to make HCO_3^-); prolonged starvation; severe infection; aspirin overdose; alcohol poisoning.
Symptoms	Hyperventilation (because the respiratory centers have told the lungs to remove excess CO_2 from the blood); increased urine output (to remove H^+ from the blood); thirst (to replace water lost as urine); drowsiness; headache; restlessness; disorientation.

Treatment   If the kidneys function, use isotonic HCO_3^- intravenously to restore HCO_3^- level, thereby neutralizing H^+ and raising pCO_2. In addition, restore water. In diabetes, use insulin therapy. If the kidneys do not function, hemodialysis must be tried.

Respiratory Acidosis Is Compensated by a Metabolic Response   In **respiratory acidosis,** either the respiratory centers or the lungs have failed, and the lungs are hypoventilating *because they cannot help it.* The blood now cannot help but retain CO_2. An increase in arterial pCO_2 is called **hypercapnia.** The retention of CO_2 functionally means the retention of acid, because CO_2 can neutralize base by the following equation.

$$HO^-(aq) + CO_2(aq) \longrightarrow HCO_3^-(aq) \qquad (23.7)$$

The decrease in the level of base lowers the pH of the blood and gives rise to respiratory acidosis.

Clinical Situations of Respiratory Acidosis

Lab results	pH ↓ (7.21); pCO_2 ↑ (70 mm Hg); $[HCO_3^-]$ ↑ (27 meq/L)
Typical patient	Chain smoker with emphysema or anyone with chronic obstructive pulmonary disease.
Range of causes	Emphysema, severe pneumonia, asthma, anterior poliomyelitis, or any cause of shallow breathing such as an overdose of narcotics, barbiturates, or general anesthesia; severe head injury.
Symptons	Shallow breathing (which is involuntary).
Treatment	Underlying problem must be treated; possibly intravenous sodium bicarbonate; possibly hemodialysis.

The body responds metabolically as best it can to respiratory acidosis by using HCO_3^- to neutralize the acid, by making more HCO_3^- in the kidneys, and by exporting H^+ via the urine.

Metabolic Alkalosis Also Receives a Respiratory Compensation, Hypoventilation   In **metabolic alkalosis,** the system has lost acid, or it has retained base (HCO_3^-), or it has been given an overdose of base (e.g., antacids). Metabolic alkalosis can also be caused by a kidney-associated decrease in the serum levels of K^+ or Cl^-. The loss of these ions means the retention of Na^+ and HCO_3^- ions, because these work in tandem and oppositely. The loss of acid could be from prolonged vomiting, which removes the gastric acid. This is followed by an effort to borrow serum H^+ to replace it, and the pH of the blood increases as it gives up H^+.

■ Compensation by hypoventilation is obviously limited by the fundamental need of the body for some oxygen.

Whatever the cause, the respiratory centers sense an increase in the level of base in the blood (as the level of acid drops), and they instruct the lungs to retain the most readily available neutralizer of base it has, namely CO_2. It removes OH^- according to Equation 23.7. The lungs hypoventilate to help retain CO_2, which can neutralize base. Thus metabolic alkalosis leads to hypoventilation.

Notice carefully that hypoventilation alone cannot be used to tell whether someone has metabolic *alkalosis* or respiratory *acidosis.* Either condition means hypoventilation. One condition, respiratory acidosis, could be treated by intravenous sodium bicarbonate, a base. But this would aggravate metabolic alkalosis.

You can see that the lab data on pH, pCO_2, and $[HCO_3^-]$ must be obtained to determine which condition is actually present. Otherwise, the treatment used could be just the opposite of what should be done.

Clinical Situations of Metabolic Alkalosis

Lab results	pH ↑ (7.53); pCO_2 ↑ (56 mm Hg); $[HCO_3^-]$ ↑ (45 meq/L)
Typical patient	Postsurgery patient with persistent vomiting.
Range of causes	Prolonged loss of stomach contents (vomiting or nasogastric suction); overdose of bicarbonate or of medications for stomach ulcers; severe exercise, or stress, or kidney disease (with loss of K^+ and Cl^-); overuse of a diuretic.
Symptoms	Hypoventilation (to retain CO_2); numbness, headache, tingling; possibly convulsions.
Treatment	Isotonic ammonium chloride (a mild acid), intravenously with great care; replace K^+ loss.

■ An overdose of "bicarb" ($NaHCO_3$), from a too aggressive use of this home remedy for "heartburn," can cause metabolic alkalosis.

■ Ammonium ion acts as a neutralizer as follows.

$$NH_4^+ + OH^- \longrightarrow NH_3 + H_2O$$

Respiratory Alkalosis Is Compensated Metabolically by a Reduced Bicarbonate Level   In **respiratory alkalosis,** the body has lost acid, usually by involuntary hyperventilation such as occurs in hysterics, prolonged crying, overbreathing at high altitudes, or by the mismanagement of a respirator. The respiratory centers have lost control, and the body expels CO_2 too rapidly. The loss of CO_2 means the loss of a base neutralizer from the blood. Hence, the level of base rises and the pH increases. To compensate, the kidneys excrete base, HCO_3^-, so the serum level of HCO_3^- decreases.

Extreme respiratory alkalosis can occur to mountain climbers, like climbers of Mount Everest (8848 m, 29,030 ft). At its summit, the barometric pressure is 253 mm Hg and the pO_2 of the air is only 43 mm Hg (as compared to 149 mm Hg at sea level). Hyperventilation brings a climber's arterial pCO_2 down to only 7.5 mm Hg (compared to a normal of 40 mm Hg) and the blood pH is above 7.7!

■ Tissue that gets too little O_2 is in a state of *hypoxia*. If it gets none at all, it is in a state of *anoxia*.

Clinical Situations of Respiratory Alkalosis

Lab results	pH ↑ (7.56); pCO_2 ↓ (23 mm Hg); $[HCO_3^-]$ ↓ (20 meq/L)
Typical patient	Someone nearing surgery and experiencing anxiety.
Range of causes	Prolonged crying; rapid breathing at high altitudes; hysterics; fever; disease of the central nervous system; improper management of a respirator.
Symptoms	Hyperventilation (that can't be helped); numbness, headache, tingling; convulsions may occur.
Treatment	Rebreathe one's own exhaled air (by breathing into a sack); administer carbon dioxide; treat underlying causes.

Take careful notice that hyperventilation alone cannot be used to tell what the condition is. Either metabolic *acidosis* or respiratory *alkalosis* is accompanied by hyperventilation, but the treatments are opposite in nature.

Combinations of Primary Acid–Base Disorders Are Possible   We have just surveyed the *four primary acid–base disorders.* Combinations of these are often seen, and health care professionals have to be alert to the ways in which the lab data vary in such combinations. Someone with diabetes, for example, might also suffer from an obstructive pulmonary disease. Diabetes causes metabolic acidosis and a *decrease* in $[HCO_3^-]$. The pulmonary disease causes respiratory acidosis with an *increase* in $[HCO_3^-]$. In combination, then, the lab data on bicarbonate level will not be in the expected pattern for either. We will not carry the study of such complications further. We mention them only to let you know that they exist. There are standard ways to recognize them.[3]

■ People working in emergency care situations get the requisite lab data rapidly, and they must be able to interpret the data on the spot.

[3] See, for example, H. Valtin and F. J. Gennari, *Acid–Base Disorders, Basic Concepts and Clinical Management.* Little, Brown and Company, Boston: 1987.

23.5 BLOOD AND THE FUNCTIONS OF THE KIDNEYS

Both filtration and chemical reactions in the kidneys help to regulate the electrolyte balance of the blood.

■ The net urine production is 0.6 to 2.5 L/day.

Diuresis is the formation of urine in the kidneys, and it is an integral part of the body's control of the electrolyte and buffer levels in blood.

Urea Is the Chief Nitrogen Waste Exported in the Urine Huge quantities of fluids leave the blood by diffusion each day at the hundreds of thousands of filtering units in each kidney. Substances in solution but not those in colloidal dispersions (e.g., proteins) leave in these fluids. Then active transport processes in kidney cells pull all of any escaped glucose, any amino acids, and most of the fluids and electrolytes back into the blood. Most of the wastes are left in the urine that is being made.

Urea is the chief nitrogen waste (30 g/day), but creatinine (1 to 2 g/day), uric acid (0.7 g/day), and ammonia (0.5 g/day) are also excreted with the urine. If the kidneys are injured or diseased and cannot function, wastes build up in the blood, which leads to a condition known as *uremic poisoning.*

■ *Ur-,* of the urine; *-emia,* of the blood. *Uremia* means substances of the urine present in the blood.

■ The monitors in the hypophysis of the blood's osmotic pressure are called *osmoreceptors.*

The Hormone Vasopressin Helps Control Water Loss A nonapeptide hormone from the hypophysis, **vasopressin,** instructs the kidneys to retain water—it's dubbed the *antidiuretic hormone* or ADH. The hypophysis releases vasopressin when the osmotic pressure of blood increases by as little as 2%. An osmotic pressure higher than normal (hypertonicity) means a higher concentration of solutes and colloids in blood. So when vasopressin arrives at the kidneys and promotes the reabsorption of water, levels of solutes in blood are prevented from going even higher. In the meantime, the thirst mechanism is stimulated to bring in water to dilute the blood.

■ In *diabetes insipidus,* vasopressin secretion is blocked, and unchecked diuresis can make from 5 to 12 L of urine a day.

Conversely, if the osmotic pressure of blood decreases (becomes hypotonic) by as little as 2%, the hypophysis retains vasopressin. None reaches the kidneys, so the water that has left the bloodstream at the filtering units does not return as much. Remember that a low osmotic pressure means a low concentration of solutes, so the absence of vasopressin at the kidneys when the blood is hypotonic lets urine form. This reduces the amount of water in the blood and thereby raises the concentrations of its dissolved matter. You can see that with the help of vasopressin a normal individual can vary the intake of water widely and yet preserve a stable, overall concentration of substances in blood.

■ The inositol phosphate "cascade" mediates the chemical signals of vasopressin.

The Hormone Aldosterone Helps the Blood Retain Sodium Ion The adrenal cortex makes **aldosterone,** a steroid hormone that works to stabilize the sodium ion level of the blood. It's secreted when this level drops. As aldosterone arrives at the kidneys, it initiates reactions that cause sodium ions, having left the blood, to return again. The return requires the return of water as well to keep serum concentrations isotonic.

Conversely, when the sodium ion level of the blood increases, aldosterone is not secreted, so sodium ions that have migrated out of the blood at a filtering unit are permitted to stay out. They remain in the urine being made, together with some extra water.

The Kidneys Make HCO_3^- for the Blood's Buffer We have seen that breathing is the body's most direct means of controlling acid as it removes or retains CO_2. *The kidneys are the body's means of controlling base,* as they make or remove HCO_3^-.

The kidneys also adjust the blood's levels of HPO_4^{2-} and $H_2PO_4^-$, the anions of the phosphate buffer. Moreover, when acidosis develops, the kidneys can put H^+ ions into the urine. Some neutralization of these ions by HPO_4^{2-} and by NH_3 takes place, but the urine becomes definitely more acidic as acidosis continues.

The Kidneys Excrete Organic Anions   When acidosis has a *metabolic* origin, the serum levels of the anions of organic acids increase. Organic acids are made at accelerated rates in metabolic disorders, like diabetes or starvation, and are the chief cause of the pH change in metabolic acidosis. The base in the blood buffer has to neutralize them, so they end up as anions.

The kidneys let organic anions stay in the urine, but only by letting increasing quantities of water stay, too. This is because there is a limit to how concentrated urine can become, so as solutes stay in the urine, water must also stay. Someone with metabolic acidosis, therefore, can experience a general dehydration as the system borrows water from other fluids to make urine. The thirst mechanism normally brings in replacement water, so the individual drinks copious amounts of fluids.

The Kidneys Can Export HCO_3^-   In alkalosis, the kidneys can put bicarbonate ion into the urine, and it no longer uses HPO_4^{2-} to neutralize H^+. Both actions increase the pH of the urine, and in severe alkalosis it can go over a value of 8.

The Kidneys Also Help to Regulate Blood Pressure   If the blood pressure drops, as in hemorrhaging, the kidneys secrete a trace of *renin* into the blood. Renin is an enzyme that acts on one of the zymogens in blood, angiotensinogen, to convert it to the enzyme, angiotensin I. This, in turn, helps to convert still another protein in blood to angiotensin II, a neurotransmitter.

Angiotensin II is the most potent vasoconstrictor known. When it makes blood capillaries constrict, the heart has to work harder, which makes the blood pressure increase. This helps to ensure that some semblance of proper filtration continues at the kidneys.

Angiotensin II also triggers the release of aldosterone, which we've already learned helps the blood to retain water. This is important because the maintenance of the overall blood volume is needed to sustain a proper blood pressure.

■ When taken after several hours of fasting, urine normally has a pH of 5.5 to 6.5.

■ In severe acidosis, the pH of urine can go as low as 4.

SUMMARY

Digestion   The release of digestive juices is under the control of nerve signals and such digestive hormones as gastrin, cholecystokinin, and secretin. α-Amylase in saliva begins the digestion of starch. Pepsin in gastric juice starts the digestion of proteins. In the duodenum, trypsinogen (from the pancreas) is activated by enteropeptidase (from the intestinal juice) and becomes trypsin, which helps to digest proteins. Enteropeptidase also activates chymotrypsin (from chymotrypsinogen) and carboxypeptidase (from procarboxypeptidase). These also help to digest proteins. The pancreas supplies an important lipase, which, with the help of the bile salts, catalyzes the digestion of hydrolyzable lipids. The bile salts also aid in the absorption of the fat-soluble vitamins, A, D, E, and K.

Intestinal juice supplies enzymes for the digestion of disaccharides, nucleic acids, small polypeptides, and lipids.

The products of the digestion of proteins are amino acids. The digestion of carbohydrates gives glucose, fructose, and galactose; and that of the triacylglycerols gives anions of fatty acids and monoacylglycerols (and some diacylglycerols). Complex lipids and nucleic acids are also hydrolyzed.

Blood   Proteins in blood give it a colloidal osmotic pressure that assists in the exchange of nutrients at capillary loops. Albumins are carriers for hydrophobic molecules and serum-soluble metallic ions. Some globulins help to defend the body against bacterial infections. Fibrinogen is the precursor of fibrin, the protein of a blood clot.

Among the electrolytes, anions of carbonic and phosphoric acid are involved in buffers, and all ions are involved in regulating the osmotic pressure of the blood. The chief cation in blood is Na^+, and the chief cation inside cells is K^+. The balances between Na^+ and K^+ as well as between Ca^{2+} and Mg^{2+} are tightly regulated. Ca^{2+} is vital to the operation of muscles, including heart muscle. Mg^{2+} is involved in a number of enzyme systems.

The blood transports oxygen and products of digestion to all tissues. It carries nitrogen wastes to the kidneys. It unloads cholesterol and heme breakdown products at the gallbladder. And it transports hormones to their target cells. Lymph, another fluid, helps to return some substances to the blood from tissues.

Sudden failure to retain the protein in blood leads to an equally sudden loss in blood volume and a condition of shock. Slower losses of protein, as in kidney disease or starvation, lead to edema.

Respiration   The relatively high pO_2 in the lungs helps to force O_2 into HHb. This creates HbO_2^- and H^+. In an isohydric shift, the H^+ is neutralized by HCO_3^-, which is returning from working tissues that make CO_2. The resulting CO_2 leaves during exhaling. Some of the H^+ also converts $HbCO_2^-$ to HHb and CO_2.

In deoxygenating HbO_2^- at cells that need oxygen, the influx of CO_2 makes HCO_3^- and H^+. The H^+ then moves (isohydric shift) to HbO_2^- and breaks it down to HHb and O_2. Both Cl^- and BPG help to ease the last of the four O_2 molecules out of oxyhemoglobin. The low oxygen affinity of blood in tissues where pCO_2 is high helps in the release of oxygen also.

In red muscle tissue, myoglobin's superior oxygen affinity ensures that such tissue can obtain oxygen from the deoxygenation of oxyhemoglobin. Fetal hemoglobin also has an oxygen affinity superior to that of adult hemoglobin.

Acid–base balance   The body uses the bicarbonate ion of the carbonate buffer to inhibit acidosis by irreversibly removing H^+ when the lungs release CO_2. HCO_3^- is replaced by the kidneys, which can also put excess H^+ into the urine. Dissolved CO_2 in the blood's carbonate buffer works to control alkalosis by neutralizing OH^-. Metabolic acidosis, with hyperventilation, and metabolic alkalosis, with hypoventilation, arise from dysfunctions in metabolism. Respiratory acidosis, with hypoventilation, and metabolic alkalosis, with hyperventilation, occur when the respiratory centers or the lungs are not working.

Diuresis   The kidneys, with the help of hormones and changes in blood pressure, blood osmotic pressure, and concentrations of ions, monitor and control the concentrations of solutes in blood. Vasopressin tells the kidneys to keep water in the bloodstream. Aldosterone tells the kidneys to keep sodium ion (and therefore water also) in the bloodstream. A drop in blood pressure tells the kidneys to release renin, which activates a vasoconstrictor, and aldosterone, which helps to raise blood pressure and retain water. In acidosis, the kidneys transfer H^+ to the urine and replace some of the HCO_3^- lost from the blood. In alkalosis the kidneys put some HCO_3^- into urine.

REVIEW EXERCISES

The answers to Review Exercises whose numbers are in color are found in Appendix E. The answers to the other Review Ecercises are found in the Study Guide that accompanies this book. The more challenging questions are marked with asterisks.

Digestion

23.1   Name the two chief extracellular fluids.

23.2   Name the fluids with digestive enzymes or digestive zymogens.

23.3   Describe how the flow of gastric juice is regulated.

23.4   What stimulates of the Na^+–K^+ pump to work in the stomach?

23.5   What is the result of the work of cholecystokinin? Of secretin?

23.6   What enzymes or zymogens are there, if any, in each of the following?
(a) Saliva   (b) Gastric juice
(c) Pancreatic juice   (d) Bile
(e) Intestinal juice

23.7   Name the enzymes and the digestive juices that supply them (or their zymogens) that catalyze the digestion of each of the following.
(a) Large polypeptides   (b) Triacylglycerols
(c) Amylose   (d) Sucrose
(e) Di- and tripeptides   (f) Nucleic acids

23.8   What are the end products of the digestion of each of the following?
(a) Proteins
(b) Carbohydrates
(c) Triacylglycerols

*23.9   What functional groups are hydrolyzed when each of the substances in Review Exercise 23.8 is digested? (Refer back to earlier chapters if necessary.)

23.10   In what way does enteropeptidase function as a "master switch" in digestion?

23.11   What would happen if the pancreatic zymogens were activated within the pancreas?

23.12   What services do the bile salts render in digestion?

23.13   What does mucin do (a) for food in the mouth and (b) for the stomach?

23.14 What is the catalyst for each of the following reactions?
(a) Pepsinogen → pepsin
(b) Trypsinogen → trypsin
(c) Chymotrypsinogen → chymotrypsin
(d) Procarboxypeptidase → carboxypeptidase
(e) Proelastase → elastase

23.15 What is the difference between rennin and renin?

23.16 Why is gastric lipase unimportant to digestive processes in the adult stomach but useful in the infant stomach?

23.17 In terms of where they work, what is different about intestinal juice compared to pancreatic juice?

23.18 What secretion neutralizes chyme, and why is this work important?

23.19 What happens to the molecules of the monoacylglycerols and fatty acids that form from digestion?

23.20 In a patient with a severe obstruction of the bile duct the feces appear clay colored. Explain why the color is light.

23.21 The cholesterol in the diet undergoes no reactions of digestion. Explain.

Substances in Blood

23.22 In terms of their general composition, what is the greatest difference between blood plasma and interstitial fluid?

23.23 What is the largest contributor to the net osmotic pressure of the blood as compared to the interstitial fluid?

23.24 What is fibrinogen? Fibrin?

23.25 What services are performed by albumins in blood?

23.26 In what two different regions are Na^+ and K^+ ions mostly found? What are the chief functions of these ions?

23.27 What causes the hyperkalemia in crushing injuries?

23.28 Excessive drinking of water tends to cause what condition, hyponatremia or hypernatremia?

23.29 The overuse of milk of magnesia can lead to what condition that involves Mg^{2+}?

23.30 Inside cells, what is a function that Mg^{2+} serves?

23.31 Where is most of the calcium ion in the body?

23.32 What does Ca^{2+} do in cells?

23.33 What condition is brought on by an overdose of vitamin D; hypercalcemia or hypocalcemia?

23.34 What is the normal range of concentration of Cl^- in blood? Explain how hypochloremia leads to alkalosis.

Exchange of Nutrients at Capillary Loops

23.35 What two opposing forces are at work on the arterial side of a capillary loop? What is the net result of these forces, and what does the net force do?

23.36 On the venous side of a capillary loop there are two opposing forces. What are they, what is the net result, and what does this cause?

23.37 Explain how a sudden change in the permeability of the capillaries can lead to shock.

Exchange of Respiratory Gases

23.38 The binding of oxygen to hemoglobin is said to be *allosteric*. What does this mean, and why is it important?

*23.39 Write the equilibrium expression for the oxygenation of hemoglobin. In what direction does this equilibrium shift when:
(a) The pH decreases?
(b) The pO_2 decreases?
(c) The red cell is in the lungs?
(d) The red cell is in a capillary loop of an actively metabolizing tissue?
(e) CO_2 comes into the red cell?
(f) HCO_3^- ions flood into the red cell?

*23.40 Using chemical equations, describe the isohydric shift when a red cell is (a) in actively metabolizing tissues and (b) in the lungs.

*23.41 In what two ways does the oxygenation of hemoglobin in red cells in alveoli help to release CO_2?

*23.42 In what way does waste CO_2 at active tissues help to release oxygen from the red cell?

*23.43 In what way does extra H^+ at active tissue help release oxygen from the red cell?

23.44 Where is carbonic anhydrase found in the blood, and what function does it have in the management of the respiratory gases in (a) an alveolus and (b) actively metabolizing tissues?

23.45 How does BPG help in the process of oxygenating hemoglobin?

23.46 In what way is BPG involved in helping to deoxygenate HbO_2^-?

23.47 What are some changes involving the blood that occur when the body remains at a high altitude for a period of time, and how do these changes help the individual?

23.48 How would the net equations for the oxygenation and the deoxygenation of blood be changed to include the function of BPG?

23.49 What are the two main forms in which waste CO_2 moves to the lungs?

23.50 How is oxygen affinity affected by pCO_2, and how is this beneficial?

23.51 What is the chloride shift and how does it aid in the exchange of respiratory gases?

23.52 In what way is the superior oxygen affinity of myoglobin over hemoglobin important?

23.53 Fetal hemoglobin has a higher oxygen affinity than adult hemoglobin. Why is this important to the fetus?

Acid–Base Balance of the Blood

23.54   Construct a table using arrows (↑) or (↓) and typical lab data that summarize the changes observed in respiratory and metabolic acidosis and alkalosis. Include a row of data showing normal values. The column headings should be as follows.

| Condition | pH | pCO_2 | $[HCO_3^-]$ |

*23.55   With respect to the *directions* of the changes in the values of pH, pCO_2, and $[HCO_3^-]$ in both respiratory acidosis and metabolic acidosis, in what way are the two types of acidosis the same? In what way are they different?

23.56   Hyperventilation is observed in what two conditions that relate to the acid–base balance of the blood? In one, giving carbon dioxide is sometimes used, and in the other, giving isotonic HCO_3^- can be a form of treatment. Which treatment goes with which condition, and why?

23.57   In what two conditions that relate to the acid–base balance of the blood is hypoventilation observed? Isotonic ammonium chloride or isotonic sodium bicarbonate are possible treatments. Which treatment goes with which condition, and how do they work?

23.58   In which condition relating to acid–base balance does hyperventilation have a beneficial effect? Explain.

*23.59   Hyperventilation is part of the *cause* of the problem in which condition relating to the acid–base balance of the blood?

*23.60   Hypoventilation is the body's way of helping itself in which condition that relates to the acid–base balance of the blood?

*23.61   In which condition that concerns the acid–base balance of the blood is hypoventilation part of the *problem* rather than the cure?

23.62   How can a general dehydration develop in metabolic acidosis?

*23.63   Which condition, metabolic or respiratory acidosis or alkalosis, results from each of the following situations?
(a) Hysterics
(b) Overdose of bicarbonate
(c) Emphysema
(d) Narcotic overdose
(e) Diabetes
(f) Overbreathing at a high altitude
(g) Severe diarrhea
(h) Prolonged vomiting
(i) Cardiopulmonary disease
(j) Barbiturate overdose

*23.64   Referring to Review Exercise 23.63, which is happening in each situation; hyperventilation or hypoventilation?

23.65   Why does hyperventilation in hysterics cause alkalosis?

23.66   Explain how emphysema leads to acidosis.

23.67   Prolonged vomiting leads to alkalosis. Explain.

23.68   Uncontrolled diarrhea can cause acidosis. Explain.

23.69   Respiratory alkalosis causes which, hypocapnia or hypercapnia?

23.70   For each 1 °C above normal human body temperature, the rate of CO_2 production increases by 13%. If the rate of breathing does not increase, what results, hypocapnia or hypercapnia?

Blood Chemistry and the Kidneys

23.71   If the osmotic pressure of the blood has increased, what, in general terms, has changed to cause this?

23.72   How does the body respond to an increase in the osmotic pressure of the blood?

23.73   If the sodium ion level of the blood falls, how does the body respond?

23.74   What is the response of the kidneys to a decrease in blood pressure?

23.75   Alcohol in the blood suppresses the secretion of vasopressin. How does this affect diuresis?

23.76   In what ways do the kidneys help to reduce acidosis?

Gastric Juice and Ulcers (Interaction 23.1)

23.77   Cimetidine, Zantac, and Prilosec are described as competitive enzyme inhibitors. What does this mean? How do they work to aid in the healing of an ulcer?

Additional Exercises

23.78   Monoacylglycerols are able to migrate through membranes of the cells of the intestinal tract that absorb them. Glycerol, however, is unable to accomplish this movement. How might we explain these relative abilities to migrate through a cell membrane?

*23.79   When the gallbladder is surgically removed, lipids of low formula mass are the only kinds that can be easily digested. Explain.

*23.80   It has been reported that some long-distance Olympic runners have trained at high altitudes and then had some of their blood withdrawn and frozen. Days or weeks after returning to lower altitudes and just prior to a long race, they have used some of this blood to replace an equal volume of what they are carrying. This is supposed to help them in the race. How would it work?

*23.81   Aquatic diving animals are known to have much larger concentrations of myoglobin in their red muscle tissue than humans. How is this important to their lives?

NUCLEIC ACIDS

24

The little guy doesn't know it, but he's going to grow up to be a polar bear and not a tulip or an Emperor penguin. The molecular basis for such important matters in nature is described in this chapter.

THIS CHAPTER IN CONTEXT

The last half of the twentieth century saw the birth and flowering of a new discipline, *molecular biology.* It explained how genes work and produced fruits of genetic engineering undreamed of a few decades ago. At the heart of the chemistry of heredity are molecular geometry, complementarity, and simple familiar forces, like hydrogen bonds and hydrophobic or hydrophilic interactions. It's amazing how so much of what we value depends on these familiar properties. The structures we'll meet may be complicated. Don't be put off by that, because the fundamental ideas are easy. They build on what we've already studied.

24.1   HEREDITY AND THE CELL

The instructions for making the chemicals in cells are carried on molecules of nucleic acids.

A requirement of reproduction is that each organism transmits to its offspring the capacity to possess the set of enzymes unique to it. The chemicals of heredity handle this assignment in a most beautiful way.

Nucleic Acids Carry Instructions for Making Enzymes   Successful reproduction does not demand that a cell or an organism duplicate the enzymes themselves and pass them directly to its offspring. Instead, the *instructions* for making its unique enzymes are reproduced and delivered. This is done by duplicating *nucleic acids,* passing on the copies to offspring, and letting these compounds direct the synthesis of enzymes. It will help us to know something of the biological context of this activity, the cell, so let's start there.

Every Cell Carries Nucleic Acids   The cell is the smallest unit of an organism that has life and can duplicate itself. Although the cells of different tissues or animals vary widely in shape and size, they generally have similar parts (Figure 24.1). All animal cells have a cell membrane, the boundary. Everything it encloses is called *protoplasm,* which contains several discrete particles or cellular bodies. Prominent among them are **mitochondria,** tiny *organelles* (subcellular bodies) where adenosine triphosphate (ATP) is made for the cell's chemical energy needs. All cells have a central body called the *nucleus.* It has its own membrane, and inside it is a weblike network of protein.

The part of the cell outside of the nucleus is called the *cytoplasm.* **Ribosomes** are one of the kinds of cytoplasmic particles. They consist mostly of nucleic acids and polypeptides and have essential functions in the synthesis of enzymes. Such synthesis, however, is under the primary control of other nucleic acid molecules kept *inside* the nucleus.

■   The liquid portion of the cytoplasm is called the *cytosol.*

Genes Are the Fundamental Units of Heredity   Besides the weblike protein, the nucleus contains twisted and intertwined filaments of nucleoprotein called *chromatin.* Chromatin has been likened to a strand of pearls. Each pearl, called a *nucleosome,* is made of barrel-shaped proteins called *histones* around which long molecules of DNA,

■   The histones are not just spools for wrapping DNA strands; they contribute to the regulation of DNA activity.

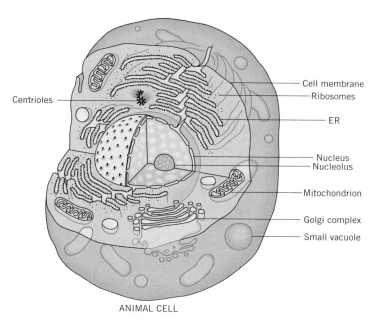

Centrioles

Cell membrane
Ribosomes
ER
Nucleus
Nucleolus
Mitochondrion
Golgi complex
Small vacuole

ANIMAL CELL

FIGURE 24.1 Model of a generalized animal cell. (ER stands for endoplasmic reticulum.) Although cells differ greatly from tissue to tissue, most have the features shown here. (From G. C. Stephens and B. B. North, *Biology, A Contemporary Perspective.* John Wiley and Sons, New York, 1974. Used by permission.)

one of the two kinds of nucleic acid, are tightly coiled. DNA strands also make up the "string" that links the "pearls."

DNA molecules carry individual **genes,** the fundamental units of heredity. A single gene contains the information for making a single polypeptide. Unique genes thus translate into unique enzymes. A major goal of this chapter is to learn how genetic information can be encoded in a molecule.

Prior to Cell Division, Genes Replicate   When cell division begins, the chromatin strands thicken and become rodlike bodies called **chromosomes.** They accept staining agents and so can be seen under a microscope. The thickening of chromatin into chromosomes is caused by the synthesis of new DNA and histones. The new chromosomes, including their DNA, are exact copies of the old, if all goes well (as it does to a remarkable degree). The reproductive duplication of DNA is called **replication,** so by the replication of DNA, exact copies of the genetic message of the first cell are made and passed on to the two new cells, one copy to each.

When two germ cells, a sperm and an ovum, unite at conception to form the single cell from which the entire organism will grow, DNA from both germ cells combine. Genetic characteristics of both parents thus are incorporated into the offspring.

Every cell produced from the first cell has the entire set of genes, but obviously most genes in the older organism are turned off most of the time. Genes that are behind the formation of fingernails, for example, must not operate in heart muscle cells! Another goal of our study is to learn how genes may be switched off and on.

■ There are 23 matched pairs of chromosomes in the human cell, for a total of 46 chromosomes.

■ Every human cell has between 50,000 and 100,000 genes.

24.2　STRUCTURE OF NUCLEIC ACIDS

Genetic information is carried by the side chains of a double-stranded polymer called the DNA double helix.

The nucleic acids that store and direct the transmission of genetic information are polymers nicknamed DNA and RNA. **DNA is deoxyribonucleic acid,** and **RNA is ribonucleic acid.** The monomer molecules for the nucleic acids are called **nucleotides.**

Nucleotides Are Monophosphate Esters of Pentoses Joined to Heterocyclic Bases Unlike the monomers of proteins, the nucleotides can be further hydrolyzed (Figure 24.2). From a representative mixture of nucleotides three kinds of products form, inorganic phosphate, a pentose sugar, and a group of heterocyclic amines called the **bases,** which have single-letter symbols.

Bases from DNA		Bases from RNA	
Adenine	A	Adenine	A
Thymine	T	Uracil	U
Guanine	G	Guanine	G
Cytosine	C	Cytosine	C

Three bases are thus common to both DNA and RNA, and one base is different. The pentose sugar unit is also different, being ribose in RNA and deoxyribose in DNA, hence the R and the D.

■ Ring carbon number 1′ is the carbon of the aldehyde group when the pentose ring is open.

Adenosine monophosphate or AMP is a typical nucleotide of RNA (Figure 24.3). The base is adenine (A). It's joined to the hemiacetal carbon (C-1′) of ribose. The C-5′ OH group of ribose has been changed into an ester of phosphoric acid. (The

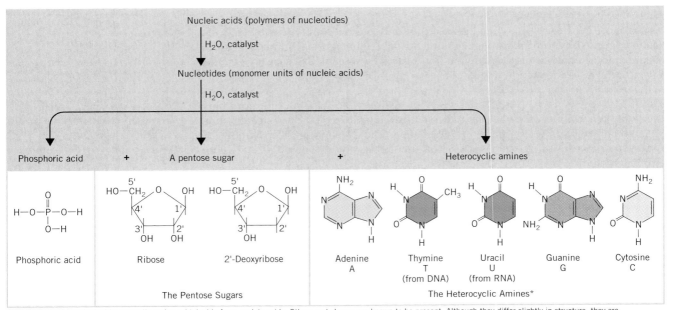

*These are the five principal heterocyclic amines obtainable from nucleic acids. Others, not shown, are known to be present. Although they differ slightly in structure, they are informationally equivalent to one or another of the five shown here.

FIGURE 24.2　The hydrolysis products of nucleic acids.

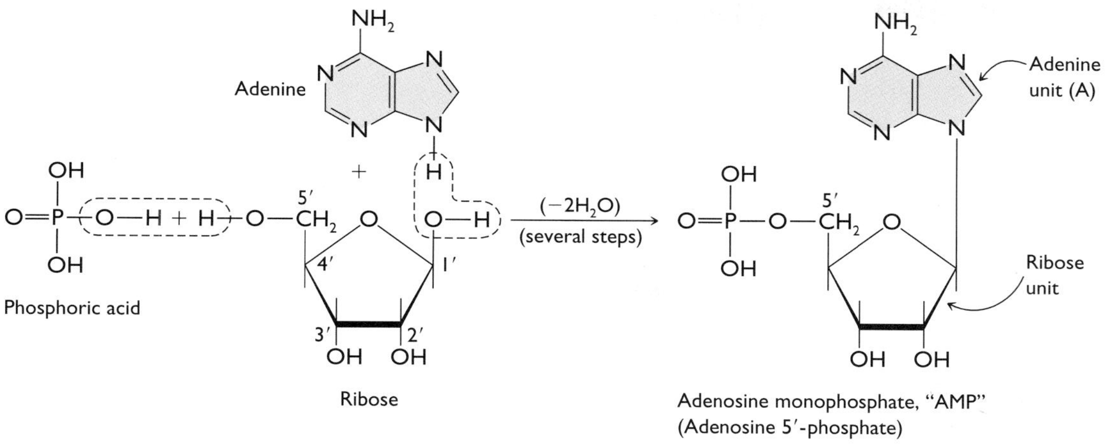

Phosphoric acid

Ribose

Adenosine monophosphate, "AMP"
(Adenosine 5'-phosphate)

FIGURE 24.3 A typical nucleotide, AMP, and the smaller units from which it is assembled. The phosphate ester group forms by the splitting out of water between an OH group of phosphoric acid and an H atom at the 5' OH group of the ribose unit. The splitting out of water between the 1' OH group of the ribose unit and the H atom on a ring nitrogen of adenine (A) joins adenine to the sugar unit. Similar structures could be drawn with the other bases of Figure 24.2.

prime, as in 1' or 5', refers to the numbering of the pentose ring; unprimed numbers are used for ring positions on the bases.) All nucleotides of RNA are monophosphate esters with structures like that of adenosine monophosphate, only the others have different bases.

When the OH at C-2' in AMP is replaced by H, and the pentose is deoxyribose, the resulting nucleotide is one for DNA. The pattern for all nucleotides, therefore, is as follows.

$$HO\text{—phosphate—pentose—OH}$$
$$\overset{\displaystyle base}{\underset{\displaystyle |}{}}$$

The Bases Project from the Backbones of Nucleic Acids Many steps and many enzymes are required to link nucleotides together to make nucleic acids. The overall effect, however, is simple (Figure 24.4). Water splits out between an OH group at a C-3' ring position of one nucleotide and the phosphate unit of the next nucleotide to create a *phosphodiester* system. When repeated thousands of times, a nucleic acid forms. Thus a nucleic acid is a copolymer for which there are four monomers, each differing in the base.

The molecule in Figure 24.4 is a portion of a DNA copolymer. If the top end of the chain were the real end of the molecule, it would, by convention, be regarded as the chain's *starting* point. It's called the 5' end after the number of the last carbon of the pentose. At the opposite end of the molecule, the chain ends at an unesterified C-3' OH group. Thus, again by convention, the chain in Figure 24.4 is said to be running, top to bottom (beginning to end), in the $5' \rightarrow 3'$ direction. The bases are in the sequence of A to T to G to C, $5' \rightarrow 3'$, so chemists write the *condensed structural formula* of the portion of the DNA molecule in Figure 24.4 simply as ATGC.

What are shown in Figure 24.4 as phosphate OH groups are moderately strong proton-donating groups, like those in phosphoric acid. In the slightly basic medium of a cell, these protons have been neutralized. Thus nucleic acids exist *in vivo* as multiply charged anions.

The repeating pattern for any nucleic acid backbone, DNA or RNA, is alternating units of phosphate and pentose, joined as phosphate diesters. Projecting from each pentose is one of the bases. The backbone holds the system together, and *the sequence of the bases carries the genetic information.* The distinctiveness of any one nucleic acid is in the sequence of its side-chain bases. Genes are carried on DNA molecules, and many involve thousands of bases (actually thousands of *pairs* of

■ Some 20 enzymes are required, the chief being DNA polymerase, discovered in 1958 by Arthur Kornberg (Nobel Prize in Physiology or Medicine, 1959).

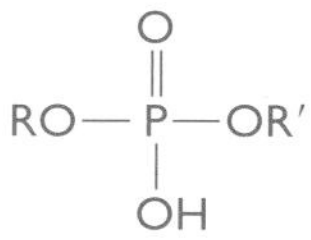

■ $RO\text{—P—}OR'$

phosphodiester

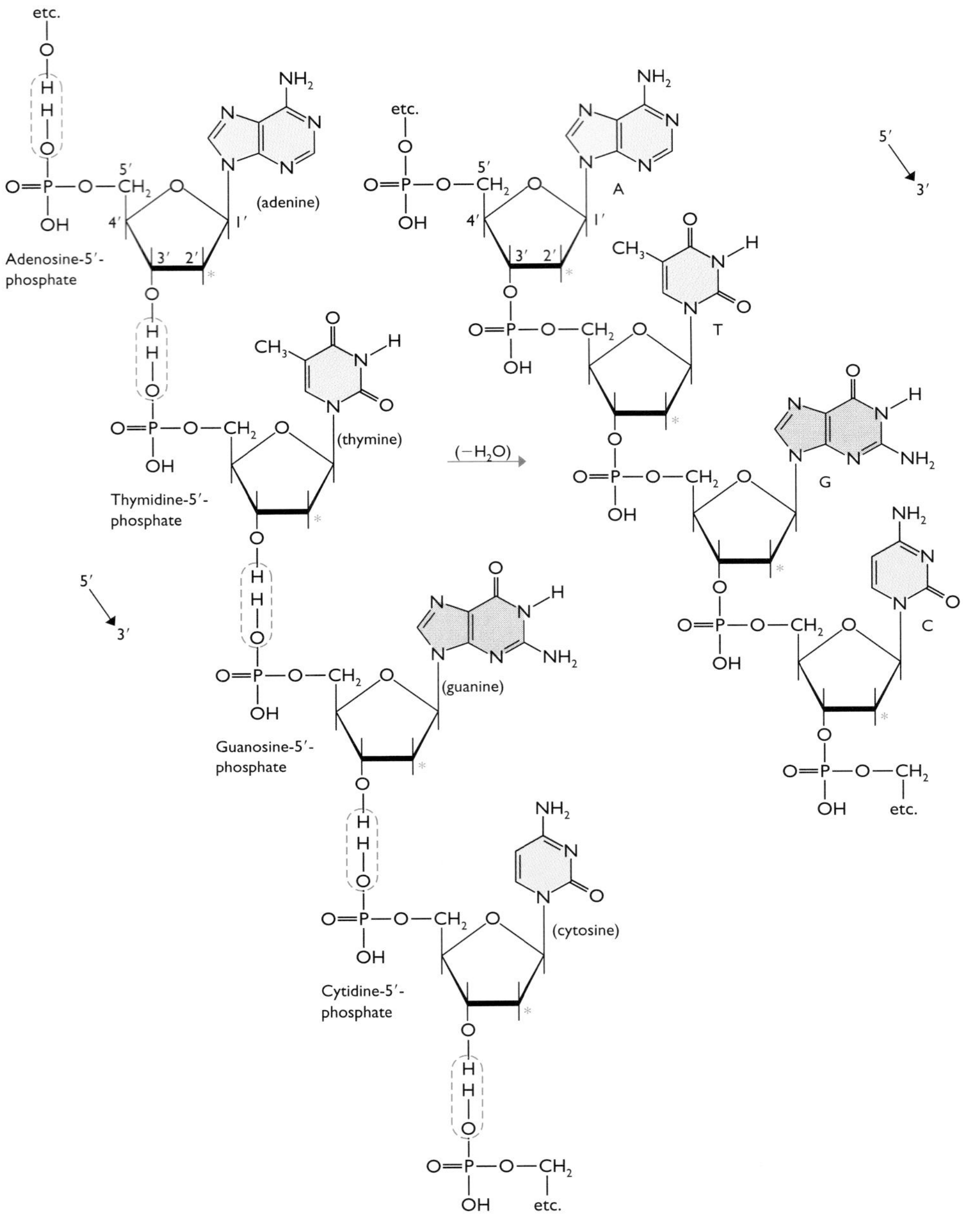

FIGURE 24.4 The relationship of a nucleic acid to its nucleotide monomers. On the right is a short section of a DNA strand. On the left are the nucleotide monomers from which it is made (after many steps). The colored asterisks by the pentose units identify the 2′ positions of these rings, where there would be another OH group if the nucleic acid were RNA (assuming that uracil also replaced thymine).

bases, as we will see), each base obviously used many times. The number of possible combinations is astronomic. There are 24 different sequences possible for just the four different bases of DNA, each base used only once.

ATGC	TAGC	GCAT	CATG
ATCG	TACG	GCTA	CAGT
AGTC	TGAC	GATC	CTAG
AGCT	TGCA	GACT	CTGA
ACTG	TCAG	GTAC	CGAT
ACGT	TCGA	GTCA	CGTA

You can begin see how it's possible for every species and every individual member of a species to have unique genes.

Pairs of Bases Are Attracted to Each Other by Hydrogen Bonds   The bases have functional groups that are located so that two bases can "fit" to each other in pairs by means of hydrogen bonds. We call the phenomenon **base pairing** (Figure 24.5). *The locations and geometries of the functional groups of the bases permit only certain base pairs to form.* In DNA, for example, G and C always form a pair, and A and T form another pair. In RNA, G and C always pair, and U and A always pair. Neither G nor C ever pairs with A, T, or U.

DNA Occurs as Paired Strands Twisted into a Double Helix   In 1953, Francis Crick of England and James Watson of the United States, working together, proposed a structure for DNA, the **DNA double helix**, also called **duplex DNA.** From X-ray data on DNA, obtained by Rosalind Franklin, they knew that DNA must be helical. Using molecular models, they deduced that two complementary DNA molecules line up side by side to form a double strand, the bases on the inside, and that the two strands twist into a right-handed helix. Crick and Watson also proposed how this kind of structure might explain in molecular terms how heredity works. Their work opened floodgates of research and marked one of the greatest advances in the history of biology. Many Nobel Prizes have been earned by those who have fleshed out important details of the Crick-Watson theory. It has been called the genetic "revolution," but true revolutions always overthrow the past. Crick and Watson didn't do that; they built on the past.

■   Crick and Watson shared the 1962 Nobel Prize in Physiology or Medicine with Maurice Wilkins.

■   Two irregular objects are *complementary* when one fits to the other, as your right hand would fit to its impression in clay.

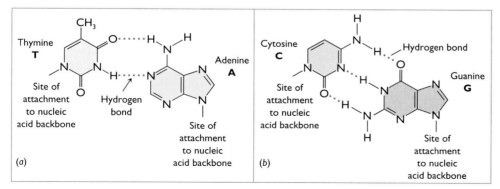

FIGURE 24.5   Hydrogen bonding between base pairs. (*a*) Thymine (T) and adenine (A) form one base pair between which are two hydrogen bonds. (*b*) Cytosine (C) and guanine (G) form another base pair between which are three hydrogen bonds. Adenine can also base pair to uracil (U).

The idea of complementary strands is a key feature of the double helix. In DNA chemistry, *strand complementarity* refers to the kind of fitting of one whole strand to the other. It occurs by base pairing *between the strands,* an idea suggested by two significant 1:1 mole ratios discovered in the late 1940s by Erwin Chargaff and now known as *Chargaff's rules.* In the DNA of *all* species there are equal numbers of A and T; there are equal numbers of G and C. What is different among species are the relative numbers of these pairs. Crick and Watson made sense of Chargaff's rules by proposing that A pairs to T *between two complementary strands,* and that G pairs to C, also between the strands.

Whenever adenine (A) is attached to one strand, then thymine (T) is attached opposite it on the complementary strand. And whenever guanine (G) projects from one strand, then cytosine (C) is opposite it on the other. A molecular model of the DNA double helix discovered by Crick and Watson is shown in Figure 24.6. It's considered to be the native form and is now called B-DNA; there are a few other forms. The system resembles a spiral staircase in which the steps, which are nearly perpendicular to the long axis of the spiral, consist of the base pairs. What may not be immediately obvious from Figure 24.6 is that the two DNA strands run in opposite directions.

■ The mole ratios of (A + T) to (C + G) vary between species.

■ Other forms are A-DNA and Z-DNA. They differ in coiling geometries.

Largely Hydrophobic Forces Stabilize the DNA Double Helix   Hydrogen bonds occur between the base pairs, but these bonds are no longer regarded as the primary source of the stability of the double helix. Hydrogen bonds do determine which bases can pair, but *hydrophobic interactions stabilize duplex DNA.* The rings of the bases themselves are planar, unsaturated, and relatively nonpolar despite the ring N atoms. In an aqueous medium there is a natural tendency for these rings to stack closely together because they cannot offer much hydrogen-bonding alternatives to water molecules. Water excludes them, so they stack together, which is exactly what the DNA duplex structure portrays. Around the edge of each spiraling backbone project the negatively charged, anionic sites of the phosphate units, giving these strongly hydrophilic groups full access to the cellular fluid.

Duplex DNA Has a Major and a Minor Groove   The structure of duplex DNA in Figure 24.6, as we said, is the native or *in vivo* form of DNA. Notice that it has two grooves, one major and one minor. These are binding sites for molecules of proteins required in several gene functions, including gene repression and gene activation. They are also sites for the initial molecular interactions of certain drugs, antibiotics, carcinogens (cancer-causing agents), and poisons. Polypeptides with net positive charges would be attracted to the spiraling duplex DNA grooves, and complementary fitting to the DNA becomes a factor in interactions between protein and DNA.

Figure 24.6 shows only a short segment of duplex DNA, a sequence too short to show that duplex DNA is further twisted and coiled into superhelices. The looping and coiling are necessary to fit the cell's DNA into its nucleus. A typical human cell nucleus, for example, is only about 10^{-7} m across, but if all of its DNA double helices were stretched out, they would measure a little over 1 m, end to end.

■ There are an estimated three billion base pairs in one nucleus of a human cell but only about 5% are a part of genes. The functions of the remaining 95% are not yet fully known.

In DNA Replication, the Bases Guide the Formation of New, Complementary Strands   When DNA replicates, the cell makes an exact complementary strand for each of the two original strands, so two identical double helices are made. The built-in guarantee that each new strand will be the complement of one of the old is the requirement that A pairs to T and C to G. A very general picture that shows how this works is in Figure 24.7. Realize that this figure explains only one of the many aspects of replication, how base pairing ensures two new complementary strands. Many of the details of how replication occurs without everything becoming hope-

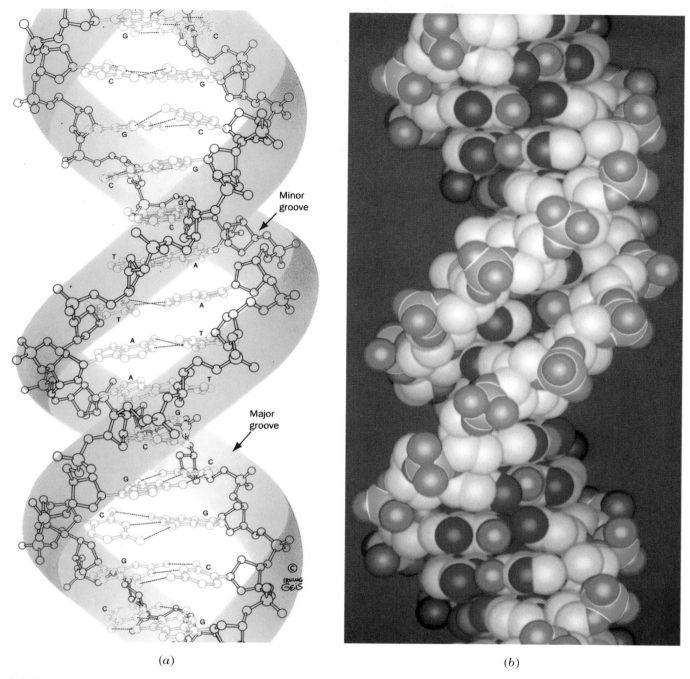

Minor
groove

Major
groove

(a)

(b)

FIGURE 24.6 The structure of duplex DNA—the native form (B-DNA)—seen here in (a) a ball-and-stick drawing and (b) a computer-generated space-filling model. (Drawing copyright © by Irving Geis.)

lessly tangled in the nucleus are understood, but we must leave them to more advanced references. One feature of DNA replication correctly shown by Figure 24.7 is that the synthesis of replica DNA strands occurs as the parent strand unwinds. The formation of the replicas does not await the complete unwinding of the parent before the replication commences.

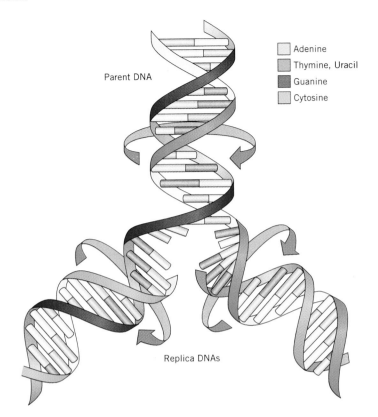

Parent DNA

Adenine
Thymine, Uracil
Guanine
Cytosine

Replica DNAs

FIGURE 24.7 The replication of DNA. (Reproduced with permission from D. Voet and J. G. Voet, *Biochemistry,* 2nd edition, John Wiley & Sons, Inc., 1995.)

■ *Exon* refers to the part that is expressed and *intron* to the segments that *interrupt* the exons.

■ There are no introns in human genes that code for histones or for α-interferon.

Sections of DNA Molecules Carry the Message of One Gene   Although a single gene can be visualized as a continuous sequence of nucleotide units, in higher organisms *a gene does not occur as a continuous sequence within chromatin DNA.* In chromatin DNA, virtually all individual genes are divided or split. This means that a gene is made up just of *sections* of a DNA chain called **exons.** A gene is thus something like the notes of a musical chord; only some keys out of an entire scale are used. Other, usually longer sections of the DNA molecule, called **introns,** interrupt the exons and separate them. The gene for the β-subunit of human hemoglobin, for example, consists of 990 bases, but two intron units, 120 and 550 bases long, interrupt the gene (Figure 24.8). Some introns are as short as 50 base pairs in length and others as long as 200,000. Introns make up an average of 80% of a human gene.

A single gene can have as many as 50 intron segments, but not all DNA molecules in higher organisms have them. A few human genes do exist with completely continuous sequences of bases. Just what purposes are served by the introns is currently under investigation. They were once called "junk DNA," but this may be inappropriate.

We'll see later in this chapter how the exons of one gene manage to get linked together and force the introns out. For the present, we can consider that a gene is

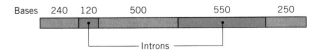

Bases   240   120   500   550   250

Introns

FIGURE 24.8 The gene for the β-subunit of hemoglobin is a split gene with two long intron sections.

just a particular section of a DNA strand minus all the intervening introns. This still leaves us with the concept that a **gene** is a specific series of bases in a definite sequence on a DNA backbone.

24.3 RIBONUCLEIC ACIDS

Triplets of bases on messenger RNA correlate with individual amino acids through the *genetic code.*

The general scheme that relates DNA to polypeptides is illustrated in Figure 24.9. You can see that varieties of RNA participate, so we'll now learn more about them.

Ribosomal RNA (rRNA) Is the Most Abundant RNA We have already mentioned the small particles in the cytoplasm, called ribosomes (Figure 24.1). Each forms from two subunits, as shown in Figure 24.9. They come together with *messenger RNA,* one of the RNA types, to form a complex.

■ American chemists Richard Roberts and Phillip Sharp shared the 1993 Nobel Prize in Physiology or Medicine for their independent discovery in 1977 of "split genes."

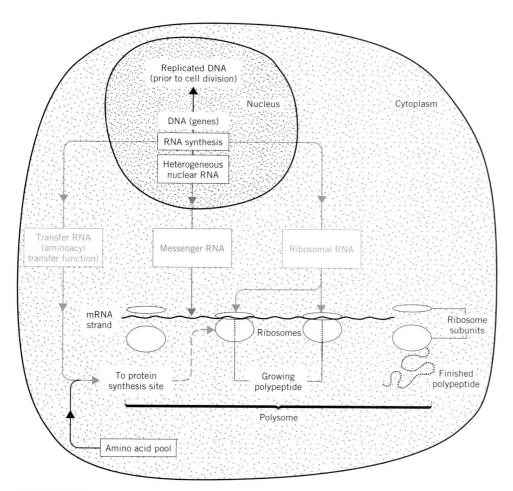

FIGURE 24.9 The relationships of nuclear DNA to the various RNAs and to the synthesis of polypeptides.

■ Over 50 kinds of polypeptides exist in one ribosome.

■ T. R. Cech and S. Altman shared the 1989 Nobel Prize in Chemistry for discovering ribozymes.

Ribosomes contain both proteins and a type of ribonucleic acid called **ribosomal RNA,** abbreviated **rRNA.** Except in a few viruses, rRNA is single stranded, but its molecules often have hairpin loops in which base pairing occurs. The rRNA of a ribosome is itself made from longer versions by losing some of its chain pieces. This loss is catalyzed by a region of the very RNA being groomed, so such RNA regions are actually working as enzymes and are called **ribozymes.** (It came as a huge surprise in the 1980s to learn that the body uses anything other than proteins as enzyme catalysts.)

Ribosomes are the sites of polypeptide synthesis. However, a ribosome does not know how to arrange the amino acid residues of a polypeptide in the right order. The rRNA of a ribosome cannot itself direct this. Another kind of RNA, *messenger RNA,* does the directing, and it's made from hnRNA, still another kind of RNA.

Heterogeneous Nuclear RNA (hnRNA) Is Complementary to DNA's Exons and Introns   When a cell calls on a gene to help make a polypeptide, the first step is to use a single DNA strand, one bearing the polypeptide's gene, to guide the assemblage of a complementary molecule of RNA, called **heterogeneous nuclear RNA,** abbreviated **hnRNA.**[1] Figure 24.10 gives a broad picture of how side-chain bases on DNA guide

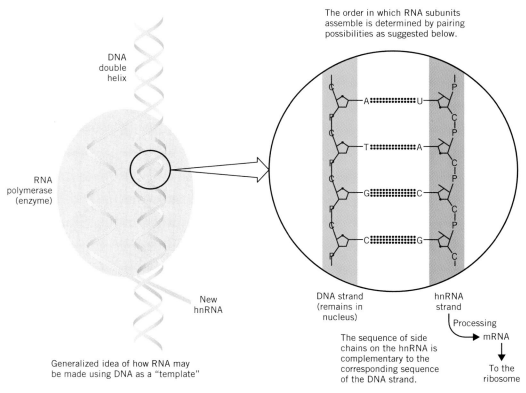

The order in which RNA subunits assemble is determined by pairing possibilities as suggested below.

DNA double helix

RNA polymerase (enzyme)

New hnRNA

Generalized idea of how RNA may be made using DNA as a "template"

DNA strand (remains in nucleus)

hnRNA strand

The sequence of side chains on the hnRNA is complementary to the corresponding sequence of the DNA strand.

Processing
mRNA

To the ribosome

FIGURE 24.10   DNA-directed synthesis of hnRNA in the nucleus of a cell in a higher organism. The shaded oval on the left represents a complex of enzymes that catalyze this step. mRNA is messenger RNA. (Notice that the direction of the hnRNA strand is opposite to that of the DNA strand.)

[1] Heterogenous nuclear RNA has been called *primary transcript RNA (ptRNA),* or simply *primary transcript,* and sometimes *pre-messenger RNA (pre-mRNA).*

the correct lineup of RNA nucleotides to make hnRNA. Uracil (U) is now used instead of thymine (T), so when a DNA strand has an adenine (A) side chain, then uracil, not thymine, takes the position opposite it on the complementary RNA. The enzyme that catalyzes DNA-directed RNA synthesis is *RNA polymerase.* Variations of it occur in all animal organisms. In certain one-celled organisms, the rate of hnRNA formation has been found to be 20 to 50 nucleotides per second.

The next general step in gene-directed polypeptide synthesis is the processing of hnRNA to make messenger RNA.

Each Messenger RNA (mRNA) Is Complementary to One Gene   Molecules of hnRNA have large sections complementary to the introns of the DNA, and these sections must be deleted. Special enzymes catalyze reactions that snip introns out and splice together the remaining units, those corresponding to the exons of the gene (Figure 24.11). The result is a shorter RNA molecule called **messenger RNA,** or **mRNA.**

In mRNA we have a sequence of bases complementary just to the gene's exons, so this mRNA now carries the unsplit genetic message. The name of the overall process that moves the genetic message from a split gene on DNA to a molecule of mRNA is **transcription.**

Triplets of Bases on mRNA Are Genetic Codons   Each group of three adjacent bases on a molecule of mRNA constitutes a unit of genetic information called a **codon.** Thus it is a sequence of *codons* on the mRNA backbone more than a sequence of individual bases that now carries the genetic message. (We'll explain shortly why *three* bases per codon are necessary.)

Once made, an mRNA molecule moves from the nucleus to the cytoplasm and attaches to a ribosome. Many ribosomes eventually string out, like beads, along one mRNA chain, and such a collection is called a *polysome* (short for *polyribosome*).

Ribosomes are traveling packages of enzymes intimately associated with rRNA. Each ribosome ratchets along the mRNA chain while mRNA codons guide the synthesis of a polypeptide. To build the polypeptide, we need to bring its individual amino acids, in order, to the assembly sites. The cell uses still another type of RNA to handle amino acid transport.

Transfer RNA (tRNA) Molecules Can Recognize both Codons and Amino Acids
The substances that carry aminoacyl units to mRNA *in the right order for a particular polypeptide* are a collection of similar compounds called **transfer RNA or tRNA.** Their molecules are small, each typically having only 75 nucleotides. They are single stranded but with hairpin loops (Figure 24.12).

We're now at the level of the molecular basis of *information,* so we can use language analogies. On a human level, language uses words, *built from a common alphabet,* to convey information. Many languages exist and the world knows several alphabets. To communicate between languages, we have to translate. The same need for translation occurs at the level of genes and polypeptides. tRNA is the master translator in cells.

tRNA is bilingual. It's able to work with two "languages," the nucleic acid or genetic language and the polypeptide language. The former is expressed in an alphabet of 4 letters, the four bases, A, T (or U), G, and C. The polypeptide language has an alphabet of 20 letters, the 20 amino acids. To translate from a 4-letter language to a 20-letter language requires that the 4 nucleic acid letters be used in groups of a minimum of 3. Then there will be enough combinations of genetic letters to handle the

■ Part of the grooming of hnRNA molecules installs at their 3′ ends a long poly-A tail (about 200 adenosine units long), and at the other end a nucleotide triphosphate "cap."

■ A small family of nuclear ribonucleoproteins, snRNPs (or "snurps"), helps this resplicing process.

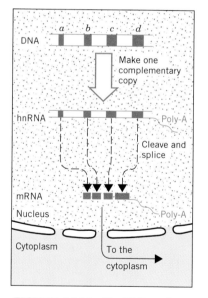

FIGURE 24.11   The RNA made directly at a DNA strand is hnRNA. Only the segments made at sites *a, b, c,* and *d*—the exons—are needed to carry the genetic message to the cytoplasm. The hnRNA is processed, therefore, and its segments that match the introns of the gene are snipped out. Then the segments that match the exons are rejoined to make the mRNA strand. (Poly-A is a section of the hnRNA molecule made entirely of repeating A nucleotides.)

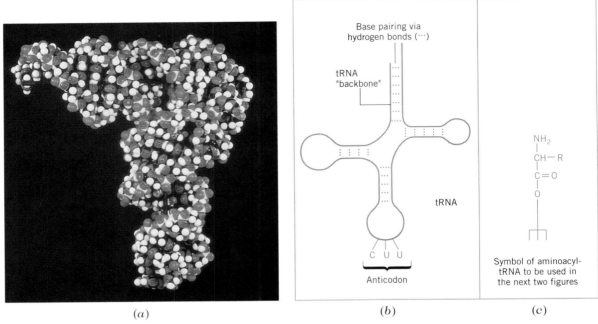

FIGURE 24.12   Transfer RNA (tRNA). (*a*) The tRNA for phenylalanine. Its anticodon occurs at the tip of the base, and the place where the phenylalanyl residue can be attached is at the upper left point. (*b*) Highly schematic representation of the model to highlight the occurrence of double-stranded regions. (*c*) Symbol of the aminoacyl—tRNA unit that will be used in succeeding figures. (Molecular model courtesy of Academic Press/Molecular Design Incorporated.)

aminoacyl unit

larger alphabet of amino acids. In this way, at least one nucleic acid "word," built of 3 letters, can exist for each of the 20 amino acids. This is exactly how tRNA molecules are structured for the work of translating between the two languages.

Each different tRNA molecule has a triplet of bases called an **anticodon,** a three-base sequence that fits, by base pairing, only to a particular three-base codon of a mRNA molecule. The tRNA in Figure 24.12, for example, carries CUU as its anticodon; its C pairs only to G on mRNA and its U's pair only to A's. Another functional group of each tRNA, the free OH group of a ribose unit, serves as the binding site for a *particular* aminoacyl unit. Each tRNA "recognizes" the amino acid belonging to it by means of special enzymes. We'll use tRNA—aa to represent this new compound, aa standing for the aminoacyl group. Because of its anticodon, a given tRNA—aa, as we said, can align with only one codon of mRNA at a polysome. This ensures that its attached aminoacyl unit is brought into position in a growing polypeptide only when it's supposed to.

When tRNA—aa molecule docks at its unique location at the mRNA chain, the growing polypeptide tethered "next door" on the same mRNA is transferred to it. You can see that *a unique series of codons can allow the polypeptide chain to grow only with an equally unique sequence of amino acid residues.* The pairing of the triplets of bases between the codons and anticodons can permit only one sequence.

The Genetic Code Is the Correlation between Codons and Amino Acids   Table 24.1 displays the known correlations of codons with amino acids, the "standard" **genetic code.** Most amino acids are associated with more than one codon, which apparently minimizes the harmful effects of genetic mutations. (These, in molecular terms, are changes in the structure of DNA.) Phenylalanine, for example, is coded ei-

TABLE 24.1   Standard Genetic Code

First	Second				Third
	U	C	A	G	
U	Phenylalanine	Serine	Tyrosine	Cysteine	U
	Phenylalanine	Serine	Tyrosine	Cysteine	C
	Leucine	Serine	CT[a]	CT	A
	Leucine	Serine	CT	Tryptophan	G
C	Leucine	Proline	Histidine	Arginine	U
	Leucine	Proline	Histidine	Arginine	C
	Leucine	Proline	Glutamine	Arginine	A
	Leucine	Proline	Glutamine	Arginine	G
A	Isoleucine	Threonine	Asparagine	Serine	U
	Isoleucine	Threonine	Asparagine	Serine	C
	Isoleucine	Threonine	Lysine	Arginine	A
	Methionine[b]	Threonine	Lysine	Arginine	G
G	Valine	Alanine	Aspartic acid	Glycine	U
	Valine	Alanine	Aspartic acid	Glycine	C
	Valine	Alanine	Glutamic acid	Glycine	A
	Valine	Alanine	Glutamic acid	Glycine	G

[a] The codon CT is a signal codon for chain termination.
[b] The codon for methionine, AUG, serves also as the codon for *N*-formylmethionine, the chain-initiating unit in polypeptide synthesis in bacteria and mitochondria.

ther by UUU or by UUC; the two are called "synonyms." Alanine is coded by any one of four synonyms, namely, GCU, GCC, GCA, or GCG. Only two amino acids go with single codons, tryptophan (Trp) and methionine (Met).

The Genetic Code Is Widely Used by Plants and Animals A few single-celled species have some codon assignments not among the standard set. Moreover, some genes within the mitochondria of mammals have unique codons. Apart from these exceptions, the genetic code of Table 24.1 is shared from the lowest to the highest forms of life in both the plant and animal kingdoms. Once again we see a remarkable kinship with nature.

mRNA Codons Are Specified by Triplets on DNA It's important to remember that a codon cannot appear on a strand of mRNA unless a complementary triplet of bases was on an exon of the original DNA strand. For example, there could not be the UUC codon on mRNA unless the DNA strand had the triplet AAG, because G pairs with C and A of DNA pairs with U of mRNA.

■ Notice that it's just the third letter in a set of synonyms that is different.

■ Some (but not all) of the enzymes present in a mitochondrion are made *within* this organelle under the direction of mitochondrial nucleic acids. According to one theory, mitochondria evolved from single-celled organisms.

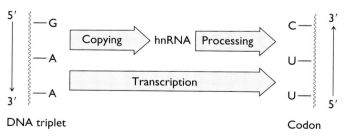

DNA triplet

Triplet of bases on
an exon of a DNA
strand

Codon

This sequence on mRNA
is complementary to the
DNA triplet

The Direction in Which a Codon Triplet Is Written Has Structural Meaning   As shown here, a DNA strand and the RNA strand made directly from it run in opposite directions. To avoid confusion in writing codons on a horizontal line, scientists use the following conventions. The 5′ end of a codon is written on the left end of three-letter symbol, and the direction, left to right, is 5′ to 3′. (See also Figure 24.4.) This is why the codon given above is written as UUC, not as CUU. Opposite this codon is the triplet GAA on the DNA strand, which is also written from the 5′ to 3′ end. To give another example, the complement to the mRNA codon, AAG, is the DNA triplet, CTT.

■ PRACTICE EXERCISE 1   Using Table 24.1, what amino acids are specified by each of the following codons on an mRNA molecule?

(a)   CCU   (b)   AGA   (c)   GAA   (d)   AAG

■ PRACTICE EXERCISE 2   What amino acids are specified by the following base triads on DNA?

(a)   GGA   (b)   TCA   (c)   TTC   (d)   GAT

24.4 mRNA-DIRECTED POLYPEPTIDE SYNTHESIS

tRNA molecules carry aminoacyl groups to a ribosome where amino acid residues are joined to make a polypeptide.

In the previous section we saw how a particular genetic message can be transcribed from the exons of a split gene to a series of codons on mRNA. In this section we will learn in broad terms how the next general step, *translation*, is accomplished. **Translation** is the mRNA-directed assemblage of a specific polypeptide.

Genetic Translation Follows Transcription   We will now use the forklike symbol shown in Figure 24.12c for tRNA—aa, with the fork's three tines standing for the anticodon triplet. We will assume that all the needed tRNA—aa combinations have been assembled and are waiting like so many spare parts to be used at the polypeptide assembly line. The cell must invest considerable amounts of chemical energy, like that of ATP, to carry out all of these steps, and we also assume that this energy is available as needed.

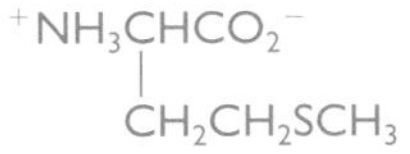

A polypeptide is started with an N-terminal methionine residue. After the polypeptide has been made, this residue is left in place only if methionine is supposed to be the N-terminal unit. Otherwise, methionine will be removed, and the second aminoacyl group will be the N-terminal unit. By anchoring the start of polypeptide synthesis in one triplet, AUG for methionine, the remaining triplets on the mRNA strand cannot overlap each other. Thus a sequence such as AUGCCGAGU . . . must have CCG as the second triplet and AGU as the third, and so forth. Although there seems to a GCC "triplet" in this sequence, it cannot be read as a triplet because the starting triplet for methionine preempts the first G in the series.

Genetic Translation, the Principal Steps

1. **An Elongation Complex Forms** The elongation complex is made of several pieces, starting with two subunits of the ribosome, the mRNA molecule, and the $tRNA_1$—Met molecule (Figure 24.13). $tRNA_1$—Met comes to rest with its anti-

■ $^+NH_3CHCO_2^-$
$|$
$CH_2CH_2SCH_3$

methionine, Met

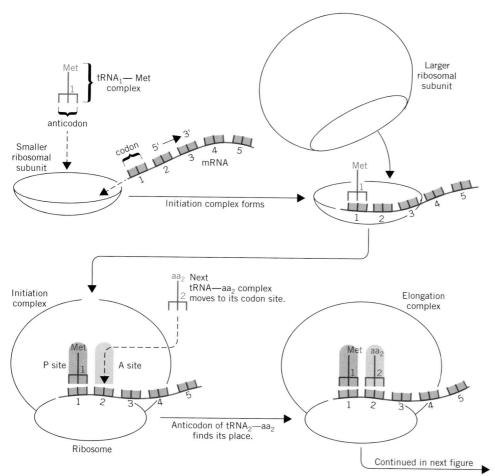

FIGURE 24.13 Formation of the elongation complex at the beginning of the synthesis of a polypeptide.

codon matched to the first codon on mRNA whose starting end rests on the ribosome. The bulk of the tRNA₁—Met molecule is in contact with what is called the P site of the ribosome's surface.

Now the second tRNA unit, tRNA₂—aa₂, holding the second aminoacyl group, aa₂, has to find the mRNA codon that matches its own anticodon. We assume that this codon is adjacent to the first and that tRNA₂—aa₂ comes to rest against a site on the ribosome called the A site, which is adjacent to the P site. The elongation complex is now complete, and chain lengthening can begin.

2. **The Polypeptide Chain Elongates** There now occurs a series of repeating steps, which you should follow in Figure 24.14. The amino group of the *second* aminoacyl unit plucks the methionine residue from its tRNA, making the first peptide bond by acyl transfer.

■ The "P" in P site refers to the peptidyl transfer site. The A site is the aminoacyl binding site. (A third site, not shown, is the E site, which temporarily holds the departing tRNA.)

■ This kind of translocation of the growing polypeptide to the next amino acid is what is blocked by the toxin of the diphtheria bacillus.

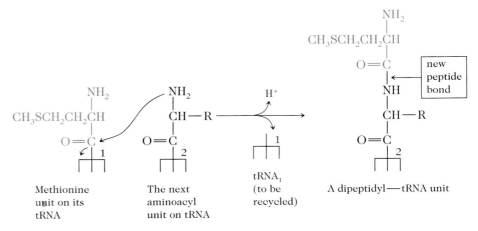

Methionine unit on its tRNA

The next aminoacyl unit on tRNA

tRNA₁ (to be recycled)

A dipeptidyl—tRNA unit

The mRNA unit now shifts one codon over, leftward as we have drawn it. (It's actually a relative motion of the mRNA and the ribosome.) This movement positions what is now a *di*peptidyl—tRNA unit over the P site. The third aminoacyl—tRNA, tRNA₃—aa₃, using base pairing, now finds the match for its anticodon at the third codon of the mRNA strand, which is over the recently vacated A site.

Elongation now occurs again. The amino group of the third amino acid (at an A site) plucks the dipeptide unit at the P site, latching onto it by a new peptide bond. A tripeptidyl system is made.

The cycle of steps can now occur again, starting with a movement of the mRNA chain relative to the ribosome that shifts the tripeptidyl—RNA unit to the P site. The fourth amino acid residue is borne on tRNA—aa₄ to the mRNA at codon 4. The amino group of residue 4 takes the tripeptidyl unit away from the P site, making still another peptide bond and a tetrapeptidyl unit. You can see how this cycle of steps can repeat, making a polypeptide, until a chain-terminating codon is reached.

3. **Polypeptide Synthesis Terminates** Once a ribosome has moved to one of the chain-terminating codons (UAA, UAG, or UGA), the synthesis is complete, and the polypeptide is released. It now acquires its higher levels of structure with the help of folding catalysts, called *chaperonins*, enzymes that guide polypeptide molecules toward their final, native shapes.

■ Polypeptide synthesis can occur at several ribosomes moving along the mRNA strand at the same time.

■ In mammals, it takes only about 1 second to move each amino acid residue into place in a growing polypeptide.

A *Repressor* Can Keep a Gene Switched Off Until an *Inducer* Acts As we have said, every cell in the body carries the entire set of genes. In any given tissue, therefore, most genes must be permanently switched off. Our aim now is only to illustrate in broad outline one of the most important kinds of activities at the molecular level of life, the control of gene expression.

Because so many steps occur between the divided gene and the finished polypeptide, there are a large number of points at which the cell can control the overall process. We'll briefly discuss only one. It's a classic example of gene regulation, done by the bacterium *Escherichia coli,* or *E. coli,* a one-celled organism in our intestinal tracts.

E. coli are able to obtain all their needed carbon atoms from the metabolism of lactose, milk sugar, a disaccharide. They must first hydrolyze it to galactose and glucose, and the enzyme β-galactosidase is required. The enzyme, of course, is not needed until lactose is present, so the gene for making the enzyme is switched off until lactose arrives.

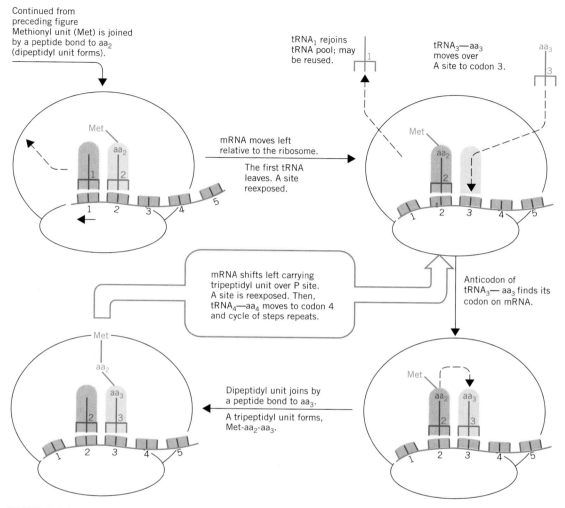

FIGURE 24.14 The elongation steps in the synthesis of a polypeptide. The dipeptidyl unit, Met—aa$_2$, formed by the process of Figure 24.13, now has a third amino acid residue, aa$_3$, added to it.

The DNA segment responsible for β-galactosidase is a region of the DNA strand called the *structural gene* (Figure 24.15). Next to it on the same DNA strand is a segment to which a **repressor** molecule can bind. It's named the *operator site* of the DNA because it's part of the "switch" that controls the operation of the structural gene. Immediately next to the operator site is a small DNA segment called the *promoter site*. It's the parking lot for an enzyme, DNA polymerase, needed to transcribe the structural gene. But the enzyme cannot work until the operator site is switched on. The structural gene plus the repressor and operator sites, taken together, make up a DNA unit called the *lac operon*.

The repressor is a polypeptide made at the direction of its own gene, a *regulator gene*, located just above the promoter site (see Figure 24.15). When repressor molecules are present, they bind to the DNA of the operator site. *The binding of repressor to operator prevents gene transcription and translation.*

Now suppose that some molecules of lactose appear. β-Galactosidase is now needed. The first of the arriving lactose molecules are altered slightly (by steps we

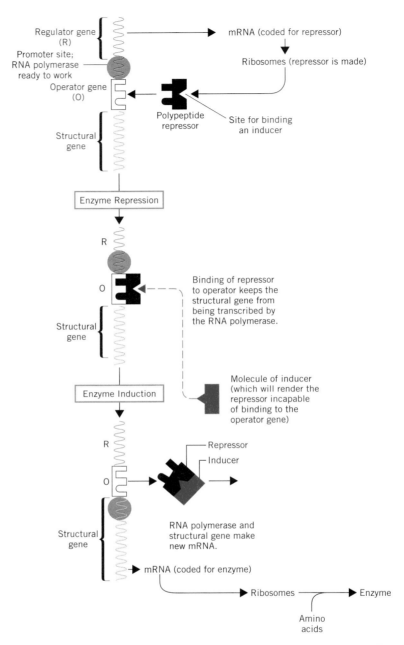

FIGURE 24.15 Repression and induction of the enzyme β-galactosidase in *E. coli*. The inducer is lactose, whose hydrolysis requires this enzyme. The first lactose molecules to arrive bind to and remove the repressor from the operator gene. Now the structural gene is free to direct the synthesis of mRNA coded to make the enzyme. The repressor is made at the direction of another gene (top), the regulator gene.

will not discuss) into **inducer** molecules. These have shapes that enable them to fit to the repressor molecule at the operator site. As they attach to the repressor, the shape of the repressor changes so that it no longer can stick to the operator gene, *and it drops off.* DNA polymerase is now released to work with the structural gene to help to make the mRNA needed for the synthesis of β-galactosidase. The overall process is called **enzyme induction,** and β-galactosidase is one of many *inducible enzymes.*

Don't let the beauty of enzyme induction be smothered by the details. Notice that if no substrate (here, lactose) is in the cell, an enzyme isn't needed, and the genetic machinery for making it stays switched off. Only when the enzyme is needed, signaled by the arrival of its substrate, is it manufactured. Many human enzymes no doubt work in the same way. A multitude of events at the molecular level of life thus happen *automatically,* leaving our minds free enough for other matters.

Many Antibiotics Kill Bacteria by Interfering with Genetic Translation or Transmission Bacteria, like us, have to manufacture polypeptides to stay alive and multiply. Inhibiting the synthesis of bacterial polypeptides at any one of several steps in the overall process kills bacteria. *Streptomycin,* for example, inhibits the initiation of polypeptide synthesis. *Chloramphenicol* inhibits the ability of newly arrived aminoacyl units to pluck the elongating strand from the P site. *Tetracycline* inhibits the binding of tRNA—aa units when they arrive at a ribosome. *Actinomycin* binds tightly to DNA and prevents transcription. *Erythromycin, puromycin,* and *cyclohex-imide* interfere with elongation. *Ricin,* a plant protein poisonous to humans, inactivates the larger subunits of our ribosomes.

■ These inhibiting activities render the *enzymes* for the various steps inactive.

X Rays and Atomic Radiations Can Damage Genes Atomic radiations, particularly X rays and gamma rays, go right through soft tissue where they create unstable ions and radicals (particles with unpaired electrons). New covalent bonds can form from radicals, making unwanted molecules. Even side-chain bases might be annealed together. If such events happen to DNA molecules, the polypeptide eventually made at their direction might be faulty. The DNA made by replication might then be seriously altered.

If the initial damage to the DNA is severe enough, replication won't be possible and the cell involved is reproductively dead. This, in fact, is the *intent,* to kill cancer cells, when massive radiation doses are used in cancer therapy.

Chemicals that are used in cancer therapy often mimic radiations by interfering with the genetic apparatus of cancer cells. Such chemicals are called **radiomimetic substances.**

24.5 VIRUSES

Viruses take over the genetic machinery of host cells to make more viruses.

Viruses are at the borderline of living systems and are generally regarded only as unique packages of dead chemicals, except when they get inside their host cells. They then seem to be living things, because they reproduce.

■ Viruses are intracellular parasites.

Viruses Consist of Nucleic Acids and Proteins **Viruses** are parasitic agents of infection made of nucleic acid surrounded by overcoats of protein. Unlike a cell, a virus has either RNA or DNA, but not both. Viruses must use host cells to reproduce because they can neither synthesize polypeptides nor generate their own energy for metabolism. The simplest virus has only four genes and the most complex has about 250.

Each kind of virus has something on its surface that is complementary to something on the surface of the host cell. The glycoproteins of membranes are involved. Thus each virus has one particular kind of host cell and does not attack all kinds. This is why viruses are so unusually selective. A virus that attacks the nerve cells in the spinal cord, for example, has no effect on heart muscle cells. A large number of viruses exist that do not affect any kind of human cell. Many viruses attack only specific plants.

The protein overcoat of some viruses includes an enzyme that catalyzes the breakdown of the cell membrane of the host cell, opening a hole into the cell. Then the viral nucleic acid squirts into the cell, or the whole virus might move in. Each virus that works this way evidently has its own unique membrane-dissolving enzyme.

Once a virus particle or the parts of one get inside the host cell, one of two possible fates awaits it. Viral nucleic acid may become turned off and changed into a *silent gene*; or the virus may take over the genetic machinery of the cell and reproduce so much of itself that it bursts the host cell walls. The new virus particles spill out and infect neighboring host cells, and the infection spreads. A silent gene might later be activated. Some cancer-causing agents, including ultraviolet light, may initiate cancer by activating silent genes.

Most viruses contain RNA, not DNA, so the manufacture of more of their RNA must somehow be managed without the direction of DNA.

RNA Viruses either Carry or Make Enzymes for Synthesizing More RNA   RNA viruses have to solve a major problem if they are to infect a host cell. Host cells normally (in health) have no enzyme that can direct the synthesis of a copy of an RNA molecule *from the instructions of another RNA.* In healthy host cells, copies of RNA molecules are made by the direction of DNA, not RNA, like the synthesis of hnRNA directed by DNA.

Two basic solutions to this problem occur involving two different enzymes. One is called *RNA replicase,* and it can catalyze the manufacture of RNA *from the directions encoded on RNA.* Some viruses carry this enzyme. Others direct the host cell to synthesize RNA replicase. Either way, once RNA replicase is inside the host cell, it handles the manufacture of the mRNA needed to make more viral RNA and protein, so new virus particles can form.

The second way by which RNA directs the synthesis of RNA occurs with viruses that carry *reverse transcriptase,* a DNA polymerase enzyme. Reverse transcriptase can use RNA information to make DNA, so it directs the synthesis of *viral* DNA, which then is used to make more viral RNA.

Four Basic Strategies Are Used by RNA Viruses to Make More RNA   Figure 24.16 outlines the strategies used by four kinds of RNA viruses to manufacture more of their own RNA. The (+) and (−) signs denote single-stranded RNA molecules of opposite complementarity. A (±) sign denotes a double-stranded nucleic acid. By convention, the messenger RNA that the virus must make is designated (+)mRNA.

The *polio virus* (Figure 24.16) contains a single-stranded RNA molecule. Inside its host cell, it functions as a messenger RNA at the host's ribosomes for the synthesis of both overcoat proteins and molecules of RNA replicase. This enzyme then synthesizes (−)RNA molecules which, in turn, direct the synthesis of the (+)mRNA that now takes over the host cell's genetic machinery.

The *rabies virus* has (−)RNA molecules, but they are not messengers in their host cells. This virus carries its own RNA replicase into the host cells, where it di-

■ A complete virus particle *outside* its host cell is called a *virion*.

■ David Baltimore and Howard Temin shared the 1975 Nobel Prize in Physiology or Medicine (together with Renato Dulbecco) for the discovery of reverse transcriptase.

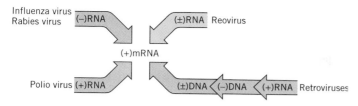

FIGURE 24.16 Overall strategies used by RNA viruses to make their messenger RNA.

rects the synthesis of (+)RNA. This is then used to make the new (−)RNA molecules required for additional virus particles.

The *influenza virus* (Figure 24.16) also has (−)RNA, each molecule bearing 10 genes. During infection, some segments of the RNA can become re-sorted even as new (−)RNA forms. In this way the influenza virus is able to change into new strains that make the job of immunizing large populations against influenza for long periods particularly difficult.

The *reovirus* of Figure 24.16 is present in the intestinal and respiratory tracts of mammals without causing disease. Its RNA is double-stranded RNA, or (±)RNA, and it carries its own RNA polymerase. It can use this on both (+) and (−) strands to make more viral (±)RNA.

The *retroviruses* (Figure 24.16), a family of viruses with (+)RNA, cannot make more RNA without first making double-stranded DNA. Retroviruses carry reverse transcriptase to do this. Thus, in retroviruses, the flow of information goes from RNA to DNA. (This explains the *retro-* prefix; it suggests a reversal or retrograde action.) Reverse transcriptase uses retroviral (+)RNA to direct the formation of (−)DNA, which directs the formation of (±)DNA. This DNA is incorporated into the host cell's collection of genes and then directs the synthesis of the (+)mRNA needed to make more retrovirus particles.

Cancer-Causing Viruses Transform Normal Genes in Host Cells The retroviruses include the only known cancer-causing RNA viruses, technically termed the *onco-genic RNA viruses*. (Several DNA-based viruses also cause cancer.) They transform host cells, causing them to grow chaotically and continuously, and doing this by changing normal genes to *oncogenes*. These cause the cancerous growth to continue.

The Host Cell of the AIDS Virus Is Part of the Human Immune System The acquired immunodeficiency syndrome, AIDS, is also caused by a retrovirus, the human immunodeficiency virus or HIV. One reason this virus is so dangerous is that its host cell, the T4 lymphocyte, is a vital part of the human immune system. By destroying T4 lymphocytes, HIV exposes the body to other infectious diseases, like pneumonia, or to rare types of cancer. The outlook for successfully *treating* HIV infections is good; the *prevention* of such infections by a vaccine seems to be far off, chiefly because HIV easily mutates.

HIV has its own multistep "life cycle," in the host cell. Being a retrovirus, it has its own reverse transcriptase to help make copies of its nucleic acid for new viral particles. So one strategy to retard the development of AIDS, if not cure it, has been to offer the HIV virus a nucleotide that can bind to its reverse transcriptase but, once bound, inhibits the further work of the enzyme. Zidovudine (AZT), a nucleotide with a modified sugar unit, has been commonly used as a reverse transcriptase inhibitor. Used by itself, however, and not as part of a combination of drugs, AZT does not delay the onset of AIDS in patients known to be infected with HIV. Lamivudine (3TC) is also a reverse transcriptase inhibitor.

■ In the 1919 worldwide influenza epidemic, 20 million people died, nearly 1% of the world's population.

■ *Reo-* in reovirus is from *respiratory, enteric orphan*—a virus in search of a disease.

■ *Oncogenic* means cancer inducing.

AZT

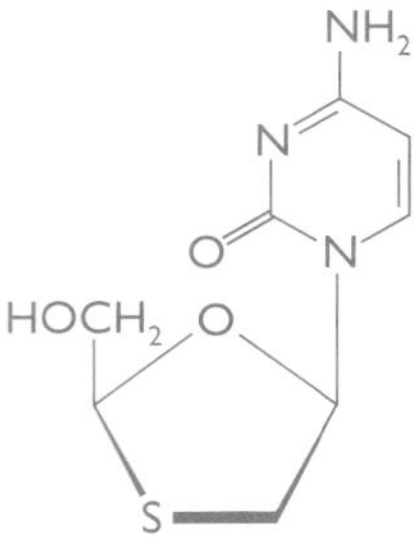

lamivudine (3TC)

Another strategy being studied is the use of special ribozymes, RNA enzymes, that would destroy the RNA in HIV.

An early step in the HIV life cycle is the splitting of two large polypeptides into smaller polypeptides. So another strategy is to administer protease inhibitors to victims of AIDS. The problem has been that HIV mutates easily, quickly giving the variants resistance to protease inhibitors. However, researchers have found that a "drug cocktail," a medley of medications, works well. For example, a combination of a protease inhibitor with a reverse transcriptase inhibitor is several times more effective than either alone. A triple drug medley consisting of two protease inhibitors and a reverse transcriptase inhibitor has also been found to reduce the population of HIV particles in patients dramatically, possibly because the particles do not develop resistance simultaneously to more than one drug. However, AIDS patients must not skip even one day of taking medication. The virus will start to replicate again, and may develop resistance to the drug cocktail. Even with these results, the immune systems of AIDS patients have been permanently damaged.

Some Viral Infections Produce Interferons, Which Fight Further Infection   One of the many features of the body's defense against some viral infections is a small family of polypeptides called the *interferons,* described further in Interaction 24.1. Supplies of interferons are now available by genetic engineering, which we will study next.

24.6   RECOMBINANT DNA TECHNOLOGY AND GENETIC ENGINEERING

Single-celled organisms can be made to manufacture the proteins of higher organisms.

Human insulin, human growth hormone, and human interferons are among the many substances now being manufactured outside the body by *recombinant DNA* technology. It represents one of the most important advances in scientific technology in history. It has permitted the *cloning,* the synthesis of identical copies, of a number of genes. The use of recombinant DNA to make genes and the products of such genes is called **genetic engineering.** With the aid of Figure 24.17, we'll learn how it works.

Bacterial Cloning of Genes   DNA intentionally made from different living sources is called **recombinant DNA.** Two kinds of enzymes, *restriction enzymes* and *DNA ligase,* are needed for its synthesis. Restriction enzymes act like scissors, snipping apart a DNA double helix but only at certain sites, where a particular short nucleotide sequence occurs. Molecular biologists have a repertory of nearly 200 such enzymes, isolated from various microorganisms, each useful for snipping at a different site. For example, one restriction enzyme from a variety of *E. coli* splits DNA between nucleotides bearing guanine (G) and adenine (A) bases, but only when they occur in the sequence GAATTC (5′ → 3′) on one strand and its "mate" on the opposite strand, CTTAAG (3′ → 5′). This takes out a short DNA double helix section, but it carries the desired gene (see the "detailed view" of restriction enzyme's work in Figure 24.17). Note in the figure that two open ends of the excised gene are "naked," that is, not paired to an opposite strand. They're called "sticky ends," and we'll appreciate why in the next step. We'll see shortly what's done with the isolated gene. But before passing on to the next step, we want to mention another major applica-

■ Bacteria use their restriction enzymes to fight off viruses that attack them. (Yes, bacteria get diseases, too!)

INTERACTION 24.1
INTERFERONS

The *interferons* are a family of glycoproteins secreted from vertebrate cells infected by viruses. The glycoproteins then bind to other host cells and change them so that they oppose viral infections of almost any sort. At least three types occur in humans, α-, β-, and γ-interferon. α-Interferon is from white blood cells; β-interferon is from certain connective tissue (fibroblasts); and γ-interferon is from cells of the immune system (lymphocytes).

Effective at concentrations as low as $3 \times 10^{-14} M$, the interferons are among the most powerful biological materials known. What an invading virus does is induce interferon synthesis in host cells. When, for example, an invading virus particle first enters a white blood cell, particularly the type made in lymph tissue, the cellular mechanism for making the mRNA coded for interferon is switched on. In a short time, the cell manufactures and releases interferon, which then acts as a signal to other white blood cells to make interferon, too. When potential host cells contain interferon, two enzymes are made, both of which inhibit the synthesis of protein that the virus must make to prepare copies of itself. Thus, interferon inhibits the replication of the virus and so prevents their spread.

The discovery of interferons in 1957 came about when two scientists, Alick Isaacs and Jean Lindenmann, wondered why victims of one viral disease never seemed to come down with a second viral disease at the same time. It seemed to them that something in the first viral attack triggered a mechanism that provided protection against an attack by a different virus. The search for this "something" led to the discovery of interferon.

α-Interferon is used in medicine more than the other two. It's been approved for use against certain viral diseases and against cancers caused by viruses. For example, α-interferon is used against chronic hepatitis B and C, genital warts (those caused by the papilloma virus), a rare cancer called hairy-cell leukemia, and Kaposi's sarcoma (a secondary disease of AIDS victims). Clinical studies are being made for the use of α-interferon in treating non-Hodgkin's lymphoma, chronic myelogenous leukemia, throat warts (from papilloma virus), colon tumors, kidney tumors, bladder cancer, and malignant melanoma.

Large-scale clinical trials using any of the interferons would have been prohibitively expensive without recombinant DNA technology. If supplies were limited to those available from blood, 25,000 pints would be needed to make 100 mg of interferon, at a cost of a few billion dollars.

---

tion of restriction enzymes, namely, their use in solving crimes by means of genetic "fingerprinting" (see Interaction 24.2). Now let's return to bacterial cloning.

Bacteria generally make polypeptides using the same genetic code as humans, but there are some differences, particularly in "machinery." An *E. coli* bacterium, for example, has DNA not only in its single chromosome but also in large, circular, supercoiled DNA double helix molecules called **plasmids.** Each plasmid carries only a few genes, but several copies of a plasmid can exist in one bacterial cell. Each plasmid is able to replicate independently of the chromosome.

The plasmids of *E. coli* can be removed from their bacteria and made to take up the desired DNA from another system. It may be the human gene that directs the synthesis of, say, human growth hormone. The desired gene, as we described, is snipped out by a selected restriction enzyme. Plasmid DNA is similarly snipped open by the same restriction enzyme, also leaving sticky ends (Figure 24.17). Then, with the aid of *DNA ligase,* a DNA-knitting enzyme, the sticky ends of the gene's DNA combine with the sticky ends of the plasmid DNA to reclose the plasmid loops. The ends are "sticky" because base pairing between open ends of the plasmid DNA and the gene DNA attracts the end of one to that of the other, as you can see in Figure 24.17. It's the altered DNA of the plasmid that is the actual recombinant DNA.

Fresh bacterial cells now take up the new plasmids. The remarkable feature of bacteria with recombinant DNA is that when they multiply, the plasmids in the offspring also have the new DNA. Still more altered DNA is made as bacterial multiplication continues.

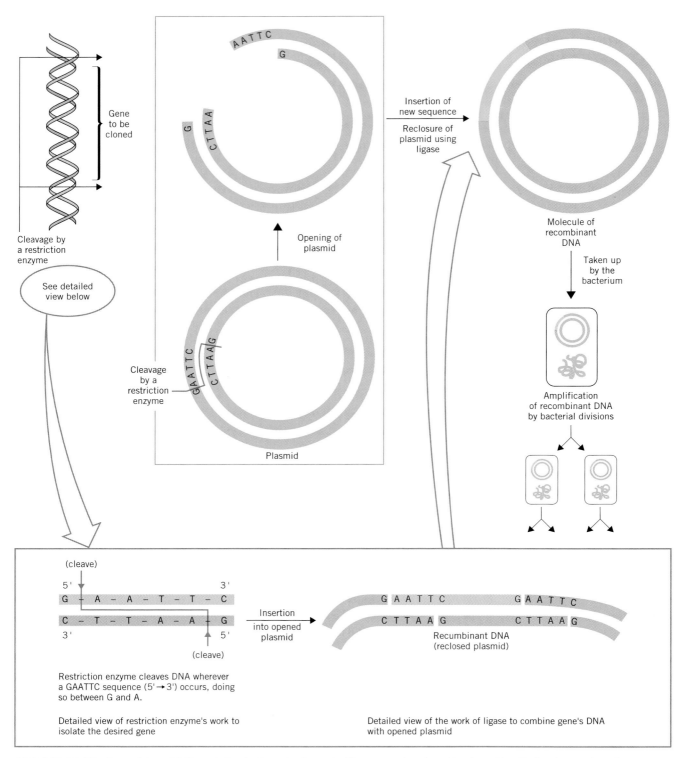

FIGURE 24.17 Recombinant DNA made by cloning, using bacteria. The gene to be cloned has been identified and its nucleotide sequence has been determined. Using this knowledge, a restriction enzyme is selected and used to snip out the DNA segment containing the gene (top left and bottom left). The restriction enzyme shown here has left "sticky ends" on the excised gene (top center) so that its DNA can join matching sticky ends on the plasmid (bottom right), the plasmid having been opened up by the same restriction enzyme. A ligase enzyme now closes the loops, making a molecule of recombinant DNA, an altered plasmid (top right). The bacteria now multiply, each new bacterium containing the recombinant DNA.

Between cell divisions, the bacteria manufacture the proteins for which they are genetically programmed, *including the proteins specified by the recombinant DNA.* In this way, bacteria can be tricked into making the *human* proteins we have mentioned. The technology isn't limited to bacteria; yeast cells work, too, and it's not confined to making human substances.

People with diabetes who once relied on the insulin of animals, like cows, pigs, or sheep, sometimes experienced allergic responses. They were also vulnerable to the availability of the pancreases of these animals, because they once were the *only* sources of insulin. The ability to make *human* insulin, therefore, has been a welcome development for diabetics.

The ability to make interferons, unusually rare and costly substances, in large quantities at low cost by recombinant DNA technology is a major technological advance. The clinical tests alone would be too costly without the availability of manufactured interferon.

Cloning Using the Polymerase Chain Reaction   The *polymerase chain reaction,* catalyzed by DNA polymerase, offers another way to clone genes (see Figure 24.18). One of the properties of duplex DNA is that heat separates the strands. So the desired DNA sample, obtained by the use of a restriction enzyme, is heated in the presence of DNA *primers.* These are only a few nucleotides long and are available by synthetic means. Primers are chosen so as to attach to ends of the heat-separated DNA strands, where they enable DNA polymerase to launch the formation of new DNA strands, those complementary to the original strands. Two clones are thus made in the first round. Their strands are separated by heat, more primer is added,

■ Kary B. Mullis (United States) was a cowinner of the 1993 Nobel Prize in Chemistry for discovering the polymerase chain reaction. The prize was shared with Michael Smith (Canada) who discovered how to make specific changes in the molecular coding of DNA so that custom-made proteins can be synthesized by living organisms.

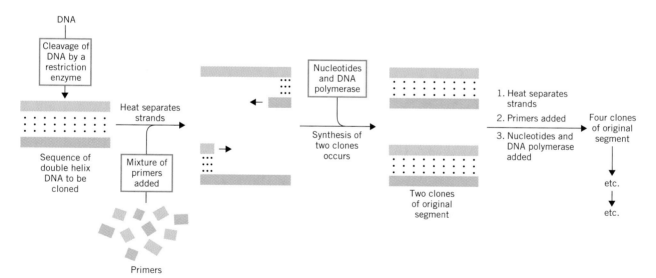

FIGURE 24.18   The cloning of DNA using the polymerase chain reaction. When the sequence of DNA to be cloned has been isolated, it is heated to separate the two strands of its double helix. (This is actually a denaturation of the double helix.) Each strand must now have a short "primer" DNA molecule (up to eight nucleotides long) join to it. (The primers are prepared using an automated DNA synthesizing apparatus.) The primer is needed to enable the DNA polymerase to get a start. This enzyme now accepts previously prepared nucleotides and joins them to the primed strands, the order of joining being determined by base pairing. Two clones are thus made. By repeating these steps, four clones are made, then eight, then 16, and so on, until sufficient DNA is available for DNA "fingerprinting" analyses or other work.

INTERACTION 24.2
DNA TYPING—"GENETIC FINGERPRINTING"—AND CRIME PROSECUTION

The genes of one human being have to be similar to those of another for the two to be members of the same species. However, every individual human being that has ever lived or ever will live, except for identical twins, has genes unique in some ways.

In the entire set of human genes, there are many regions of DNA molecules that consist of nucleotide sequences repeated in tandem. These regions all have a common core sequence, but the *number* and the *lengths* of the repeated sequences vary widely from person to person. The variations are so considerable between individuals that they are the basis of what is variously called DNA profiling, DNA typing, or DNA fingerprinting, the most important advance in forensic science (the science of crime evidence) since the development of fingerprints.

We'll first study the technique used when the sample is relatively large, for example, a spot of blood the size of a dime. The sample to be studied is isolated and its DNA is extracted. Using Figure 1, let's see how a DNA profile analysis now proceeds.

In step 1, the DNA is cut into pieces called *restriction fragments,* by the use of *restriction enzymes.* The restriction fragments obtained from any particular restriction enzyme would all have the same "sticky ends," but otherwise they vary widely in length from individual to individual. In other words, humans have restriction fragments that, in *length,* are *polymorphic* (from the Greek word *polymorphism* for "multiform"). In short(!), humans display *restriction fragment length polymorphism,* so the fragments, mercifully, are called RFLPs. *The variations in length and relative amounts of an individual's set of RFLPs constitute that person's genetic profile* or "DNA fingerprint." After chopping up the person's DNA, the technique continues with separating the RFLPs and displaying them.

Being polymorphic in length means that the RFLPs are polymorphic in *molecular mass.* This makes their separation by mass possible, using a technique called *gel electrophoresis.* "Gel" refers to a semisolid, semiliquid material that lets ions move through it. It is the supporting medium, coated on an

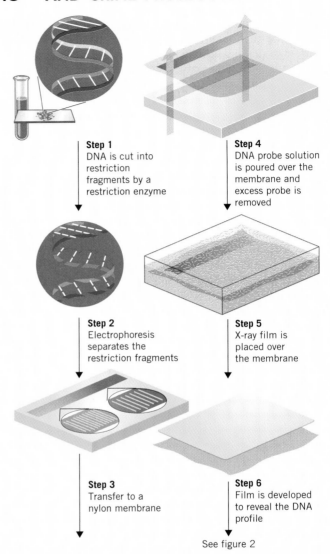

Step 1
DNA is cut into restriction fragments by a restriction enzyme

Step 2
Electrophoresis separates the restriction fragments

Step 3
Transfer to a nylon membrane

Step 4
DNA probe solution is poured over the membrane and excess probe is removed

Step 5
X-ray film is placed over the membrane

Step 6
Film is developed to reveal the DNA profile

See figure 2

FIGURE 1 Several steps (discussed in the text) are used to create a profile of the RFLPs from the DNA obtained from a crime scene sample or a defendant. (Courtesy of Cellmark Diagnostics, Germantown, Maryland.)

inert surface, for the sample of RFLPs. *Electrophoresis,* from the Greek *-phoresis,* "being carried," refers to the movement of ions in an electrical field, the ions being in the gel. It's like electrolysis, where ions also move.

In step 2 (Figure 1), a solution of RFLPs is spread in a narrow band at one end of the gel-coated surface, a direct electric current is turned on, and the ions move out. *The lighter ions move more rapidly than the heavier ions,* assuming that the charges on the ions are equal. Several samples can be run side by side at the same time on the same gel-coated plate and thus under identical conditions. (The different samples of RFLPs move in their own lanes, like runners in a sprint race.) The procedure thus separates the individual RFLPs of each sample, each sample in its own lane, spreading each RFLP (with its own unique molecular mass) into its individual narrow zone in the gel. These zones are parallel, like unevenly spaced rungs of a ladder. What remains is to locate the zones and make them visible.

A blotting technique transfers the (invisible) RFLPs to a nylon membrane (step 3), which is then immersed in a solution containing a *DNA probe* (step 4). The DNA probe consists of radiolabeled DNA molecules of short length, these having been made by standard methods. One or more atoms in the probe molecules are from a beta-emitting isotope. By base pairing to sticky ends of RFLPs, the probe molecules bind to the RFLPs wherever they are on the nylon. Any excess probe is now washed away. The binding of probe molecules to restriction fragments makes each zone of RFLPs beta emitting, but the zones are still invisible. To see the zones, a sheet of X-ray film, which is also affected by beta radiation, is placed over the nylon surface (step 5). Now the beta radiation acts on the film. After a suitable time, the X-ray film is developed (step 6), and dark bands appear wherever bands of separated RFLPs are present.

The final result is called a *DNA profile,* and it looks much like a section of a grocery store's bar code. The profile is a series of parallel lines on a piece of paper that differ from person to person in number, thickness, and spatial separation (see Figure 2). When the DNA profile of a suspect matches that of a

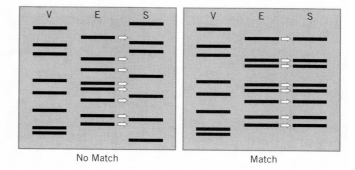

FIGURE 2 DNA analysis in crime lab testing. The DNA profiles of the victim (V) and the suspect (S) are compared to the evidence (E) obtained at the crime scene. On the left, there is no match between S and E, but on the right, the match occurs. In neither can the evidence be attributed to the victim. (Courtesy of Cellmark Diagnostics, Germantown, Maryland.)

crime scene sample of blood or semen, it's similar to having a match of fingerprints. Without a match, the suspect is innocent.

Usually, to increase the precision of the method, four or more different restriction enzymes are allowed to work on samples, each enzyme chopping up the DNA into a unique set of RFLPs to give several "bar codes." It's like using the prints of several fingers, each one unique.

If two DNA profiles do not match, another suspect is sought. Many rape suspects have been cleared by failures of DNA profile matches. When a match occurs, however, the suspect is almost certainly guilty (many experts say, *certainly*), assuming that all the standards of handling evidence and doing the profiling have been followed.

Cloning with the Polymerase Chain Reaction When the crime scene specimen is very small, there's only an ultrasmall amount of DNA. But this quantity can be "amplified" or cloned so that there will be enough DNA fragments for the rest of the analysis. The polymerase chain reaction, catalyzed by a polymerase enzyme, does the cloning work, as we discuss in connection with Figure 24.18.

and a second round of DNA formation occurs to give four clones. You can see that repetitions of this cycle can produce relatively large quantities of clones of the original DNA, enough, for example, to carry out a DNA profile analysis (see also Interaction 24.2).

Genetic Engineering Offers Major Advances in Medicine   A number of undesirable conditions are caused by flawed or absent genes, as we will study in the next section. One of the hopes of genetic engineering research is to have ways to correct such conditions. The list of potential applications of genetic engineering to health problems grows yearly. We'll mention only a few.

Dwarfism, for example, is caused by a lack of growth hormone, a relatively small polypeptide, now being made by genetic engineering. It has been successfully introduced into mice, with dramatic effects on mouse size. The hormone is now available for human use. It's also sneaking into use to make people of normal height taller, an application that specialists regard as risky.

A polypeptide hormone made by the heart, which reduces blood pressure, can be manufactured by genetic engineering and used by victims of high blood pressure.

Tissue plasminogen activator, or TPA, a clot-dissolving enzyme, is made by genetic engineering for use in reducing the damage to heart tissue following a heart attack.

Mosquitoes responsible for the transmission of dengue fever virus can be genetically altered to prevent their making the virus in their saliva, the route from which humans receive dengue fever virus. It's still in experimental stages, but this development and others like it could help control some of our greatest scourges.

Kidney dialysis patients will need fewer blood transfusions if a blood cell–producing substance, erythropoietin, can be made by this technology. Hemophiliacs who lack a blood-clotting factor may have it available.

■ There is neither a vaccine nor a cure for dengue fever. In the tropics and subtropics it strikes 50 million each year; half a million are hospitalized; and 25,000 die.

Agriculture Is Affected by Genetic Engineering   Bovine somatotropin or BST is a hormone involved in milk production in cows. When BST, made using recombinant DNA, is given to milk cows, their milk production is higher. (The practice has aroused considerable controversy.)

Genetic engineering is being tried for developing pest-resistant strains of plants, even plants that manufacture their own fertilizer. *Bacillus thuringiensis,* or BT, a common bacterium, is a natural killer of many agricultural pests. A gene derived from BT has been put into corn, making it resistant to the corn borer. Another BT gene has been incorporated into cotton, making it kill bollworms and budworms.

A gene from the petunia has been inserted into soybeans, protecting soybeans from a powerful weed killer. The weed killer can thus be used against weeds in soybean fields without killing the soybeans. Similarly, wheat can be made resistant to a potent herbicide (phosphinothricin or Basta), enabling the herbicide to be used against weeds in wheat fields without killing the wheat. Gene transfer methods are also being studied to make trees used for lumber or paper faster growing, disease resistant, and better able to handle very cold weather.

Gene Transfers Have Also Led to the Cloning of Animals   The ultimate of gene transfers involves the cloning of animals. In the 1970s, British biologist John Gurdon took the DNA from an older albino frog tadpole and inserted it into single-celled frog eggs. A number of identical albino frogs were produced, all clones of the original, but they all died once feeding started.

Would cloning work with mammals? Most scientists thought it to be impossible, but not members of one Scottish group. In early 1997, a team of biologists headed by

Ian Wilmut and Keith Campbell of Scotland's Roslin Institute unveiled to the press a lamb named Dolly that had been produced by cloning. It was the first animal "prepared" from an *adult* mammal cell. A mature cell was taken from mammary tissue of a six-year-old ewe. The host egg cell came from the same ewe. The entire nucleus of the host cell was removed, however, thus eliminating all of its genetic material. (The egg cell, in other words, would *not* form double helix DNA using DNA from a sperm donor.) Next, the mammary cell with its DNA was fused (inserted) into the egg cell with the aid of a tiny electrical charge. Thus double helix DNA belonging entirely to one ewe became a part of the egg cell; in effect, the egg was now fertilized. The electrical charge also activated the process whereby the fertilized egg began cell division. After six days, the embryo was implanted into a surrogate mother sheep. Of 277 cells that were fused, only 30 began cell division. Of 29 implanted embryos, one sheep was born.

Dolly made headlines and feature articles in news magazines around the world. A very substantial amount of unease surfaced as well. What would be the moral and ethical implications of cloning humans, assuming it to be technically possible? Such questions are being debated, and, as always, how they are resolved will be a function of basic values. You'll no doubt read much of these matters. Although Dolly represented a significant development, almost certainly the more beneficial outcomes of genetic engineering will flow from medical advances, as we'll see in the next section and in its following Interaction on gene chips.

24.7 HEREDITARY DISEASES AND GENETIC ENGINEERING

About 4000 inherited disorders in humans are caused directly or indirectly by flawed genes.

Gene Therapy Has Worked Against Enzyme Deficiencies   The introduction of a proper gene into the cells of someone with a defective gene is called **gene therapy.** The very first effort to cure a human disease using gene therapy began in 1990. It involved two girls, ages 4 and 9, born with a defective *ADA* gene that left them with almost no natural immunity. The children were injected with white blood cells that had been given the correct gene. Two years later, the girls had functioning immune systems and, instead of having to lead very isolated lives, were in public school.

■ The *ADA* gene makes the enzyme *adenosine deaminase,* which is vital to the immune system.

In Cystic Fibrosis, a Defective Gene Makes a Defective Membrane Protein   The victims of cystic fibrosis overproduce a thick mucus in the lungs and the digestive tract. It clogs these systems and often leads to death in children. Roughly one person in 20 carries the recessive cystic fibrosis gene, and the disease hits about one in every 1000 newborns.

The normal gene directs the synthesis of a transmembrane protein that lets chloride ion pass out of a cell. For those with the cystic fibrosis gene, this protein is defective, and the passage of chloride ion is impaired. When Cl^- moves unimpaired out of a cell *it also takes water along.* But this is an impaired function in cystic fibrosis, and water is thus in reduced supply outside of lung and airway cells. The mucus thickens and does not flow properly. The thickened mucus makes breathing more difficult and provides a breeding ground for bacteria that cause a certain form of pneumonia.

■ The membrane protein is called the CFTR protein after cystic fibrosis transmembrane conductance regulator.

Gene Therapy for Cystic Fibrosis Is Being Tried   One strategy against cystic fibrosis takes advantage of the ability of retroviruses, which normally carry no DNA, to accept the DNA responsible for the normal transmembrane protein. In host cells, retroviruses do make duplex DNA (from their RNA), and it becomes part of the host's chromosomes. In other words, retroviruses can pass DNA into the genetic machinery of their host cells. When the invading retrovirus already carries piggybacked, transmembrane DNA, the host cell acquires the DNA needed to counter the inherited defect.

What has to be done, of course, is to disable the virus enough so that it cannot cause infection but not enough to prevent its interacting at the host cell's surface and carrying the DNA inside. To do this, the *surface* of the retrovirus particle is left the same; it still has the original proteins by means of which the particle "recognizes" and sticks to the host cell. Only the retroviral particle *interior* is altered to prevent its being an infectious agent.

To reach lung and airway tissue in victims of cystic fibrosis, one of the common cold viruses is the delivery vehicle for the transmembrane gene. A solution containing the altered virus is simply squirted into the airways. If corrected genes become installed in enough numbers in airway cells, the defects in the movements of Cl^- and water will be removed. Much testing is underway; in the mid-1990s several patients were successfully treated by this technique.

The *Human Genome Project* Aims to Map All Human Genes and Determine Their DNA Sequences   The entire complement of genetic information of a species is called its **genome.** The Human Genome Project, formally launched in 1990, intends to discover the "map" of every human chromosome and to determine the sequence of bases in every human gene. It's perhaps the largest single project ever attempted in molecular biology, and its total cost may come within range of (but will probably be less than) what it cost to put an astronaut on the moon.

■ The gene for Huntington's disease, a deadly neurological disorder, has been found near the tip of chromosome 4.

If the genome is a library, if the chromosomes are book sections, and if the genes are individual books, then the Human Genome Project's goal in *mapping* chromosomes is to create the card catalog. "In which section of the library is the gene for cystic fibrosis?" Answer: on chromosome 7. "In which particular part (shelf) of chromosome 7 is the gene located?" Answer: in region q31.

Now comes the *sequencing* question; it's not the same as the mapping question. "What is the sequence of bases in the cystic fibrosis gene?" The answer is known, but we can't give it here; the gene involves over 6000 bases. "What's wrong with this gene to make it cause cystic fibrosis?" For about 70% of cystic fibrosis mutations, there is a deletion (an absence) of three bases in exon number 10. A phenylalanine unit, therefore, is not incorporated into the polypeptide made under the direction of the gene. So much human suffering from so small a molecular defect! By identifying the locations of human genes and determining the sequences of these genes, it is expected that specific gene therapies can be implemented to correct defects.

One of the uses of the information about human genes that is emerging from the Human Genome Project is the ability to diagnose present diseases with an unprecedented speed or to tell if a disease is likely to occur later in life. These capabilities have become possible through the development of *biochip technology* (see Interaction 24.3).

BIOCHIPS, BREAST CANCER, AND THE NO-NAME PROTEIN

Let's dream ahead some ten to thirty years. Suppose that you have a sore throat and finally decide to go to your clinic; enough is enough. The nurse takes a throat swab and sends it to the clinic's lab for *biochip diagnosis*. In a very short time—much briefer than the usual culture plate technology—the physician has a report that tells not only which specific microorganism is responsible but also whether it is in a mutant form. If a bacterium is responsible, the same technology would tell if the one bothering you is resistant to certain antibiotics but not to others. Thus, you would be given the specific antibiotic for dealing with the problem, and you'd not waste time or money on others.

Genetic information on the same swab could also tell you whether you've inherited a defective gene that disposes you toward breast cancer, prostate cancer, or other maladies. Simple changes in life style might then help you significantly delay, if not forestall, the threat. If that's not possible, then the kind of aggressive treatment that will work best can be planned, not as firing a shotgun in the dark at a noise, hoping to hit an enemy without harming friends, but as precise sniping at an enemy clearly seen.

Biochip diagnosis could tell someone who is HIV positive if his or her particular viral strain is resistant to the medications commonly used. The exact nature of the mutation could then be used to select the treatment.

A Specific DNA Probe in Each Zone of a Biochip
Actually none of the above is dreaming. The technology using the biochip is exploding. Several new companies have formed to develop biochips, and every pharmaceutical concern is working in the field. A biochip is like a computer chip, only it carries information not as sequences of zeros and ones but as fragments of DNA molecules, in other words as sequences of nucleotide units A, T, G, and C.

Imagine a glass or silicon plate; it might be as large as 10 cm^2 or as small as 20 μm^2. The surface can be divided into tiny squared zones, hundreds, thousands, even millions in number. All but one zone is covered to enable the uncovered zone to receive identical copies of the molecules having one particular DNA segment. Each such molecule is called a *probe,* and each probe is generally on the order of 20 nucleotides long. Roughly 100 million molecules of a given probe are loaded onto one zone.

The development of automated machines for synthesizing any desired nucleotide sequence has made possible the preparation of an unlimited variety of probes. Which sequences are actually desired is learned from genome projects. The sequences are those of either a healthy or a mutant human gene or are those in the genes of any microorganism chosen. The probe chosen, for example, might be a telltale sequence of part of a gene, one signifying a potential cancer danger.

The probe molecules become attached to the surface of their assigned zone. Then another zone is exposed and given a batch of the identical molecules of another DNA probe, one with a different sequence of nucleotides. (The methods of attaching the probes to the plates differ with the various corporations; we'll not describe any details.) Zone by zone, different sets of probe molecules, each set with its unique DNA sequence, are loaded onto the chip until all zones are filled. One chip might focus entirely on breast cancer in all of its mutant varieties; another chip, on HIV, and so on. A chip might, by using all of its zones, carry one single gene in its entirety, short segment by short segment, one in each zone. (The Human Genome Project is rapidly providing the nucleotide sequences of human genes.)

When a sample of genetic material, say, from the throat swab, is to be tested, its DNA molecules are chemically joined to a fluorescent marker molecule. The chip is then treated with a solution of the test material, and base pairing—A to T and G to C—leads gene molecules to "matching" zones. Which zone has fully accepted test material shows up by intense fluorescence. Sometimes, couplings happen between test material molecules and those of a probe that do not exactly match. The binding is then not as tight, and the fluorescence is less intense.

Biochips, Breast Cancer, and the No-Name Protein
When something goes chemically haywire with a cell's DNA and the cell cannot repair the damage, it's supposed to self-destruct. (Cell suicide is called *apoptosis,* and it involves several steps.) By cellular self-destruction, the dangers of the proliferation of such a cell into a lethal tumor are avoided.

What the cell mobilizes when its DNA has been damaged are extra quantities of a protein not yet fully named, the p53 protein ("p" for protein and "53" for its formula mass of 53,000). The gene that directs its synthesis is (note the italics) the *p53* gene. The protein, p53, activates another gene that makes a protein that arrests the cell cycle so that DNA repair might occur. If this fails, the cell dies.

It's been estimated that about 6.5 million of the cancer cases that arise each year throughout the world involve tumors having a mutant form of the *p53* gene. Roughly half of all lung, colon, and rectal cancers in the United States as well as about 40% of lymphomas, stomach, and pancreatic cancers involve *p53* mutations. In many human tumors, all that is wrong with the molecules of the p53 proteins is that they are improperly folded!

Mutations of the *p53* gene also occur in breast cancer. Certain of them cause cancerous breast cells to be resistant to a widely used chemotherapeutic agent, doxorubicin. Biochip technology applied to an individual's *p53* gene (Figure 1) will thus enable a breast cancer specialist to select medications (or radiation therapy) according to that individual's situation.

Additional Readings

1. J. Travis, "Chips Ahoy," *Science News,* March 8, 1997, page 144.

2. D. Stipp, "Gene Chip Breakthrough," *Fortune,* March 31, 1997, page 54.

3. R. L. Rawls, "Keeping Cancer in Check with p53," *Chemical and Engineering News,* February 17, 1997, page 38. (See also, S. Begley, "The Cancer Killer," *Newsweek,* December 23, 1996, page 42.)

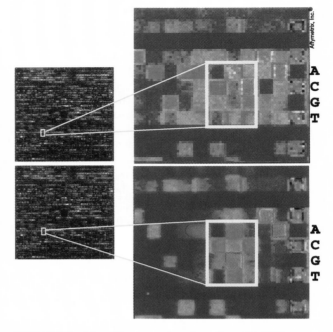

FIGURE 1 The *p53* gene probe using a GeneChip™. The top image shows the pattern for a normal *p53* gene. The bottom image is that of a mutant *p53* gene. The area of the mutation is enlarged.

SUMMARY

Hereditary information The genetic apparatus of a cell is mostly in its nucleus and consists of chromatin, a complex of DNA and proteins (histones). Strands of DNA, a polymer, carry segments that are individual genes. Chromatin replicates prior to cell division, and the duplicates segregate as the cell divides. Each new cell thereby inherits exact copies of the chromatin of the parent cell. If copying errors are made, the daughter cells (if they form at all) are mutants. They may be reproductively dead—incapable of dividing themselves—or they may transmit the mutant character to succeeding cells. The expression of this might be as a cancer, a tumor, or a birth defect. Atomic radiations, particularly X rays and gamma rays, are potent mutagens, but many chemicals, those that are radiomimetic, mimic these rays.

DNA Complete hydrolysis of DNA gives phosphoric acid, deoxyribose, and a set of four heterocyclic amines, the bases adenine (A), thymine (T), guanine (G), and cytosine (C). The molecular backbone of the DNA polymer is a series of deoxyribose units joined by phosphodiester groups. Attached to each deoxyribose is one of the four bases. The order in which triplets of bases occur is the cell's way of storing genetic information.

In higher organisms, a gene consists of successive groups of triplets, the exons, separated by introns. Thus the gene is a split system, not a continuous series of nucleotide units. DNA exists in cell nuclei as duplex DNA, double helices held to this geometry largely by hydrophobic forces. The base A always pairs with the base T and C always pairs with G. Using this structure and the faithfulness of base pairing, Crick and Watson explained the accuracy of replication. After replication, each new double helix has one of the parent DNA strands and one new, complementary strand.

RNA RNA is similar to DNA except that in RNA ribose replaces deoxyribose and uracil (U) replaces thymine (T). Four main types of RNA are involved in polypeptide synthesis. One is rRNA, which is in ribosomes. A ribosome contains both rRNA and proteins that have enzyme activity needed during polypeptide synthesis.

mRNA is the carrier of the genetic message from the nucleus to the site where a polypeptide is assembled. mRNA results from a chemical processing of the longer RNA strand, hnRNA, which is made directly under the supervision of DNA.

tRNA molecules are the smallest RNAs to participate directly in polypeptide synthesis. Their function is to convey aminoacyl units to the polypeptide assembly site. tRNAs recognize where they are to go by base pairing between an anticodon on the tRNA molecule and its complementary codon on mRNA. Both codon and anticodon consist of a triplet of bases.

Polypeptide synthesis Genetic information is first transcribed when DNA directs the synthesis of mRNA. Each base triplet on the exons of DNA specifies a codon on mRNA. The mRNA moves to the cytoplasm to form an elongation complex with subunits of a ribosome and the first and second tRNA—aa unit to become part of the developing polypeptide.

The ribosome then rolls down the mRNA as tRNA—aa units come to the mRNA codons during the moment when the latter are aligned over the proper enzyme site of a ribosome. Elongation of the polypeptide then proceeds to the end of the mRNA strand or to a chain-terminating codon. After chain termination, the polypeptide strand leaves, and it may be further modified to give it its final N-terminal amino acid residue. Chaperonin enzymes guide the polypeptides into their final geometries.

The whole operation can be controlled by a feedback mechanism in which an inducer molecule removes a repressor of the gene, thus letting the gene work.

Several antibiotics inhibit bacterial polypeptide synthesis, which causes the bacteria to die.

Viruses Viruses are packages of DNA or RNA encapsulated by protein. Once they get inside their host cell, virus particles take over the cell's genetic machinery, make enough new virus particles to burst the cell, and then repeat this in neighboring cells. Viruses are implicated in human cancer. Some viruses make new RNA under the direction of existing RNA, using RNA replicase. Others, the retroviruses, use RNA and reverse transcriptase to make DNA, which then directs the synthesis of new RNA.

Recombinant DNA Recombinant DNA is DNA made from different species, such as from bacterial plasmids and humans, using restriction enzymes and DNA ligase. It's encoded to direct the synthesis of some desired polypeptide. The altered plasmids are reintroduced into the bacteria, where they become machinery for synthesizing the polypeptide (e.g., human insulin, or growth hormone, or interferon).

Gene cloning can also be done using the polymerase chain reaction on genes removed by restriction enzymes, heated to separate the strands, and set up for the polymerase enzyme by the use of short primers.

Gene therapy Hereditary diseases stem from defects in DNA molecules that either prevent the synthesis of necessary enzymes or that make the enzymes in forms that won't work. The identification of the chromosomes bearing the defective genes has enabled the analyses of the genes themselves. In gene therapy, it is hoped that healthy genes can be substituted for those that are defective. Gene therapy for cystic fibrosis entails using a modified retrovirus to convey correct DNA material into the host cells without causing a viral infection.

REVIEW EXERCISES

The answers to Review Exercises whose numbers are in color are found in Appendix E. The answers to the other Review Exercises are found in the Study Guide that accompanies this book. The more challenging questions are marked with asterisks.

The Cell

24.1   What is the term used for each of the following?
(a) The *liquid* inside the cell but outside the nucleus
(b) The entire contents of the cell
(c) The region of the cell outside the nucleus
(d) The particle at which polypeptide synthesis occurs
(e) The particle in which ATP is synthesized
(f) The name of the chemical that makes up a gene
(g) The nucleoprotein material inside a cell nucleus
(h) The protein around which DNA strands are wound
(i) The fundamental unit of heredity

24.2   What is the relationship between a chromosome and chromatin?

24.3   The duplication of a gene occurs in what part of the cell?

24.4   In a broad, overall sense, what happens when DNA replicates?

Structural Features of Nucleic Acids

24.5   What is the general name for the chemicals that are most intimately involved in the storage and the transmission of genetic information?

24.6   The monomer units for the nucleic acids have what *general* name?

24.7   What are the names of the two sugars produced by the complete hydrolysis of all the nucleic acids in a cell?

24.8   What are the names and symbols of the four bases that are liberated by the complete hydrolysis of (a) DNA and (b) RNA?

24.9   How are all DNA molecules structurally alike?

24.10   How do different DNAs differ structurally?

24.11   How are all RNA molecules structurally alike?

24.12   What are the principal structural differences between DNA and RNA?

24.13   When DNA is hydrolyzed, the ratios of A to T and of G to C are each very close to 1:1, *regardless of the species investigated*. Explain.

24.14   What does base pairing mean, in general terms?

24.15   What is the chief stabilizing factor for the geometrical form taken by duplex DNA?

*****24.16**   If the TCAGCCT sequence appeared on a DNA strand, what would be the sequence on the DNA strand opposite it in a double helix?

24.17   The *accuracy* of replication is assured by the operation of what factors?

24.18   What is the relationship between a single molecule of single-stranded DNA and a single gene?

24.19   Suppose that a certain DNA strand has the following groups of nucleotides, where each lowercase letter represents a group several nucleotides long.

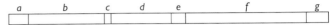

Which sections are likelier to be the introns? Why?

24.20   In general terms only, what particular contribution does a gene make to the structure of a polypeptide?

Ribonucleic Acids

24.21   What is the general composition of a ribosome, and what function does this particle have?

24.22   What is hnRNA, and what role does it have?

24.23   To which kind of RNA does the term "primary transcript" refer?

24.24   What is a codon, and what kind of nucleic acid is a continuous, uninterrupted series of codons?

24.25   What is an anticodon, and on what kind of RNA is it found?

24.26   Which triplet, ATA or CGC, cannot be a codon? Explain.

24.27   Which amino acids are specified by the following codons?
(a) UUU   (b) UCC   (c) ACA   (d) GAU

*****24.28**   What are the anticodons for the codons of Review Exercise 24.27?

*****24.29**   Suppose that sections *x, y,* and *z* of the following hypothetical DNA strand are the exons of one gene.

What is the structure of each of the following substances made under its direction?
(a) The hnRNA
(b) The mRNA
(c) The polypeptide that is made using the given genetic information. (Use the three-letter symbol format for the tripeptide structure, referring as needed to Table 24.1 for these symbols.)
(d) Suppose that the following mutation occurred. What would be the structure of the polypeptide?

Polypeptide Synthesis

24.30 Use the identifying letters to arrange the following symbols or terms in the correct order in which they are synthesized in going from a gene to an enzyme. (Place the letter of the first material to be involved on the left.)

polypeptide   duplex DNA   hnRNA   mRNA

    A   B   C   D

           $<$ $<$ $<$

earliest to be involved   last to appear

24.31 What is meant by translation, as used in this chapter? And what is meant by transcription?

*24.32 To make the pentapeptide, Met-Ala-His-Ser-Tyr, (a) What do the sequences of bases on the mRNA strand have to be? (b) What is the anticodon on the first tRNA to move into place?

24.33 In general terms, explain how the overlapping of triplets on mRNA is avoided.

24.34 The discussion about the *lac operon* was included to illustrate what aspect of the chemistry of polypeptide synthesis?

24.35 Not all of the DNA of the *lac operon* codes for the synthesis of β-galactosidase. What is the name given to the part that is so coded? What names are given to other segments of the DNA of the *lac* operon and what is their function?

24.36 In general terms, what is a repressor, and what does it do?

24.37 What does an inducer do (in general terms)?

24.38 How do some of the antibiotics work at the molecular level?

24.39 The genetic code is the key to translating between what two "languages"?

24.40 What would it mean to say that the genetic code is universal? And is the code strictly universal?

24.41 In general terms, how do X rays cause cancer?

24.42 How do X rays and gamma rays work in cancer therapy?

24.43 What is a radiomimetic substance?

Viruses

24.44 What is a virus made of?

24.45 In general terms, how does a virus discriminate among all possible host cells and "find" just one kind of host cell?

24.46 In general terms, once a virus particle has joined to the membrane of its host cell, what must occur next if the viral infection is to advance?

24.47 In general terms, what does RNA replicase do in the systems having it? What is true about a normal host cell that requires a virus to have RNA replicase?

24.48 In general terms and in connection with the work of (some) viruses, what does reverse transcriptase do?

24.49 Which kind of mRNA enables a virus to use the host cell to make more virus particles, (−)mRNA or (+)mRNA?

24.50 If the mRNA in a given virus is (−)mRNA, what must the virus–host cell system accomplish if the virus is to multiply?

24.51 If the mRNA in a given virus is (−)mRNA, what enzyme is carried into the host cell by the virus?

24.52 What does the prefix *retro-* signify in retrovirus?

24.53 What is meant by a *silent gene* and where does it come from?

24.54 Some viruses are called *oncogenic RNA viruses*. What does "oncogenic" mean here?

24.55 What is the full name of the HIV system?

24.56 What is the host cell of HIV and how does this fact make AIDS so dangerous?

24.57 What is the theory concerning the action of AZT?

24.58 How do protease inhibitors work against HIV?

24.59 What advantage is expected from the use of a two-drug or a three-drug attack on HIV?

Recombinant DNA

24.60 What is recombinant DNA?

24.61 What is a plasmid, and what is it made of?

24.62 What kinds of enzymes can snip DNA apart at specific sites?

24.63 Recombinant DNA technology is carried out to accomplish the synthesis of what kind of substance (in general terms)?

24.64 What does genetic engineering refer to?

24.65 What is the polymerase chain reaction and what kind of enzyme catalyzes it?

24.66 What property of duplex DNA makes the polymerase chain reaction possible?

24.67 Suppose that a shepherd had a large flock of sheep, all clones of each other. What is one of the risks he faces when one disease-bearing microorganism enters the flock? Why is this risk lessened if the flock is the result of the usual sexual reproduction?

Hereditary Diseases

24.68 At the molecular level of life, what kind of defect is the fundamental cause of a hereditary disease?

24.69 The defective gene in cystic fibrosis leads to an impairment of what specific activity of the cells of the affected tissues? Why does this activity result in a problem with mucus?

24.70 Only in the broadest terms, what is gene therapy meant to accomplish?

24.71 A retrovirus instead of some other kind of virus is being used in gene therapy for cystic fibrosis. Why?

24.72 In altering the retrovirus used in gene therapy for cystic fibrosis, only the interior of the virus is changed. Why is the surface of the virus particle left alone?

24.73 What is the Human Genome Project attempting?

24.74 Describe a practical application of the Human Genome Project.

Interferon (Interaction 24.1)

24.75 What kinds of compounds are the interferons and what are they able to do that protects us from disease?

24.76 What scientific observation led scientists to look for something that would do what the interferons do?

24.77 What kinds of infectious agents stimulate the production of interferons?

24.78 How are interferons made for clinical trials?

DNA Typing (Interaction 24.2)

24.79 What fact about cells makes it possible to use cells from any part of the body of a suspect for DNA fingerprinting in a rape case for which a semen sample has been obtained?

24.80 What is meant by RFLP? What does each letter mean?

24.81 How are RFLPs used to give the genetic "bar code"?

24.82 Why is the *polymerase chain reaction* part of the technology of DNA typing?

Biochips (Interaction 24.3)

24.83 What use is made of automated DNA synthesis in biochip technology?

24.84 What property of single-stranded DNA is involved when test DNA is added to a biochip?

Additional Exercises

24.85 Compare the structures of uracil and thymine (Figure 24.2).
(a) How do the structures differ?
(b) Can the structural difference affect the hydrogen-bonding capabilities of these two bases?
(c) How does the known behavior of uracil and thymine toward adenine bear on the answer to part (b)?

24.86 Write the structures of nucleotides that involve deoxyribose and each of the following bases.
(a) adenine   (b) cytosine

24.87 Consider the following compound.

(a) Is it a mononucleotide, a dinucleotide, or a higher nucleotide? How can you tell?
(b) Could it be obtained by the partial hydrolysis of DNA or RNA? How can you tell?
(c) Where is the 5′ end, at the bottom or the top of the structure, as written?
(d) In terms of the single-letter symbols for bases, how is the structure of this compound written?

BIOCHEMICAL ENERGETICS

Even at Mesa Arch, Canyonlands National Park, Utah, nature is unyielding about bio-chemical energy. It takes a one-mile run to work off the calories in only one slice of un-buttered bread.

THIS CHAPTER IN CONTEXT

Our principal *outside sources* of chemical energy are dietary carbohydrates and fatty acids, and we studied their structures partly to make this chapter possible. Our *inside sources* are other high-energy compounds, particularly triphosphates like ATP. Our *uses* of chemical energy are many, one being to make polypeptides, as we saw in the previous chapter. In this chapter we'll see how cells transfer the chemical energy of outside molecules to triphosphates and to other inside, high-energy compounds. We'll also learn how oxygen from the air we breathe oxidizes metabolites.

25.1 ENERGY FOR LIVING—AN OVERVIEW

High-energy phosphates, such as ATP, are the body's means of trapping the energy from the oxidation of the products of digestion.

■ The breaking down of molecules is called *catabolism,* from the Greek, *kata-*, down; *ballein,* to throw or cast.

The energy in a mole of glucose or a mole of a fatty acid, like palmitic acid, can be released by their oxidation to CO_2 and H_2O. We could do this in the lab simply by burning them, but this won't do in the body. The body doesn't run on heat energy, like some steam locomotive. Instead, the body oxidizes glucose or palmitic acid by a number of small steps, *some of which also make other high-energy compounds.*

When we burn 1 mol of palmitic acid, $CH_3(CH_2)_{14}CO_2H$, we get 2400 kcal (1.00×10^4 kJ) of energy. The energy yield must also be the same in the body, because the starting and ending products are identical to those from burning. Some of the energy, however, pays the energy costs of making high-energy phosphates. Only a fraction is released as heat.

Each Organophosphate Has a Potential for Transferring Its Phosphate Unit to an Acceptor   Table 25.1 gives the names and structures of the chief phosphates that we'll meet. They are arranged in their order of **phosphate group transfer potentials,** meaning their relative abilities to transfer a phosphate group to an acceptor, as in the following equation.[1]

$$RO-PO_3{}^{2-} + R'O-H \longrightarrow RO-H + R'O-PO_3{}^{2-}$$

Donor   **Acceptor**
(higher   (lower
potential)   potential)

■ The bond in red denotes the P—O bond that breaks in an energy-releasing reaction of a high-energy phosphate.

Each number in the last column of Table 25.1 is a measure of the relative potential of the compound to transfer a phosphate group to the same acceptor, water. The numbers are given as *negative* values, the convention used when energy is *released* by (subtracted from) the molecule during its reaction. A compound with a higher (more negative) phosphate group transfer potential can, in principle, always be used to make one with a lower (less negative) potential, assuming that the right enzyme and any other needed reactants are available. Thus, we can use the positions of compounds in Table 25.1 to predict whether a particular transfer is possible in principle.

■ By using a *common* acceptor, water, the numbers of Table 25.1 offer a valid *comparison* of transfer potentials.

[1] We will usually write the phosphate unit as $-OPO_3{}^{2-}$, but its state of ionization varies with the pH of the solution. At physiological pH, the unit is mostly in the singly and doubly ionized forms, $-OPO_3H^-$ and $-OPO_3{}^{2-}$.

TABLE 25.1   Some Organophosphates in Metabolism

Organophosphate	Structure	Phosphate Group Transfer Potential
Phosphoenolpyruvate	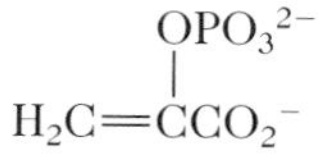 $H_2C{=}CCO_2^-$ with OPO_3^{2-}	-25.8
1,3-Bisphosphoglycerate	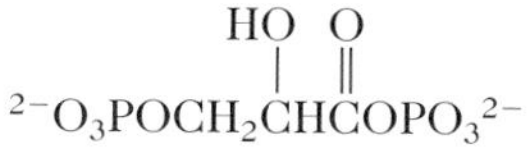 $^{2-}O_3POCH_2CHCOPO_3^{2-}$ with HO, O	-11.8
Creatine phosphate	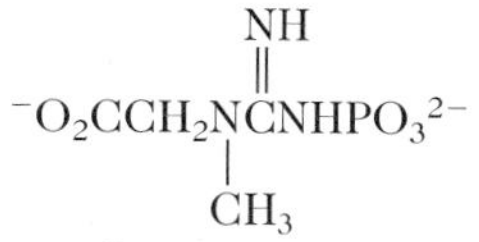 $^-O_2CCH_2NCNHPO_3^{2-}$ with NH, CH$_3$	-10.3
Acetyl phosphate	 $CH_3COPO_3^{2-}$	-10.1
Adenosine triphosphate, ATP	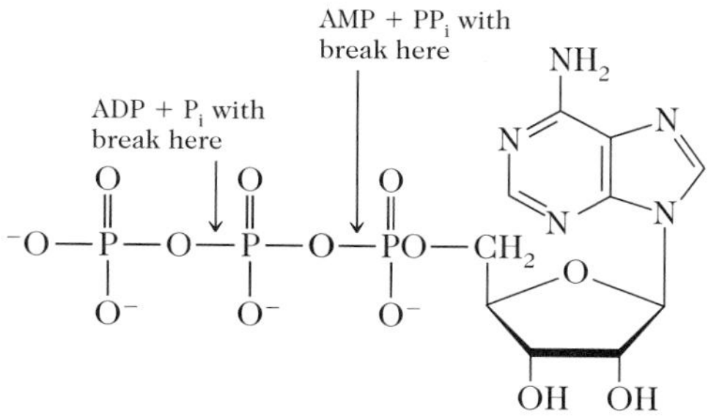	-7.3^a
Glucose-1-phosphate	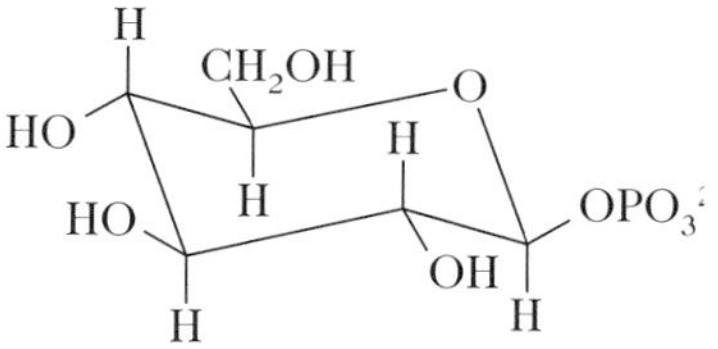	-5.0
Fructose-6-phosphate	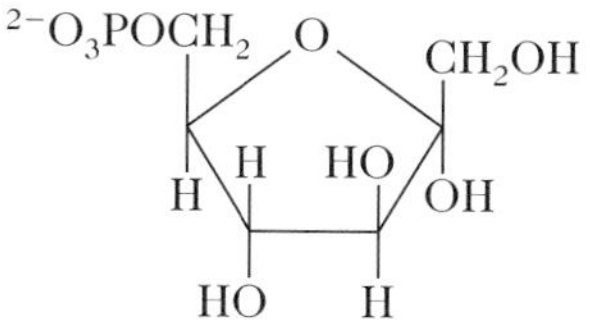	-3.8
Glucose-6-phosphate	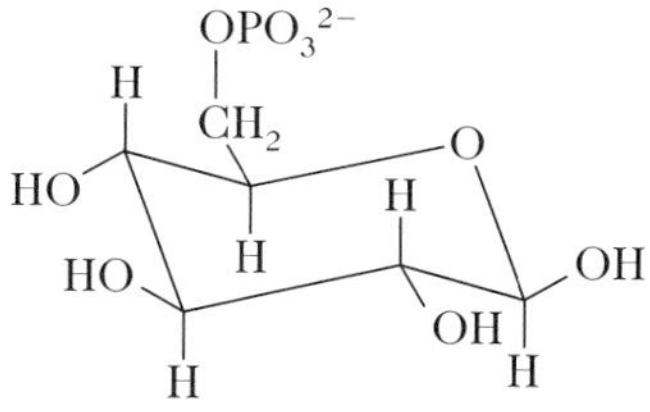	-3.3
Glycerol-1-phosphate	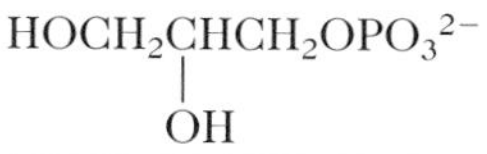 $HOCH_2CHCH_2OPO_3^{2-}$ with OH	-2.2

^{a} This value applies whether ADP or AMP forms.

■ PRACTICE EXERCISE 1   Using Table 25.1, tell whether each reaction can occur.

(a) ATP + glycerol 3-phosphate → ADP + glycerol-1,3-bisphosphate

(b) ATP + glucose → ADP + glucose-1-phosphate

(c) Glucose-1-phosphate + creatine → glucose + creatine phosphate

Adenosine Triphosphate (ATP) Is the Body's Chief Energy Broker   In Table 25.1, **adenosine triphosphate** or **ATP** is particularly important. From the lowest to the highest forms of life, it is used universally as the chief carrier of energy for living functions. Its synthesis in the body is the way that we trap some of the energy released by the oxidations of nutrients, like glucose and fatty acids. ATP can be used, for example, to power muscle contraction, and the following equation expresses the overall result.

$$\text{"Relaxed" muscle} + \text{ATP} \longrightarrow \text{"contracted" muscle} + \text{ADP} + \text{P}_i$$

The reaction produces changes in the tertiary structures of muscle proteins that cause the protein fibers to contract. Simultaneously, ATP changes to **adenosine diphosphate,** or **ADP,** plus inorganic phosphate, P_i.

ATP has two phosphate bonds either of whose rupture can release much useful energy. Occasionally the second bond (see Table 25.1) is broken in some transfer of chemical energy, and the products then are **adenosine monophosphate, AMP,** and the inorganic diphosphate ion, PP_i.

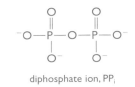

diphosphate ion, PP_i

adenosine diphosphate, ADP
(fully ionized form)

adenosine monophosphate, AMP
(fully ionized form)

Occasionally triphosphates other than ATP, like guanosine triphosphate, GTP, are the carriers of chemical energy. We'll see an example later in this chapter.

guanosine triphosphate, GTP
(fully ionized form)

ATP and the phosphates higher than it in Table 25.1 are, by definition, **high-energy phosphates.** Notice, however, that ATP is not the *highest* energy phosphate. Thus it can be made from those above it in the table by phosphate transfer to ADP. Once made, ATP can, in turn, make any of the phosphates lower on the list. The intermediate position of the ATP–ADP system is one reason why these two are so useful as energy brokers. ADP can *accept* chemical energy from the phosphates with higher (more negative) potential and be changed to ATP. Then, newly made ATP can transfer chemical energy by phosphate transfer to make compounds lower on the list.

The Resynthesis of ATP Is a Major Goal of Catabolism   Almost any energy-demanding activity of the body consumes ATP. Even at rest, the adult human consumes about 40 kg (about 80 mol) of ATP per day. Yet, at any one instant, there is less than 50 g (0.1 mol) of ATP in existence everywhere in the body! When we exercise, our rate of consumption of ATP can go as high as 0.5 kg per minute! Obviously, the rapid resynthesis of ATP is one of the highest priority activities of the body, and it occurs continuously. *Virtually all biochemical energetics come down to the synthesis and uses of ATP.*

■ Between the completely resting and the vigorously exercising states, the rate of ATP consumption can vary by a factor of 100.

Two broad routes to ATP occur. One is called **substrate phosphorylation.** It's the *direct* transfer of phosphate from a very high-energy phosphate to ADP (the substrate). The phosphate group, in other words, doesn't come from a phosphate *ion* in solution but from an organic phosphate. The other route is **oxidative phosphorylation.** This is the direct use of phosphate ion from the surrounding fluid, done by a series of reactions called the **respiratory chain.**

In Muscles, Creatine Phosphate Converts ADP to ATP   Muscular work uses ATP directly, but there isn't enough of it to sustain muscle activity for more than a fraction of a second! The *immediate* regeneration of ATP must happen or we'd give out in an instant. To handle this task, muscles, when at rest, make and store a phosphate of even higher energy than ADP or ATP, namely, creatine phosphate. Then, as soon as some ATP is used, and ADP plus P_i are made, creatine phosphate regenerates ATP.

■ The enzyme is *creatine kinase,* or CK.

What actually happens is that an increase in the supply of ADP shifts the following equilibrium to the right to raise the concentration of ATP. The high value of K_{eq} tells us that the forward reaction is favored.

$$\text{Creatine phosphate } + \text{ ADP} \xrightleftharpoons{\text{creatine kinase, CK}} \text{creatine } + \text{ ATP} \qquad K_{eq} = 162$$

Then, during periods of rest, when the ATP level is substantially increased by other reactions, the equilibrium is forced back to the left to resupply the reserves of creatine phosphate. It's interesting that the formation of creatine phosphate depends on the synthesis of ATP and that the synthesis of ATP depends on creatine phosphate. What actually happens hinges on the above equilibrium and how it's made to shift according to the supply of ADP.

Although muscle tissue has three to four times as much creatine phosphate as ATP, even this reserve cannot supply the high-energy phosphate needs of muscle for more than a 100- to 200-meter sprint. For a long sustained period of work, the body must have other methods to make ATP. We'll now take an overview of all of the routes to ATP and then look closely at individual pathways.

All Bioenergetic Pathways Converge on the Citric Acid Cycle and the Respiratory Chain   Our first interest is in what *initiates* ATP synthesis. This is everywhere, not just in muscles, under feedback control, so when the supply of ATP is high, no more is made. Only as ATP is used up is first one mechanism and then another thrown

■ When the level of ATP drops and the levels of ADP + P_i rise, the rate of breathing is also accelerated.

into action. Generally, *it's an increase in the concentration of ADP that triggers the synthesis of ATP*. When ADP appears, having been formed by some energy-using function of the body, the machinery for reconverting it to ATP is thrown into action.

Of the several metabolic pathways that can generate ATP, the most productive is the respiratory chain (Figure 25.1). It uses chemicals it does not have, so as soon as it starts, another pathway is activated to make them. It's the **citric acid cycle**, second from the bottom in Figure 25.1. *The chief purpose of the citric acid cycle is to supply the chemical needs of the respiratory chain.*

The citric acid cycle cannot run either without a unit that it does not itself make, an acetyl group. This unit enters the cycle bound to an enzyme cofactor called coenzyme A. **Acetyl coenzyme A,** or acetyl CoA, is the "fuel" for the citric acid cycle. *The catabolism of molecules from all three major foods—carbohydrates, lipids, and proteins—produces acetyl coenzyme A.*

■ Hans Krebs won a share of the 1953 Nobel Prize in Physiology or Medicine for his work on the citric acid cycle.

Fatty acids are a major source of acetyl CoA, particularly during periods of sustained activities. A series of reactions called the **β-oxidation pathway** breaks fatty acids into acetyl units, as we'll study in Chapter 26.

Most amino acids can also be catabolized to acetyl units or to intermediates of the citric acid cycle itself, as we will study in Chapter 27.

Glycolysis is the chief route that starts the extraction of the chemical energy in carbohydrates for making ATP (see Figure 25.1). The end product of glycolysis is the pyruvate ion. Then a short pathway converts pyruvate to acetyl CoA, and more ATP is then made by oxidative phosphorylation.

Glycolysis Makes ATP Even When a Cell Is Deficient in Oxygen   Glycolysis also makes some ATP by substrate phosphorylation *under low oxygen conditions* and independently of the respiratory chain. Glycolysis is thus a safety valve, a backup source of ATP for cells (temporarily) running low on oxygen.

When glycolysis is run with insufficient oxygen, its end product is lactate ion, not pyruvate ion. Once the cell is resupplied with oxygen, however, lactate is converted back to pyruvate.

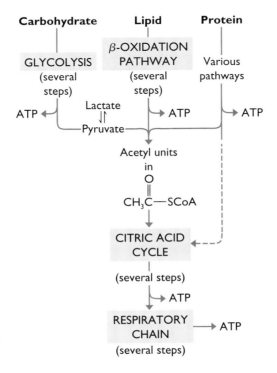

FIGURE 25.1   The major pathways for making ATP.

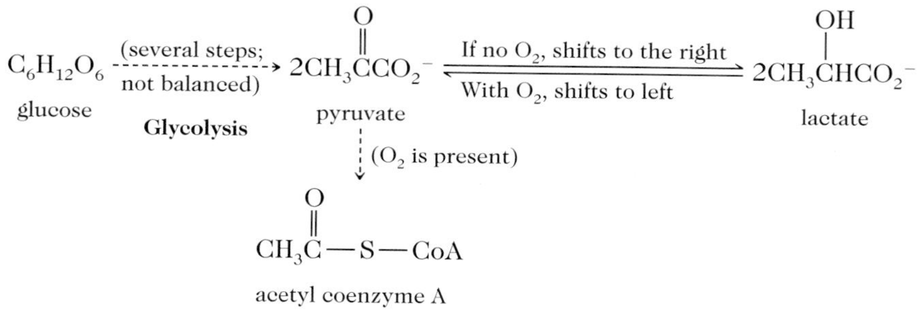

$$
\underset{\text{glucose}}{C_6H_{12}O_6} \xrightarrow[\substack{\text{(several steps;}\\\text{not balanced)}}]{} \underset{\substack{\text{pyruvate}}}{2CH_3\overset{\overset{\text{O}}{\|}}{C}CO_2^-} \underset{\text{With } O_2, \text{ shifts to left}}{\overset{\text{If no } O_2, \text{ shifts to the right}}{\rightleftharpoons}} \underset{\text{lactate}}{2CH_3\overset{\overset{\text{OH}}{|}}{C}HCO_2^-}
$$

Glycolysis

$\downarrow$ (O_2 is present)

$$
\underset{\text{acetyl coenzyme A}}{CH_3\overset{\overset{\text{O}}{\|}}{C}-S-CoA}
$$

Thus glycolysis can end either with lactate or with pyruvate, depending on the oxygen supply. We'll study the details of glycolysis in the next chapter.

The full sequence of oxygen consuming reactions from glucose to pyruvate ion to acetyl CoA and on through the citric acid cycle and the respiratory chain is the **aerobic sequence** of glucose catabolism. When glycolysis is run without oxygen, from glucose only as far as lactate ion, the series of reactions is called the **anaerobic sequence** of glucose catabolism.

The citric acid cycle and the respiratory chain are the pathways that we'll study next. They can accept breakdown products from *any* food.

■ *Aerobic* signifies the use of air. *Anaerobic,* stemming from "not air," means in the absence of the use of oxygen.

25.2 Citric Acid Cycle

Citrate ion is broken down to CO_2, bit by bit, as units of $(H:^- + H^+)$ are sent into the respiratory chain.

The *citric acid cycle*[2] is a series of reactions that dismantles acetyl groups, sending their two carbon atoms into molecules of CO_2, and their hydrogen, as $H:^-$ and H^+, into the respiratory chain. The hydrogens eventually end up in H_2O molecules, the oxygen coming from the air we breathe. The cycle occurs in the innermost part of a mitochondrion.

■ The mitochondrion is an organelle (small body) in the cytoplasm.

Coenzyme A Is the Common Carrier of Acetyl Units Before an acetyl group can enter the citric acid cycle, it must be attached to coenzyme A, CoASH. In acetyl CoA, the acetyl residue replaces the H on SH of CoASH.

■ Pantothenic acid is another of the B vitamins.

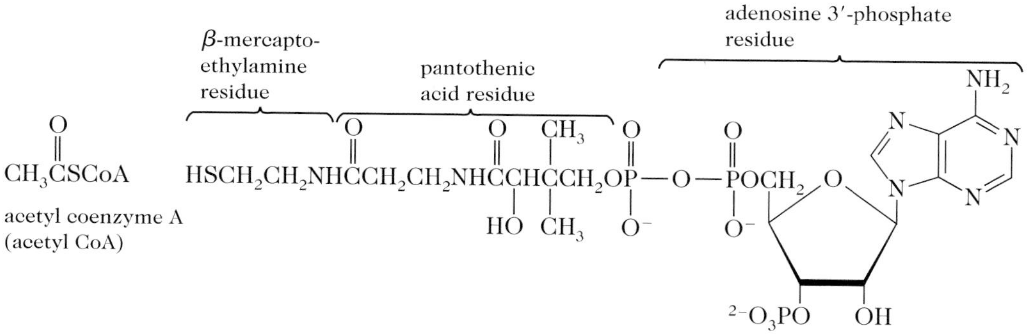

coenzyme A (CoASH)

---

[2] You should be aware that the citric acid cycle goes by two other names as well, namely, the *tricarboxylic acid cycle* and the *Krebs' cycle.* You might encounter these names in other references.

The conversion of pyruvate ion to acetyl coenzyme A involves both a decarboxylation and an oxidation. The overall equation for this complicated and irreversible change is as follows.

$$\underset{\text{pyruvate}}{CH_3\overset{\overset{\displaystyle O}{\|}}{C}CO_2^-} + \underset{\text{coenzyme A}}{HSCoA} + NAD^+ \xrightarrow{\text{5 steps}} \underset{\text{acetyl coenzyme A}}{CH_3\overset{\overset{\displaystyle O}{\|}}{C}SCoA} + CO_2 + NADH$$

ester   thioester

We leave the details to more advanced treatments, but the five steps are catalyzed by a large, multienzyme complex called *pyruvate dehydrogenase.* Three B vitamins—thiamine, nicotinic acid (niacin), and riboflavin—are among the compounds needed to make the coenzymes for the complex.

Acetyl coenzyme A is a *thioester,* which means an ester in which an oxygen atom has been replaced by a sulfur atom. A thioester happens to be far more reactive in transferring its acyl group than an ordinary ester. Acetyl CoA, therefore, is a particularly active acetyl group transfer agent. Such a transfer launches one turn of the citric acid cycle.

The Citric Acid Cycle Feeds the Pieces of H₂ to the Respiratory Enzymes The enzymes for the citric acid cycle exist in mitochondria in close proximity to those of the respiratory chain. In the cycle's first step, the acetyl group of acetyl CoA transfers to oxaloacetate ion, an ion that has two carboxylate groups and a keto group. The enzyme is *citrate synthase.* The acetyl unit *adds* across the keto group in a type of addition reaction that we have not studied before. (See Appendix D for the mechanism.)

oxaloacetate ion

The product of the addition is the citrate ion. Now the cycle itself begins (Figure 25.2). A series of reactions degrades the citrate ion, bit by bit, until another oxaloacetate ion is regenerated. The numbers of the following steps match those in Figure 25.2. Four steps involve a dehydrogenation, another name for oxidation. The dehydrogenation steps provide fuel for the respiratory chain and the synthesis of ATP. The other steps simply set molecules up for dehydrogenations. Many reactions are familiar from our study of organic compounds.

■ At physiological pH, the acids in the cycle exist as their anions.

■ The fluoroacetate ion, $FCH_2CO_2^-$, one of the most toxic of small-molecule poisons, inhibits aconitase.

1. **Dehydration** Citrate is dehydrated to give the double bond of *cis*-aconitate. This is the dehydration of an alcohol; the enzyme is *aconitase.*

2. **Hydration** Water adds to the double bond of *cis*-aconitate to give an isomer of citrate called isocitrate; aconitase also catalyzes this step.

    The net effect of steps 1 and 2 is to move the OH group in citrate to a neighboring carbon, changing the alcohol from tertiary to secondary, from one that cannot be oxidized (dehydrogenated) to one that can. (This addition of water defies Markovnikov's rule, but enzymes can handle such an assignment.)

3. **Dehydrogenation** The 2° OH group of isocitrate is dehydrogenated (oxidized) by *isocitrate dehydrogenase* to give the keto group of oxalosuccinate. The alcohol system gives up both H:⁻ and H⁺.

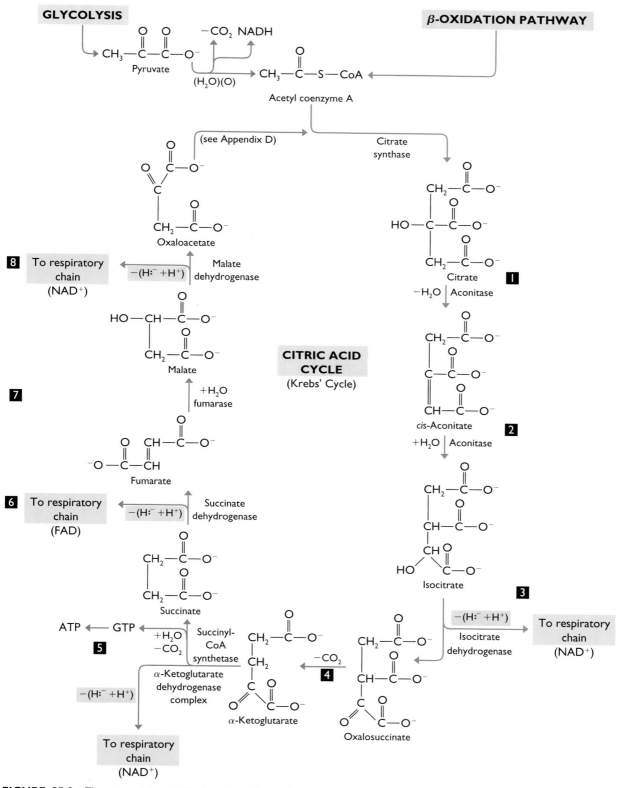

FIGURE 25.2 The citric acid cycle. The boxed numbers refer to the text discussion. The names of the enzymes for each step are given by the arrows.

$$
\begin{array}{c}
\text{(moved across inner} \\
\text{mitochondrial} \\
\text{membrane)}
\end{array}
$$

H $\cdots\to$ H$^+$

$$
\underset{\text{2° alcohol}}{\overset{\displaystyle\mathrm{O}}{\underset{\displaystyle\mathrm{H}}{\mathrm{RCR'}}}} \longrightarrow \overset{\displaystyle\overset{\mathrm{O}}{\|}}{\underset{\text{ketone}}{\mathrm{RCR'}}}
$$

H $\cdots\to$ H:$^-$ (accepted by NAD$^+$)

NAD$^+$ NADH

(reactions within
the respiratory chain)

An NAD$^+$ coenzyme accepts H:$^-$ while H$^+$ is forced across a membrane within the mitochondrion. The NAD$^+$ thereby changes to NAD:H (usually written simply as NADH). Note that the electron pair in H:$^-$ comes from a C—H bond, not a O—H bond. It's always the case; C—H bonds are storehouses of chemical energy far more than O—H bonds.

Overall, the fates of NADH and H$^+$ are intimately tied to the synthesis of three molecules of ATP by the respiratory chain, so this step is actually the first delivery of chemical energy from the citric acid cycle to a triphosphate.

4. **Decarboxylation** Oxalosuccinate, still bound to isocitrate dehydrogenase, loses a carboxyl group—it *decarboxylates*—to give α-ketoglutarate.

5. **Decarboxylative dehydrogenation** α-Ketoglutarate now undergoes a very complicated series of reactions, all catalyzed by one team of enzymes called *α-ketoglutarate dehydrogenase,* a system resembling pyruvate dehydrogenase mentioned earlier. A carboxyl group is lost and another dehydrogenation occurs. The same kind of fate occurs to H:$^-$ and H$^+$ here as happened in step 3, so three more ATPs can now be made by the respiratory chain.

   The product (not shown in Figure 25.2) is the coenzyme A derivative of succinic acid. It's converted by the action of the enzyme *succinyl-CoA synthetase* to the succinate ion, which is shown in Figure 25.2. The reaction also generates guanosine triphosphate, GTP, from its diphosphate. GTP is another high-energy triphosphate, mentioned earlier as being similar to ATP. But GTP is able to pin one of its phosphate units to ADP and so make one ATP. Thus the citric acid cycle includes one substrate phosphorylation.

6. **Dehydrogenation** Succinate donates hydrogen to FAD (not to NAD$^+$) in a reaction catalyzed by *fumarate dehydrogenase,* and the fumarate ion forms. *Both* H:$^-$ and H$^+$ are accepted by FAD, which becomes FADH$_2$. FADH$_2$ also intersects with the respiratory chain, and its further involvement in this chain produces two ATPs (not three, which are possible from NADH).

7. **Hydration** Fumarate adds water to its double bond, and malate forms, which has an oxidizable 2° alcohol group. The enzyme is *fumarase.*

8. **Dehydrogenation** The 2° alcohol group in malate is oxidized by *malate dehydrogenase* to make oxaloacetate in a reaction resembling step 3. NADH forms as the alcohol group gives up H:$^-$ to NAD$^+$. The chemical energy in NADH will be used to make three more ATP molecules by the respiratory chain.

One turn of the citric acid cycle is now complete and one molecule of the carrier, oxaloacetate, has been remade to enable another turn of the cycle. It can accept another acetyl group from acetyl coenzyme A.

■

succinate ion

■

fumarate ion
(trans geometry)

■

malate ion

25.3 RESPIRATORY CHAIN

Electron flow to oxygen in the respiratory chain creates a proton gradient in mitochondria that drives the synthesis of ATP.

The term *respiration* refers to more than just breathing. It includes all oxygen-consuming reactions in cells. We are now ready to learn specifically how oxygen is reduced to water and how the chemical energy of this event helps to drive the entire respiratory chain. The most basic statement we can write for what happens when water forms emphasizes how electrons and protons combine with an oxygen atom.

$$(\,\overset{..}{:}\,) \quad + 2H^+ \quad + \cdot \overset{..}{\underset{..}{O}} \cdot \quad \longrightarrow \quad H - \overset{..}{\underset{..}{O}} - H + \text{energy}$$

| pair of electrons | pair of protons | atom of oxygen | molecule of water |

Remember that when any particle, like the oxygen atom, gains electrons it is reduced, and the donor of electrons is oxidized. The electrons and protons for the reduction of oxygen come mostly from intermediates in the citric acid cycle. As you saw in the discussion of steps 3, 5, 6, and 8 of this cycle, the electrons of C—H bonds are used, not O—H bonds, and the hydride ion $H:^-$ is often (but not always) the vehicle for carrying the electrons from one species to another.

■ The gain of e^- or of e^- carriers such as $H:^-$ is *reduction*; the loss of e^- or of $H:^-$ is *oxidation*.

The Mitochondrion Is the Cell's "Powerhouse" The respiratory chain is one long series of oxidation–reduction reactions. The electron flow is down an energy hill all the way to oxygen, and it's irreversible. Other events simultaneously take place to make ATP from ADP and P_i.

The cell's principal site of respiratory chain activity and ATP synthesis, a *mitochondrion,* is often dubbed the powerhouse of the cell (Figure 25.3). Some tissues

■ A cell in the flight muscle of a wasp has about a million mitochondria.

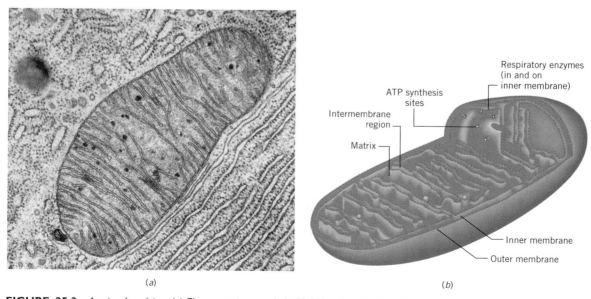

(a)   (b)

FIGURE 25.3 A mitochondrion. (*a*) Electron micrograph ($\times 53{,}000$) of a mitochondrion in a pancreas cell of a bat. (*b*) Perspective showing the interior. The respiratory enzymes are incorporated into the inner membrane. On the matrix side of this membrane are enzymes that catalyze the synthesis of ATP.

have thousands of mitochondria in the cytoplasm of a single cell. A mitochondrion has two important membranes, one outer and one inner. The inner membrane is very convoluted and has a surface area many times that of the outer membrane. The space between the two membranes is called the *intermembrane space.* The region deep inside the inner membrane is filled with a gel-like material called the *matrix.* It's less than 50% water and rich in soluble enzymes, cofactors, inorganic ions, and substrates.

The Respiratory Enzymes Are the Chief Agents of Electron Transfer   Built into the inner membrane itself are numerous immobile enzyme complexes designated as complexes I, II, III, and IV. Other parts of the inner membrane are enzymes capable of moving from one complex to another and named coenzyme Q (CoQ) and cytochrome c. Taken together, these are the **respiratory enzymes,** and they operate the respiratory chain.

The Operation of the Respiratory Chain Moves H$^+$ Ions from the Matrix to the Intermembrane Space   One important property of the inner membrane is that *only at particular channels is it permeable to chemical species, particularly ions like* H$^+$. The operation of the respiratory chain transfers H$^+$ ions from deep within the matrix to the intermembrane compartment. *This establishes an unstable proton gradient between the intermembrane region (more* H$^+$*) and the matrix (less* H$^+$*).* Eventually the fluid of the intermembrane space will have a pH 1.4 units lower (more acidic) than the matrix.

Remember, given the random motions of species in any fluid, that nature eventually destroys a gradient. However, the gradient-destroying flow of H$^+$ ion back into the matrix is permitted only at certain channels within the inner membrane. The protons, therefore, can return to the matrix only at certain locations, *places where the returning protons activate an enzyme system that makes and releases ATP.* This is basically how the cell connects reactions of the respiratory chain with the synthesis of ATP. An outline of this understanding was first formulated by English scientist Peter Mitchell, and it is called the **chemiosmotic theory.**

■ Peter Mitchell received the 1978 Nobel Prize in Chemistry.

One Entry to the Respiratory Chain Is the Synthesis of NADH   If we let *MH$_2$* represent any metabolite that can donate H:$^-$ to NAD$^+$, we can write the following equation for the dehydrogenation (oxidation) of a metabolite.

$$MH_2 + NAD^+ \longrightarrow M + NADH + H^+$$

An electron pair in *MH$_2$* has moved to NADH, which now interacts with enzyme complex I (Figure 25.4).

■ Think of *MH$_2$* as *M*:H which becomes *M* when H:$^-$ and H$^+$ leave it.

Enzyme Complex I Oxidizes NADH and Reduces CoQ   As indicated in Figure 25.4, enzyme complex I accepts an electron pair (carried by H:$^-$) from NADH. *The electron flow has now started toward oxygen.*

The acceptor of H:$^-$ in complex I is FMN. The hydride unit, H:$^-$, in NADH transfers to FMN according to the following equation.

$$NADH + H^+ + FMN \longrightarrow FMNH_2 + NAD^+$$

This restores the NAD$^+$ enzyme and moves the electron pair one more step along the respiratory chain.

What happens next is the transfer of just a pair of electrons, not H:$^-$, from FMNH$_2$ while the hydrogen *nucleus* of H:$^-$ leaves the carrier as H$^+$. *This transfer induces conformational changes in an inner membrane protein that cause protons to transfer from the matrix into the intermembrane region.* Thus the buildup of the H$^+$

■ FMN carries the B vitamin riboflavin as part of its coenzyme.

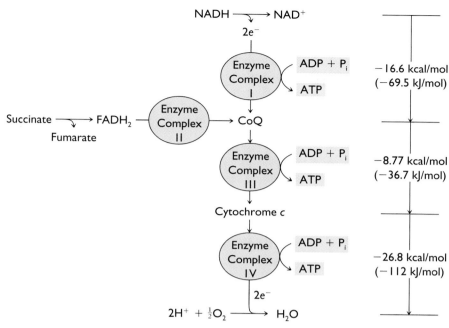

FIGURE 25.4 The enzyme complexes of the respiratory chain and the energy drop from NADH to reduced oxygen (water). The figures for energy refer to the portion of the energy change that is not unavoidably dissipated as heat. The symbols for the enzymes are as follows.

NADH = enzyme with reduced nicotinamide cofactor

FADH$_2$ = enzyme with reduced riboflavin cofactor

CoQ = enzyme with coenzyme Q cofactor

gradient across the inner mitochondrial membrane begins. The two electrons are accepted by an iron–sulfur protein, which we'll symbolize by FeS—P. The iron in FeS—P occurs as Fe^{3+}. By accepting one electron, it is reduced to the Fe^{2+} state, so we need *two* units of FeS—P to handle the pair of electrons now about to leave FMNH$_2$. This step can be written as

$$\text{FMNH}_2 + 2\text{FeS—P} \longrightarrow \text{FMN} + 2\text{FeS—P·} + 2\text{H}^+$$

where we use FeS—P· to represent the reduced form of the iron–sulfur protein in which Fe^{2+} occurs.

Before we go on, we will introduce a way to display these reactions that helps to emphasize the restoration of each enzyme to its original condition. For example, the last three reactions we have studied can be represented as follows.

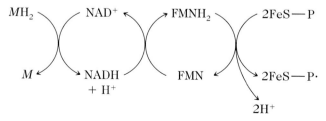

Will promote the migration of protons from the matrix to the intermembrane space

■ There are actually three types of iron–sulfur proteins having different atom ratios of iron to sulfur. The actual net charges vary.

Let's read this from left to right. At the point where the first pair of curved arrows touch, we see that as MH_2 changes to M, it passes $H\!:\!^-$ to NAD^+, changing it to NADH. H^+ is also released. The touching of the second pair of curved arrows tells us that NADH and H^+ are changed back to NAD^+ as the system passes one unit of $H\!:\!^-$ plus H^+ across to FMN to change it to $FMNH_2$. At the third pair of curved arrows, the display shows $FMNH_2$ changing back to FMN as it passes two electrons to two FeS—P molecules and also as it expels two protons (H^+). The last work of enzyme complex I is to pass electrons from the reduced iron–sulfur complex to coenzyme Q.

For each pair of H^+ ions released by respiratory chain oxidation an average of three to four protons are moved out of the matrix, across the inner membrane, and into the intermembrane space.

We've looked at the steps so far in much detail to show that when spontaneous chemical reactions (of the chain) occur in the right context (a mitochondrial membrane), a proton gradient is forced into existence. The proton gradient is now, in effect, the holder of chemical energy for making ATP. We'll now drastically reduce the attention to specific detail.

Respiratory Enzyme Complexes III and IV Continue the Respiratory Chain   The respiratory chain now becomes even more complicated, but the overall result is not hard to understand. Let's look at Figure 25.5, showing the main branch of the respiratory chain and the flow of electrons through complexes I, III, and IV. What happens can be expressed by the following overall equation. The action begins in complex I, starting with MH_2 and its reaction with NAD^+, discussed above.

$$\underbrace{MH_2 + nH^+ + \tfrac{1}{2}O_2}_{\substack{\text{in the matrix (inside} \\ \text{the inner membrane)}}} \xrightarrow[\substack{\text{respiratory chain} \\ \text{(parts of the inner} \\ \text{membrane)}}]{\substack{\text{enzyme complexes} \\ \text{I, II, and IV}}} M + H_2O + \underbrace{nH^+}_{\substack{\text{These protons are} \\ \text{released into the inter-} \\ \text{membrane space.}}}$$

■ *Cyto-*, cell; *-chrome*, pigment. Cytochromes are colored compounds.

The electrons move from complex I into a series of *cytochrome* enzymes designated, in the order in which they participate, as cytochromes b, c_1, and c. Together with an

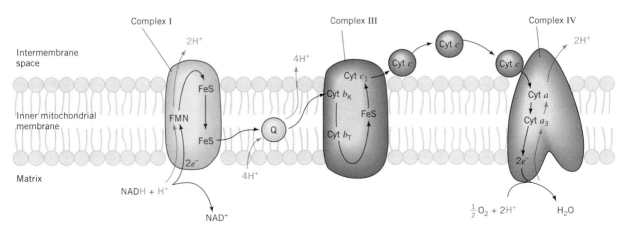

FIGURE 25.5   The electron flow (black) and the proton transfers (red) from NADH through enzyme complexes I, III, and IV of the respiratory chain. Q represents coenzyme Q, CoQ. Cyt c is cytochrome c; etc. Complex II is not shown, but it transfers electrons from succinate ion to coenzyme Q. (Reproduced by permission from D. Voet and J. G. Voet, *Biochemistry,* 2nd edition. New York, John Wiley & Sons, Inc., 1996.)

iron—sulfur protein, b and c_1 make up enzyme complex III. Moving through cytochrome c, the electrons enter cytochromes a and a_3, parts of enzyme complex IV and often called *cytochrome c oxidase.*

Complex IV is the complex that catalyzes the reduction of oxygen to water, which gives the electrons their final home. Complex IV carries a copper ion that alternates between the Cu^{2+} and the Cu^{+} states.

Complex II, indicated in Figure 25.4 (but not in Figure 25.5), is just another way to enter the respiratory chain at coenzyme Q. The succinate ion of step 6 of the citric acid cycle is one substrate for complex II.

■ Complex II includes the riboflavin-based coenzyme, FAD.

Two Gradients Are Established, a Proton and a Plus Charge Gradient We don't want all of the above details to obscure a vital result. The operation of the respiratory chain pumps protons from the mitochondrial matrix to the intermembrane region *without also putting the equivalent of negatively charged ions there as well.* Thus, *two* gradients are set up, one of H^{+} ions and one of positive charge. The latter could be erased either by the migration of negative ions to the intermembrane region *or by the migration of any kind of positive ion from this region back into the matrix.* Calcium ions, for example, might move, *and precisely such movements of* Ca^{2+} *ions are intimately involved with nerve signals and muscular work.* Thus the chemiosmotic theory also helps us understand how the catabolism of energy-rich nutrients can force the movements of any kind of ion, including those involved in nerve tissue or muscle contraction.

■ When a gradient of electrical charge is established by chemical species, it is called an *electrochemical gradient.*

■ Because chemical reactions create the gradients, we have the *chemi-* part of the term *chemiosmotic.*

Proton Channels of the Inner Mitochondrial Membrane Are Part of an Enzyme for Making ATP from ADP and P_i Embedded in the inner mitochondrial membrane are complexes of proteins that form a tube through it. At least one tube exists for every unit of respiratory enzymes. The tube itself is called the F_0 component of a complex enzyme called **proton-pumping ATPase.** Another component, the F_1 component, occurs where the proton channel ends on the matrix side of the inner membrane. The F_1 components project like lollipops from the matrix-side surface (Figure 25.6). The F_1 components consist of nine polypeptides, one of which is the proton gate and three of which are the actual sites where ADP and P_i are put together to make ATP.

In a theory proposed by Paul Boyer, the linkage between the flow of protons through F_0 and the synthesis and release of ATP from F_1 is made by a series of steps involving allosteric interactions (Figure 25.7). Each of the ATP-synthesizing F_1 subunits is capable of existing in any one of three configurations, namely, open (O), loosely closed (L), and tightly closed (T). The O configuration has little ability to bind substrates, whether ADP plus P_i or ATP. The L configuration can bind ADP and P_i *but can do nothing catalytically to change them to ATP until protons arrive to cause a change in the configuration of the protein from the L to the T state.* Now the formation of ATP is catalyzed, *but it is tightly held to the T-state subunit.* ATP cannot be released until the proton gradient builds up sufficiently to drive protons through the F_0 channel and into F_1. The arriving protons cause conformational changes in all of the F_1 subunits. The T subunit with its bound ATP changes into an O subunit, which permits the ATP to drop off. The L subunit (with loosely bound ADP plus P_i) changes to the T form as a result of which another ATP is put together to await release by further surges of protons.

It is because a series of *oxidations* creates the proton gradient that the overall synthesis is called **oxidative phosphorylation,** or sometimes *respiratory chain phosphorylation.* Each molecule of NADH that enters the chain at enzyme complex I can lead to a maximum of three ATPs. Each $FADH_2$ that enters at complex II can cause just two ATP molecules to form. Let's now go back over the major aspects of oxidative phosphorylation.

■ The proton-pumping ATPase is also called the *proton-translocating ATPase.*

■ This migration through a semipermeable membrane explains the *osmotic* part of the term *chemiosmotic.*

(a)

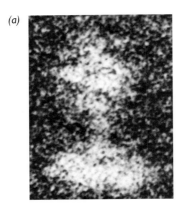

(b)

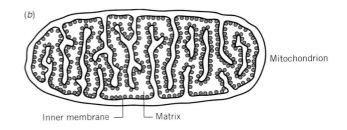

Mitochondrion

Inner membrane ⎯⏋   ⎿⎯ Matrix

(c)

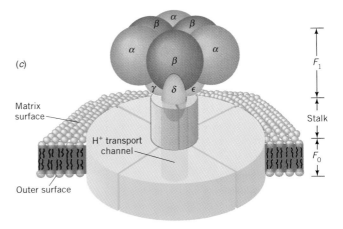

FIGURE 25.6   Proton-pumping ATPase. (a) Electron micrograph of an intact inner mitochondrial membrane showing the "lollipop" projections of the enzyme on the matrix side. [From Parsons, D. F., *Science* **140**, 985 (1963). Copyright © 1963 American Association for the Advancement of Science. Used by permission.] (b) Interpretive drawing of the mitochondrion with the same projections shown in part (a). (c) Interpretive drawing of one proton-pumping ATPase complex. The unit labeled δ is the proton gate between the F_0 proton channel (beneath) and the other subunits of the F_1 complex (above). The three subunits labeled β are ATP synthesis sites. (Adapted with permission from D. Voet and J. G. Voet, *Biochemistry*, 2nd edition. New York, John Wiley & Sons, Inc., 1996.)

Oxidative Phosphorylation

1. The synthesis of ATP occurs at an enzyme located on the matrix side of the inner mitochondrial membrane.

2. ATP synthesis is driven by a flow of protons that occurs from the intermembrane side to the matrix side of the inner membrane.

3. The flow of protons is through special channels in the membrane and down a concentration gradient of protons that exists across the inner membrane.

4. The energy that creates the proton gradient and the (+) charge gradient is provided by the flow of electrons between the respiratory chain enzymes that make up integral packages of the inner membrane.

5. The inner mitochondrial membrane is a closed envelope except for the special channels for the flow of protons and for special transport systems that let needed solutes move into or out of the innermost mitochondrial compartment.

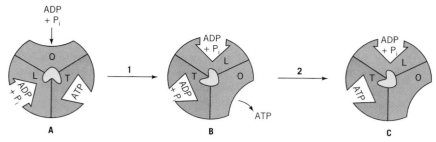

FIGURE 25.7 The synthesis and release of ATP from subunits of the F_i portion of the proton-pumping ATPase complex. Each of the three β-subunits shown in Figure 25.6 is capable of being in an O (open), L (loose), or T (tight) configuration. In set A, an ATP molecule, made in the previous cycle, is tightly held in a subunit in its T configuration. The subunit labeled L is loosely holding a molecule of ADP and P_i. The third subunit, in the O configuration, accepts ADP plus P_i. In step **1**, the central unit (gold) now changes its configuration in response to a proton flow. This forces subunits to change shape as the B form emerges. The shape of the ATP-binding subunit (T) changes to the open shape, and ATP is released. At the same time, what was an L-shaped subunit in A, and loosely holding ADP plus P_i, changes shape to T in B and, in step **2**, this forces ADP and P_i to form ATP (but it's still held tightly). C forms, which offers another open subunit that can accept another ADP–P_i pair. Another flow of protons will cause the cycle to repeat and release more ATP. (Adapted by permission from R. L. Cross, *Nature*, August 25, 1994, page 594. All rights reserved.)

A Transport Protein in the Mitochondrial Membrane Moves ADP Inside as It Carries ATP Out Both ATP and ADP are highly charged species and so cannot easily get through a lipid bilayer membrane. A transport protein of an inner mitochondrial membrane "pump" called *ATP–ADP translocase* solves this problem.

The migration of newly made ATP to the intermembrane region and thence to the cytosol is coupled to the movement of ADP in the opposite direction. The expression "is coupled" means that the membrane pump that moves ADP in one direction (in) simultaneously moves ATP in the other (out).

The Rate of ATP Resynthesis Is Sensitive to the Molar Concentrations of ADP and ATP With considerable simplification of a complex phenomenon, it is as if ADP, P_i, and ATP are involved in the following equilibrium.

$$ADP + P_i \rightleftharpoons ATP \qquad (25.1)$$

When the concentrations of ADP and P_i are high, stress thereby exists that causes ATP to be made; Equilibrium 25.1 shifts to the right, the direction predicted by Le Châtelier's principle. When the supply of ATP is relatively high, on the other hand, little more is (or need be) made. During our quiet times, when the least amounts of ADP and P_i are being made, the machinery for making ATP operates slowly, enough to maintain basal metabolism. Then, when our lives become more active, the rate of use of ATP accelerates, and ADP plus P_i are made. Their appearance shifts Equilibrium 25.1 to the right thus using them up.

Some Antibiotics and Poisons Inhibit Oxidative Phosphorylation One of the barbiturates, amytal sodium, and the powerful insecticide, rotenone, block the respiratory chain by interfering at enzyme complex I. The antibiotic antimycin A stops the chain at complex III. The cyanide ion blocks the chain at its very end, at cytochrome a_3 in complex IV. These agents, by inhibiting the chain, work to inhibit the creation of the proton gradient and in this manner inhibit ATP synthesis.

■ Rotenone is a naturally occurring insecticide.
■ Remember that the respiratory chain also occurs in bacteria, so antibiotics that interfere with the chain can kill bacteria.

■ A natural and healthy mechanism for uncoupling the chain from ATP synthesis exists in the heat-generating brown fat tissue that cold-adapted animals have.

Other substances interfere with the *use* of the gradient. They do not prevent the respiratory chain from operating, but they cancel the effect of the gradient. They *uncouple* the chain from ATP synthesis. For example, 2,4-dinitrophenol increases the permeability of the inner membrane to protons so that they can use random routes to reenter the matrix. When the gradient is destroyed without the use of the F_0 proton channel, little if any ATP synthesis can occur. The chemical energy released by the operation of the chain is then converted to heat.

Oxidative Phosphorylation Makes a Maximum of Twelve ATP Molecules Using Intermediates of the Citric Acid Cycle   The data in Table 25.2 show where ATP is generated as the citric acid cycle runs. A maximum of 12 ATPs can be made this way, and if we add the 3 ATPs made possible by the conversion of pyruvate to acetyl CoA, then a maximum of 15 ATPs can be made by the degradation of one pyruvate ion. Remember that the energy of the respiratory chain is used for other operations besides making ATP, for example, the operation of nerves. Thus Table 25.2 has to be viewed as giving the upper limits to ATP production by the respiratory chain, starting from pyruvate.

TABLE 25.2   ATP Production by Oxidative Phosphorylation

Step	Receiver of $(H:^- + H^+)$ in the Respiratory Chain	Molecules of ATP Formed
Isocitrate $\rightarrow$ α-ketoglutarate	NAD^+	3
α-Ketoglutarate $\rightarrow$ succinyl CoA	NAD^+	3
Succinyl CoA $\rightarrow$ succinate (via GTP)	—	1
Succinate $\rightarrow$ fumarate	FAD	2
Malate $\rightarrow$ oxaloacetate	NAD^+	3
Total ATP via citric acid cycle		12
Pyruvate $\rightarrow$ acetyl CoA	NAD^+	3
Total ATP from pyruvate via the citric acid cycle		15

SUMMARY

High-energy compounds   Organophosphates whose phosphate group transfer potentials equal or are higher than that of ATP are classified as high-energy phosphates. A lower energy phosphate can be made by phosphate transfer from a higher energy phosphate, a process called substrate phosphorylation.

Citric acid cycle   Acetyl groups from acetyl coenzyme A are joined to a four-carbon carrier, oxaloacetate, to make citrate. This six-carbon salt of a tri-carboxylic acid then is degraded bit by bit as pieces of hydrogen, $(H:^- + H^+)$, are fed to the respiratory chain. Each acetyl unit leads to a maximum of 12 ATP molecules.

Fatty acids and glucose are important suppliers of acetyl groups for the citric acid cycle. In aerobic glycolysis, glucose units are broken to pyruvate units. The ox-

idative decarboxylation of pyruvate leads to the synthesis of one NADH (which eventually leads to the formation of 3 ATPs), and it supplies acetyl units to the citric acid cycle from which 12 more ATPs are made.

Respiratory chain   A series of electron transfer enzymes called the respiratory enzymes occur together as groups called respiratory assemblies in the inner membranes of mitochondria. These enzymes process NADH or $FADH_2$, made by reduction reactions from metabolites (MH_2) and NAD^+ or FAD. NADH and $FADH_2$ then pass on electrons until cytochrome oxidase uses them (together with H^+) to reduce oxygen to water.

Oxidative phosphorylation   As electrons flow from NADH to oxygen in the respiratory chain, protons are released.

They cause conformational changes in proteins imbedded in the inner membrane of a mitochondrion. These changes cause protons to move from the mitochondrial matrix to the intermembrane side of the inner membrane, making a proton gradient and a positive charge gradient across the inner membrane. As protons flow back at the allowed F_0 conduits of the proton-pumping ATPase complex of this membrane, ATP is made and released by the F_1 complex of

this enzyme. In some systems, other kinds of positive ions migrate, as in the operation of nerves. Various drugs and antibiotics can block the respiratory chain. Other chemicals are able to uncouple the work of the respiratory chain from the synthesis of ATP so that the operation of the chain converts chemical energy to heat. The rate at which ATP is made increases as the concentration of ADP increases.

REVIEW EXERCISES

The answers to Review Exercises whose numbers are in color are found in Appendix E. The answers to the other Review Exercises are found in the Study Guide that accompanies this book. The more challenging questions are marked with asterisks.

Energy Sources

25.1 What products of the digestion of carbohydrates, triacylglycerols, and the polypeptides can be used as sources of biochemical energy to make ATP?

25.2 The complete catabolism of glucose gives what?

25.3 The identical products form when glucose is burned in open air as when it is fully catabolized in the body. How, then, do these two processes differ overall?

25.4 What are the end products of the complete catabolism of fatty acids?

High-energy Phosphates

25.5 Complete the following structure of ATP.

$$\text{Adenosine—O—}\overset{\overset{\textstyle O}{\|}}{\underset{\underset{\textstyle O^-}{|}}{P}}\text{—}$$

25.6 Write the structures of ADP and of AMP in the manner started by Review Exercise 25.5.

25.7 At physiological pH, what does the term *inorganic phosphate* stand for? (Give formulas and names.)

25.8 Phosphate X has a phosphate group transfer potential of -13 kcal/mol. What does this quantity refer to?

25.9 Why are creatine phosphate and ATP called high-energy phosphates but glycerol-3-phosphate is not?

*25.10 If the phosphate of M has a *higher* (more negative) phosphate group transfer potential than that of N, in which direction does the following reaction tend to go spontaneously, to the left or to the right?

$$M\text{—OPO}_3{}^{2-} + N \overset{?}{\rightleftharpoons} N\text{—OPO}_3{}^{2-} + M$$

25.11 Using data in Table 25.1, tell (yes or no) whether ATP readily transfers a phosphate group to each of the following possible compounds.
(a) Pyruvate (enol form)   (b) Fructose
(c) Glyceric acid   (d) Acetic acid

25.12 All of the possible phosphate transfers in Review Exercise 25.11 are classified as *substrate* phosphorylations. What does this mean?

25.13 What does creatine phosphate do in muscles?

25.14 Whether or not creatine phosphate is used in muscle tissue is under *feedback control*. Explain.

Overview of Metabolic Pathways

25.15 In the general area of biochemical energetics, what is the purpose of each of the following pathways?
(a) Respiratory chain   (b) Anaerobic glycolysis
(c) Citric acid cycle   (d) Fatty acid cycle

25.16 What prompts the respiratory chain to start?

25.17 Which tends to increase the rate of breathing, an increase in the body's supply of ATP or an increase in its supply of ADP? Explain.

25.18 In general terms, the intermediates that send electrons down the respiratory chain come from what metabolic pathway that consumes acetyl groups?

25.19 Arrange the following sets of terms in sequence in the order in which they occur or take place. Place the identifying letter of the first sequence of a set to occur on the left of the row of letters.

(a) Citric acid cycle   Glycolysis   Acetyl CoA
      A   B   C
   Respiratory chain   Pyruvate
      D   E

(b) Citric acid cycle   Respiratory chain   Acetyl CoA
      A   B   C
   β-Oxidation
      D

25.20 The *aerobic sequence* begins with what metabolic pathway and ends with which pathway?

25.21 The β-oxidation pathway occurs to what kind of compound?

25.22 The anaerobic sequence begins and ends with what compounds?

25.23 What does it mean for the cell that anaerobic glycolysis is a "backup"?

Citric Acid Cycle

25.24 What makes the citric acid cycle start up?

25.25 What chemical unit is degraded by the citric acid cycle? Give its name and structure.

25.26 How many times is a secondary alcohol group oxidized in the citric acid cycle?

25.27 Water adds to a carbon–carbon double bond how many times in one turn of the citric acid cycle?

*25.28 The enzyme for the conversion of isocitrate to oxalosuccinate (Figure 25.2) is stimulated by one of these two substances, ATP or ADP. Which one is the more likely activator? Explain.

25.29 What is the maximum number of ATP molecules that can be made from the use of respiratory chain phosphorylation to break down each of the following?
(a) Pyruvate   (b) An acetyl group in acetyl CoA

*25.30 Glutamic acid, one of the amino acids, can be converted to α-ketoglutarate (Figure 25.2). How many ATP molecules can be made from the entry of α-ketoglutarate into the citric acid cycle?

Respiratory Chain

25.31 Is the following (unbalanced) change a reduction or an oxidation? How can you tell?

$$\underset{\overset{|}{OH}}{CH_3CHCH_2CO_2^-} \longrightarrow \underset{\overset{\|}{O}}{CH_3CCH_2CO_2^-}$$

25.32 Which kind of enzyme would be more likely to cause the change given in Review Exercise 25.31, an enzyme like aconitase or an enzyme like isocitrate dehydrogenase? Explain.

25.33 What is missing in the following basic expression for what must happen in the respiratory chain?

$$\tfrac{1}{2}O_2 + 2H^+ \longrightarrow H_2O$$

25.34 What general name is given to the set of enzymes involved in electron transport?

*25.35 Rewrite the following as an equation.

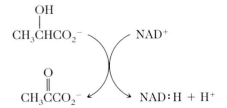

(a) Which species is oxidized? (Write its structure.)
(b) Which species is reduced?

*25.36 Write the following equation in the form of a display like that shown in Review Exercise 25.35.

$$^-O_2CCH_2CH_2CO_2^- + FAD \longrightarrow$$
$$^-O_2CCH{=}CHCO_2^- + FADH_2$$

25.37 Arrange the following in the order in which they receive and pass on electrons.

$$FMN \quad NAD^+ \quad CoQ \quad FeS{-}P$$

25.38 What does respiratory enzyme complex IV do?

25.39 What is FAD, and where is it involved in the respiratory chain?

25.40 Across which cellular membrane does the respiratory chain establish a gradient of H^+ ions? On which side of this membrane is the value of the pH lower?

25.41 According to the chemiosmotic theory, the flow of what particles most directly leads to the synthesis of ATP?

25.42 If the inner mitochondrial membrane is broken, the respiratory chain can still operate, but the phosphorylation of ADP that normally results stops. Explain.

25.43 Complete and balance the following equation. (Use 1/2 as the coefficient of oxygen as shown.)

$$MH_2 + H^+ \; + \; \tfrac{1}{2}O_2 \longrightarrow$$
$$\text{from the}$$
$$\text{matrix}$$

25.44 What is the difference between substrate and oxidative phosphorylation?

25.45 Besides a gradient of H^+ ions, what other gradient exists in mitochondria as a result of the operation of the respiratory chain? Of what significance is this gradient?

25.46 Briefly describe the theory that explains how a flow of protons across the inner mitochondrial membrane initiates the synthesis of ATP from ADP and P_i.

Additional Exercise

*25.47 The conversion of pyruvate to an acetyl unit is both an oxidation and a decarboxylation.
(a) If *only* decarboxylation occurred, what would form? Write the structure of the other product in the following.

$$\underset{\overset{\|}{O}}{CH_3CCO_2^-} + H^+ \longrightarrow \underline{\hspace{2cm}} + O{=}C{=}O$$

(b) If this product is oxidized, what is the name and the structure of the product of such oxidation?
(c) Referring to Figure 25.2, which specific compound undergoes an oxidative decarboxylation similar to that of pyruvate? (Give its name.)

METABOLISM OF CARBOHYDRATES

26

Humans and hummingbirds alike tap into the chemical energy in sugar. We'll study the metabolism of sugar and other carbohydrates in this chapter.

THIS CHAPTER IN CONTEXT

This chapter is almost entirely about the metabolism of glucose, beginning with information about circulating glucose. We'll learn about hormones such as insulin and glucagon, and that bad things result when their work goes awry. We will go into details about the catabolism of glucose both by glycolysis and by the pentose phosphate pathway. Finally, we'll learn how glucose is synthesized in the body from smaller molecules, which is a good thing for the brain.

26.1 GLYCOGEN METABOLISM

Much of the body's control over the blood sugar level is handled by the regulation of the synthesis and breakdown of glycogen.

The digestion of starch and disaccharides gives glucose, fructose, and galactose, but their catabolic pathways all converge very quickly to that of glucose itself. Galactose, for example, is changed by a few steps in the liver to glucose-1-phosphate. Fructose is changed to a compound that occurs early in glycolysis.

A Special Vocabulary Is Used to Describe the Blood Sugar Level Glucose is essential to life, particularly to the work of the nervous system. Yet both too little or too much in circulation has serious results. Specialists in diabetes even consider it a toxic compound when there is too much. No wonder that a special vocabulary has sprung up to describe its level in blood.

The concentration of monosaccharides in whole blood, expressed in milligrams per deciliter (mg/dL), is called the **blood sugar level.** This is very nearly the same as the glucose level, because glucose is overwhelmingly the major monosaccharide in blood. After several hours of fasting, the plasma sugar level, now called the **normal fasting level,** is 70 to 110 mg/dL (3.9 to 6.1 mmol/L).[1] In **hypoglycemia,** the plasma sugar level is *below* normal, and in **hyperglycemia** it is *above* the normal fasting level. **Normoglycemia** is the term used for blood sugar levels in the normal range.

When the blood sugar level becomes too hyperglycemic (becomes too high), the kidneys are unable to put back into the blood all of the glucose that leaves them in their filtering units. Glucose then appears in the urine, a condition called **glucosuria.** The blood sugar level above which this happens is called the **renal threshold** for glucose, and it is in the range of roughly 160 to 180 mg/dL, higher in some individuals.

Excess Blood Glucose Normally Is Withdrawn from Circulation When more than enough glucose is in circulation to meet energy needs, the body does not eliminate the excess but conserves its chemical energy. Two pathways handle this. One converts glucose to fat (next chapter). The other synthesizes glycogen from glucose by a series of steps, largely in the liver and muscles, called **glycogenesis** ("glycogen creation"). The liver holds 70 to 110 g of glycogen, and the muscles, taken as a whole, contain 170 to 250 g.

When muscles need glucose, they take it back out of their own glycogen. When the blood needs glucose, the liver hydrolyzes as much of its glycogen reserves as

■ -glyc-, sugar
-emia, in blood
hypo-, under, below
hyper-, above, over
renal, of the kidneys
-uria, in urine

■ Glycogen, recall, is a polymer of glucose with both $\alpha(1 \rightarrow 4)$ and $\alpha(1 \rightarrow 6)$ bridges and resembles amylopectin, a component of starch.

[1] The normal reference laboratory values or "normals" in this text are those used by the Massachusetts General Hospital and published in *The New England Journal of Medicine,* Vol. 327, September 3, 1992, page 718.

needed, the resulting glucose going into circulation. The overall series of reactions in either tissue that hydrolyzes glycogen is called **glycogenolysis** (*lysis* or hydrolysis of glycogen), a process controlled by several hormones.

Epinephrine Launches a Multiple Enzyme Glycogenolysis Cascade   When muscular work starts, the adrenal medulla secretes **epinephrine,** a hormone that initiates glycogenolysis. At its target cells, epinephrine activates the enzyme *adenylyl cyclase,* which catalyzes the conversion of some ATP to cyclic AMP.

Cyclic AMP then activates another enzyme, and this still another, and so on in a cascade of events (Figure 26.1). The effect of *each* molecule of epinephrine is multiplied by every step so that thousands of glucose units are mobilized.

While the epinephrine "cascade" releases glucose from glycogen, the affected tissue simultaneously shuts down a team of enzymes called *glycogen synthetase* that otherwise would do the opposite, change glucose to glycogen. This occurs in Figure 26.1 where the enzyme called *active protein kinase* catalyzes the phosphorylation of *two* enzymes. One is inactive phosphorylase kinase and the other is active glycogen synthetase. The first action continues the epinephrine cascade, releasing glucose units from glycogen. The second action *shuts down an enzyme that would help to remake glycogen.* The potential competition, therefore, cannot develop. It's a remarkable aspect of this system.

The end product of glycogenolysis isn't actually glucose but glucose-1-phosphate. Cells that can do glycogenolysis also have an enzyme called *phosphoglucomutase,* which catalyzes the conversion of glucose-1-phosphate to its isomer, glucose-6-phosphate.

■ Epinephrine (adrenaline) and norepinephrine (noradrenaline) are called the "fight or flight" hormones because the brain causes their release in sudden emergencies.

■ An estimated 30,000 molecules of glucose are released from glycogen for each molecule of epinephrine that initiates glycogenolysis.

■ American biochemists Edwin Krebs and Edmond Fischer shared the 1992 Nobel Prize in Physiology or Medicine for their discovery of kinases.

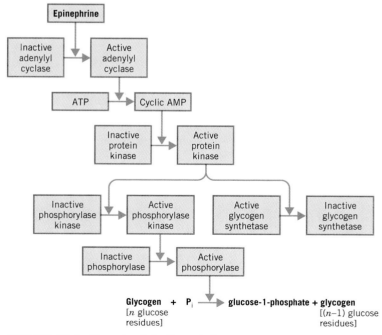

FIGURE 26.1   The epinephrine "cascade."

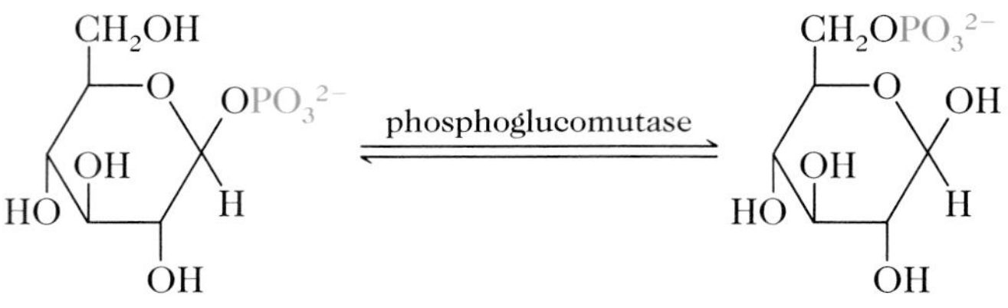

glucose-1-phosphate   glucose-6-phosphate

Glucose-6-Phosphate Traps Glucose in Muscle Cells   Glucose-6-phosphate, rather than glucose, is the form in which a glucose unit must be to enter a pathway that produces ATP. *It is also in a form that cannot migrate out of muscle cells.* Glucose cannot be lost, therefore, from muscle tissue needing it during exercise. When the supply of muscle glycogen is low, muscle cells take glucose from circulation, *trap it as glucose-6-phosphate,* and then convert this to glycogen to resupply their glycogen reserves.

Glucagon Activates Liver Glycogenolysis   The α-cells of the pancreas make a polypeptide hormone, **glucagon,** which helps to maintain a normal blood sugar level. When the blood sugar level drops, the α-cells release glucagon. Its target tissue is the liver, where it is a strong activator of glycogenolysis. Glucagon is thus an important regulator of the blood sugar level.

■ The glucagon molecule has 29 amino acid residues.

Glucagon initiates a cascade similar to that begun by epinephrine. Like epinephrine, glucagon activates adenylyl cyclase. Unlike epinephrine, glucagon *inhibits* glycolysis, which helps to maintain the glucose supply by not letting it be used for energy. Glucagon, also unlike epinephrine, does not cause an increase in blood pressure or pulse rate, and it is longer acting than epinephrine.

Liver Cells Can Release Glucose to Circulation   In the liver, glucose-6-phosphate (glucose-6-P), made by glycogenolysis, is hydrolyzed to glucose and inorganic phosphate.

■ The letter P is often used to represent the whole phosphate group in the structures of phosphate ester intermediates.

$$\text{glucose-6-P} + \text{H}_2\text{O} \xrightarrow{\text{glucose-6-phosphatase}} \text{glucose} + \text{P}_i$$

Glucose, now free, can leave the liver and so help increase the blood sugar level. During periods of fasting, the liver is actually a major supplier of glucose for the blood.

Because the brain depends on blood glucose for energy, the above changes in the liver help maintain this supply. When taken up, glucose is promptly trapped in a brain cell as glucose-6-phosphate.

Several hereditary diseases involve the glucose–glycogen interconversion, and some are discussed in Interaction 26.1.

Human Growth Hormone Stimulates the Release of Glucagon   Growth requires energy, so the action of glucagon (which helps to supply blood sugar) assists the work of human growth hormone. In some situations, such as a disfiguring condition known as *acromegaly,* there is an excessive secretion of human growth hormone that promotes too high a level of glucose in the blood. This is undesirable because a prolonged state of hyperglycemia from any cause can lead to some of the same blood capillary–related complications observed when diabetes is poorly controlled.

■ Acromegaly is sometimes called *giantism* because the bone structures of victims are enlarged.

Insulin Strongly Lowers the Blood Sugar Level   The β-cells of the pancreas make and release **insulin,** a polypeptide hormone. Its release is stimulated by an increase

INTERACTION 26.1
GLYCOGEN STORAGE DISEASES

A number of inherited diseases involve the storage of glycogen. For example, in **Von Gierke's disease,** the liver lacks the enzyme glucose-6-phosphatase, which catalyzes the hydrolysis of glucose-6-phosphate. Unless this hydrolysis occurs, glucose units cannot leave the liver. They remain as glycogen in such quantities that the liver becomes very large. At the same time, the blood sugar level falls, the catabolism of glucose accelerates, and the liver releases more and more pyruvate and lactate.

In **Cori's disease,** the liver lacks an enzyme needed to catalyze the hydrolysis of 1,6-glycosidic bonds, the bonds that give rise to the many branches

of a glycogen molecule. Without this enzyme, only a partial utilization of the glucose in glycogen is possible. The clinical symptoms resemble those of Von Gierke's disease, but they are less severe.

In **McArdle's disease,** the muscles lack phosphorylase, the enzyme needed to obtain glucose-1-phosphate from glycogen (see Figure 26.6). Although the individual is not capable of much physical activity, physical development is otherwise relatively normal.

In **Andersen's disease,** both the liver and the spleen lack the enzyme for putting together the branches in glycogen. Liver failure from cirrhosis usually causes death by age 2.

---

in the blood sugar level, which normally occurs after a carbohydrate-rich meal. As insulin moves into action, it finds its receptors at the membranes of muscle and adipose tissue cells. The insulin–receptor complexes make it possible for glucose molecules to move easily into the cells, and this lowers the blood sugar level.

In one form of diabetes mellitus, type I or **insulin-dependent diabetes mellitus** (IDDM), the β-cells have been destroyed, and the pancreas is unable to make insulin. Such individuals must receive insulin, usually by intravenous injection. If more insulin is put into circulation than needed, the blood sugar level falls too low, causing *insulin shock.*

In type II diabetes or **noninsulin-dependent diabetes mellitus** (NIDDM), *insulin resistance* develops in those tissues that normally use insulin. Insulin resistant cells fail to use insulin. Hence they do not take glucose out of circulation, with hyperglycemia resulting. The causes are complex, but they are associated with obesity, hypertension, a sedentary lifestyle, and elevated blood lipid levels. In some victims, there are genetic factors. Interaction 26.2 discusses diabetes further.

Hypoglycemia Can Make You Faint As we said, your brain relies almost entirely on glucose for its chemical energy, so if hypoglycemia develops rapidly, you may become dizzy and even faint. The brain consumes about 120 g/day of glucose, and a quick onset of hypoglycemia starves the brain cells. They do have the ability to switch over to other nutrients but can't do this rapidly.

Somatostatin Inhibits Glucagon and Slows the Release of Insulin The hypothalamus, a specific region in the brain, makes **somatostatin,** another hormone that participates in the regulation of the blood sugar level. When the β-cells of the pancreas secrete insulin, which helps to *lower* the blood sugar level, the α-cells should not at the same time release glucagon, which helps to *raise* this level. Somatostatin acts at the pancreas to inhibit the release of glucagon as well as to slow down the release of insulin. It thus helps to prevent a wild swing in the blood sugar level that insulin alone might cause.

Glucose Is Toxic in Hyperglycemia When hyperglycemia persists, something is wrong with the mechanisms for taking glucose out of circulation. Diabetes is the most common cause, but there are others. Sustained hyperglycemia particularly

■ Adipose tissue is the fatty tissue that surrounds internal organs.

■ Life-saving first aid for someone in insulin shock is sugared fruit juice or candy to counter the hypoglycemia.

■ Frederick Banting (Canada) and J. J. R. MacLeod (Scotland) shared the 1923 Nobel Prize in Physiology or Medicine for discovering insulin.

INTERACTION 26.2
DIABETES MELLITUS

What most people mean by "diabetes" is actually *diabetes mellitus*, from the Greek *diabetes*, to pass through a siphon, and *mellitus*, honey-sweet, meaning to pass urine that contains sugar. In severe, untreated diabetes, the victim's body wastes away despite efforts to satisfy a powerful thirst and hunger. To the ancients, it seemed as if the body were dissolving from within. In the United States, diabetes ranks third behind heart disease and cancer as a cause of death.

Roughly 5% of the United States population has diabetes, and almost as many others likely have it. Most victims have *type II* or *noninsulin-dependent diabetes mellitus, NIDDM.* They generally contract it when over the age of 30, so it's also called *adult-onset diabetes.* The majority of NIDDM victims manage their blood sugar levels by diet and exercise alone, without insulin injections. Their problem is actually not a lack of insulin but an *insulin resistance* by target cells. A high sugar intake calls forth such a continuous presence of insulin that insulin receptors literally wear out faster than they can be replaced. Weight reduction, particularly with physical exercise, lets the receptors rebound.

Roughly 20% of all cases of diabetes are of *type I, insulin-dependent diabetes mellitus* or *IDDM.* In those with IDDM, the β-cells of the pancreas are unable to make and secrete insulin. Most victims develop it before the age of 30, often as adolescents, so IDDM is also called *juvenile-onset diabetes.*

Type I Diabetes Develops in Stages Type I diabetes is now known to be an autoimmune disease. Those who get it very likely have a genetic defect in their immune system, probably involving more than one gene. The immune system stops recognizing the membrane proteins of its own pancreatic β-cells and sets out to destroy them.

In many victims, type I diabetes seems to start when a triggering incident, like a viral infection (for example mumps), launches a slow process. The virus changes substances on the membranes of the β-cells so that the immune system sees these cells as foreign antigens and *makes antibodies against them.* (*Antibodies* are substances made by the immune system to counteract the effects of invading substances, called *antigens*, that are alien to the body.) As β-cells are destroyed, the pancreas loses its ability to secrete insulin. Persistent hyperglycemia and type I diabetes develop. When not detected in time, and thus untreated, the immediate complication is metabolic acidosis, which is fatal.

Glucosylation of Proteins Causes Long-Term Complications of Diabetes The continuous presence of a high level of blood glucose in both IDDM and NIDDM shifts certain chemical equilibria in favor of glucosylated compounds. The aldehyde group of the open form of glucose, for example, can react with amino groups, like those on side chains of lysine residues, to couple the glucose molecule to the protein.

damages those tissues that are freely permeable to glucose, such as the retina, the kidneys, and the nerves. Therefore, those with persistent hyperglycemia face increased risks of blindness, kidney failure, and disorders of the nervous system.

In sustained high levels, glucose is toxic because it promotes unwanted increases in the activities of certain enzymes (which explains why enzyme inhibitors for treating diabetes are being studied). Glucose also reacts with and damages some of the proteins of smooth muscles, for example, those of blood vessels. This increases the risks of amputation, stroke, and heart attacks. Glucose also reacts with hemoglobin, giving glucohemoglobin (or glucosylated hemoglobin). For hyperglycemic individuals, the measurement of one's glucohemoglobin level has become the best way to determine the average blood sugar level. Given all that we've described about the blood sugar level, we have ample reason to study in more detail how the body maintains a somewhat steady concentration of blood glucose.

$$\text{glucose—CH} {=} \text{O} + \text{H}_2\text{N—protein} \longrightarrow$$
$$\text{glucose—CH} {=} \text{N—protein} + \text{H}_2\text{O}$$

When this reaction occurs in the protein support structures of blood capillaries, the *basement membranes,* they swell and thicken; the condition is called *microangiopathy.* It's believed to lead to some of the other complications, like kidney problems and gangrene of the lower limbs.

Diabetes is the leading cause of new cases of blindness in the United States, and it is the second most common cause of blindness, overall. A high blood glucose level increases the activity of an enzyme, *aldose reductase,* that reduces the aldehyde group in glucose to make sorbitol. Sorbitol, unlike glucose, tends to be trapped in lens cells, and as the sorbitol concentration increases so does the osmotic pressure in the fluid. *This draws water into the lens cells, which generates pressure and leads to cataracts and glaucoma.*

Diet Control Is Mandatory The best single treatment of diabetes is any effort that keeps a strict control on the blood sugar level to avoid the episodes of upward surges followed by precipitous declines. In 1992, the Diabetes Control and Complications Trial reported on a 10-year study of people with Type I diabetes that showed intensive insulin therapy (meaning three or more insulin injections per day or use of an insulin pump) retards the onset or progress of diabetic complications.

The insulin pump (see Figure 1), when suitable for the patient, closely approximates the behavior of a normal pancreas and gives the patient better flexi-

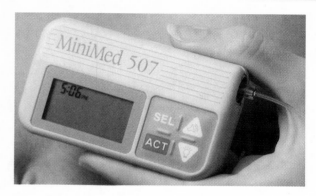

FIGURE 1   An insulin pump.

bility in life style. The pump is programmed to provide a steady infusion of regular insulin plus enhanced quantities prior to meals or intensive activities. For diabetics living in remote areas, far from diabetes specialists, an insulin pump is now available (Disetronic Medical Systems, Minnesota) that can be linked by a computer modem and a phone line to a central diabetes clinic. The pump's memory holds data on the previous several "events." These data and the latest blood sugar level reading (phoned in) guide any necessary pump adjustments, which can be made through the same modem hookup.

A Reference
Barbara Savinetti-Rose and Linda Bolmer, "Understanding Continuous Subcutaneous Insulin Infusion Therapy," *American Journal of Nursing,* March, 1997, page 42. (See also *The New England Journal of Medicine,* September 30, 1993, page 977.)

26.2 GLUCOSE TOLERANCE

The ability of the body to tolerate swings in the blood sugar level is essential to health.

Glucose tolerance is the ability of the body to manage its blood sugar level within the normal range. Many factors contribute to it.

The Cori Cycle Describes the Distributions and Uses of Glucose   The strategies used by the body to maintain its blood sugar level within the normal range form a cycle of events called the **Cori cycle** (Figure 26.2).

We begin at the bottom of Figure 26.2 where glucose enters the blood from the intestines. Some molecules stay in circulation. Others are removed, particularly by

■   Carl Cori and Gerty Cori shared the 1947 Nobel Prize in Physiology or Medicine.

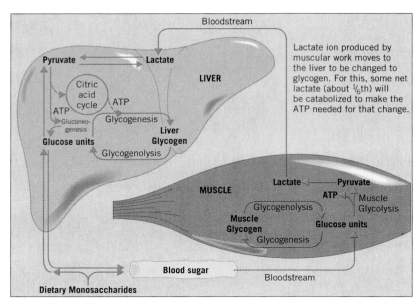

FIGURE 26.2   The Cori cycle.

muscle and liver tissue. Glucose trapped in these tissues is used to replenish their glycogen reserves or to make ATP by *glycolysis.* As we've seen, the liver is able to release glucose back into the bloodstream when the blood sugar level must be increased.

Figure 26.2 also shows glucose being used in glycolysis, the end product being either pyruvate or lactate, depending on the oxygen supply. When extensive anaerobic glycolysis occurs in a tissue, its lactate level increases. However, with a sufficient oxygen supply, either pyruvate or lactate can be used by the citric acid cycle and respiratory chain to make ATP.

Because lactate still has C—H bonds, it continues to have useful chemical energy, so instead of simply excreting excess lactate at the kidneys, the body recycles it. It reconverts a fraction of it, about five-sixths, to glucose. The synthesis of glucose from smaller molecules is called **gluconeogenesis** (to be studied later), and it requires ATP energy. The remaining one-sixth of the lactate is catabolized to supply the needed energy as ATP. The recycling of lactate completes the Cori cycle. Many of its processes, as you can see, affect the blood sugar level. Some tend to increase it and some do the opposite (Figure 26.3).

■ *Neo-,* new; *-neogenesis,* new creation; *gluconeogenesis,* the synthesis of new glucose.

Glucose Tolerance Is Measured by the Glucose Tolerance Test   In the **glucose tolerance test,** the individual is given a drink that contains glucose, generally 75 g for an adult and 1.75 g per kilogram of body weight for children, and then the blood sugar level is checked at regular intervals.

Figure 26.4 gives typical plots of the blood sugar level versus time for two types of individuals. The lower curve is that of someone with normal glucose tolerance, and the upper curve is of one whose glucose tolerance is typical of a diabetic. At the start of the test, the blood sugar level increases sharply in both types. The healthy person, however, soon manages the high level and brings it down with the help of a normal flow of insulin and somatostatin. In the diabetic, the level declines only slowly and remains mostly in the hyperglycemic range throughout.

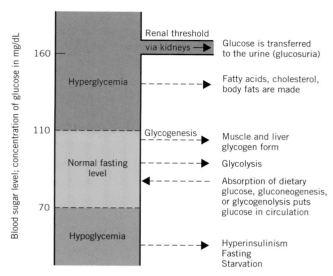

FIGURE 26.3 Factors that affect the blood sugar level.

Notice that the blood sugar level of the normal individual can sometimes drop to a mildly hypoglycemic level. It might even happen to you following a breakfast overloaded with sugar. With glucose pouring into the bloodstream, the release of a bit more insulin than needed can occur, leading to the overwithdrawal of glucose from circulation. The brain suffers from a reduced supply of glucose, and midmorning brings dizziness and maybe even fainting (or falling asleep in class). The prevention is simply a balanced breakfast. (Mother was right.)

■ Unhappily, most people with midmorning sag go into another round of sugared coffee and sugared rolls. The glucose gives a short lift, but then an oversupply of insulin restores the mild hypoglycemia of the sag.

Glucose Tolerance Is Poor in Those with Diabetes The subject of glucose tolerance is nowhere of more concern than with diabetes. Clinically (that is, for purposes of diagnosis), **diabetes** is a disease in which the blood sugar level persists in being much higher than warranted by the dietary and nutritional status of the individual. Invariably, a person with untreated diabetes has glucosuria, and the discovery of this condition often triggers the clinical work needed to diagnose diabetes.

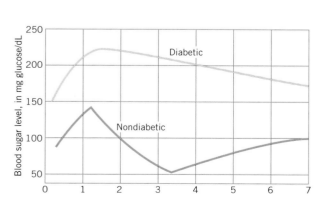

FIGURE 26.4 Glucose tolerance curves.

26.3   CATABOLISM OF GLUCOSE

Glycolysis and the pentose phosphate sequence are the chief catabolic pathways open to glucose.

■   Greek, *glykys*, sugar or sweet; *-lysis*, dissolution.

Glycolysis is a series of reactions that change glucose to pyruvate (or lactate), while making a small amount of ATP. Other monosaccharides enter the same glycolysis pathway as glucose (Figure 26.5), so when we study glycolysis we cover most monosaccharide catabolism.

Aerobic Glycolysis Ends in Pyruvate   Except during extensive exercise, glycolysis is operated with sufficient oxygen and so is *aerobic*. Its overall equation is

$$C_6H_{12}O_6 + 2ADP + 2P_i + 2NAD^+ \longrightarrow 2CH_3\overset{\overset{\displaystyle O}{\|}}{C}CO_2^- + 2ATP + 2NADH + 2H^+ + 2H_2O$$

glucose   pyruvate

Aerobic glycolysis thus makes some ATP. However, the NAD$^+$ shown here is reduced to NADH, and it delivers H:$^-$ to the respiratory chain. Thus, the generation of NADH by aerobic glycolysis drives the synthesis of even more ATP by the chain.

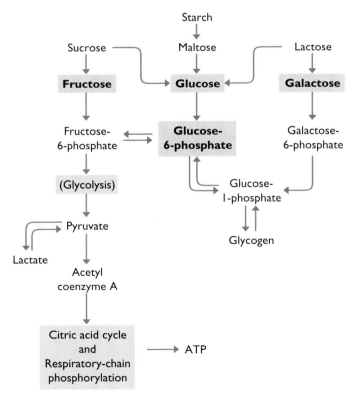

FIGURE 26.5   Convergence of the pathways in the metabolism of dietary carbohydrates.

When a cell receives oxygen at a rate slower than needed, glycolysis can still operate, but it ends in lactate, not pyruvate. The following equation, therefore, is for the *anaerobic sequence* of glucose catabolism.

$$C_6H_{12}O_6 + 2ADP + 2P_i \xrightarrow[\text{glycolysis}]{\text{anaerobic}} 2CH_3\overset{\overset{\displaystyle OH}{|}}{C}HCO_2^- + 2H^+ + 2ATP$$

glucose   lactate

Notice that ATP is still made *even though the cell is anaerobic* and not able to use the respiratory chain. It's an important backup for the cell.

Glycolysis Begins with ATP Consumption but Then Generates More Figure 26.6 outlines the steps to pyruvate (or lactate) that can begin with either glucose or glycogen. Those that lead to fructose-1,6-bisphosphate are actually up an energy hill, because they consume ATP. But this is like pushing a sled or bike up the short backside of a long hill, because the investment in energy is more than repaid by the long, downhill slide to pyruvate (or lactate) and more ATP. When glycogen is the source of glucose for glycolysis, the initial investment in ATP is slightly smaller than when glucose is the starting material.

■ The numbered steps in the discussion of glycolysis, below, refer to Figure 26.6.

1. Glucose is phosphorylated by ATP under catalysis by *hexokinase* to give glucose-6-phosphate.

■ A *kinase*, recall, is an enzyme that handles transfers of phosphate units.

2. Glucose-6-phosphate changes to its isomer, fructose-6-phosphate. The enzyme is *phosphoglucose isomerase*. This may seem to be a major structural change, but it involves little more than some shifts of bonds and hydrogens, as the arrows in the following sequence show.

■ These equilibria shift constantly to the right as long as later reactions continuously remove products as they form.

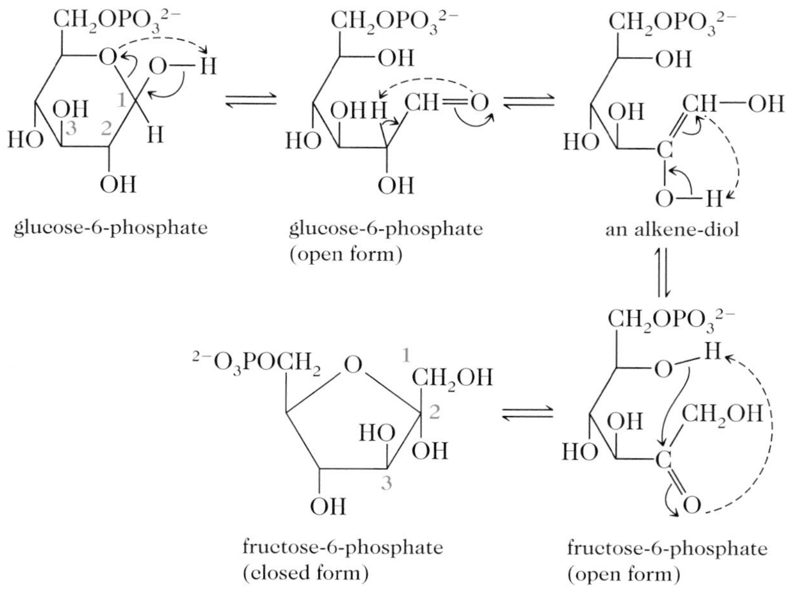

glucose-6-phosphate   glucose-6-phosphate   an alkene-diol
                           (open form)

fructose-6-phosphate   fructose-6-phosphate
(closed form)   (open form)

3. ATP phosphorylates fructose-6-phosphate to make fructose-1,6-bisphosphate. The enzyme is *phosphofructokinase*. This step, essentially irreversible, ends the energy-consuming phase of glycolysis.

■ All kinases require Mg^{2+} as a cofactor.

4. Fructose-1,6-bisphosphate breaks apart into two triose monophosphates. *Aldolase* is the catalyst. *This is actually a reverse aldol condensation (see Ap-*

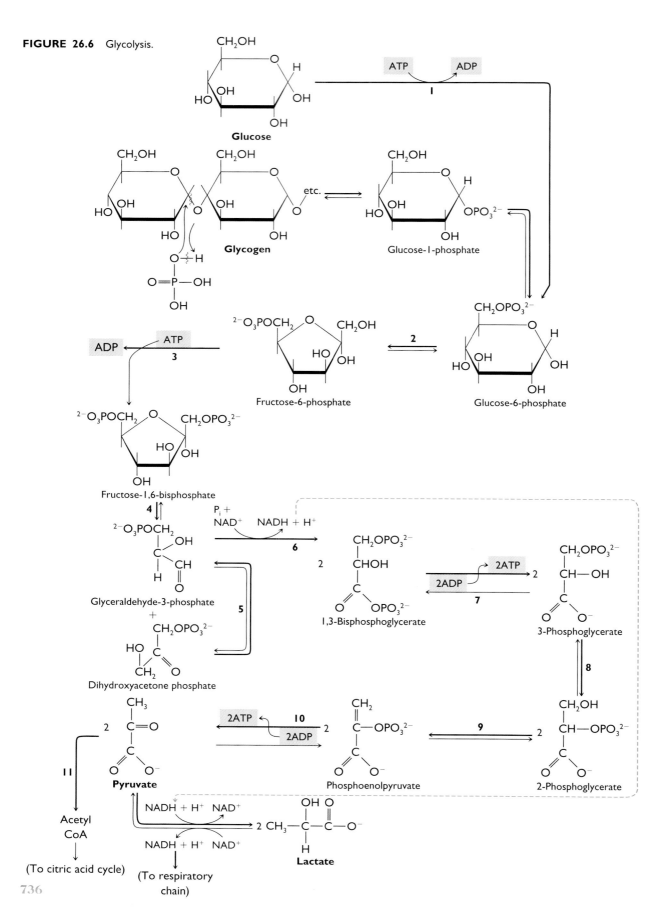

FIGURE 26.6 Glycolysis.

736

pendix D). We can visualize step 4 as a few simple and reasonable shifts of electrons and protons. The single bond between carbons 3 and 4 breaks, its electrons going to carbon 3 as they form a bond to the H being plucked from the HO group at carbon 4. This frees up an electron pair to form the new carbonyl group at carbon 4.

■ When aldolase is defective (from a defective gene), disorders such as hemolytic anemia, abnormal nerve function, and intolerance of exercise may occur.

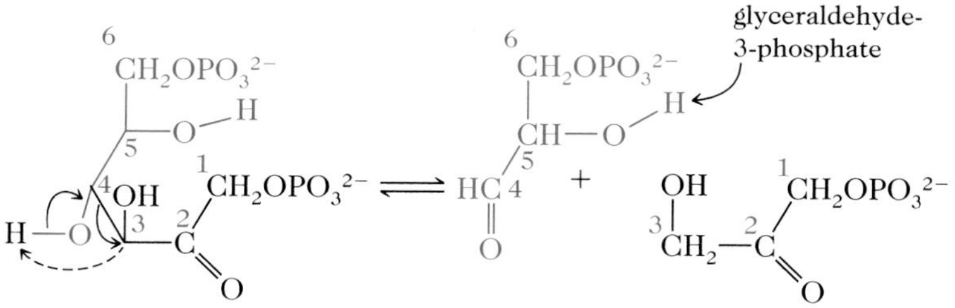

$$\underset{\substack{\text{fructose-1,6-} \\ \text{bisphosphate} \\ \text{(open form)}}}{} \quad \rightleftharpoons \quad \underset{\substack{\text{glyceraldehyde-} \\ \text{3-phosphate}}}{} + \underset{\substack{\text{dihydroxyacetone} \\ \text{phosphate}}}{}$$

5. Dihydroxyacetone phosphate, in the presence of *triose phosphate isomerase*, changes to its isomer, glyceraldehyde-3-phosphate.

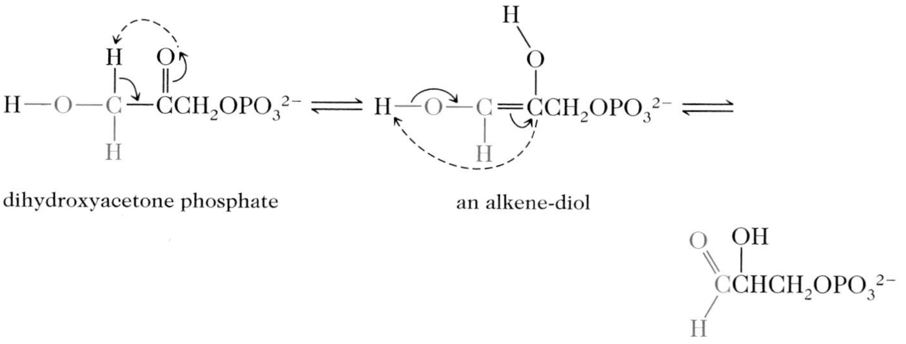

$$\underset{\text{dihydroxyacetone phosphate}}{} \rightleftharpoons \underset{\text{an alkene-diol}}{} \rightleftharpoons \underset{\substack{\text{glyceraldehyde-} \\ \text{3-phosphate}}}{}$$

This change ensures that all the chemical energy in glucose will be obtained, because the main path of glycolysis continues with glyceraldehyde-3-phosphate. All dihydroxyacetone shuttles through glyceraldehyde-3-phosphate.

6. Glyceraldehyde-3-phosphate is simultaneously oxidized and phosphorylated. The enzyme *glyceraldehyde-3-phosphate dehydrogenase* has NAD^+ as a cofactor, and an SH group on the enzyme participates. Inorganic phosphate is the source of the new phosphate group. We can visualize how it happens as follows.

■ The continuous removal of glyceraldehyde-3-phosphate by its subsequent reaction shifts the dihydroxyacetone phosphate equilibrium to the right.

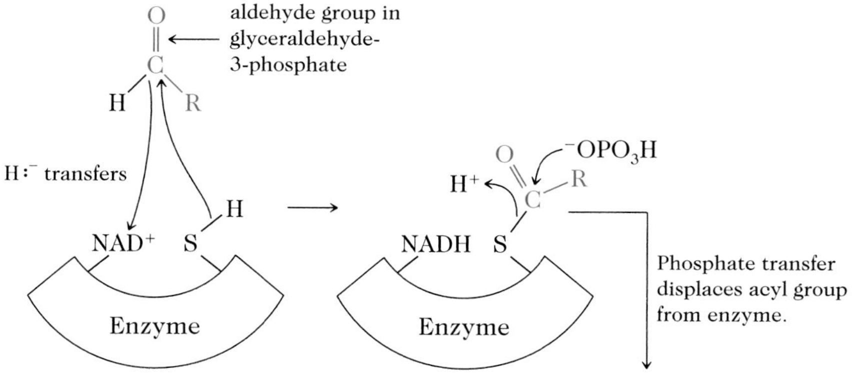

aldehyde group in glyceraldehyde-3-phosphate

$H:^-$ transfers

Phosphate transfer displaces acyl group from enzyme.

$$\text{NADH} \quad S \overset{H}{\diagdown} \quad + \quad R-\overset{\displaystyle O}{\overset{\|}{C}}-OPO_3{}^{2-} + H^+$$

mixed anyhydride
system in 1,3-bis-
phosphoglycerate

Enzyme

Notice that the step 6 enzyme now holds NADH, not NAD^+. It's a problem that we'll return to.

7. 1,3-Bisphosphoglycerate has a higher phosphate group transfer potential than ATP. With the help of *phosphoglycerate kinase,* it transfers a phosphate to ADP, so this gives us back the original investment of ATP. (Remember that each glucose molecule with six carbons is processed through *two* three-carbon molecules, so *two* ATPs are made here for each original glucose molecule.)

8. The phosphate group in 3-phosphoglycerate shifts to the 2-position, catalyzed by *phosphoglycerate mutase,* setting the molecule up for a key reaction, next.

9. The dehydration of 2-phosphoglycerate to phosphoenolpyruvate is catalyzed by *enolase.* This step is like cocking a huge bioenergetic gun. A simple, low-energy reaction, dehydration of an alcohol, converts a low-energy phosphate into the highest energy phosphate in all of metabolism, the phosphate ester of the enol form of pyruvate, phosphoenolpyruvate.[2]

10. The phosphate group in phosphoenolpyruvate transfers to ADP, and more ATP is made. The enzyme is *pyruvate kinase.* The enol form of pyruvate left behind promptly and irreversibly rearranges to the keto form of the pyruvate ion. The instability of enols (see footnote 2) is the driving force for this entire step.

11. If the mitochondrion is running aerobically when pyruvate is made, the pyruvate ion changes to acetyl coenzyme A, from which an acetyl group can enter the citric acid cycle (or be used in another way).

Two Mechanisms Exist to Restore the Step 6 Enzyme   Once used, the enzyme at step 6 has its coenzyme in its reduced form, NADH. *As long as* $H\colon^-$ *is on NADH, the enzyme is "plugged."* *Glycolysis, therefore, could not happen again unless something is done.* When the cell has sufficient oxygen, the block is continuously removed because NADH simply gives $H\colon^-$ to the respiratory chain. This regenerates the step 6 enzyme for another run of the glycolysis series.

When the cell is deficient in oxygen, NADH is changed back to NAD^+ by unloading its $H\colon^-$ onto the keto group of pyruvate, making it the 2° alcohol group in lactate. *This is why lactate is the end product of anaerobic glycolysis.* Lactate serves to store $H\colon^-$ made at step 6 until the cell again becomes aerobic. The overall change from glucose to lactate by anaerobic glycolysis can be represented by the following equation. *Note the generation of hydrogen ions.*

■ No reaction in the body is truly complete until its enzyme is fully restored.

---

[2] When a carbon atom of an alkene group holds an OH group, the compound is called an *enol* ("ene" + "−ol"). Enols are unstable alcohols that spontaneously rearrange into carbonyl compounds:

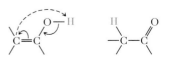

enol form   carbonyl form

$$\text{glucose} + 2ADP + 2P_i \longrightarrow 2 \text{ lactate} + 2ATP + 2H^+$$

The importance of anaerobic glycolysis is that the cell can continue to make some ATP even when insufficient oxygen is available to run the respiratory chain. Of course, there are limits. The longer the cell operates anaerobically and the more that lactate accumulates, the more the cell runs an **oxygen debt** *and the more acid is made.* Eventually the system has to slow down to let respiration bring back oxygen to metabolize lactate and buffer the acid.

Excessive Exercise Causes Lactic Acid Acidosis Anaerobic glycolysis generates acid, as we said, so extensive physical exercise that forces tissues to operate anaerobically can overtax the blood buffer. A form of metabolic acidosis called *lactic acid acidosis* results. The muscles become tired and sore. The lungs respond by hyperventilation, which blows out carbon dioxide and thus helps to remove acid from the body. During a cool down period, the body reestablishes the acid–base balance of the blood, and excess lactate ion is shuttled into the Cori cycle.

■ Hunters know that meat from animals run to exhaustion is sour.

The *Pentose Phosphate Pathway* Makes NADPH The biosyntheses of some substances in the body require a reducing agent. Fatty acids, for example, are almost entirely alkane-like, and alkanes are the most reduced types of organic compounds. The reducing agent for fatty acid biosynthesis is NADPH, the reduced form of NADH$^+$.

■ NADP$^+$ is a phosphate derivative of NAD$^+$.

The **pentose phosphate pathway** of glucose catabolism is the body's principal route to NADPH, being very active in adipose tissue, where fatty acid synthesis occurs. There are two broad sequences, one oxidative and the other nonoxidative. The oxidative reactions convert a hexose phosphate into a pentose phosphate (hence the name of the series), the overall equation being

■ The pentose phosphate pathway also goes by the names *hexose monophosphate shunt* and *phosphogluconate pathway*.

$$\text{glucose-6-phosphate} + 2NADP^+ + H_2O \longrightarrow$$
$$\text{ribose-5-phosphate} + 2NADPH + 2H^+ + CO_2$$

One product, ribose-5-phosphate, could now be used to make the pentose systems in the nucleic acids, if they are needed. If not, it undergoes the nonoxidative phase of the pentose phosphate pathway. These have the net effect of converting three pentose units into two molecules of glucose and one of glyceraldehyde. The glucose can be catabolized by glycolysis or recycled into the pentose phosphate pathway. Glyceraldehyde, we'll soon see, can be converted to glucose and recycled as well. Overall, with the recycling mentioned, the equation for the complete oxidation of one glucose molecule via the pentose phosphate pathway is as follows.

$$\text{glucose-6-phosphate} + 12NADP^+ + 6H_2O \xrightarrow{\text{pentose phosphate pathway}}$$
$$6CO_2 + 12NADPH + 12H^+ + P_i$$

26.4 GLUCONEOGENESIS

Some of the steps in gluconeogenesis are the reverse of steps in glycolysis.

Gluconeogenesis is a series of reactions by which glucose is made from smaller molecules (Figure 26.7). Excess lactate can be used as the starting material, but so can the breakdown products of several amino acids. The ability of the body to make glucose from noncarbohydrate sources, even amino acids from the body's own proteins, is an important backup during times when glucose either isn't in the diet (starvation) or is not effectively used (untreated diabetes).

■ Most of our glucose needs are met by gluconeogenesis during periods of fasting.

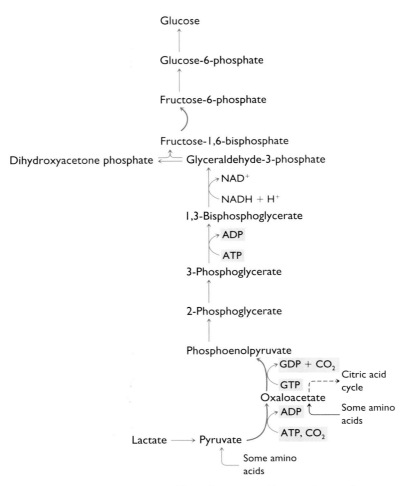

FIGURE 26.7 Gluconeogenesis. The red arrows signify steps that are the reverse of corresponding steps in glycolysis. The blue arrows denote steps that are unique to gluconeogenesis.

Gluconeogenesis Is Not the Exact Reverse of Glycolysis It might seem that the body could make glucose from pyruvate simply by running glycolysis backward. However, three steps in glycolysis cannot be directly reversed, namely, steps 1, 3, and 10 discussed earlier (see Figure 26.6). Special enzymes exist in the liver and the kidneys that create bypasses around these irreversible steps.

In the bypass that gets around step 10 of glycolysis, the synthesis of phosphoenolpyruvate from pyruvate, bicarbonate ion is used as a reactant. It's a source of CO_2 that becomes inserted into pyruvate. ATP energy is used to drive the steeply uphill reaction, and *pyruvate carboxylase* is the enzyme. Pyruvate is changed to oxaloacetate.

■ The enzyme for this step requires the vitamin biotin.

$$CH_3\overset{O}{\overset{\|}{C}}CO_2^- + HCO_3^- + ATP \xrightarrow[\text{pyruvate carboxylase}]{\text{(in mitochondria)}} {}^-O_2CCH_2\overset{O}{\overset{\|}{C}}CO_2^- + ADP + P_i$$

pyruvate   oxaloacetate

This reaction occurs in mitochondria, but the gluconeogenesis that uses the product, oxaloacetate, occurs in the cytosol. *By segregating anabolism reactions from their opposite, catabolism, the body is better able to control which pathways to operate.* After a complex shuttle mechanism gets oxaloacetate through the mitochondrial membranes, it reacts in the cytosol with another triphosphate, guanosine triphosphate (GTP), as carbon dioxide splits back out and bonds rearrange.

$$\underset{\text{oxaloacetate}}{:\ddot{O}{-}\overset{\overset{O}{\|}}{C}{-}CH_2{-}\overset{\overset{O}{\|}}{C}{-}CO_2^-} + {}^-O{-}\overset{\overset{O}{\|}}{\underset{O^-}{P}}{-}O{-}\overset{\overset{O}{\|}}{\underset{O^-}{P}}{-}O{-}\overset{\overset{O}{\|}}{\underset{O^-}{P}}{-}O{-}\text{guanosine} \longrightarrow$$

$$CO_2 + \underset{\text{phosphoenolpyruvate}}{CH_2{=}\overset{\overset{O{-}PO_3^{2-}}{|}}{C}{-}CO_2^-} + GDP$$

Because the synthesis of one glucose requires two pyruvates, and because each pyruvate requires two high-energy phosphates in gluconeogenesis, this bypass costs the equivalent of four ATPs (2ATP + 2GTP) per glucose molecule. The bypass is expensive.

At the reversal of step 7 (see Figure 26.6), two more ATPs are used per molecule of glucose made. Thus a total of six ATPs are needed to make one glucose molecule by gluconeogenesis, starting from pyruvate. We compare this cost with the two ATPs made by anaerobic glycolysis, beginning with glucose.

The bypasses to the reverses of steps 3 and 1 of glycolysis require only specific enzymes that catalyze the hydrolysis of phosphate ester groups, not high-energy boosters. Such enzymes are integral parts of the enzyme team for gluconeogenesis.

The other steps in gluconeogenesis are run as reverse shifts in equilibria for which the opposite direction is favored in glycolysis. The reverse of step 4 is an aldol condensation (Appendix D), and here is where glyceraldehyde (as the 3-phosphate), made in the pentose phosphate pathway, can be changed back to glucose.

SUMMARY

Glycogen metabolism   The regulation of glycogenesis and glycogenolysis is a part of the machinery for glucose tolerance in the body. Hyperglycemia stimulates the secretion of insulin and somatostatin, and insulin helps cells of adipose tissue to take glucose from the blood. Somatostatin helps to suppress the release of glucagon (which otherwise stimulates glycogenolysis and leads to an increase in the blood sugar level).

When glucose is abundant, the body either replenishes its glycogen reserves or makes fat. In muscular work, epinephrine stimulates a cascade of enzyme activations that begins with the activation of adenylyl cyclase and ends with the release of many glucose molecules from glycogen.

When glucose is in short supply, the body makes its own by gluconeogenesis from noncarbohydrate molecules, including several amino acids. In diabetes, some cells that are starved for glucose make their own, also. Such cells are unable to obtain glucose from circulation, so the blood sugar level is hyperglycemic to a glucosuric level. The glucose tolerance test is used to see how well the body handles an overload of glucose. In the management of diabetes, the maintenance of a reasonably steady blood sugar level in the normal range is vital. The measurement of the glucohemoglobin in circulation serves to monitor how well the normal range is kept over a long period of time.

Following strenuous exercise, when lactate is plentiful, the liver makes glucose from lactate. The many pathways that involve glycogen and glucose form a cycle of events called the Cori cycle.

Glycolysis   Under anaerobic conditions, glucose can be catabolized to lactate ion. (Galactose and fructose enter this pathway, too.) When lactate is used to store H:$^-$, one of the enzymes in glycolysis can be regenerated in the absence of oxygen, and glycolysis can be run to make some ATP without the involvement of the respiratory chain. Then, when the cell is aerobic again, the H:⁻ that this enzyme must shed to work again is given directly into the respiratory chain, and pyruvate instead of lactate becomes the end product of glycolysis.

Pentose phosphate pathway   The body's need for NADPH to make fatty acids is met by catabolizing glucose through the pentose phosphate pathway.

Gluconeogenesis   Most of the steps in gluconeogenesis are simply the reverse of steps in glycolysis, but there are a few that require rather elaborate bypasses. Special teams of enzymes and supplies of high-energy phosphate are used for these. Many amino acids can be used to make glucose by gluconeogenesis.

REVIEW EXERCISES

The answers to Review Exercises whose numbers are in color are found in Appendix E. The answers to the other Review Exercises are found in the Study Guide that accompanies this book. The more challenging questions are marked with asterisks.

Blood Sugar

26.1   What are the end products of the complete digestion of the carbohydrates in the diet?

26.2   Why can we treat the catabolism of carbohydrates as almost entirely that of glucose?

26.3   What is meant by *blood sugar level*? By *normal fasting level*?

26.4   What is the range of concentrations in mg/dL for the normal fasting level of whole blood?

*26.5   A level of 5.5 mmol/L for glucose in the blood corresponds to how many milligrams of glucose per deciliter?

26.6   What characterizes the following conditions?
(a) Glucosuria   (b) Hypoglycemia
(c) Hyperglycemia   (d) Glycogenolysis
(e) Renal threshold   (f) Glycogenesis

26.7   What is gluconeogenesis? Which condition, hypoglycemia or hyperglycemia, would activate gluconeogenesis?

26.8   Explain how severe hypoglycemia can lead to disorders of the central nervous system.

Hormones and the Blood Sugar Level

26.9   When epinephrine is secreted, what soon happens to the blood sugar level?

26.10   At which one tissue is epinephrine the most effective?

26.11   What does epinephrine activate at its target cell?

26.12   Arrange the following in the correct order in which they work in epinephrine-initiated glycogenolysis. Place the identifying letters in their correct order, left to right, in the series.

| Phosphorylase kinase | Protein kinase | Cyclic AMP |
| A | B | C |

| Adenylyl cyclase | Phosphorylase |
| D | E |

26.13   One epinephrine molecule triggers the ultimate formation of roughly how many glucose units, 10^1, 10^2, 10^3, or 10^4?

26.14   What might be the result if phosphorylase and glycogen synthetase were both activated at the same time?

26.15   What switches glycogen synthetase off when glycogenolysis is activated?

26.16   What is the end product of glycogenolysis, and what does phosphoglucomutase do to it?

26.17   Why can liver glycogen but not muscle glycogen be used to resupply blood sugar?

26.18   What is glucagon, what does it do, and what is its chief target tissue?

26.19   Which is probably better at increasing the blood sugar level, glucagon or epinephrine? Explain.

26.20   How does human growth hormone manage to promote the supply of the energy needed for growth?

26.21   What is insulin, where is it released, and what is its chief target tissue?

26.22   What triggers the release of insulin into circulation?

*26.23   If brain cells are not insulin-dependent cells, how can too much insulin cause insulin shock?

26.24   What is somatostatin, where is it released, and what kind of effect does it have on the pancreas?

Glucose Tolerance and the Cori Cycle

26.25   What is meant by *glucose tolerance*?

26.26   How is glucose trapped in muscle cells and what happens to it?

26.27   What are the main steps in the Cori cycle?

26.28 In general terms, what happens to excess lactate produced in muscles during exercise?

26.29 What is the purpose of the glucose tolerance test? How is it conducted?

26.30 Describe what happens when each of the following persons takes a glucose tolerance test.
(a) A nondiabetic individual   (b) A diabetic individual

26.31 Describe a circumstance in which hyperglycemia might arise in a nondiabetic individual.

Catabolism of Glucose

26.32 Complete and balance the following equation.

$$C_6H_{12}O_6 + 2ADP + \underline{\quad} \longrightarrow 2C_3H_5O_3^- + 2H^+ + \underline{\quad}$$
glucose   lactate

26.33 What particular significance does glycolysis have when a tissue is running an oxygen debt?

26.34 Why is the rearrangement of dihydroxyacetone phosphate into glyceraldehyde-3-phosphate important?

26.35 What happens to pyruvate (a) under aerobic conditions and (b) under anaerobic conditions?

26.36 What happens to lactate when an oxygen debt is repaid?

*26.37 What is the maximum number of ATPs that can be made by the complete catabolism of (a) one molecule of glucose and (b) one glucose residue in glycogen?

26.38 The pentose phosphate pathway uses $NADP^+$, not NAD^+. What forms *from* $NADP^+$, and how does the body use what forms (in general terms)?

Gluconeogenesis

26.39 In a period of prolonged fasting or starvation, what does the system do to try to maintain its blood sugar level?

*26.40 Amino acids are not excreted, and they are not stored in the same way that glucose residues are stored in a polysaccharide. What probably happens to the excess amino acids in a high-protein diet of an individual who does not exercise much?

26.41 The amino groups of amino acids can be replaced by keto groups. Which amino acids could give the following keto acids that participate in carbohydrate metabolism?
(a) Pyruvic acid   (b) Oxaloacetic acid

*26.42 The amino acids glutamic acid, arginine, histidine, and proline can all be catabolized to α-ketoglutarate. What metabolic cycle in the body makes it possible for α-ketoglutarate to be used eventually in gluconeogenesis? Explain.

*26.43 The carbon atoms of succinyl CoA can wind up in glucose molecules. What metabolic pathway in the body enables succinyl CoA to connect to gluconeogenesis?

Glycogen Storage Diseases (Interaction 26.1)

26.44 For each of the following diseases, name the defective enzyme, and state the biochemical and physiological consequences.
(a) Von Gierke's disease   (b) Cori's disease
(c) McArdle's disease   (d) Andersen's disease

Diabetes Mellitus (Interaction 26.2)

26.45 What is the biochemical distinction between type I and type II diabetes?

26.46 What type of diabetes usually starts at a relatively early age?

26.47 Which is the more common type of diabetes?

26.48 Viruses that cause diabetes attack which target cells?

26.49 Type I diabetes is an autoimmune disease. What does this mean?

26.50 What is one explanation for the lack of glucose uptake in NIDDM?

26.51 Sustained hyperglycemia causes damage to which specific tissue, causing damage that may be responsible for other complications?

26.52 What kind of reaction does glucose, at a high level, give with certain proteins?

26.53 Describe a theory that explains how the hydrogenation of glucose may contribute to blindness.

Additional Exercises

*26.54 Suppose that a sample of glucose is made using some atoms of carbon-13 in place of the common isotope, carbon-12. Suppose, further, that this is fed to a healthy, adult volunteer and that all of it is taken up by the muscles.
(a) Will some of the original molecules be able to go back out into circulation? Explain.
(b) Can we expect any carbon-13 compounds to end up in the liver? Explain.
(c) Can we ever expect to see carbon-13 labeled glucose molecules in circulation again? Explain.

*26.55 Referring to data in the previous chapter, (a) how much ATP can be made from the chemical energy in one pyruvate? (b) The conversion of one lactate to one pyruvate transfers one $H:^-$ to NAD^+. How many ATP molecules can be made just from the operation of this step? (c) If all the possible ATP that can be obtained from the complete catabolism of lactate were made available for gluconeogenesis, how many molecules of glucose could be made? (Assume that ATP can substitute for GTP.)

27

METABOLISM OF LIPIDS

Before these marathoners complete the Bay Bridge Run, they'll be drawing almost exclusively on fat reserves for chemical energy. The metabolism of lipids provides the reactions that make this possible.

This Chapter in Context

Our study of energy metabolism continues. We saw in the previous chapter that acetyl groups fuel the citric acid cycle and respiratory chain, making ATP. One glucose molecule supplies two acetyl groups; one molecule of palmitic acid yields eight! But if we take in more of either than we need for energy, we don't excrete the excess. Where does it go? (You probably already know.) When we eat too much sugar, acetyl groups are made and assembled into fatty acids, which we store in the same place where we put excess dietary fatty acids. Acetyl groups are also used to make cholesterol, which isn't all bad. Every cell membrane of the body needs it. We're in trouble only when our machinery for getting rid of the excess doesn't work. Acetyl groups also end up in ketone bodies, and making too many of them is very bad. Synthesizing them generates acid, causing acidosis and all manner of trouble. Clearly, our study of the molecular basis of life would be incomplete unless we looked into these matters.

27.1 Absorption and Distribution of Lipids

Lipoprotein complexes in blood transport triacylglycerols, fatty acids, cholesterol, and other lipids.

The digestion of triacylglycerols produces a mixture of the anions of long-chain fatty acids and monoacylglycerols (plus some diacylglycerols). As these move into the cells of the intestinal membrane, they become reconstituted into triacylglycerols. They, plus the cholesterol in the diet, make up most of the *exogenous lipid* material delivered to circulation.

■ *Exogenous* means "from the outside" (e.g., from the diet).

Chylomicrons Carry Dietary Lipids and Cholesterol Lipids are virtually insoluble in water, which creates a problem because they have to be transported in blood. Hence, the body combines them with proteins so they can be transported as tiny particles, more easily dispersed in blood, called **lipoprotein complexes.** There are several kinds, classified according to their densities, which range from 0.95 to 1.21 g/cm³ (see Table 27.1).

■ The solubility in water of a long-chain fatty acid is typically less than 10^{-6} mol/L, but that of a complex of the fatty acid with a plasma albumin is about 2×10^{-3} mol/L.

TABLE 27.1 Composition of Plasma Lipoproteins

Type of Complex[a]	Chief Constituents (in Order of Amount)	Density (in g/cm³)
Chylomicrons and chylomicron remnants	Triacylglycerols and cholesterol from the diet	<0.95
VLDL	Triacylglycerols from the liver, cholesterol esters, cholesterol	0.95–1.006
IDL	Cholesteryl esters, cholesterol, triacylglycerols (trace)	1.006–1.019
LDL	Cholesteryl esters, cholesterol, triacylglycerols (trace)	1.019–1.063
HDL	Cholesteryl esters, cholesterol	1.063–1.210

[a] VLDL = very low density lipoprotein complex; IDL = intermediate density lipoprotein complex; LDL = low density lipoprotein complex; and HDL = high density lipoprotein complex.

Data from M. S. Brown and J. L. Goldstein, in *Harrison's Principles of Internal Medicine,* 11th ed. Edited by E. Braunwald, K. J. Isselbacher, R. G. Petersdorf, J. D. Wilson, J. B. Martin, and A. S. Fauci. New York, McGraw-Hill, 1987, page 1651.

Lipoprotein complexes with the lowest density have only 2% protein or less and are called *chylomicrons*. They are put together from exogenous lipids within the cells of the intestinal membrane and delivered to the lymph, which carries them to the bloodstream. When in the capillaries of muscle and adipose tissue, chylomicrons are detained by binding sites where an enzyme, *lipoprotein lipase,* hydrolyzes their triacylglycerols to fatty acids and monoacylglycerols. These are promptly absorbed by the nearby tissue. Left over are *chylomicron remnants,* which still contain dietary cholesterol and so have a higher density. These break loose from the capillary surfaces and circulate to the liver, where they are absorbed (Figure 27.1). *The role of chylomicrons is thus to take exogenous fatty acids to muscle and adipose tissue and dietary cholesterol to the liver.*

■ *Chyle* is the lymph from the intestines.

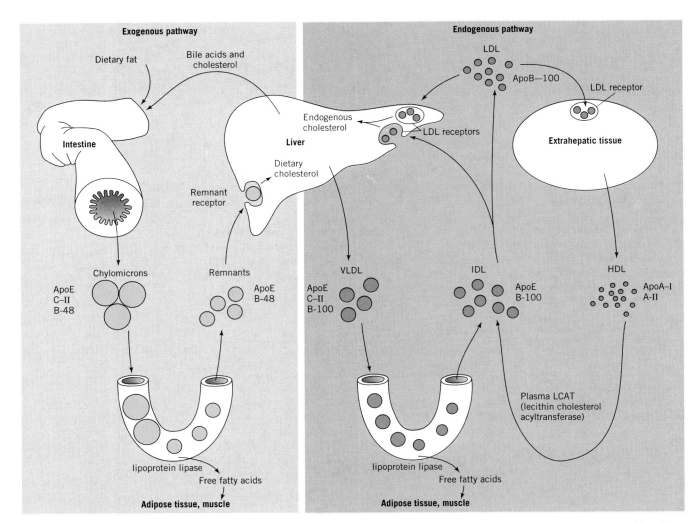

FIGURE 27.1   The transport of cholesterol and triacylglycerols by lipoprotein complexes. The abbreviations ApoE, C-II, B-48, and so forth, refer to specific proteins involved with the lipoprotein complexes. [After M. S. Brown and J. L. Goldstein, in E. Brunwald, K. J. Isselbacher, R. G. Petersdorf, J. B. Martin, and A. S. Fauci (Editors). *Harrison's Principles of Internal Medicine,* 11th edition. New York, McGraw-Hill, 1987, p. 1652.]

Lipoprotein Complexes Transport Endogenous Lipids The liver makes fatty acids and then triacylglycerols from carbohydrates. Lipids made in the liver are *endogenous* lipids ("generated from within"). Three similar lipoprotein complexes are used to transport them in the bloodstream, namely, *very low density lipoproteins (VLDL), intermediate density lipoproteins (IDL),* and *low density lipoproteins (LDL).* All three carry cholesterol, especially the LDL. Another lipoprotein complex, *high density lipoprotein (HDL),* carries cholesterol released from tissues, no longer needed by them, and ready for being exported via the bile.

■ Cholesterol (density, $d = 1.05$ g/cm³) is more dense than triacylglycerols ($d \approx 0.9$ g/cm³).

VLDL Changes to IDL and Then to LDL The liver packages triacylglycerols and cholesterol into VLDL and releases them into circulation. In circulation, VLDL undergoes continuous changes as its triacylglycerols are hydrolyzed in capillaries, thus delivering fatty acids to adipose tissue and muscles. (The changes strongly resemble those that occur to chylomicrons.) By these processes, the net density increases and so VLDL particles change to IDL (see Figure 27.1).

With continued loss of low-density triacylglycerols, the IDL change to LDL (see Figure 27.1). Some LDL is reabsorbed by the liver, but *the main purpose of LDL is to deliver cholesterol to extrahepatic tissue* (tissue other than liver tissue) *to be used to make cell membranes and, in specialized tissues, steroid hormones.* LDL carries more cholesterol in the bloodstream than any other particle.

HDL Transports Cholesterol from Extrahepatic Tissue to the Liver High density lipoprotein complexes, HDL, are cholesterol scavengers. When cells of extrahepatic tissue break down, their cholesterol molecules are picked up by HDL particles and changed to cholesteryl esters. En route to the liver, HDL undergoes some changes of its own. For example, HDL transfers some cholesteryl esters to the adrenal glands and the ovaries to make steroid hormones from cholesterol. By one means or another, some of the HDL becomes more like LDL before entering liver cells by means of LDL receptors (see Figure 27.1). Specific HDL receptors also occur on liver cells. When receptors for cholesterol-bearing lipoprotein complexes are absent, defective, or in too few numbers, either at the liver or at extrahepatic tissue, there are serious consequences, which are discussed in Interaction 27.1. You can read there about "good" and "bad" cholesterol.

■ American scientists J. L. Goldstein and M. S. Brown shared the 1985 Nobel Prize in Physiology or Medicine for their work on LDL receptor proteins and how they help to control blood cholesterol levels.

27.2 STORAGE AND MOBILIZATION OF LIPIDS

Stored triacylglycerol has a high energy density.

The energy stored per gram of tissue or per gram of a solution is called the *energy density* of the material. Isotonic glucose solution, for example, carries only about 0.2 kcal per gram. When the glucose is changed into glycogen, however, *and is no longer in solution,* there is little associated water. The energy density of wet glycogen is about 1.7 kcal/g. Triacylglycerol, in sharp contrast, has an energy density of about 7.7 kcal/g.

A 70-kg adult male has about 12 kg of triacylglycerol in storage. It would supply his caloric needs for 43 days if he needed 2500 kcal/day and if he ate no food, just had water and a vitamin–mineral supplement. Of course, during this time the body proteins would also be wasting away, and metabolic acidosis would be a problem of growing urgency.

■ Because lipids are water insoluble, they attract the least amount of associated water in storage.

■ These data are for information; they're certainly not recommendations!

INTERACTION 27.1
"GOOD" AND "BAD" CHOLESTEROL AND LIPOPROTEIN RECEPTORS

The receptor proteins for LDL, whether on the liver or on extrahepatic tissue (see Figure 27.1), have a crucial function in removing cholesterol from circulation. When the receptors are defective or absent, the level of cholesterol in the blood increases, resulting in *atherosclerosis*. This is a progressive disease in which several substances, but mostly cholesterol esters, form thickenings (atheromas, after the Greek *anthera,* mush) in the arterial wall (see Figure 1). The thickenings slowly change into fibrous, calcified plaques, which are the chief cause of arteriosclerosis (hardening of the arteries).

Coronary heart disease exists when these developments occur in heart tissue. The plaques can fissure or they can be sites where clots form. A heart attack (myocardial infarction) happens when they block blood flow in heart tissue. A number of factors contribute to these developments, such as one's genetic makeup, high blood pressure, abdominal obesity, diabetes, smoking, and diets rich in saturated fats and cholesterol.

"Good" and "Bad" Cholesterol LDL particles are the richest of all lipoproteins in cholesterol, being 35% to 40% cholesteryl esters plus 7% to 10% unesterified cholesterol. Because elevated levels of LDL are so often involved in atherosclerosis, LDL cholesterol is often called "bad cholesterol." HDL complexes perform the desirable service of helping to carry cholesterol back to the liver, so HDL cholesterol is sometimes called "good cholesterol." (HDL carries 12% cholesteryl esters and 3% to 4% free cholesterol.)

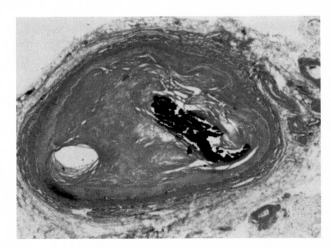

FIGURE 1 The buildup of plaque in this artery in the heart has almost closed it.

Cholesterol and Risks The Adult Treatment Panel of the National Cholesterol Education Program has assessed the risks of coronary heart disease according to two measurements, namely, the level of LDL cholesterol in the blood and the level of non-HDL cholesterol (see Table 1). The latter includes cholesterol not only in LDL but also in other lipoprotein complexes. ("Cholesterol" here means both the esterified and nonesterified forms.) The Panel suggests that, even without other risk factors, men over 35 years of age and postmenopausal women who are in the "very high" risk category be treated with cholesterol-lowering drugs. Where there are two or more other risk factors in men over 35 years of age and postmenopausal women, then the "high" risk category people probably should be given such drugs. The Panel also recommends that all adults 20 years old or more should have the total plasma cholesterol and the HDL cholesterol measured every 5 years. The total plasma cholesterol is a good predictor of an individual's eventually having coronary heart disease.

Familial Hypercholesterolemia Some people have a genetic defect that bears specifically on the LDL receptors at the liver and causes particularly high levels of LDL cholesterol. Two genes are involved. Those who carry two mutant genes have *familial hypercholesterolemia*, a genetically caused high level of cholesterol in the blood, 3 to 5 times higher than average. Even on a zero cholesterol diet, victims have very high cholesterol levels.

Victims of familial hypercholesterolemia generally have their first heart attacks as children and are dead by their early twenties. People with one defective gene and one normal gene for the LDL receptor proteins have blood cholesterol levels two or three times higher than normal. Although they number only about 0.5% of all adults, they account for 5% of all heart attacks among those younger than 60.

TABLE 1   Risk of Coronary Heart Disease[a]

Risk	LDL Cholesterol	Non-HDL Cholesterol
Very high	≥190	≥220
High	≥160	≥190
Desirable category	<130	<160
Best levels for coronary heart disease victims	<100	<130

[a] Units are in mg/dL. Data from R. J. Havel and E. Rapaport, *The New England Journal of Medicine,* Vol. 332, June 1, 1995, page 1494.

INTERACTION 27.2
BROWN FAT AND THERMOGENESIS

Both kinds of adipose tissue, white and brown, store triacylglycerols, but white adipose tissue does not metabolize them except to break them down to free fatty acids and glycerol. The fatty acids from white adipose tissue are then exported to other tissues for catabolism.

Respiratory Chain Oxidation and ATP Synthesis Are Uncoupled in Brown Adipose Tissue The triacylglycerols in brown adipose tissue are catabolized within this tissue for little other use than to generate heat. This is possible because cells of brown adipose tissue can switch off the capability of the inner mitochondrial membranes to maintain a proton gradient. Recall that the respiratory chain normally establishes a proton gradient across the inner mitochondrial membrane, and that as protons flow back through selected channels, they trigger the synthesis of ATP from ADP and P_i.

When the respiratory chain runs but cannot set up the proton gradient, the chemical energy released by the chain emerges only as heat. Such generation of heat is called *thermogenesis.* Two stimuli, both mediated by the neurotransmitter norepinephrine, trigger this heat-generating activity: exposure to cold (and the resulting shivering) and the ingestion of food.

The advantage to the body of thermogenesis induced by a cold outside temperature is that the body can oxidize its own fat to help keep itself warm.

The advantage of food-induced thermogenesis is that the individual is protected from getting fat by too much eating in relationship to exercise. Thermogenesis in brown adipose tissue removes fat by catabolism, rather than by exercise, as new calories (in the food) are imported.

Adipose Tissue Is the Principal Lipid Storage Depot Fatty acids are stored chiefly in adipose tissue of which there are two kinds, brown and white. Both are associated with internal organs, where they cushion the organs against mechanical bumps and shocks, and insulate them from swings in temperature. For a discussion of how brown adipose tissue uses the respiratory chain to generate heat, but not ATP, see Interaction 27.2. Interaction 27.3 describes a newly discovered hormone that participates in the regulation of the appetite and so has a part to play in obesity.

■ The relatively high concentration of mitochondria, which hold iron-containing enzymes (cytochromes), causes the color of brown fat tissue.

A Body in Dietary Balance Uses Fatty Acids as Well as Glucose for Energy The metabolic activities of white adipose tissue will concern us here. It stores energy as triacylglycerols chiefly on behalf of the energy budgets of other tissues. Some tissues, like heart muscle and the renal cortex, use breakdown products of fatty acids in preference to glucose for energy. Skeletal muscles can use both fatty acids and short molecules made from them for energy. In fact, most of the energy needs of resting muscle tissue are met by intermediates of fatty acid catabolism, not from glucose. Given enough time for adjustment, during starvation, for example, even brain cells can obtain some energy from fatty acid breakdown products.

Fatty acids, consequently, come and go from adipose tissue, and the balance between this tissue's receiving or releasing them is struck by the energy requirements elsewhere. In extreme distress, when either glucose is in very low supply (as in starvation) or what is available can't be used (as in diabetes), the body must turn almost entirely to its fatty acids for energy.

Figure 27.2 outlines the many steps involved in tapping the lipid reserves for their energy. (The figure also reminds us of the connection between glucose and ATP.) Triacylglycerol molecules are first hydrolyzed to free fatty acids and glycerol, the lipase being activated by epinephrine and cyclic AMP. The fatty acids are carried as albumin complexes to the liver, the chief site for their catabolism.

■ Epinephrine is also involved in glucose metabolism.

Insulin suppresses the lipase that releases fatty acids from adipose tissue. Thus when insulin is in circulation, and its presence is linked to the presence of blood glucose, the fatty acids are less required for energy.

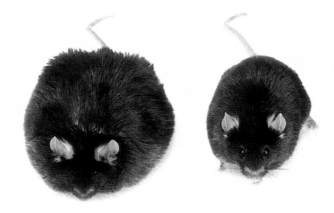
The glycerol produced when the fatty acids are released is changed to dihydroxyacetone phosphate, and it enters the pathway of glycolysis.

27.3 OXIDATION OF FATTY ACIDS

Acetyl groups, made from fatty acids, enter the citric acid cycle.

The degradation of fatty acids takes place inside mitochondria by a repeating series of steps known as the **β-oxidation pathway** (Figure 27.3).

A Fatty Acid Is First Joined to Coenzyme A   The β-oxidation pathway occurs in the mitochondrial matrix, but before a fatty acid can enter β-oxidation, has to be joined to coenzyme A in the cytosol. The cost is one ATP, which changes to AMP and PP_i. *The subsequent hydrolysis of the diphosphate is the driving force for the overall change.* So the actual cost in high-energy phosphate to launch β-oxidation is *two* high-energy phosphate bonds.

■ The β-oxidation pathway has also been called the *fatty acid cycle,* but it isn't exactly a true *cycle,* like the citric acid cycle.

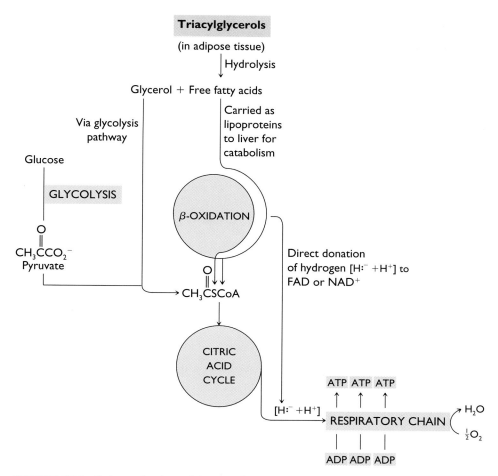

FIGURE 27.2 Pathways for the mobilization of energy reserves in the triacylglycerols of adipose tissue.

The fatty acyl unit of fatty acyl CoA is passed through the mitochondrial membrane to be rejoined inside to coenzyme A. This sequesters the fatty acyl unit in a place of *catabolism,* not anabolism.

Fatty Acyl CoA Is Catabolized by Two Carbons at a Time   The repeating sequence of the β-oxidation pathway consists of four steps. One "turn" of the sequence is used to make one molecule of $FADH_2$, one of NADH, one of acetyl coenzyme A, and a fatty acyl unit with two fewer carbons. The shortened fatty acyl unit goes through the four steps again, and the process is repeated until no more two-carbon acetyl units can be made.

The $FADH_2$ and NADH produced by β-oxidation fuel the respiratory chain directly. The acetyl groups do so indirectly, passing into the citric acid cycle, or they enter the general pool of acetyl coenzyme A that the body uses to make other substances, like cholesterol.

Let's now look at the four steps of β-oxidation in greater detail. We do so partly to demonstrate that although the molecules are complex, the individual changes are not, and we've studied most of them with simple systems. The numbers that follow refer to Figure 27.3.

■ Franz Knoop directed much of the research on the fatty acid cycle, so this pathway is sometimes called *Knoop oxidation.*

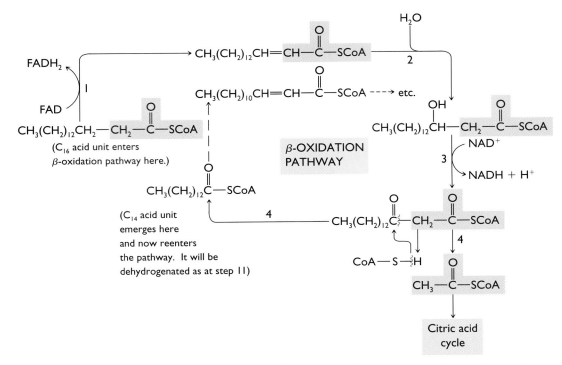

FIGURE 27.3 The β-oxidation pathway. The numbers refer to the numbered steps discussed in the text.

1. **Dehydrogenation.** FAD accepts ($H:^- + H^+$) from the α- and β-carbons of the fatty acyl unit of palmityl coenzyme A. (We've not studied an example of this kind of reaction.)

■ The enzyme is *acyl-CoA dehydrogenase.*

$$CH_3(CH_2)_{12}CH_2\!-\!CH_2\overset{\displaystyle O}{\overset{\|}{C}}SCoA + FAD \xrightarrow{\boxed{1}}$$

palmityl coenzyme A

$$CH_3(CH_2)_{12}CH\!=\!CH\overset{\displaystyle O}{\overset{\|}{C}}SCoA + FADH_2 \xrightarrow[FAD]{} [H:^- + H^+] \rightarrow$$

an α,β-unsaturated acyl derivative of coenzyme A

transfers to respiratory chain

$FADH_2$ delivers hydrogen to the respiratory chain at enzyme complex II.

2. **Hydration.** Water adds to an alkene double bond, and a 2° alcohol forms.

■ The enzyme is *enoyl-CoA hydratase.*

$$CH_3(CH_2)_{12}CH\!=\!CH\overset{\displaystyle O}{\overset{\|}{C}}SCoA + H_2O \xrightarrow{\boxed{2}} CH_3(CH_2)_{12}\overset{\displaystyle OH}{\overset{|}{C}}HCH_2\overset{\displaystyle O}{\overset{\|}{C}}SCoA$$

a β-hydroxyacyl derivative of coenzyme A

3. **Dehydrogenation.** The 2° alcohol is dehydrogenated (oxidized) to a keto group. These three steps accomplish the oxidation of the β-position of the fatty acyl

group to a keto group, which is why the pathway is called *beta*-oxidation. *The purpose is to weaken the C-2 to C-3 bond,* setting it up for being broken, next.

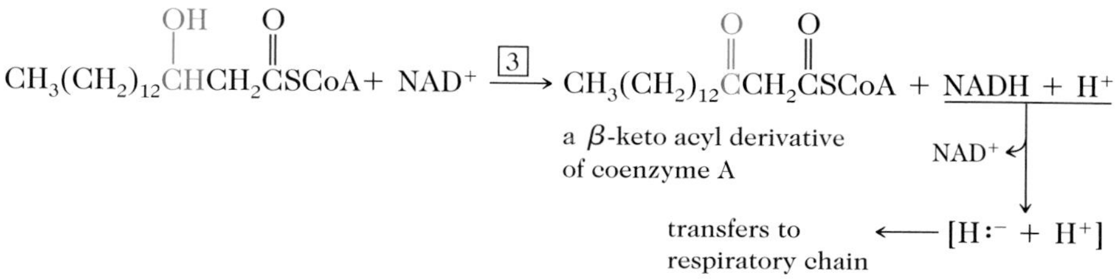

$$CH_3(CH_2)_{12}\overset{\overset{OH}{|}}{C}HCH_2\overset{\overset{O}{\|}}{C}SCoA + NAD^+ \xrightarrow{\boxed{3}} CH_3(CH_2)_{12}\overset{\overset{O}{\|}}{C}CH_2\overset{\overset{O}{\|}}{C}SCoA + \underline{NADH + H^+}$$

a β-keto acyl derivative
of coenzyme A

$$NAD^+$$

transfers to   ⟵ [H:⁻ + H⁺]
respiratory chain

■ The enzyme is 3-L-*hydroxyacyl-CoA dehydrogenase.*

4. **Bond Cleavage.** The bond between the α-carbon and the β-carbon now breaks. (This reaction is little more than a *reverse* Claisen condensation. The forward Claisen reaction is presented in Appendix D.)

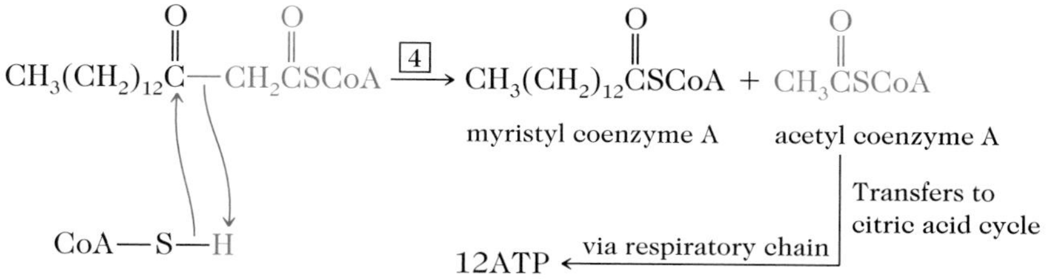

$$CH_3(CH_2)_{12}\overset{\overset{O}{\|}}{C}-CH_2\overset{\overset{O}{\|}}{C}SCoA \xrightarrow{\boxed{4}} CH_3(CH_2)_{12}\overset{\overset{O}{\|}}{C}SCoA + CH_3\overset{\overset{O}{\|}}{C}SCoA$$

myristyl coenzyme A   acetyl coenzyme A

Transfers to
citric acid cycle

CoA—S—H

12ATP ⟵ via respiratory chain

■ The enzyme is *thiolase* (or *β-ketoacyl-CoA thiolase*).

The remaining acyl unit, the original shortened by two carbons, now goes through the cycle of steps again: dehydrogenation, hydration, dehydrogenation, and bond cleavage. After seven such cycles, one molecule of palmityl coenzyme A is broken into eight molecules of acetyl coenzyme A.

One Palmityl Unit Yields 129 ATP Molecules The maximum yield of ATP from the oxidation of one unit of palmityl coenzyme A adds up to 131 ATPs (see Table 27.2). The net from palmitic acid is two ATPs fewer, or 129 ATPs, because the activation of the palmityl unit requires this initial investment, as we mentioned earlier.

27.4 BIOSYNTHESIS OF FATTY ACIDS

Acetyl CoA molecules can be made into fatty acids.

In this section we'll look at what superficially is the reverse of β-oxidation. We'll see how acetyl CoA can be made into long-chain fatty acids. Thus, acetyl CoA stands at

TABLE 27.2   Maximum Yield of ATP from Palmityl CoA

Intermediates Produced by Seven Turns of the Fatty Acid Cycle	ATP Yield per Intermediate	Total ATP Yield
7 FADH₂	2	14
7 NADH	3	21
8 acetyl coenzyme A	12	96
		131 ATP
Deduct two high-energy phosphate bonds for activating the acyl unit		− 2
Net ATP yield per palmityl unit		129 ATP

a major metabolic crossroads. It can be made from any dietary monosaccharide, from virtually all amino acids, and from fatty acids. Once made from any starting material, it can be shunted into the citric acid cycle, where its chemical energy can be used to make ATP; or its acetyl group can be made into other compounds that the body needs.

Fatty Acid Synthesis Begins with the Activation of Acetyl CoA Whenever acetyl CoA molecules are made within mitochondria but aren't needed for *catabolism* (the citric acid cycle and respiratory chain), they are exported to the cytosol for *anabolism,* the synthesis of other species.

The cell must invest some ATP to make fatty acids from smaller molecules, and the first payment occurs in the first step. Bicarbonate ion reacts with acetyl CoA to form malonyl CoA.

- The enzyme for this step, *acetyl-CoA carboxylase*, requires the vitamin biotin.

$$CH_3\overset{O}{\overset{\|}{C}}SCoA + HCO_3^- + ATP \longrightarrow {}^-O\overset{O}{\overset{\|}{C}}CH_2\overset{O}{\overset{\|}{C}}SCoA + 2H^+ + ADP + P_i$$

acetyl coenzyme A   malonyl coenzyme A

The malonyl CH_2 unit is between two carbonyl groups and so is activated for fatty acid synthesis.

A Molecular "Construction Boom" Moves the Growing Fatty Acyl Unit from Enzyme to Enzyme A huge complex of seven enzymes called *fatty acid synthase* (Figure 27.4) now builds a fatty acid. A long molecular unit, the *acyl carrier protein,* or ACP, in the center of this complex, serves as a swinging arm carrier. It's like the boom of a construction crane. Thus the arm first picks up the malonyl unit and brings it over the first enzyme. A change occurs, and the boom carries the product over the second enzyme. At each succeeding stop a reaction is catalyzed that contributes to chain lengthening.

Figure 27.4 outlines the process. The enzymes themselves are represented as colored circles keyed by numbered boxes to their names.

The transfer of the malonyl unit of malonyl CoA to the swinging arm of ACP occurs at enzyme $\boxed{1}$.

$$ {}^-O\overset{O}{\overset{\|}{C}}CH_2\overset{O}{\overset{\|}{C}}S-CoA + ACP \longrightarrow {}^-O\overset{O}{\overset{\|}{C}}CH_2\overset{O}{\overset{\|}{C}}S-ACP + CoA$$

malonyl coenzyme A   malonyl ACP

Construction booms are moved around a central axis to deliver materials to various sites of the project.

A similar reaction also occurs over enzyme $\boxed{6}$ where another molecule of acetyl CoA ties to a different unit of the synthase, *E*. The product is acetyl *E*.

$$CH_3\overset{O}{\overset{\|}{C}}S-CoA + E \longrightarrow CH_3\overset{O}{\overset{\|}{C}}S-E + CoA$$

acetyl *E*

Next, enzyme $\boxed{1}$ transfers the acetyl group of acetyl *E* to the malonyl group of malonyl ACP. Carbon dioxide, the initial activator of the acetyl unit, is ejected.

$$CH_3\overset{O}{\overset{\|}{C}}S-E + {}^-O\overset{O}{\overset{\|}{C}}CH_2\overset{O}{\overset{\|}{C}}S-ACP \longrightarrow CH_3\overset{O}{\overset{\|}{C}}CH_2\overset{O}{\overset{\|}{C}}S-ACP + CO_2 + E$$

acetyl *E*   malonyl ACP   acetoacetyl ACP

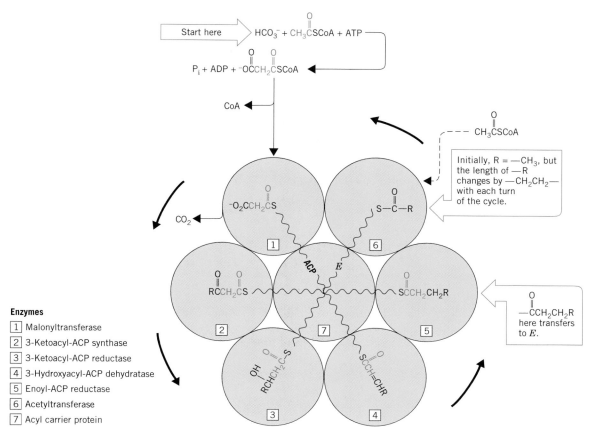

FIGURE 27.4 The synthesis of fatty acids. At the top, an acetyl group is activated and joined as a malonyl unit to an arm of the acyl carrier protein, ACP. Another acetyl group transfers from acetyl CoA to site E. In a second transfer, this acetyl group is then joined to the malonyl unit as CO_2 splits back out. This gives a β-ketoacyl system whose keto group is reduced to CH_2 by the next series of steps. One turn of the "cycle" adds a CH_2CH_2 unit to the growing acyl chain.

The loss of CO_2 is, in fact, the driving force for this transfer, a force embedded in malonyl ACP by energy from ATP. The E unit is vacated. Acetoacetyl ACP forms; it's now over enzyme 2. The chain has grown by two carbons.

The keto group in acetoacetyl ACP is next reduced by enzyme 3 to a 2° alcohol and passed to enzyme 4. The alcohol group is dehydrated by enzyme 4 to introduce a double bond, putting the unit over enzyme 5. The double bond is next reduced over enzyme 5 to give butyryl ACP. *The overall effect of these steps is to reduce the keto group to* CH_2. Notice that NADPH, the reducing agent manufactured by the pentose phosphate pathway of glucose catabolism, is used here, not NADH.

■ Glucagon, epinephrine, and cyclic AMP—all stimulators of the use of glucose to make ATP—depress the synthesis of fatty acids in the liver. Insulin, however, promotes it.

$$
\underset{\text{acetoacetyl ACP}}{CH_3\overset{O}{\overset{\|}{C}}CH_2\overset{O}{\overset{\|}{C}}S-ACP} \xrightarrow[\substack{\text{(reduction of the keto} \\ \text{group by enzyme 3)}}]{\text{NADPH}+\text{H}^+ \quad \text{NADP}^+} \underset{\substack{\text{(dehydration} \\ \text{by enzyme 4)}}}{CH_3\overset{OH}{\overset{|}{C}}HCH_2\overset{O}{\overset{\|}{C}}S-ACP} \Big\downarrow H_2O
$$

$$
\underset{\text{butyryl ACP}}{CH_3CH_2CH_2\overset{O}{\overset{\|}{C}}S-ACP} \xleftarrow[\substack{\text{(reduction of the double} \\ \text{bond at enzyme 5)}}]{\text{NADP}^+ \quad \text{NADPH}+\text{H}^+} CH_3CH=CH\overset{O}{\overset{\|}{C}}S-ACP
$$

The butyryl group is now transferred to the *vacant E* unit of the synthase, the unit that initially held an acetyl group. Butyryl *E* instead of acetyl *E* is now over enzyme $\boxed{6}$. This ends one complete cycle of the β-oxidation pathway. To recapitulate, we have gone from two-carbon acetyl units to one four-carbon butyryl unit.

The steps now repeat. A new malonyl unit is joined to the ACP. The newly made *butyryl* group is transferred to the malonyl unit as CO_2 is again ejected. This elongates the chain to six carbons, positions it on the swinging arm, and gets it ready for the multistep reduction of the keto group to CH_2. The chain will be that of the six-carbon acyl group, the hexanoyl group.

In the next repetition of this series, the six-carbon acyl group will be elongated to an eight-carbon group, and the process will repeat until the chain is 16 carbons long. Overall, the net equation for the synthesis of the palmitate ion from acetyl CoA is

■ $CH_3CH_2CH_2CH_2CH_2C-$

hexanoyl group

■ Because the symbols ATP, ADP, and P_i are given without their electrical charges, we can't provide an electrical balance to equations such as this.

■ The pentose phosphate pathway for the catabolism of glucose is the body's supplier of NADPH.

$$8CH_3CSCoA + 7ATP + 14NADPH \longrightarrow$$
$$CH_3(CH_2)_{14}CO_2^- + 7ADP + 7P_i + 8CoA + 14NADP^+ + 6H_2O$$

palmitate ion

It takes 14 NADPH to make one palmitate ion, a heavy demand. If acids with chains longer than that of the palmitate ion are needed, or if acids with double bonds are to be made, additional steps or different pathways using different enzymes are taken.

27.5 BIOSYNTHESIS OF CHOLESTEROL

Excessive cholesterol can inhibit the formation of a key enzyme required in the multistep synthesis of cholesterol.

■

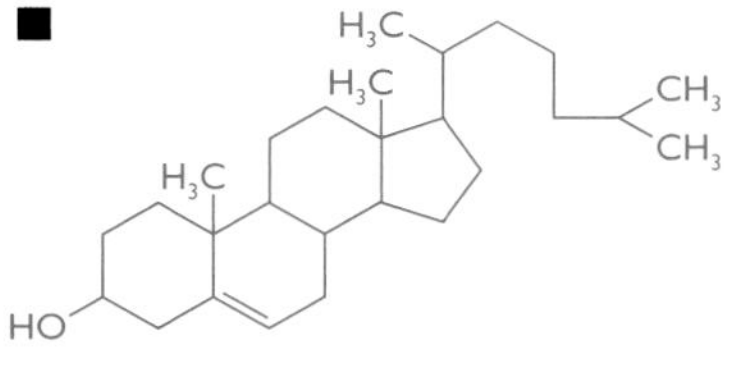

cholesterol

Cholesterol, the end product of a long, multistep process starting with acetyl CoA, is used by the body to make various bile salts, sex hormones, and cell membranes. About 80% to 95% of all cholesterol synthesis in mammals takes place in cells of the liver and the intestines. We won't go into all the details, but far enough to learn how the body controls the process. If sufficient cholesterol is in the diet, then the body's synthesis should be shut down. Let's see how this is done.

Cholesterol Is Made from Acetyl Units   When the level of acetyl CoA builds up in the liver, the following equilibrium shifts to the right (in accordance with Le Châtelier's principle).

■ The forward reaction is an example of the Claisen ester condensation (Appendix D), the joining of an acyl group of one ester molecule to the α-position of a second ester molecule.

$$2CH_3CSCoA \rightleftharpoons CH_3CCH_2CSCoA + CoASH$$

acetyl CoA   acetoacetyl CoA

When cholesterol synthesis is switched on, acetoacetyl CoA combines with another acetyl CoA.

$$CH_3\overset{O}{\overset{\|}{C}}CH_2\overset{O}{\overset{\|}{C}}-SCoA + CH_3\overset{O}{\overset{\|}{C}}SCoA \underset{}{\overset{HMG-CoA\ synthase}{\rightleftharpoons}} {}^-O\overset{O}{\overset{\|}{C}}CH_2\underset{CH_3}{\overset{OH}{\overset{|}{C}}}CH_2\overset{O}{\overset{\|}{C}}SCoA + CoASH \qquad (27.1)$$

HMG–CoA
(β-hydroxy-β-methyl-
glutaryl CoA)

■ The reaction is the addition of one carbonyl compound (acetyl CoA) to the keto group of another, so it strongly resembles the aldol condensation (Appendix D).

The Reduction of HMG–CoA Commits the Cell to Cholesterol Synthesis Both a reduction and a hydrolysis occur in the next step, which is a complex change catalyzed by *HMG–CoA reductase,* a key enzyme.

$$HMG-CoA + 2NADPH + 2H^+ \overset{HMG-CoA}{\underset{reductase}{\longrightarrow}}$$

$$HOCH_2CH_2\underset{CH_3}{\overset{OH}{\overset{|}{\underset{|}{C}}}}CH_2CO_2{}^- + 2NADP^+ + CoASH$$

mevalonate

■ Fifteen Nobel Prizes have gone to scientists who devoted the better parts of their careers to various aspects of cholesterol and its uses in the body.

Mevalonate is next carried through a long series of reactions until cholesterol is made. As we said, we'll not take it that far, but consider, instead, how cholesterol synthesis is controlled.

Cholesterol Is a Natural Inhibitor of HMG–CoA Reductase The control of HMG–CoA reductase is the major factor in the overall control of the biosynthesis of cholesterol. Cholesterol itself is one inhibitor, and it works by inhibiting both the *synthesis* of the enzyme and the enzymatic activity of any of its existing molecules. In the presence of cholesterol, the enzyme isn't totally deactivated. There is just *less* of it free to do catalytic work. Thus if the diet is relatively rich in cholesterol, the body tends to make less. If the diet is very low in cholesterol, the body makes more.

■ On a low-cholesterol diet, an adult makes about 800 mg of cholesterol per day.

Lovastatin and Compactin Lower Blood Cholesterol Levels Drugs in the *statin* family have appeared since the mid 1980s that dramatically lower the rate at which cholesterol is made. Lovastatin (Mevacor), pravastatin (Pravachol), simvastatin (Zocor), and mevastatin (Compactin) are examples. HMG–CoA reductase is a key choke point in cholesterol synthesis, and the statins work partly as competitive inhibitors of the enzyme. The statin molecules carry a part that resembles mevalonate and thus are able to compete with mevalonate for a position on the enzyme.

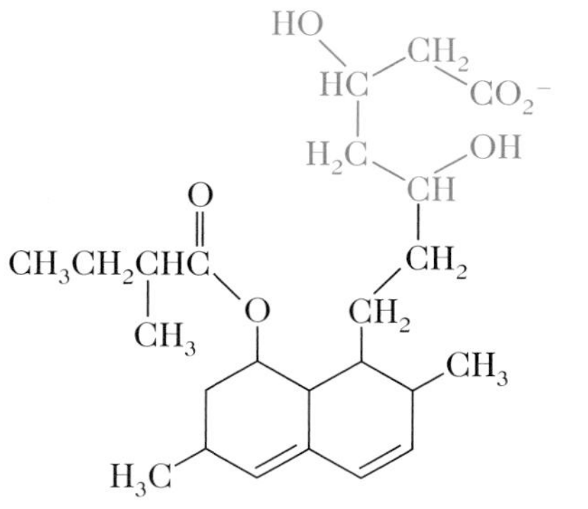

lovastatin

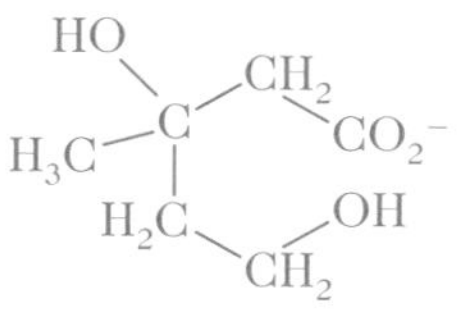

mevalonate

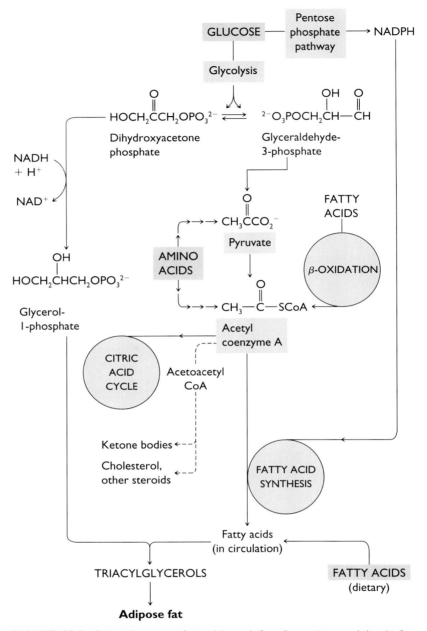

FIGURE 27.5   Principal sources of triacylglycerols for adipose tissue and the chief uses of acetyl CoA.

Some statins also increase the synthesis of the mRNA responsible for making the liver LDL receptors. With more of these, the liver is better able to reabsorb cholesterol, recycle it, or export it in the bile.

Carbohydrate and Lipid Metabolisms Are Intertwined   Figure 27.5 summarizes much of what we have covered in this and the previous two chapters about the uses of acetyl CoA and its relationship to carbohydrate and lipid catabolism. Notice particularly that triacylglycerols can be made from any of the three dietary components: carbohydrates, lipids, and proteins.

27.6 KETOACIDOSIS

An acceleration of the fatty acid cycle tips some equilibria toward ketoacidosis.

Cells of certain tissues have to engage in gluconeogenesis in two serious conditions, starvation and uncontrolled diabetes mellitus. The consequences are fatal unless the underlying causes are treated. In starvation, the blood sugar level drops because of nutritional deficiencies, so the body (principally the liver) tries to compensate by making glucose.

The Level of Acetyl CoA Increases when Gluconeogenesis Is Accelerated   Gluconeogenesis consumes oxaloacetate, the carrier of acetyl units in the citric acid cycle. Thus, when oxaloacetate is diverted from the citric acid cycle, acetyl coenzyme A is less able to unload its acetyl group into the cycle. Yet acetyl coenzyme A continues to be made by the fatty acid cycle, so *acetyl CoA levels build up.*

As the supply of acetyl CoA increases in the liver, the following equilibrium shifts to the right to make acetoacetyl CoA.

$$2CH_3\overset{O}{\overset{\|}{C}}SCoA \rightleftharpoons CH_3\overset{O}{\overset{\|}{C}}CH_2\overset{O}{\overset{\|}{C}}-SCoA + CoASH$$

acetyl CoA   acetoacetyl CoA

As the level of acetoacetyl CoA increases, equilibrium 27.1 shifts to the right to make more HMG–SCoA. As the level of HMG–SCoA increases (and little if any is being diverted to the synthesis of cholesterol), a liver enzyme splits it to acetoacetate ion and coenzyme A.

$$HMG\text{–}SCoA \longrightarrow CH_3\overset{O}{\overset{\|}{C}}CH_2\overset{O}{\overset{\|}{C}}O^- + CoASH$$

acetoacetate

The net effect of these steps, which gets us to the point, is that *excessive* amounts of acetyl CoA generate acetoacetate and hydrogen ion. The overall effect is

$$2CH_3\overset{O}{\overset{\|}{C}}SCoA + H_2O \longrightarrow CH_3\overset{O}{\overset{\|}{C}}CH_2\overset{O}{\overset{\|}{C}}O^- + 2CoASH + H^+$$

acetoacetate

The formation of hydrogen ion makes the situation dangerous, and *an increased synthesis of "new" glucose was the cause.* As we said, conditions of starvation or untreated diabetes mellitus can be instigators of accelerated gluconeogenesis. Figure 27.6 outlines the chain of events that occurs in untreated diabetes, and you may wish to refer to this figure as you continue with the following discussion.

Accelerated Acetoacetate Production Leads to Acidosis   The acid produced by the formation of acetoacetate must be neutralized by the buffer. Under an increasingly rapid production of acetoacetate and hydrogen ion, the blood buffer slowly loses

■ Again we see applications of Le Châtelier's principle; the increase in the level of acetyl CoA is the stress and the equilibrium, by shifting to the right, relieves the stress.

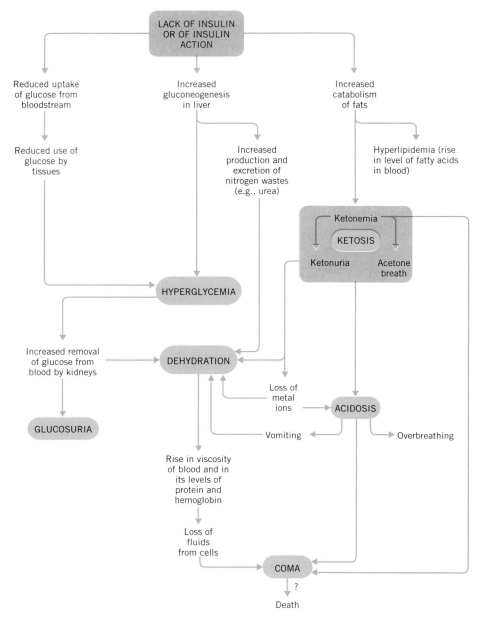

FIGURE 27.6   The principal sequence of events in untreated diabetes.

ground. A condition of *acidosis* sets in. It's *metabolic* acidosis, because the cause lies in a disorder of metabolism. Because the chief species responsible for this acidosis has a keto group, the condition is often called **ketoacidosis.**

Blood Levels of the Ketone Bodies Increase in Starvation and Diabetes   The *acetoacetate ion* is called one of three **ketone bodies.** The other two are *acetone* and the *β-hydroxybutyrate ion*. Both are produced from the acetoacetate ion. Acetone arises from acetoacetate by the loss of the carboxyl group.

$$CH_3CCH_2CO^- + H_2O \longrightarrow CH_3CCH_3 + HCO_3^-$$

acetoacetate   acetone

β-Hydroxybutyrate is produced when the keto group of acetoacetate is reduced by NADH.

$$CH_3CCH_2CO^- + NADH + H^+ \longrightarrow CH_3CHCH_2CO^- + NAD^+$$

acetoacetate β-hydroxybutyrate

The ketone bodies enter circulation. Acetone is volatile, so most of it leaves the body via the lungs. Individuals with severe ketoacidosis have "acetone breath," the noticeable odor of acetone on the breath.

The ketone bodies are not in themselves abnormal constituents of blood. Acetoacetate and β-hydroxybutyrate can be used in skeletal muscles to make ATP. Heart muscle uses these two for energy in preference to glucose. Even the brain, given time, can adapt to using these ions for energy when the blood sugar level drops in starvation or prolonged fasting. Only when the ketone bodies are produced at a rate faster than the blood buffer can handle the associated acid are they a problem.

The Conditions of Ketonemia, Ketonuria, and "Acetone Breath" Collectively Constitute Ketosis Normally, the levels of acetoacetate and β-hydroxybutyrate in the blood are, respectively, 2 μmol/dL and 4 μmol/dL. In prolonged, undetected, and untreated diabetes, these values can increase as much as 200-fold. The condition of excessive levels of ketone bodies in the blood is called **ketonemia.**

As ketonemia becomes more and more advanced, the ketone bodies begin to appear in the urine, a condition called **ketonuria.** When there is a combination of ketonemia, ketonuria, and acetone breath, the overall state is called **ketosis.** The individual will be described as *ketotic.* As unchecked ketosis becomes more severe, the associated ketoacidosis worsens and the pH of the blood continues its fatal descent.

The Urinary Removal of Organic Anions Means the Loss of Base from the Blood To leave the ketone body anions in the urine, the kidneys have to leave positive ions with them to keep everything electrically neutral. Na^+ ions, the most abundant cations, are used. One Na^+ ion has to leave with each acetoacetate ion, for example. This loss of Na^+ is often referred to as the "loss of base" from the blood, although Na^+ is not itself a base. But the loss of one Na^+ stems from the appearance of one acetoacetate ion *plus one H^+ ion* that the blood had to neutralize using HCO_3^-. Thus each Na^+ that leaves the body corresponds to the loss of one HCO_3^- ion, the true base, consumed in neutralizing one H^+. This is why the loss of Na^+ is an indicator of the loss of base.

Another way to understand the urinary loss of Na^+ as the loss of base from the blood is that a Na^+ ion has to accompany a bicarbonate ion when it goes from the kidneys into the blood. The kidneys manufacture HCO_3^- ions normally in order to replenish the blood buffer system. The greater the number of Na^+ ions that have to be left in the *urine* in order to clear ketone bodies from the blood, the less is the amount of true base, HCO_3^-, that can be put into the *blood.*

Diuresis Must Accelerate to Handle Ketosis The solutes that are leaving the body in the urine cannot, of course, be allowed to make the urine too concentrated. Otherwise, osmotic pressure balances are upset. Therefore increasing quantities of water

■ β-Hydroxybutyrate is a *ketone* body not because it has a keto group but because it is made from and is found together with a species that does.

■ The vapor pressure of acetone at body temperature is nearly 400 mm Hg (about 0.53 atm), so it readily evaporates from the blood in the lungs.

■ 1 μmol = 1 micromole = 10^{-6} mol.

■

Condition	[HCO₃⁻]$_{blood}$ in mmol/L
Normal	22–30
Mild acidosis	16–22
Moderate acidosis	10–16
Severe acidosis	<10

■ **Polyuria** is the technical name for the overproduction of urine.

must be excreted. To satisfy this need, the individual has a powerful thirst. Other wastes, such as urea, are also being produced at higher than normal rates, because amino acids are being sacrificed in gluconeogenesis. These wastes add to the demand for water to make urine.

Internal Water Shortages in Ketosis Spell Dehydration of Critical Tissues   If, during a state of ketosis, insufficient water is drunk, then water is simply taken from extracellular fluids. The blood volume therefore tends to drop, and the blood becomes more concentrated. It also thickens and becomes more viscous, which makes the delivery of blood more difficult.

Because the brain has the highest priority for blood flow, some of this flow is diverted from the kidneys to try to ensure that the brain gets what it needs. This only worsens the situation in the kidneys, and they have an increasingly difficult time clearing wastes. As the water shortage worsens, some water is borrowed from the intracellular supply. This, in addition to a combination of other developments, leads to coma and, eventually, death.

SUMMARY

Lipid absorption and distribution   As fatty acids and monoacylglycerol move out of the digestive tract they become reconstituted as triacylglycerols. These, in addition to cholesterol and proteins, are packaged as chylomicrons. As chylomicrons move in the bloodstream, they unload some of their triacylglycerols and change to chylomicron remnants, which are taken up by the liver. The liver organizes cholesterol, triacylglycerols available to it, and proteins into VLDL and then sends these very low density lipoprotein complexes into circulation. Triacylglycerols are again unloaded where they are needed, and the VLDL become more dense, changing into intermediate density complexes, IDL. Much of these are reabsorbed by the liver. Those that aren't become more dense and change over to low density lipoprotein complexes, LDL. Most of the LDL finds its way to extrahepatic tissues to deliver cholesterol for cell membranes. Endocrine glands need cholesterol to make steroid hormones. LDL can be reabsorbed by the liver to recycle its lipids. Cholesterol not needed in extrahepatic tissue is carried back to the liver by high density lipoprotein complexes, HDL. The liver can excrete excess cholesterol via the bile, or it makes bile salts.

Storage and mobilization of lipids   The favorable energy density of triacylglycerol means that more energy is stored per gram of this material than can be stored by any other chemical system. The adipose tissue is the principal storage site, and fatty material comes and goes from this tissue according to the energy budget of the body. When the energy of fatty acids is needed, they are liberated from triacylglycerols and carried to the liver, the chief site of fatty acid catabolism.

Catabolism of fatty acids   Fatty acyl groups, after being pinned to coenzyme A inside mitochondria, are catabolized

by the β-oxidation pathway. By a succession of four steps—dehydrogenation, hydration of a double bond, oxidation of the resulting alcohol, and cleavage of the bond from the α- to the β-carbon—one cycle of the β-oxidation pathway removes one two-carbon acetyl group. The pathway then repeats another cycle as the shortened fatty acyl group continues to be degraded. Each cycle of reactions produces one $FADH_2$ and one NADH, which pass $H:^-$ to the respiratory chain for the synthesis of ATP. Each cycle of β-oxidation also sends one acetyl group into the citric acid cycle which, via the respiratory chain, leads to several more ATPs. The net ATP production is 129 ATPs per palmityl residue.

Biosynthesis of fatty acids   Fatty acids can be made by a repetitive cycle of steps. It begins by building one butyryl group from two acetyl groups. The four-carbon butyryl group is attached to an acyl carrier protein that acts as a swinging arm on the enzyme complex. This arm moves the growing fatty acyl unit first over one enzyme and then another as additional two carbon units are added. The process consumes ATP and NADPH.

Biosynthesis of cholesterol   Cholesterol is made from acetyl groups by a long series of reactions. The synthesis of one of the enzymes is inhibited by excess cholesterol, which gives the system a mechanism for keeping its own cholesterol synthesis under control. Drugs in the statin family are competitive inhibitors of a key enzyme in the biosynthesis of cholesterol.

Ketoacidosis   Acetoacetate, β-hydroxybutyrate, and acetone build up in the blood—ketonemia—in starvation or in diabetes. The first two are normal sources of energy in some tissues. When they are made faster than they can be metabolized, however, there is an accompanying increased loss of bicarbonate ion from the blood's carbonate buffer. This leads to a form of metabolic acidosis called keto-acidosis.

The kidneys try to leave the anionic ketone bodies in the urine, but this requires both Na^+ (for electrical neutrality) and water (for osmotic pressure balances). The excessive loss of Na^+ in the urine is interpreted as a loss of "base." With Na^+ leaving the body in the urine, less is available to accompany replacement HCO_3^-, made by the kid-neys, when this base should be going into the blood. Under developing ketoacidosis, the kidneys have extra nitrogen wastes and, in diabetes, extra glucose that needs to leave via the kidneys in the urine. This also demands an increased volume of water. Unless this is brought in by the thirst mechanism, it has to be sought from within. However, the brain has first call on blood flowage, so the kidneys suffer more. Eventually, if these events continue unchecked, the victim goes into a coma and dies.

REVIEW EXERCISES

The answers to Review Exercises whose numbers are in color are found in Appendix E. The answers to the other Review Exercises are found in the Study Guide that accompanies this book. The more challenging questions are marked with asterisks.

Absorption and Distribution of Lipids

27.1 What are the end products of the digestion of triacylglycerols?

27.2 What happens to the products of the digestion of triacylglycerols as they migrate out of the intestinal tract?

27.3 What are chylomicrons and what is their function?

27.4 What happens to chylomicrons as they move through capillaries of, say, adipose tissue?

27.5 What happens to chylomicron remnants when they reach the liver?

27.6 What are the two chief sources of cholesterol that the liver exports?

27.7 What do the following symbols stand for?
(a) VLDL   (b) IDL   (c) LDL   (d) HDL

27.8 The loss of what kind of substance from the VLDL converts it into IDL?

27.9 What happens to cause the increase in density between VLDL and LDL?

27.10 What tissue can reabsorb IDL complexes?

27.11 What is the chief constitutent of LDL?

27.12 In extrahepatic tissue, what two general uses await delivered cholesterol?

27.13 If the liver lacks the key receptor proteins, which specific lipoprotein complexes can't be reabsorbed?

27.14 Explain the relationship between the liver's receptor proteins for lipoprotein complexes and the control of the cholesterol level of the blood.

27.15 What is the chief job of the HDL?

*27.16 Why does HDL have a higher density than chylomicrons?

Storage and Mobilization of Lipids

27.17 With reference to the storage of chemical energy in the body, what is meant by *energy density*?

27.18 Arrange the following in their order of increasing quantity of energy that they store per gram.

Wet glycogen	Isotonic glucose	Adipose lipids
1	2	3

27.19 Briefly describe two conditions in which the body would have to turn to the fatty acids for energy.

27.20 How does insulin suppress the mobilization of fatty acids from adipose tissue?

27.21 Arrange the following processes in the order in which they occur when the energy in storage in triacylglycerols is mobilized.

β-Oxidation	Lipolysis in adipose tissue	Citric acid cycle
A	B	C

Lipoprotein formation	Oxidative phosphorylation	
D	E	

27.22 What specific function does β-oxidation have in obtaining energy from fatty acids?

27.23 What specific function does the citric acid cycle have in the use of fatty acids for energy?

*27.24 Name two hormones that activate the lipase in adipose tissue. Referring to the previous chapter, what do these hormones do for the blood sugar level?

*27.25 Explain how an increase in the blood sugar level inhibits the mobilization of energy from adipose tissue.

27.26 When lipolysis occurs in adipose tissue, what happens to the glycerol?

Catabolism of Fatty Acids

27.27 How are fatty acids activated for β-oxidation?

*27.28 Write the equations for the four steps of β-oxidation as it operates on butyryl CoA. How many more cycles are possible after this one?

27.29 How is the FAD enzyme recovered from its reduced form, $FADH_2$, when β-oxidation operates?

27.30 How is the reduced form of the NAD^+ enzyme that is used in β-oxidation restored to its oxidized form?

27.31 Why is fatty acid catabolism called *beta*-oxidation?

Biosynthesis of Fatty Acids

27.32 Where are the principal sites for each activity in a liver cell?
(a) Fatty acid catabolism   (b) Fatty acid synthesis

27.33 Outline the steps that make butyryl ACP out of acetyl CoA.

27.34 What metabolic pathway in the body is the chief supplier of NADPH for fatty acid synthesis?

Biosynthesis of Cholesterol

27.35 The enzyme for the formation of which intermediate in cholesterol synthesis is the major control point in this pathway?

27.36 How does cholesterol itself work to inhibit the activity of the enzyme referred to in Review Exercise 27.35?

Ketoacidosis

27.37 What species is diverted from the citric acid cycle to gluconeogenesis?

*27.38 Why does the diversion of Review Exercise 27.37 lead to an increase in the level of acetyl CoA?

27.39 Two molecules of acetyl CoA combine to give the CoA derivative of what keto acid? Give its structure.

*27.40 In two steps, the compound of Review Exercise 27.39 gives one unit of a ketone body and one other significant species. What is it? Why is it a problem?

27.41 Give the names and structures of the ketone bodies.

27.42 What is ketonemia?

27.43 What is ketonuria?

27.44 What is meant by acetone breath?

27.45 Ketosis consists of what collection of conditions?

27.46 What is ketoacidosis? What form of acidosis is it, metabolic or respiratory?

27.47 The formation of which particular compound most lowers the supply of HCO_3^- in ketoacidosis?

27.48 Why does urine volume increase in untreated, insulin-dependent diabetes?

*27.49 If the ketone bodies (other than acetone) can normally be used by heart and skeletal muscle, what makes them dangerous in starvation or in diabetes?

*27.50 Why does the rate of urea production increase in untreated, insulin-dependent diabetes?

27.51 When a physician refers to the loss of Na^+ as the loss of *base,* what is actually meant?

Good and Bad Cholesterol (Interaction 27.1)

27.52 What is atherosclerosis? What is its relationship to arteriosclerosis?

27.53 Which of the lipoprotein complexes is sometimes called "bad cholesterol"? Explain why it is so designated.

27.54 What lipoprotein complex has "good cholesterol?"

27.55 What are some risk factors for atherosclerosis?

27.56 What is familial hypercholesterolemia and how do people get it?

Brown Adipose Tissue (Interaction 27.2)

27.57 With respect to the catabolism of fatty acids, how do white and brown adipose tissue differ?

27.58 In the ordinary operation of the respiratory chain, the chain is coupled to the synthesis of ATP because of what condition concerning mitochondria?

27.59 What happens to the energy released by the respiratory chain when it isn't used to make ATP?

27.60 What is meant by *thermogenesis*?

Leptin (Interaction 27.3)

27.61 What is leptin and how does it work in mice?

27.62 In experiments with obese mice, leptin was given by injection. Why wasn't it given orally (by mouth)?

27.63 How does the serum level of leptin in humans correlate with body mass?

27.64 Leptin causes obese mice to lose weight. Why doesn't this work with obese humans?

Additional Exercises

*27.65 Complete the following equations for one cycle of the β-oxidation of a six-carbon fatty acyl group.

(a) $$CH_3CH_2CH_2CH_2CH_2\overset{\overset{\displaystyle O}{\|}}{C}SCoA + FAD \longrightarrow$$
$$\underline{\qquad} + \underline{\qquad}$$
(b) $\underline{\qquad} + H_2O \longrightarrow \underline{\qquad}$
(c) $\underline{\qquad} + NAD^+ \longrightarrow \underline{\qquad} + NADH + H^+$
(d) $\underline{\qquad} + CoASH \longrightarrow \underline{\qquad} + \underline{\qquad}$

*27.66 Myristic acid, $CH_3(CH_2)_{12}CO_2H$, can be catabolized by β-oxidation just like palmitic acid. Calculate the maximum net number of molecules of ATP that can be made by means of the catabolism of one molecule of myristic acid. (Table 27.2 provides clues about this calculation.)

METABOLISM OF NITROGEN COMPOUNDS

28

It's quite a welcoming committee facing the salmon, but bears need amino acids. The best sources are the proteins of other animals. Most humans think so, too, but we'd rather not stand barefoot in icy water to get them.

THIS CHAPTER IN CONTEXT

This chapter will close our study of metabolism. We'll see again how much the pathways of both anabolism and catabolism among carbohydrates, lipids, and proteins are related. The body can make many (but not all) amino acids from bits and pieces of other molecules. The breakdown of amino acids gives products that can be used to make glucose or lipids. When amino acids are excessively catabolized, acidosis raises its threatening head again. Each of the standard set of amino acids has its own metabolic pathways, too many for us to study individually. The pathways, however, share a few common types of reactions, which we can describe.

28.1 SYNTHESIS OF AMINO ACIDS IN THE BODY

The body can make many amino acids from breakdown products of glucose, fatty acids, and other amino acids.

Amino acids, the end products of protein digestion, are rapidly transported across the walls of the small intestine. Some very small, simple peptides can also be absorbed. Once amino acids enter circulation, they become part of what is called the **nitrogen pool.**

The *Nitrogen Pool* Consists of All Nitrogenous Compounds Anywhere in the Body Figure 28.1 illustrates the various compartments of the nitrogen pool and how they are interrelated. Amino acids enter it not only as digestive products but also as products of the breakdown of proteins in body fluids and tissues. Our own proteins undergo constant turnover, fairly rapidly among those of the liver and blood and quite slowly among muscle proteins.

■ The polypeptides in proteins that serve as enzymes have a particularly rapid turnover.

Individual amino acids are used in any one of the following ways, depending on the body's needs of the moment.

1. The synthesis of new or replacement proteins.
2. The synthesis of nonprotein nitrogen compounds such as heme, nucleic acids, and certain hormones and neurotransmitters.

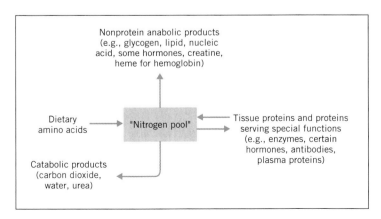

FIGURE 28.1 The nitrogen pool.

3. The production of ATP or of substances with the potential for making it, such as glycogen and fatty acids.

4. The synthesis of nonessential amino acids.

The Body Uses Essential and Nonessential Amino Acids to Make Proteins   We do not have to have *all* 20 amino acids in the diet because we can make roughly half of them ourselves from other substances in food. Those that *must* be in the diet are called the *essential amino acids*. The others are the *nonessential amino acids.* "Nonessential" in this context refers only to a *dietary* need because, one way or another, the body must have on hand all of the amino acids whenever it needs to make polypeptides.

The Ammonium Ion Supplies Nitrogen for Nonessential Amino Acids   Figure 28.2 gives an overview of how the body makes nonessential amino acids. Many of the syntheses require glutamate ion, which is made from α-ketoglutarate by **reductive amination.** In the equation below notice that ammonium ion is the source of nitrogen. A reduction also occurs because the keto group of α-ketoglutarate is in a higher oxidation state than what replaces it, namely, the $CHNH_3^+$ unit in the product, glutamate. So a reducing agent is needed, which is NADPH in most tissues.

$$^-O_2CCH_2CH_2\overset{\overset{\displaystyle O}{\|}}{C}CO_2^- + NH_4^+ + NADPH + H^+ \rightleftharpoons$$

α-ketoglutarate

$$^-O_2CCH_2CH_2\overset{\overset{\displaystyle NH_3^+}{|}}{C}HCO_2^- + NADP^+ + H_2O$$

glutamate

■ In both starvation and diabetes the body draws from its amino acid pool to make glucose.

■ This reaction is the body's principal means for incorporating inorganic nitrogen (as NH_3) into organic compounds.

■ The enzyme is *glutamate dehydrogenase.*

Glutamate Supplies the Amino Group for Many Nonessential Amino Acids   In one route to nonessential amino acids, **transamination,** the amino group of glutamate transfers to an α-keto acid, changing the keto group to an α-amino group. Aminotransferases (or transaminases) are the enzymes. The compounds with keto groups can be made from the catabolism of glucose (see Figure 28.2).

$$R\overset{\overset{\displaystyle O}{\|}}{C}CO_2^- + {}^-O_2CCH_2CH_2\overset{\overset{\displaystyle NH_3^+}{|}}{C}HCO_2^- \rightleftharpoons R\overset{\overset{\displaystyle NH_3^+}{|}}{C}HCO_2^- + {}^-O_2CCH_2CH_2\overset{\overset{\displaystyle O}{\|}}{C}CO_2^-$$

an α-keto acid   glutamate   an α-amino acid α-ketoglutarate

■ The coenzymes for aminotransferases are compounds that collectively are called vitamin B_6.

28.2   CATABOLISM OF AMINO ACIDS

Amino acid breakdown products enter catabolism pathways of carbohydrates or lipids.

Unlike our ability to store glucose as glycogen or fatty acids in adipose tissue, we have no special mechanism for the storage of amino acids. Those that happen to be in the nitrogen pool as free amino acids and that are not needed to make other amino acids or other nitrogen compounds are soon catabolized. Most of this is done in the liver.

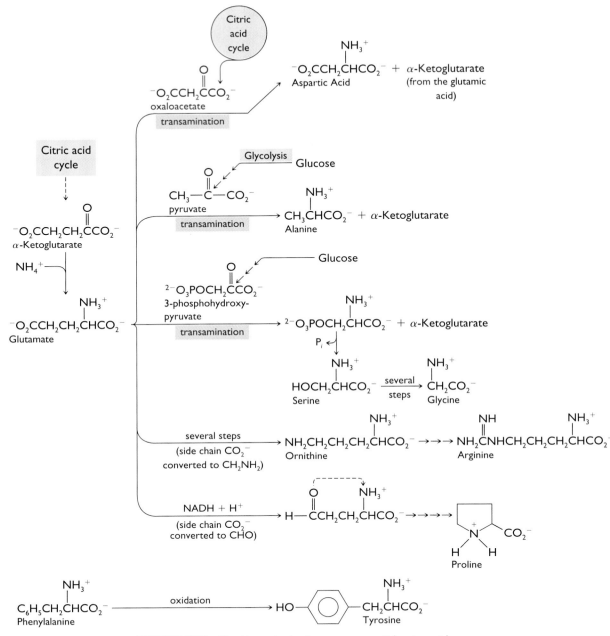

FIGURE 28.2 The biosynthesis of some nonessential amino acids.

Amino Acids Can Be Used to Make Virtually Anything Else the Body Needs The ultimate end products of the complete catabolism of amino acids are urea, carbon dioxide, and water. On the way to these, however, several intermediates form that can enter other pathways. In fact, all the pathways for the use of carbohydrates, lipids, and proteins are interconnected in one way or another (Figure 28.3).

urea

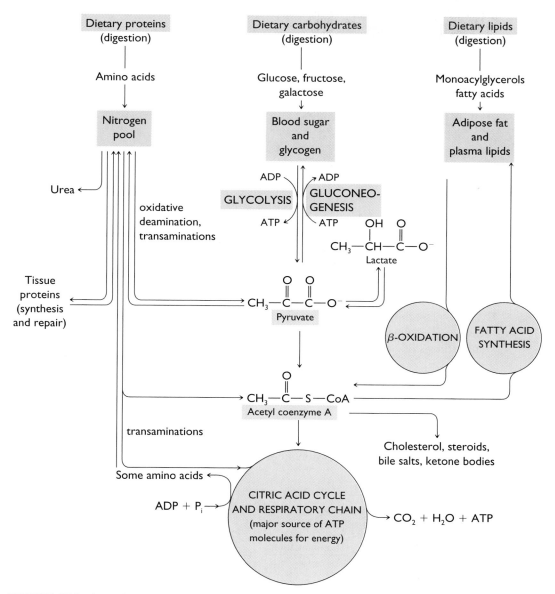

FIGURE 28.3 Interrelationships of major metabolic pathways.

Notice in Figure 28.3 the central importance of two small molecules, acetyl CoA and pyruvate. Notice also that there is no route from acetyl CoA to pyruvate, which means that (at least in all animals) *glucose cannot be made from fatty acids.*

Early in its breakdown, an amino acid gets rid of its nitrogen, which is shuttled into the synthesis of urea, our chief nitrogen waste. The nonnitrogen fragment eventually enters a pathway we have already studied. Three kinds of reactions, besides transamination, occur often, namely *oxidative deamination, direct deamination,* and *decarboxylation.* We'll study these and how they apply to certain selected amino acids.

■ Most amino acids are deaminated by this route.

The α-Ketoglutarate–Glutamate Switch Shuttles Amino Groups toward Urea   The reverse of reductive amination, namely, **oxidative deamination,** removes amino groups of amino acids. In the display that follows, transamination occurs at the first pair of curved arrows, and oxidative deamination occurs at the second. Catabolism is from left to right.

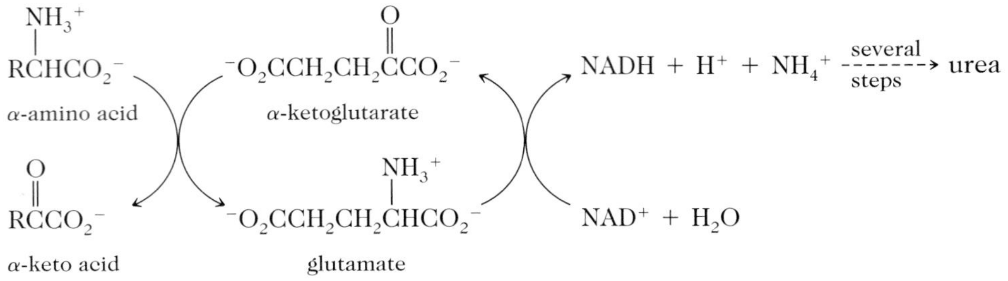

Notice that the nitrogen of the amino acid on the upper left ends up in urea. The α-ketoglutarate–glutamate pair, other words, switches amino nitrogen toward urea nitrogen. The catabolism of alanine and aspartic acid, for example, takes advantage of this switch.

Alanine Catabolizes to Pyruvate   The transamination of alanine gives pyruvate. Pyruvate can enter the citric acid cycle or be used in gluconeogenesis or in the biosynthesis of fatty acids.

■ This display shows how excess alanine from the diet, alanine not needed that day to make proteins or other nitrogen compounds, can be used to make glucose, glycogen, or fatty acids according to other needs.

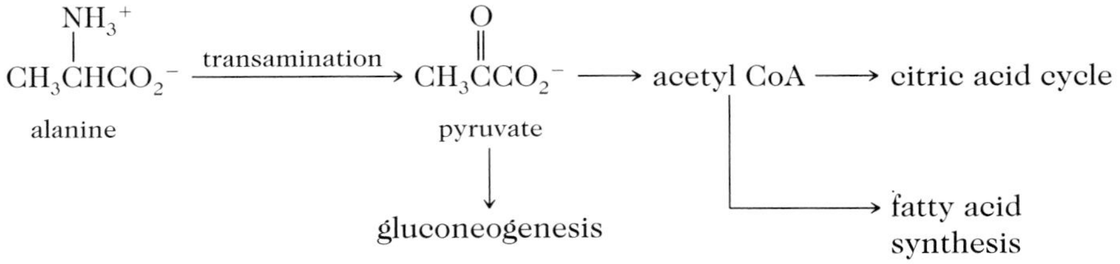

Aspartic Acid Catabolizes to Oxaloacetate   Transamination of aspartic acid gives oxaloacetate, an intermediate in both gluconeogenesis and the citric acid cycle.

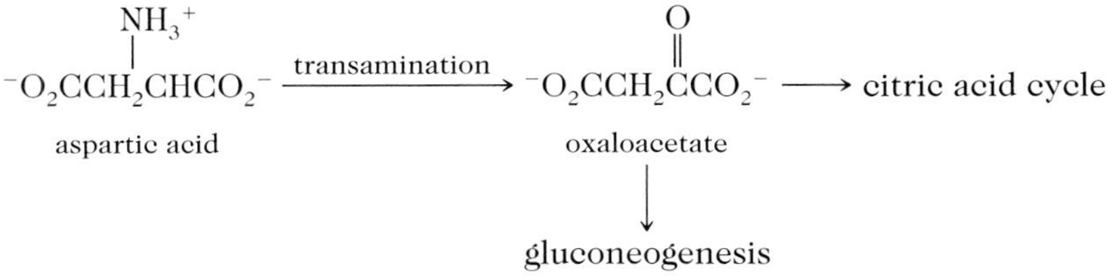

■ During prolonged fasting or starvation, the body degrades its own proteins to make the brain's favorite source of energy, glucose, via gluconeogenesis.

Direct Deamination Removes Amino Groups without Oxidation   Two amino acids, serine and threonine, have OH groups that make possible a nonoxidative loss of NH_3, called **direct deamination.** These amino acids can lose both water and ammonia directly because their OH groups are strategically located on the carbon adjacent to the one that holds an amino group. Here's how direct deamination happens with serine.

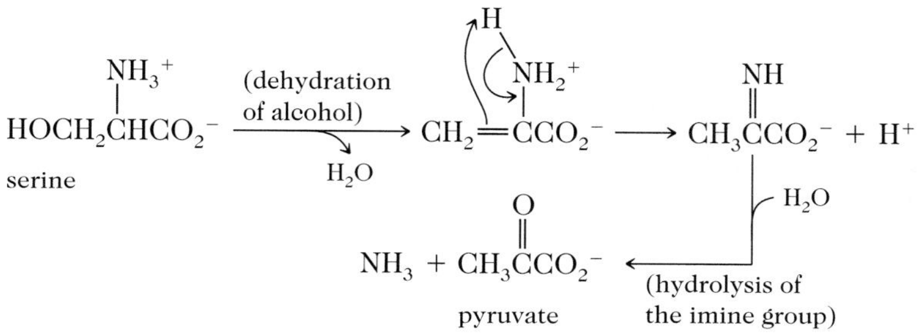

$$\underset{\text{serine}}{\text{HOCH}_2\overset{\overset{\displaystyle \text{NH}_3^+}{|}}{\text{CHCO}_2^-}} \xrightarrow[\text{H}_2\text{O}]{\overset{\text{(dehydration}}{\text{of alcohol)}}} \text{CH}_2{=}\overset{\overset{\displaystyle \text{NH}_2^+}{|}}{\text{CCO}_2^-} \longrightarrow \underset{}{\text{CH}_3\overset{\overset{\displaystyle \text{NH}}{\|}}{\text{CCO}_2^-}} + \text{H}^+$$

$$\underset{\text{pyruvate}}{\text{NH}_3 + \text{CH}_3\overset{\overset{\displaystyle O}{\|}}{\text{CCO}_2^-}} \longleftarrow \underset{\substack{\text{(hydrolysis of} \\ \text{the imine group)}}}{\overset{\text{H}_2\text{O}}{}}$$

The first step is dehydration of the alcohol group to give an unsaturated amine. This spontaneously rearranges into an *imine,* a compound with a carbon–nitrogen double bond. Water then adds to this double bond, but the product (structure not shown) spontaneously breaks up. The net effect is the hydrolysis of the imine group to give ammonia and the keto group of pyruvate. The acetyl group from the further breakdown of pyruvate, as we now well know, can go into the citric acid cycle or fatty acid synthesis. Or pyruvate could be used to make glucose.

■ Imine groups easily hydrolyze because they can add water and then split out ammonia.

Sustained Gluconeogenesis Necessarily Consumes Body Proteins   The reactions just surveyed are involved in the catabolism of several amino acids (Figure 28.4). Notice particularly that oxaloacetate occurs in two places, as an intermediate in the citric acid cycle and as the product of the oxidative deamination of aspartate. The oxaloacetate made from aspartate has two options, namely, to be used to make ATP or to make glucose. Oxaloacetate thus connects the catabolism of amino acids to carbohydrate synthesis.

Because of the occurrence of oxaloacetate in the citric acid cycle, and because its carbons can originate in fatty acids (by way of acetyl coenzyme A), we might think that fatty acids could also be used to make glucose. Not so, at least *not for a net gain of glucose.* The removal of oxaloacetate from the citric acid cycle for gluconeogenesis means the removal of what carries acetyl units (of any origin) into the cycle. This leads to a backup of acetyl CoA and a buildup of the ketone bodies.

For a *net gain* of glucose molecules via gluconeogenesis, the oxaloacetate of the citric acid cycle cannot be counted as available. *Only oxaloacetate made from amino acids can give a net gain of glucose this way.* This is why gluconeogenesis under conditions of starvation or diabetes necessarily breaks down body proteins. It needs some of their amino acids to make oxaloacetate to be able to make glucose.

The Decarboxylation of Some Amino Acids Leads to Neurotransmitters   Some special enzymes can split out just the carboxyl groups from amino acids and so make amines. The reaction, called **decarboxylation,** is used to make some neurotransmitters and hormones. Dopamine, norepinephrine, and epinephrine are all made by steps that begin with the decarboxylation of dihydroxyphenylalanine, which the body makes from the amino acid tyrosine. (The equations are on page 773.)

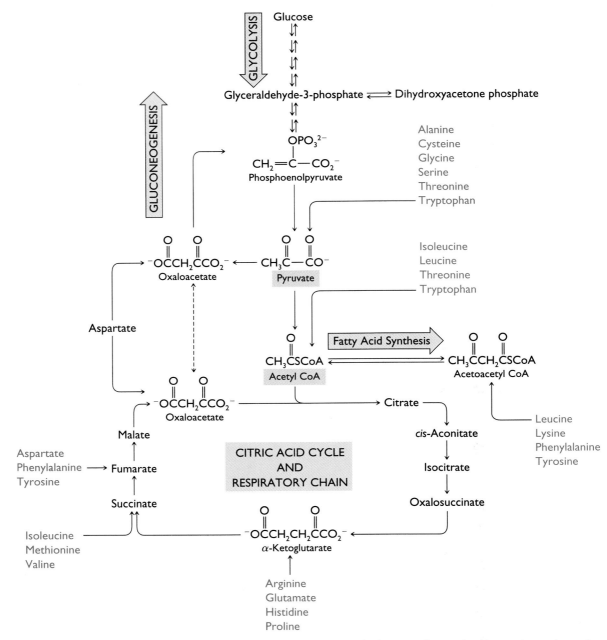

FIGURE 28.4   The catabolism of some amino acids. Amino acids named in blue can be used to make glucose; those named in red can be used to make only fatty acids. Several can be used either way.

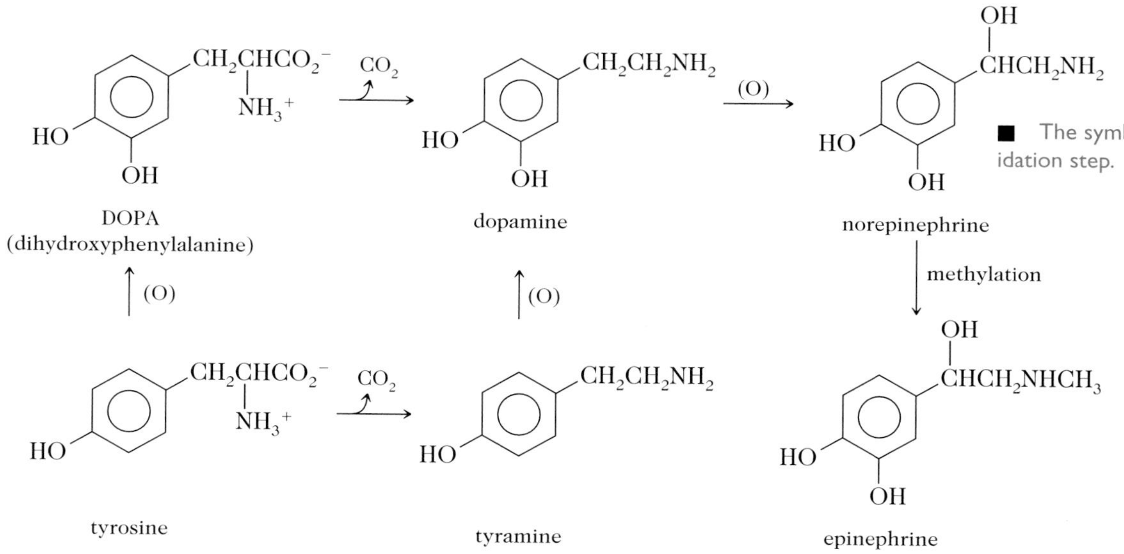

DOPA
(dihydroxyphenylalanine)
dopamine
norepinephrine
(O)
CO_2
methylation
tyrosine
tyramine
epinephrine

■ The symbol (O) signifies an oxidation step.

28.3 FORMATION OF UREA

Ammonia and amino groups are converted to urea by the urea cycle.

Urea, as we have learned, is the chief nitrogen waste made by the body. Most of its nitrogen indirectly comes from amino acids, but some comes (also indirectly) from two of the side-chain bases of nucleic acids.

Oxaloacetate Shuttles Nitrogen from Glutamate to Aspartate and Then into the Synthesis of Urea There are two direct sources for the nitrogen atoms in urea. One is the ammonium ion produced by the oxidative deamination of glutamate. Aspartate is the other nitrogen source. But aspartate can be made from glutamate by transamination, so glutamate is close to being the direct source of both nitrogens in urea. Here is the shuttle from glutamate to aspartate.

■ Recall that the amino group of glutamate can come from the amino group of any other amino acid by way of another shuttle.

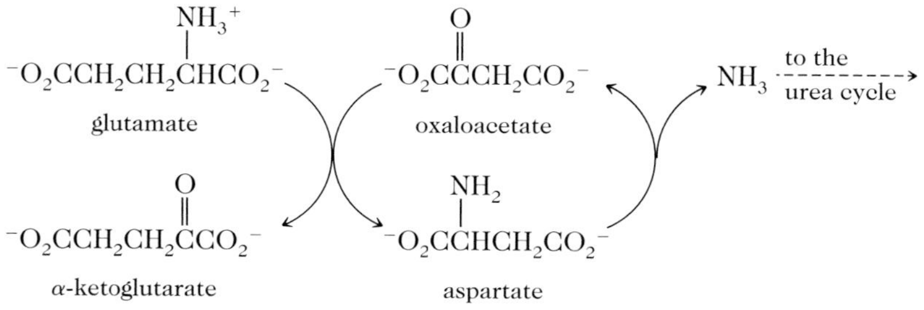

glutamate
oxaloacetate
α-ketoglutarate
aspartate
NH_3
to the urea cycle

The Urea Cycle Is the Only Way the Body Has to Make Urea A series of reactions called the **urea cycle** manufactures urea from ammonium ion, carbon dioxide, and aspartate (Figure 28.5). The boxed numbers in the figure refer to the following discussion.

■ The urea cycle is sometimes called the *Krebs ornithine cycle*.

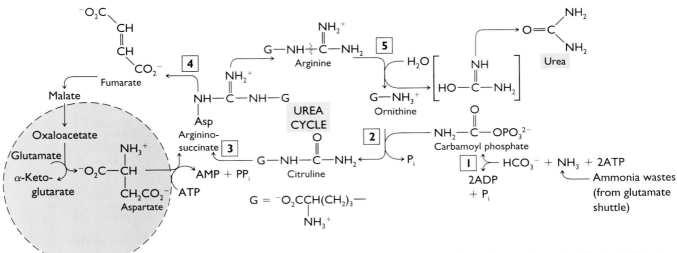

FIGURE 28.5 The urea cycle. The boxed numbers refer to the text discussion. The dashed line circle is the aspartate–oxaloacetate shuttle, also discussed in the text.

■ This step occurs inside a mitochondrion.

■ Steps 2 through 5 occur in the cell's cytosol.

1. Ammonia, with the help of ATP, reacts with CO_2 to form carbamoyl phosphate, a high-energy phosphate. In a sense, this is an activation of ammonia that launches it into the next step, which takes it into the cycle.

2. The carbamoyl group transfers to the carrier unit, ornithine, as P_i is ejected. This consumes energy from high-energy phosphate. Citrulline forms.

3. Citrulline condenses with the α-amino group of aspartate to give argininosuccinate.

4. Fumarate forms from the original aspartate as the amino group stays with the arginine that emerges. Fumarate is an intermediate in the citric acid cycle. By a transamination that involves glutamate, it is reconverted to aspartate.

5. Arginine is hydrolyzed. Urea forms and ornithine is regenerated to start another turn of the cycle.

The overall result of the urea cycle is given by the following equation (which, as Figure 28.5 makes clear, is extremely simplified).

$$2NH_3 + H_2CO_3 \longrightarrow NH_2\overset{\overset{\displaystyle O}{\|}}{C}NH_2 + 2H_2O$$
$$\text{urea}$$

To do justice to the overall event, we must factor in the ATP consumption.

$$NH_4^+ + CO_2 + 3ATP + \text{aspartate} \longrightarrow \text{urea} + 2ADP + \text{fumarate} + 4P_i + AMP$$

At Elevated Levels, the Ammonium Ion Is Toxic If an infant is born without any one of the enzymes needed for the five steps of the urea cycle, it will die soon after birth. It will be unable, on its own, to clear ammonia from its blood. (Before birth, the mother's metabolism handles this.) Ammonia and the ammonium ion are toxic at sufficiently high levels.

■ There are many genetic defects involving enzymes for the body's utilization of amino acids.

Some inherited genetic defects result in deficient urea cycle enzymes. One result is **hyperammonemia**, an elevated level of NH_3 in the blood. Infants with this genetic defect improve on low-protein diets. If the level is not high enough to cause death, it can be expected to cause mental retardation. Periodic but unremembered episodes of bizarre behavior by one man—babbling, pacing, crying, glassy eyes— were not understood until it was found that he had a rare deficiency of the enzyme for step 2 of the urea cycle (ornithine transcarbamylase).

28.4 CATABOLISM OF OTHER NITROGEN COMPOUNDS

Uric acid and bile pigments are other breakdown products of nitrogen compounds.

The nitrogen of the purine bases of nucleic acids, adenine (A) and guanine (G), is excreted as uric acid, which also has the purine nucleus. After studying how uric acid forms, we'll see how defects in this pathway can lead to gout or to a particularly difficult disease of children, the Lesch-Nyhan syndrome.

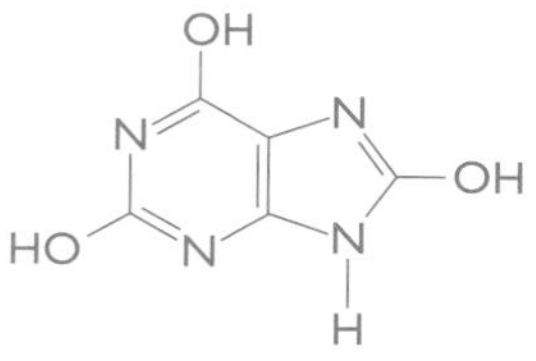

The Catabolism of AMP Gives the Urate Ion The numbered steps in Figure 28.6, discussed next, show how the adenine unit of AMP can be used to make uric acid.

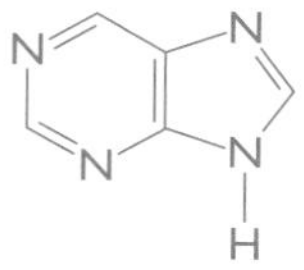

1. Transamination removes the amino group of the adenine side chain of AMP.

2. Ribose phosphate is removed and delivered to the pentose phosphate pathway of carbohydrate catabolism. The other product is hypoxanthine.

3. Hypoxanthine is oxidized to xanthine. The steps from guanine lead to xanthine, too.

4. Another oxidation produces the keto form of uric acid, which exists partly in the form of a phenol. Actually, it's the salt of uric acid, sodium urate, that forms because the acid is neutralized by base in the buffer system.

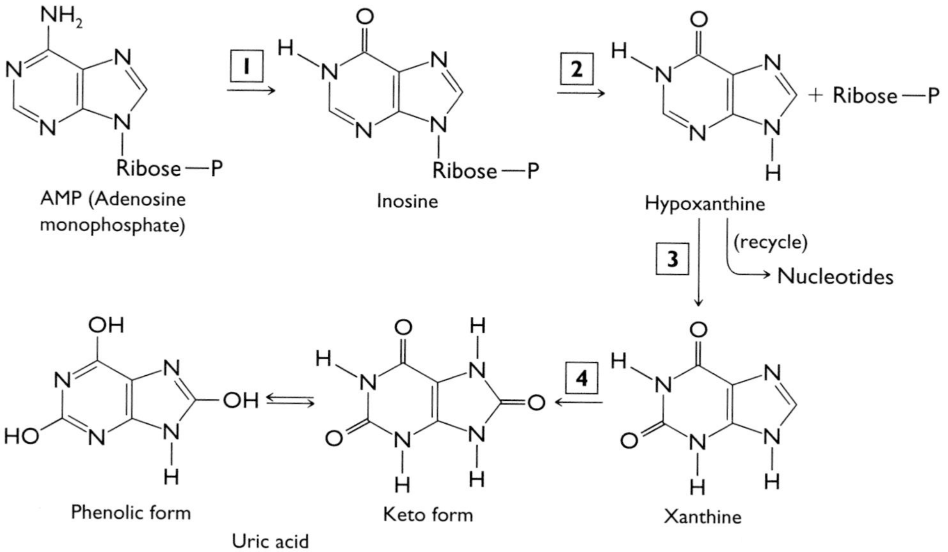

uric acid

purine

FIGURE 28.6 The catabolism of adenosine monophosphate, AMP. The boxed numbers refer to the discussion in the text.

■ Ethyl alcohol accelerates the synthesis of urate ion by increasing the rate of catabolism of adenosine monophosphate.

■ The lack of one enzyme usually means great personal and family suffering.

Overproduction of the Urate Ion Causes Gout In *gout*, the rate of formation of sodium urate is more rapid than its rate of elimination. Crystals of the salt precipitate in joints, where they cause painful inflammations and lead to a form of arthritis. Kidney stones form if the salt comes out of solution in this organ.

Just why the formation of sodium urate accelerates isn't well understood, but genetic factors are involved. Normally, some of the hypoxanthine made in step 2 is recycled back to nucleotide bases that are needed to make nucleic acids or high-energy phosphates. Some individuals with gout are known to have a partial deficiency of the enzyme system required for this recycling of hypoxanthine. Hence, most if not all of their hypoxanthine ends up as more sodium urate than normal.

The Absence of One Enzyme Needed for Hypoxanthine Recycle Leads to Self-Mutilating Behavior in Infants In the Lesch-Nyhan syndrome, the enzyme for recycling hypoxanthine is totally lacking. The result is both bizarre and traumatic. Infants with this syndrome develop compulsive, self-destructive behavior at age 2 or 3. Unless their hands are wrapped in cloth, they will bite themselves to the point of mutilation. They act with dangerous aggression toward others. Some become spastic and mentally retarded. Kidney stones develop early, and gout comes later.

The Catabolism of Heme Produces the Bile Pigments Erythrocytes have life spans of only about 120 days. Eventually they split open. Their hemoglobin spills out and then is degraded. Its breakdown products are eliminated via the feces and, to some

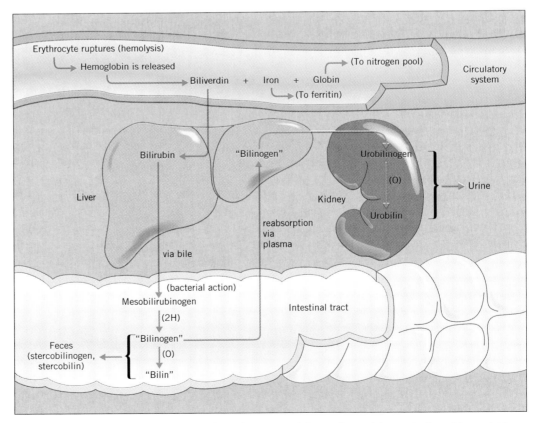

FIGURE 28.7 The formation and the elimination of the products of the catabolism of hemoglobin.

INTERACTION 28.1
JAUNDICE AND THE TETRAPYRROLE PIGMENTS

Jaundice (French, *jaune,* yellow) is a condition that is symptomatic of a malfunction somewhere along the pathway of heme metabolism. If bile pigments accumulate in the plasma in concentrations high enough to impart a yellowish coloration to the skin, the condition of *jaundice* is said to exist. Jaundice can result from one of three kinds of malfunctions.

Hemolytic jaundice results when hemolysis takes place at an abnormally fast rate. Bile pigments, particularly bilirubin, form faster than the liver can clear them. Hepatic diseases such as infectious hepatitis and cirrhosis sometimes prevent the liver from re-

moving bilirubin from circulation. The stools are usually clay colored, because the pyrrole pigments do not reach the intestinal tract.

Obstructions of bile ducts can prevent release of bile into the intestinal tract, and the tetrapyrrole pigments in bile cannot be eliminated. Under these circumstances, they tend to reenter general circulation. The kidneys remove large amounts of bilirubin, but the stools are usually clay colored. As the liver works harder and harder to handle its task of removing excess bilirubin, it can weaken and become permanently damaged.

---

extent, in the urine. In fact, the characteristic colors of feces and urine are caused by partially degraded heme molecules called the **bile pigments** (see Figure 28.7).

The degradation of heme begins before the globin portion breaks away. The heme molecule partly opens up to give a system that has a chain of four small rings, called pyrrole rings. (This is why the bile pigments are sometimes called the *tetrapyrrole pigments.*)

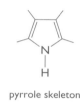

pyrrole skeleton

carbon skeleton of the bile pigments

The rings have varying numbers of double bonds according to the state of oxidation of the pigment.

The slightly broken hemoglobin molecule, now called verdohemoglobin, then splits into globin, iron(II) ion, and a greenish pigment called **biliverdin** (Latin *bilis,* bile, + *verdus,* green). Globin enters the nitrogen pool. Iron is conserved in a storage protein called *ferritin* and is reused. Biliverdin is changed in the liver to a reddish-orange pigment called **bilirubin** (Latin *bilis,* bile + *ruber,* red) the principal bile pigment in humans. Bilirubin is not only made by the liver but is also removed from circulation by the liver, which transfers it to the bile. In this fluid, bilirubin finally enters the intestinal tract, where bacterial enzymes convert it to *mesobilirubinogen,* a colorless substance. **Bilinogen** is made from it, but it usually goes by other names according to its destination rather than its structure. Thus bilinogen that leaves the body in the feces is called *stercobilinogen* (Latin, *stercus,* dung). Some bilinogen is reabsorbed via the bloodstream, comes to the liver, and finally leaves the body in the urine. Now it is called *urobilinogen.* Some bilinogen is reoxidized to give **bilin,** a brownish pigment. Depending on its destination, bilin is called *stercobilin* or *urobilin.* It all seems complicated, but tests for some of these substances are used in diagnosis. For example, Interaction 28.1 describes how the bile pigments are involved in jaundice.

■ Bile pigments are also responsible for the color of bile.

SUMMARY

Amino acid distribution The nitrogen pool receives amino acids from the diet, from the breakdown of proteins in body fluids or tissues, and from any synthesis of nonessential amino acids that occurs. Amino acids are used to build and repair tissue, replace proteins of body fluids, make nonprotein nitrogen compounds, provide chemical energy if needed, and supply molecular parts for gluconeogenesis or fatty acid synthesis.

Amino acid metabolism By the reactions of transamination, oxidative deamination, direct deamination, and decarboxylation, the α-amino acids shuttle amino groups between themselves and intermediates of the citric acid cycle. Amino acids also participate in the synthesis of urea and nonprotein nitrogen compounds.

Deaminated amino acids eventually become acetyl CoA, acetoacetyl CoA, pyruvate, or an intermediate in the citric acid cycle. The skeletons of most amino acids can be used to make glucose, or fatty acids, or the ketone bodies. Their nitrogen atoms become part of urea.

Metabolism of other nitrogen compounds The nitrogen atoms in some of the side-chain bases of nucleic acids end up in urea and those of the others are excreted as sodium urate. Urea is made by a complex cycle of reactions—the urea cycle.

Heme is catabolized to bile pigments and its iron is reused. The pigments—first, biliverdin (green), then bilirubin (red), then mesobilirubinogen (colorless), and finally bilinogen and bilin (brown)—become stercobilin or stercobilinogen, urobilin or urobilinogen, depending on the route of elimination.

REVIEW EXERCISES

The answers to Review Exercises whose numbers are in color are found in Appendix E. The answers to the other Review Exercises are found in the Study Guide that accompanies this book. The more challenging questions are marked with asterisks.

Nitrogen Pool

28.1 What is the nitrogen pool?

28.2 What are four ways in which amino acids are used in the body?

28.3 When the body retains more nitrogen than it excretes in all forms, the system is said to be in a *positive nitrogen balance*. Would this state characterize infancy or old age?

28.4 What happens to amino acids that are obtained in the diet but aren't needed to make any nitrogeneous compounds?

Biosynthesis of Amino Acids

28.5 Which amino acid is the major supplier of amino groups for the synthesis of the nonessential amino acids?

28.6 Write the equation for the reductive amination that produces glutamate. (Use NADPH as the reducing agent.)

*28.7 Write the structure of the keto acid that forms when phenylalanine undergoes transamination with α-ketoglutarate.

*28.8 When valine and α-ketoglutarate undergo transamination, what new keto acid forms? Write its structure.

The Catabolism of Amino Acids

28.9 Can cysteine be used to make glucose?

28.10 Can lysine be used to make fatty acids? To make glucose?

28.11 Can an amino acid that breaks down to acetyl CoA without going through pyruvate be used to make glucose?

*28.12 By means of two successive equations, one a transamination and the other an oxidative deamination, write the reactions that illustrate how the amino group of alanine can be removed as NH_4^+.

*28.13 Arrange the following compounds in the order in which they would be produced if the carbon skeleton of alanine were to appear in one of the ketone bodies.

Pyruvate	Acetoacetyl CoA	Acetoacetate
1	2	3
Alanine	Acetyl CoA	
4	5	

*28.14 In what order would the following compounds appear if some of the carbon atoms in glutamate were to become part of glycogen?

Oxaloacetate	α-Ketoglutarate	Glucose
1	2	3
Glycogen	Glutamate	
4	5	

28.15 Can any of the carbon atoms of glucose become part of alanine? If so, explain (in general terms).

28.16 Write the structure of the keto acid that forms by the direct deamination of threonine.

*28.17   When tyrosine undergoes decarboxylation, what forms? Write its structure.

*28.18   Write the structure of the product of the decarboxylation of tryptophan.

28.19   In the conditions of starvation or diabetes, what can the amino acids be used for?

The Formation of Urea

28.20   In the biosynthesis of urea, what are the sources of (a) the two NH_2 groups and (b) the $C=O$ group?

28.21   What is hyperammonemia and, in general terms, how does it arise and how can it be handled in infants?

28.22   What is the overall equation for the synthesis of urea?

28.23   Would a deficiency in ornithine transcarbamylase (the enzyme for step 2 of the urea cycle, Figure 28.5) cause hypoammonemia or hyperammonemia? Explain.

The Catabolism of Nonprotein Nitrogen Compounds

28.24   What compounds are catabolized to make uric acid?

28.25   What product of catabolism accumulates in the joints in gout?

*28.26   Arrange the names of the following substances in the order in which they appear during the catabolism of heme.

Biliverdin	Heme	Hemoglobin	Bilirubin
1	2	3	4

Mesobilirubinogen	Bilin	Bilinogen
5	6	7

Tetrapyrrole Pigments and Jaundice (Interaction 28.1)

28.27   Briefly describe what jaundice does to the body.

28.28   Describe how the following kinds of jaundice arise.
(a) Hemolytic jaundice
(b) The jaundice of hepatic diseases

28.29   Why should an obstruction of the bile ducts cause jaundice?

Additional Exercises

*28.30   From a study of the figures in this chapter, can the carbon atoms of serine become a part of a molecule of palmitic acid? If so, write the names of the compounds, beginning with serine, in the sequence to the start of the fatty acid synthesis.

*28.31   The carbon atoms of an amino acid that breaks down to acetyl CoA without going through pyruvate eventually are present among intermediates in the citric acid cycle, including oxaloacetate, a starting material for gluconeogenesis. Why, then, aren't all such amino acids also able to contribute to the synthesis of a net extra amount of glucose?

*28.32   In the direct deamination of serine, an imine forms as an intermediate, as we discussed. This imine adds a water molecule to give another intermediate whose structure was not shown. What is the likeliest structure of this intermediate, given that it changes to pyruvate by the loss of NH_3? Offer an explanation for the *direction* with which H_2O adds to the imine.

29 NUTRITION

Specialists in nutrition urge us to obtain our nutrients from a variety of foods, and nature obliges. This chapter describes the kinds and quantities of nutrients we should have for health.

THIS CHAPTER IN CONTEXT

Nutrition comes at the end of our study because any excursions into its *molecular* basis requires a vocabulary, which we now have. Some terms, like carbohydrates, proteins, and vitamins, are probably familiar even to those who have taken no science course. A few others, like enzyme, antioxidant, and DNA, aren't so commonly used, except by health food enthusiasts. And then there are terms like LDL, oxidative process, molecule, peroxide, phenol, double bond—the list is really quite long—that cannot be avoided if we are to be true to our goal of studying the molecular basis of life.

29.1 GENERAL NUTRITIONAL REQUIREMENTS

The recommended dietary allowances (RDAs) help meal planners meet the nutritional needs of nearly all healthy people.

Nutrition can be considered on two levels. On the personal level, we speak of our own nutrition, the sum of the foods we eat in proper proportions at the best moments as we try to maintain a state of well-being and avoid diet-related diseases and infirmities. Technically, which is the meaning we want, **nutrition** is a field of science that looks into the identities, quantities, and sources of *nutrients*.

Nutrients are chemicals that take part in any nourishing, health supporting, or health promoting metabolic activity. Carbohydrates, lipids, proteins, vitamins, minerals, and trace elements are all nutrients. Oxygen and water are usually not called nutrients, although they formally qualify under the definition. **Foods** are materials that supply one or more nutrients. **Dietetics** is the science of applying the findings of nutrition to feeding individual humans, whether they are ill or well.

The *Recommended Dietary Allowances* Describe What Nutrients Are Adequate for Healthy People   For several decades, the Food and Nutrition Board of the National Research Council of the National Academy of Sciences has established **recommended dietary allowances, or RDAs**. They are the intake levels of essential nutrients that are "adequate to meet the known nutritional needs of practically all healthy persons" (Table 29.1).

It's important to learn not only what the RDAs are but also what they are not. They are not the same as the U.S. Recommended Daily Allowances (USRDAs). These, although based on the RDAs, are standards for nutritional information on food labels and set by the U.S. Food and Drug Administration.

The RDAs are not the same as the Minimum Daily Requirements (MDRs) for any one individual. The MDRs are minimums, only that, and are set very close to the levels at which actual signs of deficiencies occur. The RDAs are two to six times the MDRs. Just as individuals differ greatly in height, weight, and appearance, they also differ greatly in specific biochemical needs. Therefore the RDAs are set far enough above average requirements so that "practically all healthy people" will thrive. Most will receive more than they need; a few will not receive enough.

The RDAs do not define *therapeutic* nutritional needs for people with chronic diseases, like prolonged infections or metabolic disorders; or for people taking certain medications on a continuing basis; or for prematurely born infants. These all require special diets. In some areas of the United States and in many parts of the world, intestinal parasites are common. These daily rob infected people of some of their food intake, and such people also need special diets.

■ Among the great scientific triumphs of the nineteenth century were the germ theory of disease and the births of the sciences of nutrition and organic chemistry.

■ It's the job of hospital dieticians to devise the large variety of diets needed by the patient population.

TABLE 29.1   Recommended Daily Dietary Allowances*a* **of the Food and Nutrition Board, National Academy of Sciences, National Research Council, Revised 1989**

Persons	Age (years)	Weight kg	Weight lb	Height cm	Height in	Protein (g)	A (μg)*b*	D (μg)*c*	E (mg)*d*	K (μg)
Infants	0.0–0.5	6	13	60	24	13	375	7.5	3	5
	0.5–1	9	20	71	28	14	375	10	4	10
Children	1–3	13	29	90	35	16	400	10	6	15
	4–6	20	44	112	44	24	500	10	7	20
	7–10	28	62	132	52	28	700	10	7	30
Males	11–14	45	99	157	62	45	1000	10	10	45
	15–18	66	145	176	69	59	1000	10	10	65
	19–24	72	160	177	70	58	1000	10	10	70
	25–50	79	174	176	70	63	1000	5	10	80
	51+	77	170	173	68	63	1000	5	10	80
Females	11–14	46	101	157	62	46	800	10	8	45
	15–18	55	120	163	64	44	800	10	8	55
	19–24	58	128	164	65	46	800	10	8	65
	25–50	63	138	163	64	50	800	5	8	65
	51+	65	143	160	63	50	800	5	8	65
Pregnant						60	800	10	10	65
Lactating	First 6 months					65	1300	10	12	65
	Second 6 months					62	1200	10	11	65

The "Fat-Soluble Vitamins" heading spans columns A, D, E, and K.

a The allowances are intended to provide for individual variations among most normal persons as they live in the United States under usual environmental stresses. Diets should be based on a variety of common foods in order to provide other nutrients for which human requirements have been less well defined.
b Retinol equivalents. 1 retinol equivalent = 1 μg retinol or 6 μg β-carotene.
c As cholecalciferol. 10 μg cholecalciferol = 400 IU of vitamin D.

■ Many scientists urge women planning pregnancies to begin supplementing their diets with certain vitamins, like folate, *before* pregnancies.

The RDAs do cover people according to age, sex, and size, and they indicate special needs for pregnant and lactating women. Special needs for extra water and salt also arise during strenuous physical activity and prolonged exposure to high temperatures.

Fad Diets are Dangerous   No single food contains all nutrients. The RDAs can and ought to be provided from a number of foods. People dangerously risk their health when they go on fad diets limited to one particular food, like brown rice, gelatin, yogurt, or liquid protein. The ancient wisdom of a varied diet that includes meat, fruit, vegetables, grains, nuts, pulses (e.g., beans), and dairy products may seem to be supported solely by cultural and aesthetic factors. But a varied diet also assures our getting any trace and needed nutrients yet to be discovered.

The U.S. Department of Agriculture has devised a "food pyramid" for educating the public about organizing personal nutrition to obtain nutrients in healthy proportions through a varied diet (Figure 29.1). The pyramid has been criticized for suggesting too much meat per day, for ignoring differences in types of lipids, and for not promoting a low-fat and low-cholesterol diet.

Adults Have a Water Budget of 2.5 to 3 L per Day   We take in water in three ways, by the fluids we drink, the foods we eat, and by the oxidations of nutrients. Most water intake is in response to the thirst mechanism. Water leaves the body in four ways, via the urine, the feces, exhaled air, and normal perspiration (Figure 29.2).

| Water-Soluble Vitamins | | | | | | | Minerals | | | | | | |
Ascorbic Acid (mg)	Folate (µg)	Niacin[e] (mg)	Riboflavin (mg)	Thiamine (mg)	Vitamin B_6 (mg)	Vitamin B_{12} (mg)	Calcium (mg)	Phosphorus (mg)	Iodine (µg)	Iron (mg)	Magnesium (mg)	Selenium (µg)	Zinc (mg)
30	25	5	0.4	0.3	0.3	0.3	400	300	40	6	40	10	5
35	35	6	0.5	0.4	0.6	0.5	600	600	50	10	60	15	5
40	50	9	0.7	0.7	1.0	0.7	800	800	70	10	80	20	10
45	75	12	0.9	0.9	1.1	1.0	800	800	90	10	120	20	10
45	100	13	1.0	1.0	1.4	1.4	800	800	120	10	170	30	10
50	150	17	1.5	1.3	1.7	2.0	1200	1200	150	12	270	40	15
60	200	20	1.8	1.5	2.0	2.0	1200	1200	150	12	400	55	15
60	200	19	1.7	1.5	2.0	2.0	1200	1200	150	10	350	70	15
60	200	19	1.7	1.5	2.0	2.0	800	800	150	10	350	70	15
60	200	15	1.4	1.2	2.0	2.0	800	800	150	10	350	70	15
50	150	15	1.3	1.1	1.4	2.0	1200	1200	150	15	280	45	12
60	180	15	1.3	1.1	1.5	2.0	1200	1200	150	15	300	50	12
60	180	15	1.3	1.1	1.6	2.0	1200	1200	150	15	280	55	12
60	180	15	1.3	1.0	1.6	2.0	800	800	150	15	280	55	12
60	180	13	1.2	1.1	1.6	2.0	800	800	150	10	280	55	12
70	400	17	1.6	1.5	2.2	2.2	1200	1200	175	30	320	65	15
95	280	20	1.9	1.6	2.1	2.6	1200	1200	200	15	355	75	19
90	280	20	1.7	1.6	2.1	2.6	1200	1200	200	15	340	75	16

[d] α-Tocopherol equivalents. 1 mg (+)-α-tocopherol = 1 α-tocopherol equivalent.
[e] 1 niacin equivalent (NE) equals 1 mg of niacin or 60 mg of dietary tryptophan.

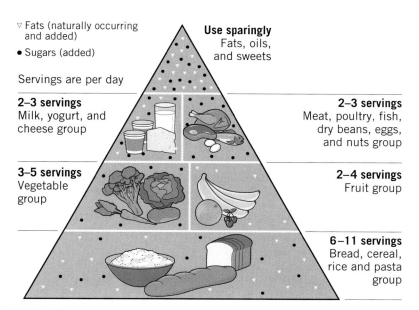

▽ Fats (naturally occurring and added)
● Sugars (added)

Servings are per day

Use sparingly
Fats, oils, and sweets

2–3 servings
Milk, yogurt, and cheese group

2–3 servings
Meat, poultry, fish, dry beans, eggs, and nuts group

3–5 servings
Vegetable group

2–4 servings
Fruit group

6–11 servings
Bread, cereal, rice and pasta group

FIGURE 29.1 The food group pyramid—a device for educating the public about nutrition. (Source: U.S. Department of Agriculture.)

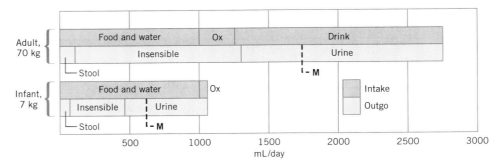

FIGURE 29.2   Water intake and outgo by the principal routes (excluding sensible perspiration or sweating). The dashed lines at **M** are the minimal volumes of urine at the maximal concentrations of solutes. "Ox" refers to water formed by the oxidation of foods. (From *Recommended Dietary Allowances*, 9th ed. National Academy of Sciences, Washington, D.C., 1980).

■ No amount of physical training or willpower can condition anyone to go without water.

Activity in the hot sun at low humidity can cause the loss of as much as 10 L of water per day along with body salts. If these losses aren't made up at a reasonable rate, heat exhaustion, heat stroke, heat cramps or even death can result.

Energy Needs Determine Oxygen Requirements   Our daily oxygen needs depend on the energy expenditures of our activities (Table 29.2). In general, when energy demands are in the range of 2.0×10^3 to 3.0×10^3 kcal/day, we need roughly 18 to 25 moles of oxygen per day.

At rest, the brain is the largest user of oxygen. Although its mass is only about 2% of body mass, it consumes, at rest, 20% of our oxygen intake. Much of the energy goes to maintain concentration gradients of ions across the billions of brain cell membranes.

■ High-protein diets, if eaten for an extended period, can make the kidneys overwork and become damaged.

Our calories should not be exclusively from either carbohydrates or fats. On a zero-carbohydrate diet, for example, the body has to make glucose to meet its needs. (Remember that the brain preferably derives nearly all its energy from the breakdown of glucose.) If the body has to make glucose over a long period of time, say weeks, a slow buildup of toxic wastes will cause harm.

On a zero-fat diet, some needed fatty acids are unavailable, and fat-soluble vitamins are not absorbed very well. A daily diet that includes at least 15 to 25 g of food fat (the equivalent of two to four pats of margarine or butter) and 50 to 100 g of digestible carbohydrate avoids these problems.

■ *Hydrogenated* vegetable oils are, like animal fats, deficient in the important fatty acids because hydrogenation removes the carbon—carbon double bonds of linoleic, linolenic, and arachidonic acids.

Linoleic Acid Must Be in the Diet   Linoleic acid is classified as an **essential fatty acid,** and it is particularly needed for the health and growth of infants.[1] Linoleic acid is so poorly supplied by animal fats, including milk fat, that we depend almost totally on vegetable oils for it. Thanks to the widespread presence of linoleic acid in all edible vegetable oils, it is virtually impossible for an adult not to obtain enough.

---

[1] All sources agree that linoleic acid is an *essential* fatty acid. Some add arachidonic acid to the list, but the body is able to make arachidonic acid from linoleic acid. Other lists include linolenic acid as an essential fatty acid, but the body can also make this acid from linoleic acid. No doubt in feeding studies, marked improvements in essential fatty acid deficiency occur with giving any of the three—linoleic, linolenic, and arachidonic acids. It is arachidonic acid that is finally crucial, because many prostaglandins are made from it, and it is possible that infants (but not adults) lack the ability to make arachidonic acid from linoleic acid.

TABLE 29.2 Daily Energy Expenditures of Adult Men and Women in Light Occupations

Activity Category	Time (hr)	Rate, Man (70 kg)		Rate, Woman (58 kg)	
		kcal/min	Total (kcal)	kcal/min	Total (kcal)
Sleeping, reclining	8	1.0–1.2	540	0.9–1.1	440
Very light[a]	12	up to 2.5	1300	up to 2.0	900
Light[b]	3	2.5–4.9	600	2.0–3.9	450
Moderate[c]	1	5.0–7.4	300	4.0–5.9	240
Heavy[d]	0	7.5–12.0		6.0–10.0	
Total	24		2740		2030

Source: J. V. G. A. Durin and R. Passmore, 1967. In *Recommended Dietary Allowances,* 9th ed. Washington, D.C., National Academy of Sciences, 1980.

[a] Seated and standing activities; driving cars and trucks; secretarial work; laboratory work; sewing; ironing; playing musical instruments.
[b] Walking (on the level, 2.5 to 3 mph), tailoring and pressing, carpentry, electrical trades, restaurant work, washing clothes, light recreation such as golf, table tennis, volleyball, sailing.
[c] Walking at 3.5 to 4 mph; garden work, scrubbing floors, shopping with heavy load, moderate sports such as skiing, tennis, dancing, bicycling.
[d] Uphill walking with a load, pick-and-shovel work, lumbering, heavy sports such as swimming, climbing, football, basketball.

Animals on a diet free of linoleic acid show poor growth, impaired healing of wounds, and skin problems and have shortened lives. Linoleic acid lowers the serum level of low density lipoprotein or LDL, the "bad cholesterol" (see Interaction 27.1). We also need linoleic acid to make the arachidonic acid required for prostaglandin synthesis (see Interaction 19.1).

29.2 PROTEIN REQUIREMENTS

A protein's biological value is determined largely by its digestibility and its ability to supply essential amino acids.

The total protein requirement is 56 g/day for adult men and 46 g/day for adult women. Small as these numbers are, they are still more than the bare minimum. Their intent is to allow about 30% extra to cover a wide range of variations in the protein needs of individuals. Moreover, the values assume that we digest and absorb into the bloodstream only 75% of dietary protein.

Our Protein Requirement Is Actually One for Essential Amino Acids Proteins supply α-amino acids, and these are what protein requirements are all about. We particularly need the **essential amino acids,** those that the body cannot synthesize, at least not at rates rapid enough to make much difference. There are eight of them (for adults), listed in the margin. The remaining 12 on the standard list of amino acids can be made in the body from parts and pieces of other molecules, including other amino acids.

Our total daily needs for the essential amino acids add up to only about 13 g, less than half an ounce. But when a cell is actually making a protein, all of the amino acids needed for it must be available *at the same time.* The absence of lysine, for example, when a protein that includes it has to be made, prevents the synthesis of the protein altogether.

■ This is about 2 ounces per day.

■

Essential Amino Acids[a]	Daily Needs (g/day)
Isoleucine	1.4
Leucine	2.2
Lysine	1.6
Methionine	2.2
Phenylalanine	2.2
Threonine	1.0
Tryptophan	0.5
Valine	1.6

[a] Histidine is essential to infants.

Data from F. E. Deatherage, *Food for Life,* Plenum Press, New York, 1975.

The Availability of Amino Acids Varies with the Digestibility of the Protein   A number of factors are involved in assessing the value of a particular source of protein. Is the protein in a particular source fully digestible? How much must be eaten to obtain all of the essential amino acids? How many calories accompany this quantity? These questions are important not just to individuals but also to policy makers dealing with starvation, famine, and undernutrition. Let's consider digestibility first.

Not all proteins are completely digested, and each is assigned a **coefficient of digestibility**, defined by Equation 29.1:

$$\text{coefficient of digestibility} = \frac{(\text{N in food eaten}) - (\text{N in feces})}{(\text{N in food eaten})} \qquad (29.1)$$

where N, of course, is nitrogen. The difference of terms in the numerator, namely,

$$(\text{N in food eaten}) - (\text{N in feces})$$

is *the nitrogen that is actually absorbed by the bloodstream.*

Animal proteins have higher coefficients of digestibility than those of fruits, 0.97 versus about 0.85 (meaning 85% digestible). In legumes and nuts, the average protein digestibility coefficient is 0.78, and in vegetables the range is 0.65 to 0.74.

The value for the proteins in whole wheat flour is 0.79, and for whole rye flour, it's 0.67. Milling improves the digestibility coefficients, raising them to 0.83 for all-purpose bread flour and 0.89 for cake flour. Milling does the same to brown rice, raising it from 0.75 to 0.84 (polished white rice). Milling, unfortunately, also reduces the quantities of vitamins, minerals, and fiber, but these are usually put back to make "enriched flours."

Proteins Low in Essential Amino Acids Have Low Nutritional Value   If a protein is lacking or is low in one or more of the essential amino acids, nutritionists rate it a *low-quality protein.* The body cannot efficiently use it to make its own proteins. Thus, when a low-quality protein fails to supply one essential amino acid, like lysine, the body still gets a load of other amino acids that it can't use to make proteins requiring lysine. There is no way to store the excess amino acids until lysine arrives, and they cannot be excreted as amino acids. To get rid of them, the body has to break them down so their nitrogen can be excreted as urea. This leaves nitrogen-free breakdown products, which the body can use for energy. But if the body isn't demanding this energy, it converts the breakdown products into fat.

One Measure of Protein Quality Is Called the Protein's *Biological Value*   Measured when the body operates under the stress of receiving not quite enough protein overall, the percentage of the nitrogen retained out of the total nitrogen eaten is called the **biological value** of the protein. It is a measure of the efficiency with which the body uses the nitrogen *of the actually absorbed amino acids* furnished by the protein.

The largest single factor in a protein's biological value is its amino acid composition, with emphasis on the essential amino acids. Human milk protein is the best of all proteins in terms of digestibility and biological value. Whole egg protein is very close, and it is often taken as the reference for experimental work.

■ Infants and schoolchildren must have a positive nitrogen balance, meaning that they must ingest more nitrogen than they excrete, in order to grow.

When the body is in **nitrogen balance**, the amount of nitrogen ingested equals the amount excreted, and the diet is well balanced in absorbable amino acids. Most 70-kg men could be in nitrogen balance by ingesting 35 g/day of the proteins in human milk, the value used as a reference standard in rating other proteins.

The Essential Amino Acids Most Poorly Supplied Put an Upper Limit on a Protein's Biological Value   Table 29.3 summarizes some information about most of the

TABLE 29.3   Comparisons of Food Proteins with Human Milk Protein

Food	Limiting Amino Acid	Food's Protein Equivalent to 35 g Human Milk Protein (g)	Digestibility Coefficient of Food's Protein	Amount of Food's Protein[a] (g)	Percentage of Protein in the Food	Amount of Food Needed[b]	Kilo-calories of Food Received
Wheat	Lysine	80.6	79.0	102	13.3%	767 g (1.7 lb)	2560
Corn	Tryptophan and lysine	72.4	60.0	120	7.8%	1540 g (3.4 lb)	5660
Rice	Lysine	51.7	75.0	68.9	7.5%	919 g (2.0 lb)	3310
Beans	Valine	50.5	78.0	64.8	24.0%	270 g (0.59 lb)	913
Soybeans	Methionine and cysteine	43.8	78.0	56.2	34.0%	165 g (0.36 lb)	665
Potatoes	Leucine	71.6	74.0	96.7	2.1%	4600 g (10.1 lb)	3500
Cassava	Methionine and cysteine	82.4	60.0	137	1.1%	12,500 g (27.5 lb)	16,400
Eggs	Leucine	36.6	97.0	37.8	12.8%	295 g (0.65 lb)	477
Meat	Tryptophan	43.1	97.0	44.4	21.5%	206 g (0.45 lb)	295
Cow's milk	Methionine and cysteine	43.8	97.0	45.2	3.2%	1410 g (3.1 lb)	903

Source: Data from F. E. Deatherage, *Food for Life,* Plenum Press, New York, 1975. Used by permission.

[a] The grams of protein that have to be obtained from each food source in order that the protein be nutritionally equivalent (with respect to essential amino acids) to 35 g of human milk protein, allowing for the poorer digestibility of that food's protein (i.e., its digestibility coefficient).

[b] The grams of each food that are equivalent in nutritional value (with respect to essential amino acids) to 35 g human milk protein allowing for the digestibility coefficient and the percent protein in the food.

proteins, named in the first column, that are prominent in various diets of the world's peoples. We now study what the data in the other columns mean.

Column 2   The essential amino acid most poorly supplied by a given protein, its **limiting amino acid,** is named in this column.

Column 3   This column gives the number of grams of each food that a 70-kg man would have to digest *and absorb* per day to get the same amount of its limiting amino acid that is available from 35 g of human milk protein. Such an intake, of course, would also supply all the other needed but nonlimiting amino acids. For example, 80.6 g of wheat protein—not wheat, but wheat protein—would have to be digested and absorbed to obtain the lysine in 35 g of human milk protein. But this figure, 80.6 g, assumes 100% digestion, so we have to adjust for a lower percentage digestion.

Column 4   This column gives the digestibility coefficients of all the proteins listed. Wheat protein has a digestibility coefficient of 0.790 so, by a factor of 100/79.0, we need more wheat protein than 80.6 g in order to get the necessary lysine.

Column 5   Here is the result of multiplying 80.6 by (100/79.0). We need 102 g of wheat protein to get the necessary limiting amino acid, lysine. But remember, we're talking about wheat *protein,* not actual wheat, so we move on to the next column.

Column 6   Wheat, on the average, is only 13.3% protein, so we have to multiply our 102 g of wheat protein by the factor 100/13.3. The grim result, 767 g, is in the next column.

Column 7   A 70-kg man would have to eat 767 g (1.69 lb) of wheat if his daily needs for all amino acids are to be met by wheat alone. But all this wheat also has calories. The implication of a daily diet of 767 g of wheat is in the last column.

Column 8   Anyone eating 767 g of wheat also gets 2560 kcal of food energy. This is nearly as much total energy per day as a 70-kg adult should have, so that isn't prohibitive, but can you imagine a diet this boring? Just imagine his problems if he had to get all his amino acids from potatoes, 10.1 lb/day, or cassava, 27.5 lb/day!

The data in Table 29.3 are important to those concerned about both the protein and total calorie needs of the world's burgeoning population. Neither children nor adults can eat enough corn, rice, potatoes, or cassava per day to meet both their protein and energy needs.

Adequate Proteins Provide a Balanced Supply of Essential Amino Acids   Proteins of eggs and meat, as you can see in Table 29.3, are particularly good. They are classified as **adequate proteins,** because they provide all the essential amino acids in suitable proportions to meet amino acid and total nitrogen needs without excessive intake of calories.

Soybeans are the best of the nonanimal sources of proteins. Cassava, a root, is an especially inadequate protein, and corn (or maize) is also poor. Unhappily, huge numbers of the world's peoples, especially those in Africa, Central and South America, India, and the countries of the Middle and Far East, rely heavily on these two foods. Although they adequately provide energy needs, not enough of them can be eaten per day to give the essential amino acids, so dietary deficiency diseases are widespread in these regions.

Variety in the Diet Offers Many Nutritional Advantages   The data in Table 29.3 also provide scientific support for the longstanding practices in all major cultures of including a wide variety of foods in the daily diet. Meat and eggs can ensure adequate protein while they leave room for an attractive variety that is important to good eating habits. Milk, soybeans, and other beans also leave room for variety.

With Planning and Good Timing, Vegetarians Obtain All Essential Amino Acids   When both rice (low in lysine) and beans (low in valine) are included in equal proportions, about 43 g of a rice–bean combination is equal in protein value to 35 g of human milk protein. Less of this combined diet is needed than rice alone or beans alone, which means that room is left for other foods that our taste buds crave.

The rice and beans should, of course, be eaten fairly closely together to ensure that all essential amino acids are simultaneously available during protein synthesis. Eating just the rice early in the morning and the beans in the evening works against efficient protein synthesis. All of us, including vegetarians, must also include vitamins, studied next.

29.3 VITAMINS

The body cannot make vitamins; they must be in the diet.

The term **vitamin** applies to any compound or a closely related group of compounds satisfying the following criteria.

1. It is organic rather than inorganic or an element.
2. It cannot be synthesized at all (or at least in sufficient amounts) by the body and thus must be in the diet.
3. Its absence causes a specific **vitamin deficiency disease.**
4. Its presence is essential to normal growth and health.
5. It is present in foods in *small* concentrations, and it is not a carbohydrate, a saponifiable lipid, an amino acid, or a protein.

Vitamins function in the body either as precursors for coenzymes or as coenzymes themselves. Most of the nutritionally required minerals function as cofactors to enzymes.

For Some Vitamins, Any Member of a Set of Compounds Works The members of a set of related compounds that prevent a specific vitamin deficiency disease are called *vitamers.* The body can convert any one of them into the active forms needed. Our needs for vitamin A, for example, are satisfied by several structurally related compounds, including a family of plant pigments called the *carotenoids.*

Genetic Defects Can Burden Some with Greater Daily Vitamin Needs In addition to vitamin deficiency diseases, there are at least 25 disorders classified as *vitamin-responsive inborn errors of metabolism.* These arise from genetic faults that can be partly and sometimes entirely overcome by the daily ingestion of 10 to 1000 times as much of a particular vitamin as normally should be present in the diet.

The Fat-Soluble Vitamins are A, D, E, and K *Fat-soluble vitamins* are largely hydrocarbon-like, and *water-soluble vitamins* are polar or ionic. We'll use these categories as we continue.

The fat-soluble vitamins occur in the lipid fractions of their sources. Their molecules have double bonds or phenol rings, so oxidizing agents readily attack them. Hence, these vitamins are destroyed by prolonged exposures to air or to the organic peroxides that develop in fats and oils turning rancid. From our point of view, these are actually good properties; because the fat-soluble vitamins are easily oxidized, they destroy oxidizing agents. Some of the latter, when inside cells or the bloodstream, are on our list of "bad things," because they are involved in the development of coronary heart disease, genetic mutations, and cancer. Interaction 29.1 is a toe-dipping excursion into the realm of poisons, antioxidants, cancer, and the diet.

As a group, the fat-soluble vitamins pose dangers when taken in excess. Our fatty tissues "dissolve" any excess of them, and such accumulations can be dangerous. Let's now take a closer look at each.

Vitamin A activity is given by several vitamers. In the body, the active forms—retinol, retinal, and retinoic acid—all have five alkene groups. β-Carotene is a source for all three because the liver has enzymes that can cleave its molecules to give the retinol skeleton. Thus β-carotene, the yellow pigment in many plants and available in liver and egg yolk, is an important source of vitamin A activity. It is an example of a *provitamin,* a compound that the body can change into a vitamin.

■ Vitamin = "vital amine," from an early belief that vitamins might all be amines.

■ Dietary surpluses of *water-soluble* vitamins leave the body in the urine.

retinol, $G = CH_2OH$
retinal, $G = CH{=\!=}O$
retinoic acid, $G = CO_2H$

β-carotene
(The dashed line divides the molecule into two retinol skeletons.)

■ Notice the alkene groups, which are easily oxidized and so can trap undesirable oxidizing agents, like peroxides.

The deficiency disease for vitamin A is nyctalopia, or night blindness, meaning impaired vision in dim light. We also need vitamin A for healthy mucous membranes.

Several studies have shown that a high intake of *β-carotene* in a diet rich in vegetables and fruit is associated with a reduced risk of cancer of the epithelial cells in several locations. (Epithelial cell cancers, like colon cancer and lung cancer, account for over 90% of cancer deaths in the United States.) The ease of oxidation of the multiply unsaturated skeleton of *β*-carotene may account for this. Two kinds of oxidizing agents, excited (activated) oxygen and free radicals, have been implicated in the onset of cancer, and *β*-carotene may serve to trap and destroy them.

■ Epithelial cells make up the tissue that lines tubes and cavities.

Chemically, *radicals* are species having at least one unpaired electron, like the hydroxyl radical, HO·, where the dot signifies the unpaired electron. Radicals, lacking outer octets, are very reactive and attack almost any organic substance with double bonds, including genes, hormones, enzymes, and vitamins.

Radicals also attack the lipids in LDL, or low density lipoprotein complexes. Partially oxidized lipids attract white blood cells, which can lead to a buildup of plaque in a blood capillary. Thus *anything that suppresses free radicals helps to protect against heart disease.*

■ *Morbidity* is the ratio of diseased to healthy individuals in a population.

Carefully controlled studies in Africa and India have shown that vitamin A supplements among peoples with chronically low availability of vitamin A significantly reduce infant mortality and morbidity. When vitamin A was given to children with severe measles, the mortality rate dropped appreciably.

Excess doses of vitamin A must not be taken because it is toxic to adults. The livers of polar bears and seals are particularly rich in vitamin A, and Eskimos are careful about eating these otherwise desirable foods. Siberian huskies also have very high levels of vitamin A in their livers, and early explorers, when driven to eating their sled dogs to survive, risked serious harm and death by eating too much husky liver. In early 1913 the Antarctic explorer X. Mertz almost certainly died in just this way, and his companion, Douglas Mawson, barely survived.

■ Those suffering from excess vitamin A recover quite quickly when they stop taking vitamin A.

Some and maybe all forms of vitamin A, when taken in excess, are also known to be teratogenic (they cause birth defects), so pregnant women are warned not to take huge doses of vitamin A.

Vitamin D exists in a number of forms. Cholecalciferol (D_3) and ergocalciferol (D_2) are two that occur naturally.

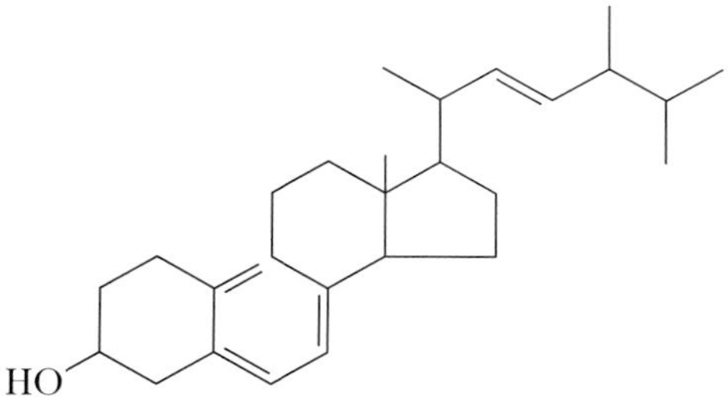

ergocalciferol (D$_2$) 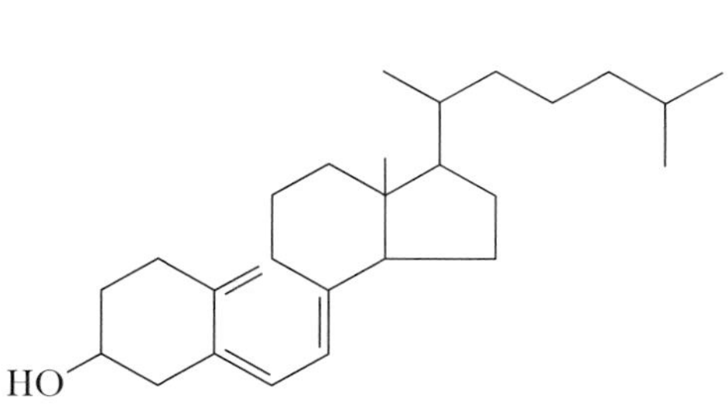   cholecalciferol (D$_3$)

These two are equally useful. The body changes either to the forms employed as hormones to stimulate the absorption and uses of calcium ions and phosphate ions. Vitamin D is especially important during the years of early growth when bones and teeth are developing. Lack of vitamin D causes the deficiency disease known as *rickets,* a bone disorder.

Our bodies can make some vitamin D from certain steroids, provided we get enough direct sunlight on the skin. Youngsters who worked from dawn to dusk in dingy factories or mines during the early years of the industrial revolution were particularly prone to rickets simply because they saw little if any sun.

Eggs, butter, liver, fatty fish, and fish oils such as cod liver oil are good natural sources of vitamin D precursors. The most common source today is vitamin D–fortified milk.

No advantage is gained by taking large doses of vitamin D, and in sufficient excess it is dangerous. Excesses promote an increase in the calcium ion level of the blood, and this damages the kidneys and causes soft tissue to calcify and harden.

Vitamin E needs are satisfied by any member of the tocopherol family. All are phenols, and the most active member is α-tocopherol.

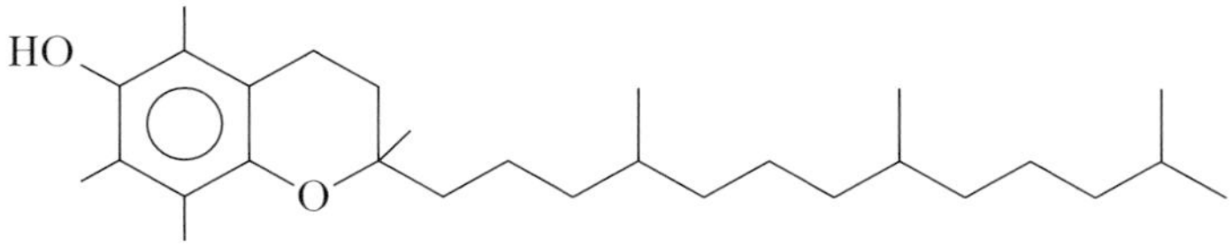

α-tocopherol

In the absence of vitamin E, the activities of certain enzymes are reduced, and red blood cells hemolyze more readily. Anemia and edema are reported in infants whose feeding formulas are low in vitamin E.

Vitamin E occurs so widely in vegetable oils that it is almost impossible not to obtain enough of it. Because vitamin E, being a phenol, has an easily oxidized ring, it is an *antioxidant.* Antioxidants destroy oxidizing agents. Vitamin E particularly detoxifies peroxides, compounds of the general formula R—O—O—H, which form when oxygen attacks unsaturated fatty acids at the CH$_2$ groups adjacent to double bonds. Considering that such acids are in the phospholipids of all cell membranes, you can see that vitamin E helps to protect membranes from oxidation. So does β-carotene.

Because vitamin E is a fat-soluble vitamin, some of it is carried in circulation inside lipoprotein complexes, particularly the low density lipoprotein complexes (LDL). Vitamin E molecules are thus strategically positioned to inhibit an important step, the oxidation of LDL, in the onsets of atherosclerosis and coronary disease. In two huge studies reported in 1993, both men and women who regularly took vitamin E supplements for more than 2 years experienced a significantly reduced risk of coronary disease. Whether the long-term intake (many years, up to a lifetime) of vit-

■ Poor management of Ca^{2+} metabolism contributes to a disfiguring bone condition of old age called *osteoporosis.*

amin E supplements will cause toxic results is not known. Until this question is answered, specialists in coronary disease caution against taking large doses of vitamin E.

Vitamin K is the antihemorrhagic vitamin because it is a cofactor in the blood-clotting mechanism. It is supplied by green, leafy vegetables. Deficiencies are rare. Sometimes women about to give birth as well as their newborn infants are given vitamin K to provide an extra measure of protection against possible hemorrhaging. Vitamin K also aids in the absorption of calcium ion into bone, so it slows the onset of osteoporosis.

The Water-Soluble Vitamins Are Parts of the Cofactors of Enzymes   The water-soluble vitamins recognized by the Food and Nutrition Board are *vitamin C, choline, thiamine, riboflavin, nicotinamide, folate, vitamin B_6, vitamin B_{12}, pantothenic acid,* and *biotin.* It is widely believed that they are among the safest substances known, but this generalization is too broad, as we will see.

Vitamin C, or ascorbic acid, prevents scurvy, a sometimes fatal disease in which collagen is not well made. It is present in citrus fruits, potatoes, leafy vegetables, and tomatoes. Vitamin C is destroyed by extended cooking, heating over steam tables, or prolonged exposure to air or to the ions of iron or copper. Even when vitamin C is kept in a refrigerator in well-capped bottles, it slowly deteriorates.

ascorbic acid
(vitamin C)

choline

Vitamin C is an antioxidant, much like vitamin E, and is a destroyer of free radicals. Because it is water soluble, it mingles within a cell's cytosol. Where low levels of vitamin C exist in seminal fluid, high levels of oxidative damage to sperm are observed, which may be a factor either in reduced fertility or in birth defects.

Whether vitamin C prevents other diseases besides scurvy is the subject of enormous controversy, speculation, and research. A variety of studies have found that vitamin C is involved in the metabolism of amino acids, in the synthesis of some adrenal hormones, and in the healing of wounds. Probably millions of people believe that vitamin C in sufficient dosages—up to several grams per day—acts to prevent or to reduce the severity of the common cold. The vitamin appears to be nontoxic at these high levels.

Choline is needed to make complex lipids. Acetylcholine, which is made from choline, is one of several substances that carry nerve signals from one nerve cell to another. The body can make choline, but the Food and Nutrition Board calls it a vitamin because there are 10 species of higher animals that have dietary requirements for it. We make it too slowly to meet all our daily needs, so having it in the diet provides protection. It occurs widely in meats, egg yolk, cereals, and legumes. No choline deficiency disease has been demonstrated in humans, but in animals the lack of dietary choline leads to fatty livers and to hemorrhagic kidney disease.

Thiamine is needed for the breakdown of carbohydrates. Its deficiency disease is beriberi, a disorder of the nervous system. Good sources are lean meats, legumes, and whole (or enriched) grains. It is stable when dry but is destroyed by alkaline conditions or prolonged cooking. Thiamine is not stored, and excesses are excreted in the urine. One's daily thiamine requirement is proportional to the number of calories that are represented by the diet.

■ Much of the vitamin K that we need is made for us by our own intestinal bacteria.

■ Collagen is the protein in bone that holds the minerals together, much as steel rods work in reinforced concrete.

■ Acetylcholine is one of several neurotransmitters.

■ In rice, most of the thiamine is in the husk, which is lost when raw rice is milled.

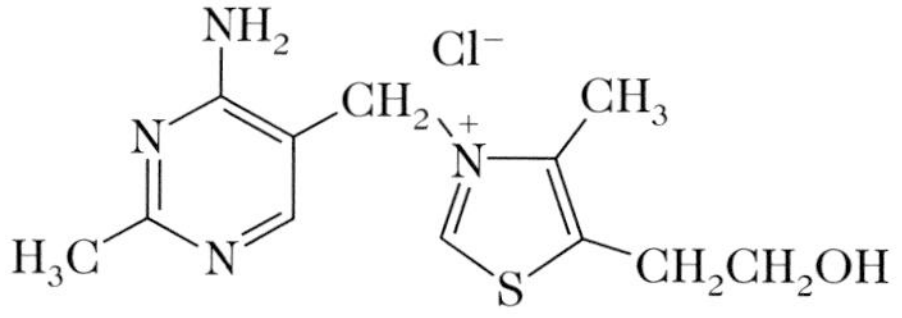

thiamine

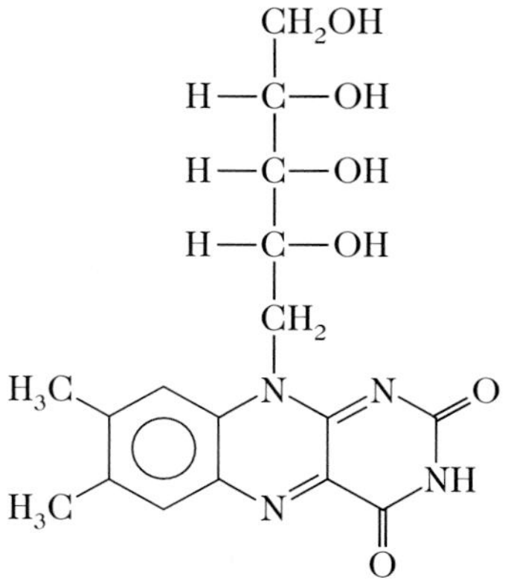

riboflavin

Riboflavin is required by a number of oxidative processes in metabolism. Deficiencies lead to the inflammation and breakdown of tissue around the mouth and nose, as well as the tongue, a scaliness of the skin, and burning, itching eyes. Wound healing is impaired. The best source is milk, but certain meats (e.g., liver, kidney, and heart) also supply it. Cereals are poor sources unless they have been enriched. Little if any riboflavin is stored, and excesses in the diet are excreted. Alkaline substances, prolonged cooking, and irradiation by light destroy this vitamin.

Niacin, meaning both nicotinic acid and nicotinamide, is essential for nearly all biological oxidations, its molecules being built into respiratory chain coenzymes, NAD^+, $NADP^+$, and their reduced forms. Niacin is needed by every cell of the body every day. Its deficiency disease is pellagra, a deterioration of the nervous system and the skin. Pellagra is particularly a problem where corn (or maize) is the major item of the diet. Corn (maize) is low in niacin and the essential amino acid tryptophan, from which we are able to make some of our own niacin. Where the diet is low in tryptophan, niacin must be provided in other foods, such as enriched grains. Prolonged cooking destroys niacin.

■ Severe niacin deficiency can cause delirium and dementia.

■ We can make about 1 mg of niacin for every 60 mg of dietary tryptophan—nearly all we need.

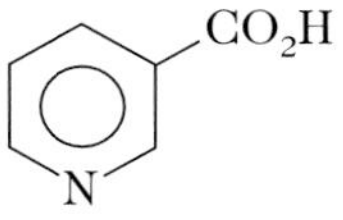

nicotinic acid
(niacin)

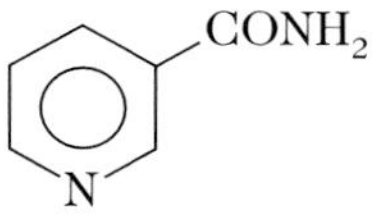

nicotinamide
(niacinamide)

Folate is the name used by the Food and Nutrition Board for folic acid and related compounds. Its deficiency disease is megaloblastic anemia. Several drugs, including alcohol, promote folic acid deficiency. The enzymes that use folic acid as a cofactor are largely involved in reactions such as the syntheses of nucleic acids and heme that transfer one-carbon units, like $-CH_3$ and $-CH=O$. Good sources of folate are fresh, leafy green vegetables, asparagus, liver, and kidney. Folate is relatively unstable to heat, air, and ultraviolet light, and its activity is often lost in both cooking and food storage.

■ Chronic alcoholism causes folate deficiency.

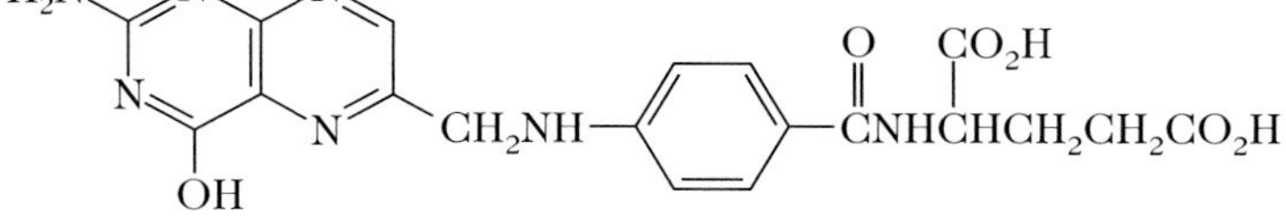

folic acid

A study sponsored by the British Medical Research Council found that supplementary folic acid (4 mg/day) prevented the recurrence of neural-tube defects (e.g., spina bifida) in 72% of the women who had a history of delivering babies with such defects. Its use during the first 4 weeks of pregnancy was particularly important. A large controlled study in Hungary showed that vitamin–mineral supplements, including folic acid, resulted in 50% fewer congenital malformations of all kinds, when taken prior to pregnancy.

As a result of such studies, the Food and Drug Administration, in 1996, issued its first fortification order since 1943. Effective in 1998, most flours, corn meal, pastas, and rice must be fortified with folate at a level of 43 to 140 μg/lb of food. The amount varies with the food, but the intent is to ensure that pregnant women will obtain 400 μg/day of folate. This is expected to reduce the risk of babies being born with neural-tube defects by as much as 70%.

Vitamin B$_6$ activity is supplied by pyridoxine, pyridoxal, or pyridoxamine. All can be changed in the body to the active form, pyridoxal phosphate. The activities of at least 60 enzymes involved in the metabolism of various amino acids depend on pyridoxal phosphate. One deficiency disease is hypochromic microcytic anemia, and disturbances in the central nervous system also occur. The vitamin is present in meat, wheat, yeast, and corn. It is relatively stable to heat, light, and alkali.

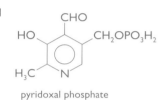

pyridoxal phosphate

pyridoxine   pyridoxal   pyridoxamine

Pyridoxine is widely used as a component of body-building diets and in the treatment of premenstrual syndrome. *Massive doses ("megavitamin doses") at levels of 500 mg/day and higher, however, can severely disable parts of the nervous system and should be avoided.* Pyridoxine is clearly not "among the safest substances known."

Vitamin B$_{12}$, or cobalamin, is a controlling factor for pernicious anemia. This deficiency disease, however, is very rare, because it is very difficult to design a diet that lacks this vitamin. Enzymes that use vitamin B$_{12}$ are, like folate, involved in the transfers of one-carbon units. The synthesis of DNA, for example, depends on vitamin B$_{12}$.

Animal products such as liver, kidney, and lean meats as well as milk products and eggs are good sources—and virtually the only sources. Thus most people who develop B$_{12}$ deficiency are true vegetarians (or are infants born of true vegetarian mothers). The onset of any symptoms of B$_{12}$ deficiency occurs slowly because the body stores it fairly well and because such minute traces are needed.

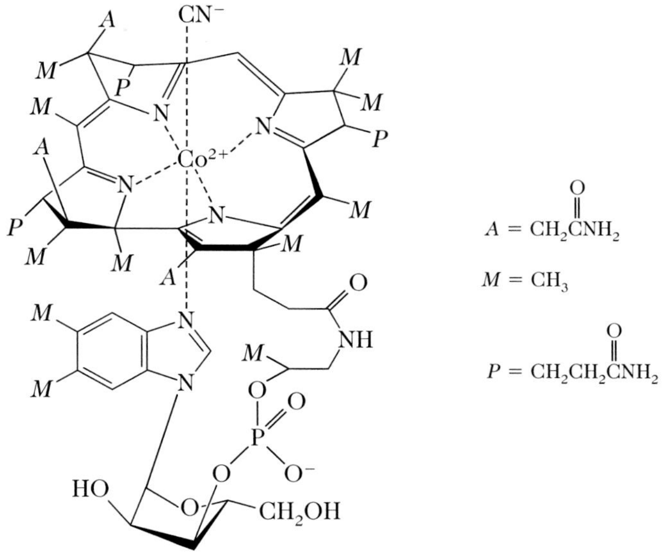

cyanocobalamin

$A = CH_2\overset{\displaystyle O}{\overset{\|}{C}}NH_2$

$M = CH_3$

$P = CH_2CH_2\overset{\displaystyle O}{\overset{\|}{C}}NH_2$

- The replacement of the CN group by other groups (e.g., OH or CH$_3$ and a few others) gives the members of a small family of compounds that function as cofactors for enzymes.

- Dorothy Crowfoot Hodgkin, a British chemist, won the 1964 Nobel Prize in Chemistry for determining the structure of cyanocobalamin.

Pantothenic acid is used to make coenzyme A, which is needed to metabolize fatty acids. Signs of a deficiency disease for pantothenic acid have not been observed clinically in humans, but the deliberate administration of compounds that lower its availability in the body causes cellular damage in vital organs. It is supplied by many foods, and especially by liver, kidney, egg yolks, and skim milk.

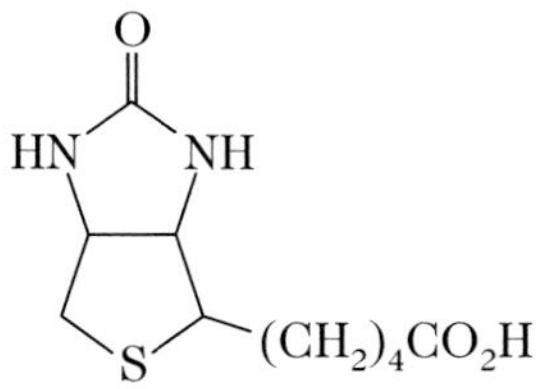

biotin

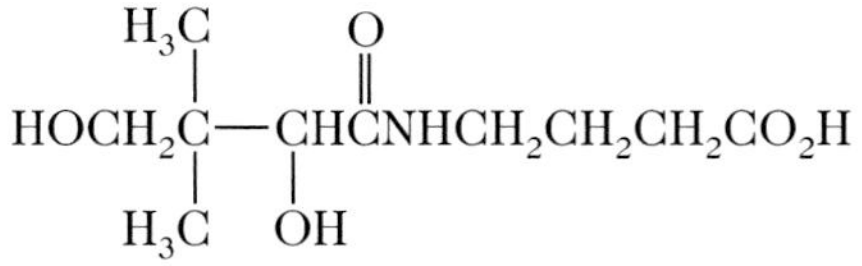

pantothenic acid

Table 29.4 describes safe and adequate daily dietary intakes of biotin and pantothenic acid.

Biotin is required for all pathways in which carbon dioxide (or bicarbonate ion) is temporarily used as a reactant, as in the synthesis of fatty acids. Signs of biotin deficiency are hard to find, but when it is deliberately induced, the individual experiences nausea, pallor, dermatitis, anorexia, and depression. When biotin is given again, the symptoms disappear. Our own intestinal microorganisms probably make biotin for us. Egg yolks, liver, tomatoes, and yeast are good sources.

29.4 MINERALS AND TRACE ELEMENTS IN NUTRITION

Many metal ions are necessary for the activities of enzymes.

Most of the minerals and trace elements needed in the diet are metal ions, but some anions are also required. The distinction between a *mineral* and a *trace element* is a matter of quantity.

TABLE 29.4   Estimated Safe and Adequate Daily Dietary Intakes of Biotin and Pantothenic Acid

Category	Age (years)	Biotin (μg)	Pantothenic Acid (mg)
Infants	0–0.5	10	2
	0.5–1	15	3
Children and	1–3	20	3
adolescents	4–6	25	3–4
	7–10	30	4–5
	11+	30–100	4–7
Adults		30–100	4–7

Data from Food and Nutrition Board, National Academy of Sciences, National Research Council, Revised 1989. Because less information is available for basing allowances, these are not in Table 29.1.

Minerals Are Needed at Levels of 100 mg/day or More   The dietary **minerals** are the chief electrolytes in the body, namely, calcium (Ca^{2+}), phosphorus (phosphate ion, P_i), magnesium (Mg^{2+}), sodium (Na^+), potassium (K^+), and chlorine (Cl^-). The calcium ion is an essential not only for bones and teeth but also to operate the nervous system and to activate enzymes.

Trace Elements Are Needed at Levels of 20 mg/day or Less   The Food and Nutrition Board recognizes 17 **trace elements**, including fluorine (as F^-) and iodine (as I^-), and the ions of chromium, manganese, iron, cobalt, copper, zinc, molybdenum, and selenium (covalently in selenocysteine). All have various functions in animals and so are quite likely needed by humans. Ten are *known* to be needed. All are toxic in excessive amounts. They occur widely in food and drink, but some are removed by food refining and processing. For trace elements not in Table 29.1, see Table 29.5 for estimated safe and adequate daily dietary intakes.

■ Selenocysteine is an amino acid with a CH_2SeH side chain in place of the CH_2SH side chain of normal cysteine.

■ 1 ppm F^- = 1 mg/L

Fluorine (as F^-) is essential to the growth and development of sound teeth, and the Food and Nutrition Board recommends that public water supplies be fluoridated at a level of about 1 ppm wherever natural fluoride levels are too low. Both the medical and the dental associations in the United States strongly support this.

TABLE 29.5   Estimated Safe and Adequate Daily Dietary Intakes of Selected Minerals[a]

Category	Age (years)	Copper (mg)	Manganese (mg)	Fluoride (mg)	Chromium (μg)	Molybdenum (μg)
Infants	0–0.5	0.4–0.6	0.3–0.6	0.1–0.5	10–40	15–30
	0.5–1	0.6–0.7	0.6–1.0	0.2–1.0	20–60	20–40
Children and	1–3	0.7–1.0	1.0–1.5	0.5–1.5	20–80	25–50
adolescents	4–6	1.0–1.5	1.5–2.0	1.0–2.5	30–120	30–75
	7–10	1.0–2.0	2.0–3.0	1.5–2.5	50–200	50–150
	11+	1.5–2.5	2.0–5.0	1.5–2.5	50–200	75–250
Adults		1.5–3.0	2.0–5.0	1.5–4.0	50–200	75–250

[a] The upper levels in this table should not be habitually exceeded because the toxic levels for many trace elements may be only several times the usual intakes.

Data from Food and Nutrition Board, National Academy of Sciences, National Research Council, Revised 1989. Because less information is available for basing allowances, these are not in Table 29.1.

Iodine (as I^-) is essential in the synthesis of thyroid hormones. Diets deficient in iodide ion result in an enlargement of the thyroid gland known as a goiter. It takes only about 1 μg (1×10^{-6} g) of I^- per kilogram of body weight each day to prevent a goiter.

Seafood is an excellent source of iodide ion, but iodized salt with 75 to 80 μg of iodide ion equivalent per gram of salt is the surest way to obtain the iodine we need. Before the days of iodized salt, goiters were quite common, especially in regions where little if any fish or other seafood was in the diet.

Goiter is not the only problem when iodine is lacking in the diets of pregnant women. In the mid-1990s an iodine-deficiency catastrophe of immense proportions in rural, inland regions of China became known in the West. Without iodine during fetal brain development, mental retardation in the offspring is probably the mildest result. Millions of cases exist in China, and hundreds of thousands are cretins, all this having happened despite China's impressive record of fighting disease under difficult conditions.

Chromium (Cr^{3+}) is required for the work of insulin and normal glucose metabolism. Most animal proteins and whole grains supply it. The daily intake should be 0.05 to 0.2 mg. Chromium occurring naturally in foods is absorbed significantly more easily than chromium in salts added to vitamin–mineral supplements.

Manganese (Mn^{2+}) is required for normal nerve function, for the development of sound bones, for reproduction, and for the activities of certain enzymes in carbohydrate metabolism. Some nutritionists recommend a daily intake of 2.5 to 5.0 mg/day. Nuts, whole grains, fruits, and vegetables supply this element, but a recommended daily allowance has yet to be set by the Food and Nutrition Board.

Iron (Fe^{2+}) is an essential cofactor for heme, certain cytochromes, and the iron–sulfur proteins of the respiratory chain. A proper level of iron in circulation is maintained by regulatory mechanisms in the intestines. A high serum iron level apparently renders an individual more susceptible to infections. Pregnant women must have larger than usual amounts of iron to keep pace with the needs of fetal blood.

Cobalt (Co^{2+}) is part of the vitamin B_{12} molecule (cyanocobalamin). Apparently there is no other use for it in humans. Without it, there can be anemia and growth retardation.

Copper (Cu^{2+}) occurs in a number of proteins and enzymes, including certain cytochromes of the respiratory chain. If deficient in copper, the individual synthesizes lower strength collagen and elastin and tends to suffer anemia, skeletal defects, and degeneration of the myelin sheaths of nerve cells. Ruptures and aneurysms of the aorta become more likely. The structure of hair is affected, and reproduction tends to fail.

Fortunately, copper occurs widely in foods, particularly in nuts, raisins, liver, kidney, certain shellfish, and legumes. An intake of just 2 mg/day assures a copper balance for nearly all people. Unfortunately, however, the copper contents of the foods in a typical American diet have declined over the last four decades. An increase in the incidence of heart disease in the United States during this period may have been caused partly by diets low in copper. In one study that involved women, the lower the copper level in the blood the higher were their blood cholesterol levels.

Zinc (Zn^{2+}) is required for the activities of several enzymes. The zinc ion is also important to stabilizing the folded conformations of a number of proteins. Without sufficient zinc in the diet, an individual will experience loss of appetite and poor wound healing. Insufficient zinc in an infant's diet, which is a chronic problem in the Middle East, causes dwarfism and poor development of the gonads. If too much zinc is in the diet, and too little copper is present, the extra zinc acts to inhibit the absorption of copper. The ratio of zinc to copper probably doesn't matter as long as sufficient copper is available. When enough zinc is present, it acts to inhibit the absorp-

■ High-fiber diets can work against the absorption of some of the trace elements.

tion of a rather toxic pollutant, cadmium (as Cd^{2+}), which is just below zinc in the periodic table.

Selenium is needed for thyroid hormone action. How much we need is not known, but it is probably 0.05 to 0.07 mg/day. *Too much is very toxic.* Selenium reduces the occurrence of certain cancers in animals, and it may possibly provide a similar benefit in humans. At excessive levels, it causes cancer in experimental animals.

A strong statistical correlation exists between high levels of selenium in livestock crops and low incidences of human deaths by heart disease. In the United States, those who live where selenium levels are high—the Great Plains between the Mississippi River and the eastern Rocky Mountains—have one-third the chance of dying from heart attack and strokes as those who live where levels are very low—the northeastern quarter of the United States, Florida, and the Pacific Northwest. Rats, lambs, and piglets on selenium-poor diets develop damage to heart tissue and abnormal electrocardiograms.

Molybdenum is needed for an enzyme required for the metabolism of nucleic acids as well as for some enzymes that catalyze oxidations. Deficiencies in humans are unknown, meaning that almost any reasonable diet furnishes enough.

Nickel, silicon, tin, vanadium, and boron are possibly trace elements for humans, because deficiency diseases for these elements have been induced in experimental animals.

SUMMARY

Nutrition Good nutrition entails the ingestion of all the substances needed for health—water, oxygen, food energy, essential amino acids and fatty acids, vitamins, minerals, and trace elements. Foods that best supply various nutrients have been identified. The Food and Nutrition Board of the National Academy of Sciences regularly publishes *recommended dietary allowances* that are intended to be amounts that will meet the nutritional needs of practically all healthy people. Some people need more, most need less. But neither more nor less of anything is necessarily better. The RDAs should be obtained by a varied diet because it promotes good eating habits and because such a diet may supply a nutrient no one yet knows is essential.

Water needs The thirst mechanism leads to our chief source of water. Our principal routes of exporting water are the urine, perspiration (both sensible and insensible), and exhaled air. When excessive water losses occur during vigorous exercise, the body also loses electrolytes.

Energy Both carbohydrates and lipids should be in the diet as sources of energy. Without carbohydrates for several days, certain poisons build up in the blood as the body works to make glucose internally from amino acids. A zero-lipid diet means zero ingestion of essential fatty acids. Our oxygen requirements adjust to the caloric demands of our activities.

Protein in the diet What we need most from proteins are certain essential amino acids, and we need some extra nitrogen if we have to make the nonessential amino acids. When a diet excretes as much nitrogen as it takes in, the individual is in nitrogen balance. The most superior, balanced proteins—those that are highly digestible and that supply the essential amino acids in the right proportions—are proteins associated with animals such as the proteins in milk and whole eggs. (Human milk protein is the standard of excellence, and whole-egg protein is very close to it.)

The most important factor in the biological value of a protein is its limiting amino acid—the essential amino acid that it supplies in the lowest quantity. Another factor is the digestibility of the protein. For several reasons—poor source of an essential amino acid; low digestibility coefficients; low concentration of the protein—several foods cannot be used as exclusive or even major components of a healthy diet. Foods that have inadequate proteins include corn (maize), rice, potatoes, and cassava.

Vitamins Organic compounds called vitamins or sets of closely related compounds that satisfy the same need, must be in the diet in at least trace amounts or an individual will suffer from a vitamin deficiency disease. Vitamins can't be sufficiently made by the body. (Essential amino acids and essential fatty acids are generally not classified as vitamins, nor are carbohydrates, proteins, or triacylglycerols and other saponifiable lipids.) Excessive amounts of the fat-soluble vitamins (A, D, E, and K)—especially A and D—must be avoided. These vitamins accumulate in fatty tissue, and in excess they can cause serious trouble. Any excesses of the water-soluble vitamins are eliminated.

Minerals and trace elements   The minerals are inorganic cations and anions that are needed in the diet in amounts in excess of 100 mg/day. The trace elements are needed at

levels of 20 mg/day or less. The whole body quantities of the minerals are large relative to the trace elements. The trace elements that are metal ions are essential to several enzyme systems. Two anions are trace elements, F^- and I^-. Fluoride ion is needed to make strong teeth, and iodide ion is needed to make thyroid hormones and to prevent goiter. It is needed during pregnancy to prevent mental retardation in the baby.

REVIEW EXERCISES

The answers to Review Exercises whose numbers are in color are found in Appendix E. The answers to the other Review Exercises are found in the Study Guide that accompanies this book. The more challenging questions are marked with asterisks.

Nutrition

29.1   What does the science of nutrition study?

29.2   What is meant by the term *nutrient*?

29.3   What is the relationship of nutrients to *foods*?

29.4   What is the relationship of the science of dietetics to nutrition?

29.5   Why are the recommended dietary allowances higher than minimum daily requirements?

29.6   What are seven situations that require special therapeutic diets?

29.7   Why should our diets be drawn from a variety of foods?

29.8   What is potentially dangerous about the following diets?
(a) A carbohydrate-free diet
(b) A lipid-free diet

29.9   All agree that one fatty acid is *essential*. What is its name and what does "essential" mean in this context?

29.10   What are two other fatty acids often cited as "essential" besides the one of Review Exercise 29.9?

29.11   Which fatty acid is needed to make prostaglandins and stands closest to prostaglandins in the synthetic pathway?

Protein and Amino Acid Requirements

29.12   Because the full complement of 20 amino acids is required to make all the body's proteins, why are fewer than half of this number considered essential amino acids?

29.13   If we are able to make all the nonessential amino acids, why should our daily protein intake include more than what is represented solely by the essential amino acids?

29.14   What does the body do with the amino acids that it absorbs from the bloodstream but doesn't use?

29.15   What is the equation that defines the coefficient of digestibility of a protein? What does the numerator in this equation stand for?

29.16   Which kind of protein is generally more fully digested by the body, the protein from an animal or a plant source?

29.17   What can be done to whole grains to improve the digestibility of their proteins? What also happens in this process that reduces the food value of the grains?

29.18   What is the most important factor in determining the biological value of a given protein?

29.19   Which specific protein has the highest coefficient of digestibility and the highest biological value for humans? Which protein comes so close to this on both counts that it is possible to use it as a substitute for research purposes?

29.20   What is meant by the *limiting* amino acid of a protein?

29.21   Why does the protein in corn have a lower biological value than the protein in whole eggs?

*****29.22**   In one variety of hybrid corn, the limiting amino acids are lysine and tryptophan. It takes 75 g of the protein of this corn to be equivalent to 35 g of human milk protein. The coefficient of digestibility of the protein in this corn is 0.59.
(a) In order to match the nutritional value of human milk protein, how many grams of the protein in this corn must be ingested?
(b) To get this much protein from this variety of corn, how many grams of the corn must be eaten? The corn is only 7.6% protein.
(c) How many kilocalories are also consumed with the quantity of corn calculated in part (b) if the corn has 360 kcal/100 g?
(d) Could a child eat enough corn per day to satisfy all its protein needs and still have room for any other food?

Vitamins

29.23 Make a table that lists each vitamin, at least one good source of each, and a serious consequence of a deficiency of each. (Use only the information available in this book.) Set up the table with the following column heads.

Vitamin	Source(s)	Problem(s) if Deficient

29.24 Why aren't the essential amino acids listed as vitamins?

29.25 On a strict vegetarian diet—no meat, eggs, and dairy products of any sort—which one vitamin is hardest to obtain?

29.26 Why should strict vegetarians use two or more different sources of proteins?

29.27 Which are the fat-soluble vitamins?

29.28 Which vitamin is activated when the skin is exposed to sunlight?

29.29 Which vitamin acts as a hormone in its active forms?

29.30 Which vitamin has been shown to be teratogenic when used in excess?

29.31 The yellow-orange pigment in carrots, β-carotene, can serve as a source of the activity of which vitamin?

29.32 The vitamin needed to participate in the blood-clotting mechanism is which one?

29.33 What is a free radical and why is this species dangerous in cells and their membranes?

29.34 What vitamins provide protection against free radicals within membranes of cells?

29.35 How might the attack of free radicals on the cholesterol in low density lipoprotein complexes (LDL) accelerate the formation of a plaque in a capillary?

29.36 Name a vitamin that helps to destroy free radicals within the cytosol.

29.37 The Food and Nutrition Board of the National Academy of Sciences recognizes which substances as the water-soluble vitamins?

29.38 Scurvy is prevented by which vitamin?

29.39 Name at least two vitamins that tend to be destroyed by prolonged cooking.

29.40 Which vitamin prevents beriberi?

29.41 When corn (maize) is the chief food in the diet, which vitamin is likely to be in short supply because a raw material for making it is in short supply?

29.42 Which vitamin, when taken during (or before) pregnancy, offers protection against neural-tube defects?

Minerals and Trace Elements

29.43 What criterion makes the distinction between *minerals* and *trace elements*?

29.44 Name and give the chemical forms of the six minerals.

29.45 Make a table of the 10 trace elements and at least one particular function of each.

29.46 What can be the body's response to an iodine-deficient diet?

Additional Exercises

*29.47 If an individual must take in 20.0 mol O_2 in 1 day to satisfy all needs, how many liters of *air* must be inhaled to supply this? (Air is 21.0% oxygen, on a volume/volume basis, and assume that the measurement is at STP.)

*29.48 The limiting amino acid in peanuts is lysine, and 62 g of protein obtained from peanuts is equivalent to 35 g of human milk protein. The coefficient of digestibility of peanut protein is 0.78. In order to match human milk protein in nutritional value,
(a) How many grams of peanut protein must be eaten?
(b) How many grams of peanuts must be eaten if peanuts are 26.2% protein?
(c) How many kilocalories are also ingested with this many grams of peanuts if peanuts have 282 kcal/100 g?
(d) Could a child eat enough peanuts per day to satisfy its protein needs and still have room for other foods?

Mathematical Concepts

A.1 EXPONENTIALS

When numbers are either very large or very small, it's often more convenient to express them in what is called *exponential notation*. Several examples are given in Table A.1, which shows how multiples of 10, such as 10,000, and submultiples of 10, such as 0.0001, can be expressed in exponential notation.

Exponential notation restates a number as the product of two numbers. The first is a digit between 1 and 10 that is multiplied by the second, namely, 10 raised to an exponent. For example, 55,000,000 is expressed in exponential notation as 5.5×10^7, in which 7 (meaning +7) is the exponent.

Exponents can be negative numbers, too, as the -3 in 3.4×10^{-3}. This appendix will teach you how to move back and forth between exponential expressions and their expanded forms.

Positive Exponents   A positive exponent tells us how many times the number given before the 10 has to be multiplied by 10 to give the same number in its expanded form. For example,

$$5.5 \times 10^7 = \underbrace{10 \times 10 \times 10 \times 10 \times 10 \times 10 \times 10}_{10^7}$$

$$6 \times 10^3 = 6 \times 10 \times 10 \times 10 = 6000$$

$$8.576 \times 10^2 = 8.576 \times 10 \times 10 = 857.6$$

The number before the 10 need not always be between 1 and 10. This is just a convention, which is sometimes ignored. But we may not ignore the rules of arithmetic in converting one form to the other. For example, 10^5 means multiplying by 10 five times; 10^3 tells us to multiply by 10 three times. Thus,

$$0.00045 \times 10^5 = 0.00045 \times 10 \times 10 \times 10 \times 10 \times 10$$
$$= 45$$
$$87.5 \times 10^3 = 87.5 \times 10 \times 10 \times 10$$
$$= 87,500$$

The problem, however, is usually given the other way around, namely that of changing a large number to its exponential form. This is very easy to do. Simply count the number of places that you have to move the decimal point to the *left* to put it right after the first digit of the given number. For example, you might have to work with a number such as 1500 (as in 1500 mL). You'd have to move the decimal point three places leftward from where it is (or is understood to be) to put it immediately after the first digit.

$$1\underset{3}{5}\underset{2}{0}\underset{1}{0}.$$

Each of these leftward moves counts as one unit for the exponent. Three leftward moves means an exponent of +3. Therefore 1500 can be written as 1.500×10^3. A really large number, and one that you'll certainly meet somewhere during the course, is

TABLE A.1

Number	Exponential Form
1	1×10^0
10	1×10^1
100	1×10^2
1000	1×10^3
10,000	1×10^4
100,000	1×10^5
1,000,000	1×10^6
0.1	1×10^{-1}
0.01	1×10^{-2}
0.001	1×10^{-3}
0.0001	1×10^{-4}
0.00001	1×10^{-5}
0.000001	1×10^{-6}

602,000,000,000,000,000,000,000. (It's called Avogadro's number, and you can see that manipulating it "as is" would be awkward and tedious. How in the world is it even pronounced?) In exponential notation, it's written simply as 6.02×10^{23}. Check it out. Do you move the decimal point leftward 23 places? (And now you could pronounce it: "six point oh two times ten to the twenty-third"; but saying "Avogadro's number" is easier.) Do the following exercises for a quick practice.

EXERCISE A.1
Write each of these numbers in exponential form.
  (a) 422,045   (b) 24,000,000,000,000,000,000   (c) 24.32

EXERCISE A.2
Expand each of these exponential numbers.
  (a) 5.050×10^6 (b) 0.0000344×10^8 (c) 324.4×10^3

Answers to Exercises A.1 and A.2

A.1 (a) 4.22045×10^5 (b) 2.4×10^{19} (c) 2.432×10^1

A.2 (b) 5,050,000   (b) 3,440   (c) 324,400

Negative Exponents   A negative exponent tells us how many times the number before the 10 must be *divided* by 10 to give the number in its expanded form. For example, an exponent of −4 means to divide by 10 four times. Thus,

$$1 \times 10^{-4} = 1 \div 10 \div 10 \div 10 \div 10$$

$$= \frac{1}{10 \times 10 \times 10 \times 10} = \frac{1}{10000} = \frac{1}{10^4}$$

$$= 0.0001$$

$$6 \times 10^{-3} = 6 \div 10 \div 10 \div 10$$

$$= \frac{6}{10 \times 10 \times 10} = \frac{6}{1000}$$

$$= 0.006$$

$$8.576 \times 10^{-2} = \frac{8.576}{10 \times 10} = \frac{8.576}{100} = 0.08576$$

Sometimes, you'll want to convert a very small number into its equivalent in exponential notation. This is also easy. This time we count *rightward* the number of times that you have to move the decimal point, one digit at a time, to place the decimal immediately to the right of the first nonzero digit in the number. For example, if the number is 0.00045, you have to move the decimal four times to the right to place it after the 4.

$$0 \underset{1 \quad 2 \quad 3 \quad 4}{.\,0\,0\,0\,4\,5}$$

Therefore we can write $0.00045 = 4.5 \times 10^{-4}$. Similarly, we can write $0.0012 = 1.2 \times 10^{-3}$. And $0.0000000000000011 = 1.1 \times 10^{-16}$. Now try these exercises.

EXERCISE A.3
Write each number in expanded form.
(a) 4.3×10^{-2} (b) 5.6×10^{-10} (c) 0.00034×10^{-2} (d) 4523.34×10^{-4}

EXERCISE A.4
Write the following numbers in exponential forms.
(a) 0.115   (b) 0.00005000041   (c) 0.000000000000345

Answers to Exercises A.3 and A.4

A.3   (a) 0.043   (b) 0.00000000056   (c) 0.0000034   (d) 0.452334
A.4   (b) 1.15×10^{-1} (b) 5.000041×10^{-5} (c) 3.45×10^{-13}

The next step is to learn how to add, subtract, multiply, or divide numbers when they are expressed in exponential form.

How to Add and Subtract Numbers in Exponential Form   We'll not spend too much time on this, because it doesn't come up very often. The only rule is that *when you add or subtract exponentials, all of the numbers must have the same exponents of 10.* If they don't, we have to reexpress them to achieve this condition. Suppose you want to add 4.41×10^3 and 2.20×10^3. The result is simply 6.61×10^3.

$$(4.41 \times 10^3) + (2.20 \times 10^3) = (4.41 + 2.20) \times 10^3$$
$$= 6.61 \times 10^3$$

However, we can't add 4.41×10^3 to 2.20×10^4 without first making the exponents equal. We can do this in either of the following ways. In one, we notice that $2.20 \times 10^4 = 2.20 \times 10 \times 10^3 = 22.0 \times 10^3$, so we have:

$$(4.41 \times 10^3) + (22.0 \times 10^3) = 26.41 \times 10^3 = 2.641 \times 10^4$$

Alternatively, we could notice that $4.41 \times 10^3 = 4.41 \times 10^{-1} \times 10^4 = 0.441 \times 10^4$, so we can do the addition as follows:

$$(0.441 \times 10^4) + (2.20 \times 10^4) = 2.641 \times 10^4$$

The result is the same both ways. The extension of this to subtraction should be obvious.[1]

---

[1] Whenever these operations are with pure numbers and not with physical quantities obtained by measurements, we are not concerned about the numbers of significant figures in the answers.

How to Multiply Numbers in Exponential Form Use the following two steps to multiply numbers that are expressed in exponential forms.

Step 1. Multiply the numbers in front of the 10s.

Step 2. *Add* the exponents of the 10s algebraically.

EXAMPLE A.1

$$(2 \times 10^4) \times (3 \times 10^5) = 2 \times 3 \times 10^{(4+5)}$$
$$= 6 \times 10^9$$

Usually, the problem you want to solve involves very large or very small numbers not yet expressed exponentially. When this happens, first convert them into their exponential forms, and then do the arithmetic. You'll see that exponentials make calculations easier.

EXAMPLE A.2

$$6576 \times 2000 = (6.576 \times 10^3) \times (2 \times 10^3)$$
$$= 13.152 \times 10^6$$
$$= 1.3152 \times 10^7$$

EXERCISE A.5

Calculate the following products after you have converted large or small numbers to exponential forms.

(a) $6{,}000{,}000 \times 0.0000002$ (b) $10^6 \times 10^{-7} \times 10^8 \times 10^{-7}$

(c) $0.003 \times 0.002 \times 0.000001$ (d) $1{,}500 \times 3{,}000{,}000{,}000{,}000$

Answers

(a) 1.2   (b) 1   (c) 6×10^{-12} (d) 4.5×10^{15}

How to Divide Numbers in Exponential Form To divide numbers expressed in exponential forms, use the following two steps.

Step 1. Divide the numbers that stand in front of the 10s.

Step 2. *Subtract* the exponents of the 10s algebraically.

EXAMPLE A.3

$$(8 \times 10^4) \div (2 \times 10^3) = (8 \div 2) \times 10^{(4-3)}$$
$$= 4 \times 10^1$$

EXAMPLE A.4

$$(8 \times 10^4) \div (2 \times 10^{-3}) = (8 \div 2) \times 10^{[4-(-3)]}$$
$$= 4 \times 10^7$$

EXERCISE A.6

Do the following calculations using exponential forms of the numbers.

(a) $6{,}000{,}000 \div 1500$ (b) $7460 \div 0.0005$

(c) $\dfrac{3\,000\,000 \times 6\,000\,000\,000}{20\,000}$ (d) $\dfrac{0.016 \times 0.0006}{0.000008}$

(e) $\dfrac{400 \times 500 \times 0.002 \times 500}{2\,500\,000}$

Answers

(a) 4×10^3 (b) 1.492×10^7 (c) 9×10^{11} (d) 1.2 (e) 8×10^{-2}

The Pocket Calculator and Exponentials   The foregoing was meant to refresh your memory about exponentials, because you have almost certainly studied them before. You probably own a pocket calculator that can take numbers in exponential form. Go ahead and use it, but be sure that you understand exponentials well, first. Otherwise, there are many pitfalls.

Most pocket calculators have a key marked $\boxed{EE}$ or $\boxed{EXP}$ used to enter exponentials. Here is where an ability to *read* exponentials comes in handy. For example, the number 2.1×10^4 reads "two point one times ten to the fourth." The $\boxed{EE}$ or $\boxed{EXP}$ key on most calculators stands for ". . . times ten to the . . . ." Therefore to enter 2.1×10^4, punch the following keys.

$$\boxed{2}\ \boxed{\cdot}\ \boxed{1}\ \boxed{EE}\ \boxed{4}$$

Try this on your own calculator, and be sure to see that the display is correct. If it isn't, you may have a calculator that works differently than most, so recheck your operations of entering and then check the owner's manual.

To enter a negative exponential you have to use one more key, the $\boxed{+/-}$ key. This switches a positive number to its negative, and you *must* use it rather than the $\boxed{-}$ key in this situation. Thus the number 2.1×10^{-5} enters as follows.

$$\boxed{2}\ \boxed{\cdot}\ \boxed{1}\ \boxed{EE}\ \boxed{+/-}\ \boxed{5}$$

Try it and check the display. To see what happens if you use the $\boxed{-}$ key instead of the $\boxed{+/-}$ key, clear the display and enter this number only using the $\boxed{-}$ key instead of the $\boxed{+/-}$ key.

A.2 CROSS-MULTIPLICATION

In this section we will learn how to solve for x in such expressions as

$$\frac{12}{x} = \frac{16}{25} \qquad \text{or} \qquad \frac{32.0}{11.2} = \frac{6.15x}{13.1}$$

The operation is called *cross-multiplication,* and its object is to get x to stand alone, all by itself, on one side of the equals sign, and above any real or understood divisor line.

> To cross-multiply, move a number or a symbol or both across the = sign and across a divisor line, and then multiply.

EXAMPLE A.5

PROBLEM: Solve for x in $\dfrac{25}{x} = 5$.

SOLUTION: Notice first the divisor lines; one of them is understood.

$$\text{Divisor line} \longrightarrow \frac{25}{x} = 5 \longleftarrow \begin{array}{l}\text{The divisor line here}\\ \text{is understood because}\end{array}$$

$$5 = \frac{5}{1}$$

Remember, we want x to stand alone, on top of a divisor line (even if this line is understood). To make this happen, we carry out cross-multiplication as indicated:

$$\frac{25}{x} \quad 5$$

Notice that the arrows show moves that carry the quantities not only across the $=$ sign but also across their respective divisor lines. *It is essential that both crossings over be done.* Now we have x standing alone above its (understood) divisor line.

$$\frac{25}{5} = x$$

Now we can do the arithmetic: $x = 5$.

EXERCISE A.7

Solve for x in the following.

(a) $\dfrac{12}{x} = \dfrac{16}{25}$ (b) $\dfrac{32.0}{11.2} = \dfrac{6.15x}{13.1}$

Answers

(a) $x = 18.75$ (b) $x = 6.085946574$

In Chapter 2, Example 2.3, the problem was to solve for Δt in the equation:

$$\frac{0.12 \text{ cal}}{\text{g} \,^{\circ}\text{C}} = \frac{115 \text{ cal}}{25.4 \text{ g} \times \Delta t}$$

Here is a problem with both units and numbers, so now we have to add one more and very important principle. We *cross-multiply units as well as numbers.*

$$\frac{\boxed{0.12 \text{ cal}}}{\boxed{\text{g} \,^{\circ}\text{C}}} \quad \frac{115 \text{ cal}}{\Delta t \times \boxed{25.4 \text{ g}}}$$

The result is the following. Draw in the cancel lines yourself to show how the units cancel correctly, leaving only $^{\circ}$C.

$$\Delta t = \frac{(115 \text{ cal}) \times (\text{g} \,^{\circ}\text{C})}{(25.4 \text{ g}) \times (0.12 \text{ cal})} = 38 \,^{\circ}\text{C}$$

How to Do Chain Calculations with the Pocket Calculator   Sometimes the steps in solving a problem lead to something like the following:

$$x = \frac{24.2 \times 30.2 \times 55.6}{2.30 \times 18.2 \times 4.44}$$

Many people will first calculate the value of the numerator and write it down. Then they'll compute the denominator and write it down. Finally, they'll divide the two results to get the final answer. There's no need to do this much work. All you have to do is enter the first number you see in the numerator, 24.2 in our example. Then use the $\boxed{\times}$ key for any number in the numerator and the $\boxed{\div}$ key for any number in the denominator. *Each number in the denominator is entered with the* $\boxed{\div}$ *key.* Any of the following sequences work. Try them.

$$24.2 \times 30.2 \times 55.6 \div 2.30 \div 18.2 \div 4.44 = 218.632 \ldots$$

Or

$$24.2 \div 2.30 \times 30.2 \div 18.2 \times 55.6 \div 4.44 = 218.632 \ldots$$

A.3 LOGARITHMS

A logarithm is an exponent. When it's specifically an exponent of 10, it's called a **common logarithm** or a **log**. The logarithm of a number N to the base 10 is the exponent to which 10 must be raised to give N. In other words, when

$$N = 10^x$$

then the log of N is simply x. For example, when

$N = 10$	$\log N = 1$	because	$10 = 10^1$
$= 100$	$= 2$		$100 = 10^2$
$= 1000$	$= 3$		$1000 = 10^3$
$= 0.1$	$= -1$		$0.1 = 10^{-1}$
$= 0.01$	$= -2$		$0.01 = 10^{-2}$
$= 0.001$	$= -3$		$0.001 = 10^{-3}$

Usually, it's not this simple. For example, suppose that $N = 7.35$ rather than some simple multiple or submultiple of 10. By our definitions, to find the log of 7.35 (usually written as log 7.35), we have to find the value of x in

$$7.35 = 10^x$$

The exponent x cannot now be a simple whole number. Here's how we proceed. If

$$a = b$$

then it must be true that

$$\log a = \log b$$

So if 7.35 takes the place of a and 10^x takes the place of b,

$$7.35 = 10^x$$

then, taking the logarithms of both sides must give us

$$\log 7.35 = \log 10^x$$

But the *definition* of a log tells us that $\log 10^x = x$, so we can now write

$$\log 7.35 = x$$

With a nod of appreciation to mathematicians who did the basic work, there are two ways to determine x. One is to use a table of logarithms and the other is to use a pocket calculator. A pocket calculator with a $\boxed{\log}$ key has, in effect, a built-in table of logarithms. Just enter the number, 7.35, and hit the $\boxed{\log}$ key, and you will see 0.866287339 come up on the screen. This number has far more digits than we need. Let's round it to 0.866. Thus

$$\log 7.35 = 0.866$$

This result means that

$$7.35 = 10^{0.866}$$

Note that this makes some rough sense. If $10 = 10^1$ and log 10 is therefore 1, the log of a number slightly smaller than 10, like 7.35, should correspond to 10 raised to a power somewhat smaller than 1, like 0.866.

The only situation in this book in which logarithms must be used is in connection with pH or pK_a problems. For pH problems, we use the pocket calculator to find the log of a number. By definition, pH is given as follows.

$$pH = -\log [H^+]$$

If $[H^+] = 3.5 \times 10^{-8}$ mol/L, for example, then the log of this value is (correctly rounded) -7.46. The pH is the negative of this number, so the pH = 7.46.

If the pH of a solution is given and the question asks you to find the value of $[H^+]$ in the solution, the alternative definition of pH is useful.

$$[H^+] = 1 \times 10^{-pH}$$

Thus if an aqueous solution has a pH of 4.66, we enter 4.66 and change sign with the $\boxed{+/-}$ key. We have now entered x for the equation

$$N = 10^x$$

where N is $[H^+]$ and $-x$ is the pH.
We now use the $\boxed{10^x}$ key to find N. Correctly rounded,

$$N = 2.2 \times 10^{-5}$$

So

$$[H^+] = 2.2 \times 10^{-5} \text{ mol/L}$$

Electron Configurations of the Elements

Atomic Number	Element	Configuration	Atomic Number	Element	Configuration	Atomic Number	Element	Configuration
1	H	$1s^1$	38	Sr	[Kr] $5s^2$	75	Re	[Xe] $6s^24f^{14}5d^5$
2	He	$1s^2$	39	Y	[Kr] $5s^24d^1$	76	Os	[Xe] $6s^24f^{14}5d^6$
3	Li	[He] $2s^1$	40	Zr	[Kr] $5s^24d^2$	77	Ir	[Xe] $6s^24f^{14}5d^7$
4	Be	[He] $2s^2$	41	Nb	[Kr] $5s^14d^4$	78	Pt	[Xe] $6s^14f^{14}5d^9$
5	B	[He] $2s^22p^1$	42	Mo	[Kr] $5s^14d^5$	79	Au	[Xe] $6s^14f^{14}5d^{10}$
6	C	[He] $2s^22p^2$	43	Tc	[Kr] $5s^24d^5$	80	Hg	[Xe] $6s^24f^{14}5d^{10}$
7	N	[He] $2s^22p^3$	44	Ru	[Kr] $5s^14d^7$	81	Tl	[Xe] $6s^24f^{14}5d^{10}6p^1$
8	O	[He] $2s^22p^4$	45	Rh	[Kr] $5s^14d^8$	82	Pb	[Xe] $6s^24f^{14}5d^{10}6p^2$
9	F	[He] $2s^22p^5$	46	Pd	[Kr] $4d^{10}$	83	Bi	[Xe] $6s^24f^{14}5d^{10}6p^3$
10	Ne	[He] $2s^22p^6$	47	Ag	[Kr] $5s^14d^{10}$	84	Po	[Xe] $6s^24f^{14}5d^{10}6p^4$
11	Na	[Ne] $3s^1$	48	Cd	[Kr] $5s^24d^{10}$	85	At	[Xe] $6s^24f^{14}5d^{10}6p^5$
12	Mg	[Ne] $3s^2$	49	In	[Kr] $5s^24d^{10}5p^1$	86	Rn	[Xe] $6s^24f^{14}5d^{10}6p^6$
13	Al	[Ne] $3s^23p^1$	50	Sn	[Kr] $5s^24d^{10}5p^2$	87	Fr	[Rn] $7s^1$
14	Si	[Ne] $3s^23p^2$	51	Sb	[Kr] $5s^24d^{10}5p^3$	88	Ra	[Rn] $7s^2$
15	P	[Ne] $3s^23p^3$	52	Te	[Kr] $5s^24d^{10}5p^4$	89	Ac	[Rn] $7s^26d^1$
16	S	[Ne] $3s^23p^4$	53	I	[Kr] $5s^24d^{10}5p^5$	90	Th	[Rn] $7s^26d^2$
17	Cl	[Ne] $3s^23p^5$	54	Xe	[Kr] $5s^24d^{10}5p^6$	91	Pa	[Rn] $7s^25f^26d^1$
18	Ar	[Ne] $3s^23p^6$	55	Cs	[Xe] $6s^1$	92	U	[Rn] $7s^25f^36d^1$
19	K	[Ar] $4s^1$	56	Ba	[Xe] $6s^2$	93	Np	[Rn] $7s^25f^46d^1$
20	Ca	[Ar] $4s^2$	57	La	[Xe] $6s^25d^1$	94	Pu	[Rn] $7s^25f^6$
21	Sc	[Ar] $4s^23d^1$	58	Ce	[Xe] $6s^24f^15d^1$	95	Am	[Rn] $7s^25f^7$
22	Ti	[Ar] $4s^23d^2$	59	Pr	[Xe] $6s^24f^3$	96	Cm	[Rn] $7s^25f^76d^1$
23	V	[Ar] $4s^23d^3$	60	Nd	[Xe] $6s^24f^4$	97	Bk	[Rn] $7s^25f^9$
24	Cr	[Ar] $4s^13d^5$	61	Pm	[Xe] $6s^24f^5$	98	Cf	[Rn] $7s^25f^{10}$
25	Mn	[Ar] $4s^23d^5$	62	Sm	[Xe] $6s^24f^6$	99	Es	[Rn] $7s^25f^{11}$
26	Fe	[Ar] $4s^23d^6$	63	Eu	[Xe] $6s^24f^7$	100	Fm	[Rn] $7s^25f^{12}$
27	Co	[Ar] $4s^23d^7$	64	Gd	[Xe] $6s^24f^75d^1$	101	Md	[Rn] $7s^25f^{13}$
28	Ni	[Ar] $4s^23d^8$	65	Tb	[Xe] $6s^24f^9$	102	No	[Rn] $7s^25f^{14}$
29	Cu	[Ar] $4s^13d^{10}$	66	Dy	[Xe] $6s^24f^{10}$	103	Lr	[Rn] $7s^25f^{14}6d^1$
30	Zn	[Ar] $4s^23d^{10}$	67	Ho	[Xe] $6s^24f^{11}$	104	Rf	[Rn] $7s^25f^{14}6d^2$
31	Ga	[Ar] $4s^23d^{10}4p^1$	68	Er	[Xe] $6s^24f^{12}$	105	Db	[Rn] $7s^25f^{14}6d^3$
32	Ge	[Ar] $4s^23d^{10}4p^2$	69	Tm	[Xe] $6s^24f^{13}$	106	Sg	[Rn] $7s^25f^{14}6d^4$
33	As	[Ar] $4s^23d^{10}4p^3$	70	Yb	[Xe] $6s^24f^{14}$	107	Bh	[Rn] $7s^25f^{14}6d^5$
34	Se	[Ar] $4s^23d^{10}4p^4$	71	Lu	[Xe] $6s^24f^{14}5d^1$	108	Hs	[Rn] $7s^25f^{14}6d^6$
35	Br	[Ar] $4s^23d^{10}4p^5$	72	Hf	[Xe] $6s^24f^{14}5d^2$	109	Mt	[Rn] $7s^25f^{14}6d^7$
36	Kr	[Ar] $4s^23d^{10}4p^6$	73	Ta	[Xe] $6s^24f^{14}5d^3$			
37	Rb	[Kr] $5s^1$	74	W	[Xe] $6s^24f^{14}5d^4$			

SOME RULES FOR NAMING INORGANIC COMPOUNDS

Only rules considered sufficient to meet most of the needs of the users of this text are in this appendix. The latest edition of the *Handbook of Chemistry and Physics,* published annually by the CRC Press, Boca Raton, FL, under the general editorship of R. C. Weast, has a section on all of the rules. Virtually all college libraries have this reference.

I. Binary Compounds—those made from only two elements

  A. One element is a metal and the other is a nonmetal.

    1. The name of the metal is written first in the name of the compound, and its symbol is placed first in the formula.

    2. The name ending of the nonmetal is changed to *-ide.* Thus the names of the simple ions of groups VIA and VIIA of the periodic table are:

Group VIIA	Group VIA
Fluoride	Oxide
Chloride	Sulfide
Bromide	Selenide
Iodide	Telluride

    3. If the metal and the nonmetal each have just one oxidation number, a binary compound of the two is named simply by writing the name of the metal and then that of the nonmetal with its ending modified by *-ide,* as shown above. Greek prefixes such as mono-, di-, tri-, etc., are not necessary. Examples are:

Some Compounds between Elements of Groups IA and VIIA		Some Compounds between Elements of Groups IA and VIA	
NaF	Sodium fluoride	Na_2O	Sodium oxide[a]
KCl	Potassium chloride	K_2S	Potassium sulfide
LiBr	Lithium bromide	Li_2O	Lithium oxide
RbI	Rubidium iodide	Cs_2S	Cesium sulfide
CsCl	Cesium chloride	Rb_2O	Rubidium oxide

[a] Not disodium oxide.

Some Compounds between Elements of Groups IIA and VIIA		Some Compounds between Elements of Groups IIA and VIA	
$BeCl_2$	Beryllium chloride	BeO	Beryllium oxide
$MgBr_2$	Magnesium bromide	MgS	Magnesium sulfide
CaF_2	Calcium fluoride	CaO	Calcium oxide
SrI_2	Strontium iodide	SrS	Strontium sulfide
$BaCl_2$	Barium chloride	BaO	Barium oxide

Some Compounds between Elements of Groups IIIA and VIIA		Some Compounds between Elements of Groups IIIA and VIA	
$AlCl_3$	Aluminum chloride	Al_2O_3	Aluminum oxide
AlF_3	Aluminum fluoride	Al_2S_3	Aluminum sulfide

4. If the metal has more than one oxidation number, but the nonmetal has just one, the formal name of the compound includes a roman numeral in parentheses following the name of the metal. This numeral stands for the oxidation number of the metal. Greek prefixes such as mono-, di-, etc., are not needed. The following compounds of iron and copper illustrate this rule.

Compounds of Iron in Oxidation States of 2+ or 3+

Formula	Formal Name	Common Name
$FeCl_2$	Iron(II) chloride[a]	Ferrous chloride
FeO	Iron(II) oxide	Ferrous oxide
Fe_2O_3	Iron(III) oxide	Ferric oxide
$FeCl_3$	Iron(III) chloride	Ferric chloride

[a] Pronounced "iron two chloride."

Compounds of Copper in Oxidation States of 1+ and 2+

Formula	Formal Name	Common Name
Cu_2O	Copper(I) oxide	Cuprous oxide
CuBr	Copper(I) bromide	Cuprous bromide
$CuCl_2$	Copper(II) chloride	Cupric chloride
CuS	Copper(II) sulfide	Cupric sulfide

B. Both elements are nonmetals, forming molecular compounds. Greek prefixes such as mono-, di-, etc., are used, sometimes for *both* elements.
   (a) Oxides of Nonmetals
       (1) Oxides of carbon
           CO   Carbon monoxide
           CO_2 Carbon dioxide
       (2) Oxides of sulfur
           SO_2 Sulfur dioxide
           SO_3 Sulfur trioxide
       (3) Oxides of nitrogen (older names in parentheses)
           N_2O Dinitrogen monoxide (nitrous oxide)
           NO   Nitrogen oxide (nitric oxide)
           N_2O_3 Dinitrogen trioxide
           NO_2 Nitrogen dioxide
           N_2O_4 Dinitrogen tetroxide
           N_2O_5 Dinitrogen pentoxide
       (4) Oxides of some halogens
           F_2O Difluorine monoxide
           Cl_2O Dichlorine monoxide
           Cl_2O_7 Dichlorine heptoxide
   (b) Some halides of carbon
       CCl_4 Carbon tetrachloride
       CBr_4 Carbon tetrabromide
   (c) Some exceptions
       H_2O Water
       NH_3 Ammonia
       CH_4 Methane

II. Compounds of Three or More Elements

A. A positive and a negative ion are combined.

1. The name of the positive ion is first followed by the name of the negative ion, just as with binary compounds between metals and nonmetals. Greek prefixes are not needed except where they occur in the name of an ion. (Older names are shown in parentheses.)

Formula	Name	Formula	Name
Li_2SO_4	Lithium sulfate	NaH_2PO_4	Sodium dihydrogen phosphate
Na_2SO_4	Sodium sulfate		
K_2SO_4	Potassium sulfate	K_2HPO_4	Potassium monohydrogen phosphate
$LiHCO_3$	Lithium hydrogen carbonate (lithium bicarbonate)[a]	$MgHPO_4$	Magnesium monohydrogen phosphate
$NaHCO_3$	Sodium hydrogen carbonate (sodium bicarbonate)	$(NH_4)_2HPO_4$	Ammonium monohydrogen phosphate
		Na_3PO_4	Sodium phosphate
Li_2CO_3	Lithium carbonate	$Ca_3(PO_4)_2$	Calcium phosphate
Na_2CO_3	Sodium carbonate	$MgSO_4$	Magnesium sulfate
$CaCO_3$	Calcium carbonate	$CaSO_4$	Calcium sulfate
$Al_2(CO_3)_2$	Aluminum carbonate	$Al_2(SO_4)_3$	Aluminum sulfate
$NaHSO_4$	Sodium hydrogen sulfate (sodium bisulfate)	$KMnO_4$	Potassium permanganate
		Na_2CrO_4	Sodium chromate
		$Mg(NO_3)_2$	Magnesium nitrate
		$NaNO_2$	Sodium nitrite

[a] *Bicarbonate* instead of *hydrogen carbonate* is used in this text because it is judged to be the more commonly used name for this ion, particularly among health scientists.

B. Molecular compounds of two or more elements. Most are organic compounds, so their rules of nomenclature are given in the chapters on organic compounds.

III. Important Inorganic Acids and Their Anions

Formula	Name	Formula	Name
H_2CO_3	Carbonic acid	HCO_3^-	Hydrogen carbonate ion (bicarbonate ion)
		CO_3^{2-}	Carbonate ion
HNO_3	Nitric acid	NO_3^-	Nitrate ion
HNO_2	Nitrous acid	NO_2^-	Nitrite ion
H_2SO_4	Sulfuric acid	HSO_4^-	Hydrogen sulfate ion (bisulfate ion)
		SO_4^{2-}	Sulfate ion
H_2SO_3	Sulfurous acid	HSO_3^-	Hydrogen sulfite ion (bisulfite ion)
		SO_3^{2-}	Sulfite ion
H_3PO_4	Phosphoric acid (orthophosphoric acid)	$H_2PO_4^-$	Dihydrogen phosphate ion
		HPO_4^{2-}	Monohydrogen phosphate ion
		PO_4^{3-}	Phosphate ion
$HClO_4$	Perchloric acid	ClO_4^-	Perchlorate ion
$HClO_3$	Chloric acid	ClO_3^-	Chlorate ion
$HClO_2$	Chlorous acid	ClO_2^-	Chlorite ion
$HClO$	Hypochlorous acid	ClO^-	Hypochlorite ion
HCl	Hydrochloric acid[a]	Cl^-	Chloride ion

[a] The name of the aqueous solution of gaseous HCl.

SOME GENERALIZATIONS ABOUT NAMES OF ACIDS AND THEIR ANIONS

1. The names of ions from acids whose names end in *-ic* all end in *-ate*.

2. When a nonmetal that forms an oxoacid whose name ends in *-ic* also forms an acid with one fewer oxygen atom, the name of the latter acid ends in *-ous*. (Compare nitric acid, HNO_3, and nitrous acid, HNO_2.)

3. When a nonmetal forms an oxoacid with one fewer oxygen atoms than are in an *-ous* acid, then the prefix *hypo-* is used. (Compare chlorous acid, $HClO_2$, and hypochlorous acid, $HClO$.)

4. The binary hydrohalogen acids are called hydrogen halides when they occur as pure gases but are called hydrohalic acids when they occur as aqueous solutions. Thus, hydrogen fluoride in water becomes hydrofluoric acid; hydrogen chloride in water becomes hydrochloric acid; and so on.

THREE ORGANIC REACTION MECHANISMS

This Appendix gathers together in one place three mechanisms that illuminate some organic reactions that otherwise would appear to occur as if by magic. These mechanisms could easily have been included within their appropriate chapters, but doing so would have interrupted the flow of topics that many instructors want.

D.1 HOW ACIDS CATALYZE THE DEHYDRATION OF ALCOHOLS

When a strong acid is added to an alcohol, the first chemical event is the ionization of the acid. The reaction is exactly analogous to the ionization of a strong acid when it is added to water; a proton transfers from the acid to the oxygen atom of a solvent molecule. Thus, in ethyl alcohol, sulfuric acid ionizes as follows to give the protonated form of ethyl alcohol.

$$HO_3SO—H + :\ddot{O} \overset{CH_2CH_3}{\underset{H}{<}} \rightleftharpoons HO_3SO^- + H—\overset{+}{\ddot{O}} \overset{CH_2CH_3}{\underset{H}{<}}$$

protonated form
of ethyl alcohol

All the covalent bonds to the oxygen atom in the protonated form of the alcohol are weak, including *the bond to carbon.* As ions and molecules bump into one another, some ions of the protonated form break up.

$$H—\overset{+}{\overset{CH_2CH_3}{\ddot{O}}} \underset{H}{<} \rightleftharpoons CH_3CH_2^+ + H—\ddot{O}: \underset{H}{}$$

protonated form   ethyl
of ethyl alcohol   carbocation

The octet for carbon in the unstable ethyl carbocation is restored when a proton from the carbon adjacent to the site of the positive charge transfers to a proton acceptor. As it transfers, the electron pair of its covalent bond pivots to form the second bond of the emerging double bond. It is a smooth, synchronous operation. One acceptor for the proton is the hydrogen sulfate ion, and its acceptance of a proton restores the catalyst, as follows.

$$HO_3SO^- + H—CH_2—\overset{+}{CH_2} \longrightarrow CH_2{=}CH_2 + HO_3SO—H$$

ethyl carbocation

Once water starts to appear as another product, its molecules can also accept the proton from the ethyl carbocation (to give H_3O^+), but this is also equivalent to the recovery of the catalyst.

D.2 THE ALDOL CONDENSATION AND REVERSE ALDOL CONDENSATION

Reactions that make new carbon–carbon bonds or break them have special places in both organic chemistry and biological chemistry. The *aldol condensation* is one such reaction, "aldol" signifying that the product has both an *ald*ehyde group and an alco*hol* group. The aldol condensation joins aldehydes or ketones together and it can be reversed. In gluconeogenesis (Figure 26.7), glyceraldehyde-3-phosphate and dihydroxyacetone phosphate are joined by an aldol condensation to give fructose-1,6-bisphosphate. In glycolysis, a reverse aldol breaks fructose-1,6-bisphosphate apart to give glyceraldehyde-3-phosphate and dihydroxyacetone phosphate.

Let's see how the aldol condensation occurs in the molecule-building or forward direction. Each step is reversible, however, so we'll write each step as an equilibrium. We'll also show the reaction using a simple aldehyde, but the same principles apply to the condensation of ketones or to a mixed reaction involving an aldehyde and a ketone.

The reaction whose mechanism we will examine is

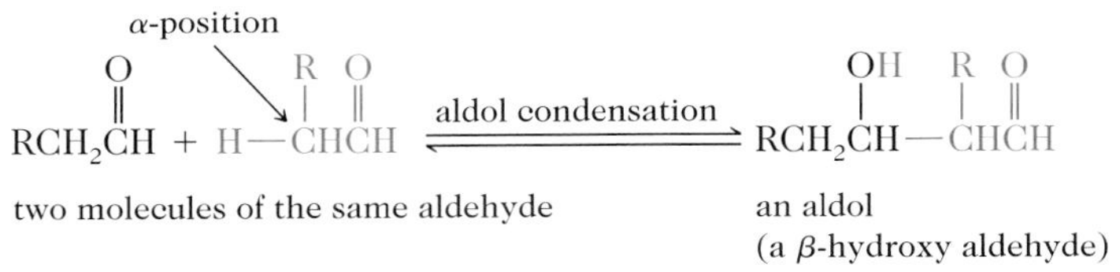

Notice that the reaction is simply the addition of one molecule of the aldehyde (the second) to another molecule (the first). *It always involves the H atom of an alpha position of the adding molecule, never a hydrogen attached at any other position on the chain.* By the same token, *the reverse aldol condensation requires that the* OH *group be beta to the carbonyl group.*

Step 1.   Proton transfer occurs from the aldehyde to a proton acceptor, $B:^-$. (*In vivo*, side chains of amino acid residues serve as carriers and transfer agents in proton shuttles.)

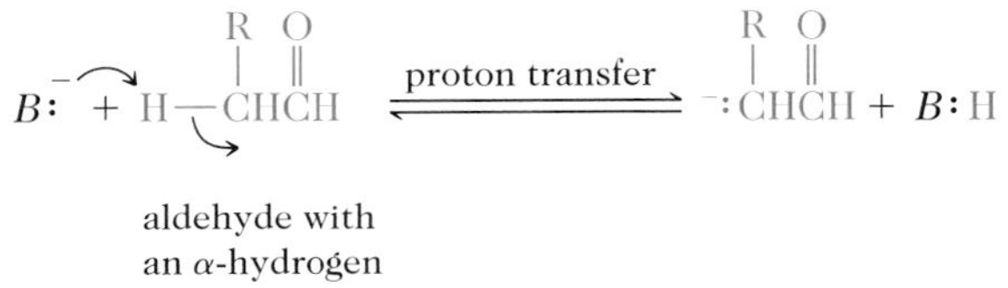

We see here why only a hydrogen attached to the *alpha* position of the aldehyde can be involved. It's the only H atom on the chain with any acidity whatsoever, and it is only *very* weakly acidic. That it has any acidity at all is caused by the nearby carbonyl group, an electronegative group that attracts electron density in the anion of the aldehyde and so stabilizes it.

Step 2.   The anion of the aldehyde is attracted to the δ+ charge on the carbonyl carbon of the other aldehyde molecule. The new carbon–carbon bond forms in this step. The product is the anion of the aldol.

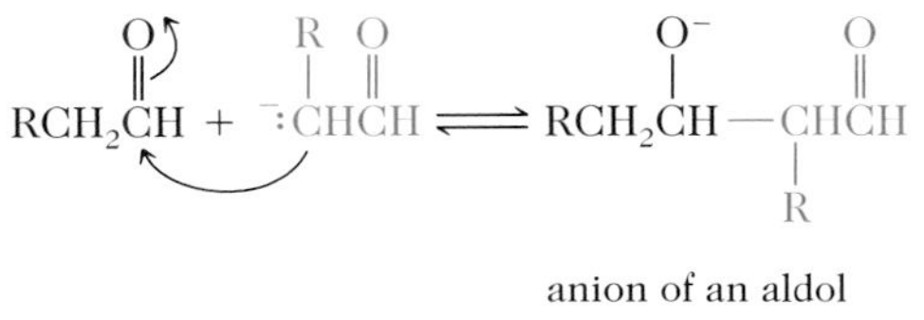

Step 3. The anion of the aldol is the conjugate base of an alcohol and therefore is a very strong base. It recovers a proton in the last step.

$$B{:}H + RCH_2\overset{\overset{\displaystyle O^-}{|}}{CH}{-}\overset{\overset{\displaystyle R}{|}}{CH}\overset{\overset{\displaystyle O}{\parallel}}{CH} \xrightarrow{\text{proton transfer}} RCH_2\overset{\overset{\displaystyle OH}{|}}{CH}{-}\overset{\overset{\displaystyle R}{|}}{CH}\overset{\overset{\displaystyle O}{\parallel}}{CH} + B{:}^-$$

anion of an aldol   aldol product

Ketones are able to engage in the same kind of reaction. All that would change in the mechanism given here is that instead of H on the carbonyl carbon there would be another R group.

In a *reverse aldol condensation,* the above steps are run in reverse. The aldol product first transfers H from its OH group to an acceptor to give the anion of the aldol. Next, the carbon–carbon bond breaks as the two smaller fragments form. *Precisely this kind of reaction occurs in step 4 of glycolysis,* because in fructose-1,6-bisphosphate there is an HO group *beta* to the keto group (see position 4 of the original ring). Thus a reverse aldol condensation can break the molecule into two half-size fragments.

D.3 CONDENSING ESTERS, A MAJOR C—C BOND-MAKING REACTION

The reaction by which an acetyl group enters the citric acid cycle is just one example of many reactions of two carbonyl compounds in which an acyl group of one becomes joined to an alpha position of another. The product is a much larger molecule. We'll first explain how this happens under laboratory conditions with a very simple system: two molecules of the same ester, ethyl acetate. The reaction is named the *Claisen ester condensation.*

The Alpha Hydrogen of a Carbonyl Compound Is "Mobile" The acid ionization constant of an alkane is estimated to be about 10^{-40}; clearly, no one would call an alkane an acid! The ionization constant of the H atom attached to the α-position of the acetyl unit (CH_3CO-) in ethyl acetate is about 10^{-18}, making it a billion trillion times as strong an acid (but still nothing we'd call an acid in water).

The reason for this greater acidity is the presence of the two nearby electronegative oxygen atoms. This greater acidity (mobility) of a hydrogen on a carbon alpha to oxygens is all we need to understand how the Claisen condensation and similar reactions in the body can occur.

With ethyl acetate, the reaction is as follows, where $B{:}^-$ is a powerful base, so powerful that the reaction cannot be run in water. (Usually it is run in ethyl alcohol.) In living systems, reactions like this use enzymes to handle necessary proton exchanges.

$$CH_3\overset{\overset{\displaystyle O}{\parallel}}{C}{-}OCH_2CH_3 + CH_3\overset{\overset{\displaystyle O}{\parallel}}{C}OCH_2CH_3 \xrightarrow{B{:}^-} CH_3\overset{\overset{\displaystyle O}{\parallel}}{C}{-}CH_2\overset{\overset{\displaystyle O}{\parallel}}{C}{-}OCH_2CH_3 + HOCH_2CH_3$$

Notice that an acetyl group (in red) has been joined to the alpha carbon of the second ester molecule. A new C—C bond has been made. Let's see how a strong base handles this reaction.

Step 1. The base takes H^+ from the alpha H—C position of one ethyl acetate.

$$B{:}^- + H{-}CH_2\overset{\overset{\displaystyle O}{\parallel}}{C}OCH_2CH_3 \longrightarrow {}^-{:}CH_2\overset{\overset{\displaystyle O}{\parallel}}{C}OCH_2CH_3 + B{-}H$$

ethyl acetate   anion of ethyl acetate

Step 2. The new anion attacks the carbonyl carbon of another ester molecule. (This carbon has a partial positive charge on it.)

Step 3. The product of this reaction expels $CH_3CH_2O^-$, which takes H^+ from $B\!:\!H$ (and the base is thus regenerated).

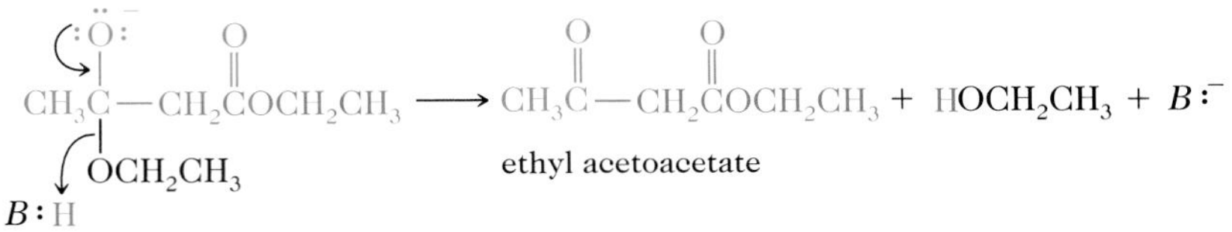

ethyl acetoacetate

This overall reaction is formally very similar to the initial reactions the body uses for several biosyntheses. These include making long hydrocarbon chains from acetate units, and making cholesterol and the sex hormones.

Citrate Synthase Manages a Claisen-like Condensation of Acetyl CoA with Oxaloacetate   If we take acetyl CoA and oxaloacetate through the same steps, we have the following reaction. The enzyme citrate synthase provides the *services* of a base. But we'll still represent the base by $B\!:\!^-$. Realize that what follows is a considerable simplification of the steps.

Step 1. Acetyl CoA gives up a proton.

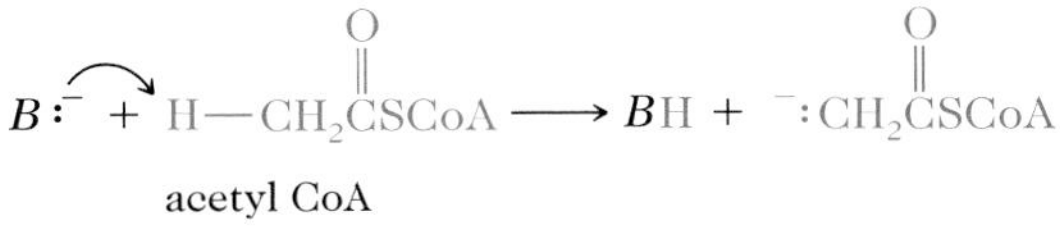

acetyl CoA

Step 2. The new anion attacks the keto group in oxaloacetate, forming a new C—C bond.

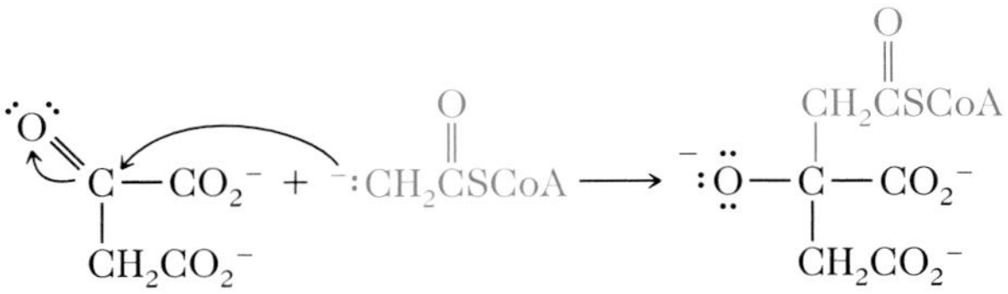

Step 3. A proton is donated so that the 3° alcohol group can form, and the CoAS group is removed by hydrolysis. The citrate ion forms.

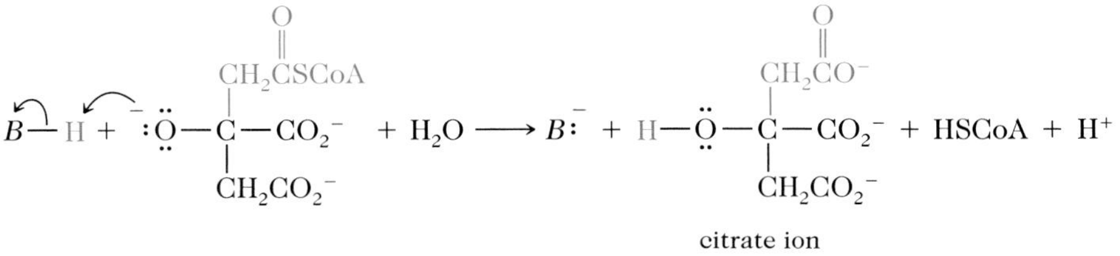

citrate ion

Thus a Claisen-like condensation launches the acetyl group of acetyl CoA into the citric acid cycle.

CHAPTER 1

Practice Exercises, Chapter 1

1. 310 K

2. (a) 5.45×10^8 (b) 5.67×10^{12} (c) 6.454×10^3
   (d) 2.5×10^1 (e) 3.98×10^{-5} (f) 4.26×10^{-3}
   (g) 1.68×10^{-1} (h) 9.87×10^{-12}

3. (a) 10^{-6} (b) 10^{-9} (c) 10^{-6} (d) 10^3

4. (a) mL   (b) μL   (c) dL   (d) mm   (e) cm   (f) kg
   (g) μg   (h) mg

5. (a) Kilogram   (b) Centimeter   (c) Deciliter
   (d) Microgram   (e) Milliliter   (f) Milligram
   (g) Millimeter   (h) Microliter

6. (a) 1.5 Mg   (b) 3.45 μL   (c) 3.6 mg   (d) 6.2 mL
   (e) 1.68 kg   (f) 5.4 dm

7. (a) 275 kg   (b) 62.5 μL   (c) 82 nm or 0.082 μm

8. (a) 95   (b) 11.36   (c) 0.0263   (d) 1.3000
   (e) 16.1   (f) 3.8×10^2 (g) 9.31   (h) 9.1×10^2

9. (a) $\dfrac{1\ \text{g}}{1000\ \text{mg}}$ or $\dfrac{1000\ \text{mg}}{1\ \text{g}}$

   (b) $\dfrac{1\ \text{kg}}{2.205\ \text{lb}}$ or $\dfrac{2.205\ \text{lb}}{1\ \text{kg}}$

10. 0.324 g of aspirin

11. (a) 324 mg of aspirin   (b) 3.28×10^4 ft
    (c) 18.5 mL   (d) 17.72 g
    (e) 4.78×10^3 μL

12. 40 °C

13. 59 °F (Quite cool.)

14. 20.7 mL

15. 32.1 g

Review Exercises, Chapter 1

1.8   The better question is (b). Choice (a) begins with a bias, namely that the hypothesis is true, that it cannot be false.

1.14   The observation of a *chemical* property necessarily converts the substance into a different substance. The observation or measurement of a physical property does not do this.

1.16   Measurement

1.20   When we use a two-pan balance, we have both pans at almost identical locations so the gravitational attractions are the same for the objects on both pans when balanced. Therefore, their masses are the same, too.

1.24   It can be defined (or *derived*) from a base quantity, length.

$$\text{Volume} = (\text{length})^3$$

1.37   The kelvin, by a factor of 9/5 (180/100)

1.41   41 °F (5 °C); too cold without a coat or heavy sweater for most people.

1.43   104 °F (The patient has a fever.)

1.44   (a) They are accurate; the average is 59.84. They are close to the true value.
      (b) Evidently the numbers could be read to within one unit of the second decimal place, so the uncertainty is small ($\pm$0.01 kg).

1.46   (a) 6   (b) 5   (c) 4 (by our rule)   (d) 3

1.47   (a) 2×10^5 mi   (b) 1.6×10^5 mi
      (c) 1.61×10^5 mi   (d) 1.605×10^5 mi
      (e) 1.6054×10^5 mi

1.49   (a) 6.324×10^3 g   (b) 6.78×10^{-9} L
      (c) 8.746000×10^6 m   (d) 1.00493×10^{-3} g

1.51   (a) 6.324 kg   (b) 6.78 nL   (c) 8.746000 Mm
      (d) 1.00493 mg

1.53 1.1×10^3 autos

1.55   (a) 2×10^7 (b) 1.7×10^7 (c) 1.66×10^7
      (d) 1.656×10^7 (e) 1.6560×10^7
      (f) 1.65600×10^7

1.58   (a) 1.5×10^{-4} (b) 3.5×10^4 (c) 2.50×10^{24}
      (d) 6.65×10^1 (e) 2.8   (f) 4.5025×10^1
      (g) 3.0×10^1 (h) 1×10^{-1} (i) 2.00

1.60   (a) 20 in.   (b) 0.23 kg   (c) 0.25 oz   (d) 4.5 g

1.62   (a) 50.6 kg   (b) 75.6 in.

1.64   10.0 liquid ounces

1.66   2 tablets

1.68   29028 ft

1.70   0.500 g

1.72   (a) Density = 0.787 g/cm^3 (on a mass of 1.70×10^3 g and a volume of 2.16×10^3 cm^3).
      (b) 24.8 kg (if lead) or 54.7 lb. (Answers can vary slightly depending on the timing of rounding.)

1.74   (a) 0.910 g/mL   (b) 2.75×10^2 mL

1.76   Its density approaches that of water itself, so the urine contains little if any dissolved substances.

1.78   1.03

1.82   Living systems from a huge variety of species, both plant and animal, can either nourish or receive nourishment from each other.

1.86   (a) A, B, D   (b) E, F, G   (c) C, H, I

1.87   (a) 3.00000×10^1 (b) 3.00×10^1 (c) 4.65×10^1 (d) 6.75×10^1 (e) 4.35×10^1 (f) 1×10^{-5}

1.89   No. The vehicle has a mass of 2.0×10^3 kg, too much for the bridge.

1.91   7.00 mL

1.93   675 g, 1.49 lb

1.97   (a) 6×10^3 g   (b) 6.1×10^3 g   (c) 6.06×10^3 g   (d) 6.060×10^3 g

1.98   (a) 1.1×10^4 mg/6.0 pt; 3.9×10^3 mg/L   (b) 2.4×10^3 mg/6.0 pt; 8.5×10^2 mg/L

CHAPTER 2

Practice Exercises, Chapter 2

1. Na_2S

2.   Potassium, carbon, and oxygen in a ratio of $2:1:3$

3.   (a) $\dfrac{1 \text{ cal}}{4.184 \text{ J}}$ or $\dfrac{4.184 \text{ J}}{1 \text{ cal}}$ (b) 333 J or 0.333 kJ

4. $\Delta t = 4.53$ °C, so the final temperature is 24.5 °C.

Review Exercises, Chapter 2

2.8   In *chemical* changes, substances change into other substances. This inevitably is accompanied by changes in physical appearances and properties, because different substances differ from each other in at least one physical way.

2.10   To the water only. In chemistry we restrict the word *substance* to elements and compounds, not to mixtures like sugar in water.

2.11   Compounds obey the law of definite proportions; mixtures do not.

2.15   Substance

2.27   Yes, the ratio, $\dfrac{7.94454}{3.97265} = 1.99981$, is extremely close to the whole number 2.

2.28   No is the symbol of an element, nobelium, because the second letter is lower case. NO is the symbol of a compound, nitric oxide, made of the elements nitrogen (N) and oxygen (O).

2.35   One atom of iron

2.38   (a) A discrete "package" of atoms made from two H atoms and one O atom exists and is capable of moving around independently.

(b) Yes. The formula unit of water has the composition H_2O.

(c) Yes. It shows the kinds of atoms and their ratio in terms of the smallest whole numbers.

(d) A package made from only one Na atom and one Cl atom that is capable of moving around independently does not exist.

2.44   The kinetic energy is multiplied by four. Notice that the velocity is doubled, and KE is directly proportional to the *square* of the velocity, so if the velocity is multiplied by 2, the KE is multiplied by 2^2, or 4.

2.46   (a) $kg \times \dfrac{m^2}{s^2}$ (b) 6.25×10^5 J   (c) 1.49×10^5 cal or 1.49×10^2 kcal   (d) 35.4 m/s (about 80 mi/hr) (Notice that the velocity does not have to double to double the associated kinetic energy.)

2.48   As reactive chemicals within the battery, so it is in the form of chemical energy (a potential energy)

2.51   (a) From A to B.   (b) B is either melting or boiling.

2.57   The substance's temperature.

2.68   According to the law of multiple proportions, the answer has to be 15.8731, because only this number, when divided by 7.93655, gives a whole number, 2.

2.70   (a) 67.1 mi/hr   (b) 22.4 times the speed of walking   (c) 31.5 kJ in the vehicle versus 6.28×10^{-2} kJ for walking

2.72 3.3×10^2 kcal (rounded from 332 kcal)

2.74   (a) 3.9×10^2 kcal (rounded from 385 kcal)   (b) 1.3 hr; 4.6 mi

CHAPTER 3

Practice Exercises, Chapter 3

1.   (a) 15   (b) 24   (c) 24

2.   (a) 5   (b) 18   (c) 20

3.   52.5

4.   (a) Sn   (b) Cl   (c) Rb   (d) Mg   (e) Ar

5.   (a) $1s^2 2s^2 2p^6 3s^2 3p_x^1$ (b) $1s^2 2s^2 2p^6 3s^2 3p_x^2 3p_y^2 3p_z^1$ (c) $1s^2 2s^2 2p^6 3s^2 3p_x^1 3p_y^1$ (d) $1s^2 2s^2 2p^6 3s^2 3p^6 4s^2$

6.   (a) 1   (b) 6   (c) 5   (d) 7

Review Exercises, Chapter 3

3.4   1.007 g; almost identical, numerically, to the atomic mass of H.

3.7   Its *electron configuration*.

3.10   Pair 1, because both have atomic number 10 so they both are isotopes of the same element.

3.12   These two isotopes have identical chemical properties, so no specification that distinguishes each isotope has to be used.

3.16 (a) NaH   (b) SeH_2 (usually written as H_2Se)
(c) GaH_3 (d) GeH_4 (e) AsH_3 (f) CaH_2
(g) BrH (usually written as HBr)

3.19 (a) VIIA; halogens   (b) Acid
(c) $HX + KOH \rightarrow KX + H_2O$
(d) Acid–base neutralization

3.26 Principal energy level or shell

3.29

Principal Energy Level Number	Number of Sublevels	Number of Orbitals
1	1	1
2	2	4
3	3	9
4	4	16

3.31 $3s^1$

3.38 (a) 7, nitrogen   (b) 14, silicon

3.39 (a) 15, phosphorus   (b) 20, calcium

3.40 (a) $1s^22s^22p^63s^23p_x^23p_y^13p_z^1$
(b) $[Ne]\,3s^23p_x^23p_y^13p_z^1$

3.42 (a) 30; its number of electrons, 30, equals its number of protons, which is its atomic number.
(b) Yes, it has 18 electrons.
(c) No, all occupied orbitals are *filled* orbitals (each with 2 electrons), and the Pauli exclusion principle tells us that orbitals can hold 2 electrons only if their spins are opposite or paired.
(d) $[Ar]\,4s^23d^{10}$. (Argon has no $3d$ electrons so the $3d^{10}$ electrons must be shown as part of the condensed electron configuration.)
(e) $[Ar]\,4s^23d^{10}4p_x^24p_y^14p_z^1$ Atomic number = 34. Atomic symbol = Se
(f) Three of the four $4p$ electrons spread out among the three $4p$ orbitals.
(g) The fourth $4p$ electron pairs up with another electron in the $4p_x$ orbital, and we assume that the spins of these two are opposite.

3.44 (a) $1s^22s^1$ or $[He]\,2s^1$ (b) $1s^22s^22p_x^22p_y^12p_z^1$ or $[He]\,2s^22p_x^22p_y^12p_z^1$ (or $[He]\,2s^22p^4$)
(c) $1s^22s^22p^63s^2$ or $[Ne]\,3s^2$
(d) $1s^22s^22p^63s^23p_x^23p_y^13p_z^1$ or $[Ne]\,3s^23p_x^23p_y^13p_z^1$ (or $[Ne]\,3s^23p^4$)

3.46 (a) 1+   (b) 1−   (c) Attract; they have opposite charges.   (d) X, 23; Y, 35

3.48 (a) Density = 3×10^{15} g/cm^3 (b) 3×10^9 metric ton/cm^3 (or 3 billion metric tons per cubic centimeter)

3.53 (a) The same element; they have identical numbers of electrons and so identical atomic numbers.
(b) Configuration **1**; its electrons are in the lowest available energy levels.
(c) Configuration **2**. Configuration **2** has an electron in the $4s$ orbital while having empty $3p$ orbitals.
(d) The atom has to be given energy to change it

from **1** to **2**. (The change from **2** to **1** would release the same amount of energy.)

3.55 (a) IVA. All inner levels, 1–3, are full and the outer level (5) has 4 electrons, like all elements in group IVA.
(b) $[Kr]\,5s^24d^{10}5p^2$
(c) $[Kr]\,5s^24d^{10}5p^3$
(d) $[Kr]\,5s^24d^{10}5p^1$
(e) $[Ar]\,4s^23d^{10}4p^2$ (Notice that only the numbers of the principal energy levels change in shifting from element 50 to the element immediately above it in the periodic table.)

CHAPTER 4

Practice Exercises, Chapter 4

1. (a) AgBr   (b) Na_2O (c) Fe_2O_3 (d) $CuCl_2$

2. (a) Copper(II) sulfide   (b) Sodium fluoride
(c) Iron(II) iodide   (d) Zinc bromide
(e) Copper(I) oxide

3. (a) 3+   (b) 2+   (c) 2+

4. (a) $1s^22s^22p^63s^23p^64s^1$. 1+ charge on the ion
(b) $1s^22s^22p^63s^23p_x^23p_y^13p_z^1$. 2− charge on the ion
(c) $1s^22s^22p^63s^23p_x^13p_y^1$. No ion is predicted (or exists).

5. (a) $1s^22s^22p^63s^23p^6$ (b) $1s^22s^22p^63s^23p_x^23p_y^23p_z^2$
(c) No ion exists

6. (a) Cs^+ (b) F^- (c) No ion is predicted.   (d) Sr^{2+}

7. (a) Mg is oxidized; S is reduced. Mg is the reducing agent; S is the oxidizing agent.
(b) Zn is oxidized; Cu^{2+} is reduced. Zn is the reducing agent; Cu^{2+} is the oxidizing agent.

8. $Na\cdot$ $\cdot Mg\cdot$ $\cdot \overset{\cdot}{Al}\cdot$ $\cdot \overset{\cdot}{\underset{\cdot}{Si}}\cdot$ $\cdot \overset{\cdot\cdot}{\underset{\cdot}{P}}\cdot$ $\cdot \overset{\cdot\cdot}{\underset{\cdot}{S}}\cdot$ $:\overset{\cdot\cdot}{\underset{\cdot\cdot}{Cl}}\cdot$ $:\overset{\cdot\cdot}{\underset{\cdot\cdot}{Ar}}:$

9. $\cdot \overset{\cdot}{\underset{\cdot}{Sb}}\cdot$

10. $\cdot Ca\cdot\; +\; \cdot\overset{\cdot\cdot}{\underset{\cdot\cdot}{O}}\cdot \;\rightarrow\; Ca^{2+} + \left[:\overset{\cdot\cdot}{\underset{\cdot\cdot}{O}}:\right]^{2-}$

11. $H-\underset{\underset{\displaystyle H}{|}}{\overset{\overset{\displaystyle H}{|}}{Si}}-H$

12. (a) $KHCO_3$ (b) Na_2HPO_4 (c) $(NH_4)_3PO_4$

13. (a) Sodium cyanide   (b) Potassium nitrate
(c) Sodium hydrogen sulfite   (d) Ammonium carbonate   (e) Sodium acetate

14. $H\;\; O\;\; \underset{\displaystyle O}{\overset{\displaystyle O}{S}}\;\; O\;\; H \qquad H\;\; O\;\; \underset{\displaystyle O}{\overset{\displaystyle O}{P}}\;\; O\;\; H$
$\qquad\qquad\qquad\qquad\qquad\qquad\qquad\qquad H$

15. 32

16.

$$O$$

H O Cl O   26 valence electrons

$$H—\ddot{\underset{..}{O}}—\ddot{\underset{..}{Cl}}—\ddot{\underset{..}{O}}:$$

with $:\ddot{\underset{..}{O}}:$ above Cl

17.

$$:\ddot{\underset{..}{O}}:\overset{\cdot\cdot}{\underset{..}{S}}:\ddot{\underset{..}{O}}:$$

18. (a) $\left[\begin{array}{c} H \\ H:\ddot{O}:H \end{array}\right]^{+}$ (b) $\left[H:\ddot{\underset{..}{O}}: \right]^{-}$

19. (a) $\left[:\ddot{\underset{..}{O}}:\overset{..}{\underset{..}{C}}:\ddot{\underset{..}{O}}: \right]^{2-}$ (b) $\left[:\ddot{\underset{..}{O}}:\overset{..}{N}:\ddot{\underset{..}{O}}: \right]^{-}$

20.

The molecule is polar.

Review Exercises, Chapter 4

4.6

potassium atom

fluorine atom

potassium ion

fluoride ion

4.10 (a) NaI (b) $FeBr_2$
(c) Al_2O_3 (d) $CaCl_2$
(e) BaO (f) Cu_2S

4.11 (a) Potassium iodide (b) Barium chloride
(c) Calcium sulfide (d) Iron(II) chloride
(e) Aluminum oxide (f) Silver iodide

4.15 IIIA

4.18 Yes, 1−

4.20 Yes, 2−

4.22 (a) M^+ $1s^2$
(b) Q^+ $1s^2 2s^2 2p^6 3s^2 3p^6$
(c) Z^{2-} $1s^2 2s^2 2p^6 3s^2 3p^6$

4.24 (a) 4+ (b) 5+ (c) 3+
(d) 3+ (e) 4+ (f) 3+

4.26 (a) O_2 (b) Al (c) Al (d) O_2

4.28 Both atoms and molecules are small particles and both are electrically neutral. A molecule, however, has two or more atomic nuclei and an atom has just one.

4.39 (a) Rb· (b) ·Sr· (c) ·Ga· (d) ·Te·

4.44 (a) Sodium carbonate (b) Ammonium nitrate
(c) Magnesium hydroxide (d) Calcium acetate
(e) Potassium bicarbonate (f) Barium sulfate
(g) Sodium nitrite (h) Ammonium phosphate

4.45 (a) 14 (b) 22 (c) 15

4.55

(a) $:\ddot{\underset{..}{Cl}}—Pb—\ddot{\underset{..}{Cl}}:$ with $:\ddot{\underset{..}{Cl}}:$ above and $:\ddot{Cl}:$ below

(b) $:\ddot{\underset{..}{F}}—\ddot{\underset{..}{O}}—\ddot{\underset{..}{F}}:$

(c) $:\ddot{\underset{..}{Cl}}—N—\ddot{\underset{..}{Cl}}:$ with $:\ddot{Cl}:$ above

(d) $H—N—H$ with H above

(e) $\left[H—\ddot{\underset{..}{O}}—S—\ddot{\underset{..}{O}}: \right]^{-}$ with $:\ddot{O}:$ above and $:\ddot{O}:$ below

(f) $\left[H—\ddot{\underset{..}{O}}—S—\ddot{\underset{..}{O}}: \right]^{-}$ with $:\ddot{O}:$ above

4.58 (a) The four axes point to the corners of a regular tetrahedron and so make angles of 109.5° with respect to each other.
(b) There are three axes, they lie in the same plane, they point to the corners of a regular (equilateral) triangle, and they make angles of 120°.
(c) There are two axes, they are colinear, and so they make an angle of 180°.

4.62 (a) Yes (b) δ+ is near X and δ− is near Y.

4.79 (a) Zinc oxide (b) Lithium oxide
(c) Iron(III) bromide (d) Magnesium chloride
(e) Sodium fluoride (f) Copper(II) bromide

4.81 (a) X^{2+}, $1s^2 2s^2 2p^6 3s^2 3p^6$
(b) Y^-, $1s^2 2s^2 2p^6 3s^2 3p^6$
(c) Z has no ions.

4.83 VIA

4.85 No. The atom has an outer octet.

4.87 W_2O_5

4.89 The $3p_z$ atomic orbitals, each with one electron and one such atomic orbital from each of the two Cl atoms

4.93 In ethylene. There is twice as much electron density between the nuclei, so they are more strongly attracted toward this region and thus closer together.

4.96 (a) $(NH_4)_2CO_3$ (b) Na_2HPO_4
(c) $Al(OH)_3$ (d) $NaHSO_4$
(e) $LiHCO_3$ (f) $Ca(NO_3)_2$
(g) $NH_4H_2PO_4$ (h) $NaMnO_4$

4.97 (a) Potassium hydrogen sulfate
(b) Lithium monohydrogen phosphate
(c) Calcium cyanide   (d) Sodium dichromate
(e) Sodium sulfite   (f) Barium chromate
(g) Aluminum sulfate
(h) Potassium permanganate

4.99 49

4.103 (a) NO_2
(b)
$$:\overset{..}{O}::\overset{..}{O}:$$
$$:\overset{..}{O}:\overset{..}{N}:\overset{..}{N}:\overset{..}{O}:$$

4.105 The same as in methane, CH_4, 109.5°.

CHAPTER 5[1]

Practice Exercises, Chapter 5

1. 8.68×10^{22} atoms of gold per ounce
2. (a) 180   (b) 58.3   (c) 858.6
3. 408 g of NH_3
4. 0.0380 mol or 38.0 mmol of aspirin
5. $3O_2 \rightarrow 2O_3$
6. $4Al + 3O_2 \rightarrow 2Al_2O_3$
7. (a) $2Ca + O_2 \rightarrow 2CaO$
   (b) $2KOH + H_2SO_4 \rightarrow 2H_2O + K_2SO_4$
   (c) $Cu(NO_3)_2 + Na_2S \rightarrow CuS + 2NaNO_3$
   (d) $2AgNO_3 + CaCl_2 \rightarrow 2AgCl + Ca(NO_3)_2$
   (e) $2Al + 3H_2SO_4 \rightarrow Al_2(SO_4)_3 + 3H_2$
   (f) $CH_4 + 2O_2 \rightarrow 2H_2O + CO_2$
8. 0.500 mol of H_2O. 500 mmol of H_2O
9. 4.20 mol of N_2 and 4.20 mol of O_2
10. 450 mol of H_2 and 150 mol of N_2
11. 11.5 g of O_2
12. 18.4 g of Na is used up and 46.8 g NaCl forms.
13. (a) 2.45 g of H_2SO_4 (b) 9.01 g of $C_6H_{12}O_6$
14. 156 mL of 0.800 M Na_2CO_3 solution
15. 9.82 mL of 0.112 M H_2SO_4, when calculated step by step with rounding after each step.

[1]If your calculated answers differ slightly from the answers given here, it may be caused by a conceptually unimportant difference in handling the calculation. All answers to computational problems given here were obtained by *chain* calculations. The formula masses needed for these calculations were first computed in the usual way from atomic masses rounded to the first decimal place (H to 1.01), and then they were rounded to the number of significant figures allowed by the given data *before* they were used in the calculations.

9.84 mL of 0.112 M H_2SO_4, when found by a chain calculation.

16. Dilute 20.0 mL of 0.200 M $K_2Cr_2O_7$ to a final volume of 100 mL.
17. Dilute 14 mL of 18 M H_2SO_4 to a final volume of 250 mL.

Review Exercises, Chapter 5

5.9 6.02×10^{22} atoms of cobalt

5.11 (a) 36.5   (b) 56.1   (c) 106.0
(d) 98.1   (e) 84.0   (f) 261.3
(g) 132.1   (h) 158.2   (i) 180.1

5.15 (a) 4.56 g HCl   (b) 7.01 g KOH
(c) 13.3 g Na_2CO_3 (d) 12.3 g H_2SO_4
(e) 10.5 g $NaHCO_3$ (f) 32.7 g $Ba(NO_3)_2$
(g) 16.5 g $(NH_4)_2HPO_4$ (h) 19.8 g $Ca(C_2H_3O_2)_2$
(i) 22.5 g $C_6H_{12}O_6$

5.16 (a) 1.37 mol HCl   (b) 0.891 mol KOH
(c) 0.472 mol Na_2CO_3 (d) 0.510 mol H_2SO_4
(e) 0.595 mol $NaHCO_3$ (f) 0.191 mol $Ba(NO_3)_2$
(g) 0.379 mol $(NH_4)_2HPO_4$
(h) 0.316 mol $Ca(C_2H_3O_2)_2$
(i) 0.278 mol $C_6H_{12}O_6$

5.18 2.09×10^{21} molecules H_2O

5.21 Two molecules of nitrogen monoxide react with one molecule of oxygen to give two molecules of nitrogen dioxide.

5.23 (a) $2Mg + O_2 \rightarrow 2MgO$
(b) $CaO + 2HCl \rightarrow CaCl_2 + H_2O$
(c) $MgCl_2 + 2AgNO_3 \rightarrow Ca(NO_3)_2 + 2AgCl$
(d) $2HBr + Ca(OH)_2 \rightarrow CaBr_2 + 2H_2O$
(e) $Na_2CO_3 + 2HCl \rightarrow 2NaCl + CO_2 + H_2O$

5.25 $$\dfrac{1 \text{ mol } Ca(OH)_2}{2 \text{ mol } HCl} \qquad \dfrac{2 \text{ mol } HCl}{1 \text{ mol } Ca(OH)_2}$$

5.27 (a) $C_6H_{12}O_6 + 6O_2 \rightarrow 6CO_2 + 6H_2O$
(b) $$\dfrac{1 \text{ mol } C_6H_{12}O_6}{6 \text{ mol } O_2} \qquad \dfrac{6 \text{ mol } O_2}{1 \text{ mol } C_6H_{12}O_6}$$

5.29 (a) 0.111 mol O_2 (b) 0.0740 mol Fe_2O_3

5.31 (a) $2Al_2O_3 \rightarrow 4Al + 3O_2$
(b) 52.9 g Al   (c) 47.1 g O_2
(d) 100.0 g (Al + O_2), which is identical to the initial mass of Al_2O_3 used. Law of conservation of mass in chemical reactions.

5.33 48.6 g Na_2CO_3

5.36 (a) Saturated   (b) Concentrated

5.42 (a) 0.0625 mol or 3.66 g NaCl
(b) 0.0250 mol or 4.50 g $C_6H_{12}O_6$
(c) 0.0250 mol or 2.45 g H_2SO_4
(d) 0.0625 mol or 6.62 g Na_2CO_3

5.44 600 mL

5.46 0.100 mol Na_2CO_3

5.48 19.8 mL

5.51 36.7 mL of 0.350 M Na_2SO_4 solution and 32.1 mL of 0.400 M $Ba(NO_3)_2$ solution

5.53 (a) 32.6 mL of 0.450 M Na_2CO_3
(b) 2.13 g of Na_2CO_3

5.55 8.33 mL of 15 M H_3PO_4. You would place some water into a 250-mL volumetric flask, slowly add the 15 M H_3PO_4, and then make up the final volume to the 250-mL mark.

5.59 4.79 g Na, 0.210 g H, 10.0 g O

5.67 6×10^{15} molecules O_3

5.69 (a) 1.5 mol N_2/mol TNT   (b) 3.5 mol CO/mol TNT
(c) 2.5 mol H_2O vapor/mol TNT
(d) 7.5 mol gases/mol TNT

(e) $1.7 \times 10^3 \dfrac{\text{L gases}}{\text{L TNT}}$ (Note that this is nearly a 2000-fold expansion.)

5.71 (a) 15.9 g $Mn(NO_3)_2$ (b) 62.2 g $NaBiO_3$
(c) 39.2 g HNO_3 (d) 87.7 g $Bi(NO_3)_3$

5.74 223 mL

5.76 1.67 L of 0.120 M Na_2CO_3

CHAPTER 6

Practice Exercises, Chapter 6

1. 2.6 L
2. 547 mL
3. 484 mm Hg
4. 214 mL
5. 98.1 g O_2
6. 52 mm Hg

Review Exercises, Chapter 6

6.9 The weight doubles; the pressure stays the same because pressure is the *ratio* of weight to area.

6.14 A 25-kg force acting on 5 cm². (It exerts a pressure of 5 kg/cm² whereas the other system exerts a pressure of 4 kg/cm².)

6.16 (a) 250 mm Hg   (b) 350 mm Hg

6.18 1.5×10^2 mm Hg or 1.5×10^2 torr. 20 kPa

6.21 1.06×10^3 mm Hg

6.23 5.73 L

6.25 715 mm Hg

6.27 4.16 atm

6.29 No. (P = 228 atm, well below 500 atm)

6.31 30.2 mmol

6.33 (a) 9.82×10^{-3} mol CO_2
(b) 9.82×10^{-3} mol $CaCO_3$
(c) 0.983 g $CaCO_3$
(d) 0.551 g CaO

6.37 0.0112 mol Zn; 0.732 g Zn

6.48 Octane; its molecules are larger than those of butane but both are compounds of carbon and hydrogen and so have roughly the same (very small) permanent polarity.

6.51 (a) B   (b) B   (c) A
(d) C   (e) A, B   (f) 100 °C
(g) Their rates are equal.   (h) Equilibrium

6.55 (a) Condensation of alcohol vapor to alcohol liquid occurs.
(b) Forward change
(c) Shift to the left
(d) Shift to the right; this supplies vapor to compensate for the loss of vapor and tends to restore equilibrium.
(e) Yes, the equilibrium will shift to the left because this releases heat and so tends to compensate for the heat loss caused by the cooling.

6.79 (a) 33.9 ft
(b) One can use a much shorter glass tube.

6.81 1.08 atm

6.83 6.02×10^{23} (1 mol)

6.85 195 atm

6.87 317 mL

6.89 Yes. The sum of the partial pressures is only 725 mm Hg, so something else must be present to account for the "missing" 20 mm Hg of partial pressure.

6.91 (a) Forward. $2NO_2 \rightleftharpoons N_2O_4$ + heat
(b) Forward. The forward reaction is the conversion of 2 mol of gas (NO_2) into 1 mol (N_2O_4). Increasing the pressure favors this volume-reducing effect.

6.93

6.95 0.759 mol

6.96 1.89 g $CaCO_3$; 4.73 mL HCl

6.97 789 mm Hg. 105 kPa

CHAPTER 7

Practice Exercises, Chapter 7

1. 1.47 mg N_2/100 g H_2O
2. 10.2 g of 96% H_2SO_4
3. 1.25 g of glucose and 499 g of water (rounded from 498.75, and assuming that the density of water is 1.00 g/mL)
4. 1.67×10^4 mm Hg
5. (a) 0.020 Osm   (b) 0.015 Osm   (c) 0.100 Osm
(d) 0.150 Osm

Review Exercises, Chapter 7

7.5 (a) Stirred suspensions
(b) Colloidal dispersions
(c) Colloidal dispersions
(d) Solutions
(e) Unstirred suspensions

7.13 The $\delta+$ ends of polar water molecules

7.17 The dissolving of a solid and a liquid is usually an endothermic change, so heat shifts the following equilibrium to the right, in favor of more solute being in solution.

$$NH_4Cl(s) + heat \rightleftharpoons NH_4^+(aq) + Cl^-(aq)$$

The removal of heat (by lowering the temperature) shifts this equilibrium to the left and more solid NH_4Cl forms.

7.20 $MgSO_4(s) + 7H_2O \rightarrow MgSO_4 \cdot 7H_2O(s)$

7.23 $y = 10$. The formula is $Y \cdot 10H_2O$

7.24 0.040 g/L

7.29 The first region (where the gas tension is 79 mm Hg).

7.30 $\dfrac{0.915 \text{ g NaOH}}{100 \text{ g NaOH soln}}$ and $\dfrac{100 \text{ g NaOH soln}}{0.915 \text{ g NaOH}}$

7.32 $\dfrac{30 \text{ mL alcohol}}{100 \text{ mL alcohol soln}}$ and $\dfrac{100 \text{ mL alcohol soln}}{30 \text{ mL alcohol}}$

7.34 (a) 31.5 g NaBr   (b) 8.75 g NH$_4$Cl
(c) 1.00 g Mg(NO$_3$)$_2$ (d) 2.25 g NaCl

7.36 (a) 100 g 3.00% NaCl solution
(b) 6.50 g 5.00% NaHCO$_3$ solution
(c) 7.29 mL 3.50% glucose solution
(d) 224 g 3.00% NaCl solution
(e) 514 mL 3.50% glucose solution

7.38 125 g 10.0% NaOH solution

7.40 (a) 35.9% (w/w) HCl
(b) 88.1 mL (104 g) of 35.9% (w/w) HCl. (88.5 mL by a chain calculation)

7.42 $-1.86 \,°C$

7.49 0.080 M Na$_2$SO$_4$ solution, because it is $3 \times 0.080\,M = 0.240\,M$ in its osmolarity.

7.50 Solution A. Whereas the osmolarities of A and B are identical with respect to their dissolved salts and sugars, A has the higher concentration of the colloidally dispersed starch.

7.52 457 mm Hg

7.54 (a) Hypotonic (b) (1) Crenation (2) Hemolysis

7.68 Make sure that the solution has undissolved NaNO$_3$ present.

7.70 0.0194 g/L

7.72 (a) 1.50 g NaCl   (b) 0.575 g NaI
(c) 1.00 g glucose   (d) 3.75 g H$_2$SO$_4$

7.73 37.5 mL methyl alcohol

7.75 7.14 M H$_2$SO$_4$

7.77 228 mm Hg

7.79 0.33 osmol. (0.12 osmol NaCl + 0.060 osmol NaHCO$_3$ + 0.040 osmol KCl + 0.11 osmol glucose)

CHAPTER 8

Practice Exercises, Chapter 8

1.

2. $H_2SO_3(aq) \rightleftharpoons H^+(aq) + HSO_3^-(aq)$
$HSO_3^-(aq) \rightleftharpoons H^+(aq) + SO_3^{2-}(aq)$

3. $HNO_3(aq) + KOH(aq) \rightarrow KNO_3(aq) + H_2O$
$H^+(aq) + NO_3^-(aq) + K^+(aq) + OH^-(aq) \rightarrow$
$\qquad\qquad\qquad K^+(aq) + NO_3^-(aq) + H_2O$
$H^+(aq) + OH^-(aq) \rightarrow H_2O$

4. $Mg(OH)_2(s) + 2HCl(aq) \rightarrow MgCl_2(aq) + 2H_2O$
$Mg(OH)_2(s) + 2H^+(aq) \rightarrow Mg^{2+}(aq) + 2H_2O$

5. K_2SO_4

6. $2NaHCO_3(aq) + H_2SO_4(aq) \rightarrow$
$\qquad\qquad Na_2SO_4(aq) + 2CO_2(g) + 2H_2O$
$2Na^+(aq) + 2HCO_3^-(aq) + 2H^+(aq) + SO_4^{2-}(aq) \rightarrow$
$\qquad 2Na^+(aq) + SO_4^{2-}(aq) + 2CO_2(g) + 2H_2O$
$HCO_3^-(aq) + H^+(aq) \rightarrow CO_2(g) + H_2O$

7. $K_2CO_3(aq) + H_2SO_4(aq) \rightarrow$
$\qquad\qquad K_2SO_4(aq) + CO_2(g) + H_2O$
$2K^+(aq) + CO_3^{2-}(aq) + 2H^+(aq) + SO_4^{2-}(aq) \rightarrow$
$\qquad 2K^+(aq) + SO_4^{2-}(aq) + CO_2(g) + H_2O$
$CO_3^{2-}(aq) + 2H^+(aq) \rightarrow CO_2(g) + H_2O$

8. $MgCO_3(s) + 2HNO_3(aq) \rightarrow$
$\qquad\qquad Mg(NO_3)_2(aq) + CO_2(g) + H_2O$
$MgCO_3(s) + 2H^+(aq) + 2NO_3^-(aq) \rightarrow$
$\qquad\qquad Mg^{2+}(aq) + 2NO_3^-(aq) + CO_2(g) + H_2O$
$MgCo_3(s) + 2H^+(aq) \rightarrow Mg^{2+}(aq) + CO_2(g) + H_2O$

9. (a) $NH_3(aq) + HBr(aq) \rightarrow NH_4Br(aq)$
$NH_3(aq) + H^+(aq) \rightarrow NH_4^+(aq)$
(b) $2NH_3(aq) + H_2SO_4(aq) \rightarrow (NH_4)_2SO_4(aq)$
$NH_3(aq) + H^+(aq) \rightarrow NH_4^+(aq)$

10. $Mg(s) + 2HCl(aq) \rightarrow MgCl_2(aq) + H_2(g)$
$Mg(s) + 2H^+(aq) \rightarrow Mg^{2+}(aq) + H_2(g)$

11. (a) HNO_3 (b) HSO_3^- (c) HCO_3^- (d) HSO_4^-
    (e) HCl (f) H_3O^+ (g) H_2O

12. (a) CO_3^{2-} (b) PO_4^{3-} (c) HSO_4^- (d) SO_4^{2-}
    (e) Br^- (f) H_2O (g) OH^-

13. (a) Weak acid (b) Weak acid (c) Weak acid

14. (a) Weak base (b) Weak base
    (c) Strong base (d) Strong base

15. Yes. $NO_2^-(aq) + H^+(aq) \rightleftharpoons HNO_2(aq)$. The product is favored because we know that HNO_2 is a weak acid (not being on the list of strong acids).

16. $Na_2S(aq) + Cu(NO_3)_2(aq) \rightarrow CuS(s) + 2NaNO_3(aq)$
    $S^{2-}(aq) + Cu^{2+}(aq) \rightarrow CuS(s)$

17. The acetate ion combines with (neutralizes) the hydrogen ion. In the following equilibrium, the product is favored.

$$H^+(aq) + C_2H_3O_2^-(aq) \rightleftharpoons HC_2H_3O_2(aq)$$

18. (a) AgCl precipitates. $Ag^+(aq) + Cl^-(aq) \rightarrow AgCl(s)$
    (b) CO_2 evolves. $CaCO_3(s) + 2H^+(aq) \rightarrow$
        $Ca^{2+}(aq) + CO_2(g) + H_2O$
    (c) No reaction occurs.

Review Exercises, Chapter 8

8.1 (a) $Mg(NO_3)_2$
    (b) Magnesium ion, Mg^{2+}, and nitrate ions, NO_3^-.
    (c) The compound forms separated ions when it dissolves in water and the solution conducts electricity.
    (d) Dissociation

8.3 Ionization. The fact that SO_3 is a *gas* tells us that SO_3 contains no preformed ions, so it must react with water as it dissolves to generate ions.

8.7 Cations

8.9 Methyl alcohol does not furnish any ions of any kind in water. The formula units of methyl alcohol do not break up in water but those of NaOH do.

8.11 It is a molecular compound. It does not consist of ions in the liquid state.

8.14 (a) No (b) Yes, any polyatomic ion such as SO_4^{2-}
    (c) No (d) No

8.16 (a) Acidic (b) Neutral (c) Basic

8.20 Proton and hydronium ion. The nucleus of the hydrogen atom.

8.27 Both H's in H_2SO_4 can react with a base, but only one H in $HC_2H_3O_2$ can neutralize a base.

8.29 With greater difficulty. The second H^+ ion has to pull away from a particle that has more negative charge than the neutral molecule from which the first H^+ ion must leave. Unlike charges attract, so the more unlike the charges are, the more energy is needed to separate particles bearing them.

8.32 Sulfuric acid. The extra oxygen atom provides more electron withdrawing action, and this weakens the H—O bonds more.

8.37 What little does dissolve in water still ionizes 100%.

8.43 (a) Yes (b) No; it's not balanced. (c) No

8.45 (a) $OH^-(aq) + H^+(aq) \rightarrow H_2O$
    (b) $HCO_3^-(aq) + H^+(aq) \rightarrow CO_2(g) + H_2O$
    (c) $CO_3^{2-}(aq) + 2H^+(aq) \rightarrow CO_2(g) + H_2O$
    (d) $NH_3(aq) + H^+(aq) \rightarrow NH_4^+(aq)$

8.47 $M(OH)_2(s) + 2H^+(aq) \rightarrow M^{2+}(aq) + 2H_2O$

8.50 0.457 mol $NaHCO_3$

8.52 6.84 g NaOH

8.54 0.989 g K_2CO_3

8.56 40.9 mL KOH solution

8.58 Any acid stronger than this gives up a proton to H_2O, and thus it is replaced by H_3O^+ and the conjugate base of the stronger acid.

8.61 (a) NH_2^- (b) OH^- (c) S^{2-}

8.63 The reactants are favored because they include the weaker acid and the weaker base.

8.65 An aqueous solution of NaOH or KOH, when added to the solution to be tested, releases ammonia, which has a characteristic odor, if the solution in the test tube is ammonium chloride.

$$NH_4^+(aq) + OH^-(aq) \rightarrow NH_3(aq) + H_2O$$

8.67 Potassium hydroxide, KOH
    $KOH(aq) + HCl(aq) \rightarrow KCl(aq) + H_2O$
    Potassium bicarbonate, $KHCO_3$
    $KHCO_3(aq) + HCl(aq) \rightarrow KCl(aq) + CO_2(g) + H_2O$
    Potassium carbonate, K_2CO_3
    $K_2CO_3(aq) + 2HCl(aq) \rightarrow$
    $$2KCl(aq) + CO_2(g) + H_2O$$

8.69 Compounds c, d, and f are insoluble in water.

8.71 (a) $Mg^{2+}(aq) + 2OH^-(aq) \rightarrow Mg(OH)_2(s)$
    (b) No reaction
    (c) $OH^-(aq) + H^+(aq) \rightarrow H_2O$
    (d) $Pb^{2+}(aq) + 2Cl^-(aq) \rightarrow PbCl_2(s)$
    (e) No reaction
    (f) $S^{2-}(aq) + Ni^{2+}(aq) \rightarrow NiS(s)$
    (g) $SO_4^{2-}(aq) + Ba^{2+}(aq) \rightarrow BaSO_4(s)$
    (h) $OH^-(aq) + H^+(aq) \rightarrow H_2O$
    (i) $S^{2-}(aq) + Cu^{2+}(aq) \rightarrow CuS(s)$
    (j) $Ag^+(aq) + Br^-(aq) \rightarrow AgBr(s)$
    (k) $HCO_3^-(aq) + H^+(aq) \rightarrow CO_2(g) + H_2O$
    (l) $Cl^-(aq) + Ag^+(aq) \rightarrow AgCl(s)$

8.73 3.5–5.0 meq K^+/L

8.75 5.01 meq of K^+

8.90 In H_3O^+, because the weaker bond in H—Br(g) breaks to form the stronger bond in the hydronium ion by the following reaction.

$$HBr(g) + H_2O \longrightarrow H_3O^+(aq) + Br^-(aq)$$

8.93 (a) $HNO_3(aq) + KOH(aq) \rightarrow KNO_3(aq) + H_2O$
$H^+(aq) + OH^-(aq) \rightarrow H_2O$
(b) $K_2CO_3(aq) + 2HNO_3(aq) \rightarrow$
$\qquad\qquad 2KNO_3(aq) + CO_2(g) + H_2O$
$CO_3^{2-}(aq) + 2H^+(aq) \rightarrow CO_2(g) + H_2O$
(c) $2HBr(aq) + MgCO_3(s) \rightarrow$
$\qquad\qquad MgBr_2(aq) + CO_2(g) + H_2O$
$2H^+(aq) + MgCO_3(s) \rightarrow$
$\qquad\qquad Mg^{2+}(aq) + CO_2(g) + H_2O$
(d) $CaCO_3(s) + 2HI(aq) \rightarrow$
$\qquad\qquad CaI_2(aq) + CO_2(g) + H_2O$
$CaCO_3(s) + 2H^+(aq) \longrightarrow$
$\qquad\qquad Ca^{2+}(aq) + CO_2(g) + H_2O$
(e) $NH_3(aq) + HI(aq) \rightarrow NH_4I(aq)$
$NH_3(aq) + H^+(aq) \rightarrow NH_4^+(aq)$
(f) $Mg(OH)_2(s) + 2HBr(aq) \rightarrow MgBr_2(aq) + 2H_2O$
$Mg(OH)_2(s) + 2H^+(aq) \rightarrow Mg^{2+}(aq) + 2H_2O$
(g) $2Al(s) + 6HCl(aq) \rightarrow 2AlCl_3(aq) + 3H_2(g)$
$2Al(s) + 6H^+(aq) \rightarrow 2Al^{3+}(aq) + 3H_2(g)$

8.95 3.13 g $MgCO_3$

8.97 (a) $CaCO_3(s) + 2HCl(aq) \rightarrow$
$\qquad\qquad CaCl_2(aq) + CO_2(g) + H_2O$
$CaCO_3(s) + 2H^+(aq) \rightarrow$
$\qquad\qquad Ca^{2+}(aq) + CO_2(g) + H_2O$
(b) 54.3 g $CaCO_3$; 235 mL HCl solution

8.103 Compounds a, d, and f are insoluble in water.

8.105 (a) We look for the salt with the anion having the highest charge to be least soluble. This salt would involve the strongest attraction between cation and anion and so would be least able to release the ions from the crystal.
(b) $Ca(H_2PO_4)_2$
(c) $Ca_3(PO_4)_2$

8.107 A weak base. It turns red litmus blue, as bases do. Its aqueous solution is a poor conductor, which means that it generates very low concentrations of ions. This makes it a weak base.

8.110 (a) $NaHCO_3(aq) + HC_2H_3O_2(aq) \rightarrow$
$\qquad\qquad H_2CO_3(aq) + NaC_2H_3O_2(aq)$
or $NaHCO_3(aq) + HC_2H_3O_2(aq) \rightarrow$
$\qquad\qquad H_2O + CO_2(g) + NaC_2H_3O_2(aq)$
$HCO_3^-(aq) + HC_2H_3O_2(aq) \rightarrow$
$\qquad\qquad H_2CO_3(aq) + C_2H_3O_2^-(aq)$
(b) Carbonic acid is weaker than acetic acid; "the stronger (acetic acid) gives way to the weaker (carbonic acid)."

8.112 (a) $H_2S(aq) + Cu^{2+}(aq) \rightarrow CuS(s) + 2H^+(aq)$
(b) $OH^-(aq) + H^+(aq) \rightarrow H_2O$
(c) $SO_4^{2-}(aq) + Ba^{2+}(aq) \rightarrow BaSO_4(s)$
(d) $SO_4^{2-}(aq) + Pb^{2+}(aq) \rightarrow PbSO_4(s)$
(e) No reaction
(f) $HCO_3^-(aq) + H^+(aq) \rightarrow CO_2(g) + H_2O$
(g) $S^{2-}(aq) + Cd^{2+}(aq) \rightarrow CdS(s)$
(h) $OH^-(aq) + H^+(aq) \rightarrow H_2O$
(i) No reaction
(j) $HCO_3^-(aq) + H^+(aq) \rightarrow CO_2(g) + H_2O$
(k) No reaction
(l) $Pb^{2+}(aq) + CrO_4^{2-}(aq) \rightarrow PbCrO_4(s)$

8.114 Make a solution of the unknown. If it is K_2O, you will notice that the temperature increases, because K_2O reacts exothermically with water to give KOH ($K_2O + 2H_2O \rightarrow 2KOH$). The solution can be tested with litmus. If the unknown is K_2O, red litmus will turn blue. KNO_3 will not do this.

8.116 0.232 g K_2CO_3

8.118 3.11 g or 3.11×10^3 mg of Na^+

CHAPTER 9

Practice Exercises, Chapter 9

1. (a) 2.5×10^{-6} mol OH^-/L. Basic
   (b) 9.1×10^{-8} mol OH^-/L. Acidic
   (c) 1.1×10^{-7} mol OH^-/L. Basic
2. (a) 1.60 (b) 10.40 (c) 10.70
3. 7.14. Basic
4. (a) 4.6×10^{-7} mol/L, acidic (b) 1.3×10^{-8} mol/L, basic
5. 5.2×10^{-8} mol/L
6. 3.6×10^{-10} mol/L
7. $HC_2H_3O_2(aq) \rightleftharpoons H^+(aq) + C_2H_3O_2^-(aq)$
   $$K_a = \frac{[H^+][C_2H_3O_2^-]}{[HC_2H_3O_2]}$$
8. $HCO_3^-(aq) \rightleftharpoons H^+(aq) + CO_3^{2-}(aq)$
   $$K_a = \frac{[H^+][CO_3^{2-}]}{[HCO_3^-]}$$
9. $NH_4^+(aq) \rightleftharpoons H^+(aq) + NH_3(aq)$
   $$K_a = \frac{[H^+][NH_3]}{[NH_4^+]}$$
10. Ascorbic acid
11. Cyanide ion
12. (a) $CO_3^{2-}(aq) + H_2O \rightleftharpoons HCO_3^-(aq) + OH^-(aq)$
    $$K_b = \frac{[HCO_3^-][OH^-]}{[CO_3^{2-}]}$$
    (b) $C_2H_3O_2^-(aq) + H_2O \rightleftharpoons HC_2H_3O_2(aq) + OH^-(aq)$
    $$K_b = \frac{[HC_2H_3O_2][OH^-]}{[C_2H_3O_2^-]}$$
    (c) $NH_3(aq) + H_2O \rightleftharpoons NH_4^+(aq) + OH^-(aq)$
    $$K_b = \frac{[NH_4^+][OH^-]}{[NH_3]}$$
13. (a) Yes, basic (b) Yes, basic
    (c) Yes, basic (d) Yes, acidic
    (e) Yes, basic (f) Yes, basic

14. Basic
15. Acidic
16. Yes, decrease the pH because NH_4^+ hydrolyzes to give some H^+.
17. 10.33
18. NH_3, $K_b = 1.8 \times 10^{-5}$
19. 4.8
$$CN^-(aq) + H_2O \rightleftharpoons HCN(aq) + OH^-(aq)$$
$$K_b = \frac{[HCN][OH^-]}{[CN^-]}$$
20. 3.89
21. 3.86
22. (a) $pH = pK_a + 1$ (b) $pH = pK_a - 1$
23. 0.105 M NaOH
24. 0.125 M H_2SO_4

Review Exercises, Chapter 9

9.6   A and D
9.11   (a) No effect   (b) No effect   (c) Reduces it
    (d) Increases
9.14   A catalyst can also make a reaction go at the same rate as in its absence, but do so at a lower temperature and so under milder conditions.
9.17   By increasing the collision frequency and by increasing the *fraction* of successful collisions.
9.19   (a) E_{act}
    (b) Heat of reaction
    (c) If the reaction has both a large E_{act} and a large (evolving) heat of reaction.
    (d) The reaction is instantaneous (an explosion).
    (e) E_{act}
9.21   (a) CO and O_2 (b) CO_2
    (c) C   (d) B
    (e) Exothermic. The energy of the product at D is less than that of the reactants at A, and the difference in energy *evolves* as the heat of reaction.
    (f) E
9.23

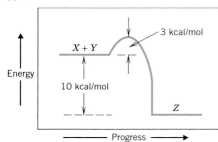

9.26   (a) $K_{eq} = \dfrac{[NH_4^+][OH^-]}{[NH_3][H_2O]}$

(b) $K_{eq} = \dfrac{[H_2O]^2}{[H_2]^2[O_2]}$

(c) $K_{eq} = \dfrac{[NH_3]^2}{[N_2][H_2]^3}$

9.30 2.43×10^{-14}. Neutral, because $[H^+] = [OH^-]$.
9.32   Acidic
9.34   4.0
9.36 4.5×10^{-6} mol/L. Slightly acidic
9.38   Weak. At a pH of 5.72, the value of $[H^+]$ is between 1×10^{-5} to 1×10^{-6} mol/L, but the initial concentration of the acid is much higher $(1 \times 10^{-3}$ mol/L). Therefore, only a small percentage of the acid molecules can be ionized.
9.41 $HSO_3^-(aq) \rightleftharpoons H^+(aq) + SO_3^{2-}(aq)$
$$K_{eq} = \frac{[H^+][SO_3^{2-}]}{[HSO_3^-]}$$
9.43   Ammonium ion
9.45 $NO_2^-(aq) + H_2O \rightleftharpoons HNO_2(aq) + OH^-(aq)$
$$K_b = \frac{[HNO_2][OH^-]}{[NO_2^-]}$$
9.46   Ammonia
9.49   (a) Neutral   (b) Acidic   (c) Basic
    (d) Acidic   (e) Basic
9.51   Basic.
9.53   (a) 10.85   (b) 10.25
9.55   (a) 7.1×10^{-4}, 3.15   (b) 1.8×10^{-4}, 3.75
9.57 N is the stronger base and it has the weaker conjugate acid.
9.61   More alkaline. Alkalosis
9.65 $H_2PO_4^-(aq) + OH^-(aq) \rightarrow HPO_4^{2-}(aq) + H_2O$
9.66 $HPO_4^{2-}(aq) + H^+(aq) \rightarrow H_2PO_4^-(aq)$
9.68   Acid is neutralized by:
$$H^+(aq) + HCO_3^-(aq) \rightarrow CO_2(aq) + H_2O$$
    Base is neutralized by:
$$OH^-(aq) + CO_2(aq) \rightarrow HCO_3^-(aq)$$
9.73   A decrease in the blood's pH (acidosis). The hyperventilation helps to remove acid.
9.75   pH = 4.83
9.78   (a) 24 mmol/L   (b) 1.2 mmol/L
9.79   (a) pH = 6.1
    (b) pH = 7.1
    (c) Hyperventilation expels additional CO_2.
    (d) A net synthesis of HCO_3^- for the blood is performed by the kidneys.
9.83   The titration of any strong acid (e.g., HCl) with any strong base (e.g., NaOH) produces a salt whose ions do not hydrolyze.
9.85   (a) 0.3781 M HCl   (b) 0.4000 M HBr
9.87   64.00

9.88 (a) 7.292 g HCl   (b) 4.845 g HNO_3
(c) 0.5518 g H_2SO_4

9.89 (a) 0.2239 M NaOH   (b) 8.956 g NaOH/L

9.95 (a) **B**, because the product level is shown at a lower energy than the reactant level, as it must in an exothermic reaction.
(b) The O—O bond. The only way that energy can be released (as heat) in this reaction is for the stronger bond not to break but to *form*.

9.98 9.0×10^{-16}

9.100 pH = 12(pOH = 2)

9.101 1.5×10^{-5} mol/L

9.103 $NH_3(aq) \rightleftharpoons NH_2^-(aq) + H^+(aq)$

$$K_a = \frac{[NH_2^-][H^+]}{[NH_3]}$$

9.104 A strong acid. We can calculate from the pH that $[H^+]$ equals the molarity of the acid. This is possible only if the acid is 100% ionized.

9.109 (a) Alkalosis
(b) For each CO_2 lost at the lungs, one H^+ ion is permanently neutralized. The loss of H^+ ion, of course, means an increase in the value of the pH of the blood.

9.111 3.92

9.113 (a) pH = 5.10   (b) pH = 4.96   (c) pH = 5.24
(d) After the addition of acid to 500 mL water, pH = 1.40
After the addition of base to 500 mL water, pH = 12.60

9.115 (a) The ammonium ion:
$NH_4^+(aq) + OH^-(aq) \rightarrow NH_3(aq) + H_2O$
(b) Ammonia: $NH_3(aq) + H^+(aq) \rightarrow NH_4^+(aq)$
(c) 9.24

9.118 (a) Yes
(b) NH_4^+ and NH_3
(c) $NH_4^+ + OH^- \rightarrow NH_3 + H_2O$

CHAPTER 10

Practice Exercises, Chapter 10

1. $^{131}_{53}I \rightarrow ^{131}_{54}Xe + ^{0}_{-1}e + ^{0}_{0}\gamma$

2. $^{239}_{94}Pu \rightarrow ^{235}_{92}U + ^{4}_{2}He + ^{0}_{0}\gamma$

3. 10,000 units

4. 8.5 m

Review Exercises, Chapter 10

10.5 It is the most massive of the particles and it carries the largest charge. Therefore, it collides very quickly with a molecule in the air or other matter

that it enters and so travels the shorter distance (is least penetrating).

10.9 A neutron changes into a proton as an electron is ejected.

10.10 (a) $^{211}_{83}Bi$ (b) $^{216}_{84}Po$ (c) $^{140}_{57}La$

10.11 (a) $^{144}_{60}Nd \rightarrow ^{140}_{58}Ce + ^{4}_{2}He$
(b) $^{40}_{19}K \rightarrow ^{40}_{20}Ca + ^{0}_{-1}e$
(c) $^{149}_{62}Sm \rightarrow ^{149}_{63}Eu + ^{0}_{-1}e$
(d) $^{251}_{98}Cf \rightarrow ^{247}_{96}Cm + ^{4}_{2}He + ^{0}_{0}\gamma$

10.14 1.500 ng

10.26 15.1 m

10.40 Gallium-67 $^{66}_{30}Zn + ^{1}_{1}p \rightarrow ^{67}_{31}Ga$

10.42 $^{14}_{7}N + ^{2}_{1}H \rightarrow ^{15}_{8}O + ^{1}_{0}n$

10.51 It emits only gamma radiation, and it has a shorter half-life.

10.82 Isotope. A radionuclide is a radioactive *isotope* that consists of *atoms*.

10.84 (a) $^{187}_{75}Re \rightarrow ^{187}_{76}Os + ^{0}_{-1}e$
(b) $^{242}_{94}Pu \rightarrow ^{238}_{92}U + ^{4}_{2}He + ^{0}_{0}\gamma$
(c) $^{131}_{53}I \rightarrow ^{131}_{54}Xe + ^{0}_{-1}e$
(d) $^{243}_{95}Am \rightarrow ^{239}_{93}Np + ^{4}_{2}He$

10.86 2.0×10^3 millirad

10.90 Neutrons $^{113}_{49}In \rightarrow ^{111}_{49}In + 2^{1}_{0}n$

10.91 $^{67}_{31}Ga \xrightarrow{\text{electron capture}} ^{67}_{30}Zn$

10.93 A lithium-6 nucleus. $^{27}_{13}Al + ^{6}_{3}Li \rightarrow ^{32}_{15}P + ^{1}_{1}p$

CHAPTER 11

Practice Exercises, Chapter 11

1. (a) CH_3—CH_2—CH_3 (b) CH_3—$\overset{\overset{\displaystyle |}{CH_3}}{CH}$—$CH_3$

(c) CH_3—$\overset{\overset{\displaystyle CH_3}{|}}{\underset{\underset{\displaystyle CH_3}{|}}{C}}$——$\overset{\overset{\displaystyle CH_3}{|}}{CH}$—$\overset{\overset{\displaystyle CH_3}{|}}{CH}$—$CH_3$

2. (a) $CH_3CH_2CH_3$ (b) $CH_3\underset{\underset{\displaystyle CH_3}{|}}{CH}CH_3$

(c) $CH_3\overset{\overset{\displaystyle CH_3}{|}}{\underset{\underset{\displaystyle CH_3}{|}}{C}}$——$\overset{\overset{\displaystyle CH_3}{|}}{CH}$—$CHCH_3$ (Sometimes horizontal bonds are retained to make room for substituents.)

3.

(a) H—$\overset{\overset{\displaystyle H}{|}}{\underset{\underset{\displaystyle H}{|}}{C}}$—$\overset{\overset{\displaystyle H}{|}}{\underset{\underset{\displaystyle H}{|}}{C}}$—$H$

(b) 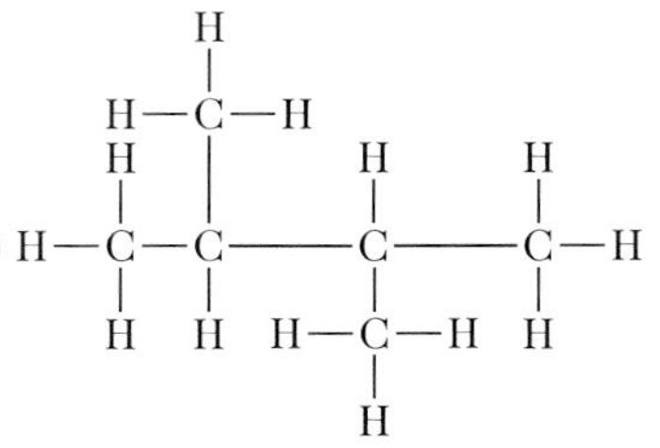
$$CH_3CH_2C(CH_3)(CH_2CH_3)CH_3$$

(c)

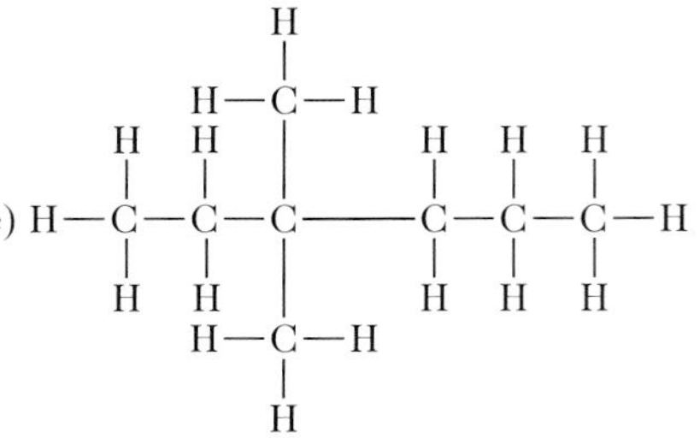

4. Structures (b) and (c) violate the tetravalences of carbon at one point.

5. (a)

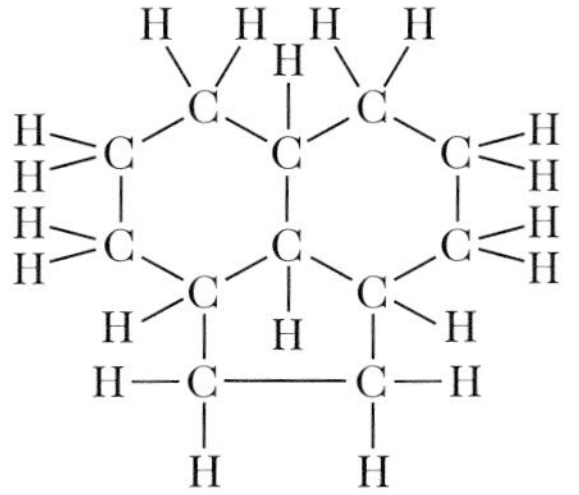

(b)

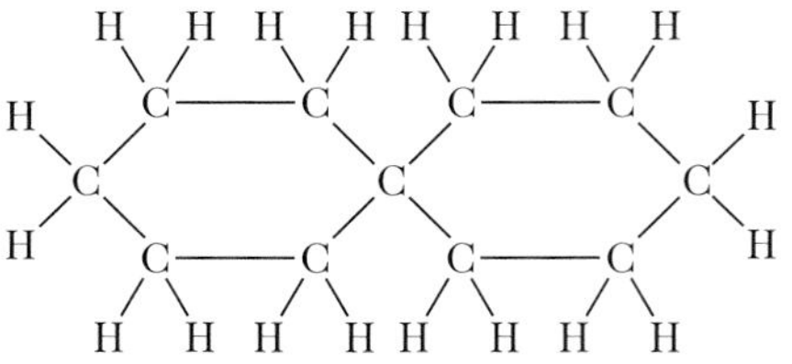

6.  CH_3CH_2—⬡—⬠

7. (a) Identical   (b) Isomers
   (c) Identical   (d) Isomers
   (e) Different in another way

8. 2-Methyl-1-butanol (less polar)

9. (a) 3-Methylhexane
   (b) 4-Ethyl-2,3-dimethylheptane
   (c) 5-Ethyl-2,4,6-trimethyloctane

10. (a) 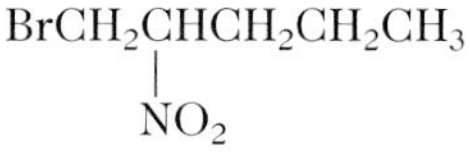 $BrCH_2CHCH_2CH_2CH_3$
             |
            NO_2

(b) 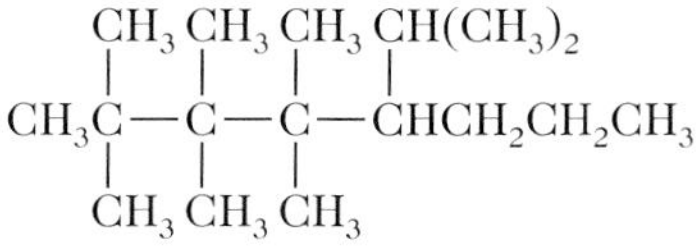
$$CH_3C(CH_3)_2C(CH_3)_2CHCH(CH_3)CH_2CH_2CH_3$$

(c) 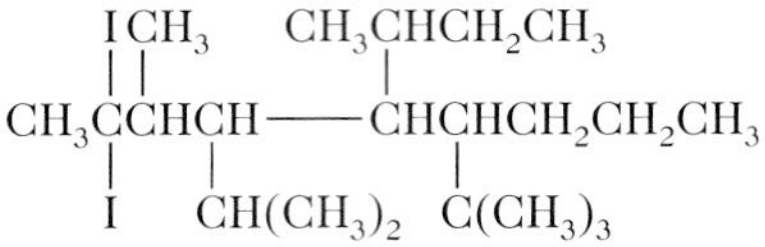
$$CH_3CC(I)CHCH(CH_3)CH_2CH_2CH_3$$

(d) 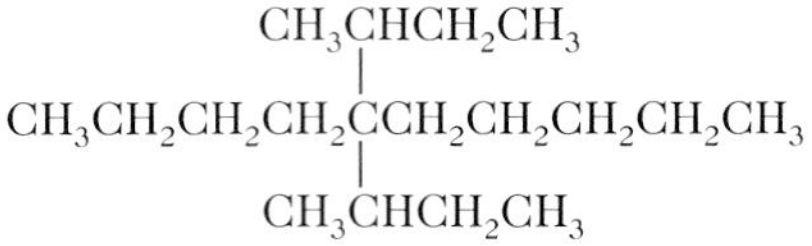
$$CH_3CH_2CH_2CH_2CCH_2CH_2CH_2CH_2CH_3$$

11.

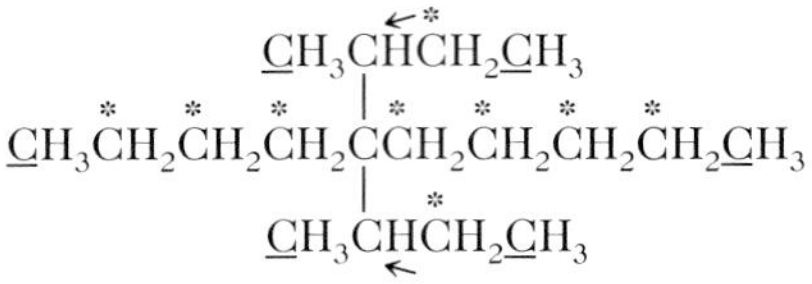

12. (a) Ethyl chloride   (b) Butyl bromide
     (c) Isobutyl chloride (d) *t*-Butyl bromide

13. (a) Two. Butyl chloride, 1-chlorobutane
       sec-Butyl chloride, 2-chlorobutane
     (b) Two. Isobutyl chloride,
         1-chloro-2-methylpropane
         t-Butyl chloride, 2-chloro-2-methylpropane

Review Exercises, Chapter 11

11.6 Compounds b, d, and e are considered to be inorganic.

11.9 (a) Molecular; low melting and flammable.
    (b) Ionic; water soluble (and likely a carbonate or a bicarbonate).
    (c) Molecular; no ionic compound is a gas (or a liquid) at room temperature.
    (d) Ionic; high melting and nonflammable.
    (e) Molecular; most liquid organic compounds are insoluble in water but will burn.
    (f) Molecular; no ionic compound is a liquid at room temperature.

11.10 Straight chain. The hexane molecule has no tertiary carbons as do all branched chain, open chain compounds.

11.12 (a) H—$C(H)(H)$—O—H (b) H—$C(Cl)(Cl)$—H

     (c) H—$N(H)$—$N(H)$—H (d) H—$C(H)(H)$—$C(H)(H)$—H

     (e) H—$C(=O)$—H (f) H—$C(=O)$—O—H

     (g) H—$N(H)$—O—H (h) H—$C(H)=C(H)$—H

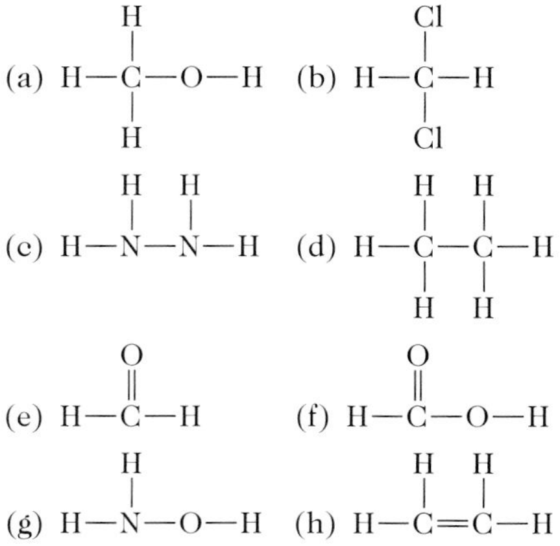

(i) $H-\underset{\underset{Cl}{|}}{\overset{\overset{Cl}{|}}{C}}-Cl$ (j) $H-C\equiv N$

(k) $H-\underset{\underset{H}{|}}{\overset{\overset{H}{|}}{C}}-C\equiv N$ or $H-\overset{\overset{H}{|}}{C}=C=N-H$

(l) $H-\underset{\underset{H}{|}}{\overset{\overset{H\quad H}{|\quad|}}{C}}-N-H$

11.15 (a) $CH_3CH_2\underset{\underset{OH}{|}}{CH}CH_2CH_2CH_3$

(b)

(c) H_3C-

11.18 The rotation does not reduce the attractive force between the shared electrons and the two carbon nuclei they attract.

11.20 Compounds (a) and (b) are saturated. Compound (d) can only be ethene.

11.21 Identical compounds: a, b, c, f, d, i, m, n
Isomers: e, g, h, k
Unrelated: j, l

11.23 (a) Aldehyde   (b) Ketone
(c) Alkane (or cycloalkane)   (d) Ester

11.24 (a) Alkane   (b) Alkane
(c) Amine   (d) Alcohol
(e) Cycloalkane, alkene
(f) ketone   (g) Ketone
(h) Alcohol–ketone, carboxylic acid
(i) Carboxylic acid   (j) Ether–ketone, ester
(k) Carboxylic acid–alcohol, ester–alcohol
(m) Aldehyde (aromatic)   (n) Ether–amine

11.26 Compound B; it has two polar, water-like OH groups. Compound A is much more hydrocarbon-like and has only one functional group.

11.28 Add a drop of the liquid to water. If it dissolves, it is methyl alcohol.

11.30 $CH_3CH_2CH_2CH_2CH_2CH_3$ hexane

11.32 (a) CH_3CH_2Cl (b) $CH_3CH_2CH_2Br$
(c) $CH_3\underset{\underset{I}{|}}{CH}CH_3$ (d) $CH_3\underset{\underset{Cl}{|}}{CH}CH_2CH_3$

11.34 (a)

1,2-dimethyl-cyclohexane

(b) $CH_3\underset{\underset{CH_3}{|}}{CH}CH_2\underset{\underset{CH_3}{|}}{CH}CHCH_3$

2,3,5-trimethylhexane

(c) $CH_3CH_2CH_2CH_2Cl$ (d) $\underset{\underset{propane}{}}{CH_3CH_2}\overset{\overset{CH_3}{|}}{}$

1-chlorobutane

11.36 $CH_3CH_2OH + 3O_2 \rightarrow 2CO_2 + 3H_2O$

11.38 CH_3CHCl_2, 1,1-dichloroethane
$ClCH_2CH_2Cl$, 1,2-dichloroethane

11.50 **A**, **B**, and **C** are identical; **D** is an isomer of any one of them.

11.52 (a) $CH_3CH_2CH_2CH_2CH_2CH_3$

(b) $CH_3\overset{\overset{CH_3}{|}}{CH}CH_2CH_2CH_2CH_3$

11.54 5-sec-Butyl-4-t-butyl-6-propyldecane

11.57 $CH_3CH_2CHCl_2$ $ClCH_2CH_2CH_2Cl$

1,1-dichloropropane   1,3-dichloropropane

$CH_3\overset{\overset{Cl}{|}}{CH}CH_2Cl$ $CH_3\underset{\underset{Cl}{|}}{\overset{\overset{Cl}{|}}{C}}CH_3$

1,2-dichloropropane

2,2-dichloropropane

11.59 No reaction with any of the given reactants.

11.61 11.1 g chlorocyclopentane

CHAPTER 12

Practice Exercises, Chapter 12

1. (a) 2-Methylpropene
(b) 4-Isobutyl-3,6-dimethyl-3-heptene
(c) 1-Chloropropene
(d) 3-Bromopropene
(e) 4-Methyl-1-hexene
(f) 4-Methylcyclohexene

2.
(a) $CH_3CH=CH\overset{\overset{CH_3}{|}}{CH}CH_3$

(b) $H_2C=CH\underset{\underset{CH_2CH_2CH_3}{|}}{CH}CH_2CH_2CH_3$

(c) $H_2C=CH\underset{\underset{CH_3}{|}}{C}CH_2Cl$

(d) H_3C CH_3 on $CH_3C{=}CCH_3$

3. Cis–trans isomerism is possible for a and b. For part (a), cis versus trans is based on the way the main chain passes through the double bond.

(a) 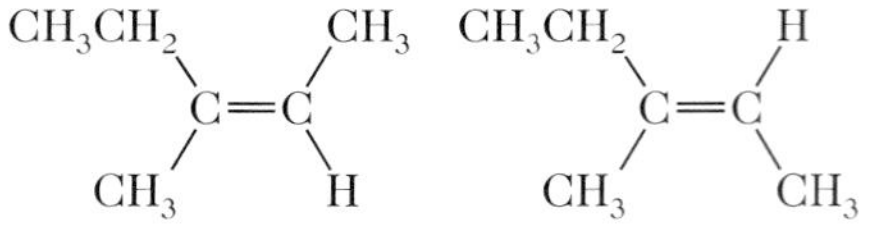

cis isomer trans isomer

(b) 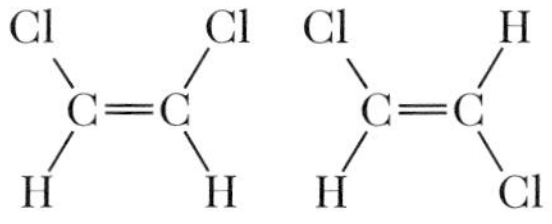

cis isomer trans isomer

4. (a) $CH_3CH_2CH_3$
(b) No reaction
(c) 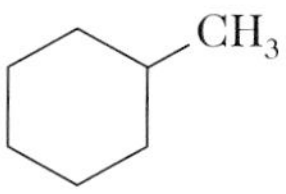
(d) $CH_3(CH_2)_{16}CO_2H$

5. (a) CH_3 / CH_3CCH_2Br / Br
(b) No reaction (in the absence of heat or UV radiation)
(c) $ClCH_2CHCH_2CH_3$ / Cl
(d) $CH_3CH_2CH_2CH_3$

6. (a) $CH_3CHCH_2CH_3$ / Cl
(b) CH_3 / CH_3CCH_3 / Br
(c) 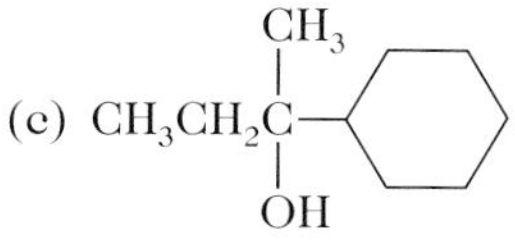
(d) 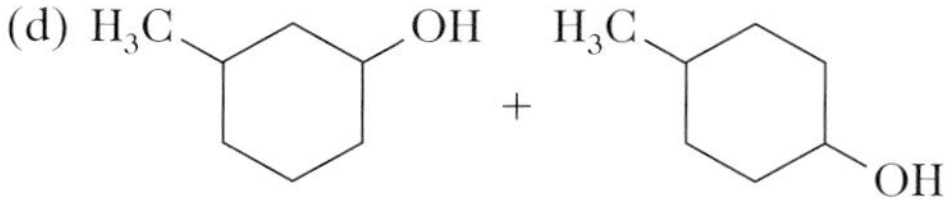
(e) No reaction

7. (a) $CH_3CH_2CH_2\overset{+}{C}H_2$ and $CH_3CH_2\overset{+}{C}HCH_3$ (preferred) — $CH_3CH_2CHCH_3$ / Cl

(b) CH_3 / $CH_3CHCH_2^+$ and CH_3 / $CH_3\underset{+}{C}CH_3$ (preferred) — CH_3 / CH_3CCH_3 / Cl

(c) 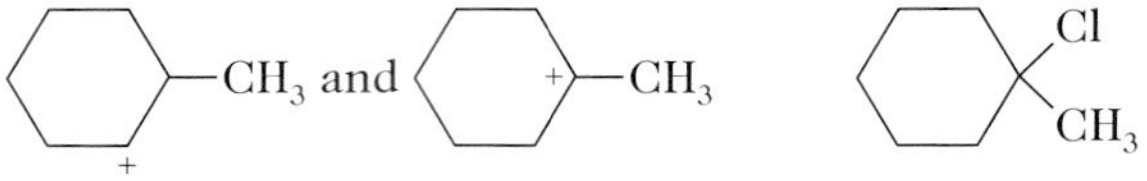

preferred

(d) Only one carbocation is possible: $CH_3CH_2{}^+CHCH_3$ — Cl / $CH_3CH_2CHCH_3$

(e) $CH_3{}^+CHCH_2CH_2CH_3$ — $CH_3CHCH_2CH_2CH_3$ / Cl
and
$CH_3CH_2{}^+CHCH_2CH_3$ — $CH_3CH_2CHCH_2CH_3$ / Cl

(Both carbocations are 2°, so both are equally possible and both form. Thus two isomeric chloropentanes form.)

(f) $CH_3CHCH_2CH_2CH_3$ / OH and $CH_3CH_2CHCH_2CH_3$ / OH

Two 2° carbocations can form, so two isomeric pentanols form. Only one carbocation can form from propene, namely, the more stable isopropyl carbocation.

Review Exercises, Chapter 12

12.1 (a) C, D, E, (b) A (c) B, C, E (d) all

12.4 (a) 3-Heptene
(b) 4-Methyl-2-pentene
(c) 4-Methyl-2-propyl-1-hexene
(d) *cis*-2-Pentene
(e) 1,3-Dimethylcyclohexene
(f) 2,3-Dimethylcyclohexene

12.6 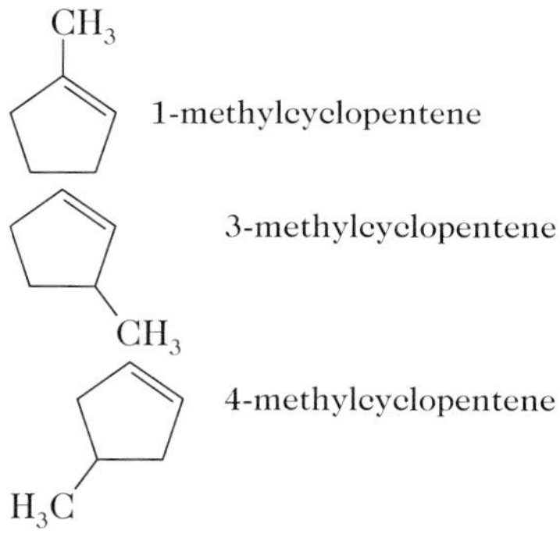

12.8 (a) Identical (b) Identical
(c) Isomers (d) Identical
(e) Identical

12.10 (a)

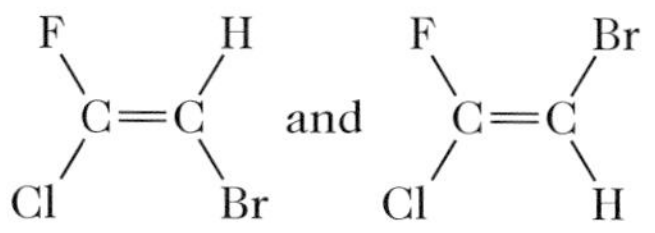

and

(b) No geometric isomers

(c)

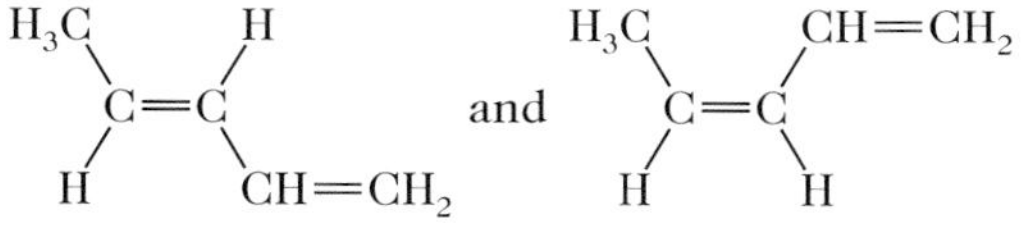

and

12.13 (a)

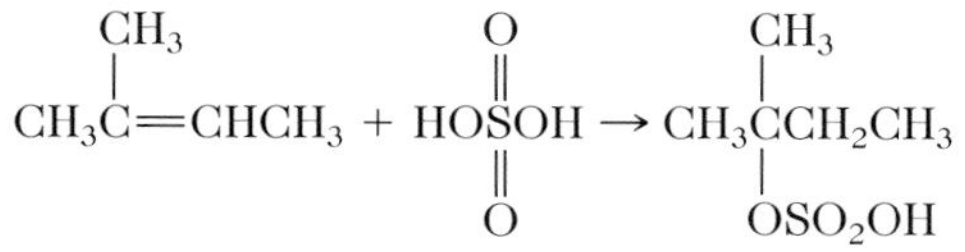

$$CH_3\overset{\underset{\displaystyle |}{CH_3}}{C}=CHCH_3 + HO\overset{\underset{\displaystyle \|}{\overset{\displaystyle O}{}}}{S}OH \rightarrow CH_3\overset{\underset{\displaystyle |}{CH_3}}{C}CH_2CH_3$$
$$\qquad\qquad\qquad\qquad\qquad\qquad OSO_2OH$$

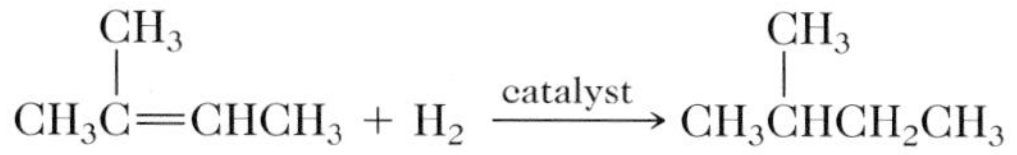

(b) $CH_3\overset{\underset{\displaystyle |}{CH_3}}{C}=CHCH_3 + H_2 \xrightarrow{catalyst} CH_3\overset{\underset{\displaystyle |}{CH_3}}{C}HCH_2CH_3$

(c) $CH_3\overset{\underset{\displaystyle |}{CH_3}}{C}=CHCH_3 + H_2O \xrightarrow{H^+} CH_3\overset{\underset{\displaystyle |}{CH_3}}{C}CH_2CH_3$
$\qquad\qquad\qquad\qquad\qquad\qquad\quad OH$

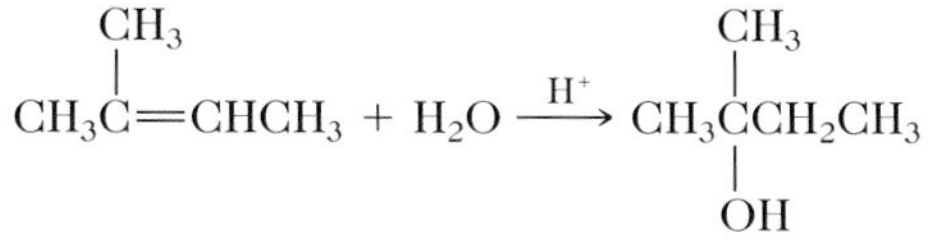

(d) 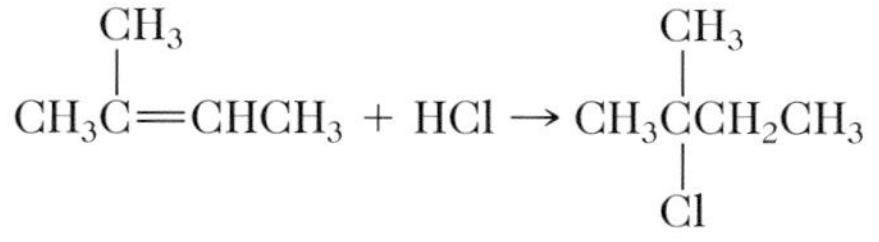 $CH_3\overset{\underset{\displaystyle |}{CH_3}}{C}=CHCH_3 + HCl \rightarrow CH_3\overset{\underset{\displaystyle |}{CH_3}}{C}CH_2CH_3$
$\qquad\qquad\qquad\qquad\qquad\qquad\quad Cl$

(e) 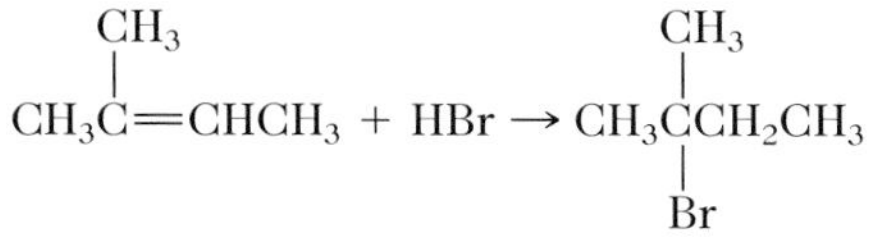 $CH_3\overset{\underset{\displaystyle |}{CH_3}}{C}=CHCH_3 + HBr \rightarrow CH_3\overset{\underset{\displaystyle |}{CH_3}}{C}CH_2CH_3$
$\qquad\qquad\qquad\qquad\qquad\qquad\quad Br$

(f) 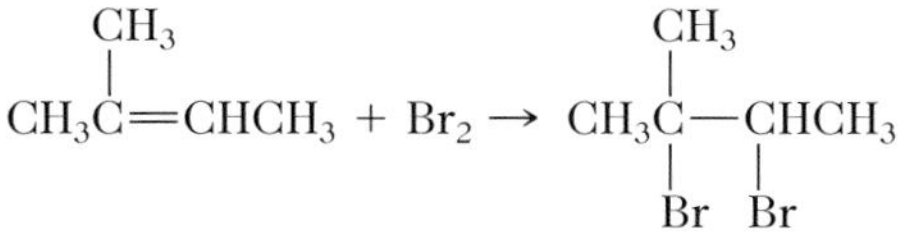 $CH_3\overset{\underset{\displaystyle |}{CH_3}}{C}=CHCH_3 + Br_2 \rightarrow CH_3\overset{\underset{\displaystyle |}{CH_3}}{C}-CHCH_3$
$\qquad\qquad\qquad\qquad\qquad\qquad\quad Br \;\; Br$

12.15 Cycloalkanes with the formula C_5H_{10} have no alkene groups and so cannot react with the given reactants. For example,

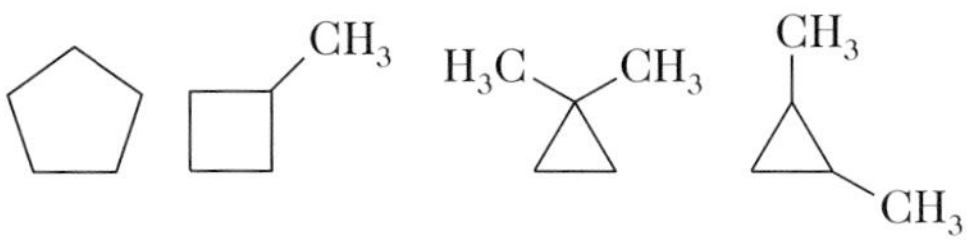

12.16 (a) 2KOH
(b) 63.1 g $KMnO_4$
(c) 33.3 g $K_2C_6H_8O_4$

12.18 (a)

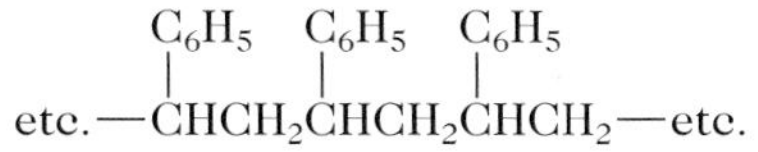

$$\text{etc.} -\overset{\underset{\displaystyle |}{C_6H_5}}{C}HCH_2\overset{\underset{\displaystyle |}{C_6H_5}}{C}HCH_2\overset{\underset{\displaystyle |}{C_6H_5}}{C}HCH_2- \text{etc.}$$

(b)

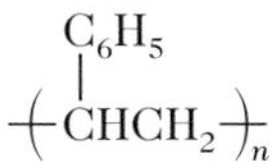

12.20 Some traces of alkenes are also present.

12.22 Sulfanilamide has a benzene ring.

12.24 Any monoalkylbenzene, like toluene, $C_6H_5CH_3$, or ethylbenzene, $C_6H_5CH_2CH_3$

12.40 (a) Cycloalkanes   (b) Cycloalkenes and open chain dienes

12.42 No cis–trans isomers are possible.

12.44 $HC{\equiv}CCH_2CH_3$ 1-butyne
$CH_3C{\equiv}CCH_3$ 2-butyne

12.47 Compound A could be either of the following:

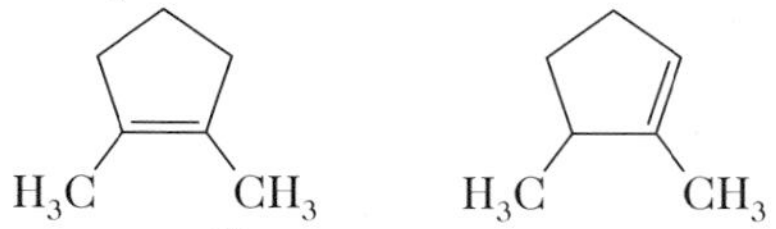

12.49

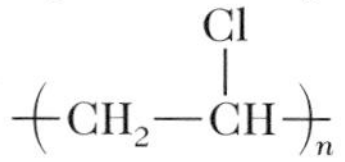

12.51

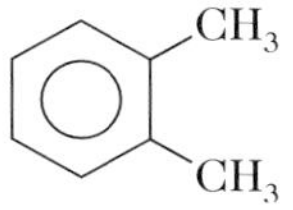

12.52 (a) 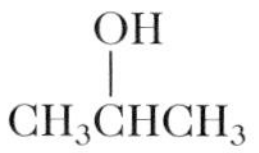 $CH_3\overset{\underset{\displaystyle |}{OH}}{C}HCH_3$

(b) 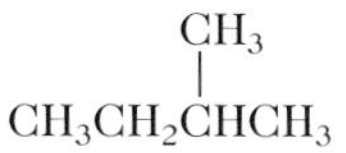 $CH_3CH_2\overset{\underset{\displaystyle |}{CH_3}}{C}HCH_3$

(c) $C_6H_5Br + HBr$

(d) No reaction

(e)

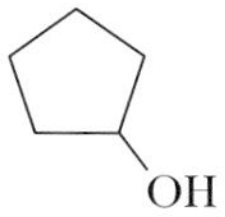

(f) $CH_3CH_2CH_2CH_2CH_3$

(g) No reaction

(h)

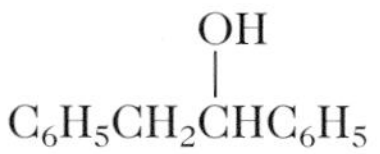

$$C_6H_5CH_2\overset{\underset{\displaystyle |}{OH}}{C}HC_6H_5$$

(i) No reaction

(j) $C_5H_2 + 8O_2 \rightarrow 5CO_2 + 6H_2O$

(k) No reaction

CHAPTER 13

Practice Exercises, Chapter 13

1. (a) Monohydric, secondary
   (b) Monohydric, secondary
   (c) Dihydric, unstable (two OH groups on the same carbon)
   (d) Dihydric, both are secondary
   (e) Monohydric, primary
   (f) Monohydric, primary
   (g) Monohydric, tertiary
   (h) Monohydric, secondary
   (i) Unstable (three OH groups on the same carbon)

2. (a) Alcohol   (b) Phenol
(c) Carboxylic acid   (d) Alcohol (but unstable;
                         an enol)
(e) Alcohol   (f) Alcohol

3. (a) 4-Methyl-1-pentanol
(b) 2-Methyl-2-propanol
(c) 2-Ethyl-2-methyl-1-pentanol
(d) 2-Methyl-1,3-propanediol

4. In 1,2-propanediol. Its boiling point is 189 °C, much higher than that of 1-butanol (bp 117 °C). 1,2-Propanediol is more soluble in water.

5. (a) $CH_3CH=CH_2$ (b) $CH_3CH=CH_2$

(c) $CH_2=\overset{\underset{\displaystyle |}{CH_3}}{C}CH_3$ (d)

6. (a) $CH_3\overset{\underset{\displaystyle |}{CH_3}}{C}HCH=O$ then $CH_3\overset{\underset{\displaystyle |}{CH_3}}{C}HCO_2H$
(b) $C_6H_5CH=O$ then $C_6H_5CO_2H$

7. (a) $C_6H_5\overset{\overset{\displaystyle O}{\|}}{C}CH_3$ (b)

8. (a) $CH_3\overset{\overset{\displaystyle O}{\|}}{C}CH_2CHO$ and $CH_3\overset{\overset{\displaystyle O}{\|}}{C}CH_2CO_2H$
(b) No reaction
(c) $CH_3\overset{\underset{\displaystyle |}{CH_3}}{C}CH=O$ and $CH_3\overset{\underset{\displaystyle |}{CH_3}}{C}CO_2H$
(d) $CH_3\overset{\underset{\displaystyle |}{CH_3}}{C}H\overset{\overset{\displaystyle O}{\|}}{C}CH_3$

9. (a) CH_3OCH_3 (b) $CH_3CH_2CH_2OCH_2CH_2CH_3$

(c)

10. (a) $2CH_3SH$ (b) $CH_3\overset{\underset{\displaystyle |}{CH_3}}{C}HS\overset{\underset{\displaystyle |}{CH_3}}{C}HCH_3$

(c)   (d)

Review Exercises, Chapter 13

13.2   1, 1° alcohol; 2, ketone; 3, 3° alcohol; 4, alkene; 5, ketone

13.5   (a) Isopropyl alcohol   (b) Ethyl alcohol

(c) *sec*-Butyl alcohol   (d) *t*-Butyl alcohol

13.7 $HOCH_2\overset{\underset{\displaystyle OH}{|}}{CH}CH_2OH$, glycerol (1,2,3,propanetriol)

13.10   3-Ethyl-1-hexanol

13.13 $B < D < A < C$

13.14

(a) $CH_3\overset{\underset{\displaystyle |}{CH_3}}{C}=CH_2$ (b)

(c) $CH_3\overset{\underset{\displaystyle |}{CH_3}}{C}=CHCH_2CH_3$ (d) $CH_3\overset{\overset{\displaystyle H_3C}{|}}{C}=\overset{\underset{\displaystyle CH_3}{|}}{C}CH_3$

  plus some   plus some

$CH_2=\overset{\underset{\displaystyle |}{CH_3}}{C}CH_2CH_2CH_3$ $CH_2=\overset{\overset{\displaystyle H_3C}{|}}{C}-\overset{\underset{\displaystyle CH_3}{|}}{CH}CH_3$

(e) H_3C CH_3 (f)

13.15

(a) $CH_3\overset{\overset{\displaystyle O}{\|}}{C}\overset{\underset{\displaystyle CH_3}{|}}{CH}CH$ $CH_3\overset{\overset{\displaystyle O}{\|}}{C}\overset{\underset{\displaystyle CH_3}{|}}{CH}COH$ (b) No reaction

(c) No reaction   (d) No reaction

(e) H_3C CH_3 (f) $\overset{\overset{\displaystyle O}{\|}}{C}CH(CH_3)_2$

13.17   (a) $HOCH_2CH_2CH_2CH_3$ (b) $CH_3CH_2\overset{\underset{\displaystyle OH}{|}}{CH}CH(CH_3)_2$

(c) $HOCH_2CH_2\overset{\underset{\displaystyle CH_3}{|}}{CH}CH_3$ (d) CH_2OH

13.19   **A** reacts with aqueous NaOH; **B** does not. **B** can be dehydrated to an alkene; **A** cannot. (Both react with oxidizing agents, but in different ways.)

13.21   (a) $CH_3CH_2OCH_2CH_3$ (b) $CH_3\overset{\overset{\displaystyle CH_3}{|}}{CH}O\overset{\overset{\displaystyle CH_3}{|}}{CH}CH_3$

(c) $CH_3OCH_2CH_2OCH_2CH_2OCH_3$

(d) CH_2OCH_2

13.23 $CH_3OCH_3 + CH_3OCH_2CH_3 + CH_3CH_2OCH_2CH_3$

13.26   (a) $CH_3CH_2CH_2SSCH_2CH_2CH_3$
(b) $(CH_3)_2CHCH_2SH$
(c) $HSCH_2CH_2CH_2SH$ (d) $(CH_3)_2CHSSCH(CH_3)_2$

13.45   3-*t*-Butyl-4-ethyl-2-isopropyl-3-methyl-1-octanol

13.48 B. A is a 3° alcohol; it cannot be oxidized by permanganate and so cannot decolorize this oxidizing agent.

13.50 B is a 1° alcohol and can be dehydrated.

13.53 (a)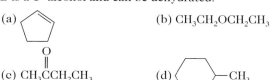

(b) $CH_3CH_2OCH_2CH_3$

(c) $CH_3\overset{\overset{O}{\|}}{C}CH_2CH_3$

(d) cyclohexyl—CH_3

(e) No reaction

(f) No reaction

(g) $CH_3CH_2CH_2\overset{\overset{OH}{|}}{C}HCH_2CH_3$

(h) No reaction

(i) 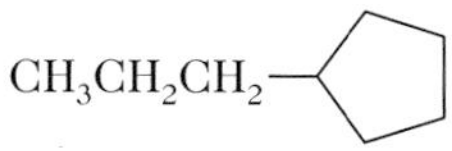 $CH_3CH_2CH_2$—cyclopentyl

(j) 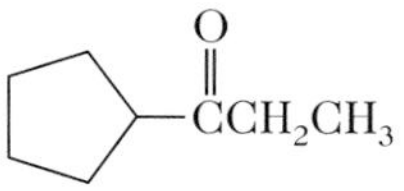 cyclopentyl—$\overset{\overset{O}{\|}}{C}CH_2CH_3$

13.55 (a) 3.00 mol acetone
(b) 0.0760 mol $KMnO_4$
(c) 27.3 g $KMnO_4$
(d) 15.1 g acetone; 15.0 g MnO_2

13.57 20.5 g cyclopentanol needed; 20.8 g $Na_2Cr_2O_7$ required.

CHAPTER 14

Practice Exercises, Chapter 14

1. (a) 2-Methylpropanal
   (b) 3-Bromobutanal
   (c) 4-Ethyl-2,4,6-trimethylheptanal

2. 2-Isopropylpropanal would have the following structure.

$$CH_3\overset{\overset{O}{\|}}{C}H\overset{\underset{\underset{CH_3CHCH_3}{|}}{}}{C}H$$

It should be named 2,3-dimethylbutanal.

3. (a) 2-Butanone
   (b) 6-Methyl-2-heptanone
   (c) 2-Methylcyclohexanone

4. (a) $CH_3CH_2\overset{\overset{O}{\|}}{C}\overset{\underset{\underset{CH_3}{|}}{}}{C}HCH_3$

   (b) $CH_3\overset{\overset{O}{\|}}{C}C_6H_5$

   (c) $CH_3CH_2CH_2\overset{\overset{O}{\|}}{C}CH_2CH_2CH_3$

   (d) $(CH_3)_3C\overset{\overset{O}{\|}}{C}C(CH_3)_3$

5. (a) $CH_3CH_2\overset{\overset{OH}{|}}{C}HCH_3$

   (b) $CH_3\overset{\underset{\underset{CH_3}{|}}{}}{C}HCH_2CH_2OH$

(c) 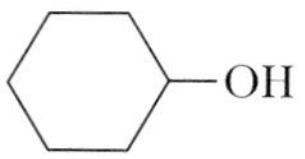 cyclohexyl—OH

6. (a) Not a hemiacetal   (b) Not a hemiacetal

(c) $HO\overset{*}{C}H_2OCH_2CH_3$ (d) 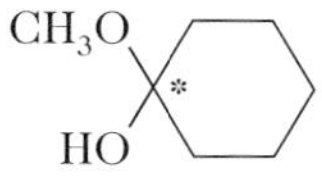 cyclohexane with CH_3O and HO substituents

7. (a) $CH_3\overset{\overset{OH}{|}}{C}HOCH_3$

   (b) $CH_3CH_2CH_2\overset{\overset{OH}{|}}{C}HOCH_2CH_3$

   (c) $C_6H_5\overset{\overset{OH}{|}}{C}HOCH_2CH_2CH_3$

   (d) $HOCH_2OCH_3$

8. (a) $CH_3CH_2\overset{\overset{O}{\|}}{C}H + HOCH_3$

   (b) $CH_3CH_2OH + H\overset{\overset{O}{\|}}{C}CH_2CH_3$

9. (a) Neither an acetal nor a ketal
   (b) A ketal. Its carbon atom that holds two oxygen atoms was a keto group carbon.

10. (a) $2CH_3OH + H\overset{\overset{O}{\|}}{C}H$ (b) No reaction

    (c) $2CH_3OH + CH_3\overset{\overset{H_3C}{|}}{C}H\overset{\overset{O}{\|}}{C}CH_3$

Review Exercises, Chapter 14

14.1 (a) Ketone   (b) Aldehyde
(c) Carboxylic acid   (d) Aldehyde
(e) Carboxylic acid   (f) Ketone

14.3 (a) $CH_3CH_2\overset{\overset{CH_3}{|}}{C}HCH_2\overset{\overset{O}{\|}}{C}H$ (b) 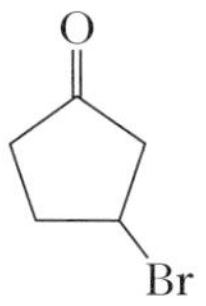 cyclopentanone

(c) 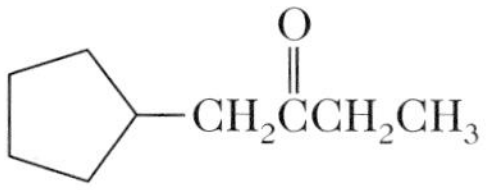 cyclopentyl—$CH_2\overset{\overset{O}{\|}}{C}CH_2CH_3$ (d) $CH_3\overset{\overset{O}{\|}}{C}H$ (with Br)

(e) $CH_3\overset{\overset{O}{\|}}{C}CH_3$ (f) $CH_3CH_2CH_2\overset{\overset{O}{\|}}{C}H$

14.5 (a) 2-Methylcyclohexanone (b) Propionic acid
(c) 2-Pentanone   (d) Propanal
(e) 2-Methylpentanal

14.7 5-Ketohexanal

14.9 Valeraldehyde

14.11 C < B < D < A

14.13 C < B < D < A

14.16   (a) $CH_3\overset{O}{\overset{\|}{C}}CH_2CH_3$ (b) $CH_3\underset{CH_3}{\overset{}{C}H}\overset{O}{\overset{\|}{C}}H$

2-butanone   2-methylpropanal

(c) 2-methylcyclopentanone

(d) $C_6H_5CH_2\underset{CH_3}{\overset{}{C}H}CHO$

2-methyl-3-phenylpropanal

14.18 C_3H_6O is $CH_3CH_2CH{=}O$; $C_3H_6O_2$ is $CH_3CH_2CO_2H$

14.20   Positive Benedict's tests are given by a, b, and c.

14.26 $CH_3\overset{O}{\overset{\|}{C}}CH_2CO_2^-$

14.29   (a) $CH_3CH_2O^-$
   (b) $CH_3CH_2O^- + H_2O \rightarrow CH_3CH_2OH + OH^-$
   (c) ethanol

14.31   (a) cyclohexanone ($=O$)   (b) $CH_3\overset{O}{\overset{\|}{C}}CH_2CH_3$

   (c) $H\overset{O}{\overset{\|}{C}}\underset{CH_3}{\overset{}{C}H}CH_2CH_3$ (d) $Cl{-}$benzene ring$-\overset{O}{\overset{\|}{C}}H$

14.33   (a) Hemiacetal
   (b) Something else (an ether–alcohol)
   (c) Hemiacetal   (d) Alcohol and acetal

14.35   (a) $CH_3\underset{OH}{\overset{CH_3}{\overset{}{C}}}OCH_3$ $CH_3\underset{OCH_3}{\overset{CH_3}{\overset{}{C}}}OCH_3$

   (b) $CH_3\underset{OH}{\overset{CH_3}{\overset{}{C}}}OCH_2CH_3$ $CH_3\underset{OCH_2CH_3}{\overset{CH_3}{\overset{}{C}}}OCH_2CH_3$

14.37   (a)   (b)

14.39   (a) $CH_3CH_2CHO + HOCH_2CH_3 + HOCH_3$
   (b) $CH_3CHO + (CH_3)_2CHOH + HOCH_3$

   (c) $OCHCH_2OCH_3 + 2\,HOCH_3$
   (d) $HOCH_2CH_2OH + OCHCH_3$

14.41   (a) 4,4-Dimethyl-2-pentanone
   (b) 3-Ethyl-3,6-dimethylheptanal
   (c) 2,4-Dimethylcycloheptanone
   (d) 4-Phenyl-2-butanone

14.43 $CH_3\underset{}{\overset{OH}{\overset{\|}{C}}H}CO_2^-$

14.45 $CH_3\underset{}{\overset{O^-}{\overset{\|}{C}}H}CH_2\overset{O}{\overset{\|}{C}}S{-}$(enzyme) $CH_3\underset{}{\overset{OH}{\overset{\|}{C}}H}CH_2\overset{O}{\overset{\|}{C}}S{-}$(enzyme)

   A   B

14.47   (a)   (b)

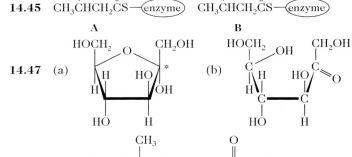

14.48   (a) $HOCH_2CH_2\underset{CH_3}{\overset{}{C}H}CH_3$ (b) $CH_3\overset{O}{\overset{\|}{C}}CH(CH_3)_2$
   (c) No reaction   (d) $CH_3CH_2CH_2CH_3$

   (e) $CH_3CH_2\underset{}{\overset{OH}{\overset{\|}{C}}H}OCH_3$ (f) $CH_3CH_2OH + Mtb^+$

   (g) $CH_3\underset{OCH_2CH_3}{\overset{}{C}H}OCH_2CH_3$ (h) $CH_3CHO + 2CH_3OH$

   (i) $CH_3CH_2CH_2\overset{O}{\overset{\|}{C}}OH$ (j) No reaction

14.50 $CH_3CH_2\overset{O}{\overset{\|}{C}}H$ $CH_3CH_2CH_2OH$ $CH_3CH{=}CH_2$
   A   B   C

   $CH_3\underset{}{\overset{OH}{\overset{\|}{C}}H}CH_3$ $CH_3\overset{O}{\overset{\|}{C}}CH_3$
   D   E

14.52   (a) 0.173 mol butanal
   (b) 1.23 mol CH_3OH
   (c) Yes, 11.1 g CH_3OH needed but 39.4 g taken
   (d) 3.11 g H_2O obtained
   (e) To ensure that the equilibria involved in the reaction are all shifted as much as possible to the right, in favor of the products.

CHAPTER 15

Practice Exercises, Chapter 15

1.   (a) 2,2-Dimethylpropanoic acid
   (b) 5-Ethyl-5-isopropyl-3-methyloctanoic acid

(c) Sodium ethanoate

(d) 5-Chloro-3-methylheptanoic acid

2. Pentanedioic acid

3. 9-Octadecenoic acid

4. (a) $CH_3CH_2CO_2^-$ (b) $CH_3O-\langle\bigcirc\rangle-CO_2^-$

(c) $CH_3CH{=}CHCO_2^-$

5. (a) $CH_3O-\langle\bigcirc\rangle-CO_2H$ (b) $CH_3CH_2CO_2H$

(c) $CH_3CH{=}CHCO_2H$

6. (a) $CH_3\overset{\overset{\displaystyle O}{\|}}{C}OCH_3$ (b) $CH_3\overset{\overset{\displaystyle O}{\|}}{C}OCH_2CH_2CH_3$

(c) $CH_3\overset{\overset{\displaystyle O}{\|}}{C}O\overset{\overset{\displaystyle CH_3}{|}}{C}HCH_3$

7. (a) $H\overset{\overset{\displaystyle O}{\|}}{C}OCH_2CH_3$ (b) $CH_3CH_2\overset{\overset{\displaystyle O}{\|}}{C}OCH_2CH_3$

(c) $C_6H_5\overset{\overset{\displaystyle O}{\|}}{C}OCH_2CH_3$

8. (a) Methyl propanoate

(b) Propyl 3-methylpentanoate

9. (a) t-Butyl acetate   (b) Ethyl butyrate

10. (a) $CH_3OH + CH_3CO_2H$

(b) $(CH_3)_2CHOH + CH_3CH_2CO_2H$

(c) $CH_3CH_2CH_2OH + (CH_3)_2CHCO_2H$

11. (a) $C_6H_5OH + CH_3CO_2^-$ (Actually, the salt of phenol would form, because phenol neutralizes base.)

(b) $CH_3OH + {}^-O_2C-\langle\bigcirc\rangle-OCH_3$

Review Exercises, Chapter 15

15.1 (a) B, E   (b) A, C, D

(c) B   (d) None (E is an alkene and an alcohol)

15.5 (a) $CH_3CO_2^-$ (b) $CH_3CH_2CH_2\overset{\overset{\displaystyle CH_3}{|}}{C}HCH_2CHCO_2H$ with $\overset{}{\underset{\displaystyle Cl}{|}}$

(c) $HO_2CCH_2CH_2CH_2CO_2H$ (d) $CH_3\overset{\overset{\displaystyle CH_3}{|}}{C}H\overset{\overset{\displaystyle CH_3}{|}}{C}HCO_2^-$

15.7 (a) 2,2-Dimethylpropanoic acid

(b) 3-Methylpentanoic acid

(c) Potassium butanoate

(d) Sodium benzoate

15.9 *trans*-Butenedioic acid

15.13 B < C < D < A

15.15 C < A < D < D

15.16 (a) $\langle\bigcirc\rangle-OH + OH^-(aq) \rightarrow$

$\langle\bigcirc\rangle-O^- + H_2O$

(b) $H^+(aq) + OH^-(aq) \rightarrow H_2O$

(c) No reaction

(d) $CH_3CO_2H(aq) + OH^-(aq) \rightarrow$

$CH_3CO_2^-(aq) + H_2O$

15.19 (a) $CH_3CH_2CO_2^- + H^+ \rightarrow CH_3CH_2CO_2H$

(b) $^-O_2CCH_2CH_2CH_2CO_2^- + H^+ \rightarrow$

$HO_2CCH_2CH_2CH_2CO_2^-$

(c) $NH_3 + H^+ \rightarrow NH_4^+$

15.21 (a) $C_6H_5\overset{\overset{\displaystyle O}{\|}}{C}OH + HOCH_3$ (b) $C_6H_5\overset{\overset{\displaystyle O}{\|}}{C}Cl + HOCH_3$

(c) $C_6H_5\overset{\overset{\displaystyle O}{\|}}{C}O\overset{\overset{\displaystyle O}{\|}}{C}C_6H_5 + HOCH_3$

15.23 (a) $CH_3CH_2CO_2CH_3$ (b) $CH_3CH_2\overset{\overset{\displaystyle CH_3}{|}}{C}HCO_2CH_3$

(c) $Br-\langle\bigcirc\rangle-CO_2CH_3$ (d) $\langle\bigcirc\rangle\overset{\displaystyle CO_2CH_3}{\underset{\displaystyle CO_2CH_3}{}}$

15.26 The acid chloride has a more stable leaving group (the weakly basic Cl^- ion) than the ester, for which the leaving group is a very strongly basic anion of an alcohol.

15.28 (a) $CH_3CH_2\overset{\overset{\displaystyle O}{\|}}{C}OC(CH_3)_3$

(b) $CH_3CH_2\overset{\overset{\displaystyle H_3C}{|}}{C}H\overset{\overset{\displaystyle O}{\|}}{C}OCH(CH_3)_2$

15.30 (a) $CH_3CH_2\overset{\overset{\displaystyle O}{\|}}{C}OCH(CH_3)_2 + H_2O \xrightarrow{H^+}$

$CH_3CH_2\overset{\overset{\displaystyle O}{\|}}{C}OH + HOCH(CH_3)_2$

(b) $CH_3CH_2\overset{\overset{\displaystyle O}{\|}}{C}O-\langle\bigcirc\rangle + H_2O \xrightarrow{H^+}$

$CH_3CH_2\overset{\overset{\displaystyle O}{\|}}{C}OH + HO-\langle\bigcirc\rangle$

(c) No reaction   (d) No reaction

15.32 $HOCH_2\overset{\overset{}{\underset{\displaystyle OH}{|}}}{C}HCH_2OH + 3CH_3(CH_2)_{12}CO_2H$

15.33 (a) $CH_3CH_2\overset{\overset{\displaystyle O}{\|}}{C}ONa + HOCH(CH_3)_2$

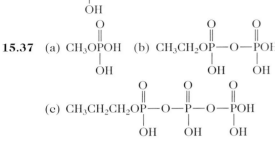

(b) CH_3CH_2CONa + HO—⬡

(c) No reaction   (d) No reaction

15.35 $HOCH_2CHCH_2OH$ + $3CH_3(CH_2)_{12}CO_2Na$
         |
         OH

15.37   (a) CH_3OPOH (b) CH_3CH_2OP—O—POH
                                      ‖   ‖   ‖
                              O   O   O   O
                                      | | |
                             OH   OH   OH

(c) $CH_3CH_2CH_2OP$—O—P—O—POH
                    ‖   ‖   ‖
                    O   O   O
                    | | |
                   OH   OH   OH

15.49   (a) $HCO_2H + H_2O \rightleftharpoons HCO_2^- + H_3O^+$
         (b) Toward the formate ion. The added OH^- (from NaOH) neutralizes H_3O^+ and so reduces the concentration of H_3O^+ in the equilibrium. The equilibrium thus must shift to the right to replace the lost H_3O^+.
         (c) $K_a = \dfrac{[HCO_2^-][H^+]}{[HCO_2H]}$
         (d) Stronger

15.52   (a) $HOCH_2CH_2CO_2^- + H^+ \rightarrow HOCH_2CH_2CO_2H$
         (b) No reaction
         (c) $C_6H_5O^- + H^+ \rightarrow C_6H_5OH$

15.55   Shift it to the right, because the stress in the equilibrium is the loss of a *product*. In accordance with Le Châtelier's principle, the equilibrium must shift in the direction that tries to replace this loss.

15.57 $HOCH_2CH_2CH_2CH_2CO_2H$

15.60   (a) 6.29 g methyl benzoate
         (b) 1.48 g CH_3OH; 1.88 ml CH_3OH
         (c) The reaction involves an equilibrium. When a large excess of methyl alcohol is used, the equilibrium shifts in accordance with Le Châtelier's principle so that essentially all of the benzoic acid is converted to the ester.

15.61   (a) CH_3OH + $HOCCH_2CH_3$
                            ‖
                            O
         (b) CH_3CONa + H_2O
                  ‖
                  O
         (c) $C_6H_5CO_2H$
         (d) CH_3—⬡
         (e) $CH_3CHCH_2CO_2Na$ + $HOCH_3$
                |
                CH_3
         (f) $CH_3CCH_2CH_3$ + H_2O
                  ‖
                  O

(g) $CH_3COCH_2CH_2CH_3$ + HCl
         ‖
         O
(h) CH_3CHO + $2HOCH_3$
(i) No reaction
(j) $C_6H_5CO_2CH_2CH_3$ + H_2O
(k) $CH_3CH_2CHCH_3$
                 |
                 Cl
(l) No reaction

CHAPTER 16

Practice Exercises, Chapter 16

1.   (a) Isopropyldimethylamine
     (b) Cyclohexylamine
     (c) *t*-Butylisobutylamine

2.   (a) $(CH_3)_3CNHCHCH_2CH_3$
                              |
                              CH_3
     (b) NO_2—⬡—NH_2
     (c) NH_2—⬡—CO_2H

3.   (a) $C_6H_5NH_3^+$ (b) $(CH_3)_3NH^+$ (c) $^+NH_3CH_2CH_2NH_3^+$

4.   (a) HO—⬡—$\overset{\overset{\displaystyle OH}{|}}{\underset{\underset{\displaystyle H}{|}}{C}}$—$CH_2$—$NHCH_3$
            (HO at top left)
     (b) CH_3O—⬡—$CH_2CH_2NH_2$
            (OCH_3 top and OCH_3 bottom)

5.   (a) 4-Methylhexanamide
     (b) 2-Ethylbutanamide

6.   (a) $(CH_3)_2CHCNHCH_3$ (b) $CH_3CNHC_6H_5$
                  ‖   ‖
                  O   O
     (c) No amide forms.   (d) No amide forms.

7.   (a) $C_6H_5CO_2H + NH_2CH_3$ (b) No hydrolysis occurs.
     (c) $C_6H_5NH_2 + HO_2CCH_3$
     (d) $NH_2CH_2CH_2NH_2 + 2CH_3CO_2H$

8. $NH_2CH_2CO_2H + NH_2CHCO_2H +$
                              |
                              CH_3
         $NH_2CHCO_2H + NH_2CHCO_2H$
               | |
             CH_3CH   CH_2SH
               |
              CH_3

Review Exercises, Chapter 16

16.2 (a) Although this is an aromatic *compound* (it has the benzene ring), it is an aliphatic *amine* because the amino group is not attached directly to the ring.
(b) Aromatic amine
(c) Amide, aliphatic, heterocyclic
(d) Ketone, amine, aliphatic, heterocyclic

16.4 (a) 1, alkene; 2, ester; 3, heterocyclic amine
(b) 1, heterocyclic amine; 2, alcohol; 3, ester
(c) 1, alkene; 2, alcohol; 3, ether; 4, heterocyclic amine
(d) 1, amide; 2, heterocyclic amine; 3, alkene; 4, (aliphatic, heterocyclic) amine

16.6 (a) Triethylammonium chloride
(b) Cyclohexylisopropylamine
(c) *t*-Butylamine   (d) Triisopropylamine

16.7 (a) $CH_3CH_2CH_2NH_3^+$ (b) $CH_3CH_2CH_2NH_2$
(c) No reaction   (d) No reaction

16.9 A is the stronger base; it is an amine (plus a ketone). B is an amide, and amides are neutral.

16.11 (a) Butanamide   (b) 3-Methylbutanamide

16.13 $C_6H_5CON(CH_3)_2$

16.15
$$CH_3\overset{O}{\overset{\|}{C}}Cl + 2NH_3 \rightarrow CH_3\overset{O}{\overset{\|}{C}}NH_2 + NH_4Cl$$

$$CH_3\overset{O}{\overset{\|}{C}}O\overset{O}{\overset{\|}{C}}CH_3 + 2NH_3 \rightarrow CH_3\overset{O}{\overset{\|}{C}}NH_2 + NH_4O\overset{O}{\overset{\|}{C}}CH_3$$

16.17 (a) $NH_2CH_2\overset{O}{\overset{\|}{C}}NHCH\overset{O}{\overset{\|}{C}}{-}$ (b) two
with CH_3 below

16.19 (a) $NH_2\underset{CH_3}{CH}CH_2CO_2H + NH_2CH_2CO_2H$
(b) $2NH_3 + HO_2CCH_2\underset{CH_3}{CH}CO_2H$
(c) $CH_3CHCH_2CH_2CH_2CO_2H$ with NH_2 below
(d) $2NH_3 + (H_2CO_3)$. The latter breaks up into $CO_2 + H_2O$.

16.31

16.34 Acetal (or ketal) group in carbohydrates; ester group in fats and oils; and the amide group in proteins.

16.35 (a) 1.66 g benzoic acid
(b) 28.3 mL 0.482 M HCl

16.36 (a) $CH_3CO_2H + CH_3OH$
(b) $CH_3\underset{OH}{CH}CH_2CH_3$
(c) No reaction
(d) $CH_3CH_2\overset{O}{\overset{\|}{C}}NH_2$
(e) No reaction
(f) No reaction
(g) $CH_3CHO + 2HOCH_2CH_3$
(h) $CH_3CH_2CO_2H$
(i) No reaction
(j) $CH_3CH_2\overset{O}{\overset{\|}{C}}OCH_3$
(k) $CH_3CH_2\overset{O}{\overset{\|}{C}}ONa + NH_3$
(l) $CH_3CH_2\underset{OCH_3}{CH}OCH_3$
(m) $CH_3CH_2SSCH_2CH_3$
(n) $C_6H_5CO_2Na + HOCH_2CH(CH_3)_2$
(o) $Cl^{-+}NH_3CH_2CH_2CH(CH_3)_2$
(p) $CH_3CH_2CH_2CH_2OCH_3$

16.38 (a) Methyl butanoate
(b) 1-Bromo-2-methylpropane
(c) 3-Methylbutanal
(d) 2-Pentene
(e) 2,2,4-Trimethylpentane
(f) 2-Methyl-3-hexanone
(g) 2-Methyl-2-propanol
(h) Pentanoic acid
(i) Sodium ethanoate

16.39 B. It is a carboxylic acid that will become an anion at the basic pH and so more soluble in water. (A is an ester and C is an amine.)

CHAPTER 17

Practice Exercises, Chapter 17

1. (a)

(b) $CH_3\overset{*}{C}H\underset{OH}{C}O_2H$

(c) $CH_3\underset{HO}{\overset{*}{C}H}\underset{NH_3^+}{\overset{*}{C}H}CO_2^-$

(d) $HOCH_2\underset{OH}{\overset{*}{C}H}{-}\underset{OH}{\overset{*}{C}H}{-}\underset{OH}{\overset{*}{C}H}\overset{O}{\overset{\|}{C}}H$

2. (a) 3   (b) 8   (c) 4

3. CH₃C̊HC̊HCH₃
         | |
        HO OH

    The two tetrahedral stereocenters are identical; they hold identical sets of four different groups, CH_3, OH, H, and $CH_3CH(OH)$.

Review Exercises, Chapter 17

17.2   (a) $CH_3CH_2CH_2OH$ and CH_3CHCH_3
                                           |
                                          OH

                     O   O
                     ‖   ‖
       (b) $CH_3CCH_2CH_2CH_3$ and $CH_3CH_2CCH_2CH_3$
       (c) $CH_2{=}CHCH_2CH_3$ and $CH_3CH{=}CHCH_3$

17.5   (a) Identical   (b) Identical
       (c) Stereoisomers   (d) Constitutional isomers

17.7   (a) Constitutional isomers

                                           O
                                         ‖
       (b) HOCH₂C̊H—C̊H—C̊H—C̈CH₂OH
                    | | |
                    OH   OH   OH

          fructose (open form)

                                        O
                                       ‖
      HOCH₂C̊H—C̊H—C̊H—C̊H—CH
                | | | |
               OH OH OH OH

          Glucose (open form)
       (c) In fructose, three different tetrahedral stereocenters. In glucose, four different such centers.
       (d) For fructose, 8 (2^3); for glucose, 16 (2^4)
       (e) For fructose, 4 pairs; for glucose, 8 pairs

17.9   (a) No; citric acid has no tetrahedral stereo-center.

              CH₂CO₂CH₃
                    |
       (b) HOC̊CO₂H
                    |
              CH₂CO₂H
       (The bottom carboxyl group, but not the middle one, might also have been chosen for showing it as a methyl ester.)

17.10   148.5 °C. The designations (+) and (−) placed before otherwise identical names tell us that the two compounds are enantiomers, and enantiomers have identical physical properties.

17.14   −0.375°

17.15   3.01 g/100 mL

17.17   Strychnine. The calculated specific rotation for the sample is −139°, which corresponds to the value for strychnine, not for brucine.

CHAPTER 18

Practice Exercises, Chapter 18

1. (a) a = b
        c = e
    (b) Compounds a and d are enantiomers.
    (c) Compound c (or e) is a meso compound.

2.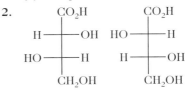

           a pair of enantiomers

    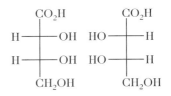

           a pair of enantiomers

Review Exercises, Chapter 18

18.5   (a) C   (b) D   (c) A
        (d) B and D

                         O
                         ‖
18.7   HOCH₂CHCH
                 |
               OH

        glyceraldehyde

18.9   Polysaccharide

18.11   (a) 2   (b) Trisaccharide

18.13   A is ruled out because it has one carbon holding two OH groups, an unstable system.

18.15   (a) 2 enantiomers

                  CHO   CHO
                         | |
     (b) H——OH   HO——H
                 CH₂OCH₃   CH₂OCH₃
     (c)   D   L

18.17   CH=O
             |
     HO—C—H
             |
      H—C—OH
             |
     HO—C—H
             |
     HO—C—H
             |
          CH₂OH

18.19

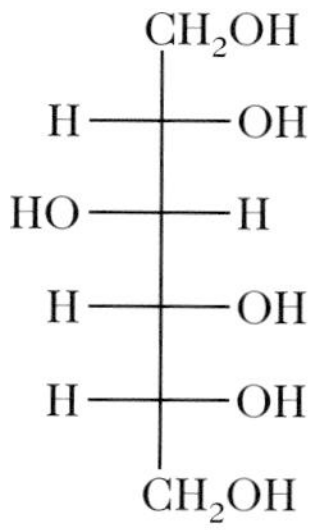

$$
\begin{array}{c}
CH_2OH \\
H{-}{-}OH \\
HO{-}{-}H \\
H{-}{-}OH \\
H{-}{-}OH \\
CH_2OH
\end{array}
$$

18.21 (a)

$$
\begin{array}{c}
CH_2OH \\
C{-}OH \\
C\ \ H\ \ H\ \ CH{=}O \\
HO\ \ C{-}C \\
OH\ \ OH
\end{array}
$$

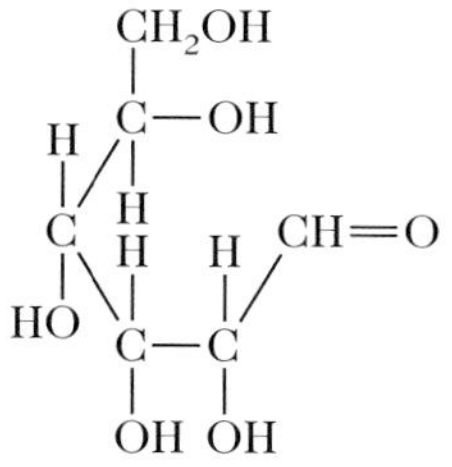

(b) At carbon 3.

(c) D-Family. The relative positions of the CH₂OH group and the O atom of the ring tell us that the compound is in the D-family.

(d) D-Allose

18.23

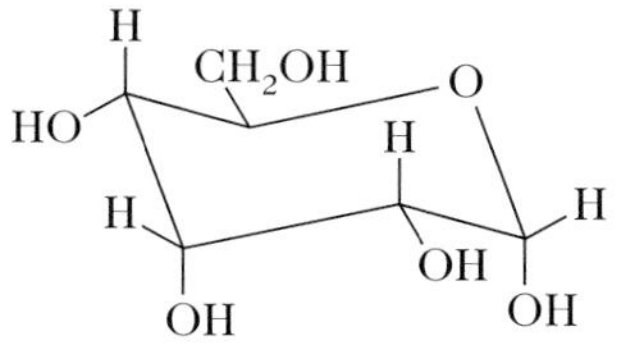

α-allose

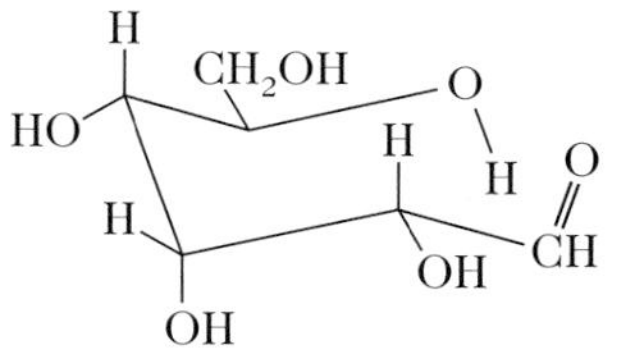

open form of allose

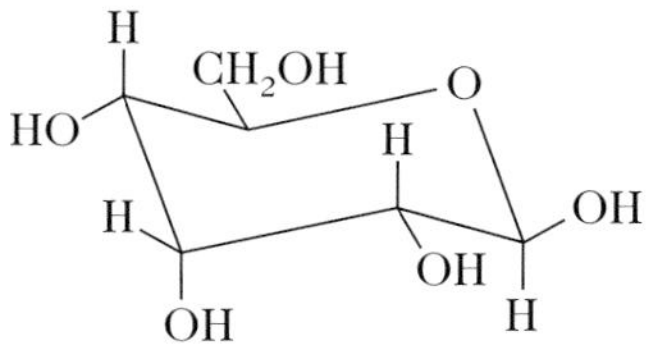

β-allose

18.26 HOCH₂ ... (furanose ring structure with H, H, OH, OH, OH)

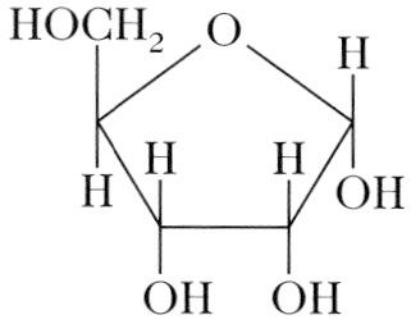

18.28 No, an OH group is required at position 4 to make possible the formation of the five-membered ring.

18.30

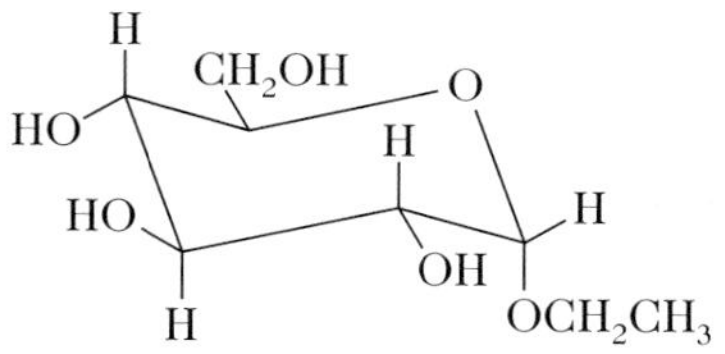

ethyl α-glucoside

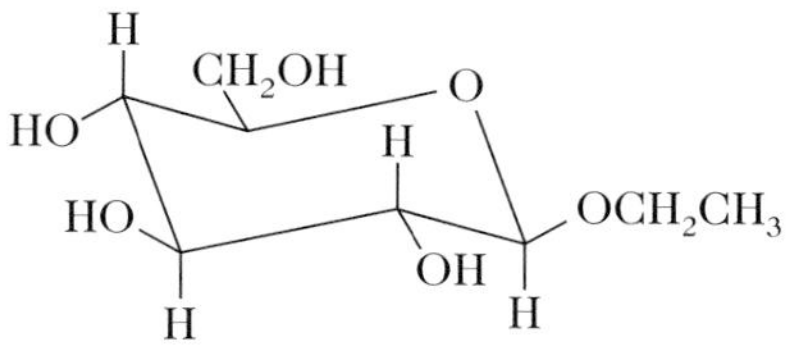

ethyl β-glucoside

Another kind (diastereomers, Interaction 17.1).

18.35 (a) Yes, see arrow

(b) Yes, see enclosure

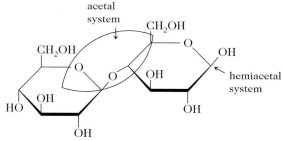

(c) A β(1 → 4) bridge

(d) Yes, it has the hemiacetal system so the open form of the corresponding ring (on the right) has an aldehyde group.

(e) Maltose has an α(1 → 4) bridge between the two rings.

(f) Two glucose molecules

18.37

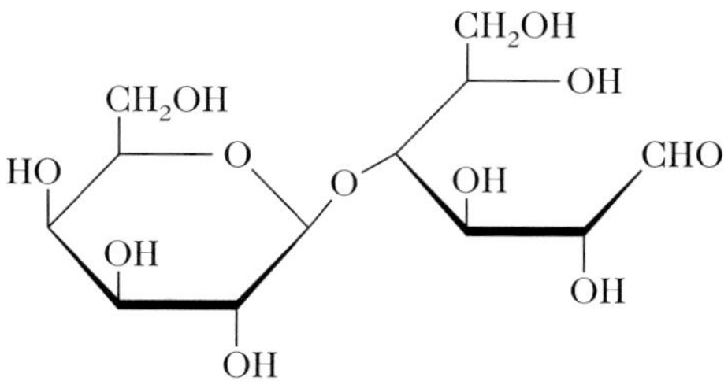

(disaccharide ring structure)

18.56

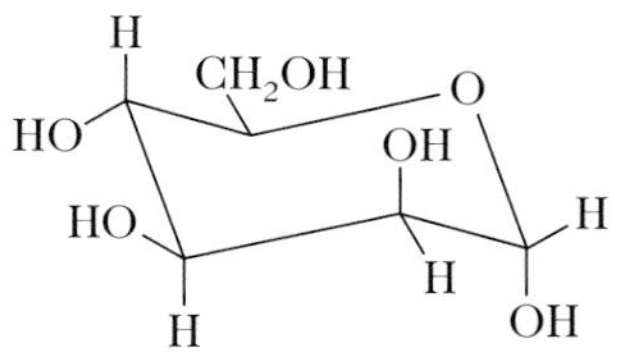

α-mannose

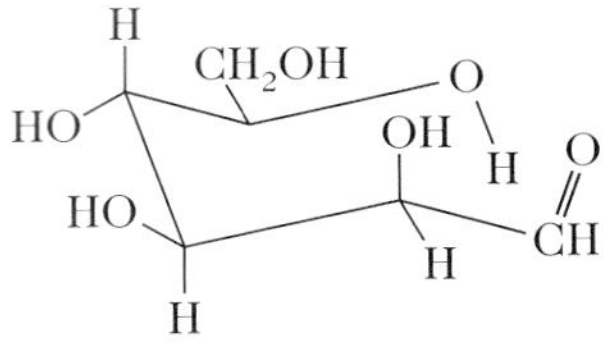

open form of mannose

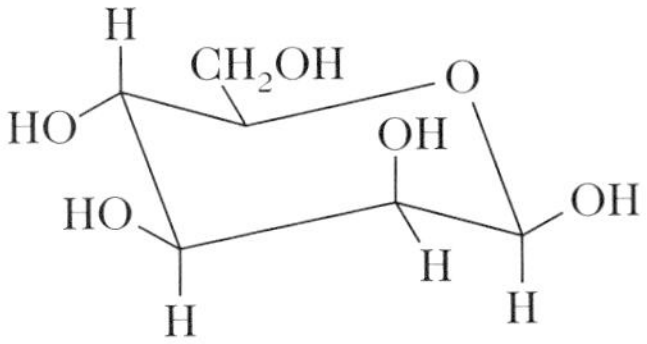

β-mannose

18.58 (a) No, it has no hemiacetal or hemiketal system and so cannot give a Tollens' or a Benedict's test.
(b) No, for the same reason given in (a).
(c) Two molecules of glucose

18.61 5

CHAPTER 19

Practice Exercises, Chapter 19

1. 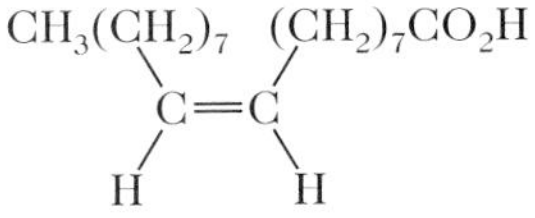 $CH_3(CH_2)_7$ $(CH_2)_7CO_2H$
    $\underset{H}{\overset{}{C}}=\underset{H}{\overset{}{C}}$

2. $CH_3(CH_2)_{26}CO_2(CH_2)_{25}CH_3$

3. $1 + 3NaOH \longrightarrow$

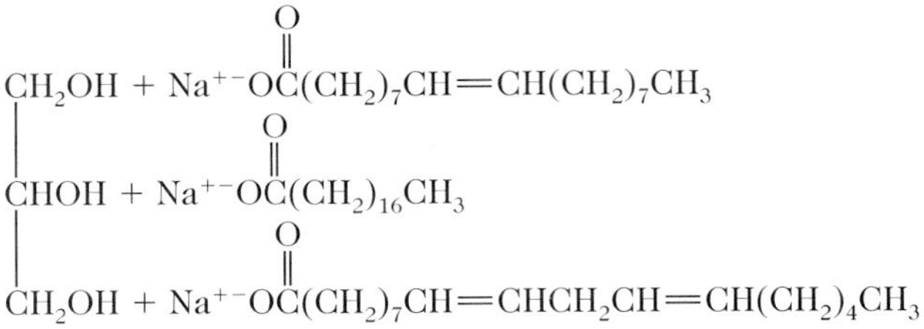

    $CH_2OH + Na^{+-}OC(CH_2)_7CH=CH(CH_2)_7CH_3$
    | O
    $CHOH + Na^{+-}OC(CH_2)_{16}CH_3$
    | O
    $CH_2OH + Na^{+-}OC(CH_2)_7CH=CHCH_2CH=CH(CH_2)_4CH_3$

4. 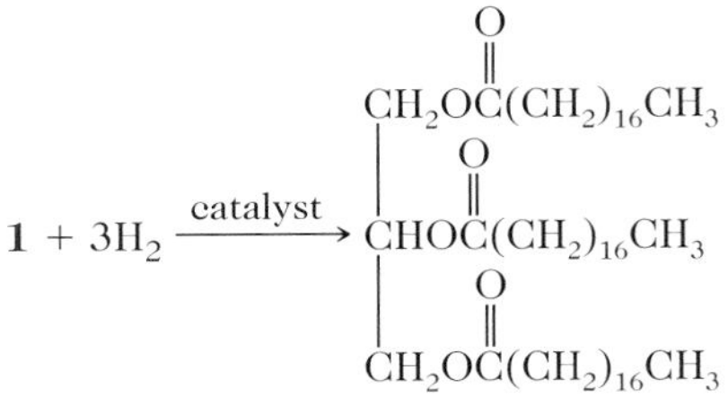 $1 + 3H_2 \xrightarrow{\text{catalyst}}$
    $CH_2OC(CH_2)_{16}CH_3$
    | O
    $CHOC(CH_2)_{16}CH_3$
    | O
    $CH_2OC(CH_2)_{16}CH_3$

Review Exercises, Chapter 19

19.1 It is not obtainable from living plants or animals.

19.3 It is soluble in water, and it isn't present in plant or animal sources.

19.5 Palmitic acid, $CH_3(CH_2)_{14}CO_2H$
Stearic acid, $CH_3(CH_2)_{16}CO_2H$

19.7 (a) $CH_3(CH_2)_{12}CO_2H + NaOH \longrightarrow$
$CH_3(CH_2)_{12}CO_2^-Na^+ + H_2O$
(b) $CH_3(CH_2)_{12}CO_2H + CH_3OH \xrightarrow{H^+}$
$CH_3(CH_2)_{12}CO_2CH_3 + H_2O$

19.8 The organic products of the reactions are the following.
(a) $CH_3(CH_2)_7CH-CH(CH_2)_7CO_2H$
              | |
              Br   Br
(b) $CH_3(CH_2)_7CH=CH(CH_2)_7CO_2^-Na^+$
(c) $CH_3(CH_2)_{16}CO_2H$
(d) $CH_3(CH_2)_7CH=CH(CH_2)_7CO_2CH_3$

19.11
              O
              ‖
$CH_2OC(CH_2)_7CH=CHCH_2CH=CHCH_2CH=CHCH_2CH_3$
| O
| ‖
$CHOC(CH_2)_7CH=CHCH_2CH=CH(CH_2)_4CH_3$
| O
| ‖
$CH_2OC(CH_2)_{14}CH_3$

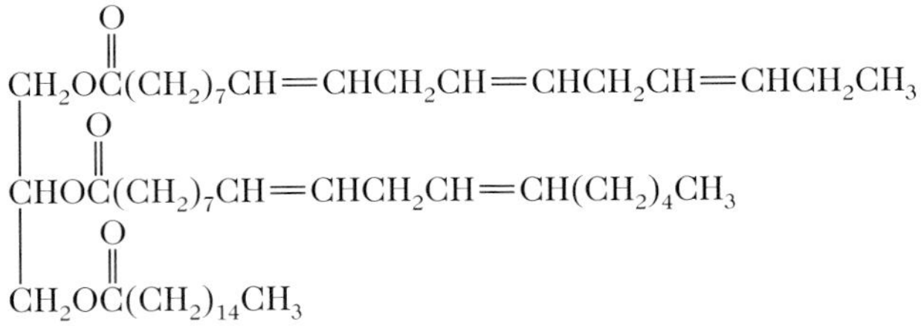

19.13 $HOCH_2CHCH_2OH$
                |
                OH
+ $HO_2C(CH_2)_7CH=CHCH_2CH=CH(CH_2)_4CH_3$
+ $HO_2C(CH_2)_{14}CH_3$ + $HO_2C(CH_2)_7CH=CH(CH_2)_7CH_3$

19.14 $HOCH_2CHCH_2OH$
                |
                OH
+ $NaO_2C(CH_2)_7CH=CHCH_2CH=CH(CH_2)_4CH_3$
+ $NaO_2C(CH_2)_{14}CH_3$ + $NaO_2C(CH_2)_7CH=CH(CH_2)_7CH_3$

19.16 Only one structure is possible if the molecule is to be chiral (have a tetrahedral stereocenter indicated by the asterisk).
              O
              ‖
$CH_2OC(CH_2)_7CH=CH(CH_2)_7CH_3$
| O
| ‖
$*CHOC(CH_2)_{16}CH_3$
| O
| ‖
$CH_2OC(CH_2)_{16}CH_3$

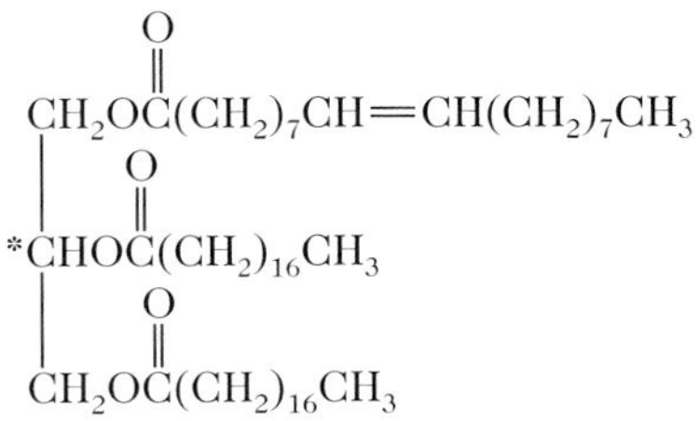

19.21 $CH_3(CH_2)_{16}CO_2(CH_2)_{17}CH_3$

19.30 Glycosidic links. The linkage involves the hemiacetal carbon of the sugar ring. The glycosidic link is more easily hydrolyzed (being an acetal, not an ordinary ether).

19.31 (a)

$$CH_2OC(CH_2)_7CH=CHCH_2CH=CHCH_2CH=CHCH_2CH_3$$

*$CHOC(CH_2)_7CH=CH(CH_2)_7CH_3$

$CH_2OPOCH_2CH_2\overset{+}{N}(CH_3)_3$

O^-

   (b) A glycerophospholipid, because it is based on glycerol, not sphingosine.
   (c) Yes, the asterisk in the structure of part (a) marks the tetrahedral stereocenter.
   (d) A lecithin, because its hydrolysis would give 2-(trimethylamino)ethanol.

19.42 The water-avoiding properties of the hydrophobic units and the water-attracting properties of the hydrophilic units

19.55 (a) Yes; it has both hydrophobic sections (hydrocarbon like) and polar groups (OH).
   (b) No, steroids have *three* six-membered rings plus a five-membered ring.

CHAPTER 20

Practice Exercises, Chapter 20

1. Glycine $^+NH_3CH_2CO_2^-$
   Alanine $^+NH_3CHCO_2^-$
                 CH_3
   Lysine $^+NH_3CHCO_2^-$
                 $CH_2CH_2CH_2CH_2NH_2$
   Glutamic acid $^+NH_3CHCO_2^-$
                 $CH_2CH_2CO_2H$

2. (a) $^+NH_3CHCO_2^-$ (b) $^+NH_3CHCO_2^-$
      $CH_2CO_2^-$ CH_2CONH_2

3. $^+NH_3CHCO_2^-$ NH_2^+
      $CH_2CH_2CH_2NHCNH_2$

4. Hydrophilic; neutral. (The side chain has an amide group, not an amino group.)

5. $^+NH_3CHC-NHCHCO_2^-$ $^+NH_3CHC-NHCHCO_2^-$
      CH_3 CH_2 CH_2 CH_3
             CH_2 CH_2
             CO_2H CO_2H

   Ala-Glu   Glu-Ala

Review Exercises, Chapter 20

20.1 B. Its NH_3^+ group is not on the same carbon that holds the CO_2^- group.

20.2 (a) $-NHCHC-$
             $CH_2CH(CH_3)_2$
   (b) Leucine, Leu
   (c) Hydrophobic

20.4 $NH_2CHCO_2^-$
         CH_3

20.7 A. It has amine-like groups that can both donate hydrogen bonds to water molecules and accept them. (B has an alkyl group side chain, which is hydrophobic.)

20.9 (a) At a pH of 10. At the more basic pH, all protons that can be donated to the base from the amino acid have left the molecule leaving it with a net charge of 1−.
   (b) To the anode

20.11 Oxidizing agent

20.13 CO_2^-
         $^+NH_3$——H
            CH_2OH

20.14 A has an amide bond not to the amino group of the α-position of an amino acid unit but to an amino group of a side chain (that of lysine). B has a proper peptide bond.

20.16 $^+NH_3CHC-NHCHCO^-$
             H $(CH_2)_2CO_2H$
          Gly-Glu

       $^+NH_3CHC——NHCHCO^-$
          $(CH_2)_2CO_2H$ H
          Glu-Gly

20.18 Gly-Cys-Ala   Cys-Ala-Gly   Ala-Gly-Cys
       Gly-Ala-Cys   Cys-Gly-Ala   Ala-Cys-Gly

20.20
$^+NH_3CHC-NHCHC-NHCHC-NHCHC-NHCHCO^-$
CH_3CHCH_3 $CH_2C_6H_5$ CH_3 H $CH_2CH(CH_3)_2$

20.22 (a) A
   (b) B; it has only hydrophilic side chains. All those in A are hydrophobic.

20.24   Gly-Cys-Ala
|
Gly-Cys-Ala

20.27   Primary structure

20.29   Tertiary

20.31   Quaternary

20.33   It is a right-handed helix stabilized by hydrogen bonds between carbonyl oxygen atoms and H atoms on N atoms farther down the helix. The side chains project to the outside of the helix.

20.35   It aids in the hydroxylation of proline and lysine residues, without which collagen is not adequately made.

20.37   Covalent linkages fashioned from lysine side chains.

20.39   No, they represent portions of the secondary structure of a polypeptide and often both features are present.

20.41   The force of attraction between a site bearing a full negative charge (e.g., a CO_2^- group on a glutamic acid or aspartic acid side chain) and a site with a full positive charge (e.g., a NH_3^+ group on a lysine side chain).

20.43   It consists of more than one polypeptide associated together in a specific way, each with primary, secondary, and tertiary structure.

20.45 $^+NH_3CHCO_2^-$ + $^+NH_3CHCO_2^-$ + $^+NH_3CHCO_2^-$
| | |
CH_2OH CH_3 CH_3CHCH_3

+ $^+NH_3CHCO_2^-$ + $^+NH_3CH_2CO_2^-$
|
$(CH_2)_4NH_2$

20.49   A change in the value of something, like concentration, from one place to another.

20.51   Cell fluid

20.53   It moves sodium and potassium ions through membranes against their concentration gradients in order to reestablish these gradients.

20.55   Gap junctions are tubules made of proteins and fastened between cells that provide avenues for the direct movements of ions and molecules from one cell to another.

20.57   A unit of a monosaccharide molecule.

20.59   It is the generic name of all polysaccharides.

20.61   A shock-absorbing gel-like material made of glycosaminoglycans and found in cartilage and other extracellular spaces that hold fibrous proteins

20.63   The resiliency of ground substance depends on the hydrogen bonds increasing the "stickiness" of the molecules of ground substance and their abilities to hold large amounts of water as water of hydration.

20.65   Fibrous proteins are insoluble in water; globular proteins are more soluble.

20.67   They both have strengthening functions in tissue; both are fibrous proteins. The action of hot water on collagen turns it to gelatin, but elastin is unaffected in this way.

20.69   Fibrin is the protein that forms a blood clot. Fibrinogen is changed to fibrin by the clotting mechanism.

20.78

$$^+NH_3CHC-NHCHC-NHCHC-NHCHC-NHCHCO_2^-$$
with O double bonds above each C, and CH_3 below each CH

CHAPTER 21

Practice Exercises, Chapter 21

1.   (a) Sucrose   (b) Glucose   (c) Protein   (d) An ester

2.   Feedback inhibition

Review Exercises, Chapter 21

21.3   (a) An apoenzyme is the wholly polypeptide part of the enzyme.
(b) A cofactor is a nonpolypeptide molecule or ion needed to make the complete enzyme.
(c) A coenzyme is one kind of cofactor, an organic ion or molecule.

21.5   It catalyzes the rapid reestablishment of the equilibrium after it has been disturbed.

21.8 CH_3CHCH_3 (with OH) + $NAD^+ \rightarrow CH_3CCH_3$ (with O) + $NAD:H + H^+$

21.9 $H^+ + NADH + FAD \rightarrow NAD^+ + FADH_2$

21.12   Lactose is a disaccharide and the substrate for the enzyme, lactase.

21.17   By the necessity of the fitting of the substrate molecule to the surface of the enzyme much as a key must fit to a particular lock.

21.20   (a) $V \propto [E_o]$ (b) $V \propto [S]$

21.21   The value of $[S]$ at which the reaction rate is one-half of the maximum rate.

21.22   At another site. *Allosteric* describes an action induced at a site on an enzyme molecule at some distance from the active site.

21.23   The enzyme has more than one active site and that the (slower) activation of one site automatically causes the activation of the other(s).

21.25   An effector binds allosterically to the enzyme (by binding at a place other than any of the catalytically active sites) and induces changes in shape that activate these sites.

21.28   At higher Ca^{2+} concentration, $Ca_3(PO_4)_2$ would precipitate.

21.30 Ca^{2+} converts them to activated effectors.

21.32   When a zymogen is cleaved properly, an active enzyme emerges. Trypsinogen is the zymogen for trypsin.

21.34   A proteolytic, blood clot–dissolving enzyme. It normally circulates in its inactive form, plasminogen.

21.36   The inhibitor is a nonsubstrate molecule resembling the true substrate enough to enable the binding of the inhibitor to the enzyme. By thus occupying the active site, it inhibits the enzyme's work.

21.38   Because it shuts down a pathway when it is no longer needed but lets the pathway occur when it is needed.

21.40   (a) It binds to a metal ion cofactor and so deactivates the enzyme.
(b) It denatures enzymes by combining with their SH groups.
(c) They deactivate enzymes of the nervous system.

21.42   It inhibits an enzyme needed for the growth of bacteria.

21.44   The CK(*MB*) band originates in the leakage of this isoenzyme only from damaged heart muscle.

21.47   To help to diagnose mild heart attacks. When a troponin subunit is found in the blood it indicates that cells of heart muscle have been injured or are dying.

21.49   Streptokinase and tissue plasminogen activator (TPA). TPA occurs in human blood.

21.51   TPA initiates the dissolving of a blood clot whether the clot is responsible for a heart attack or for an ischemic stroke.

21.63   Competitive inhibition.

CHAPTER 22

Review Exercises, Chapter 22

22.1   (a) Endocrine glands   (b) Nerve cells

22.3   They are primary chemical messengers.

22.5   G-protein (actually one of the three subunits of the G-protein)

22.7   It is an enzyme activator.

22.9   It is started in the same general way, up to and including the step in which the G-protein does its work.

22.11   Steroids, polypeptides, simple amino compounds, and local hormones (prostaglandins).

22.13   (a) Glucose   (b) amino acids   (c) metal ions

22.15   It is hydrolyzed back to acetic acid and choline. The enzyme is acetylcholine esterase. Nerve poisons inactivate this enzyme.

22.17   It prevents the release of acetylcholine from the presynaptic neuron.

22.19   Iproniazid inhibits the monoamine oxidases and thus lets norepinephrine work at a higher level of activity.

22.21   It inhibits the reabsorption of serotonin.

22.23   They bind to dopamine receptors in the postsynaptic nerve and inhibit the action of dopamine.

22.25   Degenerated neurons can use L-DOPA to make dopamine.

22.27   The signal-inhibiting activity of GABA is enhanced by Valium and Librium.

22.29   When the flow of Ca^{2+} into cells of heart muscles is reduced, the heart beats with reduced vigor.

22.31   It consists of very tiny, gaseous molecules that easily slip through cell membranes.

22.33   It binds to Fe^{2+} ion in a heme unit that is part of the enzyme.

CHAPTER 23

Review Exercises, Chapter 23

23.2   Saliva, gastric juice, pancreatic juice, and intestinal juice.

23.4   Histamine.

23.6   (a) α-Amylase.
(b) Pepsinogen and gastric lipase.
(c) α-Amylase, lipase, nuclease, trypsinogen, chymotrypsinogen, procarboxypeptidase, and proelastase.
(d) No enzymes.
(e) Amylase, aminopeptidase, sucrase, lactase, maltase, lipase, nucleases, enteropeptidase.

23.8   (a) Amino acids.
(b) Glucose, fructose, and galactose.
(c) Fatty acids and monoacylglycerols (plus some diacylglycerols).

23.10   It catalyzes the conversion of trypsinogen to trypsin. Then trypsin catalyzes the conversion of other zymogens to chymotrypsin, carboxypeptidase, and elastase. Thus enteropeptidase turns on enzyme activity for three major protein-digesting enzymes.

23.12   They are surface-active agents that help to break up lipid globules, wash lipids from the particles of food, and aid in the absorption of fat-soluble vitamins.

23.14   (a) HCl   (b) Enteropeptidase
(c) Trypsin   (d) Trypsin
(e) Trypsin

23.16   This enzyme is inactive at the high acidity of the digesting mixture in the adult stomach, but the acidity of this mixture in the infant's stomach is less. Thus, infants start lipid digestion in their stomachs.

23.18 Dilute sodium bicarbonate released from the pancreas. This raises the pH of the chyme to the optimum pH for the action of the enzymes that will function in the duodenum.

23.20 The flow of bile normally delivers colored breakdown products from hemoglobin in the blood, and these products give the normal color to feces. When no bile flows, no colored products are available to the feces.

23.22 The concentration of soluble proteins is greater in blood.

23.24 Fibrinogen is a protein in blood that is changed to fibrin, the insoluble protein of a blood clot, by the clotting mechanism.

23.26 Na^+ is in blood plasma and other extracellular fluids; K^+ is chiefly in intracellular fluids. The two ions help to maintain osmotic pressure relationships; are part of the regulatory mechanisms for acid-base balance; and participate in the smooth working of the muscles and the nervous system.
23.27 ~ 145 meq/L

23.29 Hypermagnesemia and cardiac arrest.

23.31 In bones and teeth

23.33 Hypercalcemia.

23.35 Blood pressure and osmotic pressure. Blood pressure tends to force blood fluids out of the blood vessel and osmotic pressure tends to force fluids back. The return of fluids to the blood from the interstitial compartment on the arterial side is overbalanced by the blood pressure. The net effect on the arterial side is a diffusion of fluids from the blood.

23.37 Serum proteins are lost from the blood, which upsets the osmotic pressure of the blood. Water leaves the blood for the interstitial compartment, and the blood volume drops. Loss of blood delivery to the brain leads to the symptoms of shock.

23.38 The first oxygen molecule to bind changes the shapes of other parts of the hemoglobin molecule and makes it much easier for the remaining three oxygen molecules to bind. This ensures that all four oxygen-binding sites of each hemoglobin molecule will leave the lungs fully loaded with oxygen.

23.39 $HHb + O_2 \rightleftharpoons HbO_2^- + H^+$
(a) To the left   (b) To the left
(c) To the right   (d) To the left
(e) To the left   (f) To the right

23.40
(a) $HHb + O_2 \longleftarrow HbO_2^- + H^+$ ⎫ isohydric shift in
   $CO_2 + H_2O \longrightarrow HCO_3^- + H^+$ ⎭ metabolizing tissue
(b) $HHb + O_2 \longrightarrow HbO_2^- + H^+$ ⎫ isohydric shift
   $CO_2 + H_2O \longleftarrow HCO_3^- + H^+$ ⎭ in alveolus

23.42 Waste CO_2 combines with water to give HCO_3^- and

the H^+ that is needed to react with HbO_2^- to form HHb and release O_2.

23.44 In red cells.
(a) It catalyzes the conversion of HCO_3^- and H^+ to CO_2 and H_2O.
(b) It catalyzes the conversion of CO_2 and H_2O to HCO_3^- and H^+. It can do both because it accelerates *both* the forward and the reverse reactions in the following equilibrium.

$$CO_2 + H_2O \rightleftharpoons HCO_3^- + H^+$$

Other factors, such as the value of pCO_2, determine whether the forward or the reverse reaction is favored.

23.46 It migrates into a cavity within the hemoglobin molecule and helps to change the shapes of subunits so that oxygen molecules are more easily ejected.

23.48 For oxygenation:

$$HHbBPG + O_2 + HCO_3^- \longrightarrow$$
$$HbO_2^- + BPG + CO_2 + H_2O$$

For deoxygenation:

$$HbO_2^- + BPG + CO_2 + H_2O \longrightarrow$$
$$HHbBPG + O_2 + HCO_3^-$$

23.50 Oxygen affinity is lowered. Where the partial pressure of CO_2 is relatively high (as in actively metabolizing tissue) there is a need for oxygen, so the lowering effect of CO_2 on oxygen affinity helps to release O_2 precisely where O_2 is most needed.

23.52 Myoglobin can take oxygen from oxyhemoglobin and thus ensure that the oxygen needs of myoglobin-containing tissue are met.

23.55 The pH of the blood decreases in both but both pCO_2 and $[HCO_3^-]$ increase in respiratory acidosis and both decrease in metabolic acidosis.

23.56 Hyperventilation is observed in metabolic acidosis, and HCO_3^- can be given intravenously to neutralize excess acid. Hyperventilation is also observed in respiratory alkalosis (because the patient can't help hyperventilating), and CO_2 is given (by rebreathing one's own air) to keep up the supply of H_2CO_3, which can neutralize excess base.

23.57 Hypoventilation is observed in metabolic alkalosis, and isotonic ammonium chloride can be given to neutralize the excess base. Involuntary hypoventilation is observed in respiratory acidosis, and isotonic sodium bicarbonate might be given to neutralize excess acid.

23.58 In metabolic acidosis, because it helps to blow out CO_2 and thereby to reduce the level of H^+ in the blood and simultaneously raise the pH.

23.59 In respiratory alkalosis. The involuntary loss of

CO_2 reduces the level of H_2CO_3 in the blood and thereby reduces the level of H^+.

23.61 In respiratory acidosis.

23.63 (a) Respiratory alkalosis   (b) Metabolic alkalosis
(c) Respiratory acidosis   (d) Respiratory acidosis
(e) Metabolic acidosis   (f) Respiratory alkalosis
(g) Metabolic acidosis   (h) Metabolic alkalosis
(i) Respiratory acidosis   (j) Respiratory acidosis

23.64 (a) Hyperventilation   (b) Hypoventilation
(c) Hypoventilation   (d) Hypoventilation
(e) Hyperventilation   (f) Hyperventilation
(g) Hyperventilation   (h) Hypoventilation
(i) Hypoventilation   (j) Hypoventilation

23.66 Hypoventilation in emphysema lets the blood retain carbon dioxide, and the pH decreases.

23.69 Hypocapnia

23.71 The blood has become more concentrated in solutes.

23.73 Aldosterone is secreted from the adrenal cortex, and it instructs the kidneys to retain sodium ion in the blood.

23.75 The rate of diuresis increases.

23.80 The blood produced while at a high altitude has a higher concentration of hemoglobin and of BPG. This aids in their ability to use oxygen during a race.

CHAPTER 24

Practice Exercises, Chapter 24

1. (a) Proline   (b) Arginine
(c) Glutamic acid   (d) Lysine

2. (a) Serine   (b) CT (chain termination)
(c) Glutamic acid   (d) Isoleucine

Review Exercises, Chapter 24

24.1 (a) Cytosol   (b) Protoplasm
(c) Cytoplasm   (d) Ribosome
(e) Mitochondrion   (f) Deoxyribonucleic acid
(g) Chromatin   (h) Histone
(i) Gene

24.3 The cell nucleus.

24.10 In the sequence of bases attached to the deoxyribose units of the main chain.

24.11 The main chains all have the same phosphate–ribose–phosphate–ribose repeating system.

24.13 A and T pair to each other, so they must be in a 1 : 1 ratio regardless of the species. Similarly, G and C pair to each other and must be in a 1 : 1 ratio.

24.16
$$5' \rightarrow 3'$$
Given segment: TCAGCCT
Opposite segment: AGTCGGA
$$3' \leftarrow 5'$$

24.19 The introns are b, d, and f, because they are the longer segments.

24.26 ATA. A codon is RNA material, and T does not occur in RNA.

24.27 (a) Phenylalanine   (b) Serine
(c) Threonine   (d) Aspartic acid

24.28 Writing them in the 5′ to 3′ direction:
(a) AAA   (b) GGA
(c) UGU   (d) AUC

24.29 (a) UAUCUUAUAGAGUCCCCAACAGAU
$$5' \rightarrow 3'$$
(b) UAUUCCACAGAU
(c) Tyr-Ser-Thr-Asp
(d) Phe-Ser-Thr-Asp

24.32 (a) A large number of sequences are possible because three of the specific amino acid residues are coded by more than one codon. The possibilities are indicated by:
Met—Ala—His—Ser—Tyr

AUG	GCU	CAU	UCU	UAU	$5' \rightarrow 3'$
	GCC	CAC	UCC	UAC	
	GCA		UCA		
	GCG		UCG		
			AGU		
			AGC		

(b) CAU ($5' \rightarrow 3'$) or if ($3' \rightarrow 5'$), then UAC

24.34 How polypeptide synthesis can be controlled by the use of repressors and inducers.

24.37 It cancels the effect of a repressor.

24.45 A molecular feature on the surface of a virus particle fits by a flexible lock-and-key mechanism to a specific glycoprotein on the membrane of one specific kind of host cell.

24.48 Reverse transcriptase, a DNA polymerase enzyme, directs the synthesis of viral DNA that subsequently is used to code for the synthesis of more viral RNA. Reverse transcriptase is able to use RNA information to make DNA.

24.50 It must not only make (+)RNA to direct multiplication, it must also make the (−)RNA that new virus particles require.

24.52 That the cell is able to move information backward, from RNA to DNA. (Normally information always flows from DNA to RNA.)

24.57 The theory is that AZT molecules bind to reverse transcriptase and inhibit the work of this enzyme in HIV.

24.58 They inhibit the breaking up of large proteins, a process in the normal life cycle of HIV.

24.59 The HIV in a specific infected person will not be able to develop resistance to more than one drug simultaneously.

24.60 DNA made by joining DNA segments from two different species.

24.62 Restriction enzymes.

24.64 The cloning of genes or the products of genes and their use.

24.66 Heat separates the strands of duplex DNA.

24.70 To insert correct DNA by some mechanism into cells lacking such DNA and so to repair the defect.

24.72 The surface carries the recognition molecules needed to find the appropriate host cells.

24.74 Specific gene therapies can be implemented to correct genetic defects.

24.85 (a) By one methyl group
(b) No
(c) Both uracil and thymine can form a base pair with adenine.

24.87 (a) A dinucleotide; it has two side-chain bases and two ribose units.
(b) Of RNA, because the sugar units are those of ribose
(c) At the top
(d) AC

CHAPTER 25

Practice Exerxise, Chapter 25

1. (a) Yes   (b) Yes   (c) No

Review Exercises, Chapter 25

25.2 Carbon dioxide and water

25.4 Carbon dioxide and water

25.6
$$
\text{Adenosine}-\text{O}-\overset{\overset{\text{O}}{\|}}{\underset{\underset{\text{O}^-}{|}}{\text{P}}}-\text{O}-\overset{\overset{\text{O}}{\|}}{\underset{\underset{\text{O}^-}{|}}{\text{P}}}-\text{O}^-
$$

ADP

$$
\text{Adenosine}-\text{O}-\overset{\overset{\text{O}}{\|}}{\underset{\underset{\text{O}^-}{|}}{\text{P}}}-\text{O}^-
$$

AMP

25.8 The relative potential that phosphate X has for donating a phosphate unit in the synthesis of another organophosphate compound. The higher the number is in a negative sense, the higher is this potential.

25.10 The right

25.11 (a) No
(b) Yes
(c) No
(d) No

25.13 It stores phosphate group energy and transfers phosphate to ADP to remake the ATP consumed by muscular work.

25.15 (a) The aerobic synthesis of ATP
(b) The synthesis of ATP when a tissue operates anaerobically
(c) The supply of metabolites for the respiratory chain
(d) The supply of metabolites for the citric acid cycle and the respiratory chain.

25.17 An increase in its supply of ADP. The need to convert ADP back to ATP is met by metabolism, which requires oxygen.

25.19 (a) $B < E < C < A < D$
(b) $D < C < A < B$

25.22 It starts with glucose and ends with lactate.

25.26 Two

25.28 ADP. The cycle helps the cell make ATP from ADP, so activation by ADP is logical.

25.30 9

25.31 An oxidation; the reactant loses hydrogen.

25.33 A pair of electrons on the left of the arrow.

25.35
$$
\underset{\underset{}{}}{\overset{\overset{\text{OH}}{|}}{\text{CH}_3\text{CHCO}_2^-}} + \text{NAD}^+ \longrightarrow
$$
$$
\overset{\overset{\text{O}}{\|}}{\text{CH}_3\text{CCO}_2^-} + \text{NAD:H} + \text{H}^+
$$
(a) $\overset{\overset{\text{O}}{\|}}{\text{CH}_3\text{CCO}_2^-}$
(b) NAD^+

25.37 $\text{NAD}^+ < \text{FMN} < \text{FeSP} < \text{CoQ}$

25.39 It is a riboflavin-containing coenzyme that in its reduced form, FADH_2, passes electrons and H^+ into the respiratory chain.

25.41 The flow of protons across the inner mitochondrial membrane.

25.43 $MH_2 + nH^+ + \frac{1}{2}O_2 \rightarrow M + H_2O + nH^+$ (outside the inner mitochondrial membrane)

25.45 A gradient of positive charge. The migration of *any* cation away from the region of higher positive charge density or the migration of *any* anion toward this region will be spontaneous.

25.47 (a) $\text{CH}_3\text{CH}{=}\text{O}$
(b) Acetic acid, $\text{CH}_3\text{CO}_2\text{H}$ (or acetate ion, CH_3CO_2^-)
(c) α-Ketoglutarate

CHAPTER 26

Review Exercises, Chapter 26

26.2   After a few steps, the metabolic pathways of galactose and fructose merge with the pathway of glucose.

26.4   70–110 mg/dL (3.9–6.1 mmol/L). (Note. Various references give slightly different ranges of values.)

26.6   (a) Glucose in urine.
(b) A low blood sugar level.
(c) A high blood sugar level.
(d) The conversion of glycogen to glucose.
(e) The concentration of something (e.g., glucose) in blood above which that solute appears in the urine.
(f) The synthesis of glycogen from glucose.

26.8   The lack of glucose means the lack of the one nutrient most needed by the brain.

26.9   It increases.

26.11   The enzyme adenylyl cyclase.

26.12   D < C < B < A < E

26.14   Glucose might be changed back to glycogen as rapidly as it is released from glycogen, and no glucose would be made available to the cell.

26.16   Glucose-1-phosphate is the end product, and phosphoglucomutase catalyzes its change to glucose-6-phosphate.

26.18   It is a polypeptide hormone made in the alpha cells of the pancreas and released into circulation when the blood sugar level drops. At the liver it activates adenylyl cyclase, which leads to glycogenolysis and the release of glucose into circulation.

26.20   It stimulates the release of glucagon, which leads to the release of glucose into circulation.

26.22   An increase in the blood sugar level.

26.24   Somatostatin is a polypeptide hormone released by the hypothalamus, and it acts at the pancreas to inhibit the release of glucagon and to slow down the release of insulin.

26.25   The body's ability to manage dietary glucose without letting the blood sugar level swing too widely from its normal fasting level.

26.28   Some is catabolized in the liver to provide the energy to convert the rest, via gluconeogenesis, to glucose.

26.32 $C_6H_{12}O_6 + 2ADP + 2P_i \rightarrow 2C_3H_5O_3^- + 2H^+ + 2ATP$

26.35   (a) It undergoes oxidative decarboxylation and becomes the acetyl group in acetyl CoA.
(b) Its keto group is reduced by NADH to a 2° alcohol group in lactate, which enables the NADH to be reoxidized to NAD^+ and then reused for more glycolysis.

26.37   (a) 17 ATP   (b) 18 ATP

26.39   It makes glucose out of smaller molecules obtained by the catabolism of fatty acids and amino acids.

26.41   (a) Alanine   (b) Aspartic acid

26.43   Succinyl units are in the citric acid cycle, which ends with the formation of oxaloacetate, and the latter can be used in gluconeogenesis.

26.54   (a) Yes, either pyruvate or lactate containing carbon-13 may reenter circulation.
(b) Yes, either pyruvate or lactate containing carbon-13 might be absorbed by the liver from the bloodstream.
(c) Yes, glucose with carbon-13 atoms might be made via gluconeogenesis from either pyruvate or lactate containing carbon-13.

CHAPTER 27

Review Exercises, Chapter 27

27.4   They unload some of their triacylglycerol.

27.8   Triacylglycerol.

27.9   The loss of the less dense triacylglycerol leaves a higher concentration of the more dense cholesterol.

27.11   Cholesterol.

27.13   IDL and LDL.

27.15   Return to the liver any cholesterol that extrahepatic tissue cannot use.

27.18   2 < 1 < 3

27.20   Insulin suppresses the lipase needed to hydrolyze triacylglycerols in storage prior to the release of their fatty acids into circulation.

27.21   B < D < A < C < E

27.23   The citric acid cycle processes the acetyl units manufactured by the β-oxidation pathway and so fuels the respiratory chain.

27.25   An increase in the blood sugar level triggers the release of insulin which inhibits the release of fatty acids from adipose fat.

27.28
$$CH_3CH_2CH_2\overset{\overset{O}{\|}}{C}SCoA + FAD \rightarrow$$
$$CH_3CH{=}CH\overset{\overset{O}{\|}}{C}SCoA + FADH_2$$
$$CH_3CH{=}CH\overset{\overset{O}{\|}}{C}SCoA + H_2O \rightarrow CH_3\overset{\overset{OH}{|}}{C}HCH_2\overset{\overset{O}{\|}}{C}SCoA$$

$$
\underset{\text{OH}}{\overset{\text{O}}{\text{CH}_3\text{CHCH}_2\text{CSCoA}}} + \text{NAD}^+ \rightarrow
$$

$$
\underset{\text{O}}{\overset{\text{O}}{\text{CH}_3\text{CCH}_2\text{CSCoA}}} + \text{NAD:H} + \text{H}^+
$$

$$
\overset{\text{O}\quad\text{O}}{\text{CH}_3\text{CCH}_2\text{CSCoA}} + \text{CoASH} \rightarrow 2\,\overset{\text{O}}{\text{CH}_3\text{CSCoA}}
$$

No more turns of the β-oxidation pathway are possible.

27.30 NADH passes its hydrogen into the respiratory chain and is changed back to NAD$^+$.

27.32 (a) Inside mitochondria.
(b) Cytosol.

27.34 The pentose phosphate pathway of glucose catabolism.

27.36 Cholesterol inhibits the synthesis of HMG-CoA reductase.

27.38 Oxaloacetate is the carrier of acetyl groups in the citric acid cycle, so its loss means that the acetyl CoA level increases.

27.40 A proton or hydrogen ion, H$^+$. If the level of hydrogen ion increases, the problem is acidosis.

27.47 Acetoacetate.

27.49 Their *over*production leads to acidosis.

27.66 112

CHAPTER 28

Review Exercises, Chapter 28

28.3 Infancy

28.5 Glutamic acid (glutamate)

28.7 $\text{C}_6\text{H}_5\text{CH}_2\overset{\text{O}}{\text{C}}\text{CO}_2\text{H}$

28.9 Yes

28.11 No

28.13 $4 < 1 < 5 < 2 < 3$

28.15 Yes: glucose $\xrightarrow[\text{glycolysis}]{\text{aerobic}}$ pyruvate $\xrightarrow{\text{transamination}}$ alanine

28.17 HO—⬡—CH$_2$CH$_2$NH$_2$

 tyramine

28.19 To synthesize glucose by means of gluconeogenesis

28.20 (a) Originally, the amino groups of amino acids.
(b) Carbon dioxide.

28.23 Hyperammonemia. Step 2 consumes carbamoyl phosphate, which is made using ammonia. If car-

bamoyl phosphate levels rise, a backup occurs to cause ammonia levels to increase.

28.26 $3 < 2 < 1 < 4 < 5 < 7 < 6$

28.30 Serine → pyruvate → acetyl CoA → (fatty acid synthesis) → palmitic acid

28.32 $\underset{\text{OH}}{\overset{\text{NH}_2}{\text{CH}_3\text{CCO}_2^-}}$

The carbon holding the NH group of the imine carries a partial positive charge because of the electronegativity of the NH group. Therefore, the partially negatively charged oxygen of a water molecule would go to this carbon rather than to the N at the other end of the imine double bond (which itself has a partial negative charge, too).

CHAPTER 29

Review Exercises, Chapter 29

29.2 It is any compound needed for health.

29.4 Dietetics is the application of the findings of nutrition to the feeding of individuals, whether ill or well.

29.6 1. People with chronic diseases
2. People who must take special medications
3. Prematurely born infants
4. Pregnant women
5. Lactating women
6. People involved in strenuous physical activity
7. People exposed for prolonged periods to high temperatures

29.8 (a) The body must make its own glucose, which can lead to a buildup of harmful substances.
(b) It lacks the essential fatty acids and it makes the absorption of the fat-soluble vitamins more difficult.

29.10 Linolenic acid and arachidonic acid.

29.12 The body can make several amino acids itself.

29.14 It breaks them down, eliminates the nitrogen, and converts other parts to fatty acids.

29.16 From an animal source

29.18 The proportions of essential amino acids available from it

29.20 The essential amino acid most poorly supplied by the protein

29.22 (a) 1.3×10^2 g
(b) 1.7×10^3 g
(c) 6.1×10^3 kcal
(d) Very likely not, since 1.7×10^3 g is nearly 4 lb.

29.24 They are needed in much more than trace amounts, and they come from proteins.

29.26 No single vegetable source has a balanced supply of essential amino acids.

29.28 Vitamin D

29.30 Vitamin A

29.32 Vitamin K

29.34 Vitamin E. (Vitamin C works mostly in fluids.)

29.36 Vitamin C

29.38 Vitamin C

29.40 Thiamine

29.42 Folate

29.44 Calcium, Ca^{2+}; phosphorus, P_i (chiefly, the mix of HPO_4^{2-} and $H_2PO_4^-$ plus some PO_4^{3-} that exists at body pH and in bone); magnesium, Mg^{2+}; sodium, Na^+; potassium, K^+; and chloride, Cl^-.

29.46 Goiter; mental retardation

29.48 (a) 79 g
(b) 3.0×10^2 g
(c) 8.5×10^2 kcal
(d) Because 3.0×10^2 g of peanuts is about 2/3 lb, a child could probably get this much down.

GLOSSARY[1]

Absolute configuration The actual arrangement in space about each tetrahedral stereocenter in a molecule. (18.3)

Absolute zero The coldest temperature attainable; 0 K or $-273.15\,°C$. (1.4)

Accuracy In science, the degree of conformity to some accepted standard or reference; freedom from error or mistake; correctness. (1.4)

Acetal Any organic compound in which two ether-like linkages extend from one CH unit. (14.5)

Acetyl coenzyme A The molecule from which acetyl groups are transferred into the citric acid cycle or into the synthesis of fatty acids. (25.1)

Achiral Not possessing chirality; that quality of a molecule (or other object) that allows it to be superimposed on its mirror image. (17.2)

Acid *Brønsted theory:* Any substance that can donate a proton (H^+). (3.2, 4.6, 8.2)

Acid anhydride In organic chemistry, a compound formed by splitting water out between two OH groups of the acid functions of two organic acids. (14.3)

Acid–base indicator (See *Indicator.*)

Acid–base neutralization The reaction of an acid with a base. (3.2)

Acid chloride A derivative of a carboxylic acid in which the OH group of the acid has been replaced by Cl. (14.3)

Acid derivative Any organic compound that can be made from an organic acid or that can be changed back to the acid by hydrolysis. (Examples are acid chlorides, acid anhydrides, esters, and amides.) (14.3)

Acidic solution A solution in which the molar concentration of hydronium ions is greater than that of hydroxide ions. (8.2, 9.5)

Acid ionization constant (K_a) A modified equilibrium constant for the following equilibrium:

$$HA + H_2O \rightleftharpoons H_3O^+ + A^-$$

$$K_a = \frac{[H_3O^+][A^-]}{[HA]} \qquad (9.6)$$

Acidosis A condition in which the pH of the blood is below normal. *Metabolic acidosis* is brought on by a defect in

some metabolic pathway. *Respiratory acidosis* is caused by a defect in the respiratory centers or in the mechanisms of breathing. (9.9, 23.4, 27.6)

Acid rain Rain made acidic by the presence of air pollutants such as oxides of sulfur and nitrogen. (Interaction 9.2)

Active transport The movement of a substance through a biological membrane against a concentration gradient and caused by energy consuming chemical changes that involve parts of the membrane. (20.8)

Activity series A list of elements (or other substances) in the order of the ease with which they release electrons under standard conditions and become oxidized. (8.3)

Acyl group

$$R\!-\!\overset{\overset{\displaystyle O}{\|}}{C}\!- \qquad (15.3)$$

acyl group

Acyl group transfer reaction Any reaction in which an acyl group transfers from a donor to an acceptor. (15.3)

Addition reaction Any reaction in which two parts of a reactant molecule add to a double or a triple bond. (11.4)

Adenosine diphosphate (ADP) A high-energy diphosphate ester obtained from adenosine triphosphate (ATP) when part of the chemical energy in ATP is tapped for some purpose in a cell. (25.1)

Adenosine monophosphate (AMP) A low-energy phosphate ester that can be obtained by the hydrolysis of ATP or ADP; a monomer for the biosynthesis of nucleic acids. (25.1)

Adenosine triphosphate (ATP) A high-energy triphosphate ester used in living systems to provide chemical energy for metabolic needs. (25.1)

Adequate protein A protein that, when digested, makes available all of the essential amino acids in suitable proportions to satisfy both the amino acid and total nitrogen requirements of good nutrition without providing excessive calories. (29.2)

ADP (See *Adenosine diphosphate.*)

Aerobic sequence An oxygen-consuming sequence of catabolism that starts with glucose and proceeds through glycolysis, the citric acid cycle, and the respiratory chain. (25.1, 26.3)

Agonist A compound whose molecules can bind to a recep-

---

[1] The entries in this Glossary include the terms that appear in boldface within the chapters, including the margin comments, as well as several additional entries. The numbers in parentheses following the definitions are the section numbers (or Interactions) where the entry is introduced or discussed.

tor on a cell membrane and cause a response by the cell. (22.2)

Albumin One of a family of globular proteins that tend to dissolve in water, and that in blood contribute to the blood's colloidal osmotic pressure and aid in the transport of metal ions, fatty acids, cholesterol, triacylglycerols, and other water-insoluble substances. (20.9, 23.2)

Alcohol Any organic compound whose molecules have the OH group attached to a saturated carbon; ROH. (12.1)

Alcohol group The OH group when it is joined to a saturated carbon. (12.1)

Aldehyde An organic compound that has a carbonyl group joined to H on one side and C on the other. (14.1)

Aldehyde group $-CH{=}O$ (14.1)

Aldohexose A monosaccharide whose molecules have six carbon atoms and an aldehyde group. (18.2)

Aldose A monosaccharide whose molecules have an aldehyde group. (18.2)

Aldosterone A steroid hormone, made in the adrenal cortex, secreted into the bloodstream when the sodium ion level is low, and that signals the kidneys to leave sodium ions in the bloodstream. (23.5)

Aliphatic compound Any organic compound whose molecules lack a benzene ring or a similar structural feature. (11.5)

Alkali metals The elements of group IA of the periodic table—lithium, sodium, potassium, rubidium, cesium, and francium. (3.2)

Alkaline earth metals The elements of group IIA of the periodic table—beryllium, magnesium, calcium, strontium, barium, and radium. (3.2)

Alkaloid A physiologically active, heterocyclic amine isolated from plants. (16.2)

Alkalosis A condition in which the pH of the blood is above normal. *Metabolic alkalosis* is caused by a defect in metabolism. *Respiratory alkalosis* is caused by a defect in the respiratory centers of the brain or in the apparatus of breathing. (9.9, 23.4)

Alkane A saturated hydrocarbon, one that has only single bonds. A *normal* alkane is any whose molecules have straight chains. (11.5)

Alkene A hydrocarbon whose molecules have one or more double bonds. (11.5, 12.1)

Alkyl group A substituent group that is an alkane minus one H atom. (11.6)

Alkyne A hydrocarbon whose molecules have triple bonds. (11.5, 12.1)

Allosteric activation The activation of an enzyme's catalytic site by the binding of some molecule at a position elsewhere on the enzyme. (22.3)

Allosteric inhibition The inhibition of the activity of an enzyme caused by the binding of an inhibitor molecule at some site other than the enzyme's catalytic site. (21.4)

Alloy A mixture of two or more metals made by mixing them in their molten states. (2.1)

Alpha (α) particle The nucleus of a helium atom; ^4_2He. (10.1)

Alpha (α) radiation A stream of high-energy alpha particles. (10.1)

Amide An organic compound whose molecules have a carbonyl-to-nitrogen single bond. (15.3, 16.3)

Amide bond The single bond that holds the carbonyl group to the nitrogen atom in an amide. (16.3)

Amine An organic compound whose molecules have a trivalent nitrogen atom, as in $R-NH_2$, $R-NH-R$, or R_3N. (16.1)

Amine salt An organic compound whose molecules have a positively charged, tetravalent, protonated nitrogen atom, as in RNH_3^+, $R_2NH_2^+$, or R_3NH^+. (16.2)

Amino acid Any organic compound whose molecules have both an amino group and a carboxyl group. (20.1)

Amino acid residue A structural unit in a polypeptide,

$$-NH-\underset{\overset{|}{R}}{CH}-\overset{\overset{O}{\|}}{C}-$$

furnished by an α-amino acid, where R is the side chain group of a particular amino acid. (20.1)

AMP (See *Adenosine monophosphate*.)

Amphipathic compound A substance whose molecules have both hydrophilic and hydrophobic groups. (19.5)

Anaerobic sequence The oxygen-independent catabolism of glucose to lactate ion. (25.1, 26.3)

Anhydrous Without water. (7.2)

Anion A negatively charged ion. (4.1)

Anode The positive electrode to which negatively charged ions (anions) are attracted during electrolysis. (8.1)

Anoxia A condition of a tissue in which it receives no oxygen. (23.4)

Antagonist A compound that can bind to a membrane receptor but not cause any response by the cell. (22.2)

Antibiotics Antimetabolites made by bacteria and fungi. (21.4)

Anticodon A sequence of three adjacent side-chain bases on a molecule of tRNA that is complementary to a codon and that fits to its codon on an mRNA chain during polypeptide synthesis. (24.3)

Antimetabolite A substance that inhibits the growth of bacteria. (21.4)

Apoenzyme The wholly polypeptide part of an enzyme. (21.1)

Aromatic compound Any organic compound whose molecules have a benzene ring (or a feature very similar to this). (11.5, 12.1)

Atmosphere, standard (See *Standard atmosphere.*)

Atom A small particle with one nucleus and zero charge; the smallest particle of a given element that bears the chemical properties of the element. (2.2, 3.1)

Atomic mass The average mass, in atomic mass units (u), of the atoms of the isotopes of a given element as they occur naturally. (3.1, 5.2)

Atomic mass number (See *Mass number.*)

Atomic mass unit (u) $1.6605665 \times 10^{-24}$ g. A mass very close to that of a proton or a neutron. (3.1)

Atomic number The positive charge on an atom's nucleus; the number of protons in an atom's nucleus. (3.1)

Atomic orbital A region in space close to an atom's nucleus in which one or two electrons can reside. (3.4)

Atomic symbol A one- or two-letter symbol for an element or one of its atoms. (2.2)

Atomic weight (See *Atomic mass.*)

ATP (See *Adenosine triphosphate.*)

Aufbau principle A principle regarding the construction of electron configurations: As each additional proton is located in an atomic nucleus, an electron enters whichever of the available orbitals corresponds to the lowest energy. Hund's rule and the Pauli exclusion principle govern the term "available orbitals of lowest energy." (3.4)

Avogadro's number 6.02×10^{23}. The number of formula units in one mole of any element or compound. (5.1)

Avogadro's principle Equal volumes of gases contain equal numbers of moles when they are compared at identical temperatures and pressures. (6.3)

Background radiation Cosmic rays plus the natural atomic radiation emitted by the traces of radioactive isotopes in soils and rocks plus any radiation that escapes from the operations of nuclear facilities. (10.2)

Balanced equation (See *Equation, balanced.*)

Barometer An instrument for measuring atmospheric pressure. (6.1)

Basal activities The minimum activities of the body needed to maintain muscle tone, control body temperature, circulate the blood, handle wastes, breathe, and carry out other essential activities. (Interaction 2.1)

Basal metabolic rate The rate at which energy is expended to maintain basal activities. (Interaction 2.1)

Basal metabolism The total of all of the chemical reactions that support basal activities. (Interaction 2.1)

Base Any acid neutralizer. *Brønsted theory:* A proton acceptor; a compound that neutralizes hydrogen ions. (3.2, 4.6, 8.2)

Base, heterocyclic A heterocyclic amine obtained from the hydrolysis of nucleic acids: adenine, thymine, guanine, cytosine, or uracil. (24.2)

Base ionization constant (K_b) For the equilibrium (where B is some base)

$$B + H_2O \rightleftharpoons BH^+ + OH^-$$

$$K_b = \frac{[BH^+][OH^-]}{[B]} \quad (9.7, 15.2)$$

Base pairing In nucleic acid chemistry, the association by means of hydrogen bonds of two heterocyclic, side-chain bases—adenine with thymine (or uracil) and guanine with cytosine. (24.2)

Base quantity A fundamental quantity of physical measurement such as mass, length, and time; a quantity used to define derived quantities such as mass/volume for density. (1.4)

Base unit A fundamental unit of measurement for a base quantity—such as the kilogram for mass, the meter for length, the second for time, the kelvin for temperature degree, and the mole for quantity of chemical substance; a unit to which derived units of measurement are related. (1.4)

Basic solution A solution in which the molar concentration of hydroxide ions is greater than that of hydronium ions. (8.2, 9.5)

Becquerel (Bq) The SI unit for the activity of a radioactive source; one nuclear disintegration (or other transformation) per second. (10.3)

Benedict's reagent A solution of copper(II) sulfate, sodium citrate, and sodium carbonate that is used in the Benedict's test. (14.3)

Benedict's test The use of Benedict's reagent to detect the presence of any compound whose molecules have easily oxidized functional groups—α-hydroxyaldehydes and α-hydroxyketones—such as those present in monosaccharides. In a positive test the intensely blue color of the reagent disappears and a reddish precipitate of copper(I) oxide separates. (14.3)

Beta oxidation The catabolism of a fatty acid by a series of repeating steps that produce acetyl units (in acetyl CoA); the fatty acid cycle of catabolism. (25.1, 27.3)

Beta (β) particle A high-energy electron emitted from a nucleus, $_{-1}^{0}e$. (10.1)

Beta (β) radiation A stream of high-energy electrons. (10.1)

Bile A secretion of the gall bladder that empties into the upper intestine and furnishes bile salts; a route of excretion for cholesterol and bile pigments. (23.1)

Bile pigment Colored products of the partial catabolism of heme that are transferred from the liver to the gall bladder for secretion via the bile.   (28.4)

Bile salts Steroid-based detergents in bile that emulsify fats and oils during digestion.   (23.1)

Bilin The brownish pigment that is the end product of the catabolism of heme and that contributes to the characteristic colors of feces and urine.   (28.4)

Bilinogen A product of the catabolism of heme that contributes to the characteristic colors of feces and urine; some of it is oxidized to bilin.   (28.4)

Bilirubin A reddish-orange substance that forms from biliverdin during the catabolism of heme and that enters the intestinal tract via the bile and is eventually changed into bilinogen and bilin.   (28.4)

Biliverdin A greenish pigment that forms when partly catabolized hemoglobin (as verdohemoglobin) is further broken down and that is changed in the liver to bilirubin.   (28.4)

Binary compound A compound made from two elements.   (3.2)

Biochemistry The study of the structures and properties of substances found in living systems.   (18.1)

Biological value In nutrition, the percentage of the nitrogen of ingested protein that is absorbed from the digestive tract and retained by the body when the total protein intake is less than normally required.   (29.2)

Biotin A water-soluble vitamin needed to make enzymes used in fatty acid synthesis.   (29.3)

2,3-Bisphosphoglycerate (BPG) An organic ion that nestles within the hemoglobin molecule in deoxygenated blood but is expelled from the hemoglobin molecule during oxygenation.   (23.3)

Blood sugar The carbohydrates—mostly glucose—that are present in blood.   (18.2)

Blood sugar level The concentration of carbohydrate—mostly glucose—in the blood; usually stated in units of mg/dL.   (26.1)

Boat form A conformation of a six-membered ring that resembles a boat.   (Interaction 18.2)

Bohr effect The stimulation of hemoglobin to bind oxygen caused by the removal (neutralization) of the hydrogen ion released by oxygen binding.   (23.3)

Bohr model of the atom The solar system model of the structure of an atom, proposed by Niels Bohr, that pictures the electrons circling the nucleus in discrete energy states called orbits.   (3.3)

Boiling The turbulent behavior in a liquid when its vapor pressure equals the atmospheric pressure and when the liquid absorbs heat while experiencing no increase in temperature.   (6.7)

Boiling point, normal The temperature at which a substance boils when the atmospheric pressure is 760 mm Hg (1 atm).   (2.4, 6.7)

Bond, chemical A net electrical force of attraction that holds atomic nuclei near each other within compounds.   (4.1)

Boron family Group IIIA of the periodic table: boron, aluminum, gallium, indium, and thallium.   (3.2)

Boyle's law (See *Pressure–volume law.*)

BPG (See *2,3-Bisphosphoglycerate.*)

Branched chain A sequence of atoms to which additional atoms are attached at points other than the ends.   (11.2)

Brønsted theory An acid is a proton donor and a base is a proton acceptor.   (8.2)

Brownian movement The random, chaotic movements of particles in a colloidal dispersion that can be seen with a microscope.   (7.1)

Buffer A combination of solutes that holds the pH of a solution relatively constant even if small amounts of acids or bases are added.   (9.9)

Butyl group $CH_3CH_2CH_2CH_2$—   (11.6)

sec-Butyl group $CH_3CH_2CH(CH_3)$—   (11.6)

t-Butyl group $(CH_3)_3C$—   (11.6)

Calorie The amount of heat that raises the temperature of 1 g of water by 1 degree Celsius from 14.5 °C to 15.5 °C.   (2.4)

Carbaminohemoglobin Hemoglobin that carries chemically bound carbon dioxide.   (23.3)

Carbocation Any cation in which a carbon atom has just six outer-level electrons.   (12.5)

Carbohydrate Any naturally occurring substance whose molecules are polyhydroxyaldehydes or polyhydroxyketones or can be hydrolyzed to such compounds.   (18.2)

Carbon family The group IVA elements in the periodic table—carbon, silicon, germanium, tin, and lead.   (3.2)

Carbonate buffer A mixture or a solution that includes bicarbonate ions and dissolved carbon dioxide in which the bicarbonate ion can neutralize added acid and carbon dioxide can neutralize added base.   (9.9)

Carbonyl group The atoms carbon and oxygen joined by a double bond, C=O.   (14.1)

Carboxylic acid A compound whose molecules have the carboxyl group, CO_2H.   (15.1)

Carcinogen A chemical or physical agent that induces the onset of cancer or the formation of a tumor that may or may not become cancerous.   (10.2)

Cardiovascular compartment The entire blood-carrying network of the body, including the heart.   (23.2)

Catabolism The reactions of metabolism that break molecules down.   (25.1)

Catalysis The phenomenon of an increase in the rate of a chemical reaction brought about by a relatively small amount of a chemical—the catalyst—that is not permanently changed by the reaction.   (9.1)

Catalyst A substance that is able, in relatively low concentrations, to accelerate the rate of a chemical reaction without itself being permanently changed. (In living systems, the catalysts are called enzymes.)   (9.1)

Cathode The negative electrode to which positively charged ions—cations—are attracted during electrolysis.   (8.1)

Cation A positively charged ion.   (4.1)

Centimeter (cm) A length equal to one-hundredth of the meter.

$$1 \text{ cm} = 0.01 \text{ m} = 0.394 \text{ in.}   (1.4)$$

Chair form A conformation of a six-membered ring that resembles a chair.   (Interaction 18.2)

Charles' law (See *Temperature–volume law.*)

Chemical bond (See *Bond, chemical.*)

Chemical energy The potential energy that substances have because their arrangements of electrons and atomic nuclei are not as stable as are alternative arrangements that become possible in chemical reactions.   (2.3)

Chemical equation A shorthand representation of a chemical reaction that uses formulas instead of names for reactants and products, that separates reactant formulas from product formulas by an arrow, that separates formulas on either side of the arrow by plus signs, and that expresses the mole proportions of the chemicals by simple numbers (coefficients) placed before the formulas.   (2.2, 5.3)

Chemical property Any chemical reaction that a substance can undergo and the ability to undergo such a reaction.   (1.3)

Chemical reaction Any event in which substances change into different chemical substances.   (1.3, 2.1)

Chemiosmotic theory An explanation of how oxidative phosphorylation is related to a flow of protons in a proton gradient established by the respiratory chain, a gradient that extends across the inner membrane of a mitochondrion.   (25.3)

Chemistry The study of the compositions and structures of substances and their ability to change into other substances.   (1.3)

Chiral Having handedness in a molecular structure. (See also *Chirality.*)   (17.2)

Chiral carbon (See *Tetrahedral stereocenter.*)

Chirality The quality of handedness that a molecular structure has that prevents this structure from being superimposable on its mirror image.   (17.2)

Chloride shift An interchange of chloride ions and bicar-

bonate ions between a red blood cell and the surrounding blood serum.   (23.3)

Choline A compound needed to make complex lipids and acetylcholine; classified as a vitamin.   (29.3)

Chromosome Small threadlike bodies in a cell nucleus that carry genes in a linear array and that are microscopically visible during cell division.   (24.1)

Citric acid cycle A series of reactions that dismantles acetyl units and sends electrons (and protons) into the respiratory chain; a major source of metabolites for the respiratory chain.   (25.1, 25.2)

Codon A sequence of three adjacent side chain bases in a molecule of mRNA that codes for a specific amino acid residue when the mRNA participates in polypeptide synthesis.   (24.3)

Coefficient of digestibility The proportion of an ingested protein's nitrogen that enters circulation rather than elimination (in feces); the difference between the nitrogen ingested and the nitrogen in the feces divided by the nitrogen ingested.   (29.2)

Coefficients Numbers placed before formulas in chemical equations to indicate the mole proportions of reactants and products.   (2.2, 5.3)

Coenzyme An organic compound needed to make a complete enzyme from an apoenzyme.   (21.1)

Cofactor A nonprotein compound or ion that is an essential part of an enzyme.   (21.1)

Collagen The fibrous protein of connective tissue that changes to gelatin in boiling water.   (20.6, 20.9)

Colligative property A property of a solution that depends only on the concentrations of the solute and the solvent and not on their chemical identities (e.g., osmotic pressure).   (7.5)

Collision theory A theory about the rates of chemical reactions that postulates collisions between reacting particles.   (9.2)

Colloidal dispersion A relatively stable, uniform distribution in some dispersing medium of colloidal particles—those with at least one dimension between 1 and 1000 nm.   (7.1)

Colloidal osmotic pressure The contribution made to the osmotic pressure of a solution by substances colloidally dispersed in it.   (7.5)

Combined gas law (See *General gas law.*)

Common ion effect The reduction in the solubility of a salt in some solution by the addition of another solute that furnishes one of the ions of this salt.   (8.5)

Competitive inhibition The inhibition of an enzyme by the binding of a molecule that can compete with the substrate for the occupation of the catalytic site.   (21.4)

Complex (See *Complex ion.*)

Complex ion A combination of a metal ion with one or

more ligands—negatively charged or neutral electron-rich species.   (14.3)

Compound A substance made from the atoms of two or more elements that are present in a definite proportion by mass and by atoms.   (2.1)

Concentration The quantity of some component of a mixture in a unit of volume or a unit of mass of the mixture.   (5.5)

Condensation The physical change of a substance from its gaseous state to its liquid state.   (6.7)

Condensed structure (See *Structural formula.*)

Conformation One of the infinite number of contortions of a molecule that are permitted by free rotations around single bonds.   (11.2)

Conjugate acid–base pair Two particles whose formulas differ by only one H^+, such as NH_4^+ and NH_3, or HCl and Cl^-.   (8.4)

Constitutional isomerism The existence of two or more compounds with identical molecular formulas but different atom-to-atom sequences.   (11.3, 17.1)

Constitutional isomers Compounds with identical molecular formulas but different atom-to-atom sequences.   (17.1)

Conversion factor A fraction that expresses a relationship between quantities that have different units, such as 2.54 cm/in.   (1.4)

Coordinate covalent bond A covalent bond in which both of the electrons of the shared pair have originated from one of the atoms involved in the bond.   (4.6)

Cori cycle The sequence of chemical events and transfers of substances in the body that describes the distribution, storage, and mobilization of blood sugar, including the reconversion of lactate to glycogen.   (26.2)

Cosmic radiation A stream of ionizing radiations from the sun and outer space that consists mostly of protons but also includes alpha particles, electrons, and the nuclei of atoms up to atomic number 26.   (Interaction 10.1)

Covalence number The number of covalent bonds that an atom can have in a molecule.   (4.5)

Covalent bond The net force of attraction that arises as two atomic nuclei share a pair of electrons. One pair is shared in a single bond, two pairs in a double bond, and three electron pairs are shared in a triple bond.   (4.4, 7.2)

Crenation The shrinkage of red blood cells when they are in contact with a hypertonic solution.   (7.6)

Curie (Ci) A unit of activity of a radioactive source.

$$1 \text{ Ci} = 3.70 \times 10^{10} \text{ disintegrations/s}   (10.3)$$

Dalton (D) A unit for a formula mass; one atomic mass unit.   (5.2)

Dalton's law (See *Law of partial pressures.*)

Dalton's theory A theory that accounts for the laws of chemical combination by postulating that matter consists of indestructible atoms, that all atoms of the same element are identical in mass and other properties, that the atoms of different elements are different in mass and other properties, and that in the formation of a compound atoms join together in definite, whole-number ratios.   (2.2)

Deamination The removal of an amino group from an amino acid.   (28.2)

Decarboxylation The removal of a carboxyl group.   (28.2)

Degree Celsius One-one hundredth (1/100) of the interval on a thermometer between the freezing point and the boiling point of water.   (1.4)

Degree Fahrenheit One-one hundred and eightieth (1/180) of the interval on a thermometer between the freezing point and the boiling point of water.   (1.4)

Deliquescence The ability of a substance to attract water vapor to itself to form a concentrated solution.   (7.2)

Denaturation The conversion of a native protein into a denatured protein.   (20.2)

Denatured protein A protein whose molecules have suffered the loss of their native shape and form as well as their ability to function biologically.   (20.2)

Density The ratio of the mass of an object to its volume; the mass per unit volume. Density = mass/volume (usually expressed in g/mL).   (1.7)

Deoxyribonucleic acid (DNA) The chemical of a gene; one of a large number of polymers of deoxyribonucleotides and whose sequences of side chain bases constitute the genetic messages of genes.   (24.2)

Derived quantity A quantity based on a relationship that involves one or more base quantities of measurement such as volume (length^3) or density (mass/volume).   (1.4)

Desiccant A substance that combines with water vapor to form a hydrate and thereby reduces the concentration of water vapor in the air space around the substance.   (7.2)

Detergent A surface-active agent; a soap.   (Interaction 19.4)

Dextrorotatory That property of an optically active substance by which it can cause the plane of plane-polarized light to rotate clockwise.   (17.3)

D-Family; L-Family The names of the two optically active families to which substances can belong when they are considered solely according to one kind of molecular chirality (molecular handedness) or the other.   (18.3)

Diabetes mellitus A disease in which there is an insufficiency of effective insulin and an impairment of glucose tolerance.   (26.2)

Dialysis The passage through a dialyzing membrane of water and particles in solution, but not of particles that have colloidal size. (7.5)

Dialyzing membrane A membrane permeable to solvent and small ions or molecules but impermeable to colloidal sized particles. (7.5)

Diastereomers Stereoisomers whose molecules are not related as an object is to its mirror image. (17.2)

Dietetics The application of the findings of the science of nutrition to the feeding of individual humans, whether well or ill. (29.1)

Diffusion A physical process whereby particles, by random motions, intermingle and spread out so as to erase concentration gradients. (6.5)

Digestive juice A secretion into the digestive tract that consists of a dilute aqueous solution of digestive enzymes (or their zymogens) and inorganic ions. (23.1)

Dihydric alcohol An alcohol with two OH groups; a glycol. (13.1)

Dipeptide A compound whose molecules have two α-amino acid residues joined by a peptide (amide) bond. (20.3)

Dipolar ion A molecule that carries one plus charge and one minus charge, such as an α-amino acid. (20.1)

Dipole, electrical A pair of equal but opposite (and usually partial) electrical charges separated by a small distance in a molecule. (4.9)

Dipole–dipole attraction The electrical force of attraction between $\delta+$ and $\delta-$ sites of polar molecules. (7.2)

Diprotic acid An acid with two protons available per molecule to neutralize a base, e.g., H_2SO_4. (8.2)

Disaccharide A carbohydrate that can be hydrolyzed into two monosaccharides. (18.5)

Dissociation The separation of preexisting ions from one another as an ionic compound dissolves or melts. (7.2)

Disulfide link The sulfur–sulfur covalent bond in polypeptides. (20.1)

Disulfide system S—S as in R—S—S—R. (13.6)

DNA (See *Deoxyribonucleic acid;* see also *Double helix DNA model.*)

Double bond A covalent bond in which two pairs of electrons are shared. (4.5)

Double displacement reaction A reaction in which a compound is made by the exchange of partner ions between two salts. (8.5)

Double helix DNA model A spiral arrangement of two intertwining DNA molecules held together by hydrogen bonds between side-chain bases. (24.2)

Duplex DNA DNA in its double stranded form. (24.2)

Dynamic equilibrium (See *Equilibrium, dynamic.*)

Edema The swelling of tissue caused by the retention of water. (23.2)

Effector A chemical other than a substrate that can allosterically activate an enzyme. (21.4)

Elastin The fibrous protein of tendons and arteries. (20.9)

Electrical balance The condition of a net ionic equation wherein the algebraic sum of the positive and negative charges of the reactants equals that of the products. (8.3)

Electrode A metal object, usually a wire, suspended in an electrically conducting medium through which electricity passes to or from an external circuit. (8.1)

Electrolysis A procedure in which an electrical current is passed through a solution that contains ions, or through a molten salt, for the purpose of bringing about a chemical change. (8.1)

Electrolyte Any substance whose solution in water conducts electricity; or the solution itself of such a substance. (8.1)

Electrolytes, blood The ionic substances dissolved in the blood. (23.2)

Electron A subatomic particle that bears one unit of negative charge and has a mass that is 1/1836 the mass of a proton. (3.1)

Electron cloud A mental model that views the one or two rapidly moving electrons of an orbital as creating a cloudlike distribution of negative charge. (3.4)

Electron configuration The most stable arrangement (that is, the arrangement of lowest energy) of the electrons of an atom, ion, or molecule. (3.3)

Electron dot structure A Lewis structure of a molecule in which all valence shell electrons, whether shared or unshared, are shown either by dots or by lines. (4.5)

Electronegativity The ability of an atom joined to another by a covalent bond to attract the electrons of the bond toward itself. (4.9)

Electron sharing The joint attraction of two atomic nuclei toward a pair of electrons situated between the nuclei and between which, therefore, a covalent bond exists. (4.4)

Electron shell An alternative name for *principal energy level.* (3.3)

Electron volt (eV) A very small unit of energy used to describe the energy of a radiation; 1.6×10^{-19} joule; 3.8×10^{-20} calorie. (10.3)

Element A substance that cannot be broken down into anything that is both stable and more simple; a substance in which all of the atoms have the same atomic number and the same electron configuration; one of the three broad kinds of matter, the others being compounds and mixtures. (2.1)

Emulsion A colloidal dispersion of tiny microdroplets of one liquid in another liquid.   (7.1)

Enantiomers Stereoisomers whose molecules are related as an object is related to its mirror image but that cannot be superimposed.   (17.2)

Endothermic Describing a change that needs a constant supply of heat energy to happen.   (2.4, 9.2)

End point The stage in a titration when the operation is stopped.   (9.11)

Energy A capacity to cause a change that can, in principle, be harnessed for useful work.   (2.3)

Energy level A principal energy state in which electrons of an atom can be.   (3.3)

Energy of activation The minimum energy that must be provided by the collision between reactant particles to initiate the rearrangement of electrons relative to nuclei that must happen if the reaction is to occur.   (9.2)

Enzyme A catalyst in a living system.   (9.1, 21.1)

Enzyme induction The chemical process whereby the synthesis of an enzyme is prompted.   (24.4)

Enzyme–substrate complex The temporary combination that an enzyme must form with its substrate before catalysis can occur.   (21.2)

Epinephrine A hormone of the adrenal medulla that activates the enzymes needed to release glucose from glycogen.   (26.1)

Equation, balanced A chemical equation in which all of the atoms represented in the formulas of the reactants are present in identical numbers among the products, and in which any net electrical charge provided by the reactants equals the same charge indicated by the products. (See also *Chemical equation.*)   (2.2, 5.3)

Equation of state for an ideal gas (See *Ideal gas law.*)

Equilibrium, dynamic A situation in which two opposing events occur at identical rates so that no net change happens.   (6.7)

Equilibrium constant The value that the mass action expression has when a chemical system is at equilibrium. (9.3)

Equilibrium equation A chemical equation in which oppositely pointing arrows separate reactants and products that are in equilibrium.   (6.7)

Equilibrium law The mathematical equation that describes the interrelationships among the molar concentrations of reactants and products when equilibrium exists. (9.3)

Equivalence point The stage in a titration when the reactants have been mixed in the exact molar proportions represented by the balanced equation; in an acid–base titration, the stage when the moles of hydrogen ions furnished by the acid matches the moles of hydroxide ions (or other proton acceptor) supplied by the base. (9.11)

Equivalent (eq) For an ion, usually its mass in grams divided by the amount of its electrical charge.   (8.5)

Error In a measurement, the difference between the measured value and the correct value of a physical quantity.   (1.4)

Erythrocyte A red blood cell.   (23.2)

Essential amino acid An α-amino acid that the body cannot make from other amino acids and that must be supplied by the diet.   (29.2)

Essential fatty acid A fatty acid that must be supplied by the diet.   (29.1)

Ester A derivative of an acid and an alcohol that can be hydrolyzed to these parent compounds. Esters of carboxylic acids and phosphoric acid occur in living systems.   (15.3)

$$(H)R-\overset{\displaystyle O}{\overset{\|}{C}}-O-R' \qquad RO-\overset{\displaystyle O}{\overset{\|}{\underset{\underset{\displaystyle OH}{|}}{P}}}-OH$$

carboxylic acid ester   phosphoric acid ester

Esterification The formation of an ester.   (15.3)

Ether An organic compound whose molecules have an oxygen attached by single bonds to separate carbon atoms neither of which is a carbonyl carbon atom: R—O—R′.   (13.5)

Ethyl group CH_3CH_2- (11.6)

Evaporation The conversion of a substance from its liquid to its vapor state.   (6.7)

Exon A segment of a DNA strand that eventually becomes expressed as a corresponding sequence of aminoacyl residues in a polypeptide.   (24.2)

Exothermic Describing a change by which heat energy is released from the system.   (2.4, 9.2)

Extensive property Any property whose value is directly proportional to the size of the sample, such as volume or mass.   (1.7)

Extracellular fluids Body fluids that are outside of cells. (23.1)

Factor-label method A strategy for solving computational problems that uses conversion factors and the cancellation of the units of physical quantities as an aid in working toward the solution.   (1.6)

Fatty acid Any carboxylic acid that can be obtained by the hydrolysis of animal fats or vegetable oils. (15.1, 19.1)

Fatty acid cycle (See *Beta oxidation.*)

Feedback inhibition The competitive inhibition of an enzyme by a product of its own action.   (21.4)

Fibrin The fibrous protein of a blood clot that forms from fibrinogen during clotting.   (20.9, 23.2)

Fibrinogen A protein in blood that is changed to fibrin during clotting.   (23.2)

Fibrous proteins Water-insoluble proteins found in fibrous tissues.   (20.9)

Fischer projection structure A two-dimensional representation, prepared according to rules, of the configuration at a tetrahedral stereocenter.   (18.3)

Fission The splitting of the nucleus of a heavy atom approximately in half, which is accompanied by the release of one or a few neutrons and energy.   (10.7)

Folate A vitamin supplied by folic acid or pteroylglutamic acid and that is needed to prevent megaloblastic anemia and neural tube defects.   (29.3)

Food A material that supplies one or more nutrients without contributing materials that, either in kind or quantity, would be harmful to most healthy people.   (29.1)

Formula, chemical A shorthand representation of a substance that uses atomic symbols and following subscripts to describe the elemental composition and the mole ratios in which the atoms of the elements are combined.   (2.2)

Formula, empirical A chemical symbol for a compound that gives just the ratios of the atoms and not necessarily the composition of a complete molecule.   (2.2, 4.1)

Formula, molecular A chemical symbol for a substance that gives the composition of a complete molecule.   (4.5)

Formula, structural A chemical symbol for a substance that uses atomic symbols and lines to describe the pattern in which the atoms are joined together in a molecule.   (4.5)

Formula mass The sum of the atomic masses of the atoms represented in a chemical formula.   (5.2)

Formula unit A small particle—an atom, a molecule, or a set of ions—that has the composition given by the chemical formula of the substance.   (2.2)

Forward reaction In a chemical equilibrium, the reaction whereby substances to the left of the double arrows are changed to the products shown on the right-hand side of the arrows.   (6.7)

Free rotation The absence of a barrier to the rotation of two groups with respect to each other when they are joined by a single, covalent bond.   (11.2)

Functional group An atom or a group of atoms in a molecule that is responsible for the particular set of reactions that all compounds with this group have.   (11.4)

Gamma radiation A natural radiation similar to but more powerful than X rays.   (10.1)

Gap junctions Tubules made of membrane-bound proteins that interconnect one cell to neighboring cells and through which materials can pass directly.   (20.8)

Gas Any substance that must be contained in a wholly closed space and whose shape and volume are determined entirely by the shape and volume of its container; a state of matter.   (1.3, 6.1)

Gas constant, universal (R) The ratio of PV to nT for a gas, where P = the gas pressure, V = volume, n = number of moles, and T = the Kelvin temperature. Depending on the units of P and V,

$$R = 0.0821 \text{ L atm/mol K}$$
$$R = 6.24 \times 10^4 \text{ mL mm Hg/mol K}   (6.3)$$

Gas tension The partial pressure of a gas over its solution in some liquid when the system is in equilibrium.   (7.3)

Gastric juice The digestive juice secreted into the stomach and that contains pepsinogen, hydrochloric acid, and gastric lipase.   (23.1)

Gay-Lussac's law (See *Pressure–temperature law.*)

Gel A colloidal dispersion of a solid in a liquid that has adopted a semisolid form.   (7.1)

Gene A unit of heredity carried on a cell's chromosomes and consisting of DNA.   (24.1, 24.2)

General gas law $P_1V_1/T_1 = P_2V_2/T_2$.   (6.2)

Gene therapy Procedures for inserting correct genetic material into defective tissue.   (24.6)

Genetic code The set of correlations that specify which codons on mRNA chains are responsible for which amino acyl residues when the latter are steered into place during the mRNA-directed synthesis of polypeptides.   (24.3)

Genetic engineering The use of recombinant DNA to manufacture substances or to repair genetic defects.   (24.6)

Genome The entire complement of genetic information of a species; all the genes of an individual.   (24.7)

Geometric isomerism Stereoisomerism caused by restricted rotation that gives different geometries to the same structural organization; cis–trans isomerism.   (12.3)

Geometric isomers Stereoisomers whose molecules have identical atomic organizations but different geometries; cis–trans isomers.   (12.3)

Globular proteins Proteins that are soluble in water or in water that contains certain dissolved salts.   (20.9)

Globulins Globular proteins in the blood that include γ-globulin, an agent in the body's defense against infectious diseases.   (20.9, 23.2)

Glucagon A hormone, secreted by the α-cells of the pancreas in response to a decrease in the blood sugar level, that stimulates the liver to release glucose from its glycogen stores.   (26.1)

Gluconeogenesis The synthesis of glucose from compounds with smaller molecules or ions.   (26.2, 26.4)

Glucose tolerance The ability of the body to manage the intake of dietary glucose while keeping the blood sugar level from fluctuating widely. (26.2)

Glucose tolerance test A series of measurements of the blood sugar level after the ingestion of a considerable amount of glucose; used to obtain information about an individual's glucose tolerance. (26.2)

Glucoside An acetal formed from glucose (in its cyclic, hemiacetal form) and an alcohol. (18.4)

Glucosuria The presence of glucose in urine. (26.1)

Glycerophospholipid A hydrolyzable lipid that has an ester linkage between glycerol and one phosphoric acid unit (this, in turn, forming another ester link to a small molecule). In *phosphatides,* the remaining two OH units of glycerol are esterified with fatty acids. In *plasmalogens,* one OH is esterified with a fatty acid and the other is joined by an ether link to a long-chain unsaturated alcohol. (19.3)

Glycogenesis The synthesis of glycogen. (26.1)

Glycogenolysis The breakdown of glycogen to glucose. (26.1)

Glycol A dihydric alcohol. (13.1)

Glycolipid A lipid whose molecules include a glucose unit, a galactose unit, or some other carbohydrate unit. (19.3, 20.8)

Glycolysis A series of chemical reactions that breaks down glucose or glucose units in glycogen until pyruvate remains (when the series is operated aerobically) or lactate forms (when the conditions are anaerobic). (25.1, 26.3)

Glycoprotein A protein, often membrane bound, that is joined to a carbohydrate unit. (20.8)

Glycoside An acetal or a ketal formed from the cyclic form of a monosaccharide and an alcohol. (18.4)

Glycosidic link An oxygen bridge from one monosaccharide unit to another. (18.4)

Gradient The presence of a change in value of some physical quantity with distance, as in a *concentration* gradient in which the concentration of a solute is different in different parts of the system. (7.3, 20.8)

Gram (g) A mass equal to one-thousandth of the kilogram mass, the SI standard mass.

    1 g = 0.001 kg = 1000 mg; 1 lb = 453.6 g   (1.4)

Gray (Gy) The SI unit of absorbed dose of radiation equal to one joule of energy absorbed per kilogram of tissue. (10.3)

Greenhouse effect The entrapment of heat radiating from the Earth by the greenhouse gases in the atmosphere (principally carbon dioxide, water, and methane). (Interaction 8.4)

Ground substance A gellike material present in cartilage and other extracellular spaces that gives flexibility to collagen and other fibrous proteins. (20.8)

Group A vertical column in the periodic table; a family of elements. (3.2)

Half-life The time needed for half of the atoms in a sample of a particular radioactive isotope to undergo radioactive decay. (10.1)

Halogens The elements of group VIIA of the periodic table—fluorine, chlorine, bromine, iodine, and astatine. (3.2)

Hard water Water that contains one or more of the metallic ions Mg^{2+}, Ca^{2+}, Fe^{2+}, or Fe^{3+}. The negative ions present are usually Cl^- and $SO_4{}^{2-}$. If $HCO_3{}^-$ is the chief negative ion, the water is said to be *temporary hard water;* otherwise it is *permanent hard water.* (Interaction 8.3)

Heat The form of energy that transfers between two objects in contact that have initially different temperatures. (2.4)

Heat of fusion The quantity of heat that one gram of a substance absorbs when it changes from its solid to its liquid state at its melting point. (6.9)

Heat of reaction The net energy difference between the reactants and the products of a reaction. (9.2)

Heat of vaporization The quantity of heat that one gram of a substance absorbs when it changes from its liquid to its gaseous state. (6.7)

Heisenberg uncertainty principle It is impossible simultaneously to determine with precision and accuracy both the position and the velocity of an electron. (3.3)

α-Helix One kind of secondary structure of a polypeptide in which its molecules are coiled. (20.4)

Heme The deep-red, iron-containing prosthetic group in hemoglobin and myoglobin. (20.6)

Hemiacetal Any compound whose molecules have both an OH and an OR group coming to a CH unit. (14.5)

Hemiketal Any compound whose molecules have both an OH and an OR group coming to a carbon that otherwise bears no H atoms. (14.5)

Hemoglobin The oxygen-carrying protein in red blood cells. (20.6, 23.3)

Hemolysis The bursting of a red blood cell. (7.6)

Henderson-Hasselbalch equation An equation used in buffer calculations when the buffer is prepared using a weak acid of some known pK_a and the anion of the weak acid:

$$pH = pK_a + \log \frac{[\text{anion}]}{[\text{acid}]}$$

where [anion] is the *initial* molar concentration of the anion component of a buffer pair and [acid] is the *ini-*

tial molar concentration of the acid component. (9.10)

Henry's law (See *Pressure–solubility law.*)

Heterocyclic compound An organic compound with a ring in which an atom other than carbon takes up at least one position in the ring. (11.2)

Heterogeneous mixture A mixture in which the composition of one small portion is not identical with that of another. (7.1)

Heterogeneous nuclear RNA (hnRNA) RNA made directly at the guidance of DNA and from which messenger RNA (mRNA) is made. (24.3)

High-energy phosphate An organophosphate with a phosphate group transfer potential equal to or higher than that of ADP or ATP. (25.1)

Homeostasis The response of an organism to a stimulus such that the organism is restored to its prestimulated state. (21.4)

Homogeneous mixture A mixture in which the composition and properties are uniform throughout. (7.1)

Hormone A primary chemical messenger made by an endocrine gland and carried by the bloodstream to a target organ where a particular chemical response is initiated. (22.1)

Hund's rule Electrons become evenly distributed among *different* orbitals of the same sublevel insofar as there is room. (3.4)

Hydrate A compound in which intact molecules of water are held in a definite molar proportion to the other components. (7.2)

Hydrated ion An ion around which molecules of water have been drawn by ion–dipole attractions. (7.2)

Hydration The association of water molecules with dissolved ions or polar molecules. (7.2)

Hydrocarbon An organic compound that consists entirely of carbon and hydrogen. (11.5)

Hydrogen bond The force of attraction between $\delta+$ on a hydrogen held by a covalent bond to oxygen or nitrogen (or fluorine) and a $\delta-$ charge on a nearby atom of oxygen or nitrogen (or fluorine). (6.8, 7.2)

Hydrolase An enzyme that catalyzes a hydrolysis reaction. (21.1)

Hydrolysis of anions Reactions in which anions (other than OH^-) react with water and increase the pH of a solution. (9.7)

Hydrolysis of cations Reactions in which cations (other than H_3O^+) react with water and decrease the pH of the solution. (9.6)

Hydrolyzable lipid A lipid that can be hydrolyzed or saponified. (Formerly called a saponifiable lipid.) (19.1)

Hydronium ion H_3O^+ (8.2)

Hydrophilic group Any part of a molecular structure that attracts water molecules; a polar or ionic group such as $-OH$, $-CO_2^-$, $-NH_3^+$, or $-NH_2$. (19.5)

Hydrophobic group Any part of a molecular structure that has no attraction for water molecules; a nonpolar group such as any alkyl group. (19.5)

Hydrophobic interaction The water avoidance of nonpolar groups or side chains that is partly responsible for the shape adopted by a polypeptide or nucleic acid molecule in an aqueous environment. (20.1, 24.2)

Hydroxide ion OH^- (8.2)

Hygroscopic Describing a substance that can reduce the concentration of water vapor in the surrounding air by forming a hydrate. (7.2)

Hyperammonemia An elevated level of ammonium ion in the blood. (28.3)

Hypercapnia An elevated level of carbon dioxide in the blood as indicated by a partial pressure of CO_2 in venous blood above 50 mm Hg. (23.4)

Hyperglycemia An elevated level of glucose in the blood—above 110 mg/dL in whole blood. (26.1)

Hyperkalemia An elevated level of potassium ion in blood—above 5.0 meq/L. (23.2)

Hypernatremia An elevated level of sodium ion in blood—above 145 meq/L. (23.2)

Hyperthermia A condition of a core body temperature above normal. (Interaction 2.1)

Hypertonic Having an osmotic pressure greater than some reference; having a total concentration of all solute particles higher than that of some reference. (7.6)

Hyperventilation Breathing considerably faster and deeper than normal. (9.9)

Hypocapnia A condition of a below-normal concentration of carbon dioxide in the blood as indicated by a partial pressure of CO_2 in venous blood of less than 35 mm Hg. (23.4)

Hypoglycemia A low level of glucose in blood—below 65 mg/dL of whole blood. (26.1)

Hypokalemia A low level of potassium ion in blood—below 3.5 meq/L. (23.2)

Hyponatremia A low level of sodium ion in blood—below 135 meq/L. (23.2)

Hypothermia A low body temperature. (Interaction 2.1)

Hypothesis A conjecture, subject to being disproved, that explains a set of facts in terms of a common cause and that serves as the basis for the design of additional tests or experiments. (1.2)

Hypotonic Having an osmotic pressure less than some reference; having a total concentration of dissolved solute particles less than that of some reference. (7.6)

Hypoventilation Breathing more slowly and less deeply than normal; shallow breathing. (9.9)

Hypoxia A condition of a low supply of oxygen. (23.4)

Ideal gas A hypothetical gas that obeys the gas laws exactly. (6.2)

Ideal gas law $PV = nRT$ (6.3)

Indicator A dye that has one color in solution below a measured pH range and a different color above this range. (8.2, 9.5)

Induced fit model Many enzymes are induced by their substrate molecules to modify their shapes to accommodate the substrate. (21.2)

Inducer A substance whose molecules remove repressor molecules from operator genes and so open the way for structural genes to direct the overall syntheses of particular polypeptides. (24.4)

Inertia The resistance of an object to a change in its position or its motion. (1.4)

Inhibitor A substance that interacts with an enzyme to prevent its acting as a catalyst. (21.4)

Inner transition elements The elements of the lanthanide and actinide series of the periodic table. (3.2)

Inorganic compound Any compound that is not an organic compound. (8.2, 11.1)

Insulin A protein hormone made by the pancreas, released in response to an increase in the blood sugar level, and used by certain tissues to help them take up glucose from circulation. (26.1)

Intensive property Any property whose value is independent of the size of the sample, such as temperature and density. (1.7)

Internal environment Everything enclosed within an organism. (23.1)

International System of Units (SI) The successor to the metric system with new reference standards for the base units but with the same names for the units and the same decimal relationships. (1.4)

International Union of Pure and Applied Chemistry system (IUPAC system) A set of systematic rules for naming compounds and designed to give each compound one unique name and for which only one structure can be drawn. (11.6)

Interstitial fluids Fluids in tissues but not inside cells or the blood. (23.1)

Intestinal juice The digestive juice that empties into the duodenum from the intestinal mucosa and whose enzymes also work within the intestinal mucosa as molecules migrate through. (23.1)

Intracellular fluids Fluids inside cells. (23.1)

Intron A segment of a DNA strand that separates exons and that does not become expressed as a segment of a polypeptide. (24.2)

Inverse square law The intensity of radiation varies inversely with the square of the distance from its source. (10.2)

Invert sugar A 1:1 mixture of glucose and fructose. (18.5)

In Vitro Occurring in laboratory vessels. (13.3)

In Vivo Occurring within a living system. (13.3)

Iodine test A test for starch by which a drop of iodine reagent produces an intensely purple color if starch is present. (18.6)

Ion An electrically charged, atomic- or molecule-sized particle; a particle that has one or a few atomic nuclei and either one or two (seldom, three) too many or too few electrons to render the particle electrically neutral. (4.1)

Ion–dipole attraction The attraction between a partially charged site of a polar molecule and a fully charged ion. (7.2)

Ionic bond The force of attraction between oppositely charged ions in an ionic compound. (4.1, 7.2)

Ionic compound A compound that consists of an orderly aggregation of oppositely charged ions that assemble in whatever ratio ensures overall electrical neutrality. (4.1)

Ionic equation A chemical equation that explicitly shows all of the particles—ions, atoms, or molecules—that are involved in a reaction even if some are only spectator particles. (See also *Net ionic equation; Equation, balanced.*) (8.3)

Ion–Ion Attraction The electrical force at work between oppositely charged ions, particularly in an ionic crystal. (7.2)

Ionization A reaction, usually involving water molecules, whereby molecules change into ions. (8.1)

Ionizing radiation Any radiation, such as alpha, beta, gamma, X, and cosmic radiation, that can create ions from molecules within the medium that it enters. (10.2)

Ion product constant of water (K_w) The product of the molar concentrations of hydrogen ions and hydroxide ions in water at a given temperature.

$$K_w = [H^+][OH^-]$$
$$= 1.0 \times 10^{-14} \text{ (at 25 °C)} \quad (9.4)$$

Isobutyl group $(CH_3)_2CHCH_2—$ (11.6)

Isoelectric molecule A molecule that has an equal number of positive and negative sites. (20.1)

Isoelectric point (pI) The pH of a solution in which a specified amino acid or a protein is in an isoelectric condition; the pH at which there is no net migration of the amino acid or protein in an electric field. (20.1)

Isoenzymes Enzymes that have identical catalytic functions but that are made of slightly different polypeptides.   (21.1)

Isohydric shift In actively metabolizing tissue, the use of a hydrogen ion released from newly formed carbonic acid to react with and liberate oxygen from oxyhemoglobin; in the lungs, the use of hydrogen ion released when hemoglobin oxygenates to combine with bicarbonate ion and liberate carbon dioxide for exhaling. (23.3)

Isomerase An enzyme that catalyzes the conversion of a compound into one of its isomers.   (21.1)

Isomerism The phenomenon of the existence of two or more compounds with identical molecular formulas but different structures.   (11.3)

Isomers Compounds with identical molecular formulas but different structures.   (11.3)

Isopropyl group $(CH_3)_2CH—$ (11.6)

Isotonic Having an osmotic pressure identical to that of a reference; having a concentration equivalent to the reference with respect to the ability to undergo osmosis. (7.6)

Isotope A substance in which all of the atoms are identical in atomic number, mass number, and electron configuration.   (3.1)

Isozyme (See *Isoenzymes.*)

IUPAC system (See *International Union of Pure and Applied Chemistry system.*)

Joule (J) The SI derived unit of energy.   (2.3)

K_a (See *Acid ionization constant.*)

K_b (See *Base ionization constant.*)

K_w (See *Ion product constant of water.*)

Kelvin The SI unit of temperature degree and equal to 1/100th of the interval between the freezing point and the boiling point of water when measured under standard conditions.   (1.4)

Kelvin scale The scale of absolute temperatures expressed in kelvins beginning with 0 K for the coldest temperature attainable.   (1.4)

Keratin The fibrous protein of hair, fur, fingernails, and hooves.   (20.9)

Ketal A substance whose molecules have two OR groups joined to a carbon that also holds two hydrocarbon groups.   (14.5)

Ketoacidosis The acidosis caused by untreated ketonemia. (27.6)

Keto group The carbonyl group when it is joined on each side to carbon atoms.   (14.1)

Ketohexose A monosaccharide whose molecules contain six carbon atoms and have a keto group.   (18.2)

Ketone Any compound with a carbonyl group attached to two carbon atoms, as in $R_2C{=}O$.   (14.1)

Ketone bodies Acetoacetate, β-hydroxybutyrate—or their parent acids—and acetone.   (27.6)

Ketonemia An elevated concentration of ketone bodies in the blood.   (27.6)

Ketonuria An elevated concentration of ketone bodies in the urine.   (27.6)

Ketose A monosaccharide whose molecules have a ketone group.   (18.2)

Ketosis The combination of ketonemia, ketonuria, and acetone breath.   (27.6)

Kilocalorie (kcal) The quantity of heat equal to 1000 calories.   (2.4)

Kilojoule (kJ) The quantity of heat equal to 1000 joules. (2.3)

Kilogram (kg) The SI base unit of mass; 1000 g; 2.205 lb. (1.4)

Kinase An enzyme that catalyzes the transfer of a phosphate group.   (21.1)

Kinetic energy (KE) The energy of an object by virtue of its motion.

$$KE = \tfrac{1}{2} \text{ mass} \times \text{velocity}^2 \quad (2.3)$$

Kinetics The field of chemistry that deals with the rates of chemical reactions.   (9.1)

Kinetic theory of gases A set of postulates about the nature of an ideal gas: that it consists of a large number of very small particles in constant, random motion; that in their collisions the particles lose no frictional energy; that between collisions the particles neither attract nor repel each other; and that the motions and collisions of the particles obey all the laws of motion. (6.5)

Kreb's cycle (See *Citric acid cycle.*)

Law of conservation of energy Energy can be neither created nor destroyed but only transformed from one form to another.   (2.3)

Law of conservation of mass Matter is neither created nor destroyed in chemical reactions; the masses of all products equals the masses of all reactants.   (2.1)

Law of definite proportions The elements in a compound occur in definite proportions by mass.   (2.1)

Law of mass action (law of Guldberg and Waage) The molar proportions of the interacting substances in a chemical equilibrium are related by the following equation (in which the ratio on the right is called the *mass action expression* for the system).

$$K_{eq} = \frac{[C]^c[D]^d}{[A]^a[B]^b}$$

The symbols refer to the following generalized equilibrium:

$$aA + bB \rightleftharpoons cC + dD$$

and the brackets, [ ], denote molar concentrations. When an equilibrium involves additional substances, the equation for the equilibrium constant is adjusted accordingly.   (9.3)

Law of multiple proportions When two elements can combine to form more than one compound, the different masses of the first that can combine with the same mass of the second are in the ratio of small whole numbers.   (2.2)

Law of partial pressures (Dalton's law) The total pressure of a mixture of gases is the sum of their individual partial pressures.   (6.4)

Le Châtelier's principle If a system is in equilibrium and a change is made in its conditions, the system will change in whichever way most directly restores equilibrium.   (6.7)

Length The base quantity for expressing distances or how long something is.   (1.4)

Levorotatory The property of an optically active substance that causes a counterclockwise rotation of the plane of plane-polarized light.   (17.3)

Lewis structure A structural formula of a particle (atom, ion, or molecule) that shows all valence shell electrons in their correct places.   (4.5)

Ligand An electron-rich species, either negatively charged or electrically neutral, that binds with a metal ion to form a complex ion.   (14.3)

Ligase An enzyme that catalyzes the formation of bonds at the expense of triphosphate energy.   (21.1)

Like-dissolves-like rule Polar solvents dissolve polar or ionic solutes and nonpolar solvents dissolve nonpolar or weakly polar solutes.   (11.5)

Limiting amino acid The essential amino acid most poorly provided by a dietary protein.   (29.2)

Limiting reactant The reactant that is completely consumed while one or more other reactants is not used up.   (5.3)

Lipid A plant or animal product that tends to dissolve in such nonpolar solvents as ether, carbon tetrachloride, and benzene.   (19.1)

Lipid bilayer The sheetlike array of two layers of lipid molecules, interspersed with molecules of cholesterol and proteins, that make up the membranes of cells in animals.   (19.5)

Lipoprotein complex A combination of lipid and protein molecules that serves as the vehicle for carrying the lipid in the bloodstream.   (27.1)

Liquid A state of matter in which a substance's volume but not its shape is independent of the shape of its container.   (1.3, 6.6)

Liter (L) A volume equal to 1000 cm³ or 1000 mL or 1.057 liquid quart.   (1.4)

Lock and key theory The specificity of an enzyme for its substrate is caused by the need for the substrate molecule to fit to the enzyme's surface much as a key fits to and turns only one tumbler lock.   (21.2)

London force A net force between molecules that arises from temporary polarities induced in the molecules by collisions or near collisions with neighboring molecules.   (7.2)

Lyase An enzyme that catalyzes an elimination reaction to form a double bond.   (21.1)

Macromolecule Any molecule with a very high formula mass, generally several thousand or more.   (7.1)

Markovnikov's rule In the addition of an unsymmetrical reactant to an unsymmetrical double bond of a simple alkene, the positive part of the reactant molecule (usually H^+) goes to the carbon that has the greater number of hydrogen atoms and the negative part goes to the other carbon of the double bond.   (12.4)

Mass A quantitative measure of inertia based on an artifact at Sèvres, France, called the standard kilogram mass; a measure of the quantity of matter in an object relative to this reference standard.   (1.4)

Mass number The sum of the numbers of protons and neutrons in one atom of an isotope.   (3.1)

Material balance The condition of a chemical equation in which all of the atoms present among the reactants are also found in the products.   (8.3)

Matter Anything that occupies space and has mass.   (2.1)

Measurement An operation that obtains a value for a physical quantity by the use of an instrument.   (1.4)

Melting point The temperature at which a solid changes into its liquid form; the temperature at which equilibrium exists between the solid and liquid forms of a substance.   (2.4, 6.9)

Mercaptan A thioalcohol; R—S—H.   (13.6)

Meso compound One of a set of optical isomers whose own molecules are not chiral and which, therefore, is optically inactive.   (Interaction 17.1)

Messenger RNA (mRNA) RNA that carries the genetic code in the form of a specific series of codons for a specific polypeptide from the cell's nucleus to the cytoplasm.   (24.3)

Metabolism The sum total of all of the chemical reactions that occur in an organism.   (2.4)

Metal Any substance, usually an element, that is shiny, conducts electricity well, and (if a solid) can be hammered into sheets and drawn into wires.   (2.1)

Metalloids Elements that have some metallic and some nonmetallic properties. (3.2)

Metathesis reaction (See *Double displacement reaction.*)

Meter (m) The base unit of length in the International System of Units (SI)

$$1 \text{ m} = 100 \text{ cm} = 39.37 \text{ in.} = 3.280 \text{ ft} = 1.093 \text{ yd} \quad (1.4)$$

Methyl group CH_3— (11.6)

Micelle A globular arrangement of the molecules of an amphipathic compound in water in which their hydrophobic parts intermingle inside the globule and their hydrophilic parts are exposed to the water. (19.5)

Microgram (μg) A mass equal to one-thousandth of a milligram.

$$1 \text{ } \mu\text{g} = 0.001 \text{ mg} = 1 \times 10^{-6} \text{ g} \quad (1.4)$$

Microliter (μL) A volume equal to one-thousandth of a milliliter.

$$1 \text{ } \mu\text{L} = 0.001 \text{ mL} = 1 \times 10^{-6} \text{ L} \quad (1.4)$$

Milliequivalent (meq) A quantity of substance equal to one-thousandth of an equivalent. (8.5)

Milligram (mg) A mass equal to one-thousandth of a gram.

$$1 \text{ mg} = 0.001 \text{ g}; \quad 1000 \text{ mg} = 1 \text{ g}. \quad (1.4)$$

Milliliter (mL) A volume equal to one-thousandth of a liter.

$$1 \text{ mL} = 0.001 \text{ L} \quad (1.4)$$

Millimeter (mm) A length equal to one-thousandth of a meter.

$$1 \text{ mm} = 0.001 \text{ m} = 0.0394 \text{ in.} \quad (1.4)$$

Millimeter of mercury (mm Hg) A unit of pressure equal to 1/760 atm; also called the *torr*. (6.1)

Millimole (mmol) One-thousandth of a mole.

$$1000 \text{ mmol} = 1 \text{ mol} \quad (5.2)$$

Minerals, dietary Ions that must be provided in the diet at levels of 100 mg/day or more; Ca^{2+}, Mg^{2+}, Na^+, K^+, Cl^-, and phosphate. (29.4)

Mitochondrion A unit inside a plant or animal cell in which the machinery for making high-energy phosphates by oxidative phosphorylation is located. (24.1)

Mixture One of three kinds of matter (together with elements and compounds); any substance made up of two or more elements or compounds combined physically in no particular proportion by mass and separable into its component parts by physical means. (2.1, 7.1)

Model, scientific A mental construction, often involving pictures or diagrams, that is used to explain a number of facts. (3.2)

Moderate acid An acid with a K_a in the range of 1 to 10^{-3}. (9.6)

Molar concentration (M) A solution's concentration in units of moles of solute per liter of solution; molarity. (5.5)

Molarity (See *Molar concentration.*)

Molar volume, standard The volume occupied by one mole of a gas under standard conditions of temperature and pressure; 22.4 L at 273 K and 1 atm. (6.3)

Mole (mol) A mass of a compound or element that equals its formula mass in grams; Avogadro's number of a substance's formula units. (5.1, 5.2)

Molecular compound A compound whose smallest representative particle is a molecule; a covalent compound. (4.4)

Molecular equation An equation that shows the complete formulas of all of the substances present in a mixture undergoing a reaction. (See also *Net ionic equation; Equation, balanced.*) (8.3)

Molecular formula (See *Formula, molecular.*)

Molecular kinetic energy The energy of motion of individual atoms, ions, or molecules. (2.5)

Molecular mass The formula mass of a substance. (5.2)

Molecular orbital A region in the space that envelopes two (or sometimes more) atomic nuclei where a shared pair of electrons of a covalent bond resides. (4.4)

Molecular weight (See *Molecular mass.*)

Molecule An electrically neutral (but often polar) particle made up of the nuclei and electrons of two or more atoms and held together by covalent bonds; the smallest representative sample of a molecular compound. (2.2, 4.4)

Monatomic ion An ion possessing only one atomic nucleus. (4.4)

Monoamine oxidase An enzyme that catalyzes the inactivation of neurotransmitters or other amino compounds of the nervous system. (22.2)

Monohydric alcohol An alcohol whose molecules have one OH group. (13.1)

Monomer A compound that can be used to make a polymer. (12.6)

Monoprotic acid An acid with one proton per molecule that can neutralize a base. (8.2)

Monosaccharide A carbohydrate that cannot be hydrolyzed. (18.2)

Mucin A viscous glycoprotein released in the mouth and the stomach that coats and lubricates food particles and protects the stomach from the acid and pepsin of gastric juice. (23.1)

Mutagen A chemical or physical agent that can induce the mutation of a gene without preventing the gene from replicating. (10.2)

Mutarotation The gradual change in the specific rotation of a substance in solution but without a permanent, irreversible chemical change occurring. (18.4)

Myosins Proteins in contractile muscle. (20.9)

Native protein A protein whose molecules are in the configuration and shape they normally have within a living system.   (20.2)

Net ionic equation A chemical equation in which all spectator particles are omitted so that only the particles that participate directly are represented.   (8.3)

Neurotransmitter A substance released by one nerve cell to carry a signal to the next nerve cell.   (22.1)

Neutralization, acid–base A reaction between an acid and a base.   (3.2)

Neutralizing capacity The capacity of a solution or a substance to neutralize an acid or a base—expressed as a molar concentration.   (9.11)

Neutral solution A solution in which the molar concentration of hydronium ions exactly equals the molar concentration of hydroxide ions.   (8.2, 9.5)

Neutron An electrically neutral subatomic particle with a mass of 1 u.   (3.1)

Niacin A water-soluble vitamin needed to prevent pellagra and essential to the coenzymes in NAD^+ and $NADP^+$; nicotinic acid or nicotinamide.   (29.3)

Nitrogen balance A condition of the body in which it excretes as much nitrogen as it receives in the diet.   (29.2)

Nitrogen family The elements of group VA of the periodic table: nitrogen, phosphorus, arsenic, antimony, and bismuth.   (3.2)

Nitrogen pool The sum total of all nitrogen compounds in the body.   (28.1)

Noble gases The elements of group 0 of the periodic table: helium, neon, argon, krypton, xenon, and radon.   (3.2)

Noble gas rule (See *Octet rule.*)

Nomenclature The system of names and the rules for devising such names, given structures, or for writing structures, given names.   (11.6)

Nonelectrolyte Any substance that cannot furnish ions when dissolved in water or when melted.   (8.1)

Nonfunctional group A section of an organic molecule that remains unchanged during a chemical reaction at a functional group.   (11.4)

Nonhydrolyzable lipid Any lipid, such as the steroids, that cannot be hydrolyzed or similarly broken down by aqueous alkali.   (19.1)

Nonmetal Any element that is not a metal. (See *Metal.*)   (2.1)

Nonvolatile liquid Any liquid with a very low vapor pressure at room temperature and that does not readily evaporate.   (6.7)

Normal fasting level The normal concentration of something in the blood, such as blood sugar, after about 4 hours without food.   (26.1)

Normoglycemia The condition of the blood in a healthy person in which the concentration of glucose is normal.   (26.1)

Nuclear chain reaction The mechanism of nuclear fission by which one fission event makes enough fission initiators (neutrons) to cause more than one additional fission event.   (10.7)

Nuclear equation A representation of a nuclear transformation in which the chemical symbols of the reactants and products include mass numbers and atomic numbers.   (10.1)

Nucleic acid A polymer of nucleotides in which the repeating units are pentose phosphate diesters, each pentose unit bearing a side-chain base (one of five heterocyclic amines); polymeric compounds that are involved in the storage, transmission, and expression of genetic messages.   (24.2)

Nucleotide A monomer of a nucleic acid that consists of a pentose phosphate ester in which the pentose unit carries one of five heterocyclic amines as a side-chain base.   (24.2)

Nucleus In chemistry and physics, the subatomic particle that serves as the core of an atom and that is made up of protons and neutrons.   (3.1) In biology, the organelle in a cell that houses DNA.   (24.1)

Nutrient Any one of a large number of substances in food and drink that is needed to sustain growth and health.   (29.1)

Nutrition The science of the substances of the diet that are necessary for growth, operation, energy, and repair of body tissues.   (29.1)

Octet, outer A condition of an atom or ion in which its highest occupied energy level has eight electrons—a condition of stability.   (4.2)

Octet rule (noble gas rule) The atoms of a reactive element tend to undergo those chemical reactions that most directly give them the electron configuration of the noble gas that stands nearest the element in the periodic table (all but one of which, He, have outer octets).   (4.2)

Olefin An alkene.   (12.6)

One substance–one structure rule If two samples of matter have identical physical and chemical properties, they have identical molecules.   (17.1)

Optical isomer One of a set of compounds whose molecules differ only in their chiralities.   (17.3)

Optically active The ability of a substance to rotate the plane of polarization of plane-polarized light.   (17.3)

Optical rotation The degrees of rotation of the plane of plane-polarized light caused by an optically active solution; the observed rotation of such a solution.   (17.3)

Orbital (See *Atomic orbital.*)

Orbital overlap The interpenetration of one atomic orbital by another from an adjacent atom to form a molecular orbital. (4.4)

Organic chemistry The chemistry of carbon compounds. (11.1)

Organic compounds Compounds of carbon other than those related to carbonic acid and its salts, or to the oxides of carbon, or to the cyanides. (8.2, 11.1)

Osmolarity The molar concentration of all osmotically active solute particles in a solution. (7.5)

Osmosis The passage of water only, without any solute, from a less concentrated solution (or pure water) to a more concentrated solution when the two solutions are separated by a semipermeable membrane. (7.5)

Osmotic membrane A semipermeable membrane that permits only osmosis, not dialysis. (7.5)

Osmotic pressure The pressure that would have to be applied to a solution to prevent osmosis if the solution were separated from water by an osmotic membrane. (7.5)

Outer octet (See *Octet, outer.*)

Outside shell electrons Electrons occupying any of the available orbitals at the highest occupied principal energy level. (3.4)

Oxidase (See *Oxidoreductase.*)

Oxidation A reaction in which the oxidation number of one of the atoms of a reactant becomes more positive; in organic chemistry, the loss of hydrogen or the gain of oxygen. (4.3)

Oxidation number For simple monatomic ions, the quantity and sign of the electrical charge on the ion. (4.3)

Oxidation–reduction reaction A reaction in which oxidation numbers change. (4.3)

Oxidative deamination The change of an amino group to a keto group with loss of nitrogen. (28.2)

Oxidative phosphorylation The synthesis of high-energy phosphates such as ATP from lower energy phosphates and inorganic phosphate by the reactions that involve the respiratory chain. (25.1, 25.3)

Oxidizing agent A substance that can cause an oxidation. (4.3)

Oxidoreductase An enzyme that catalyzes the formation of an oxidation–reduction equilibrium. (21.1)

Oxygen affinity The percentage to which all of the hemoglobin molecules in the blood are saturated with oxygen molecules. (23.3)

Oxygen debt The condition in a tissue when anaerobic glycolysis has operated and lactate has been excessively produced. (26.3)

Oxygen family The elements in group VIA of the periodic table: oxygen, sulfur, selenium, tellurium, and polonium. (3.2)

Oxyhemoglobin Hemoglobin carrying its capacity of oxygen. (23.3)

P_i Inorganic phosphate ion(s) of whatever mix of PO_4^{3-}, HPO_4^{2-}, $H_2PO_4^{-}$, and even traces of H_3PO_4 that is possible at the particular pH of the system, but almost entirely $HPO_4^{2-} + H_2PO_4^{-}$. (15.6)

Pancreatic juice The digestive juice that empties into the duodenum from the pancreas. (23.1)

Pantothenic acid A water-soluble vitamin needed to make coenzyme A. (29.3)

Partial pressure The pressure contributed to the total pressure by an individual gas in a mixture of gases. (6.4)

Parts per billion (ppb) The number of parts in a billion parts. (Two drops of water in a railway tank car that holds 34,000 gallons of water correspond roughly to 1 ppb.) (7.4)

Parts per million (ppm) The number of parts in a million parts. (Two drops of water in a large 32-gallon trash can correspond roughly to 1 ppm.) (7.4)

Pascal (Pa) The SI derived unit of pressure.

$$133.3224 \text{ Pa} = 1 \text{ mm Hg} \quad (6.1)$$

Pauli exclusion principle No more than two electrons can occupy the same orbital at the same time, and two can be present only if they have opposite spin. (3.4)

Pentose phosphate pathway The synthesis of NADPH that uses chemical energy in glucose-6-phosphate and that involves pentoses as intermediates. (26.3)

Peptide bond The amide linkage in a protein; a carbonyl-to-nitrogen bond. (20.3)

Peptide unit An amino acid residue as it occurs within a polypeptide molecule. (20.1)

peptide unit

Percent (%) A measure of concentration.
Vol/vol (v/v) percent: The number of volumes of solute in 100 volumes of solution.
Wt/wt (w/w) percent: The number of grams of solute in 100 g of the solution.
Wt/vol (w/v) percent: The number of grams of solute in 100 mL of the solution. (7.4)

Period A horizontal row in the periodic table. (3.2)

Periodic law Many properties of the elements are periodic functions of their atomic numbers. (3.2)

Periodic table A display of the elements that emphasizes the family relationships. (3.2)

pH The negative power to which the base 10 must be

raised to express the molar concentration of hydrogen ions in an aqueous solution.

$$[H^+] = 1 \times 10^{-pH}$$

$$-\log [H^+] = pH \quad (9.5)$$

Phenol Any organic compound whose molecules have an OH group attached to a benzene ring. (13.4)

Phenyl group The benzene ring minus one H atom; C_6H_5. (12.7)

Phosphate buffer Usually a mixture or a solution that contains dihydrogen phosphate ions ($H_2PO_4^-$) to neutralize OH^- and monohydrogen phosphate ions (HPO_4^{2-}) to neutralize H^+. (9.9)

Phosphate group transfer potential The relative ability of an organophosphate to transfer a phosphate group to some acceptor. (25.1)

Phosphatide A glycerophospholipid whose molecules are esters between glycerol, two fatty acids, phosphoric acid, and a small alcohol. (19.3)

Phosphoglyceride (See *Glycerophospholipid.*)

Phospholipids Lipids such as the glycerophospholipids (phosphatides and plasmalogens) and the sphingomyelins whose molecules include phosphate ester units. (19.3)

Photon A package of energy released when an electron in an atom moves from a higher to a lower energy state; a unit of light energy. (3.3)

Photosynthesis The synthesis in plants of complex compounds from carbon dioxide, water, and minerals with the aid of sunlight captured by the plant's green pigment, chlorophyll. (2.4, 18.2)

Physical property Any observable characteristic of a substance other than a chemical property, such as color, density, melting point, boiling point, temperature, and quantity. (1.3)

Physical quantity A property of something to which we assign both a numerical value and a unit, such as mass, volume, or temperature;

physical quantity = number × unit. (1.4)

Physical states Solid, liquid, and gas. (1.3)

Physiological saline solution A solution of sodium chloride with an osmotic pressure equal to that of blood. (7.6)

pI (See *Isoelectric point.*)

pK_a $pK_a = -\log K_a$ (9.8)

pK_b $pK_b = -\log K_b$ (9.8)

Plane-polarized light Light whose electrical field vibrations are all in the same plane. (17.3)

Plasmalogens Glycerophospholipids whose molecules include an unsaturated fatty alcohol unit. (19.3)

Plasmid A circular molecule of supercoiled DNA in a bacterial cell. (24.6)

β-Pleated sheet A secondary structure for a polypeptide in which the molecules are aligned side by side in a sheetlike array with the sheet partially pleated. (20.4)

pOH The negative power to which the base 10 must be raised to express the concentration of hydroxide ions in an aqueous solution in mol/L.

$$[OH^-] = 1 \times 10^{-pOH}$$

At 25 °C, pH + pOH = 14.00 (9.5)

Poison A substance that reacts in some way in the body to cause changes in metabolism that threaten health or life. (21.4)

Polar bond A bond at which we can write $\delta+$ at one end and $\delta-$ at the other end, the end that has the more electronegative atom. (4.9)

Polarimeter An instrument for detecting and measuring optical activity. (17.3)

Polar molecule A molecule that has sites of partial positive and partial negative charge and a permanent electrical dipole. (4.9)

Polyatomic ion An ion made from two or more atoms, such as OH^-, SO_4^{2-}, and CO_3^{2-}. (4.6)

Polymer Any substance with a very high formula mass whose molecules have a repeating structural unit. (12.6)

Polymerization A chemical reaction that makes a polymer from a monomer. (12.6)

Polypeptide A polymer with repeating α-aminoacyl units joined by peptide (amide) bonds. (20.1)

Polysaccharide A carbohydrate whose molecules are polymers of monosaccharides. (18.2)

Potential energy Stored or inactive energy. (2.3)

PP$_i$ Inorganic diphosphate ion(s). (25.1)

ppb (See *Parts per billion.*)

ppm (See *Parts per million.*)

Precipitate A solid that separates from a solution as the result of a chemical reaction. (5.4)

Precipitation The formation and separation of a precipitate. (5.4)

Precision The fineness of a measurement or the degree to which successive measurements agree with each other when several are taken one after the other. (See also *Accuracy.*) (1.4)

Pressure Force per unit area. (6.1)

Pressure–solubility law (Henry's law) The concentration of a gas in a liquid at any given temperature is directly proportional to the partial pressure of the gas on the solution. (7.3)

Pressure–temperature law (Gay-Lussac's law) The pressure of a gas is directly proportional to its Kelvin temperature when the gas volume is constant. (6.2)

Pressure–volume law (Boyle's law) The volume of a gas is inversely proportional to its pressure when the temperature is constant.   (6.2)

Primary alcohol An alcohol in whose molecules an OH group is attached to a primary carbon, as in RCH_2OH. (13.1)

Primary carbon In a molecule, a carbon atom that is joined directly to just one other carbon, such as the end carbons in $CH_3CH_2CH_3$.   (11.6)

Primary structure The sequence of amino acyl residues held together by peptide bonds in a polypeptide. (20.2)

Primary transcript RNA (ptRNA) (See *Heterogeneous nuclear RNA.*)

Principal energy level A space near an atomic nucleus where there are one or more sublevels and orbitals in which electrons can reside; an electron shell.   (3.3)

Product A substance that forms in a chemical reaction. (2.1)

Proenzyme (See *Zymogen.*)

Property A characteristic of something by means of which we can identify it.   (1.3)

Propyl group $CH_3CH_2CH_2$—   (11.6)

Prosthetic group A nonprotein molecule joined to a polypeptide to make a biologically active protein. (20.5)

Protein A naturally occurring polymeric substance made up wholly or mostly of polypeptide molecules.   (20.1)

Proton A subatomic particle that bears one unit of positive charge and has a mass of 1 u.   (3.1)

Proton-pumping ATPase The enzyme on the matrix side of the inner mitochondrial membrane that catalyzes the formation of ATP from ADP and P_i under the influence of a flow of protons across this membrane.   (25.3)

Quantum A quantity of energy possessed by a photon. (3.3)

Quaternary structure An aggregation of two or more polypeptide strands each with its own primary, secondary, and tertiary structure.   (20.2, 20.6)

Racemic mixture A 1:1 mixture of enantiomers and therefore optically inactive.   (17.3)

Rad One rad equals 10^{-2} gray $(1 \times 10^{-5}$ J) of energy absorbed per gram of tissue as a result of ionizing radiation.   (10.3)

Radiation In atomic physics, the emission of some ray such as an alpha, beta, or gamma ray; any of the rays themselves.   (10.1)

Radiation sickness The set of symptoms that develops following exposure to heavy doses of ionizing radiation. (10.2)

Radical A particle with one or more unpaired electrons. (10.2)

Radioactive The property of unstable atomic nuclei whereby they emit alpna, beta, or gamma rays.   (10.1)

Radioactive decay The change occurring to a radioactive isotope by which it emits alpha rays, beta rays, or gamma rays.   (10.1)

Radioactive disintegration series A series of isotopes selected and arranged such that each isotope except the first is produced by the radioactive decay of the preceding isotope and the last isotope is nonradioactive. (10.1)

Radioactivity The ability to emit atomic radiation.   (2.1, 10.1)

Radiomimetic substance A substance whose chemical effect in a cell mimics the effect of ionizing radiation. (24.4)

Radionuclide A radioactive isotope.   (10.1)

Rate of reaction The number of successful (product-forming) collisions that occur each second in each unit of volume of a reacting mixture.   (9.2)

Reactant One of the substances that reacts in a chemical reaction.   (2.1)

Reaction, chemical An event in which chemical substances change into other substances.   (2.1)

Reagent Any mixture of chemicals, usually a solution, that is used to carry out a chemical test.   (7.4)

Receptor molecule A molecule of a protein built into a cell membrane that can accept a molecule of a hormone or a neurotransmitter.   (20.8, 22.1)

Recombinant DNA DNA made by combining the natural DNA of plasmids in bacteria or the natural DNA in yeasts with DNA from external sources, such as the DNA for human insulin, and made as a step in a process that uses altered bacteria or yeasts to make specific proteins (e.g., interferons, human growth hormone, and insulin).   (24.6)

Recommended dietary allowance (RDA) The level of intake of a particular nutrient as determined by the Food and Nutrition Board of the National Research Council of the National Academy of Sciences to meet the known nutritional needs of most healthy individuals. (29.1)

Redox reaction Abbreviation of *reduction–oxidation;* a reaction in which oxidation numbers change.   (4.3)

Reducing agent A substance that can cause another to be reduced.   (4.3)

Reducing carbohydrate A carbohydrate that gives a positive Benedict's test.   (18.2)

Reductase An enzyme that catalyzes a reduction. (See *Oxidoreductase.*)   (21.1)

Reduction A reaction in which the oxidation number of an atom of one reactant becomes less positive or more negative; in organic chemistry, the gain of hydrogen or the loss of oxygen.   (4.3)

Reductive amination The conversion of a keto group to an amino group by the action of ammonia and a reducing agent. (28.1)

Rem One rem is the quantity of a radiation that produces the same effect in humans as one roentgen of X rays or gamma rays. (10.3)

Renal threshold That concentration of a substance in blood above which it appears in the urine. (26.1)

Replication The reproductive duplication of DNA double helix. (24.1)

Representative element Any element in any A-group of the periodic table; any element in groups IA–VIIA and those in group 0. (3.2)

Repressor A substance whose molecules can bind to a gene and prevent the gene from directing the synthesis of a polypeptide. (24.4)

Respiration The intake and chemical use of oxygen by the body and the release of carbon dioxide. (25.3)

Respiratory chain The reactions that transfer electrons from the intermediates made by other pathways to oxygen; the mechanism that creates a proton gradient across the inner membrane of a mitochondrion and that leads to ATP synthesis; the enzymes that handle these reactions. (25.1, 25.2)

Respiratory enzymes The enzymes of the respiratory chain. (25.3)

Respiratory gases Oxygen and carbon dioxide. (25.3)

Reverse reaction The reaction that undoes the effect of the forward reaction of an equilibrium. (See *Forward reaction.*) (6.7)

Riboflavin A B vitamin needed to give protection against the breakdown of tissue around the mouth, the nose, and the tongue, as well as to aid in wound healing. (29.3)

Ribonucleic acids (RNA) Polymers of nucleotides made using ribose that participate in the transcription and the translation of the genetic messages into polypeptides. (See also *Heterogeneous nuclear RNA, Messenger RNA, Ribosomal RNA,* and *Transfer RNA.*) (24.3)

Ribosomal RNA (rRNA) RNA that is incorporated into cytoplasmic bodies called ribosomes. (24.3)

Ribosome A granular complex of rRNA that becomes attached to an mRNA strand and that supplies some of the enzymes for mRNA-directed polypeptide synthesis. (24.1)

Ribozyme An enzyme whose molecules consist of ribonucleic acid, not polypeptide. (24.3)

Ring compound A compound whose molecules contain three or more atoms joined in a ring. (11.2)

RNA (See *Ribonucleic acid.*)

Roentgen One roentgen is the quantity of X rays or gamma radiation that generates ions with an aggregate 2.1 ×

10^9 units of charge in 1 mL of dry air at normal pressure and temperature. (10.3)

Saliva The digestive juice secreted in the mouth whose enzyme, amylase, catalyzes the partial digestion of starch. (23.1)

Salt Any crystalline compound that consists of oppositely charged ions (other than H^+, OH^-, or O^{2-}). (4.6, 8.5)

Salt bridge A force of attraction between $(+)$ and $(-)$ sites on polypeptide molecules. (20.5)

Saponifiable lipid (See *Hydrolyzable lipid.*)

Saponification The reaction of an ester with a base to give an alcohol and the salt of an acid. (15.5)

Saturated compound A compound whose molecules have only single bonds. (11.2)

Scientific method A method of solving a problem that uses facts to devise a hypothesis to explain the facts and to suggest further tests or experiments designed to discover whether the hypothesis is true or false. (1.2)

Scientific notation The method of writing a number as the product of two numbers, one being 10^x, where x is some positive or negative whole number. (1.5)

Second (s) The SI unit of time; 1/60th minute. (1.4)

Secondary alcohol An alcohol in whose molecules an OH group is attached to a secondary carbon atom; R_2CHOH. (13.1)

Secondary carbon Any carbon atom in an organic molecule that has two and only two bonds to other carbon atoms, such as the middle carbon atom in $CH_3CH_2CH_3$. (11.6)

Secondary structure A shape, such as the α-helix or a unit in a β-pleated sheet, that all or a large part of a polypeptide molecule adopts under the influence of hydrogen bonds, salt bridges, and hydrophobic interactions after its peptide bonds have been made. (20.2, 20.4)

Semipermeable Descriptive of a membrane that permits only certain kinds of molecules to pass through and not others. (7.5)

Shock, traumatic A medical emergency in which relatively large volumes of blood fluid leave the vascular compartment and enter the interstitial spaces. (23.2)

Side chain, amino acid The organic group attached at the α-carbon of an α-amino acid. (20.1)

Sievert (Sv) The SI unit of radiation dose equivalent. (10.3)

Significant figures The number of digits in a numerical measurement or in the result of a calculation that are known with certainty to be accurate plus one more digit. (1.4)

Simple lipid (See *Triacylglycerol.*)

Simple salt A salt that consists of only one kind of cation and one kind of anion. (8.5)

Simple sugar Any monosaccharide. (18.2)

Single bond A covalent bond involving one shared pair of electrons. (4.5)

Soap A detergent that consists of the salts of long-chain fatty acids. (Interaction 19.4)

Soft water Water with little if any of the hardness ions— Mg^{2+}, Ca^{2+}, Fe^{2+}, or Fe^{3+}. (Interaction 19.4)

Sol A colloidal dispersion of tiny particles of a solid in a liquid. (7.1)

Solid A state of matter in which the visible particles of the substance have both definite shapes and definite volumes. (1.3, 6.9)

Solubility The extent to which a substance dissolves in a fixed volume or mass of a solvent at a given temperature. (5.4)

Solute The component of a solution that is understood to be dissolved in or dispersed in a continuous solvent. (5.4)

Solution A homogeneous mixture of two or more substances that are at the smallest levels of their states of subdivision—at the ion, atom, or molecule level. (5.4, 7.1, 7.2)

Solution, aqueous A solution in which water is the solvent. (5.4)

Solution, concentrated A solution with a high ratio of solute to solvent. (5.4)

Solution, dilute A solution with a low ratio of solute to solvent. (5.4)

Solution, saturated A solution into which no more solute can be dissolved at the given temperature; a solution in which dynamic equilibrium exists between the dissolved and the undissolved solute. (5.4, 7.2)

Solution, supersaturated An unstable solution that has a higher concentration of solute than that of the saturated solution. (5.4)

Solution, unsaturated A solution into which more solute could be dissolved without changing the temperature. (5.4)

Solvent That component of a solution into which the solutes are considered to have dissolved; the component that is present as a continuous phase. (5.4)

Somatostatin A hormone of the hypothalamus that inhibits or slows the release of glucagon and insulin from the pancreas. (26.1)

Specific gravity The ratio of the density of an object to the density of water. (1.7)

Specific heat The amount of heat that one gram of a substance can absorb per degree Celsius increase in temperature:

$$\text{Specific heat} = \frac{\text{heat}}{g\,\Delta t}$$

where Δt = the change in temperature. (2.4)

Specific rotation $[\alpha]$ The optical rotation of a solution per unit of concentration per unit of path length in decimeters:

$$[\alpha] = \frac{\alpha}{cl}$$

where α = observed rotation; c = concentration in g/mL, and l = path length (in dm). (17.3)

Sphingolipid A lipid that, when hydrolyzed, gives sphingosine instead of glycerol, plus fatty acids, phosphoric acid, and a small alcohol or a monosaccharide; sphingomyelins and cerebrosides. (19.3)

Standard, reference A physical description or embodiment of a base unit of measurement, such as the standard meter or the standard kilogram mass. (1.4)

Standard atmosphere (atm) The pressure that supports a column of mercury 760 mm high when the mercury has a temperature of 0 °C. (6.1)

Standard conditions of temperature and pressure (STP) 0 °C (or 273 K) and 1 atm (or 760 mm Hg). (6.3)

Standard solution Any solution for which the concentration is accurately known. (9.11)

States of matter The three possible physical conditions of aggregation of matter—solid, liquid, and gas. (1.3)

Stereoisomers Isomers whose molecules have the same atom-to-atom sequences but different geometric arrangements; geometric (cis–trans) or optical isomers. (17.1)

Stereoisomerism The existence of stereoisomers. (17.1)

Steroids Nonhydrolyzable lipids such as cholesterol and several sex hormones whose molecules have the four fused rings of the steroid nucleus. (19.4)

Stoichiometry The branch of chemistry that deals with the mole proportions of chemicals in reactions. (3.1)

Straight chain A continuous, open sequence of covalently bound carbon atoms from which no additional carbon atoms are attached at interior locations of the sequence. (11.2)

Stress In equilibrium chemistry, anything that upsets an equilibrium. (6.7)

Strong acid An acid with a high percentage ionization and a high value of acid ionization constant, K_a. A *strong Brønsted acid* is any species, molecule or ion, that has a strong tendency to donate a proton to some acceptor. (8.2, 9.6)

Strong base A metal hydroxide with a high percentage ionization in solution. A *strong Brønsted base* is any species, molecular or ionic, that binds an accepted proton strongly. (8.2, 9.7)

Strong electrolyte Any substance that has a high percentage ionization in solution. (8.1)

Structural formula A formula that uses lines representing covalent bonds to connect the atomic symbols in the pattern that occurs in one molecule of a compound. (4.5, 11.2)

Structure (See *Structural formula*.)

Subatomic particle An electron, a proton, or a neutron; the atomic nucleus as a whole is also a subatomic particle.   (3.1)

Sublimation The change from the solid state directly to the gaseous state.   (6.9)

Subscripts Numbers placed to the right and a half space below the atomic symbols in a chemical formula. (2.2)

Subshell An energy sublevel; a region that makes up part (sometimes all) of a principal energy level and that can itself be subdivided into individual orbitals.   (3.4)

Substance, pure An element or a compound but not a mixture.   (2.1)

Substitution reaction A reaction in which one atom or group replaces another atom or group in a molecule. (11.7)

Substrate The substance on which an enzyme performs its catalytic work.   (17.2)

Substrate phosphorylation The direct transfer of a phosphate unit from an organophosphate to a receptor molecule.   (25.1)

Superimposition A chirality testing operation to see whether one molecular model can be made to blend simultaneously at exactly every point with another model.   (17.2)

Supersaturated Describing an unstable condition of a solution in which more solute is in solution than could be if there were equilibrium between the undissolved and dissolved states of the solute.   (5.4)

Surface-active agent (See *Surfactant*.)

Surface tension The quality of a liquid's surface by which it behaves as if it were a thin, invisible, elastic membrane.   (6.8)

Surfactant A substance, such as a detergent, that reduces the surface tension of water.   (6.8)

Suspension A mixture in which the particles of at least one component have average diameters greater than 1000 nm.   (7.1)

Synapse The fluid-filled gap between the end of the axon of one nerve cell and the next nerve cell.   (22.2)

Target cell A cell at which a hormone molecule finds a site where it can become attached and then causes some action that is associated with the hormone.   (22.1)

Target tissue The organ whose cells are recognizable by the molecules of a particular hormone.   (22.1)

Temperature The measure of the hotness or coldness of an object. *Degrees* of temperature, such as those of the Celsius, Fahrenheit, or Kelvin scales, are intervals of equal separation on a thermometer.   (1.4)

Temperature–pressure law (Gay-Lussac's law) The pressure of a gas is directly proportional to its Kelvin temperature when the volume is kept constant.   (6.2)

Temperature–volume law (Charles' law) The volume of a gas is directly proportional to its Kelvin temperature when the pressure is kept constant.   (6.2)

Teratogen A chemical or physical agent that can cause birth defects in a fetus other than inherited defects. (10.2)

Tertiary alcohol An alcohol in whose molecules an OH group is held by a carbon from which three bonds extend to other carbon atoms; R_3COH.   (13.1)

Tertiary carbon A carbon in an organic molecule that has three and only three bonds to adjacent *carbon* atoms. (11.6)

Tertiary structure The shape of a polypeptide molecule that arises from further folding or coiling of secondary structures.   (20.2, 20.5)

Tetrahedral Descriptive of the geometry of bonds at a central atom in which the bonds project to the corners of a regular tetrahedron.   (4.8)

Tetrahedral stereocenter An atom in a molecule with four single bonds arranged tetrahedrally and holding four different atoms or groups.   (17.2)

Theory An explanation for a large number of facts, observations, and hypotheses in terms of one or a few fundamental assumptions of what the world (or some small part of the world) is like.   (1.2)

Thermal property A physical property of a substance in response to a temperature change, such as specific heat. (2.4)

Thiamin A B vitamin needed to prevent beriberi.   (29.3)

Thioalcohol A compound whose molecules have the SH group attached to a saturated carbon atom; a mercaptan.   (13.6)

Threshold exposure The level of exposure to some toxic agent below which no harm is done.   (10.2)

Time A period during which something endures, exists, or continues.   (1.4)

Titration An experimental procedure for mixing two solutions using a buret in order to compare the concentration of one of the solutions with that of the other, the standard solution.   (9.11)

Tollens' reagent A slightly alkaline solution of the diammine complex of the silver ion, $Ag(NH_3)_2^+$, in water. (14.3)

Tollens' test The use of Tollen's reagent to detect an easily oxidized group such as the aldehyde group.   (14.3)

Torr A unit of pressure; 1 torr = 1 mm Hg; 1 atm = 760 torr.   (6.1)

Trace element, dietary Any element that the body needs each day in an amount of no more than 20 mg.   (29.4)

Transamination The transfer of an amino group from an amino acid to a receiver with a keto group such that the keto group changes to an amino group. (28.1)

Transcription The synthesis of messenger RNA under the direction of DNA. (24.3)

Transferase An enzyme that catalyzes the transfer of some group. (21.1)

Transfer RNA (tRNA) RNA that serves to carry an aminoacyl group to a specific acceptor site of an mRNA molecule at a ribosome where the aminoacyl group is placed into a growing polypeptide chain. (24.3)

Transition elements The elements between those of group IIA and group IIIA in the long periods of the periodic table; a metallic element other than one in group IA or IIA or in the actinide or lanthanide families. (3.2)

Translation The synthesis of a polypeptide under the direction of messenger RNA. (24.4)

Transmutation The change of an isotope of one element into an isotope of a different element. (10.1)

Triacylglycerol A lipid that can be hydrolyzed to glycerol and fatty acids; a triglyceride; sometimes, simply called a glyceride or a simple lipid. (19.1)

Tricarboxylic acid cycle (See *Citric acid cycle.*)

Triglyceride (See *Triacylglycerol.*)

Trihydric alcohol An alcohol with three OH groups per molecule. (13.1)

Triple bond A covalent bond involving the sharing of three pairs of electrons. (4.5)

Triple helix The quaternary structure of tropocollagen in which three polypeptide chains are coiled together. (20.6)

Triprotic acid An acid that can supply three protons per molecule. (8.2)

Tyndall effect The scattering of light by colloidal sized particles in a colloidal dispersion. (7.1)

Uncertainty The estimate of how finely a number can be read from a measuring instrument. (1.4)

Universal gas law (See *Ideal gas law.*)

Unsaturated compound Any compound whose molecules have a double or a triple bond. (11.2)

Unshared pairs Pairs of valence-shell electrons not involved in covalent bonds. (4.5)

Urea cycle The reactions by which urea is made from amino acids. (28.3)

Vacuum An enclosed space in which there is no matter. (6.1)

Valence shell The highest energy level of an atom that is occupied by electrons; the outside shell. (4.2)

Valence-shell electron-pair repulsion theory (VSEPR) Bond angles at a central atom are caused by the repulsions of the electron clouds of valence-shell electron pairs. (4.8)

Vaporization The change of a liquid into its vapor. (6.7)

Vapor pressure The pressure exerted by the vapor that is in equilibrium with its liquid state at a given temperature. (6.7)

Vascular compartment The entire network of blood vessels and their contents. (23.2)

Vasopressin A hypophysis hormone that acts at the kidneys to help regulate the concentrations of solutes in the blood by instructing the kidneys to retain water (if the blood is too concentrated) or to excrete water (if the blood is too dilute). (23.5)

Ventilation The movement of air into and out of the lungs by breathing. (9.9)

Virus One of a large number of substances that consist of nucleic acid surrounded by a protein overcoat and that can enter host cells, multiply, and destroy the host. (24.5)

Vital force theory A discarded theory that organic compounds could be made in the laboratory only if the chemicals possessed a vital force contributed by some living thing. (11.1)

Vitamin An organic substance that must be in the diet, whose absence causes a deficiency disease, that is present in foods in trace concentrations, and that isn't a carbohydrate, lipid, protein, or amino acid. (29.3)

Vitamin A Retinol; a fat-soluble vitamin in yellow colored foods and needed to prevent night blindness and certain conditions of the mucous membranes. (29.3)

Vitamin B_6 Pyridoxine, pyridoxal, or pyridoxamine; a vitamin needed to prevent hypochromic microcytic anemia and used in enzymes of amino acid catabolism. (29.3)

Vitamin C Ascorbic acid; a vitamin needed to prevent scurvy. (29.3)

Vitamin D Cholecalciferol (D_3) or ergocalciferol (D_2); a fat-soluble vitamin needed to prevent rickets and to ensure the formation of healthy bones and teeth. (29.3)

Vitamin deficiency diseases Diseases caused not by bacteria or viruses but by the absence of specific vitamins, such as pernicious anemia (B_{12}), hypochromic microcytic anemia (B_6), pellagra (niacin), the breakdown of certain tissues (riboflavin), megaloblastic anemia (folate), beriberi (thiamin), scurvy (C), hemorrhagic disease (K), rickets (D), and night blindness (A). (29.3)

Vitamin E A mixture of tocopherols; a fat-soluble vitamin apparently needed for protection against edema and anemia (in infants) and possibly against dystrophy, paralysis, and heart attacks. (29.3)

Vitamin K The antihemorrhagic vitamin that serves as a cofactor in the formation of a blood clot. (29.3)

Volatile liquid A liquid that has a high vapor pressure and readily evaporates at room temperature.   (6.7)

Volt (V) The SI unit of electrical potential, the force that drives a flow of electrons when a current of electricity flows.   (10.3)

Volume The capacity of an object to occupy space.   (1.4)

VSEPR theory (See *Valence-shell electron-pair repulsion theory.*)

Water of hydration Water molecules held in a hydrate in some definite mole ratio to the rest of the compound. (7.2)

Wax A lipid whose molecules are esters of long-chain monohydric alcohols and long-chain fatty acids. (19.1)

Weak acid An acid with a low percentage ionization in solution and with a small acid ionization constant, K_a. A *weak Brønsted acid* is any species, molecule or ion, that has a weak tendency to donate a proton and poorly serves as a proton donor.   (8.2, 8.4, 9.6)

Weak base A base with a low percentage ionization in solution and a small base ionization constant. A *weak Brønsted base* is any species, molecule or ion, that weakly holds an accepted proton and poorly serves as a proton acceptor.   (8.2, 8.4, 9.7)

Weak electrolyte Any electrolyte that has a low percentage ionization in solution.   (8.1)

Weight The gravitational force of attraction on an object as compared to that of some reference.   (1.4)

Zwitterion (See *Dipolar ion.*)

Zymogen A polypeptide that is changed into an enzyme by the loss of a few amino acid residues or by some other change in its structure; a proenzyme.   (21.4)

PHOTO CREDITS

Chapter 1 *Page 1:* Pal Hermansen/Tony Stone Images/ New York, Inc. *Page 5:* Joel Glenn/The Image Bank. *Page 7 (bottom left):* Michael Watson. *Page 7 (bottom right):* Courtesy Central Scientific Co. *Page 7 (top):* Marty Loken/ Tony Stone Images/New York, Inc. *Page 8:* Michael Watson. *Page 9 (top):* Michael Watson. *Page 9 (center):* Courtesy Bureau International des Poids et Mesures, France. *Pages 9 (bottom) and 10:* Ken Karp.

Chapter 2 *Page 29:* Studio-7/The Stock Market. *Page 31:* Michael Watson. *Page 32:* Ken Karp. *Page 33:* Courtesy AIP Neils Bohr Library. *Page 34:* OPC, Inc. *Page 39:* Ken Karp. *Page 40:* Mike Malyszko/FPG International.

Chapter 3 *Page 49:* Philip Long/Tony Stone Images/New York, Inc. *Page 50:* Mark Burnett/Stock, Boston. *Page 51:* Drawing by William Numeroff. *Page 55:* Courtesy New York Public Library Picture Collection. *Page 58:* Courtesy Nobel Foundation, Stockholm.

Chapter 4 *Page 74:* Hans Pfletschinger/Peter Arnold, Inc. *Page 76:* M. Claye/Jacana/Photo Researchers. *Page 78:* Photo by George Chin, DOTS, Inc. *Page 96:* Courtesy NASA. *Page 105:* Ken Karp.

Chapter 5 *Page 117:* Diane Schiumo/Fundamental Photographs. *Pages 120 and 123:* Michael Watson. *Page 132:* Andy Washnik. *Page 134:* Michael Watson. *Page 140:* OPC, Inc.

Chapter 6 *Page 146:* Tom Till/Tony Stone Images/ New York, Inc. *Page 154:* Kairos/Latin Stock/Science Photo Library/Photo Researchers. *Page 157:* Eric Reynolds/Adventure Photo. *Page 166:* John Kelly/ The Image Bank. *Page 170:* St. Bartholomew's Hospital/Science Photo Library/Photo Researchers. *Page 171:* Michael Watson.

Chapter 7 *Page 177:* Stephen Studd/Tony Stone Images/ New York, Inc. *Page 179:* OPC, Inc. *Page 184:* Michael Watson. *Page 186:* Peter Lerman.

Chapter 8 *Page 202:* Arnulf Husmo/Tony Stone Images/ New York, Inc. *Page 205:* OPC, Inc. *Page 206:* Robert Capece. *Page 208:* Andy Washnik. *Page 216:* OPC, Inc. *Page 219:* Courtesy Miles Laboratories. *Pages 220–221 and 232 (bottom):* OPC, Inc. *Page 222:* Ken Karp. *Page 232 (top):* Index Stock. *Page 235:* Courtesy of Betz Company. *Page 236:* Andy Washnik.

Chapter 9 *Page 246:* Tom Smart/Gamma Liaison. *Page 247:* OPC, Inc. *Page 248:* Courtesy USDA. *Page 263:* OPC, Inc. *Page 264 (left):* Andy Washnik. *Page 264 (right):* Courtesy Fisher Scientific. *Page 275:* Richard Megna/ Fundamental Photographs. *Page 282:* Matt Meadows/ Peter Arnold, Inc. *Page 284:* Michael Watson.

Chapter 10 *Page 293:* Ron Sanford/Tony Stone Images/ New York, Inc. *Page 308:* Courtesy U.S. Council for Energy Awareness, Washington, D.C. *Page 311 (left):* Courtesy GE Medical Systems. *Page 311 (right):* Courtesy Dimensional Medicine Inc., Minnetonka, MN. *Page 312:* Hank Morgan/ Rainbow. *Page 313:* Courtesy E.D. London, National Institute on Drug Abuse. *Page 314:* Howard Sochurek/ Woodfin Camp & Associates. *Page 317:* Shone/Gamma Liaison.

Chapter 11 *Pages 323, 326, 330–332, 346–347, and 351:* Tripos Associates. *Page 340 (top):* Kent & Donna Dannen. *Page 340 (bottom):* Bob Anderson/Masterfile. *Page 353:* J.B. Diederich/The Stock Market. *Page 355:* A. Bradshaw/Sipa Press.

Chapter 12 *Page 361:* Richard Megna/Fundamental Photographs. *Pages 367–368:* Tripos Associates. *Page 375 and 378:* Andy Washnik. *Page 379:* Jim Mendenhall. *Page 386 (top):* Russ Schleipman/The Stock Shop. *Page 386 (bottom):* Tim Davis/Photo Researchers. *Page 390:* Andy Washnik.

Chapter 13 *Page 399:* Jerry Alexander/Tony Stone Images/New York, Inc. *Page 402:* J. Pickerell/FPG International. *Page 418:* Tony Page/Tony Stone Images/ New York, Inc.

Chapter 14 *Page 428:* Tony Stone Images/New York, Inc. *Page 435:* Andy Washnik. *Page 436:* Michael Watson.

Chapter 15 *Page 452:* James Andrew Bareham/Tony Stone Images/New York, Inc. *Page 458:* Courtesy Lisa Passmore. *Page 466:* Harry J. Przekop, Jr./Medichrome/ The Stock Shop.

Chapter 16 *Page 479:* Jean François Causse/Tony Stone Images/New York, Inc. *Page 488:* LaFoto/H. Armstrong Roberts, Inc.

Chapter 17 *Page 498:* Renee Lynn/Tony Stone Images/ New York, Inc. *Pages 503–504:* Michael Watson. *Page 510:* Diane Schiumo/Fundamental Photographs.

Chapter 18 *Page 516:* Andy Sacks/Tony Stone Images/ New York, Inc. *Page 518:* Paul Barton/The Stock Market. *Page 539:* Andy Washnik.

Chapter 19   *Page 545:* Comstock, Inc. *Page 547:* Tripos Associates. *Page 549:* Courtesy Bayer AG, Leverkusen, Germany. *Page 552:* Tripos Associates. *Page 555:* Nino Mascardi/The Image Bank. *Pages 557–559:* Courtesy Richard Pastor, FDA.

Chapter 20   *Page 567:* David Parker/Science Photo Library/Photo Researchers. *Page 584:* Courtesy Dr. Jerome Gross, Massachusetts General Hospital. *Page 585:* Bill Longcore/Photo Researchers.

Chapter 21   *Page 601:* John M. Roberts/The Stock Market. *Page 607:* Courtesy T.A. Steitz, Department of Molecular Biophysics and Biochemistry, Yale University. *Page 618:* Courtesy Boehringer Mannheim Diagnostics.

Chapter 22   *Page 622:* Jess Stock/Tony Stone Images/ New York, Inc. *Page 628:* P. Gontier/Eurelios/Science Photo Library/Photo Researchers.

Chapter 23   *Page 640:* Chris Noble/Tony Stone Images/ New York, Inc.

Chapter 24   *Page 667:* T. Davis and W. Bilenduke/ Tony Stone Images/New York, Inc. *Page 675:* Courtesy Robert Stodola, Fox Chase Cancer Center. *Page 680:* Courtesy S. H. Kim, Duke University Medical Center. *Page 700:* Courtesy Affymetrix.

Chapter 25   *Page 705:* John Kelly/The Image Bank. *Page 715:* Courtesy Keith R. Porter. *Page 720:* From Parsons, D.F., *Science,* 140, 985 (1973).

Chapter 26   *Page 725:* Sanford/Agliolo/Index Stock. *Page 731:* Courtesy MiniMed, Sylmar, CA.

Chapter 27   *Page 744:* David Madison/Tony Stone Images/ New York, Inc. *Page 748:* W. Ober/Visuals Unlimited. *Page 750:* Courtesy John Sholtis, The Rockefeller University, New York. *Page 754:* Jeff Smith/The Image Bank.

Chapter 28   *Page 765:* Sanford/Agliolo/The Stock Market.

Chapter 29
Page 780: Otto Rogge/The Stock Market.

INDEX[a]

RELATIONSHIP OF UNITS
(Values in boldface are exact.)

Length
1 in. = **2.54** cm
1 ft. = **30.48** cm
1 yd. = **91.44** cm

Volume
1 liq oz = **29.57353** mL
1 liq qt = **946.352946** mL
1 gallon = **3.785411784** L

Mass
1 oz = **28.349523125** g
1 lb = **453.59237** g

Pressure

$$1 \text{ mm Hg} = 1 \text{ torr} = \frac{\textbf{1}}{\textbf{760}} \text{ atm}$$

Energy
1 cal = **4.184** joule

STRONG AQUEOUS ACIDS

Hydrochloric acid, HCl
Hydrobromic acid, HBr
Hydroiodic acid, HI
Nitric acid, HNO_3
Sulfuric acid, H_2SO_4

STRONG, HIGHLY SOLUBLE AQUEOUS BASES

Sodium hydroxide, NaOH
Potassium hydroxide, KOH

STRONG, SLIGHTLY SOLUBLE AQUEOUS BASES

Calcium hydroxide, $Ca(OH)_2$
Magnesium hydroxide, $Mg(OH)_2$

PHYSICAL CONSTANTS

Atomic mass unit (u) = $1.6605665 \times 10^{-24}$ g
Avogadro's number = 6.0221367×10^{23}

Gas constant, R = $6.24 \times 10^4 \dfrac{\text{mm Hg mL}}{\text{mol K}}$

= $0.0821 \dfrac{\text{L atm}}{\text{mol K}}$

Molar volume = 22.41383 L/mol (ideal gas, at 273.15 K and 760 mm Hg)

ABBREVIATIONS OF UNITS

atm	atmosphere of pressure
Bq	becquerel
°C	degree Celsius
cal	calorie
cc	cubic centimeter
Ci	curie
cm	centimeter
D	rad
dL	deciliter
eq	equivalent
eV	electron-volt
°F	degree Fahrenheit
ft	foot
g	gram
gal	gallon
GeV	gigaelectron-volt
Gy	gray
in.	inch
J	joule
K	kelvin
kcal	kilocalorie
keV	kiloelectron-volt
kg	kilogram
km	kilometer
kPa	kilopascal
L	liter
lb	pound
m	meter
M	mol/L (molarity)
meq	milliequivalent
MeV	megaelectron-volt
mg	milligram
μg	microgram
mi	mile
mL	milliliter
μL	microliter
mm	millimeter
μm	micrometer
mm Hg	millimeter of mercury
mmol	millimole
mol	mole
mOsm	milliosmole
Osm	osmole
oz	ounce
Pa	pascal
ppb	parts per billion
ppm	parts per million
pt	pint
qt	quart
s	second
Sv	sievert
T	Kelvin temperature
t_c	Celsius temperature
t_f	Fahrenheit temperature
u	atomic mass unit
yd	yard